RELATED TITLES OF INTEREST FROM PRENTICE HALL

Numerical Methods Using MATLAB, 4/E (0-13-065248-2)
John H. Mathews and Kurtis Fink

Introduction to Numerical Methods and MATLAB: Implements and Applications (0-201-30860-6)
Gerald Recktenwald

Applied Numerical Methods for Engineers and Scientists (0-13-089480-X)
Singiresu S. Rao

Introduction to Scientific Computing: A Matrix-Vector Approach to Using MATLAB, 2/E (0-13-949157-0)
Charles F. Van Loan

Applied Numerical Analysis Using MATLAB (0-13-319849-9)
Laurene Fausett

Numerical Methods: Algorithms and Applications (0-13-031400-5)
Laurene Fausett

Numerical Methods Using MATLAB, 2/E (0-13-012641-1)
George Lindfield and John Penny

Calculus of Variations: Mechanics, Control, and Other Applications (0-13-142383-5)
Charles MacCluer

An Introduction to Dynamical Systems: Continuous and Discrete (0-13-143140-4)
R. Clark Robinson

Modeling Differential Equations in Biology (0-13-017325-8)
Clifford Taubes

The Theory of Differential Equations: Classical and Qualitative (0-13-102026-9)
Walter Kelley and Allan Peterson

A First Course in the Qualitative Theory of Differential Equations (0-13-008380-1)
James Liu

Partial Differential Equations and Boundary Value Problems, 2/E (0-13-148096-0)
Nakhle H. Asmar

Applied Partial Differential Equations with Fourier Series and Boundary Value Problems, 4/E (0-13-065243-1)
Richard Haberman

A Very Applied First Course in Partial Differential Equations (0-13-030417-4)
Michael Keane

Partial Differential Equations (0-13-674359-5)
Arthur David Snider

Partial Differential Equations, 2/E (0-13-009335-1)
Robert McOwen

A First Course in Fourier Analysis (0-13-578782-3)
David Kammler

A First Course in Wavelets with Fourier Analysis (0-13-022809-5)
Albert Boggess and Francis Narcowich

Classical and Modern Fourier Analysis (0-13-035399-X)
Loukas Grafakos

Introduction to the Mathematics of Medical Imaging (0-13-067548-2)
Charles Epstein

Advanced Engineering Mathematics (0-13-321431-1)
Michael Greenberg

Applied Complex Analysis with Partial Differential Equations (0-13-089239-4)
Nakhle H. Asmar

Real Analysis with Real Applications (0-13-041647-9)
Ken Davidson and Alan Donsig

Differential Equations and Linear Algebra (0-13-086250-9)
Jerry Farlow, James Hall, Jean McDill, Beverly West

Differential Equations and Linear Algebra, 2/E (0-13-263757-X)
Stephen Goode

Differential Equations and Linear Algebra (0-13-011118-X)
Michael Greenberg

Multivariable Mathematics, 4/E (0-13-067276-9)
Williamson/Trotter

Differential Equations: An Applied Approach (0-13-044930-X)
J.M. Cushing

Continued on inside backcover...

A Friendly Introduction to Numerical Analysis

A Friendly Introduction to Numerical Analysis

Brian Bradie

Christopher Newport University
Department of Mathematics

PEARSON
Prentice Hall

Upper Saddle River, New Jersey 07458

Original edition, entitled FRIENDLY INTRODUCTION TO NUMERICAL ANALYSIS, A, 1st Edition, by BRADIE, BRIAN, published by Pearson Education, Inc, publishing as Pearson, Copyright © 2006.

This edition is manufactured in KOREA, and is authorized for sale only in KOREA.

ISBN 978-89-450-4522-5
ISBN 89-450-4522-8

Printed in the United States of America
10 9 8 7 6 5 4 3 2

ISBN 0-13-013054-0

Pearson Education LTD., *London*
Pearson Education Australia PTY, Limited, *Sydney*
Pearson Education Singapore, Pte. Ltd
Pearson Education North Asia Ltd, *Hong Kong*
Pearson Education Canada, Ltd., *Toronto*
Pearson Educacion de Mexico, S.A. de C.V.
Pearson Education - Japan, *Tokyo*
Pearson Education Malaysia, Pte. Ltd

Contents

APPENDICES

***Answers to Selected Exercises are available to text users at the web address** of `www.pcs.cnu.edu/~bbradie/textbookanswers.html`

Preface

This is an introduction to the fundamental concepts and techniques of numerical analysis and numerical methods for undergraduates, as well as for graduate engineers and applied scientists receiving their first exposure to numerical analysis. Applications drawn from the literature of many different fields will prepare students to use the techniques covered to solve a wide variety of practical problems. There is also sufficient mathematical detail to prepare students to embark upon an investigation of more advanced topics, especially in PDEs. The presentation style is what I like to call tell and show. This means that the concepts and techniques are first developed in a clear, concise, and easy-to-read manner, and then illustrated with at least one fully worked example. In total, nearly 250 fully worked examples are presented to help the students grasp the sequence of calculations associated with a particular method and gain better insight into algorithm operation.

The text is organized around mathematical problems, with each chapter devoted to a single type of problem (e.g., rootfinding, numerical calculus: differentiation and integration, the matrix eigenvalue problem, and elliptic partial differential equations). Within each chapter the presentation begins with the simplest and most basic methods, progressing gradually to more advanced topics. Early chapters generally contain easier material, while later chapters proceed at increasing levels of difficulty and complexity. Throughout, emphasis is placed on understanding and being able to work with the key concepts of rate/order of convergence and stability, and assessing the accuracy of numerical results. This emphasis helps students develop skill in numerically verifying theoretical convergence speed. More importantly, the text emphasizes that it is not sufficient to obtain the correct answer from a numerical algorithm; one must also check that convergence toward the correct answer is happening at the correct speed. I have always felt very strongly that a textbook must provide students with some means of checking their understanding and honing their skills, some means of making the knowledge their own. This is invariably accomplished through the exercises. This text features more than 1200 numbered exercises (many with multiple parts) organized into exercise sets at the end of each section. Each exercise set contains problems designed to provide students with the opportunity to practice (with paper, pencil, and calculator) the sequence of calculations associated with a particular method. The exercises usually also require the verification of theoretical error bounds and/or theoretical rates of convergence. Additional exercises may require the derivation of a method, an examination of conditions under which methods perform better or worse than predicted by theory, or extension of material presented in the section. Many exercises require students to code a numerical method on the computer and then use that computer code, and many exercises are application problems that require interpretation of results.

Distinctive Features

A quick scan of the table of contents will reveal that certain topics typically found in a book of this nature, such as approximation (orthogonal least-squares, FFT, rational function approximation) and optimization, have been omitted. In place of these topics is an extensive coverage of material not usually found, or only briefly discussed, in other texts. This extensive coverage includes treatment of non-Dirichlet boundary conditions, and artificial singularities for one-dimensional boundary value problems; treatment of non-Dirichlet boundary conditions, the multigrid method and irregular domains for elliptic partial differential equations; treatment of source and decay terms, non-Dirichlet boundary conditions, polar coordinates and problems in two space dimensions for parabolic partial differential equations; and treatment of the advection and convection-diffusion equations. Why did I select such non-standard topics for inclusion? My primary objective in writing this text was to create a book that would allow students to immediately apply the numerical techniques they have learned to real-world problems. After reviewing technical journals and textbooks to determine the most commonly used basic numerical techniques and discussing topics with my colleagues and non-academic scientists/engineers, I felt that students would benefit more from an expanded coverage of boundary value problems and partial differential equations than they would from a superficial coverage of these same topics and inclusion of those topics which have been omitted.

In keeping with the objective of preparing students to apply numerical techniques, an extensive set of application problems has been compiled from the literature of many different fields. Physics, biology, chemistry, chemical engineering, thermodynamics, heat transfer, electrostatics, ecology, manufacturing and sociology are among the fields represented. Each chapter opens with outlines of several real-world problems which serve to motivate the study and to demonstrate the broad applicability of the class of methods which will be treated in that chapter. Application problems then appear throughout the chapter as both worked examples and exercises. An added benefit of the application problems is that they afford the opportunity to discuss practical issues, such as introducing nondimensional variables, treating singularities, and manipulating problems into the form required by a particular method. Perhaps the most distinctive feature of this book is the minimal amount of pseudocode which appears. This feature is in marked contrast to other introductory textbooks on numerical analysis, which tend to have a lot of pseudocode, and usually some *Maple* and/or MATLAB code fragments, too. Unfortunately, it has been my experience that most students don't use pseudocode properly. What is intended by an author as a teaching tool, more often than not, is used by the students just to expedite the completion of an assignment. Instead of digging through each line of code to develop a deeper understanding of how and why each method works, the student simply translates the code into whatever happens to be the language of choice. When this happens, little or no transfer of knowledge takes place. The end result in such a case is that the presence of pseudocode hinders, rather than promotes, student learning. For this reason, I have chosen to include pseudocode only when the natural language description of an algorithm became too cumbersome or when the pseudocode was needed to develop some other essential idea. Although pseudocode has largely been removed, students have most certainly not been left without guidance in the production of efficient, working code. Where appropriate, programming hints have been provided, and important implementation details have been discussed. Then, of course, there are the worked examples. These provide dynamic demonstrations for each of the algorithms being developed and contain sufficient detail to suggest an overall structure for the implementation of the algorithm. Furthermore, to recognize that structure, the student will have to become actively involved with the details of the example. Thus, when compared with pseudocode, which is a static representation of an algorithm, worked examples are, in my opinion, the superior alternative.

Supplements and Software

This text is accompanied by an Instructors Solutions Manual that can be obtained (by instructors only) by contacting either the local Prentice Hall sales rep or `george_lobell@prenhall.com`. There are also 70 plus pages of Answers to Selected Exercises for students, found on the website

`www.pcs.cnu.edu/~bbradie/textbookanswers.html`

To accommodate differing viewpoints on the pseudocode issue, implementations for all of the methods developed in the text can be downloaded via the Internet. Each method is available in MATLAB and C++ formats. Depending upon demand, *Maple*, Mathematica, MathCad and Fortran implementations may be added. Instructions for using the MATLAB functions are contained in the header of the corresponding m-file and are accessible through the standard MATLAB help ⟨function name⟩ facility. Each C++ function is described in the comments at the beginning of the code. The main page for obtaining the software is located at `www.pcs.cnu.edu/~bbradie/textbookcode.html`.

To the Student

The best advice that I can give for working with this textbook is to be an active reader. This means that each time you come to a worked example, you should verify the results of all calculations on your own and attempt to fill in all of the missing details. A similar procedure should be employed for each proof that you read. Working in this fashion will not only hone your general mechanical and analytical skills, but will also significantly improve your understanding of how and why each numerical method works, and will stimulate the process by which you assimilate new knowledge and make it your own. The most common stumbling block encountered by numerical analysis students is difficulty in translating the natural language description of an algorithm into working computer code. Here is a scheme that you may find helpful in overcoming this problem. Start by identifying the inputs. The inputs should include every item that must be known for the code to perform its intended task. Don't forget values that are needed to control the termination of an iterative process. Next, identify the outputs, which are, of course, the values which the code is supposed to compute. Once the inputs and the outputs have been clearly identified, focus on the construction of a logical and well-defined sequence of steps that will produce the outputs from the inputs. The worked examples should prove extremely useful at this point. Finally, convert each step into the appropriate set of assignment statements, conditional/branching statements, loop structures and function calls. As with any new skill, the more you practice, the better you will become.

Acknowledgments

At the top of the list of those to whom I owe a debt of gratitude are George Lobell, Jennifer Urban, and Debbie Ryan at Prentice Hall. I doubt that working with a novice author is ever easy, but they led me through the process to publication with unbelievable patience and kindness. Next comes George Webb, the Dean of the College of Science and Technology at Christopher Newport University, who was a constant source of encouragement and enthusiasm. The reviewers of earlier drafts supplied numerous important comments/suggestions and criticisms/corrections to make what was a good book a much better one. They are: Herman Gollwitzer, Drexel University; Jan S. Hesthaven, Brown University; Carl Gardner, Arizona State University; Scott R. Fulton, Clarkson University; Charles Odion, Prairie View A&M University; Mark S. Gockenbach, Michigan Technological University; Jeffrey S. Scroggs, North Carolina State University; Stephen E. Stuckwisch, Auburn University; Chi-Ming Tang, SUNY Geneseo; Alejandro Engel, Rochester Institute of Technology; Mark Arnold, University of Arkansas; Kuiyuan Li, University of West Florida; Richard A. Zalik, Auburn University; and Rakhim Aitbayev, New Mexico Institute of Mining and Technology. The students who served as guinea pigs for the rough drafts—Brad Abbott, Marchele Bachini, Neal Cheney, Chandon Davis, Brent Goodwin, James Laverty, Peter Parker, Nathan Phillips, Sonny Price, Michelle Reaves, and Kenneth Surles-Law—also made many valuable suggestions. Peter Parker and Michelle Reaves deserve special recognition for detecting more typographical errors than I would like to admit. Finally, I wish to thank Vernon Mason, Melissa Waugh and Debbie Decker, whose workouts helped me to relieve a tremendous amount of stress and to break through several rough bouts of writers block.

If you find any errors, please email me as soon as possible. I will make corrections immediately and post on the website errata.

Brian Bradie
bbradie@pcs.cnu.edu

CHAPTER 1

Getting Started

AN OVERVIEW

The diagram shown below provides a greatly oversimplified view of applied mathematics. The starting point is almost always some real-world problem or real-world phenomena that needs to be studied. The axioms and postulates of the appropriate discipline(s)—be they from the physical, natural, or social sciences—are then used to develop a set of assumptions and a set of equations, known as a mathematical model, which will be used for subsequent analysis. The type of equations that arise can range from simple algebraic equations to extremely complicated coupled systems of nonlinear partial differential equations. Once the model has been set, the next step is to solve the equations and interpret the results in the context of the original problem. If the predictions of the model are in agreement with experimental data, the model is accepted and can be used to make predictions regarding situations for which experimental data is unavailable. On the other hand, if the model fails to accurately reflect some desired aspect of the dynamical behavior of the system, it is necessary to return to the model building phase and reexamine the validity of the model's basic assumptions.

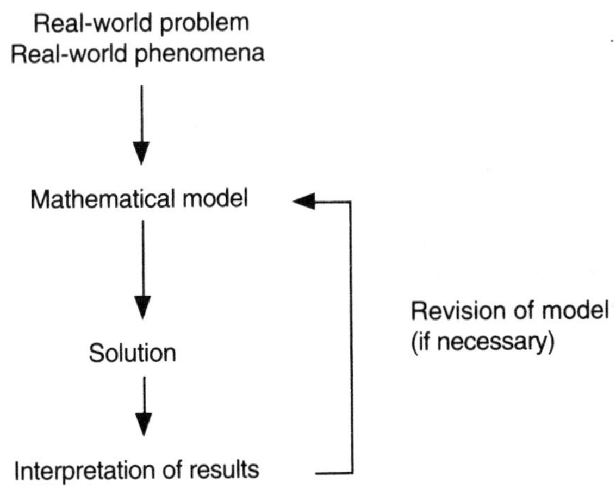

During the solution phase, ideally, an analytical solution is obtained. Unfortunately, an analytical solution is generally available for the simplest cases only. The vast majority of situations require the use of approximate solution techniques.

The objective of this text is to develop methods for determining approximate solutions for several classes of mathematical problems that commonly arise during the modeling of real-world phenomena. These will include such tasks as locating the roots of a function, determining the value of a definite integral, finding the solution of a two-point boundary value problem, and so on.

When dealing with approximation methods, there is an essential separation into what could be referred to as the engineering side of the matter and the mathematical side. There are, of course, the issues of which methods can be applied to which problems and what is the best way to implement a particular method (the engineering side); however, there are also the theoretical issues of how the methods work, how well the methods work and under what circumstances the methods can be expected to work (the mathematical side). This book will routinely address both sets of issues.

An Example of the Modeling Process

As an example, consider the motion of a simple pendulum. The rigid rod that forms the arm of the pendulum has length L and is assumed to be of negligible mass. If it is further assumed that the pendulum will undergo small amplitude oscillations and that energy losses due to air resistance will be negligible, then the second-order differential equation

$$\ddot{\theta} + \omega^2\theta = 0$$

provides a reasonable model. Here θ denotes the angle made by the pendulum arm with the vertical position, dots denote differentiation with respect to time, $\omega = \sqrt{L/g}$ is the natural frequency of the oscillations, and g is the acceleration due to gravity. Since the model equation has constant coefficients, an analytical solution is possible in this case. If it turns out that the amplitude of oscillations is not small and that air resistance cannot be neglected, a more appropriate model would be

$$\ddot{\theta} + b\dot{\theta} + \omega^2 \sin\theta = 0,$$

where b is a drag coefficient. The presence of the sine term makes this a nonlinear equation for which an analytical solution is no longer possible. Approximation techniques will have to be used to carry the analysis further.

The remainder of this section presents a variety of different problems that require the use of some numerical approximation technique. In each case, reference is made to the type(s) of techniques that will be needed and to the chapter in which those techniques will be presented.

Solving a Crime

Commissioner Gordon has been found dead in his office. At 8:00 PM, the county coroner determined the core temperature of the corpse to be 90° F. One hour later, the core temperature had dropped to 85° F. Maintenance reported that the building's air conditioning unit broke down at 4:00 PM. The temperature in the commissioner's office was 68° F at that time. The computerized climate control system recorded that the office temperature rose at a rate of 1° F per hour after the air conditioning stopped working.

Captain Furillo believes that the infamous Doc B killed the commissioner. Doc B, however, claims that he has an alibi. Lois Lane was interviewing him at the Daily Planet Building, just across the street from the commissioner's office. The receptionist at the Daily Planet Building checked Doc B into the building at 6:35 PM, and the interview tapes confirm that Doc B was occupied from 6:40 PM until 7:15 PM. Could Doc B have killed the commissioner?

To answer this question, we need to determine the time of death from the information we have at hand. We will assume the core temperature of the corpse was 98.6° F at the time of death and began decreasing immediately following death. We will further assume that the decrease in core temperature proceeded according to Newton's Law of Cooling. This principle states that the temperature of an object will change at a rate proportional to the difference between the temperature of the object and that of its surroundings.

To explicitly formulate our model, let $T(t)$ denote the core temperature of the corpse as a function of time, with time measured in hours. Take $t = 0$ to correspond to 8:00 PM. Using this coordinate system, we know

$$T(0) = 90 \quad \text{and} \quad T(1) = 85. \tag{1}$$

Furthermore, the office temperature is given by $T_{\text{office}} = 72 + t$. Applying Newton's Law of Cooling, we obtain

$$\frac{dT}{dt} = -k(T - 72 - t), \tag{2}$$

where k is a positive constant of proportionality. To complete our analysis, we must first determine the solution of (2) that satisfies the conditions stated in (1). Then, using this solution, we must determine the time when the core temperature of the corpse was 98.6° F.

Fortunately, (2) is a linear, first-order differential equation that may be solved by the method of integrating factors. The solution obtained in this manner is

$$T(t) = \left(72 + t - \frac{1}{k}\right) + ce^{kt}, \tag{3}$$

where c is a constant of integration. If we substitute $t = 0$ and use the fact that $T(0) = 90$, we find

$$c = 18 + \frac{1}{k}. \tag{4}$$

To determine the value of k, substitute (4) and $t = 1$ into (3) and use $T(1) = 85$ to obtain

$$73 - \frac{1}{k} + \left(18 + \frac{1}{k}\right) e^{-k} = 85. \tag{5}$$

Equation (5) cannot be solved explicitly for k, so we will have to settle for an approximate solution. Once a value for k has been obtained, the time of death, t_d, is determined by the equation

$$72 + t_d - \frac{1}{k} + \left(18 + \frac{1}{k}\right) e^{-kt_d} = 98.6. \tag{6}$$

Note that this equation also cannot be solved explicitly for t_d.

Equations (5) and (6) are examples of a class of problems known as rootfinding problems. In Chapter 2, we will develop several techniques for computing approximate solutions to rootfinding problems.

Steady-State Distribution of the British Workforce

In a study of class mobility in modern Great Britain, Goldthorpe and Llewellyn [1] classified workers into seven occupational classes:

1. higher-grade professionals/administrators;
2. lower-grade professionals/administrators and higher-grade technicians;
3. routine nonmanual employees;
4. small proprietors;
5. lower-grade technicians and supervisors of manual laborers;
6. skilled manual laborers; and
7. semiskilled and unskilled manual laborers.

Using data from 10,309 men aged 20–64 living in England and Wales, the proportion of children, p_{ij}, born to parents in occupational class j who eventually became members of occupational class i was estimated. The collection of values obtained by Goldthorpe and Llewellyn is summarized in the matrix

$$P = [p_{ij}] = \begin{bmatrix} 0.452 & 0.291 & 0.184 & 0.126 & 0.142 & 0.078 & 0.065 \\ 0.189 & 0.231 & 0.157 & 0.114 & 0.136 & 0.088 & 0.078 \\ 0.115 & 0.119 & 0.128 & 0.080 & 0.101 & 0.083 & 0.082 \\ 0.077 & 0.070 & 0.078 & 0.244 & 0.077 & 0.065 & 0.066 \\ 0.048 & 0.096 & 0.128 & 0.087 & 0.157 & 0.123 & 0.125 \\ 0.054 & 0.106 & 0.156 & 0.144 & 0.212 & 0.304 & 0.235 \\ 0.065 & 0.087 & 0.169 & 0.205 & 0.175 & 0.259 & 0.349 \end{bmatrix}.$$

In the vocabulary of discrete dynamical systems, each p_{ij} is called a transition probability, and the matrix P is called a transition matrix. Note that the sum of the entries in each column of P is equal to 1.

Suppose the vector $\boldsymbol{\pi}_0$ denotes an initial distribution of workers across the indicated occupational classes. Given the definition of the entries in the transition matrix P, it follows that the distribution of workers in the next generation is given by the vector

$$\boldsymbol{\pi}_1 = P\boldsymbol{\pi}_0.$$

The distribution of workers in subsequent generations is then given by

$$\begin{aligned}\pi_2 &= P\pi_1 = P^2\pi_0,\\ \pi_3 &= P\pi_2 = P^3\pi_0,\\ \pi_4 &= P\pi_3 = P^4\pi_0,\end{aligned}$$

and so on. If the transition matrix P remains valid over time, will the vectors $\pi_n = P^n\pi_0$ approach a fixed, steady-state distribution vector π for any initial vector π_0?

To answer this question, we need to identify another property of the matrix P. Observe that every element in P is strictly greater than zero. Any transition matrix for which some integer power of the matrix has all positive entries is called regular; hence, P is a regular transition matrix. Because P is a regular transition matrix, it can be shown that as $n \to \infty$ and for any initial distribution π_0, the sequence of vectors $P^n\pi_0$ approaches the fixed vector π that satisfies

$$P\pi = \pi \tag{7}$$

subject to the constraint that the sum of the entries in π is equal to 1 (see Anton and Rorres [2]).

A vector that satisfies equation (7) is called an eigenvector of the matrix P. Since $P\pi$ is specifically equal to $1\cdot\pi$, we say that π is an eigenvector of P associated with the eigenvalue 1. We will develop techniques for approximating eigenvalues and eigenvectors of matrices in Chapter 4.

Estimating a Coefficient of Friction

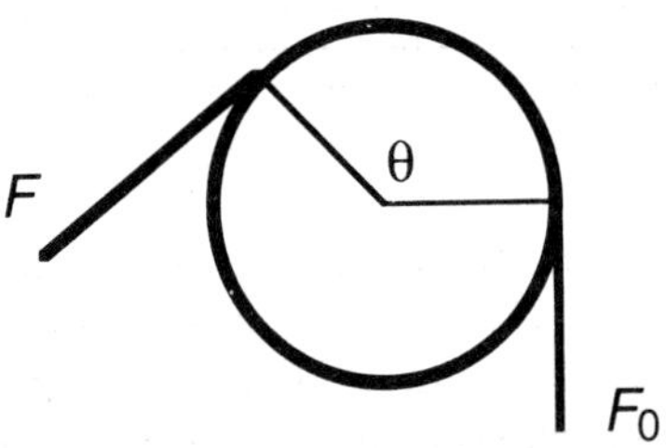

When a flexible rope is wrapped around a rough cylinder, a small "restraining" force of magnitude F_0 applied at one end of the rope can withstand a force of magnitude $F > F_0$ applied at the other end. The magnitude of the force that can be withstood depends on the angle θ through which the rope is wrapped around the cylinder (see figure above) and on the coefficient of friction, μ, between the rope and the cylinder. The coefficient of friction is defined by the relation

$$\frac{dF}{d\theta} = \mu F.$$

As part of a physics lab, students need to estimate the coefficient of friction between a given rope and cylinder. For a restraining force of $F_0 = 5$ lb, they have obtained the following data for F as a function of θ.

θ	0	$\pi/2$	π	$3\pi/2$	2π	$5\pi/2$	3π	$7\pi/2$	4π	$9\pi/2$	5π
F(θ)	5.00	7.83	12.27	19.22	30.10	47.15	73.86	115.70	181.24	283.90	444.71

To complete the lab, the students must be able to take this data and estimate $dF/d\theta$. This problem requires techniques for numerical differentiation, which are considered in Chapter 6.

Projection Printing

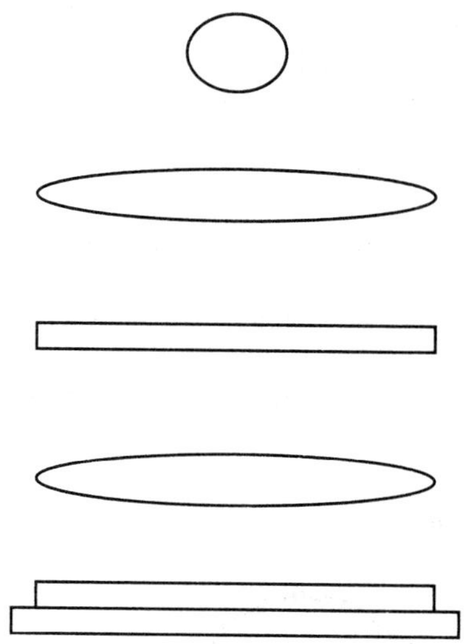

Projection printing is one of the major processes involved in the fabrication of semiconductor devices. A simple schematic is shown in the figure above. The light source is used to project a pattern from the mask onto the surface of the photoresist film. The pattern may, for example, indicate the locations where wires are to be placed or where metal contacts are to be acid etched.

The photoresist film contains a chemical called a photoactive compound, or PAC. The absorption of light by the PAC causes a reaction in which the PAC breaks down. At the end of the exposure process, the contour lines of PAC concentration form a latent image in the resist film, much like a photographic negative.

Let $I(z,t)$ denote the light intensity and $M(z,t)$ denote the normalized concentration of PAC within the photoresist film during the exposure process. Here, z measures depth into the film, with $z = 0$ corresponding to the air-film interface, and t denotes time. Our objective is to determine $M(z, t_{\text{exp}})$, where t_{exp} is the duration of the exposure.

Assuming that the substrate absorbs the incident light and effectively eliminates reflection, then I and M satisfy Dill's equations [3], which are given by

$$\frac{\partial I}{\partial z} = -I(AM + B) \tag{8}$$

$$\frac{\partial M}{\partial t} = -IMC. \tag{9}$$

The constants A, B and C are material properties of the photoresist film whose values can be measured experimentally. The auxiliary conditions associated with (8) and (9) are

$$M(z, 0) = 1 \tag{10}$$

and

$$I(0, t) = I_0. \tag{11}$$

Following Babu and Barouch [4], we can reduce the system of partial differential equations (8) and (9) to an initial value problem involving an ordinary differential equation for $M(z, t_{\text{exp}})$.

If we divide (9) by MC, substitute the resulting expression for $-I$ into (8) and then rearrange terms, we obtain

$$C\frac{\partial I}{\partial z} = \frac{AM + B}{M}\frac{\partial M}{\partial t} = \frac{\partial}{\partial t}\left[AM + B \ln M\right]. \tag{12}$$

Alternatively, if we divide (9) by $-M$ and differentiate with respect to z, we find

$$C\frac{\partial I}{\partial z} = \frac{\partial}{\partial t}\left[-\frac{1}{M}\frac{\partial M}{\partial z}\right]. \tag{13}$$

Combining (12) and (13) then yields

$$\frac{\partial}{\partial t}\left[AM + B \ln M + \frac{1}{M}\frac{\partial M}{\partial z}\right] = 0,$$

which, upon integration with respect to t, gives

$$AM + B \ln M + \frac{1}{M}\frac{\partial M}{\partial z} = f(z) \tag{14}$$

for some arbitrary function f. Substituting $t = 0$ into (14) and applying (10), we find $f(z) = A$. Using this expression in (14) and evaluating at $t = t_{\text{exp}}$, it follows that

$$\frac{dM(z, t_{\text{exp}})}{dz} = M(z, t_{\text{exp}})\left[A\left(1 - M(z, t_{\text{exp}})\right) - B \ln M(z, t_{\text{exp}})\right]. \tag{15}$$

To obtain the initial condition associated with (15), first substitute $z = 0$ into (9). Using (11), this yields

$$\frac{\partial M(0, t)}{\partial t} = -I_0 CM(0, t).$$

The solution of this equation, taking into account (10), is $M(0,t) = \exp(-I_0Ct)$; therefore,

$$M(0, t_{\text{exp}}) = \exp(-I_0Ct_{\text{exp}}). \tag{16}$$

Equations (15) and (16) form the desired initial value problem. Techniques for approximating the solution of initial value problems are developed in Chapter 7.

Rise in the Water Table due to the Spring Thaw

An aquifer is a geological formation through which groundwater flows easily to supply a well or a spring. When studying an aquifer the primary quantity of interest is the water table. The water table is defined to be the location where the relative pressure (i.e., the pressure due to the water only) is zero and is generally a function of both time and position. The diagram following shows an idealized one-dimensional aquifer and its water table. Note the presence of wells at which the height of the water table can be monitored.

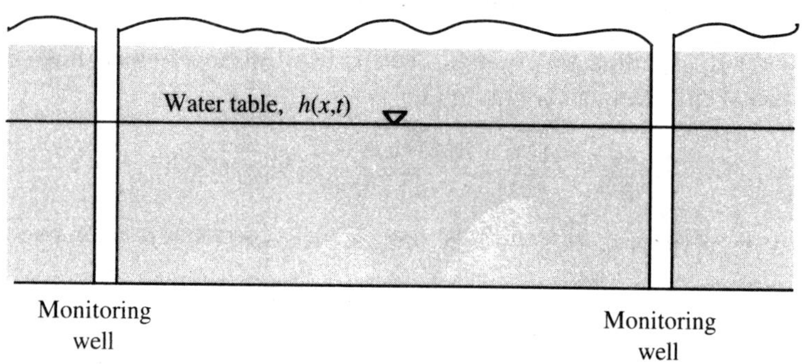

Suppose the monitoring wells in the following aquifer are separated by a distance of 800 meters. During the spring thaw, the water level in each well is recorded on a regular basis. Using this information, we want to determine the behavior of the water table over time. Let $h(x,t)$ denote the water table at a distance of x meters from the left monitoring well and at time t. Take $t = 0$ to correspond to the start of the spring thaw. (See figure at top of next page.)

To model the change in h, consider the representative section of the aquifer shown above. The section has a length of one unit in the direction perpendicular to the page. Assuming that the density (mass per unit volume) of the water is constant, conservation of mass requires that the change in the volume of water within this representative section be equal to the net volume of water that flows into the section. The change in the volume of water within the section is given by

$$S\Delta h\Delta x \cdot 1, \tag{17}$$

where S is the hydraulic storativity, which measures the ability of the aquifer to

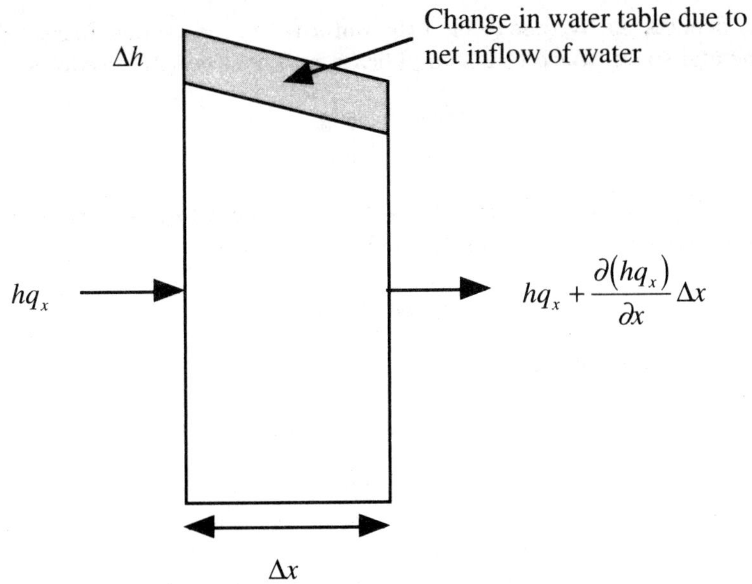

absorb and release water. The net volume of water that flows into the section is

$$hq_x \Delta t \cdot 1 - \left(hq_x + \frac{\partial (hq_x)}{\partial x} \Delta x \right) \Delta t \cdot 1 = -\frac{\partial (hq_x)}{\partial x} \Delta x \Delta t \cdot 1. \tag{18}$$

Here, q_x is the specific flow rate (volumetric flow rate per unit area) of water in the x direction. Equating (17) and (18), dividing by the product $\Delta x \Delta t$ and taking the limit as $\Delta t \to 0$, we obtain

$$S\frac{\partial h}{\partial t} = -\frac{\partial (hq_x)}{\partial x}. \tag{19}$$

The specific flow rate is related to the slope of the water table by Darcy's law, which states

$$q_x = -K\frac{\partial h}{\partial x}, \tag{20}$$

where K is the hydraulic conductivity of the aquifer. Substituting (20) into (19) yields

$$S\frac{\partial h}{\partial t} = \frac{\partial}{\partial x}\left(hK\frac{\partial h}{\partial x} \right). \tag{21}$$

Next, we assume that K is constant. Further, we make what is known as the Boussinesq approximation, which involves replacing the expression

$$hK\frac{\partial h}{\partial x} \quad \text{by} \quad h_{\text{av}}K\frac{\partial h}{\partial x}.$$

Here, h_{av} denotes the average level of the water table, the average being taken over both time and space. With these simplifications, equation (21) becomes

$$\frac{\partial h}{\partial t} = a\frac{\partial^2 h}{\partial x^2}, \tag{22}$$

where $a = \frac{Kh_{\text{av}}}{S}$ is the hydraulic diffusivity. To completely determine the water table, we must supplement (22) with the initial condition

$$h(x,0) = h_0(x) \tag{23}$$

and the boundary conditions

$$h(0,t) = h_L(t) \quad \text{and} \quad h(800,t) = h_R(t). \tag{24}$$

The function $h_0(x)$ specifies the water table at the start of the spring thaw, and the functions $h_L(t)$ and $h_R(t)$ are obtained from the measurements made at the left and right monitoring wells, respectively.

Equation (22) is an example of a parabolic partial differential equation, and the combination of equations (22), (23) and (24) is called an initial boundary value problem. We will develop techniques for approximating the solution of an initial boundary value problem involving a parabolic partial differential equation in Chapter 10.

References

1. J. H. Goldthorpe, *Social Mobility and Class Structure in Modern Britain*, Clarendon Press, Oxford, 1987.
2. H. Anton and C. Rorres, *Elementary Linear Algebra with Applications*, John Wiley and Sons, New York, 1987.
3. F. H. Dill, W. P. Hornberger, P. S. Hauge and J. M. Shaw, "Characterization of Positive Photoresist," *IEEE Trans. Electron Devices*, **ED-22**, 445–452, 1975.
4. S. V. Babu and E. Barouch, "Exact Solution of Dill's Model Equations for Positive Photoresist Kinetics," *IEEE Electron Device Lett.*, **EDL-7**, 252–253, 1986.

1.1 ALGORITHMS

At the heart of numerical analysis is the concept of an *algorithm.*

Definition. An ALGORITHM is a precisely defined sequence of steps for performing a specified task.

Throughout this text, we will design and implement (either by hand or with the aid of a calculator or computer) algorithms for computing approximate solutions to certain classes of mathematical problems. We will also critically examine the performance of these algorithms. Our objectives will include, but not be limited to, determining the conditions under which an algorithm is expected to work, how accurately the solution produced by an algorithm approximates the exact solution of the underlying problem, and how various parameter values affect algorithm performance.

Though the algorithms we develop will vary in terms of the number of steps involved, overall complexity, and general objective, they will all consist of three basic components. The first is a list of the input parameters. These are the quantities that must be supplied for the algorithm to be able to carry out its designated task. The second component then states specifically what operations need to be performed and in what order they need to be performed. The final component identifies the output of the algorithm, the information that is reported back to the user.

An Example from Statistics

A common task in data analysis is the calculation of the mean, $\bar{x}$, and the standard deviation, s, of a collection of data. The mean is the most commonly used value to locate the "middle" of a data set, and the standard deviation measures the variation of the data about the mean. Let n denote the number of elements in the data set, and let x_i denote an individual element from the data set, where i ranges from 1 through n. The formulas for the mean and standard deviation, which may be found in any elementary statistics textbook, are

$$\bar{x} = \frac{\sum_{i=1}^{n} x_i}{n} \quad \text{and} \quad s = \sqrt{\frac{n\sum_{i=1}^{n} x_i^2 - \left(\sum_{i=1}^{n} x_i\right)^2}{n(n-1)}}. \tag{1}$$

To construct an algorithm to compute the mean and standard deviation, we first identify the inputs. Examining the formulas in (1), it is clear that the only values that must be supplied are the x_i and n. Next, we focus on the steps that need to be performed. The key expressions in (1) are the two summations

$$\sum_{i=1}^{n} x_i \quad \text{and} \quad \sum_{i=1}^{n} x_i^2.$$

Once these values are known, the remaining calculations can be performed. Finally, we note that the algorithm must report both the mean and the standard deviation. Bringing all of this information together, we obtain the following algorithm.

GIVEN: an array of real numbers, x_i
the number of elements in the array, n

STEP 1: initialize *xsum* and *x2sum* to 0
STEP 2: for i from 1 to n
add x_i to *xsum*
add x_i^2 to *x2sum*
end
STEP 3: calculate *xbar* = *xsum* / n

STEP 4:	calculate s = sqrt($(n(x2sum)$ - $(xsum)^2)$ / $(n(n-1))$)
OUTPUT:	$xbar$ and s

Observe that in STEP 1 the variables that will be used to accumulate the two important summations are initialized to zero. It is always good practice to initialize any variable before it is used later in the algorithm.

EXAMPLE 1.1 The Statistics Algorithm in Action

Let's apply the above algorithm to calculate the mean and standard deviation of the data set

$$1, \quad 3, \quad 5, \quad 7, \quad 9.$$

The inputs are thus

$$x_1 = 1, \quad x_2 = 3, \quad x_3 = 5, \quad x_4 = 7, \quad x_5 = 9, \quad \text{and} \quad n = 5.$$

Working sequentially through the steps of the algorithm then produces the following results.

STEP 1:	$xsum = 0$; $x2sum = 0$		
STEP 2:	$i = 1:$	$xsum = 0 + 1 = 1$;	$x2sum = 0 + 1^2 = 1$
	$i = 2:$	$xsum = 1 + 3 = 4$;	$x2sum = 1 + 3^2 = 10$
	$i = 3:$	$xsum = 4 + 5 = 9$;	$x2sum = 10 + 5^2 = 35$
	$i = 4:$	$xsum = 9 + 7 = 16$;	$x2sum = 35 + 7^2 = 84$
	$i = 5:$	$xsum = 16 + 9 = 25$;	$x2sum = 84 + 9^2 = 165$
STEP 3:	$xbar = 25 / 5 = 5$		
STEP 4:	$s = \sqrt{(5(165) - (25)^2)/(5 \cdot 4)} = 3.162$		
OUTPUT:	$xbar = 5$ and $s = 3.162$		

Thus, for the data set consisting of the five numbers 1, 3, 5, 7, and 9, the mean is $\bar{x} = 5$ and the standard deviation is $s = 3.162$.

An Algorithm for Approximating a Definite Integral

As a second example, consider the trapezoidal rule. This is one of the most basic schemes for approximating the value of the definite integral

$$\int_a^b f(x)\,dx$$

for an arbitrary integrand, f. For a fixed positive integer n, introduce the partition

$$a = x_0 < x_1 < x_2 < \cdots < x_{n-1} < x_n = b$$

over the integration interval $[a, b]$, where $x_i = a + ih$ for each i and $h = (b - a)/n$. Using this partition, the trapezoidal rule approximation is then given by

$$\int_a^b f(x)\,dx \approx \frac{h}{2}\left[f(a) + 2\sum_{i=1}^{n-1} f(x_i) + f(b)\right]. \tag{2}$$

Clearly, the input to an algorithm that computes the trapezoidal rule approximation must include the integrand, f, and the limits of integration, a and b. Without these items, the underlying mathematical problem is not even defined. The algorithm also requires the positive integer n, which defines the partition. In terms of calculations, note that (2) contains a summation, so we will have to initialize a variable to accumulate that value. We also need to compute h so we can later compute each x_i. To apply the formula in (2), we work from inside the square brackets outward. First, calculate the summation. Next, multiply by two, and add in the function values $f(a)$ and $f(b)$. Finally, multiply by $h/2$ and output the result.

Here is the final algorithm.

GIVEN:	the limits of integration a and b the integrand f the number of subintervals n
STEP 1:	compute $h = (b - a)/n$; and initialize *sum* to 0
STEP 2:	for i from 1 to $n - 1$ add $f(a + ih)$ to *sum*
STEP 3:	multiply *sum* by 2
STEP 4:	add $f(a)$ and $f(b)$ to *sum*
OUTPUT:	$(h/2)$*sum*

EXAMPLE 1.2 Trapezoidal Rule in Action

We will now demonstrate the trapezoidal rule algorithm by approximating the value of the definite integral

$$\int_1^2 \frac{dx}{x}.$$

Matching this specific problem to the general pattern, $\int_a^b f(x)\,dx$, we see that $a = 1$, $b = 2$ and $f(x) = 1/x$. For the last input parameter, let's take $n = 4$.

Working sequentially through the steps yields

STEP 1:	$h = (2 - 1)/4 = 1/4;\quad sum = 0$
STEP 2:	$i = 1:\quad sum = 0 + 1 / (1 + 1/4) = 4 / 5$ $i = 2:\quad sum = 4/5 + 1 / (1 + 1/2) = 22 / 15$ $i = 3:\quad sum = 22/15 + 1 / (1 + 3/4) = 214 / 105$
STEP 3:	$sum = 2 (214 / 105) = 428 / 105$
STEP 4:	$sum = 428/105 + 1 / 1 + 1 / 2 = 1171 / 210$
OUTPUT:	$(1 / 8) (1171 / 210) = 1171 / 1680 = 0.697023809$

Therefore,

$$\int_1^2 \frac{dx}{x} \approx 0.697023809.$$

The exact value of the integral is, of course, $\ln 2$, so the absolute error in the trapezoidal rule approximation is $|\ln 2 - 0.697023809| \approx 3.877 \times 10^{-3}$.

As a prelude of things to come, let's investigate the effect of the input parameter n on the performance of the trapezoidal rule. For several different values of n, Table 1.1 displays the trapezoidal rule approximation, T_n, to $\int_1^2 \frac{1}{x}\,dx$ and the corresponding absolute error, $|e_n|$, where

$$e_n = T_n - \int_1^2 \frac{1}{x}\,dx.$$

Note that on successive rows of the table, the value of n doubles. Clearly, $|e_n|$ is a decreasing function of n. Can we now use the data to determine a specific functional form for the relationship between $|e_n|$ and n?

n	Approximation, T_n	Absolute Error, $\lvert e_n\rvert$	Error Ratio, $\lvert e_n\rvert/\lvert e_{n/2}\rvert$
2	0.708333333	0.015186152	
4	0.697023810	0.003876628	0.255273883
8	0.694121850	0.000974670	0.251422112
16	0.693391202	0.000244022	0.250363712
32	0.693208208	0.000061028	0.250092204

TABLE 1.1: Trapezoidal Rule Approximation to $\int_1^2 \frac{1}{x}\,dx$ as a Function of n

This is where the fourth column of Table 1.1 comes into play. This column lists the ratio $|e_n|/|e_{n/2}|$. Observe that each time n is doubled, the absolute error is reduced by a factor of roughly one-quarter. Thus, the numerical evidence from this problem suggests that, for the trapezoidal rule,

$$|e_n| \approx c/n^2,$$

where c is independent of n. When we study the trapezoidal rule in more detail in Chapter 6, we will show that this is, in fact, the correct functional relationship.

Approximating a Square Root

Many of the algorithms developed in later chapters will work in much the same way as the trapezoidal rule. They will compute a single approximation to the solution of a particular mathematical problem. In contrast, other algorithms will generate a sequence of approximations which hopefully converge toward the desired solution. Algorithms of this type are called *iterative*.

When constructing an iterative algorithm, there are a few important details to keep in mind. First, every iterative algorithm must contain a *stopping condition.* A stopping condition is just a test that is used to decide when to terminate the iterative process and to accept the most recently computed term in the sequence as a final approximation. Second, to accommodate those times when the sequence either does not converge or converges very slowly, it is advisable to impose an upper limit on the number of iterations that will be allowed. Third, it is generally not necessary to save every term from the sequence. Typically, an algorithm can perform its task with knowledge of only a few of the terms.

To illustrate these points, consider the following. Let a be a nonnegative real number. For any positive real number x_0, the sequence generated by the rule

$$x_{n+1} = \frac{1}{2}\left(x_n + \frac{a}{x_n}\right) \tag{3}$$

for $n = 0, 1, 2, \ldots$ converges to $\sqrt{a}$. In the next section, we will investigate this sequence in more detail and establish that the quantity $|x_{n+1} - x_n|$ provides an estimate for the difference $|x_{n+1} - \sqrt{a}|$. An appropriate stopping condition would therefore be to terminate iteration as soon as $|x_{n+1} - x_n|$ falls below an input parameter known as the convergence tolerance, which we shall denote by ϵ. To prevent the algorithm from getting caught in an infinite loop, we limit the number of iterations to *Nmax*, which is another input parameter. Finally, note that during any iteration, we have to know the value of x_n in order to compute x_{n+1}. Once x_{n+1} has been calculated, we still need to have x_n available so that we may determine whether the stopping condition has been met. If it is determined that another iteration needs to be performed, however, we no longer need x_n. Accordingly, only two terms from the sequence need to be saved at any given time.

Bringing all of this information together, we arrive at the algorithm

GIVEN:	nonnegative real number a starting approximation x_0 convergence parameter ϵ maximum number of iterations *Nmax*
STEP 1:	for *iter* from 1 to *Nmax*
STEP 2:	compute $x_1 = (x_0 + a/x_0)/2$
STEP 3:	if $|x_1 - x_0| < \epsilon$, OUTPUT x_1
STEP 4:	copy the value of x_1 to x_0
	end
OUTPUT:	"maximum number of iterations has been exceeded"

The variable x_0 holds the value of x_n, while x_1 holds the value of x_{n+1}. STEP 4 is required so that x_0 contains the correct value for the next iteration.

EXAMPLE 1.3 Square Root Algorithm in Action

Suppose we wish to approximate the value of $\sqrt{2}$. Since the algorithm we just constructed generates an approximation to $\sqrt{a}$, this fixes $a = 2$. For the remaining input parameters, let's use $x_0 = 2$, $\epsilon = 0.005$ and $Nmax = 10$. With the input parameters set, the first iteration of the algorithm proceeds as follows:

STEP 1:	$iter = 1$
STEP 2:	$x_1 = (2 + 2/2)/2 = 3/2$
STEP 3:	$\lvert x_1 - x_0 \rvert = 1/2 > 0.005$, so
STEP 4:	$x_0 = 3/2$.

In the second iteration, we find

STEP 1:	$iter = 2$
STEP 2:	$x_1 = (3/2 + 4/3)/2 = 17/12$
STEP 3:	$\lvert x_1 - x_0 \rvert = 1/12 > 0.005$, so
STEP 4:	$x_0 = 17/12$.

The third iteration then yields

STEP 1:	$iter = 3$
STEP 2:	$x_1 = (17/12 + 24/17)/2 = 577/408$
STEP 3:	$\lvert x_1 - x_0 \rvert = 1/408 \approx 0.00245 < 0.005$, so
	OUTPUT $x_1 = 577/408 \approx 1.414215686$.

Thus $\sqrt{2} \approx 1.414215686$, which is in error by roughly 2.124×10^{-6}.

EXERCISES

1. Use the statistics algorithm from the text to compute the mean, $\bar{x}$, and the standard deviation, s, of the data set: $-5, -3, 2, -2, 1$.

2. With $n = 4$, use the trapezoidal rule algorithm from the text to approximate the value of the definite integral

$$\int_0^1 \frac{1}{1+x^2}\,dx.$$

3. Use the square root algorithm from the text to approximate $\sqrt{5}$. Take $x_0 = 5$, $\epsilon = 5 \times 10^{-4}$ and $Nmax = 10$.

4. A different scheme for approximating the square root of a positive real number a is based on the recursive formula

$$x_{n+1} = \frac{x_n^3 + 3x_n a}{3x_n^2 + a}.$$

   (a) Construct an algorithm for approximating the square root of a given positive real number a using this formula.

(b) Test your algorithm using $a = 2$ and $x_0 = 2$. Allow a maximum of 10 iterations and use a convergence tolerance of $\epsilon = 5 \times 10^{-5}$. Compare the performance of this algorithm with the one presented in the text.

5. Let A be an $n \times m$ matrix and B be an $m \times p$ matrix. The $n \times p$ matrix $C = AB$ has elements defined by

$$c_{ik} = \sum_{j=1}^{m} a_{ij} b_{jk}$$

for each $i = 1, 2, 3, \ldots, n$ and each $k = 1, 2, 3, \ldots, p$. Construct an algorithm to compute the product of two matrices.

6. Consider the computation of the following sum,

$$\sum_{i=1}^{n} \sum_{j=1}^{n} a_i b_j,$$

where the a_i and b_j are real numbers.

(a) How many multiplications and how many additions are required to compute the sum? Each answer should be a function of n.

(b) Modify the summation to an equivalent form that reduces the number of operations needed. How many multiplications and how many additions are required to compute the sum in its revised form?

7. Let a be a nonzero real number. For any x_0 satisfying $0 < x_0 < 2/a$, the recursive sequence defined by

$$x_{n+1} = x_n(2 - ax_n)$$

converges to $1/a$.

(a) Construct an algorithm for approximating the reciprocal of a given nonzero real number a using this formula.

(b) Test your algorithm using $a = 37$ and $x_0 = 0.01$. Allow a maximum of 10 iterations and use a convergence tolerance of $\epsilon = 5 \times 10^{-4}$.

8. Given two positive integers a and b, the greatest common divisor of a and b is the largest integer which divides both a and b (i.e., the largest integer n for which both a/n and b/n are integers).

(a) Construct an algorithm to compute the greatest common divisor of two positive integers.

(b) How many divisions does your algorithm require?

9. The inner product, or dot product, of two n-vectors $\mathbf{x}$ and $\mathbf{y}$ is given by

$$\mathbf{x} \cdot \mathbf{y} = x_1 y_1 + x_2 y_2 + x_3 y_3 + \cdots + x_n y_n.$$

(a) Construct an algorithm to compute the inner product of two n-vectors.

(b) Apply your algorithm to calculate the inner product of the vectors

$$\mathbf{x} = \begin{bmatrix} -3 & 4 & 1 & 2 \end{bmatrix}^T \quad \text{and} \quad \mathbf{y} = \begin{bmatrix} 1 & -3 & 2 & 5 \end{bmatrix}^T.$$

10. The linear correlation coefficient for n ordered pairs (x_i, y_i) is given by the formula

$$r = \frac{n\sum_{i=1}^{n} x_i y_i - \left(\sum_{i=1}^{n} x_i\right)\left(\sum_{i=1}^{n} y_i\right)}{\sqrt{n\sum_{i=1}^{n} x_i^2 - \left(\sum_{i=1}^{n} x_i\right)^2}\sqrt{n\sum_{i=1}^{n} y_i^2 - \left(\sum_{i=1}^{n} y_i\right)^2}}.$$

(a) Construct an algorithm to compute the linear correlation coefficient for a given set of ordered pairs.

(b) Apply your algorithm to compute the linear correlation coefficient for the following set of ordered pairs:

x_i	3	7	9	2	7	0	3
y_i	−5	10	15	−8	11	−10	−4

11. The midpoint rule approximates the value of a definite integral using the formula

$$\int_a^b f(x)\,dx \approx 2h\sum_{j=1}^{n} f(x_j),$$

where $h = (b-a)/2n$ and $x_j = a + (2j-1)h$.

(a) Construct an algorithm to approximate the value of a definite integral using the midpoint rule.

(b) Apply your algorithm to approximate the value of $\int_1^2 dx/x$. Take $n = 4$. Compare the approximation obtained from the midpoint rule with the approximation obtained from the trapezoidal rule.

12. Consider the following algorithm for the trapezoidal rule:

GIVEN:	the limits of integration a and b
	the integrand f
	the number of subintervals n
STEP 1:	compute $h = (b-a)/n$; and initialize *sum* to 0
STEP 2:	for i from 1 to $n-1$
	add $2f(a+ih)$ to *sum*
STEP 3:	add $f(a)$ and $f(b)$ to *sum*
OUTPUT:	$(h/2)sum$

Compare the number of arithmetic operations required by this algorithm to the number of operations required by the algorithm presented in the text.

13. Rewrite the algorithm for the trapezoidal rule that was presented in the text to reduce both the number of additions and the number of multiplications/divisions by one.

14. Let $P(x) = a_n x^n + a_{n-1}x^{n-1} + \cdots + a_1 x + a_0$ be an nth-degree polynomial with all real coefficients, and let x_0 be a given real number.

(a) Treating integer powers as repeated multiplication, how many multiplications and how many additions are required to evaluate $P(x)$ at $x = x_0$?

(b) Devise an algorithm for computing the value of an nth-degree polynomial that reduces the required number of arithmetic operations. How many multiplications and how many additions are required by your algorithm?

For Exercises 15–18, make use of the fact that when the sum of a convergent alternating series is approximated using the sum of the first n terms, the error in this approximation is smaller than the magnitude of the $(n+1)$st term; that is, if $\sum(-1)^n a_n$ is an alternating series with sum S, then

$$\left| S - \sum_{k=0}^{n-1} (-1)^k a_k \right| < a_n.$$

15. The value of π is given by

$$\pi = 4\sum_{n=0}^{\infty} \frac{(-1)^n}{2n+1} = 4\left(1 - \frac{1}{3} + \frac{1}{5} - \frac{1}{7} + - \cdots\right).$$

(a) Construct an algorithm to approximate the value of π to within a specified tolerance, ϵ.

(b) Test your algorithm with a tolerance value of $\epsilon = 5 \times 10^{-3}$.

16. The value of $1/e$ is given by

$$1/e = \sum_{n=0}^{\infty} \frac{(-1)^n}{n!} = 1 - \frac{1}{1!} + \frac{1}{2!} - \frac{1}{3!} + - \cdots.$$

(a) Construct an algorithm to approximate the value of $1/e$ to within a specified tolerance, ϵ.

(b) Test your algorithm with a tolerance value of $\epsilon = 5 \times 10^{-7}$.

17. The value of $\sin(\pi/10)$ is given by

$$\begin{aligned}\sin\left(\frac{\pi}{10}\right) &= \sum_{n=0}^{\infty} \frac{(-1)^n}{(2n+1)!}\left(\frac{\pi}{10}\right)^{2n+1} \\ &= \frac{\pi}{10} - \frac{1}{3!}\left(\frac{\pi}{10}\right)^3 + \frac{1}{5!}\left(\frac{\pi}{10}\right)^5 - \frac{1}{7!}\left(\frac{\pi}{10}\right)^7 + - \cdots.\end{aligned}$$

(a) Construct an algorithm to approximate the value of $\sin(\pi/10)$ to within a specified tolerance, ϵ.

(b) Test your algorithm with a tolerance value of $\epsilon = 5 \times 10^{-7}$. Note that the exact value of $\sin(\pi/10)$ is $\frac{1}{4}(\sqrt{5} - 1)$.

18. The value of $\cos(\pi/5)$ is given by

$$\cos\left(\frac{\pi}{5}\right) = \sum_{n=0}^{\infty} \frac{(-1)^n}{(2n)!}\left(\frac{\pi}{5}\right)^{2n} = 1 - \frac{1}{2!}\left(\frac{\pi}{5}\right)^2 + \frac{1}{4!}\left(\frac{\pi}{5}\right)^4 - \frac{1}{6!}\left(\frac{\pi}{5}\right)^6 + - \cdots.$$

(a) Construct an algorithm to approximate the value of $\cos(\pi/5)$ to within a specified tolerance, ϵ.

(b) Test your algorithm with a tolerance value of $\epsilon = 5 \times 10^{-7}$. Note that the exact value of $\cos(\pi/5)$ is $\frac{1}{4}(\sqrt{5} + 1)$.

1.2 CONVERGENCE

Many of the algorithms that will be developed in later chapters will be iterative in nature. These algorithms will generate a sequence of approximations that converge toward the desired solution. When several techniques are available for solving a particular problem, we would generally like to choose a technique whose sequence converges as rapidly as possible. To facilitate a comparison between competing methods, we will introduce in this section two quantitative measures of convergence speed.

Rate of Convergence

For completeness, remember that convergence of a sequence is defined as follows.

Definition. The sequence $\{x_n\}$ CONVERGES to the value L provided

$$\lim_{n\to\infty} x_n = L,$$

or, equivalently,

$$\lim_{n\to\infty} |x_n - L| = 0.$$

The value to which the sequence converges, L, is called the *limit* of the sequence. A sequence for which $\lim_{n\to\infty} x_n$ does not exist is said to *diverge.*

The two principal measures of convergence speed are known as rate of convergence and order of convergence. Let's consider rate of convergence first.

Definition. Let $\{p_n\}$ be a sequence that converges to a number p. If there exists a sequence $\{\beta_n\}$ that converges to zero and a positive constant λ, independent of n, such that

$$|p_n - p| \le \lambda |\beta_n|$$

for all sufficiently large values of n, then $\{p_n\}$ is said to converge to p with RATE OF CONVERGENCE $O(\beta_n)$.

The expression $O(\beta_n)$ is read "big-O of β_n" and is referred to as big-O notation. When $\{p_n\}$ converges to p with rate of convergence $O(\beta_n)$, it is common to express this in shorthand by writing $p_n = p + O(\beta_n)$; hence, the big-O term provides a reference for how quickly the error approaches zero.

The sequence $\{\beta_n\}$, which is typically taken to be of the form $1/n^a$ or $1/a^n$ for some positive constant a, serves as a benchmark and allows for ease of comparison between different sequences. For example, a sequence with rate of convergence $O(1/n^2)$ converges more slowly than one with a rate of convergence $O(1/n^{10})$, which in turn converges more slowly than a sequence with rate of convergence $O(1/2^n)$.

EXAMPLE 1.4 Comparing Rates of Convergence

Consider the sequences

$$\left\{\frac{n+3}{n+7}\right\} \quad \text{and}\textit{and} \quad \left\{\frac{2^n+3}{2^n+7}\right\}.$$

n	$(n+3)/(n+7)$	$(2^n+3)/(2^n+7)$
1	0.5000000000	0.5555555556
2	0.5555555556	0.6363636364
3	0.6000000000	0.7333333333
4	0.6363636364	0.8260869565
5	0.6666666667	0.8974358974
6	0.6923076923	0.9436619718
7	0.7142857143	0.9703703704
8	0.7333333333	0.9847908745
9	0.7500000000	0.9922928709

TABLE 1.2: Corresponding Terms in Two Sequences that Converge to 1

Since

$$\lim_{n\to\infty} \frac{n+3}{n+7} = 1 \qquad \text{and} \qquad \lim_{n\to\infty} \frac{2^n+3}{2^n+7} = 1,$$

it follows that both sequences converge to the limit 1. Although these two sequences have the same limit value, as seen in Table 1.2, the terms in the sequence $(2^n + 3)/(2^n + 7)$ appear to be approaching 1 much more rapidly than the terms in the sequence $(n+3)/(n+7)$.

We now determine the rate of convergence of each sequence. After some basic algebra, we find

$$\left|\frac{n+3}{n+7} - 1\right| = \frac{4}{n+7} < 4 \cdot \frac{1}{n}.$$

Hence, we may take $\lambda = 4$ and $\beta_n = 1/n$ in the definition of rate of convergence. It follows that the sequence

$$\left\{\frac{n+3}{n+7}\right\}$$

converges to 1 with rate of convergence $O(1/n)$. Working in a similar manner, we find that

$$\left|\frac{2^n+3}{2^n+7} - 1\right| = \frac{4}{2^n+7} < 4 \cdot \frac{1}{2^n}.$$

Hence, we may take $\lambda = 4$ and $\beta_n = 1/2^n$ in the definition of rate of convergence, so the sequence

$$\left\{\frac{2^n+3}{2^n+7}\right\}$$

converges to 1 with rate of convergence $O(1/2^n)$. These results confirm our numerical evidence since $1/2^n$ approaches zero faster than $1/n$ as $n \to \infty$.

In later chapters, many of the methods that will be developed will have theoretical error bounds that are expressed as a function of a method parameter. For example, most of the numerical integration techniques covered in Chapter 6 will have error bounds expressed in terms of the parameter h, the spacing between the points at which the integrand is sampled. To facilitate the comparison between different techniques, it will be useful to have big-O notation defined for functions.

Definition. Let f be a function defined on the interval (a, b) that contains $x = 0$, and suppose $\lim_{x\to 0} f(x) = L$. If there exists a function g for which $\lim_{x\to 0} g(x) = 0$ and a positive constant K such that

$$|f(x) - L| \leq K|g(x)|$$

for all sufficiently small values of x, then $f(x)$ is said to converge to L with RATE OF CONVERGENCE $O(g(x))$.

In these instances, the benchmark function $g(x)$ will tend to be of the form x^a for some positive exponent a. An error term with rate of convergence $O(x)$ then approaches zero more slowly than an error term with rate of convergence $O(x^4)$, say, as the value of x approaches zero.

EXAMPLE 1.5 Determining Rate of Convergence for a Function

Consider the function

$$f(x) = \frac{\cos x - 1 + \frac{1}{2}x^2}{x^4}.$$

What is the limit of f as $x \to 0$? Furthermore, at what rate does f converge to this limit?

We can actually answer both of these questions simultaneously by using Taylor's Theorem. From Taylor's Theorem (which will be reviewed at the end of this section), we know that

$$\cos x = 1 - \frac{1}{2}x^2 + \frac{1}{24}x^4 - \frac{1}{720}x^6 \cos \xi,$$

for some ξ between 0 and x. Hence,

$$\frac{\cos x - 1 + \frac{1}{2}x^2}{x^4} = \frac{1}{24} - \frac{1}{720}x^2 \cos \xi.$$

Finally, because

$$\left| \frac{\cos x - 1 + \frac{1}{2}x^2}{x^4} - \frac{1}{24} \right| = \frac{1}{720}|x^2 \cos \xi| \leq \frac{1}{720}|x^2|,$$

it follows that $\lim_{x\to 0} f(x) = \frac{1}{24}$ and the rate of convergence is $O(x^2)$.

Order of Convergence

Order of convergence provides a different measure of convergence speed than rate of convergence. Whereas rate of convergence examines individually the terms in the sequence of error values, $e_n = p_n - p$, order of convergence examines the relationship between successive error values, measuring the effectiveness with which each iteration reduces the approximation error.

Definition. Let $\{p_n\}$ be a sequence that converges to a number p. Let $e_n = p_n - p$ for $n \geq 0$. If there exist positive constants λ and α such that

$$\lim_{n\to\infty} \frac{|p_{n+1} - p|}{|p_n - p|^\alpha} = \lim_{n\to\infty} \frac{|e_{n+1}|}{|e_n|^\alpha} = \lambda,$$

then $\{p_n\}$ is said to converge to p of ORDER α with ASYMPTOTIC ERROR CONSTANT λ.

It follows that for a sequence that converges of order α, the error satisfies the asymptotic relation $|e_{n+1}| \approx \lambda |e_n|^\alpha$.

An iterative method is said to be of order α if the sequence it generates converges of order α. The most common values of α in practice are: $\alpha = 1$ (also known as linear convergence), $\alpha = 2$ (quadratic convergence) and $\alpha = 3$ (cubic convergence). Noninteger values for α are possible. Note that when $\alpha = 1$, the sequence of error values satisfies

$$|e_{n+1}| \approx \lambda |e_n| \approx \lambda^2 |e_{n-1}| \approx \lambda^3 |e_{n-2}| \approx \cdots \approx \lambda^n |e_0|.$$

Hence, a linearly convergent sequence converges with rate of convergence $O(\lambda^n)$.

To demonstrate the difference between the various orders of convergence, suppose there are three methods, one linear, one quadratic, and one cubic, all being applied to the same problem. Each method has an asymptotic error constant of $\lambda = 0.5$, and there is unit error in the initial approximation (i.e., $|e_0| = 1$). The chart below displays the error associated with each method through several iterations.

	LINEAR $\Rightarrow \|e_{n+1}\| \approx 0.5\|e_n\|$	QUADRATIC $\Rightarrow \|e_{n+1}\| \approx 0.5\|e_n\|^2$	CUBIC $\Rightarrow \|e_{n+1}\| \approx 0.5\|e_n\|^3$
$\|e_1\|$	0.5	0.5	0.5
$\|e_2\|$	0.25	0.125	0.0625
$\|e_3\|$	0.125	7.8125×10^{-3}	1.2207×10^{-4}
$\|e_4\|$	0.0625	3.0518×10^{-5}	9.0949×10^{-13}
$\|e_5\|$	0.03125	4.6566×10^{-10}	3.7616×10^{-37}
$\|e_6\|$	0.015625	1.0842×10^{-19}	
$\|e_7\|$	7.8125×10^{-3}	5.8775×10^{-39}	

Note the dramatic difference between the linear and quadratic methods. The linear method would take more than 100 iterations to achieve the accuracy attained by the quadratic method in just seven iterations. Even the more modest accuracy achieved by the quadratic method in five iterations would take the linear method 31 iterations. Unless each iteration of the quadratic method requires significantly more work than each iteration of the linear method, the linear method will never compete with the quadratic. On the other hand, there is only a slight difference (two or three iterations) between the quadratic and cubic methods. In practice, the extra work needed to achieve cubic convergence would therefore not be justified.

EXAMPLE 1.6 Determining Order of Convergence

In Section 1.1, we used the recursive scheme

$$x_{n+1} = \frac{1}{2}\left(x_n + \frac{a}{x_n}\right) \tag{1}$$

to compute an approximation to the square root of a positive real number a. Here, we would like to determine the order of convergence of the generated sequence. To accomplish this, we must be able to compare the error in the $(n+1)$st term in the sequence, $x_{n+1} - \sqrt{a}$, with the error in the nth term, $x_n - \sqrt{a}$.

We start by subtracting $\sqrt{a}$ from both sides of (1) and performing some basic algebra. This yields

$$\begin{aligned} x_{n+1} - \sqrt{a} &= \frac{1}{2}\left(x_n + \frac{a}{x_n}\right) - \sqrt{a} \\ &= \frac{x_n^2 - 2x_n\sqrt{a} + a}{2x_n} \\ &= \frac{(x_n - \sqrt{a})^2}{2x_n}. \end{aligned}$$

Accordingly,

$$\lim_{n\to\infty} \frac{|x_{n+1} - \sqrt{a}|}{|x_n - \sqrt{a}|^2} = \lim_{n\to\infty} \frac{1}{2x_n} = \frac{1}{2\sqrt{a}}.$$

Hence, the sequence generated by this scheme has order of convergence equal to 2 and asymptotic error constant $1/(2\sqrt{a})$.

A common task throughout the remainder of the text will be confirming a theoretical order of convergence using numerical data. For example, we have just established, theoretically, that the sequence generated by equation (1) should converge of order 2. Does the sequence actually achieve quadratic convergence in practice? To answer this question we need to select an a, generate the resulting sequence and examine the ratio $|e_n|/|e_{n-1}|^2$. If this ratio approaches a constant as n increases (the ratio should, in particular, approach the asymptotic error constant $1/2\sqrt{a}$), then we have numerical evidence of quadratic convergence.

EXAMPLE 1.7 Numerical Verification of Quadratic Convergence

With $a = 9$ and $x_0 = 9$, the first five terms of the sequence generated by equation (1) are listed in the second column of the following table. The absolute error in each term in the sequence (remember that the sequence is supposed to converge to $\sqrt{a} = \sqrt{9} = 3$) is given in the third column of the table. For our present purposes, the most important information in the table is found in the fourth column, which shows the ratio $|e_n|/|e_{n-1}|^2$.

n	x_n	$\lvert e_n\rvert = \lvert x_n - 3\rvert$	$\lvert e_n\rvert/\lvert e_{n-1}\rvert^2$
0	9	6	
1	5	2	0.055556
2	3.4	0.4	0.100000
3	3.02352941176471	2.35294×10^{-2}	0.147059
4	3.00009155413138	9.15541×10^{-5}	0.165370
5	3.00000000139698	1.39698×10^{-9}	0.166661

Note that the ratio $|e_n|/|e_{n-1}|^2$ approaches a constant, thereby providing numerical confirmation of the quadratic convergence of the sequence. Further, the error ratio appears to be approaching $1/6 = 1/(2\sqrt{9})$, providing numerical confirmation that the asymptotic error constant for equation (1) is $\lambda = 1/(2\sqrt{a})$.

Review of Taylor's Theorem

Taylor's Theorem is an important tool in many branches of mathematics, including numerical analysis. The theorem indicates how to construct a polynomial approximation for a sufficiently differentiable function.

> **Theorem.** Suppose f is continuous on $[a, b]$, has n continuous derivatives on (a, b) and $f^{(n+1)}$ exists on $[a, b]$. Let $x_0 \in [a, b]$. For every $x \in [a, b]$ there exists a number $\xi(x)$ between x and x_0 such that
>
> $$f(x) = P_n(x) + R_n(x),$$
>
> where
>
> $$P_n(x) = \sum_{k=0}^{n} \frac{f^{(k)}(x_0)}{k!}(x - x_0)^k \text{ and } R_n(x) = \frac{f^{(n+1)}(\xi(x))}{(n+1)!}(x - x_0)^{n+1}.$$

Here, P_n is called the *nth-degree Taylor polynomial* for f about $x = x_0$. In practice, $P_n(x)$ is used as an approximation to $f(x)$ for values of x near $x = x_0$. The term $R_n(x)$ is called the *remainder term* associated with P_n. For each x, the remainder term gives the error incurred by using $P_n(x)$ to approximate $f(x)$. In practice, the exact value of ξ is rarely known, so the remainder term is used to determine an error bound rather than the actual approximation error.

EXAMPLE 1.8 A Taylor Polynomial and Its Remainder Term

We will now use Taylor's theorem to obtain the second-degree Taylor polynomial and its associated remainder term for the function $f(x) = \sqrt{x}$ about $x_0 = 16$. To determine the coefficients in the second-degree Taylor polynomial, we need f and its first two derivatives. For the remainder term, we will need the third derivative of f. Starting from $f(x) = \sqrt{x}$, we compute

$$f'(x) = \frac{1}{2}x^{-1/2}, \quad f''(x) = -\frac{1}{4}x^{-3/2} \quad \text{and} \quad f'''(x) = \frac{3}{8}x^{-5/2}.$$

Therefore,

$$\begin{aligned} f(x_0) &= \sqrt{16} = 4 \\ f'(x_0) &= \frac{1}{2\sqrt{16}} = \frac{1}{8} \\ f''(x_0) &= -\frac{1}{4(\sqrt{16})^3} = -\frac{1}{256} \quad \text{and} \\ f'''(\xi) &= \frac{3}{8\xi^{5/2}}. \end{aligned}$$

Finally,

$$\begin{aligned} f(x) &= P_2(x) + R_2(x) \\ &= 4 + \frac{1}{8}(x-16) - \frac{1}{512}(x-16)^2 + \frac{1}{16\xi^{5/2}}(x-16)^3. \end{aligned}$$

Suppose we now take $x = 17$. Using the Taylor polynomial and remainder term we just calculated, we find

$$\sqrt{17} = f(17) \approx P_2(17) = 4 + \frac{1}{8} - \frac{1}{512} = 4.123046875$$

with an absolute error given by

$$|R_2(17)| = \left| \frac{1}{16\xi^{5/2}} \right|,$$

where $16 < \xi < 17$. Because ξ must be larger than 16, it follows that

$$|R_2(17)| < \frac{1}{16 \cdot 16^{5/2}} = \frac{1}{16384} \approx 6.10 \times 10^{-5}.$$

This last inequality provides what is called a *theoretical error bound.* The value of $P_2(17)$ can differ from $\sqrt{17}$ by no more than this amount. In fact, the actual difference between $P_2(17)$ and $\sqrt{17}$ is roughly 5.88×10^{-5}.

We conclude this section by stating the nth-degree Taylor polynomial and associated remainder term for several common functions. The derivations of these expressions are left for the exercises.

$$e^x = 1 + x + \frac{x^2}{2} + \cdots + \frac{x^n}{n!} + \frac{x^{n+1}}{(n+1)!}e^\xi$$

$$\sin x = x - \frac{x^3}{3!} + \frac{x^5}{5!} - + \cdots + \frac{(-1)^n x^{2n+1}}{(2n+1)!} + \frac{(-1)^{n+1}x^{2n+3}}{(2n+3)!}\sin\xi$$

$$\cos x = 1 - \frac{x^2}{2!} + \frac{x^4}{4!} - + \cdots + \frac{(-1)^n x^{2n}}{(2n)!} + \frac{(-1)^{n+1}x^{2n+2}}{(2n+2)!}\cos\xi$$

$$\frac{1}{1+x} = 1 - x + x^2 - + \cdots + (-1)^n x^n + \frac{(-1)^{n+1}x^{n+1}}{(1+\xi)^{n+2}}$$

EXERCISES

1. Compute each of the following limits and determine the corresponding rate of convergence.
 (a) $\lim_{n\to\infty} \frac{n-1}{n^3+2}$
 (b) $\lim_{n\to\infty} \left(\sqrt{n+1} - \sqrt{n}\right)$
 (c) $\lim_{n\to\infty} \frac{\sin n}{n}$
 (d) $\lim_{n\to\infty} \frac{3n^2-1}{7n^2+n+2}$

2. Compute each of the following limits and determine the corresponding rate of convergence.
 (a) $\lim_{x\to 0} \frac{e^x-1}{x}$
 (b) $\lim_{x\to 0} \frac{\sin x}{x}$
 (c) $\lim_{x\to 0} \frac{e^x - \cos x - x}{x^2}$
 (d) $\lim_{x\to 0} \frac{\cos x - 1 + x^2/2 - x^4/24}{x^6}$

3. Numerically determine which of the following sequences approaches 1 faster, and then confirm the numerical evidence by determining the rate of convergence of each sequence.

$$\lim_{x\to 0} \frac{\sin x^2}{x^2} \quad \text{versus} \quad \lim_{x\to 0} \frac{(\sin x)^2}{x^2}$$

4. Suppose that $0 < a < b$.
 (a) Show that if $\alpha_n = \alpha + O(1/n^b)$, then $\alpha_n = \alpha + O(1/n^a)$.
 (b) Show that if $f(x) = L + O(x^b)$, then $f(x) = L + O(x^a)$.

5. Suppose that $f_1(x) = L_1 + O(x^a)$ and $f_2(x) = L_2 + O(x^b)$. Show that

$$c_1 f_1(x) + c_2 f_2(x) = c_1 L_1 + c_2 L_2 + O(x^c),$$

where $c = \min(a, b)$.

6. The table below lists the errors of successive iterates for three different methods for approximating $\sqrt[3]{5}$. Estimate the order of convergence of each method, and explain how you arrived at your conclusions.

Method 1	Method 2	Method 3
4.0×10^{-2}	3.7×10^{-4}	4.3×10^{-3}
9.1×10^{-4}	1.2×10^{-15}	1.8×10^{-8}
4.8×10^{-7}	1.5×10^{-60}	1.4×10^{-24}

7. Let $\{p_n\}$ be a sequence that converges to the limit p.

(a) If

$$\lim_{n\to\infty} \frac{|p_{n+1} - p|}{|p_n - p|^\alpha} = 0,$$

what can be said about the order of convergence of $\{p_n\}$ to p?

(b) If

$$\lim_{n\to\infty} \frac{|p_{n+1} - p|}{|p_n - p|^\alpha} \to \infty,$$

what can be said about the order of convergence of $\{p_n\}$ to p?

8. Suppose theory indicates that the sequence $\{p_n\}$ converges to p of order 1.5. Explain how you would numerically verify this order of convergence.

9. Theory indicates that the following sequence should converge to $\sqrt{3}$ of order 1.618. Does the sequence actually achieve an order of convergence of 1.618? If not, what is the actual order?

n	p_n
0	2.000000000000000
1	1.666666666666667
2	1.727272727272727
3	1.732142857142857
4	1.732050680431722
5	1.732050807565499

10. Theory indicates that the following sequence should converge to 4/3 of order 1.618. Does the sequence actually achieve an order of convergence of 1.618? If not, what is the actual order?

n	p_n
0	1.498664098580016
1	1.497353997792205
2	1.428801977335339
3	1.401092915389552
4	1.376493676051456
5	1.361345745573130
6	1.351034482500881
7	1.344479850695066

11. Show that the convergence of the sequence generated by the formula

$$x_{n+1} = \frac{x_n^3 + 3x_n a}{3x_n^2 + a}$$

toward $\sqrt{a}$ is third order. What is the asymptotic error constant?

12. Let a be a nonzero real number. For any x_0 satisfying $0 < x_0 < 2/a$, the recursive sequence defined by

$$x_{n+1} = x_n(2 - ax_n)$$

converges to $1/a$. What are the order of convergence and the asymptotic error constant?

13. Suppose that the sequence $\{p_n\}$ converges linearly to the limit p with asymptotic error constant λ. Further suppose that $p_{n+1} - p$, $p_n - p$ and $p_{n-1} - p$ are all of the same sign. Show that

$$\frac{p_{n+1} - p_n}{p_n - p_{n-1}} \approx \lambda.$$

14. A sequence $\{p_n\}$ converges *superlinearly* to p provided

$$\lim_{n\to\infty} \frac{|p_{n+1} - p|}{|p_n - p|} = 0.$$

Show that if $p_n \to p$ of order α for $\alpha > 1$, then $\{p_n\}$ converges superlinearly to p.

15. Suppose that $\{p_n\}$ converges superlinearly to p (see Exercise 14). Show that

$$\lim_{n\to\infty} \frac{|p_{n+1} - p_n|}{|p_n - p|} = 1.$$

16. **(a)** Determine the third-degree Taylor polynomial and associated remainder term for the function $f(x) = \ln(1 - x)$. Use $x_0 = 0$.

(b) Using the results of part (a), approximate $\ln(0.25)$ and compute the theoretical error bound associated with this approximation. Compare the theoretical error bound with the actual error.

(c) Compute the following limit and determine the corresponding rate of convergence:

$$\lim_{x\to 0} \frac{\ln(1 - x) + x + \frac{1}{2}x^2}{x^3}.$$

17. **(a)** Determine the third-degree Taylor polynomial and associated remainder term for the function $f(x) = \sqrt{1 + x}$. Use $x_0 = 0$.

(b) Using the results of part (a), approximate $\sqrt{1.5}$ and compute the theoretical error bound associated with this approximation. Compare the theoretical error bound with the actual error.

(c) Compute the following limit and determine the corresponding rate of convergence:

$$\lim_{x\to 0} \frac{\sqrt{1 + x} - 1 - \frac{1}{2}x}{x^2}.$$

In Exercises 18–21, verify that Taylor's theorem produces the indicated formula, where ξ is between 0 and x.

18.

$$e^x = 1 + x + \frac{x^2}{2} + \cdots + \frac{x^n}{n!} + \frac{x^{n+1}}{(n+1)!}e^{\xi}$$

19.

$$\sin x = x - \frac{x^3}{3!} + \frac{x^5}{5!} - + \cdots + \frac{(-1)^n x^{2n+1}}{(2n+1)!} + \frac{(-1)^{n+1} x^{2n+3}}{(2n+3)!} \sin \xi$$

20.

$$\cos x = 1 - \frac{x^2}{2!} + \frac{x^4}{4!} - + \cdots + \frac{(-1)^n x^{2n}}{(2n)!} + \frac{(-1)^{n+1} x^{2n+2}}{(2n+2)!} \cos \xi$$

21.

$$\frac{1}{1+x} = 1 - x + x^2 - + \cdots + (-1)^n x^n + \frac{(-1)^{n+1} x^{n+1}}{(1+\xi)^{n+2}}$$

1.3 MATHEMATICS ON THE COMPUTER FLOATING POINT NUMBER SYSTEMS

Any meaningful discourse on numerical methods/numerical analysis must include a discussion of errors. After all, numerical methods are generally designed to determine approximate solutions. Sources of error can be broadly grouped into four categories:

- modeling error;
- discretization and truncation error;
- roundoff and data error;
- human error.

The assumptions that are made during the model building phase (as described in the overview to this chapter) give rise to equations which are at best approximations to the system being studied. A quantification and classification of these approximation errors can be found in most textbooks on mathematical modeling but is beyond the scope of this text.

Many of the techniques that are developed in the forthcoming chapters involve the conversion of a continuous problem into a discrete one. This conversion process introduces what are referred to as discretization errors. Still other techniques involve the truncation of an infinite series, giving rise to truncation errors in the approximation solution. As methods are developed, these types of error will be examined in detail. Although no one likes to admit it, we all make programming errors and computation errors. Great care must be taken to ensure that all human errors are detected and corrected. Checking programs with test problems whose exact solution is known to verify that theoretical error bounds and rates of convergence are satisfied is a powerful technique for achieving this goal.

Unlike discretization and truncation errors, which arise due to the formulation of a numerical method, roundoff and data errors are inherent in the way that computers and data acquisition hardware represent real numbers. The objectives of this section are to examine the representation of real numbers on computers and to make a precise definition of roundoff error. The important concept of conditioning is also introduced. Discussions of floating point arithmetic, the accumulation of roundoff error through a sequence of calculations, and the types of operations that should be avoided in the construction of a numerical algorithm will be deferred to the next section.

An Example to Set the Stage

As part of a laboratory experiment, a group of students needs to calculate the modulus of elasticity, E, of a steel beam. An object of mass $m = 0.491$ kg is suspended from one end of a beam whose length is $l = 0.451$ m, whose width is $a = 0.021$ m and whose thickness is $b = 0.003$ m. The resulting deflection of the tip of the beam is measured to be $d = 0.142$ m. Substituting these values into the formula

$$E = \frac{4mgl^3}{dab^3},$$

where $g = 9.81$ m/s^2 is the acceleration due to gravity, the students calculate

$$E = \frac{4(0.491)(9.81)(0.451)^3}{(0.142)(0.021)(0.003)^3} = 21.952 \times 10^9 \text{ N/m}^2.$$

A standard table of the properties of steel, however, indicates that the actual value should be $E = 30 \times 10^9$ N/m^2. Is the value calculated by the students within acceptable limits of the tabulated value?

To answer this question, we must recognize that all physical measurements are made with finite precision and hence include some amount of error. For instance, if it is assumed that all of the measured values given above have been rounded to the digits shown, then the mass of the object that has been suspended from the beam can actually be anywhere between 0.4905 kg and 0.4915 kg. Similarly, the length of the beam is between 0.4505 m and 0.4515 m, the width of the beam is between 0.0205 m and 0.0215 m, the thickness of the beam is between 0.0025 m and 0.0035 m, and the deflection of the tip of the beam is between 0.1415 m and 0.1425 m. Since each measured value is really an interval, the equation for the modulus of elasticity should be used to determine a range of possible values:

$$\frac{4(0.4905)(9.81)(0.4505)^3}{(0.1425)(0.0215)(0.0035)^3} < E < \frac{4(0.4915)(9.81)(0.4515)^3}{(0.1415)(0.0205)(0.0025)^3},$$

or

$$13.397 \times 10^9 \text{ N/m}^2 < E < 39.165 \times 10^9 \text{ N/m}^2.$$

Note this range includes the tabulated value of $E = 30 \times 10^9$ N/m^2. Therefore, taking into account the accuracy of the measurements, the value calculated by the students is within acceptable limits of the tabulated value.

Floating Point Number Systems

Although practical problems deal with real valued quantities and the theorems upon which numerical methods are based are written in terms of real valued functions, computers, like measurement devices, have no concept of the real numbers. Instead, computers represent real numbers in what is known as a *floating point number system.*

Definition. A FLOATING POINT NUMBER SYSTEM, $\mathbf{F}(\beta, k, m, M)$, is a subset of the real number system characterized by the parameters

β:	the base
k:	the number of digits in the base β expansion
m:	the minimum exponent
M:	the maximum exponent

Elements of $\mathbf{F}(\beta, k, m, M)$ are those real numbers that can be expressed exactly as

$$\pm (0.d_1 d_2 d_3 \cdots d_k)_\beta \times \beta^e,$$

where $m \leq e \leq M$. The first base β digit, d_1, must be nonzero, except when the number being represented is 0, in which case $d_1 = 0$.

The restriction is made on the first digit, d_1, to guarantee that each element in the set has a unique representation. Computers primarily use $\beta = 2$, or a binary number system, though some computers use $\beta = 16$, or a hexidecimal number system. Handheld calculators typically use a decimal number system; *i.e.*, $\beta = 10$. Commonly used values for k, m and M will be noted later in the section and in the exercises.

There are three important ways in which a floating point number system differs from the real number system. First, a floating point number system is a discrete set. In contrast, the real number system is continuous, meaning that between any two real numbers there are infinitely many other real numbers. Second, a floating point number system is a finite set, whereas the real number system is an infinite set. Here, finite refers not only to the number of elements in the set, but also to the range of values. A floating point number system contains both a smallest and a largest positive element, as well as a smallest and a largest negative element. The real number system has no such elements. Third, real numbers are uniformly distributed while finite precision imposes a nonuniform distribution upon the elements of a floating point number system. In particular, the elements near zero are more closely spaced than the elements at the extremes of the representable values.

As an example, consider the system

$$\begin{aligned}\mathbf{F}(10, 2, 0, 2) = \{0,& \\ &\pm 0.10, \pm 0.11, \pm 0.12, \ldots, \pm 0.19, \\ &\vdots \\ &\pm 0.90, \pm 0.91, \pm 0.92, \ldots, \pm 0.99, \\ &\pm 1.0, \pm 1.1, \pm 1.2, \ldots, \pm 1.9, \\ &\vdots \\ &\pm 9.0, \pm 9.1, \pm 9.2, \ldots, \pm 9.9,\end{aligned}$$

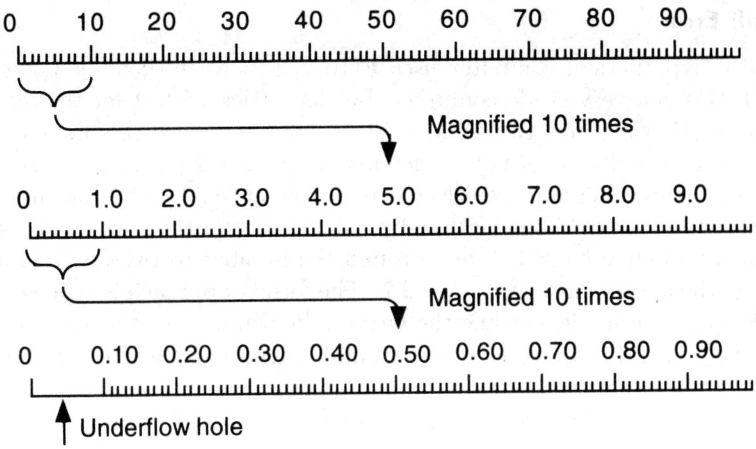

Figure 1.1 The positive elements from the floating point number system $\mathbf{F}(10, 2, 0, 2)$.

$$\pm 10, \pm 11, \pm 12, \ldots, \pm 19,$$

$$\vdots$$

$$\pm 90, \pm 91, \pm 92, \ldots, \pm 99\}.$$

Figure 1.1, which displays the positive elements from $\mathbf{F}(10, 2, 0, 2)$, clearly illustrates the discrete nature of the number system. Whether we examine the list of elements given above or the figure, we see that the smallest nonzero elements in $\mathbf{F}(10, 2, 0, 2)$ are ± 0.10, while the elements of largest magnitude are ± 99. Taking into account both positive and negative values, the system contains only 541 elements. Finally, Figure 1.1 also makes it clear that the elements in this system are not uniformly distributed along the number line. In particular, there is a gap of one-tenth between 0 and the smallest nonzero numbers, elements are separated by a distance of one-hundredth in the range from 0.1 to 1.0, by a distance of one-tenth in the range from 1.0 to 10.0 and by a distance of 1 from 10 through 99.

The fact that a floating point number system has smallest nonzero elements, as well as largest elements, carries certain practical implications. Suppose, for instance, that a calculation produces a number that falls between zero and one of the smallest nonzero elements of the system [e.g., the operation 0.20/87 performed in $\mathbf{F}(10, 2, 0, 2)$]. Because the operation has resulted in a number that is too small to be represented in the system, we say that an *underflow* has occurred. Typically, underflow is handled by replacing the number with zero. On the other hand, a calculation that produces a number that is too large to be represented in the system generates an *overflow* exception. An example of an overflow exception would be the calculation 5.7×43 performed in $\mathbf{F}(10, 2, 0, 2)$. Overflow usually causes a process to halt execution.

Roundoff Error

Suppose a hypothetical computer uses $\mathbf{F}(10,2,0,2)$ as its floating point system. Granted, this is a very crude computer, but let's stick with it for the moment. A given calculation requires the value $\sqrt{7.1} \approx 2.66458$. Since this number requires more than two digits, it is not a member of the set $\mathbf{F}(10,2,0,2)$. An element from the set must therefore be selected to represent $\sqrt{7.1}$. The two most natural approaches to take would be to either drop all digits after the second one, producing the approximation $\sqrt{7.1} \approx 2.6$, or to round the number to two digits of accuracy, producing the approximation $\sqrt{7.1} \approx 2.7$. The former approach is typically referred to as *chopping* the number, while the latter is known as *rounding* the number.

In the general case, let y be a real number whose expansion is given by

$$y = \pm\,(0.d_1d_2d_3\cdots d_kd_{k+1}\cdots)_\beta \times \beta^e$$

with $d_1 \neq 0$ and $m \leq e \leq M$. Denote by $fl(y) \in \mathbf{F}(\beta,k,m,M)$ the *floating point equivalent* of y (i.e., that element from the floating point system that will be used to represent y). When the number is chopped, the floating point equivalent is given by

$$fl_{\text{chop}}(y) = \pm\,(0.d_1d_2d_3\cdots d_k)_\beta \times \beta^e;$$

when the number is rounded,

$$fl_{\text{round}}(y) = \begin{cases} \pm\,(0.d_1d_2d_3\cdots d_k)_\beta \times \beta^e, & d_{k+1} < \beta/2 \\ \pm\left[(0.d_1d_2d_3\cdots d_k)_\beta + \beta^{-k}\right] \times \beta^e, & d_{k+1} \geq \beta/2 \end{cases}.$$

Regardless of whether the number is chopped or rounded, the conversion of a number into its floating point equivalent introduces some amount of error. This type of error is known as *roundoff error*.

Definition. The error introduced by converting a real number to its floating point equivalent is called ROUNDOFF ERROR.

The standard metrics of absolute and relative error, as defined below, are used to quantify the effect of roundoff.

Definition. Let p^* denote any approximation to the value p. The ABSOLUTE ERROR in p^* is given by

$$|p^* - p|.$$

The RELATIVE ERROR in p^* is

$$|p^* - p|/|p|,$$

provided that $p \neq 0$, and is usually, though not always, expressed as a percentage.

For the example involving $\sqrt{7.1}$, the absolute and relative errors associated with chopping the number are

$$|2.6 - 2.66458| = 6.458 \times 10^{-2}$$

and

$$\frac{|2.6 - 2.66458|}{2.66458} = 0.0242 = 2.42\%,$$

respectively. With rounding, the corresponding errors are 3.542×10^{-2} and 1.33%.

In the general case, consider chopping first. A bound on the absolute size of the roundoff error is

$$\begin{aligned} |fl_{\text{chop}}(y) - y| &= (0.d_{k+1}d_{k+2}d_{k+3}\cdots)_\beta \times \beta^{e-k} \\ &\leq (1.0)_\beta \times \beta^{e-k} \\ &= \beta^{e-k}. \end{aligned}$$

To bound the relative error, a lower bound for $|y|$ is needed. Provided $y \neq 0$, given the restriction on the value of d_1,

$$\begin{aligned} |y| &= (0.d_1d_2d_3\cdots)_\beta \times \beta^{e} \\ &\geq (0.1)_\beta \times \beta^{e} \\ &= \beta^{e-1}. \end{aligned}$$

Therefore,

$$\frac{|fl_{\text{chop}}(y) - y|}{|y|} \leq \frac{\beta^{e-k}}{\beta^{e-1}} = \beta^{1-k}.$$

By proceeding in a similar manner, it can be shown that when a number is rounded, the bounds on both the absolute and relative error due to roundoff are one-half the bounds obtained when a number is chopped. That is,

$$|fl_{\text{round}}(y) - y| \leq \frac{1}{2}\beta^{e-k}$$

and

$$\frac{|fl_{\text{round}}(y) - y|}{|y|} \leq \frac{1}{2}\beta^{1-k}.$$

Note that the bound on the relative error due to roundoff, regardless of whether the number has been chopped or rounded, is independent of the number. The bound depends only on the base, β, and the number of digits, k, of the floating point system in use. This bound is therefore a function of the hardware implementation; accordingly, the phrase *machine precision*, or *machine epsilon*, is often used.

Definition. The MACHINE PRECISION, u, is given by

$$u = \begin{cases} \beta^{1-k}, & \text{chopping} \\ \frac{1}{2}\beta^{1-k}, & \text{rounding,} \end{cases}$$

where β is the base and k the number of digits in the implemented floating point number system.

Suppose that x and y are two nearly equal numbers. Because floating point numbers are represented using only a finite number of digits, it is natural to measure the "closeness" of two numbers not only in terms of absolute and relative difference, but also in terms of the number of *significant digits* which they have in common.

Definition. Suppose that $x \neq 0$ and that

$$\beta^{-(t+1)} < \left| \frac{x - y}{x} \right| \leq \beta^{-t}$$

for some positive integer t. Then we say that x and y agree to at least t and at most $t + 1$ SIGNIFICANT base β DIGITS.

Clearly, if two numbers agree to at least t significant base β digits, these numbers will be indistinguishable in any floating point number system with base β and $k \leq t$.

EXAMPLE 1.9 How Close Are Two Numbers

Consider the numbers $\cos(0.1°) = 0.999998476$ and $\cos(0.11°) = 0.999998157$. Since

$$\left| \frac{\cos(0.1°) - \cos(0.11°)}{\cos(0.1°)} \right| = 3.198 \times 10^{-7}$$

and

$$10^{-7} < 3.198 \times 10^{-7} \leq 10^{-6},$$

it follows that $\cos(0.1°)$ and $\cos(0.11°)$ agree to at least 6 and at most 7 decimal digits. To how many significant binary digits do these two cosine values agree? Since

$$2^{-22} = 2.384 \times 10^{-7} < 3.198 \times 10^{-7} < 4.768 \times 10^{-7} = 2^{-21},$$

we see that $\cos(0.1°)$ and $\cos(0.11°)$ agree to at least 21 and at most 22 binary digits.

The IEEE Standard

To this point, we've been dealing with floating point number systems in the abstract. But what systems are we likely to encounter in practice? In the 1970s, work was begun to develop standards for the representation and arithmetic of binary ($\beta = 2$) floating point numbers on microprocessors. One of the major objectives was to eliminate inconsistencies when moving code from one machine to another. The culmination of this effort came in 1985 with the publication of the report *Binary Floating Point Arithmetic Standard 754-1985* by the American IEEE (Institute for Electrical and Electronics Engineers) computer society. This report contained specifications for the representation of floating point numbers, the elementary operations and rounding rules available, the rules for converting between number systems, and the handling of exceptional cases. In 1989, the International Electrotechnical Commission let the IEEE standards become international standards. Today, these standards are generally adhered to by microprocessor manufacturers.

The IEEE standard actually specifies two different floating point formats: a basic format and an extended format. Each format contains both a single precision and a double precision number system. The basic format includes the single precision floating point number system $\mathbf{F}(2, 24, -125, 128)$ and the double precision system $\mathbf{F}(2, 53, -1021, 1024)$. For the extended format, only lower or upper bounds on number system parameters are provided. In particular, an extended single precision system must have $k \geq 32$, $m \leq -1021$, and $M \geq 1024$. An extended double precision system must have $k \geq 64$, $m \leq -16381$, and $M \geq 16384$.

Let's examine the IEEE standard single precision number system in more detail. With $\beta = 2$ and $k = 24$, machine precision with rounding is

$$u = \frac{1}{2}2^{1-24} = 2^{-24} \approx 5.96 \times 10^{-8}.$$

Accordingly, there are between seven and eight significant decimal digits available in single precision. The smallest positive number in single precision is

$$(0.1)_2 \times 2^{-125} = 2^{-126} \approx 1.18 \times 10^{-38},$$

while the largest positive number is

$$(0.111\cdots1)_2 \times 2^{128} = \left(1 - 2^{-24}\right) 2^{128} \approx 3.40 \times 10^{38}.$$

An examination of IEEE standard double precision, as well as several other number systems that can be found on specific processors, is left for the exercises.

Conditioning

In the next section, several examples will be used to demonstrate that certain operations can lead to a dramatic and devastating accumulation of roundoff error when a sequence of finite precision calculations are performed. There are also, however, mathematical problems for which a small change in input data, such as brought about by roundoff, leads to large changes in the analytical solution to the problem. Here, input data can refer to, among other things, the coefficients of a polynomial whose roots are being computed, the initial and/or boundary values associated with a differential equation, and the tabulated values from which a derivative and/or an integral needs to be approximated.

Consider the initial value problem

$$x' - x = e^{-2t}, \qquad x(0) = -1/3,$$

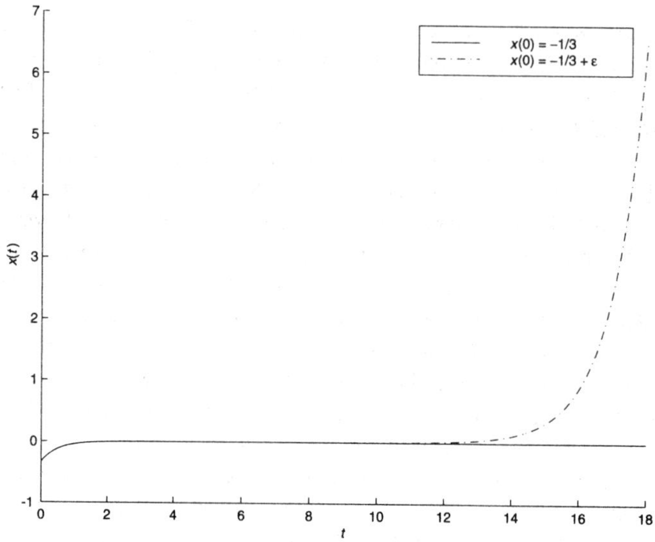

Figure 1.2 Comparison between the solution of $x' - x = e^{-2t}$ subject to the initial condition $x(0) = -1/3$ and $x(0) = -1/3 + \epsilon$ with $\epsilon = 10^{-7}$.

whose exact solution is $x(t) = -e^{-2t}/3$. If the initial condition is changed to $x(0) = -1/3 + \epsilon$, the exact solution of the problem becomes $x(t) = -\epsilon e^t - e^{-2t}/3$. No matter how small the perturbation, ϵ, to the initial condition, the difference between the two solutions, ϵe^t, grows without bound. Figure 1.2 plots both solutions with $\epsilon = 10^{-7}$.

As another example, take the polynomial

$$P(x) = (x-1)(x-2)(x-3)(x-4)(x-5)(x-6)(x-7)(x-8)(x-9)(x-10).$$

The roots of this tenth-degree polynomial are clearly the consecutive integers from 1 through 10, inclusive. Consider the perturbed polynomial $\tilde{P}(x) = P(x) + x^5$. The coefficient of x^5 in $P(x)$ is -902055, so the relative change in this coefficient in $\tilde{P}(x)$ is roughly one-thousandth of one percent. Figure 1.3 displays the roots of both $P(x)$ and $\tilde{P}(x)$. The relative change in the first two roots is on the order of the change in the coefficient of x^5, while the changes in the third and fourth roots are roughly 1% and 27%, respectively. The final six roots have been transformed into three complex conjugate pairs with definite nonzero imaginary parts.

Problems like these two, for which a small change in input data results in a large change in the output, are said to be *ill conditioned.* It is important to note that this ill conditioning is inherent in the mathematical problem itself and is not an artifact of any numerical computation scheme. In forthcoming chapters, the conditioning of the various mathematical problems being investigated will be discussed where appropriate.

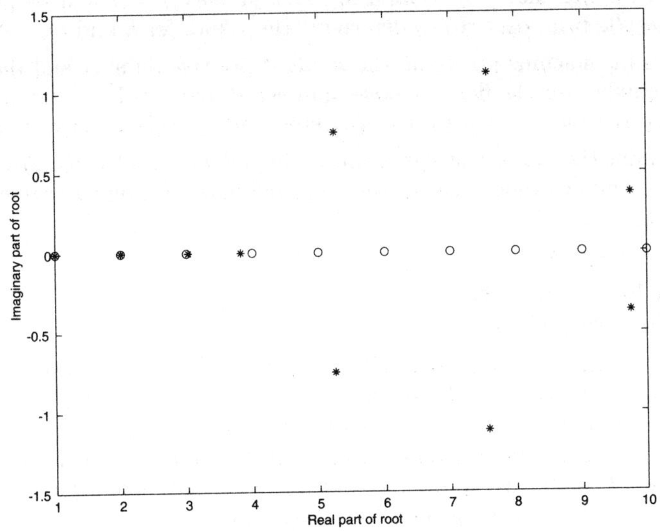

Figure 1.3 Roots of the polynomial $P(x)$, $\circ$, versus those of the polynomial $\tilde{P}(x)$, $*$, in the complex plane.

EXERCISES

1. Provide the floating point equivalent for each of the following numbers from the floating point number system $\mathbf{F}(10, 4, 0, 4)$. Consider both chopping and rounding. Compute the absolute and relative error in each floating point equivalent.

(a) π **(b)** e

(c) $\sqrt{2}$ **(d)** $1/7$

(e) $\cos 22°$ **(f)** $\ln 10$

(g) $\sqrt[3]{9}$

2. Prove the bounds on the absolute and relative roundoff error associated with rounding:

$$|fl_{\text{round}}(y) - y| \leq \frac{1}{2}\beta^{e-k} \quad \text{and} \quad \frac{|fl_{\text{round}}(y) - y|}{|y|} \leq \frac{1}{2}\beta^{1-k}.$$

3. Show that machine precision is the smallest floating point number, v, such that $fl(1 + v) > 1$.

4. **(a)** Construct an algorithm to determine machine precision and another algorithm to determine the smallest positive number of a floating point number system.

(b) Implement the algorithms from part (a) to determine machine precision and the smallest positive number on your computing system. Consider both single and double precision.

(c) Assuming that your computing system uses $\beta = 2$ and rounding, use the results from part (b) to determine the values for k and m.

5. Determine machine precision, the smallest positive number and the largest positive number for the floating point number system used by your calculator. Assuming the calculator uses $\beta = 10$, determine the values for k, m, and M.

6. Determine the number of significant decimal digits and the number of significant binary digits to which each of the following pairs of numbers agree.
   (a) $355/113$ and π
   (b) $685/252$ and e
   (c) $\sqrt{10002}$ and $\sqrt{10001}$
   (d) $103/280$ and $1/e$

7. The ideal gas law states that $PV = nRT$, where P is the pressure of the gas, V is the volume, n is the number of moles, T is the temperature, and $R = 0.08206$ atm $\cdot$ m^3/moles $\cdot$ K is the universal gas constant.
   (a) Experimentally, it has been determined that $P = 0.750$ atm, $V = 1.15$ m^3, and $T = 294.1$K. Assuming that all values have been rounded to the digits shown, in what range of values does n fall?
   (b) Experimentally, it has been determined that $V = 0.331$ m^3, $n = 0.00712$ moles, and $T = 264.7$K. Assuming that all values have been rounded to the digits shown, in what range of values does P fall?

8. In a physics laboratory, students measure the mass of a rectangular block to be 243.27 ± 0.005 grams. The length, width, and depth of the block are measured to be 7.8 ± 0.05 cm, 3.1 ± 0.05 cm, and 4.2 ± 0.05 cm, respectively.
   (a) In what range of values does the volume of the block fall?
   (b) In what range of values does the density of the block fall? Density is mass per unit volume.

9. Students are using a pendulum to experimentally determine the acceleration due to gravity, g. They measure the period, T, of the pendulum to be 2.2 seconds, and the length, l, of the pendulum to be 1.15 meters. Assuming that all values are correct to the digits shown, in what range of values does g fall? The variables in this problem are related by the formula $T = 2\pi\sqrt{l/g}$.

10. Determine machine precision, the smallest positive number and the largest positive number in the IEEE standard double precision system. Approximately how many significant decimal digits does the double precision standard supply?

11. In addition to the standard single and double precision floating point systems, Intel microprocessors also have an extended precision system $\mathbf{F}(2, 64, -16381, 16384)$. Determine machine precision, the smallest positive number and the largest positive number for this extended precision system.

12. IBM System/390 mainframes provide three floating point number systems: short precision $\mathbf{F}(16, 6, -64, 63)$, long precision $\mathbf{F}(16, 14, -64, 63)$, and extended precision $\mathbf{F}(16, 28, -64, 63)$. Compare machine precision, the smallest positive number, and the largest positive number for each of these number systems.

13. A common floating point number system used on modern calculators is $\mathbf{F}(10, 10, -98, 100)$. Determine machine precision, the smallest positive number and the largest positive number for this extended precision system.

14. **(a)** Show that the number of elements in the set $\mathbf{F}(\beta, k, m, M)$ is given by $1 + 2(\beta - 1)\beta^{k-1}(M - m + 1)$.

(b) How many elements are in the IEEE standard single precision number system?

(c) How many elements are in the IEEE standard double precision number system?

15. Consider the function $f(x) = x^2 - 4x + 4$.

(a) What are the zeros of f?

(b) Suppose we were to change the constant term to $4 - 10^{-8}$. What are the zeros of this new function? Relative to the size of the change in the constant term, how big is the change in the zeros of the function?

(c) Now, suppose we were to change the constant term to $4 + 10^{-8}$. What are the zeros of this new function? Relative to the size of the change in the constant term, how big is the change in the zeros of the function?

16. Consider the linear, first-order differential equation

$$\frac{dx}{dt} + \frac{1}{t}x = \frac{\sin t}{t}.$$

(a) Solve this equation subject to the initial condition $x(\pi/2) = x_0$.

(b) Solve this equation subject to the perturbed initial condition $x(\pi/2) = x_0 + \epsilon$.

(c) By considering the difference between the solutions obtained in parts (a) and (b), comment on the conditioning of this problem.

17. Consider the linear, first-order differential equation

$$\frac{dx}{dt} - \frac{1}{t}x = t \sin t.$$

(a) Solve this equation subject to the initial condition $x(\pi/2) = x_0$.

(b) Solve this equation subject to the perturbed initial condition $x(\pi/2) = x_0 + \epsilon$.

(c) By considering the difference between the solutions obtained in parts (a) and (b), comment on the conditioning of this problem.

18. Consider the linear system of equations

$$\begin{bmatrix} 1.1 & 2.1 \\ 2 & 3.8 \end{bmatrix} \begin{bmatrix} x \\ y \end{bmatrix} = \mathbf{b}.$$

(a) Solve the system for the right-hand side vector $\mathbf{b} = \begin{bmatrix} 3.2 & 5.8 \end{bmatrix}^T$.

(b) Solve the system for the right-hand side vector $\mathbf{b} = \begin{bmatrix} 3.21 & 5.79 \end{bmatrix}^T$.

(c) Solve the system for the right-hand side vector $\mathbf{b} = \begin{bmatrix} 3.1 & 5.7 \end{bmatrix}^T$.

(d) By considering the difference between the solutions obtained in parts (a), (b), and (c), comment on the conditioning of this problem.

1.4 MATHEMATICS ON THE COMPUTER: FLOATING POINT ARITHMETIC

The objective of numerical analysis is not just to input numbers to the computer, but rather to perform a sequence of calculations to produce a desired result. As each floating point operation is performed, a new amount of roundoff error will be introduced. It is therefore reasonable to expect a slow accumulation of roundoff error as calculations proceed. If positive and negative deviations are randomly distributed, it is possible for roundoff error to remain level or even to decrease. The opposite extreme is, unfortunately, also possible. Certain mathematical operations, when performed in finite precision, can lead to a dramatic accumulation of roundoff. A theoretical and experimental investigation of this issue will be presented in this section.

Floating Point Arithmetic

Since computers have no concept of the real numbers, they also have no concept of real arithmetic. Instead, computers perform calculations within their floating point number system. The scheme for floating point arithmetic that is adopted below glosses over many of the precise implementation details, which would of course be machine dependent but is sufficient for illustration purposes.

Let x and y be any real numbers, and let @ denote one of the binary arithmetic operators of addition (+), subtraction (−), multiplication (×), or division (÷). The corresponding floating point equivalent operator, $@_{fl}$, is defined by the relation

$$x @_{fl} y = fl(fl(x) @ fl(y)). \tag{1}$$

In other words, floating point arithmetic will be assumed to consist of three steps. First, each operand is replaced by its floating point equivalent. Next, the exact arithmetic is performed. Finally, the result is replaced by its floating point equivalent. For example, using 4 decimal digit rounding arithmetic,

$$\begin{aligned}(5/3) \times_{fl} \sqrt{3} &= fl(fl(5/3) \times fl(\sqrt{3})) \\ &= fl(1.667 \times 1.732) \\ &= fl(2.887244) \\ &= 2.887.\end{aligned}$$

Although this scheme glosses over any implementation specific details, the IEEE standard requires that relation (1) hold provided that neither underflow nor overflow occurs, so it is completely realistic.

Before turning to a discussion of roundoff error accumulation, it is important to note that floating point arithmetic does not satisfy many of the properties of real arithmetic that are taken for granted. When considered as real quantities, the values 0.1329, 1.543, and 23.21 can be added in any order to obtain the result 24.8859. However, in 4 decimal digit rounding arithmetic,

$$(0.1329 + 1.543) + 23.21 = 1.676 + 23.21 = 24.89,$$

but

$$0.1329 + (1.543 + 23.21) = 0.1329 + 24.75 = 24.88.$$

Note that all intermediate results have been rounded. Because the two final values are not equal, we see that floating point arithmetic is not associative. Note the more accurate result is obtained by adding the values in ascending order. In general, this will produce the most accurate result. The reasoning behind this is straightforward. By starting with the smallest values, information in the least significant bits can accumulate and influence the final result. When the numbers above were added in descending order, the two least significant digits in 0.1329 were simply discarded.

The distributive laws also do not hold in floating point arithmetic. Take, for example,

$$(0.1351 + 23.21) \times 1.543 = 23.35 \times 1.543 = 36.03.$$

When computed as the sum of two products, the result is

$$0.1351 \times 1.543 + 23.21 \times 1.543 = 0.2085 + 35.81 = 36.02.$$

The exact value in this case is 36.0214893.

Accumulation of Roundoff Error

Now that a framework for performing floating point operations has been established, we are in a position to discuss the mechanisms by which roundoff error accumulates in floating point operations. Again, let x and y be any real numbers, let @ denote an exact binary operator and let $@_{fl}$ denote the corresponding floating point operator. The difference between the exact value $x@y$ and the floating point value $x@_{fl}y$ can be broken into two components:

$$\begin{aligned} x@_{fl}y - x@y &= fl(fl(x)@fl(y)) - x@y \\ &= [fl(fl(x)@fl(y)) - fl(x)@fl(y)] + [fl(x)@fl(y) - x@y]. \end{aligned}$$

The first term, $fl(fl(x)@fl(y)) - fl(x)@fl(y)$, is known as the *introduced error.* It represents the roundoff error associated with the final step in the operation; [i.e., replacing the value $fl(x)@fl(y)$ by its floating point equivalent]. The second component, $fl(x)@fl(y) - x@y$, measures the effect that the roundoff error in the two operands has on the result of the exact arithmetic operation. This component is referred to as the *propagated error.* Whereas the introduced error is small, bounded in relative terms by machine precision, the propagated error can be large.

To examine the propagated error, let δ_x denote the absolute deviation and ϵ_x denote the relative deviation between x and $fl(x)$; that is,

$$\delta_x = fl(x) - x \qquad \epsilon_x = \frac{fl(x) - x}{x} = \frac{\delta_x}{x}.$$

Note that from these definitions, $fl(x)$ can be written as either $x + \delta_x$ or $x + x\epsilon_x = x(1 + \epsilon_x)$. Similar definitions apply for δ_y and ϵ_y, and similar expressions can be used to express $fl(y)$.

Starting with multiplication, it is found that

$$
\begin{aligned}
fl(x) \times fl(y) &= [x(1+\epsilon_x)] \times [y(1+\epsilon_y)] \\
&= xy(1+\epsilon_x+\epsilon_y+\epsilon_x\epsilon_y) \\
&= xy(1+\epsilon_{xy}),
\end{aligned} \tag{2}
$$

where $\epsilon_{xy} = \epsilon_x + \epsilon_y + \epsilon_x\epsilon_y \approx \epsilon_x + \epsilon_y$ provided $|\epsilon_x|, |\epsilon_y| \ll 1$. For division,

$$
\begin{aligned}
fl(x) \div fl(y) &= [x(1+\epsilon_x)] \div [y(1+\epsilon_y)] \\
&= \frac{x}{y}(1+\epsilon_x)(1-\epsilon_y+\epsilon_y^2-\epsilon_y^3+-\cdots) \\
&= \frac{x}{y}(1+\epsilon_{x/y}),
\end{aligned} \tag{3}
$$

where $\epsilon_{x/y} \approx \epsilon_x - \epsilon_y$. The geometric series expansion

$$(1+\epsilon_y)^{-1} = 1-\epsilon_y+\epsilon_y^2-\epsilon_y^3+-\cdots$$

was used in the second line of (3). From expressions (2) and (3), it is seen that with multiplication and division, the relative error propagates slowly in the sense that, if the relative error in the operands is small, the relative error in the product or quotient will also be small. It should be noted, though, that absolute error can grow rapidly when multiplying by a large number or dividing by a small number. Whenever possible, algorithms should therefore be written to avoid these situations.

When working with addition and subtraction, the behavior of absolute and relative error is very different from that associated with multiplication and division. Consider

$$
\begin{aligned}
fl(x) \pm fl(y) &= [x+\delta_x] \pm [y+\delta_y] \\
&= (x \pm y) + (\delta_x + \delta_y) \\
&= (x \pm y)\left(1 + \frac{x\epsilon_x \pm y\epsilon_y}{x \pm y}\right).
\end{aligned} \tag{4}
$$

From the second line of (4), it is clear that if the absolute error in the operands is small, the absolute error in the result will also be small. However, even when the relative error in the operands is small, the relative error in the result, $\frac{x\epsilon_x \pm y\epsilon_y}{x \pm y}$, can still be large if $x \pm y$ is close to zero. This will happen when two nearly equal numbers are subtracted or when numbers of nearly equal magnitude but different sign are added. Drastic propagation of error due to this situation is known as *cancellation error.* This type of error is directly linked to the loss of significant digits in the calculation. Algorithms should therefore also avoid the subtraction of nearly equal numbers whenever possible.

We now give three examples in which cancellation error can arise if care is not exercised.

EXAMPLE 1.10 The Quadratic Formula

A common mathematical procedure is the solution of a quadratic equation using the quadratic formula:

$$\frac{-b \pm \sqrt{b^2 - 4ac}}{2a}.$$

In most circumstances, this formula will produce entirely reasonable and accurate results. However, whenever the quantity b^2 is much, much greater than the product $4ac$, the discriminant will have a value very close to b^2. The numerator in this case will then be roughly equal to $-b \pm |b|$. Cancellation error will then occur either in the calculation of $-b + \sqrt{b^2 - 4ac}$ or in the calculation of $-b - \sqrt{b^2 - 4ac}$, depending on the sign of the parameter b.

As an illustration, consider the quadratic equation $0.2x^2 - 47.91x + 6 = 0$. To ten digits, the roots of this equation are 0.1253003555 and 239.4246996. Let's now calculate the roots using 4 decimal digit rounding arithmetic. The computation proceeds as follows:

$$\begin{aligned}\frac{47.91 \pm \sqrt{47.91^2 - 4 \cdot 0.2 \cdot 6}}{2 \cdot 0.2} &= \frac{47.91 \pm \sqrt{2295 - 4.8}}{0.4} \\ &= \frac{47.91 \pm \sqrt{2290}}{0.4} \\ &= \frac{47.91 \pm 47.85}{0.4}.\end{aligned}$$

One root of the quadratic is thus computed to be $\frac{47.91+47.85}{0.4} = \frac{95.76}{0.4} = 239.4$, while the second is computed to be $\frac{47.91-47.85}{0.4} = \frac{0.06}{0.4} = 0.15$. This problem encompasses both extremes of roundoff error accumulation. Despite individual roundoff errors in the computation of b^2, the difference between b^2 and $4ac$ and the square root of the discriminant, the larger root is correct to all digits shown—239.4 is the floating point equivalent of 239.4246996 in a 4 decimal digit number system with rounding. The smaller root, on the other hand, is in error by nearly 20%. The subtraction of the nearly equal numbers 47.91 and 47.85 resulted in the loss of three significant digits—from four in each of the operands to one in the result—which, in turn, produced a large relative error.

Is there some way to reformulate the calculation of the smaller root so as to obtain a more accurate result? The answer is yes. Because $b = -47.91 < 0$ for this problem, cancellation error occurred for the

$$\frac{-b - \sqrt{b^2 - 4ac}}{2a}$$

portion of the quadratic formula. Rationalizing the numerator of this expression yields

$$\frac{-b - \sqrt{b^2 - 4ac}}{2a} = \frac{2c}{-b + \sqrt{b^2 - 4ac}}$$

as an alternative formula for calculating the smaller root. Note that in the denominator of this new formula, two nearly equal numbers are added, rather than subtracted, so the possibility of cancellation error has been eliminated. Substituting the coefficient values into the new formula produces

$$\frac{2 \cdot 6}{47.91 + 47.85} = \frac{12}{95.76} = 0.1253,$$

which is the floating point equivalent of 0.1253003555 in a 4 decimal digit number system with rounding.

EXAMPLE 1.11 A Linear System of Equations

Consider the solution of the linear system of three equations in three unknowns, given in augmented matrix form by

$$\left[\begin{array}{ccc|c} 6 & -2 & 3 & 5 \\ 1 & -1/3 & 1/3 & 2 \\ 1 & 3 & -1 & 5 \end{array}\right].$$

One pass of Gaussian elimination using exact arithmetic produces the matrix

$$\left[\begin{array}{ccc|c} 6 & -2 & 3 & 5 \\ 0 & 0 & -1/6 & 7/6 \\ 0 & 10/3 & -3/2 & 25/6 \end{array}\right],$$

from which the exact solution is computed to be $x_1 = 3.7$, $x_2 = -1.9$, and $x_3 = -7$. To arrive at these values, interchange the second and third rows of the reduced matrix and perform back substitution.

To examine the effect of working in 4 decimal digit rounding arithmetic, first express the original matrix in floating point:

$$\left[\begin{array}{ccc|c} 6 & -2 & 3 & 5 \\ 1 & -0.3333 & 0.3333 & 2 \\ 1 & 3 & -1 & 5 \end{array}\right].$$

For the first pass of Gaussian elimination, the pivot is placed in the first row, first column, and the multiplier needed to eliminate the 1's in the first column of the second and third rows is $fl(-1/6) = -0.1667$. The computations for the non-zero entries in the new second row produce

$$-0.3333 + (-2) \times (-0.1667) = -0.3333 + 0.3334 = 0.0001;$$
$$0.3333 + (3) \times (-0.1667) = 0.3333 - 0.5001 = -0.1668; \quad \text{and}$$
$$2 + (5) \times (-0.1667) = 2 - 0.8335 = 1.167.$$

Note the loss of significant figures as the result of the subtraction of nearly equal numbers in the computation of the element in row 2, column 2. The computations for the new third row produce

$$
\begin{aligned}
&3 + (-2) \times (-0.1667) = 3 + 0.3334 = 3.333; \\
&-1 + (3) \times (-0.1667) = -1 - 0.5001 = -1.500; \text{and} \\
&5 + (5) \times (-0.1667) = 5 - 0.8335 = 4.167.
\end{aligned}
$$

The matrix for the next pass of Gaussian elimination is then

$$
\left[\begin{array}{ccc|c} 6 & -2 & 3 & 5 \\ 0 & 0.0001 & -0.1668 & 1.167 \\ 0 & 3.333 & -1.500 & 4.167 \end{array}\right].
$$

The multiplier needed to eliminate the entry in the second column of the third row is $fl(-3.333/0.0001) = -33330$, resulting in the computations

$$
\begin{aligned}
&-1.500 + (-0.1668) \times (-33330) = -1.500 + 5559 = 5558; \text{and} \\
&4.167 + (1.167) \times (-33330) = 4.167 - 38900 = -38900.
\end{aligned}
$$

A cascade of effects has occurred here. Cancellation error led to a small pivot element, which then led to a large multiplier. The use of the large multiplier then resulted in the subtraction of numbers with drastically different magnitudes with a concomitant loss of significant digits. For example, the value 4.167 originally in the last row, last column of the matrix had no effect on the calculation of the new element at that location.

The final reduced matrix now takes the form

$$
\left[\begin{array}{ccc|c} 6 & -2 & 3 & 5 \\ 0 & 0.0001 & -0.1668 & 1.167 \\ 0 & 0 & 5558 & -38900 \end{array}\right].
$$

From the last row, it follows that $x_3 = -6.999$, which is a pretty good approximation for the exact value of x_3. Substituting $x_3 = -6.999$ into the second row generates the equation $0.0001x_2 - 0.1668(-6.999) = 1.167$, which is equivalent to $0.0001x_2 + 1.167 = 1.167$ in 4 decimal digit rounding arithmetic. The solution of this last equation is $x_2 = 0.0000$. Finally, substituting the values for x_2 and x_3 into the first row gives the equation $6x_1 + (-6.999) = 5$, whose solution is $x_1 = 4.333$. The error in x_2 is 100%, while that in x_1 is slightly more than 17%. These large errors can be directly traced to the loss of significant figures caused ultimately by the subtraction of nearly equal numbers.

One possible strategy for alleviating the accumulation of roundoff error during Gaussian elimination will be explored in the exercises. More will be said on this subject in Chapter 3.

EXAMPLE 1.12 Values of a Function

Even the seemingly straightforward task of evaluating a function can prove difficult. Suppose we need to evaluate the function $f(x) = e^x - \cos x - x$ for values of x very near zero. Because both e^x and $\cos x$ approach one as x approaches zero, it is possible that cancellation error will be a problem. Before we start calculating function values, however, let's do some analysis to determine what we should expect.

To start, we find that at $x = 0$, $f(0) = 0$. Next, we turn to the first derivative of f, $f'(x) = e^x + \sin x - 1$. For $0 < x < \pi$, $e^x > 1$ and $\sin x > 0$, so $f'(x) > 0$ and f is a strictly increasing function. On the other hand, for $-\pi < x < 0$, $e^x < 1$ and $\sin x < 0$, so $f'(x) < 0$ and f is a strictly decreasing function. Therefore, $x = 0$ is the only zero of f on the interval $(-\pi, \pi)$. Lastly, examine the second derivative, $f''(x) = e^x + \cos x$. For $-\pi/2 < x < \pi/2$, $e^x > 0$ and $\cos x > 0$, so $f''(x) > 0$ and f is concave up.

Now let's calculate some function values. Figure 1.4 plots f over the interval $-5 \times 10^{-8} \le x \le 5 \times 10^{-8}$. Points were generated at 1001 uniformly spaced abscissas, and all calculations were performed in IEEE standard double precision. The overall trend in the graph is in agreement with our analysis, but the fine detail clearly is not. For instance, the graph suggests several zeros closely spaced around $x = 0$ as opposed to a unique zero at $x = 0$.

Is there some way to reformulate this problem so as to more accurately reproduce the fine details of f? Given that the objective is to evaluate f near $x = 0$, it seems natural to try replacing both e^x and $\cos x$ by their respective Taylor series about $x = 0$. This yields

$$\begin{aligned} f(x) &= \left(1 + x + \frac{x^2}{2} + \frac{x^3}{6} + \frac{x^4}{24} + O(x^5)\right) - \left(1 - \frac{x^2}{2} + \frac{x^4}{24} + O(x^6)\right) - x \\ &= x^2 + \frac{x^3}{6} + O(x^5). \end{aligned}$$

It is left to the exercises to show that the relative error incurred by using $x^2 + \frac{x^3}{6}$ to approximate $e^x - \cos x - x$ is roughly 10^{-24} for $|x| \le 5 \times 10^{-8}$. Hence, for $|x| \le 5 \times 10^{-8}$, $e^x - \cos x - x = x^2 + \frac{x^3}{6}$ to machine precision in IEEE standard double precision. Furthermore, by rearranging the polynomial as

$$x^2 + \frac{x^3}{6} = x^2\left(1 + \frac{x}{6}\right),$$

we see that calculations with x near zero will not involve the subtraction of nearly equal numbers, so there should not be any problem with cancellation error. The plot in Figure 1.5 confirms this assessment.

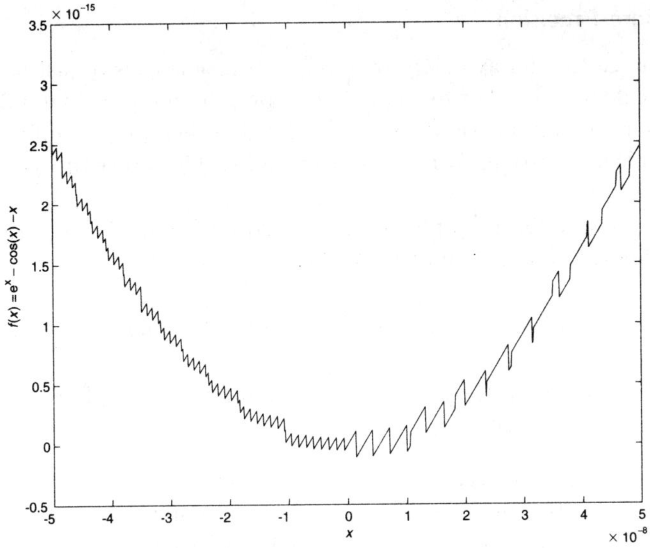

Figure 1.4 Graph of the function $f(x) = e^x - \cos x - x$ computed in IEEE standard double precision.

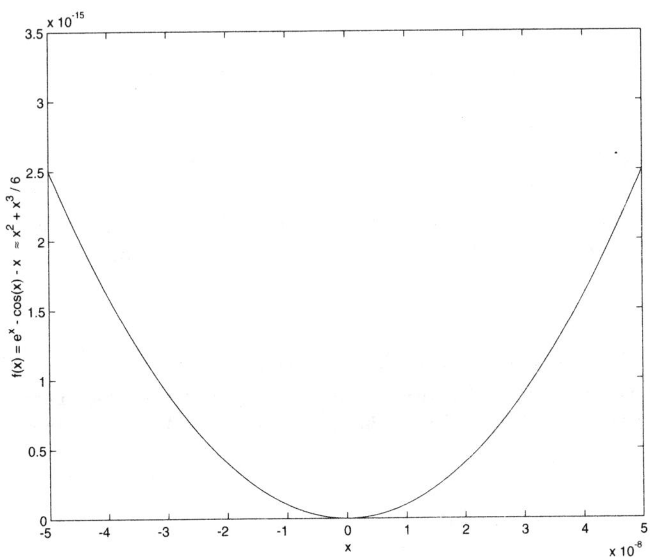

Figure 1.5 Graph of the function $f(x) = e^x - \cos x - x$ computed from the reformulated expression $f(x) \approx x^2 + \frac{x^3}{6}$ in IEEE standard double precision.

Just Add More Precision

A common suggestion for avoiding roundoff error problems is to just use higher precision in all calculations. For example, when programming in FORTRAN, change variable declarations from REAL*8 to REAL*16. When programming in C/C++, change variable declarations from `float` to `double`. This procedure does not always work, however.

Consider the following problem introduced by Rump [1] and reconsidered by Aberth [2]. Let $a = 77617.0$ and $b = 33096.0$ and compute

$$333.5b^6 + a^2\left(11a^2b^2 - b^6 - 121b^4 - 2\right) + 5.5b^8 + \frac{a}{2b}.$$

Rump reports that on an IBM System 370 mainframe, using FORTRAN, the following results were obtained:

single precision	$+1.172603\ldots$
double precision	$+1.1726039400531\ldots$
extended precision	$+1.172603940053178\ldots$

Although it would be tempting based on these values to claim that, to 7 digits, the value of this expression is $+1.172603$, the true value to 15 decimal places is -0.827396059946821. The FORTRAN values do not even generate the correct sign. In FORTRAN 77 on a Sun SPARCclassic workstation, the values

single precision	-6.33825×10^{29}
double precision	$-1.180591627174 \times 10^{21}$
extended precision	$+1.172603940053178\ldots$

were obtained. The single and double precision values have the correct sign but have dramatically different magnitude than the true value. The extended precision value agrees with that obtained by Rump. Finally, the Digits parameter in MAPLE Release 5 had to be set to at least 37 to obtain the true value.

References

1. Rump, S. M., "Algorithms for Verified Inclusions: Theory and Practice," in *Reliability in Computing*, R.E. Moore, ed., Academic Press, San Diego, 1988.
2. O. Aberth, *Precise Numerical Methods Using C++*, Academic Press, San Diego, 1998.

EXERCISES

1. Determine the value of each of the following expressions using 4-digit rounding and 4-digit chopping arithmetic. For each quantity, compute the absolute and the relative error.

   (a) $\pi + e - \cos 22^\circ$ **(b)** $e/7 + \sqrt{2}\ln\pi$

   (c) $\pi\ln 2 + \sqrt{10}\cos 22^\circ$ **(d)** $\left(\ln 2 - \sqrt{10} + \tan 22^\circ\right) / \left(7\sqrt[3]{9}\right)$

2. Identify the potential roundoff error problems in the following algorithm for calculating the roots of the quadratic equation $ax^2 + bx + c = 0$.

GIVEN:	real coefficients a, b, c
STEP 1:	calculate $disc = \sqrt{b^2 - 4ac}$
STEP 2:	calculate $root1 = (-b + disc)/(2a)$
STEP 3:	calculate $root2 = c/(a \cdot root1)$
OUTPUT:	$root1$ and $root2$

Note that this algorithm uses the fact that the product of the roots of $ax^2+bx+c=0$ is equal to c/a.

3. Identify the potential roundoff error problems in the following algorithm for calculating the roots of the quadratic equation $ax^2 + bx + c = 0$.

GIVEN:	real coefficients a, b, c
STEP 1:	calculate $disc = \sqrt{b^2 - 4ac}$
STEP 2:	calculate $root1 = -2c/(b + disc)$
STEP 3:	calculate $root2 = -(b/a) - root1$
OUTPUT:	$root1$ and $root2$

Note that this algorithm uses the fact that the sum of the roots of $ax^2+bx+c=0$ is equal to $-b/a$.

4. Construct an algorithm that computes the roots of the quadratic equation $ax^2+bx+c=0$ and that avoids as many roundoff error problems as possible. Test your algorithm by computing the roots of the quadratic equations $0.2x^2-47.91x+6=0$ and $0.025x^2+7x-0.1=0$. Use 4 decimal digit rounding arithmetic in your calculations.

5. Show that the relative error incurred by using $x^2+\frac{x^3}{6}$ to approximate $e^x-\cos x-x$ is roughly 10^{-24} for $|x| \le 5 \times 10^{-8}$.

6. In the floating point number system $\mathbf{F}(10, 10, -98, 100)$, subtract each of the following pairs of numbers. How many significant decimal digits are lost in performing the subtraction, and how does this compare with the number of significant decimal digits to which the numbers agree?

(a) $355/113$ and π

(b) $685/252$ and e

(c) $\cos(0.1°)$ and $\cos(0.11°)$

(d) $103/280$ and $1/e$

7. **(a)** To how many significant decimal digits do the numbers $\sqrt{10002}$ and $\sqrt{10001}$ agree?

(b) In the floating point number system $\mathbf{F}(10, 10, -98, 100)$, subtract $\sqrt{10001}$ from $\sqrt{10002}$. How many significant decimal digits are lost in performing the subtraction?

(c) Explain how you would rearrange your computations to obtain a more accurate answer.

8. **(a)** For what values of x does $(1 - \cos x)/x^2 = 1/2$ to full machine precision. Consider both IEEE standard single precision and double precision. (Hint: Use Taylor series.)

(b) Repeat part (a) to determine the values of x for which $e^{-x} = 1$ to full machine precision.

(c) Repeat part (a) to determine the positive values of x for which $\ln(1+x) - \cos x - x = -1$ to full machine precision.

9. (a) Plot the function $f(x) = 1 - \cos x$ over the interval $-5 \times 10^{-8} \leq x \leq 5 \times 10^{-8}$. Generate points at 1001 uniformly spaced abscissas and perform all calculations in IEEE standard double precision.

(b) Reformulate f to avoid cancellation error and then repeat part (a).

10. Repeat Exercise 9 for the function $f(x) = \tan^{-1} x - \sin x$.

11. Repeat Exercise 9 for the function $f(x) = \ln(1+x) - \cos x - x + 1$ over the interval $-5 \times 10^{-6} \leq x \leq 5 \times 10^{-6}$.

12. Near certain values of x each of the following functions cannot be accurately computed using the formula as given due to cancellation error. Identify the values of x which are involved (e.g., near $x = 0$ or large positive x) and propose a reformulation of the function (e.g., using Taylor series, rationalization, trig identities, etc.) to remedy the problem.

(a) $f(x) = 1 + \cos x$

(b) $f(x) = e^{-x} + \sin x - 1$

(c) $f(x) = \ln x - \ln(1/x)$

(d) $f(x) = \sqrt{x^2+1} - \sqrt{x^2+4}$

(e) $f(x) = 1 - 2\sin^2 x$

(f) $f(x) = \ln(x + \sqrt{x^2+1})$

(g) $f(x) = x - \sin x$

(h) $f(x) = \ln x - 1$

13. (a) Verify that

$$f(x) = 1 - \sin x \quad \text{and} \quad g(x) = \frac{\cos^2 x}{1 + \sin x}$$

are identical functions.

(b) Which function should be used for computations when x is near $\pi/2$? Why?

(c) Which function should be used for computations when x is near $3\pi/2$? Why?

14. It was noted that evaluation of the expression

$$333.5b^6 + a^2\left(11a^2b^2 - b^6 - 121b^4 - 2\right) + 5.5b^8 + \frac{a}{2b}$$

when $a = 77617.0$ and $b = 33096.0$ requires at least 37 decimal digits of precision.

(a) HP workstations have a double precision extended format that corresponds to the floating point number system $\mathbf{F}(2, 113, -16381, 16384)$. Does this system provide enough precision to evaluate the above expression?

(b) What is the smallest value for k for which the floating point number system $\mathbf{F}(2, k, m, M)$ provides 37 decimal digits of precision?

15. Consider the following linear system of equations:

$$\begin{bmatrix} 3.02 & -1.05 & 2.53 \\ 4.33 & 0.56 & -1.78 \\ -0.83 & -0.54 & 1.47 \end{bmatrix} \begin{bmatrix} x_1 \\ x_2 \\ x_3 \end{bmatrix} = \begin{bmatrix} -1.61 \\ 7.23 \\ -3.38 \end{bmatrix}.$$

(a) Determine the solution of this system using exact arithmetic during Gaussian elimination.

(b) Determine the solution of this system using 3 decimal digit rounding arithmetic during Gaussian elimination.

(c) Explain the difference between the answers found in part (a) and those found in part (b).

16. One strategy for alleviating the accumulation of roundoff error during Gaussian elimination is known as partial pivoting (a more detailed description of this process will be provided in Chapter 3). The basic idea is as follows. During the ith pass of Gaussian elimination, find the row, starting at row i and running through the last row of the matrix, that has the largest entry in column i. Interchange this row with the current row i and proceed with the elimination phase.

(a) Repeat Exercise 15(b) using this partial pivoting strategy. What is the relative error in each component of the computed solution?

(b) Repeat the Linear System of Equations problem considered in the text using the partial pivoting strategy. What is the relative error in each component of the computed solution?

CHAPTER 2

Rootfinding

AN OVERVIEW

Fundamental Mathematical Problem

In this chapter, several techniques will be developed for finding approximate solutions to the general mathematical problem

given a function f, find a value for x such that $f(x) = 0$.

Such an x is called a *zero* of the function f or a *root* of the equation $f(x) = 0$. This problem is therefore known as the *rootfinding problem.* In the most general setting of the rootfinding problem, both the function f and the independent variable x could be vector valued. In this chapter, only the scalar case will be considered. Systems of nonlinear equations will be treated in Chapter 3.

The "Solving a Crime" problem capsule in the Chapter 1 Overview (see page 2) is one application that gives rise to a rootfinding problem. Here are two more examples.

van der Waals Equation

Every student of high school chemistry has been exposed to the ideal gas law:

$$PV = nRT,$$

which relates the pressure (P), volume (V), and temperature (T) of an ideal gas. Here, n represents the number of moles of gas present, and R is the universal gas constant. Real gases satisfy this equation only approximately; under conditions of high pressure and/or low volume the approximation becomes more crude. One attempt to model the relationship among pressure, volume, and temperature for real gases is the van der Waals equation:

$$\left(P + \frac{n^2a}{V^2}\right)(V - nb) = nRT.$$

The term involving the parameter a corrects the pressure for intramolecular attractive forces (i.e., the pressure would be higher if not for the attractive forces exerted among the molecules in the gas). The term involving the parameter b is a correction for that portion of the volume of the gas that is not compressible due to the intrinsic volume of the gas molecules. Suppose that one mole of chlorine gas has a pressure of 2 atmospheres and a temperature of 313 K. For chlorine gas, $a = 6.29$ atm $\cdot$ liter^2/mole^2 and $b = 0.0562$ liter/mole. What is the volume of the gas?

Depth of Submersion

Suppose we want to determine how far a spherical object of radius R will sink into a fluid such as water or oil. According to Archimedes' principle, the object will sink to the depth at which the weight of the fluid displaced by the object equals the weight of the object. Now, the weight of the object is the product of its mass, m, and the acceleration due to gravity, g. If we assume the object has a constant mass density, ρ_o, then $m = \frac{4}{3}\pi R^3 \rho_o$ and

$$\text{weight of the object} = \frac{4}{3}\pi R^3 \rho_o g. \tag{1}$$

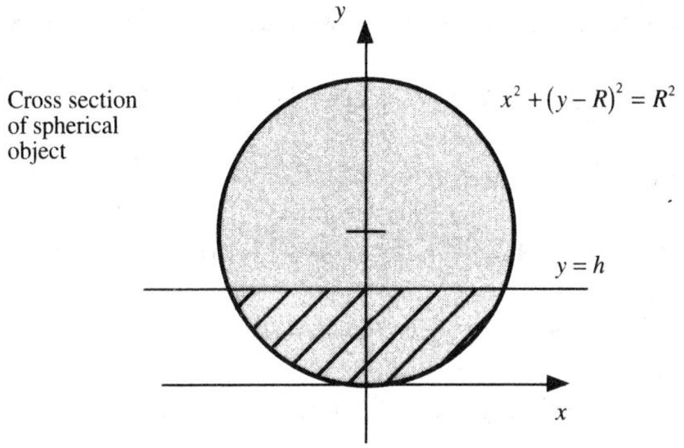

What about the weight of the displaced fluid? Assuming the fluid has density ρ_f and V_d is the volume of fluid displaced by the object, then

$$\text{weight of displaced fluid} = \rho_f V_d g. \tag{2}$$

To complete the specification of the problem, we need to determine V_d. Suppose the object sinks to a depth h. Considering the geometry shown in the diagram above and applying some basic Calculus (volume by slicing to be exact), we find that

$$V_d = \pi \int_0^h x^2 dy = \pi \int_0^h (2Ry - y^2)\, dy = \pi h^2 \left(R - \frac{h}{3}\right). \tag{3}$$

Substituting (3) into (2) and equating the resulting expression with (1) yields

$$\frac{4}{3}\pi R^3 \rho_o g = \pi h^2 \left(R - \frac{h}{3}\right) \rho_f g.$$

After some algebraic simplification, this becomes

$$\frac{\rho_f}{3} h^3 - R\rho_f h^2 + \frac{4}{3} R^3 \rho_o = 0. \tag{4}$$

Therefore, given values for R, ρ_o and ρ_f, the depth to which the object sinks is determined by solving equation (4) for h.

Multiplicity

Throughout this chapter, the concept of the *multiplicity* of a root will play an important role.

Definition. A root p of the equation $f(x) = 0$ is said to be a ROOT OF MULTIPLICITY m if f can be written in the form

$$f(x) = (x-p)^m q(x),$$

where $\lim_{x\to p} q(x) \neq 0$. A root of multiplicity one is called a SIMPLE ROOT.

For polynomials, multiplicity can be determined by first factoring the polynomial and then examining the power on each factor. For example, since

$$x^6 + x^5 - 12x^4 + 2x^3 + 41x^2 - 51x + 18 = (x-1)^3(x+3)^2(x-2),$$

it follows that the equation

$$x^6 + x^5 - 12x^4 + 2x^3 + 41x^2 - 51x + 18 = 0$$

has a root of multiplicity 3 at $x = 1$, a root of multiplicity 2 at $x = -3$, and a simple root at $x = 2$.

What about the equation $f(x) = 0$, where

$$f(x) = 2x + \ln\left(\frac{1-x}{1+x}\right)?$$

Clearly, $f(0) = 0$, so the equation has a root at $x = 0$. But what is the multiplicity of this root? For equations with non-polynomial functions, the following theorem is helpful.

Theorem. Let f be a continuous function with m continuous derivatives. The equation $f(x) = 0$ has a root of multiplicity m at $x = p$ if and only if $f(p) = f'(p) = f''(p) = \cdots = f^{(m-1)}(p) = 0$ but $f^{(m)}(p) \neq 0$.

Proof. Let f be a continuous function with m continuous derivatives. Suppose that $f(p) = f'(p) = f''(p) = \cdots = f^{(m-1)}(p) = 0$ but $f^{(m)}(p) \neq 0$. Expanding f in a Taylor polynomial of degree $m-1$ about the point $x = p$ yields

$$f(x) = \sum_{k=0}^{m-1} \frac{f^{(k)}(p)}{k!}(x-p)^k + \frac{f^{(m)}(\xi(x))}{m!}(x-p)^m,$$

where $\xi(x)$ is between x and p. Using the hypotheses regarding the value of $f^{(k)}(p)$ for $0 \leq k \leq m-1$, this last expression simplifies to

$$f(x) = \frac{f^{(m)}(\xi(x))}{m!}(x-p)^m.$$

Let's examine the factor $f^{(m)}(\xi(x))$ more closely. Since $f^{(m)}$ is continuous, it follows that

$$\lim_{x\to p} f^{(m)}(\xi(x)) = f^{(m)}(\lim_{x\to p} \xi(x)) = f^{(m)}(p) \neq 0.$$

Now, define $q(x) = f^{(m)}(\xi(x))/m!$. Then $f(x) = (x-p)^m q(x)$, where

$$\lim_{x\to p} q(x) \neq 0,$$

and the equation $f(x) = 0$ has a root of multiplicity m at $x = p$.

Conversely, suppose that $f(x) = 0$ has a root of multiplicity m at $x = p$. Then there exists a function q, with $\lim_{x\to p} q(x) \neq 0$, such that $f(x) = (x-p)^m q(x)$. By direct calculation, we find

$$f(p) = f'(p) = f''(p) = \cdots = f^{(m-1)}(p) = 0;$$

however, for the mth derivative, we find

$$f^{(m)}(p) = \lim_{x\to p} f^{(m)}(x) = m! \lim_{x\to p} q(x) \neq 0,$$

as desired. □

Returning to the problem posed before the theorem, for $f(x) = 2x + \ln\left(\frac{1-x}{1+x}\right)$, we calculate

$$f(0) = f'(0) = f''(0) = 0, \quad \text{but} \quad f'''(0) = -4 \neq 0.$$

Hence, the equation $f(x) = 0$ has a root of multiplicity 3 at $x = 0$.

Remainder of Chapter

Basic techniques for solving the rootfinding problem fall into two categories: simple enclosure methods and fixed point iteration schemes. Although the word *iteration* appears in the name of only one of these categories, both classes of methods are, in fact, iterative processes. The development of two simple enclosure methods, the bisection method and the method of false position, will be the focus of Sections 1 and 2, respectively. These techniques will be guaranteed to converge to a root of the specified function under very mild conditions; the rate of convergence will tend to be quite slow, however. The general theory of fixed point iteration schemes will be discussed in the next section. Properly constructed, such schemes will exhibit very rapid convergence; unfortunately, these techniques require stronger conditions to guarantee convergence. Section 4 will be devoted to the development of Newton's method, the classical fixed point iteration scheme. The secant method, which can be considered a variation on either Newton's method or the method of false position, will be developed in Section 5. In Section 6, general techniques for accelerating the convergence of iterative schemes will be developed. Special techniques for accelerating Newton's method will also be presented. The chapter concludes with a section dealing with the special problem of locating roots of polynomials.

A Remark about Pathological Examples

Although many powerful techniques for locating the roots of functions will be developed in this chapter, it must be kept in mind that there are problems for which it will be difficult, if not downright impossible, for even the best of techniques to find the desired solution. Consider, for example, locating the roots of

$$f(x) = 3x^2 + \frac{1}{\pi^4}\ln\left[(\pi - x)^2\right] + 1.$$

It is fairly easy to establish that f has two simple real roots. First, note that since $3x^2 + 1$ is always positive, the term involving $\ln\left[(\pi - x)^2\right]$ must be negative for f to evaluate to zero. This implies that any roots must lie on the interval $x \in (\pi - 1, \pi + 1)$. Combining the facts that $\lim_{x\to\pi} \ln\left[(\pi - x)^2\right] \to -\infty$ and f is continuous everywhere except at $x = \pi$ with the knowledge that $f(\pi - 1) > 0$ and $f(\pi+1) > 0$ guarantees the existence of a root on each of the intervals $(\pi - 1, \pi)$ and $(\pi, \pi + 1)$. Monotonicity of f on $(-\infty, \pi)$ and on (π, ∞) guarantees the uniqueness of the root in each interval.

Since the natural logarithm term must balance $3x^2 + 1$ and the coefficient on the logarithm term is roughly 0.01, it is reasonable to assume that both roots are close to π. It follows that $\ln\left[(\pi - x)^2\right]$ must be roughly $-\left(3x^2 + 1\right)\pi^4\big|_{x=\pi}$ or -2981.6. Therefore, $(\pi - x)$ must be on the order of $\pm e^{-1490.788} \approx \pm 10^{-647}$, so that $x \approx \pi \pm 10^{-647}$. The floating point number system on most machines will never be able to resolve these values. The moral of the story is a simple one: Pathological problems do exist, so always perform some basic analysis before rushing to the computer.

2.1 THE BISECTION METHOD

As noted in the overview to this chapter, rootfinding techniques are generally divided into two categories: simple enclosure methods and fixed point iteration schemes. All simple enclosure methods are based on the Intermediate Value Theorem. These methods essentially work by first finding an interval which is guaranteed to contain a root and then systematically shrinking the size of that interval. In this section, we will develop and study the performance of the most basic simple enclosure method, which is known as the bisection method.

Intermediate Value Theorem

Before we begin our development of the bisection method, let's take a moment to review the Intermediate Value Theorem. This theorem appears in probably every calculus book; for a proof, consult a textbook on advanced calculus or real analysis.

> **Theorem.** Let f be a continuous function over the closed interval $[a, b]$, and let k be any real number that lies between the values $f(a)$ and $f(b)$. Then there exists a real number c with $a < c < b$ such that $f(c) = k$.

In plain English, a function that is continuous on a closed interval is guaranteed to assume every value between the values achieved at the endpoints of the interval.

So what does this have to do with the rootfinding problem? Basically, the Intermediate Value Theorem provides a means for identifying intervals which enclose the real zeros of continuous functions. All that is needed is to find an interval such that the values of the function at the endpoints of that interval are of opposite sign. The magnitudes of these endpoint values are irrelevant. As long as one endpoint value is positive and the other negative, zero is somewhere between the values, and at least one zero of the function is guaranteed to exist on that interval.

To demonstrate this idea, consider the function $f(x) = x^3 + 2x^2 - 3x - 1$. The value of this function at a string of consecutive integers is listed below. Each change in the sign of the function value signals an interval that contains a real zero of the function. This function, therefore, clearly has three simple real zeros, one each on the intervals: $(-3, -2)$, $(-1, 0)$, and $(1, 2)$.

$$\left.\begin{array}{l} f(-3) = -1 \\ f(-2) = 5 \end{array}\right\rangle \qquad \left.\begin{array}{l} f(-1) = 3 \\ f(0) = -1 \end{array}\right\rangle \qquad \left.\begin{array}{l} f(1) = -1 \\ f(2) = 9 \end{array}\right\rangle$$

Bisection Method

Suppose we have used the Intermediate Value Theorem to locate an interval that contains a zero of a continuous function. What do we do next? Our objective will be to systematically shrink the size of that root enclosing interval. Perhaps the simplest and most natural way to accomplish a reduction in interval size is to cut the interval in half. Once this has been done, we determine which half contains a root, by once again using the Intermediate Value Theorem, and then repeat the process on that half. This technique is known as the *bisection method.*

From this very basic description of the bisection method, it should be clear that the method generates a sequence of root enclosing intervals. For notational convenience, let (a_n, b_n) be the enclosing interval during the nth iteration of the method. Furthermore, let p_n denote the midpoint of the interval $[a_n, b_n]$; that is,

$$p_n = \frac{a_n + b_n}{2}.$$

We will use p_n not only as one of the endpoints for the next enclosing interval, but also as an approximation to the location of the exact root p. If p_n is an accurate enough approximation—an issue that will be addressed shortly—the iteration is terminated; otherwise, the Intermediate Value Theorem is invoked to determine which of the two subintervals, (a_n, p_n) or (p_n, b_n), contains the root and becomes (a_{n+1}, b_{n+1}). The entire process is then repeated on that subinterval.

EXAMPLE 2.1 Bisection Method in Action

We discovered earlier that the function $f(x) = x^3 + 2x^2 - 3x - 1$ has a simple zero on the interval $(1, 2)$. Let's run through a few iterations of the bisection method to demonstrate the general procedure.

For the first iteration, we have $(a_1, b_1) = (1, 2)$ and we know that $f(a_1) < 0$ and that $f(b_1) > 0$. The midpoint of this first interval, and our first approximation to the location of the exact root, is

$$p_1 = \frac{a_1 + b_1}{2} = \frac{1+2}{2} = 1.5.$$

To determine whether the root is contained on $(a_1, p_1) = (1, 1.5)$ or on $(p_1, b_1) = (1.5, 2)$, we calculate

$$f(p_1) = 2.375 > 0.$$

Since $f(a_1)$ and $f(p_1)$ are of opposite sign, the Intermediate Value Theorem tells us the root is between a_1 and p_1. For the next iteration, we therefore take $(a_2, b_2) = (a_1, p_1) = (1, 1.5)$.

The midpoint of this new interval, and our second approximation to the location of the root, is

$$p_2 = \frac{a_2 + b_2}{2} = \frac{1+1.5}{2} = 1.25.$$

Note that

$$f(p_2) \approx 0.328 > 0,$$

which is of opposite sign from $f(a_2)$. Hence, the Intermediate Value Theorem tells us the root is between a_2 and p_2, so we take $(a_3, b_3) = (a_2, p_2) = (1, 1.25)$.

In the third iteration, we then calculate

$$p_3 = \frac{a_3 + b_3}{2} = \frac{1+1.25}{2} = 1.125$$

and

$$f(p_3) \approx -0.420 < 0.$$

Here, we find that $f(a_3)$ and $f(p_3)$ are of the same sign, which implies that the root must lie somewhere between p_3 and b_3. For the fourth iteration, we will therefore have $(a_4, b_4) = (p_3, b_3) = (1.125, 1.25)$ and

$$p_4 = \frac{a_4 + b_4}{2} = 1.1875.$$

To ten decimal places, $p = 1.1986912435$, so the absolute error in p_4 is 1.119×10^{-2}.

Even though we've developed the basic iterative process which lies at the heart of the bisection method, we're not yet ready to construct an algorithm. Since the bisection method is iterative in nature, the algorithm must contain a stopping condition. We have to have some way to decide when p_n is sufficiently accurate to terminate the iteration. However, to properly formulate a stopping condition, we need to understand the convergence properties of the sequence generated by the bisection method. We will now undertake an analysis of these properties.

Convergence Analysis

Under what circumstances will the sequence of approximations generated by the bisection method converge to a root of $f(x) = 0$? When the sequence does converge, what is the speed of convergence? Much of the information we need to answer these questions is contained in the following theorem.

Theorem. Let f be continuous on the closed interval $[a, b]$ and suppose that $f(a)f(b) < 0$. The bisection method generates a sequence of approximations $\{p_n\}$ which converges to a root $p \in (a, b)$ with the property

$$|p_n - p| \leq \frac{b-a}{2^n}.$$

Notes

1. Pay close attention to the conclusion of this theorem. It states that the bisection method converges to *a* root of f, not *the* root of f. The condition $f(a)f(b) < 0$ implies differing signs at the endpoints of the interval, which guarantees the existence of a root, but not uniqueness. There may be more than one root on the interval and there is no way to know, a priori, to which root the sequence will converge, but it will converge to one of them.
2. Since $|p_n - p|$ is the absolute error in the approximation p_n, the expression on the right-hand side of the inequality at the end of the theorem is referred to as a theoretical error bound. The error at any stage of the iterative process can never be larger than this quantity. Working with problems for which the analytical solution is known and verifying that a theoretical error bound is satisfied is a powerful tool for eliminating "human errors" in the development of computer codes.
3. The requirement that an interval $[a, b]$ be found such that $f(a)f(b) < 0$ implies that the bisection method cannot be used to locate roots of even multiplicity. For such roots, the sign of the function does not change on either side of the root. This restriction is, in fact, common to all simple enclosure techniques and is not peculiar to the bisection method.

Proof of Theorem. Since the quantity $b - a$ is constant and $2^{-n} \to 0$ as $n \to \infty$, establishing the error bound will be sufficient to prove convergence of the bisection method sequence. By construction of the bisection algorithm and using the notation introduced previously, for each n, $p \in (a_n, b_n)$ and p_n is taken as the midpoint of (a_n, b_n). This implies that p_n can differ from p by no more than half the length of (a_n, b_n); that is,

$$|p_n - p| \leq \frac{1}{2}(b_n - a_n).$$

However, again by construction,

$$b_n - a_n = \frac{1}{2}(b_{n-1} - a_{n-1}) = \frac{1}{4}(b_{n-2} - a_{n-2}) = \cdots = \frac{1}{2^{n-1}}(b_1 - a_1).$$

Recalling that $b_1 = b$ and $a_1 = a$ and combining the last two equations produces the desired error bound

$$|p_n - p| \le \frac{b-a}{2^n}. \qquad \square$$

So the sequence of approximations generated by the bisection method is always guaranteed to converge to a root of the equation $f(x) = 0$. How quickly will the sequence converge? From the theoretical error bound, we have

$$|p_n - p| \le \frac{b-a}{2^n} = (b-a)\frac{1}{2^n}.$$

Hence, if we take $\lambda = b - a$ and $\beta_n = 1/2^n$ in the definition of rate of convergence, we see that the sequence generated by the bisection method has rate of convergence $O(1/2^n)$.

What about order of convergence? Given that each time the enclosing interval is cut in half we obtain an extra base 2 digit of accuracy, we might expect to find that convergence is linear (i.e., $\alpha = 1$). Unfortunately, we run into a slight problem when we try to apply the definition. Examine the last column of Table 2.1. This table shows the results of fifteen iterations of the bisection method when applied to the function $f(x) = x^3 + 2x^2 - 3x - 1$ with a starting interval of $(1, 2)$. Observe that sometimes the error drops sharply from iteration to iteration (e.g., from iteration 10 to iteration 11), sometimes the error deceases only slightly (from iteration 6 to 7), and sometimes the error actually increases. It is therefore quite likely that the limit which appears in the definition of order of convergence won't exist.

All is not lost, however. In Section 1.2, we saw that for a linearly convergent sequence

$$|p_{n+1} - p| \approx \lambda^n |p_1 - p|,$$

where λ is the asymptotic error constant. As evidenced by the theoretical error bound, the bisection method sequence does satisfy this relationship with $\lambda = 1/2$. Furthermore, observe from Figure 2.1 that the overall relationship between $\log|e_{n+1}|$ and $\log|e_n|$ appears to be linear with slope one. From this, it then follows that the general trend between old and new errors is linear. We therefore stretch the definition of order of convegence and say that the convergence of the bisection method sequence is order $\alpha = 1$ with asymptotic error constant $\lambda = 1/2$.

Stopping Condition and Algorithm

We are now in position to select a stopping condition. In what follows, let ϵ be a specified convergence tolerance. For any rootfinding technique, there are three primary measures of convergence with which to construct the stopping condition. These are

(1) the absolute error in the location of the root

Terminate the iteration when $|p_n - p| < \epsilon$.

	Enclosing Interval	Approximation	Absolute Error
1	(1.000000,2.000000)	1.500000	0.3013087565
2	(1.000000,1.500000)	1.250000	0.0513087565
3	(1.000000,1.250000)	1.125000	0.0736912435
4	(1.125000,1.250000)	1.187500	0.0111912435
5	(1.187500,1.250000)	1.218750	0.0200587565
6	(1.187500,1.218750)	1.203125	0.0044337565
7	(1.187500,1.203125)	1.195312	0.0033787435
8	(1.195312,1.203125)	1.199219	0.0005275065
9	(1.195312,1.199219)	1.197266	0.0014256185
10	(1.197266,1.199219)	1.198242	0.0004490560
11	(1.198242,1.199219)	1.198730	0.0000392252
12	(1.198242,1.198730)	1.198486	0.0002049154
13	(1.198486,1.198730)	1.198608	0.0000828451
14	(1.198608,1.198730)	1.198669	0.0000218099
15	(1.198669,1.198730)	1.198700	0.0000087077

TABLE 2.1: Fifteen Iterations of Bisection Method Applied to $f(x) = x^3 + 2x^2 - 3x - 1$ Starting from the Interval $(1, 2)$

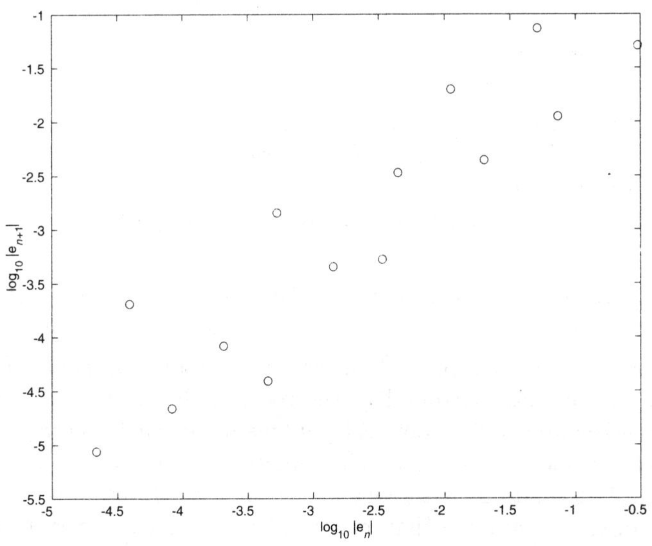

Figure 2.1 Error after $n + 1$ iterations versus error after n iterations for approximations generated by the bisection method when applied to the function $f(x) = x^3 + 2x^2 - 3x - 1$. A log-log scale has been used to accomodate the variation in the order of magnitude of the errors.

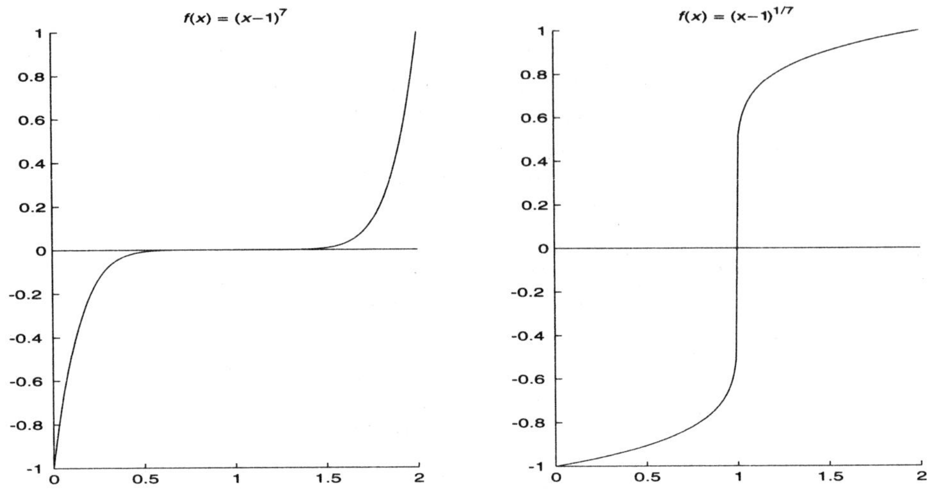

Figure 2.2 (Left graph) The error conditions based on the location of the root will provide more reliable termination of the iterative process for this function. Simply requiring the value of the function to be small could lead to large errors in the approximation of the location of the root. (Right graph) The test for root condition will provide more reliable termination of the iterative process for this function. Simply requiring that p_n be "close" to p could lead to $f(p_n)$ being far from zero.

(2) the relative error in the location of the root

Terminate the iteration when $|p_n - p| < \epsilon|p_n|$.

(3) the test for a root

Terminate the iteration when $|f(p_n)| < \epsilon$.

There is no general rule of thumb for selecting one stopping condition over another, and it is worth noting that none of these conditions works well in all cases.

Consider the function plotted on the left in Figure 2.2, $f(x) = (x-1)^7$, which has a wide, flat plateau surrounding the root. Either of the stopping conditions based on the location of the root will produce more reliable termination than the test for root condition in the sense that a function value near zero will not guarantee a small error in the approximate location of the root. In particular, even if $f(x) \approx 10^{-7}$, we can only guarantee that $x \approx 1 \pm 10^{-1}$. A small error in the location of the root, on the other hand, will always lead to a function value near zero. The reliability of the stopping conditions is reversed for the function plotted on the right in Figure 2.2, $f(x) = (x-1)^{1/7}$, which has a nearly vertical portion surrounding the root. For this function, a function value near zero guarantees that x must be close to 1, but having x close to 1 does not imply that $f(x)$ will be close to 0. In particular, if $x \approx 1 \pm 10^{-7}$, then $f(x) \approx 10^{-1}$.

Throughout our development of rootfinding techniques we will implement a stopping condition based on the absolute error in the location of the root. The careful reader is now asking how we are going to accomplish this since we don't know the value of p. In the case of the bisection method, though, we do have some theory to fall back on. From the proof of the bisection method convergence theorem, we know that

$$|p_n - p| \leq \frac{b_n - a_n}{2}.$$

We can therefore terminate the bisection method when

$$\frac{b_n - a_n}{2} < \epsilon.$$

Bringing together the basic iterative structure we developed earlier and the stopping condition we just selected, we can now construct an algorithm for the bisection method.

```
GIVEN:    function whose zero is to be located, f
          left endpoint of interval, a
          right endpoint of interval, b
          convergence tolerance, ε
          maximum number of iterations, Nmax

STEP 1:   save sfa = sign ( f(a) )
STEP 2:   for i from 1 to Nmax do
STEP 3:        p = a + (b − a)/2
STEP 4:        if ( (b − a) < 2ε ) then OUTPUT p
STEP 5:        save sfp = sign ( f(p) )
STEP 6:        if ( sfa * sfp < 0 ) then
                    assign the value of p to b
               else
                    assign the value of p to a
                    assign the value of sfp to sfa
               end
          end
STEP 7:   OUTPUT a message that the maximum number
               of iterations has been exceeded prior to
               achieving convergence
```

There are a few important remarks that need to be made regarding this algorithm. First, although we introduced the bisection method in terms of a sequence of enclosing intervals and a sequence of approximations to the location of the zero, a close examination of the method indicates that we only need to know the current enclosing interval and the current approximation to ensure proper execution. This is why the variables a, b, and p in the above algorithm are scalars and not arrays.

Second, the standard measure for the amount of work performed by a rootfinding technique is the number of times the function f is evaluated, not the number of iterations. In general, the function may be very complicated, and each evaluation may require many floating point operations. We therefore want to avoid any unnecessary function evaluations. This is the reason that the signs of $f(a)$ and $f(p)$ are saved in STEPS 1 and 5. Had these signs not been saved, we would have had to re-evaluate the function at both a and p to perform the test in STEP 6. By saving the signs, the algorithm, as written, requires only one new function evaluation per iteration.

Finally, observe that in STEP 6, we worked with the signs of the function values, rather than checking the sign of the product $f(a)f(p)$. By construction, both a and p will be converging toward a zero of f. Hence, both $f(a)$ and $f(p)$ will be approaching zero. Multiplying these values together could then result in underflow.

EXAMPLE 2.2 A Second Demonstration Problem

As we develop additional rootfinding techniques in subsequent sections, we will want to have at least a couple of problems on hand with which to compare performance. One of the problems we will use is locating the root of

$$x^3 + 2x^2 - 3x - 1 = 0$$

on the interval $(1, 2)$. The values contained in the first three columns of Table 2.1 (see page 63) were obtained by applying the bisection method algorithm to this problem with a convergence tolerance of 5×10^{-5}.

As a second example, we will use the equation

$$\tan(\pi x) - x - 6 = 0.$$

This equation actually has an infinite number of roots. Here, we want to approximate the smallest positive root, which Figure 2.3 suggests lies on the interval $(0.40, 0.48)$. Applying the bisection method algorithm to the function $f(x) = \tan(\pi x) - x - 6$ with a convergence tolerance of 5×10^{-5} produces the results

	Enclosing Interval	Approximation
1	(0.400000,0.480000)	0.4400000000
2	(0.440000,0.480000)	0.4600000000
3	(0.440000,0.460000)	0.4500000000
4	(0.450000,0.460000)	0.4550000000
5	(0.450000,0.455000)	0.4525000000
6	(0.450000,0.452500)	0.4512500000
7	(0.450000,0.451250)	0.4506250000
8	(0.450625,0.451250)	0.4509375000
9	(0.450937,0.451250)	0.4510937500
10	(0.450937,0.451094)	0.4510156250
11	(0.451016,0.451094)	0.4510546875

To ten decimal places, $p = 0.4510472588$, so the absolute error in the final bisection method approximation is roughly 7.426×10^{-6}.

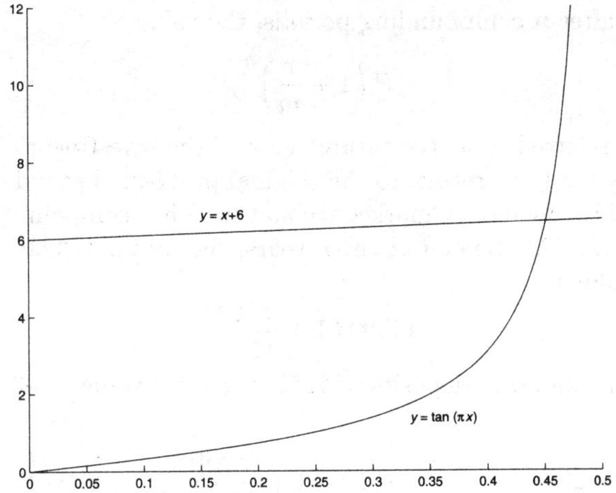

Figure 2.3 The point of intersection between the graph of $y = x + 6$ and the graph of $y = \tan(\pi x)$ is the location of the smallest positive root of $\tan(\pi x) - x - 6 = 0$.

An Application Problem: Saving for a Down Payment

A couple plans to open a money market account in which they will save the down payment for the purchase of a home. They have \$13,500 from the sale of some stock with which to open the account. After examining their budget, they feel they can comfortably deposit an extra \$250 into the account each month. What is the minimum interest rate, compounded on a monthly basis, that the couple must earn on their investment to reach their goal of accumulating \$25,000 within three years?

To answer this question, we need to determine how money that earns compound interest grows over time. Suppose that P dollars are invested at an annual interest rate r, compounded m times per year. At the end of the first compounding period, interest in the amount of $P\frac{r}{m}$ is credited to the account. The total value of the investment is then

$$P + P\frac{r}{m} = P\left(1 + \frac{r}{m}\right).$$

The interest earned at the end of the next compounding period is $P\left(1 + \frac{r}{m}\right)\frac{r}{m}$, so the value of the account grows to

$$P\left(1 + \frac{r}{m}\right) + P\left(1 + \frac{r}{m}\right)\frac{r}{m} = P\left(1 + \frac{r}{m}\right)\left(1 + \frac{r}{m}\right) = P\left(1 + \frac{r}{m}\right)^2.$$

After the third compounding period, the account value becomes

$$P\left(1 + \frac{r}{m}\right)^2 + P\left(1 + \frac{r}{m}\right)^2\frac{r}{m} = P\left(1 + \frac{r}{m}\right)^2\left(1 + \frac{r}{m}\right) = P\left(1 + \frac{r}{m}\right)^3,$$

and, in general, after n compounding periods, the value is

$$P\left(1+\frac{r}{m}\right)^{n}.$$

This quantity is referred to as the future value of the investment.

We are now ready to return to the original problem. Let r denote the annual interest rate paid by the money market account, which is compounded on a monthly basis (i.e., $m = 12$). At the end of three years, the initial \$13,500 investment will have grown in value to

$$13500\left(1+\frac{r}{12}\right)^{36}$$

dollars. As for the monthly deposits of \$250, the total value of all 36 deposits will be

$$250\left(1+\frac{r}{12}\right)^{35}+250\left(1+\frac{r}{12}\right)^{34}+250\left(1+\frac{r}{12}\right)^{33}+\cdots+250\left(1+\frac{r}{12}\right)+250. \quad (1)$$

Here, we have used the fact that the first monthly deposit earns 35 months of interest, the second earns 34 months of interest, the third earns 33 months of interest, and so on. The sum of the geometric progression given by (1) may be expressed in closed form as

$$250\frac{\left(1+\frac{r}{12}\right)^{36}-1}{r/12}.$$

If no other deposits are made to and no withdrawals are taken from the account, the couple will therefore have saved

$$13500\left(1+\frac{r}{12}\right)^{36}+250\frac{\left(1+\frac{r}{12}\right)^{36}-1}{r/12}$$

dollars for the down payment by the end of three years. The minimum interest rate that the money market account must pay for the couple to reach their goal is, accordingly, the solution of the equation

$$13500\left(1+\frac{r}{12}\right)^{36}+250\frac{\left(1+\frac{r}{12}\right)^{36}-1}{r/12}=25000.$$

Let's define

$$f(r)=13500\left(1+\frac{r}{12}\right)^{36}+250\frac{\left(1+\frac{r}{12}\right)^{36}-1}{r/12}-25000.$$

Note that $f(0.01) = -1956.54$, but $f(0.10) = 3645.91$, so the desired interest rate is somewhere between 1% and 10%. Using the bisection method with a convergence tolerance of 5×10^{-6}, we find, after 15 iterations,

$$r = 0.0439395.$$

The couple therefore needs to find an account paying roughly 4.40%, compounded monthly.

EXERCISES

1. Verify that each of the following equations has a root on the interval $(0, 1)$. Next, perform the bisection method to determine p_3, the third approximation to the location of the root, and to determine (a_4, b_4), the next enclosing interval.
   (a) $\ln(1+x) - \cos x = 0$ **(b)** $x^5 + 2x - 1 = 0$
   (c) $e^{-x} - x = 0$ **(d)** $\cos x - x = 0$

In Exercises 2–5, verify that the given function has a zero on the indicated interval. Next, perform the first five (5) iterations of the bisection method and verify that each approximation satisfies the theoretical error bound of the bisection method, but that the actual errors do not steadily decrease. The exact location of the zero is indicated by the value of p.

2. $f(x) = x^3 + x^2 - 3x - 3, \quad (1, 2), \quad p = \sqrt{3}$
3. $f(x) = \sin x, \quad (3, 4), \quad p = \pi$
4. $f(x) = 1 - \ln x, \quad (2, 3), \quad p = e$
5. $f(x) = x^6 - 3, \quad (1, 2), \quad p = \sqrt[6]{3}$
6. Determine a formula which relates the number of iterations, n, required by the bisection method to converge to within an absolute error tolerance of ϵ, starting from the initial interval (a, b).
7. Modify the algorithm for the bisection method as follows. Remove the input *Nmax*, and calculate the number of iterations needed to achieve the specified convergence tolerance using the results of Exercise 6.
8. Suppose that an equation is known to have a root on the interval $(0, 1)$. How many iterations of the bisection method are needed to achieve full machine precision in the approximation to the location of the root assuming calculations are performed in IEEE standard double precision? What if the root were known to be contained in the interval $(8, 9)$? (*Hint:* Consider the number of base 2 digits already known in the location of the root and how many base 2 digits are available in the indicated floating point system.)
9. By construction, the endpoints of the enclosing intervals produced by the bisection method satisfy $a_1 \le a_2 \le a_3 \le \cdots \le b_3 \le b_2 \le b_1$. Prove that the sequences $\{a_n\}$ and $\{b_n\}$ converge and that
$$\lim_{n\to\infty} a_n = \lim_{n\to\infty} b_n = \lim_{n\to\infty} p_n = p.$$
10. It was noted that the function $f(x) = x^3 + 2x^2 - 3x - 1$ has a zero on the interval $(-3, -2)$ and another on the interval $(-1, 0)$. Approximate both of these zeroes to within an absolute tolerance of 5×10^{-5}.
11. Approximate $\sqrt[3]{13}$ to three decimal places by applying the bisection method to the equation $x^3 - 13 = 0$.
12. Approximate $1/37$ to five decimal places by applying the bisection method to the equation $1/x - 37 = 0$.
13. In one of the worked examples of this section, the smallest positive root of the equation $\tan(\pi x) - x - 6 = 0$ was approximated. Graphically determine an interval which contains the next smallest positive root of this equation, and then approximate the root to within an absolute tolerance of 5×10^{-5}.

14. The equation $(x-0.5)(x+1)^3(x-2) = 0$ clearly has roots at $x = -1$, $x = 0.5$, and $x = 2$. Each of the intervals listed below encompasses all of these roots. Determine to which root the bisection method converges when each of the intervals below is used as the starting interval.

(a) $(-3, 3)$ **(b)** $(-1.5, 3)$
(c) $(-2, 4)$ **(d)** $(-2, 3)$
(e) $(-1.5, 2.2)$ **(f)** $(-7, 3)$

15. It can be shown that the equation

$$\frac{3}{2}x - 6 - \frac{1}{2}\sin(2x) = 0$$

has a unique real root.

(a) Find an interval on which this unique real root is guaranteed to exist.

(b) Using the interval found in part (a) and the bisection method, approximate the root to within an absolute tolerance of 10^{-5}.

16. For each of the functions given below, use the bisection method to approximate all real zeros. Use an absolute tolerance of 10^{-6} as a stopping criterion.

(a) $f(x) = e^x + x^2 - x - 4$
(b) $f(x) = x^3 - x^2 - 10x + 7$
(c) $f(x) = 1.05 - 1.04x + \ln x$

17. Peters ("Optimum Spring-Damper Design for Mass Impact," *SIAM Review*, **39**(1), pp. 118–122, 1997) models the impact of an object on a spring-damper system. If the displacement of the object following impact is limited, then the maximum force exerted on the object is minimized when the nondimensional damping coefficient, ζ, is the solution of the equation

$$\cos\left[4\zeta\sqrt{1-\zeta^2}\right] = -1 + 8\zeta^2 - 8\zeta^4$$

on the interval $0 < \zeta < 1/2$. The maximum (nondimensional) force is then given by

$$F_m = \exp\left[-\zeta(\tau_f + \tau_m)\right],$$

where

$$\tau_f = \cos^{-1}\zeta/\sqrt{1-\zeta^2}$$

is the time of the end of the stroke and

$$\tau_m = \cos^{-1}\left[\zeta(3-4\zeta^2)\right]/\sqrt{1-\zeta^2}$$

is the time when the maximum force occurs. Determine ζ to within an absolute tolerance of 5×10^{-7}, and then calculate τ_f, τ_m and F_m.

18. DeSantis, Gironi, and Marelli ("Vector-Liquid Equilibrium from a Hard-Sphere Equation of State," *Industrial and Engineering Chemistry Fundamentals*, **15**, 182–189, 1976) derive a relationship for the compressibility factor of real gases of the form

$$z = \frac{1+y+y^2-y^3}{(1-y)^3},$$

where y is related to the van der Waals volume correction factor. If $z = 0.892$, what is the value of y?

19. Reconsider the "Saving for a Down Payment" application problem. Which of the following scenarios requires a smaller compounded monthly interest rate to achieve a goal of \$25,000 after three years:

(a) a \$14,000 initial investment with \$250 per month thereafter; or

(b) a \$12,500 initial investment with \$300 per month thereafter?

2.2 THE METHOD OF FALSE POSITION

In Section 2.1, we developed the bisection method for approximating the zeros of continuous functions. On the plus side, the bisection method is straightforward to implement. In terms of computational cost, only one new function evaluation is needed per iteration, so the method is as inexpensive as one can expect. Most important, the sequence of approximations generated by the method is guaranteed to converge.

On the minus side, the sequence of approximations generated by the bisection method converges only linearly, with a rate of convergence of $O(1/2^n)$. Furthermore, even though there is a theoretical bound available for the error in each approximation, the bound can be overly pessimistic. As a result, it is possible that an approximation which is accurate to within the specified convergence tolerance will fail to terminate the iteration.

In this section we will develop a second simple enclosure method, one which is known as the *method of false position.* We will show that the sequence of approximations obtained from this new method is still guaranteed to converge and that the convergence of the sequence is still only linear. However, for the method of false position, we will be able to compute a fairly accurate estimate for the error in each approximation, not just a theoretical error bound. This error estimate will allow the formulation of a stopping condition which should greatly reduce the possibility that a sufficiently accurate approximation will fail to terminate the iteration.

Method of False Position

Being a simple enclosure method, the method of false position iteratively determines a sequence of root enclosing intervals, (a_n, b_n), and a sequence of approximations, which we shall denote by p_n. During each iteration, a single point is selected from (a_n, b_n) to approximate the location of the root and serve as p_n. If p_n is an accurate enough approximation, the iterative process is terminated. Otherwise, the Intermediate Value Theorem is used to determine whether the root lies on the subinterval (a_n, p_n) or on the subinterval (p_n, b_n). The entire process is then repeated on that subinterval.

The method of false position differs from other simple enclosure methods in the procedure used to select p_n. Whereas the bisection method simply chooses the midpoint of the enclosing interval, the method of false position uses the x-intercept of the line which passes through the points $(a_n, f(a_n))$ and $(b_n, f(b_n))$ as p_n (see Figure 2.4). The equation of the line which passes through $(a_n, f(a_n))$

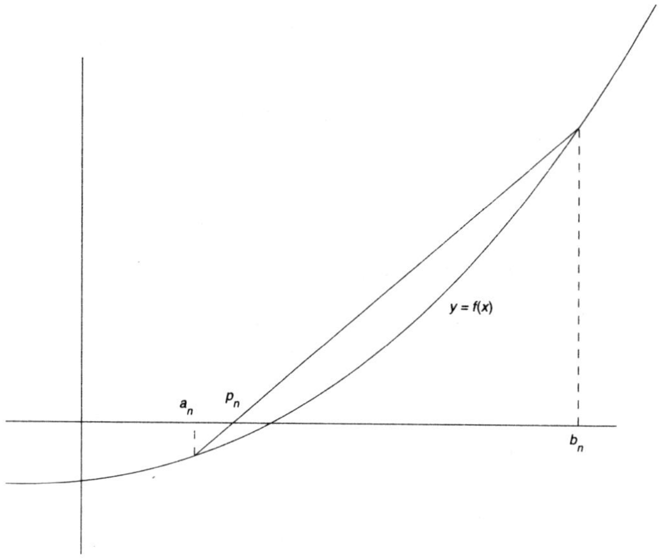

Figure 2.4 Schematic for the selection of p_n for the method of false position.

and $(b_n, f(b_n))$ is given by

$$y - f(b_n) = \frac{f(b_n) - f(a_n)}{b_n - a_n}(x - b_n).$$

Setting $y = 0$ and then solving for $x = p_n$ yields the formula

$$p_n = b_n - f(b_n)\frac{b_n - a_n}{f(b_n) - f(a_n)}.$$

EXAMPLE 2.3 Method of False Position in Action

We know that the function $f(x) = x^3 + 2x^2 - 3x - 1$ has a simple zero on the interval $(1, 2)$. Let's run through a few iterations of the method of false position to demonstrate the general procedure.

For the first iteration, we have $(a_1, b_1) = (1, 2)$ and we know that $f(a_1) = -1 < 0$ and that $f(b_1) = 9 > 0$. Our first approximation to the location of the zero is then

$$p_1 = b_1 - f(b_1)\frac{b_1 - a_1}{f(b_1) - f(a_1)} = 2 - 9\frac{2 - 1}{9 - (-1)} = 1.1.$$

To determine whether the zero is contained on $(a_1, p_1) = (1, 1.1)$ or on $(p_1, b_1) = (1.1, 2)$, we calculate

$$f(p_1) = -0.549 < 0.$$

Since $f(a_1)$ and $f(p_1)$ are of the same sign, the Intermediate Value Theorem tells us the zero is between p_1 and b_1. For the next iteration, we therefore take $(a_2, b_2) = (p_1, b_1) = (1.1, 2)$.

Our second approximation to the location of the zero is

$$p_2 = 2 - 9\frac{2 - 1.1}{9 - (-0.549)} = 1.151743638.$$

Note that

$$f(p_2) \approx -0.274 < 0,$$

which is of the same sign as $f(a_2)$. Hence, the Intermediate Value Theorem tells us the zero is now between p_2 and b_2, so we take $(a_3, b_3) = (p_2, b_2) = (1.151743638, 2)$.

In the third iteration, we calculate

$$p_3 = b_3 - f(b_3)\frac{b_3 - a_3}{f(b_3) - f(a_3)} = 1.17684091$$

and

$$f(p_3) \approx -0.131 < 0.$$

Hence, we once again find that $f(a_3)$ and $f(p_3)$ are of the same sign, which implies that the zero must lie somewhere between p_3 and b_3. For the fourth iteration, we will therefore have $(a_4, b_4) = (p_3, b_3) = (1.17684091, 2)$. Recall that to ten decimal places, $p = 1.1986912435$, so the absolute error in p_3 is 2.185×10^{-2}.

Having seen how the method of false position works, the next issue to address is the performance of this new method relative to that of the bisection method. Recall that when comparing the performance of rootfinding techniques, the fundamental measure of work is the number of function evaluations. For the bisection method, we have seen that n iterations require n function evaluations. For the method of false position, note that four function evaluations—$f(a_1)$, $f(b_1)$, $f(p_1)$ and $f(p_2)$—were needed in the preceding example to obtain p_3. In particular, $f(a_1)$ and $f(b_1)$ were used to calculate p_1. Each additional iteration then required one new function value: $f(p_1)$ for the second iteration and $f(p_2)$ for the third. Thus, a properly constructed false position algorithm, one that saves function values that will be needed for subsequent iterations, will cost $n + 1$ function evaluations for n iterations.

Returning to the preceding example, it should be clear that the appropriate comparison to make is between the approximation p_3 from the method of false position and the approximation p_4 from the bisection method. With $f(x) = x^3 + 2x^2 - 3x - 1$ and an initial interval of $(1, 2)$, the bisection method produces $p_4 = 1.1875$ (see the first example in Section 2.1). The error in this approximation is roughly 1.119×10^{-2}, which is about half the error in the value p_3 obtained above. On the other hand, with one more iteration each, the error from the bisection approximation is roughly 2.006×10^{-2}, which is nearly twice the false position error of 1.043×10^{-2}. Hence, for this problem, the two methods perform equally well.

There are, however, problems for which bisection clearly outperforms false position. As a case in point, consider the function $f(x) = \tan(\pi x) - x - 6$, which has a zero on $(0.4, 0.48)$. To guarantee an absolute error of less than 5×10^{-5}, the bisection method uses 11 function evaluations. With these 11 evaluations, the actual error is approximately 7.43×10^{-6}. The method of false position needs 15 function evaluations to produce an error less than 5×10^{-5} and 18 evaluations to achieve an error of roughly 7.43×10^{-6}.

As one might expect, there are other problems for which false position outperforms bisection. As an example, consider $f(x) = x^3 + 2x^2 - 3x - 1$, which has a zero on $(-3, -2)$. To guarantee an absolute error of less than 5×10^{-5}, the bisection method uses 15 function evaluations, and the actual error is roughly 2.83×10^{-5}. The method of false position achieves both of these error levels with only 5 function evaluations.

So what should we make from all of this? From the outset, we might have expected false position to always outperform bisection. After all, false position uses more information about the function. However, our examples have shown that this is not the case. There is also no general theory to indicate which method will be better for a given problem. The main advantage which the method of false position has over the bisection method is the existence of a computable error estimate. We will derive this error estimate toward the end of the section when we discuss an appropriate stopping condition. In addition to providing more reliable termination, the existence of an error estimate will allow us, in Section 2.6, to accelerate the convergence of the false position sequence. Such acceleration is not possible for the bisection method.

Convergence Analysis

Does the sequence of approximations generated by the method of false position, $\{p_n\}$, converge to a root p? If so, what is the order of convergence? To answer these questions, we need to examine the associated sequence of errors, $\{e_n\}$, where $e_n = p_n - p$. The sequence $\{p_n\}$ converges if and only if $|e_n| \to 0$ as $n \to \infty$, and the order of convergence is determined by the asymptotic relationship between $|e_n|$ and $|e_{n-1}|$.

The error sequence $\{e_n\}$ is governed by what is known as the *error evolution equation.* To construct the error evolution equation, take the equation for p_n and subtract p from both sides. This yields

$$p_n - p = b_n - p - f(b_n)\frac{b_n - a_n}{f(b_n) - f(a_n)}. \tag{1}$$

Next, approximate the function values $f(a_n)$ and $f(b_n)$ by the second degree Taylor polynomials

$$\begin{aligned} f(a_n) &\approx f'(p)(a_n - p) + \frac{f''(p)}{2}(a_n - p)^2, \\ f(b_n) &\approx f'(p)(b_n - p) + \frac{f''(p)}{2}(b_n - p)^2, \end{aligned} \tag{2}$$

where the fact that $f(p) = 0$ has been taken into account. The term $f(b_n) - f(a_n)$ is then approximately

$$
\begin{aligned}
f(b_n) - f(a_n) &\approx f'(p)(b_n - a_n) + \frac{f''(p)}{2}\left[(b_n - p)^2 - (a_n - p)^2\right] \\
&= (b_n - a_n)\left[f'(p) + \frac{f''(p)}{2}(b_n + a_n - 2p)\right].
\end{aligned} \tag{3}
$$

Substituting (2) and (3) into (1), factoring the term $b_n - p$, and dividing out the term $b_n - a_n$ yields

$$
\begin{aligned}
p_n - p &\approx (b_n - p)\left[1 - \frac{f'(p) + \frac{f''(p)}{2}(b_n - p)}{f'(p) + \frac{f''(p)}{2}(b_n + a_n - 2p)}\right] \\
&\approx (b_n - p)(a_n - p)\frac{f''(p)}{2f'(p) + f''(p)(b_n + a_n - 2p)}.
\end{aligned} \tag{4}
$$

Before we can proceed any further, we have to make one very important observation. Table 2.2 displays the results of ten iterations of the method of false position applied to three different test problems. Focus on the middle column, which lists the enclosing interval for each iteration. Notice that in each case one of the endpoints remains fixed, while the other endpoint is just the previous approximation to the location of the root. Now, not every problem will have one endpoint fixed for all iterations; however, when using the method of false position, as the iteration proceeds, one of the endpoints will always eventually become fixed. In fact, the method of false position will always eventually settle into one of the configurations shown in Figure 2.5.

Let's now return to the error evolution equation. Because one endpoint of the enclosing interval remains fixed while the other is just p_{n-1}, (4) becomes

$$e_n \approx \lambda e_{n-1},$$

where

$$\lambda = \frac{lf''(p)}{2f'(p) + lf''(p)}$$

and

$$l = \begin{cases} a_n - p, & a_n \text{ remains fixed} \\ b_n - p, & b_n \text{ remains fixed} \end{cases}.$$

Provided $|\lambda| < 1$, it then follows that the sequence generated by the method of false position converges, and the convergence is linear with asymptotic error constant $|\lambda|$.

The only question that remains is whether $|\lambda|$ really is less than 1. Remember that the method of false position eventually settles into one of the configurations shown in Figure 2.5. Let's consider the configuration in the upper left corner in more detail. Because a_n is fixed, $l = a_n - p$. Now, $a_n - p < 0$ and $f''(p) <$

$f(x) = x^3 + 2x^2 - 3x - 1$ on $(1, 2)$

n	Enclosing Interval	Approximation
1	(1.0000000000,2.0000000000)	1.1000000000
2	(1.1000000000,2.0000000000)	1.1517436381
3	(1.1517436381,2.0000000000)	1.1768409100
4	(1.1768409100,2.0000000000)	1.1886276733
5	(1.1886276733,2.0000000000)	1.1940789113
6	(1.1940789113,2.0000000000)	1.1965820882
7	(1.1965820882,2.0000000000)	1.1977277544
8	(1.1977277544,2.0000000000)	1.1982513178
9	(1.1982513178,2.0000000000)	1.1984904185
10	(1.1984904185,2.0000000000)	1.1985995764

$f(x) = \tan(\pi x) - x - 6$ on $(0.4, 0.48)$

n	Enclosing Interval	Approximation
1	(0.4000000000,0.4800000000)	0.4208674108
2	(0.4208674108,0.4800000000)	0.4332027501
3	(0.4332027501,0.4800000000)	0.4404957388
4	(0.4404957388,0.4800000000)	0.4448079249
5	(0.4448079249,0.4800000000)	0.4473577484
6	(0.4473577484,0.4800000000)	0.4488655162
7	(0.4488655162,0.4800000000)	0.4497571072
8	(0.4497571072,0.4800000000)	0.4502843380
9	(0.4502843380,0.4800000000)	0.4505961108
10	(0.4505961108,0.4800000000)	0.4507804752

$f(x) = x^3 + 2x^2 - 3x - 1$ on $(-3, -2)$

n	Enclosing Interval	Approximation
1	$(-3.0000000000, -2.0000000000)$	-2.8333333333
2	$(-3.0000000000, -2.8333333333)$	-2.9079283887
3	$(-3.0000000000, -2.9079283887)$	-2.9120026293
4	$(-3.0000000000, -2.9120026293)$	-2.9122172667
5	$(-3.0000000000, -2.9122172667)$	-2.9122285522
6	$(-3.0000000000, -2.9122285522)$	-2.9122291456
7	$(-3.0000000000, -2.9122291456)$	-2.9122291768
8	$(-3.0000000000, -2.9122291768)$	-2.9122291784
9	$(-3.0000000000, -2.9122291784)$	-2.9122291785
10	$(-3.0000000000, -2.9122291785)$	-2.9122291785

TABLE 2.2: Ten Iterations of the Method of False Position Applied to Three Test Problems

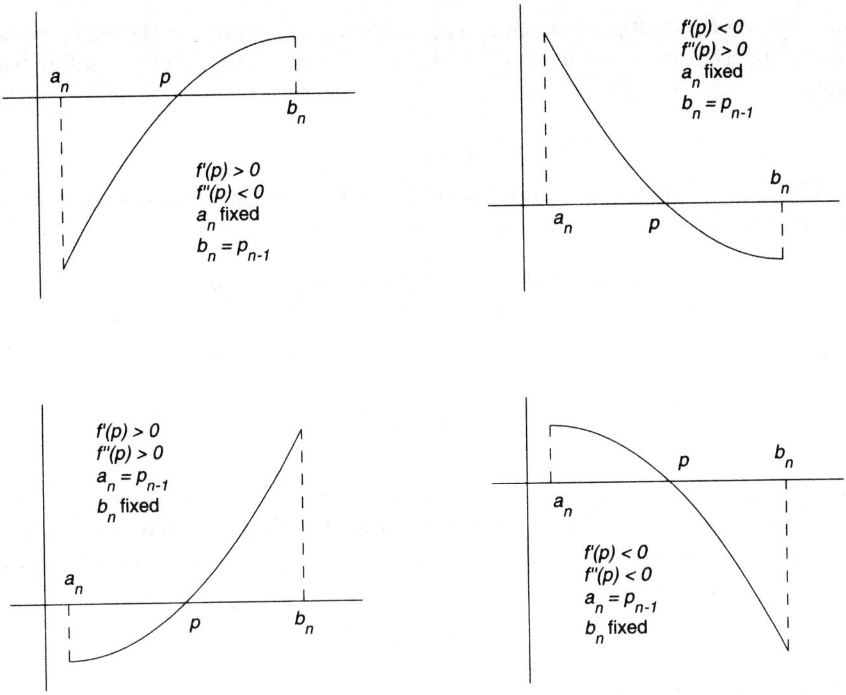

Figure 2.5 Eventual configurations for the method of false position.

0, so $(a_n - p)f''(p) > 0$. Since $f'(p)$ is also greater than zero, it follows that $2f'(p) + (a_n - p)f''(p) > (a_n - p)f''(p)$ and

$$0 < \frac{(a_n - p)f''(p)}{2f'(p) + (a_n - p)f''(p)} = \lambda < 1.$$

Hence $|\lambda| < 1$ and the method of false position converges. In a similar fashion, the remaining three configurations can be shown to lead to $|\lambda| < 1$. The details are left as an exercise.

Stopping Condition

We would like to terminate the method of false position when $|e_n|$ falls below a specified convergence tolerance, ϵ. To implement this idea, we must have a formula for estimating $|e_n|$ which involves only quantities which can be calculated during the course of the iteration. As a starting point, note that

$$\begin{aligned} e_n &= p_n - p \\ &= p_n - p_{n-1} + p_{n-1} - p \\ &= p_n - p_{n-1} + e_{n-1}. \end{aligned} \tag{5}$$

From the error evolution equation, $e_n \approx \lambda e_{n-1}$, or, equivalently, $e_{n-1} \approx e_n/\lambda$. Substituting this expression into (5), solving for e_n, and taking absolute values yields

$$|e_n| \approx \left|\frac{\lambda}{\lambda-1}\right| |p_n - p_{n-1}|. \tag{6}$$

Next, we focus on λ. The value of λ can be estimated using terms from the sequence $\{p_n\}$ as follows. Consider the ratio

$$\frac{p_n - p_{n-1}}{p_{n-1} - p_{n-2}} = \frac{(p_n - p) - (p_{n-1} - p)}{(p_{n-1} - p) - (p_{n-2} - p)} = \frac{e_n - e_{n-1}}{e_{n-1} - e_{n-2}}.$$

Using the relations $e_n \approx \lambda e_{n-1}$ and $e_{n-2} \approx e_{n-1}/\lambda$, we find

$$\frac{p_n - p_{n-1}}{p_{n-1} - p_{n-2}} \approx \frac{(\lambda - 1)e_{n-1}}{(1 - \frac{1}{\lambda})e_{n-1}} = \lambda. \tag{7}$$

Equations (6) and (7) together constitute a computable estimate for $|e_n|$. The accuracy of this estimate is demonstrated in Table 2.3. Consequently, an appropriate stopping condition for the method of false position is to terminate the iteration when

$$\left|\frac{\lambda}{\lambda-1}\right| |p_n - p_{n-1}| < \epsilon,$$

where λ is obtained from (7).

An Application Problem: Depth of Submersion

In the Chapter 2 Overview (see page 55), we found that a spherical object of radius R and density ρ_o placed on the surface of a fluid of density ρ_f would sink to a depth h which is a root of the equation

$$\frac{\rho_f}{3}h^3 - R\rho_f h^2 + \frac{4}{3}R^3\rho_o = 0.$$

In deriving this equation it was assumed that the object was not fully submerged in the fluid.

Suppose we place a spherical ball of cork with a radius of $R = 5$ cm and a density of $\rho_o = 0.120$ g/cm^3 into motor oil with a density of $\rho_f = 0.890$ g/cm^3. Five iterations of the method of false position with an initial interval of $(0, 10)$ and a convergence tolerance of $\epsilon = 5 \times 10^{-5}$ yield $p_5 = 2.3043353119$. The estimate for the error in p_5 is roughly 4.378×10^{-5}. Thus, the ball of cork sinks to a depth of roughly 2.304 cm in the motor oil.

EXERCISES

1. Each of the following equations has a root on the interval $(0, 1)$. Perform the method of false position to determine p_3, the third approximation to the location of the root, and to determine (a_4, b_4), the next enclosing interval.
   (a) $\ln(1 + x) - \cos x = 0$ **(b)** $x^5 + 2x - 1 = 0$
   (c) $e^{-x} - x = 0$ **(d)** $\cos x - x = 0$

$$f(x) = x^3 + 2x^2 - 3x - 1 \text{ on } (1, 2)$$

n	Absolute Error, $\|e_n\|$	Error Estimate, $\left\|\frac{\lambda}{\lambda-1}\right\| \|p_n - p_{n-1}\|$
1	0.0986912435	
2	0.0469476054	
3	0.0218503335	0.0236382347
4	0.0100635702	0.0104374516
5	0.0046123322	0.0046903760
6	0.0021091553	0.0021254290
7	0.0009634891	0.0009668808
8	0.0004399257	0.0004406324
9	0.0002008251	0.0002009723
10	0.0000916671	0.0000916978

$$f(x) = \tan(\pi x) - x - 6 \text{ on } (0.4, 0.48)$$

n	Absolute Error, $\|e_n\|$	Error Estimate, $\left\|\frac{\lambda}{\lambda-1}\right\| \|p_n - p_{n-1}\|$
1	0.0301798480	
2	0.0178445088	
3	0.0105515200	0.0105481928
4	0.0062393339	0.0062382353
5	0.0036895105	0.0036891385
6	0.0021817426	0.0021816149
7	0.0012901516	0.0012901074
8	0.0007629208	0.0007629055
9	0.0004511480	0.0004511426
10	0.0002667836	0.0002667817

$$f(x) = x^3 + 2x^2 - 3x - 1 \text{ on } (-3, -2)$$

n	Absolute Error, $\|e_n\|$	Error Estimate, $\left\|\frac{\lambda}{\lambda-1}\right\| \|p_n - p_{n-1}\|$
1	7.8895845e-02	
2	4.3007897e-03	
3	2.2654917e-04	2.3538350e-04
4	1.1911748e-05	1.1936259e-05
5	6.2624802e-07	6.2631578e-07
6	3.2924183e-08	3.2924370e-08
7	1.7309461e-09	1.7309465e-09
8	9.1002317e-11	9.1002226e-11
9	4.7841731e-12	4.7843508e-12
10	2.5091040e-13	2.5153164e-13

TABLE 2.3: Confirmation of Error Estimate

2. Construct an algorithm for the method of false position. Remember to save function values which will be needed for later iterations and to implement a stopping condition based on equations (6) and (7).

3. Confirm that $|\lambda| < 1$ for the remaining configurations in Figure 2.5.

In Exercises 4–7, an equation, an interval on which the equation has a root, and the exact value of the root are specified.

(1) Perform the first five (5) iterations of the method of false position.

(2) Verify that the absolute error in the third, fourth and fifth approximations satisfies the error estimate

$$|p_n - p| \approx \left|\frac{\lambda}{\lambda - 1}\right| |p_n - p_{n-1}|.$$

(3) How does the error in the fifth false position approximation compare to the maximum error which would result from six iterations of the bisection method?

4. The equation $x^3 + x^2 - 3x - 3 = 0$ has a root on the interval $(1, 2)$, namely $x = \sqrt{3}$.

5. The equation $x^7 = 3$ has a root on the interval $(1, 2)$, namely $x = \sqrt[7]{3}$.

6. The equation $x^3 - 13 = 0$ has a root on the interval $(2, 3)$, namely $\sqrt[3]{13}$.

7. The equation $1/x - 37 = 0$ has a zero on the interval $(0.01, 0.1)$, namely $x = 1/37$.

8. The function $f(x) = \sin x$ has a zero on the interval $(3, 4)$, namely $x = \pi$. Perform three iterations of the method of false position to approximate this zero. Determine the absolute error in each of the three computed approximations. What is the apparent order of convergence? What explanation can you provide for this behavior?

9. (a) Verify that the equation $x^4 - 18x^2 + 45 = 0$ has a root on the interval $(1, 2)$. Next, perform three iterations of the method of false position. Given that the exact value of the root is $x = \sqrt{3}$, compute the absolute error in the three approximations just obtained. What is the apparent order of convergence? What explanation can you provide for this behavior?
   (b) Verify that the equation $x^4 - 18x^2 + 45 = 0$ also has a root on the interval $(3, 4)$. Perform five iterations of the method of false position, and compute the absolute error in each approximation. The exact value of the root is $x = \sqrt{15}$. What is the apparent order of convergence in this case?
   (c) What explanation can you provide for the different convergence behavior between parts (a) and (b)?

10. The function $f(x) = x^3 + 2x^2 - 3x - 1$ has a zero on the interval $(-1, 0)$. Approximate this zero to within an absolute tolerance of 5×10^{-5}.

11. For each of the functions given below, use the method of false position to approximate all real roots. Use an absolute tolerance of 10^{-6} as a stopping condition.
   (a) $f(x) = e^x + x^2 - x - 4$
   (b) $f(x) = x^3 - x^2 - 10x + 7$
   (c) $f(x) = 1.05 - 1.04x + \ln x$

12. In the literature, it is not uncommon to find the method of false position terminated when $|p_n - p_{n-1}| < \epsilon$. Comment on the accuracy of this stopping condition. Consider the cases $\lambda \approx 0$, $\lambda \approx 1/2$ and $\lambda \approx 1$.

13. A storage tank is in the shape of a horizontal cylinder with length L and radius r. The volume V of fluid in the tank is related to the depth h of the fluid by the equation

$$V = \left[r^2 \cos^{-1}\left(\frac{r-h}{r}\right) - (r-h)\sqrt{2rh - h^2} \right] L.$$

If $r = 1$ meter, $L = 3$ meters, and $V = 7$ cubic meters, determine h.

14. The equation $x^2 = 1 - \cos(\sqrt{2}x) + \sqrt{2}\sin(\sqrt{2}x)$ has two real roots. One of them is at $x = 0$. Determine an interval that contains the other root, and then approximate this root to three decimal places. This problem arises in the calculation of the amplitude of the solution to a nonlinear third-order differential equation. See Gottlieb ("Simple Nonlinear Jerk Functions with Periodic Solutions," *American Journal of Physics*, **66** (10), 903–906, 1998) for details.

15. Rework the "Depth of Submersion" problem to determine the depth to which a glass marble of radius 2 cm and density 0.040 g/cm^3 sinks in water of density 0.998 g/cm^3.

2.3 FIXED POINT ITERATION SCHEMES

In the previous two sections, we studied simple enclosure methods for solving rootfinding problems. We found that simple enclosure methods are guaranteed to converge to a root of the equation being studied. Unfortunately, the rate of convergence (i.e., the number of iterations required to achieve a given level of precision in the approximate root) tends to be slow. In this section, the concept of fixed point iteration schemes will be presented. When properly constructed, these schemes will exhibit rapid convergence. The price paid for this accelerated convergence, however, is the loss of guaranteed convergence.

Review of Mean Value Theorem

The theorems which we prove in this section will make extensive use of the Mean Value Theorem. Therefore, rather than jumping straight into a discussion of fixed points and fixed point iteration schemes, we'll start with a review of the statement and consequences of this important theorem.

> **Theorem.** If the function f is continuous on the closed interval $[a, b]$ and differentiable on the open interval (a, b), then there exists a real number $\xi \in (a, b)$ such that
> $$f'(\xi) = \frac{f(b) - f(a)}{b - a}.$$

The Mean Value Theorem has an interesting geometric interpretation (see Figure 2.6). The expression $(f(b) - f(a))/(b - a)$ gives the slope of the line which passes through the points $(a, f(a))$ and $(b, f(b))$. The line through these two points

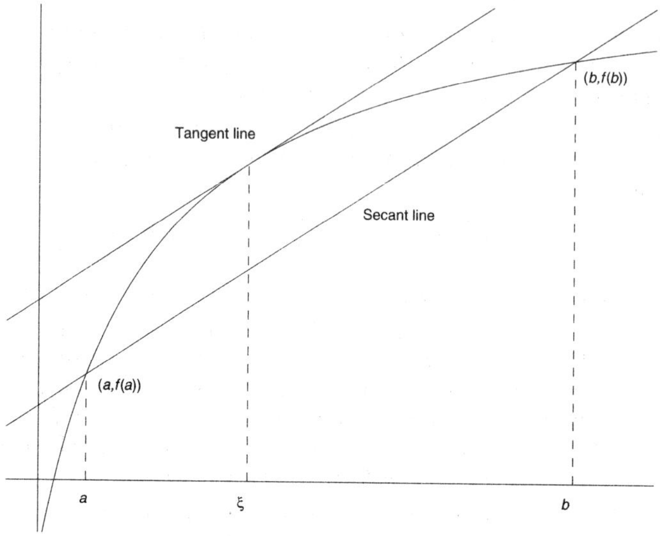

Figure 2.6 Geometric interpretation of the Mean Value Theorem.

is often called a secant line. Of course, $f'(\xi)$ gives the slope of the line tangent to the graph of f at the location $x = \xi$. Hence, for a continuous and differentiable function, the Mean Value Theorem guarantees there is at least one point on the graph of the function at which the tangent line is parallel to the secant line.

When we actually apply the Mean Value Theorem later in the section, we will use a slightly different formulation. Taking the equation

$$f'(\xi) = \frac{f(b) - f(a)}{b - a}$$

and mutliplying through by $b - a$ yields

$$f(b) - f(a) = f'(\xi)(b - a).$$

Note this latter equation relates the difference of two function values to the difference of the corresponding input values. Throughout the remainder of this section, be on the lookout for expressions with this property. An application of the Mean Value Theorem is likely to follow soon after.

EXAMPLE 2.4 An Inequality Involving the Sine Function

Consider the inequality

$$|\sin b - \sin a| \leq |b - a|,$$

where a and b are any real numbers. Here is one of those expressions alluded to above. Note that the left side of the inequality involves the difference of two function values and the right side involves the difference of the corresponding input values. This is a clear indication that we should use the Mean Value Theorem.

So let a and b be any two real numbers. If $a = b$, then the inequality is trivially satisfied. Suppose then that $a \neq b$. Because the sine function is both continuous and differentiable everywhere, the function is certainly continuous and differentiable on the interval between $x = a$ and $x = b$. Applying the Mean Value Theorem to the function $f(x) = \sin x$, it follows that

$$\sin b - \sin a = \cos \xi (b - a),$$

for some ξ between a and b. Taking the absolute value of both sides of this last expression and using the fact that the magnitude of the cosine function is always less than or equal to one produces the desired result.

Background for Fixed Points

Consider the function $\sin x$. Since the sine function maps $x = \pi/4$ to the value $\sqrt{2}/2$, the sine function may be thought of as moving the input value of $\pi/4$ to the output value of $\sqrt{2}/2$. On the other hand, the sine function maps zero to zero; *i.e.*, $\sin 0 = 0$. In keeping with the analogy just established, the sine function fixes the location of 0. For this reason $x = 0$ is said to be a fixed point of the function $\sin x$. In general, the following definition is made:

> **Definition.** A FIXED POINT of the function g is any real number, p, for which $g(p) = p$; that is, whose location is fixed by g.

This definition provides a direct analytical means for determining fixed points, which can be used in simple cases.

EXAMPLE 2.5 Fixed Points of the Logistic Equation

One of the most popular mathematical models for the generation-by-generation growth of a population is the logistic equation:

$$p_{n+1} = cp_n(1 - p_n),$$

where $0 < c < 4$ is a constant and p_n denotes the normalized size of the population in the n-th generation, measured relative to the maximum population which the environment can support. The fixed points of the function on the right-hand side of the logistic equation play an important role in the dynamics of the long-term behavior of the population. Using the definition, the fixed points for the logistic equation are the solutions of

$$p = cp(1 - p).$$

Solving this quadratic equation produces $p = 0$ and $p = (c - 1)/c$.

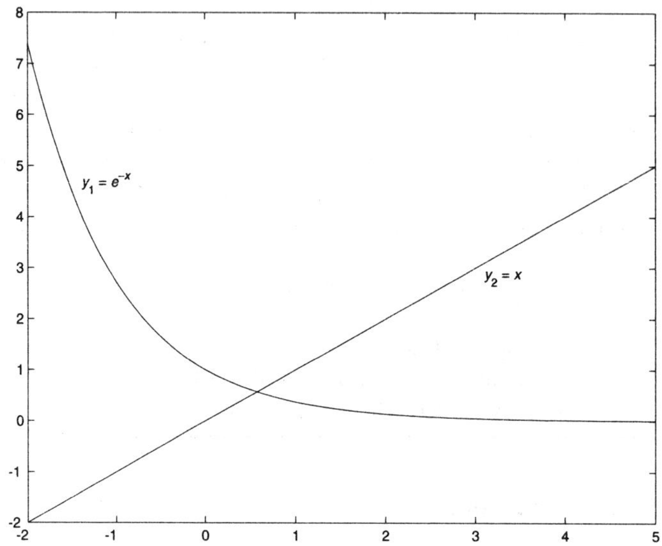

Figure 2.7 Graphical determination of the existence of a fixed point for the function $g(x) = e^{-x}$.

For those functions for which the analytical approach fails, it is possible to investigate the existence of fixed points using a graphical approach. Begin by graphing $y_1 = g(x)$ and $y_2 = x$ on the same set of coordinate axes. At any point of intersection between the two graphs, we are guaranteed that $y_1 = y_2$, and hence, $g(x) = x$. Therefore, points of intersection between the graph of g and the graph of the identity function represent the fixed points of g. To illustrate this process, consider $g(x) = e^{-x}$. The equation $e^{-x} = x$ cannot be solved by analytical means, but Figure 2.7 indicates the existence of a unique fixed point somewhere in the vicinity of 0.6.

We have so far encountered a function with two fixed points and a function with a unique fixed point. There are also functions which do not have any fixed points. The graphical approach described in the previous paragraph is sufficient to establish that e^x and $\ln x$ have no fixed points (their graphs never intersect the graph of the identity function). Either the analytical approach or the graphical approach can be used to establish that $x^2 + 1$ also has no fixed points. To conclude the introductory material on fixed points, here is a theorem which states conditions under which a function is guaranteed to have a unique fixed point.

Theorem. Let g be continuous on the closed interval $[a, b]$ with $g : [a, b] \to [a, b]$. Then g has a fixed point $p \in [a, b]$. Furthermore, if g is differentiable on the open interval (a, b) and there exists a positive constant $k < 1$ such that $|g'(x)| \le k < 1$ for all $x \in (a, b)$, then the fixed point in $[a, b]$ is unique.

Proof. (1) Existence

Assume that g is continuous on the closed interval $[a,b]$ with $g:[a,b]\to[a,b]$. Define the auxiliary function $h(x) = g(x) - x$. Note that h satisfies two important properties. First, since h is the difference of two functions that are continuous on $[a,b]$, h is also continuous on that interval. Second, by construction, the roots of h are precisely the fixed points of g.

Now, since $\min_{x\in[a,b]} g(x) \geq a$ and $\max_{x\in[a,b]} g(x) \leq b$, it follows that

$$h(a) = g(a) - a \geq 0 \qquad \text{and} \qquad h(b) = g(b) - b \leq 0.$$

If either $h(a) = 0$ or $h(b) = 0$, then we have found a root of h, which, by construction, is a fixed point of g, and we are done. If neither $h(a) = 0$ nor $h(b) = 0$, then $h(b) < 0 < h(a)$. Since h is continuous on $[a,b]$, the Intermediate Value Theorem may be invoked to guarantee the existence of $p \in [a,b]$ such that $h(p) = 0$, which implies $g(p) = p$.

(2) Uniqueness

This part of the proof will proceed by contradiction. Suppose that p and q are both fixed points of the function g on the interval $[a,b]$, with $p \neq q$. By the definition of a fixed point, $g(p) = p$ and $g(q) = q$. Then

$$\begin{aligned}
|p-q| &= |g(p) - g(q)| \\
&= |g'(\xi)(p-q)| && \text{by the Mean Value Theorem} \\
&= |g'(\xi)||p-q| \\
&\leq k|p-q| < |p-q|,
\end{aligned}$$

which is a contradiction. Hence, $p = q$, and the fixed point is unique. □

It should be noted that the hypotheses of this theorem are sufficient conditions. By themselves, these conditions guarantee the existence and uniqueness of a fixed point. However, the hypotheses are not necessary conditions, meaning that it is possible for a function to violate one or more of the hypotheses, yet still have a (possibly unique) fixed point. For example, consider the function $g(x) = 4x(1-x)$ on the interval $[0.1, \infty)$. Since $\lim_{x\to\infty} g(x) \to -\infty$, g clearly does not map $[0.1,\infty)$ onto itself. Furthermore, $\lim_{x\to\infty} |g'(x)| \to +\infty$, so that g also violates the hypothesis regarding the magnitude of the first derivative. However, g has fixed points at $x = 0$ and $x = 3/4$, so that g does in fact have a unique fixed point on the interval $[0.1, \infty)$.

Fixed Point Iteration

If it is known that a function g has a fixed point, one way to approximate the value of that fixed point is to use what is known as a *fixed point iteration scheme.* These can be defined as follows:

Definition. A FIXED POINT ITERATION SCHEME (also known as a FUNCTIONAL iteration scheme) to approximate the fixed point, p, of a function g, generates the sequence $\{p_n\}$ by the rule $p_n = g(p_{n-1})$ for all $n \geq 1$, given a starting approximation, p_0.

Within a fixed point iteration scheme, the function g is often referred to as the *iteration function.*

EXAMPLE 2.6 Fixed Point Iteration in Action

In Figure 2.7, we saw that the function $g(x) = e^{-x}$ has a unique fixed point somewhere near $x = 0.6$. To locate this fixed point more precisely, we will now perform fixed point iteration with g as the iteration function and $p_0 = 0$. The first ten iterations yield

$$
\begin{aligned}
p_0 = 0 \qquad p_1 &= g(p_0) = 1.0000000000 \\
p_2 &= g(p_1) = 0.3678794412 \\
p_3 &= g(p_2) = 0.6922006276 \\
p_4 &= g(p_3) = 0.5004735006 \\
p_5 &= g(p_4) = 0.6062435351 \\
p_6 &= g(p_5) = 0.5453957860 \\
p_7 &= g(p_6) = 0.5796123355 \\
p_8 &= g(p_7) = 0.5601154614 \\
p_9 &= g(p_8) = 0.5711431151 \\
p_{10} &= g(p_9) = 0.5648793474\,.
\end{aligned}
$$

The sequence appears to be converging, albeit, very slowly. In fact, it takes more than 20 iterations for p_n to agree with the exact fixed point to at least 5 significant decimal digits.

Although the study of fixed points is an important subject in its own right, the objective of this chapter is still the rootfinding problem—some connection between fixed point problems and rootfinding problems therefore has to be established. Fortunately, or perhaps unfortunately, every rootfinding problem can be transformed into any number of different fixed point problems. Some of these fixed point problems will converge rapidly, some will converge slowly and some will not converge at all. The conversion process is actually quite simple. Take the rootfinding equation, $f(x) = 0$, and algebraically transform it into an equation of the form $x = \ldots$; the expression on the right-hand side of the resulting equation is a corresponding iteration function $g(x)$.

To demonstrate the process, consider the function $f(x) = x^3 + x^2 - 3x - 3$, which has a unique zero on the interval $(1, 2)$. The objective here will be to approximate this root using fixed point iteration. Starting from the equation $x^3 + x^2 - 3x - 3 = 0$, one possible iteration function arises by transposing the linear term to the right-hand side and dividing by 3. Let $g_1(x)$ denote the resulting iteration function, $g_1(x) = \left(x^3 + x^2 - 3\right)/3$. Alternatively, both the linear and constant terms could be transposed to the right-hand side. Dividing the resulting equation by x^2 and subtracting 1 produces the iteration function $g_2(x) = -1 + (3x+3)/x^2$. Three additional iteration functions that will be examined are $g_3(x) = \sqrt[3]{3+3x-x^2}$, $g_4(x) = \sqrt{(3+3x-x^2)/x}$, and $g_5(x) = x-(x^3+x^2-3x-3)/(3x^2+2x-3)$. The function $g_3(x)$ is derived by isolating the cubic term and taking the cube root, while $g_4(x)$ is derived by isolating the cubic term, dividing by x, and then taking the square root. An explanation for $g_5(x)$ will be deferred to the next section. The results of applying each of these five functions, with a starting approximation of $p_0 = 1$ and a stopping criterion of $|p_n - p_{n-1}| < 10^{-7}$, are shown in Table 2.4.

The results in Table 2.4 display a wide range of convergence behaviors. The sequence generated by the first iteration function converges, but to a fixed point outside the interval $(1, 2)$. The sequence generated by $g_2(x)$ fails to converge despite attaining values quite close to the fixed point determined by $g_1(x)$—see the values of p_9 and p_{10}. Convergence of the sequences for the third and fifth iteration functions is rapid, but is achieved in very different manners. Each iteration with $g_3(x)$ produces roughly one additional decimal place of accuracy, whereas $g_5(x)$ roughly doubles the number of correct decimal places with each iteration. Had greater precision been requested, the fifth sequence would have converged much faster than the third. Finally, the fourth sequence converges to the desired fixed point, but does so very slowly.

Convergence

Based on the results of this experiment, the following basic problem must be tackled whenever fixed point iteration is to be used to solve a rootfinding problem:

> Given a function f whose roots are to be determined, can an iteration function, g, be constructed such that the fixed points of g are the roots of f AND for some starting approximation p_0, the sequence $p_n = g(p_{n-1})$ converges to p, a root of f?

To have any hope of answering this question in the affirmative, conditions on g that will guarantee convergence of the iteration scheme must first be known. The following theorem provides the needed information.

Theorem. Let g be continuous on the closed interval $[a, b]$ with $g : [a, b] \to [a, b]$. Furthermore, suppose that g is differentiable on the open interval (a, b) and there exists a positive constant $k < 1$ such that $|g'(x)| \leq k < 1$ for all $x \in (a, b)$. Then
(1) the sequence $\{p_n\}$ generated by $p_n = g(p_{n-1})$ converges to the fixed point p for any $p_0 \in [a, b]$;
(2) $|p_n - p_{n-1}| \leq k^n \max(p_0 - a, b - p_0)$; and,
(3) $|p_n - p| \leq \frac{k^n}{1-k}|p_1 - p_0|$.

Note that the hypotheses of this theorem are precisely those that were sufficient to guarantee that a function has a unique fixed point, so the reference in conclusion (1) to THE fixed point of g is justified. As with the previous theorem

n	$g_1(x)$	(b) $g_2(x)$	(c) $g_3(x)$	(d) $g_4(x)$	(e) $g_5(x)$
0	+1.0	+1.0	+1.0	+1.0	+1.0
1	−.3333333333	5.0000000000	1.709975947	2.236067978	3.000000000
2	−.9753086420	−.2800000000	1.733134316	1.451059202	2.200000000
3	−.9921709716	26.5510204100	1.731994802	1.901682432	1.830150754
4	−.9974310264	−.8827544117	1.732053695	1.635808067	1.737795453
5	−.9991480696	−.5486245114	1.732050659	1.788336635	1.732072292
6	−.9997165068	3.4989256110	1.732050815	1.699764653	1.732050808
7	−.9999055558	.1024544340	1.732050807	1.750767137	1.732050808
8	−.9999685246	314.0796731		1.721269132	
9	−.9999895088	−.9904178717		1.738283891	
10	−.9999965030	−.9706946913		1.728454854	
11	−.9999988343	−.9066955739		1.734127847	
12	−.9999996114	−.6595130205		1.730851932	
13	−.9999998704	1.3484159170		1.732743080	
14	−.9999999568	2.8747932030		1.731651158	
15		.4065545020		1.732281557	
16		24.52938017		1.731917588	
17		−.8727117320		1.732127723	
18		−.4986188487		1.732006401	
19		5.0499512450		1.732076446	
20		−.2882970612		1.732036005	
25		Does not converge		1.732051758	
30				1.732050747	
31				1.732050842	

$$g_1(x) = \frac{x^3 + x^2 - 3}{3}, \quad g_2(x) = -1 + \frac{3x + 3}{x^2}, \quad g_3(x) = \sqrt[3]{3 + 3x - x^2},$$

$$g_4(x) = \sqrt{\frac{3 + 3x - x^2}{x}}, \quad g_5(x) = x - \frac{x^3 + x^2 - 3x - 3}{3x^2 + 2x - 3}$$

TABLE 2.4: Comparison of Fixed Point Iteration Schemes

in this section, the hypotheses of this theorem are sufficient conditions for convergence of the iteration scheme, though not necessary. Direct calculation shows that for the fifth iteration function examined above, $\max_{x\in[1,2]} |g_5'(x)| = 8$, so that the derivative bound is violated, yet the sequence generated by $g_5(x)$ converged very rapidly.

Proof. (1) To establish the first part of the theorem, it must be shown that $|p_n - p| \to 0$ as $n \to \infty$ for any starting value $p_0 \in [a, b]$. Therefore, let $p_0 \in [a, b]$. Since $g : [a, b] \to [a, b]$, we are guaranteed that $p_n = g(p_{n-1})$ is well-defined and that $p_n \in [a, b]$ for all n. Furthermore,

$$\begin{aligned}
|p_n - p| &= |g(p_{n-1}) - g(p)| && \text{definition of } p_n \text{ and } p \\
&= |g'(\xi)||p_{n-1} - p| && \text{Mean Value Theorem} \\
&\le k|p_{n-1} - p| && \text{hypothesis on } g' \\
&\le k^2|p_{n-2} - p| && \text{repeat previous 3 steps} \\
&\ \ \vdots \\
&\le k^n|p_0 - p|.
\end{aligned}$$

Now, since $k < 1$,

$$\lim_{n\to\infty} |p_n - p| \le \lim_{n\to\infty} k^n|p_0 - p| = |p_0 - p| \lim_{n\to\infty} k^n = 0.$$

(2) Combining the bound

$$|p_n - p| \le k^n|p_0 - p|$$

obtained in the proof of part (1) with the bound

$$|p_0 - p| \le \max(p_0 - a, b - p_0)$$

establishes the second conclusion of the theorem.

(3) Proceeding in exactly the same manner as in part (1), it can be shown that

$$|p_{n+1} - p_n| \le k^n|p_1 - p_0|.$$

Now, let $m > n$. Then

$$\begin{aligned}
|p_m - p_n| &= |p_m - p_{m-1} + p_{m-1} - p_{m-2} + \cdots + p_{n+1} - p_n| \\
&\le |p_m - p_{m-1}| + |p_{m-1} - p_{m-2}| + \cdots + |p_{n+1} - p_n| \\
&\le k^{m-1}|p_1 - p_0| + k^{m-2}|p_1 - p_0| + \cdots + k^n|p_1 - p_0| \\
&= k^n|p_1 - p_0|(1 + k + k^2 + \cdots + k^{m-n-1}).
\end{aligned}$$

In part (1), it was established that $p_m \to p$ as $m \to \infty$, so

$$|p - p_n| \le k^n|p_1 - p_0| \sum_{i=0}^{\infty} k^i = \frac{k^n}{1-k}|p_1 - p_0|,$$

where the formula for the sum of a convergent geometric series was used to obtain the final result. □

The theoretical error bound established in part (3) of this theorem clearly demonstrates the importance of the parameter k on the convergence behavior of a fixed point iteration scheme. When k is "small," the error in the approximation to the fixed point will be reduced rapidly, but as $k \to 1$, the rate of convergence of the approximation sequence should decrease dramatically. Notice that when $k = 1/2$, the rate of convergence should be roughly the same as the bisection method.

Order of Convergence for Fixed Point Iteration Schemes

Analytically determining the order of convergence for an arbitrary iteration scheme can be very difficult. Fortunately, for fixed point iteration schemes of the form $p_n = g(p_{n-1})$, the order of convergence can be completely characterized in terms of the value of the derivatives of the iteration function at the fixed point; that is, in terms of the values of $g^{(k)}(p)$.

Theorem. Let g be a continuous function on the closed interval $[a, b]$ with $g : [a, b] \to [a, b]$ and suppose that g' is continuous on the open interval (a, b) with $|g'(x)| \leq k < 1$ for all $x \in (a, b)$. If $g'(p) \neq 0$, then for any $p_0 \in [a, b]$, the sequence $p_n = (p_{n-1})$ converges only linearly to the fixed point p.

Proof. First note that the hypotheses stated in the first sentence of the theorem are precisely those that guarantee that g has a unique fixed point, p, on the interval $[a, b]$ and, that for any starting value $p_0 \in [a, b]$, the sequence generated by $p_n = g(p_{n-1})$ will converge to p. Therefore, the only thing that needs to be proven here is that if $g'(p) \neq 0$, then the convergence is only linear. In other words, it must be shown that

$$\lim_{n\to\infty} \frac{|p_{n+1} - p|}{|p_n - p|} = \lambda \in (0, 1)$$

for some λ.

So consider $|p_{n+1} - p|$. Using the definition of the sequence $\{p_n\}$, the definition of a fixed point and the Mean Value Theorem, it can be shown that

$$\begin{aligned} |p_{n+1} - p| &= |g(p_n) - g(p)| \\ &= |g'(c_n)||p_n - p|, \end{aligned}$$

where c_n is between p_n and p. Since $p_n \to p$, it follows by the squeeze theorem that $c_n \to p$. Furthermore, because g' is continuous on (a, b),

$$\lim_{n\to\infty} |g'(c_n)| = |g'(\lim_{n\to\infty} c_n)| = |g'(p)|.$$

Hence,

$$\lim_{n\to\infty} \frac{|p_{n+1} - p|}{|p_n - p|} = |g'(p)|.$$

Finally, because $|g'(p)| \in (0,1)$ by hypothesis, the order of convergence of the sequence $\{p_n\}$ is one (linear convergence) with asymptotic error constant $|g'(p)|$. □

To obtain higher-order convergence, it is clear that the iteration function must have a zero derivative at the fixed point. The next theorem indicates that the more derivatives of the iteration function which are zero at the fixed point, the higher will be the order of convergence of the generated sequence.

Theorem. Let g be a continuous function on the closed interval $[a,b]$ with $\alpha > 1$ continuous derivatives on the open interval (a,b). Further, let $p \in (a,b)$ be a fixed point of g. If

$$g'(p) = g''(p) = \cdots = g^{(\alpha-1)}(p) = 0,$$

but $g^{(\alpha)}(p) \neq 0$, then there exists a $\delta > 0$ such that for any $p_0 \in [p-\delta, p+\delta]$, the sequence $p_n = g(p_{n-1})$ converges to the fixed point p of order α with asymptotic error constant

$$\lim_{n\to\infty} \frac{|e_{n+1}|}{|e_n|^\alpha} = \frac{|g^{(\alpha)}(p)|}{\alpha!}.$$

Proof. Let's start by establishing the existence of a $\delta > 0$ such that for any $p_0 \in [p-\delta, p+\delta]$, the sequence $p_n = g(p_{n-1})$ converges to the fixed point p. We'll tackle the question of order of convergence later.

Let $k < 1$. Since $g'(p) = 0$ and g' is continuous, it follows that there exists a $\delta > 0$ such that $|g'(x)| \le k < 1$ for all $x \in I \equiv [p-\delta, p+\delta]$. From this it follows that $g : I \to I$; for if $x \in I$ then,

$$\begin{aligned} |g(x) - p| &= |g(x) - g(p)| \\ &= |g'(\xi)||x-p| \\ &\le k|x-p| < |x-p| \le \delta. \end{aligned}$$

Therefore, by the general fixed point iteration theorem established earlier, the sequence $p_n = g(p_{n-1})$ converges to the fixed point p for any $p_0 \in [p-\delta, p+\delta]$.

To establish the order of convergence, let $x \in I$ and expand the iteration function g into a Taylor series about $x = p$:

$$g(x) = g(p) + g'(p)(x-p) + \cdots + \frac{g^{(\alpha-1)}(p)}{(\alpha-1)!}(x-p)^{\alpha-1} + \frac{g^{(\alpha)}(\xi)}{\alpha!}(x-p)^\alpha,$$

where ξ is between x and p. Using the hypotheses regarding the value of $g^{(k)}(p)$ for $1 \le k \le \alpha - 1$ and letting $x = p_n$, the Taylor series expansion simplifies to

$$p_{n+1} - p = \frac{g^{(\alpha)}(\xi)}{\alpha!}(p_n - p)^\alpha,$$

where ξ is now between p_n and p. The definitions of the fixed point iteration scheme and of a fixed point have been used to replace $g(p_n)$ with p_{n+1} and $g(p)$ with p. Finally, let $n \to \infty$. Then $p_n \to p$, forcing $\xi \to p$ also. Hence

$$\lim_{n\to\infty} \frac{|e_{n+1}|}{|e_n|^\alpha} = \frac{|g^{(\alpha)}(p)|}{\alpha!},$$

or $p_n \to p$ of order α. □

Stopping Condition

For a fixed point iteration scheme that produces a linearly convergent sequence (i.e., $g'(p) \neq 0$), a stopping condition can be formulated in much the same manner as one was formulated for the method of false position. In particular, an estimate for $|e_n|$ can be constructed from terms in the sequence $\{p_n\}$. In the case of fixed point iteration, the relevant formulas are

$$|e_n| \approx \left| \frac{g'(p)}{g'(p) - 1} \right| |p_n - p_{n-1}| \tag{1}$$

and

$$g'(p) \approx \frac{p_n - p_{n-1}}{p_{n-1} - p_{n-2}}. \tag{2}$$

The details are left as an exercise. Thus an appropriate stopping condition would involve estimating $g'(p)$ and $|e_n|$ using the formulas given above and terminating the iteration when $|e_n|$ falls below the convergence tolerance ϵ.

EXAMPLE 2.7 Error Estimate and Stopping Condition

We know that the function $g(x) = e^{-x}$ has a unique fixed point somewhere near $x = 0.6$. To ten decimal places, the fixed point happens to be $p = 0.5671432904$. The absolute error in the first ten approximations obtained from fixed point iteration with $p_0 = 0$ is listed below, along with the error estimate as obtained from equations (1) and (2).

		Error Estimate
$p_1 = 1.0000000000$	$\lvert e_1\rvert = 0.4328567096$	
$p_2 = 0.3678794412$	$\lvert e_2\rvert = 0.1992638492$	
$p_3 = 0.6922006276$	$\lvert e_3\rvert = 0.1250573371$	0.1099745306
$p_4 = 0.5004735006$	$\lvert e_4\rvert = 0.0666697898$	0.0712322670
$p_5 = 0.6062435351$	$\lvert e_5\rvert = 0.0391002447$	0.0376047292
$p_6 = 0.5453957860$	$\lvert e_6\rvert = 0.0217475044$	0.0222212089
$p_7 = 0.5796123355$	$\lvert e_7\rvert = 0.0124690451$	0.0123155830
$p_8 = 0.5601154614$	$\lvert e_8\rvert = 0.0070278290$	0.0070769665
$p_9 = 0.5711431151$	$\lvert e_9\rvert = 0.0039998247$	0.0039839812
$p_{10} = 0.5648793474$	$\lvert e_{10}\rvert = 0.0022639430$	0.0022690318

The first few error estimates differ from the actual errors by as much as about 12%. After p_7, however, the error estimates differ from the actual errors by less than 1%.

If iterations are continued until the error estimate falls below $\epsilon = 5\times 10^{-6}$, the final approximation to p is $p_{21} = 0.5671477143$. The estimate of the error in this approximation is 4.42383×10^{-6}, which is in excellent agreement with the actual error of 4.42385×10^{-6}.

When the sequence produced by fixed point iteration has order of convergence $\alpha > 1$, a simpler stopping condition can be used. Recall that any order of convergence $\alpha > 1$ corresponds to superlinear convergence and that all superlinearly convergent sequences satisfy the limit

$$\lim_{n\to\infty} \frac{|p_n - p_{n-1}|}{|p_{n-1} - p|} = 1$$

(see Exercises 14 and 15 in Section 1.2). This limit implies that

$$|p_{n-1} - p| \approx |p_n - p_{n-1}|$$

for any superlinearly convergent sequence. Since p_n is supposed to be a better approximation to p than p_{n-1}, it follows that $|p_n - p_{n-1}|$ should be a conservative estimate of the error $|e_n| = |p_n - p|$. Consequently, whenever the order of convergence is greater than one, an appropriate stopping condition would be to terminate the iteration as soon as $|p_n - p_{n-1}|$ falls below the convergence tolerance ϵ.

EXERCISES

1. Suppose the sequence $\{p_n\}$ is generated by the fixed point iteration scheme $p_n = g(p_{n-1})$. Further, suppose that the sequence converges linearly to the fixed point p.

(a) Show that

$$g'(p) \approx \frac{p_n - p_{n-1}}{p_{n-1} - p_{n-2}}.$$

(b) Show that

$$|e_n| \approx \left|\frac{g'(p)}{g'(p) - 1}\right| |p_n - p_{n-1}|.$$

2. Construct an algorithm for fixed point iteration when the order of convergence is linear.

3. Construct an algorithm for fixed point iteration when the order of convergence is superlinear.

4. In the literature, it is not uncommon to find fixed point iteration terminated when $|p_n - p_{n-1}| < \epsilon$, even when convergence is only linear. Comment on the accuracy of this stopping condition when convergence is linear. Consider the cases $g'(p) \approx 0$, $g'(p) \approx 1/2$ and $g'(p) \approx 1$.

5. Consider the function $g(x) = \cos x$.
   (a) Graphically verify that this function has a unique fixed point on the real line.
   (b) Can we prove that the fixed point is unique using the theorems of this section? Why or why not?
   (c) What order of convergence do we expect from the fixed point iteration scheme $p_n = g(p_{n-1}) = \cos(p_{n-1})$? Why?
   (d) Perform seven iterations starting from $p_0 = 0$. Verify that the appropriate error estimate is valid. To ten decimal places, the fixed point is $x \approx 0.7390851332$.

6. Consider the function $g(x) = 1 + x - \frac{1}{8}x^3$.
   (a) Analytically verify that this function has a unique fixed point on the real line.
   (b) Can we prove that the fixed point is unique using the theorems of this section? Why or why not?
   (c) What order of convergence do we expect from the fixed point iteration scheme $p_n = g(p_{n-1})$? Why?
   (d) Perform seven iterations starting from $p_0 = 0$. Verify that the appropriate error estimate is valid.

7. Consider the function $g(x) = 2x(1 - x)$, which has fixed points at $x = 0$ and at $x = 1/2$.
   (a) Why should we expect that fixed point iteration, starting even with a value very close to zero, will fail to converge toward $x = 0$?
   (b) Why should we expect that fixed point iteration, starting with $p_0 \in (0, 1)$ will converge toward $x = 1/2$? What order of convergence should we expect?
   (c) Perform seven iterations starting from an arbitrary $p_0 \in (0, 1)$ and numerically confirm the order of convergence.

8. Verify that $x = \sqrt{a}$ is a fixed point of the function

$$g(x) = \frac{1}{2}\left(x + \frac{a}{x}\right).$$

Use the techniques of this section to determine the order of convergence and the asymptotic error constant of the sequence $p_n = g(p_{n-1})$ toward $x = \sqrt{a}$.

9. Verify that $x = \sqrt{a}$ is a fixed point of the function

$$g(x) = \frac{x^3 + 3xa}{3x^2 + a}.$$

Use the techniques of the this section to determine the order of convergence and the asymptotic error constant of the sequence $p_n = g(p_{n-1})$ toward $x = \sqrt{a}$.

10. Verify that $x = 1/a$ is a fixed point of the function $g(x) = x(2 - ax)$. Use the techniques of the this section to determine the order of convergence and the asymptotic error constant of the sequence $p_n = g(p_{n-1})$ toward $x = 1/a$.

11. Consider the function $g(x) = e^{-x^2}$.

(a) Prove that g has a unique fixed point on the interval $[0, 1]$.

(b) With a starting approximation of $p_0 = 0$, use the iteration scheme $p_n = e^{-p_{n-1}^2}$ to approximate the fixed point on $[0, 1]$ to within 5×10^{-7}.

(c) Use the theoretical error bound $|p_n - p| \le \frac{k^n}{1-k}|p_1 - p_0|$ to obtain a theoretical bound on the number of iterations needed to approximate the fixed point to within 5×10^{-7}. How does the number of iterations performed in part (b) compare with the theoretical bound?

12. Repeat Exercise 11 for the function $g(x) = \frac{1}{2}\cos x$.

13. Repeat Exercise 11 for the function $g(x) = \frac{1}{3}(2 - e^x + x^2)$.

14. The function $f(x) = e^x + x^2 - x - 4$ has a unique zero on the interval $(1, 2)$. Create three different iteration functions corresponding to this function, and compare their convergence properties for approximating the zero on $(1, 2)$. Use the same starting approximation, p_0, for each iteration function.

15. Repeat Exercise 14 for the function $f(x) = x^3 - x^2 - 10x + 7$ on the interval $(0, 1)$.

16. Repeat Exercise 14 for the function $f(x) = 1.05 - 1.04x + \ln x$ on the interval $(1, 2)$.

2.4 NEWTON'S METHOD

The fundamental concepts of fixed point iteration schemes were developed in the previous section, and the connection between the fixed point problem and the rootfinding problem was made. In this section, we will present Newton's method, which is perhaps the most well known fixed point iteration scheme for approximating the roots of an arbitrary function. Many students are first introduced to this technique when discussing applications of the derivative in a first course on calculus.

Newton's Method

The basic idea behind Newton's method is quite straightforward. Let p_n denote the most recent approximation to a zero, p, of the function f. Replace f by its tangent line approximation based at the location $x = p_n$, and take the x-intercept of the tangent line as the next approximation, p_{n+1}, to the root (see Figure 2.8). Since the tangent line approximation based at $x = p_n$ is given by

$$y - f(p_n) = f'(p_n)(x - p_n),$$

the explicit expression for p_{n+1} is

$$p_{n+1} = p_n - \frac{f(p_n)}{f'(p_n)}.$$

This last equation provides the definition for the iteration function of Newton's method.

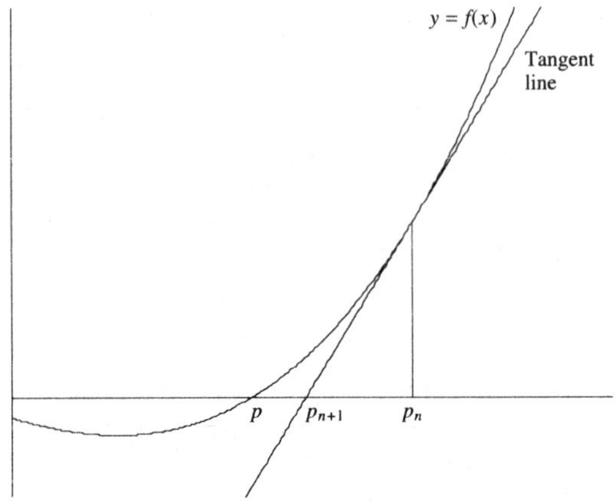

Figure 2.8 Newton's method for approximating the zero of a function: schematic for a single iteration.

Definition. Newton's Method is the fixed point iteration scheme based on the iteration function

$$g(x) = x - \frac{f(x)}{f'(x)};$$

that is, starting from an initial approximation, p_0, the sequence $\{p_n\}$ is generated via $p_n = g(p_{n-1})$.

Note that each iteration of Newton's method requires two separate function evaluations: one evaluation of the function and one of its derivative. This number should be compared to the single function evaluation needed per iteration for both the bisection method and the method of false position.

EXAMPLE 2.8 Newton's Method in Action

Recall the primary demonstration problem from previous sections: Locate the unique zero of the function $f(x) = x^3 + 2x^2 - 3x - 1$ on the interval $(1, 2)$. To apply Newton's method, the derivative of f is needed. For this problem, $f'(x) = 3x^2 + 4x - 3$. With a starting approximation of $p_0 = 1$, four iterations of Newton's method produce the results

$$p_1 = p_0 - \frac{f(p_0)}{f'(p_0)} = 1 - \frac{-1}{4} = 1.25$$

$$p_2 = p_1 - \frac{f(p_1)}{f'(p_1)} = 1.2009345794;$$

$$p_3 = p_2 - \frac{f(p_2)}{f'(p_2)} = 1.1986958411; \text{ and}$$

$$p_4 = p_3 - \frac{f(p_3)}{f'(p_3)} = 1.1986912435.$$

The approximation p_4 is correct to the digits shown and has an absolute error of roughly 1.937×10^{-11}.

In the preceding example, Newton's method achieved an accuracy of 1.937×10^{-11} with only eight function evaluations—four evaluations of f and four evaluations of f'. For comparison, starting from the interval $(1, 2)$, the bisection method needs 36 function evaluations and the method of false position needs 31 evaluations to produce similar accuracy. The information summarized in Table 2.5 indicates that Newton's method also outperforms both the bisection method and false position on the other two standard test problems; that is, locating the zero of $f(x) = \tan(\pi x) - x - 6$ on $(0.4, 0.48)$ and locating the zero of $f(x) = x^3 + 2x^2 - 3x - 1$ on $(-3, -2)$. Note that false position performs exceptionally well on the latter problem, yet still uses 50% more function evaluations than Newton's method.

	$\tan(\pi x) - x - 6$ $p = 0.4510472588$	$x^3 + 2x^2 - 3x - 1$ $p = -2.9122291785$
Newton's Method	$p_0 = 0.48$ $p_5 = 0.4510472613$ $\lvert p_5 - p\rvert \approx 2.448 \times 10^{-9}$ 10 function evaluations	$p_0 = -3$ $p_3 = -2.9122291786$ $\lvert p_3 - p\rvert \approx 9.346 \times 10^{-11}$ 6 function evaluations
Bisection Method	$(a_1, b_1) = (0.4, 0.48)$ $p_{25} = 0.4510472608$ $\lvert p_{25} - p\rvert \approx 1.931 \times 10^{-9}$ 25 function evaluations	$(a_1, b_1) = (-3, -2)$ $p_{34} = -2.9122291785$ $\lvert p_{34} - p\rvert \approx 4.716 \times 10^{-11}$ 34 function evaluations
False Position	$(a_1, b_1) = (0.4, 0.48)$ $p_{32} = 0.4510472563$ $\lvert p_{32} - p\rvert \approx 2.551 \times 10^{-9}$ 33 function evaluations	$(a_1, b_1) = (-3, -2)$ $p_8 = -2.9122291784$ $\lvert p_8 - p\rvert \approx 9.100 \times 10^{-11}$ 9 function evaluations

TABLE 2.5: Comparison of Newton's Method, the Bisection Method and the Method of False Position

Before delving into an analysis of the convergence of Newton's method, let's investigate the influence of the initial approximation, p_0. Consider the function $f(x) = x^3 + 2x^2 - 3x - 1$. Table 2.6 displays the sequences generated by Newton's method for $p_0 = 1$, $p_0 = 2$, and $p_0 = 3$. Here, changing the value of p_0 results only in a variation in the number of iterations needed to achieve convergence. A similar

n	$p_0 = 1$	$p_0 = 2$	$p_0 = 3$
1	1.2500000000	1.4705882353	2.0277777778
2	1.2009345794	1.2471326788	1.4845011523
3	1.1986958411	1.2006987324	1.2514517238
4	1.1986912435	1.1986949265	1.2010586170
5		1.1986912435	1.1986963626
6			1.1986912435

TABLE 2.6: Influence of p_0 on Newton's Method Sequence

observation can be made when Newton's method is initialized with $p_0 = -3$, $p_0 = -2.5$, and $p_0 = -2$. All three sequences converge to -2.9122291785, just in a different number of iterations.

Much more substantial changes in performance are noted when working with the function $f(x) = \tan(\pi x) - x - 6$. Recall that when $p_0 = 0.48$, Newton's method converges to 0.4510472613 in five iterations. With a starting approximation of $p_0 = 0.4$, however, the sequence generated by Newton's method fails to converge, even after 5000 iterations. Thus, unlike simple enclosure methods, Newton's method is not guaranteed to converge for an arbitrary starting approximation. With $p_0 = 0$, we observe another interesting phenomenon. The sequence converges after 42 iterations to 697.4995475. This is, in fact, one of the many zeroes of $f(x) = \tan(\pi x) - x - 6$. Hence, even when Newton's method converges, it may converge to a value very far from p_0.

Clearly, these examples demonstrate that the convergence of the Newton's method sequence is heavily dependent upon the choice of p_0—but with a "good" choice of starting approximation, the sequence will converge very rapidly.

Convergence Analysis for Newton's Method

Let's take a more formal look into the convergence properties of Newton's method, in an attempt to quantify the dependence on the choice of p_0. The simplest plan of attack is to apply the general fixed point iteration convergence theorem which was proven in Section 2.3. To do this, it must be shown that there exists an interval, I, which contains the root, p, for which

1. g is continuous on the interval I;
2. g maps I into I; and
3. $|g'(x)| \leq k < 1$ for all $x \in I$,

where g is the Newton's method iteration function. If all three of these conditions can be established, then by the theorem of Section 2.3, it can be concluded that Newton's method will converge for any starting approximation $p_0 \in I$.

Theorem. Let f be a twice continuously differentiable function on the interval $[a, b]$ with $p \in (a, b)$ and $f(p) = 0$. Further suppose that $f'(p) \neq 0$. Then there exists a $\delta > 0$ such that for any $p_0 \in I = [p - \delta, p + \delta]$, the sequence $\{p_n\}$ generated by Newton's method converges to p.

Proof. First recall that the Newton iteration function is given by

$$g(x) = x - \frac{f(x)}{f'(x)}.$$

The condition $f(p) = 0$ then implies that $g(p) = p$. The proof of the theorem will now proceed in three steps.

Step 1. Show that g is continuous "near" p.
Given the definition of the iteration function g and the continuity assumptions on the function f, the only possible discontinuity in g would be from division by zero with f'. However, the continuity of f' and the assumption that $f'(p) \neq 0$ imply the existence of a positive constant δ_1 such that $f'(x) \neq 0$ for all $x \in I_1 = [p - \delta_1, p + \delta_1] \subset [a, b]$. The reasoning behind this conclusion is that, although f' could have a zero somewhere in the vicinity of p, continuity requires that the distance between p and that zero be of finite, not infinitesimal, size. Therefore, g is well defined, and hence continuous, on I_1.

Step 2. Show that $|g'(x)|$ is "small" near p.
A straightforward calculation shows that

$$g'(x) = \frac{f(x)f''(x)}{[f'(x)]^2}.$$

Having already established that $f'(x) \neq 0$ for all $x \in I_1$, it follows that g' is continuous on I_1. Furthermore, $g'(p) = 0$—which follows from $f(p) = 0$. Choose any k that satisfies $0 < k < 1$. By an argument similar to the one applied in Step 1, it follows that there exists a positive constant δ, with $\delta \leq \delta_1$, such that

$$|g'(x)| \leq k < 1$$

for all $x \in I = [p - \delta, p + \delta]$.

Step 3. Show that g maps the interval I into itself.
Let $x \in I$. Then

$$\begin{aligned} |g(x) - p| &= |g(x) - g(p)| \\ &= |g'(\xi)||x - p| \qquad \text{for some } \xi \text{ between } x \text{ and } p \\ &\leq k|x - p| < |x - p| \leq \delta, \end{aligned}$$

since $x \in [p - \delta, p + \delta]$. In the second line, the mean value theorem was applied. Therefore, $g(x) \in [p - \delta, p + \delta] = I$.

Summary: Having established that

1. g is continuous on the interval I;
2. g maps I into I; and
3. $|g'(x)| \leq k < 1$ for all $x \in I$,

the convergence theorem from Section 2.3 guarantees that the sequence generated by Newton's method will converge to the root p for any starting approximation $p_0 \in I$. □

Although this theorem guarantees that δ exists, it may be very small, implying the need for a very good starting approximation to ensure convergence of the sequence. For instance, in test problem 2, locating the root of $f(x) = \tan(\pi x) - x - 6$ on the interval $[0, 0.48]$, it can be shown that $\delta \approx 0.02$. It is therefore not uncommon, in practice, to find Newton's method combined with a simple enclosure method. Several iterations of the simple enclosure method are performed to obtain the starting approximation for Newton's method. The interval on which the root has been localized can then be used to test the approximations generated by Newton's method. If one of those approximations is found to fall outside the localizing interval, additional iterations of the simple enclosure method are performed. Newton's method is then restarted with a refined initial estimate of the root. This procedure is repeated as necessary until convergence is obtained.

During the course of the above proof, we established that, provided $f'(p) \neq 0$,

$$g(p) = p \quad \text{and} \quad g'(p) = 0,$$

where g is the iteration function for Newton's method. Therefore, the order of convergence for a sequence generated by Newton's method is at least quadratic. To determine the exact order of convergence, we need to continue calculating derivatives of g and evaluating them at $x = p$ until we find a derivative which does not evaluate to zero. For the second derivative, we find

$$g''(x) = \frac{f''(x)}{f'(x)} + \frac{f(x)f'''(x)}{[f'(x)]^2} - 2\frac{f(x)\,[f''(x)]^2}{[f'(x)]^3},$$

from which it follows that $g''(p) = \frac{f''(p)}{f'(p)}$. Since this will not be zero in general, Newton's method is of order two, with asymptotic error constant $\lambda = f''(p)/2f'(p)$, provided $f'(p) \neq 0$. Because the convergence of Newton's method is superlinear, our work at the end of the previous section suggests that an appropriate stopping condition for Newton's method is to terminate the iteration when $|p_n - p_{n-1}|$ falls below a specified convergence tolerance ϵ.

EXAMPLE 2.9 Demonstration of Order of Convergence and Asymptotic Error Constant

Consider Newton's method applied to the function $f(x) = x^3 + 2x^2 - 3x - 1$ with a starting approximation of $p_0 = 2$. The absolute error in p_0 and the first five

Newton's method approximations is listed in the table below. Also listed in the table is the ratio $|e_n|/|e_{n-1}|^2$.

n	Absolute Error, $\lvert e_n\rvert$	$\lvert e_n\rvert/\lvert e_{n-1}\rvert^2$
0	8.0130876×10^{-1}	
1	2.7189699×10^{-1}	0.42345
2	4.8441435×10^{-2}	0.65525
3	2.0074889×10^{-3}	0.85549
4	3.6829405×10^{-6}	0.91387
5	$1.2432499 \times 10^{-11}$	0.91657

Note that the ratio $|e_n|/|e_{n-1}|^2$ approaches a constant, thereby providing numerical confirmation of the quadratic convergence of the sequence. Further, the error ratio appears to be approaching

$$\frac{f''(p)}{2f'(p)} \approx 0.916586,$$

providing numerical confirmation that the asymptotic error constant for Newton's method is $\lambda = f''(p)/2f'(p)$.

Newton's Method with Roots of Multiplicty > 1

The final issue that we will discuss at this time regarding Newton's method is the performance of the method when $f'(p) = 0$. If $f(p) = f'(p) = 0$, then f must have a zero of multiplicity $m \geq 2$ at $x = p$. This implies that f can be written in the form

$$f(x) = (x-p)^m q(x),$$

where $\lim_{x\to p} q(x) \neq 0$. Substituting this expression for f into the Newton's method iteration function

$$g(x) = x - \frac{f(x)}{f'(x)},$$

we find

$$g(x) = x - \frac{(x-p)q(x)}{(x-p)q'(x) + mq(x)}$$

and

$$g'(x) = \frac{\left[m(m-1)q(x) + 2m(x-p)q'(x) + (x-p)^2 q''(x)\right] q(x)}{\left[(x-p)q'(x) + mq(x)\right]^2}.$$

Therefore, $g(p) = p$, but $g'(p) = 1 - 1/m$, which is nonzero for any root of multiplicity greater than one. Accordingly, Newton's method provides only linear convergence for roots of multiplicity greater than one. Since the asymptotic error constant is given by $g'(p)$, the rate of convergence in these cases for Newton's method will be $O\left((1-1/m)^n\right)$. Note that for a root of multiplicity greater than two, this rate of convergence is slower than that of the bisection method.

EXAMPLE 2.10 Newton's Method for a Problem with a Root of Multiplicity > 1

Consider the function $f(x) = x(1 - \cos x)$, which has a root of multiplicity three at $x = 0$. The following table shows the results of ten iterations of Newton's method applied to this problem with a starting value of $p_0 = 1$. For comparison, the results of the bisection method, starting from the interval $[-2, 1]$ are shown in the third column.

	Newton's Method	Bisection Method
1	0.6467039965	−0.5000000000
2	0.4259712109	0.2500000000
3	0.2825304410	−0.1250000000
4	0.1879335654	0.0625000000
5	0.1251658102	−0.0312500000
6	0.0834075192	0.0156250000
7	0.0555942620	−0.0078125000
8	0.0370596587	0.0039062500
9	0.0247054965	−0.0019531250
10	0.0164700517	0.0009765625

For this problem, the bisection method significantly outperforms Newton's method, especially considering that Newton's method uses two function evaluations per iteration while the bisection method uses just one.

Furthermore, notice that with Newton's method, the error in each approximation is roughly two-thirds of the error in the previous approximation. This is due to the root being of multiplicity three. With a root of multiplicity three, the analysis given above implies that the rate of convergence for Newton's method will be $O\left((2/3)^n\right)$.

Application Problem 1: Volume of Chlorine Gas

The van der Waals equation,

$$\left(P + \frac{n^2 a}{V^2}\right)(V - nb) = nRT,$$

which relates the pressure (P), volume (V) and temperature (T) of a gas, was introduced in the Chapter 2 Overview (see page 54). Here, n represents the number of moles of gas present and R is the universal gas constant. The term involving the parameter a corrects the pressure for intermolecular attractive forces, while the term involving the parameter b is a correction for that portion of the volume of the gas that is not compressible due to the intrinsic volume of the gas molecules.

Suppose that one mole of chlorine gas has a pressure of 2 atmospheres and a temperature of 313 K. For chlorine gas, $a = 6.29$ atm · liter^2/mole^2 and $b = 0.0562$ liter/mole. What is the volume of the gas?

In the units of this problem, the universal gas constant has the value $R = 0.08206$ atm $\cdot$ liter/mole $\cdot$ K. We will solve for the volume using Newton's method. The van der Waals equation is first rewritten as the function

$$f(V) = \left(P + \frac{n^2 a}{V^2}\right)(V - nb) - nRT.$$

The derivative of this function is

$$f'(V) = P + \frac{n^2 a}{V^2} - \frac{2n^2 a}{V^3}(V - nb).$$

A convergence tolerance of 5×10^{-7} is used, and a maximum of 10 iterations are allowed. The initial approximation for the volume is taken from the ideal gas law:

$$V_0 = \frac{nRT}{P} = \frac{(1 \text{ mole})(0.08206 \text{ atm} \cdot \text{liter/mole} \cdot \text{K})(313 \text{ K})}{2 \text{ atm}} = 12.84239 \text{ liter}.$$

The actual volume is found to be $V = 12.6510993$ liter, which is roughly 1.5% below the ideal gas law value. Three iterations of Newton's method were needed to achieve convergence.

Application Problem 2: Location of Maximum in an Energy Distribution

The energy density ψ within an isothermal blackbody enclosure is given by Planck's radiation law

$$\psi = \frac{8\pi ch\lambda^{-5}}{e^{ch/\lambda kT} - 1},$$

where λ is the wavelength of the radiation, t is the absolute temperature of the blackbody, h is Planck's constant, k is Boltzmann's constant and C is the speed of light. To determine the wavelength which maximizes the energy density, we first calculate

$$\frac{d\psi}{d\lambda} = \frac{8\pi ch\lambda^{-6}}{e^{ch/\lambda kT} - 1}\left(-5 + \frac{(ch/\lambda kT)e^{ch/\lambda kT}}{e^{ch/\lambda kT} - 1}\right).$$

The term in front of the parentheses is zero in the limits as $\lambda \to 0$ and as $\lambda \to \infty$; however, both of these situations give rise to minima in the energy density. The maximum we are seeking arises when the term inside the parentheses is zero. This happens when

$$1 - \frac{ch}{5\lambda_{\text{max}} kT} = e^{-ch/\lambda_{\text{max}} kT},$$

where λ_{max} is the wavelength that maximizes the energy density. If we let $x = ch/\lambda_{\text{max}} kT$, then the equation for the maximum becomes

$$1 - \frac{x}{5} = e^{-x}.$$

Let's define

$$f(x) = e^{-x} - 1 + \frac{x}{5}$$

and calculate

$$f'(x) = -e^{-x} + \frac{1}{5}.$$

Note that f has a zero at $x = 0$, but we know that we don't want that root. Since the line $1 - \frac{x}{5}$ has an intercept at $x = 5$ and $e^{-5} \approx 6.74 \times 10^{-3}$, it is likely that f has a zero near $x = 5$. Applying Newton's method with $p_0 = 5$ and $\epsilon = 5 \times 10^{-3}$, two iterations produce the approximation $x \approx 4.965$. Therefore,

$$\lambda_{\max} \approx \frac{ch}{4.965kT}.$$

EXERCISES

1. Each of the following equations has a root on the interval $(0, 1)$. Perform Newton's method to determine p_4, the fourth approximation to the location of the root.

(a) $\ln(1 + x) - \cos x = 0$ (b) $x^5 + 2x - 1 = 0$
(c) $e^{-x} - x = 0$ (d) $\cos x - x = 0$

2. Construct an algorithm for Newton's method. Is it necessary to save all calculated terms in the sequence $\{p_n\}$?

In Exercises 3–6, an equation, an interval on which the equation has a root, and the exact value of the root are specified.

(1) Perform five (5) iterations of Newton's method.
(2) For $n \geq 1$, compare $|p_n - p_{n-1}|$ with $|p_{n-1} - p|$ and $|p_n - p|$.
(3) For $n \geq 1$, compute the ratio $|p_n - p|/|p_{n-1} - p|^2$ and show that this value approaches $|f''(p)/2f'(p)|$.

3. The equation $x^3 + x^2 - 3x - 3 = 0$ has a root on the interval $(1, 2)$, namely $x = \sqrt{3}$.

4. The equation $x^7 = 3$ has a root on the interval $(1, 2)$, namely $x = \sqrt[7]{3}$.

5. The equation $x^3 - 13 = 0$ has a root on the interval $(2, 3)$, namely $\sqrt[3]{13}$.

6. The equation $1/x - 37 = 0$ has a zero on the interval $(0.01, 0.1)$, namely $x = 1/37$.

7. Show that when Newton's method is applied to the equation $x^2 - a = 0$, the resulting iteration function is $g(x) = \frac{1}{2}\left(x + \frac{a}{x}\right)$.

8. Show that when Newton's method is applied to the equation $1/x - a = 0$, the resulting iteration function is $g(x) = x(2 - ax)$.

9. The function $f(x) = \sin x$ has a zero on the interval $(3, 4)$, namely $x = \pi$. Perform three iterations of Newton's method to approximate this zero, using $p_0 = 4$. Determine the absolute error in each of the computed approximations. What is the apparent order of convergence? What explanation can you provide for this behavior? (Note: If you have access to *Maple*, perform five iterations with the `Digits` parameter set to at least 100.)

10. **(a)** Verify that the equation $x^4 - 18x^2 + 45 = 0$ has a root on the interval $(1, 2)$. Next, perform three iterations of Newton's method, with $p_0 = 1$. Given that the exact value of the root is $x = \sqrt{3}$, compute the absolute error in the approximations just obtained. What is the apparent order of convergence? What explanation can you provide for this behavior? (Note: If you have access to *Maple*, perform five iterations with the `Digits` parameter set to at least 100.)

(b) Verify that the equation $x^4 - 18x^2 + 45 = 0$ also has a root on the interval $(3, 4)$. Perform five iterations of Newton's method, and compute the absolute error in each approximation. The exact value of the root is $x = \sqrt{15}$. What is the apparent order of convergence in this case?

(c) What explanation can you provide for the different convergence behavior between parts (a) and (b)?

11. The function $f(x) = 27x^4 + 162x^3 - 180x^2 + 62x - 7$ has a zero at $x = 1/3$. Perform ten iterations of Newton's method on this function, starting with $p_0 = 0$. What is the apparent order of convergence of the sequence of approximations? What is the multiplicity of the zero at $x = 1/3$? Would the sequence generated by the bisection method converge faster?

12. Repeat Exercise 11 for the function

$$f(x) = \frac{x}{1+x^2} - \frac{500}{841}\left(1 - \frac{21x}{125}\right),$$

which has a zero at $x = 2.5$. Start Newton's method with $p_0 = 2$.

13. The function $f(x) = x^3 + 2x^2 - 3x - 1$ has a zero on the interval $(-1, 0)$. Approximate this zero to within an absolute tolerance of 5×10^{-5}.

14. For each of the functions given below, use Newton's method to approximate all real roots. Use an absolute tolerance of 10^{-6} as a stopping condition.

(a) $f(x) = e^x + x^2 - x - 4$

(b) $f(x) = x^3 - x^2 - 10x + 7$

(c) $f(x) = 1.05 - 1.04x + \ln x$

15. An equation of state relates the volume V occupied by one mole of a gas to the instantaneous pressure p and the Kelvin absolute temperature t of the gas. The Redlich-Kwong equation of state is given by

$$P = \frac{RT}{V-b} - \frac{a}{V(V+b)\sqrt{T}},$$

where a and b are related to the critical temperature T_c and the critical pressure P_c by the equations

$$a = 0.42747\left(\frac{R^2T_c^{5/2}}{P_c}\right) \quad \text{and} \quad b = 0.08664\left(\frac{RT_c}{P_c}\right).$$

The coefficient R is a universal constant equal to 0.08206.

(a) Determine the volume of one mole of carbon dioxide at a temperature of $T = 323.15$K and a pressure of one atmosphere. For carbon dioxide, $T_c = 304.2$K and $P_c = 72.9$ atmospheres.

(b) Determine the volume of one mole of ammonia at a temperature of T = 450K and a pressure of 56 atmospheres. For ammonia, T_c = 405.5K and P_c = 111.3 atmospheres.

16. In determining the minimum cushion pressure needed to break a given thickness of ice using an air cushion vehicle, Muller ("Ice Breaking with an Air Cushion Vehicle," in *Mathematical Modeling: Classroom Notes in Applied Mathematics*, M. S. Klamkin, editor, SIAM, 1987) derived the equation

$$p^3(1-\beta^2) + \left(0.4h\beta^2 - \frac{\sigma h^2}{r^2}\right)p^2 + \frac{\sigma^2 h^4}{3r^4}p - \left(\frac{\sigma h^2}{3r^2}\right)^3 = 0,$$

where p denotes the cushion pressure, h the thickness of the ice field, R the size of the air cushion, σ the tensile strength of the ice, and β is related to the width of the ice wedge. Take $\beta = 0.5$, $r = 40$ feet, and $\sigma = 150$ pounds per square inch (psi). Determine p for $h = 0.6$, 1.2, 1.8, 2.4, 3.0, 3.6, and 4.2 feet.

17. A frame structure is composed of two vertical columns and one horizontal beam, as shown below. The vertical columns are of length L and have modulus of elasticity E and moment of inertia I. The horizontal beam connecting the tops of the columns is of length L_1 with modulus of elasticity E and moment of inertia I_1. The structure is pinned at the bottom and free to displace laterally at the top. The buckling load, p, for the structure is given by

$$P = (kL)^2\frac{EI}{L^2},$$

where kL is the smallest positive solution of

$$kL\tan kL = 6\frac{I_1 L}{IL_1}.$$

Suppose $E = 30 \times 10^6$ lb/in^2, $I = 15.2$ in^4, $L = 144$ in, $I_1 = 9.7$ in^4 and $L_1 = 120$ in. Determine the buckling load of the structure.

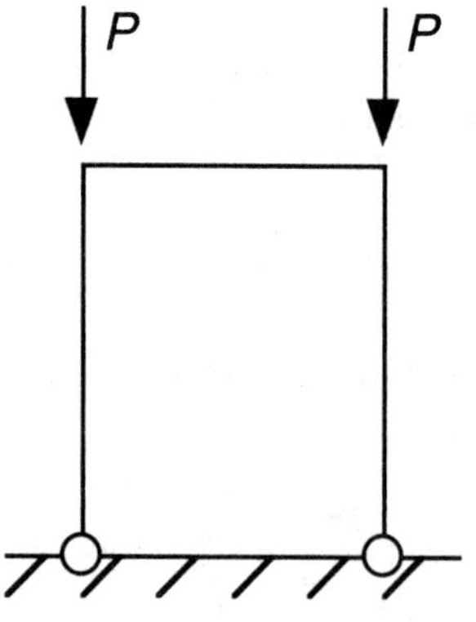

Figure 2.9

2.5 SECANT METHOD

Newton's method is an extremely powerful rootfinding technique. With a "good" starting approximation, the sequence generated by Newton's method converges very rapidly—quadratically, in fact. However, Newton's method does require two new function evaluations per iteration, as well as knowledge of the derivative of the function whose zero is being approximated. In this section, we will develop a rootfinding technique known as the *secant method*, which addresses both of these negative aspects associated with Newton's method.

Secant Method

The secant method can actually be viewed as a variation on either the method of false position or Newton's method. Like the method of false position, the secant method computes the next approximation, p_{n+1}, as the x-intercept of a line that passes through two points on the graph of f. The distinguishing features of the secant method are as follows: First, no attempt is made to maintain an interval that contains the root; and, second, the line from which p_{n+1} is calculated is passed through the points associated with the current and previous approximations, p_n and p_{n-1} (see Figure 2.10). The equation of this line is

$$y - f(p_n) = \frac{f(p_n) - f(p_{n-1})}{p_n - p_{n-1}}(x - p_n),$$

so p_{n+1} is given by

$$p_{n+1} = p_n - f(p_n)\frac{p_n - p_{n-1}}{f(p_n) - f(p_{n-1})}.$$

Definition. The SECANT METHOD is the rootfinding scheme based on the recurrence relation

$$p_{n+1} = p_n - f(p_n)\frac{p_n - p_{n-1}}{f(p_n) - f(p_{n-1})}. \tag{1}$$

From this definition, it is clear that the secant method does not require the derivative of f. Recognizing the similarity between the formula for p_{n+1} given in (1) and the formula for p_{n+1} from the method of false position, it should also be clear that a properly constructed secant method algorithm will use only one new function evaluation per iteration.

It is worth noting that equation (1) can also be derived by approximating the derivative term in Newton's method by

$$f'(p_n) \approx \frac{f(p_n) - f(p_{n-1})}{p_n - p_{n-1}};$$

in other words, the slope of the tangent line at $x = p_n$ is replaced by the slope of the secant line formed between $x = p_n$ and $x = p_{n-1}$. Since the calculation of p_{n+1} requires both p_n and p_{n-1}, the secant method needs two starting values, p_0 and p_1,

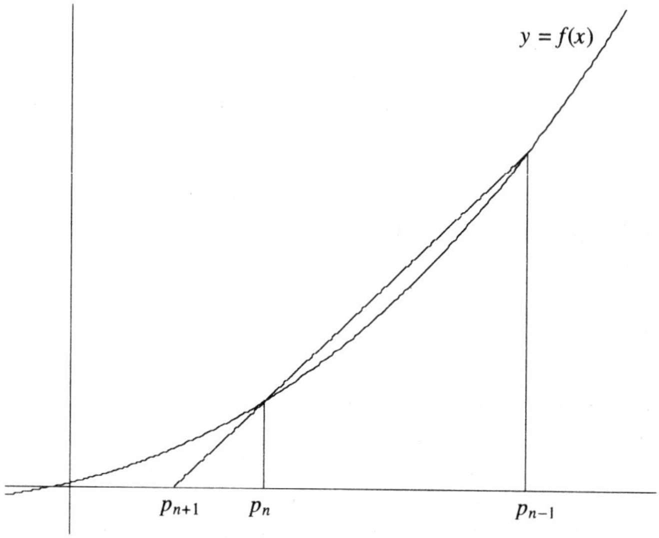

Figure 2.10 The secant method for approximating the zero of a function: schematic for a single iteration.

to initiate the iteration. One obvious choice for starting values is the endpoints of an interval which contain the root.

EXAMPLE 2.11 The Secant Method in Action

Consider the function $f(x) = x^3 + 2x^2 - 3x - 1$, which we know has a unique zero on the interval $(1, 2)$. Taking $p_0 = 2$ and $p_1 = 1$, the secant method produces

$$p_2 = p_1 - f(p_1)\frac{p_1 - p_0}{f(p_1) - f(p_0)} = 1 - (-1)\frac{1-2}{-1-9} = 1.1.$$

With $p_1 = 1$ and $p_2 = 1.1$, the next secant method approximation is

$$\begin{aligned} p_3 &= p_2 - f(p_2)\frac{p_2 - p_1}{f(p_2) - f(p_1)} \\ &= 1.1 - (-0.549)\frac{1.1 - 1}{-0.549 - (-1)} = 1.2217294900. \end{aligned}$$

The next four iterations produce

$$p_4 = p_3 - f(p_3)\frac{p_3 - p_2}{f(p_3) - f(p_2)} = 1.1964853266;$$

$$p_5 = p_4 - f(p_4)\frac{p_4 - p_3}{f(p_4) - f(p_3)} = 1.1986453684;$$

$$p_6 = p_5 - f(p_5)\frac{p_5 - p_4}{f(p_5) - f(p_4)} = 1.1986913364; \quad \text{and}$$

$$p_7 = p_6 - f(p_6)\frac{p_6 - p_5}{f(p_6) - f(p_5)} = 1.1986912435.$$

The approximation p_7 is correct to the digits shown and has an absolute error of roughly 3.907×10^{-12}.

To obtain the approximation p_7, the secant method went through six iterations (the first iteration produced p_2, the second p_3, etc.) and evaluated the function f at $x = p_0, p_1, p_2, p_3, p_4, p_5$, and p_6. So an absolute error of roughly 3.907×10^{-12} was achieved for a total cost of seven function evaluations. For comparison, Newton's method achieved similar accuracy in only four iterations; however, those four iterations were performed at a cost of eight function evaluations. Thus, even though the secant method took more iterations than Newton's method to achieve a given level of accuracy, the secant method used fewer function evaluations.

Does this phenomenon occur for other rootfinding problems? Consider the function $f(x) = \tan(\pi x) - x - 6$. We know that with $p_0 = 0.48$, Newton's method produces an absolute error of roughly 2.448×10^{-9} with five iterations and, therefore, ten function evaluations. Starting from $p_0 = 0.4$ and $p_1 = 0.48$, the secant method reaches similar accuracy (an absolute error of 2.022×10^{-9}) in eight iterations, but this amounts to only nine function evaluations. Thus, once again, the secant method requires more iterations but fewer function evaluations than Newton's method to achieve a given level of accuracy. An examination of our third standard test problem, locating the zero of $f(x) = x^3 + 2x^2 - 3x - 1$ on $(-3, -2)$, is left for the exercises.

The influence of the starting approximations p_0 and p_1 on the performance of the secant method will also be explored in the exercises.

Order of Convergence

To determine the order of convergence for the secant method, we need to derive the corresponding error evolution equation. The first step is to subtract the true root, p, from both sides of the recurrence formula for p_{n+1}, yielding

$$p_{n+1} - p = p_n - p - f(p_n)\frac{p_n - p_{n-1}}{f(p_n) - f(p_{n-1})}.$$

The remaining steps are nearly identical to those used to derive the error evolution equation for the method of false position. The details are therefore left as an exercise. The end result is

$$p_{n+1} - p \approx (p_n - p)(p_{n-1} - p)\frac{f''(p)}{2f'(p) + f''(p)(p_n + p_{n-1} - 2p)}.$$

As p_n and p_{n-1} approach p, the term in the denominator involving the second derivative can be dropped and the leading term in the error is given by

$$|e_{n+1}| \approx C|e_n||e_{n-1}|, \quad \text{where } C = f''(p)/2f'(p). \tag{2}$$

Now, suppose that the Secant method is of order α with asymptotic error constant λ; that is, successive errors are related by the asymptotic formula $|e_{n+1}| \approx \lambda |e_n|^\alpha$. This relationship can also be written as $|e_n| \approx \lambda |e_{n-1}|^\alpha$, which, when solved for $|e_{n-1}|$, yields $|e_{n-1}| \approx \lambda^{-1/\alpha}|e_n|^{1/\alpha}$. Substituting for $|e_{n+1}|$ and $|e_{n-1}|$ in (2) leads to

$$\lambda |e_n|^\alpha \approx C|e_n|\lambda^{-1/\alpha}|e_n|^{1/\alpha}. \tag{3}$$

Equating powers on $|e_n|$ in (3), it follows that α must satisfy the algebraic equation $\alpha = 1 + 1/\alpha$. The single positive root of this equation is $\alpha = (1+\sqrt{5})/2$. Hence, the secant method is of order $(1+\sqrt{5})/2 \approx 1.618$. Furthermore, equating the coefficients of $|e_n|$ yields

$$\lambda \approx C^{1/\alpha} = \left(\frac{f''(p)}{2f'(p)}\right)^{\alpha-1}.$$

EXAMPLE 2.12 Demonstration of Order of Convergence and Asymptotic Error Constant

Consider the secant method applied to the function $f(x) = x^3 + 2x^2 - 3x - 1$ with starting approximations of $p_0 = 2$ and $p_1 = 1$. The absolute error in p_0, p_1 and the first six secant method approximations is listed in the following table. Also listed in the table is the ratio $|e_n|/|e_{n-1}|^{1.618}$.

n	Absolute Error, $\lvert e_n\rvert$	$\lvert e_n\rvert/\lvert e_{n-1}\rvert^{1.618}$
0	8.0130876×10^{-1}	
1	1.9869124×10^{-1}	0.28434
2	9.8691244×10^{-2}	1.34842
3	2.3038247×10^{-2}	0.97658
4	2.2059169×10^{-3}	0.98434
5	4.5875100×10^{-5}	0.91128
6	9.2909772×10^{-8}	0.97192
7	$3.9070969 \times 10^{-12}$	0.93225

Note that the ratio $|e_n|/|e_{n-1}|^{1.618}$ approaches a constant, thereby providing numerical confirmation that the order of convergence of the sequence is $\alpha = 1.618$. Further, the error ratio appears to be approaching

$$\left(\frac{f''(p)}{2f'(p)}\right)^{0.618} \approx 0.94759,$$

providing numerical confirmation that the asymptotic error constant for the secant method is $\lambda = (f''(p)/2f'(p))^{0.618}$.

The analysis we've just completed is based on the assumption that $f'(p) \neq 0$; that is, p is a simple zero of f. We saw in the previous section that the order of convergence of Newton's method drops to linear when approximating a zero of multiplicity greater than one. In the exercises we will explore whether the same fate befalls the secant method.

Application Problem: Solving a Crime

In the Chapter 1 Overview (see the problem capsule "Solving a Crime" on page 2), we were presented with the problem of determining time of death from core temperature measurements. To summarize, Commissioner Gordon had been found dead in his office. At 8:00 PM, the county coroner determined the core temperature of the corpse to be 90°F. One hour later, the core temperature had dropped to 85°F. Captain Furillo believed that the infamous Doc B had killed the commissioner. Doc B, however, claimed to have an alibi. Lois Lane was interviewing him at the Daily Planet Building, just across the street from the commissioner's office. The receptionist at the Daily Planet Building checked Doc B into the building at 6:35 PM, and the interview tapes confirmed that Doc B was occupied from 6:40 PM until 7:15 PM.

To determine time of death, we used Newton's Law of Cooling to model the temperature of the corpse as a function of time. Taking $t = 0$ to correspond to 8:00 PM, the time when the first core temperature was taken, we were able to reduce the problem to two nonlinear algebraic equations. We will now solve those equations with the help of the secant method.

The first equation that we have to solve is

$$73 - \frac{1}{k} + \left(18 + \frac{1}{k}\right) e^{-k} = 85,$$

where k is the constant of proportionality from Newton's Law of Cooling and measures the rate at which the corpse loses heat to its surroundings. To prepare for using the secant method, we rewrite the above equation as the function

$$f(k) = -12 - \frac{1}{k} + \left(18 + \frac{1}{k}\right) e^{-k}.$$

With $p_0 = 0.1$ and $p_1 = 1$, six iterations of the secant method produce the value $k = 0.337114$. Iterations were terminated when $|p_n - p_{n-1}|$ fell below $\epsilon = 5 \times 10^{-7}$. (Why is this an appropriate stopping condition for use with the secant method?)

With the value of k determined, we can now turn to solving the second equation,

$$72 + t_d - \frac{1}{k} + \left(18 + \frac{1}{k}\right) e^{-kt_d} = 98.6,$$

where t_d is the time of death, measured in hours, and 98.6 is the assumed temperature of the corpse at the time of death. To prepare for using the secant method here, we substitute the value of k and rearrange the equation as

$$f(t_d) = -26.6 + t_d - \frac{1}{0.337114} + \left(18 + \frac{1}{0.337114}\right) e^{-0.337114 t_d}.$$

With $p_0 = -2$, $p_1 = 0$ and the same stopping condition noted above, six iterations yield $t_d = -1.130939$. Thus, the time of death was roughly 1 hour and 8 minutes prior to 8:00 PM, or 6:52 PM. This is right in the middle of Doc B's interview with Lois Lane, so Doc B could not have killed the commissioner.

EXERCISES

1. Each of the following equations has a root on the interval $(0, 1)$. Perform the secant method to determine p_4, the fourth approximation to the location of the root.
   (a) $\ln(1 + x) - \cos x = 0$ (b) $x^5 + 2x - 1 = 0$
   (c) $e^{-x} - x = 0$ (d) $\cos x - x = 0$
2. Construct an algorithm for the secant method.
3. Show that the equation for the secant method can be rewritten as
$$p_{n+1} = \frac{f(p_n)p_{n-1} - f(p_{n-1})p_n}{f(p_n) - f(p_{n-1})}.$$
   Explain why this formula is inferior to the one used in the text.
4. Fill in the missing details in the derivation of the error evolution equation
$$p_{n+1} - p \approx (p_n - p)(p_{n-1} - p)\frac{f''(p)}{2f'(p) + f''(p)(p_n + p_{n-1} - 2p)}.$$

In Exercises 5–8, an equation, an interval on which the equation has a root, and the exact value of the root are specified.

(a) Perform seven (7) iterations of the secant method.
(b) For $n \geq 2$, compare $|p_n - p_{n-1}|$ with $|p_{n-1} - p|$ and $|p_n - p|$.
(c) For $n \geq 2$, compute the ratio $|p_n - p|/|p_{n-1} - p|^{1.618}$ and show that this value approaches $\left(|f''(p)/2f'(p)|\right)^{0.618}$.

5. The equation $x^3 + x^2 - 3x - 3 = 0$ has a root on the interval $(1, 2)$, namely $x = \sqrt{3}$.
6. The equation $x^7 = 3$ has a root on the interval $(1, 2)$, namely $x = \sqrt[7]{3}$.
7. The equation $x^3 - 13 = 0$ has a root on the interval $(2, 3)$, namely $\sqrt[3]{13}$.
8. The equation $1/x - 37 = 0$ has a zero on the interval $(0.01, 0.1)$, namely $x = 1/37$.
9. The function $f(x) = \sin x$ has a zero on the interval $(3, 4)$, namely $x = \pi$. Perform five iterations of the secant method to approximate this zero, using $p_0 = 3$ and $p_1 = 4$. Determine the absolute error in each of the computed approximations. What is the apparent order of convergence? What explanation can you provide for this behavior? (Note: If you have access to *Maple*, perform seveitem n iterations with the `Digits` parameter set to at least 100.)
10. (a) Verify that the equation $x^4 - 18x^2 + 45 = 0$ has a root on the interval $(1, 2)$. Next, perform five iterations of the secant method, using $p_0 = 1$ and $p_1 = 2$. Given that the exact value of the root is $x = \sqrt{3}$, compute the absolute error in the approximations just obtained. What is the apparent order of convergence? What explanation can you provide for this behavior? (Note: If you have access to *Maple*, perform seven iterations with the `Digits` parameter set to at least 100.)
    (b) Verify that the equation $x^4 - 18x^2 + 45 = 0$ also has a root on the interval $(3, 4)$. Perform seven iterations of the secant method, and compute the

absolute error in each approximation. The exact value of the root is $\sqrt{15}$. What is the apparent order of convergence in this case? What explanation can you provide for the different convergence behavior between parts (a) and (b)?

11. It was observed that Newton's method provides only linear convergence towards roots of multiplicity greater than one. How does the secant method perform under such circumstances? Each of the following functions has a zero at the specified location. Perform ten iterations of the secant method to locate these zeros. Does the sequence generated by the secant method converge with order $\alpha \approx 1.618$ or has the order dropped to $\alpha = 1$?

(a) $f(x) = x(1 - \cos x)$ has a zero at $x = 0$ – use $p_0 = -1$ and $p_1 = 2$

(b) $f(x) = 27x^4 + 162x^3 - 180x^2 + 62x - 7$ has a zero at $x = 1/3$

(c) $f(x) = \frac{x}{1+x^2} - \frac{500}{841}\left(1 - \frac{21x}{125}\right)$ has a zero at $x = 2.5$

12. Newton's method approximates the zero of $f(x) = x^3+2x^2-3x-1$ on the interval $(-3, -2)$ to within 9.436×10^{-11} in 3 iterations and 6 function evaluations. How many iterations and how many function evaluations are needed by the secant method to approximate this zero to a similar accuracy? Take $p_0 = -2$ and $p_1 = -3$.

In Exercises 13–16 we will investigate the influence of the starting approximations p_0 and p_1 on the performance of the secant method. In each exercise, apply the secant method to the indicated function using the indicated values for p_0 and p_1. Iterate until $|p_n - p_{n-1}| < 5 \times 10^{-7}$. Record and compare the final approximation and the number of iterations in each case.

13. $f(x) = x^3 + 2x^2 - 3x - 1$

(a) $p_0 = -3, \quad p_1 = -2$

(b) $p_0 = -2, \quad p_1 = -3$

(c) $p_0 = -4, \quad p_1 = -2$

(d) $p_0 = -2, \quad p_1 = -4$

14. $f(x) = x^3 + 2x^2 - 3x - 1$

(a) $p_0 = 1, \quad p_1 = 2$

(b) $p_0 = 2, \quad p_1 = 1$

(c) $p_0 = 3, \quad p_1 = 2$

(d) $p_0 = 2, \quad p_1 = 3$

15. $f(x) = \tan(\pi x) - x - 6$

(a) $p_0 = 0, \quad p_1 = 0.48$

(b) $p_0 = 0.24, \quad p_1 = 0.48$

(c) $p_0 = 0.4, \quad p_1 = 0.48$

16. $f(x) = x^3 - 2x - 5$

(a) $p_0 = 1, \quad p_1 = 3$

(b) $p_0 = 1, \quad p_1 = 2$

(c) $p_0 = 3, \quad p_1 = 2$

17. The function $f(x) = x^3 + 2x^2 - 3x - 1$ has a simple zero on the interval $(-1, 0)$. Approximate this zero to within an absolute tolerance of 5×10^{-5}.

18. For each of the functions given below, use the secant method to approximate all real roots. Use an absolute tolerance of 10^{-6} as a stopping condition.

(a) $f(x) = e^x + x^2 - x - 4$

(b) $f(x) = x^3 - x^2 - 10x + 7$

(c) $f(x) = 1.05 - 1.04x + \ln x$

19. Keller ("Probability of a Shutout in Racquetball," *SIAM Review*, **26**, 267–268, 1984) showed that the probability that Player A will shut out Player B in a game of racquetball is given by

$$P = \frac{1+w}{2}\left(\frac{w}{1-w+w^2}\right)^{21},$$

where w denotes the probability that Player A will win any specific rally, independent of the server. Determine the minimal value of w that will guarantee that Player A will shut out Player B in at least one-quarter of the games they play. Repeat your calculations for at least half the games being shutouts and at least three-quarters of the games being shutouts.

20. A couple wishes to open a money market account in which they will save the down payment for purchasing a house. The couple has \$13,000 from the sale of some stock with which to open the account and plans to deposit an additional \$200 each month thereafter. By the end of three years, the couple hopes to have saved \$20,000. If the money market account pays an annual interest of $R\%$, compounded monthly, then at the end of three years, the balance of the account will be

$$13000\left(1+\frac{r}{12}\right)^{36} + 200\frac{\left(1+\frac{r}{12}\right)^{36}-1}{\frac{r}{12}}.$$

What is the lowest interest rate which will achieve the couple's goal of saving \$20,000? What is the lowest interest rate if the couple can raise their monthly deposit to \$250?

21. Suppose it was discovered that Commissioner Gordon had the flu when he died, and his core temperature at the time of his death was 103°F. With $k = 0.337114$, solve the equation

$$72 + t_d - \frac{1}{k} + \left(18 + \frac{1}{k}\right)e^{-kt_d} = 103$$

to determine the time of death based on this new information. Does Doc B's alibi still hold?

2.6 ACCELERATING CONVERGENCE

Having spent so much time discussing speed of convergence, a natural question to ask would be whether it is possible to accelerate the convergence speed of a sequence. For example, can anything be done to speed up the convergence of a linearly convergent sequence? Also, can anything be done to restore quadratic convergence to Newton's method when attempting to approximate a root of multiplicity greater than one? These questions will be addressed in this section.

Aitken's Δ^2-Method

Let's start by accelerating the convergence of a linearly convergent sequence. Thus far, the only truly linearly convergent sequences we've encountered have been generated by either the method of false position or fixed point iteration. Remember that we had to stretch the definition of linear convergence to make the bisection method fit.

During our development of the method of false position, we found that the error associated with the n-th term in the false position sequence, p_n, can be estimated by the formula

$$p - p_n \approx \frac{\lambda}{1-\lambda}(p_n - p_{n-1}). \tag{1}$$

Here, p denotes the limit value of the sequence, and

$$\lambda \approx \frac{p_n - p_{n-1}}{p_{n-1} - p_{n-2}}. \tag{2}$$

Similarly, we found that the error associated with the n-th term in a sequence generated by fixed point iteration can be estimated by the formula

$$p - p_n \approx \frac{g'(p)}{1-g'(p)}(p_n - p_{n-1}), \tag{3}$$

where g is the iteration function and

$$g'(p) \approx \frac{p_n - p_{n-1}}{p_{n-1} - p_{n-2}}. \tag{4}$$

Substituting (2) into (1) or (4) into (3) and solving the resulting expression for p yields

$$p \approx p_n - \frac{(p_n - p_{n-1})^2}{p_n + p_{n-2} - 2p_{n-1}}. \tag{5}$$

Given the approximate nature of (1), (2), (3), and (4), the value given by (5) is not likely to be the exact limit of the sequence, but it should, at least, be a better approximation to that limit than p_n. This is the fundamental idea behind what is known as *Aitken's Δ^2-method.*

From a linearly convergent sequence $\{p_n\}$, Aitken's Δ^2-method constructs the sequence $\{\hat{p}_n\}$ according to the rule

$$\hat{p}_n = p_n - \frac{(p_n - p_{n-1})^2}{p_n + p_{n-2} - 2p_{n-1}}.$$

The formula for $\hat{p}_n$ is usually written in a more compact notation—from which the method derives its name. Let Δ denote the differencing operator that is defined by the relation $\Delta p_n \equiv p_n - p_{n-1}$. The numerator of the second term on the right-hand side of the formula for $\hat{p}_n$ can then be expressed as $(\Delta p_n)^2$. As for the denominator, note that

$$\begin{aligned} p_n - 2p_{n-1} + p_{n-2} &= (p_n - p_{n-1}) - (p_{n-1} - p_{n-2}) \\ &= \Delta p_n - \Delta p_{n-1} \\ &= \Delta(\Delta p_n) = \Delta^2 p_n. \end{aligned}$$

The formula for $\hat{p}_n$ can therefore be written as

$$\hat{p}_n = p_n - \frac{(\Delta p_n)^2}{\Delta^2 p_n}.$$

The sequence $\{\hat{p}_n\}$ is guaranteed to converge more rapidly than the sequence $\{p_n\}$ in the sense that

$$\lim_{n\to\infty} \frac{|\hat{p}_n - p|}{|p_n - p|} = 0.$$

A proof of this statement will be developed in Exercise 16.

EXAMPLE 2.13 Accelerating the Method of False Position

In Section 2.2, the method of false position was used to approximate the zero of $f(x) = x^3 + 2x^2 - 3x - 1$ on the interval $(1, 2)$. Here, we will accelerate the convergence of the false position sequence by applying Aitken's Δ^2-method.

Note that the formula for $\hat{p}_n$ requires three consecutive terms from the p_n sequence. We therefore start by using the method of false position to calculate $p_1 = 1.1$, $p_2 = 1.1517436381$ and $p_3 = 1.1768409100$. Substituting these values into the $\hat{p}_n$ formula for $n = 3$, we find

$$\begin{aligned}
\hat{p}_3 &= p_3 - \frac{(p_3 - p_2)^2}{p_3 + p_1 - 2p_2} \\
&= 1.1768409100 - \frac{(1.1768409100 - 1.1517436381)^2}{1.1768409100 + 1.1 - 2(1.1517436381)} \\
&= 1.2004791447.
\end{aligned}$$

We now return to false position and calculate $p_4 = 1.1886276733$. Substituting the values for p_2, p_3 and p_4 into the $\hat{p}_n$ formula with $n = 4$ then yields

$$\begin{aligned}
\hat{p}_4 &= 1.1886276733 - \frac{(1.1886276733 - 1.1768409100)^2}{1.1886276733 + 1.1517436381 - 2(1.1768409100)} \\
&= 1.1990651249.
\end{aligned}$$

Continuing in this fashion, alternating between the method of false position and Aitken's Δ^2-method, we obtain the values listed below. Recall that to ten decimal places $p = 1.1986912435$.

n	False Position, p_n	Aitken's Δ^2, $\hat{p}_n$
1	1.1000000000	
2	1.1517436381	
3	1.1768409100	1.2004791447
4	1.1886276733	1.1990651249
5	1.1940789113	1.1987692873
6	1.1965820882	1.1987075172
7	1.1977277544	1.1986946351
8	1.1982513178	1.1986919502

Clearly the sequence generated by Aitken's Δ^2-method is converging faster—p_8 is accurate to three decimal places, while $\hat{p}_8$ is accurate to six decimal places. The next table demonstrates that both sequences are converging linearly, but the asymptotic error constant for the $\hat{p}_n$ sequence is less than half the asymptotic error constant for the p_n sequence. Thus, Aitken's Δ^2-method accelerates convergence not by increasing the order of convergence, α, but by reducing the asymptotic error constant, λ.

	False Position		Aitken's Δ^2	
n	Absolute Error, $\lvert e_n\rvert$	$\lvert e_n\rvert/\lvert e_{n-1}\rvert$	Absolute Error, $\lvert e_n\rvert$	$\lvert e_n\rvert/\lvert e_{n-1}\rvert$
1	9.8691244×10^{-2}			
2	4.6947605×10^{-2}	0.4757		
3	2.1850334×10^{-2}	0.4654	1.7879012×10^{-3}	
4	1.0063570×10^{-2}	0.4606	3.7388139×10^{-4}	0.2091
5	4.6123322×10^{-3}	0.4583	7.8043755×10^{-5}	0.2087
6	2.1091553×10^{-3}	0.4573	1.6273731×10^{-5}	0.2085
7	9.6348913×10^{-4}	0.4568	3.3916255×10^{-6}	0.2084
8	4.3992572×10^{-4}	0.4566	7.0667577×10^{-7}	0.2084

EXAMPLE 2.14 Accelerating Fixed Point Iteration

Next, let's accelerate the convergence of fixed point iteration when applied to approximate the fixed point of the function $g(x) = e^{-x}$. Here, we proceed exactly as we did above. With $p_0 = 0$, we first use fixed point iteration to calculate $p_1 = 1$, $p_2 = 0.3678794412$, and $p_3 = 0.6922006276$. Then, substituting these values into the $\hat{p}_n$ formula for $n = 3$, we find

$$\hat{p}_3 = p_3 - \frac{(p_3 - p_2)^2}{p_3 + p_1 - 2p_2} = 0.6922006276 - \frac{(0.6922006276 - 0.3678794412)^2}{0.6922006276 + 1 - 2(0.3678794412)}$$
$$= 0.5822260970.$$

We continue by calculating one term in the p_n sequence followed by one term in the $\hat{p}_n$ sequence. After ten iterations of fixed point iteration, we have the values listed below. To ten decimal places, the fixed point of g is $x = 0.5671432904$.

n	Fixed Point, p_n	Aitken's Δ^2, $\hat{p}_n$
1	1.0000000000	
2	0.3678794412	
3	0.6922006276	0.5822260970
4	0.5004735006	0.5717057675
5	0.6062435351	0.5686388059
6	0.5453957860	0.5676169948
7	0.5796123355	0.5672967525
8	0.5601154614	0.5671924279
9	0.5711431151	0.5671591338
10	0.5648793474	0.5671483792

Again, it is clear that the sequence generated by Aitken's Δ^2-method is converging faster—p_{10} is accurate to only two decimal places, while $\hat{p}_{10}$ is accurate to five decimal places. The next table shows that both sequences are converging linearly, but the asymptotic error constant for the $\hat{p}_n$ sequence is less than 60% of the asymptotic error constant for the p_n sequence.

	Fixed Point		Aitken's Δ^2	
n	Absolute Error, $\lvert e_n\rvert$	$\lvert e_n\rvert/\lvert e_{n-1}\rvert$	Absolute Error, $\lvert e_n\rvert$	$\lvert e_n\rvert/\lvert e_{n-1}\rvert$
1	0.4328567096			
2	0.1992638492	0.4603		
3	1.2505734×10^{-1}	0.6276	1.5082807×10^{-2}	
4	6.6669790×10^{-2}	0.5331	4.5624771×10^{-3}	0.3025
5	3.9100245×10^{-2}	0.5865	1.4955155×10^{-3}	0.3278
6	2.1747504×10^{-2}	0.5562	4.7370444×10^{-4}	0.3167
7	1.2469045×10^{-2}	0.5734	1.5346208×10^{-4}	0.3240
8	7.0278290×10^{-3}	0.5636	4.9137477×10^{-5}	0.3202
9	3.9998247×10^{-3}	0.5691	1.5843424×10^{-5}	0.3224
10	2.2639430×10^{-3}	0.5660	5.0888172×10^{-6}	0.3212

Steffensen's Method

For linearly convergent fixed point iteration schemes of the form $p_{n+1} = g(p_n)$, it is possible to accelerate convergence even further by applying a variation of the Aitken's Δ^2-method. The basic idea can be explained as follows. Suppose the starting approximation p_0 is given and the values $p_1 = g(p_0)$ and $p_2 = g(p_1)$ are calculated. Aitken's Δ^2-method is then applied to compute $\hat{p}$. Since $\hat{p}$ is supposed to be a better approximation to the fixed point than p_2, it seems counterproductive to continue the iteration using p_2. Why not reinitialize the iteration function using $\hat{p}$? This three-step process is then repeated: From the current approximation, perform two fixed point iterations and then combine the current approximation and the two intermediate values according to the Aitken's Δ^2 formula to form the next approximation.

The scheme just described is known as *Steffensen's method.* The sequence of calculations is depicted in Figure 2.11. The sequence of approximations is denoted by $\{\hat{p}_n\}$, and for consistency, the initial approximation is denoted by $\hat{p}_0$. Finally, $p_{1,n}$ and $p_{2,n}$ denote the intermediate values calculated using the iteration function starting with $x = \hat{p}_n$. With this notation the Aitken's Δ^2 formula takes the form

$$\hat{p}_{n+1} = p_{2,n} - \frac{(p_{2,n} - p_{1,n})^2}{p_{2,n} - 2p_{1,n} + \hat{p}_n}.$$

Let's do an example to determine how much acceleration this technique produces over the basic strategy of Aitken's Δ^2-method.

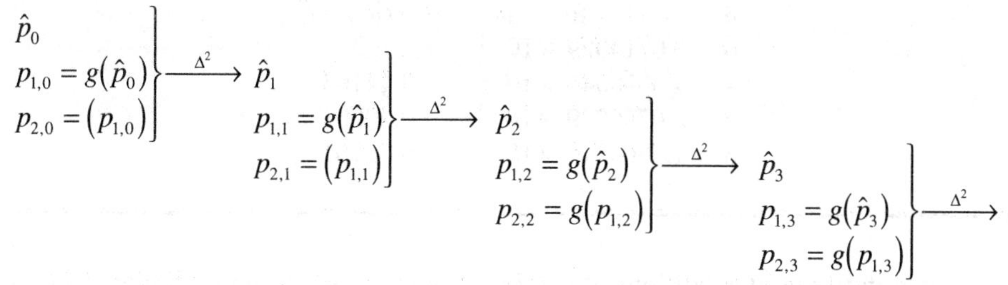

Figure 2.11 Graphical depiction of the sequence of calculations in Steffensen's method. The sequence of approximations is denoted by the $\hat{p}$ values. The values $p_{1,n}$ and $p_{2,n}$ are intermediate values used in the Aitken's Δ^2 formula.

EXAMPLE 2.15 Steffensen's Method in Action

Let's reconsider the fixed point iteration problem examined above—this time using Steffensen's method to accelerate convergence. With a starting approximation of $\hat{p}_0 = 0$, we calculate $p_{1,0} = g(\hat{p}_0) = 1$, $p_{2,0} = g(p_{1,0}) = 0.3678794412$, and

$$\begin{aligned}\hat{p}_1 &= 0.3678794412 - \frac{(0.3678794412 - 1)^2}{0.3678794412 + 0 - 2(1)} \\ &= 0.6126998368.\end{aligned}$$

Reinitializing the iteration with $\hat{p}_1$, we obtain $p_{1,1} = g(\hat{p}_1) = 0.541885888$, $p_{2,1} = g(p_{1,1}) = 0.5816502896$, and

$$\begin{aligned}\hat{p}_2 &= 0.5816502896 - \frac{(0.5816502896 - 0.541885888)^2}{0.5816502896 + 0.6126998368 - 2(0.541885888)} \\ &= 0.5673508577.\end{aligned}$$

At this point, note that $\hat{p}_2$ is more accurate than the tenth term in the sequence generated by fixed point iteration.

The third iteration of Steffensen's method produces $\hat{p}_3 = 0.5671432948$. This value is correct to eight decimals places and has an absolute error of roughly 4.421×10^{-9}. Thus, at a cost of only six evaluations of the function g, Steffensen's method has produced a significantly more accurate approximation to the fixed point than was obtained with Aitken's Δ^2-method at a cost of ten function evaluations.

Having obtained such a small absolute error in so few iterations suggests that Steffensen's method has done more than just reduce the asymptotic error constant for a linearly convergent sequence. Examining the final column in the following table provides evidence that the sequence generated by Steffensen's method is converging quadratically.

n	$\lvert e_n\rvert = \lvert\hat{p}_n - p\rvert$	$\lvert e_n\rvert/\lvert e_{n-1}\rvert^2$
0	5.6714329×10^{-1}	
1	4.5556546×10^{-2}	0.14163
2	2.0756729×10^{-4}	0.10001
3	4.4209310×10^{-9}	0.10261

Under fairly mild conditions, it can be shown (see Isaacson and Keller [1]) that starting from an iteration function that produces linear convergence, Steffensen's method will produce quadratic convergence. This is accomplished with two function evaluations per iteration, the same as Newton's method, but does not require knowledge of the derivative.

Restoring Quadratic Convergence to Newton's Method

This brings us to the problem of restoring quadratic convergence to Newton's method when a root of multiplicity greater than one is being approximated. There are two different approaches that can be taken. In the first approach, the function to which Newton's method is applied is modified in such a way that the root being approximated is guaranteed to be a simple root. In the second approach, the iteration function of Newton's method is modified so that roots of a particular multiplicity can be approximated with a quadratically convergent sequence. Both techniques will now be developed, and the merits and shortcomings of each will be discussed.

Let's start with the approach that modifies the function to that Newton's method is applied. Suppose f has a root of multiplicity m at $x = p$, and consider the function f defined by $F(x) = f(x)/f'(x)$. Since f can be written in the form $f(x) = (x-p)^m q(x)$, where $\lim_{x\to p} q(x) \neq 0$, it follows that $f'(x) = (x-p)^{m-1}\left[(x-p)q'(x) + mq(x)\right]$ and

$$F(x) = \frac{(x-p)q(x)}{(x-p)q'(x) + mq(x)} = (x-p)\tilde{q}(x),$$

where $\lim_{x\to p} \tilde{q}(x) = 1/m \neq 0$. Therefore, f has a simple root at $x = p$, which implies that Newton's method applied to f is guaranteed to converge quadratically.

Substituting $F(x) = f(x)/f'(x)$ into the iteration function for Newton's method yields

$$\begin{aligned} g(x) &= x - \frac{F(x)}{F'(x)} \\ &= x - \frac{f(x)/f'(x)}{\frac{[f'(x)]^2 - f(x)f''(x)}{[f'(x)]^2}} \\ &= x - \frac{f(x)f'(x)}{[f'(x)]^2 - f(x)f''(x)}. \end{aligned}$$

This last formula points out the two main disadvantages of this approach. First, both the first and the second derivatives of f are needed. Second, each iteration requires three function evaluations. On the other hand, this approach will work regardless of the multiplicity of the root and requires no prior knowledge of that multiplicity.

EXAMPLE 2.16 Restoring Quadratic Convergence to Newton's Method—Approach 1

Consider the function $f(x) = 1 + \ln x - x$. It is clear that this function has a root at $x = 1$. The multiplicity of this root happens to be two. The results of applying Newton's method to $f(x)$—standard Newton's method implementation—and to $f(x)/f'(x)$—modified implementation—with a starting approximation of $p_0 = 2$ and a convergence tolerance of 10^{-5} are given below.

	$f(x)$	$f(x)/f'(x)$
1	1.3862943611	1.1146099182
2	1.1721921890	1.0036656204
3	1.0815404027	1.0000044517
4	1.0397051441	1.0000000001
5	1.0195949175	
6	1.0097340850	
7	1.0048513268	
8	1.0024217503	
9	1.0012098989	
10	1.0006047056	
11	1.0003022919	
12	1.0001511307	
13	1.0000755615	
14	1.0000377798	
15	1.0000188897	
16	1.0000094448	

The sequence generated by the standard Newton's method implementation clearly converges only linearly. Note that the error is cut by one-half with each iteration, exactly what is to be expected when approximating a root of multiplicity two. The modified implementation reduced the number of iterations by a factor of four. Even with the extra work required for each iteration, the modified approach has reduced the overall amount of work dramatically—from 32 function evaluations down to only 12.

The second approach to restoring quadratic convergence to Newton's method makes a modification to the method's iteration function, g. Recall that it was shown in Section 2.4 that Newton's method converges only linearly to roots of multiplicity

$m > 1$ because $g'(p) = 1 - 1/m$, which is nonzero for $m \neq 1$. The term $1/m$ comes from the value of the derivative of the term $f(x)/f'(x)$ in the iteration function. This suggests multiplying the term $f(x)/f'(x)$ by m and replacing the standard Newton's method iteration function with

$$\tilde{g}(x) = x - m\frac{f(x)}{f'(x)}.$$

The derivative of this iteration function, evaluated at $x = p$, would then be equal to $1 - (1/m) \times m = 0$, implying quadratic convergence of the generated sequence.

This approach to restoring quadratic convergence to Newton's method has the advantage over the previous approach of requiring no new function evaluations. Unfortunately, this approach does require a priori knowledge of the multiplicity of the root. When the multiplicity is known, however, it can be supplied as an input parameter to the rootfinding routine. In Exercise 13 we will explore a procedure for estimating the multiplicity of a root.

EXAMPLE 2.17 Restoring Quadratic Convergence to Newton's Method—Approach 2

Reconsider the function $f(x) = 1 + \ln x - x$ that has a root of multiplicity two at $x = 1$. With a starting approximation of $p_0 = 2$, a convergence tolerance of 10^{-5}, and using $m = 2$ in the iteration function $\tilde{g}$, we obtain the results

1	0.7725887222
2	0.9804852866
3	0.9998718053
4	0.9999999945
5	0.9999999945

Even with one additional iteration needed to achieve convergence, fewer function evaluations (10 versus 12) were performed with this approach than Approach 1, above—for this example, at least.

References

1. E. Isaacson and H.B. Keller, *Analysis of Numerical Methods*, John Wiley and Sons, New York, 1966.

EXERCISES

1. Show that the equation for Aitken's Δ^2-method can be rewritten as

$$\hat{p}_n = \frac{p_n p_{n-2} - p_{n-1}^2}{p_n - 2p_{n-1} + p_{n-2}}.$$

Explain why this formula is inferior to the one used in the text.

2. Should Aitken's Δ^2-method be applied to a sequence generated by the bisection method? Explain.

3. The sequence listed below was obtained from the method of false position applied to the function $f(x) = \tan(\pi x) - x - 6$ over the interval $(0.40, 0.48)$.

1	0.420867411
2	0.433202750
3	0.440495739
4	0.444807925
5	0.447357748
6	0.448865516
7	0.449757107

(a) Apply Aitken's Δ^2-method to the given sequence.

(b) To nine digits, the zero of f on $(0.40, 0.48)$ is $x = 0.451047259$. Use this to show that both the original sequence and the output from Aitken's Δ^2-method are linearly convergent and estimate the corresponding asymptotic error constants. By how much has Aitken's Δ^2-method reduced the asymptotic error constant?

4. The sequence listed below was obtained from Newton's method applied to the function $f(x) = x(1 - \cos x)$ to approximate the zero at $x = 0$.

1	0.646703997
2	0.425971211
3	0.282530441
4	0.187933565
5	0.125165810
6	0.083407519
7	0.055594262

(a) Apply Aitken's Δ^2-method to the given sequence.

(b) Verify that both the original sequence and the output from Aitken's Δ^2-method are linearly convergent and estimate the corresponding asymptotic error constants. By how much has Aitken's Δ^2-method reduced the asymptotic error constant?

5. The sequence listed below was obtained from fixed point iteration applied to the function $g(x) = \sqrt{10/(2 + x)}$, which has a unique fixed point.

1	2.236067977
2	1.536450382
3	1.681574897
4	1.648098560
5	1.655643081
6	1.653933739
7	1.654320556

(a) Apply Aitken's Δ^2-method to the given sequence.

(b) To ten digits, the fixed point of g is $x = 1.654249158$. Use this to show that both the original sequence and the output from Aitken's Δ^2-method are linearly convergent and estimate the corresponding asymptotic error constant. By how much has Aitken's Δ^2-method reduced the asymptotic error constant?

6. Apply Steffensen's method to the iteration function $g(x) = \frac{1}{2}\sqrt{10 - x^3}$ using a starting value of $p_0 = 1$. Perform four iterations, compute the absolute error in each approximation and confirm quadratic convergence. To twenty digits, the fixed point of g nearest $x = 1$ is $x = 1.3652300134140968458$.

7. (a) Perform ten iterations to approximate the fixed point of $g(x) = \cos x$ using $p_0 = 0$. Verify that the sequence converges linearly and estimate the asymptotic error constant. To 20 digits, the fixed point is

$$x = 0.73908513321516064166.$$

   (b) Accelerate the convergence of the sequence obtained in part (a) using Aitken's Δ^2-method. By how much has Aitken's Δ^2-method reduced the asymptotic error constant?

   (c) Apply Steffensen's method to $g(x) = \cos x$ using the same starting approximation specified in part (a). Perform four iterations, and verify that convergence is quadratic.

8. (a) Perform ten iterations to approximate the fixed point of $g(x) = \ln(4 + x - x^2)$ using $p_0 = 2$. Verify that the sequence converges linearly and estimate the asymptotic error constant. To 20 digits, the fixed point is $x = 1.2886779668238684115$.

   (b) Accelerate the convergence of the sequence obtained in part (a) using Aitken's Δ^2-method. By how much has Aitken's Δ^2-method reduced the asymptotic error constant?

   (c) Apply Steffensen's method to $g(x) = \ln(4 + x - x^2)$ using the same starting approximation specified in part (a). Perform four iterations, and verify that convergence is quadratic.

9. (a) Perform ten iterations to approximate the fixed point of $g(x) = (1.05 + \ln x)/1.04$ using $p_0 = 1$. Verify that the sequence converges linearly and estimate the asymptotic error constant. To 20 digits, the fixed point is $x = 1.1097123038867133005$.

   (b) Accelerate the convergence of the sequence obtained in part (a) using Aitken's Δ^2-method. By how much has Aitken's Δ^2-method reduced the asymptotic error constant?

   (c) Apply Steffensen's method to $g(x) = (1.05 + \ln x)/1.04$ using the same starting approximation specified in part (a). Perform five iterations, and verify that convergence is quadratic.

10. The function $f(x) = 27x^4 + 162x^3 - 180x^2 + 62x - 7$ has a zero of multiplicity 3 at $x = 1/3$. Apply both techniques for restoring quadratic convergence to Newton's method to this problem. Use $p_0 = 0$, and verify that both resulting sequences converge quadratically.

11. The function $f(x) = \frac{x}{1+x^2} - \frac{500}{841}\left(1 - \frac{21x}{125}\right)$ has a zero of multiplicity 2 at $x = 2.5$. Apply both techniques for restoring quadratic convergence to Newton's method to this problem. Use $p_0 = 2$, and verify that both resulting sequences converge quadratically.

12. The function $f(x) = x(1 - \cos x)$ has a zero of multiplicity 3 at $x = 0$. Apply both techniques for restoring quadratic convergence to Newton's method to this problem, using $p_0 = 1$. You should observe that the resulting sequences appear

to converge faster than quadratically. What apparent order of convergence do you observe? Why is convergence faster than quadratic for this problem?

13. Suppose Newton's method is applied to a function with a zero of multiplicity $m > 1$. Show that the multiplicity of the zero can be estimated as the integer nearest to

$$m \approx \frac{1}{1 - \frac{p_n - p_{n-1}}{p_{n-1} - p_{n-2}}}.$$

Verify that this formula produces an accurate estimate when applied to the sequence listed in Exercise 4 and when applied to the sequence generated when Newton's method was applied to the function $f(x) = 1 + \ln x - x$ in the text.

14. Each of the following functions has a zero of multiplicity greater than one at the specified location. In each case, apply the secant method to the function $f(x)/f'(x)$ to approximate the indicated zero. Has the order of convergence been restored to $\alpha \approx 1.618$?

(a) $f(x) = 1 + \ln x - x$ has a zero at $x = 1$ – use $p_0 = -1$ and $p_1 = 2$.

(b) $f(x) = 27x^4 + 162x^3 - 180x^2 + 62x - 7$ has a zero at $x = 1/3$.

(c) $f(x) = \frac{x}{1+x^2} - \frac{500}{841}\left(1 - \frac{21x}{125}\right)$ has a zero at $x = 2.5$.

15. Repeat Exercise 14, but this time replace the standard secant method formula for p_{n+1} by the formula

$$p_{n+1} = p_n - mf(p_n)\frac{p_n - p_{n-1}}{f(p_n) - f(p_{n-1})},$$

where m is the multiplicity of the zero being approximated. The functions in (a) and (c) have $m = 2$, and the function in (b) has $m = 3$.

16. The method of false position and fixed point iteration generate linearly convergent sequences for which

$$\lim_{n\to\infty} \frac{p_n - p}{p_{n-1} - p} \tag{6}$$

exists. Note that this limit does not involve absolute values. Let λ denote the value of this limit. This exercise will lead us through the proof that the sequence produced by Aitken's Δ^2-method converges more rapidly than linearly convergent sequences for which (6) exists.

(a) Let

$$\epsilon_n = \frac{p_n - p}{p_{n-1} - p} - \lambda.$$

Show that $\epsilon_n \to 0$ as $n \to \infty$.

(b) Show that

$$\Delta p_n = (p_n - p)\left(1 - \frac{1}{\epsilon_n + \lambda}\right).$$

(c) Show that

$$\Delta^2 p_n = (p_n - p)\left[1 - \frac{1}{\epsilon_n + \lambda} - \frac{1}{\epsilon_n + \lambda}\left(1 - \frac{1}{\epsilon_{n-1} + \lambda}\right)\right]$$
$$= \frac{p_n - p}{(\epsilon_n + \lambda)(\epsilon_{n-1} + \lambda)}\left[(\lambda - 1)^2 + \epsilon_n'\right],$$

where $\epsilon_n' = \epsilon_n\epsilon_{n-1} + \lambda(\epsilon_n + \epsilon_{n-1}) - 2\epsilon_{n-1}$. Further, show that $\epsilon_n' \to 0$ as $n \to \infty$.

(d) Show that

$$\frac{\hat{p}_n - p}{p_n - p} = 1 - \frac{\epsilon_{n-1} + \lambda}{\epsilon_n + \lambda} \cdot \frac{(\epsilon_n + \lambda - 1)^2}{(\lambda - 1)^2 + \epsilon_n'};$$

hence, as $n \to \infty$,

$$\frac{\hat{p}_n - p}{p_n - p} \to 0.$$

2.7 ROOTS OF POLYNOMIALS

The previous sections have been devoted to the development of techniques for solving the rootfinding problem with an arbitrary nonlinear function. We conclude the chapter by examining the special problem of locating the roots of polynomial functions. Such polynomial rootfinding problems arise in a variety of situations, including the solution of constant coefficient differential equations and the derivation of numerical integration techniques known as Gaussian quadrature rules (a topic which will be covered in Chapter 6).

Working with Polynomials

Anyone who has worked with MATLAB, *Maple*, or *Mathematica* has no doubt recognized that these systems treat polynomials differently from other functions when it comes to rootfinding. For instance, MATLAB provides two built-in m-files: `roots`, which is designed for polynomials and computes all roots, and `fzero`, which can be used with any type of function but determines one root only. The *Maple* and *Mathematica* rootfinding commands can take any type of function as input but internally distinguish between polynomials, for which all roots are computed, and non-polynomials, for which only one root is determined.

Why are polynomials treated differently from other functions? The answer to this question comes in two parts. First, although there is no general theory for the number of roots that an arbitrary nonlinear equation has, as a consequence of the Fundamental Theorem of Algebra, every nth-degree polynomial has precisely n roots in the complex plane, counting multiplicities. When working with polynomials with real coefficients (as we will do exclusively in this section), there is the additional fact that complex roots can occur only in conjugate pairs. Therefore, if $3 + i$ is found to be a root of a given polynomial with real coefficients, $3 - i$ must also be a root.

The second reason that polynomials are treated differently with regard to rootfinding is the following. Let p be an nth-degree polynomial, and suppose that $x = x^*$ is a root of p. Then the monomial $x - x^*$ can be factored from p, leaving $p(x) = (x - x^*)q(x)$, where q is a polynomial of degree $n - 1$. Hence, polynomial rootfinding possesses a natural reduction of order. This process of removing a previously determined root and reducing the size of the remaining problem is known as *deflation.* We will encounter deflation again in Chapter 4 when working with the algebraic eigenvalue problem.

Polynomial Deflation

Suppose the nth-degree polynomial p is given by

$$p(x) = a_n x^n + a_{n-1}x^{n-1} + a_{n-2}x^{n-2} + \cdots + a_1 x + a_0.$$

Further, suppose that $x = x^*$ is a root of p. Then, as noted above, p can be written in the form $p(x) = (x - x^*)q(x)$, where

$$q(x) = b_{n-1}x^{n-1} + b_{n-2}x^{n-2} + b_{n-3}x^{n-3} + \cdots + b_1 x + b_0$$

is a polynomial of degree $n - 1$. To determine the relationship between the coefficients of the deflated polynomial q and the original polynomial p, expand the product $(x - x^*)q(x)$ to obtain

$$b_{n-1}x^n + (b_{n-2} - b_{n-1}x^*)\,x^{n-1} + (b_{n-3} - b_{n-2}x^*)\,x^{n-2} + \cdots + (b_0 - b_1 x^*)\,x - b_0 x^*.$$

Equating coefficients on likes powers of x between this last expression and the polynomial p yields

$$b_{n-1} = a_n \qquad \text{and} \tag{1}$$

$$b_k = a_{k+1} + b_{k+1}x^*, \qquad k = n-2, n-3, n-4, ..., 0. \tag{2}$$

The algorithm given by equations (1) and (2) is commonly known as *synthetic division.*

EXAMPLE 2.18 Polynomial Deflation in Action

To demonstrate both the synthetic division algorithm and the role of deflation in the polynomial rootfinding process, consider the fourth-degree polynomial

$$p(x) = x^4 + 2x^3 + 4x^2 - 2x - 5.$$

Upon inspection, we note that the sum of the coefficients of p is zero, meaning that $p(1) = 0$. Thus, $x^* = 1$ is a root. Applying synthetic division to p with $x^* = 1$ yields

$$\begin{aligned} b_3 &= a_4 = 1; \\ b_2 &= a_3 + b_3 x^* = 2 + 1 \cdot 1 = 3; \\ b_1 &= a_2 + b_2 x^* = 4 + 3 \cdot 1 = 7; \quad \text{and} \\ b_0 &= a_1 + b_1 x^* = -2 + 7 \cdot 1 = 5. \end{aligned}$$

Therefore, $p(x) = (x - 1)q(x)$, where $q(x) = x^3 + 3x^2 + 7x + 5$.

By trial and error, we find that $q(-1) = 0$, so $x^* = -1$ is a root of q, and hence also of p. Applying synthetic division to q with $x^* = -1$ yields

$$\begin{aligned} b_2 &= a_3 = 1; \\ b_1 &= a_2 + b_2 x^* = 3 + 1(-1) = 2; \quad \text{and} \\ b_0 &= a_1 + b_1 x^* = 7 + 2(-1) = 5. \end{aligned}$$

Therefore, $q(x) = (x+1)r(x)$, where $r(x) = x^2 + 2x + 5$.

Finally, because r is a quadratic polynomial, we can use the quadratic formula. This produces the complex conjugate roots $-1 \pm 2i$. Bringing all of this information together, we see that the polynomial $p(x) = x^4 + 2x^3 + 4x^2 - 2x - 5$ has a pair of distinct real roots, ± 1, and a complex conjugate pair of roots, $-1 \pm 2i$.

Unlike the previous example, in which exact roots were determined at each stage, the roots in practical problems will only be calculated to finite precision. The use of an approximate root in the synthetic division procedure will then introduce inaccuracies into the coefficients of the deflated polynomial. For many polynomials, this will not present a serious problem; however, there are polynomials that are extremely ill conditioned. The roots of these polynomials can be very sensitive to changes in the coefficients. Consequently, deflation based on an approximate root can lead to degraded accuracy in subsequently determined roots.

To reduce the effect of deflation induced errors, we can treat the roots obtained from deflated polynomials as merely initial estimates for the roots of the original polynomial. These initial estimates can then be refined, or polished, by applying a rootfinding technique to the original polynomial. This refinement process is not without its pitfalls, though. It is possible for two roots obtained from different deflated polynomials to converge to the same root of the original polynomial, thereby producing a spurious multiplicity. For a more detailed analysis and discussion of the deflation and refinement processes, consult Wilkinson [1] and Peters and Wilkinson [2].

Laguerre's Method

Any of the basic rootfinding techniques discussed earlier in this chapter could be used at the heart of a polynomial rootfinding algorithm. We will, however, base our algorithm on a technique known as Laguerre's method, which is specifically designed for use with polynomials. This technique requires only one starting value and is guaranteed to converge to a root from any starting value. For simple roots, convergence is cubic. Laguerre's method can also produce an approximation to a complex root from a real starting value.

So how does Laguerre's method work? The basic process is iterative in nature. Let $\tilde{x}$ denote the current approximation to a root of the polynomial

$$p(x) = c(x - x_1)(x - x_2)\cdots(x - x_n).$$

Here, c is some constant. The starting approximation is often taken to be zero so as to favor convergence toward the root of smallest magnitude. Next, consider the functions

$$\begin{aligned} G(x) &= \frac{d\ln|p(x)|}{dx} = \frac{p'}{p} \\ &= \frac{1}{x-x_1} + \frac{1}{x-x_2} + \cdots + \frac{1}{x-x_n} \end{aligned}$$

and

$$\begin{aligned} H(x) &= -\frac{d^2\ln|p(x)|}{dx^2} = \left(\frac{p'}{p}\right)^2 - \frac{p''}{p} \\ &= \frac{1}{(x-x_1)^2} + \frac{1}{(x-x_2)^2} + \cdots + \frac{1}{(x-x_n)^2}. \end{aligned}$$

Now, make the following set of assumptions. First, assume that the current approximation is some distance a from the root x_i. Second, assume that *all other roots* are a distance b from the approximation. In other words, assume

$$\tilde{x} - x_i = a \quad \text{but} \quad \tilde{x} - x_j = b \quad \text{for all } j \neq i.$$

Evaluating G and H at $\tilde{x}$ and taking into account the above expressions then yields

$$G = \frac{1}{a} + \frac{n-1}{b} \quad \text{and} \quad H = \frac{1}{a^2} + \frac{n-1}{b^2}.$$

The solution of these equations for a is

$$a = \frac{n}{G \pm \sqrt{(n-1)(nH - G^2)}},$$

where the sign in front of the radical should be chosen to make the magnitude of the denominator as large as possible. Finally, replace $\tilde{x}$ by $\tilde{x} - a$ and repeat. The iteration is terminated when the magnitude of a falls below a specified convergence tolerance.

Each iteration of Laguerre's method requires the evaluation of p, p' and p''. Fortunately, these three function values can be calculated simultaneously, without ever having to explicitly compute either of the derivative functions. The key observation is that with one extra calculation, synthetic division will produce the value of p at $x = x^*$. In particular, consider the quantity $a_0 + b_0x^*$. Working backward through the coefficients computed by the synthetic division algorithm, it follows that

$$\begin{aligned} a_0 + b_0x^* &= a_0 + (a_1 + b_1x^*)x^* = a_0 + a_1x^* + b_1(x^*)^2 \\ &= a_0 + a_1x^* + (a_2 + b_2x^*)(x^*)^2 = a_0 + a_1x^* + a_2(x^*)^2 + b_2(x^*)^3 \\ &= \cdots \\ &= a_0 + a_1x^* + a_2(x^*)^2 + \cdots + a_{n-1}(x^*)^{n-1} + b_{n-1}(x^*)^n \\ &= a_0 + a_1x^* + a_2(x^*)^2 + \cdots + a_{n-1}(x^*)^{n-1} + a_n(x^*)^n = p(x^*). \end{aligned}$$

Now let's focus on the evaluation of the derivatives of p. While computing $p(x^*)$, synthetic division produces the coefficients of a polynomial $q(x)$ such that

$$p(x) = (x - x^*)q(x) + p(x^*).$$

Taking the first derivative of this equation and evaluating at $x = x^*$ gives $p'(x^*) = q(x^*)$. Hence, the value of $p'(x^*)$ can be obtained by applying synthetic division to the coefficients of q as they are computed during the evaluation of p. Of course, this process will not only determine the value of $q(x^*)$, it will also determine the coefficients of a polynomial $r(x)$ for which

$$q(x) = (x - x^*)r(x) + q(x^*),$$

or, equivalently, for which

$$p(x) = (x - x^*)^2 r(x) + (x - x^*)q(x^*) + p(x^*).$$

Taking two derivatives of this latter equation and substituting $x = x^*$ shows that $p''(x^*) = 2r(x^*)$. Therefore, the second derivative of p can be evaluated by applying synthetic division to the coefficients of r. Bringing all of this information together suggests the following algorithm for simultaneously computing $p(x^*)$, $p'(x^*)$ and $p''(x^*)$:

$$\begin{array}{l}
p := a_n, q := r := 0 \\
\text{for } I \text{ from } n - 1 \text{ downto } 0 \text{ do} \\
\qquad r := r \cdot x + q \\
\qquad q := q \cdot x + p \\
\qquad p := p \cdot x + a_i \\
p' := q, p'' := 2r
\end{array}$$

EXAMPLE 2.19 Laguerre's Method in Action

To demonstrate Laguerre's method, consider the polynomial

$$p(x) = 16x^4 + 70x^3 - 169x^2 - 580x + 75.$$

Let's take $x_0 = 0$ as an initial approximation and $\epsilon = 5 \times 10^{-11}$ as a convergence tolerance. Evaluating p, p' and p'' at $x = x_0$ yields

$$p(x_0) = 75, \quad p'(x_0) = -580 \quad \text{and} \quad p''(x_0) = -338.$$

With these values we now calculate

$$G = \frac{p'(x_0)}{p(x_0)} = -\frac{580}{75} = -7.7333333333; \quad \text{and}$$

$$H = \left(\frac{p'(x_0)}{p(x_0)}\right)^2 - \frac{p''(x_0)}{p(x_0)} = 64.3111111111.$$

Since G is negative, we choose the negative sign in front of the radical in the denominator of the formula for a. With $n = 4$. we find

$$a = \frac{4}{G - \sqrt{3(4H - G^2)}} = -0.12472343153714;$$

therefore,

$$x_1 = x_0 - a = 0.12472343153714.$$

Because $|a| > \epsilon$, we perform another iteration. Evaluating p, p', and p'' at $x = x_1$ yields

$$p(x_1) = 0.1711418588, \quad p'(x_1) = -618.1876560151, \quad \text{and}$$

$$p''(x_1) = -282.6294193545.$$

These values then lead to

$$G = -3.6155129190 \times 10^3, \qquad H = 1.3073585101 \times 10^7$$

and

$$a = -2.7656845983 \times 10^{-4}.$$

Thus,

$$x_2 = x_1 - a = 0.12499999999697.$$

Once again, $|a| > \epsilon$, so we perform a third iteration. Evaluating p, p', and p'' at $x = x_2$ yields

$$p(x_2) = 1.8773818056 \times 10^{-9}, \quad p'(x_2) = -618.8437499991, \quad \text{and}$$

$$p''(x_2) = -282.5000000014.$$

From here, we calculate

$$G = -3.2963127061 \times 10^{11}, \qquad H = 1.0865677457 \times 10^{23}$$

and

$$a = -3.0336927626 \times 10^{-12}.$$

Therefore,

$$x_3 = x_2 - a = 0.12500000000000.$$

Since $|a|$ is now less than ϵ, we terminate the iteration and accept x_3 as an approximation to one of the roots of p. The value of x_3 is correct to all digits shown.

Laguerre's method will locate one root of a polynomial. To approximate all of the roots of a polynomial, we proceed as follows. After a root is determined, that root is used to deflate the polynomial. If a complex root is found, then the polynomial is deflated by both that root and its complex conjugate. This process (locate a root and deflate) is repeated until either one or two roots remain to be found. If one root remains, then the original polynomial has been reduced to a linear function, whose root may be found directly; if two roots remain, then the original polynomial has been reduced to a quadratic function, to which the quadratic formula can be applied.

EXAMPLE 2.20 Finding All of the Roots

Above, we found that $x_1^* = 0.12500000000000$ was a root of the fourth-degree polynomial

$$p(x) = 16x^4 + 70x^3 - 169x^2 - 580x + 75.$$

Using this root to deflate p yields the third-degree polynomial

$$q(x) = 16x^3 + 72x^2 - 160x - 600.$$

If we now apply Laguerre's method to q with the same starting approximation and convergence tolerance as used earlier, five iterations produce the root $x_2^* = -2.50000000000000$. Deflating q with this root leaves the quadratic

$$r(x) = 16x^2 + 32x - 240.$$

From the quadratic formula we obtain the final two roots: $x_3^* = 3.00000000000000$ and $x_4^* = -5.00000000000000$. All roots are correct to the digits shown.

EXAMPLE 2.21 Two More Polynomials

Consider the fifth-degree polynomial

$$p(x) = x^5 - 3.4x^4 + 5.4531x^3 - 4.20772x^2 + 1.50924x - 0.20304.$$

Using a convergence tolerance of 5×10^{-11} and a starting approximation of 0 for each application of Laguerre's method, the roots of this polynomial were found to be

$$\begin{aligned}
x_1^* &= 0.44999999975510 \\
x_2^* &= 0.47000000070755 \\
x_3^* &= 0.47999999953778 \\
x_4^* &= 0.99999999999978 + 1.00000000000002i \\
x_5^* &= 0.99999999999978 - 1.00000000000002i
\end{aligned}$$

The first two roots were obtained after six iterations each, and the third root was obtained after four iterations. The complex conjugate pair was obtained from the quadratic formula. The exact roots of this polynomial are 0.45, 0.47, 0.48, and $1 \pm i$.

Finally, consider the polynomial

$$p(x) = 4x^4 - 9x^3 + 3x^2 + 5x - 3,$$

which has a simple root at $x = -0.75$ and a root of multplicity 3 at $x = 1$. With the same convergence tolerance and starting approximation as used above, the following estimates for the roots were obtained:

$$\begin{aligned}
x_1^* &= 0.99988146132218 \\
x_2^* &= 1.00005926933960 + 0.00010265287051i \\
x_3^* &= 1.00005926933960 - 0.00010265287051i \\
x_4^* &= -0.75000000000139.
\end{aligned}$$

Ten iterations were needed to achieve convergence for the first root and eleven for the complex conjugate pair. As might have been expected, the algorithm has no problem estimating the simple root at $x = -0.75$, but has difficulty estimating the triple root at $x = 1$. In fact, the algorithm does not find three real roots, but rather one real root and a complex conjugate pair tightly clustered around $x = 1$.

An Application Problem: Chemical Equilibrium

One mole of nitrogen gas and one mole of hydrogen gas are injected into a one liter reaction chamber. The temperature within the chamber is maintained at 1000 K, and the reversible reaction

$$N_2 + 3H_2 \rightleftharpoons 2NH_3$$

is allowed to proceed to equilibrium. If the equilibrium constant for the indicated reaction is $k = 2.37 \times 10^{-3}$ at 1000 K, how much ammonia (NH_3) is present at equilibrium?

The equilibrium constant for a reversible reaction is given by the product of the concentrations of the substances that appear on the right side of the reaction equation divided by the concentrations of the substances that appear on the left side. Each of these concentrations is raised to the power of the coefficient of that substance in the reaction equation. For this reaction, then,

$$k = 2.37 \times 10^{-3} = \frac{[NH_3]^2}{[N_2][H_2]^3},$$

where $[\ \cdot\ ]$ denotes the concentration of the indicated substance.

Let's assume that at equilibrium there are x moles/liter of NH_3; that is, $[NH_3] = x$. Since the reaction chamber has a volume of one liter, this means that x moles of NH_3 are present. From the reaction equation, it is seen that for every two moles of NH_3 produced, one mole of N_2 is used. Thus, to produce x moles of NH_3, $x/2$ moles of N_2 must have reacted; hence, at equilibrium, $1 - \frac{x}{2}$ moles of N_2 remain. In other words, $[N_2] = 1 - \frac{x}{2}$ at equilibrium. By a similar argument, it follows that $[H_2] = 1 - \frac{3x}{2}$ at equilibrium. Substituting these concentrations into the equilibrium constant expression yields

$$2.37 \times 10^{-3} = \frac{x^2}{\left(1 - \frac{x}{2}\right)\left(1 - \frac{3x}{2}\right)^3}.$$

This last equation can be rearranged into the form

$$3.999375x^4 - 15.9975x^3 - 978.67x^2 - 11.85x + 2.37 = 0.$$

The roots of this equation are $x = 0.04351$, $x = -0.05566$, $x = -13.76348$ and $x = 17.77563$. The two negative roots must be discarded since the concentration of ammonia cannot be negative. The root $x = 17.77563$ must also be discarded since this would correspond to a negative concentration of both nitrogen and hydrogen. Therefore, at equilibrium, there are 0.04351 moles/liter of NH_3 present.

Other Polynomial Rootfinding Schemes

The Jenkins-Traub method [3,4,5] is widely used in software libraries for polynomial rootfinding. This method is best described as a polyalgorithm, combining multiple schemes to produce a robust and efficient computational procedure. Details can be found in Householder [6] and Ralston and Rabinowitz [7]. The Lehmer-Schur algorithm generalizes the one-dimensional notion of bracketing and isolates roots inside circles in the complex plane. Consult Acton [8] for an introduction and Householder [6] for further details. For polynomials with real coefficients, Bairstow's method seeks quadratic factors. This avoids the need for complex arithmetic. A completely different approach to polynomial rootfinding is to formulate the rootfinding problem as a matrix eigenvalue problem. Here, a matrix is first constructed whose eigenvalues are the roots of the given polynomial. Then numerical techniques to locate the eigenvalues of the matrix are applied.

References

1. J. Wilkinson, "The perfidious polynomial," in G. Golub, ed., *Studies in Numerical Analysis*, Mathematical Association of America, Washington, D.C., 1984.
2. G. Peters and J. Wilkinson, "Practical Problems Arising in the Solution of Polynomial Equations," *Journal of the Institute for Mathematics and Its Applications*, **8**, 16–35, 1971.
3. M. A. Jenkins and J. F. Traub, "A Three-Stage Algorithm for Real Polynomials using Quadratic Iteration," *SIAM Journal on Numerical Analysis*, **7**, 545–566, 1970.
4. M. A. Jenkins and J. F. Traub, "Algorithm 419—Zeros of a Complex Polynomial," *Communications of the ACM*, **15**, 97–99, 1972.
5. M. A. Jenkins, "Algorithm 493—Zeros of a Real Polynomial," *ACM Transactions on Mathematical Software*, **1**, 178–189, 1975.

6. A. S. Householder, *The Numerical Treatment of a Single Nonlinear Equation*, McGraw-Hill, New York, 1970.

7. A. Ralston and P. Rabinowitz, *A First Course in Numerical Analysis*, 2nd edition, McGraw-Hill, New York, 1978.

8. F. S. Acton, *Numerical Methods That Work*, Harper and Row, New York, 1970.

EXERCISES

1. Use synthetic division to deflate the given polynomial by the indicated root.

(a) $p(x) = x^4 - 2.25x^3 - 25.75x^2 + 28.5x + 126, \qquad x^* = 3$

(b) $p(x) = x^4 + 1.83x^3 - 0.081x^2 + 1.83x - 1.081, \qquad x^* = -2.3$

(c) $p(x) = x^4 + 20.5x^3 + 129.5x^2 + 230x - 150, \qquad x^* = 0.5$

2. Apply Laguerre's method to each of the following polynomials with a starting approximation of $x_0 = 0$ and a convergence tolerance of 5×10^{-11}.

(a) $p(x) = x^3 - 4x^2 - 3x + 5$

(b) $p(x) = x^3 - 7x^2 + 14x - 6$

(c) $p(x) = x^4 + 20.5x^3 + 129.5x^2 + 230x - 150$

(d) $p(x) = x^4 - 2x^3 - 5x^2 + 12x - 5$

3. Construct an algorithm to deflate the nth-degree polynomial

$$p(x) = a_n x^n + a_{n-1}x^{n-1} + a_{n-2}x^{n-2} + \cdots + a_1 x + a_0$$

by the quadratic factor $x^2 + \alpha x + \beta$; that is, find the polynomial

$$q(x) = b_{n-2}x^{n-2} + b_{n-3}x^{n-3} + b_{n-4}x^{n-4} + \cdots + b_1 x + b_0$$

such that $p(x) = (x^2 + \alpha x + \beta)q(x)$.

4. Determine all roots for each of the following polynomials. Use a convergence tolerance of 5×10^{-11}.

(a) $p(x) = 2x^5 - 6x^4 + 5x^3 + x^2 + 2$

(b) $p(x) = -3x^6 + x^3 + 10x - 1$

(c) $p(x) = x^6 + x^5 - 9x^4 - 8x^3 + 29x^2 - 4x + 4$

(d) $p(x) = x^4 + 5x^3 + 7x^2 + 1$

(e) $p(x) = 16x^4 - 40x^3 + 5x^2 + 20x + 6$

(f) $p(x) = 10x^3 - 8.3x^2 + 2.295x - 0.21141$

5. The Chebyshev polynomials, $T_i(x)$, are a special class of functions. They satisfy the two-term recurrence relation

$$T_{i+1}(x) = 2xT_i(x) - T_{i-1}(x)$$

with $T_0(x) = 1$ and $T_1(x) = x$.

(a) Using the recurrence relation, determine the formula for $T_6(x)$.

(b) Locate all roots of $T_6(x)$.

6. The Hermite polynomials, $H_i(x)$, are a special class of functions. They satisfy the two-term recurrence relation

$$H_{i+1}(x) = 2xH_i(x) - 2iH_{i-1}(x)$$

with $H_0(x) = 1$ and $H_1(x) = x$.
(a) Using the recurrence relation, determine the formula for $H_5(x)$.
(b) Locate all roots of $H_5(x)$.

7. The Laguerre polynomials, $\mathcal{L}_i(x)$, are a special class of functions. They satisfy the two-term recurrence relation

$$\mathcal{L}_{i+1}(x) = (1 + 2i - x)\mathcal{L}_i(x) - i^2\mathcal{L}_{i-1}(x)$$

with $\mathcal{L}_0(x) = 1$ and $\mathcal{L}_1(x) = 1 - x$.
(a) Using the recurrence relation, determine the formula for $\mathcal{L}_4(x)$.
(b) Locate all roots of $\mathcal{L}_4(x)$.

8. The Legendre polynomials, $P_i(x)$, are a special class of functions. They satisfy the two-term recurrence relation

$$P_{i+1}(x) = \frac{2i+1}{i+1}xP_i(x) - \frac{i}{i+1}P_{i-1}(x)$$

with $P_0(x) = 1$ and $P_1(x) = x$.
(a) Using the recurrence relation, determine the formula for $P_5(x)$.
(b) Locate all roots of $P_5(x)$.

9. The concentration, C, of a certain chemical in the bloodstream t hours after injection into muscle tissue is given by

$$C = \frac{3t^2 + t}{50 + t^3}.$$

At what time is the concentration greatest?

10. DeSanti ("A Model for Predicting Aircraft Altitude Loss in a Pull-Up from a Dive," *SIAM Review*, **30** (4), 625–628, 1988) develops the following relationship for the ratio between the final velocity, V_f, and the initial velocity, V_0, for an aircraft executing a pull-up from a dive:

$$\frac{1}{3}\left(\frac{V_f}{V_0}\right)^3 - B\frac{V_f}{V_0} - \frac{1}{3} + B\cos\gamma_0 = 0.$$

γ_0 is the initial flight path angle and $B = g/(kV_0^2)$, where g is the acceleration due to gravity and k is related to the coefficient of lift. The altitude loss during the pull-up can be determined from the ratio V_f/V_0 using the equation

$$\Delta y = \frac{1 - (V_f/V_0)^2}{2kB}.$$

Determine the altitude loss associated with each of the following sets of system parameters (take $g = 9.8$ m/s^2):
(a) $V_0 = 100$ m/s, $\gamma_0 = -30°$, $k = 0.00196$ m^{-1}

(b) $V_0 = 150$ m/s, $\gamma_0 = -10°$, $k = 0.00145$ m^{-1}
(c) $V_0 = 200$ m/s, $\gamma_0 = -45°$, $k = 0.00128$ m^{-1}
(d) $V_0 = 250$ m/s, $\gamma_0 = -30°$, $k = 0.00112$ m^{-1}

11. In determining the minimum cushion pressure needed to break a given thickness of ice using an air cushion vehicle, Muller ("Ice Breaking with an Air Cushion Vehicle," in *Mathematical Modeling: Classroom Notes in Applied Mathematics*, M. S. Klamkin, editor, SIAM, 1987) derived the equation

$$p^3(1-\beta^2) + \left(0.4h\beta^2 - \frac{\sigma h^2}{r^2}\right)p^2 + \frac{\sigma^2 h^4}{3r^4}p - \left(\frac{\sigma h^2}{3r^2}\right)^3 = 0,$$

where p denotes the cushion pressure, h the thickness of the ice field, r the size of the air cushion, σ the tensile strength of the ice, and β is related to the width of the ice wedge. Taking $\beta = 0.5$, $r = 40$ feet, and $\sigma = 150$ pounds per square inch (psi), determine the cushion pressure needed to break a sheet of ice 6 feet thick.

12. Determine the roots of the polynomials

$$P(x) = (x-1)(x-2)(x-3)(x-4)(x-5)(x-6)(x-7)(x-8)(x-9)(x-10)$$

and

$$\tilde{P}(x) = (x-1)(x-2)(x-3)(x-4)(x-5)(x-6)(x-7)(x-8)(x-9)(x-10)+x^5$$

with Laguerre's method as the central rootfinding scheme. Apply a convergence tolerance of 5×10^{-11}, and take 0 as the initial approximation.

13. One mole of H_2S is injected into a two liter reaction chamber, and the reversible reaction

$$2H_2S \rightleftharpoons 2H_2 + S_2$$

is allowed to proceed to equilibrium. If the equilibrium constant for the indicated reaction is $k = 0.016$, how much H_2 and S_2 are present at equilibrium?

14. The reversible reaction

$$2SO_2 + O_2 \rightleftharpoons 2SO_3$$

is allowed to proceed to equilibrium in a one liter reaction chamber. If 0.012 moles of SO_2 and 0.0076 moles of O_2 are initially present and the equilibrium constant for the indicated reaction is $k = 44.643$, how much SO_3 is present at equilibrium?

CHAPTER 3

Systems of Equations

AN OVERVIEW

Fundamental Mathematical Problems

In this chapter we will discuss the general mathematical problems associated with the simultaneous solution of systems of n algebraic equations in n unknowns. These problems take one of two possible forms:

> *Linear Systems of Equations*
> Given a nonsingular $n \times n$ matrix A and an n-vector $\mathbf{b}$, determine the n-vector $\mathbf{x}$ that satisfies the equation $A\mathbf{x} = \mathbf{b}$.
> *Nonlinear Systems of Equations*
> Given a function $\mathbf{F} : \mathbf{R}^n \to \mathbf{R}^n$, find an n-vector $\mathbf{x}$ such that $\mathbf{F}(\mathbf{x}) = \mathbf{0}$.

Techniques for solving linear systems of equations separate into two classes—direct techniques and iterative techniques—whereas nonlinear systems are solved exclusively with iterative methods.

Direct techniques produce a solution to the system of equations in a fixed number of steps. The solution obtained by these techniques would be exact except for roundoff error. Iterative techniques, on the other hand, generate a sequence of approximations which converge toward the true solution. The amount of work required by these techniques therefore depends on the specific problem being attacked and on the choice of a starting approximation. When working with iterative methods, both the conditions under which and the speed with which the resulting sequence converges must be explored.

Here are a few applications which give rise to a system of algebraic equations.

Forces in a Plane Truss

Consider the statically determinate plane truss shown in Figure 3.1. The structure is pinned to a stationary support at the upper right and supported by a roller at the lower left. Furthermore, the structure is subjected to a 3-kN (kilo-Newton) horizontal force at the lower right joint and to a 2-kN force acting at a 45° angle to the horizontal at the upper left joint. Our objective is to determine the resulting forces within the members and the reaction forces at the stationary support and the roller.

Assuming each force acts in the direction indicated in the diagram, balancing the horizontal and vertical components at each joint provides eight simultaneous

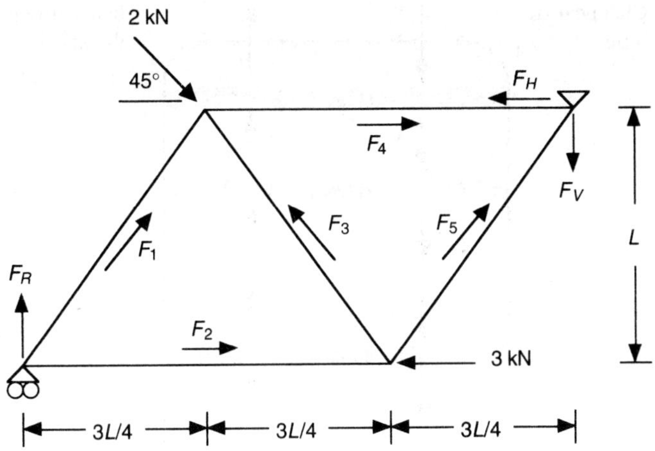

Figure 3.1 Forces acting on a statically determinate truss

linear equations for the eight unknowns. The table below summarizes the equations in this system.

Joint	*Horizontal Component*	*Vertical Component*
lower left	$\frac{3}{5}F_1 + F_2 = 0$	$\frac{4}{5}F_1 + F_R = 0$
lower right	$F_2 - \frac{3}{5}F_3 + \frac{3}{5}F_5 - 3 = 0$	$\frac{4}{5}F_3 + \frac{4}{5}F_5 = 0$
upper left	$\frac{3}{5}F_1 - \frac{3}{5}F_3 + F_4 + \sqrt{2} = 0$	$\frac{4}{5}F_1 + \frac{4}{5}F_3 - \sqrt{2} = 0$
upper right	$F_4 + \frac{3}{5}F_5 - F_H = 0$	$\frac{4}{5}F_5 - F_V = 0$

Multistage Chemical Extraction

Figure 3.2 shows a schematic for an n-stage countercurrent chemical extraction reactor. Water containing a mass fraction x_{in} of a certain chemical enters at one end of the reactor, while a solvent containing a mass fraction y_{in} of the same chemical enters at the other end. The water stream has a mass flow rate of W, and the solvent stream has a mass flow rate of S. As the streams move from stage to stage, the chemical is extracted from the water and transferred to the solvent. Given values for x_{in}, y_{in}, W, S, and n, find the mass fraction of the chemical leaving each stage of the reactor in both the water and solvent streams.

A material balance around stage i (for $i = 2, 3, 4, \ldots, n-1$) yields the equation

$$Wx_{i-1} + Sy_{i+1} = Wx_i + Sy_i. \tag{1}$$

At equilibrium, the linear relationship

$$y_i = mx_i \tag{2}$$

is assumed to hold at each stage of the reactor, where m is a constant that depends on the chemical being extracted and the solvent. Substituting (2) into (1) and

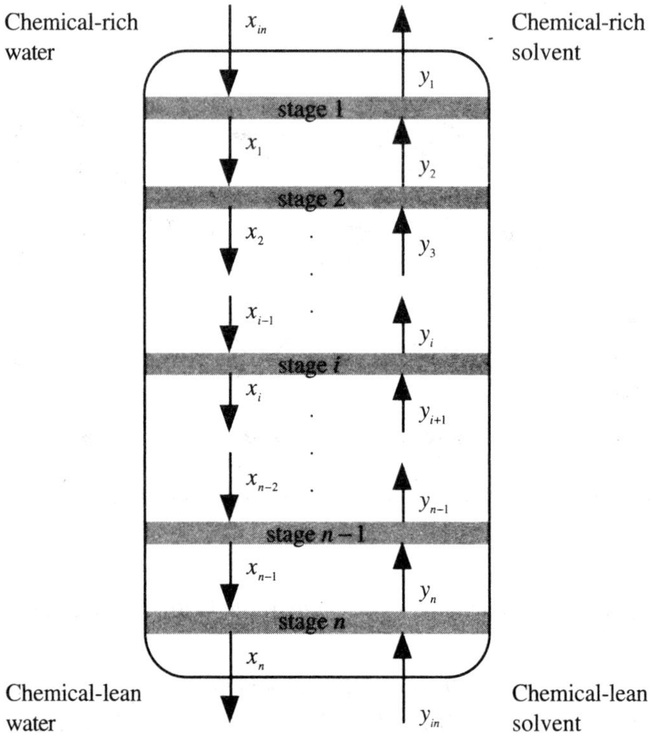

Figure 3.2 Schematic of an n-stage countercurrent chemical extraction reactor.

rearranging terms gives

$$Wx_{i-1} - (W + Sm)x_i + Smx_{i+1} = 0. \tag{3}$$

Performing material balances around stage 1 and stage n produces the equations

$$-(W + Sm)x_1 + Smx_2 = -Wx_{in} \tag{4}$$

and

$$Wx_{n-1} - (W + Sm)x_n = -Sy_{in}, \tag{5}$$

respectively. Equations (3), (4), and (5) form a complete system of n linear equations for the x_i, the mass fractions in the water stream. After solving this system, the mass fractions in the solvent stream, the y_i, are obtained from (2).

Coupled Reversible Chemical Reactions

Consider the coupled pair of reversible chemical reactions

$$2A + B \rightleftharpoons C$$
$$A + D \rightleftharpoons C,$$

where A, B, C, and D represent certain chemical compounds. Suppose A_0 moles of chemical A, B_0, moles of B and D_0 moles of D are injected into a 1-liter reaction chamber and the above reactions are allowed to proceed to equilibrium. If k_1 and k_2 are the equilibrium constants of the first and second reactions, respectively, then how many moles of C are present at equilibrium?

Let c_1 denote the number of moles of C produced by the first reaction and c_2 denote the number of moles of C produced by the second reaction. From the equation for the first reaction, we see that to produce c_1 moles of C, c_1 moles of B, and $2c_1$ moles of A must have reacted. For the second reaction to produce c_2 moles of C, c_2 moles of both A and D must have reacted. Therefore, at equlibrium, there will be $A_0 - 2c_1 - c_2$ moles of A, $B_0 - c_1$ moles of B, $c_1 + c_2$ moles of C, and $D_0 - c_2$ moles of D present.

An equilibrium constant measures the ratio of the concentrations of products to reactants, each raised to the power of their respective coefficients in the chemical equation. Thus

$$k_1 = \frac{[C]}{[A]^2[B]} \quad \text{and} \quad k_2 = \frac{[C]}{[A][D]},$$

where $[\ \cdot\ ]$ denotes the concentration of the indicated chemical. Since the reaction chamber has a volume of one liter, it follows that at equilibrium $[A] = A_0 - 2c_1 - c_2$ moles/liter, $[B] = B_0 - c_1$ moles/liter, $[C] = c_1 + c_2$ moles/liter, and $[D] = D_0 - c_2$ moles/liter. Substituting these concentrations into the expressions for the two equilibrium constants yields the system of nonlinear equations

$$k_1 = \frac{c_1 + c_2}{(A_0 - 2c_1 - c_2)^2(B_0 - c_1)} \quad \text{and} \quad k_2 = \frac{c_1 + c_2}{(A_0 - 2c_1 - c_2)(D_0 - c_2)}.$$

For given values of the parameters A_0, B_0, D_0, k_1, and k_2, we need to solve this system for c_1 and c_2.

Remainder of Chapter

In Section 1 the basic algorithm that will be used for the solution of systems of linear algebraic equations, Gaussian elimination, will be introduced. A careful count of the number of operations required by this algorithm will also be presented. To reduce the effect of roundoff error on the computed solution, pivoting strategies will be considered in the next section. Section 3 then introduces the concepts of vector and matrix norms, while Section 4 presents estimates for the error in the computed solution. The concept of an LU decomposition and its place in the practical implementation of a solution algorithm will be discussed in Section 5.

Coverage of direct techniques for linear systems concludes with a discussion of direct factorization techniques in Section 6 and a discussion of solution algorithms for matrices of special structure in Section 7. Sections 8 and 9 then present iterative techniques for linear systems, with the basic concepts and the classical methods covered in Section 8 and the conjugate gradient method discussed in Section 9. The solution of nonlinear systems of equations is treated in the final section.

3.0 LINEAR ALGEBRA REVIEW

In this section, we will review several definitions and concepts of linear algebra. The focus will be on topics needed throughout the remainder of the chapter. For a more detailed review of this material, consult a standard linear algebra textbook, such as Lay [1], Leon [2], or Shifrin and Adams [3].

Matrices

A matrix is one of the most important tools in linear algebra.

Definition. A MATRIX is a rectangular collection of numbers arranged in rows and columns.

A matrix with n rows and m columns is said to be of dimension $n \times m$, which is read "n by m." Standard notation is to use a capital letter, such as A, to denote a matrix, and the corresponding lowercase letter with two subscripts, as in a_{ij}, to denote the elements in the matrix. The first subscript indicates the row, and the second indicates the column. For example,

$$A = \begin{bmatrix} 2 & -4 & 1 \\ -6 & 0 & 5 \end{bmatrix}$$

is a 2×3 matrix, with

$$\begin{array}{lll} a_{11} = 2; & a_{12} = -4; & a_{13} = 1; \\ a_{21} = -6; & a_{22} = 0; & a_{23} = 5. \end{array}$$

A matrix with one column is called a *column vector*, while a matrix with a single row is called a *row vector*. Vectors will be denoted by a lowercase letter in boldface, such as $\mathbf{x}$. The elements in a vector will be denoted by the same lowercase letter, not in boldface, with a single subscript, as in x_i. When we wish to indicate that a specific vector has n elements, we will refer to the vector as an n-vector.

A matrix that has the same number of rows as columns is called a *square* matrix. Suppose A is an $n \times n$ square matrix. The elements a_{11}, a_{22}, a_{33}, ..., a_{nn} are called the *diagonal* elements of A. All other elements are referred to as *off-diagonal* elements. If all of the off-digonal elements of the square matrix A are zero (i.e., $a_{ij} = 0$ for $i \neq j$), then A is called a *diagonal matrix*. An $n \times n$ diagonal matrix whose diagonal elements are all equal to 1 is called the *identity matrix*, which is denoted by I_n. When the context makes the dimension of the matrix clear, the subscript n is generally omitted and the identity matrix is simply denoted as I.

Operations on Matrices

In order to work with equations involving matrices, we need to define some basic matrix concepts and operations. The first topic we consider is matrix equality.

Definition. Two $n \times m$ matrices A and B are EQUAL if $a_{ij} = b_{ij}$ for each i and j.

Note the role that the dimension of the matrices plays in this definition. Thus, even though every element in the matrix

$$A = \begin{bmatrix} -2 & 6 \\ 1 & 3 \end{bmatrix}$$

is equal to the corresponding element in the matrix

$$B = \begin{bmatrix} -2 & 6 & 5 & 0 \\ 1 & 3 & 0 & -3 \\ 0 & 1 & 2 & 3 \end{bmatrix},$$

these two matrices are not equal because they have different dimensions.

The two most basic algebraic operations involving matrices are matrix addition and scalar multiplication. These are defined as follows.

Definition. The SUM of two $n \times m$ matrices A and B is an $n \times m$ matrix $C = A + B$ whose elements are given by $c_{ij} = a_{ij} + b_{ij}$ for each i and j.

Definition. Let A be an $n \times m$ matrix and α be a real number. The SCALAR MULTIPLICATION of α and A is an $n \times m$ matrix $C = \alpha A$ whose elements are given by $c_{ij} = \alpha a_{ij}$.

EXAMPLE 3.1 Matrix Addition and Scalar Multiplication

Let

$$A = \begin{bmatrix} 2 & 1 \\ 1 & 1 \\ 2 & 3 \end{bmatrix} \quad \text{and} \quad B = \begin{bmatrix} -4 & 1 \\ 3 & -1 \\ -2 & 1 \end{bmatrix}.$$

Then

$$2A = \begin{bmatrix} 2(2) & 2(1) \\ 2(1) & 2(1) \\ 2(2) & 2(3) \end{bmatrix} = \begin{bmatrix} 4 & 2 \\ 2 & 2 \\ 4 & 6 \end{bmatrix} \quad \text{and} \quad -3B = \begin{bmatrix} 12 & -3 \\ -9 & 3 \\ 6 & -3 \end{bmatrix}.$$

Further,

$$A + B = \begin{bmatrix} 2+(-4) & 1+1 \\ 1+3 & 1+(-1) \\ 2+(-2) & 3+1 \end{bmatrix} = \begin{bmatrix} -2 & 2 \\ 4 & 0 \\ 0 & 4 \end{bmatrix}$$

and

$$2A - 3B = 2A + (-3)B = \begin{bmatrix} 16 & -1 \\ -7 & 5 \\ 10 & 3 \end{bmatrix}.$$

Another important matrix operation is the transpose.

Definition. The TRANSPOSE of an $n \times m$ matrix A, denoted by A^T, is an $m \times n$ matrix whose ith row is the ith column of A for each i.

For example, if

$$A = \begin{bmatrix} 2 & 1 \\ 1 & 1 \\ 2 & 3 \end{bmatrix},$$

then

$$A^T = \begin{bmatrix} 2 & 1 & 2 \\ 1 & 1 & 3 \end{bmatrix}.$$

For the matrix

$$A = \begin{bmatrix} 4 & -1 & 0 \\ -1 & 4 & -2 \\ 0 & -2 & 4 \end{bmatrix},$$

note that

$$A^T = \begin{bmatrix} 4 & -1 & 0 \\ -1 & 4 & -2 \\ 0 & -2 & 4 \end{bmatrix} = A.$$

A square matrix A is called *symmetric* if $A^T = A$.

The final basic matrix operation that we will discuss is matrix multiplication.

Definition. Let A be an $n \times m$ matrix and B be a $m \times p$ matrix. The PRODUCT of A and B is an $n \times p$ matrix C whose elements are defined by

$$c_{ij} = \sum_{k=1}^{m} a_{ik} b_{kj},$$

for each i and j.

Note that the number of columns in the first matrix must equal the number of rows in the second matrix for the matrix product to be defined.

EXAMPLE 3.2 Matrix Multiplication

Let

$$A = \begin{bmatrix} 2 & -6 \\ -4 & 0 \\ 1 & 5 \end{bmatrix}, \quad B = \begin{bmatrix} 1 & 1 \\ 2 & 3 \end{bmatrix}, \quad \text{and} \quad C = \begin{bmatrix} 3 & -1 \\ -2 & 1 \end{bmatrix}.$$

Then

$$\begin{aligned} AB &= \begin{bmatrix} (2)(1) + (-6)(2) & (2)(1) + (-6)(3) \\ (-4)(1) + (0)(2) & (-4)(1) + (0)(3) \\ (1)(1) + (5)(2) & (1)(1) + (5)(3) \end{bmatrix} \\ &= \begin{bmatrix} -10 & -16 \\ -4 & -4 \\ 11 & 16 \end{bmatrix}, \end{aligned}$$

and

$$AC = \begin{bmatrix} 18 & -8 \\ -12 & 4 \\ -7 & 4 \end{bmatrix},$$

but the products BA and CA are not defined because the number of columns in B and C is not equal to the number of rows in A.

Finally,

$$BC = CB = \begin{bmatrix} 1 & 0 \\ 0 & 1 \end{bmatrix} = I.$$

The Inverse Matrix

In the last example, we saw two square matrices whose product was the identity matrix. Square matrices with this property are called *inverses*.

Definition. Let A be an $n \times n$ matrix. If there exists an $n \times n$ matrix B such that $BA = AB = I$, then the matrix B is called the INVERSE of A and is denoted by A^{-1}.

Not all square matrices have inverses. For example, consider the 2×2 matrix

$$A = \begin{bmatrix} 1 & 0 \\ 0 & 0 \end{bmatrix}.$$

If

$$B = \begin{bmatrix} b_{11} & b_{12} \\ b_{21} & b_{22} \end{bmatrix}, \quad \text{then} \quad AB = \begin{bmatrix} b_{11} & b_{12} \\ 0 & 0 \end{bmatrix}.$$

From here, we see that no choice of the elements b_{11}, b_{12}, b_{21} and b_{22} will result in the product AB being equal to I. Hence, A does not have an inverse. Matrices that do not have an inverse are called *singular*, whereas matrices that do have an inverse are called *nonsingular*.

The following theorem lists several important results regarding inverses. The proofs will be considered in the exercises.

Theorem. Let A be a nonsingular matrix. Then

1. A^{-1} is unique;
2. A^{-1} is nonsingular and $(A^{-1})^{-1} = A$;
3. A^T is nonsingular and $(A^T)^{-1} = (A^{-1})^T$; and
4. If B is nonsingular, then AB is nonsingular and $(AB)^{-1} = B^{-1}A^{-1}$.

The Determinant

Associated with every square matrix is a real number called the *determinant* of A, which we shall denote by $\det(A)$.

Definition. The DETERMINANT of a square matrix A is defined recursively as follows.

1. If A is the 1×1 matrix $[a_{11}]$, then $\det(A) = a_{11}$.
2. If A is an $n \times n$ matrix with $n > 1$, then

$$\det(A) = \sum_{j=1}^{n} a_{ij}(-1)^{i+j} m_{ij}$$

for any choice of the row i, or

$$\det(A) = \sum_{i=1}^{n} a_{ij}(-1)^{i+j} m_{ij}$$

for any choice of the column j, where m_{ij} is the determinant of the $(n-1) \times (n-1)$ matrix obtained by deleting the ith row and the jth column from A.

Each m_{ij} is called a *minor* of A, and the expression $(-1)^{i+j} m_{ij}$ is called the *cofactor* associated with a_{ij}. The procedure for calculating determinants given by this definition is therefore known as expansion by cofactors.

The second statement in this definition provides $2n$ different ways to calculate a determinant, depending on the row or column chosen for the cofactor expansion. However, all expansions will lead to the same numerical value. We can therefore use the flexibility of the definition to our advantage by choosing to expand along the row or column with the most zero elements.

EXAMPLE 3.3 Calculating a Determinant

Consider the 4×4 matrix

$$A = \begin{bmatrix} 1 & 0 & 4 & 1 \\ -2 & 1 & -3 & 2 \\ 0 & 0 & 0 & 2 \\ 3 & 2 & 1 & -1 \end{bmatrix}.$$

Because the third row contains only one non-zero element, we choose to expand along the third row. Since $a_{31} = a_{32} = a_{33} = 0$,

$$\det(A) = a_{34}(-1)^{3+4} m_{34} = -2 \det\left(\begin{bmatrix} 1 & 0 & 4 \\ -2 & 1 & -3 \\ 3 & 2 & 1 \end{bmatrix}\right).$$

Expanding along the second column of this new matrix, we find

$$\begin{aligned}
\det\left(\begin{bmatrix} 1 & 0 & 4 \\ -2 & 1 & -3 \\ 3 & 2 & 1 \end{bmatrix}\right)
&= 1(-1)^{2+2}\det\left(\begin{bmatrix} 1 & 4 \\ 3 & 1 \end{bmatrix}\right) + 2(-1)^{3+2}\det\left(\begin{bmatrix} 1 & 4 \\ -2 & -3 \end{bmatrix}\right) \\
&= [(1)(1) - (3)(4)] - 2\,[(1)(-3) - (-2)(4)] \\
&= -11 - 2(5) = -21,
\end{aligned}$$

where we expanded along the first column of each of the 2×2 matrices. Therefore, $\det(A) = -2(-21) = 42$.

The following theorem presents several important properties of determinants. The proof of this theorem can be found in the linear algebra texts cited in the references below.

Theorem. Let A be an $n \times n$ matrix.

1. If A has a row or column consisting only of zero entries, then $\det(A) = 0$;
2. If A has two rows the same or two columns the same, then $\det(A) = 0$;
3. $\det(A^T) = \det(A)$;
4. If A is nonsingular, then $\det(A^{-1}) = (\det(A))^{-1}$;
5. If B is an $n \times n$ matrix, then $\det(AB) = \det(A)\det(B)$.

We close our linear algebra review with a theorem which links the concepts of nonsingular matrices, determinants and solutions of linear systems of equations. This theorem will be extremely important throughout the remainder of this chapter. For a proof, consult one of the texts listed below.

Theorem. For any $n \times n$ matrix A, the following statements are equivalent:

1. A is nonsingular;
2. $\det(A) \neq 0$;
3. The equation $A\mathbf{x} = \mathbf{0}$ has the unique solution $\mathbf{x} = \mathbf{0}$; and
4. The equation $A\mathbf{x} = \mathbf{b}$ has a unique solution for any right-hand side vector $\mathbf{b}$.

References

1. D. C. Lay, *Linear Algebra and Its Applications,* 2nd edition, Addison-Wesley, Reading, MA., 1997.
2. S. J. Leon, *Linear Algebra with Applications, Sixth Edition,* Prentice Hall, Upper Saddle River, NJ, 2002.
3. T. Shifrin and M. R. Adams, *Linear Algebra: A Geometric Approach,* W. H. Freeman, New York, 2002.

EXERCISES

In Exercises 1–9, compute the indicated matrices given

$$A = \begin{bmatrix} 1 & -1 & 3 \\ 2 & 0 & 5 \end{bmatrix}, \quad B = \begin{bmatrix} 2 & 1 & 0 \\ -3 & -1 & 5 \\ 1 & 3 & 4 \end{bmatrix}, \quad C = \begin{bmatrix} 4 & 2 \\ 3 & -1 \\ 2 & -4 \end{bmatrix},$$

$$\text{and } D = \begin{bmatrix} 1 & -1 & 4 \\ 0 & 2 & -2 \\ 0 & 0 & 3 \end{bmatrix}.$$

If an operation cannot be performed, indicate why not.

1. **(a)** $2A + C^T$ **(b)** $C - 3B$

2. **(a)** AB **(b)** AD

3. **(a)** CA **(b)** AC

4. **(a)** BD **(b)** DB

5. **(a)** BC **(b)** CB

6. **(a)** $3B - 2D$ **(b)** $2D^T + B$

7. **(a)** $\det(D)$ **(b)** $\det(A)$

8. **(a)** $C^T D$ **(b)** BA^T

9. **(a)** $-2A^T + 5C$ **(b)** $B^T + D$

10. Let A be a nonsingular matrix.

(a) Show that A^{-1} is unique.

(b) Show that A^{-1} is nonsingular and $(A^{-1})^{-1} = A$.

(c) Show that A^T is nonsingular and $(A^T)^{-1} = (A^{-1})^T$.

(d) If B is nonsingular, show that AB is nonsingular and $(AB)^{-1} = B^{-1}A^{-1}$.

11. Can an $n \times m$ matrix with $n \neq m$ be symmetric? Explain.

12. Recalculate the determinant of the matrix

$$A = \begin{bmatrix} 1 & 0 & 4 & 1 \\ -2 & 1 & -3 & 2 \\ 0 & 0 & 0 & 2 \\ 3 & 2 & 1 & -1 \end{bmatrix}$$

by first expanding along the second column.

13. Show that

$$\det\left(\begin{bmatrix} a_{11} & a_{12} \\ a_{21} & a_{22} \end{bmatrix}\right) = a_{11}a_{22} - a_{12}a_{21}.$$

14. Let

$$A = \begin{bmatrix} a_{11} & a_{12} \\ a_{21} & a_{22} \end{bmatrix}.$$

(a) Show that A is nonsingular provided $a_{11}a_{22} - a_{12}a_{21} \neq 0$.

(b) If $a_{11}a_{22} - a_{12}a_{21} \neq 0$, show that

$$A^{-1} = \frac{1}{a_{11}a_{22} - a_{12}a_{21}} \begin{bmatrix} a_{22} & -a_{12} \\ -a_{21} & a_{11} \end{bmatrix}.$$

15. Let D be an $n \times n$ diagonal matrix. Show that $\det(D) = d_{11}d_{22}d_{33}\cdots d_{nn}$.

16. Let α be a real number and let

$$A = \begin{bmatrix} \alpha & 4 \\ 1 & \alpha \end{bmatrix} \quad \text{and} \quad B = \begin{bmatrix} 2 & \alpha & 0 \\ -3 & -1 & 5 \\ 1 & 3 & \alpha \end{bmatrix}.$$

(a) For what value(s) of α is A singular?
(b) For what value(s) of α is B singular?

3.1 GAUSSIAN ELIMINATION

In this chapter we study techniques for the solution of systems of linear algebraic equations. The most general system of n linear equations in n unknowns can be written as

$$\begin{array}{ccccccccccc} a_{11}x_1 & + & a_{12}x_2 & + & a_{13}x_3 & + & \cdots & + & a_{1n}x_n & = & b_1 \\ a_{21}x_1 & + & a_{22}x_2 & + & a_{23}x_3 & + & \cdots & + & a_{2n}x_n & = & b_2 \\ a_{31}x_1 & + & a_{32}x_2 & + & a_{33}x_3 & + & \cdots & + & a_{3n}x_n & = & b_3 \\ \vdots & & \vdots & & \vdots & & & & \vdots & & \vdots \\ a_{n1}x_1 & + & a_{n2}x_2 & + & a_{n3}x_3 & + & \cdots & + & a_{nn}x_n & = & b_n. \end{array}$$

The a_{ij} and the b_i are known constants, and the x_i are the variables. This system can be expressed very compactly in matrix notation as $A\mathbf{x} = \mathbf{b}$, where A is the $n \times n$ matrix

$$\begin{bmatrix} a_{11} & a_{12} & a_{13} & \cdots & a_{1n} \\ a_{21} & a_{22} & a_{23} & \cdots & a_{2n} \\ a_{31} & a_{32} & a_{33} & \cdots & a_{3n} \\ & \vdots & & & \vdots \\ a_{n1} & a_{n2} & a_{n3} & \cdots & a_{nn} \end{bmatrix}$$

and $\mathbf{x}$ and $\mathbf{b}$ are the n-dimensional column vectors $\begin{bmatrix} x_1 & x_2 & x_3 & \cdots & x_n \end{bmatrix}^T$ and $\begin{bmatrix} b_1 & b_2 & b_3 & \cdots & b_n \end{bmatrix}^T$, respectively. A is called the *coefficient matrix*, $\mathbf{x}$ the *solution vector* and $\mathbf{b}$ the *right-hand side vector* for the system.

We will focus on the solution technique known as Gaussian elimination with back substitution. After a review of the basic algorithm, several examples will be presented to demonstrate the technique. Finally, a detailed account of the number of operations required to compute the solution will be given, and a comparison with other possible solution strategies will be made.

Reviewing the Basics

The first step in the solution of a linear system of equations is to gather all the information needed to compute the solution (that is, the coefficients and the right-hand sides) into one structure, known as the *augmented matrix* for the system. For a system of n equations in n unknowns, the augmented matrix will have dimensions $n \times (n+1)$. The first n columns are the coefficient matrix, A, for the system. The right-hand side vector, $\mathbf{b}$, forms the last column. For the general system of linear equations given earlier, the augmented matrix is

$$\left[\begin{array}{cccccc|c} a_{11} & a_{12} & a_{13} & \cdot & \cdot & \cdot & a_{1n} & b_1 \\ a_{21} & a_{22} & a_{23} & \cdot & \cdot & \cdot & a_{2n} & b_2 \\ a_{31} & a_{32} & a_{33} & \cdot & \cdot & \cdot & a_{3n} & b_3 \\ & \cdot & & & & & \cdot & \cdot \\ & \cdot & & & & & \cdot & \cdot \\ & \cdot & & & & & \cdot & \cdot \\ a_{n1} & a_{n2} & a_{n3} & \cdot & \cdot & \cdot & a_{nn} & b_n \end{array}\right].$$

It is customary to use a vertical line to separate the two portions, coefficient and right-hand side, of the augmented matrix.

The objective of Gaussian elimination is to transform the coefficient portion of the augmented matrix into *upper triangular* form.

> **Definition.** A matrix U is called UPPER TRIANGULAR if all elements below the main diagonal are zero; that is, if $u_{ij} = 0$ whenever $i > j$.

In this definition, and throughout our discussions involving matrices, we will assume the conventional interpretation for subscripts on matrix elements: The first subscript refers to the row and the second refers to the column.

The transformation of the coefficient portion of the augmented matrix is carried out through the systematic application of three *elementary row operations* (EROs). The three operations, and the notation we will use to signify each, are

ERO$_1$: Any two rows can be interchanged. The notation $R_i \leftrightarrow R_j$ indicates that row i was interchanged with row j.

ERO$_2$: Any row can be multiplied by a nonzero constant. The notation $r_i \leftarrow mR_i$ indicates that row i was multiplied by m.

ERO$_3$: Any multiple of one row can be added to another row. The notation $r_i \leftarrow R_i + mR_j$ indicates that m times row j was added to row i.

The system of equations corresponding to the matrix which results after any sequence of these operations is performed is equivalent to the original system in the sense that it has the same solution set. As we will see below, the majority of the work in Gaussian elimination consists of repeatedly applying the third operation.

To illustrate the Gaussian elimination process, consider the system

$$\begin{array}{rcrcrcrcl} x_1 & + & x_2 & + & x_3 & + & x_4 & = & 1 \\ x_1 & + & x_2 & + & 2x_3 & + & 3x_4 & = & 2 \\ -x_1 & & & + & 2x_3 & + & x_4 & = & 1 \\ 3x_1 & + & 2x_2 & - & x_3 & & & = & 1. \end{array}$$

We begin by placing the *pivot* in the first row, first column of the augmented matrix. In the matrices shown below, the location of the pivot is indicated by angled braces, $\langle\,\rangle$. The pivot serves as a reference location for organizing subsequent calculations. The goal is to replace each element below the pivot, within the pivot column, with a zero. This can be done by performing ERO_3 on the rows below the pivot row, each time adding an appropriate multiple of the pivot row. The required multiple, m, is determined by the formula

$$m = -\frac{\text{element to be replaced by zero}}{\text{element in pivot}}.$$

For the problem at hand, the multipliers for the second, third, and fourth rows are -1, $+1$, and -3, respectively. The result of carrying out the corresponding row operations is

$$\left[\begin{array}{cccc|c} \langle 1\rangle & 1 & 1 & 1 & 1 \\ 1 & 1 & 2 & 3 & 2 \\ -1 & 0 & 2 & 1 & 1 \\ 3 & 2 & -1 & 0 & 1 \end{array}\right] \xrightarrow{\substack{r_2 \leftarrow R_2 - R_1 \\ r_3 \leftarrow R_3 + R_1 \\ r_4 \leftarrow R_4 - 3R_1}} \left[\begin{array}{cccc|c} 1 & 1 & 1 & 1 & 1 \\ 0 & 0 & 1 & 2 & 1 \\ 0 & 1 & 3 & 2 & 2 \\ 0 & -1 & -4 & -3 & -2 \end{array}\right].$$

Having completed one elimination pass through the matrix (generating zeros in one column), the pivot is moved down one row and to the right one column to set up for the next pass. At this point, we have a slight problem—the pivot element is zero. This problem can be bypassed by locating a row below the pivot row which has a nonzero entry in the pivot column. Provided the original coefficient matrix was nonsingular, it will always be possible to find such a row. The current pivot row and the selected row are then interchanged. Here, we choose to interchange rows 2 and 3. Adding the new pivot row to the fourth row completes the second elimination pass.

$$\left[\begin{array}{cccc|c} 1 & 1 & 1 & 1 & 1 \\ 0 & \langle 0\rangle & 1 & 2 & 1 \\ 0 & 1 & 3 & 2 & 2 \\ 0 & -1 & -4 & -3 & -2 \end{array}\right] \xrightarrow{R_2 \leftrightarrow R_3} \left[\begin{array}{cccc|c} 1 & 1 & 1 & 1 & 1 \\ 0 & \langle 1\rangle & 3 & 2 & 2 \\ 0 & 0 & 1 & 2 & 1 \\ 0 & -1 & -4 & -3 & -2 \end{array}\right]$$

$$\xrightarrow{r_4 \leftarrow R_4 + R_2} \left[\begin{array}{cccc|c} 1 & 1 & 1 & 1 & 1 \\ 0 & 1 & 3 & 2 & 2 \\ 0 & 0 & 1 & 2 & 1 \\ 0 & 0 & -1 & -1 & 0 \end{array}\right]$$

For the third, and in this case final, pass through the matrix, the pivot is moved to the third row, third column. As a general rule, the number of elimination passes is always one less than the number of equations. By adding the third row to the fourth row, the transformation of the coefficient portion of the augmented matrix to upper triangular form is complete:

$$\left[\begin{array}{cccc|c} 1 & 1 & 1 & 1 & 1 \\ 0 & 1 & 3 & 2 & 2 \\ 0 & 0 & \langle 1 \rangle & 2 & 1 \\ 0 & 0 & -1 & -1 & 0 \end{array}\right] \xrightarrow{r_4 \leftarrow R_4 + R_3} \left[\begin{array}{cccc|c} 1 & 1 & 1 & 1 & 1 \\ 0 & 1 & 3 & 2 & 2 \\ 0 & 0 & 1 & 2 & 1 \\ 0 & 0 & 0 & 1 & 1 \end{array}\right].$$

To obtain the solution to the system, we are now in position to perform back substitution. The equation corresponding to the bottom row of the transformed augmented matrix contains just one variable and can be solved directly. Here, we find $x_4 = 1$. This value is then substituted into the equation corresponding to the next to last row to give $x_3 + 2(1) = 1$, or $x_3 = -1$. Continuing to work back up the matrix, the values for x_4 and x_3 are substituted into the second equation, yielding $x_2 + 3(-1) + 2(1) = 2$, or $x_2 = 3$. Finally, from the first equation we find $x_1 + 3 - 1 + 1 = 1$, or $x_1 = -2$. Collecting these four values, the solution vector is

$$\mathbf{x} = \begin{bmatrix} -2 & 3 & -1 & 1 \end{bmatrix}^T.$$

Application Problem 1: An Electrical Circuit

Consider the electrical circuit shown in Figure 3.3. We would like to determine the currents flowing through the different branches. To deal with such circuits, we apply Kirchoff's current equation and Kirchoff's voltage equation. The current equation states that, at any junction, the sum of the current flowing into the junction must be equal to the sum of the current flowing out from the junction. The voltage equation states that the sum of the changes in voltage around any closed loop must be equal to zero. To apply the voltage equation for this problem, we will also need to use Ohm's law, which states that the voltage drop across a resistor is equal to the product of the current and the resistance.

Applying Kirchoff's current equation to the junction on the right side of the circuit yields

$$I_1 = I_2 + I_3 \quad \text{or} \quad I_1 - I_2 - I_3 = 0.$$

Balancing the currents flowing into and out from the junction between the 1-Ω resistor and the 6-volt source and the junction at the bottom of the circuit gives the equations

$$I_2 - I_4 - I_5 = 0$$

and

$$I_3 + I_4 - I_6 = 0,$$

respectively. The equation obtained from the junction on the left side of the circuit is just the sum of the three previous equations, so we do not include it in the system.

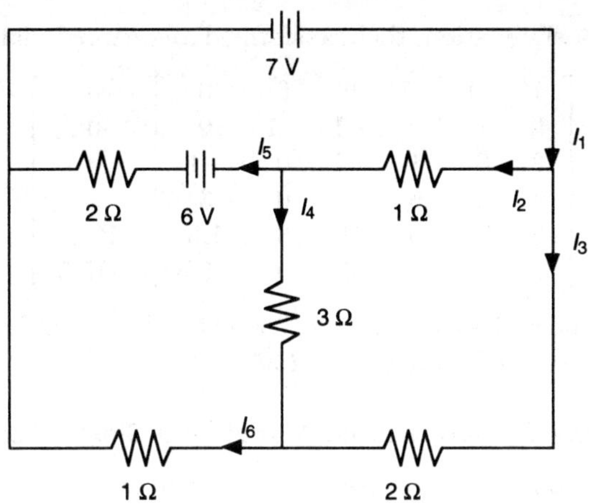

Figure 3.3

To obtain three more equations, we now turn to Kirchoff's voltage equation. Traveling clockwise around the outermost loop of the circuit, we find that the current I_3 flows through a 2-Ω resistor and the current I_6 flows through a 1-Ω resistor. This produces a total voltage drop of $2I_3 + I_6$, which must balance the 7-volt increase produced by the voltage source at the top of the circuit. Hence, we have

$$2I_3 + I_6 = 7.$$

Traveling clockwise around the loop at the top of the circuit and clockwise around the loop at the lower right yields

$$I_2 + 2I_5 = 13$$

and

$$-I_2 + 2I_3 - 3I_4 = 0,$$

respectively. The negative signs appear in the last equation because a clockwise loop travels through the 3-Ω and 1-Ω resistors in the opposite direction of the indicated current.

The augmented matrix corresponding to this system of six equations in the six unknown currents is

$$\left[\begin{array}{cccccc|c} 1 & -1 & -1 & 0 & 0 & 0 & 0 \\ 0 & 1 & 0 & -1 & -1 & 0 & 0 \\ 0 & 0 & 1 & 1 & 0 & -1 & 0 \\ 0 & 0 & 2 & 0 & 0 & 1 & 7 \\ 0 & 1 & 0 & 0 & 2 & 0 & 13 \\ 0 & -1 & 2 & -3 & 0 & 0 & 0 \end{array}\right].$$

Following Gaussian elimination, the transformed augmented matrix takes the form

$$\left[\begin{array}{cccccc|c} 1 & -1 & -1 & 0 & 0 & 0 & 0 \\ 0 & 1 & 0 & -1 & -1 & 0 & 0 \\ 0 & 0 & 1 & -1 & 0 & -1 & 0 \\ 0 & 0 & 0 & -2 & 0 & 3 & 7 \\ 0 & 0 & 0 & 0 & 3 & 1.5 & 16.5 \\ 0 & 0 & 0 & 0 & 0 & -6.5 & -15.5 \end{array}\right].$$

Back substitution now gives the desired currents: $I_1 = 6.692$, $I_2 = 4.385$, $I_3 = 2.308$, $I_4 = 0.0769$, $I_5 = 4.308$, and $I_6 = 2.385$.

Application Problem 2: Input-Output Model for a Simple Economy

Consider a simple economy that consists of four sectors: (1) agriculture; (2) energy; (3) manufacturing; and (4) labor. Producing output in one sector generally requires input from all four sectors. For example, to produce agricultural output, we likely will need seeds and fertilizer from the agriculture sector, farming equipment from the manufacturing sector, fuel to run the equipment from the energy sector, and farm hands from the labor sector. The relationships between the various sectors can be represented by an *input-output matrix* A, where the element a_{ij} is defined as the input required from sector i to produce one unit of output from sector j.

Let the vector $\mathbf{x}$ denote the total output of the economy. The vector $A\mathbf{x}$ then gives the total input needed from each sector of the economy to produce the output $\mathbf{x}$. This quantity is known as the *internal demand.* If there is additionally an external consumer demand on each sector, denoted by the vector $\mathbf{d}$, then the total output must be sufficient to cover both the internal and the external demand. In other words, $\mathbf{x}$ must satisfy

$$\mathbf{x} = A\mathbf{x} + \mathbf{d} \quad \text{or} \quad (I - A)\mathbf{x} = \mathbf{d}.$$

Suppose that the input-output matrix for our hypothetical four-sector economy is given by

$$A = \begin{bmatrix} 0.05 & 0.09 & 0.09 & 0.19 \\ 0.16 & 0.15 & 0.28 & 0.21 \\ 0.19 & 0.21 & 0.22 & 0.27 \\ 0.27 & 0.04 & 0.35 & 0.02 \end{bmatrix}$$

and that there is consumer demand of \$23 billion for agriculture, \$45 billion for energy, \$39 for manufacturing and \$12 billion for labor. The augmented matrix for determining $\mathbf{x}$ is then

$$\left[\begin{array}{cccc|c} 0.95 & -0.09 & -0.09 & -0.19 & 23 \\ -0.16 & 0.85 & -0.28 & -0.21 & 45 \\ -0.19 & -0.21 & 0.78 & -0.27 & 39 \\ -0.27 & -0.04 & -0.35 & 0.98 & 12 \end{array}\right].$$

Applying Gaussian elimination with back substitution, we find that an output of \$64.59 billion from the agriculture sector, \$127.27 billion from the energy sector,

$128.02 billion from the manufacturing sector and $80.96 billion from the labor sector will meet the indicated consumer demand.

Operation Counts

How much work does it take to perform Gaussian elimination with back substitution to obtain the solution to an arbitrary system of n equations? As a measure of work, we will use the number of arithmetic operations being performed. Let's start with Gaussian elimination. The segment of pseudocode given below captures the major components of Gaussian elimination.

```
for pass from 1 to n − 1
        for row from pass + 1 to n
                m = −a_{row,pass}/a_{pass,pass}
                set a_{row,pass} = 0
                for col from pass + 1 to n + 1
                        a_{row,col} ← a_{row,col} + m a_{pass,col}
```

Note that we set $a_{row,pass} = 0$ to avoid an unnecessary calculation. The innermost loop has $n+1$ as its final index because each row in the augmented matrix also has the entry from the right-hand side vector.

Traditionally, the number of additions and subtractions have been counted separately from the number of multiplications and divisions. On older computers, multiplication and division were significantly more time-consuming than addition and subtraction. On many modern architectures, however, multiplication is no more expensive than either addition or subtraction, and division is not even twice as expensive. We will therefore break from tradition and just count the total number of arithmetic operations.

A scan of the pseudocode indicates that two arithmetic operations are performed each time the innermost loop is executed and one more operation is performed inside the middle loop. Therefore, the total number of arithmetic operations for Gaussian elimination is

$$\begin{aligned}\sum_{pass=1}^{n-1}\sum_{row=pass+1}^{n}\left[1+\sum_{col=pass+1}^{n+1}2\right] &= \sum_{pass=1}^{n-1}\sum_{row=pass+1}^{n}(2n-2pass+3)\\ &= \sum_{pass=1}^{n-1}(2n-2pass+3)(n-pass)\\ &= \frac{2}{3}n^3+\frac{1}{2}n^2-\frac{7}{6}n.\end{aligned}$$

To determine the number of operations required by back substitution, examine the pseudocode

```
x_n = b_n/a_{n,n}
for row from n − 1 to 1 by −1
```

$$
\begin{aligned}
&sum = b_{row} \\
&\text{for } col \text{ from } row + 1 \text{ to } n \\
&\qquad\qquad sum = sum - a_{row,col}x_{col} \\
&x_{row} = sum/a_{row,row}
\end{aligned}
$$

One operation is performed before the loops, one inside the outer loop and two inside the inner loop. The total number of operations for back substitution is then

$$1 + \sum_{row=1}^{n-1}\left[1 + \sum_{col=row+1}^{n} 2\right] = n^2.$$

In summary, solving a system of n linear equations in n unknowns by Gaussian elimination with back substitution requires

$$\frac{2}{3}n^3 + \frac{3}{2}n^2 - \frac{7}{6}n$$

arithmetic operations.

Why Not Gauss-Jordan Elimination or Multiplication by the Inverse Matrix?

Gaussian elimination with back substitution is not the only direct procedure available for solving linear systems of equations. Two alternative strategies are Gauss-Jordan elimination and multiplication by the inverse of the coefficient matrix. An examination of the operation counts for these two techniques will make it clear why Gaussian elimination is preferred.

Gauss-Jordan elimination replaces the elements both above and below the pivot with zeros and generates ones along the main diagonal (by using ERO_2), thereby reducing the coefficient portion of the augmented matrix to the identity matrix. This process removes the need to perform back substitution and saves n^2 arithmetic operations. The elimination phase, however, requires more operations. The number of operations needed to reduce the coefficient portion of the augmented matrix to the identity matrix is

$$
\begin{aligned}
\sum_{pass=1}^{n-1}\sum_{row=pass+1}^{n}\left[1 + \sum_{col=pass+1}^{n+1} 2\right] + \sum_{pass=2}^{n}\sum_{row=1}^{pass-1}\left[1 + \sum_{col=pass+1}^{n+1} 2\right] + n \\
= n^3 + n^2 - n.
\end{aligned}
$$

The two triple sums take into account the generation of zeros below the pivot and above the pivot, respectively, and the term following the summations corresponds to the n divisions needed to place ones along the main diagonal. Although the operation count still has a leading term of n^3, the leading coefficient for Gauss-Jordan elimination is half again larger than that for Gaussian elimination. For any n larger than 2, the added cost of Gauss-Jordan elimination outweighs the savings achieved by not needing to perform back substitution.

Multiplying by the inverse of the coefficient matrix is even more expensive. First, it can be shown (see Exercise 9) that computation of the inverse matrix requires $2n^3 - 2n^2 + n$ total arithmetic operations. The multiplication $A^{-1}\mathbf{b}$ then requires $2n^2 - n$ additional operations, which is nearly twice the work needed to perform back substitution.

EXERCISES

In Exercises 1–5, write out the augmented matrix for the indicated linear system of equations and then obtain the solution using Gaussian elimination with back substitution.

1.
$$\begin{aligned} 2x_1 - x_2 + x_3 &= -1 \\ 4x_1 + 2x_2 + x_3 &= 4 \\ 6x_1 - 4x_2 + 2x_3 &= -2 \end{aligned}$$

2.
$$\begin{aligned} \tfrac{1}{3}x_1 + \tfrac{2}{3}x_2 + 2x_3 &= -1 \\ x_1 + 2x_2 + \tfrac{3}{2}x_3 &= \tfrac{3}{2} \\ \tfrac{1}{2}x_1 + 2x_2 + \tfrac{12}{5}x_3 &= \tfrac{1}{10} \end{aligned}$$

3.
$$\begin{aligned} x_1 + 2x_2 - x_3 &= 1 \\ 2x_1 - x_2 + x_3 &= 3 \\ -x_1 + 2x_2 + 3x_3 &= 7 \end{aligned}$$

4.
$$\begin{aligned} x_2 + x_3 + x_4 &= 0 \\ 3x_1 + 3x_3 - 4x_4 &= 7 \\ x_1 + x_2 + x_3 + 2x_4 &= 6 \\ 2x_1 + 3x_2 + x_3 + 3x_4 &= 6 \end{aligned}$$

5.
$$\begin{aligned} 3x_1 - x_2 + 3x_3 + x_4 &= 6 \\ 6x_1 + 9x_3 - 2x_4 &= 13 \\ -12x_1 - 10x_3 + 5x_4 &= -17 \\ 72x_1 - 8x_2 + 48x_3 - 19x_4 &= 93 \end{aligned}$$

6. Let U be an $n \times n$ upper triangular matrix. Show that

$$\det(U) = u_{11}u_{22}u_{33}\cdots u_{nn}.$$

7. Suppose we had not assigned the value 0 to the element $a_{row,pass}$ in our Gaussian elimination pseudocode and had instead computed the value inside the innermost loop. How many arithmetic operations would that have added to the operation count for the elimination phase?

8. **(a)** Construct an algorithm to carry out Gauss-Jordan elimination; that is, during each pass through the matrix, generate zeros both above and below the pivot element; after all n passes, place ones along the diagonal.

(b) Show that the total number of arithmetic operations needed to solve a system of n equations in n unknowns using Gauss-Jordan elimination is $n^3 + n^2 - n$.

9. The inverse of an $n \times n$ matrix can be computed by performing Gauss-Jordan elimination on an $n \times 2n$ augmented matrix, where the last n columns are the $n \times n$ identity matrix.

(a) Show that if one naively applies Gauss-Jordan elimination without taking into account the structure of the identity matrix, then computation of the inverse requires $3n^3 - 2n^2$ arithmetic operations.

(b) Show that if one takes into account the structure of the identity matrix (and does not perform multiplication when the matrix element is a one and does not perform addition/subtraction when one of the elements is known to be zero), then computation of the inverse can be reduced to $2n^3 - 2n^2 + n$ operations.

10. (a) Solve the system

$$\begin{aligned} 3.02x_1 - 1.05x_2 + 2.53x_3 &= -1.61 \\ 4.33x_1 + 0.56x_2 - 1.78x_3 &= 7.23 \\ -0.83x_1 - 0.54x_2 + 1.47x_3 &= -3.38 \end{aligned}$$

using Gaussian elimination with back substitution.

(b) Change the coefficient of x_1 in the first equation to 3.01 and solve the resulting system. By what percentage have the three components of the solution vector changed?

(c) Return the coefficient of x_1 in the first equation to 3.02, but change the right-hand side of the last equation to -3.39 and solve the resulting system. By what percentage have the three components of the solution vector changed from their values in part (a)?

11. (a) Solve the system

$$\begin{aligned} 6x_1 - 2x_2 + 3x_3 &= 5 \\ x_1 - \tfrac{1}{3}x_2 + \tfrac{1}{3}x_3 &= 2 \\ x_1 + 3x_2 - x_3 &= 5 \end{aligned}$$

using Gaussian elimination with back substitution.

(b) Change the coefficient of x_1 in the first equation to 6.01 and solve the resulting system. By what percentage have the three components of the solution vector changed?

(c) Return the coefficient of x_1 in the first equation to 6, but change the right-hand side of the second equation to 1.99 and solve the resulting system. By what percentage have the three components of the solution vector changed from their values in part (a)?

12. Let A be the $n \times n$ matrix whose entries are given by $a_{ij} = 1/(i + j - 1)$ for $1 \leq i, j \leq n$.

(a) For $n = 5, 6$ and 7, solve the system $A\mathbf{x} = \mathbf{b}$ using single precision arithmetic. In each case, take $\mathbf{b}$ as the vector that corresponds to an exact solution of $x_i = 1$ for each $i = 1, 2, 3, \ldots, n$. Calculate the maximum component-wise error between the computed solution and the exact solution for each n.

(b) For $n = 11, 12$ and 13, solve the system $A\mathbf{x} = \mathbf{b}$ using double precision arithmetic. In each case, take $\mathbf{b}$ as the vector that corresponds to an exact solution of $x_i = 1$ for each $i = 1, 2, 3, \ldots, n$. Calculate the maximum component-wise error between the computed solution and the exact solution for each n.

13. The circuit shown in Figure 3.4 could be used as part of a system for charging a car battery. Assuming that the internal resistance of the generator and the battery are negligible and applying Kirchoff's loop equation around the left and right loops of the circuit (traveling counterclockwise about the left loop and clockwise around the right loop) produces the equations

$$-4I_2 + 15I_3 = 12$$

and

$$10I_1 + 15I_3 = 100.$$

Balancing the current flowing into and out from the junction between the 4-Ω and 10-Ω resistors yields the equation $I_1 = I_2 + I_3$. Determine the current flowing through each branch of the circuit.

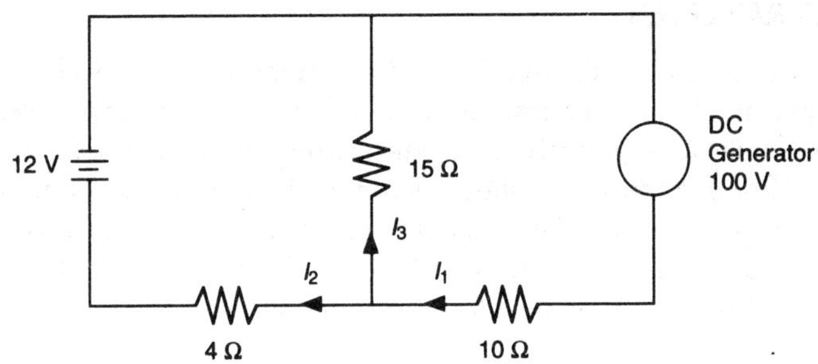

Figure 3.4 Figure for Exercise 13.

14. Consider a simple economy that consists of three sectors: food, clothing, and shelter. The production of one unit of food requires 0.43 units of food, 0.17 units of clothing and 0.18 units of shelter. The production of one unit of clothing requires 0.08 units of food, 0.23 units of clothing, and 0.28 units of shelter. The production of one unit of shelter requires 0.23 units of food, 0.16 units of clothing, and 0.14 units of shelter. If consumer demand is for \$90 million worth of food, \$32 million worth of clothing, and \$245 million worth of shelter, what total output from each sector is needed?

15. Suppose the coefficient matrix and the control vector for the longitudinal dynamics of an aircraft are given by

$$A = \begin{bmatrix} -0.0507 & -3.861 & 0 & -32.17 \\ -0.00117 & -0.5164 & 1 & 0 \\ -0.000129 & 1.4168 & -0.4932 & 0 \\ 0 & 0 & 1 & 0 \end{bmatrix}$$

and $\mathbf{b} = \begin{bmatrix} 0 & -0.0717 & -1.645 & 0 \end{bmatrix}^T$, respectively. In order to change the open loop coefficient vector $\mathbf{a} = \begin{bmatrix} 1.0603 & -1.115 & -0.0565 & -0.0512 \end{bmatrix}^T$ into the closed loop coefficient vector $\hat{\mathbf{a}} = \begin{bmatrix} 2.52 & 6.31 & 0.150 & 0.0625 \end{bmatrix}^T$, the gain vector, $\mathbf{g}$, in the feedback control law must satisfy the equation

$$(QW)^T\mathbf{g} = \hat{\mathbf{a}} - \mathbf{a}.$$

The matrix Q takes the form $Q = \begin{bmatrix} \mathbf{b} & A\mathbf{b} & A^2\mathbf{b} & A^3\mathbf{b} \end{bmatrix}$, and

$$W = \begin{bmatrix} 1 & 1.0603 & -1.115 & -0.0565 \\ 0 & 1 & 1.0603 & -1.115 \\ 0 & 0 & 1 & 1.0603 \\ 0 & 0 & 0 & 1 \end{bmatrix}.$$

Compute $\mathbf{g}$.

16. Solve the system of equations associated with the "Forces in a Plane Truss" problem capsule presented in the Chapter 3 Overview (see page 138).

3.2 PIVOTING STRATEGIES

It is sometimes necessary during Gaussian elimination to interchange rows while solving a system of linear equations so as to avoid a zero pivot element. When performing calculations in finite precision arithmetic, it may also be necessary to interchange rows to reduce the effect of roundoff error on the computed solution. In this section, we will first illustrate the problems which can arise when solving linear systems using finite precision arithmetic. We will then discuss two strategies which can be used to reduce the effects of roundoff error.

An Example to Motivate the Discussion

Consider the system of three equations in three unknowns

$$\begin{array}{rcrcrcr} \frac{2}{3}x_1 & + & \frac{2}{7}x_2 & + & \frac{1}{5}x_3 & = & \frac{43}{15} \\ \frac{1}{3}x_1 & + & \frac{1}{7}x_2 & - & \frac{1}{2}x_3 & = & \frac{5}{6} \\ \frac{1}{5}x_1 & - & \frac{3}{7}x_2 & + & \frac{2}{5}x_3 & = & -\frac{12}{5}. \end{array}$$

Working in exact arithmetic, the first pass of Gaussian elimination requires that the elementary row operations $r_2 \leftarrow R_2 - \frac{1}{2}R_1$ and $r_3 \leftarrow R_3 - \frac{3}{10}R_1$ be carried out. This yields the equivalent system

$$\begin{array}{rcrcrcr} \frac{2}{3}x_1 & + & \frac{2}{7}x_2 & + & \frac{1}{5}x_3 & = & \frac{43}{15} \\ & & & & -\frac{3}{5}x_3 & = & -\frac{3}{5} \\ & & -\frac{36}{70}x_2 & + & \frac{17}{50}x_3 & = & -\frac{163}{50}. \end{array}$$

Interchanging the second and third equations produces an upper triangular system, from which the exact solution $x_1 = 1$, $x_2 = 7$, and $x_3 = 1$ is obtained by back substitution.

What happens when we try to solve this system using four decimal digit rounding arithmetic? First, we replace each of the rational numbers in the original system by its floating point equivalent. This transforms the original system to

$$\begin{aligned} 0.6667x_1 &+ 0.2857x_2 + 0.2000x_3 = 2.867 \\ 0.3333x_1 &+ 0.1429x_2 - 0.5000x_3 = 0.8333 \\ 0.2000x_1 &- 0.4286x_2 + 0.4000x_3 = -2.400. \end{aligned}$$

The first pass of Gaussian Elimination produces

$$\begin{aligned} 0.6667x_1 + 0.2857x_2 &+ 0.2000x_3 = 2.867 \\ 0.0001x_2 &- 0.6000x_3 = -0.5997 \\ -0.5143x_2 &+ 0.3400x_3 = -3.260. \end{aligned}$$

The coefficient on x_2 in the second equation should have been zero, but cancellation error has left us with a small nonzero coefficient, which becomes the pivot for the second elimination pass. This second pass then produces the upper triangular system

$$\begin{aligned} 0.6667x_1 + 0.2857x_2 &+ 0.2000x_3 = 2.867 \\ 0.0001x_2 &- 0.6000x_3 = -0.5997 \\ &-3086x_3 = -3087. \end{aligned}$$

Turning to back substitution, we find $x_3 = (-3087)/(-3086) = 1.000$. Substituting this value into the second equation and solving for x_2 yields

$$x_2 = \frac{-0.5997 - (1.000)(-0.6000)}{0.0001} = \frac{0.0003}{0.0001} = 3.000.$$

Note the cancellation error in the calculation of the numerator. In fact, three of the four significant digits have been lost. This cancellation error is then magnified when we divide by the small pivot element, producing a value which is in error by more than 57%. Substituting x_3 and x_2 into the first equation leads to $x_1 = 2.715$, which is in error by nearly 200%.

Why is the computed solution so inaccurate? We can ultimately trace the problem back to the cancellation error introduced in the first elimination pass. This generated the small pivot element, 0.0001, for the second pass. Additional cancellation error was introduced during back substitution, and this error was then magnified when we divided by that same small pivot element. Now, during Gaussian elimination, there is always going to be the possibility of cancellation error when the third elementary row operation is applied. There isn't much we can do about this. However, we can and, as this example clearly demonstrates, should avoid using small pivot elements.

Partial Pivoting

To avoid small pivot elements, we can employ a *pivoting strategy*. In general, a pivoting strategy is any systematic scheme for interchanging the rows (and possibly the columns) of the coefficient matrix to place a selected element in the pivot position. The simplest such scheme is known as partial pivoting.

PARTIAL PIVOTING During the ith elimination pass of Gaussian elimination, let

$$M_i = \max_{i \le j \le n} |a_{j,i}|,$$

and let j_0 be the smallest value of j for which this maximum occurs. If $j_0 > i$, then interchange rows i and j_0.

In other words, we find the element in the pivot column, starting from the i-th row and continuing to the bottom of the matrix, which is of largest magnitude, and then make that element the pivot element.

EXAMPLE 3.4 Partial Pivoting in Action

Reconsider the system from above, whose representation in a four decimal digit floating point system with rounding was

$$\begin{aligned} 0.6667x_1 + 0.2857x_2 + 0.2000x_3 &= 2.867 \\ 0.3333x_1 + 0.1429x_2 - 0.5000x_3 &= 0.8333 \\ 0.2000x_1 - 0.4286x_2 + 0.4000x_3 &= -2.400. \end{aligned}$$

The first pass of Gaussian elimination proceeds exactly as before because the largest element in the first column ($i = 1$) is initially in the first equation ($j_0 = 1$). Since $j_0 = i$, no interchange of equations is required. Hence, the second pass starts from

$$\begin{aligned} 0.6667x_1 + 0.2857x_2 + 0.2000x_3 &= 2.867 \\ 0.0001x_2 - 0.6000x_3 &= -0.5997 \\ -0.5143x_2 + 0.3400x_3 &= -3.260. \end{aligned}$$

Here, note that the largest element in the second column ($i = 2$) is located in the third equation ($j_0 = 3$). The partial pivoting strategy therefore requires that the second and third equations be interchanged. This yields

$$\begin{aligned} 0.6667x_1 + 0.2857x_2 + 0.2000x_3 &= 2.867 \\ -0.5143x_2 + 0.3400x_3 &= -3.260 \\ 0.0001x_2 - 0.6000x_3 &= -0.5997, \end{aligned}$$

from which the elimination of x_2 from the last equation leaves us with

$$\begin{aligned} 0.6667x_1 + 0.2857x_2 + 0.2000x_3 &= 2.867 \\ -0.5143x_2 + 0.3400x_3 &= -3.260 \\ -0.5999x_3 &= -0.6003. \end{aligned}$$

Back substitution produces the solution $x_3 = 1.001$, $x_2 = 7.000$, and $x_1 = 1.000$. To four digits, the values of x_1 and x_2 are exact, while the value of x_3 is in error by only one-tenth of one percent.

In the preceding example, the necessary row interchange was carried out explicitly so as not to draw attention away from the action of the pivoting strategy. In practice, it is usually more efficient to handle row interchanges implicitly. This is accomplihsed by maintaining a vector of n elements, such that the ith element of the vector indicates the row within the matrix that contains the coefficients for the ith equation. Let's denote this *row vector* by $\mathbf{r}$. The vector is initialized to

$$\mathbf{r} = \begin{bmatrix} 1 & 2 & 3 & \cdots & n \end{bmatrix}^T.$$

With this vector, each time a row interchange is required, we need only swap the corresponding elements of the vector.

Handling row interchanges in an implicit manner does require one important change to our Gaussian elimination and back subsitution algorithms. Every reference to a row of the coefficient matrix or to a row of the right-hand side vector must be made through the row vector $\mathbf{r}$. For instance, the coefficient of the fifth variable in the seventh equation must be accessed as $a_{r_7,5}$, while the right-hand side of the third equation must be accessed as b_{r_3}.

EXAMPLE 3.5 A System with Four Equations

Consider the system whose augmented matrix is

$$\left[\begin{array}{cccc|c} 3 & 1 & 4 & -1 & 7 \\ 2 & -2 & -1 & 2 & 1 \\ 5 & 7 & 14 & -8 & 20 \\ 1 & 3 & 2 & 4 & -4 \end{array}\right].$$

The exact solution of this system is

$$\mathbf{x} = \begin{bmatrix} 1 & -1 & 1 & -1 \end{bmatrix}^T.$$

Solving this system using Gaussian elimination without pivoting in four decimal digit rounding arithmetic produces the solution

$$\mathbf{x} = \begin{bmatrix} 1.131 & -0.7928 & 0.8500 & -0.9987 \end{bmatrix}^T$$

(see Exercise 13).

Initialize the row vector to

$$\mathbf{r} = \begin{bmatrix} 1 & 2 & 3 & 4 \end{bmatrix}^T.$$

To determine the location of the pivot, examine the values

$$|a_{r_1,1}| = 3, \quad |a_{r_2,1}| = 2, \quad |a_{r_3,1}| = 5, \quad \text{and} \quad |a_{r_4,1}| = 1.$$

The largest value in this list corresponds to row r_3, so $j_0 = 3$. Since $j_0 = 3 > 1 = i$, we need to swap the first and third elements in the row vector. Thus, for the first pass, we have

$$\mathbf{r} = \begin{bmatrix} 3 & 2 & 1 & 4 \end{bmatrix}^T.$$

After the elimination pass, the matrix becomes

$$\left[\begin{array}{cccc|c} 0 & -3.200 & -4.400 & 3.800 & -5.000 \\ 0 & -4.800 & -6.600 & 5.200 & -7.000 \\ 5.000 & 7.000 & 14.00 & -8.000 & 20.00 \\ 0 & 1.600 & -0.8000 & 5.600 & -8.000 \end{array}\right].$$

To determine the location of the next pivot, examine the values

$$|a_{r_2,2}| = 4.800, \quad |a_{r_3,2}| = 3.200, \quad \text{and} \quad |a_{r_4,2}| = 1.600.$$

The largest value in this list corresponds to row r_2, so $j_0 = 2$. Since $j_0 = 2 = i$, no row interchange is needed for the second pass.

The second elimination pass produces the matrix

$$\left[\begin{array}{cccc|c} 0 & 0 & 0 & 0.3330 & -0.333 \\ 0 & -4.800 & -6.600 & 5.200 & -7.000 \\ 5.000 & 7.000 & 14.00 & -8.000 & 20.00 \\ 0 & 0 & -3.000 & 7.333 & -10.33 \end{array}\right].$$

The location of the final pivot is determined by examining the values

$$|a_{r_3,3}| = 0.0000 \quad \text{and} \quad |a_{r_4,3}| = 3.000.$$

The largest value here corresponds to row r_4, so $j_0 = 4$. Since $j_0 = 4 > 3 = i$, we need to swap the third and fourth elements in the row vector, giving

$$\mathbf{r} = \left[\begin{array}{cccc} 3 & 2 & 4 & 1 \end{array}\right]^T.$$

Since the element in the $a_{r_4,3} = a_{1,3}$ position is already zero, the third elimination pass makes no changes to the matrix. The final contents of the matrix and the row vector are therefore as shown above. Back substitution now yields

$$\begin{aligned} x_4 &= \frac{b_{r_4}}{a_{r_4,4}} = \frac{b_1}{a_{1,4}} = \frac{-0.333}{0.333} = -1.000 \\ x_3 &= \frac{b_{r_3} - a_{r_3,4}x_4}{a_{r_3,3}} = \frac{b_4 - a_{4,4}x_4}{a_{4,3}} = 0.9990 \\ x_2 &= \frac{b_{r_2} - a_{r_2,3}x_3 - a_{r_2,4}x_4}{a_{r_2,2}} = \frac{b_2 - a_{2,3}x_3 - a_{2,4}x_4}{a_{2,2}} = -0.9985 \\ x_1 &= \frac{b_{r_1} - a_{r_1,2}x_2 - a_{r_1,3}x_3 - a_{r_1,4}x_4}{a_{r_1,1}} = 1.000. \end{aligned}$$

The maximum error in any component of this solution is only 0.15%, which is a dramatic improvement over the solution obtained without pivoting.

Scaled Partial Pivoting

Partial pivoting works well in many instances but does not reduce the effects of roundoff error for all problems. Consider the system

$$\begin{aligned} 0.7x_1 &+ 1725x_2 &= 1739 \\ 0.4352x_1 &- 5.433x_2 &= 3.271, \end{aligned}$$

whose exact solution is $x_1 = 20$ and $x_2 = 1$. If we were to solve this system using four decimal digit rounding arithmetic with partial pivoting, we would leave the equations in the order listed since the coefficient on x_1 in the first equation (0.7) is larger than the coefficient on x_1 in the second equation (0.4352). Eliminating x_1 from the second equation, we obtain the equivalent system

$$\begin{aligned} 0.7x_1 + 1725x_2 &= 1739 \\ -1077x_2 &= -1078. \end{aligned}$$

Back substitution produces the solution $x_2 = 1.001$ and $x_1 = 17.14$. The value of x_2 is in excellent agreement with the exact value, but the value of x_1 is in error by more than 14%.

Partial pivoting performed poorly on this system because it did not take into account the sizes of the coefficients in locations other than the pivot column when selecting the pivot. In this case, although 0.7 is larger than 0.4352, when measured relative to the other coefficients in each equation, 0.7 is actually smaller than 0.4352; that is,

$$\frac{0.7}{1725} < \frac{0.4352}{5.433},$$

where 1725 and 5.433 are the absolute values of the coefficients of greatest magnitude in the first and second equations, respectively. Had we decided to choose the element in the pivot column which is largest in magnitude relative to the other coefficients in its equation, then we would have switched the order of the equations in this system prior to eliminating variables:

$$\begin{aligned} 0.4352x_1 &- 5.433x_2 &= 3.271 \\ 0.7x_1 &+ 1725x_2 &= 1739. \end{aligned}$$

Now eliminating x_1 from the second equation yields the equivalent system

$$\begin{aligned} 0.4352x_1 - 5.433x_2 &= 3.271 \\ 1734x_2 &= 1734. \end{aligned}$$

Back substitution from this set of equations yields $x_2 = 1.000$ and $x_1 = 20.00$, which are exact to four digits.

The pivoting strategy we have just applied is known as *scaled partial pivoting.*

SCALED PARTIAL PIVOTING Before starting Gaussian elimination, construct a scale vector **s** as follows. For each $1 \le i \le n$, let

$$s_i = \max_{1 \le j \le n} |a_{i,j}|.$$

Also, initialize the row vector to

$$\mathbf{r} = \begin{bmatrix} 1 & 2 & 3 & \cdots & n \end{bmatrix}^T.$$

During the ith elimination pass, let

$$M_i = \max_{i \le j \le n} \left(\frac{|a_{r_j,i}|}{s_{r_j}} \right),$$

and let j_0 be the smallest value of j for which this maximum occurs. If $j_0 > i$, then interchange rows i and j_0.

Note that while the row vector will generally change from pass to pass, the scale vector is set at the beginning of the process and is not changed thereafter. We could modify the scale vector after each pass if we wanted, but for the majority of linear systems the added calculations produce negligible benefit.

EXAMPLE 3.6 Scaled Partial Pivoting in Action

Reconsider the system whose augmented matrix is

$$\left[\begin{array}{cccc|c} 3 & 1 & 4 & -1 & 7 \\ 2 & -2 & -1 & 2 & 1 \\ 5 & 7 & 14 & -8 & 20 \\ 1 & 3 & 2 & 4 & -4 \end{array}\right]$$

and whose exact solution is

$$\mathbf{x} = \begin{bmatrix} 1 & -1 & 1 & -1 \end{bmatrix}^T.$$

Earlier, we noted dramatic improvement in the computed solution of this system when using partial pivoting. Let's now examine the effect of using scaled partial pivoting.

The first step is to construct the scale vector. Since

$$\max_{1 \le j \le 4} |a_{1,j}| = 4, \quad \max_{1 \le j \le 4} |a_{2,j}| = 2, \quad \max_{1 \le j \le 4} |a_{3,j}| = 14 \text{ and } \max_{1 \le j \le 4} |a_{4,j}| = 4,$$

we find

$$\mathbf{s} = \begin{bmatrix} 4 & 2 & 14 & 4 \end{bmatrix}^T.$$

Next, we initialize the row vector to

$$\mathbf{r} = \begin{bmatrix} 1 & 2 & 3 & 4 \end{bmatrix}^T.$$

To determine the location of the pivot, we examine the values

$$\frac{|a_{r_1,1}|}{s_{r_1}} = \frac{3}{4}, \quad \frac{|a_{r_2,1}|}{s_{r_2}} = \frac{2}{2}, \quad \frac{|a_{r_3,1}|}{s_{r_3}} = \frac{5}{14}, \quad \text{and} \quad \frac{|a_{r_4,1}|}{s_{r_4}} = \frac{1}{4}.$$

The largest value in this list corresponds to row r_2, so $j_0 = 2$. Since $j_0 = 2 > 1 = i$, we need to swap the first and second elements in the row vector. Thus, for the first elimination pass, we have

$$\mathbf{r} = \begin{bmatrix} 2 & 1 & 3 & 4 \end{bmatrix}^T.$$

After the elimination pass, the matrix becomes

$$\left[\begin{array}{cccc|c} 0 & 4.000 & 5.500 & -4.000 & 5.500 \\ 2.000 & -2.000 & -1.000 & 2.000 & 1.000 \\ 0 & 12.00 & 16.50 & -13.00 & 17.50 \\ 0 & 4.000 & 2.500 & 3.000 & -4.500 \end{array}\right].$$

To determine the location of the next pivot, examine the values

$$\frac{|a_{r_2,2}|}{s_{r_2}} = \frac{4.000}{4}, \quad \frac{|a_{r_3,2}|}{s_{r_3}} = \frac{12.00}{14}, \quad \text{and} \quad \frac{|a_{r_4,2}|}{s_{r_4}} = \frac{4.000}{4}.$$

The largest value in this list is 1, which occurs for both row r_2 and row r_4. Choosing the first occurrence of the maximum value, we have $j_0 = 2 = i$. Hence, no row interchange is needed for the second pass.

The second elimination pass produces the matrix

$$\left[\begin{array}{cccc|c} 0 & 4.000 & 5.500 & -4.000 & 5.500 \\ 2.000 & -2.000 & -1.000 & 2.000 & 1.000 \\ 0 & 0 & 0 & -1.000 & 1.000 \\ 0 & 0 & -3.000 & 7.000 & -10.00 \end{array}\right].$$

The location of the final pivot is determined by examining the values

$$\frac{|a_{r_3,3}|}{s_{r_3}} = \frac{0}{14} \quad \text{and} \quad \frac{|a_{r_4,3}|}{s_{r_4}} = \frac{3.000}{4}.$$

The largest value here corresponds to row r_4, so $j_0 = 4 > 3 = i$. We therefore need to swap the third and fourth elements in the row vector, giving

$$\mathbf{r} = \begin{bmatrix} 2 & 1 & 4 & 3 \end{bmatrix}^T.$$

Since the element in the $a_{r_4,3} = a_{3,3}$ position is already zero, the third elimination pass makes no changes to the matrix. Back substitution now yields

$$\begin{aligned}
x_4 &= \frac{b_{r_4}}{a_{r_4,4}} = \frac{b_3}{a_{3,4}} = \frac{1.000}{-1.000} = -1.000 \\
x_3 &= \frac{b_{r_3} - a_{r_3,4}x_4}{a_{r_3,3}} = \frac{b_4 - a_{4,4}x_4}{a_{4,3}} = 1.000 \\
x_2 &= \frac{b_{r_2} - a_{r_2,3}x_3 - a_{r_2,4}x_4}{a_{r_2,2}} = \frac{b_1 - a_{1,3}x_3 - a_{1,4}x_4}{a_{1,2}} = -1.000 \\
x_1 &= \frac{b_{r_1} - a_{r_1,2}x_2 - a_{r_1,3}x_3 - a_{r_1,4}x_4}{a_{r_1,1}} = 1.000.
\end{aligned}$$

This solution is exact to four decimal digits.

EXERCISES

1. For each of the following augmented matrices, identify the entry that would serve as the first pivot element for

(i) Gaussian elimination with no pivoting;
(ii) Gaussian elimination with partial pivoting; and
(iii) Gaussian elimination with scaled partial pivoting.

(a) $\left[\begin{array}{cccc|c} 0 & 1 & 1 & 1 & 0 \\ 3 & 0 & 3 & -4 & 7 \\ 1 & 1 & 1 & 2 & 6 \\ 2 & 3 & 1 & 3 & 6 \end{array}\right]$

(b) $\left[\begin{array}{cccc|c} 3 & -1 & 3 & 1 & 6 \\ 6 & 0 & 9 & -2 & 13 \\ -12 & 0 & -10 & 5 & -17 \\ 72 & -8 & 48 & -19 & 93 \end{array}\right]$

(c) $\left[\begin{array}{ccc|c} -1.78 & 0.56 & 4.33 & 7.23 \\ 2.53 & -1.05 & 3.02 & -1.61 \\ 1.47 & -0.54 & -0.83 & -3.38 \end{array}\right]$

(d) $\left[\begin{array}{ccc|c} 0.25 & 0.35 & 0.15 & 0.60 \\ 0.20 & 0.20 & 0.25 & 0.90 \\ 0.15 & 0.20 & 0.25 & 0.70 \end{array}\right]$

(e) $\left[\begin{array}{cccc|c} 0.2115 & 2.296 & 2.715 & 3.215 & 8.438 \\ 0.4371 & 3.916 & 1.683 & 2.852 & 8.888 \\ 6.099 & 4.324 & 23.20 & 1.578 & 35.20 \\ 4.623 & 0.8926 & 15.32 & 5.305 & 26.14 \end{array}\right]$

For the augmented matrices indicated in Exercises 2–6, show the contents of the matrix after one pass of

(i) Gaussian elimination with no pivoting;
(ii) Gaussian elimination with partial pivoting; and
(iii) Gaussian elimination with scaled partial pivoting.

For (ii) and (iii), show the contents of the row vector, and for (iii), show the contents of the scale vector.

2. The augmented matrix from Exercise 1(a).

3. The augmented matrix from Exercise 1(b).

4. The augmented matrix from Exercise 1(c).

5. The augmented matrix from Exercise 1(d).

6. The augmented matrix from Exercise 1(e).

In Exercises 7–12:

(a) Solve the indicated system using Gaussian elimination with partial pivoting. Show all intermediate matrices and the row vector at each step.

(b) Repeat part (a) using Gaussian elimination with scaled partial pivoting. Show the contents of the scale vector.

7.
$$\begin{aligned} 2x_1 + 3x_2 + x_3 &= -4 \\ 4x_1 + x_2 + 4x_3 &= 9 \\ 3x_1 + 4x_2 + 6x_3 &= 0 \end{aligned}$$

8.
$$\begin{aligned} 2x_1 - x_2 + x_3 &= 2 \\ 4x_1 + 2x_2 + x_3 &= 7 \\ 6x_1 - 4x_2 + 2x_3 &= 4 \end{aligned}$$

9.
$$\begin{aligned} 3x_2 + x_3 &= 1 \\ x_1 + 2x_2 - 2x_3 &= 7 \\ 2x_1 + 5x_2 + 4x_3 &= -1 \end{aligned}$$

10.
$$\begin{aligned} x_1 + 8x_2 + 6x_3 &= -1 \\ -3x_1 - 4x_2 + 5x_3 &= 6 \\ 2x_1 + 4x_2 - 6x_3 &= -8 \end{aligned}$$

11.
$$\begin{aligned} x_1 - 3x_2 + 7x_3 &= 2 \\ 2x_1 + 4x_2 - 3x_3 &= -1 \\ -3x_1 + 7x_2 + 2x_3 &= 3 \end{aligned}$$

12.
$$\begin{aligned} x_1 - 2x_2 + x_3 - x_4 &= -5 \\ x_1 + 5x_2 - 7x_3 + 2x_4 &= 2 \\ 3x_1 + x_2 - 5x_3 + 3x_4 &= 1 \\ 2x_1 + 3x_2 - 5x_3 &= 17 \end{aligned}$$

13. Show that when the system

$$\left[\begin{array}{cccc|c} 3 & 1 & 4 & -1 & 7 \\ 2 & -2 & -1 & 2 & 1 \\ 5 & 7 & 14 & -8 & 20 \\ 1 & 3 & 2 & 4 & -4 \end{array}\right]$$

is solved using Gaussian elimination with no pivoting and four decimal digit rounding arithmetic, the resulting solution is

$$\mathbf{x} = \begin{bmatrix} 1.131 & -0.7928 & 0.8500 & -0.9987 \end{bmatrix}^T.$$

In Exercises 14–17, solve the given system in the indicated finite precision arithmetic using

(i) Gaussian elimination with no pivoting;
(ii) Gaussian elimination with partial pivoting; and
(iii) Gaussian elimination with scaled partial pivoting.

Compare the results obtained from each technique with the exact solution of the system.

14. 3 decimal digit rounding arithmetic

$$\begin{aligned} 0.5x_1 + 1.1x_2 + 3.1x_3 &= 6.0 \\ 2.0x_1 + 4.5x_2 + 0.36x_3 &= 0.02 \\ 5.0x_1 + 0.96x_2 + 6.5x_3 &= 0.96 \end{aligned}$$

15. 3 decimal digit rounding arithmetic

$$\begin{aligned} 3.41x_1 + 1.23x_2 - 1.09x_3 &= 4.72 \\ 2.71x_1 + 2.14x_2 + 1.29x_3 &= 3.10 \\ 1.89x_1 - 1.91x_2 - 1.89x_3 &= 2.91 \end{aligned}$$

16. 4 decimal digit rounding arithmetic

$$\left[\begin{array}{cccc|c} 3 & 1 & 4 & -1 & 7 \\ 2 & -2 & -1 & 2 & 1 \\ 5 & 7 & 14 & -9 & 21 \\ 1 & 3 & 2 & 4 & -4 \end{array}\right]$$

17. 4 decimal digit rounding arithmetic

$$\begin{array}{rcrcrcr} 1.985x_1 & - & 1.358x_2 & + & 2.113x_3 & = & -5.56 \\ 0.953x_1 & - & 0.6522x_2 & - & 1.815x_3 & = & 0.1592 \\ 2.607x_1 & + & 0.2065x_2 & + & 3.79x_3 & = & -0.357 \end{array}$$

18. Let A be the $n \times n$ matrix whose entries are given by $a_{ij} = 1/(i+j-1)$ for $1 \le i, j \le n$.

(a) For $n = 5, 6$ and 7, solve the system $A\mathbf{x} = \mathbf{b}$ using single precision arithmetic. In each case, take $\mathbf{b}$ as the vector that corresponds to an exact solution of $x_i = 1$ for each $i = 1, 2, 3, \dots, n$. Compare the solutions obtained using Gaussian elimination without pivoting, with partial pivoting and with scaled partial pivoting.

(b) For $n = 11, 12$, and 13, solve the system $A\mathbf{x} = \mathbf{b}$ using double precision arithmetic. In each case, take $\mathbf{b}$ as the vector that corresponds to an exact solution of $x_i = 1$ for each $i = 1, 2, 3, \dots, n$. Compare the solutions obtained using Gaussian elimination without pivoting, with partial pivoting and with scaled partial pivoting.

19. Solve the following system in single precision arithmetic.

$$\begin{array}{rcrcrcr} -149x_1 & - & 50x_2 & - & 154x_3 & = & 353 \\ 537x_1 & + & 180x_2 & + & 546x_3 & = & -1263 \\ -27x_1 & - & 9x_2 & - & 25x_3 & = & 61 \end{array}$$

Use Gaussian elimination without pivoting, with partial pivoting and with scaled partial pivoting. Which technique provided the most accurate solution? The exact solution for this problem is $\mathbf{x} = \begin{bmatrix} -1 & -1 & -1 \end{bmatrix}^T$.

20. Solve the following system in double precision arithmetic.

$$\left[\begin{array}{ccccc|c} -9 & 11 & -21 & 63 & -252 & -356 \\ 70 & -69 & 141 & -421 & 1684 & 2385 \\ -575 & 575 & -1149 & 3451 & -13801 & -19551 \\ 3891 & -3891 & 7782 & -23345 & 93365 & 132274 \\ 1024 & -1024 & 2048 & -6144 & 24572 & 34812 \end{array}\right]$$

Use Gaussian elimination without pivoting, with partial pivoting and with scaled partial pivoting. Which technique provided the most accurate solution? The exact solution for this problem is $\mathbf{x} = \begin{bmatrix} 1 & -1 & 1 & -1 & 1 \end{bmatrix}^T$.

3.3 VECTOR AND MATRIX NORMS

Partial pivoting and scaled partial pivoting were introduced as strategies for reducing the impact of roundoff error during the Gaussian elimination process. But how much error can we expect when solving a system of linear equations, and how does the error depend on the properties of the coefficient matrix and the right-hand side vector? Further, when A and $\mathbf{b}$ are known only approximately, how is the error in the solution related to the errors in the data?

To provide meaningful answers to these questions, tools are needed for measuring the "size" of a vector and the "size" of a matrix. The development of such tools will be the focus of this section.

Vector Norms

When working with scalar quantities as in Chapter 2, it was natural to measure size using the absolute value function. To measure size when working with n-dimensional vectors, a generalization of absolute value, referred to as a vector norm, is needed.

Definition. The function $\|\cdot\| : \mathbf{R}^n \to \mathbf{R}$ is called a VECTOR NORM if, for all $\mathbf{x}, \mathbf{y} \in \mathbf{R}^n$ and all $\alpha \in \mathbf{R}$, the following properties hold:

(i) $\|\mathbf{x}\| \geq 0$;
(ii) $\|\mathbf{x}\| = 0$ if and only if $\mathbf{x} = \mathbf{0}$;
(iii) $\|\alpha\mathbf{x}\| = |\alpha|\,\|\mathbf{x}\|$; and
(iv) $\|\mathbf{x}+\mathbf{y}\| \leq \|\mathbf{x}\| + \|\mathbf{y}\|$.

Note that the standard absolute value function satisfies each of these properties for $n = 1$. Property (iv) is often called the triangle inequality.

There are infinitely many vector norms that can be constructed. Here, we will restrict our attention to the two most commonly used in practice. A third will be considered in the exercises. The two norms on which we will concentrate are the l_2, or Euclidean, norm and the l_∞, or maximum, norm.

Definition. Let $\mathbf{x} \in \mathbf{R}^n$. The l_2-norm of $\mathbf{x}$, which is denoted by $\|\mathbf{x}\|_2$, is defined by

$$\|\mathbf{x}\|_2 = \left(\sum_{i=1}^{n} x_i^2\right)^{1/2}.$$

The l_∞-norm of $\mathbf{x}$, which is denoted by $\|\mathbf{x}\|_\infty$, is defined by

$$\|\mathbf{x}\|_\infty = \max_{1\leq i\leq n} |x_i|.$$

We will now show that $\|\cdot\|_2$ satisfies the properties of a vector norm. The verification that $\|\cdot\|_\infty$ satisfies the required properties is left as an exercise.

EXAMPLE 3.7 $\|\cdot\|_2$ **Is a Vector Norm**

To establish that $\|\cdot\|_2$ is a vector norm, we must show that $\|\cdot\|_2$ satisfies each of the four properties of the definition. In what follows, let $\mathbf{x}$ and $\mathbf{y}$ be arbitrary n-vectors, and let α be an arbitrary real number.

(i): $\|\mathbf{x}\|_2 \geq 0$

Since $x_i^2 \geq 0$ for any real number x_i, it follows that

$$\|\mathbf{x}\|_2 = \left(\sum_{i=1}^{n} x_i^2\right)^{1/2} \geq 0.$$

(ii): $\|\mathbf{x}\|_2 = 0$ if and only if $\mathbf{x} = \mathbf{0}$

If $\mathbf{x} = \mathbf{0}$, then $x_i = 0$ for each i. Therefore, $\sum x_i^2 = 0$ and $\|\mathbf{x}\|_2 = 0$. Conversely, if $\|\mathbf{x}\|_2 = 0$, then $\sum x_i^2 = 0$. This can happen only if $x_i = 0$ for each i, so $\mathbf{x} = \mathbf{0}$.

(iii): $\|\alpha\mathbf{x}\|_2 = |\alpha| \, \|\mathbf{x}\|_2$

$$\|\alpha\mathbf{x}\|_2 = \left(\sum (\alpha x_i)^2\right)^{1/2} = \left(\alpha^2 \sum x_i^2\right)^{1/2} = |\alpha| \left(\sum x_i^2\right)^{1/2} = |\alpha| \, \|\mathbf{x}\|_2$$

(iv): $\|\mathbf{x}+\mathbf{y}\|_2 \leq \|\mathbf{x}\|_2 + \|\mathbf{y}\|_2$

To show that $\|\cdot\|_2$ satisfies this property requires the Cauchy-Buniakowski-Schwarz inequality, which states that for any $\mathbf{x}, \mathbf{y} \in \mathbf{R}^n$,

$$\left|\sum_{i=1}^{n} x_i y_i\right| \leq \|\mathbf{x}\|_2 \|\mathbf{y}\|_2.$$

We will simply apply this result here and provide a proof below.

$$\begin{aligned}
\|\mathbf{x}+\mathbf{y}\|_2^2 &= \sum (x_i + y_i)^2 \\
&= \sum x_i^2 + 2\sum x_i y_i + \sum y_i^2 \\
&\leq \|\mathbf{x}\|_2^2 + 2\left|\sum x_i y_i\right| + \|\mathbf{y}\|_2^2 \\
&\leq \|\mathbf{x}\|_2^2 + 2\|\mathbf{x}\|_2\|\mathbf{y}\|_2 + \|\mathbf{y}\|_2^2 \\
&= (\|\mathbf{x}\|_2 + \|\mathbf{y}\|_2)^2
\end{aligned}$$

Upon taking the square root of both sides, the triangle inequality results.

For completeness, we now restate and prove the Cauchy-Buniakowski-Schwarz inequality.

Theorem. Let $\mathbf{x}, \mathbf{y} \in \mathbf{R}^n$. Then

$$\left|\sum_{i=1}^{n} x_i y_i\right| \leq \|\mathbf{x}\|_2 \|\mathbf{y}\|_2.$$

Proof. The inequality is trivially satisfied if $\mathbf{x} = \mathbf{0}$ or $\mathbf{y} = \mathbf{0}$. Therefore, suppose $\mathbf{x}$ and $\mathbf{y}$ are both nonzero. Let λ be any real number. Then

$$\begin{aligned} 0 \le \|\mathbf{x} + \lambda \mathbf{y}\|_2^2 &= \sum (x_i + \lambda y_i)^2 \\ &= \sum x_i^2 + 2\lambda \sum x_i y_i + \lambda^2 \sum y_i^2 \\ &= \|\mathbf{x}\|_2^2 + 2\lambda \sum x_i y_i + \lambda^2 \|\mathbf{y}\|_2^2. \end{aligned}$$

Let $a = \|\mathbf{y}\|_2^2$, $b = \sum x_i y_i$ and $c = \|\mathbf{x}\|_2^2$. The preceding inequality then becomes

$$a\lambda^2 + 2b\lambda + c \ge 0$$

for all $\lambda \in \mathbf{R}$. This can happen if and only if the discriminant, $(2b)^2 - 4ac$, is non-positive. Hence $b^2 < ac$. Substituting the values for a, b, and c gives

$$\left(\sum_{i=1}^{n} x_i y_i \right)^2 \le \|\mathbf{x}\|_2^2 \|\mathbf{y}\|_2^2,$$

from which the required inequality follows upon taking the square root of both sides. □

EXAMPLE 3.8 Calculating Vector Norms

Consider the three vectors

$$\begin{aligned} \mathbf{x}_1 &= \begin{bmatrix} 1 & -2 & 3 \end{bmatrix}^T; \\ \mathbf{x}_2 &= \begin{bmatrix} 2 & 0 & -1 & 2 \end{bmatrix}^T; \\ \mathbf{x}_3 &= \begin{bmatrix} 0 & 1 & -4 & 2 & -1 \end{bmatrix}^T. \end{aligned}$$

The maximum norm of each of these vectors is computed as follows:

$$\begin{aligned} \|\mathbf{x}_1\|_\infty &= \max\{|1|, |-2|, |3|\} = 3; \\ \|\mathbf{x}_2\|_\infty &= \max\{|2|, |0|, |-1|, |2|\} = 2; \\ \|\mathbf{x}_3\|_\infty &= \max\{|0|, |1|, |-4|, |2|, |-1|\} = 4. \end{aligned}$$

The Euclidean norm of each vector is

$$\begin{aligned} \|\mathbf{x}_1\|_2 &= \sqrt{1^2 + (-2)^2 + 3^2} = \sqrt{14} \approx 3.74; \\ \|\mathbf{x}_2\|_2 &= \sqrt{2^2 + 0^2 + (-1)^2 + 2^2} = \sqrt{9} = 3; \\ \|\mathbf{x}_3\|_2 &= \sqrt{0^2 + 1^2 + (-4)^2 + 2^2 + (-1)^2} = \sqrt{22} \approx 4.69. \end{aligned}$$

We will, of course, use vector norms for more than just assigning a size to a vector. One of the primary uses of vector norms will be to establish the convergence of a sequence of vectors, say $\{\mathbf{x}^{(k)}\}$. This is done by showing that $\|\mathbf{x}^{(k)} - \mathbf{x}\| \to 0$ as $k \to \infty$ for some limit vector $\mathbf{x}$. Given that different norms can be used to establish convergence, two important questions naturally arise. First, is it possible for a sequence to converge in one norm but to diverge in another norm? Second, is it possible for a sequence to converge to different limit values in different norms? Fortunately, the answer to both of these questions is no!

The reason that the choice of vector norm is irrelevant when considering convergence is that, on $\mathbf{R}^n$, all vector norms are equivalent.

> **Definition.** Let $\|\cdot\|$ and $\|\cdot\|'$ be vector norms on $\mathbf{R}^n$. If there exist positive constants c_1 and c_2 such that
>
> $$c_1\|\mathbf{x}\| \le \|\mathbf{x}\|' \le c_2\|\mathbf{x}\|$$
>
> for all $\mathbf{x} \in \mathbf{R}^n$, the two norms are said to be EQUIVALENT.

The connection between equivalence and convergence can be explained as follows. Suppose the sequence $\{\mathbf{x}^{(k)}\}$ converges to $\mathbf{x}$ in the $\|\cdot\|$-norm; that is, $\|\mathbf{x}^{(k)} - \mathbf{x}\| \to 0$ as $k \to \infty$. The right side inequality in the equivalence definition then guarantees that $\|\mathbf{x}^{(k)} - \mathbf{x}\|' \to 0$ as $k \to \infty$. Hence, the sequence converges to the same limit value in the $\|\cdot\|'$-norm. Similarly, if $\{\mathbf{x}^{(k)}\}$ converges to $\mathbf{x}$ in the $\|\cdot\|'$-norm, the left side inequality in the definition guarantees convergence of the sequence to the same limit value in the $\|\cdot\|$-norm.

Establishing that the l_2-norm and the l_∞-norm are equivalent is straightforward. Let $\mathbf{x} \in \mathbf{R}^n$ and suppose that x_j is a component for which $\|\mathbf{x}\|_\infty = |x_j|$. Then

$$\begin{aligned}\|\mathbf{x}\|_\infty^2 = |x_j|^2 = x_j^2 &\le \sum_{i=1}^n x_i^2 \\ &\le \sum_{i=1}^n x_j^2 = n x_j^2 = n\|\mathbf{x}\|_\infty^2.\end{aligned}$$

Thus,

$$\|\mathbf{x}\|_\infty \le \left(\sum_{i=1}^n x_i^2\right)^{1/2} = \|\mathbf{x}\|_2 \le \sqrt{n}\|\mathbf{x}\|_\infty,$$

so $c_1 = 1$ and $c_2 = \sqrt{n}$ in the above definition. The proof that any pair of vector norms on $\mathbf{R}^n$ are equivalent can be found in Ortega [1] or Ortega and Rheinboldt [2].

Matrix Norms

To measure the errors introduced during the solution of a linear system, it will be necessary to have a means for quantifying the "size" of a matrix. This is done using matrix norms.

Definition. A MATRIX NORM is a function $\|\cdot\| : \mathbf{R}^{n \times n} \to \mathbf{R}$ that, for all $A, B \in \mathbf{R}^{n\times n}$ and all $\alpha \in \mathbf{R}$, satisfies

(i) $\|A\| \geq 0$;
(ii) $\|A\| = 0$ if and only if $A = 0$;
(iii) $\|\alpha A\| = |\alpha| \, \|A\|$;
(iv) $\|A + B\| \leq \|A\| + \|B\|$; and
(v) $\|AB\| \leq \|A\| \, \|B\|$.

Although the same symbol is used to denote both vector and matrix norms, the type of norm being used should be clear from the specific context.

As with vector norms, there are various ways to obtain matrix norms. We will, however, restrict attention to those matrix norms which are related to vector norms. These are referred to as the *natural matrix norms.*

Definition. Let $\|\cdot\|_v$ be a vector norm. The real-valued function $\|\cdot\|$ that is defined for all $A \in \mathbf{R}^{n\times n}$ by

$$\|A\| = \max_{\|\mathbf{x}\|_v \neq \mathbf{0}} \frac{\|A\mathbf{x}\|_v}{\|\mathbf{x}\|_v}$$

is called the NATURAL, or OPERATOR, NORM associated with (generated by, induced by) the vector norm $\|\cdot\|_v$.

All natural matrix norms possess an important *consistency property.* Since $\|A\|$ is defined as the maximum of the ratio $\|A\mathbf{x}\|_v/\|\mathbf{x}\|_v$, it follows that for any nonzero n-vector $\mathbf{x}$

$$\|A\| \geq \frac{\|A\mathbf{x}\|_v}{\|\mathbf{x}\|_v},$$

or, equivalently, $\|A\mathbf{x}\|_v \leq \|A\| \, \|\mathbf{x}\|_v$. This inequality is often used to provide a bound on the value of $\|A\mathbf{x}\|_v$ and plays a central role in proving that a natural norm possesses the properties in the definition of a matrix norm.

Theorem. Let $\|\cdot\|_v$ be a vector norm. The natural norm associated with $\|\cdot\|_v$ is a matrix norm.

Proof. Properties (i) and (iii) of a matrix norm follow directly from the corresponding properties of a vector norm. To establish property (ii), note that

$$\begin{aligned} \|A\| = 0 &\Leftrightarrow \|A\mathbf{x}\|_v = 0 \text{ for all } \mathbf{x} \neq \mathbf{0} \\ &\Leftrightarrow A\mathbf{x} = \mathbf{0} \text{ for all } \mathbf{x} \neq \mathbf{0} \\ &\Leftrightarrow A = 0. \end{aligned}$$

For property (iv), let $\mathbf{x}$ be any nonzero n-vector. Then

$$\|(A + B)\mathbf{x}\|_v = \|A\mathbf{x} + B\mathbf{x}\|_v \leq \|A\mathbf{x}\|_v + \|B\mathbf{x}\|_v$$

by the triangle inequality for the vector norm. By the consistency property of the natural norm, $\|A\mathbf{x}\|_v \leq \|A\| \, \|\mathbf{x}\|_v$ and $\|B\mathbf{x}\|_v \leq \|B\| \, \|\mathbf{x}\|_v$. Thus

$$\|(A+B)\mathbf{x}\|_v \leq \|A\| \, \|\mathbf{x}\|_v + \|B\| \, \|\mathbf{x}\|_v,$$

or

$$\frac{\|(A+B)\mathbf{x}\|_v}{\|\mathbf{x}\|_v} \leq \|A\| + \|B\| \quad \Rightarrow \quad \|A+B\| \leq \|A\| + \|B\|.$$

Property (v) can be established in a similar manner and is left as an exercise. □

The maximum matrix norm, $\|A\|_\infty$, is fairly easy to calculate in terms of the entries in A. For, suppose that $\mathbf{x}$ is any nonzero n-vector. Recall that the ith component of the product $A\mathbf{x}$ is given by

$$(A\mathbf{x})_i = \sum_{j=1}^{n} a_{ij} x_j.$$

Therefore,

$$\begin{aligned} \|A\mathbf{x}\|_\infty &= \max_i \left| \sum_{j=1}^{n} a_{ij} x_j \right| \leq \max_i \sum_{j=1}^{n} |a_{ij}||x_j| \\ &\leq \max_j |x_j| \max_i \sum_{j=1}^{n} |a_{ij}| = \|\mathbf{x}\|_\infty \max_i \sum_{j=1}^{n} |a_{ij}|, \end{aligned}$$

from which it follows that

$$\|A\|_\infty \leq \max_i \sum_{j=1}^{n} |a_{ij}|. \tag{1}$$

Now, let k be an index for which

$$\sum_{j=1}^{n} |a_{kj}| = \max_i \sum_{j=1}^{n} |a_{ij}|,$$

and define the vector $\mathbf{x}$ by

$$x_j = \begin{cases} 1, & a_{kj} \geq 0 \\ -1, & a_{kj} < 0 \end{cases}.$$

Then $a_{kj} x_j = |a_{kj}|$ for each j. For this $\mathbf{x}$,

$$\begin{aligned} \|A\mathbf{x}\|_\infty &= \max_i \left| \sum_{j=1}^{n} a_{ij} x_j \right| \geq \left| \sum_{j=1}^{n} a_{kj} x_j \right| \\ &= \sum_{j=1}^{n} |a_{kj}| = \|\mathbf{x}\|_\infty \max_i \sum_{j=1}^{n} |a_{ij}|, \end{aligned}$$

where we have used the fact that $\|\mathbf{x}\|_\infty = 1$. Therefore,

$$\|A\|_\infty \geq \max_i \sum_{j=1}^{n} |a_{ij}|. \tag{2}$$

Combining equations (1) and (2) yields

$$\|A\|_\infty = \max_i \sum_{j=1}^{n} |a_{ij}|.$$

Since $\|A\|_\infty$ is based on sums of absolute values of the entries along each row of A, the l_∞ natural matrix norm is also referred to as the row norm of A.

The l_2 natural matrix norm, unfortunately, is not as straightforward to calculate and requires knowledge of the eigenvalues of the matrix.

Definition. Let $A \in \mathbf{R}^{n\times n}$. If for some number λ (which may be complex) there exists a nonzero vector $\mathbf{x}$ such that $A\mathbf{x} = \lambda\mathbf{x}$, then λ is an EIGENVALUE of A and $\mathbf{x}$ is an EIGENVECTOR corresponding to λ.

The eigenvalue relation $A\mathbf{x} = \lambda\mathbf{x}$ is equivalent to the linear system $(A - \lambda I)\mathbf{x} = \mathbf{0}$. For this system to have a nonzero solution for $\mathbf{x}$, the matrix $A - \lambda I$ must be singular. Thus, the eigenvalues of A are those values of λ for which $\det(A - \lambda I) = 0$. As a function of λ, $\det(A - \lambda I)$ is an nth-degree polynomial, known as the *characteristic polynomial* of A. So, counting multiplicities, an $n \times n$ matrix has precisely n eigenvalues. The set of all eigenvalues for a given matrix A is called the *spectrum* of A and is denoted by $\sigma(A)$.

EXAMPLE 3.9 Calculating Eigenvalues

Consider the matrix

$$A = \begin{bmatrix} 18 & 10 \\ 10 & 13 \end{bmatrix}.$$

The characteristic polynomial associated with this matrix is

$$\begin{aligned} p(\lambda) &= \det(A - \lambda I) \\ &= \det\left(\begin{bmatrix} 18-\lambda & 10 \\ 10 & 13-\lambda \end{bmatrix}\right) \\ &= (18-\lambda)(13-\lambda) - 100 = \lambda^2 - 31\lambda + 134. \end{aligned}$$

The eigenvalues of A are the roots of this polynomial:

$$\lambda = \frac{31 \pm \sqrt{31^2 - 4(134)}}{2} = \frac{31 \pm \sqrt{425}}{2} = \frac{31 \pm 5\sqrt{17}}{2}.$$

As a second example, consider the 3×3 matrix

$$A = \begin{bmatrix} 2 & -1 & 1 \\ -1 & 2 & 0 \\ 1 & 0 & 6 \end{bmatrix}.$$

The characteristic polynomial associated with this matrix is

$$\begin{aligned} p(\lambda) &= \det(A - \lambda I) \\ &= (2-\lambda)(2-\lambda)(6-\lambda) - (2-\lambda) - (6-\lambda) \\ &= -\lambda^3 + 10\lambda^2 - 26\lambda + 16. \end{aligned}$$

To five decimal places, the roots of this polynomial, and hence the eigenvalues of A, are

$$\lambda_1 = 0.89722, \quad \lambda_2 = 2.85363 \quad \text{and} \quad \lambda_3 = 6.24914.$$

One of the most important quantities related to the eigenvalues of a matrix is the spectral radius.

Definition. The SPECTRAL RADIUS $\rho(A)$ of the matrix A is defined by

$$\rho(A) = \max_{\lambda \in \sigma(A)} |\lambda|.$$

The relationship between the spectral radius and the norm of the matrix is provided by the following theorem.

Theorem. Let A be an $n \times n$ matrix. Then

(i) $\|A\|_2 = \sqrt{\rho(A^T A)}$;
(ii) $\rho(A) \le \|A\|$ for any natural norm; and
(iii) for any $\epsilon > 0$, there exists a natural norm $\|\cdot\|$ for which $\|A\| \le \rho(A) + \epsilon$.

Proof. (i) and (iii): See Isaacson and Keller [3] or Ortega [1].
(ii) Let $\lambda \in \sigma(A)$ with associated eigenvector $\mathbf{x}$. Taking the norm of both sides of the eigenvalue relation $A\mathbf{x} = \lambda\mathbf{x}$ then yields

$$|\lambda|\,\|\mathbf{x}\| = \|\lambda\mathbf{x}\| = \|A\mathbf{x}\| \le \|A\|\,\|\mathbf{x}\|,$$

or

$$|\lambda| \le \|A\|.$$

Therefore,

$$\rho(A) = \max_{\lambda \in \sigma(A)} |\lambda| \le \|A\|. \qquad \square$$

Note that conclusions (ii) and (iii) of this theorem indicate that the spectral radius is the greatest lower bound for the natural norms of a matrix. Since $\|A\|_2$ is based on the spectral radius, the l_2 natural matrix norm is also referred to as the spectral norm.

EXAMPLE 3.10 Calculating the l_2 and l_∞ Norms of a Matrix

Let's calculate both the l_2 and the l_∞ norms of the matrices

$$A_1 = \begin{bmatrix} 1 & -2 \\ 4 & 3 \end{bmatrix} \quad \text{and} \quad A_2 = \begin{bmatrix} 1 & 0 & 2 \\ 0 & 1 & -1 \\ -1 & 1 & 1 \end{bmatrix}.$$

Starting with the matrix A_1,

$$\|A_1\|_\infty = \max\{|1| + |-2|, |4| + |3|\} = \max\{3, 7\} = 7.$$

To determine the l_2 norm, we first compute

$$A_1^T A_1 = \begin{bmatrix} 1 & 4 \\ -2 & 3 \end{bmatrix} \begin{bmatrix} 1 & -2 \\ 4 & 3 \end{bmatrix} = \begin{bmatrix} 18 & 10 \\ 10 & 13 \end{bmatrix}.$$

The eigenvalues of this matrix were previously found to be $\frac{1}{2}(31 \pm 5\sqrt{17})$. Hence,

$$\rho(A_1^T A_1) = \frac{1}{2}(31 + 5\sqrt{17}) \quad \text{and} \quad \|A_1\|_2 = \sqrt{\frac{1}{2}(31 + 5\sqrt{17})} \approx 5.08013.$$

For the matrix A_2,

$$\begin{aligned} \|A_2\|_\infty &= \max\{|1| + |0| + |2|, |0| + |1| + |-1|, |-1| + |1| + |1|\} \\ &= \max\{3, 2, 3\} = 3, \end{aligned}$$

and

$$A_2^T A_2 = \begin{bmatrix} 2 & -1 & 1 \\ -1 & 2 & 0 \\ 1 & 0 & 6 \end{bmatrix}.$$

The eigenvalues of $A_2^T A_2$ were previously found to be 0.89722, 2.85363, and 6.24914. Hence,

$$\rho(A_2^T A_2) = 6.24914 \quad \text{and} \quad \|A_2\|_2 = \sqrt{6.24914} \approx 2.49983.$$

References

1. J. M. Ortega, *Numerical Analysis—A Second Course,* Academic Press, New York, 1972.
2. J. M. Ortega and W. C. Rheinboldt, *Iterative Solution of Nonlinear Equations in Several Variables,* Academic Press, New York, 1970.
3. E. Isaacson and H. Keller, *Analysis of Numerical Methods,* John Wiley & Sons, New York, 1966.

EXERCISES

1. Verify that the l_∞-norm,

$$\|\mathbf{x}\|_\infty = \max_{1\le i\le n} |x_i|,$$

satisfies the properties of a vector norm.

2. Compute the l_2-norm and the l_∞-norm for each of the following vectors.

(a) $\mathbf{x} = \begin{bmatrix} 3 & -5 & \sqrt{2} \end{bmatrix}^T$

(b) $\mathbf{x} = \begin{bmatrix} 2 & 1 & -3 & 4 \end{bmatrix}^T$

(c) $\mathbf{x} = \begin{bmatrix} 4 & -8 & 1 \end{bmatrix}^T$

(d) $\mathbf{x} = \begin{bmatrix} -2\sqrt{3} & -6 & 4 & 2 \end{bmatrix}^T$

(e) $\mathbf{x} = \begin{bmatrix} e & \pi & -1 \end{bmatrix}^T$

3. (a) Show that the function $\|\cdot\|_1 : \mathbf{R}^n \to \mathbf{R}$ defined by

$$\|\mathbf{x}\|_1 = \sum_{i=1}^{n} |x_i|$$

is a vector norm. The operator $\|\cdot\|_1$ is known as the l_1-norm.

(b) Compute the l_1-norm for each of the vectors in Exercise 2.

(c) Show that $\|\mathbf{x}\|_\infty \le \|\mathbf{x}\|_1 \le n\|\mathbf{x}\|_\infty$ for all $\mathbf{x} \in \mathbf{R}^n$.

(d) Show that $\|\mathbf{x}\|_2 \le \|\mathbf{x}\|_1 \le \sqrt{n}\|\mathbf{x}\|_2$ for all $\mathbf{x} \in \mathbf{R}^n$.

4. Let $\|\cdot\|_v$ be a vector norm. Show that the natural norm associated with $\|\cdot\|_v$ satisfies $\|AB\| \le \|A\|\,\|B\|$ for all $A, B \in \mathbf{R}^{n\times n}$.

5. Compute the spectrum of each of the following matrices.

(a) $A = \begin{bmatrix} 4 & -2 \\ 1 & 1 \end{bmatrix}$

(b) $A = \begin{bmatrix} 0.7 & 0.2 \\ 0.3 & 0.8 \end{bmatrix}$

(c) $A = \begin{bmatrix} 2 & -3 & 1 \\ 1 & -2 & 1 \\ 1 & -3 & 2 \end{bmatrix}$

(d) $A = \begin{bmatrix} 1 & 2 & 1 \\ 0 & 3 & 1 \\ 0 & 5 & -1 \end{bmatrix}$

6. Compute the l_2-norm and the l_∞-norm for each of the following matrices.

(a) $A = \begin{bmatrix} 5 & -4 \\ -1 & 7 \end{bmatrix}$

(b) $A = \begin{bmatrix} 4 & 2 \\ 1 & 3 \end{bmatrix}$

(c) $A = \begin{bmatrix} 4 & -1 & -2 \\ 1 & 2 & -3 \\ 0 & 0 & 4 \end{bmatrix}$

(d) $A = \begin{bmatrix} 2 & 1 & 0 \\ -1 & 2 & -1 \\ -3 & 4 & -4 \end{bmatrix}$

7. **(a)** Prove that the natural matrix norm associated with the l_1 vector norm (see Exercise 3) is given by

$$\|A\|_1 = \max_{1 \le j \le n} \sum_{i=1}^{n} |a_{ij}|$$

for all $A \in \mathbf{R}^{n \times n}$. This is also known as the *column norm* of A.

(b) Compute $\|\cdot\|_1$ for each of the matrices in Exercise 6.

8. The *Frobenius norm* (which is not a natural matrix norm) is defined by

$$\|A\|_F = \left(\sum_{i=1}^{n} \sum_{j=1}^{n} |a_{ij}|^2 \right)^{1/2}$$

for all $A \in \mathbf{R}^{n \times n}$.

(a) Show that $\|\cdot\|_F$ is a matrix norm.

(b) Compute the Frobenius norm for each of the matrices in Exercise 6.

9. **(a)** Let λ be an eigenvalue of the matrix A with associated eigenvector $\mathbf{x}$. For any integer $k \ge 1$, show that λ^k is an eigenvalue of A^k with eigenvector $\mathbf{x}$.

(b) Let A be a symmetric matrix. Show that $\|A\|_2 = \rho(A)$.

10. Show that if A is a matrix with $\rho(A) < 1$, then the matrix $I - A$ is nonsingular. (*Hint*: Assume that $I - A$ is singular and show this leads to the conclusion that $\lambda = 1$ is an eignevalue of A.)

11. **(a)** Let D be an $n \times n$ diagonal matrix. Show that the eigenvalues of D are the diagonal elements $d_{11}, d_{22}, d_{33}, \ldots, d_{nn}$.

(b) Let U be an $n \times n$ upper triangular matrix. Show that the eigenvalues of U are the diagonal elements $u_{11}, u_{22}, u_{33}, \ldots, u_{nn}$.

3.4 ERROR ESTIMATES AND CONDITION NUMBER

Having developed the appropriate tools (i.e., vector and matrix norms), we now address the questions raised at the beginning of the last section. In particular, how much error can we expect when solving a system of linear equations using Gaussian elimination, and how does the error depend on the properties of the coefficient matrix and the right-hand side vector? Further, when A and $\mathbf{b}$ are known only approximately, how is the error in the solution related to the errors in the data?

Error Estimates

Suppose $\tilde{\mathbf{x}}$ is an approximate solution to the linear system $A\mathbf{x} = \mathbf{b}$, whose exact solution is the vector $\mathbf{x}$. In practice, the exact solution to the system is unknown, so the error in $\tilde{\mathbf{x}}$, $\mathbf{e} = \tilde{\mathbf{x}} - \mathbf{x}$, cannot be directly computed. However, the *residual vector*, which is defined as $\mathbf{r} = A\tilde{\mathbf{x}} - \mathbf{b}$, can be easily computed. Note that the residual measures the amount by which the approximate solution fails to satisfy the linear system. When $\mathbf{r} = \mathbf{0}$, it follows that $\tilde{\mathbf{x}}$ is the exact solution, so $\mathbf{e} = \mathbf{0}$. It seems reasonable to expect, therefore, that whenever $\|\mathbf{r}\|$ is small, $\|\mathbf{e}\|$ will be small as well. Unfortunately, this need not always be the case.

EXAMPLE 3.11 A Small Residual But A Large Error

The linear system

$$\begin{bmatrix} 1 & -2 \\ -0.99 & 1.99 \end{bmatrix} \mathbf{x} = \begin{bmatrix} -1 \\ 1 \end{bmatrix}$$

has $\mathbf{x} = \begin{bmatrix} 1 & 1 \end{bmatrix}^T$ as its exact solution. The vector $\tilde{\mathbf{x}} = \begin{bmatrix} -1 & 0 \end{bmatrix}^T$ is an obviously poor approximation to $\mathbf{x}$:

$$\mathbf{e} = \begin{bmatrix} -2 & -1 \end{bmatrix}^T \quad \Rightarrow \quad \|\mathbf{e}\|_\infty = 2.$$

However, the residual associated with $\tilde{\mathbf{x}}$ is

$$\begin{aligned} \mathbf{r} = A\tilde{\mathbf{x}} - \mathbf{b} &= \begin{bmatrix} 1 & -2 \\ -0.99 & 1.99 \end{bmatrix} \begin{bmatrix} -1 \\ 0 \end{bmatrix} - \begin{bmatrix} -1 \\ 1 \end{bmatrix} \\ &= \begin{bmatrix} -1 \\ 0.99 \end{bmatrix} - \begin{bmatrix} -1 \\ 1 \end{bmatrix} = \begin{bmatrix} 0 \\ -0.01 \end{bmatrix}, \end{aligned}$$

so $\|\mathbf{r}\|_\infty = 0.01$. Thus, the error is 200 times larger than the residual.

The next theorem shows that the norm of both the coefficient matrix and its inverse play an important role in the reliability of the residual as a predictor of error.

Theorem. Let A be a nonsingular matrix, $\tilde{\mathbf{x}}$ be an approximate solution to the linear system $A\mathbf{x} = \mathbf{b}$, $\mathbf{r} = A\tilde{\mathbf{x}} - \mathbf{b}$ and $\mathbf{e} = \tilde{\mathbf{x}} - \mathbf{x}$. Then, for any natural matrix norm $\|\cdot\|$,

$$\frac{1}{\|A\|}\|\mathbf{r}\| \leq \|\mathbf{e}\| \leq \|A^{-1}\|\,\|\mathbf{r}\|$$

and

$$\frac{1}{\|A\|\,\|A^{-1}\|}\frac{\|\mathbf{r}\|}{\|\mathbf{b}\|} \leq \frac{\|\mathbf{e}\|}{\|\mathbf{x}\|} \leq \|A\|\,\|A^{-1}\|\frac{\|\mathbf{r}\|}{\|\mathbf{b}\|}, \tag{1}$$

provided $\mathbf{x} \neq \mathbf{0}$ and $\mathbf{b} \neq \mathbf{0}$.

Proof. First we need a relationship between $\mathbf{e}$ and $\mathbf{r}$. Combining $\mathbf{r} = A\tilde{\mathbf{x}} - \mathbf{b}$, $\mathbf{b} = A\mathbf{x}$ and $\mathbf{e} = \tilde{\mathbf{x}} - \mathbf{x}$, it follows that

$$\mathbf{r} = A\tilde{\mathbf{x}} - \mathbf{b} = A\tilde{\mathbf{x}} - A\mathbf{x} = A(\tilde{\mathbf{x}} - \mathbf{x}) = A\mathbf{e}.$$

Equivalently, $\mathbf{e} = A^{-1}\mathbf{r}$. Now, let $\|\cdot\|$ be any natural matrix norm. An immediate consequence of $\mathbf{e} = A^{-1}\mathbf{r}$ is

$$\|\mathbf{e}\| = \|A^{-1}\mathbf{r}\| \le \|A^{-1}\|\,\|\mathbf{r}\|.$$

From $\mathbf{r} = A\mathbf{e}$, we obtain

$$\|\mathbf{r}\| = \|A\mathbf{e}\| \le \|A\|\,\|\mathbf{e}\|,$$

or $\|\mathbf{e}\| \ge \|\mathbf{r}\|/\|A\|$. Thus,

$$\frac{1}{\|A\|}\|\mathbf{r}\| \le \|\mathbf{e}\| \le \|A^{-1}\|\,\|\mathbf{r}\|. \tag{2}$$

Next, suppose that $\mathbf{x} \ne \mathbf{0}$ and $\mathbf{b} \ne \mathbf{0}$. Taking the norm of both sides of $A\mathbf{x} = \mathbf{b}$ yields

$$\|\mathbf{b}\| = \|A\mathbf{x}\| \le \|A\|\,\|\mathbf{x}\| \quad \Rightarrow \quad \frac{1}{\|\mathbf{x}\|} \le \frac{\|A\|}{\|\mathbf{b}\|}.$$

Similarly, from $\mathbf{x} = A^{-1}\mathbf{b}$, we obtain

$$\frac{1}{\|A^{-1}\|\,\|\mathbf{b}\|} \le \frac{1}{\|\mathbf{x}\|},$$

so

$$\frac{1}{\|A^{-1}\|\,\|\mathbf{b}\|} \le \frac{1}{\|\mathbf{x}\|} \le \frac{\|A\|}{\|\mathbf{b}\|}. \tag{3}$$

Finally, combining (2) and (3) yields

$$\frac{1}{\|A\|\,\|A^{-1}\|}\frac{\|\mathbf{r}\|}{\|\mathbf{b}\|} \le \frac{\|\mathbf{e}\|}{\|\mathbf{x}\|} \le \|A\|\,\|A^{-1}\|\frac{\|\mathbf{r}\|}{\|\mathbf{b}\|}. \qquad \square$$

The inequalities in (1) provide lower and upper bounds on the relative error in an approximate solution to $A\mathbf{x} = \mathbf{b}$ in terms of the relative residual, $\|\mathbf{r}\|/\|\mathbf{b}\|$, and the norms of A and its inverse. The quantity $\kappa(A) = \|A\|\,\|A^{-1}\|$ occurs frequently in the analysis of linear systems and is known as the *condition number* of A. The value of $\kappa(A)$ depends heavily upon the matrix norm being used; however, for any nonsingular matrix and any natural matrix norm, the following bound applies:

$$1 = \|I\| = \|A \cdot A^{-1}\| \le \|A\|\,\|A^{-1}\| = \kappa(A).$$

When $\kappa(A)$ is small (i.e., ≈ 1), the relative residual provides a good measure for the error in the approximate solution. On the other hand, when $\kappa(A)$ is large, the relative residual can be a very poor indicator of the accuracy of the approximate solution.

EXAMPLE 3.12 A Small Residual But A Large Error, Continued

The inverse of the coefficient matrix from the previous example,

$$A = \begin{bmatrix} 1 & -2 \\ -0.99 & 1.99 \end{bmatrix},$$

is

$$A^{-1} = \begin{bmatrix} 199 & 200 \\ 99 & 100 \end{bmatrix}.$$

Hence, $\|A\|_\infty = 3$, $\|A^{-1}\|_\infty = 399$, and $\kappa_\infty(A) = 3(399) = 1197$. The relative error in an approximate solution to a system with A as its coefficient matrix can therefore be as small as 1/1197 times, or as large as 1197 times, the relative residual.

Perturbations to A and b

What if the entries in A and $\mathbf{b}$ are known only approximately, due perhaps to data errors or roundoff errors or both? How does the error in the computed solution depend on the errors in A and $\mathbf{b}$? Let δA and $\delta\mathbf{b}$ denote the perturbations to A and $\mathbf{b}$, respectively, and let $\mathbf{x}+\delta\mathbf{x}$ denote the solution to the system with perturbed coefficient matrix and right-side vector. That is,

$$(A + \delta A)(\mathbf{x} + \delta\mathbf{x}) = \mathbf{b} + \delta\mathbf{b}. \tag{4}$$

Further, suppose that $\|\delta A\| < 1/\|A^{-1}\|$, which guarantees that $A + \delta A$ remains nonsingular (see Exercise 3).

Now, expand the product on the left side of (4), cancel the term $A\mathbf{x}$ on the left with the term $\mathbf{b}$ on the right (since $A\mathbf{x} = \mathbf{b}$) and rearrange the remaining terms to produce

$$\delta\mathbf{x} = A^{-1}\left[\delta\mathbf{b} - (\delta A)\mathbf{x} - (\delta A)(\delta\mathbf{x})\right].$$

Next, take norms, repeatedly apply the consistency property of the natural matrix norm and the triangle inequality and solve for $\|\delta\mathbf{x}\|$. The result is

$$\begin{aligned}
\|\delta\mathbf{x}\| &\le \frac{\|A^{-1}\|}{1 - \|A^{-1}\|\,\|\delta A\|}(\|\delta\mathbf{b}\| + \|\delta A\|\,\|\mathbf{x}\|) \\
&= \frac{\|A\|\,\|A^{-1}\|}{1 - \|A\|\,\|A^{-1}\|(\|\delta A\|/\|A\|)}\left(\frac{\|\delta\mathbf{b}\|}{\|A\|} + \frac{\|\delta A\|}{\|A\|}\|\mathbf{x}\|\right) \\
&= \frac{\kappa(A)}{1 - \kappa(A)(\|\delta A\|/\|A\|)}\left(\frac{\|\delta\mathbf{b}\|}{\|A\|} + \frac{\|\delta A\|}{\|A\|}\|\mathbf{x}\|\right).
\end{aligned}$$

Finally, divide through by $\|\mathbf{x}\|$ and use the relation $\|A\|\|\mathbf{x}\| \ge \|\mathbf{b}\|$ to obtain

$$\frac{\|\delta\mathbf{x}\|}{\|\mathbf{x}\|} \le \frac{\kappa(A)}{1 - \kappa(A)(\|\delta A\|/\|A\|)}\left(\frac{\|\delta\mathbf{b}\|}{\|\mathbf{b}\|} + \frac{\|\delta A\|}{\|A\|}\right). \tag{5}$$

Note the presence of the condition number in (5). Even if the relative errors in A and $\mathbf{b}$ are small, the relative error in the approximate solution may be signficant if $\kappa(A)$ is large.

EXAMPLE 3.13 A Perturbed System

Let

$$A = \begin{bmatrix} 1 & -2 \\ -0.99 & 1.99 \end{bmatrix} \quad \text{and} \quad \mathbf{b} = \begin{bmatrix} -1 \\ 1 \end{bmatrix}.$$

Recall that the true solution to $A\mathbf{x} = \mathbf{b}$ is $\mathbf{x} = \begin{bmatrix} 1 & 1 \end{bmatrix}^T$ and that $\kappa_\infty(A) = 1197$.

As an experiment, let's first change the right side vector to $\mathbf{b} + \delta\mathbf{b}$, where $\delta\mathbf{b} = \begin{bmatrix} 0.01 & 0.01 \end{bmatrix}^T$. The true solution of $A\tilde{\mathbf{x}} = (\mathbf{b} + \delta\mathbf{b})$ is $\tilde{\mathbf{x}} = \begin{bmatrix} 4.99 & 2.99 \end{bmatrix}^T$. Thus $\delta\mathbf{x} = \begin{bmatrix} 3.99 & 1.99 \end{bmatrix}^T$, and

$$\frac{\|\delta\mathbf{x}\|_\infty}{\|\mathbf{x}\|_\infty} = \frac{3.99}{1} = 3.99.$$

Though this constitutes a substantial relative change in the solution, it is quite a bit less than the maximum possible change, which for $\kappa_\infty(A) = 1197$, $\|\delta\mathbf{b}\|_\infty = 0.01$ and $\|\mathbf{b}\|_\infty = 1$ is

$$\frac{1197}{1 - 1197 \cdot 0}\left(\frac{0.01}{1} + 0\right) = 11.97.$$

Next, suppose we perturb both A and $\mathbf{b}$ with

$$\delta A = \begin{bmatrix} -0.001 & -0.001 \\ -0.001 & -0.001 \end{bmatrix} \quad \text{and} \quad \delta\mathbf{b} = \begin{bmatrix} 0.01 \\ 0.01 \end{bmatrix}.$$

The true solution of the system $(A + \delta A)\tilde{\mathbf{x}} = (\mathbf{b} + \delta\mathbf{b})$ is $\tilde{\mathbf{x}} \approx \begin{bmatrix} 12.910 & 6.940 \end{bmatrix}^T$. Hence,

$$\frac{\|\delta\mathbf{x}\|_\infty}{\|\mathbf{x}\|_\infty} \approx \frac{11.91}{1} = 11.91.$$

Once again, though quite large, this relative change in the solution is significantly less than the maximum possible change, which for $\kappa_\infty(A) = 1197$, $\|\delta A\|_\infty = 0.002$, $\|A\|_\infty = 3$, $\|\delta\mathbf{b}\|_\infty = 0.01$, and $\|\mathbf{b}\|_\infty = 1$ is

$$\frac{1197}{1 - 1197 \cdot \frac{0.002}{3}}\left(\frac{0.01}{1} + \frac{0.002}{3}\right) = 63.21.$$

Rounding Errors Introduced by Gaussian Elimination

Equation (5) was derived without reference to any numerical method for solving the linear system; hence, this result represents a fundamental property of the mathematical problem. Using a process known as *backward error analysis* (see Wilkinson

[1,2] or Atkinson [3]), it can be shown that the approximate solution to the system $A\mathbf{x} = \mathbf{b}$, obtained by applying Gaussian elimination with pivoting in t-digit decimal floating point arithmetic, is the exact solution of the perturbed system $(A + \delta A)\tilde{\mathbf{x}} = \mathbf{b}$, where

$$\frac{\|\delta A\|_\infty}{\|A\|_\infty} \leq f(n) \cdot 10^{1-t} \cdot \frac{\max_{i,j,k} |a_{ij}^{(k)}|}{\|A\|_\infty}.$$

Here, n denotes the size of the system, and the $a_{ij}^{(k)}$ are the elements in the coefficient matrix during the kth elimination pass. In practice, $f(n) \approx n$, and $f(n) \leq 1.01(n^3 + 3n^2)$ in the worst case. Wilkinson has observed that the empirical bound

$$\frac{\|\delta A\|_\infty}{\|A\|_\infty} \leq n \cdot 10^{1-t}$$

is seldom exceeded when pivoting is used. Hence, the rounding errors introduced by Gaussian elimination would be bounded by

$$\frac{\|\delta \mathbf{x}\|_\infty}{\|\mathbf{x}\|_\infty} \leq \frac{\kappa_\infty(A) \cdot n \cdot 10^{1-t}}{1 - \kappa_\infty(A) \cdot n \cdot 10^{1-t}}.$$

From here, it is clear that to obtain a "good" solution we must have $\kappa_\infty(A)$ much smaller than $n \cdot 10^{1-t}$. Further, if $\kappa_\infty(A) \approx 10^r$ for some $r \geq 0$, then we can expect to lose roughly r decimal digits of precision in computing an approximate solution.

Summary

We've discovered that the condition number of a matrix A, $\kappa(A)$, is central to the error analysis of the linear system $A\mathbf{x} = \mathbf{b}$. From a numerical analysis standpoint, the order of magnitude of $\kappa(A)$ provides an indicator for the number of significant decimal digits that likely will be lost when computing a solution of $A\mathbf{x} = \mathbf{b}$ using Gaussian elimination. More fundamentally, the condition number measures the sensitivity of the exact solution of $A\mathbf{x} = \mathbf{b}$ to changes in the coefficient matrix and the right-hand side vector. The larger the condition number, the more sensitive the solution.

Recall that polynomials whose roots are sensitive to changes in the coefficients are called ill conditioned. Similarly, a matrix with a large condition number is said to be ill conditioned. But what constitutes a large condition number? The answer depends on the floating point number system being used. Specifically, the order of magnitude of the condition number needs to be compared to the number of significant digits available in the given system. Thus, a condition number of 10^6 would be large in IEEE standard single precision which provides only 7 decimal digits of accuracy. However, in IEEE standard double precision, which provides 16 decimal digits of accuracy, a condition number of 10^6 would not be considered large.

References

1. J. H. Wilkinson, *Rounding Errors in Algebraic Processes*, Prentice Hall, Englewood Cliffs, NJ, 1963.
2. J. H. Wilkinson, *The Algebraic Eigenvalue Problem*, Oxford University Press, Oxford, 1965.
3. K. E. Atkinson, *An Introduction to Numerical Analysis*, John Wiley & Sons, New York, 1978.

EXERCISES

1. Let A and B be $n \times n$ matrices, and let α be a nonzero real number.
   (a) Show that $\kappa(AB) \leq \kappa(A)\kappa(B)$.
   (b) Show that $\kappa(\alpha A) = \kappa(A)$.

2. Let A be an $n \times n$ matrix, and suppose that $A\mathbf{x} = \mathbf{y}$ for some vectors $\mathbf{x}$ and $\mathbf{y}$. Show that

$$\kappa(A) \geq \frac{\|A\|\|\mathbf{x}\|}{\|\mathbf{y}\|}.$$

3. **(a)** Let A be a nonsingular matrix. Show that if

$$\|A - B\| < 1/\|A^{-1}\|,$$

then B is nonsingular. (*Hint*: Write $B = A-(A-B) = A(I-A^{-1}(A-B))$, and focus on the matrix $A^{-1}(A-B)$. You will need to use Exercise 10 from Section 3.3.)
   (b) Let A be a nonsingular matrix and suppose that $\|\delta A\| < 1/\|A^{-1}\|$. Show that $A + \delta A$ is nonsingular.

4. For each of the following floating point number systems, what is roughly the largest condition number for which the solution to the system $A\mathbf{x} = \mathbf{b}$, computed in that number system using Gaussian elimination with pivoting, would be accurate to ten (10) decimal digits? See Section 1.3 for an explanation of the notation.
   (a) IEEE standard double precision, $\mathbf{F}(2, 53, -1021, 1024)$
   (b) Intel extended precision, $\mathbf{F}(2, 64, -16381, 16384)$
   (c) HP double extended precision, $\mathbf{F}(2, 113, -16381, 16384)$
   (d) IBM System/390 long precision, $\mathbf{F}(16, 14, -64, 63)$
   (e) IBM System/390 extended precision, $\mathbf{F}(16, 28, -64, 63)$

5. Suppose the matrix A has a condition number of $\approx 10^5$. If the system of equations $A\mathbf{x} = \mathbf{b}$ is solved using Gaussian elimination with pivoting in each of the following floating point number systems, how many decimal digits of precision can be expected in the approximate solution?
   (a) IEEE standard double precision, $\mathbf{F}(2, 53, -1021, 1024)$
   (b) Intel extended precision, $\mathbf{F}(2, 64, -16381, 16384)$
   (c) HP double extended precision, $\mathbf{F}(2, 113, -16381, 16384)$
   (d) IBM System/390 long precision, $\mathbf{F}(16, 14, -64, 63)$
   (e) IBM System/390 extended precision, $\mathbf{F}(16, 28, -64, 63)$

(f) IEEE standard single precision, $\mathbf{F}(2, 24, -125, 128)$

(g) IBM System/390 short precision, $\mathbf{F}(16, 6, -64, 63)$

6. Repeat Exercise 5 if the matrix A has a condition number of $\approx 10^{12}$.

7. Compute κ_∞ for each of the following matrices.

(a) $A = \begin{bmatrix} 1 & 2 \\ 1.001 & 2 \end{bmatrix}$

(b) $A = \begin{bmatrix} 2.01 & 1.99 \\ 1.99 & 2.01 \end{bmatrix}$

(c) $A = \begin{bmatrix} 1 & -1 & -1 \\ 0 & 1 & -1 \\ 0 & 0 & 1 \end{bmatrix}$

(d) $A = \begin{bmatrix} 1 & \frac{1}{2} & \frac{1}{3} \\ \frac{1}{2} & \frac{1}{3} & \frac{1}{4} \\ \frac{1}{3} & \frac{1}{4} & \frac{1}{5} \end{bmatrix}$

8. In each of the following problems, a linear system $A\mathbf{x} = \mathbf{b}$ is given, along with the exact solution, $\mathbf{x}$, and an approximate solution, $\tilde{\mathbf{x}}$. Compute the error $\mathbf{e} = \tilde{\mathbf{x}} - \mathbf{x}$ and the residual $\mathbf{r} = A\tilde{\mathbf{x}} - \mathbf{b}$ and then compare the relative error to the condition number times the relative residual. Use the l_∞ norm in all cases. Note that the coefficient matrices in these problems are the same matrices from Exercise 7.

(a) $\begin{bmatrix} 1 & 2 \\ 1.001 & 2 \end{bmatrix} \mathbf{x} = \begin{bmatrix} 3 \\ 3.001 \end{bmatrix}$

$\mathbf{x} = \begin{bmatrix} 1 & 1 \end{bmatrix}^T$

$\tilde{\mathbf{x}} = \begin{bmatrix} 3 & 0 \end{bmatrix}^T$

(b) $\begin{bmatrix} 2.01 & 1.99 \\ 1.99 & 2.01 \end{bmatrix} \mathbf{x} = \begin{bmatrix} 4 \\ 4 \end{bmatrix}$

$\mathbf{x} = \begin{bmatrix} 1 & 1 \end{bmatrix}^T$

$\tilde{\mathbf{x}} = \begin{bmatrix} 2 & 0 \end{bmatrix}^T$

(c) $\begin{bmatrix} 1 & -1 & -1 \\ 0 & 1 & -1 \\ 0 & 0 & 1 \end{bmatrix} \mathbf{x} = \begin{bmatrix} 0 \\ 2 \\ 0 \end{bmatrix}$

$\mathbf{x} = \begin{bmatrix} 2 & 2 & 0 \end{bmatrix}^T$

$\tilde{\mathbf{x}} = \begin{bmatrix} 1.9 & 2.1 & -0.1 \end{bmatrix}^T$

(d) $\begin{bmatrix} 1 & \frac{1}{2} & \frac{1}{3} \\ \frac{1}{2} & \frac{1}{3} & \frac{1}{4} \\ \frac{1}{3} & \frac{1}{4} & \frac{1}{5} \end{bmatrix} \mathbf{x} = \begin{bmatrix} 1 \\ \frac{7}{12} \\ \frac{13}{30} \end{bmatrix}$

$\mathbf{x} = \begin{bmatrix} 1 & -2 & 3 \end{bmatrix}^T$

$\tilde{\mathbf{x}} = \begin{bmatrix} 1.02 & -1.96 & 2.94 \end{bmatrix}^T$

9. Let

$$A = \begin{bmatrix} 3 & 1.5 & 1 \\ 1.5 & 1 & 0.75 \\ 1 & 0.75 & 0.6 \end{bmatrix}.$$

(a) Compute $\kappa_\infty(A)$.

(b) Let $\mathbf{b} = \begin{bmatrix} 0.2 & 1 & 1 \end{bmatrix}^T$, and solve the system $A\mathbf{x} = \mathbf{b}$. Now perturb $\mathbf{b}$ by $\delta\mathbf{b} = \begin{bmatrix} 0.01 & -0.01 & 0.01 \end{bmatrix}^T$ and solve the resulting perturbed system. Compare the actual value of $\|\delta\mathbf{x}\|_\infty / \|\mathbf{x}\|_\infty$ with the theoretical upper bound predicted by equation (5).

(c) Repeat part (b), but start with $\mathbf{b} = \begin{bmatrix} 5.5 & 3.25 & 2.35 \end{bmatrix}^T$.

10. Let

$$A = \begin{bmatrix} 25 & 19 \\ 21 & 16 \end{bmatrix}.$$

(a) Compute $\kappa_\infty(A)$.

(b) Let $\mathbf{b} = \begin{bmatrix} 6 & 5 \end{bmatrix}^T$, and solve the system $A\mathbf{x} = \mathbf{b}$. Now perturb $\mathbf{b}$ by $\delta\mathbf{b} = \begin{bmatrix} 0.01 & -0.01 \end{bmatrix}^T$ and solve the resulting perturbed system. Compare the actual value of $\|\delta\mathbf{x}\|_\infty / \|\mathbf{x}\|_\infty$ with the theoretical upper bound predicted by equation (5).

(c) Repeat part (b), but start with $\mathbf{b} = \begin{bmatrix} 1 & 1 \end{bmatrix}^T$.

11. Let

$$A = \begin{bmatrix} 0.25 & 0.35 & 0.15 \\ 0.20 & 0.20 & 0.25 \\ 0.15 & 0.20 & 0.25 \end{bmatrix} \quad \text{and} \quad \mathbf{b} = \begin{bmatrix} 0.60 \\ 0.90 \\ 0.70 \end{bmatrix}.$$

(a) Compute $\kappa_\infty(A)$.

(b) Solve the system $A\mathbf{x} = \mathbf{b}$.

(c) Perturb the coefficient matrix and right-side vector by

$$\delta A = \begin{bmatrix} 0.01 & 0 & 0 \\ 0 & 0 & -0.01 \\ 0 & -0.01 & 0 \end{bmatrix} \quad \text{and} \quad \delta\mathbf{b} = \begin{bmatrix} 0.01 \\ 0.02 \\ -0.03 \end{bmatrix}$$

and solve the resulting perturbed system. Compare the actual value of $\|\delta\mathbf{x}\|_\infty / \|\mathbf{x}\|_\infty$ with the theoretical upper bound predicted by equation (5).

(d) Perturb the original coefficient matrix and right-side vector by

$$\delta A = \begin{bmatrix} 0 & -0.01 & 0.01 \\ -0.01 & 0.01 & 0 \\ 0.01 & 0 & 0.01 \end{bmatrix} \quad \text{and} \quad \delta\mathbf{b} = \begin{bmatrix} 0.02 \\ 0.01 \\ -0.01 \end{bmatrix}$$

and solve the resulting perturbed system. Compare the actual value of $\|\delta\mathbf{x}\|_\infty / \|\mathbf{x}\|_\infty$ with the theoretical upper bound predicted by equation (5).

12. Let

$$A = \begin{bmatrix} 5.1 & 8.7 \\ 2.4 & 4.1 \end{bmatrix} \quad \text{and} \quad \mathbf{b} = \begin{bmatrix} 9.48 \\ 4.48 \end{bmatrix}.$$

(a) Compute $\kappa_\infty(A)$.

(b) Solve the system $A\mathbf{x} = \mathbf{b}$.

(c) Perturb the coefficient matrix and right-side vector by

$$\delta A = \begin{bmatrix} -0.001 & 0 \\ 0.001 & 0 \end{bmatrix} \quad \text{and} \quad \delta\mathbf{b} = \begin{bmatrix} 0.05 \\ -0.05 \end{bmatrix}$$

and solve the resulting perturbed system. Compare the actual value of $\|\delta\mathbf{x}\|_\infty / \|\mathbf{x}\|_\infty$ with the theoretical upper bound predicted by equation (5).

(d) Perturb the original coefficient matrix and right-side vector by

$$\delta A = \begin{bmatrix} 0.001 & -0.001 \\ -0.001 & 0.001 \end{bmatrix} \quad \text{and} \quad \delta\mathbf{b} = \begin{bmatrix} -0.1 \\ 0.1 \end{bmatrix}$$

and solve the resulting perturbed system. Compare the actual value of $\|\delta\mathbf{x}\|_\infty / \|\mathbf{x}\|_\infty$ with the theoretical upper bound predicted by equation (5).

13. Let A be the $n \times n$ matrix whose entries are given by $a_{ij} = 1/(i + j - 1)$ for $1 \le i, j \le n$.

(a) For $n = 5$, solve the system $A\mathbf{x} = \mathbf{b}$ using Gaussian elimination with scaled partial pivoting in single precision arithmetic. Take $\mathbf{b}$ as the vector that corresponds to an exact solution of $x_i = 1$ for each $i = 1, 2, 3, ..., n$. Estimate $\kappa(A)$ based on the results of this experiment.

(b) Repeat part (a) with $n = 11$ and double precision arithmetic.

14. Solve the following system in single precision arithmetic.

$$\begin{array}{rcrcrcr} -149x_1 & - & 50x_2 & - & 154x_3 & = & 353 \\ 537x_1 & + & 180x_2 & + & 546x_3 & = & -1263 \\ -27x_1 & - & 9x_2 & - & 25x_3 & = & 61 \end{array}$$

Use Gaussian elimination with scaled partial pivoting. The exact solution for this problem is $\mathbf{x} = \begin{bmatrix} -1 & -1 & -1 \end{bmatrix}^T$. Estimate the condition number of the coefficient matrix based on the outcome of this experiment.

15. Solve the following system in double precision arithmetic.

$$\left[\begin{array}{rrrrr|r} -9 & 11 & -21 & 63 & -252 & -356 \\ 70 & -69 & 141 & -421 & 1684 & 2385 \\ -575 & 575 & -1149 & 3451 & -13801 & -19551 \\ 3891 & -3891 & 7782 & -23345 & 93365 & 132274 \\ 1024 & -1024 & 2048 & -6144 & 24572 & 34812 \end{array}\right]$$

Use Gaussian elimination with scaled partial pivoting. The exact solution for this problem is $\mathbf{x} = \begin{bmatrix} 1 & -1 & 1 & -1 & 1 \end{bmatrix}^T$. Estimate the condition number of the coefficient matrix based on the outcome of this experiment.

3.5 *LU* DECOMPOSITION

Suppose we need to solve several linear systems, all with the same coefficient matrix, but each with a different right-hand-side vector. If all of the right-hand-side vectors are known from the outset, we can place the coefficient matrix and all of the vectors into a large augmented matrix. Gaussian elimination with back substitution applied to this large augmented matrix would then produce a simultaneous solution to all of the systems.

But what if the right-hand-side vectors are not all known from the outset? For example, the solution vector for one system may be the right-hand-side vector for the next system. Several methods developed later in the text will work in precisely this manner. Although it is the elements in the coefficient matrix which dictate the operations to perform during Gaussian elimination, these operations are also carried out on the right-hand-side vector. As a result, each time we change the right-hand-side vector, exactly the same sequence of operations has to be repeated on the new augmented matrix. That's $O(n^3)$ operations repeated again and again. From an efficiency standpoint, it would be better to have a solution scheme that treats the coefficient matrix and the right-hand-side vector separately, thereby reducing the effort which must be expended when the right-hand-side vector is changed. The objective of this section is to develop such a scheme.

LU Decomposition

Suppose we needed to solve the following cubic equation: $x^3 - 5x^2 + 4x = 0$. We would start by factoring the cubic polynomial into $x(x-4)(x-1)$ and then reducing the original problem into, in this case, three simpler problems: $x = 0$ or $x - 4 = 0$ or $x - 1 = 0$. Given the success of this approach, it is natural to ask whether it is possible to factor a matrix in such a way that the original problem $A\mathbf{x} = \mathbf{b}$ can be reduced to solving simpler problems. The answer to this question is yes. Fortunately, the resulting solution scheme will also provide us with an efficient scheme for handling multiple right-hand sides.

What structure should we seek for the matrix factors in our factorization of the coefficient matrix? In Section 3.1, we saw that a system of equations with an upper triangular coefficient matrix is easily solved using back substitution. In the solution of systems of linear equations, lower triangular matrices are equally as important as upper triangular matrices.

Definition. The matrix L is called LOWER TRIANGULAR if all elements above the main diagonal are zero; that is, if $l_{ij} = 0$ whenever $i < j$.

A system with a coefficient matrix that is lower triangular can also be easily solved. For lower triangular matrices the solution technique is known as forward substitution, which is identical to back substitution except that we work from the top of the matrix to the bottom. Based on these considerations, we will try to factor the original coefficient matrix into the product of a lower triangular matrix and an upper triangular matrix—in that order.

Given a matrix A, a lower triangular matrix L and an upper triangular matrix U for which $LU = A$ are said to form an LU *decomposition* of A. For example, because

$$\begin{bmatrix} 1 & 0 & 0 \\ 2 & 1 & 0 \\ 5 & 12 & 1 \end{bmatrix} \begin{bmatrix} 1 & 4 & 3 \\ 0 & -1 & 3 \\ 0 & 0 & -53 \end{bmatrix} = \begin{bmatrix} 1 & 4 & 3 \\ 2 & 7 & 9 \\ 5 & 8 & -2 \end{bmatrix},$$

the matrices

$$L = \begin{bmatrix} 1 & 0 & 0 \\ 2 & 1 & 0 \\ 5 & 12 & 1 \end{bmatrix} \quad \text{and} \quad U = \begin{bmatrix} 1 & 4 & 3 \\ 0 & -1 & 3 \\ 0 & 0 & -53 \end{bmatrix}$$

form an LU decomposition for the matrix

$$A = \begin{bmatrix} 1 & 4 & 3 \\ 2 & 7 & 9 \\ 5 & 8 & -2 \end{bmatrix}.$$

Not every matrix has an LU decomposition (see Exercises 7, 8, and 9), but it is possible to rearrange the rows of any nonsingular matrix so that the resulting matrix does have an LU decomposition. The algorithm we develop below will automatically perform the needed row interchanges.

When a matrix has an LU decomposition, that decomposition is not unique. In fact, when the matrix A has an LU decomposition, there are an infinite number of different choices available for the matrices L and U. This situation should not be surprising considering that between them, the factor matrices have $n^2 + n$ elements to be determined (each matrix has $(n^2 + n)/2$ nonzero elements), but the matrix A has only n^2 elements. The problem of computing L and U is therefore underdetermined.

Though the LU decomposition process does not uniquely determine L and U, the pairs of matrices that form different decompositions of the same matrix are related. For if $A = L_1U_1 = L_2U_2$, where L_1 and L_2 are lower triangular matrices and U_1 and U_2 are upper triangular matrices, then it follows that

$$L_2^{-1}L_1 = U_2U_1^{-1}.$$

The matrix on the left-hand side of this equation is lower triangular, while the matrix on the right-hand side is upper triangular. For these two matrices to be equal, they must be equal to a diagonal matrix, call it D. Thus the matrices that form different LU decompositions for the same matrix must be related by

$$L_1 = L_2D \quad \text{and} \quad U_2 = DU_1$$

for some diagonal matrix D. Hence, we say an LU decomposition is unique up to scaling by a diagonal matrix.

EXAMPLE 3.14 Multiple *LU* Decompositions

We've already established that the matrix

$$A = \begin{bmatrix} 1 & 4 & 3 \\ 2 & 7 & 9 \\ 5 & 8 & -2 \end{bmatrix}$$

has an LU decomposition that consists of the matrices

$$L_1 = \begin{bmatrix} 1 & 0 & 0 \\ 2 & 1 & 0 \\ 5 & 12 & 1 \end{bmatrix} \quad \text{and} \quad U_1 = \begin{bmatrix} 1 & 4 & 3 \\ 0 & -1 & 3 \\ 0 & 0 & -53 \end{bmatrix}.$$

Another LU decomposition of A consists of the pair

$$L_2 = \begin{bmatrix} 1 & 0 & 0 \\ 2 & -1 & 0 \\ 5 & -12 & -53 \end{bmatrix} \quad \text{and} \quad U_2 = \begin{bmatrix} 1 & 4 & 3 \\ 0 & 1 & -3 \\ 0 & 0 & 1 \end{bmatrix},$$

which can be verified directly by multiplication. In this case,

$$U_1 = DU_2 \quad \text{and} \quad L_2 = L_1 D,$$

where the diagonal matrix

$$D = \begin{bmatrix} 1 & 0 & 0 \\ 0 & -1 & 0 \\ 0 & 0 & -53 \end{bmatrix}.$$

Yet another LU decomposition of A consists of the pair

$$L_3 = \begin{bmatrix} 1/2 & 0 & 0 \\ 1 & -1/3 & 0 \\ 5/2 & -4 & 53 \end{bmatrix} \quad \text{and} \quad U_3 = \begin{bmatrix} 2 & 8 & 6 \\ 0 & 3 & -9 \\ 0 & 0 & -1 \end{bmatrix}.$$

Here, $U_3 = DU_2$ and $L_2 = L_3 D$, where the diagonal matrix

$$D = \begin{bmatrix} 2 & 0 & 0 \\ 0 & 3 & 0 \\ 0 & 0 & -1 \end{bmatrix}.$$

What is the diagonal matrix which relates the pair L_1 and U_1 to the pair L_3 and U_3?

Obtaining an *LU* Decomposition

Different approaches can be taken for computing an LU decomposition. One general approach is known as *direct factorization*. In this approach, we would write out the n^2 equations for the n^2+n entries in L and U implied by the matrix equation $LU = A$. These n^2 equations would then be supplemented with n auxiliary conditions (such as requiring all of the diagonal entries of the matrix U be equal to one) so that the problem will be well defined. Finally, the calculations would be organized so that the system could be solved as efficiently as possible. We will defer a detailed discussion of direct factorization until the next section.

Here, we will focus on modifying Gaussian elimination to produce an LU decomposition. The key to making this modification is recognizing that Gaussian elimination can be represented as a sequence of matrix multiplications. To see how this comes about, consider the following matrix multiplication:

$$\begin{bmatrix} 1 & 0 & 0 & 0 \\ 0 & 1 & 0 & 0 \\ 0 & 2 & 1 & 0 \\ 0 & 3 & 0 & 1 \end{bmatrix} \begin{bmatrix} a_1 & a_2 & a_3 & a_4 \\ b_1 & b_2 & b_3 & b_4 \\ c_1 & c_2 & c_3 & c_4 \\ d_1 & d_2 & d_3 & d_4 \end{bmatrix}$$
$$= \begin{bmatrix} a_1 & a_2 & a_3 & a_4 \\ b_1 & b_2 & b_3 & b_4 \\ c_1+2b_1 & c_2+2b_2 & c_3+2b_3 & c_4+2b_4 \\ d_1+3b_1 & d_2+3b_2 & d_3+3b_3 & d_4+3b_4 \end{bmatrix}.$$

Careful examination of the product matrix reveals that premultiplication by the matrix

$$\begin{bmatrix} 1 & 0 & 0 & 0 \\ 0 & 1 & 0 & 0 \\ 0 & 2 & 1 & 0 \\ 0 & 3 & 0 & 1 \end{bmatrix}$$

has carried out the elementary row operations $r_3 \leftarrow R_3 + 2R_2$ and $r_4 \leftarrow R_4 + 3R_2$. Note that the multiples of row 2 that have been added to rows 3 and 4 are exactly the entries in row 3, column 2 and row 4, column 2 of the premultiplying matrix, respectively. Generalizing this result, it follows that the ith pass of Gaussian elimination is equivalent to premultiplication of the coefficient matrix by the matrix M_i, which is the identity matrix with the zero entries below the diagonal in the ith column replaced by $m_{j,i}$ $(j = i+1, i+2, i+3, \dots, n)$. Recall that $m_{j,i}$ is just the multiple of row i needed to generate a zero in row j of column i. Thus, assuming that no row interchanges were necessary, the entire Gaussian elimination process is given by $M_{n-1}M_{n-2}M_{n-3}\cdots M_3M_2M_1A = U$.

Each of the matrices M_i is nonsingular, so we can solve the matrix representation of Gaussian elimination for the matrix A, yielding $A = M_1^{-1}M_2^{-1}M_3^{-1} \cdots M_{n-1}^{-1}U$. It is straightforward to show that M_i^{-1} is given by the identity matrix with the zero entries below the diagonal in the ith column replaced by $-m_{j,i}$ $(j = i+1, i+2, i+3, \dots, n)$—see Exercise 3. Carrying out the multiplication of the

M_i^{-1}, we find that

$$M_1^{-1}M_2^{-1}M_3^{-1}\cdots M_{n-1}^{-1} = \begin{bmatrix} 1 & & & & & \\ -m_{2,1} & 1 & & & & \\ -m_{3,1} & -m_{3,2} & 1 & & & \\ -m_{4,1} & -m_{4,2} & -m_{4,3} & 1 & & \\ \vdots & \vdots & & \ddots & \ddots & \\ -m_{n,1} & -m_{n,2} & -m_{n,3} & \cdots & -m_{n,n-1} & 1 \end{bmatrix},$$

which is a lower triangular matrix. Note that because of its special structure, we can obtain this matrix with no additional arithmetic.

Hence, the U matrix produced by Gaussian elimination is the upper triangular matrix of an LU decomposition, where the lower triangular matrix has ones along the main diagonal and $-m_{j,i}$ in the locations below the diagonal. Since the elements along the main diagonal of L are always equal to 1, there is no need to store these elements explicitly. We only need to record the $-m_{j,i}$, which can be conveniently and efficiently done by overwriting the elements which are being set to zero. This amounts to changing the line "set $a_{row,pass} = 0$" in the Gaussian elimination algorithm to "set $a_{row,pass} = -m$."

To this point our development has assumed that no row interchanges would be necessary. However, we know that row interchanges are sometimes necessary to avoid a zero pivot and are usually necessary as part of a pivoting strategy to reduce the effects of roundoff error. How do these row interchanges affect the LU decomposition process? Let's do an example and find out.

EXAMPLE 3.15 Determining an *LU* Decomposition

Let's determine an LU decomposition for the matrix

$$A = \begin{bmatrix} 1 & 4 & 3 \\ 2 & 7 & 9 \\ 5 & 8 & -2 \end{bmatrix}$$

using Gaussian elimination with scaled partial pivoting. The scale vector associated with the matrix A is given by

$$\mathbf{s} = \begin{bmatrix} 4 & 9 & 8 \end{bmatrix}^T,$$

and we initialize the row vector to

$$\mathbf{r} = \begin{bmatrix} 1 & 2 & 3 \end{bmatrix}^T.$$

Examining the ratios

$$\frac{|a_{r_1,1}|}{s_{r_1}} = \frac{1}{4}, \quad \frac{|a_{r_2,1}|}{s_{r_2}} = \frac{2}{9}, \quad \text{and} \quad \frac{|a_{r_3,1}|}{s_{r_3}} = \frac{5}{8},$$

we find the largest value corresponds to row r_3, so we need to swap the first and third elements in the row vector. Thus, for the first elimination pass, we have

$$\mathbf{r} = \begin{bmatrix} 3 & 2 & 1 \end{bmatrix}^T.$$

Following this elimination pass, the contents of the matrix are

$$\begin{bmatrix} (1/5) & 12/5 & 17/5 \\ (2/5) & 19/5 & 49/5 \\ 5 & 8 & -2 \end{bmatrix}.$$

Note how the opposite of each multiplier overwrites the element which is being set to zero. To distinguish the multipliers from the other elements in the matrix, the multipliers are displayed within parentheses.

To determine the location of the next pivot, we examine the ratios

$$\frac{|a_{r_2,2}|}{s_{r_2}} = \frac{19/5}{9} = \frac{19}{45}, \quad \text{and} \quad \frac{|a_{r_3,2}|}{s_{r_3}} = \frac{12/5}{4} = \frac{3}{5}.$$

The largest of these corresponds to row r_3, so we swap the second and third elements in the row vector, which becomes

$$\mathbf{r} = \begin{bmatrix} 3 & 1 & 2 \end{bmatrix}^T.$$

Following the second, and final, elimination pass, the contents of the matrix are

$$\begin{bmatrix} (1/5) & 12/5 & 17/5 \\ (2/5) & (19/12) & 265/60 \\ 5 & 8 & -2 \end{bmatrix}.$$

To identify the L and U matrices in the decomposition, we first need to read the rows of the final matrix in the order indicated by the row vector; that is, start with the third row, then the first row, and finally the second row. The upper triangular matrix in the decomposition is then obtained by setting the elements below the main diagonal to zero. The lower triangular matrix is obtained by setting the elements along the main diagonal to 1 and the elements above the diagonal to zero. Therefore, we find

$$L = \begin{bmatrix} 1 & 0 & 0 \\ 1/5 & 1 & 0 \\ 2/5 & 19/12 & 1 \end{bmatrix} \quad \text{and} \quad U = \begin{bmatrix} 5 & 8 & -2 \\ 0 & 12/5 & 17/5 \\ 0 & 0 & 265/60 \end{bmatrix}.$$

Note that if we multiply the matrices L and U, we obtain

$$LU = \begin{bmatrix} 5 & 8 & -2 \\ 1 & 4 & 3 \\ 2 & 7 & 9 \end{bmatrix},$$

which is not equal to the matrix A. The rows of LU are the rows of A, but listed in a different order. Observe, in particular, that the rows of LU are the rows of A listed in the order indicated by the final row vector.

To clarify the outcome of this last example, let

$$P = \begin{bmatrix} 0 & 0 & 1 \\ 1 & 0 & 0 \\ 0 & 1 & 0 \end{bmatrix}.$$

This matrix was obtained by taking the 3×3 identity matrix and reordering the rows according to the contents of the row vector $\mathbf{r} = \begin{bmatrix} 3 & 1 & 2 \end{bmatrix}^T$. If we now multiply P into the matrix A, we obtain

$$PA = \begin{bmatrix} 5 & 8 & -2 \\ 1 & 4 & 3 \\ 2 & 7 & 9 \end{bmatrix},$$

which is equal to the product LU calculated above. Hence, with row interchanges, we have found an LU decomposition for the matrix PA.

A matrix such as P, which is an identity matrix with its rows reordered, is called a *permutation matrix.* Thus, when row interchanges are used, the LU decomposition we calculate will not be for the original matrix A, but will be for the matrix PA, for some permutation matrix P. The specific permutation matrix can be found by reordering the rows of the $n \times n$ identity matrix according to the final contents of the row vector $\mathbf{r}$.

Solving a Linear System using an *LU* Decomposition

Suppose we need to solve the linear system $A\mathbf{x} = \mathbf{b}$, and we have already found a lower triangular matrix L and an upper triangular matrix U such that $LU = PA$ for some permutation matrix P. If we multiply the linear system by P and then substitute LU for PA, we find that solving the original system is equivalent to solving $LU\mathbf{x} = P\mathbf{b}$, or $L(U\mathbf{x}) = P\mathbf{b}$. Now, let $\mathbf{z} = U\mathbf{x}$. This transforms what had been one problem, solve $A\mathbf{x} = \mathbf{b}$ for $\mathbf{x}$, into a sequence of two problems: Solve $L\mathbf{z} = P\mathbf{b}$ for $\mathbf{z}$ and then solve $U\mathbf{x} = \mathbf{z}$ for $\mathbf{x}$. These two subproblems, however, are easy to solve as a result of the structure we imposed on the matrices L and U. Forward substitution applied to $L\mathbf{z} = P\mathbf{b}$ produces the vector $\mathbf{z}$, and then back substitution applied to $U\mathbf{x} = \mathbf{z}$ gives the solution vector $\mathbf{x}$.

It is important to note that to carry out this solution process, we don't need to explicitly construct the permutation matrix P and form the matrix-vector product $P\mathbf{b}$. The matrix P is completely determined by the final contents of the row vector $\mathbf{r}$ generated during the LU decomposition process. Furthermore, multiplication by P merely rearranges the rows of $\mathbf{b}$. Therefore, to carry out the solution process, we simply need to know the row vector $\mathbf{r}$ and access the rows of $\mathbf{b}$ through the row vector.

EXAMPLE 3.16 Demonstration of Solution Process Based on *LU* Decomposition

Consider the linear system

$$\begin{bmatrix} 1 & 4 & 3 \\ 2 & 7 & 9 \\ 5 & 8 & -2 \end{bmatrix} \mathbf{x} = \begin{bmatrix} -4 \\ -10 \\ 9 \end{bmatrix}.$$

Above, the LU decomposition process applied to the coefficient matrix for this system produced the matrix

$$\begin{bmatrix} (1/5) & 12/5 & 17/5 \\ (2/5) & (19/12) & 265/60 \\ 5 & 8 & -2 \end{bmatrix}$$

and the row vector

$$\mathbf{r} = \begin{bmatrix} 3 & 1 & 2 \end{bmatrix}^T.$$

Performing forward substitution, the intermediate vector $\mathbf{z}$ is computed as follows:

$$z_1 = b_{r_1} = b_3 = 9;$$

$$z_2 = b_{r_2} - l_{r_2,1}z_1 = b_1 - l_{1,1}z_1 = -4 - \frac{1}{5}(9) = -\frac{29}{5};$$

and

$$\begin{aligned} z_3 &= b_{r_3} - l_{r_3,1}z_1 - l_{r_3,2}z_2 = b_2 - l_{2,1}z_1 - l_{2,2}z_2 \\ &= -10 - \frac{2}{5}(9) - \frac{19}{12}\left(-\frac{29}{5}\right) = -\frac{265}{60}. \end{aligned}$$

The notation $l_{i,j}$ refers to the elements in the lower triangular matrix of the LU decomposition, as stored in the matrix shown above.

Back substitution now determines the solution to the original system of equations:

$$x_3 = \frac{z_3}{u_{r_3,3}} = \frac{z_3}{u_{2,3}} = \frac{-265/60}{265/60} = -1;$$

$$x_2 = \frac{z_2 - u_{r_2,3}x_3}{u_{r_2,2}} = \frac{z_2 - u_{1,3}x_3}{u_{1,2}} = \frac{-29/5 - (17/5)(-1)}{12/5} = -1;$$

and

$$\begin{aligned} x_1 &= \frac{z_1 - u_{r_1,2}x_2 - u_{r_1,3}x_3}{u_{r_1,1}} = \frac{z_1 - u_{3,2}x_2 - u_{3,3}x_3}{u_{3,1}} \\ &= \frac{9 - 8(-1) - (-2)(-1)}{5} = 3. \end{aligned}$$

Hence, $\mathbf{x} = \begin{bmatrix} 3 & -1 & -1 \end{bmatrix}^T$.

Summary and Comparison

In this section, we've developed a two-step algorithm for solving a linear system of equations. The first step, known as the factor step, determines an LU decomposition for the coefficient matrix of the system. The coefficient matrix is the only input, and the output consists of the LU decomposition overwritten onto the original contents of the coefficient matrix and a row vector which indicates the final ordering of the rows. The second step of the algorithm, known as the solve step, takes the LU decomposition, the row vector, and the right-hand-side vector as inputs and then performs forward substitution followed by back substitution to produce a solution vector.

For a linear system with n equations in n unknowns, the factor step has a computational cost of $\frac{2}{3}n^3 - \frac{1}{2}n^2 - \frac{1}{6}n$ arithmetic operations. This is slightly lower than the cost of Gaussian elimination because we've moved the processing of the right-hand-side vector to the solve step. The solve step then requires $2n^2 - n$ arithmetic operations, which is slightly higher than the cost of back substitution. Our two step algorithm therefore has a total cost of $\frac{2}{3}n^3+\frac{3}{2}n^2-\frac{7}{6}n$ operations, which is identical to the cost of solving a single linear system using Gaussian elimination with back substitution.

What if we have multiple systems, all of which have the same coefficient matrix? Specifically, suppose we need to solve m systems of n equations in n unknowns. Using the factor and solve algorithm, the factor step would be performed once, at a cost of $\frac{2}{3}n^3-\frac{1}{2}n^2-\frac{1}{6}n$ operations. The solve step would then be repeated for each of the m right-hand-side vectors, at a cost of $2mn^2-mn$ operations. Thus, the total cost for solving all m systems is $\frac{2}{3}n^3+\left(2m-\frac{1}{2}\right)n^2-\left(m+\frac{1}{6}\right)n$ arithmetic operations.

If all m right-hand-side vectors are known from the outset, we can construct an $n \times (n+m)$ augmented matrix and perform simultaneous Gaussian elimination with back substitution. The computational cost of this algorithm is identical to that of the factor and solve algorithm (see Exercise 2). However, if the right-hand-side vectors are not all known from the outset, then performing Gaussian elimination with back substitution sequentially on all m systems incurs a cost of $\frac{2}{3}mn^3+\frac{3}{2}mn^2-\frac{7}{6}mn$, which is substantially higher than the cost of the factor and solve algorithm. Thus, in this case, the two-step algorithm is superior to Gaussian elimination with back subsititution.

Continuing to consider the case of multiple linear systems, all with the same coefficient matrix, what about computing A^{-1} followed by m matrix-vector multiplications as a solution algorithm? Each multiplication $A^{-1}\mathbf{b}$ has a cost of $2n^2-n$ operations, which is identical to the cost of the solve step. However, computing A^{-1} is nearly three times more expensive than determining an LU decomposition. We can therefore conclude that when we have multiple systems, all with the same coefficient matrix, and the right-hand-side vectors are not all known in advance, the factor and solve algorithm is the most efficient solution scheme.

An Application Problem: The Inverse Power Method

In Chapter 4, we will study techniques for approximating the eigenvalues and eigenvectors of an arbitrary matrix. One of the techniques we will study is called the inverse power method. This is an iterative technique that requires an initial estimate for an eigenvalue, λ_0, and a nonzero vector, $\mathbf{x}^{(0)}$, as input. In each iteration, the following calculations are made:

$$\begin{aligned} \mathbf{x}^{(m)} &= (A - \lambda_0 I)^{-1}\mathbf{x}^{(m-1)} \\ \lambda_m &= x^{(m)}_{p_{m-1}} \\ \mathbf{x}^{(m)} &= \mathbf{x}^{(m)}/x^{(m)}_{p_m}. \end{aligned}$$

The quantity $\lambda_0+(1/\lambda_m)$ converges toward the eigenvalue of A that is closest to λ_0, and $\mathbf{x}^{(m)}$ converges toward a corresponding eigenvector. The integer p_m is chosen so that $\left|x^{(m)}_{p_m}\right| = \|\mathbf{x}^{(m)}\|_\infty$. In implementing this algorithm, we will not compute $(A - \lambda_0 I)^{-1}$; it is more efficient to solve the system $(A - \lambda_0 I)\mathbf{x}^{(m)} = \mathbf{x}^{(m-1)}$ for $\mathbf{x}^{(m)}$. Since the matrix $A - \lambda_0 I$ does not change from iteration to iteration, we can perform an LU decomposition once and use it in each iteration.

As an illustration, let

$$A = \begin{bmatrix} 1 & -1 & 0 \\ -2 & 4 & -2 \\ 0 & -1 & 2 \end{bmatrix}.$$

This matrix has an eigenvalue near 5. (To ten decimal places, the eigenvalue is 5.1248854198.) Take $\lambda_0 = 5$, so that

$$A - \lambda_0 I = \begin{bmatrix} -4 & -1 & 0 \\ -2 & -1 & -2 \\ 0 & -1 & -3 \end{bmatrix},$$

and let $\mathbf{x}^{(0)} = \begin{bmatrix} 1 & -4 & 1 \end{bmatrix}^T$. With this vector, note that $p_0 = 2$. An LU decomposition for $A - \lambda_0 I$ is

$$L = \begin{bmatrix} 1 & 0 & 0 \\ 1/2 & 1 & 0 \\ 0 & 2 & 1 \end{bmatrix} \quad \text{and} \quad U = \begin{bmatrix} -4 & -1 & 0 \\ 0 & -1/2 & -2 \\ 0 & 0 & 1 \end{bmatrix}.$$

For the first iteration of the inverse power method, forward substitution applied to $L\mathbf{z} = \mathbf{x}^{(0)}$ gives $\mathbf{z} = \begin{bmatrix} 1 & -9/2 & 10 \end{bmatrix}^T$. Back substitution on $U\mathbf{x}^{(1)} = \mathbf{z}$ then gives $\mathbf{x}^{(1)} = \begin{bmatrix} 7.5 & -31 & 10 \end{bmatrix}^T$. With $p_0 = 2$, we find $\lambda_1 = x_2^{(1)} = -31$ or $\lambda_0 + 1/\lambda_1 \approx 4.9677$. Finally, since $p_1 = 2$, we set

$$\mathbf{x}^{(1)} = \frac{\mathbf{x}^{(1)}}{x_2^{(1)}} = \begin{bmatrix} -15/62 & 1 & -10/31 \end{bmatrix}^T$$

to prepare for the next iteration.

Forward substitution applied to $L\mathbf{z} = \mathbf{x}^{(1)}$ gives

$$\mathbf{z} = (1/124)\begin{bmatrix} -30 & 139 & -318 \end{bmatrix}^T,$$

which leads to $\mathbf{x}^{(2)} = (1/124)\begin{bmatrix} -241 & 994 & -318 \end{bmatrix}^T$ when back substitution is applied to $U\mathbf{x}^{(2)} = \mathbf{z}$. It follows that $\lambda_2 = x_2^{(2)} = 994/124$ and $\lambda_0 + 1/\lambda_2 \approx 5.1247$, which is already correct to three decimal places. The approximate eigenvector is

$$\mathbf{x}^{(2)} = \frac{\mathbf{x}^{(2)}}{x_2^{(2)}} = \begin{bmatrix} -241/994 & 1 & -318/994 \end{bmatrix}^T.$$

We leave it as an exercise to perform the next couple of iterations.

EXERCISES

1. (a) Show that the algorithm to obtain an LU decomposition based on Gaussian elimination requires $\frac{2}{3}n^3 - \frac{1}{2}n^2 - \frac{1}{6}n$ arithmetic operations.
   (b) Show that the solve step—forward substitution followed by backward substitution—requires $2n^2 - n$ arithmetic operations.
   (c) Suppose A^{-1} has been calculated. Show that the multiplication $A^{-1}\mathbf{b}$ requires $2n^2 - n$ arithmetic operations.
2. Let A be an $n \times n$ matrix, and suppose that we need to solve m linear systems $A\mathbf{x} = \mathbf{b}_i$ for $i = 1, 2, 3, \ldots, m$. Consider constructing an $n \times (n+m)$ augmented matrix that contains all of the right-hand-side vectors and performing Gaussian elimination with back substitution on this matrix. Show that this algorithm requires $\frac{2}{3}n^3 + \left(2m - \frac{1}{2}\right)n^2 - \left(m + \frac{1}{6}\right)n$ arithmetic operations.
3. Show that

$$\begin{bmatrix} 1 & 0 & 0 & 0 & 0 \\ 0 & 1 & 0 & 0 & 0 \\ 0 & m_{3,2} & 1 & 0 & 0 \\ 0 & m_{4,2} & 0 & 1 & 0 \\ 0 & m_{5,2} & 0 & 0 & 1 \end{bmatrix}^{-1} = \begin{bmatrix} 1 & 0 & 0 & 0 & 0 \\ 0 & 1 & 0 & 0 & 0 \\ 0 & -m_{3,2} & 1 & 0 & 0 \\ 0 & -m_{4,2} & 0 & 1 & 0 \\ 0 & -m_{5,2} & 0 & 0 & 1 \end{bmatrix}.$$

4. Let

$$A = \begin{bmatrix} 1 & 2 \\ 3 & 4 \end{bmatrix}.$$

Verify that each of the following pairs forms an LU decomposition of A, and then use the decomposition to solve the system $A\mathbf{x} = \begin{bmatrix} 4 & 6 \end{bmatrix}^T$.

(a) $L_1 = \begin{bmatrix} 1 & 0 \\ 3 & 1 \end{bmatrix}$, $U_1 = \begin{bmatrix} 1 & 2 \\ 0 & -2 \end{bmatrix}$

(b) $L_2 = \begin{bmatrix} 1 & 0 \\ 3 & -2 \end{bmatrix}$, $U_2 = \begin{bmatrix} 1 & 2 \\ 0 & 1 \end{bmatrix}$

(c) $L_3 = \begin{bmatrix} -1 & 0 \\ -3 & -2 \end{bmatrix}$, $U_3 = \begin{bmatrix} -1 & -2 \\ 0 & 1 \end{bmatrix}$

5. Let

$$A = \begin{bmatrix} 2 & 7 & 5 \\ 6 & 20 & 10 \\ 4 & 3 & 0 \end{bmatrix}.$$

Verify that each of the following pairs forms an LU decomposition of A, and then use the decomposition to solve the system $A\mathbf{x} = \begin{bmatrix} 0 & 4 & 1 \end{bmatrix}^T$.

(a) $L_1 = \begin{bmatrix} 1 & 0 & 0 \\ 3 & 1 & 0 \\ 2 & 11 & 1 \end{bmatrix}, \quad U_1 = \begin{bmatrix} 2 & 7 & 5 \\ 0 & -1 & -5 \\ 0 & 0 & 45 \end{bmatrix}$

(b) $L_2 = \begin{bmatrix} 1 & 0 & 0 \\ 3 & -1 & 0 \\ 2 & -11 & 45 \end{bmatrix}, \quad U_2 = \begin{bmatrix} 2 & 7 & 5 \\ 0 & 1 & 5 \\ 0 & 0 & 1 \end{bmatrix}$

(c) $L_3 = \begin{bmatrix} 2 & 0 & 0 \\ 6 & -1 & 0 \\ 4 & -11 & 45 \end{bmatrix}, \quad U_3 = \begin{bmatrix} 1 & 7/2 & 5/2 \\ 0 & 1 & 5 \\ 0 & 0 & 1 \end{bmatrix}$

6. Let

$$A = \begin{bmatrix} 1 & 3 & 1 & -2 \\ 2 & 4 & -1 & 2 \\ 3 & 1 & 1 & 5 \\ 4 & 2 & -1 & 6 \end{bmatrix}.$$

Verify that each of the following pairs forms an LU decomposition of A, and then use the decomposition to solve the system $A\mathbf{x} = \begin{bmatrix} 3 & 7 & 10 & 11 \end{bmatrix}^T$.

(a) $L_1 = \begin{bmatrix} 1 & 0 & 0 & 0 \\ 2 & 1 & 0 & 0 \\ 3 & 4 & 1 & 0 \\ 4 & 5 & 1 & 1 \end{bmatrix}, \quad U_1 = \begin{bmatrix} 1 & 3 & 1 & -2 \\ 0 & -2 & -3 & 6 \\ 0 & 0 & 10 & -13 \\ 0 & 0 & 0 & -3 \end{bmatrix}$

(b) $L_1 = \begin{bmatrix} 1 & 0 & 0 & 0 \\ 2 & -1 & 0 & 0 \\ 3 & -4 & 10 & 0 \\ 4 & -5 & 10 & -3 \end{bmatrix}, \quad U_1 = \begin{bmatrix} 1 & 3 & 1 & -2 \\ 0 & 2 & 3 & -6 \\ 0 & 0 & 1 & -13/10 \\ 0 & 0 & 0 & 1 \end{bmatrix}$

7. (a) Show that the matrix

$$\begin{bmatrix} 0 & 1 \\ 1 & 1 \end{bmatrix}$$

has no LU decomposition. (*Hint*: Write out the equations corresponding to

$$\begin{bmatrix} l_{11} & 0 \\ l_{21} & l_{22} \end{bmatrix} \begin{bmatrix} u_{11} & u_{12} \\ 0 & l_{22} \end{bmatrix} = \begin{bmatrix} 0 & 1 \\ 1 & 1 \end{bmatrix}$$

and show that the resulting system is inconsistent.)

(b) Reverse the order of the rows of A and show that the resulting matrix does have an LU decomposition.

8. (a) Show that the matrix

$$\begin{bmatrix} 0 & 0 & 1 \\ 0 & 1 & 0 \\ 1 & 0 & 1 \end{bmatrix}$$

has no LU decomposition.

(b) Rearrange the rows of A so that the resulting matrix does have an LU decomposition.

9. Repeat Exercise 8 for the matrix

$$A = \begin{bmatrix} 1 & 1 & 1 \\ 1 & 1 & 2 \\ -1 & 0 & 2 \end{bmatrix}$$

10. Consider the matrix

$$A = \begin{bmatrix} 1 & 1 & 2 \\ -1 & 0 & 2 \\ 3 & 2 & -1 \end{bmatrix}.$$

(a) Find a lower triangular matrix L with ones along its diagonal and an upper triangular matrix U such that $A = LU$.

(b) Find matrices L, D and U such that $A = LDU$, where L is a lower triangular matrix with ones along its diagonal, D is a diagonal matrix, and U is an upper triangular matrix with ones along its diagonal.

(c) Find a lower triangular matrix L and an upper triangular matrix U with ones along its diagonal such that $A = LU$.

For Exercises 11–15,

(a) Using scaled partial pivoting during the factor step, find matrices L, U and P such that $LU = PA$.

(b) Solve the system $A\mathbf{x} = \mathbf{b}$ for each of the given right-hand-side vectors.

11. $A = \begin{bmatrix} 1 & 2 & 3 & 4 \\ -1 & 1 & 2 & 3 \\ 1 & -1 & 1 & 2 \\ -1 & 1 & -1 & 5 \end{bmatrix}$ $\mathbf{b}_1 = \begin{bmatrix} 10 \\ 5 \\ 3 \\ 4 \end{bmatrix}$, $\mathbf{b}_2 = \begin{bmatrix} -4 \\ -5 \\ -3 \\ -4 \end{bmatrix}$, $\mathbf{b}_3 = \begin{bmatrix} -2 \\ -3 \\ 1 \\ -8 \end{bmatrix}$

12. $A = \begin{bmatrix} 1 & 0 & 2 & 0 \\ -1 & 4 & 3 & 6 \\ 0 & -2 & 5 & -3 \\ 3 & 1 & 1 & 0 \end{bmatrix}$ $\mathbf{b}_1 = \begin{bmatrix} 3 \\ 12 \\ 0 \\ 5 \end{bmatrix}$, $\mathbf{b}_2 = \begin{bmatrix} -1 \\ -6 \\ -4 \\ 3 \end{bmatrix}$, $\mathbf{b}_3 = \begin{bmatrix} 3 \\ -8 \\ 10 \\ 2 \end{bmatrix}$

13. $A = \begin{bmatrix} 1 & 3 & 1 & -2 \\ 2 & 4 & -1 & 2 \\ 3 & 1 & 1 & 5 \\ 4 & 2 & 6 & -1 \end{bmatrix}$ $\mathbf{b}_1 = \begin{bmatrix} 1 \\ -5 \\ -2 \\ 9 \end{bmatrix}$, $\mathbf{b}_2 = \begin{bmatrix} -5 \\ -3 \\ 6 \\ -5 \end{bmatrix}$, $\mathbf{b}_3 = \begin{bmatrix} 5 \\ 5 \\ -2 \\ 1 \end{bmatrix}$

14. $A = \begin{bmatrix} 2 & 7 & 5 \\ 6 & 20 & 10 \\ 4 & 3 & 0 \end{bmatrix}$ $\mathbf{b}_1 = \begin{bmatrix} 14 \\ 36 \\ 7 \end{bmatrix}$, $\mathbf{b}_2 = \begin{bmatrix} -4 \\ -16 \\ -7 \end{bmatrix}$, $\mathbf{b}_3 = \begin{bmatrix} -3 \\ -12 \\ 6 \end{bmatrix}$

15. $A = \begin{bmatrix} 13 & 39 & 2 & 57 & 28 \\ -4 & -12 & 0 & -19 & -9 \\ 3 & 0 & -9 & 2 & 1 \\ 6 & 17 & 9 & 5 & 7 \\ 19 & 42 & -17 & 107 & 44 \end{bmatrix}$ $\mathbf{b}_1 = \begin{bmatrix} -53 \\ 18 \\ -7 \\ 0 \\ -103 \end{bmatrix}$, $\mathbf{b}_2 = \begin{bmatrix} 57 \\ -18 \\ -11 \\ 18 \\ 69 \end{bmatrix}$,

$$\mathbf{b}_3 = \begin{bmatrix} -145 \\ 49 \\ -27 \\ -4 \\ -286 \end{bmatrix}$$

16. In the text, the Inverse Power Method, a technique for approximating the eigenvalues and eignevectors for an arbitrary matrix, was described. Given an initial estimate for the eigenvalue, λ_0, and a nonzero vector, $\mathbf{x}^{(0)}$, the following sequence of calculations are iterated:

$$\mathbf{x}^{(m)} = (A - \lambda_0 I)^{-1}\mathbf{x}^{(m-1)}$$
$$\lambda_m = x^{(m)}_{p_{m-1}}$$
$$\mathbf{x}^{(m)} = \mathbf{x}^{(m)}/x^{(m)}_{p_m}.$$

The quantity $\lambda_0 + (1/\lambda_m)$ converges toward the eigenvalue of A that is closest to λ_0, and $\mathbf{x}^{(m)}$ converges toward a corresponding eigenvector. The integer p_m is is chosen so that $\left|x^{(m)}_{p_m}\right| = \|\mathbf{x}^{(m)}\|_\infty$.

For the remainder of this exercise, let

$$A = \begin{bmatrix} 1 & -1 & 0 \\ -2 & 4 & -2 \\ 0 & -1 & 2 \end{bmatrix}.$$

(a) For $\lambda_0 = 5$ and $\mathbf{x}^{(0)} = \begin{bmatrix} 1 & -4 & 1 \end{bmatrix}^T$, we found that $\lambda_2 = 994/124$ and $\mathbf{x}^{(2)} = \begin{bmatrix} -241/994 & 1 & -318/994 \end{bmatrix}^T$. Perform the next two iterations. How does the value $\lambda_0 + (1/\lambda_4)$ compare to the true eigenvalue 5.1248854198?

(b) For $\lambda_0 = 2$ and $\mathbf{x}^{(0)} = \begin{bmatrix} 1 & -1 & 2 \end{bmatrix}^T$, perform the first four iterations of the inverse power method. How does the value $\lambda_0 + (1/\lambda_4)$ compare to the true eigenvalue 1.6366717621?

17. Determine the member and reaction forces within the plane truss shown in Figure 3.5 when the truss is subjected to each of the following loading configurations.

(a) 500-pound forces directed vertically downward at nodes #3 and #5, and a 1000-pound force directed vertically downward at node #4.

(b) 500-pound force acting at node #3, a 1000-pound force acting at node #4, and a 1500-pound force acting at node #5, all forces acting vertically downward.

(c) 1500-pound force acting at node #3, a 1000-pound force acting at node #4, and a 500-pound force acting at node #5, all forces acting vertically downward.

(d) 500-pound force acting at node #4, and a 1000-pound force acting at node #3, both forces acting horizontally to the right.

(e) 500-pound force acting at node #4 and a 1000-pound force acting at node #5, both forces acting horizontally to the left.

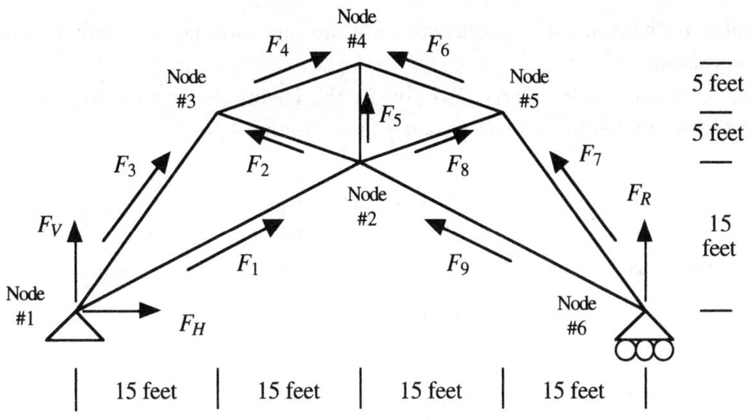

Figure 3.5 Figure for Exercise 17.

3.6 DIRECT FACTORIZATION

In the previous section we saw how Gaussian elimination could be used for determining the LU decomposition of a nonsingular matrix. Direct factorization is an alternative procedure for obtaining an LU decomposition. Why do we need another technique for calculating an LU decomposition? On the one hand, there are matrices which have special structure to them. Direct factorization will make it possible to construct schemes that take advantage of that structure. On the other hand, the formulas associated with direct factorization will allow us, on some computers, to take advantage of architecture to improve both speed and accuracy.

Direct Factorization

Given a matrix A, recall that the objective of an LU decomposition is to determine a lower triangular matrix L and an upper triangular matrix U such that $LU = A$. This matrix equation is a shorthand for a system of n^2 equations, assuming that A is an $n \times n$ matrix, for the $n^2 + n$ nonzero entries in the matrices L and U. To produce a well-posed problem, we must specify n additional equations.

Recall that the factors in an LU decomposition are determined only up to a scaling by a diagonal matrix. Therefore, different factorizations may be viewed as resulting from different choices for the diagonal elements of either L or U. The two most common choices for the diagonal entries are

$$\begin{aligned} l_{ii} &= 1 \text{ for each } i = 1, 2, 3, \ldots, n; \text{ and} \\ u_{ii} &= 1 \text{ for each } i = 1, 2, 3, \ldots, n, \end{aligned}$$

which give rise to what are known as the Doolittle decomposition and the Crout decomposition, respectively. When implementing a direct factorization algorithm, the most important issue is the efficient organization of the calculations. To demonstrate the proper sequence of calculations, let's focus on the Crout decomposition.

The Doolittle decomposition proceeds in a similar manner and will be considered in the exercises.

Let A be an $n \times n$ matrix. To obtain the Crout decomposition of A we must determine the entries l_{ij} $(i \geq j)$ and u_{ij} $(i < j)$ such that

$$\begin{bmatrix} l_{11} & & & & \\ l_{21} & l_{22} & & & \\ l_{31} & l_{32} & l_{33} & & \\ \vdots & & & \ddots & \\ l_{n1} & l_{n2} & l_{n3} & \cdots & l_{nn} \end{bmatrix} \begin{bmatrix} 1 & u_{12} & u_{13} & \cdots & u_{1n} \\ & 1 & u_{23} & \cdots & u_{2n} \\ & & 1 & & u_{3n} \\ & & & \ddots & \vdots \\ & & & & 1 \end{bmatrix}$$

$$= \begin{bmatrix} a_{11} & a_{12} & a_{13} & \cdots & a_{1n} \\ a_{21} & a_{22} & a_{23} & \cdots & a_{2n} \\ a_{31} & a_{32} & a_{33} & \cdots & a_{3n} \\ & & \vdots & & \vdots \\ a_{n1} & a_{n2} & a_{n3} & \cdots & a_{nn} \end{bmatrix}.$$

As with Gaussian elimination, we will organize our calculations into passes. In total, an $n \times n$ matrix will require n passes.

For the first pass, note that the first column of U contains a single nonzero entry, a 1 in the first row. Therefore, the product of the ith row of L (for $i = 1, 2, 3, \ldots, n$) with the first column of U is simply the element l_{i1}. The decomposition equation requires that this value be equated to a_{i1}; that is,

$$l_{i1} = a_{i1}.$$

These equations determine the first column of L. Now that the l_{11} entry is known, multiplying the first row of L with the jth column of U (for $j = 2, 3, 4, \ldots, n$) and equating the result to a_{1j} produces the equation $l_{11}u_{1j} = a_{1j}$. Dividing by l_{11}, we find

$$u_{1j} = a_{1j}/l_{11},$$

thus determining the first row of U.

Each subsequent pass of the algorithm computes one more column of L and one more row of U. In particular, the kth column of L and the kth row of U are determined during the kth pass. To compute the elements l_{ik} (for $i = k, k+1, k+2, \ldots, n$), form the product of the ith row of L with the kth column of U, and equate that to a_{ik}. The resulting equation is

$$l_{ik} + \sum_{j=1}^{k-1} l_{ij}u_{jk} = a_{ik},$$

whose solution for l_{ik} is

$$l_{ik} = a_{ik} - \sum_{j=1}^{k-1} l_{ij}u_{jk}. \tag{1}$$

With l_{kk} now known, the elements u_{kj} (for $j = k+1, k+2, k+3, \ldots, n$) are found by equating the product of the kth row of L and the j-th column of U to the element a_{kj}. The formula for u_{kj} is

$$u_{kj} = \frac{1}{l_{kk}}\left(a_{kj} - \sum_{i=1}^{k-1} l_{ki}u_{ij}\right). \tag{2}$$

Note that for the final pass, only l_{nn} is left to be determined, so the formula for u_{kj} does not need to be applied.

EXAMPLE 3.17 Crout Decomposition in Action

Consider the 3×3 matrix

$$A = \begin{bmatrix} 1 & 4 & 3 \\ 2 & 7 & 9 \\ 5 & 8 & -2 \end{bmatrix}.$$

The Crout decomposition of this matrix will consist of the matrices

$$L = \begin{bmatrix} l_{11} & 0 & 0 \\ l_{21} & l_{22} & 0 \\ l_{31} & l_{32} & l_{33} \end{bmatrix} \quad \text{and} \quad U = \begin{bmatrix} 1 & u_{12} & u_{13} \\ 0 & 1 & u_{23} \\ 0 & 0 & 1 \end{bmatrix}.$$

Since A is a 3×3 matrix, it will take three passes to compute all of the entries in L and U. Following the general description provided above, in each pass, we will compute the elements in one column of L and in one row of U.

Forming the product of each row of L with the first column of U and equating the result with the corresponding element from A determines the elements in the first column of L:

$$l_{11} = 1 \qquad l_{21} = 2 \qquad \text{and} \qquad l_{31} = 5.$$

The first row of U is obtained by multiplying the first row of L with the second and third columns of U and then equating the result with the corresponding element from A. This yields the equations

$$l_{11}u_{12} = 4 \qquad \text{and} \qquad l_{11}u_{13} = 3,$$

whose solutions are

$$u_{12} = 4 \qquad \text{and} \qquad u_{13} = 3.$$

Thus, after the first pass, the Crout decomposition looks like

$$L = \begin{bmatrix} 1 & 0 & 0 \\ 2 & l_{22} & 0 \\ 5 & l_{32} & l_{33} \end{bmatrix} \quad \text{and} \quad U = \begin{bmatrix} 1 & 4 & 3 \\ 0 & 1 & u_{23} \\ 0 & 0 & 1 \end{bmatrix}.$$

For the second pass, we multiply the second and third rows of L with the second column of U. Equating each product with the corresponding element from A generates the equations

$$l_{21}u_{12} + l_{22} = 7 \quad \text{and} \quad l_{31}u_{12} + l_{32} = 8.$$

Substituting the values determined during the first pass and solving for the elements in the second column of L gives

$$l_{22} = -1 \quad \text{and} \quad l_{32} = -12.$$

Next, we multiply the second row of L into the third column of U to derive the equation

$$l_{21}u_{13} + l_{22}u_{23} = 9 \quad \text{or} \quad (2)(3) + (-1)u_{23} = 9.$$

Solving for u_{23}, we find $u_{23} = -3$. Thus, after the second pass, the Crout decomposition looks like

$$L = \begin{bmatrix} 1 & 0 & 0 \\ 2 & -1 & 0 \\ 5 & -12 & l_{33} \end{bmatrix} \quad \text{and} \quad U = \begin{bmatrix} 1 & 4 & 3 \\ 0 & 1 & -3 \\ 0 & 0 & 1 \end{bmatrix}.$$

All that remains is the computation of l_{33}. Multiplying the third row of L and the third column of U generates the equation

$$l_{31}u_{13} + l_{32}u_{23} + l_{33} = -2.$$

Substituting the values determined from the previous passes, we find

$$l_{33} = -53.$$

Thus, the complete Crout decomposition is

$$L = \begin{bmatrix} 1 & 0 & 0 \\ 2 & -1 & 0 \\ 5 & -12 & -53 \end{bmatrix} \quad \text{and} \quad U = \begin{bmatrix} 1 & 4 & 3 \\ 0 & 1 & -3 \\ 0 & 0 & 1 \end{bmatrix}.$$

Observe that each element a_{ij} from the original matrix appears in only one factorization equation. In particular, a_{ij} appears in the equation for the element l_{ij} whenever $i \geq j$ and in the equation for the element u_{ij} whenever $i < j$. Hence, as each new factorization element l_{ij} or u_{ij} is computed, it can be stored in the location previously occupied by a_{ij}. The same will be true for Doolittle decomposition.

Direct Factorization versus Decomposition via Gaussian Elimination

In terms of general performance, direct factorization and decomposition via Gaussian elimination are identical. First, both schemes can be carried out "in place." By overwriting the entries in A, no additional storage is needed to carry out either algorithm, an important benefit when the A matrix is large. Second, both schemes have exactly the same computational cost of

$$\frac{2}{3}n^3 - \frac{1}{2}n^2 - \frac{1}{6}n$$

arithmetic operations for an $n \times n$ matrix.

Direct factorization does offer certain advantages over Gaussian elimination when the matrix has special structure. This issue will be explored in detail in the next section. Furthermore, note that the direct factorization equations, (1) and (2), for computing l_{ik} and u_{kj} involve inner products of vectors (i.e., multiplication of a row vector into a column vector). These calculations can be carried out in a separate routine that employs higher precision to accumulate the inner product (thereby improving accuracy) and/or that exploits machine architecture, such as vectorization (thereby improving speed).

EXERCISES

In Exercises 1–6, determine the Crout decomposition of the given matrix, and then solve the system $A\mathbf{x} = \mathbf{b}$ for each of the given right-hand-side vectors.

1. $A = \begin{bmatrix} 2 & 7 & 5 \\ 6 & 20 & 10 \\ 4 & 3 & 0 \end{bmatrix}$ $\mathbf{b}_1 = \begin{bmatrix} 0 \\ 4 \\ 1 \end{bmatrix}$, $\mathbf{b}_2 = \begin{bmatrix} -4 \\ -16 \\ -7 \end{bmatrix}$, $\mathbf{b}_3 = \begin{bmatrix} -3 \\ -12 \\ 6 \end{bmatrix}$

2. $A = \begin{bmatrix} 1 & 1 & 2 \\ -1 & 0 & 2 \\ 3 & 2 & -1 \end{bmatrix}$ $\mathbf{b}_1 = \begin{bmatrix} 3 \\ -1 \\ 4 \end{bmatrix}$, $\mathbf{b}_2 = \begin{bmatrix} -9 \\ -10 \\ 7 \end{bmatrix}$, $\mathbf{b}_3 = \begin{bmatrix} -2 \\ -1 \\ 0 \end{bmatrix}$

3. $A = \begin{bmatrix} -3 & 2 & -1 \\ 6 & 8 & 1 \\ 4 & 2 & 7 \end{bmatrix}$ $\mathbf{b}_1 = \begin{bmatrix} 7 \\ 3 \\ -33 \end{bmatrix}$, $\mathbf{b}_2 = \begin{bmatrix} -12 \\ 1 \\ 1 \end{bmatrix}$, $\mathbf{b}_3 = \begin{bmatrix} 17 \\ -19 \\ -35 \end{bmatrix}$

4. $A = \begin{bmatrix} 1 & 4 & 5 \\ 2 & 6 & 4 \\ -1 & -2 & 3 \end{bmatrix}$ $\mathbf{b}_1 = \begin{bmatrix} -15 \\ -14 \\ -7 \end{bmatrix}$, $\mathbf{b}_2 = \begin{bmatrix} -10 \\ -10 \\ -10 \end{bmatrix}$, $\mathbf{b}_3 = \begin{bmatrix} -21 \\ -14 \\ -17 \end{bmatrix}$

5. $A = \begin{bmatrix} 1 & 2 & 3 & 4 \\ -1 & 1 & 2 & 3 \\ 1 & -1 & 1 & 2 \\ -1 & 1 & -1 & 5 \end{bmatrix}$ $\mathbf{b}_1 = \begin{bmatrix} 10 \\ 5 \\ 3 \\ 4 \end{bmatrix}$, $\mathbf{b}_2 = \begin{bmatrix} -4 \\ -5 \\ -3 \\ -4 \end{bmatrix}$, $\mathbf{b}_3 = \begin{bmatrix} -2 \\ -3 \\ 1 \\ -8 \end{bmatrix}$

6. $A = \begin{bmatrix} 1 & 3 & 1 & -2 \\ 2 & 4 & -1 & 2 \\ 3 & 1 & 1 & 5 \\ 4 & 2 & 6 & -1 \end{bmatrix}$ $\mathbf{b}_1 = \begin{bmatrix} 1 \\ -5 \\ -2 \\ 9 \end{bmatrix}$, $\mathbf{b}_2 = \begin{bmatrix} -5 \\ -3 \\ 6 \\ -5 \end{bmatrix}$, $\mathbf{b}_3 = \begin{bmatrix} 5 \\ 5 \\ -2 \\ 1 \end{bmatrix}$

7. Show that computing the Crout decomposition of an $n \times n$ matrix requires $\frac{2}{3}n^3 - \frac{1}{2}n^2 - \frac{1}{6}n$ arithmetic operations.

8. (a) Construct an algorithm to compute the Doolittle decomposition of an $n \times n$ matrix.

(b) Show that computing the Doolittle decomposition of an $n \times n$ matrix requires $\frac{2}{3}n^3 - \frac{1}{2}n^2 - \frac{1}{6}n$ arithmetic operations.

In Exercises 9–14, determine the Doolittle decomposition (see Exercise 8) of the given matrix, and then solve the system $A\mathbf{x} = \mathbf{b}$ for each of the given right-hand side-vectors.

9. Use the matrix and right-hand-side vectors from Exercise 1.

10. Use the matrix and right-hand-side vectors from Exercise 2.

11. Use the matrix and right-hand-side vectors from Exercise 3.

12. Use the matrix and right-hand-side vectors from Exercise 4.

13. Use the matrix and right-hand-side vectors from Exercise 5.

14. Use the matrix and right-hand-side vectors from Exercise 6.

15. (a) Construct an algorithm to factor an $n \times n$ matrix into the product LDU, where L is a lower triangular matrix with ones along its diagonal, D is a diagonal matrix, and U is an upper triangular matrix with ones along its diagonal.

(b) Suppose the matrix A has been factored into the product LDU, where the matrices L, D, and U have the form specified in part (a). Construct an algorithm to use this factorization to solve the system $A\mathbf{x} = \mathbf{b}$.

(c) How many arithmetic operations are required to compute the factorization in part (a)? How does this total compare to the number of operations needed to compute an LU decomposition?

(d) How many arithmetic operations are required by the algorithm in part (b) to solve a system given an LDU decomposition of the coefficient matrix? How does this total compare to the number of operations needed by forward and backward substitution?

(e) How does the total number of arithmetic operations needed to solve a system of equations using an LDU decomposition compare to the number of operations needed to solve a system using an LU decomposition?

In Exercises 16–21, determine the LDU decomposition (see Exercise 15) of the given matrix, and then solve the system $A\mathbf{x} = \mathbf{b}$ for each of the given right-hand-side vectors.

16. Use the matrix and right-hand-side vectors from Exercise 1.

17. Use the matrix and right-hand-side vectors from Exercise 2.

18. Use the matrix and right-hand-side vectors from Exercise 3.

19. Use the matrix and right-hand-side vectors from Exercise 4.

20. Use the matrix and right-hand-side vectors from Exercise 5.

21. Use the matrix and right-hand-side vectors from Exercise 6.

3.7 SPECIAL MATRICES

Linear systems which arise in practice often have coefficient matrices that have special properties or structure. Additionally, many numerical methods involve the construction and solution of linear systems with coefficient matrices that have special properties or structure. In this section, we will discuss three important classes of special matrices which arise frequently.

Strictly Diagonally Dominant Matrices

The first class of special matrices that we will discuss is the *strictly diagonally dominant* matrices.

Definition. An $n \times n$ matrix A is STRICTLY DIAGONALLY DOMINANT if, for each row, the magnitude of the diagonal element is strictly larger than the sum of the magnitudes of the other elements on the row; that is, if, for each i,

$$|a_{ii}| > \sum_{j=1, j\neq i}^{n} |a_{ij}|.$$

As an illustration, consider the two 3×3 matrices

$$\begin{bmatrix} 3 & -1 & 1 \\ 2 & -6 & 3 \\ -9 & 7 & -20 \end{bmatrix} \quad \text{and} \quad \begin{bmatrix} 3 & -2 & 2 \\ 2 & -6 & 3 \\ -9 & 7 & -20 \end{bmatrix}.$$

The first matrix is strictly diagonally dominant since

$$\begin{aligned} |3| = 3 > 2 &= |-1| + |1|, \\ |-6| = 6 > 5 &= |2| + |3|, \text{ and} \\ |-20| = 20 > 16 &= |-9| + |7|. \end{aligned}$$

The second matrix, however, is not strictly diagonally dominant since in the very first row

$$|3| = 3 < 4 = |-2| + |2|.$$

The following theorem indicates three of the most important characteristics of strictly diagonally dominant matrices.

Theorem. Let A be a strictly diagonally dominant matrix. Then

1. A is nonsingular, so $A\mathbf{x} = \mathbf{b}$ has a unique solution for any right-hand-side vector $\mathbf{b}$;
2. Gaussian elimination and direct factorization can be performed on A without row interchanges; and
3. The calculations during Gaussian elimination and direct factorization are stable with respect to the growth of roundoff error; that is, no pivoting strategy needs to be applied.

Proof. (1) Let A be an $n \times n$ strictly diagonally dominant matrix, but suppose A is singular. Then there exists a non-zero vector $\mathbf{x}$ such that $A\mathbf{x} = \mathbf{0}$. Let i be an index for which $|x_i| = \|\mathbf{x}\|_\infty$. Now, focus on the ith component of the vector equation $A\mathbf{x} = \mathbf{0}$, which reads

$$\sum_{j=1}^{n} a_{ij}x_j = 0 \qquad \text{or} \qquad a_{ii}x_i + \sum_{i=j,j\neq i}^{n} a_{ij}x_j = 0.$$

Solving this last equation for a_{ii}, taking the absolute value and repeatedly applying the triangle inequality leads to

$$|a_{ii}| \leq \sum_{j=1,j\neq i}^{n} |a_{ij}|\frac{|x_j|}{|x_i|} < \sum_{j=1,j\neq i}^{n} |a_{ij}|.$$

This violates the hypothesis that A is strictly diagonally dominant. Hence, A must be nonsingular.

See Wendroff [1] for a proof of parts (2) and (3). □

This theorem has both theoretical and practical implications. On the theoretical side, many of the techniques developed in this text require the solution of a linear system of equations. The coefficient matrices of some of these systems will be strictly diagonally dominant. When this happens, we will be guaranteed that the systems have a unique solution and that the underlying numerical methods are well defined.

On the practical side, this theorem provides information to help us choose the most efficient solution technique available. If the coefficient matrix for a specific system of equations is strictly diagonally dominant, we don't need a solution techqniue which uses sophisticated pivoting strategies. We don't need any pivoting strategy at all.

Symmetric Positive Definite Matrices

The second class of special matrices that we will consider is the *symmetric positive definite* matrices. Recall that a matrix A is symmetric if $A^T = A$; that is, if $a_{ji} = a_{ij}$ for each i and j.

Definition. A matrix A is SYMMETRIC POSITIVE DEFINITE if it is symmetric and $\mathbf{x}^T A\mathbf{x} > 0$ for any nonzero vector $\mathbf{x}$.

Consider the 3×3 matrix

$$A_1 = \begin{bmatrix} 3 & 1 & -1 \\ 1 & 4 & 2 \\ -1 & 2 & 5 \end{bmatrix}.$$

By inspection, this matrix is symmetric. Let $\mathbf{x} = \begin{bmatrix} x_1 & x_2 & x_3 \end{bmatrix}^T$. Then

$$\begin{aligned}\mathbf{x}^T A_1\mathbf{x} &= 3x_1^2 + 4x_2^2 + 5x_3^2 + 2x_1x_2 + 4x_2x_3 - 2x_1x_3 \\ &= x_1^2 + x_2^2 + x_3^2 + (x_1 + x_2)^2 + (x_1 - x_3)^2 + 2(x_2 + x_3)^2,\end{aligned}$$

which is clearly greater than zero for any non-zero $\mathbf{x}$. Therefore, A_1 is symmetric positive definite. On the other hand, the matrix

$$A_2 = \begin{bmatrix} 3 & -2 & -1 \\ -2 & 3 & -2 \\ -1 & -2 & 3 \end{bmatrix}$$

is not symmetric positive definite. For this matrix,

$$\mathbf{x}^T A_2 \mathbf{x} = 2\left[(x_1 - x_2)^2 + (x_1 - x_3)^2 + (x_2 - x_3)^2\right],$$

which equals zero for any vector whose components satisfy $x_1 = x_2 = x_3$; for example, $\mathbf{x} = \begin{bmatrix} 1 & 1 & 1 \end{bmatrix}^T$ or $\mathbf{x} = \begin{bmatrix} \pi & \pi & \pi \end{bmatrix}^T$.

It should be clear from these two examples that determining whether or not a given matrix is symmetric positive definite using the definition is a nontrivial task. Fortunately, there are simpler conditions that can be checked. We begin by considering a set of three necessary conditions. All symmetric positive definite matrices must satisfy each of these conditions. Therefore, any matrix which violates one of these conditions cannot be symmetric positive definite.

Theorem. Let A be an $n \times n$ symmetric positive definite matrix. Then

1. $a_{ii} > 0$ for each $i = 1, 2, 3, \ldots, n$;

2. $\max_{1 \le k,j \le n} |a_{kj}| \le \max_{1 \le i \le n} |a_{ii}|$; and

3. $a_{ij}^2 < a_{ii} a_{jj}$ for $i \ne j$.

Proof. We will prove the second part of the theorem here and leave the first and third parts for the exercises. For $j \ne k$, define the vector $\mathbf{x}$ by

$$x_i = \begin{cases} 1, & i = j \\ -1, & i = k \\ 0, & \text{otherwise} \end{cases}.$$

Since $\mathbf{x} \ne \mathbf{0}$, it follows that

$$\mathbf{x}^T A \mathbf{x} = a_{jj} - a_{jk} - a_{kj} + a_{kk} > 0,$$

or

$$a_{kj} < \frac{a_{jj} + a_{kk}}{2}, \tag{1}$$

where we have used the fact that $a_{jk} = a_{kj}$ since A is symmetric. Next, consider the vector $\mathbf{x}$ defined by

$$x_i = \begin{cases} 1, & i = j \text{ or } i = k \\ 0, & \text{otherwise} . \end{cases}$$

Now, $\mathbf{x}^T A \mathbf{x} > 0$ is equivalent to

$$a_{kj} > -\frac{a_{jj} + a_{kk}}{2}. \tag{2}$$

Combining (1) and (2) yields

$$|a_{kj}| < \frac{a_{jj} + a_{kk}}{2} \le \max_{1 \le i \le n} |a_{ii}|,$$

whenever $j \neq k$. Hence,

$$\max_{1 \le k,j \le n} |a_{kj}| \le \max_{1 \le i \le n} |a_{ii}|.$$

□

EXAMPLE 3.18 Matrices that are not Symmetric Positive Definite

None of the matrices

$$\begin{bmatrix} 2 & 1 & 0 \\ 1 & -2 & 1 \\ 0 & 1 & 2 \end{bmatrix}, \quad \begin{bmatrix} 8 & -1 & 1 \\ -1 & 8 & 9 \\ 1 & 9 & 7 \end{bmatrix}, \quad \text{and} \quad \begin{bmatrix} 3 & 1 & 5 \\ 1 & 4 & 2 \\ 5 & 2 & 8 \end{bmatrix}$$

is symmetric positive definite. The first matrix cannot be symmetric positive definite because the diagonal element $a_{22} = -2$ is negative, which violates the first condition from the theorem. The second matrix cannot be symmetric positive definite because it violates the second condition. Note that

$$\max_{1 \le k,j \le 3} |a_{kj}| = 9 > 8 = \max_{1 \le i \le 3} |a_{ii}|.$$

The final matrix cannot be symmetric positive definite because

$$a_{13}^2 = 5^2 > 24 = a_{11}a_{33}.$$

Next, we present a few results which allow us to show that a matrix is symmetric positive definite. Proofs of the first two results will be deferred until the next chapter after we've developed some more theory regarding eigenvalues and eigenvectors.

Theorem. If A is a symmetric matrix and all of its eigenvalues are positive, then A is symmetric positive definite.

Corollary. If A is symmetric, strictly diagonally dominant and $a_{ii} > 0$ for each i, then A is symmetric positive definite.

A matrix can also be identified as symmetric positive definite by examining its leading principal submatrices.

Definition. For each $k = 1, 2, 3, \ldots n$, the first k rows and the first k columns of the $n \times n$ matrix A form the kth LEADING PRINCIPAL SUBMATRIX of A.

The connection between leading principal submatrices and symmetric positive definite matrices is given by the following theorem. For a proof, see Stewart [2].

Theorem. A symmetric matrix is symmetric positive definite if and only if each of its leading principal submatrices has positive determinant.

EXAMPLE 3.19 Matrices that are Symmetric Positive Definite

Consider the matrices

$$\begin{bmatrix} 6 & -2 & 3 \\ -2 & 8 & 1 \\ 3 & 1 & 7 \end{bmatrix} \quad \text{and} \quad \begin{bmatrix} 3 & -1 & 2 \\ -1 & 3 & 1 \\ 2 & 1 & 3 \end{bmatrix}.$$

The first matrix is symmetric, strictly diagonally dominant and each of its diagonal elements is positive. Hence, by the corollary given above, this matrix is symmetric positive definite. The second matrix is not strictly diagonally dominant, so the corollary does not apply. However,

$$\det([3]) = 3 > 0; \quad \det\left(\begin{bmatrix} 3 & -1 \\ -1 & 3 \end{bmatrix}\right) = 8 > 0; \quad \text{and}$$

$$\det\left(\begin{bmatrix} 3 & -1 & 2 \\ -1 & 3 & 1 \\ 2 & 1 & 3 \end{bmatrix}\right) = 5 > 0,$$

so all of the leading principal submatrices have positive determinant. Consequently, the second matrix is symmetric positive definite.

The following theorem summarizes the important properties of symmetric positive definite matrices relative to the problem of solving linear systems of equations. The proof of the first part is considered in Exercise 8. For a proof of second and third parts, see Wendroff [1].

Theorem. Let A be a symmetric positive definite matrix. Then

1. A is nonsingular, so $A\mathbf{x} = \mathbf{b}$ has a unique solution for any right-hand-side vector $\mathbf{b}$;
2. Gaussian elimination and direct factorization can be performed on A without row interchanges; and
3. The calculations during Gaussian elimination and direct factorization are stable with respect to the growth of roundoff error; that is, no pivoting strategy needs to be applied.

The implications of this theorem are the same as those of the theorem presented earlier for strictly diagonally dominant matrices. From a theoretical standpoint, we will know that a numerical method which requires the solution of a linear system with a symmetric positive definite coefficient matrix is well-defined. From a practical stnadpoint, the theorem helps us select an efficient computational scheme. In particular, no pivoting strategy is necessary.

When working with symmetric positive definite matrices, even greater efficiency can be obtained by taking into account the symmetry of the matrix. To do this, rather than factor the matrix into LU form, we factor the matrix into the form

$$A = LL^T$$
$$= \begin{bmatrix} l_{11} & & & & \\ l_{21} & l_{22} & & & \\ l_{31} & l_{32} & l_{33} & & \\ \vdots & & & \ddots & \\ l_{n1} & l_{n2} & l_{n3} & \cdots & l_{nn} \end{bmatrix} \begin{bmatrix} l_{11} & l_{21} & l_{31} & \cdots & l_{n1} \\ & l_{22} & l_{32} & \cdots & l_{n2} \\ & & l_{33} & & l_{n3} \\ & & & \ddots & \vdots \\ & & & & l_{nn} \end{bmatrix}.$$

This produces what is known as the Cholesky decomposition. The equations for the Cholesky decomposition are developed in the same manner as were the equations for the Crout decomposition: We make passes through the matrix, each time computing the elements in one column of L. In the first pass, we find

$$l_{11} = \sqrt{a_{11}} \quad \text{and} \quad l_{i1} = a_{i1}/l_{11} \text{ (for } i = 2, 3, 4, \ldots, n).$$

For $k = 2, 3, 4, \ldots, n-1$, the diagonal element of the kth column is given by

$$l_{kk} = \sqrt{a_{kk} - \sum_{j=1}^{k-1} l_{kj}^2},$$

while the remaining elements in the column satisfy

$$l_{ik} = \left(a_{ik} - \sum_{j=1}^{k-1} l_{ij} l_{kj} \right) / l_{kk},$$

for $k + 1 \le i \le n$. The final element in the L matrix is then given by

$$l_{nn} = \sqrt{a_{nn} - \sum_{j=1}^{n-1} l_{nj}^2}.$$

EXAMPLE 3.20 Cholesky Decomposition in Action

Consider the matrix

$$A = \begin{bmatrix} 4 & 2 & -1 \\ 2 & 4 & 1 \\ -1 & 1 & 4 \end{bmatrix}.$$

For the first pass through this matrix, we calculate the elements in the first column of L as follows:

$$l_{11} = \sqrt{a_{11}} = \sqrt{4} = 2;$$
$$l_{21} = \frac{a_{21}}{l_{11}} = 1; \text{ and}$$
$$l_{31} = \frac{a_{31}}{l_{11}} = -\frac{1}{2}.$$

The second and third passes then produce

$$l_{22} = \sqrt{a_{22} - l_{21}^2} = \sqrt{4-1} = \sqrt{3};$$
$$l_{32} = \frac{a_{32} - l_{31}l_{21}}{l_{22}} = \frac{1+\frac{1}{2}}{\sqrt{3}} = \frac{\sqrt{3}}{2};$$

and

$$l_{33} = \sqrt{a_{33} - l_{31}^2 - l_{32}^2} = \sqrt{4 - \frac{1}{4} - \frac{3}{4}} = \sqrt{3}.$$

Therefore,

$$\begin{bmatrix} 4 & 2 & -1 \\ 2 & 4 & 1 \\ -1 & 1 & 4 \end{bmatrix} = \begin{bmatrix} 2 & 0 & 0 \\ 1 & \sqrt{3} & 0 \\ -\frac{1}{2} & \frac{\sqrt{3}}{2} & \sqrt{3} \end{bmatrix} \begin{bmatrix} 2 & 0 & 0 \\ 1 & \sqrt{3} & 0 \\ -\frac{1}{2} & \frac{\sqrt{3}}{2} & \sqrt{3} \end{bmatrix}^T.$$

Taking the symmetry of the matrix into account when constructing the factorization algorithm provides more than just aesthetic benefits. The Cholesky decomposition requires $\frac{1}{3}n^3 + \frac{1}{2}n^2 - \frac{5}{6}n$ arithmetic operations, plus n square roots. This is roughly half the number of operations used by Gaussian elimination and general direct factorization. It is possible to eliminate the square roots by factoring the matrix into the product LDL^T, where L is a lower triangular matrix with ones along the main diagonal and D is a diagonal matrix. The problem of constructing such a factorization algorithm is considered in Exercise 17.

Tridiagonal Matrices

The final special class of matrices we will discuss is the *tridiagonal* matrices.

Definition. The matrix A is TRIDIAGONAL if $a_{ij} = 0$ whenever $|i-j| > 1$; that is, on the ith row of the matrix, the only nonzero elements are $a_{i,i-1}$, $a_{i,i}$ and $a_{i,i+1}$.

Given the structure of a tridiagonal matrix, we will modify the general Crout decomposition algorithm and seek lower and upper triangular matrices of the form

$$L = \begin{bmatrix} l_{11} & & & & & \\ l_{21} & l_{22} & & & & \\ & l_{32} & l_{33} & & & \\ & & \ddots & \ddots & & \\ & & & \ddots & \ddots & \\ & & & & l_{n,n-1} & l_{nn} \end{bmatrix} \qquad U = \begin{bmatrix} 1 & u_{12} & & & & \\ & 1 & u_{23} & & & \\ & & 1 & u_{34} & & \\ & & & \ddots & \ddots & \\ & & & & \ddots & u_{n-1,n} \\ & & & & & 1 \end{bmatrix}.$$

Working again in passes, with each pass computing the elements in one column of L and one row of U, we find that the first pass gives

$$l_{11} = a_{11}, \quad l_{21} = a_{21} \quad \text{and} \quad u_{12} = a_{12}/l_{11}.$$

For each subsequent pass (that is, for $k = 2, 3, 4, \ldots, n-1$), the relevant formulas are

$$l_{kk} = a_{kk} - l_{k,k-1}u_{k-1,k}, \quad l_{k+1,k} = a_{k+1,k} \quad \text{and} \quad u_{k,k+1} = a_{k,k+1}/l_{kk}.$$

The last element along the diagonal of L is given by $l_{nn} = a_{nn} - l_{n,n-1}u_{n-1,n}$.

EXAMPLE 3.21 Solving a System with a Tridiagonal Coefficient Matrix

$$\begin{bmatrix} 4 & -1 & & \\ 2 & 4 & -1 & \\ & -2 & 4 & -1 \\ & & -2 & 4 \end{bmatrix} \mathbf{x} = \begin{bmatrix} 6 \\ -6 \\ 3 \\ 4 \end{bmatrix}.$$

To solve this system, we start by factoring the coefficient matrix. In the first pass, we calculate

$$l_{11} = a_{11} = 4; \quad l_{21} = a_{21} = 2; \quad u_{12} = \frac{a_{12}}{l_{11}} = -\frac{1}{4}.$$

The second and third passes then calculate

$$l_{22} = a_{22} - l_{21}u_{12} = \frac{9}{2}; \quad l_{32} = a_{32} = -2; \quad u_{23} = \frac{a_{23}}{l_{22}} = -\frac{2}{9}$$

and

$$l_{33} = a_{33} - l_{32}u_{23} = \frac{32}{9}; \quad l_{43} = a_{43} = -2; \quad u_{34} = \frac{a_{34}}{l_{33}} = -\frac{9}{32}.$$

The final pass gives

$$l_{44} = a_{44} - l_{43}u_{43} = \frac{55}{16}.$$

The complete LU decomposition of the coefficient matrix is then

$$L = \begin{bmatrix} 4 & & & \\ 2 & \frac{9}{2} & & \\ & -2 & \frac{32}{9} & \\ & & -2 & \frac{55}{16} \end{bmatrix} \quad \text{and} \quad U = \begin{bmatrix} 1 & -\frac{1}{4} & & \\ & 1 & -\frac{2}{9} & \\ & & 1 & -\frac{9}{32} \\ & & & 1 \end{bmatrix}.$$

Moving on to the solution step, forward substitution applied to $L\mathbf{z} = \mathbf{b}$ yields

$$z_1 = \frac{b_1}{l_{11}} = \frac{3}{2}; \qquad z_2 = \frac{b_2 - l_{21}z_1}{l_{22}} = -2;$$
$$z_3 = \frac{b_3 - l_{32}z_2}{l_{33}} = -\frac{9}{32}; \qquad z_4 = \frac{b_4 - l_{43}z_3}{l_{44}} = 1.$$

Forward subsitution then produces

$$x_4 = z_4 = 1; \qquad x_3 = z_3 - u_{34}x_4 = 0;$$
$$x_2 = z_2 - u_{23}x_3 = -2; \qquad x_1 = z_1 - u_{12}x_2 = 1.$$

Note that the solution of a linear system with a tridiagonal coefficient matrix is very inexpensive. The factor step requires $3n - 3$ arithmetic operations, and the solve step requires $5n - 4$ operations. The total cost is thus only $8n - 7$ operations. This is a significant reduction from the roughly $\frac{2}{3}n^3$ operations needed to solve a linear system with an arbitrary coefficient matrix. Because tridiagonal matrices will arise so often in the coming chapters, a complete algorithm for solving a system with a tridiagonal coefficient matrix is given in Appendix B.

Application Problem: Multistage Chemical Extraction

In the Chapter 3 Overview (see page 139), we developed the system of equations

$$\begin{aligned} &-(W + Sm)x_1 + Smx_2 = -Wx_{in} \\ &Wx_{i-1} - (W + Sm)x_i + Smx_{i+1} = 0 \qquad (i = 2, 3, 4, \ldots, n-1) \\ &Wx_{n-1} - (W + Sm)x_n = -Sy_{in}, \end{aligned}$$

where x_i denotes the mass fraction of a chemical exiting the ith stage of a counter-current extraction reactor in the water stream. The parameters are the flow rate in the water stream W, the flow rate in the solvent stream S, the ratio of the mass fraction of the chemical in the solvent stream to the mass fraction in the water stream m, the mass fraction in the input water stream x_{in}, and the mass fraction in the input solvent stream y_{in}.

Suppose we are working with a six-stage reactor with a water stream flow rate of $W = 200$ kg/hr and a solvent stream flow rate of $S = 50$ kg/hr. The input mass fractions are $x_{in} = 0.075$ and $y_{in} = 0$, and $m = 7$. The system of equations for the x_i then becomes

$$\begin{bmatrix} -550 & 350 & & & & \\ 200 & -550 & 350 & & & \\ & 200 & -550 & 350 & & \\ & & 200 & -550 & 350 & \\ & & & 200 & -550 & 350 \\ & & & & 200 & -550 \end{bmatrix} \begin{bmatrix} x_1 \\ x_2 \\ x_3 \\ x_4 \\ x_5 \\ x_6 \end{bmatrix} = \begin{bmatrix} -15 \\ 0 \\ 0 \\ 0 \\ 0 \\ 0 \end{bmatrix}.$$

The solution of this system is

$$\mathbf{x} = \begin{bmatrix} 0.042205 & 0.023465 & 0.012756 & 0.006637 & 0.003140 & 0.001142 \end{bmatrix}^T.$$

Note that the mass fraction in the water stream as it exits the reactor is only 0.001142, which is a reduction of nearly 98.5% from the input mass fraction.

References

1. B. Wendroff, *Theoretical Numerical Analysis*, Academic Press, New York, 1966.
2. G. W. Stewart, *Introduction to Matrix Computations*, Academic Press, New York, 1973.

EXERCISES

1. Classify each of the following matrices as strictly diagonally dominant, symmetric positive definite, both, or neither.

(a) $\begin{bmatrix} 2 & -1 & 0 \\ -1 & 4 & 2 \\ 0 & 2 & 6 \end{bmatrix}$ (b) $\begin{bmatrix} 1 & 2 & 0 \\ 4 & 6 & -1 \\ -3 & 2 & 0 \end{bmatrix}$

(c) $\begin{bmatrix} 5 & -3 & 2 \\ -3 & 1 & 0 \\ 2 & 0 & 6 \end{bmatrix}$ (d) $\begin{bmatrix} 4 & -2 & 2 \\ -2 & 6 & 4 \\ 2 & 4 & 7 \end{bmatrix}$

(e) $\begin{bmatrix} 8 & 2 & 4 & 1 \\ 0 & -3 & 1 & 1 \\ 0 & 0 & 6 & 2 \\ 0 & 0 & 0 & 1 \end{bmatrix}$ (f) $\begin{bmatrix} 4 & 1 & 1 & 1 \\ 1 & 3 & 0 & -1 \\ 1 & 0 & -2 & 0 \\ 1 & -1 & 0 & 4 \end{bmatrix}$

2. Consider the 2×2 symmetric matrix

$$\begin{bmatrix} a & b \\ b & c \end{bmatrix}.$$

What conditions must the elements a, b, and c satisfy to guarantee that the matrix is positive definite?

3. Consider the matrix

$$\begin{bmatrix} a & -1 & 0 \\ -1 & 4 & 1 \\ 0 & 1 & 5 \end{bmatrix}.$$

(a) For what values of a will this matrix be positive definite?

(b) For what values of a will this matrix be strictly diagonally dominant?

4. Repeat Exercise 3 for the matrix

$$\begin{bmatrix} 5 & -2 & 2 \\ -2 & 6 & a \\ 2 & a & 7 \end{bmatrix}.$$

5. Consider the matrix

$$\begin{bmatrix} b & -1 & a \\ -1 & 3 & 0 \\ a & 0 & 4 \end{bmatrix}.$$

(a) What conditions must a and b satisfy for this matrix to be symmetric positive definite?

(b) What conditions must a and b satisfy for this matrix to be strictly diagonally dominant?

6. (a) Suppose that A is a strictly diagonally dominant matrix. Show that the matrix $-A$ is strictly diagonally dominant but that the matrix A^T need not be strictly diagonally dominant.

(b) Suppose that A and B are both strictly diagonally dominant matrices. Show that $A + B$, $A - B$, and AB need not be strictly diagonally dominant.

7. (a) Suppose that A is a symmetric positive definite matrix. Show that the matrix $-A$ is not symmetric positive definite but that the matrix A^T is symmetric positive definite.

(b) Suppose that A and B are both symmetric positive definite matrices. Show that $A + B$ is symmetric positive definite but that $A - B$ need not be symmetric positive definite.

8. Show that if the matrix A is symmetric positive definite, then A is nonsingular.

9. Let A be an $n \times n$ symmetric positive definite matrix.

(a) Show that $a_{ii} > 0$ for each $i = 1, 2, 3, \dots, n$.

(b) Show that $a_{ij}^2 < a_{ii}a_{jj}$ for $i \neq j$.

10. Compute the Cholesky decomposition for each of the following matrices.

(a) $\begin{bmatrix} 16 & -28 & 0 \\ -28 & 53 & 10 \\ 0 & 10 & 29 \end{bmatrix}$ **(b)** $\begin{bmatrix} 9/4 & 3 & 3/2 \\ 3 & 25/4 & 7/2 \\ 3/2 & 7/2 & 17/4 \end{bmatrix}$.

(c) $\begin{bmatrix} 4 & -2 & -2 & 0 \\ -2 & 5 & 1 & -2 \\ -2 & 1 & 10 & 3 \\ 0 & -2 & 3 & 18 \end{bmatrix}$ **(d)** $\begin{bmatrix} 1 & -2 & 3 & -2 \\ -2 & 20 & -2 & 8 \\ 3 & -2 & 11 & -5 \\ -2 & 8 & -5 & 9 \end{bmatrix}$

11. Show that the computation of a Cholesky decomposition for an $n \times n$ matrix requires $\frac{1}{3}n^3 + \frac{1}{2}n^2 - \frac{5}{6}n$ arithmetic operations plus n square roots.

12. (a) Construct an algorithm to perform forward and backward substitution on the system $A\mathbf{x} = \mathbf{b}$, given a Cholesky decomposition ($A = LL^T$) for the coefficient matrix?

(b) How many arithmetic operations are required by the algorithm from part (a)?

13. Solve each of the following systems by computing a Cholesky decomposition for the coefficient matrix and then performing forward and backward substitution (see Exercise 12a).

(a) A = matrix given in Exercise 10a, $\mathbf{b} = \begin{bmatrix} 8 & -2 & 38 \end{bmatrix}^T$

(b) A = matrix given in Exercise 10b, $\mathbf{b} = \begin{bmatrix} 3 & 1 & 9 \end{bmatrix}^T$

(c) A = matrix given in Exercise 10c, $\mathbf{b} = \begin{bmatrix} 4 & -4 & 4 & -13 \end{bmatrix}^T$

(d) A = matrix given in Exercise 10d, $\mathbf{b} = \begin{bmatrix} 15 & -12 & 56 & -35 \end{bmatrix}^T$

14. Solve each of the following systems of equations. Note that each system has a tridiagonal coefficient matrix.

(a)

$$\begin{array}{rcrcrcrcr} 3x_1 & - & x_2 & & & & & = & 4 \\ x_1 & + & 4x_2 & + & 2x_3 & & & = & -7 \\ & & 3x_2 & + & 5x_3 & - & x_4 & = & -15 \\ & & & & -2x_3 & + & 7x_4 & = & 18 \end{array}$$

(b)

$$\begin{array}{rcrcrcrcr} 2x_1 & - & x_2 & & & & & = & 0 \\ -x_1 & + & 2x_2 & & x_3 & & & = & 0 \\ & & -x_2 & + & 2x_3 & - & x_4 & = & 0 \\ & & & & -x_3 & + & 2x_4 & = & 5 \end{array}$$

(c)

$$\begin{array}{rcrcrcrcr} 4x_1 & - & x_2 & & & & & = & 3 \\ -x_1 & - & 5x_2 & + & 6x_3 & & & = & 0 \\ & & x_2 & - & 3x_3 & + & 2x_4 & = & -4 \\ & & & & x_3 & + & 3x_4 & = & -2 \end{array}$$

15. Repeat the "Multistage Chemical Extraction" problem with a solvent stream input mass fraction of $y_{in} = 0.02$. By what percentage is the mass fraction in the water stream reduced?

16. An absorption column works much like an extraction reactor (see page 139). A gas stream with flow rate G and input mass fraction y_{in} of a chemical is used to transfer the chemical to a liquid stream that has a flow rate L and an input mass fraction x_{in}. At equilibrium, it is assumed that $y_i = mx_i$, where x_i and y_i are the mass fractions of the chemical within the liquid and gas streams, respectively, as they exit the ith stage of the column.

(a) Set up the system of equations for an n stage absorption column.

(b) If $L = 2500$ kg/hr, $G = 4000$ kg/hr, $x_{in} = 0$, $y_{in} = 0.05$, and $m = 1.46$, what is the mass fraction in the liquid stream as it exits an eight stage column?

17. (a) Construct an algorithm to factor an $n \times n$ symmetric positive definite matrix into the form LDL^T, where L is a lower triangular matrix with ones along its diagonal and D is a diagonal matrix. How many arithmetic operations are required to compute the LDL^T decomposition? How does this compare with the number of operations needed to compute a Cholesky decomposition?

(b) Construct an algorithm to solve the system $A\mathbf{x} = \mathbf{b}$ given an LDL^T decomposition of the coefficient matrix. How many arithmetic operations does this solve step require? How does this compare with the number of operations required by the solve step associated with a Cholesky decomposition?

18. Repeat Exercise 13 using an LDL^T decomposition rather than a Cholesky decomposition.

19. A matrix A is *pentadiagonal* if $a_{ij} = 0$ whenever $|i - j| > 2$.

(a) Construct an algorithm to efficiently compute the Crout decomposition of a pentadiagonal matrix.

(b) How many operations are required by the algorithm from part (a)?

(c) How many operations are needed to carry out forward and backward substitution using the decomposition obtained from part (a)?

3.8 ITERATIVE TECHNIQUES FOR LINEAR SYSTEMS: BASIC CONCEPTS AND METHODS

Having just devoted several sections to the development of direct techniques for linear systems—techniques that produce an answer in a fixed number of operations—it is natural to ask why we would want or even need to develop iterative techniques. For systems of small dimension, there is no need. Direct techniques will perform very efficiently. However, linear systems arising from practical applications will frequently be quite large. The coefficient matrices associated with these systems also tend to be sparse, meaning that only a small percentage of the entries are nonzero. We will encounter systems of this type in Chapter 9 when we treat the solution of elliptic partial differential equations.

For systems with large, sparse coefficient matrices, direct techniques are often less efficient than iterative techniques. Even though multiple iterations need to be performed to achieve convergence, an iterative solution will typically require fewer total operations than a direct solution. It will often happen that the nonzero elements in the coefficient matrix will exhibit a well-defined pattern. In these cases, an iterative solution will not require the storage of the coefficient matrix at all—only the structure of the equations will be needed. As an added bonus, iterative techniques are generally insensitive to roundoff error.

Basic Concepts

Basic iterative techniques for the solution of linear systems of equations are analogous to the fixed-point techniques which were discussed in Chapter 2. The original linear system $A\mathbf{x} = \mathbf{b}$, which can be interpreted as the rootfinding problem

$$\text{find the } n\text{-vector } \mathbf{x} \text{ so that } A\mathbf{x} - \mathbf{b} = \mathbf{0},$$

is first converted to the fixed point problem

$$\text{find the } n\text{-vector } \mathbf{x} \text{ so that } \mathbf{x} = T\mathbf{x} + \mathbf{c},$$

for some matrix T and vector $\mathbf{c}$. Next, starting from some initial approximation to the solution of the fixed point problem, $\mathbf{x}^{(0)}$, a sequence of vectors $\{\mathbf{x}^{(k)}\}$ is computed according to the rule

$$\mathbf{x}^{(k+1)} = T\mathbf{x}^{(k)} + \mathbf{c}. \tag{1}$$

Within this context, the matrix T is called the *iteration matrix*. The functional iteration is terminated when some appropriate measure of the difference between successive vectors in the sequence, $\mathbf{x}^{(k)}$ and $\mathbf{x}^{(k+1)}$, falls below a user specified tolerance.

The analysis of the functional iteration scheme given by (1) boils down to four important questions. First, what conditions guarantee a unique solution to the fixed point problem? Second, under what conditions will the sequence generated by (1) converge to this unique fixed point? Third, when the sequence generated by (1) converges, how quickly does it converge? Fourth, what conditions must the matrix T and the vector $\mathbf{c}$ satisfy in order for the fixed point problem to be consistent with the original rootfinding problem (i.e., for the two problems to have the same solution)?

The following theorem from general matrix theory plays a major role in establishing answers to these questions.

Theorem. Let A be an $n \times n$ matrix. Then the following statements are equivalent:

1. $\rho(A) < 1$, where $\rho(A)$ denotes the spectral radius of A;
2. $A^k \to 0$ as $k \to \infty$; and
3. $A^k\mathbf{x} \to \mathbf{0}$ as $k \to \infty$ for any vector $\mathbf{x}$.

A proof of this result can be found in Isaacson and Keller [1].

Let's start with the question of the uniqueness of the solution of the fixed point problem. Manipulating the fixed-point equation, we find

$$\begin{aligned}\mathbf{x} = T\mathbf{x} + \mathbf{c} \Leftrightarrow \mathbf{x} - T\mathbf{x} &= \mathbf{c} \\ \Leftrightarrow (I - T)\mathbf{x} &= \mathbf{c}.\end{aligned}$$

From this last equation, it follows that the fixed point problem has a unique solution if and only if the matrix $I - T$ is nonsingular. A sufficient condition for $I - T$ to be nonsingular is $\rho(T) < 1$ (see Exercise 10 from Section 3.3). Hence, the fixed point problem is guaranteed to have a unique solution whenever $\rho(T) < 1$.

To establish convergence, let $\mathbf{x}^\star$ denote the solution to the fixed point problem, and define the iteration error vector by $\mathbf{e}^{(k)} = \mathbf{x}^{(k)} - \mathbf{x}^\star$. Subtracting the equation $\mathbf{x}^\star = T\mathbf{x}^\star + \mathbf{c}$ from equation (1) yields the error evolution equation

$$\mathbf{e}^{(k+1)} = T\mathbf{e}^{(k)}.$$

Working backward through this equation, we find

$$\begin{aligned}\mathbf{e}^{(k+1)} &= T\mathbf{e}^{(k)} \\ &= T(T\mathbf{e}^{(k-1)}) = T^2\mathbf{e}^{(k-1)} \\ &= T^2(T\mathbf{e}^{(k-2)}) = T^3\mathbf{e}^{(k-2)} \\ &= \cdots \\ &= T^{k+1}\mathbf{e}^{(0)}.\end{aligned}$$

Ideally, $\mathbf{e}^{(k+1)}$ should approach $\mathbf{0}$ as $k \to \infty$ for any choice of initial vector $\mathbf{x}^{(0)}$, that is, for any initial error vector $\mathbf{e}^{(0)}$. The theorem stated above indicates that this will happen if and only if $\rho(T) < 1$. Hence, the iteration scheme defined by equation (1) will converge for any choice of the initial vector $\mathbf{x}^{(0)}$ if and only if $\rho(T) < 1$.

From the error evolution equation, we find

$$\|\mathbf{e}^{(k+1)}\| \leq \|T\| \, \|\mathbf{e}^{(k)}\|$$

for any vector norm $\|\cdot\|$ and associated natural matrix norm. Provided $\|T\| < 1$, it can be shown that

$$\|\mathbf{e}^{(k)}\| \leq \frac{\|T\|^k}{1 - \|T\|} \|\mathbf{x}^{(1)} - \mathbf{x}^{(0)}\|$$

(see Exercise 14). These two inequalities imply that the sequence $\{\mathbf{x}^{(k)}\}$ converges linearly with an asymptotic error constant that is less than or equal to $\|T\|$. Carrying out a more precise analysis, it can be shown that the asymptotic error constant is equal to $\rho(T)$. The proof is based on the fact that $\rho(T)$ is the greatest lower bound for all natural matrix norms of T. See Ortega [2] for details. Thus, the smaller the spectral radius of the iteration matrix, the faster the convergence of the corresponding iterative scheme.

The final preliminary issue to discuss is that of consistency. In order for the iteration defined by (1) to be of any practical use, the solution of the fixed point problem, $\mathbf{x}^\star = (I - T)^{-1}\mathbf{c}$, must be identical to the solution of the original linear system, $\mathbf{x}^\star = A^{-1}\mathbf{b}$. Hence, when constructing the fixed point problem from the linear system, we must be certain that T and $\mathbf{c}$ satisfy the relation

$$(I - T)^{-1}\mathbf{c} = A^{-1}\mathbf{b}.$$

Splitting Methods

A broad class of consistent iterative methods, known as splitting methods, can be constructed by introducing the notion of a splitting.

> **Definition.** Let A be a given $n \times n$ matrix. If M and N are $n \times n$ matrices with M nonsingular and $A = M - N$, then the pair (M, N) is called a SPLITTING of the matrix A.

So let's suppose that (M, N) forms a splitting of the matrix A. Then

$$A\mathbf{x} = \mathbf{b} \quad \text{is equivalent to} \quad (M - N)\mathbf{x} = \mathbf{b}.$$

Clearing the parentheses and transposing the term involving the matrix N to the right-hand side of the equation yields

$$M\mathbf{x} = N\mathbf{x} + \mathbf{b}.$$

Finally, premultiplying by M^{-1} produces

$$\mathbf{x} = M^{-1}N\mathbf{x} + M^{-1}\mathbf{b}.$$

Hence the splitting $A = M - N$ determines the fixed point problem $\mathbf{x} = T\mathbf{x} + \mathbf{c}$ and associated iteration scheme $\mathbf{x}^{(k+1)} = T\mathbf{x}^{(k)} + \mathbf{c}$, where

$$T = M^{-1}N \quad \text{and} \quad \mathbf{c} = M^{-1}\mathbf{b}.$$

To establish that splitting methods are always consistent, first note that with $T = M^{-1}N$

$$\begin{aligned} I - T &= I - M^{-1}N \\ &= M^{-1}(M - N) \\ &= M^{-1}A. \end{aligned}$$

Therefore, $(I - T)^{-1} = A^{-1}M$. Finally, with $\mathbf{c} = M^{-1}\mathbf{b}$,

$$\begin{aligned} (I - T)^{-1}\mathbf{c} &= A^{-1}M \cdot M^{-1}\mathbf{b} \\ &= A^{-1}\mathbf{b}, \end{aligned}$$

as required. Descriptions of the three most commonly used splitting methods are presented below. The convergence properties for these three methods will then be discussed at the end of the section.

The Jacobi Method, Gauss-Seidel Method, and SOR Method

To identify the splittings associated with the Jacobi method, the Gauss-Seidel method, and the SOR method, first express the coefficient matrix A in the form

$$A = D - L - U.$$

Here, D is the diagonal part of A, $-L$ is the strictly lower triangular part of A, and $-U$ is the strictly upper triangular part. It is important to keep in mind that the matrices L and U used here are in no way related to the LU decomposition of the coefficient matrix. As an example, suppose

$$A = \begin{bmatrix} 5 & 1 & 2 \\ -3 & 9 & 4 \\ 1 & 2 & -7 \end{bmatrix}.$$

Then

$$D = \begin{bmatrix} 5 & 0 & 0 \\ 0 & 9 & 0 \\ 0 & 0 & -7 \end{bmatrix}, \quad L = \begin{bmatrix} 0 & 0 & 0 \\ 3 & 0 & 0 \\ -1 & -2 & 0 \end{bmatrix} \quad \text{and} \quad U = \begin{bmatrix} 0 & -1 & -2 \\ 0 & 0 & -4 \\ 0 & 0 & 0 \end{bmatrix}.$$

The Jacobi method is based on the splitting $M = D$ and $N = L + U$. In order for M to be nonsingular, it must be the case that, for each i, $d_{i,i} \equiv a_{i,i} \neq 0$. If this relationship does not hold for even a single value of i, then the equations in the system must be reordered before the Jacobi method can be applied. With the specific choice of splitting indicated above, the iteration scheme for the Jacobi method is defined by

$$\mathbf{x}^{(k+1)} = T_{jac}\mathbf{x}^{(k)} + \mathbf{c}_{jac}, \tag{2}$$

where

$$T_{jac} = D^{-1}(L+U) \quad \text{and} \quad \mathbf{c}_{jac} = D^{-1}\mathbf{b}.$$

Taking into account the structure of the iteration matrix, T_{jac}, and the vector $\mathbf{c}_{jac}$, the individual components of equation (2) can be written as

$$x_i^{(k+1)} = \frac{1}{a_{i,i}}\left[b_i - \sum_{j=1}^{i-1} a_{i,j}x_j^{(k)} - \sum_{j=i+1}^{n} a_{i,j}x_j^{(k)}\right]. \tag{3}$$

Hence, the Jacobi method is equivalent to solving the i-th equation in the system for the unknown x_i.

Since, in general, the value of $x_i^{(k)}$ will be needed to compute $x_j^{(k+1)}$ for each $j = i+1, i+2, \ldots, n$, the value of $x_i^{(k)}$ cannot be overwritten by the newly computed value of $x_i^{(k+1)}$. This implies that when implementing the Jacobi method, two storage arrays will have to be maintained, one for the old approximation vector $\mathbf{x}^{(k)}$ and one for the new approximation vector $\mathbf{x}^{(k+1)}$. It also follows that the components of $\mathbf{x}^{(k+1)}$ can be computed in any order and that, on a parallel or vector machine, all components of $\mathbf{x}^{(k+1)}$ can be computed simultaneously. For this reason, the Jacobi method is often called the method of Simultaneous Relaxation.

EXAMPLE 3.22 The Jacobi Method in Action

Consider the system of equations

$$\begin{array}{rcrcrcr} 5x_1 & + & x_2 & + & 2x_3 & = & 10 \\ -3x_1 & + & 9x_2 & + & 4x_3 & = & -14 \\ x_1 & + & 2x_2 & - & 7x_3 & = & -33. \end{array}$$

The Jacobi method, when applied to this system, will produce the sequence of approximations $\{\mathbf{x}^{(k)}\}$ according to the rules

$$\begin{aligned} x_1^{(k+1)} &= \frac{1}{5}\left[10 - x_2^{(k)} - 2x_3^{(k)}\right] \\ x_2^{(k+1)} &= \frac{1}{9}\left[-14 + 3x_1^{(k)} - 4x_3^{(k)}\right] \\ x_3^{(k+1)} &= -\frac{1}{7}\left[-33 - x_1^{(k)} - 2x_2^{(k)}\right]. \end{aligned}$$

If we start with $\mathbf{x}^{(0)} = \begin{bmatrix} 0 & 0 & 0 \end{bmatrix}^T$, then the components of $\mathbf{x}^{(1)}$ are

$$\begin{aligned} x_1^{(1)} &= \frac{1}{5}\left[10 - x_2^{(0)} - 2x_3^{(0)}\right] = 2 \\ x_2^{(1)} &= \frac{1}{9}\left[-14 + 3x_1^{(0)} - 4x_3^{(0)}\right] = -\frac{14}{9} \\ x_3^{(1)} &= -\frac{1}{7}\left[-33 - x_1^{(0)} - 2x_2^{(0)}\right] = \frac{33}{7}. \end{aligned}$$

The following table summarizes the 14 iterations of the Jacobi method that were needed for $\|\mathbf{x}^{(k+1)} - \mathbf{x}^{(k)}\|_\infty$ to fall below 5×10^{-4}. Other stopping criteria can be imposed, but this is the most common. The exact solution to this problem is

$$\mathbf{x} = \begin{bmatrix} 1 & -3 & 4 \end{bmatrix}^T.$$

Hence, $\|\mathbf{x} - \mathbf{x}^{(14)}\|_\infty = 2.43 \times 10^{-4}$.

k	$\mathbf{x}^{(k)}$
0	$[\ 0.000000 \quad 0.000000 \quad 0.000000\ ]^T$
1	$[\ 2.000000 \quad -1.555556 \quad 4.714286\ ]^T$
2	$[\ 0.425397 \quad -2.984127 \quad 4.555556\ ]^T$
3	$[\ 0.774603 \quad -3.438448 \quad 3.922449\ ]^T$
4	$[\ 1.118710 \quad -3.040665 \quad 3.842530\ ]^T$
5	$[\ 1.071121 \quad -2.890443 \quad 4.005340\ ]^T$
6	$[\ 0.975953 \quad -2.978666 \quad 4.041462\ ]^T$
7	$[\ 0.979148 \quad -3.026443 \quad 4.002660\ ]^T$
8	$[\ 1.004225 \quad -3.008133 \quad 3.989466\ ]^T$
9	$[\ 1.005840 \quad -2.993910 \quad 3.998280\ ]^T$
10	$[\ 0.999470 \quad -2.997289 \quad 4.002574\ ]^T$
11	$[\ 0.998428 \quad -3.001321 \quad 4.000699\ ]^T$
12	$[\ 0.999985 \quad -3.000835 \quad 3.999398\ ]^T$
13	$[\ 1.000408 \quad -2.999738 \quad 3.999759\ ]^T$
14	$[\ 1.000044 \quad -2.999757 \quad 4.000133\ ]^T$

An obvious improvement that can be made to the Jacobi method is to use the value of $x_i^{(k+1)}$ as soon as it has been calculated in the computation of all subsequent entries in the vector $\mathbf{x}^{(k+1)}$, rather than waiting until the next iteration. After all, $x_i^{(k+1)}$ is supposed to be a better approximation to x_i than $x_i^{(k)}$. This modification amounts to changing equation (3) to

$$x_i^{(k+1)} = \frac{1}{a_{i,i}}\left[b_i - \sum_{j=1}^{i-1} a_{i,j}x_j^{(k+1)} - \sum_{j=i+1}^{n} a_{i,j}x_j^{(k)}\right]; \tag{4}$$

the only difference between the equations is that the superscript on x in the first summation is now $k+1$. The iteration scheme corresponding to equation (4) is known as the Gauss-Seidel method. Note that the Gauss-Seidel method is not vectorizable. The entries in $\mathbf{x}^{(k+1)}$ must be computed in succession. Hence, the Gauss-Seidel method is also known as Successive Relaxation.

Working backward from equation (4), we find that the splitting upon which the Gauss-Seidel method is based is

$$M = D - L \quad \text{and} \quad N = U.$$

Thus, the iteration matrix for the Gauss-Seidel method is given by

$$T_{gs} = (D-L)^{-1}U,$$

and the vector $\mathbf{c}$ is given by

$$\mathbf{c}_{gs} = (D-L)^{-1}\mathbf{b}.$$

The necessary and sufficient condition for the matrix M to be nonsingular is the same as above: for each i, we must have $d_{i,i} \equiv a_{i,i} \neq 0$.

EXAMPLE 3.23 The Gauss-Seidel Method in Action

Reconsider the system of equations

$$\begin{array}{rcrcrcr} 5x_1 & + & x_2 & + & 2x_3 & = & 10 \\ -3x_1 & + & 9x_2 & + & 4x_3 & = & -14 \\ x_1 & + & 2x_2 & - & 7x_3 & = & -33\,. \end{array}$$

The Gauss-Seidel method, when applied to this system, will produce the sequence of approximations $\{\mathbf{x}^{(k)}\}$ according to the rules

$$\begin{aligned} x_1^{(k+1)} &= \frac{1}{5}\left[10 - x_2^{(k)} - 2x_3^{(k)}\right] \\ x_2^{(k+1)} &= \frac{1}{9}\left[-14 + 3x_1^{(k+1)} - 4x_3^{(k)}\right] \\ x_3^{(k+1)} &= -\frac{1}{7}\left[-33 - x_1^{(k+1)} - 2x_2^{(k+1)}\right]. \end{aligned}$$

If we start with $\mathbf{x}^{(0)} = \begin{bmatrix} 0 & 0 & 0 \end{bmatrix}^T$, then the components of $\mathbf{x}^{(1)}$ are

$$\begin{aligned} x_1^{(1)} &= \frac{1}{5}\left[10 - x_2^{(0)} - 2x_3^{(0)}\right] = 2 \\ x_2^{(1)} &= \frac{1}{9}\left[-14 + 3x_1^{(1)} - 4x_3^{(0)}\right] = -\frac{8}{9} \\ x_3^{(1)} &= -\frac{1}{7}\left[-33 - x_1^{(1)} - 2x_2^{(1)}\right] = \frac{299}{63}. \end{aligned}$$

The following table summarizes the 10 iterations of the Gauss-Seidel method that were needed for $\|\mathbf{x}^{(k+1)} - \mathbf{x}^{(k)}\|_\infty$ to fall below 5×10^{-4}. Recall that the exact solution to this problem is

$$\mathbf{x} = \begin{bmatrix} 1 & -3 & 4 \end{bmatrix}^T.$$

Hence, $\|\mathbf{x} - \mathbf{x}^{(10)}\|_\infty = 7.8 \times 10^{-5}$. Note that convergence is obtained with the Gauss-Seidel method in roughly 30% fewer iterations than the Jacobi method. Further, the error in the final Gauss-Seidel approximation is roughly one-third the error in the final Jacobi approximation.

k	$\mathbf{x}^{(k)}$
0	$[\ 0.000000\quad 0.000000\quad 0.000000\ ]^T$
1	$[\ 2.000000\quad -0.888889\quad 4.746032\ ]^T$
2	$[\ 0.279365\quad -3.571781\quad 3.733686\ ]^T$
3	$[\ 1.220882\quad -2.808011\quad 4.086409\ ]^T$
4	$[\ 0.927039\quad -3.062724\quad 3.971656\ ]^T$
5	$[\ 1.023883\quad -2.979442\quad 4.009286\ ]^T$
6	$[\ 0.992174\quad -3.006736\quad 3.996958\ ]^T$
7	$[\ 1.002564\quad -2.997793\quad 4.000997\ ]^T$
8	$[\ 0.999160\quad -3.000723\quad 3.999673\ ]^T$
9	$[\ 1.000275\quad -2.999763\quad 4.000107\ ]^T$
10	$[\ 0.999910\quad -3.000078\quad 3.999965\ ]^T$

The final iterative technique that we will discuss in this section is the SOR method. An explanation for the name of the method will be provided shortly. This technique attempts to improve upon the convergence of the Gauss-Seidel method by computing $x_i^{(k+1)}$ as a weighted average of $x_i^{(k)}$ and the value produced by the Gauss-Seidel method, as given in equation (4). Let the weighting parameter, also known as a relaxation parameter, be denoted by ω. Then the analogue of equations (3) and (4) for the SOR method is

$$x_i^{(k+1)} = (1-\omega)x_i^{(k)} + \frac{\omega}{a_{i,i}}\left[b_i - \sum_{j=1}^{i-1} a_{i,j}x_j^{(k+1)} - \sum_{j=i+1}^{n} a_{i,j}x_j^{(k)}\right]. \tag{5}$$

Note that when $\omega = 1$, the SOR method reduces to the Gauss-Seidel method. Typically, there exists a range of ω values for which the SOR method will converge faster than the Gauss-Seidel method. The splitting associated with the SOR method is

$$M = \frac{1}{\omega}D - L \quad \text{and} \quad N = \left(\frac{1}{\omega} - 1\right)D + U.$$

Therefore,

$$T_{sor} = \left(\frac{1}{\omega}D - L\right)^{-1}\left[\left(\frac{1}{\omega} - 1\right)D + U\right]$$

and

$$\mathbf{c}_{sor} = \left(\frac{1}{\omega}D - L\right)^{-1}\mathbf{b}.$$

EXAMPLE 3.24 The SOR Method in Action

One more time we will consider the system of equations

$$\begin{array}{rcrcrcr} 5x_1 & + & x_2 & + & 2x_3 & = & 10 \\ -3x_1 & + & 9x_2 & + & 4x_3 & = & -14 \\ x_1 & + & 2x_2 & - & 7x_3 & = & -33. \end{array}$$

With $\omega = 0.9$, the iteration equations for the SOR method become

$$\begin{aligned} x_1^{(k+1)} &= 0.1x_1^{(k)} + \frac{0.9}{5}\left[10 - x_2^{(k)} - 2x_3^{(k)}\right] \\ x_2^{(k+1)} &= 0.1x_2^{(k)} + \frac{0.9}{9}\left[-14 + 3x_1^{(k+1)} - 4x_3^{(k)}\right] \\ x_3^{(k+1)} &= 0.1x_3^{(k)} - \frac{0.9}{7}\left[-33 - x_1^{(k+1)} - 2x_2^{(k+1)}\right]. \end{aligned}$$

If we start with $\mathbf{x}^{(0)} = \begin{bmatrix} 0 & 0 & 0 \end{bmatrix}^T$, then the components of $\mathbf{x}^{(1)}$ are

$$\begin{aligned} x_1^{(1)} &= 0.1x_1^{(0)} + \frac{0.9}{5}\left[10 - x_2^{(0)} - 2x_3^{(0)}\right] = 1.8 \\ x_2^{(1)} &= 0.1x_2^{(0)} + \frac{0.9}{9}\left[-14 + 3x_1^{(1)} - 4x_3^{(0)}\right] = -0.86 \\ x_3^{(1)} &= 0.1x_3^{(0)} - \frac{0.9}{7}\left[-33 - x_1^{(1)} - 2x_2^{(1)}\right] = 4.253143. \end{aligned}$$

The table shown below summarizes the 6 iterations of the SOR method that were needed for $\|\mathbf{x}^{(k+1)} - \mathbf{x}^{(k)}\|_\infty$ to fall below 5×10^{-4}. Note that $\|\mathbf{x} - \mathbf{x}^{(6)}\|_\infty = 6.0 \times 10^{-5}$.

k	$\mathbf{x}^{(k)}$
0	$[\ 0.000000\ \ 0.000000\ \ 0.000000\ ]^T$
1	$[\ 1.800000\ \ -0.860000\ \ 4.253143\ ]^T$
2	$[\ 0.603669\ \ -3.006157\ \ 3.972774\ ]^T$
3	$[\ 0.971276\ \ -2.998342\ \ 3.994011\ ]^T$
4	$[\ 0.998985\ \ -2.997743\ \ 3.999851\ ]^T$
5	$[\ 0.999546\ \ -2.999851\ \ 3.999965\ ]^T$
6	$[\ 0.999940\ \ -2.999989\ \ 3.999992\ ]^T$

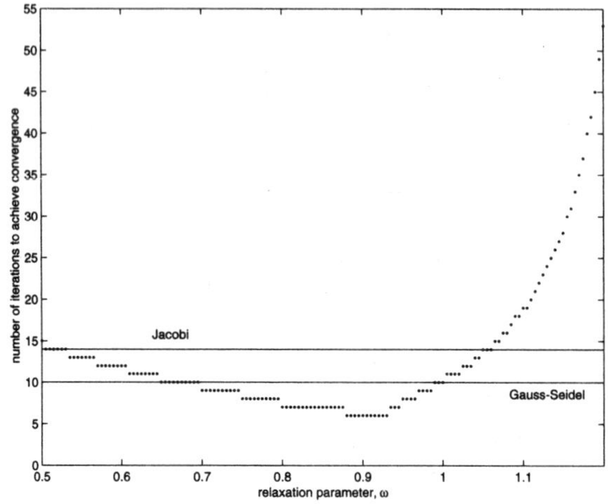

The selection of the parameter ω is crucial to the performance of the SOR method. The figure above displays the number of iterations required for the SOR method to converge to within a tolerance of 5×10^{-4} as a function of ω. The horizontal lines indicate the number of iterations required by the Gauss-Seidel method (bottom line) and the Jacobi method (top line). We see that there is a range of values, roughly $0.65 \leq \omega \leq 1.0$, for which the performance of the SOR method is as good as or better than the performance of the Gauss-Seidel method. As expected, the SOR method outperforms the Jacobi method over a broader range of ω values.

For the example problem we have been examining in this section, the SOR method performed better than the Gauss-Seidel method primarily for $\omega < 1$. However, for many of the practical problems to which the SOR method is applied (such as the systems of equations associated with the solution of elliptic partial differential equations which we will encounter in Chapter 9), performance is better than the Gauss-Seidel method for $\omega > 1$. When ω is selected greater than 1, the iterative method is referred to as an overrelaxation scheme. Since the Gauss-Seidel method is successive relaxation, this new technique is Successive overrelaxation, or the SOR method for short.

Specific Convergence Properties for the Jacobi, Gauss-Seidel, and SOR Methods

We know that the general iterative method $\mathbf{x}^{(k)} = T\mathbf{x}^{(k)} + \mathbf{c}$ converges if and only if $\rho(T) < 1$. Are there conditions that, when imposed upon the coefficient matrix A, will guarantee that the Jacobi, Gauss-Seidel, and SOR methods converge? The answer to this question is yes; unfortunately, there is no general theory, just a collection of special cases. For example, it is known that strict diagonal dominance of A is sufficient to guarantee that both the Jacobi method and the Gauss-Seidel method will converge for any choice of the initial vector $\mathbf{x}^{(0)}$. The proof for the Jacobi method is considered in Exercise 15, while the proof for the Gauss-Seidel method can be found in Ortega [2].

The following results regarding the Gauss-Seidel method are also useful in practice.

Theorem.

1. If A is real and symmetric with all positive diagonal elements, then the Gauss-Seidel method converges if and only if A is positive definite.
2. If A is positive definite, then the Gauss-Seidel method will converge for any choice of the initial vector $\mathbf{x}^{(0)}$.

A proof of the first part of this theorem can be found in Isaacson and Keller [1]. Consult Ortega [2] or Ralston and Rabinowitz [3] for a proof of the second part. For the SOR method, the most important convergence results are as follows:

Theorem.

1. If A has all nonzero diagonal elements, then $\rho(T_{sor}) \geq |\omega - 1|$. Therefore, the SOR method can converge only if $0 < \omega < 2$.
2. If A is positive definite and $0 < \omega < 2$, then the SOR method will converge for any choice of the initial vector $\mathbf{x}^{(0)}$.

See Ortega [2], Young [4], or Golub and Ortega [5] for details.

What about the speed of convergence? For the sample problem treated in this section, it was found that the Gauss-Seidel method converged in fewer iterations than the Jacobi method. Will this relative performance hold in general? The answer to this question is no. There are, in fact, coefficient matrices for which the Jacobi method will converge, but the Gauss-Seidel method will not—see Exercise 13. Furthermore, there is no general theory to indicate which method, Jacobi or Gauss-Seidel, will perform best on an arbitrary problem, just a collection of special cases. For example (see Ralston and Rabinowitz [3] or Young [4] for a proof).

Theorem. Suppose A is an $n \times n$ matrix. If $a_{i,i} > 0$ for each i and $a_{i,j} \leq 0$ whenever $i \neq j$, then one and only one of the following statements holds:

1. $0 \leq \rho(T_{gs}) = \rho(T_{jac}) < 1$;
2. $1 < \rho(T_{jac}) < \rho(T_{gs})$;
3. $\rho(T_{jac}) = \rho(T_{gs}) = 0$;
4. $\rho(T_{jac}) = \rho(T_{gs}) = 1$.

Thus, under the hypotheses of this theorem, when one method converges, so will the other method, with the Gauss-Seidel method converging faster. On the other hand, when one method diverges, so will the other, with the Gauss-Seidel method diverging faster.

The final issue we will address is that of the choice of the relaxation parameter for the SOR method. In practice, one of the most important special cases is the following.

Theorem. If A is positive definite and tridiagonal, then $\rho(T_{gs}) = [\rho(T_{jac})]^2 < 1$, and the optimal choice of the relaxation parameter, ω, for the SOR method is

$$\omega = \frac{2}{1+\sqrt{1-[\rho(T_{jac})]^2}}.$$

With this choice of ω, $\rho(T_{sor}) = \omega - 1$.

Ortega [2] provides a proof of this result. There are more general conditions under which the formula for the optimal value of ω given in the above theorem holds, but these require more advanced matrix theory concepts than we have developed here. For details consult Stoer and Bulirsch [6], Young [4], or Varga [7]. The optimal choice for ω for an arbitrary linear system remains an open question. Methods for computing the optimal value of ω during the iterative process are discussed by Hageman and Young [8].

References

1. E. Isaacson and H. B. Keller, *Analysis of Numerical Methods*, John Wiley and Sons, New York, 1966.
2. J. M. Ortega, *Numerical Analysis: A Second Course*, Academic Press, New York, 1972.
3. A. Ralston and P. Rabinowitz, *A First Course in Numerical Analysis*, 2nd edition, McGraw-Hill, New York, 1978.
4. D. M. Young, *Iterative Solution of Large Linear Systems*, Academic Press, New York, 1971.
5. G. H. Golub and J. M. Ortega, *Scientific Computing and Differential Equations: An Introduction to Numerical Methods*, Academic Press, Boston, 1992.
6. J. Stoer and R. Bulirsch, *Introduction to Numerical Analysis*, Springer-Verlag, New York, 1980.
7. R. S. Varga, *Matrix Iterative Analysis*, Prentice Hall, Englewood Cliffs, 1962.
8. L. A. Hageman and D. M. Young, *Applied Iterative Methods*, Academic Press, New York, 1981.

EXERCISES

In Exercises 1–4,

(a) Compute T_{jac} and T_{gs} for the given matrix.

(b) Determine the spectral radius of each iteration matrix from part (a).

(c) Will the Jacobi method converge for any choice of initial vector $\mathbf{x}^{(0)}$? Will the Gauss-Seidel method converge for any choice of initial vector $\mathbf{x}^{(0)}$? Explain.

1. $\begin{bmatrix} 2 & -1 \\ -1 & 3 \end{bmatrix}$

2. $\begin{bmatrix} 1 & 2 \\ 3 & 4 \end{bmatrix}$

3. $\begin{bmatrix} 4 & -1 & -2 \\ -1 & 3 & 0 \\ 0 & -1 & 3 \end{bmatrix}$

4. $\begin{bmatrix} 3 & 2 & -2 \\ -2 & -2 & 1 \\ 5 & -5 & 4 \end{bmatrix}$

5. For each of the following coefficient matrices and right-hand side vectors, write out the components of the Jacobi method iteration equation. Then, starting with the initial vector $\mathbf{x}^{(0)} = \mathbf{0}$, perform two iterations of the Jacobi method.

(a) $\begin{bmatrix} 2 & -1 & 0 \\ -1 & 4 & 2 \\ 0 & 2 & 6 \end{bmatrix}, \quad \begin{bmatrix} -1 \\ 3 \\ 5 \end{bmatrix}$

(b) $\begin{bmatrix} 3 & -1 & 1 \\ 2 & -6 & 3 \\ -9 & 7 & -20 \end{bmatrix}, \quad \begin{bmatrix} 4 \\ -13 \\ 7 \end{bmatrix}$

(c) $\begin{bmatrix} 4 & 2 & -1 \\ 2 & 4 & 1 \\ -1 & 1 & 4 \end{bmatrix}, \quad \begin{bmatrix} 1 \\ -1 \\ 1 \end{bmatrix}$

(d) $\begin{bmatrix} 4 & -1 & 0 & 0 \\ 2 & 4 & -1 & 0 \\ 0 & -2 & 4 & -1 \\ 0 & 0 & -2 & 4 \end{bmatrix}, \quad \begin{bmatrix} 0 \\ 2 \\ -3 \\ 1 \end{bmatrix}$

6. Repeat Exercise 5 for the Gauss-Seidel method.

In Exercises 7–10, use both the Jacobi method and the Gauss-Seidel method to solve the indicated linear system of equations. Take $\mathbf{x}^{(0)} = \mathbf{0}$, and terminate iteration when $\|\mathbf{x}^{(k+1)} - \mathbf{x}^{(k)}\|_\infty$ falls below 5×10^{-6}. Record the number of iterations required to achieve convergence.

7.

$$\begin{array}{rcrcrcrcr} 4x_1 & + & x_2 & + & x_3 & + & x_4 & = & -5 \\ x_1 & + & 8x_2 & + & 2x_3 & + & 3x_4 & = & 23 \\ x_1 & + & 2x_2 & - & 5x_3 & & & = & 9 \\ -x_1 & & & + & 2x_3 & + & 4x_4 & = & 4 \end{array}$$

8.

$$\begin{array}{rcrcrcr} 4x_1 & - & x_2 & & & = & 2 \\ -x_1 & + & 4x_2 & - & x_3 & = & 4 \\ & - & x_2 & + & 4x_3 & = & 10 \end{array}$$

9.

$$\begin{array}{rcrcrcrcrcr} 7x_1 & - & 3x_2 & & & & & & & = & 4 \\ -3x_1 & + & 9x_2 & + & x_3 & & & & & = & -6 \\ & & x_2 & + & 3x_3 & - & x_4 & & & = & 3 \\ & & & & -x_3 & + & 10x_4 & - & 4x_5 & = & 7 \\ & & & & & & -4x_4 & + & 6x_5 & = & 2 \end{array}$$

10.

$$\begin{array}{rcrcrcrcrcrcr}
4x_1 & - & x_2 & & & - & 2x_4 & & & & & = & -1 \\
-x_1 & + & 4x_2 & - & x_3 & & & - & 2x_5 & & & = & 0 \\
 & - & x_2 & + & 4x_3 & & & & & - & 2x_6 & = & 1 \\
-x_1 & & & & & + & 4x_4 & - & x_5 & & & = & -2 \\
 & - & x_2 & & & - & x_4 & + & 4x_5 & - & x_6 & = & 1 \\
 & & & - & x_3 & & & - & x_5 & + & 4x_6 & = & 2
\end{array}$$

11. The linear systems in Exercises 8 and 9 have positive definite, tridiagonal coefficient matrices. Determine the optimal value of the relaxation parameter for the SOR method for each system. Using the corresponding optimal value of ω, solve the systems from Exercises 8 and 9. Take $\mathbf{x}^{(0)} = \mathbf{0}$, and terminate iteration when $\|\mathbf{x}^{(k+1)} - \mathbf{x}^{(k)}\|_\infty$ falls below 5×10^{-6}.

12. For each of the linear systems in Exercises 7–10, generate a plot of the number of iterations required by the SOR method to achieve convergence as a function of the relaxation parameter ω. Take $\mathbf{x}^{(0)} = \mathbf{0}$, and terminate iteration when $\|\mathbf{x}^{(k+1)} - \mathbf{x}^{(k)}\|_\infty$ falls below 5×10^{-6}. Over roughly what range of ω values does the SOR method outperform the Gauss-Seidel method? the Jacobi method?

13. Let

$$A = \begin{bmatrix} 2 & 4 & -4 \\ 3 & 3 & 3 \\ 10 & 10 & 5 \end{bmatrix}.$$

(a) Write out the iteration matrix T_{jac} corresponding to the matrix A, and determine $\rho(T_{jac})$. Will the Jacobi method converge for any choice of the initial vector $\mathbf{x}^{(0)}$?

(b) Write out the iteration matrix T_{gs} corresponding to the matrix A, and determine $\rho(T_{gs})$. Will the Gauss-Seidel method converge for any choice of the initial vector $\mathbf{x}^{(0)}$?

14. Consider the iteration scheme $\mathbf{x}^{(k+1)} = T\mathbf{x}^{(k)} + \mathbf{c}$, and suppose that $\|T\| < 1$ for some natural matrix norm. Show that for any $\mathbf{x}^{(0)} \in \mathbf{R}^n$

$$\|\mathbf{x} - \mathbf{x}^{(k)}\| \leq \frac{\|T\|^k}{1 - \|T\|}\|\mathbf{x}^{(1)} - \mathbf{x}^{(0)}\|.$$

(*Hint*: Review the proof of the fixed point iteration convergence theorem in Section 2.3.)

15. Let A be a strictly diagonally dominant matrix and let T_{jac} be the Jacobi method iteration matrix associated with A. Show that $\rho(T_{jac}) < 1$. (*Hint*: Show that $\|T_{jac}\|_\infty < 1$ and use the fact that the spectral radius of a matrix is smaller than any natural matrix norm of that matrix.)

16. Suppose that $\rho(T) < 1$. Show that

$$(I - T)^{-1} = \sum_{k=0}^{\infty} T^k.$$

17. The variables of interest in an absorption column are the steady-state composition of solute in the liquid, x_i, on each plate and the steady-state composition of the solute in the gas, y_i, on each plate. Suppose we have a six plate absorption column, where the inlet compositions, $x_0 = 0.05$ kg solute/kg liquid and $y_7 = 0.3$ kg solute/kg inert gas, are known, as are the liquid and gas flow rates, $L = 40.8$ kg/min and $G = 66.7$ kg/min. Further, we will assume that the linear equilibrium relationship $y_i = 0.72x_i$ holds. Performing a material balance around an arbitrary plate, we find that the x_i satisfy the system

$$\begin{bmatrix} -88.824 & 48.024 & 0 & 0 & 0 & 0 \\ 40.8 & -88.824 & 48.024 & 0 & 0 & 0 \\ 0 & 40.8 & -88.824 & 48.024 & 0 & 0 \\ 0 & 0 & 40.8 & -88.824 & 48.024 & 0 \\ 0 & 0 & 0 & 40.8 & -88.824 & 48.024 \\ 0 & 0 & 0 & 0 & 40.8 & -88.824 \end{bmatrix} \begin{bmatrix} x_1 \\ x_2 \\ x_3 \\ x_4 \\ x_5 \\ x_6 \end{bmatrix} = \begin{bmatrix} -2.04 \\ 0 \\ 0 \\ 0 \\ 0 \\ -20.01 \end{bmatrix}.$$

Determine the x_i using

(a) the Jacobi method;

(b) the Gauss-Seidel method; and

(c) the SOR method with ω ranging from 1.1 through 1.9 in increments of 0.1.

3.9 ITERATIVE TECHNIQUES FOR LINEAR SYSTEMS: CONJUGATE GRADIENT METHOD

Not all iterative methods for the solution of linear systems are based on the notion of a splitting and the conversion to a fixed point problem. For example, there is a completely separate class of techniques based on the equivalence between the solution of a linear system and the minimization of an associated quadratic functional. The conjugate gradient method, which is a popular choice for the solution of large sparse systems, belongs to this latter class of techniques. For an overview of the many other methods which belong to this class of minimization methods, consult Ueberhuber [1].

Minimizing a Quadratic Functional and Solving a Linear System

Suppose that A is a symmetric and positive definite $n \times n$ matrix. This assumption will be made throughout the section. Consider $f : \mathbf{R}^n \to \mathbf{R}$ defined by

$$f(\mathbf{x}) = \frac{1}{2}\mathbf{x}^T A\mathbf{x} - \mathbf{b}^T\mathbf{x} + \mathbf{c}.$$

A mathematical object such as this, which operates on the components of a vector to produce a scalar output, is referred to as a functional. In particular, f happens to be a quadratic functional – $\mathbf{x}^T A\mathbf{x}$ consists of a collection of terms, all of which are of degree two in the components of $\mathbf{x}$. Under the assumption that A is positive definite, f behaves much like an upward opening parabola and hence has a unique global minimizer; that is, there exists a unique vector $\mathbf{x}^*$ such that $f(\mathbf{x}^*) < f(\mathbf{x})$ for all $\mathbf{x} \in \mathbf{R}^n$, $\mathbf{x} \neq \mathbf{x}^*$.

From basic calculus, we know that the location of the minimum value of f can be determined by requiring the gradient of f,

$$\nabla f = \begin{bmatrix} \frac{\partial f}{\partial x_1} & \frac{\partial f}{\partial x_2} & \frac{\partial f}{\partial x_3} & \cdots & \frac{\partial f}{\partial x_n} \end{bmatrix}^T,$$

to be identically zero when evaluated at $\mathbf{x} = \mathbf{x}^*$. In terms of the elements of A and the components of $\mathbf{x}$,

$$f(\mathbf{x}) = \frac{1}{2}\sum_{i=1}^{n} x_i \left(\sum_{j=1}^{n} a_{ij}x_j\right) - \sum_{j=1}^{n} b_j x_j + c_j.$$

Therefore, for each k,

$$\begin{aligned} \frac{\partial f}{\partial x_k} &= \frac{1}{2}\left(\sum_{i=1}^{n} x_i a_{ik} + \sum_{j=1}^{n} a_{kj}x_j\right) - b_k \\ &= \frac{1}{2}\left(\sum_{i=1}^{n} a_{ki} x_i + \sum_{j=1}^{n} a_{kj}x_j\right) - b_k \\ &= \sum_{j=1}^{n} a_{kj}x_j - b_k = (A\mathbf{x} - \mathbf{b})_k, \end{aligned}$$

where the symmetry of A was used in going from the first line to the second. Thus,

$$\nabla f = A\mathbf{x} - \mathbf{b},$$

and locating the minimum of f is equivalent to solving $A\mathbf{x} = \mathbf{b}$.

So how do we numerically approximate the global minimizer of the functional f? Most of the techniques developed for this problem start with an initial guess, $\mathbf{x}^{(0)}$, and then generate the sequence $\{\mathbf{x}^{(m)}\}$ according to the rule

$$\mathbf{x}^{(m+1)} = \mathbf{x}^{(m)} + \lambda_m \mathbf{d}^{(m)}.$$

The vector $\mathbf{d}^{(m)}$ is called the *search direction*, and the scalar λ_m is the *step size*. Of course, different minimization methods are determined by different choices of the step size and the search direction. We will now turn our attention to a discussion of the step sizes and search directions which define the conjugate gradient method.

Choosing the Step Size

Suppose that the approximate minimizer, $\mathbf{x}^{(m)}$, and the search direction, $\mathbf{d}^{(m)}$, are known. We will apply the following basic principle to determine the corresponding step size:

select λ_m so that $f(\mathbf{x})$ is minimized along the line $\mathbf{x} = \mathbf{x}^{(m)} + \lambda \mathbf{d}^{(m)}$.

Substituting $\mathbf{x} = \mathbf{x}^{(m)} + \lambda \mathbf{d}^{(m)}$ into $f(\mathbf{x})$ gives

$$\begin{aligned}
f(\mathbf{x}^{(m)} + \lambda \mathbf{d}^{(m)}) &= \frac{1}{2}\left(\mathbf{x}^{(m)} + \lambda \mathbf{d}^{(m)}\right)^T A \left(\mathbf{x}^{(m)} + \lambda \mathbf{d}^{(m)}\right) \\
&\quad - \mathbf{b}^T \left(\mathbf{x}^{(m)} + \lambda \mathbf{d}^{(m)}\right) + \mathbf{c} \\
&= \frac{1}{2}\mathbf{x}^{(m)^T} A\mathbf{x}^{(m)} + \frac{1}{2}\lambda \mathbf{d}^{(m)^T} A\mathbf{x}^{(m)} + \frac{1}{2}\lambda \mathbf{x}^{(m)^T} A\mathbf{d}^{(m)} \\
&\quad \frac{1}{2}\lambda^2 \mathbf{d}^{(m)^T} A\mathbf{d}^{(m)} - \mathbf{b}^T\mathbf{x}^{(m)} - \mathbf{b}^T\lambda\mathbf{d}^{(m)} + \mathbf{c} \\
&= \left(\frac{1}{2}\mathbf{d}^{(m)^T} A\mathbf{d}^{(m)}\right)\lambda^2 \\
&\quad + \left[\frac{1}{2}\mathbf{d}^{(m)^T} A\mathbf{x}^{(m)} + \frac{1}{2}\mathbf{x}^{(m)^T} A\mathbf{d}^{(m)} - \mathbf{b}^T\mathbf{d}^{(m)}\right]\lambda \\
&\quad + \frac{1}{2}\mathbf{x}^{(m)^T} A\mathbf{x}^{(m)} - \mathbf{b}^T\mathbf{x}^{(m)} + \mathbf{c}.
\end{aligned}$$

Since the matrix A is symmetric and, for any two vectors $\mathbf{u}$ and $\mathbf{v}$, $\mathbf{u}^T\mathbf{v} = \mathbf{v}^T\mathbf{u}$, it follows that

$$\mathbf{x}^{(m)^T} A\mathbf{d}^{(m)} = \mathbf{x}^{(m)^T} A^T\mathbf{d}^{(m)} = \left(A\mathbf{x}^{(m)}\right)^T \mathbf{d}^{(m)} = \mathbf{d}^{(m)^T} A\mathbf{x}^{(m)}$$

and $\mathbf{b}^T\mathbf{d}^{(m)} = \mathbf{d}^{(m)^T}\mathbf{b}$. Therefore,

$$\begin{aligned}
f(\mathbf{x}^{(m)} + \lambda \mathbf{d}^{(m)}) &= \left(\frac{1}{2}\mathbf{d}^{(m)^T} A\mathbf{d}^{(m)}\right)\lambda^2 + \left(\mathbf{d}^{(m)^T} A\mathbf{x}^{(m)} - \mathbf{d}^{(m)^T}\mathbf{b}\right)\lambda + \tilde{\mathbf{c}} \\
&= \left(\frac{1}{2}\mathbf{d}^{(m)^T} A\mathbf{d}^{(m)}\right)\lambda^2 + \mathbf{d}^{(m)^T}\mathbf{r}^{(m)}\lambda + \tilde{\mathbf{c}},
\end{aligned}$$

where $\tilde{\mathbf{c}} = \frac{1}{2}\mathbf{x}^{(m)^T} A\mathbf{x}^{(m)} - \mathbf{b}^T\mathbf{x}^{(m)} + \mathbf{c}$ and $\mathbf{r}^{(m)} = A\mathbf{x}^{(m)} - \mathbf{b}$. Note that $\mathbf{r}^{(m)}$ is just the residual associated with the approximation $\mathbf{x}^{(m)}$. The positive definiteness of A guarantees that $\mathbf{d}^{(m)^T} A\mathbf{d}^{(m)} > 0$ for any $\mathbf{d}^{(m)} \neq 0$, which implies that $f(\mathbf{x}^{(m)} + \lambda \mathbf{d}^{(m)})$ is an upward opening parabola in λ. $f(\mathbf{x}^{(m)} + \lambda \mathbf{d}^{(m)})$ then achieves its minimum value when

$$\frac{\partial f}{\partial \lambda} = \mathbf{d}^{(m)^T} A\mathbf{d}^{(m)}\lambda + \mathbf{d}^{(m)^T}\mathbf{r}^{(m)} = 0.$$

Hence,

$$\lambda_m = -\frac{\mathbf{d}^{(m)^T}\mathbf{r}^{(m)}}{\mathbf{d}^{(m)T} A\mathbf{d}^{(m)}}.$$

Choosing the Search Direction

As the name of the method implies, the search directions in the conjugate gradient method are based on the gradient of f. Recall it was established above that

$$\nabla f = A\mathbf{x} - \mathbf{b};$$

thus

$$\begin{aligned}\nabla f(\mathbf{x}^{(m)}) &= A\mathbf{x}^{(m)} - \mathbf{b}\\ &= \mathbf{r}^{(m)},\end{aligned}$$

the residual vector. To simplify the notation in the derivation that follows, we will therefore use $\mathbf{r}^{(m)}$ in place of $\nabla f(\mathbf{x}^{(m)})$ when referencing the gradient.

For the first iteration of the conjugate gradient method, the search direction is chosen as $\mathbf{d}^{(0)} = -\mathbf{r}^{(0)}$. That is, we start from $\mathbf{x}^{(0)}$ and travel in the direction opposite to that of the gradient—standard arguments from multivariable calculus guarantee that this is the direction of maximum decrease in the value of f. All subsequent search directions are determined by

$$\mathbf{d}^{(m+1)} = -\mathbf{r}^{(m+1)} + \alpha_m \mathbf{d}^{(m)}, \qquad m \geq 0.$$

The scalar α_m is chosen so that the search direction $\mathbf{d}^{(m+1)}$ is A-conjugate to the direction $\mathbf{d}^{(m)}$.

Definition. Let A be a symmetric and positive definite matrix. Two vectors, $\mathbf{u}$ and $\mathbf{v}$, are A-CONJUGATE if

$$\mathbf{u}^T A\mathbf{v} = 0.$$

Thus, to determine the formula for α_m, we start by computing

$$\begin{aligned}\mathbf{d}^{(m+1)^T} A\mathbf{d}^{(m)} &= \left(-\mathbf{r}^{(m+1)} + \alpha_m \mathbf{d}^{(m)}\right)^T A\mathbf{d}^{(m)}\\ &= -\mathbf{r}^{(m+1)^T} A\mathbf{d}^{(m)} + \alpha_m \mathbf{d}^{(m)^T} A\mathbf{d}^{(m)}.\end{aligned}$$

For this last expression to be zero, we see that

$$\alpha_m = \frac{\mathbf{r}^{(m+1)^T} A\mathbf{d}^{(m)}}{\mathbf{d}^{(m)T} A\mathbf{d}^{(m)}}.$$

The matrix-vector product $A\mathbf{d}^{(m)}$ and the vector-vector product $\mathbf{d}^{(m)^T}\left(A\mathbf{d}^{(m)}\right)$ would already have been computed as part of the calculation of the step size λ_m, so the calculation of α_m requires only one additional vector-vector product and one division.

Gathering together the approximate solution update equation, the formula for the step size and the information regarding the construction of the search directions, we arrive at the following basic algorithm for the conjugate gradient method:

$$
\begin{aligned}
&\mathbf{r}^{(0)} = A\mathbf{x}^{(0)} - \mathbf{b} \\
&\mathbf{d}^{(0)} = -\mathbf{r}^{(0)} \\
&\text{for } m = 0, 1, 2, \ldots \\
&\qquad \lambda_m = -\mathbf{d}^{(m)^T}\mathbf{r}^{(m)}/\mathbf{d}^{(m)^T}A\mathbf{d}^{(m)} \\
&\qquad \mathbf{x}^{(m+1)} = \mathbf{x}^{(m)} + \lambda_m\mathbf{d}^{(m)} \\
&\qquad \mathbf{r}^{(m+1)} = A\mathbf{x}^{(m+1)} - \mathbf{b} \\
&\qquad \text{if } \sqrt{\mathbf{r}^{(m+1)T}\mathbf{r}^{(m+1)}} < TOL, \text{ OUTPUT } \mathbf{x}^{(m+1)} \\
&\qquad \alpha_m = \mathbf{r}^{(m+1)^T}A\mathbf{d}^{(m)}/\mathbf{d}^{(m)^T}A\mathbf{d}^{(m)} \\
&\qquad \mathbf{d}^{(m+1)} = -\mathbf{r}^{(m+1)} + \alpha_m\mathbf{d}^{(m)} .
\end{aligned}
$$

The inputs to this routine are the coefficient matrix A, the right-hand-side vector $\mathbf{b}$, the initial vector $\mathbf{x}^{(0)}$ and the convergence tolerance TOL. The first two lines set the initial search direction. Inside the iteration loop, the step size, the updated approximate solution, and the new residual are computed, in that order. If the length of the new residual is not below the convergence tolerance, the new search direction is computed in preparation for the next iteration.

Making the Algorithm More Efficient

The pseudocode that was presented above for the conjugate gradient method is not as efficient as it could be. Notice that each iteration requires the calculation of 2 matrix-vector products, 4 vector-vector products, and 3 vector additions. By constructing a formula that allows the new residual $\mathbf{r}^{(m+1)}$ to be calculated from the old residual $\mathbf{r}^{(m)}$ and by taking advantage of the special properties of the residual vectors and search directions, the operation count for the conjugate gradient method can be reduced to 1 matrix-vector product, 2 vector-vector products, and 3 vector additions.

Let's see how $\mathbf{r}^{(m+1)}$ can be calculated from $\mathbf{r}^{(m)}$. Take the approximate solution update equation, $\mathbf{x}^{(m+1)} = \mathbf{x}^{(m)} + \lambda_m\mathbf{d}^{(m)}$, multiply through by A, and then subtract $\mathbf{b}$ from both sides. This produces the equation

$$A\mathbf{x}^{(m+1)} - \mathbf{b} = A\mathbf{x}^{(m)} - \mathbf{b} + \lambda_m A\mathbf{d}^{(m)}.$$

Applying the definition of the residual vector then yields

$$\mathbf{r}^{(m+1)} = \mathbf{r}^{(m)} + \lambda_m A\mathbf{d}^{(m)}. \tag{1}$$

Equation (1) still requires a matrix-vector product, $A\mathbf{d}^{(m)}$, but this is the same matrix vector product that is needed for the calculation of both λ_m and α_m. Using (1) instead of $A\mathbf{x}^{(m+1)} - \mathbf{b}$ to calculate $\mathbf{r}^{(m+1)}$ will therefore reduce the number of matrix-vector products performed each iteration by one.

Further efficiencies can be gained by developing different formulas for λ_m and α_m. To derive these formulas, two special relationships between the residuals and the search directions are needed. First, take equation (1) and premultiply by $\mathbf{d}^{(m)^T}$. This gives

$$\begin{aligned}\mathbf{d}^{(m)^T}\mathbf{r}^{(m+1)} &= \mathbf{d}^{(m)^T}\mathbf{r}^{(m)} + \lambda_m \mathbf{d}^{(m)^T} A\mathbf{d}^{(m)} \\ &= \mathbf{d}^{(m)^T}\mathbf{r}^{(m)} - \left(\mathbf{d}^{(m)^T}\mathbf{r}^{(m)}/\mathbf{d}^{(m)^T}A\mathbf{d}^{(m)}\right)\mathbf{d}^{(m)^T}A\mathbf{d}^{(m)} \\ &= \mathbf{d}^{(m)^T}\mathbf{r}^{(m)} - \mathbf{d}^{(m)^T}\mathbf{r}^{(m)} = 0,\end{aligned}$$

so the previous search direction and the new residual vector are orthogonal. With α_m chosen to make the search directions A-conjugate, it turns out that the residual vectors are also orthogonal to one another: that is,

$$\mathbf{r}^{(m)^T}\mathbf{r}^{(l)} = 0 \qquad \text{for } m \neq l.$$

For a proof of this result, see Golub and van Loan [2].

Solving equation (1) for $A\mathbf{d}^{(m)}$ gives

$$A\mathbf{d}^{(m)} = \frac{1}{\lambda_m}\left[\mathbf{r}^{(m+1)} - \mathbf{r}^{(m)}\right]. \tag{2}$$

Premultiplying (2) by $\mathbf{d}^{(m)^T}$ then yields

$$\begin{aligned}\mathbf{d}^{(m)^T}A\mathbf{d}^{(m)} &= \frac{1}{\lambda_m}\mathbf{d}^{(m)^T}\left[\mathbf{r}^{(m+1)} - \mathbf{r}^{(m)}\right] \\ &= -\frac{1}{\lambda_m}\mathbf{d}^{(m)^T}\mathbf{r}^{(m)} \\ &= \frac{1}{\lambda_m}\left[\mathbf{r}^{(m)} - \alpha_{m-1}\mathbf{d}^{(m-1)}\right]^T\mathbf{r}^{(m)} \\ &= \frac{1}{\lambda_m}\mathbf{r}^{(m)^T}\mathbf{r}^{(m)}.\end{aligned} \tag{3}$$

In going from the first line to the second and from the third line to the fourth, the orthogonality of the previous search direction and the new residual vector has been used. Equation (3) gives the new formula for λ_m:

$$\lambda_m = \frac{\mathbf{r}^{(m)^T}\mathbf{r}^{(m)}}{\mathbf{d}^{(m)T}A\mathbf{d}^{(m)}}.$$

The vector-vector product $\mathbf{r}^{(m)^T}\mathbf{r}^{(m)}$ will have been computed for the convergence test during the previous iteration, so the new formula for λ_m saves one vector-vector product over the original formula.

Let's now return to equation (2) and premultiply by $\mathbf{r}^{(m+1)^T}$. The result of this operation is

$$\begin{aligned}\mathbf{r}^{(m+1)^T}A\mathbf{d}^{(m)} &= \frac{1}{\lambda_m}\mathbf{r}^{(m+1)^T}\left[\mathbf{r}^{(m+1)} - \mathbf{r}^{(m)}\right] \\ &= \frac{1}{\lambda_m}\mathbf{r}^{(m+1)^T}\mathbf{r}^{(m+1)},\end{aligned} \tag{4}$$

where the term $\mathbf{r}^{(m+1)^T}\mathbf{r}^{(m)}$ has been eliminated by applying the orthogonality of the residual vectors. Dividing (4) by (3) produces

$$\alpha_m = \frac{\mathbf{r}^{(m+1)^T}\mathbf{r}^{(m+1)}}{\mathbf{r}^{(m)^T}\mathbf{r}^{(m)}}.$$

This formula reduces the operation count for each iteration of the conjugate gradient method by one more vector-vector product.

Taking into account the new formulas for $\mathbf{r}^{(m+1)}$, λ_m and α_m, we arrive at an efficient algorithm for the conjugate gradient method:

$$
\begin{array}{l}
\mathbf{r}^{(0)} = A\mathbf{x}^{(0)} - \mathbf{b} \\
\mathbf{d}^{(0)} = -\mathbf{r}^{(0)} \\
\text{set } \delta^{(0)} = \mathbf{r}^{(0)^T}\mathbf{r}^{(0)} \\
\text{for } m = 0,\ 1,\ 2,\ \ldots \\
\qquad \text{set } \mathbf{u} = A\mathbf{d}^{(m)} \\
\qquad \lambda_m = \delta^{(m)}/\mathbf{d}^{(m)^T}\mathbf{u} \\
\qquad \mathbf{x}^{(m+1)} = \mathbf{x}^{(m)} + \lambda_m\mathbf{d}^{(m)} \\
\qquad \mathbf{r}^{(m+1)} = \mathbf{r}^{(m)} + \lambda_m\mathbf{u} \\
\qquad \text{set } \delta^{(m+1)} = \mathbf{r}^{(m+1)^T}\mathbf{r}^{(m+1)} \\
\qquad \text{if } \sqrt{\delta^{(m+1)}} < TOL,\ \text{OUTPUT } \mathbf{x}^{(m+1)} \\
\qquad \alpha_m = \delta^{(m+1)}/\delta^{(m)} \\
\qquad \mathbf{d}^{(m+1)} = -\mathbf{r}^{(m+1)} + \alpha_m\mathbf{d}^{(m)}.
\end{array}
$$

This algorithm requires only one matrix-vector product, two vector-vector products, and three vector additions per iteration.

EXAMPLE 3.25 Conjugate Gradient Method in Action

Consider the linear system

$$\begin{bmatrix} 1 & -1/4 & -1/4 & 0 \\ -1/4 & 1 & 0 & -1/4 \\ -1/4 & 0 & 1 & -1/4 \\ 0 & -1/4 & -1/4 & 1 \end{bmatrix} \mathbf{x} = \begin{bmatrix} -1 \\ 0 \\ 1 \\ -1 \end{bmatrix}.$$

Let's take $\mathbf{x}^{(0)} = \begin{bmatrix} 0 & 0 & 0 & 0 \end{bmatrix}^T$. Then

$$\mathbf{r}^{(0)} = A\mathbf{x}^{(0)} - \mathbf{b} = -\mathbf{b} = \begin{bmatrix} 1 & 0 & -1 & 1 \end{bmatrix}^T$$

and

$$\mathbf{d}^{(0)} = -\mathbf{r}^{(0)} = \begin{bmatrix} -1 & 0 & 1 & -1 \end{bmatrix}^T.$$

We have three preliminary calculations to make before determining the step size λ_0:

$$
\begin{array}{l}
\delta^{(0)} = \mathbf{r}^{(0)^T}\mathbf{r}^{(0)} = 3; \\
\mathbf{u} = A\mathbf{d}^{(0)} = \begin{bmatrix} -5/4 & 1/2 & 3/2 & -5/4 \end{bmatrix}^T; \quad \text{and} \\
\mathbf{d}^{(0)^T}\mathbf{u} = 4.
\end{array}
$$

Therefore,

$$\lambda_0 = \frac{\delta^{(0)}}{\mathbf{d}^{(0)^T}\mathbf{u}} = \frac{3}{4},$$

and

$$\mathbf{x}^{(1)} = \mathbf{x}^{(0)} + \lambda_0\mathbf{d}^{(0)} = \begin{bmatrix} -3/4 & 0 & 3/4 & -3/4 \end{bmatrix}^T.$$

The residual associated with this new approximate solution is given by

$$\mathbf{r}^{(1)} = \mathbf{r}^{(0)} + \lambda_0\mathbf{u} = \begin{bmatrix} 1 \\ 0 \\ -1 \\ 1 \end{bmatrix} + \frac{3}{4}\begin{bmatrix} -5/4 \\ 1/2 \\ 3/2 \\ -5/4 \end{bmatrix} = \begin{bmatrix} 1/16 \\ 3/8 \\ 1/8 \\ 1/16 \end{bmatrix},$$

and $\delta^{(1)} = \mathbf{r}^{(1)^T}\mathbf{r}^{(1)} = 21/128$. Then

$$\alpha_0 = \frac{\delta^{(1)}}{\delta^{(0)}} = \frac{7}{128}$$

and

$$\mathbf{d}^{(1)} = -\mathbf{r}^{(1)} + \alpha_0\mathbf{d}^{(0)} = \begin{bmatrix} -1/16 \\ -3/8 \\ -1/8 \\ -1/16 \end{bmatrix} + \frac{7}{128}\begin{bmatrix} -1 \\ 0 \\ 1 \\ -1 \end{bmatrix} = \begin{bmatrix} -15/128 \\ -3/8 \\ -9/128 \\ -15/128 \end{bmatrix}.$$

This completes the first iteration.

To start the second iteration, only two preliminary calculations are needed to determine the step size λ_1:

$$\mathbf{u} = A\mathbf{d}^{(1)} = \begin{bmatrix} -3/512 & -81/256 & -3/256 & -3/512 \end{bmatrix}^T; \quad \text{and}$$
$$\mathbf{d}^{(1)^T}\mathbf{u} = 495/4096.$$

Recall that the value $\delta^{(1)} = \mathbf{r}^{(1)^T}\mathbf{r}^{(1)} = 21/128$ is available from the first iteration. Hence,

$$\lambda_1 = \frac{\delta^{(1)}}{\mathbf{d}^{(1)^T}\mathbf{u}} = \frac{224}{165},$$

$$\mathbf{x}^{(2)} = \mathbf{x}^{(1)} + \lambda_1\mathbf{d}^{(1)} = \begin{bmatrix} -3/4 \\ 0 \\ 3/4 \\ -3/4 \end{bmatrix} + \frac{224}{165}\begin{bmatrix} -15/128 \\ -3/8 \\ -9/128 \\ -15/128 \end{bmatrix} = \begin{bmatrix} -10/11 \\ -28/55 \\ 36/55 \\ -10/11 \end{bmatrix}$$

and

$$\mathbf{r}^{(2)} = \mathbf{r}^{(1)} + \lambda_1\mathbf{u} = \begin{bmatrix} 3/55 & -3/55 & 6/55 & 3/55 \end{bmatrix}^T.$$

To complete the second iteration we compute

$$\delta^{(2)} = \mathbf{r}^{(2)^T}\mathbf{r}^{(2)} = \frac{63}{3025}$$

$$\alpha_1 = \frac{\delta^{(2)}}{\delta^{(1)}} = \frac{384}{3025}$$

and

$$\mathbf{d}^{(2)} = -\mathbf{r}^{(2)} + \alpha_1 \mathbf{d}^{(1)} = \begin{bmatrix} -42/605 & 21/3025 & -357/3025 & -42/605 \end{bmatrix}^T.$$

For the third iteration, we make the following calculations:

$$\mathbf{u} = A\mathbf{d}^{(2)} = \begin{bmatrix} -126/3025 & 126/3025 & -252/3025 & -126/3025 \end{bmatrix}^T;$$
$$\mathbf{d}^{(2)^T}\mathbf{u} = 2646/166375;$$

$$\lambda_2 = \frac{\delta^{(2)}}{\mathbf{d}^{(2)T}\mathbf{u}} = \frac{55}{42};$$

$$\mathbf{x}^{(3)} = \mathbf{x}^{(2)} + \lambda_2 \mathbf{d}^{(2)} = \begin{bmatrix} -1 & -1/2 & 1/2 & -1 \end{bmatrix}^T; \quad \text{and}$$

$$\mathbf{r}^{(3)} = \mathbf{r}^{(2)} + \lambda_2 \mathbf{u} = \begin{bmatrix} 0 & 0 & 0 & 0 \end{bmatrix}^T.$$

Thus, three iterations of the conjugate gradient method have produced the exact solution. Of course, all of our calculations were performed in exact arithmetic. What if we perform the calculations in finite-precision arithmetic? The following table displays the output from an implementation of the conjugate gradient method in MATLAB.

k	$\mathbf{x}^{(k)}$
0	$\begin{bmatrix} 0.000000 & 0.000000 & 0.000000 & 0.000000 \end{bmatrix}^T$
1	$\begin{bmatrix} -0.750000 & 0.000000 & 0.750000 & -0.750000 \end{bmatrix}^T$
2	$\begin{bmatrix} -0.909091 & -0.509091 & 0.654545 & -0.909091 \end{bmatrix}^T$
3	$\begin{bmatrix} -1.000000 & -0.500000 & 0.500000 & -1.000000 \end{bmatrix}^T$

The entries in $\mathbf{x}^{(3)}$ are actually correct to sixteen decimal places. For comparison, to achieve similar accuracy, the Jacobi method requires 55 iterations, the Gauss-Seidel method 29 iterations and the SOR method (with $\omega = 1.075$) 17 iterations.

Some Final Comments

We observed in the example presented above that the conjugate gradient method produced the exact solution to the linear system in only three iterations. This was not the result of a "lucky" choice of the coefficient matrix and the right-hand-side vector. When working in exact arithmetic, the conjugate gradient method will always produce the exact solution to an $n \times n$ system in at most n iterations. To see why this happens, suppose, without loss of generality, that $\mathbf{x}^{(0)} = \mathbf{0}$. With this choice for the initial vector, it follows that

$$\mathbf{x}^{(m)} = \sum_{k=1}^{m} c_k \mathbf{d}^{(k)}$$

for some scalars c_1, c_2, c_3, $\ldots$, c_m. By construction, the A-conjugate search directions are also linearly independent (see Exercise 11), so the set

$$\left\{\mathbf{d}^{(1)}, \mathbf{d}^{(2)}, \mathbf{d}^{(3)}, \ldots, \mathbf{d}^{(n)}\right\}$$

forms a basis for $\mathbf{R}^n$. The exact solution of the linear system will therefore lie in

$$\text{span}\left\{\mathbf{d}^{(1)}, \mathbf{d}^{(2)}, \mathbf{d}^{(3)}, \ldots, \mathbf{d}^{(m)}\right\}$$

for some $m \leq n$, and the iteration will terminate.

In finite precision arithmetic, roundoff error will generally prevent the direction vectors from remaining conjugate and the residual vectors from remaining orthogonal. The conjugate gradient method will still converge since each iteration is guaranteed to reduce the value of f, but convergence can be slow, particularly when the coefficient matrix is nearly singular. To handle this situation, it is common to modify, or precondition, the linear system prior to applying the conjugate gradient method. Preconditioning is accomplished by premultiplying the original system by the inverse of a nonsingular matrix, M, known as the preconditioning matrix. This multiplication converts $A\mathbf{x} = \mathbf{b}$ into $A^P\mathbf{x}^P = \mathbf{b}^P$, where $A^P = M^{-1}AM$, $\mathbf{x}^P = M^{-1}\mathbf{x}$ and $\mathbf{b}^P = M^{-1}\mathbf{b}$. Determining an appropriate preconditioning matrix M is a nontrivial task, but a good choice for M can result in convergence after $O(\sqrt{n})$ or fewer iterations. For a detailed discussion of preconditioning of the conjugate gradient method, consult Golub and van Loan [2], Ueberhuber [1], or Khosla and Rubin [3].

What if the coefficient matrix is nonsymmetric or not positive definite or both? A vast array of variants on the conjugate gradient method have been developed to handle these more general linear systems. For example, the generalized minimal residual (GMRES) method (Saad and Schultz [4]) will converge in at most n iterations for an $n \times n$ system, but the computational cost and storage requirements increase linearly with the iteration count. The biconjugate gradient (BiCG) method (Golub and van Loan [2]) replaces the orthogonal sequence of residuals by two mutually orthogonal sequences but no longer provides a minimization of the residuals, requires two matrix-vector products per iteration, often experiences irregular convergence behavior, and may even break down. The quasi-minimal residual (QMR) method (Freund and Nachtigal [5], [6]) converges about as rapidly as the GMRES method and typically exhibits convergence behavior that is much smoother than the BiCG method. The squared conjugate gradient (CGS) method (Sonneveld [7]) applies the basic operator of the BiCG method twice, so converges roughly twice as fast, but the convergence behavior is often erratic. The stabilized biconjugate gradient (BiGSTAB) method (van den Vorst [8]) can be interpreted as a combination of the BiCG method with a repeatedly applied GMRES method. This technique generally converges about as fast as the CGS method, but without the erratic convergence behavior. The two survey articles, Golub and O'Leary [9] and Broyden [10], would be good starting points for further study.

References

1. C. Ueberhuber, *Numerical Computation 2: Methods, Software and Analysis*, Springer-Verlag, Berlin, 1997.

2. G. Golub and C. van Loan, *Matrix Computations*, 3rd edition, Johns Hopkins Press, Baltimore, 1996.

3. P. K. Khosla and S. G. Rubin, "A Conjugate Gradient Iterative Method," *Computational Fluids*, **9**, 109–121, 1981.

4. Y. Saad and M. Schultz, "GMRES: A Generalized Minimal Residual Algorithm for Solving Nonsymmetric Linear Systems," *SIAM Journal on Scientific and Statistical Computing*, **7**, 856–869, 1986.

5. R. Freund and N. Nachtigal, "QMR: A Quasi-Minimal Residual Method for Non-Hermitian Linear Systems," *Numerische Mathematik*, **60**, 315–339, 1991.

6. R. Freund and N. Nachtigal, "An Implementation of the QMR Method Based on Two Coupled Two-Term Recurrences," Technical Report 92.15, RIACS, NASA Ames, 1992.

7. P. Sonneveld, "CGS: A Fast Lanczos-type Solver for Nonsymmetric Linear Systems," *SIAM Journal on Scientific and Statistical Computing*, **10**, 36–52, 1989.

8. H. A. van den Vorst, "Bi-CGSTAB: A Fast and Smoothly Converging Variant of Bi-CG for the Solution of Nonsymmetric Linear Systems," *SIAM Journal on Scientific and Statistical Computing*, **13**, 631–644, 1992.

9. G. Golub and D. O'Leary, "Some History of the Conjugate Gradient and Lanczos Methods," *SIAM Review*, **31**, 50–102, 1989.

10. C. G. Broyden, "A New Taxonomy of Conjugate Gradient Methods," *Computers and Mathematics with Applications*, **31**, 7–17, 1996.

EXERCISES

In Exercises 1–4, solve the indicated linear system using the conjugate gradient method in exact arithmetic. Show that the exact solution is obtained in each case in three or fewer iterations.

1.
$$\begin{aligned} 3x_1 - x_2 + 2x_3 &= -6 \\ -x_1 + 3x_2 + x_3 &= 3 \\ 2x_1 + x_2 + 3x_3 &= -4 \end{aligned}$$

2.
$$\begin{aligned} 4x_1 - x_2 &= 2 \\ -x_1 + 4x_2 - x_3 &= 4 \\ - x_2 + 4x_3 &= 10 \end{aligned}$$

3.
$$\begin{aligned} 6x_1 - 2x_2 + 3x_3 &= 11 \\ -2x_1 + 8x_2 + x_3 &= -9 \\ 3x_1 + x_2 + 7x_3 &= 9 \end{aligned}$$

4.
$$\begin{aligned} 3x_1 + x_2 - x_3 &= 2 \\ x_1 + 4x_2 + 2x_3 &= 7 \\ -x_1 + 2x_2 + 5x_3 &= 6 \end{aligned}$$

In Exercises 5–10, use the conjugate gradient method to solve the indicated linear system of equations. Take $\mathbf{x}^{(0)} = \mathbf{0}$, and use a convergence tolerance of 5×10^{-7}. Compare the number of iterations required to achieve convergence with the number of iterations required by the Jacobi method and the Gauss-Seidel method using the same starting vector and convergence tolerance. For Exercises 7 and 8, also determine the number of iterations required by the SOR method. The optimal values of the relaxation parameter for Exercises 7 and 8 are $\omega = 1.0923$ and $\omega = 1.1128$, respectively.

5.

$$\begin{aligned}
4x_1 + x_2 + x_3 - x_4 &= 8\\
x_1 + 8x_2 + 2x_3 + 3x_4 &= -12\\
x_1 + 2x_2 + 5x_3 - 2x_4 &= 15\\
-x_1 + 3x_2 - 2x_3 + 4x_4 &= -20
\end{aligned}$$

6.

$$\begin{aligned}
3x_1 - x_3 - x_5 &= 3\\
4x_2 + x_3 + 2x_6 &= 7\\
-x_1 \quad x_2 + 5x_3 \quad x_6 &= 6\\
6x_4 - x_5 - 2x_6 &= 11\\
-x_1 - x_4 + 7x_5 + 2x_6 &= 1\\
2x_2 + x_3 - 2x_4 + 2x_5 + 8x_6 &= 7
\end{aligned}$$

7.

$$\begin{aligned}
7x_1 - 3x_2 &= 4\\
-3x_1 + 9x_2 + x_3 &= -6\\
x_2 + 3x_3 - x_4 &= 3\\
-x_3 + 10x_4 + 4x_5 &= 7\\
4x_4 + 6x_5 &= 2
\end{aligned}$$

8.

$$\begin{aligned}
4x_1 - x_2 - x_4 &= -1\\
-x_1 + 4x_2 - x_3 - x_5 &= 0\\
- x_2 + 4x_3 - x_6 &= 1\\
-x_1 + 4x_4 - x_5 &= -2\\
- x_2 - x_4 + 4x_5 - x_6 &= 1\\
- x_3 - x_5 + 4x_6 &= 2
\end{aligned}$$

9.

$$\begin{aligned}
4x_1 + x_2 + x_3 + x_4 &= 33/2\\
x_1 + 3x_2 - x_3 + x_4 &= 1/2\\
x_1 - x_2 + 2x_3 &= 17/2\\
x_1 + x_2 + 3x_4 &= 27/2
\end{aligned}$$

10.

$$\begin{aligned}
10x_1 + x_2 + 2x_3 + 3x_4 + 4x_5 &= 12\\
x_1 + 9x_2 - x_3 + 2x_4 - 3x_5 &= -27\\
2x_1 - x_2 + 7x_3 + 3x_4 - 5x_5 &= 14\\
3x_1 + 2x_2 + 3x_3 + 12x_4 - x_5 &= -17\\
4x_1 - 3x_2 - 5x_5 - x_4 + 15x_5 &= 12
\end{aligned}$$

11. Let A be an $n \times n$ symmetric and positive definite matrix and suppose that the nonzero vectors $\mathbf{v}_1, \mathbf{v}_2, \mathbf{v}_3, \ldots, \mathbf{v}_n$ form an A-conjugate set; that is, $\mathbf{v}_i^T A \mathbf{v}_j = 0$ whenever $i \neq j$. Show that

$$c_1\mathbf{v}_1 + c_2\mathbf{v}_2 + c_3\mathbf{v}_3 + \cdots + c_n\mathbf{v}_n = \mathbf{0}$$

requires that $c_1 = c_2 = c_3 = \cdots = c_n = 0$. Hence the set

$$\{\mathbf{v}_1, \mathbf{v}_2, \mathbf{v}_3, \ldots, \mathbf{v}_n\}$$

is linearly independent and forms a basis for $\mathbf{R}^n$.

12. A simpler choice for the search direction would be to set $\mathbf{d}^{(m)} = -\mathbf{r}^{(m)}$. This amounts to always selecting the direction in which f decreases most rapidly in the vicinity of $\mathbf{x}^{(m)}$ and produces what is known as the method of steepest descent. The resulting algorithm is summarized in the following pseudocode.

$\mathbf{r}^{(0)} = A\mathbf{x}^{(0)} - \mathbf{b}$
for $m = 0, 1, 2, \ldots$
  $\mathbf{d}^{(m)} = -\mathbf{r}^{(m)}$
  $\lambda_m = -\mathbf{d}^{(m)^T}\mathbf{r}^{(m)} / \mathbf{d}^{(m)^T} A\mathbf{d}^{(m)}$
  $\mathbf{x}^{(m+1)} = \mathbf{x}^{(m)} + \lambda_m \mathbf{d}^{(m)}$
  $\mathbf{r}^{(m+1)} = \mathbf{r}^{(m)} + \lambda_m A\mathbf{d}^{(m)}$
  if $\sqrt{\mathbf{r}^{(m+1)T}\mathbf{r}^{(m+1)}} < TOL$, OUTPUT $\mathbf{x}^{(m+1)}$

Solve the linear systems in Exercises 5–10 using the method of steepest descent with $\mathbf{x}^{(0)} = \mathbf{0}$ and a convergence tolerance of 5×10^{-7}. Compare the performance of the method of steepest descent with that of the conjugate gradient method.

13. The coefficients of the least squares cubic polynomial $a_0 + a_1 x + x_2 x^2 + a_3 x^3$ that fits the data

x	0	0.5	1.0	1.5	2.0	2.5	3.0	3.5
y	1.0	1.7	2.1	2.0	1.1	0.9	1.4	3.1

satisfy the linear system of equations

$$\begin{bmatrix} 8 & 14 & 35 & 98 \\ 14 & 35 & 98 & 292.25 \\ 35 & 98 & 292.25 & 906.5 \\ 98 & 292.25 & 906.5 & 2887.8125 \end{bmatrix} \begin{bmatrix} a_0 \\ a_1 \\ a_2 \\ a_3 \end{bmatrix} = \begin{bmatrix} 13.3 \\ 25.45 \\ 67.625 \\ 202.6375 \end{bmatrix}.$$

Determine the values of a_0, a_1, a_2, and a_3.

3.10 NONLINEAR SYSTEMS OF EQUATIONS

Suppose we need to solve the system of three nonlinear equations

$$\begin{aligned} x_1^3 - 2x_2 - 2 &= 0 \\ x_1^3 - 5x_3^2 + 7 &= 0 \\ x_2 x_3^2 - 1 &= 0. \end{aligned}$$

Although we cannot express this system in matrix notation because the equations are nonlinear, we can express the system in vector notation. First, define the functions

$$\begin{aligned} f_1(x_1, x_2, x_3) &= x_1^3 - 2x_2 - 2; \\ f_2(x_1, x_2, x_3) &= x_1^3 - 5x_3^2 + 7; \text{ and} \\ f_3(x_1, x_2, x_3) &= x_2 x_3^2 - 1. \end{aligned}$$

Note that each of these functions represents the left-hand side of one of the equations from the nonlinear system. Next, let $\mathbf{x} = \begin{bmatrix} x_1 & x_2 & x_3 \end{bmatrix}^T$, and construct the vector-valued function

$$\mathbf{F}(\mathbf{x}) = \begin{bmatrix} f_1(x_1, x_2, x_3) \\ f_2(x_1, x_2, x_3) \\ f_3(x_1, x_2, x_3) \end{bmatrix}.$$

In terms of this vector-valued function, the original system of three nonlinear equations can be expressed concisely as the single vector equation $\mathbf{F}(\mathbf{x}) = \mathbf{0}$.

The problem of finding a vector $\mathbf{x}$ for which the vector-valued function $\mathbf{F}$ evaluates to $\mathbf{0}$ (i.e., the zero vector) is a generalization of the rootfinding problem which was investigated in Chapter 2. It ought to be possible, then, to modify one of the techniques developed in that chapter to suit our present needs. We will focus our attention on Newton's method and Broyden's method, which is a modification of Newton's method.

Newton's Method for a System of Nonlinear Equations

Recall that given a scalar function, f, of a single scalar argument and given an initial approximation, x_0, for a root of that function, Newton's method computes a sequence of (hopefully) improved approximations to the root according to the rule

$$x_{n+1} = x_n - f(x_n)/f'(x_n).$$

Now, let $\mathbf{F}$ be a vector-valued function of a vector argument $\mathbf{x}$, assuming that both vectors contain m components. To apply Newton's method to the problem of approximating a solution of $\mathbf{F}(\mathbf{x}) = \mathbf{0}$, we would like to write

$$\mathbf{x}^{(n+1)} = \mathbf{x}^{(n)} - \mathbf{F}(\mathbf{x}^{(n)})/\mathbf{F}'(\mathbf{x}^{(n)}).$$

This, however, brings up the immediate question of what is meant by $\mathbf{F}'(\mathbf{x}^{(n)})$. First, $\mathbf{F}'(\mathbf{x}^{(n)})$ must include the derivative of each scalar component function with respect to each component of the argument vector. That's m^2 individual partial derivatives. These partial derivatives should be organized so that $d\mathbf{F} = \mathbf{F}'(\mathbf{x}^{(n)})\Delta\mathbf{x}$ provides an estimate for the change in $\mathbf{F}(\mathbf{x})$ when the argument changes from $\mathbf{x}$ to $\mathbf{x} + \Delta\mathbf{x}$. From multivariable calculus we know that

$$df = \frac{\partial f}{\partial x_1}\Delta x_1 + \frac{\partial f}{\partial x_2}\Delta x_2 + \frac{\partial f}{\partial x_3}\Delta x_3 + \cdots + \frac{\partial f}{\partial x_m}\Delta x_m$$

for a scalar function of m arguments, which suggests that the partial derivatives in $\mathbf{F}'(\mathbf{x})$ be organized into matrix form as follows:

$$\mathbf{F}'(\mathbf{x}) = \begin{bmatrix} \partial f_1/\partial x_1 & \partial f_1/\partial x_2 & \partial f_1/\partial x_3 & \cdots & \partial f_1/\partial x_m \\ \partial f_2/\partial x_1 & \partial f_2/\partial x_2 & \partial f_2/\partial x_3 & \cdots & \partial f_2/\partial x_m \\ \partial f_3/\partial x_1 & \partial f_3/\partial x_2 & \partial f_3/\partial x_3 & \cdots & \partial f_3/\partial x_m \\ & \vdots & & & \vdots \\ \partial f_m/\partial x_1 & \partial f_m/\partial x_2 & \partial f_m/\partial x_3 & \cdots & \partial f_m/\partial x_m \end{bmatrix}.$$

This matrix is known as the *Jacobian matrix* for the system and is typically denoted by $J(\mathbf{x})$. Having established that $\mathbf{F}'(\mathbf{x})$ is a matrix, this brings up a second question: how do we divide by a matrix? Simple. We multiply by its inverse. Thus, Newton's method for a system of equations takes the form

$$\mathbf{x}^{(n+1)} = \mathbf{x}^{(n)} - \left[J(\mathbf{x}^{(n)})\right]^{-1} \mathbf{F}(\mathbf{x}^{(n)}).$$

When implementing this scheme, we will not actually compute the inverse of the Jacobian matrix. Instead, we define

$$\mathbf{v}^{(n)} = -\left[J(\mathbf{x}^{(n)})\right]^{-1} \mathbf{F}(\mathbf{x}^{(n)}),$$

and then solve the linear system of equations

$$\left[J(\mathbf{x}^{(n)})\right] \mathbf{v}^{(n)} = -\mathbf{F}(\mathbf{x}^{(n)})$$

for $\mathbf{v}^{(n)}$. Once $\mathbf{v}^{(n)}$ is known, the next iterate is computed according to the rule $\mathbf{x}^{(n+1)} = \mathbf{x}^{(n)} + \mathbf{v}^{(n)}$.

EXAMPLE 3.26 A System of Three Nonlinear Algebraic Equations

Let's apply Newton's method to the system of three nonlinear algebraic equations

$$\begin{aligned} x_1^3 - 2x_2 - 2 &= 0 \\ x_1^3 - 5x_3^2 + 7 &= 0 \\ x_2 x_3^2 - 1 &= 0\,. \end{aligned}$$

Recall that this system is equivalent to the vector equation $\mathbf{F}(\mathbf{x}) = \mathbf{0}$, where

$$\mathbf{F}(\mathbf{x}) = \begin{bmatrix} f_1(x_1, x_2, x_3) \\ f_2(x_1, x_2, x_3) \\ f_3(x_1, x_2, x_3) \end{bmatrix} = \begin{bmatrix} x_1^3 - 2x_2 - 2 \\ x_1^3 - 5x_3^2 + 7 \\ x_2 x_3^2 - 1 \end{bmatrix}.$$

The Jacobian matrix associated with $\mathbf{F}(\mathbf{x})$ is easily found to be

$$J(\mathbf{x}) = \begin{bmatrix} 3x_1^2 & -2 & 0 \\ 3x_1^2 & 0 & -10x_3 \\ 0 & x_3^2 & 2x_2x_3 \end{bmatrix}.$$

Starting from the initial vector $\mathbf{x}^{(0)} = \begin{bmatrix} 1 & 1 & 1 \end{bmatrix}^T$, we compute

$$\mathbf{F}(\mathbf{x}^{(0)}) = \begin{bmatrix} -3 & 3 & 0 \end{bmatrix}^T$$

and

$$J(\mathbf{x}^{(0)}) = \begin{bmatrix} 3 & -2 & 0 \\ 3 & 0 & -10 \\ 0 & 1 & 2 \end{bmatrix}.$$

Solving the linear system $\left[J(\mathbf{x}^{(0)})\right]\mathbf{v}^{(0)} = -\mathbf{F}(\mathbf{x}^{(0)})$ yields the update vector $\mathbf{v}^{(0)} = \begin{bmatrix} 3/7 & -6/7 & 3/7 \end{bmatrix}^T$, and then $\mathbf{x}^{(1)} = \mathbf{x}^{(0)} + \mathbf{v}^{(0)} = \begin{bmatrix} 10/7 & 1/7 & 10/7 \end{bmatrix}^T$. Continuing to iterate until the maximum norm of $\mathbf{v}^{(n)}$ is less than 5×10^{-4}, we obtain the results listed below.

n	$\mathbf{x}^{(n)^T}$		
0	1.00000000000000	1.00000000000000	1.00000000000000
1	1.42857142857143	0.14285714285714	1.42857142857143
2	1.44011117287382	0.49305169538633	1.41331295163980
3	1.44225533875822	0.50000806218205	1.41421499021415
4	1.44224957033522	0.50000000001480	1.41421356237591

The exact solution to this system, in the neighborhood of the initial vector $\mathbf{x}^{(0)} = \begin{bmatrix} 1 & 1 & 1 \end{bmatrix}^T$, is $\mathbf{x} = \begin{bmatrix} \sqrt[3]{3} & 1/2 & \sqrt{2} \end{bmatrix}^T$. Thus, four iterations of Newton's method have produced results that are correct to eight decimal places.

Quasi-Newton Methods

Newton's method for a system of nonlinear equations provides the same quadratic convergence that was observed for scalar equations in Chapter 2. Unfortunately, the cost associated with each iteration of Newton's method for a system of equations is quite high. Not only does each iteration require the calculation of $\mathbf{F}(\mathbf{x}^{(k)})$, which is equivalent to m scalar function evaluations, it also requires the evaluation of $J(\mathbf{x}^{(k)})$. This adds m^2 scalar function evaluations to the tally, one for each of the m^2 partial derivatives that make up the Jacobian. On top of all of these scalar function evaluations, an additional $O(m^3)$ algebraic operations are needed to solve the linear system

$$J(\mathbf{x}^{(k)})\mathbf{v}^{(k)} = -\mathbf{F}(\mathbf{x}^{(k)})$$

for the update vector $\mathbf{v}^{(k)}$.

To reduce the per iteration workload, a variety of modified Newton methods, also known as inexact Newton methods and quasi-Newton methods, have been developed. These methods involve such procedures as using the Jacobian from the first iteration for all subsequent iterations, updating the Jacobian matrix only every kth iteration and performing a violent diagonalization of the Jacobian matrix. This last procedure produces a scheme that is similar in nature to the Gauss-Seidel method (Section 3.8) for linear systems. See Dennis and Schnabel [1], Eisenstat and Walker [2], and Ortega and Rheinboldt [3] for general discussions of quasi-Newton methods.

Below, we will develop one particularly popular quasi-Newton method, known as Broyden's method ([4] and [5]). Broyden's method is based on a clever procedure for constructing an approximation to the inverse of the Jacobian matrix as the iteration proceeds. The resulting scheme not only reduces the number of scalar function evaluations required each iteration, but also dramatically decreases the operation count associated with computing the update vector $\mathbf{v}^{(k)}$. On the down side, Broyden's method is only superlinearly convergent, as opposed to quadratically convergent. For most practical problems, the decrease in order of convergence is an acceptable trade-off for the reduction in computational effort. There are procedures which reduce the workload per iteration and maintain quadratic convergence, such as Brown's method ([6] and [7]) and Brent's method [8]. Moré and Cosnard [9] present a comparative study of some of the commonly used methods of this type. These methods, however, are considerably more difficult to implement efficiently than is Broyden's method.

Broyden's Method

The first iteration of Broyden's method is almost identical to that of Newton's method. Using an initial vector $\mathbf{x}^{(0)}$, $\mathbf{F}(\mathbf{x}^{(0)})$ and $J(\mathbf{x}^{(0)})$ are evaluated. For later convenience, we will denote the Jacobian matrix $J(\mathbf{x}^{(0)})$ by A_0. Next, compute A_0^{-1}. Actually compute a matrix inverse? Yes! This will turn out to be one of those rare situations in which it is more efficient to compute the inverse than to solve the corresponding system of equations. The reason for this will be made clear below. Then multiply A_0^{-1} into $\mathbf{F}(\mathbf{x}^{(0)})$ to form $\mathbf{v}^{(0)}$. Finally, update $\mathbf{x}^{(0)}$ to $\mathbf{x}^{(1)}$.

For all subsequent iterations, Broyden's method forgoes the calculation of the Jacobian. Rather, a matrix A_k is sought which approximates $J(\mathbf{x}^{(k)})$ in the sense that

$$A_k(\mathbf{x}^{(k)} - \mathbf{x}^{(k-1)}) = \mathbf{F}(\mathbf{x}^{(k)}) - \mathbf{F}(\mathbf{x}^{(k-1)}). \tag{1}$$

In one variable, the condition imposed in equation (1) is equivalent to

$$f'(x_k) \approx \frac{f(x_k) - f(x_{k-1})}{x_k - x_{k-1}};$$

hence, Broyden's method is a multivariable version of the Secant method. Unfortunately, (1) does not uniquely determine the matrix A_k – it provides only n equations for the n^2 elements of A_k. The best additional constraints (see Dennis and Schnabel [1]) are to require

$$A_k\mathbf{u} = A_{k-1}\mathbf{u} \text{ for all vectors } \mathbf{u} \text{ such that } (\mathbf{x}^{(k)} - \mathbf{x}^{(k-1)})^T\mathbf{u} = 0.$$

Combining these conditions with (1) uniquely determines (Dennis and Moré [10])

$$A_k = A_{k-1} - \frac{\mathbf{y} - A_{k-1}\mathbf{\Delta}}{\mathbf{\Delta}^T\mathbf{\Delta}}\mathbf{\Delta}^T, \tag{2}$$

where $\mathbf{y} = \mathbf{F}(\mathbf{x}^{(k)}) - \mathbf{F}(\mathbf{x}^{(k-1)})$ and $\mathbf{\Delta} = \mathbf{x}^{(k)} - \mathbf{x}^{(k-1)}$. By using the matrix A_k in place of $J(\mathbf{x}^{(k)})$, n^2 scalar function evaluations are saved each iteration.

A more significant reduction in computational effort is achieved by making use of the Sherman-Morrison formula [10]: If M is a nonsingular matrix and $\mathbf{u}$ and $\mathbf{v}$ are vectors for which $\mathbf{v}^TM^{-1}\mathbf{u} \neq -1$, then the matrix $M + \mathbf{u}\mathbf{v}^T$ is nonsingular and

$$(M + \mathbf{u}\mathbf{v}^T)^{-1} = M^{-1} - \frac{M^{-1}\mathbf{u}\mathbf{v}^TM^{-1}}{1 + \mathbf{v}^TM^{-1}\mathbf{u}}. \tag{3}$$

Identifying the matrix A_{k-1} with M, the vector $(\mathbf{y} - A_{k-1}\boldsymbol{\Delta})/(\boldsymbol{\Delta}^T\boldsymbol{\Delta})$ with $\mathbf{u}$ and the vector $\boldsymbol{\Delta}$ with $\mathbf{v}$, equations (2) and (3) imply

$$\begin{aligned} A_k^{-1} &= \left(A_{k-1} - \frac{\mathbf{y} - A_{k-1}\boldsymbol{\Delta}}{\boldsymbol{\Delta}^T\boldsymbol{\Delta}}\boldsymbol{\Delta}^T \right)^{-1} \\ &= A_{k-1}^{-1} - \frac{A_{k-1}^{-1}\left(\frac{\mathbf{y}-A_{k-1}\boldsymbol{\Delta}}{\boldsymbol{\Delta}^T\boldsymbol{\Delta}}\boldsymbol{\Delta}^T\right)A_{k-1}^{-1}}{1 + \boldsymbol{\Delta}^T A_{k-1}^{-1}\left(\frac{\mathbf{y}-A_{k-1}\boldsymbol{\Delta}}{\boldsymbol{\Delta}^T\boldsymbol{\Delta}}\right)} \\ &= A_{k-1}^{-1} + \frac{(\boldsymbol{\Delta} - A_{k-1}^{-1}\mathbf{y})\boldsymbol{\Delta}^T A_{k-1}^{-1}}{\boldsymbol{\Delta}^T A_{k-1}^{-1}\mathbf{y}}. \end{aligned} \tag{4}$$

Note that equation (4) uses only matrix-vector multiplication and hence provides a procedure for calculating A_k^{-1} directly from A_{k-1}^{-1} in only $O(n^2)$ algebraic operations. With A_k^{-1} known, the need to solve a linear system at each iteration to compute the update vector $\mathbf{v}^{(k)}$ is eliminated; instead, $\mathbf{v}^{(k)}$ is determined by forming the matrix-vector product $A_k^{-1}\mathbf{F}(\mathbf{x}^{(k)})$. Thus, each iteration of Broyden's method will require only $O(n^2)$ algebraic operations, as compared to the $O(n^3)$ algebraic operations required by Newton's method.

EXAMPLE 3.27 A System of Three Nonlinear Algebraic Equations Revisited

Let's apply Broyden's method to the same system of nonlinear algebraic equations that we investigated earlier, $\mathbf{F}(\mathbf{x}) = \mathbf{0}$, where

$$\mathbf{F}(\mathbf{x}) = \begin{bmatrix} f_1(x_1, x_2, x_3) \\ f_2(x_1, x_2, x_3) \\ f_3(x_1, x_2, x_3) \end{bmatrix} = \begin{bmatrix} x_1^3 - 2x_2 - 2 \\ x_1^3 - 5x_3^2 + 7 \\ x_2 x_3^2 - 1 \end{bmatrix}.$$

Recall that the Jacobian matrix associated with $\mathbf{F}(\mathbf{x})$ is given by

$$J(\mathbf{x}) = \begin{bmatrix} 3x_1^2 & -2 & 0 \\ 3x_1^2 & 0 & -10x_3 \\ 0 & x_3^2 & 2x_2x_3 \end{bmatrix}.$$

Starting from the initial vector $\mathbf{x}^{(0)} = \begin{bmatrix} 1 & 1 & 1 \end{bmatrix}^T$, we compute

$$\mathbf{F}(\mathbf{x}^{(0)}) = \begin{bmatrix} -3 & 3 & 0 \end{bmatrix}^T$$

and

$$A_0 = J(\mathbf{x}^{(0)}) = \begin{bmatrix} 3 & -2 & 0 \\ 3 & 0 & -10 \\ 0 & 1 & 2 \end{bmatrix}.$$

It follows that

$$A_0^{-1} = \frac{1}{42}\begin{bmatrix} 10 & 4 & 20 \\ -6 & 6 & 30 \\ 3 & -3 & 6 \end{bmatrix},$$

$$\mathbf{v}^{(0)} = -A_0^{-1}\mathbf{F}(\mathbf{x}^{(0)}) = \frac{1}{42}\begin{bmatrix} 10 & 4 & 20 \\ -6 & 6 & 30 \\ 3 & -3 & 6 \end{bmatrix}\begin{bmatrix} -3 \\ 3 \\ 0 \end{bmatrix} = \begin{bmatrix} 3/7 \\ -6/7 \\ 3/7 \end{bmatrix}$$

and $\mathbf{x}^{(1)} = \mathbf{x}^{(0)} + \mathbf{v}^{(0)} = \begin{bmatrix} 10/7 & 1/7 & 10/7 \end{bmatrix}^T$.

For the next iteration, we start by computing

$$\mathbf{F}(\mathbf{x}^{(1)}) = \begin{bmatrix} 0.62973760932945 \\ -0.28862973760933 \\ -0.70845481049563 \end{bmatrix}$$

and

$$\mathbf{y} = \mathbf{F}(\mathbf{x}^{(1)}) - \mathbf{F}(\mathbf{x}^{(0)}) = \begin{bmatrix} 3.62973760932945 \\ -3.28862973760933 \\ -0.70845481049563 \end{bmatrix}$$

and also noting that $\boldsymbol{\Delta} = \mathbf{v}^{(0)}$. To compute A_1^{-1} according to equation (4), we will need the intermediate results

$$A_0^{-1}\mathbf{y} = \begin{bmatrix} 0.21366097459392 \\ -1.49437734277384 \\ 0.39296126613911 \end{bmatrix},$$

$$\boldsymbol{\Delta}^T A_0^{-1} = \begin{bmatrix} 0.25510204081633 & -0.11224489795918 & -0.34693877551020 \end{bmatrix}$$

and $\boldsymbol{\Delta}^T A_0^{-1}\mathbf{y} = 1.54087582554888$. It then follows that

$$A_1^{-1} = \begin{bmatrix} 0.27367506516073 & 0.07958297132928 & 0.42780191138141 \\ -0.03735881841877 & 0.09643788010426 & 0.57080799304952 \\ 0.07732406602954 & -0.07402258905300 & 0.13483927019983 \end{bmatrix},$$

$$\mathbf{v}^{(1)} = -A_1^{-1}\mathbf{F}(\mathbf{x}^{(1)}) = \begin{bmatrix} 0.15370485292292 \\ 0.45575276157379 \\ 0.02546853667618 \end{bmatrix}$$

and

$$\mathbf{x}^{(2)} = \mathbf{x}^{(1)} + \mathbf{v}^{(1)} = \begin{bmatrix} 1.58227628149435 \\ 0.59860990443093 \\ 1.45403996524761 \end{bmatrix}.$$

Continuing to iterate until the maximum norm of $\mathbf{v}^{(n)}$ is less than 5×10^{-4}, we obtain the results listed below. As expected, Broyden's method requires more iterations than does Newton's method, but each iteration is cheaper in terms of computational effort. The l_∞-norm of the difference between $\mathbf{x}^{(7)}$ and the exact solution $\mathbf{x} = \begin{bmatrix} \sqrt[3]{3} & 1/2 & \sqrt{2} \end{bmatrix}^T$ is 1.038×10^{-5}.

n	$\mathbf{x}^{(n)^T}$		
0	1.00000000000000	1.00000000000000	1.00000000000000
1	1.42857142857143	0.14285714285714	1.42857142857143
2	1.58227628149435	0.59860990443093	1.45403996524761
3	1.35508942048637	0.49542848694730	1.41165011868362
4	1.44643226689446	0.50532150828807	1.40994166582347
5	1.44288531438848	0.49855503157583	1.41541277217107
6	1.44207572772478	0.50011152020649	1.41412254801611
7	1.44225995004753	0.49999540342064	1.41421807651088

Application Problem 1: Coupled Reversible Chemical Reactions

In the Chapter 3 Overview (see page 141), we developed the system of equations

$$k_1 = \frac{c_1 + c_2}{(A_0 - 2c_1 - c_2)^2(B_0 - c_1)} \quad \text{and} \quad k_2 = \frac{c_1 + c_2}{(A_0 - 2c_1 - c_2)(D_0 - c_2)}$$

for the number of moles, c_1 and c_2, of a chemical C produced at equilibrium by the coupled chemical reactions

$$\begin{aligned} 2A + B &\rightleftharpoons C \\ A + D &\rightleftharpoons C. \end{aligned}$$

The subscript 1 refers to the first of these reactions, and the subscript 2 refers to the second. The parameters A_0, B_0, and D_0 are the initial number of moles of the chemicals A, B, and D, respectively, injected into the reaction chamber. The equilibrium constants for the two reactions are denoted by k_1 and k_2.

Suppose $A_0 = 20$ moles, $B_0 = D_0 = 10$ moles, $k_1 = 1.63 \times 10^{-4}$, and $k_2 = 3.27 \times 10^{-3}$. Substituting these values into the equations for c_1 and c_2 and rearranging the terms yields

$$\begin{aligned} c_1 + c_2 - 1.63 \times 10^{-4}(20 - 2c_1 - c_2)^2(10 - c_1) &= 0 \\ c_1 + c_2 - 3.27 \times 10^{-3}(20 - 2c_1 - c_2)(10 - c_2) &= 0\,. \end{aligned}$$

Applying Newton's method to this system of nonlinear equations, with an initial guess of $\mathbf{x}^{(0)} = \begin{bmatrix} 0.5 & 0.5 \end{bmatrix}^T$ and a convergence tolerance of 5×10^{-6}, the following values were obtained after four iterations:

$$c_1 = 0.10987 \quad \text{and} \quad c_2 = 0.49001.$$

Therefore, there are $0.10987 + 0.49001 = 0.59988$ moles of C present at equilibrium.

Application Problem 2: Flow Distribution in a Pipe Flow Network

The following diagram shows a pipe network through which water at 20°C is flowing. Given that the pump produces an outlet pressure of 5.2×10^5 Pa, we would like to determine the volumetric flow rates (measured in m^3/s) through each pipe in the

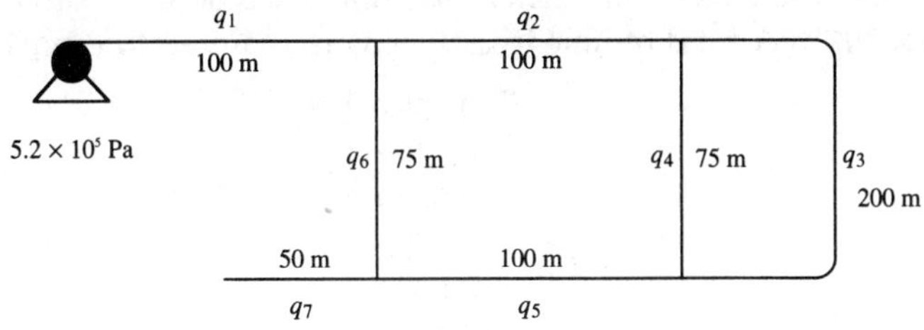

network. These quantities are labeled q_1, q_2, q_3, q_4, q_5, q_6, and q_7 in the diagram. The length of each pipe is also listed.

The analysis of a pipe network such as this is similar to the analysis of an electric circuit. We focus on junctions and loops. At each junction, the rate at which fluid enters the junction must equal the rate at which fluid leaves the junction. Starting with the junction at the upper left and proceeding clockwise about the network, we obtain the equations

$$\begin{aligned} q_1 - q_2 - q_6 &= 0; \\ q_2 - q_3 - q_4 &= 0; \\ q_3 + q_4 - q_5 &= 0; \quad \text{and} \\ q_5 + q_6 - q_7 &= 0. \end{aligned}$$

Around any loop in the network, the sum of the pressure drops around the loop must equal zero. The pressure drop along each pipe is due to friction and is given by the Darcy-Weisbach equation (see White [11])

$$\text{pressure drop} = \frac{8f\rho L}{\pi^2 d^5} q^2.$$

Here, f is the Darcy friction factor, ρ is the density of the fluid, L is the length of the pipe, q is the volumetric flow rate, and d is the inside diameter of the pipe. Suppose that all of the pipes in this network have a friction factor $f = 0.02$ and an inside diameter $d = 0.2$ m. At 20°C, water has a density of 998 kg/m^3. Traveling clockwise around the rightmost, middle, and leftmost loops in the network and dividing the resulting equations by $\frac{8f\rho}{\pi^2 d^5}$ yields

$$\begin{aligned} 200q_3^2 - 75q_4^2 &= 0; \\ 100q_2^2 + 75q_4^2 + 100q_5^2 - 75q_6^2 &= 0; \quad \text{and} \\ 100q_1^2 + 75q_6^2 + 50q_7^2 - 5.2 \times 10^5 \frac{\pi^2(0.2)^5}{8(0.02)(998)} &= 0. \end{aligned}$$

Combining the junction equations with the loop equations produces a set of seven nonlinear equations in the seven unknown flow rates. Broyden's method was applied to this system, with an initial guess of $q_i = 0.1$ for each i. Iterations were terminated when the maximum norm of the difference between successive iterates fell below 5×10^{-6}. A total of nine iterations were required to compute

$$\mathbf{q} = \begin{bmatrix} 0.2388 \\ 0.0869 \\ 0.0330 \\ 0.0539 \\ 0.0869 \\ 0.1519 \\ 0.2388 \end{bmatrix}.$$

References

1. J. E. Dennis and R. E. Schnabel, *Numerical Methods for Unconstrained Optimization and Nonlinear Equations*, Prentice Hall, Englewood Cliffs, NJ, 1983.
2. A. C. Eisenstat and H. Walker, "Globally Convergent Inexact Newton Methods," *SIAM Journal of Optimization*, **4**, 393–422, 1994.
3. J. Ortega and W. Rheinboldt, *Iterative Solution of Nonlinear Equations in Several Variables*, Academic Press, New York, 1970.
4. C. G. Broyden, "A class of methods for solving nonlinear simultaneous equations," *Mathematics of Computation*, **19**, 577–593, 1965.
5. C. G. Broyden, "Quasi-Newton Methods and Their Application to Function Minimization," *Mathematics of Computation*, **21**, 368–381, 1967.
6. K. M. Brown, "A Quadratically Convergent Newton-Like Method Based upon Gaussian Elimination," *SIAM Journal on Numerical Analysis*, **6**, 560–569, 1969.
7. G. D. Byrne and C. A. Hall, eds., *Numerical Solution of Systems of Nonlinear Algebraic Equations*, Academic Press, New York, 1973.
8. R. P. Brent, "An Algorithm with Guaranteed Convergence for Finding a Zero of a Function," *Computer Journal*, **14**, 422–425, 1971.
9. J. J. Moré and M. Y. Cosnard, "Numerical Solution of Nonlinear Equations," *ACM Transactions on Mathematical Software*, **5**, 64–85, 1979.
10. J. Dennis and J.J. Moré, "Quasi-Newton Methods: Motivation and Theory," *SIAM Review*, **19**, 46–89, 1977.
11. F. White, *Fluid Mechanics*, 2nd edition, McGraw-Hill, New York, 1986.

EXERCISES

1. For each of the following nonlinear systems, write out the vector-valued function **F** associated with the system and compute the Jacobian of **F**.

(a) $\begin{aligned} x_1 - x_2 - x_1^3 &= 0 \\ x_1 + x_2 - x_2^3 &= 0 \end{aligned}$ **(b)** $\begin{aligned} 1 + x_2 - e^{-x_1} &= 0 \\ x_1^3 - x_2 &= 0 \end{aligned}$

(c) $\begin{aligned} 2x_1 - 3x_2 + x_3 - 4 &= 0 \\ 2x_1 + x_2 - x_3 + 4 &= 0 \\ x_1^2 + x_2^2 + x_3^2 - 4 &= 0 \end{aligned}$

2. For each of the nonlinear systems in Exercise 1, carry out two iterations of Newton's method. Use the initial vector indicated below.

(a) $\mathbf{x}^{(0)} = [\ \tfrac{1}{2} \ \ \tfrac{1}{2}\ ]^T$

(a) $\mathbf{x}^{(0)} = \begin{bmatrix} \frac{1}{2} & \frac{1}{2} \end{bmatrix}^T$

(b) $\mathbf{x}^{(0)} = \begin{bmatrix} 1 & 1 \end{bmatrix}^T$

(c) $\mathbf{x}^{(0)} = \begin{bmatrix} -\frac{1}{2} & -\frac{3}{2} & \frac{3}{2} \end{bmatrix}^T$

3. For each of the nonlinear systems in Exercise 1, carry out two iterations of Broyden's method. Use the initial vectors indicated in Exercise 2.

In Exercises 4–10, solve the indicated nonlinear system of equations using both Newton's method and Broyden's method. Use the indicated initial vector, and terminate the iteration process when the maximum norm of the difference between successive iterates falls below 5×10^{-6}. Compare the number of iterations required by the two methods to achieve convergence.

4. $$\begin{aligned} 5\cos x + 6\cos(x+y) - 10 &= 0 \\ 5\sin x + 6\sin(x+y) - 4 &= 0 \end{aligned} \qquad \mathbf{x}^{(0)} = \begin{bmatrix} 0.7 & 0.7 \end{bmatrix}^T$$

5. $$\begin{aligned} x_1^2 + x_2^2 + x_3^2 - 1 &= 0 \\ x_1^2 + x_3^2 - 0.25 &= 0 \\ x_1^2 + x_2^2 - 4x_3 &= 0 \end{aligned} \qquad \mathbf{x}^{(0)} = \begin{bmatrix} 1 & 1 & 1 \end{bmatrix}^T$$

6. $$\begin{aligned} x_1^2 + x_2^2 + x_3^2 - 10 &= 0 \\ x_1 + 2x_2 - 2 &= 0 \\ x_1 + 3x_3 - 9 &= 0 \end{aligned} \qquad \mathbf{x}^{(0)} = \begin{bmatrix} 2 & 0 & 2 \end{bmatrix}^T$$

7. $$\begin{aligned} x^3 + 10x - y - 5 &= 0 \\ x + y^3 - 10y + 1 &= 0 \end{aligned} \qquad \mathbf{x}^{(0)} = \begin{bmatrix} 1 & 0 \end{bmatrix}^T$$

8. $$\begin{aligned} x_1^2 + 50x_1 + x_2^2 + x_3^2 - 200 &= 0 \\ x_1^2 + 20x_2 + x_3^2 - 50 &= 0 \\ -x_1^2 - x_2^2 + 40x_3 + 75 &= 0 \end{aligned} \qquad \mathbf{x}^{(0)} = \begin{bmatrix} 2 & 2 & 2 \end{bmatrix}^T$$

9. $$\begin{aligned} 2x - \cos y &= 0 \\ 2y - \sin x &= 0 \end{aligned} \qquad \mathbf{x}^{(0)} = \begin{bmatrix} 0 & 0 \end{bmatrix}^T$$

10. $$\begin{aligned} x_2 - \tfrac{1}{3}x_1 &= 0 \\ \tfrac{x_1^2}{1+x_1^2} - \tfrac{1}{2}x_2 &= 0 \end{aligned} \qquad \mathbf{x}^{(0)} = \begin{bmatrix} 7 & 2 \end{bmatrix}^T$$

11. The following systems have the indicated number of solutions. Approximate each of the solutions to within a convergence tolerance of 5×10^{-5}.

(a)

$$\begin{aligned} e^x - y &= 0 \\ ey^2 - 6x - 4 &= 0 \end{aligned} \qquad \text{2 solutions}$$

(b)

$$\begin{aligned} 4x_1 - x_2 + x_3 - x_1x_4 &= 0 \\ -x_1 + 3x_2 - 2x_3 - x_2x_4 &= 0 \\ x_1 - 2x_2 + 3x_3 - x_3x_4 &= 0 \\ x_1^2 + x_2^2 + x_3^2 - 1 &= 0 \end{aligned} \qquad \text{4 solutions}$$

(c)

$$\begin{aligned} x^2 + y^2 - 5 &= 0 \\ x^3 + y^3 - 2 &= 0 \end{aligned} \qquad \text{2 solutions}$$

12. The filter coefficients—h_1, h_2, h_3, and h_4—for the Daubechies wavelet of length 4 are solutions of the system

$$\begin{aligned} h_1 + h_2 + h_3 + h_4 &= \sqrt{2} \\ h_1 - h_2 + h_3 - h_4 &= 0 \\ 3h_1 - 2h_2 + h_3 &= 0 \\ h_1^2 + h_2^2 + h_3^2 + h_4^2 &= 1\,. \end{aligned}$$

Determine h_1, h_2, h_3, and h_4.

13. (a) Repeat the "Coupled Reversible Chemical Reactions" application problem from the text using Broyden's method. Use the same initial vector and convergence tolerance as were used in the example.

(b) Repeat the "Flow Distribution in a Pipe Flow Network" application problem from the text using Newton's method. Use the same initial vector and convergence tolerance as were used in the example.

14. Repeat the "Coupled Reversible Chemical Reactions" application problem changing the parameter values to $A_0 = 5$ moles, $B_0 = 2$ moles, $D_0 = 1$ mole, $k_1 = 4.25 \times 10^{-2}$, and $k_2 = 0.286$.

15. Suppose 3 moles of chemical A, 2 moles of chemical B, and 1 mole of chemical D are injected into a one-liter reaction chamber and the coupled chemical reactions

$$\begin{aligned} A + 2B &\rightleftharpoons 2C \\ 2A + D &\rightleftharpoons C \end{aligned}$$

are allowed to proceed to equilibrium. The equilibrium constants for the reactions are $k_1 = 1.00 \times 10^{-2}$ and $k_2 = 5.12 \times 10^{-2}$. How many moles of C are present at equilibrium?

16. The diagram given below shows a pipe network through which water at 20°C is flowing. Given that the pump produces an outlet pressure of 4.1×10^5 Pa and that all of the pipes have a friction factor of $f = 0.00225$ and an inside diameter of $d = 0.15$ m, determine the volumetric flow rates (measured in m^3/s) through each pipe in the network.

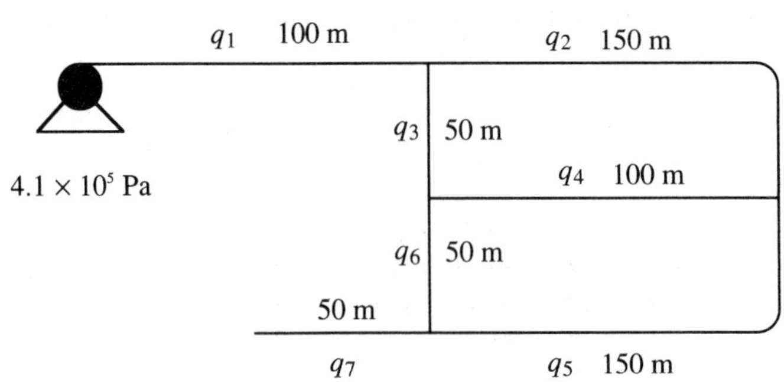

CHAPTER 4

Eigenvalues and Eigenvectors

AN OVERVIEW

Fundamental Mathematical Problem

In this chapter, we will develop a variety of techniques for approximating the eigenvalues and eigenvectors of an $n \times n$ matrix. Recall that an eigenvalue of a matrix A is any number, typically denoted by λ, for which the equation $A\mathbf{v} = \lambda\mathbf{v}$ has a nonzero solution for the vector $\mathbf{v}$. Since the equation $A\mathbf{v} = \lambda\mathbf{v}$ is equivalent to $(A - \lambda I)\mathbf{v} = \mathbf{0}$, we see that the eigenvalues of A are those values of λ for which the matrix $A - \lambda I$ is singular; that is, those values of λ for which $\det(A - \lambda I) = 0$.

As a function of λ, $\det(A - \lambda I)$ is an nth-degree polynomial, known as the *characteristic polynomial* of A. Counting multiplicities, an $n \times n$ matrix therefore has precisely n eigenvalues. Furthermore, the coefficients of the characteristic polynomial are sums and products of the elements in A. If A is a real matrix, it then follows that the eigenvalues must be real or occur in complex conjugate pairs. The collection of all eigenvalues of a matrix is known as the *spectrum* of the matrix.

The nonzero vector $\mathbf{v}$ for which $A\mathbf{v} = \lambda\mathbf{v}$ is called an *eigenvector* of the matrix A associated with the eignevalue λ. Since $\mathbf{v}$ is a solution to the matrix equation $(A - \lambda I)\mathbf{v} = 0$ when $A - \lambda I$ is singular, eigenvectors are not unique. They are, however, determined up to a multiplicative constant. In other words, if $\mathbf{v}$ is an eigenvector associated with the eigenvalue λ, then $\alpha\mathbf{v}$ is also an eigenvector associated with the same eigenvalue, for any nonzero constant α.

The "Steady-State Distribution of the British Workforce" problem capsule from the Chapter 1 Overview (see page 4) is one example that gives rise to an eigenvalue problem. Here is another example.

Measuring the Student Experience

Table 4.1 summarizes the correlations among seven different measures of the "student experience" for the four-year colleges and universities in the Commonwealth of Virginia. The measures include the percentage of first-year students who return for their second year (RET), the percentage of classes with fewer than 20 students (< 20), the percentage of classes with more than 50 students (> 50), the percentage of classes taught by full-time faculty (FTFAC), the average number of years needed to graduate in the current graduating class (GTIME), the percentage of first-time full-time students who graduate within six years (GRATE), and the donation rate for alumni (ALUM). Suppose we want to use the information in Table 4.1 to construct two or three composite measures, or indices, of the student experience, thereby making it easier to compare different institutions.

	RET	< 20	> 50	FTFAC	GTIME	GRATE	ALUM
RET	1.0000	−0.2411	0.4931	0.3009	−0.6865	0.9493	0.7538
< 20	−0.2411	1.0000	−0.5535	−0.0387	0.1256	−0.1698	0.0684
> 50	0.4931	−0.5535	1.0000	−0.2095	−0.1546	0.3972	−0.0643
FTFAC	0.3009	−0.0387	−0.2095	1.0000	−0.2357	0.3994	0.4033
GTIME	−0.68654	0.1256	−0.1546	−0.2357	1.0000	−0.77614	−0.7330
GRATE	0.9493	−0.1698	0.3972	0.3994	−0.7761	1.0000	0.7601
ALUM	0.7538	0.0684	−0.0643	0.4033	−0.7330	0.7601	1.0000

TABLE 4.1: Correlations Among Measures of the Student Experience at Four-Year Colleges and Universities in Virginia

Let R denote the 7×7 matrix of correlation values given in Table 4.1. The eigenvectors of R, scaled to unit length in the l_2-norm, are called the *principal components* of the original set of variables. Note that each principal component represents a specific linear combination of the variables. Moreover, the principal components are uncorrelated. Consequently, the principal components are ideal candidates for the composite measures we seek.

But which principal components should we use? The variation in the original data is divided among the principal components. In particular, the percentage of variation accounted for by each principal component is given by the ratio of the associated eigenvalue to the number of variables. It is therefore standard procedure to rank the principal components according to the size of the eigenvalues of R. The largest eigenvalue is associated with the *first principal component*, the next largest eigenvalue with the *second principal component*, and so on. In order to capture as much of the variation in the original data as possible, we should then choose the first few principal components as our composite measures of the student experience.

Remainder of the Chapter

Section 1 focuses on the power method, which is used to determine what is known as the dominant eigenvalue and its associated eigenvector. The inverse power method, which can be used to approximate the smallest eigenvalue of a matrix or to approximate the eigenvalue nearest to a given value, is described in Section 2. This method also produces an estimate for the associated eigenvector. The topic of deflation, which involves transforming a matrix so as to "remove" a previously determined eigenvalue from the spectrum, is discussed in Section 3. In the final two sections of the chapter, techniques for simultaneously approximating all of the eigenvalues of a symmetric matrix are presented. We first consider the reduction of symmetric matrices to tridiagonal form (Section 4) and finally discuss determining the eigenvalues for symmetric tridiagonal matrices (Section 5).

Localizing Eigenvalues

Before launching into the development of numerical techniques for approximating eigenvalues, let's consider an important analytical result. For obvious reasons, this theorem is known as the *Gerschgorin Circle Theorem.*

Theorem. Let A be an $n \times n$ matrix and define $r_i = \sum_{j=1,\ j\neq i}^{n} |a_{ij}|$ for each $i = 1, 2, 3, \ldots, n$. Further, let

$$C_i = \{z \in \mathbf{C} : |z - a_{ii}| \leq r_i\},$$

where $\mathbf{C}$ denotes the complex plane.
1. If λ is an eigenvalue of A, then λ lies in one of the circles C_i.
2. If k of the circles C_i form a connected region R in the complex plane, disjoint from the remaining $n - k$ circles, then the region R contains exactly k eigenvalues.

Proof. Let λ be an eigenvalue of A with associated eigenvector $\mathbf{x}$. Define $r_i = \sum_{j=1,\ j\neq i}^{n} |a_{ij}|$ for each $i = 1, 2, 3, \ldots, n$. Further, let k be an index for which $|x_k| = \|\mathbf{x}\|_\infty$. Equating the kth elements in the eigenvalue relation $A\mathbf{x} = \lambda\mathbf{x}$ yields

$$\sum_{j=1}^{n} a_{kj}x_j = \lambda x_k,$$

or

$$(\lambda - a_{kk})x_k = \sum_{j=1}^{k-1} a_{kj}x_j + \sum_{j=k+1}^{n} a_{kj}x_j.$$

Hence, upon taking the absolute value and repeatedly applying the triangle inequality,

$$\begin{aligned}
|\lambda - a_{kk}||x_k| &\leq \left|\sum_{j=1}^{k-1} a_{kj}x_j\right| + \left|\sum_{j=k+1}^{n} a_{kj}x_j\right| \\
&\leq \|\mathbf{x}\|_\infty \sum_{j=1}^{k-1} |a_{kj}| + \|\mathbf{x}\|_\infty \sum_{j=k+1}^{n} |a_{kj}| \\
&= r_k\|\mathbf{x}\|_\infty.
\end{aligned}$$

From here it follows that $|\lambda - a_{kk}| \leq r_k$, so $\lambda \in C_k$ and the first part of the theorem is proven. Ortega [1] contains a very readable proof of the second conclusion. □

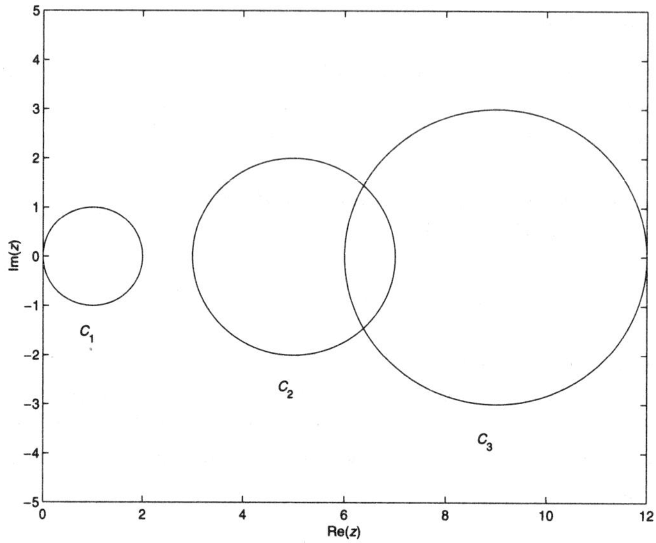

Figure 4.1 Gerschgorin circles for the matrix $A = \begin{bmatrix} 1 & -1 & 0 \\ 1 & 5 & 1 \\ -2 & -1 & 9 \end{bmatrix}$.

EXAMPLE 4.1 Localizing Eigenvalues using the Gerschgorin Circle Theorem

Consider the matrix

$$A = \begin{bmatrix} 1 & -1 & 0 \\ 1 & 5 & 1 \\ -2 & -1 & 9 \end{bmatrix}.$$

Proceeding row by row, we find that the radii of the Gerschgorin circles for this matrix are

$$\begin{aligned} r_1 &= |-1| + |0| = 1 \\ r_2 &= |1| + |1| = 2 \\ r_3 &= |-2| + |-1| = 3. \end{aligned}$$

The circles are therefore given by

$$\begin{aligned} C_1 &= \{z \in \mathbf{C} : |z-1| \le 1\} \\ C_2 &= \{z \in \mathbf{C} : |z-5| \le 2\} \\ C_3 &= \{z \in \mathbf{C} : |z-9| \le 3\} \end{aligned}$$

and are plotted in Figure 4.1. Note that C_1 is disjoint from the other circles, which implies that one of the eigenvalues must be contained in C_1. Furthermore, since A is a real matrix, the eigenvalue in C_1 must be real and therefore must lie on the closed interval $[0, 2]$. On the other hand, circles C_2 and C_3 overlap. Their union must therefore contain the two other eigenvalues of A. These final two eigenvalues could be two real eigenvalues or a complex conjugate pair.

References

1. J. Ortega, *Numerical Analysis—A Second Course*, Academic Press, New York, 1972.

4.1 THE POWER METHOD

Some matrix eigenvalue problems require the computation of a single eigenvalue, others the computation of several eigenvalues and yet others the computation of all of the eigenvalues. The corresponding eigenvectors may or may not also be needed. To handle each of these situations efficiently, we will need to develop several different solution strategies. In this section we will introduce the power method, an iterative technique for locating what is known as the dominant eigenvalue of a matrix. The power method also computes an associated eigenvector. In subsequent sections we will introduce extensions to the power method which allow for the computation of other eigenvalues.

Derivation of Method

Let A be an $n \times n$ matrix with eigenvalues $\lambda_1, \lambda_2, \lambda_3, \ldots, \lambda_n$, not necessarily distinct, that satisfy the relations $|\lambda_1| > |\lambda_2| \geq |\lambda_3| \geq \cdots \geq |\lambda_n|$. The eigenvalue λ_1, which is largest in magnitude, is known as the *dominant* eigenvalue of the matrix A. Furthermore, assume that the associated eigenvectors $\mathbf{v}_1$, $\mathbf{v}_2$, $\mathbf{v}_3, \ldots, \mathbf{v}_n$ are linearly independent, and therefore form a basis for $\mathbf{R}^n$. It should be noted at this point that not all matrices have eigenvalues and eigenvectors which satisfy the conditions we've assumed here. At the end of the section and in the exercises, we will explore what happens when these conditions are violated.

Let $\mathbf{x}^{(0)}$ be a nonzero element of $\mathbf{R}^n$. Since the eigenvectors of A form a basis for $\mathbf{R}^n$, it follows that $\mathbf{x}^{(0)}$ can be written as a linear combination of $\mathbf{v}_1$, $\mathbf{v}_2$, $\mathbf{v}_3, \ldots, \mathbf{v}_n$; that is, there exist constants $\alpha_1, \alpha_2, \alpha_3, \ldots, \alpha_n$ such that

$$\mathbf{x}^{(0)} = \alpha_1\mathbf{v}_1 + \alpha_2\mathbf{v}_2 + \alpha_3\mathbf{v}_3 + \cdots + \alpha_n\mathbf{v}_n.$$

Next, construct the sequence of vectors $\{\mathbf{x}^{(m)}\}$ according to the rule $\mathbf{x}^{(m)} = A\mathbf{x}^{(m-1)}$ for $m \geq 1$. By direct calculation we find

$$\begin{aligned}
\mathbf{x}^{(1)} &= A\mathbf{x}^{(0)} = \alpha_1(A\mathbf{v}_1) + \alpha_2(A\mathbf{v}_2) + \alpha_3(A\mathbf{v}_3) + \cdots + \alpha_n(A\mathbf{v}_n) \\
&= \alpha_1(\lambda_1\mathbf{v}_1) + \alpha_2(\lambda_2\mathbf{v}_2) + \alpha_3(\lambda_3\mathbf{v}_3) + \cdots + \alpha_n(\lambda_n\mathbf{v}_n), \\
\mathbf{x}^{(2)} &= A\mathbf{x}^{(1)} = A^2\mathbf{x}^{(0)} \\
&= \alpha_1(A^2\mathbf{v}_1) + \alpha_2(A^2\mathbf{v}_2) + \alpha_3(A^2\mathbf{v}_3) + \cdots + \alpha_n(A^2\mathbf{v}_n) \\
&= \alpha_1(\lambda_1^2\mathbf{v}_1) + \alpha_2(\lambda_2^2\mathbf{v}_2) + \alpha_3(\lambda_3^2\mathbf{v}_3) + \cdots + \alpha_n(\lambda_n^2\mathbf{v}_n)
\end{aligned}$$

and, in general,

$$\begin{aligned}\mathbf{x}^{(m)} &= A\mathbf{x}^{(m-1)} = \cdots = A^m\mathbf{x}^{(0)} \\ &= \alpha_1(A^m\mathbf{v}_1) + \alpha_2(A^m\mathbf{v}_2) + \alpha_3(A^m\mathbf{v}_3) + \cdots + \alpha_n(A^m\mathbf{v}_n) \\ &= \alpha_1(\lambda_1^m\mathbf{v}_1) + \alpha_2(\lambda_2^m\mathbf{v}_2) + \alpha_3(\lambda_3^m\mathbf{v}_3) + \cdots + \alpha_n(\lambda_n^m\mathbf{v}_n).\end{aligned}$$

In deriving these expressions we have made repeated use of the relation $A\mathbf{v}_j = \lambda_j\mathbf{v}_j$, which follows from the fact that $\mathbf{v}_j$ is an eigenvector associated with the eigenvalue λ_j.

Factoring λ_1^m from the right-hand side of the equation for $\mathbf{x}^{(m)}$ gives

$$\mathbf{x}^{(m)} = \lambda_1^m\left[\alpha_1\mathbf{v}_1 + \alpha_2\left(\frac{\lambda_2}{\lambda_1}\right)^m\mathbf{v}_2 + \alpha_3\left(\frac{\lambda_3}{\lambda_1}\right)^m\mathbf{v}_3 + \cdots + \alpha_n\left(\frac{\lambda_n}{\lambda_1}\right)^m\mathbf{v}_n\right]. \quad (1)$$

By assumption, $|\lambda_j/\lambda_1| < 1$ for each j, so $|\lambda_j/\lambda_1|^m \to 0$ as $m \to \infty$. It therefore follows that

$$\lim_{m\to\infty}\frac{\mathbf{x}^{(m)}}{\lambda_1^m} = \alpha_1\mathbf{v}_1.$$

Since any nonzero constant times an eigenvector is still an eigenvector associated with the same eigenvalue, we see that the scaled sequence $\{\mathbf{x}^{(m)}/\lambda_1^m\}$ converges to an eigenvector associated with the dominant eigenvalue provided $\alpha_1 \neq 0$. Furthermore, convergence toward the eigenvector is linear with asymptotic error constant $|\lambda_2/\lambda_1|$.

An approximation for the dominant eigenvalue of A can be obtained from the sequence $\{\mathbf{x}^{(m)}\}$ as follows. Let i be an index for which $x_i^{(m-1)} \neq 0$, and consider the ratio of the ith element from the vector $\mathbf{x}^{(m)}$ to the ith element from $\mathbf{x}^{(m-1)}$. By equation (1),

$$\frac{x_i^{(m)}}{x_i^{(m-1)}} = \frac{\lambda_1^m\alpha_1 v_{1,i}\left[1 + O\left((\lambda_2/\lambda_1)^m\right)\right]}{\lambda_1^{m-1}\alpha_1 v_{1,i}\left[1 + O\left((\lambda_2/\lambda_1)^{m-1}\right)\right]} = \lambda_1\left[1 + O\left((\lambda_2/\lambda_1)^{m-1}\right)\right],$$

provided $v_{1,i} \neq 0$, where $v_{1,i}$ denotes the ith element from the vector $\mathbf{v}_1$. Hence, the ratio $x_i^{(m)}/x_i^{(m-1)}$ converges toward the dominant eigenvalue, and the convergence is linear with asymptotic rate constant $|\lambda_2/\lambda_1|$.

To avoid overflow and underflow problems when calculating the sequence $\{\mathbf{x}^{(m)}\}$ (note that $\lim_{m\to\infty}\lambda_1^m \to \pm\infty$ when $|\lambda_1| > 1$, whereas $\lim_{m\to\infty}\lambda_1^m \to 0$ when $|\lambda_1| < 1$), it is common practice to scale the vectors $\mathbf{x}^{(m)}$ so that they are all of unit length. Here, we will use the l_∞-norm to measure vector length. Thus, in a practical implementation of the power method, the vector $\mathbf{x}^{(m)}$ would be computed in two steps: First multiply the previous vector by the matrix A and then scale the resulting vector to unit length.

To simplify the notation, let's introduce the vector $\mathbf{y}^{(m)}$ to denote the result of multiplying by the matrix A; that is, $\mathbf{y}^{(m)} = A\mathbf{x}^{(m-1)}$. $\mathbf{x}^{(m)}$ is then calculated by the formula

$$\mathbf{x}^{(m)} = \frac{\mathbf{y}^{(m)}}{y_{p_m}^{(m)}},$$

where p_m is an integer chosen so that $|y_{p_m}^{(m)}| = \|\mathbf{y}^{(m)}\|_\infty$. Note that p_m is an index into the vector $\mathbf{y}^{(m)}$. Whenever there is more than one possible choice for the index p_m, we will adopt the convention of always selecting the smallest value. The vector $\mathbf{x}^{(m)}$ now converges specifically to that multiple of $\mathbf{v}_1$ which has unit length measured in the infinity norm. As for the eigenvalue, since $\mathbf{x}^{(m-1)}$ is approximately an eigenvector associated with λ_1, $\mathbf{y}^{(m)} = A\mathbf{x}^{(m-1)} \approx \lambda_1 \mathbf{x}^{(m-1)}$. By construction $x_{p_{m-1}}^{(m-1)} = 1$, so it follows that $y_{p_{m-1}}^{(m)}$ converges to λ_1.

The power method is an iterative scheme, so a convergence tolerance must be specified and a stopping condition implemented. Three possibilities for the stopping condition immediately come to mind. Iteration could be terminated when

$$\begin{aligned} \left|\lambda^{(m)} - \lambda^{(m-1)}\right| &< TOL; \text{ when} \\ \|\mathbf{x}^{(m)} - \mathbf{x}^{(m-1)}\|_\infty &< TOL; \text{ or when} \\ \|A\mathbf{x}^{(m)} - \lambda^{(m)}\mathbf{x}^{(m)}\|_\infty &< TOL, \end{aligned}$$

where TOL denotes the specified convergence tolerance and $\lambda^{(m)}$ is used to denote the approximation to the eigenvalue during the mth iteration. These conditons represent checking for convergence of the eigenvalue, checking for convergence of the eigenvector and checking for convergence of the residual, respectively.

Checking for convergence of the residual most accurately reflects the underlying mathematical problem, as it measures how closely the eigenpair satisfies the eigenvalue equation, but using this convergence condition can be problematic. Knowing that eigenvalues are the roots of a polynomial, it should not be surprising that eigenvalues can be very ill conditioned. As with linear systems, where it was found that $A\mathbf{x}^\star - \mathbf{b}$ can be a poor estimator of the error in $\mathbf{x}^\star$ when A is ill conditioned, the residual $A\mathbf{x}^{(m)} - \lambda^{(m)}\mathbf{x}^{(m)}$ can be a poor estimator of the error in $\lambda^{(m)}$ and $\mathbf{x}^{(m)}$ when the eigenvalue λ is ill conditioned. Consult Golub and van Loan [1] or Ueberhuber [2] for details. Checking for convergence of the eigenvalue is the least computationally expensive option, but it is possible for the correct eigenvalue to be determined while $\mathbf{x}^{(m)}$ is still far from the true eigenvector (see Exercise 9). For these reasons, we choose to implement the stopping condition on convergence of the eigenvector.

EXAMPLE 4.2 A Demonstration of the Power Method

Consider the matrix

$$A = \begin{bmatrix} -2 & -2 & 3 \\ -10 & -1 & 6 \\ 10 & -2 & -9 \end{bmatrix},$$

whose eigenvalues are $\lambda_1 = -12$, $\lambda_2 = -3$, and $\lambda_3 = 3$. Let's start with the vector $\mathbf{x}^{(0)} = \begin{bmatrix} 1 & 0 & 0 \end{bmatrix}^T$, which already has an infinity norm of 1. Since the first element in $\mathbf{x}^{(0)}$ is the only element that has an absolute value of one, we set $p_0 = 1$.

For the first iteration of the power method, we compute

$$\mathbf{y}^{(1)} = A\mathbf{x}^{(0)} = \begin{bmatrix} -2 & -10 & 10 \end{bmatrix}^T,$$

from which we obtain our first estimate for the dominant eigenvalue: x$\lambda^{(1)} = y_{p_0}^{(1)} = y_1^{(1)} = -2$. Note that the infinity norm of the vector $\mathbf{y}^{(1)}$ is 10. Sticking with our convention of selecting the smallest index for which the magnitude of the vector element is equal to the infinity norm of the vector, we take $p_1 = 2$. Therefore, for the second iteration, we have

$$\mathbf{x}^{(1)} = \frac{\mathbf{y}^{(1)}}{-10} = \begin{bmatrix} 1/5 & 1 & -1 \end{bmatrix}^T.$$

The calculations for the second iteration produce the results

$$\begin{aligned} \mathbf{y}^{(2)} = A\mathbf{x}^{(1)} &= \begin{bmatrix} -27/5 & -9 & 9 \end{bmatrix}^T, \\ \lambda^{(2)} = y_{p_1}^{(2)} &= y_2^{(2)} = -9, \\ p_2 &= 2 \end{aligned}$$

and

$$\mathbf{x}^{(2)} = \frac{\mathbf{y}^{(2)}}{-9} = \begin{bmatrix} 3/5 & 1 & -1 \end{bmatrix}^T.$$

The third iteration then produces

$$\begin{aligned} \mathbf{y}^{(3)} = A\mathbf{x}^{(2)} &= \begin{bmatrix} -31/5 & -13 & 13 \end{bmatrix}^T, \\ \lambda^{(3)} = y_{p_2}^{(3)} &= y_2^{(3)} = -13, \\ p_3 &= 2 \end{aligned}$$

and

$$\mathbf{x}^{(3)} = \frac{\mathbf{y}^{(3)}}{-13} = \begin{bmatrix} 31/65 & 1 & -1 \end{bmatrix}^T.$$

The following table displays the output from the 11 iterations of the power method needed for the eigenvector to converge to within a tolerance of 5×10^{-6}. The final estimates are

$$\lambda_1 \approx -12.000014 \quad \text{and} \quad \mathbf{v}_1 \approx \begin{bmatrix} 0.500000 & 1.000000 & -1.000000 \end{bmatrix}^T.$$

The eigenvalue estimate is in error by roughly 0.0001%, while the eigenvector is correct to the digits shown.

The values in the column headed "Convergence" were computed according to the formula

$$\left| \frac{\lambda^{(j)} - \lambda^{(j-1)}}{\lambda^{(j-1)} - \lambda^{(j-2)}} \right|.$$

This quantity is an estimate for the asymptotic rate of linear convergence of the sequence $\{\lambda^{(j)}\}$ toward the value $\lambda_1 = -12$. Note the values in this column approach the value predicted by theory: $|\lambda_2/\lambda_1| = 3/12 = 0.25$.

j	$\mathbf{x}^{(j)^T}$			$\lambda^{(j)}$	Convergence
0	1.000000	0.000000	0.000000		
1	0.200000	1.000000	−1.000000	-2.000000	
2	0.600000	1.000000	−1.000000	-9.000000	
3	0.476923	1.000000	−1.000000	-13.000000	0.571429
4	0.505882	1.000000	−1.000000	-11.769231	0.307692
5	0.498537	1.000000	−1.000000	-12.058824	0.235294
6	0.500366	1.000000	−1.000000	-11.985366	0.253659
7	0.499908	1.000000	−1.000000	-12.003663	0.249084
8	0.500023	1.000000	−1.000000	-11.999085	0.250229
9	0.499994	1.000000	−1.000000	-12.000229	0.249943
10	0.500001	1.000000	−1.000000	-11.999943	0.250014
11	0.500000	1.000000	−1.000000	-12.000014	0.249996

Variation for Symmetric Matrices

When the matrix A is symmetric, a slight modification to the power method provides more rapid convergence: convergence is still linear, but the asymptotic rate constant is smaller. The modifications to the basic algorithm consist of using a different norm to scale the vectors $\mathbf{x}^{(m)}$ and using a different formula to compute the eigenvalue estimate. These changes are based on the following theorem, which is discussed in most textbooks on linear algebra (see, for example, Anton and Rorres [3], Lay [4], Leon [5], or Shifrin and Adams [6]). An elementary proof can be found in Anton and Rorres [3].

Theorem. If A is an $n \times n$ symmetric matrix, then there exists a set of n eigenvectors $\mathbf{v}_1$, $\mathbf{v}_2$, $\mathbf{v}_3, \ldots, \mathbf{v}_n$ that are orthogonal with respect to the standard inner product on $\mathbf{R}^n$; that is, whenever $i \neq j$

$$\mathbf{v}_i^T \mathbf{v}_j = 0.$$

Recall that for arbitrary vectors $\mathbf{x}, \mathbf{y} \in \mathbf{R}^n$, the standard inner product is the scalar quantity $\mathbf{x}^T\mathbf{y}$ (or $\mathbf{y}^T\mathbf{x}$), and associated with this inner product is the norm $\sqrt{\mathbf{x}^T\mathbf{x}}$. When written in component form,

$$\sqrt{\mathbf{x}^T\mathbf{x}} = \left(\sum_{i=1}^{n} x_i^2 \right)^{1/2},$$

we can readily see that this is just the l_2, or Euclidean, norm.

To exploit the orthogonality of the eigenvectors of a symmetric matrix within the power method, we will measure vector length and scale the vectors $\mathbf{x}^{(m)}$ to unit length using the Euclidean norm. Furthermore, we will compute an estimate for the dominant eigenvalue using the standard inner product as follows. Premultiplying both sides of the relation $\mathbf{y}^{(m)} \approx \lambda_1 \mathbf{x}^{(m-1)}$ by $\mathbf{x}^{(m-1)^T}$ yields $\mathbf{x}^{(m-1)^T}\mathbf{y}^{(m)} \approx \lambda_1 \mathbf{x}^{(m-1)^T}\mathbf{x}^{(m-1)} = \lambda_1$, since $\mathbf{x}^{(m-1)^T}\mathbf{x}^{(m-1)} = 1$ by construction. Putting these changes together, we arrive at the power method for symmetric matrices:

let $\mathbf{x}^{(0)}$ be a nonzero element of $\mathbf{R}^n$ with $\mathbf{x}^{(0)^T}\mathbf{x}^{(0)} = 1$. For $m = 1, 2, 3, \ldots$, calculate

$$\begin{aligned}\mathbf{y}^{(m)} &= A\mathbf{x}^{(m-1)}\\ \lambda^{(m)} &= \mathbf{x}^{(m-1)^T}\mathbf{y}^{(m)} \quad \text{and}\\ \mathbf{x}^{(m)} &= \mathbf{y}^{(m)}/\sqrt{\mathbf{y}^{(m)T}\mathbf{y}^{(m)}}.\end{aligned}$$

Then $\lambda^{(m)} \to \lambda_1$ and $\mathbf{x}^{(m)}$ converges to an eigenvector associated with λ_1 that has unit length in the Euclidean norm.

What about the convergence rate for this version of the power method? Using equation (1) and the orthogonality of the eigenvectors, it follows that

$$\begin{aligned}\mathbf{x}^{(m-1)} &= \frac{\lambda_1^{m-1}\left[\alpha_1\mathbf{v}_1 + \sum_{i=2}^n \alpha_i\left(\frac{\lambda_i}{\lambda_1}\right)^{m-1}\mathbf{v}_i\right]}{\sqrt{\lambda_1^{2m-2}\left[\alpha_1\mathbf{v}_1 + \sum_{i=2}^n \alpha_i\left(\frac{\lambda_i}{\lambda_1}\right)^{m-1}\mathbf{v}_i\right]^T\left[\alpha_1\mathbf{v}_1 + \sum_{i=2}^n \alpha_i\left(\frac{\lambda_i}{\lambda_1}\right)^{m-1}\mathbf{v}_i\right]}}\\ &= \frac{\alpha_1\mathbf{v}_1 + \sum_{i=2}^n \alpha_i(\lambda_i/\lambda_1)^{m-1}\mathbf{v}_i}{\sqrt{\alpha_1^2\mathbf{v}_1^T\mathbf{v}_1 + \sum_{i=2}^n \alpha_i^2(\lambda_i/\lambda_1)^{2m-2}\mathbf{v}_i^T\mathbf{v}_i}}.\end{aligned}$$

Then

$$\begin{aligned}\mathbf{x}^{(m-1)^T}\mathbf{y}^{(m)} &= \mathbf{x}^{(m-1)^T}A\mathbf{x}^{(m-1)}\\ &= \frac{\lambda_1\left[\alpha_1^2\mathbf{v}_1^T\mathbf{v}_1 + \sum_{i=2}^n \alpha_i^2(\lambda_i/\lambda_1)^{2m-1}\mathbf{v}_i^T\mathbf{v}_i\right]}{\left[\alpha_1^2\mathbf{v}_1^T\mathbf{v}_1 + \sum_{i=2}^n \alpha_i^2(\lambda_i/\lambda_1)^{2m-2}\mathbf{v}_i^T\mathbf{v}_i\right]}\\ &= \lambda_1\left\{1 + O\left((\lambda_2/\lambda_1)^{2(m-1)}\right)\right\},\end{aligned}$$

and $\mathbf{x}^{(m-1)^T}\mathbf{y}^{(m)} \to \lambda_1$ linearly with asymptotic rate constant $|\lambda_2/\lambda_1|^2$.

While the eigenvalues of a general matrix can be very ill conditioned, the eigenvalues of symmetric matrices are well conditioned. In particular, we have the following remarkable result.

Theorem. Let A be an $n \times n$ symmetric matrix, let $\hat{\lambda}$ be a real number, let $\hat{\mathbf{x}}$ be an n-vector with $\|\hat{\mathbf{x}}\|_2 = 1$ and define $\mathbf{r} = A\hat{\mathbf{x}} - \hat{\lambda}\hat{\mathbf{x}}$. Then A has an eigenvalue λ with

$$|\lambda - \hat{\lambda}| \leq \|\mathbf{r}\|_2.$$

Thus, for a symmetric matrix, the l_2-norm of the residual vector $\mathbf{r}$ provides an upper bound on the error in the eigenvalue estimate $\hat{\lambda}$. A proof of this theorem will be developed in Exercise 21. Unfortunately, even if $\|\mathbf{r}\|_2$ is small, the error in the eigenvector estimate $\hat{\mathbf{x}}$ may still be large (see Exercise 11). We will therefore continue to implement a stopping condition based on convergence of the eigenvector sequence.

With this variation of the power method for symmetric matrices, there is one glitch which can arise when checking for the convergence of the eigenvector sequence. When $\lambda_1 < 0$, the sequence $\{\mathbf{x}^{(m)}\}$ won't converge to a multiple of $\mathbf{v}_1$, but will alternate between a multiple of $\mathbf{v}_1$ and its opposite. To circumvent this "convergence problem," we can compare the norm of $\mathbf{x}^{(m)} - \text{sgn}(\lambda_1)\mathbf{x}^{(m-1)}$ to the convergence tolerance, rather than the norm of $\mathbf{x}^{(m)} - \mathbf{x}^{(m-1)}$.

EXAMPLE 4.3 Demonstration of Power Method for Symmetric Matrices

Consider the 4×4 symmetric matrix

$$A = \begin{bmatrix} 5.5 & -2.5 & -2.5 & -1.5 \\ -2.5 & 5.5 & 1.5 & 2.5 \\ -2.5 & 1.5 & 5.5 & 2.5 \\ -1.5 & 2.5 & 2.5 & 5.5 \end{bmatrix},$$

whose eigenvalues are $\lambda_1 = 12$, $\lambda_2 = 4$, $\lambda_3 = 4$, and $\lambda_4 = 2$. The eigenvector associated with λ_1 that has unit Euclidean norm is $\mathbf{v}_1 = \begin{bmatrix} -1/2 & 1/2 & 1/2 & 1/2 \end{bmatrix}^T$.

We will start the iteration with the vector $\mathbf{x}^{(0)} = \begin{bmatrix} 0.5 & 0.5 & 0.5 & 0.5 \end{bmatrix}^T$. Note that

$$\mathbf{x}^{(0)^T}\mathbf{x}^{(0)} = (0.5)(0.5) + (0.5)(0.5) + (0.5)(0.5) + (0.5)(0.5) = 1,$$

so this vector is already properly scaled in the Euclidean norm. For $m = 1$, we calculate

$$\begin{aligned} \mathbf{y}^{(1)} &= A\mathbf{x}^{(0)} = \begin{bmatrix} -0.5 & 3.5 & 3.5 & 4.5 \end{bmatrix}^T, \\ \lambda^{(1)} &= \mathbf{x}^{(0)^T}\mathbf{y}^{(1)} = (0.5)(-0.5) + (0.5)(3.5) + (0.5)(3.5) + (0.5)(4.5) \\ &= 5.5 \end{aligned}$$

and

$$\begin{aligned} \mathbf{x}^{(1)} &= \frac{\mathbf{y}^{(1)}}{\sqrt{\mathbf{y}^{(1)^T}\mathbf{y}^{(1)}}} = \frac{\begin{bmatrix} -0.5 & 3.5 & 3.5 & 4.5 \end{bmatrix}^T}{\sqrt{(-0.5)(-0.5) + (3.5)(3.5) + (3.5)(3.5) + (4.5)(4.5)}} \\ &= \frac{\begin{bmatrix} -0.5 & 3.5 & 3.5 & 4.5 \end{bmatrix}^T}{3\sqrt{5}}. \end{aligned}$$

Continuing on to the second iteration, we find

$$\begin{aligned} \mathbf{y}^{(2)} &= A\mathbf{x}^{(1)} = \frac{1}{3\sqrt{5}}\begin{bmatrix} -27 & 37 & 37 & 43 \end{bmatrix}^T, \\ \lambda^{(2)} &= \mathbf{x}^{(1)^T}\mathbf{y}^{(2)} = \frac{(-0.5)(-27) + (3.5)(37) + (3.5)(37) + (4.5)(43)}{(3\sqrt{5})(3\sqrt{5})} \\ &= \frac{466}{45} = 10.355556 \end{aligned}$$

and

$$\mathbf{x}^{(2)} = \frac{\mathbf{y}^{(2)}}{\sqrt{\mathbf{y}^{(2)^T}\mathbf{y}^{(2)}}} = \frac{\frac{1}{3\sqrt{5}}\begin{bmatrix} -27 & 37 & 37 & 43 \end{bmatrix}^T}{\sqrt{\frac{(-27)(-27)+(37)(37)+(37)(37)+(43)(43)}{45}}}$$
$$= \frac{\begin{bmatrix} -27 & 37 & 37 & 43 \end{bmatrix}^T}{2\sqrt{1329}}.$$

The table below displays the results of these first two iterations, together with the next eight. Calculations were terminated when the Euclidean norm of the difference between successive approximations to the eigenvector fell below 5×10^{-5}. The eigenvalue estimate is correct to the digits shown, while the eigenvector is correct to four digits in the first and fourth components and to all digits shown in the second and third components. The estimate for the asymptotic rate of linear convergence given in the last column of the table is in excellent agreement with the theoretical value: $|\lambda_2/\lambda_1|^2 = (4/12)^2 = 1/9$.

j	$\mathbf{x}^{(j)^T}$				$\lambda^{(j)}$	Convergence
0	[0.500000	0.500000	0.500000	0.500000]		
1	[−0.074536	0.521749	0.521749	0.670820]	5.500000	
2	[−0.370315	0.507469	0.507469	0.589761]	10.355556	
3	[−0.460013	0.501622	0.501622	0.533985]	11.799850	0.297452
4	[−0.487194	0.500309	0.500309	0.511882]	11.977899	0.123278
5	[−0.495812	0.500056	0.500056	0.504042]	11.997556	0.110403
6	[−0.498617	0.500010	0.500010	0.501360]	11.999729	0.110531
7	[−0.499541	0.500002	0.500002	0.500455]	11.999970	0.110923
8	[−0.499847	0.500000	0.500000	0.500152]	11.999997	0.111059
9	[−0.499949	0.500000	0.500000	0.500051]	12.000000	0.111098
10	[−0.499983	0.500000	0.500000	0.500017]	12.000000	0.111108

Application Problem 1: Age Demographics of a Female Population

The Leslie model describes the dynamics of the female portion of a population. In this model, females are divided into age classes of equal duration. Suppose there are a total of n distinct age classes. The birth and death processes which control the future evolution of the population are described by the parameters a_i $i = 1, 2, 3, \ldots, n)$ and b_i $(i = 1, 2, 3, \ldots, n-1)$, which measure the average number of daughters born to each female in the ith age class and the fraction of females in the ith age class who survive into the $(i+1)$st age class, respectively.

If $\mathbf{x}^{(k)}$ denotes the distribution of the population among the age classes at some time $t = t_k$, then the distribution at time $t = t_{k+1}$, denoted by the vector $\mathbf{x}^{(k+1)}$, is given by

$$\mathbf{x}^{(k+1)} = L\mathbf{x}^{(k)},$$

where the *Leslie matrix*, L, has the form

$$L = \begin{bmatrix} a_1 & a_2 & a_3 & \cdot & \cdot & \cdot & a_n \\ b_1 & 0 & 0 & \cdot & \cdot & \cdot & 0 \\ 0 & b_2 & 0 & \cdot & \cdot & \cdot & 0 \\ \cdot & \cdot & \cdot & \cdot & & & \cdot \\ \cdot & & \cdot & \cdot & \cdot & & \cdot \\ \cdot & & & \cdot & \cdot & \cdot & \cdot \\ 0 & \cdot & \cdot & \cdot & 0 & b_{n-1} & 0 \end{bmatrix}.$$

It is required that the elapsed time between $t = t_k$ and $t = t_{k+1}$ be equal to the width of the age classes. The dominant eigenvalue of the Leslie matrix indicates the growth rate of the female population. The coordinates of the associated eigenvector indicate the steady-state distribution of the population across the age brackets.

The following parameters for a sheep population were obtained by Caughley [7] from data collected by Hickey [8,9].

Age (years)	a_i	b_i
0	0.000	1.000
1	0.045	0.845
2	0.391	0.824
3	0.472	0.795
4	0.484	0.755
5	0.546	0.699
6	0.543	0.626
7	0.502	0.532
8	0.468	0.418
9	0.459	0.289
10	0.433	0.162
11	0.421	

Using the power method on the 12×12 Leslie matrix corresponding to this data, with $\mathbf{x}^{(0)} = \begin{bmatrix} 1 & 1 & 1 & 1 & 1 & 1 & 1 & 1 & 1 & 1 & 1 & 1 \end{bmatrix}^T$ and $TOL = 5 \times 10^{-6}$, produces

$$\lambda_1 \approx 1.08999$$

and

$$\mathbf{x} \approx \begin{bmatrix} 1.00000 & 0.91744 & 0.71123 & 0.53767 & 0.39216 & 0.27164 \\ 0.17420 & 0.10004 & 0.04883 & 0.01872 & 0.00496 & 0.00074 \end{bmatrix}^T.$$

Twenty-six iterations were required to achieve convergence. The eigenvalue indicates that the population in each age bracket will increase by 8.999% each year. Rescaling the eigenvector so that the sum of the components is one, we find that the model predicts 23.9% of the female population in the 0–1 age class, 22.0% in the 1–2 age class, 17.0% in the 2–3 age class, and so on.

Application Problem 2: Eigenvalues and Undirected Graphs

Geometrically, an undirected graph consists of a set of marked points, called *vertices*, together with a set of lines which connect pairs of vertices, called *edges*. For example, Figure 4.2 displays an undirected graph with seven vertices (labeled 1, 2, 3, ..., 7) and nine edges.

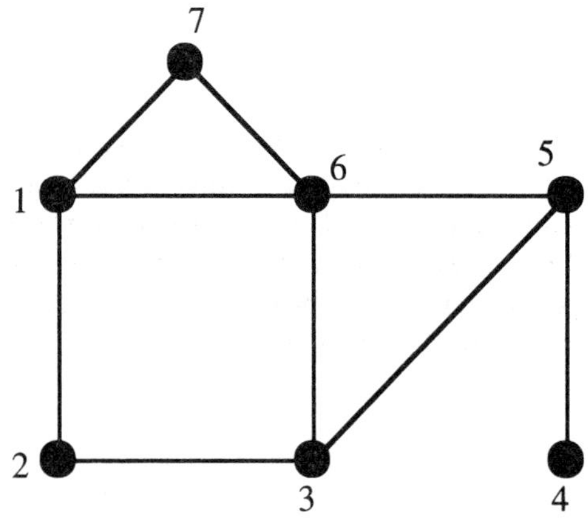

Figure 4.2 An undirected graph.

Two vertices that are connected by an edge are said to be *adjacent*. For instance, vertices 1 and 2 are adjacent in Figure 4.2, but vertices 1 and 3 are not. The overall adjacency structure of an undirected graph can be summarized in an *adjacency matrix*. For an undirected graph with n vertices, the adjacency matrix, which we shall denote by A, is the $n \times n$ matrix whose elements are defined by

$$a_{ij} = \begin{cases} 1, & \text{vertex } i \text{ is adjacent to vertex } j \\ 0, & \text{otherwise}. \end{cases}$$

The adjacency matrix for the undirected graph in Figure 4.2 is

$$A = \begin{bmatrix} 0 & 1 & 0 & 0 & 0 & 1 & 1 \\ 1 & 0 & 1 & 0 & 0 & 0 & 0 \\ 0 & 1 & 0 & 0 & 1 & 1 & 0 \\ 0 & 0 & 0 & 0 & 1 & 0 & 0 \\ 0 & 0 & 1 & 1 & 0 & 1 & 0 \\ 1 & 0 & 1 & 0 & 1 & 0 & 1 \\ 1 & 0 & 0 & 0 & 0 & 1 & 0 \end{bmatrix}.$$

Applying the power method for symmetric matrices to A produces the estimates

$$\lambda_1 \approx 2.86081$$

and

$$\mathbf{v}_1 \approx \begin{bmatrix} 0.406691 & 0.290865 & 0.425420 & 0.134554 \\ 0.384933 & 0.541244 & 0.331352 \end{bmatrix}^T.$$

What significance does the dominant eigenvalue of an adjacency matrix have? Suppose we wish to assign a color to each vertex in an undirected graph in such a way that adjacent vertices are assigned different colors. Such an assignment is called a *proper coloring* of a graph and may be used by, say, a mapmaker to determine how to color the geographic regions on a map so that regions that share a common border receive different colors. The minimum number of colors that can be used in a proper coloring of a graph is called the *chromatic number* and is denoted by χ. It can be shown that $\frac{n}{n-\lambda_1} \leq \chi \leq 1 + \lambda_1$, where n is the number of vertices in the graph and λ_1 is the dominant eigenvalue of the adjacency matrix. See Cvetkovic, Doob, and Sachs [10] for a proof of the lower bound and van Lint and Wilson [11] for a proof of the upper bound. Hence, for the graph in Figure 4.2, $1.69 \leq \chi \leq 3.86$, or since the chromatic number must be an integer, $2 \leq \chi \leq 3$.

Next, suppose the vertices in an undirected graph represent cities and an edge represents the existence of a direct traveling route between two cities. Geographers have shown that the entries in an eigenvector associated with the dominant eigenvalue of the adjacency matrix provide a measure of the accessibility of the cities (Straffin [12]). Thus, since the sixth entry in $\mathbf{v}_1$ is the largest, vertex 6 represents the most accessible city. Further, since the first, third, and fifth entries of $\mathbf{v}_1$ are roughly equal, the cities represented by vertices 1, 3, and 5 are nearly equal in terms of accessibility. Finally, as might have been expected, the city represented by vertex 4 is the least accessible.

Some Final Comments Regarding the Power Method

Throughout this section, we have made several important assumptions. First, we assumed that the eigenvalues of A satisfy the relations $|\lambda_1| > |\lambda_2| \geq |\lambda_3| \geq \cdots \geq |\lambda_n|$; that is, that A possesses a unique dominant eigenvalue of multiplicity one. Second, we assumed that the associated eigenvectors $\mathbf{v}_1, \mathbf{v}_2, \mathbf{v}_3, \ldots, \mathbf{v}_n$ are linearly independent. Finally, we assumed that when the vector $\mathbf{x}^{(0)}$ is written as a linear combination of the eigenvectors,

$$\mathbf{x}^{(0)} = \alpha_1\mathbf{v}_1 + \alpha_2\mathbf{v}_2 + \alpha_3\mathbf{v}_3 + \cdots + \alpha_n\mathbf{v}_n,$$

the coefficient α_1 is nonzero. We will now discuss what happens when these assumptions are violated. Demonstrations of these points will be provided in the exercises.

Let's start with the last of our assumptions. What happens if $\alpha_1 = 0$? This is not a major problem. Eventually, the roundoff errors produced during the iterations will generate a vector $\mathbf{x}^{(m)}$ for which $\alpha_1 \neq 0$. We can also reduce the likelihood that $\alpha_1 = 0$ from the outset by using a random number generator to select the components of $\mathbf{x}^{(0)}$.

If A has a unique dominant eigenvalue of multiplicity one but does not possess a set of n linearly independent eigenvectors, the power method will generally still converge to the dominant eigenvalue. In some cases, however, depending on the choice of $\mathbf{x}^{(0)}$, the power method may converge to a non-dominant eigenvalue. What if A does not have a unique dominant eigenvalue; for example, $|\lambda_1| = |\lambda_2|$ but $\lambda_1 \neq \lambda_2$? In some instances, the eigenvalue sequence, $\lambda^{(m)}$, will converge to one of the dominant eigenvalues. Generally, however, neither the eigenvalue sequence nor the eigenvector sequence, $\mathbf{x}^{(m)}$, will converge.

Finally, what happens when A has a unique dominant eigenvalue of multiplicity greater than 1? Suppose that $\lambda_1 = \lambda_2 = \lambda_3 = \cdots = \lambda_r$. If this eigenvalue has a set of r linearly independent eigenvectors, then the eigenvalue sequence of the power method will converge to λ_1. The eigenvector sequence will converge to a linear combination of the r linearly independent eigenvectors, with the specific linear combination dependent on the choice of $\mathbf{x}^{(0)}$. On the other hand, if λ_1 does not have a set of r linearly independent eigenvectors, then the power method will generally fail to converge. In some cases, the power method will still converge, but the convergence will be extremely slow.

Clearly, the power method has its drawbacks. However, if $|\lambda_2/\lambda_1| \ll 1$, the power method is an efficient method for obtaining the dominant eigenvalue and an associated eigenvector. It is also important to note that to implement the power method, we only need to be able to calculate the matrix-vector product $A\mathbf{x}^{(m)}$. It is not actually necessary to store the matrix A in an $n \times n$ array. Therefore, the power method can be of interest for large, sparse matrices.

References

1. G. H. Golub and C. F. van Loan, *Matrix Computations*, The Johns Hopkins University Press, Baltimore, 1989.

2. C. W. Ueberhuber, *Numerical Computation 2: Methods, Software and Analysis*, Springer-Verlag, Berlin, 1997.

3. H. Anton and C. Rorres, *Elementary Linear Algebra with Applications*, John Wiley and Sons, New York, 1987.

4. D. C. Lay, *Linear Algebra and Its Applications*, 2nd edition, Addison-Wesley, Reading, MA., 1997.

5. S. J. Leon, *Linear Algebra with Applications*, 6th edition, Prentice Hall, Upper Saddle River, NJ, 2002.

6. T. Shifrin and M. R. Adams, *Linear Algebra: A Geometric Approach*, W. H. Freeman, New York, 2002.

7. G. Caughley, "Parameters for Seasonally Breeding Populations," *Ecology*, **48 (5)**, 834–839, 1967.

8. F. Hickey, "Death and Reproductive Rate of Sheep in Relation to Flock Culling and Selection," *New Zealand J. Agricultural Research*, **3**, 332–344, 1960.

9. F. Hickey, "Sheep Mortality in New Zealand," *New Zealand Agriculturalist*, **15**, 1–3, 1963.

10. D. M. Cvetkovic, M. Doob, and H. Sachs, *Spectra of Graphs*, Academic Press, New York, 1979.
11. J. H. van Lint and R. M. Wilson, *A Course in Combinatorics*, Cambridge University Press, Cambridge, 1992.
12. P. D. Straffin, "Linear Algebra in Geography: Eigenvectors of Networks," *Mathematics Magazine*, **53**, 269–276, 1980.

EXERCISES

In Exercises 1–7, a matrix A and a vector $\mathbf{x}^{(0)}$ are given. Perform five iterations of the appropriate version of the power method.

1. $A = \begin{bmatrix} 3 & 2 & -2 \\ -3 & -1 & 3 \\ 1 & 2 & 0 \end{bmatrix}$ and $\mathbf{x}^{(0)} = \begin{bmatrix} 1 & 0 & 0 \end{bmatrix}^T$

2. $A = \begin{bmatrix} 15 & 7 & -7 \\ -1 & 1 & 1 \\ 13 & 7 & -5 \end{bmatrix}$ and $\mathbf{x}^{(0)} = \begin{bmatrix} 1 & 0 & 0 \end{bmatrix}^T$

3. $A = \begin{bmatrix} 1 & -0.4 & -0.6 \\ -0.4 & 1 & 0.4 \\ -0.6 & 0.4 & 1 \end{bmatrix}$ and $\mathbf{x}^{(0)} = \begin{bmatrix} 1 & 1 & 1 \end{bmatrix}^T$

4. $A = \begin{bmatrix} 19 & -9 & -6 \\ 25 & -11 & -9 \\ 17 & -9 & -4 \end{bmatrix}$ and $\mathbf{x}^{(0)} = \begin{bmatrix} 0 & 0 & 1 \end{bmatrix}^T$

5. $A = \begin{bmatrix} 1 & 4 & 5 \\ 4 & -3 & 0 \\ 5 & 0 & 7 \end{bmatrix}$ and $\mathbf{x}^{(0)} = \begin{bmatrix} 1 & 0 & 1 \end{bmatrix}^T$

6. $A = \begin{bmatrix} 10 & -4 & 0 & -4 \\ -4 & -5 & 0 & 1 \\ 0 & 0 & -2 & 0 \\ -4 & 1 & 0 & -5 \end{bmatrix}$ and $\mathbf{x}^{(0)} = \begin{bmatrix} 1/2 & 1/2 & 1/2 & 1/2 \end{bmatrix}^T$

7. $A = \begin{bmatrix} 1 & 0.25 & 0 & 0 & 0 & 0 \\ 0 & 0.25 & 0 & 1 & 0.25 & 0 \\ 0 & 0 & 0 & 0 & 0.25 & 0 \\ 0 & 0.25 & 0 & 0 & 0 & 0 \\ 0 & 0.25 & 1 & 0 & 0.25 & 0 \\ 0 & 0 & 0 & 0 & 0.25 & 1 \end{bmatrix}$ and $\mathbf{x}^{(0)} = \begin{bmatrix} 1 & 0 & 1 & 0 & 1 & 0 \end{bmatrix}^T$

8. Since the sequences $\{\mathbf{x}^{(m)}\}$ and $\{\lambda^{(m)}\}$ converge linearly, convergence can be accelerated by applying Aitken's Δ^2-method. Discuss how you would incorporate Aitken's Δ^2-method into the power method. Consider both the generic form and the variation for symmetric matrices.

9. For the following matrices, use the power method with a randomly selected initial vector and a convergence tolerance of 5×10^{-5} to estimate the dominant eigenvalue and its associated eigenvector. How many iterations are needed for convergence of the eigenvector? Compare this with the number of iterations required for convergence of the eigenvalue. Had convergence of the eigenvalue been used for the stopping criterion, what would the error have been for the eigenvector estimate?

(a) $A = \begin{bmatrix} 4 & 2 & -2 & 2 \\ 1 & 3 & 1 & -1 \\ 0 & 0 & 2 & 0 \\ 1 & 1 & -3 & 5 \end{bmatrix}$ (b) $A = \begin{bmatrix} 1.36 & 0.48 & 0 \\ 0.48 & 1.64 & 0 \\ 0 & 0 & 3 \end{bmatrix}$

10. In the example "Demonstration of Power Method for Symmetric Matrices," we estimated the dominant eigenvalue and an associated eigenvector for the matrix

$$A = \begin{bmatrix} 5.5 & -2.5 & -2.5 & -1.5 \\ -2.5 & 5.5 & 1.5 & 2.5 \\ -2.5 & 1.5 & 5.5 & 2.5 \\ -1.5 & 2.5 & 2.5 & 5.5 \end{bmatrix}$$

using the variation of the power method designed for symmetric matrices. Using the generic power method algorithm, recompute the dominant eigenvalue for this matrix. Take $\mathbf{x}^{(0)} = \begin{bmatrix} 1 & 1 & 1 & 1 \end{bmatrix}^T$ as the starting vector and use a convergence tolerance of 5×10^{-5}. Compare the performance of the generic power method with that of the symmetric matrix variation.

11. Consider the matrix

$$A = \begin{bmatrix} 1 & \epsilon \\ \epsilon & 1 \end{bmatrix}.$$

The eigenvalues of A are $\lambda_1 = 1 + \epsilon$ and $\lambda_2 = 1 - \epsilon$. The corresponding eigenvectors that have unit length in the l_∞-norm are $\mathbf{v}_1 = \begin{bmatrix} 1 & 1 \end{bmatrix}^T$ and $\mathbf{v}_2 = \begin{bmatrix} 1 & -1 \end{bmatrix}^T$. Let $\hat{\lambda} = 1$ and $\hat{\mathbf{x}} = \begin{bmatrix} 1 & 0 \end{bmatrix}^T$.

(a) Calculate the residual vector $\mathbf{r} = A\hat{\mathbf{x}} - \hat{\lambda}\hat{\mathbf{x}}$.

(b) Compare the norm of $\mathbf{r}$ with the norms of $\hat{\mathbf{x}} - \mathbf{v}_1$ and $\hat{\mathbf{x}} - \mathbf{v}_2$.

12. Consider the matrix

$$A = \begin{bmatrix} 1 & 2 & 3 \\ 0 & 4 & 5 \\ 0 & 0 & 4.001 \end{bmatrix},$$

whose dominant eigenvalue is clearly 4.001. Perform 20 iterations of the power method starting with a randomly selected vector $\mathbf{x}^{(0)}$. How does the l_∞-norm of the residual vector $\mathbf{r} = A\mathbf{x}^{(20)} - \lambda^{(20)}\mathbf{x}^{(20)}$ compare with the absolute error in $\lambda^{(20)}$?

13. Each of the following matrices has a unique dominant eigenvalue of multiplicity one but does not possess a complete set of linearly independent eigenvectors. Use the power method to determine the dominant eigenvalue and an associated eigenvector for each matrix.

(a) $A = \begin{bmatrix} 0 & 0 & 5 \\ 1 & 0 & 9 \\ 0 & 1 & 3 \end{bmatrix}$ (b) $A = \begin{bmatrix} 0 & 0 & 0 & 5 \\ 1 & 0 & 0 & 14 \\ 0 & 1 & 0 & 12 \\ 0 & 0 & 1 & 2 \end{bmatrix}$

14. Each of the following matrices has a unique dominant eigenvalue of multiplicity greater than one, and the eigenvalue does possess a complete set of linearly independent eigenvectors. Use the power method with several different randomly selected initial vectors and observe that the eigenvalue sequence converges to the same value each time but the eigenvector sequence converges to different vectors.

(a) $A = \begin{bmatrix} 5 & 0 & 0 \\ 0 & 5 & 0 \\ 1 & -1 & 3 \end{bmatrix}$ **(b)** $A = \begin{bmatrix} 5 & -1 & 3 & -1 \\ -1 & 5 & -3 & 1 \\ -1 & 1 & 1 & 1 \\ 1 & -1 & 3 & 3 \end{bmatrix}$

15. Each of the following matrices has a unique dominant eigenvalue of multiplicity greater than one, but the eigenvalue does not possess a complete set of linearly independent eigenvectors. Apply the power method with a randomly selected initial vector. Limit calculations to at most 20 iterations and comment on the behavior of the eigenvalue and eigenvector sequences.

(a) $A = \begin{bmatrix} 0 & 0 & 0 & \frac{1}{2} \\ 1 & 0 & 0 & \frac{1}{2} \\ 0 & 1 & 0 & -\frac{3}{2} \\ 0 & 0 & 1 & \frac{5}{2} \end{bmatrix}$ **(b)** $A = \begin{bmatrix} 1 & 1 & 1 & 0 \\ 0 & 1 & 1 & 0 \\ 0 & 0 & 1 & 0 \\ 0 & 0 & 0 & \frac{1}{2} \end{bmatrix}$

16. The following matrices do not have a unique dominant eigenvalue. Apply the power method with a randomly selected initial vector. Limit calculations to at most 20 iterations and comment on the behavior of the eigenvalue and eigenvector sequences.

(a) $A = \begin{bmatrix} 0 & 1 & 0 & 1 & 0 \\ 1 & 0 & 1 & 0 & 1 \\ 0 & 1 & 0 & 1 & 0 \\ 1 & 0 & 1 & 0 & 1 \\ 0 & 1 & 0 & 1 & 0 \end{bmatrix}$ **(b)** $A = \begin{bmatrix} 0 & 0 & 10 \\ 1 & 0 & -16 \\ 0 & 1 & 7 \end{bmatrix}$

17. Suppose that a particular insect has a lifespan of five years and that the parameters of the Leslie matrix for this insect are

Age Bracket	a_i	b_i
0–1	0.0	0.7
1–2	0.0	0.9
2–3	1.2	0.9
3–4	2.3	0.6
4–5	0.9	

Determine the annual growth rate of the female population and the steady-state distribution of the female population among the age brackets.

18. For each of the undirected graphs in Figure 4.3, write out the corresponding adjacency matrix and then determine the upper and lower bounds on the chromatic number of the graph and rank the accessibility of the vertices.

19. In studying loggerhead sea turtles, Crouse, Crowder and Caswell ("A Stage-Based Population Model for Loggerhead Sea Turtles and Implications for Conservation," *Ecology*, **68** (5), 1412–1423, 1987) developed a variation of a Leslie

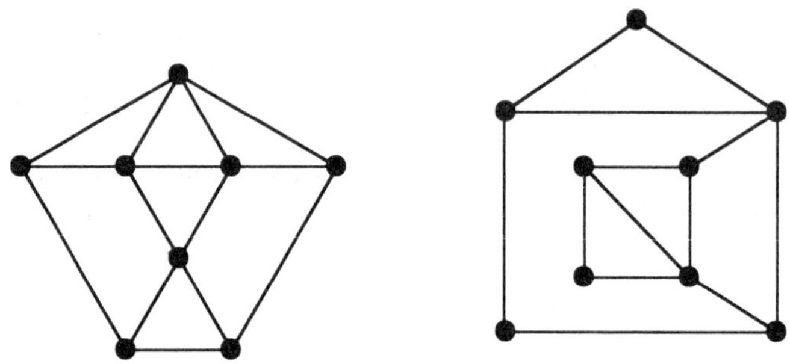

Figure 4.3 Undirected graphs for Exercise 18.

model that involved the matrix

$$M = \begin{bmatrix} 0 & 0 & 0 & 0 & 127 & 4 & 80 \\ 0.6747 & 0.7370 & 0 & 0 & 0 & 0 & 0 \\ 0 & 0.0486 & 0.6610 & 0 & 0 & 0 & 0 \\ 0 & 0 & 0.0147 & 0.6907 & 0 & 0 & 0 \\ 0 & 0 & 0 & 0.0518 & 0 & 0 & 0 \\ 0 & 0 & 0 & 0 & 0.8091 & 0 & 0 \\ 0 & 0 & 0 & 0 & 0 & 0.8091 & 0.8089 \end{bmatrix}.$$

The dominant eigenvalue and its associated eigenvector for this matrix have the same interpretation as for a Leslie matrix. Determine the annual growth rate of this population and the steady-state distribution of the population among the various classes.

20. Let A be an $n \times n$ symmetric matrix.

(a) Show that if A is symmetric positive defintite, then all of the eigenvalues of A are positive.

(b) Show that if all of the eigenvalues of A are positive, then A is symmetric positive definite.

21. Let A be an $n \times n$ symmetric matrix, let $\hat{\lambda}$ be a real number, let $\hat{\mathbf{x}}$ be an n-vector with $\|\hat{\mathbf{x}}\|_2 = 1$ and define $\mathbf{r} = A\hat{\mathbf{x}} - \hat{\lambda}\hat{\mathbf{x}}$. We will establish that A has an eigenvalue λ with

$$|\lambda - \hat{\lambda}| \leq \|\mathbf{r}\|_2.$$

Since A is symmetric, there exists a set of n eigenvectors $\mathbf{v}_1, \mathbf{v}_2, \mathbf{v}_3, \ldots, \mathbf{v}_n$ which are orthogonal in the standard inner product on $\mathbf{R}^n$. We can therefore write

$$\hat{\mathbf{x}} = \sum_{i=1}^{n} \beta_i \mathbf{v}_i,$$

for some constants $\beta_1, \beta_2, \beta_3, \ldots, \beta_n$.

(a) Show that $1 = \|\hat{\mathbf{x}}\|_2^2 = \sum_{i=1}^n \beta_i^2 \|\mathbf{v}_i\|_2^2$.

(b) Show that $\mathbf{r} = \sum_{i=1}^n \beta_i (\lambda_i - \hat{\lambda})\mathbf{v}_i$, where the λ_i are the eigenvalues of A.

(c) Show that $\|\mathbf{r}\|_2^2 \geq \left(\min_{1 \leq i \leq n} |\lambda_i - \hat{\lambda}|\right)^2$.

(d) Use (c) to deduce that A has an eigenvalue λ with $|\lambda - \hat{\lambda}| \leq \|\mathbf{r}\|_2$.

4.2 THE INVERSE POWER METHOD

The power method is designed to approximate the dominant eigenvalue (the eigenvalue that is largest in magnitude) of a matrix. There are many instances, however, in which an eigenvalue other than the dominant one is needed. For example, the buckling load of a beam, the fundamental vibrational frequency of a structure, and the ground state of a quantum operator require the eigenvalue that is smallest in magnitude. In a more general setting, we may have an estimate for an arbitrary eigenvalue, obtained perhaps using the Gerschgorin Circle Theorem, and want to determine a more accurate approximation. In this section, we present a technique, the inverse power method, for addressing these problems.

Some Theory

To derive the inverse power method, we will need the following result, which relates the eigenvalues of a matrix A to a class of matrices which can be constructed from A.

Theorem. Let A be an $n \times n$ matrix with eigenvalues λ_1, λ_2, $\lambda_3, \dots, \lambda_n$ and associated eigenvectors $\mathbf{v}_1$, $\mathbf{v}_2$, $\mathbf{v}_3, \dots, \mathbf{v}_n$.
1. If $B = a_0 I + a_1 A + a_2 A^2 + \cdots + a_m A^m = p(A)$, where p is the polynomial $p(x) = a_0 + a_1 x + a_2 x^2 + \cdots + a_m x^m$, then the eigenvalues of B are

$$p(\lambda_1), p(\lambda_2), p(\lambda_3), \dots, p(\lambda_n)$$

with associated eigenvectors $\mathbf{v}_1$, $\mathbf{v}_2$, $\mathbf{v}_3, \dots, \mathbf{v}_n$.
2. If A is nonsingular, then A^{-1} has eigenvalues

$$\frac{1}{\lambda_1}, \frac{1}{\lambda_2}, \frac{1}{\lambda_3}, \dots, \frac{1}{\lambda_n}$$

with associated eigenvectors $\mathbf{v}_1$, $\mathbf{v}_2$, $\mathbf{v}_3, \dots, \mathbf{v}_n$.

Proof. Let A be an $n \times n$ matrix with eigenvalues λ_1, λ_2, $\lambda_3, \dots, \lambda_n$ and associated eigenvectors $\mathbf{v}_1$, $\mathbf{v}_2$, $\mathbf{v}_3, \dots, \mathbf{v}_n$.

Part 1:
Note that for any positive integer k,

$$\begin{aligned} A^k \mathbf{v}_i &= A^{k-1} (A\mathbf{v}_i) = \lambda_i A^{k-1} \mathbf{v}_i \\ &= \lambda_i A^{k-2} (A\mathbf{v}_i) = \lambda_i^2 A^{k-2} \mathbf{v}_i \\ &= \cdots \\ &= \lambda_i^{k-1} (A\mathbf{v}_i) = \lambda_i^k \mathbf{v}_i. \end{aligned}$$

Now, let $B = a_0 I + a_1 A + a_2 A^2 + \cdots + a_m A^m = p(A)$, where p is the polynomial

$p(x) = a_0 + a_1x + a_2x^2 + \cdots + a_mx^m$. Then, for each $i = 1, 2, 3, \ldots, n$,

$$\begin{aligned} B\mathbf{v}_i &= (a_0I + a_1A + a_2A^2 + \cdots + a_mA^m)\mathbf{v}_i \\ &= a_0\mathbf{v}_i + a_1A\mathbf{v}_i + a_2A^2\mathbf{v}_i + \cdots + a_mA^m\mathbf{v}_i \\ &= a_0\mathbf{v}_i + a_1\lambda_i\mathbf{v}_i + a_2\lambda_i^2\mathbf{v}_i + \cdots + a_m\lambda_i^m\mathbf{v}_i \\ &= (a_0 + a_1\lambda_i + a_2\lambda_i^2 + \cdots + a_m\lambda_i^m)\mathbf{v}_i \\ &= p(\lambda_i)\mathbf{v}_i. \end{aligned}$$

Hence, the eigenvalues of B are

$$p(\lambda_1), p(\lambda_2), p(\lambda_3), \ldots, p(\lambda_n)$$

with associated eigenvectors $\mathbf{v}_1$, $\mathbf{v}_2$, $\mathbf{v}_3, \ldots, \mathbf{v}_n$.

Part 2:
Suppose A is nonsingular. Since $\mathbf{v}_i$ is an eigenvector associated with the eigenvalue λ_i, it follows that

$$A\mathbf{v}_i = \lambda_i\mathbf{v}_i.$$

Premultiplying this equation by $(1/\lambda_i)A^{-1}$ yields

$$\frac{1}{\lambda_i}A^{-1}(A\mathbf{v}_i) = \frac{1}{\lambda_i}A^{-1}(\lambda_i\mathbf{v}_i),$$

or

$$\frac{1}{\lambda_i}\mathbf{v}_i = A^{-1}\mathbf{v}_i.$$

Therefore, for each $i = 1, 2, 3, \ldots, n$, $1/\lambda_i$ is an eigenvalue of A^{-1}, with associated eigenvector $\mathbf{v}_i$. □

The Method

Once again, let A be an $n \times n$ matrix with eigenvalues λ_1, λ_2, $\lambda_3, \ldots, \lambda_n$ and associated eigenvectors $\mathbf{v}_1$, $\mathbf{v}_2$, $\mathbf{v}_3, \ldots, \mathbf{v}_n$. Let q be any constant for which $A - qI$ is nonsingular (this will hold true for any q that is not an eigenvalue of A), and consider the matrix $B = (A - qI)^{-1}$. As a consequence of the theorem we just finished proving, the eigenvalues of B are

$$\mu_1 = \frac{1}{\lambda_1 - q}, \mu_2 = \frac{1}{\lambda_2 - q}, \mu_3 = \frac{1}{\lambda_3 - q}, \ldots, \mu_n = \frac{1}{\lambda_n - q}$$

with associated eigenvectors $\mathbf{v}_1$, $\mathbf{v}_2$, $\mathbf{v}_3, \ldots, \mathbf{v}_n$.

If we apply the power method to the matrix B, the eigenvalue estimates $\lambda^{(m)}$ will converge to the dominant eigenvalue, say μ_k. Note, however, that μ_k will be the dominant eigenvalue of B if and only if λ_k is the eigenvalue of A that is closest to the number q. Hence, if by some means we determine that A has an eigenvalue in the vicinity of q, we can obtain an approximation to that eigenvalue by applying the power method to the matrix $B = (A - qI)^{-1}$. This procedure is known as the inverse power method.

An implementation of the inverse power method can be obtained from code for the power method with only a few modifications. First, an extra input value, the number q, must be included in the parameter list. Second, the operation $\mathbf{y}^{(m)} = A\mathbf{x}^{(m-1)}$ must be replaced by $\mathbf{y}^{(m)} = (A - qI)^{-1}\mathbf{x}^{(m-1)}$. Of course, in practice, we will solve the linear system $(A - qI)\mathbf{y}^{(m)} = \mathbf{x}^{(m-1)}$ for $\mathbf{y}^{(m)}$. Since the matrix $A - qI$ does not change during the iteration process, the factorization of $A - qI$ can be computed once prior to the iteration loop and only the solve step (forward and backward substitution) need be performed with each iteration. Third, remember that the sequence $\{\lambda^{(m)}\}$ converges to $\mu_k = (\lambda_k - q)^{-1}$. To obtain an approximation to λ_k, we must compute $(1/\lambda^{(m)}) + q$. The eigenvectors of A and $(A - qI)^{-1}$ are the same, so no manipulation of the sequence $\{\mathbf{x}^{(m)}\}$ is necessary.

Note that $(A - qI)^{-1}$ is symmetric whenever A is symmetric. The inverse power method can therefore be implemented with both the general version of the power method and its variation for symmetric matrices. What about the convergence of the inverse power method sequences? As above, suppose that μ_k is the dominant eigenvalue of $(A - qI)^{-1}$, and further suppose that μ_l is the second largest eigenvalue. The sequences $\{\mathbf{x}^{(m)}\}$ and $\{\lambda^{(m)}\}$ then converge linearly with asymptotic error constant

$$O(|\mu_l/\mu_k|) = O\left(\left|\frac{\lambda_k - q}{\lambda_l - q}\right|\right)$$

for general matrices and with asymptotic error constant

$$O(|\mu_l/\mu_k|^2) = O\left(\left|\frac{\lambda_k - q}{\lambda_l - q}\right|^2\right)$$

for symmetric matrices. Hence, convergence of the inverse power method depends not only on the separation of the eigenvalues, but also on the accuracy of the estimate q.

EXAMPLE 4.4 A Demonstration of the Inverse Power Method

Consider the 5×5 matrix

$$A = \begin{bmatrix} 12 & 1 & 1 & 0 & 3 \\ -1 & 3 & 0 & 1 & 0 \\ 1 & 0 & -6 & 2 & 1 \\ 0 & 2 & 1 & 9 & 0 \\ 1 & 0 & 1 & 0 & -2 \end{bmatrix}.$$

The Gerschgorin circles for A are plotted in Figure 4.4. Each circle C_i corresponds to the ith row from the matrix. Note that circle $C_2 = \{z \in \mathbf{C} : |z - 3| \le 2\}$ is

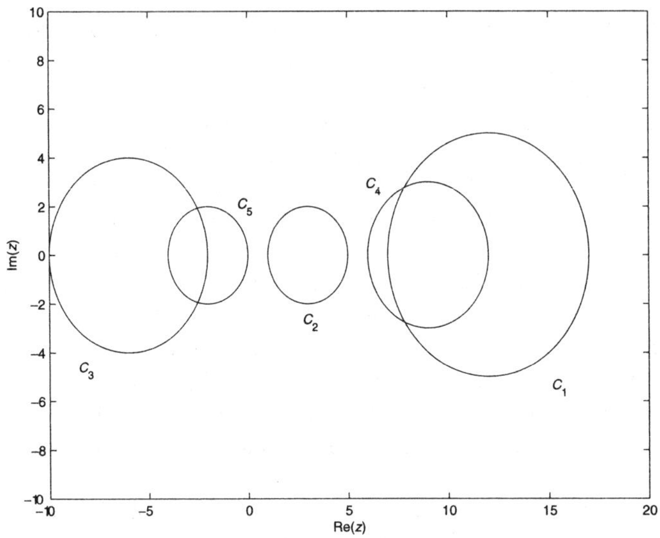

Figure 4.4 Gerschgorin circles for the matrix

$$A = \begin{bmatrix} 12 & 1 & 1 & 0 & 3 \\ -1 & 3 & 0 & 1 & 0 \\ 1 & 0 & -6 & 2 & 1 \\ 0 & 2 & 1 & 9 & 0 \\ 1 & 0 & 1 & 0 & -2 \end{bmatrix}.$$

disjoint from the other four circles and hence is guaranteed to contain one of the five eigenvalues. Furthermore, since A is a real matrix, the eigenvalue in C_2 must be real and therefore must lie in the interval $[1, 5]$.

Unfortunately, as is clear from Figure 4.4, the eigenvalue in C_2 is not the dominant eigenvalue of the matrix, so the power method will not locate it. However, the inverse power method can. Let's take $q = 3$, since this is the center of the Gerschgorin circle. With a starting vector of

$$\mathbf{x}^{(0)} = \begin{bmatrix} 1 & 1 & 1 & 1 & 1 \end{bmatrix}^T$$

and a convergence tolerance of $TOL = 5 \times 10^{-6}$, only five iterations are needed to compute

$$\lambda \approx 2.779638$$

and

$$\mathbf{v} \approx \begin{bmatrix} -0.087621 & 1.000000 & -0.084234 & -0.307983 & -0.035955 \end{bmatrix}^T.$$

Table 4.2 displays the output of each iteration.

j	$\mathbf{x}^{(j)T}$	$3+1/\lambda^{(j)}$
0	$\begin{bmatrix} 1.000000 & 1.000000 & 1.000000 & 1.000000 & 1.000000 \end{bmatrix}$	
1	$\begin{bmatrix} -0.130952 & 1.000000 & -0.068452 & -0.360119 & 0.005952 \end{bmatrix}$	4.750000
2	$\begin{bmatrix} -0.087393 & 1.000000 & -0.083210 & -0.306325 & -0.033860 \end{bmatrix}$	2.781069
3	$\begin{bmatrix} -0.087658 & 1.000000 & -0.084217 & -0.308045 & -0.035867 \end{bmatrix}$	2.779612
4	$\begin{bmatrix} -0.087622 & 1.000000 & -0.084233 & -0.307981 & -0.035952 \end{bmatrix}$	2.779641
5	$\begin{bmatrix} -0.087621 & 1.000000 & -0.084234 & -0.307983 & -0.035955 \end{bmatrix}$	2.779638

TABLE 4.2: Table for Example 4.4.

Eigenvalue Smallest in Magnitude

As noted in the introduction to this section, there are many practical applications that require approximating the eigenvalue of a matrix that is smallest in magnitude. How do we proceed in this case? Recall that each eigenvalue of A^{-1} is the reciprocal of an eigenvalue of A. It follows that the dominant eigenvalue of the matrix A^{-1} corresponds to the eigenvalue of A that is smallest in magnitude. Thus, to approximate the eigenvalue that is smallest in magnitude, we can apply the inverse power method with $q = 0$.

EXAMPLE 4.5 Approximating the Eigenvalue Smallest in Magnitude

In the previous section, we used the power method to approximate the dominant eigenvalue, $\lambda = 12$, for the matrix

$$A = \begin{bmatrix} 5.5 & -2.5 & -2.5 & -1.5 \\ -2.5 & 5.5 & 1.5 & 2.5 \\ -2.5 & 1.5 & 5.5 & 2.5 \\ -1.5 & 2.5 & 2.5 & 5.5 \end{bmatrix}.$$

Here, we will use the inverse power method, with $q = 0$, to approximate the eigenvalue that is smallest in magnitude. The table below lists the output from the 15 iterations needed to obtain convergence of the eigenvector with a convergence tolerance of $TOL = 5 \times 10^{-5}$, starting from the initial vector $\mathbf{x}^{(0)} = \begin{bmatrix} 0.5 & 0.5 & 0.5 & 0.5 \end{bmatrix}^T$. The final estimates are

$$\lambda_{\min} \approx 2.00000$$

and

$$\mathbf{v} \approx \begin{bmatrix} 0.500031 & 0.500000 & 0.500000 & -0.499969 \end{bmatrix}^T.$$

For comparison, the exact eigenvalue is $\lambda = 2$, and the exact eigenvector with unit length in the l_2 norm is $\mathbf{v} = \begin{bmatrix} 1/2 & 1/2 & 1/2 & -1/2 \end{bmatrix}^T$.

j	$\mathbf{x}^{(j)^T}$				$1/\lambda^{(j)}$
0	0.500000	0.500000	0.500000	0.500000	
1	0.741620	0.471940	0.471940	0.067420	3.692308
2	0.693774	0.484333	0.484333	−0.222531	2.435424
3	0.613172	0.494640	0.494640	−0.366991	2.118843
4	0.559931	0.498442	0.498442	−0.435417	2.030804
5	0.530668	0.499577	0.499577	−0.468229	2.007783
6	0.515488	0.499889	0.499889	−0.484246	2.001951
7	0.507780	0.499971	0.499971	−0.492156	2.000488
8	0.503898	0.499993	0.499993	−0.496086	2.000122
9	0.501951	0.499998	0.499998	−0.498045	2.000031
10	0.500976	0.500000	0.500000	−0.499023	2.000008
11	0.500488	0.500000	0.500000	−0.499512	2.000002
12	0.500244	0.500000	0.500000	−0.499756	2.000000
13	0.500122	0.500000	0.500000	−0.499878	2.000000
14	0.500061	0.500000	0.500000	−0.499939	2.000000
15	0.500031	0.500000	0.500000	−0.499969	2.000000

Application Problem 1: Steady-State Distribution of the British Workforce

In the Chapter 1 Overview (see page 4), we discussed a model of the British workforce. Recall that the workforce population was divided into the seven occupational classes

1. higher-grade professionals/administrators;
2. lower-grade professionals/administrators and higher-grade technicians;
3. routine nonmanual employees;
4. small proprietors;
5. lower-grade technicians and supervisors of manual laborers;
6. skilled manual laborers; and
7. semiskilled and unskilled manual laborers,

and the transition matrix between the classes was estimated to be

$$P = \begin{bmatrix} 0.452 & 0.291 & 0.184 & 0.126 & 0.142 & 0.078 & 0.065 \\ 0.189 & 0.231 & 0.157 & 0.114 & 0.136 & 0.088 & 0.078 \\ 0.115 & 0.119 & 0.128 & 0.080 & 0.101 & 0.083 & 0.082 \\ 0.077 & 0.070 & 0.078 & 0.244 & 0.077 & 0.065 & 0.066 \\ 0.048 & 0.096 & 0.128 & 0.087 & 0.157 & 0.123 & 0.125 \\ 0.054 & 0.106 & 0.156 & 0.144 & 0.212 & 0.304 & 0.235 \\ 0.065 & 0.087 & 0.169 & 0.205 & 0.175 & 0.259 & 0.349 \end{bmatrix}.$$

Each entry, p_{ij}, represents the proportion of children born to parents in occupational class j who became members of occupational class i.

Assuming that this transition matrix remains valid over time, the steady state distribution of the workforce among the indicated classes, denoted by $\boldsymbol{\pi}$, is the eigenvector of P that is associated with the eigenvalue 1 and whose elements sum to one. Applying the inverse power method to the matrix P with $q = 1.01$, an initial vector of $\mathbf{x}^{(0)} = \begin{bmatrix} 1 & 1 & 1 & 1 & 1 & 1 & 1 \end{bmatrix}^T$ and a convergence tolerance of $TOL = 5 \times 10^{-6}$, four iterations are required to produce the eigenvector estimate

$$\begin{bmatrix} 1.0000 & 0.6945 & 0.4952 & 0.4253 & 0.5187 & 0.8561 & 0.9351 \end{bmatrix}^T.$$

This is not quite the vector that we want, because the sum of the entries is not equal to one. Rescaling this vector, we find the steady state workforce distribution vector to be

$$\pi = \begin{bmatrix} 0.2030 & 0.1410 & 0.1005 & 0.0864 & 0.1053 & 0.1738 & 0.1899 \end{bmatrix}^T.$$

Hence, if the state transition matrix given above were to remain valid over time, we would expect roughly 20.30% of the population to be employed as higher-grade professionals/administrators, 14.10% to be employed as lower-grade professionals/administrators and higher-grade technicians, 10.05% to be employed as routine nonmanual employees, and so on.

Application Problem 2: Critical Buckling Load

Consider a long, slender rod of length L, as shown in the diagram below. The rod is clamped at one end and hinged at the other. At the hinged end, the rod is compressed axially by a force P. As long as the magnitude of P remains below a certain critical load, P_{cr}, the rod will remain straight. However, for $P > P_{\text{cr}}$, the rod will deflect from the straight configuration, or buckle. We would like to determine both P_{cr} and the shape of the buckled configuration.

Let X denote distance along the length of the rod, with $X = 0$ corresponding to the clamped end. Further, let $w(X)$ denote the deflection of the rod from its straight configuration. The differential equation for w is

$$\frac{d^2}{dX^2}\left(EI\frac{d^2w}{dX^2}\right) + \frac{d}{dX}\left(P\frac{dw}{dX}\right) = 0, \tag{1}$$

where E is Young's modulus and I is the moment of inertia of the rod's cross section. See Timoshenko and Gere [1] for a derivation of this equation. At the clamped end, the rod can neither deflect nor rotate, so

$$w(0) = \frac{dw}{dX}(0) = 0; \tag{2}$$

at the hinged end, the rod cannot deflect and no bending moment can be imparted, so

$$w(L) = \frac{d^2w}{dX^2}(L) = 0. \tag{3}$$

Introducing the nondimensional independent variable $x = X/L$, equations (1), (2), and (3) become

$$w^{(4)} + \lambda w'' = 0, \qquad w(0) = w'(0) = w(1) = w''(1) = 0, \tag{4}$$

where primes denote differentiation with respect to x and $\lambda = \frac{PL^2}{EI}$.

To approximate the solution of (4), we will start by dividing the interval $[0, 1]$ into three equal sized pieces. Then we will assume that

$$w(x) \approx \sum_{i=1}^{5} c_i \phi_i(x), \tag{5}$$

where the c's are constants to be determined and the ϕ's are known *basis functions.* For this problem, we will use the functions

$$\phi_1(x) = \begin{cases} 27x^2(1-2x), & 0 \le x < \frac{1}{3} \\ (6x-1)(3x-2)^2, & \frac{1}{3} \le x < \frac{2}{3} \\ 0, & \text{elsewhere} \end{cases}$$

$$\phi_2(x) = \begin{cases} 3x^2(3x-1), & 0 \le x < \frac{1}{3} \\ (x-\frac{1}{3})(2-3x)^2, & \frac{1}{3} \le x < \frac{2}{3} \\ 0 & \text{elsewhere} \end{cases}$$

$$\phi_3(x) = \phi_1(x - \frac{1}{3}),$$

$$\phi_4(x) = \phi_2(x - \frac{1}{3}),$$

and

$$\phi_5(x) = \begin{cases} 9(x-1)(x-\frac{2}{3})^2, & \frac{2}{3} \le x < 1 \\ 0 & \text{elsewhere.} \end{cases}$$

Graphs of these functions are shown in Figure 4.5. The function pieces that are used to construct these basis functions are called *Hermite cubics.* Hermite cubics will be studied in more detail in Section 5.7.

Next, multiply the differential equation in (4) by $\phi_j(x)$ (for $j = 1, 2, 3, 4$, and 5) and integrate the resulting expression from $x = 0$ to $x = 1$. This yields

$$0 = \int_0^1 \left(w^{(4)}(x) + \lambda w''(x)\right) \phi_j(x)\, dx = \int_0^1 w^{(4)}(x)\phi_j(x)\, dx + \lambda \int_0^1 w''(x)\phi_j(x)\, dx.$$

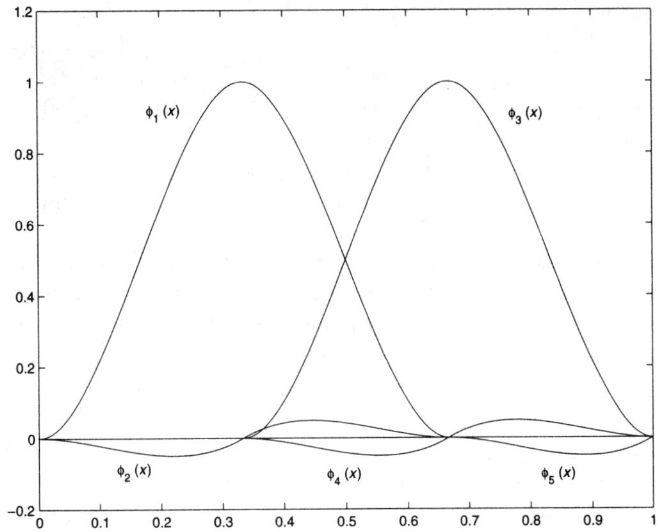

Figure 4.5 Graphs of basis functions used in Application Problem 2.

If we now integrate the integral containing $w^{(4)}$ by parts twice and the integral containing w'' by parts once, we obtain

$$\begin{aligned} 0 = & \int_0^1 (w''(x)\phi_j''(x) - w'(x)\phi_j'(x))\,dx \\ & + \left[w'''(x)\phi_j(x) + w'(x)\phi_j(x) - w''(x)\phi_j'(x)\right]\Big|_0^1. \end{aligned} \tag{6}$$

Since $\phi_j(0) = \phi_j(1) = \phi_j'(0) = 0$ for each j and $w''(1) = 0$, the second term on the right-hand side of equation (6) is identically zero. Consequently,

$$\int_0^1 (w''(x)\phi_j''(x) - w'(x)\phi_j'(x))\,dx = 0. \tag{7}$$

Equation (7) is called the *variational*, or *weak*, formulation of (4).

Substituting (5) into (7) and rearranging terms leads to the system of algebraic equations

$$A\mathbf{c} = \lambda B\mathbf{c}, \tag{8}$$

where $\mathbf{c} = \begin{bmatrix} c_1 & c_2 & c_3 & c_4 & c_5 \end{bmatrix}^T$,

$$A = \left[\int_0^1 \phi_i''(x)\phi_j''(x)\,dx\right] = \begin{bmatrix} 648 & 0 & -324 & 54 & 0 \\ 0 & 24 & -54 & 6 & 0 \\ -324 & -54 & 648 & 0 & 54 \\ 54 & 6 & 0 & 24 & 6 \\ 0 & 0 & 54 & 6 & 12 \end{bmatrix}$$

and

$$B = \left[\int_0^1 \phi_i'(x)\phi_j'(x)\,dx \right] = \begin{bmatrix} 36/5 & 0 & -18/5 & 1/10 & 0 \\ 0 & 4/45 & -1/10 & -1/90 & 0 \\ -18/5 & -1/10 & 36/5 & 0 & 1/10 \\ 1/10 & -1/90 & 0 & 4/45 & -1/90 \\ 0 & 0 & 1/10 & -1/90 & 2/45 \end{bmatrix}.$$

We are interested in nonzero solutions for the vector $\mathbf{c}$, as these correspond to deflected configurations for the rod. In particular, we want to determine the smallest value of λ (as this corresponds to the smallest axial force) for which (8) has nonzero solutions. Thus, we have transformed what was a boundary value problem into an eigenvalue problem. Because matrices appear on both sides of (8), this is called a *generalized* eigenvalue problem.

To proceed with our analysis, note that B is symmetric positive definite. If we let LL^T denote the Cholesky factorization of B and then define $C = L^{-1}A(L^T)^{-1}$ and $\mathbf{y} = L^T\mathbf{c}$, the generalized eigenvalue problem $A\mathbf{c} = \lambda B\mathbf{c}$ becomes the standard eigenvalue problem $C\mathbf{y} = \lambda\mathbf{y}$. In this instance,

$$C = \begin{bmatrix} 90.0000 & 0.0000 & 0.0000 & 57.2385 & 11.4146 \\ 0.0000 & 270.0000 & -39.3835 & 105.2214 & 29.4498 \\ 0.0000 & -39.3835 & 74.6809 & 40.9283 & 108.1915 \\ 57.2385 & 105.2214 & 40.9283 & 280.6100 & 141.9272 \\ 11.4146 & 29.4498 & 108.1915 & 141.9272 & 292.2044 \end{bmatrix}.$$

Applying the inverse power method with $q = 0$, a randomly selected initial vector and a convergence tolerance of 5×10^{-6} produces

$$\lambda \approx 20.314651 \qquad \Rightarrow \qquad P_{\text{cr}} \approx 20.314651\frac{EI}{L^2}$$

and

$$\mathbf{y} \approx [\ 0.101824 \quad 0.206840 \quad 0.903036 \quad -0.052444 \quad -0.358643\ ]^T.$$

Solving the equation $\mathbf{y} = L^T\mathbf{c}$ for $\mathbf{c}$ and normalizing the result to unit length in the l_2-norm yields

$$\mathbf{c} \approx [\ 0.117527 \quad 0.511028 \quad 0.195196 \quad -0.192401 \quad -0.806175\ ]^T.$$

Finally, using the elements of $\mathbf{c}$ in (5) generates the deflected configuration shown in Figure 4.6.

The technique we used to convert the boundary value problem in (4) into the eigenvalue problem in (8) is known as *finite element analysis.* The reader who is interested in learning more about finite element analysis should consult one of the standard texts, such as Johnson [2], Logan [3], or Wait and Mitchell [4]. For a more theoretical treatment, see Brenner and Scott [5] or Oden and Reddy [6].

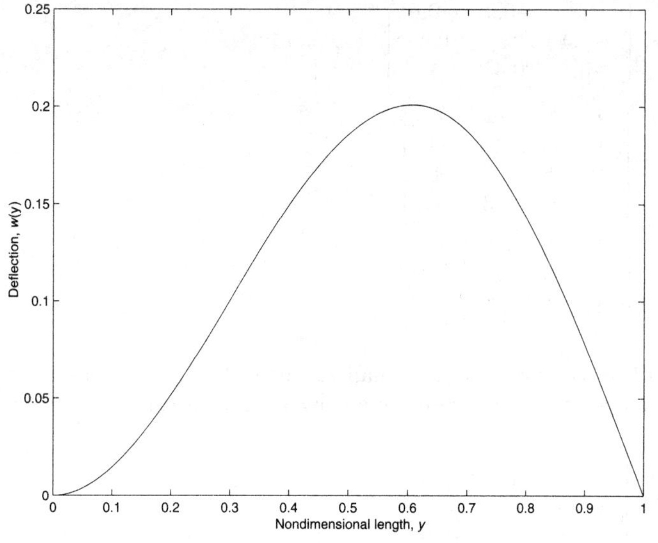

Figure 4.6 Deflection of clamped-hinged rod when loaded with its critical buckling load P_{cr}.

References

1. S. P. Timoshenko and J. M. Gere, *Theory of Elastic Stability*, McGraw-Hill, New York, 1961.
2. C. Johnson, *Numerical Solution of Partial Differential Equations by the Finite Element Method*, Cambridge University Press, Cambridge, 1987.
3. D. L. Logan, *A First Course in the Finite Element Method*, 2nd edition, PWS-Kent, Boston, 1992.
4. R. Wait and A. R. Mitchell, *Finite Element Analysis and Applications*, John Wiley and Sons, New York, 1985.
5. S. C. Brenner and L. R. Scott, *The Mathematical Theory of Finite Element Methods*, Springer-Verlag, New York, 1994.
6. J. T. Oden and J. N. Reddy, *An Introduction to the Mathematical Theory of Finite Elements*, John Wiley and Sons, New York, 1976.
7. A. J. Hoffman, "On eigenvalues and colorings of graphs," in *Graph Theory and Its Applications*, B. Harris, ed., Academic Press, 1970.

EXERCISES

In Exercises 1–4, approximate the eigenvalue of the given matrix that is nearest to the indicated value, and determine its associated eigenvector. In each case use a convergence tolerance of 5×10^{-5}.

1. $A = \begin{bmatrix} 1 & 4 & 5 \\ 4 & -3 & 0 \\ 5 & 0 & 7 \end{bmatrix} \qquad q = 1$

2. $A = \begin{bmatrix} 1 & -0.4 & -0.6 \\ -0.4 & 1 & 0.4 \\ -0.6 & 0.4 & 1 \end{bmatrix}$ $q = 0.7$

3. $A = \begin{bmatrix} 4 & 1 & -2 & 2 \\ 1 & -1 & 1 & -1 \\ 1 & 0 & 2 & 0 \\ 1 & 1 & -3 & 4 \end{bmatrix}$ $q = 2$

4. $A = \begin{bmatrix} 16 & 7 & -7 \\ -1 & 2 & 1 \\ 11 & 7 & -5 \end{bmatrix}$ $q = 3$

In Exercises 5–8, approximate the smallest eigenvalue, and its associated eigenvector, for the given matrix. In each case use a convergence tolerance of 5×10^{-5}.

5. $A = \begin{bmatrix} 4 & 2 & -2 & 2 \\ 1 & 3 & 1 & -1 \\ 1 & 0 & 2 & 0 \\ 1 & 1 & -3 & 4 \end{bmatrix}$

6. $A = \begin{bmatrix} 1 & 2 & 3 \\ 2 & -4 & -1 \\ 3 & -1 & 0 \end{bmatrix}$

7. $A = \begin{bmatrix} 5 & -2 & 2 \\ 4 & -3 & 4 \\ 3 & -6 & 7 \end{bmatrix}$

8. $A = \begin{bmatrix} -10 & -4 & 0 & -4 \\ -4 & 5 & 0 & 1 \\ 0 & 0 & 2 & 0 \\ -4 & 1 & 0 & 5 \end{bmatrix}$

In Exercises 9–12, sketch the Gerschgorin circles for the given matrix, and obtain an approximation for every eigenvalue which is contained in an isolated circle.

9. $A = \begin{bmatrix} 7 & 2 & 1 & 0 & 1 \\ 1 & -8 & 1 & 2 & -1 \\ 0 & -1 & 0 & -1 & 0 \\ 2 & 0 & 0 & 9 & 1 \\ -1 & 0 & -1 & 0 & 12 \end{bmatrix}$

10. $A = \begin{bmatrix} 16 & -8 & 2 & 1 \\ 2 & -12 & 1 & 0 \\ -1 & 1 & -4 & 1 \\ 0 & -1 & 2 & 3 \end{bmatrix}$

11. $A = \begin{bmatrix} 8 & 2 & 0 & 1 \\ 2 & 0 & 2 & 0 \\ 0 & 2 & 1 & 1 \\ 1 & 0 & 1 & -8 \end{bmatrix}$

12. $A = \begin{bmatrix} 20 & 3 & -1 & 1 \\ 3 & 7 & -2 & 2 \\ -1 & -2 & -5 & 1 \\ 1 & 2 & 1 & -8 \end{bmatrix}$

13. Each of the following matrices has two distinct eigenvalues that are equidistant from the indicated value of q. Apply the inverse power method with the indicated value of q and several randomly selected initial vectors. Limit calculations to at most 20 iterations and comment on the behavior of the eigenvalue and eigenvector sequences.

(a) $A = \begin{bmatrix} 5.5 & -2.5 & -2.5 & -1.5 \\ -2.5 & 5.5 & 1.5 & 2.5 \\ -2.5 & 1.5 & 5.5 & 2.5 \\ -1.5 & 2.5 & 2.5 & 5.5 \end{bmatrix} \qquad q = 3$

(b) $A = \begin{bmatrix} 0 & 1 & 0 \\ 0 & 0 & 1 \\ 4 & -17 & 8 \end{bmatrix} \qquad q = 2$

14. Each of the following matrices has a unique eigenvalue of multiplicity greater than one nearest to the indicated value of q, but that eigenvalue does not possess a complete set of linearly independent eigenvectors. Apply the inverse power method with the indicated value of q and several randomly selected initial vectors. Limit calculations to at most 20 iterations and comment on the behavior of the eigenvalue and eigenvector sequences.

(a) $A = \begin{bmatrix} 0 & 0 & 0 & 0 & -189 \\ 1 & 0 & 0 & 0 & 27 \\ 0 & 1 & 0 & 0 & 126 \\ 0 & 0 & 1 & 0 & -74 \\ 0 & 0 & 0 & 1 & 15 \end{bmatrix} \qquad q = 4$

(b) $A = \begin{bmatrix} 2.75 & 0.25 & -3.75 & -3.25 \\ 0.25 & 2.75 & -3.25 & -3.75 \\ -3.25 & -3.75 & 2.25 & 0.75 \\ -3.75 & -3.25 & 0.75 & 2.25 \end{bmatrix} \qquad q = 1$

15. A common problem which arises in computer graphics is the determination of the axis and angle of rotation for a composite rotation matrix, R (see S. Alessandrini, "A Motivational Example for the Numerical Solution of the Algebraic Eigenvalue Problem," *SIAM Review*, **40**(4), 935–940, 1998). Here, we will focus on determining the axis of rotation. By definition, the axis of rotation, **a**, is unchanged by R; that is, $R\mathbf{a} = \mathbf{a}$. Hence, **a** is an eigenvector of the rotation matrix associated with the eigenvalue 1. It is conventional to report the axis of rotation with unit length in the Euclidean norm. Determine the axis of rotation for each of the following rotation matrices.

(a) $R = \begin{bmatrix} 0.686237 & -0.714977 & -0.133738 \\ 0.652225 & 0.686237 & -0.321996 \\ 0.321996 & 0.133738 & 0.937247 \end{bmatrix}$

(b) $R = \begin{bmatrix} 0.230560 & -0.494104 & 0.838273 \\ 0.041493 & -0.855708 & -0.515793 \\ 0.972173 & 0.153704 & -0.176790 \end{bmatrix}$

(c) $R = \begin{bmatrix} 0.753220 & 0.041955 & 0.656429 \\ 0.550316 & 0.506441 & -0.663830 \\ -0.360293 & 0.861253 & 0.358373 \end{bmatrix}$

16. A sporting goods store sells 20-gallon aquariums. The store starts each week with at most four aquariums in stock. At the end of each week, if there are no aquariums left in stock, an order for four aquariums is placed; otherwise no order is placed.

The steady-state probability distribution for the number of aquariums that are in stock at the beginning of any given week (1, 2, 3, or 4) is given by an eigenvector of the state transition matrix

$$P = \begin{bmatrix} e^{-\mu} & 0 & 0 & 1-e^{-\mu} \\ \mu e^{-\mu} & e^{-\mu} & 0 & 1-(1+\mu)e^{-\mu} \\ \mu^2 e^{-\mu}/2 & \mu e^{-\mu} & e^{-\mu} & 1-(1+\mu+\mu^2/2)e^{-\mu} \\ \mu^3 e^{-\mu}/6 & \mu^2 e^{-\mu}/2 & \mu e^{-\mu} & 1-(\mu+\mu^2/2+\mu^3/6)e^{-\mu} \end{bmatrix}$$

associated with the eigenvalue $\lambda = 1$. To be a valid probability distribution, the eigenvector must be normalized so that the sum of its components is equal to one. Here, μ represents the average number of aquarium sales per week.

Determine the steady-state probability distribution for the number of aquariums in stock at the start of a week for $\mu = 1.0$ through $\mu = 2.0$ in increments of 0.25.

17. In a study of the distribution of red and gray squirrels among habitats in Scotland (Usher, Crawford, and Banwell, "An American Invasion of Great Britain: The Case of the Native and Alien Squirrel Species," *Conservation Biology*, **6**, 108–115, 1992), the following transition matrix was reported:

$$\begin{bmatrix} 0.874 & 0.095 & 0.077 & 0.059 \\ 0.015 & 0.722 & 0.007 & 0.193 \\ 0.109 & 0.095 & 0.906 & 0.143 \\ 0.002 & 0.087 & 0.009 & 0.605 \end{bmatrix}.$$

In constructing this matrix, each habitat was classified into one of four categories: only red squirrels present, only gray squirrels present, neither species present and both species present. Determine the steady-state distribution associated with the given transition matrix.

18. The paper cited in Exercise 17 also contains the transition matrix

$$\begin{bmatrix} 0.8797 & 0.0212 & 0.0981 & 0.0010 \\ 0.0382 & 0.8002 & 0.0273 & 0.1343 \\ 0.0525 & 0.0041 & 0.8802 & 0.0633 \\ 0.0008 & 0.0143 & 0.0527 & 0.9322 \end{bmatrix},$$

which considers habitats in all of Great Britain, not just Scotland. Determine the steady-state distribution associated with the given transition matrix.

19. Usher ("Studies on a Wood-Feeding Termite Community in Ghana, West Africa," *Biotropica*, **7**, 217–233, 1975) investigated changes in the species of termite infesting baitwood blocks over a 48-week period. Every four weeks, the blocks were inspected and classified according to the following schema: species *Ancistrotermes* present; species *Macrotermes* present; species *Pseudacanthotermes* present; species *Microtermes* present; known species other than previous four present; unknown species present; and no termite activity. The transition matrix among these classifications was found to be

$$\begin{bmatrix} 0.471 & 0.336 & 0.326 & 0.287 & 0.378 & 0.400 & 0.375 \\ 0.029 & 0.187 & 0.054 & 0.045 & 0.010 & 0.041 & 0.018 \\ 0.057 & 0.075 & 0.203 & 0.071 & 0.057 & 0.021 & 0.064 \\ 0.068 & 0.075 & 0.084 & 0.310 & 0.057 & 0.041 & 0.048 \\ 0.031 & 0.030 & 0.015 & 0.047 & 0.130 & 0.028 & 0.030 \\ 0.018 & 0.007 & 0.007 & 0.017 & 0.093 & 0.055 & 0.035 \\ 0.326 & 0.291 & 0.310 & 0.223 & 0.275 & 0.414 & 0.430 \end{bmatrix}.$$

Determine the steady-state distribution associated with the given transition matrix.

20. Let A be an $n \times n$ matrix. Suppose the eigenvalues of A satisfy the relations

$$\lambda_1 > \lambda_2 \geq \lambda_3 \geq \cdots \geq \lambda_{n-1} > \lambda_n.$$

Note the absence of absolute values. The eigenvalue λ_1 is the largest eigenvalue of A, but not necessarily the dominant eigenvalue. Similarly, λ_n is the smallest eigenvalue of A, but not necessarily the eigenvalue of smallest magnitude. Devise an algorithm to determine λ_1 and λ_n (and corresponding eigenvectors) assuming $|\lambda_1| \neq |\lambda_n|$.

21. Using the algorithm from Exercise 20, determine the largest (λ_1) and the smallest (λ_n) eigenvalues of the following matrices:

(a) the matrix from Exercise 9. **(b)** the matrix from Exercise 10.

(c) the matrix from Exercise 11. **(d)** the matrix from Exercise 12.

22. Hoffman [7] established the following lower bound for the chromatic number of a graph:

$$\chi \geq 1 + \frac{\lambda_1}{|\lambda_n|},$$

where λ_1 is the largest and λ_n the smallest eigenvalue of the adjacency matrix associated with the graph. Use this formula to obtain a lower bound for the chromatic number of the graphs in Figures 4.2 and 4.3.

Exercises 23–28 deal with the generalized eigenvalue problem

$$A\mathbf{x} = \lambda B\mathbf{x},$$

where A and B are $n \times n$ matrices.

23. Suppose that B is symmetric positive definite and has the Cholesky factorization LL^T. Show that the generalized eigenvalue problem $A\mathbf{x} = \lambda B\mathbf{x}$ is equivalent to the standard eigenvalue problem $C\mathbf{y} = \lambda\mathbf{y}$, where $C = L^{-1}A(L^T)^{-1}$ and $\mathbf{y} = L^T\mathbf{x}$.

24. In the "Critical Buckling Load" application problem, the matrix A was also symmetric positive definite. Let LL^T denote the Cholesky factorization of A and then define $C = L^{-1}B(L^T)^{-1}$, $\mathbf{y} = L^T\mathbf{c}$, and $\omega = 1/\lambda$.

(a) Show that the generalized eigenvalue problem $A\mathbf{c} = \lambda B\mathbf{c}$ is equivalent to the standard eigenvalue problem $C\mathbf{y} = \omega\mathbf{y}$.

(b) Determine the dominant eigenvalue and associated eigenvector for the problem $C\mathbf{y} = \omega\mathbf{y}$. What are the corresponding values of λ and $\mathbf{c}$?

(c) How do the results of part (b) compare to those presented in the text?

25. Suppose that B is nonsingular. Construct an algorithm to approximate the dominant eigenvalue of $A\mathbf{x} = \lambda B\mathbf{x}$ and its associated eigenvector.

26. Suppose that A is nonsingular. Construct an algorithm to approximate the eigenvalue of smallest magnitude of $A\mathbf{x} = \lambda B\mathbf{x}$ and its associated eigenvector.

27. For the given matrices A and B, use the algorithm of Exercise 25 to determine the dominant eigenvalue of $A\mathbf{x} = \lambda B\mathbf{x}$ and its associated eigenvector.

(a) $A = \begin{bmatrix} 16 & -8 & 0 & 0 \\ -8 & 14 & -8 & 1 \\ 0 & -8 & 16 & -8 \\ 0 & 1 & -8 & 7 \end{bmatrix} \qquad B = \begin{bmatrix} 16 & 2 & 0 & 0 \\ 2 & 8 & 2 & -1 \\ 0 & 2 & 16 & 2 \\ 0 & -1 & 2 & 4 \end{bmatrix}$

(b) $A = \begin{bmatrix} 4 & 0 & 1 \\ -2 & 1 & 0 \\ -2 & 0 & 1 \end{bmatrix} \qquad B = \begin{bmatrix} 0 & 1 & 0 \\ 0 & 0 & 1 \\ 4 & -17 & 8 \end{bmatrix}$

28. For the matrices A and B given in Exercise 27, use the algorithm of Exercise 26 to determine the eigenvalue of smallest magnitude of $A\mathbf{x} = \lambda B\mathbf{x}$ and its associated eigenvector.

4.3 DEFLATION

In the previous two sections, we have developed procedures for approximating the dominant eigenvalue of a matrix, the eigenvalue of a matrix that is smallest in magnitude and the eigenvalue of a matrix closest to some specified value. What if we need to approximate several of the largest eigenvalues or several of the smallest eigenvalues? For example, in a principal component analysis, we may want to determine more than the first principal component.

One approach to handling problems of this type would be to compute the entire spectrum of the matrix—approximate every eigenvalue. For a small matrix, this might not be a bad approach. For a large matrix, however, computing the entire spectrum when only a small portion of the spectrum is needed would be extremely wasteful of computational effort. When only a small portion of the spectrum is desired, it is much more efficient to employ a deflation strategy.

We first encountered the technique of deflation in Section 2-8 while investigating the polynomial rootfinding problem. Recall that the objective of deflation is to "remove" an already determined solution from the problem. Within the context of polynomial rootfinding, we removed each root as it was computed by dividing out the corresponding monomial. For the matrix eigenvalue problem, "removal" is conventionally taken to mean modifying the matrix so as to shift the previously determined eigenvalue to zero, while leaving the remainder of the spectrum unchanged.

In this section we will consider two deflation techniques: Wielandt deflation, which can be applied to any matrix, and Hotelling deflation, which is specifically designed for symmetric matrices. An additional technique will be considered in the exercises.

Eigenvalues and Eigenvectors of a Matrix and Its Transpose

To establish the key theorem behind Wielandt deflation, we must first establish the relationship among the eigenvalues and eigenvectors of a matrix A and its transpose. Recall that the eigenvalues of A^T satisfy the equation $\det(A^T - \lambda I) = 0$. Since

$$\det(A^T - \lambda I) = \det\left[(A - \lambda I)^T\right] = \det(A - \lambda I),$$

it follows that the eigenvalues of A^T are the same as those of A. The eigenvectors are generally not the same, but there is an important relationship among the eigenvectors associated with different eigenvalues.

Let $\mathbf{v}_i$ be an eigenvector for the matrix A associated with the eigenvalue λ_i, and let $\mathbf{w}_j$ be an eigenvector for the matrix A^T associated with the eigenvalue λ_j, where $\lambda_i \neq \lambda_j$. Taking the transpose of the eigenvalue equation $A\mathbf{v}_i = \lambda_i \mathbf{v}_i$ and postmultiplying the result by $\mathbf{w}_j$ yields $\mathbf{v}_i^T A^T \mathbf{w}_j = \lambda_i \mathbf{v}_i^T \mathbf{w}_j$. Using the eigenvalue equation $A^T\mathbf{w}_j = \lambda_j \mathbf{w}_j$ then leads to $\lambda_j \mathbf{v}_i^T \mathbf{w}_j = \lambda_i \mathbf{v}_i^T \mathbf{w}_j$ or $(\lambda_j - \lambda_i)\mathbf{v}_i^T \mathbf{w}_j = 0$. Since we have assumed $\lambda_i \neq \lambda_j$, it follows that $\mathbf{v}_i^T \mathbf{w}_j = 0$. In other words, eigenvectors from A and A^T that are associated with different eigenvalues are orthogonal with respect to the standard inner product on $\mathbf{R}^n$.

An Important Matrix Transformation

Let A be an $n \times n$ matrix with eigenvalues $\lambda_1, \lambda_2, \lambda_3, \ldots, \lambda_n$ and associated eigenvectors $\mathbf{v}_1, \mathbf{v}_2, \mathbf{v}_3, \ldots, \mathbf{v}_n$. Suppose that by some means (*e.g.*, the power method or the inverse power method) we have obtained approximations for λ_1 and $\mathbf{v}_1$, and we now wish to deflate the spectrum of the matrix A.

Consider the matrix

$$B = A - \lambda_1 \mathbf{v}_1 \mathbf{x}^T, \tag{1}$$

where $\mathbf{x}$ is an arbitrary n-vector. If we postmultiply equation (1) by the vector $\mathbf{v}_1$, we find

$$\begin{aligned} B\mathbf{v}_1 &= A\mathbf{v}_1 - \lambda_1 \mathbf{v}_1 \mathbf{x}^T \mathbf{v}_1 \\ &= \lambda_1 \mathbf{v}_1 - \lambda_1 \mathbf{v}_1 \mathbf{x}^T \mathbf{v}_1 \\ &= \lambda_1 \mathbf{v}_1 \left(1 - \mathbf{x}^T \mathbf{v}_1\right). \end{aligned}$$

Hence, provided $\mathbf{x}$ is chosen so that $\mathbf{x}^T\mathbf{v}_1 = 1$, zero is an eigenvalue of the matrix B with associated eigenvector $\mathbf{v}_1$.

What about the other eigenvalues of A? Have they been changed by the transformation performed in equation (1)? Taking the transpose of equation (1) and postmultiplying the result by $\mathbf{w}_i$, where $i = 2, 3, 4, \ldots, n$ and $\mathbf{w}_i$ is an eigenvector of A^T associated with the eigenvalue λ_i, gives

$$\begin{aligned} B^T\mathbf{w}_i &= A^T\mathbf{w}_i - \lambda_1\mathbf{x}\mathbf{v}_1^T\mathbf{w}_i \\ &= \lambda_i\mathbf{w}_i - 0 \\ &= \lambda_i\mathbf{w}_i. \end{aligned}$$

Note that in going from the first line to the second, we have used the orthogonality of the eigenvectors of A and A^T that are associated with different eigenvalues. This establishes that $\lambda_2, \lambda_3, \lambda_4, \ldots, \lambda_n$ are eigenvalues of B^T, which implies that they are also eigenvalues of B.

The last issue to address is the eigenvectors of the matrix B. Let $\mathbf{u}_i$ denote an eigenvector of B associated with the eigenvalue λ_i. We have already established that $\lambda_1 = 0$ and $\mathbf{u}_1 = \mathbf{v}_1$. Given the construction of the matrix B, we can assume that for $i = 2, 3, 4, \ldots, n$,

$$\mathbf{v}_i = \alpha\mathbf{u}_i + \beta\mathbf{v}_1, \tag{2}$$

where α and β are constants whose value is to be determined. If we postmultiply equation (1) by $\mathbf{v}_i$,

$$\begin{aligned} B\mathbf{v}_i &= A\mathbf{v}_i - \lambda_1\mathbf{v}_1\mathbf{x}^T\mathbf{v}_i \\ &= \lambda_i\mathbf{v}_i - \lambda_1\mathbf{v}_1\mathbf{x}^T\mathbf{v}_i \end{aligned}$$

and then substitute for $\mathbf{v}_i$ from equation (2), we find

$$B(\alpha\mathbf{u}_i + \beta\mathbf{v}_1) = \lambda_i(\alpha\mathbf{u}_i + \beta\mathbf{v}_1) - \lambda_1\mathbf{v}_1\mathbf{x}^T(\alpha\mathbf{u}_i + \beta\mathbf{v}_1).$$

Clearing parentheses and using the relations $\mathbf{x}^T\mathbf{v}_1 = 1$, $B\mathbf{v}_1 = 0$, and $B\mathbf{u}_i = \lambda_i\mathbf{u}_i$ leads to

$$\alpha\lambda_i\mathbf{u}_i = \lambda_i\alpha\mathbf{u}_i + \lambda_i\beta\mathbf{v}_1 - \alpha\lambda_1(\mathbf{x}^T\mathbf{u}_i)\mathbf{v}_1 - \beta\lambda_1\mathbf{v}_1,$$

or

$$0 = \left[\beta(\lambda_i - \lambda_1) - \alpha\lambda_1(\mathbf{x}^T\mathbf{u}_i)\right]\mathbf{v}_1. \tag{3}$$

One solution of this equation is

$$\alpha = (\lambda_i - \lambda_1) \quad \text{and} \quad \beta = \lambda_1(\mathbf{x}^T\mathbf{u}_i),$$

which yields

$$\mathbf{v}_i = (\lambda_i - \lambda_1)\mathbf{u}_i + \lambda_1(\mathbf{x}^T\mathbf{u}_i)\mathbf{v}_1. \tag{4}$$

Any other solution to equation (3) will produce a multiple of the eigenvector given by this last equation.

We can summarize these results into the following theorem:

Theorem. Let A be an $n \times n$ matrix with eigenvalues $\lambda_1, \lambda_2, \lambda_3, \ldots, \lambda_n$ and associated eigenvectors $\mathbf{v}_1, \mathbf{v}_2, \mathbf{v}_3, \ldots, \mathbf{v}_n$, and let $\mathbf{x}$ be any n-vector for which $\mathbf{x}^T\mathbf{v}_1 = 1$. Then the matrix

$$B = A - \lambda_1 \mathbf{v}_1 \mathbf{x}^T$$

has eigenvalues $0, \lambda_2, \lambda_3, \ldots, \lambda_n$ with associated eigenvectors $\mathbf{v}_1, \mathbf{u}_2, \mathbf{u}_3, \ldots, \mathbf{u}_n$ where for $i = 2, 3, 4, \ldots, n$,

$$\mathbf{v}_i = (\lambda_i - \lambda_1)\mathbf{u}_i + \lambda_1(\mathbf{x}^T\mathbf{u}_i)\mathbf{v}_1.$$

Wielandt Deflation

In Wielandt Deflation, the deflation vector $\mathbf{x}$ is chosen to be

$$\mathbf{x} = \frac{1}{\lambda_1 v_{1,k}} \begin{bmatrix} a_{k1} \\ a_{k2} \\ a_{k3} \\ \cdot \\ \cdot \\ \cdot \\ a_{kn} \end{bmatrix},$$

where $v_{1,k}$ denotes the kth element of the vector $\mathbf{v}_1$. The values $a_{k1}, a_{k2}, a_{k3}, \ldots, a_{kn}$ correspond to the kth row of the matrix A written as a column vector. The value of k can be any index for which $v_{1,k}$ is nonzero, but we will consistently choose the smallest index for which $|v_{1,k}|$ is equal to the infinity norm of the vector $\mathbf{v}_1$. With this choice for $\mathbf{x}$,

$$\begin{aligned} \mathbf{x}^T\mathbf{v}_1 &= \frac{1}{\lambda_1 v_{1,k}} [k\text{th row of } A]\, \mathbf{v}_1 \\ &= \frac{1}{\lambda_1 v_{1,k}} [k\text{th element of the product } A\mathbf{v}_1] \\ &= \frac{1}{\lambda_1 v_{1,k}} [k\text{th element of } \lambda_1 \mathbf{v}_1] \\ &= \frac{1}{\lambda_1 v_{1,k}} \lambda_1 v_{1,k} = 1. \end{aligned}$$

Therefore, the hypothesis of the deflation theorem is satisfied.

We get an extra bonus with the Wielandt deflation vector. Each row of the matrix $\lambda_1\mathbf{v}_1\mathbf{x}^T$ is a multiple of the kth row of A. In particular, the ith row of $\lambda_1\mathbf{v}_1\mathbf{x}^T$ is $v_{1,i}/v_{1,k}$ times the kth row of A. This implies that the kth row of $B = A - \lambda_1\mathbf{v}_1\mathbf{x}^T$ consists entirely of zeros. Suppose that $\mathbf{u}$ is an eigenvector of B associated with the eigenvalue $\lambda \neq 0$. Given that B has all zeros along the kth row, the kth element of the product $B\mathbf{u}$, which is just λu_k, must be zero, and therefore $u_k = 0$. This, in turn, implies that the kth column of B has no influence on the product $B\mathbf{u}$. Thus, before searching for the next eigenpair, we can reduce the size of B by deleting the kth row and the kth column.

When we take advantage of this reduction in size, an extra detail must be accounted for when equation (4) is applied to convert the eigenvectors of B into the eigenvectors of A. The vector $\mathbf{u}$ that appears on the right-hand side of the equation will be one element smaller than the other vectors. To compensate for the size difference, a zero must be placed between the $(k-1)$st and kth elements of $\mathbf{u}$ before equation (4) is used.

EXAMPLE 4.6 Wielandt Deflation in Action

Consider the 4×4 matrix

$$A = \begin{bmatrix} 11 & -6 & 4 & -2 \\ 4 & 1 & 0 & 0 \\ -9 & 9 & -6 & 5 \\ -6 & 6 & -6 & 7 \end{bmatrix}.$$

Let's determine the two largest eigenvalues and associated eigenvectors of A.

Applying the power method, we find the dominant eigenvalue of A to be $\lambda_1 = 5$, with corresponding eigenvector $\mathbf{v}_1 = \begin{bmatrix} 1 & 1 & 0 & 0 \end{bmatrix}^T$. In order to focus on the deflation process, we will ignore the effects of roundoff error in this example. With $k = 1$, the deflation vector is

$$\mathbf{x} = \frac{1}{\lambda_1 v_{1,1}} [\text{first row of } A]^T = \frac{1}{5} \begin{bmatrix} 11 & -6 & 4 & -2 \end{bmatrix}^T.$$

Forming the matrix $\lambda_1 \mathbf{v}_1 \mathbf{x}^T$ and subtracting the result from A gives the matrix B:

$$\lambda_1 \mathbf{v}_1 \mathbf{x}^T = \begin{bmatrix} 11 & -6 & 4 & -2 \\ 11 & -6 & 4 & -2 \\ 0 & 0 & 0 & 0 \\ 0 & 0 & 0 & 0 \end{bmatrix} \quad \Rightarrow \quad B = \begin{bmatrix} 0 & 0 & 0 & 0 \\ -7 & 7 & -4 & 2 \\ -9 & 9 & -6 & 5 \\ -6 & 6 & -6 & 7 \end{bmatrix}.$$

Deleting the first row and first column from this matrix produces the matrix

$$B' = \begin{bmatrix} 7 & -4 & 2 \\ 9 & -6 & 5 \\ 6 & -6 & 7 \end{bmatrix}.$$

Applying the power method to B' generates $\lambda_2 = 4$ and $\mathbf{u}_2' = \begin{bmatrix} 0 & 1/2 & 1 \end{bmatrix}^T$.

To complete the deflation process, we must convert the eigenvector $\mathbf{u}_2'$ to correspond to the original matrix. Since $k = 1$, we first prepend a zero to $\mathbf{u}_2'$ to create the 4-vector $\mathbf{u}_2$ which appears on the right-hand side of equation (4). Using equation (4), we then obtain

$$\begin{aligned}
\mathbf{v}_2 &= (\lambda_2 - \lambda_1)\mathbf{u}_2 + \lambda_1\left(\mathbf{x}^T\mathbf{u}_2\right)\mathbf{v}_1 \\
&= -\begin{bmatrix} 0 \\ 0 \\ 1/2 \\ 1 \end{bmatrix} + 5 \cdot \frac{1}{5}\left(\begin{bmatrix} 11 & -6 & 4 & -2 \end{bmatrix}\begin{bmatrix} 0 \\ 0 \\ 1/2 \\ 1 \end{bmatrix}\right)\begin{bmatrix} 1 \\ 1 \\ 0 \\ 0 \end{bmatrix} \\
&= \begin{bmatrix} 0 \\ 0 \\ -1/2 \\ -1 \end{bmatrix}.
\end{aligned}$$

We therefore have the eigenpairs

$$(5, \begin{bmatrix} 1 & 1 & 0 & 0 \end{bmatrix}^T) \quad \text{and} \quad (4, \begin{bmatrix} 0 & 0 & -1/2 & -1 \end{bmatrix}^T).$$

Variation for Symmetric Matrices

Hotelling Deflation is specifically designed for symmetric matrices. Therefore, let A be a symmetric $n \times n$ matrix with eigenvalues $\lambda_1, \lambda_2, \lambda_3, \ldots, \lambda_n$ and associated orthogonal eigenvectors $\mathbf{v}_1, \mathbf{v}_2, \mathbf{v}_3, \ldots, \mathbf{v}_n$. Suppose that by some means (e.g., the power method or the inverse power method) we have obtained approximations for λ_1 and $\mathbf{v}_1$, and consider the matrix

$$B = A - \frac{\lambda_1}{\mathbf{v}_1^T\mathbf{v}_1}\mathbf{v}_1\mathbf{v}_1^T. \tag{5}$$

To begin, note that

$$\begin{aligned}
B^T &= A^T - \frac{\lambda_1}{\mathbf{v}_1^T\mathbf{v}_1}(\mathbf{v}_1\mathbf{v}_1^T)^T \\
&= A - \frac{\lambda_1}{\mathbf{v}_1^T\mathbf{v}_1}\mathbf{v}_1\mathbf{v}_1^T = B,
\end{aligned}$$

so B is a symmetric matrix. Next, by direct calculation, we find

$$\begin{aligned}
B\mathbf{v}_1 &= A\mathbf{v}_1 - \frac{\lambda_1}{\mathbf{v}_1^T\mathbf{v}_1}\mathbf{v}_1\mathbf{v}_1^T\mathbf{v}_1 \\
&= \lambda_1\mathbf{v}_1 - \lambda_1\mathbf{v}_1 \\
&= 0,
\end{aligned}$$

so that 0 is an eigenvalue of B with associated eigenvector $\mathbf{v}_1$. Furthermore, for

$i = 2, 3, 4, \ldots, n$

$$\begin{aligned} B\mathbf{v}_i &= A\mathbf{v}_i - \frac{\lambda_1}{\mathbf{v}_1^T\mathbf{v}_1}\mathbf{v}_1\mathbf{v}_1^T\mathbf{v}_i \\ &= \lambda_i\mathbf{v}_i - 0 \\ &= \lambda_i\mathbf{v}_i, \end{aligned}$$

where, in going from the first line to the second, we have used the orthogonality of the eigenvectors of a symmetric matrix. Thus the transformation given by equation (5) shifts the eigenvalue λ_1 to zero, but preserves every other eigenvalue and every eigenvector of the matrix A.

EXAMPLE 4.7 Hotelling Deflation in Action

Consider the symmetric 4×4 matrix

$$A = \begin{bmatrix} 4 & 2/3 & -4/3 & 4/3 \\ 2/3 & 4 & 0 & 0 \\ -4/3 & 0 & 6 & 2 \\ 4/3 & 0 & 2 & 6 \end{bmatrix}.$$

We will now determine the eigenpairs with the two largest eigenvalues using Hotelling deflation.

The power method for symmetric matrices applied to A finds the dominant eigenvalue to be $\lambda_1 = 8$ with associated eigenvector, normalized in the Euclidean norm, $\mathbf{v}_1 = \begin{bmatrix} 0 & 0 & \sqrt{2}/2 & \sqrt{2}/2 \end{bmatrix}^T$. Using this eigenpair, we compute

$$\begin{aligned} \frac{\lambda_1}{\mathbf{v}_1^T\mathbf{v}_1}\mathbf{v}_1\mathbf{v}_1^T &= \frac{8}{1}\begin{bmatrix} 0 \\ 0 \\ \sqrt{2}/2 \\ \sqrt{2}/2 \end{bmatrix}\begin{bmatrix} 0 & 0 & \sqrt{2}/2 & \sqrt{2}/2 \end{bmatrix} \\ &= \begin{bmatrix} 0 & 0 & 0 & 0 \\ 0 & 0 & 0 & 0 \\ 0 & 0 & 4 & 4 \\ 0 & 0 & 4 & 4 \end{bmatrix} \end{aligned}$$

and then

$$B = \begin{bmatrix} 4 & 2/3 & -4/3 & 4/3 \\ 2/3 & 4 & 0 & 0 \\ -4/3 & 0 & 2 & -2 \\ 4/3 & 0 & -2 & 2 \end{bmatrix}.$$

The power method for symmetric matrices applied to B then gives $\lambda_2 = 6$ and $\mathbf{v}_2 = \begin{bmatrix} \sqrt{2}/2 & \sqrt{2}/6 & -\sqrt{2}/3 & \sqrt{2}/3 \end{bmatrix}^T$.

Application Problem: Measuring the Student Experience

In the Chapter 4 Overview (see page 261), the following matrix of correlations among seven measures of the "student experience" for the four-year colleges and universities in the Commonwealth of Virginia was presented:

$$R = \begin{bmatrix} 1.0000 & -0.2411 & 0.4931 & 0.3009 & -0.6865 & 0.9493 & 0.7538 \\ -0.2411 & 1.0000 & -0.5535 & -0.0387 & 0.1256 & -0.1698 & 0.0684 \\ 0.4931 & -0.5535 & 1.0000 & -0.2095 & -0.1546 & 0.3972 & -0.0643 \\ 0.3009 & -0.0387 & -0.2095 & 1.0000 & -0.2357 & 0.3994 & 0.4033 \\ -0.6865 & 0.1256 & -0.1546 & -0.2357 & 1.0000 & -0.7761 & -0.7330 \\ 0.9493 & -0.1698 & 0.3972 & 0.3994 & -0.7761 & 1.0000 & 0.7601 \\ 0.7538 & 0.0684 & -0.0643 & 0.4033 & -0.7330 & 0.7601 & 1.0000 \end{bmatrix}.$$

The measures are, in order, the percentage of first year students who return for their second year, the percentage of classes with fewer than 20 students, the percentage of classes with more than 50 students, the percentage of classes taught by full-time faculty, the average number of years needed to graduate in the current graduating class, the percentage of first-time full-time students who graduate within six years and the donation rate for alumni.

Recall that the eigenvectors of R represent uncorrelated linear combinations of the original variables known as principal components. These principal components are ranked according to the eigenvalues of R. In particular, the eigenvector associated with the largest eigenvalue of R is called the first principal component, the eigenvector associated with the next largest eigenvalue is called the second principal component, and so on. Further, the percentage of variation accounted for by each principal component is given by the ratio of the associated eigenvalue to the number of variables.

Table 4.3 displays the first two principal components for the matrix R. These were obtained by using the power method for symmetric matrices and Hotelling deflation. Observe that these two principal components account for more than 75% of the variation in the original data. The largest entries in the first principal component correspond to students returning for the second year, years to graduation, graduation rate and alumni donation rate. This component may therefore be interpreted as a measure of the overall college experience. On the other hand, the largest entries in the second principal component primarily correspond to classroom data. This component may therefore be interpreted as a measure of the classroom experiences of students.

Repeated Deflation

The process of spectrum deflation, be it Wielandt deflation or Hotelling deflation, can be repeated as each new eigenvalue-eigenvector pair is determined. Combining repeated deflation with the power method can be used to determine the first few largest eigenvalues of a matrix; alternatively, repeated deflation combined with the inverse power method can be used to determine the first few smallest eigenvalues.

	First Component	Second Component
Students returning for second year	0.4971	−0.0994
Classes with fewer than 20 students	−0.1318	0.5731
Classes with more than 50 students	0.1942	−0.6610
Classes taught by full-time faculty	0.2265	0.3422
Time to graduation	−0.4444	−0.1027
Graduation within six years	0.5073	−0.0072
Alumni donation rate	0.4378	0.3115
Eigenvalue	3.6403	1.6759
Cumulative % Variation	52.00	75.95

TABLE 4.3: Principal Components for Student Experience

In principle, deflation could be used to compute all of the eigenvalues of a matrix, but the accumulation of roundoff error makes such a scheme impractical even for matrices of only moderate size.

References

1. B. Mohar and S. Poljak, "Eigenvalues in Combinatorial Optimization," in *Combinatorial and Graph-Theoretical Problems in Linear Algebra*, R. A. Brualdi, S. Friedland, and V. Klee, eds., IMA Volumes in Mathematics and Its Applications, volume 50, Springer-Verlag, Berlin, 1993.

EXERCISES

1. For each of the following matrices, an eigenvalue-eigenvector pair is given. Determine the deflation vector $\mathbf{x}$ and the deflated matrix B corresponding to Wielandt deflation.

(a) $A = \begin{bmatrix} 3 & 0 & 0 \\ 0 & 2 & 2 \\ 0 & 1 & 3 \end{bmatrix}, \quad \lambda_1 = 4, \quad \mathbf{v}_1 = \begin{bmatrix} 0 \\ 1 \\ 1 \end{bmatrix}$

(b) $A = \begin{bmatrix} 0 & 1 & 0 \\ 0 & 0 & 1 \\ 4 & -17 & 8 \end{bmatrix}, \quad \lambda_1 = 4, \quad \mathbf{v}_1 = \begin{bmatrix} 1/16 \\ 1/4 \\ 1 \end{bmatrix}$

(c) $A = \begin{bmatrix} 19 & -9 & -6 \\ 25 & -11 & -9 \\ 17 & -9 & -4 \end{bmatrix}, \quad \lambda_1 = 2, \quad \mathbf{v}_1 = \begin{bmatrix} 3/4 \\ 3/4 \\ 1 \end{bmatrix}$

(d) $A = \begin{bmatrix} 4 & 2 & -2 & 2 \\ 1 & 3 & 1 & -1 \\ 0 & 0 & 2 & 0 \\ 1 & 1 & -3 & 5 \end{bmatrix}, \quad \lambda_1 = 6, \quad \mathbf{v}_1 = \begin{bmatrix} 1 \\ 0 \\ 0 \\ 1 \end{bmatrix}$

2. For each of the following matrices, an eigenvalue-eigenvector pair is given. Determine the deflated matrix B corresponding to Hotelling deflation.

(a) $A = \begin{bmatrix} 3 & 0 & 0 \\ 0 & 2 & 2 \\ 0 & 2 & 5 \end{bmatrix}, \quad \lambda_1 = 6, \quad \mathbf{v}_1 = \begin{bmatrix} 0 \\ 1 \\ 2 \end{bmatrix}$

(b) $A = \begin{bmatrix} 4 & -1 & 1 \\ -1 & 3 & -2 \\ 1 & -2 & -3 \end{bmatrix}, \quad \lambda_1 = 6, \quad \mathbf{v}_1 = \begin{bmatrix} 1 \\ -1 \\ 1 \end{bmatrix}$

(c) $A = \begin{bmatrix} -2 & 0 & -36 \\ 0 & -3 & 0 \\ -36 & 0 & -23 \end{bmatrix}, \quad \lambda_1 = -50, \quad \mathbf{v}_1 = \begin{bmatrix} 3/5 \\ 0 \\ 4/5 \end{bmatrix}$

(d) $A = \begin{bmatrix} 5 & -1 & -1 & -1 \\ -1 & 10 & -1 & -1 \\ -1 & -1 & 5 & -1 \\ -1 & -1 & -1 & 10 \end{bmatrix}, \quad \lambda_1 = 11, \quad \mathbf{v}_1 = \begin{bmatrix} 0 \\ -1 \\ 0 \\ 1 \end{bmatrix}$

3. Given the following information, use equation (4) to construct the eigenvector $\mathbf{v}_2$.

(a) $\mathbf{u}_2' = \begin{bmatrix} 1 \\ 1 \end{bmatrix}, \quad k = 3, \quad \mathbf{x} = \frac{1}{4}\begin{bmatrix} 6 \\ -6 \\ 7 \end{bmatrix}, \quad \mathbf{v}_1 = \begin{bmatrix} 0 \\ 1/2 \\ 1 \end{bmatrix}, \quad \lambda_1 = 4, \ \lambda_2 = 3$

(b) $\mathbf{u}_2' = \begin{bmatrix} -1 \\ -1 \\ 0 \end{bmatrix}, \quad k = 1, \quad \mathbf{x} = \frac{1}{5}\begin{bmatrix} 11 \\ -6 \\ 4 \\ -2 \end{bmatrix}, \quad \mathbf{v}_1 = \begin{bmatrix} 1 \\ 1 \\ 0 \\ 0 \end{bmatrix}, \quad \lambda_1 = 5, \ \lambda_2 = 3$

(c) $\mathbf{u}_2' = \begin{bmatrix} 1 \\ -1 \end{bmatrix}, \quad k = 1, \quad \mathbf{x} = \frac{1}{6}\begin{bmatrix} 4 \\ -1 \\ 1 \end{bmatrix}, \quad \mathbf{v}_1 = \begin{bmatrix} 1 \\ -1 \\ 1 \end{bmatrix}, \quad \lambda_1 = 6, \ \lambda_2 = 3$

For Exercises 4–9, determine the first two dominant eigenvalues and associated eigenvectors for the specified matrix.

4. $\begin{bmatrix} -101 & 51 & -12 & 0 & 0 \\ -174 & 88 & -20 & 0 & 0 \\ 136 & -68 & 19 & 0 & 0 \\ 840 & -420 & 105 & -32 & 18 \\ 2106 & -1008 & 252 & -84 & 46 \end{bmatrix}$

5. $\begin{bmatrix} 7 & 2 & 1 & 0 & 1 \\ 1 & -8 & 1 & 2 & -1 \\ 0 & -1 & 0 & -1 & 0 \\ 2 & 0 & 0 & 9 & 1 \\ -1 & 0 & -1 & 0 & 12 \end{bmatrix}$

6. $\begin{bmatrix} -10 & -4 & 0 & -4 \\ -4 & 5 & 0 & 1 \\ 0 & 0 & 2 & 0 \\ -4 & 1 & 0 & 5 \end{bmatrix}$

7. $\begin{bmatrix} 52 & 6 & 15 & 1 & 5 & 3 & 5 \\ 6 & 16 & 0 & -13 & 4 & -9 & -2 \\ 15 & 0 & 58 & -3 & 8 & 1 & 6 \\ 1 & -13 & -3 & 30 & 1 & -7 & -3 \\ 5 & 4 & 8 & 1 & 42 & 3 & 5 \\ 3 & -9 & 1 & -7 & 3 & 28 & -1 \\ 5 & -2 & 6 & -3 & 5 & -1 & 44 \end{bmatrix}$

8. $\begin{bmatrix} 12 & 1 & 1 & 0 & 3 \\ -1 & 3 & 0 & 1 & 0 \\ 1 & 0 & -6 & 2 & 1 \\ 0 & 2 & 1 & 9 & 0 \\ 1 & 0 & 1 & 0 & -2 \end{bmatrix}$

9. $\begin{bmatrix} 1.00 & 0.01 & 0.97 & 0.44 & 0.02 \\ 0.01 & 1.00 & 0.15 & 0.69 & 0.86 \\ 0.97 & 0.15 & 1.00 & 0.51 & 0.12 \\ 0.44 & 0.69 & 0.51 & 1.00 & 0.78 \\ 0.02 & 0.86 & 0.12 & 0.78 & 1.00 \end{bmatrix}$

In Exercises 10–15, a correlation matrix for a set of measured variables is given. Determine the first two principal components and the percentage of variation accounted for by those components. Provide an interpretation for each principal component.

10. Dunteman (*Principal Components Analysis*, Sage University Press Series on Quantitative Applications in the Social Sciences, 07–069, Sage Publications, Beverly Hills, 1989) presents the following matrix related to the satisfaction of married army enlisted personnel. The variables are satisfaction with job, job training, working conditions, medical care, and dental care.

$$\begin{bmatrix} 1.000 & 0.451 & 0.511 & 0.197 & 0.162 \\ 0.451 & 1.000 & 0.445 & 0.252 & 0.238 \\ 0.511 & 0.445 & 1.000 & 0.301 & 0.227 \\ 0.197 & 0.252 & 0.301 & 1.000 & 0.620 \\ 0.162 & 0.238 & 0.227 & 0.620 & 1.000 \end{bmatrix}$$

11. Harman (*Modern Factor Analysis*, The University of Chicago Press, Chicago, 1960) summarizes the correlation among eight measured variables for 305 female subjects. The measured variables are, in order, height, arm span, forearm length, lower leg length, weight, pelvic breadth, chest girth and chest width.

$$\begin{bmatrix} 1.000 & 0.846 & 0.805 & 0.859 & 0.473 & 0.398 & 0.301 & 0.382 \\ 0.846 & 1.000 & 0.881 & 0.826 & 0.376 & 0.326 & 0.277 & 0.415 \\ 0.805 & 0.881 & 1.000 & 0.801 & 0.380 & 0.319 & 0.237 & 0.345 \\ 0.859 & 0.826 & 0.801 & 1.000 & 0.436 & 0.329 & 0.327 & 0.365 \\ 0.473 & 0.376 & 0.380 & 0.436 & 1.000 & 0.762 & 0.730 & 0.629 \\ 0.398 & 0.326 & 0.319 & 0.329 & 0.762 & 1.000 & 0.583 & 0.577 \\ 0.301 & 0.277 & 0.237 & 0.327 & 0.730 & 0.583 & 1.000 & 0.539 \\ 0.382 & 0.415 & 0.345 & 0.365 & 0.629 & 0.577 & 0.539 & 1.000 \end{bmatrix}$$

12. Using data from a study of Olympic track records (D. N. Naik and R. Khattree, "Revisiting Olympic Track Records: Some Practical Considerations in the Principal Components Analysis," *The American Statistician*, **50**, 140–144, 1996), the following matrix summarizes the correlations among speeds for the 100 m, 200 m, 400 m, 800 m, 1500 m, 5000 m, 10000 m, and marathon events.

$$\begin{bmatrix} 1.000 & 0.910 & 0.829 & 0.751 & 0.692 & 0.600 & 0.610 & 0.499 \\ 0.910 & 1.000 & 0.848 & 0.803 & 0.771 & 0.686 & 0.688 & 0.588 \\ 0.829 & 0.848 & 1.000 & 0.872 & 0.831 & 0.767 & 0.780 & 0.703 \\ 0.751 & 0.803 & 0.872 & 1.000 & 0.907 & 0.851 & 0.856 & 0.796 \\ 0.692 & 0.771 & 0.831 & 0.907 & 1.000 & 0.924 & 0.931 & 0.860 \\ 0.600 & 0.686 & 0.767 & 0.851 & 0.924 & 1.000 & 0.970 & 0.927 \\ 0.610 & 0.688 & 0.780 & 0.856 & 0.931 & 0.970 & 1.000 & 0.942 \\ 0.499 & 0.588 & 0.703 & 0.796 & 0.860 & 0.927 & 0.942 & 1.000 \end{bmatrix}$$

13. In a study of vertical jump ability (I. Kollias, V. Hatzitaki, G. Papaiakovou, and G. Giatsis, "Using Principal Components Analysis to Identify Individual Differences in Vertical Jump Performance," *Research Quarterly for Exercise and Sport*, **72**, 63–66, 2001), the following correlation matrix is presented. The measured variables are peak force relative to body mass, peak power relative to body mass, maximum rate of force development, time to peak force, push-off duration and vertical displacement of center of mass.

$$\begin{bmatrix} 1.000 & 0.897 & 0.578 & -0.381 & -0.483 & -0.226 \\ 0.897 & 1.000 & 0.494 & -0.237 & -0.391 & -0.054 \\ 0.578 & 0.494 & 1.000 & -0.650 & -0.589 & -0.088 \\ -0.381 & -0.237 & -0.605 & 1.000 & 0.915 & 0.087 \\ -0.483 & -0.391 & -0.589 & 0.915 & 1.000 & 0.141 \\ -0.226 & -0.054 & -0.088 & 0.087 & 0.141 & 1.000 \end{bmatrix}$$

14. In a study of the effect of pigment levels on photosynthesis, Cassie ("Relationship Between Plant Pigments and Gross Primary Production in *Skeletonema costatum*," *Limnology and Oceanography*, **8**, 433–439, 1963) reports the following correlation matrix for chlorophyll a, chlorophyll c, carotenoids, cell count and gross production.

$$\begin{bmatrix} 1.000 & 0.989 & 0.892 & 0.972 & 0.788 \\ 0.989 & 1.000 & 0.855 & 0.958 & 0.694 \\ 0.892 & 0.855 & 1.000 & 0.874 & 0.808 \\ 0.972 & 0.958 & 0.874 & 1.000 & 0.770 \\ 0.788 & 0.694 & 0.808 & 0.770 & 1.000 \end{bmatrix}$$

15. Johnson and Wichern (*Applied Multivariate Statistical Analysis*, Prentice Hall, Englewood Cliffs, 1982) report the following correlations among the weekly rates of return for three chemical stocks (Allied Chemical, DuPont, Union Carbide) and two oil stocks (Exxon and Texaco).

$$\begin{bmatrix} 1.000 & 0.577 & 0.509 & 0.387 & 0.462 \\ 0.577 & 1.000 & 0.599 & 0.389 & 0.322 \\ 0.509 & 0.599 & 1.000 & 0.436 & 0.426 \\ 0.387 & 0.389 & 0.436 & 1.000 & 0.523 \\ 0.462 & 0.322 & 0.426 & 0.523 & 1.000 \end{bmatrix}$$

16. Let A be an $n \times n$ matrix. Suppose the eigenvalues of A satisfy the relations

$$\lambda_1 > \lambda_2 \geq \lambda_3 \geq \cdots \geq \lambda_{n-1} > \lambda_n.$$

Note the absence of absolute values. Devise an algorithm to determine λ_2. On what assumptions is your algorithm based?

17. Using the algorithm from Exercise 16, determine λ_2 for the following matrices:

(a) $A = \begin{bmatrix} 7 & 2 & 1 & 0 & 1 \\ 1 & -8 & 1 & 2 & -1 \\ 0 & -1 & 0 & -1 & 0 \\ 2 & 0 & 0 & 9 & 1 \\ -1 & 0 & -1 & 0 & 12 \end{bmatrix}$

(b)

(b) $A = \begin{bmatrix} 16 & -8 & 2 & 1 \\ 2 & -12 & 1 & 0 \\ -1 & 1 & -4 & 1 \\ 0 & -1 & 2 & 3 \end{bmatrix}$

(c) $A = \begin{bmatrix} 8 & 2 & 0 & 1 \\ 2 & 0 & 2 & 0 \\ 0 & 2 & 1 & 1 \\ 1 & 0 & 1 & -8 \end{bmatrix}$

(d) $A = \begin{bmatrix} 20 & 3 & -1 & 1 \\ 3 & 7 & -2 & 2 \\ -1 & -2 & -5 & 1 \\ 1 & 2 & 1 & -8 \end{bmatrix}$

18. The second largest eigenvalue, λ_2, of the adjacency matrix associated with a graph can be related to what are known as the diameter and mean distance of the graph. See Mohar and Poljak [1] for details. Determine the second largest eigenvalue of the adjacency matrices associated with the graphs in Figures 4.2 and 4.3.

19. Let A be an $n \times n$ matrix with eigenvalues $\lambda_1, \lambda_2, \lambda_3, \ldots, \lambda_n$ and associated eigenvectors $\mathbf{v}_1, \mathbf{v}_2, \mathbf{v}_3, \ldots, \mathbf{v}_n$. Further, let $\mathbf{w}_1$ be the eigenvector of A^T associated with λ_1 for which $\mathbf{w}_1^T\mathbf{v}_1 = 1$. Consider the matrix $B = A - \lambda_1\mathbf{v}_1\mathbf{w}_1^T$.

(a) Show that B has the same eigenvectors as A.

(b) Show that B has eigenvalues $0, \lambda_2, \lambda_3, \ldots, \lambda_n$.

20. Construct an algorithm to determine the first two dominant eigenvalues of a matrix A using the deflation strategy of Exercise 19.

21. Repeat Exercises 4–9 using the algorithm developed in Exercise 20.

4.4 REDUCTION TO SYMMETRIC TRIDIAGONAL FORM

In the final two sections of this chapter we will address the issue of computing all of the eigenvalues of a matrix. Because the eigenvalues of symmetric matrices are well-conditioned whereas the eigenvalues of nonsymmetric matrices can be poorly conditioned and because an $n \times n$ symmetric matrix always possesses n linearly independent eigenvectors whereas a nonsymmetric matrix may not, we will restrict our attention to symmetric matrices only. For discussions and detailed algorithms related to the nonsymmetric eigenvalue problem, we refer the interested reader to the classic works of Wilkinson [1] and Wilkinson and Reinsch [2], as well as the more recent texts by Golub and van Loan [3] and Saad [4].

To compute all of the eigenvalues of a symmetric matrix, we will proceed in two stages. First, the matrix will be transformed to symmetric tridiagonal form. This stage requires a fixed, finite number of operations (i.e., the procedure is not iterative). In the second stage, an iterative procedure is applied to the symmetric tridiagonal matrix produced by the first stage. The iteration generates a sequence of matrices which will converge to a diagonal matrix. The eigenvalues of this diagonal matrix are, of course, just the elements along the main diagonal (recall Exercise 11 of Section 3.3). Taking into account the cumulative effect upon the eigenvalues of all of the transformations which have been applied to the original matrix yields the eigenvalues of the original matrix.

So why do we proceed in two stages? Why don't we just perform the iterative technique on the original matrix? Simply put, the answer is efficiency. Transforming an $n \times n$ symmetric matrix to symmetric tridiagonal form requires on the order of $\frac{4}{3}n^3$ arithmetic operations for large n. The iterative reduction of the symmetric triadiagonal matrix to diagonal form then requires $O(n^2)$ arithmetic operations. On the other hand, applying the iterative technique directly to the original matrix requires on the order of $\frac{4}{3}n^3$ arithmetic operations *per iteration*. Thus, by first transforming the matrix to a simpler form we significantly reduce the computational cost.

We will focus on the reduction to symmetric tridiagonal form in this section. The second stage will be considered in the next section. To develop the reduction procedure, two special tools are needed: similarity transformations and orthogonal matrices. We will therefore start with a discussion of these items. Next an algorithm for reducing a symmetric matrix to symmetric tridiagonal form will be developed. Finally, we will discuss the steps needed to obtain the eigenvectors of a symmetric matrix.

Similarity Transformations and Orthogonal Matrices

Transforming a symmetric matrix to symmetric tridiagonal form is meaningless unless we have precise knowledge of how the eigenvalues have been affected by each transformation that has been performed. Fortunately, there is a class of transformations which does not change the spectrum of a matrix. These are known as *similarity transformations*.

> **Definition.** Let A be an $n \times n$ matrix and let M be any nonsingular $n \times n$ matrix. The matrix $B = M^{-1}AM$ is said to be SIMILAR to A. The process of converting A to B is called a SIMILARITY TRANSFORMATION.

To establish that a similarity transformation does not affect any of the eigenvalues of A, we proceed as follows. The eigenvalues of B are solutions of the equation $\det(B - \lambda I) = 0$; but

$$\begin{aligned}\det(B - \lambda I) &= \det(M^{-1}AM - \lambda I)\\ &= \det\left[M^{-1}(A - \lambda I)M\right]\\ &= \det(M^{-1})\det(A - \lambda I)\det(M)\\ &= (\det(M))^{-1}\det(A - \lambda I)\det(M)\\ &= \det(A - \lambda I).\end{aligned}$$

Thus, $\det(B - \lambda I) = 0$ if and only if $\det(A - \lambda I) = 0$, which implies that A and B have exactly the same eigenvalues.

Although any nonsingular matrix can be used to generate a similarity transformation, we would like to use matrices whose inverses are easy to compute. The class of *orthogonal matrices* will suit our needs nicely.

Definition. The $n \times n$ matrix Q is called an ORTHOGONAL MATRIX if $Q^{-1} = Q^T$.

As an example, consider the matrix

$$Q = \begin{bmatrix} 1 & 0 & 0 \\ 0 & \sqrt{2}/2 & -\sqrt{2}/2 \\ 0 & \sqrt{2}/2 & \sqrt{2}/2 \end{bmatrix}.$$

Direct multiplication shows that

$$QQ^T = \begin{bmatrix} 1 & 0 & 0 \\ 0 & \sqrt{2}/2 & -\sqrt{2}/2 \\ 0 & \sqrt{2}/2 & \sqrt{2}/2 \end{bmatrix}\begin{bmatrix} 1 & 0 & 0 \\ 0 & \sqrt{2}/2 & \sqrt{2}/2 \\ 0 & -\sqrt{2}/2 & \sqrt{2}/2 \end{bmatrix} = \begin{bmatrix} 1 & 0 & 0 \\ 0 & 1 & 0 \\ 0 & 0 & 1 \end{bmatrix} = I$$

and that $Q^TQ = I$. Hence, $Q^{-1} = Q^T$ and Q is an orthogonal matrix.

Aside from having an inverse matrix that is easy to compute, an orthogonal matrix has several other important properties, many of which will be treated in the exercises. The most important of these properties is related to the conditioning of the eigenvalue problem. As noted on several occasions, the eigenvalues of symmetric matrices are well conditioned. Suppose then that A is a symmetric matrix and $B = Q^{-1}AQ = Q^TAQ$ for some orthogonal matrix Q. It follows that

$$B^T = (Q^TAQ)^T = Q^TAQ = B.$$

Hence, a similarity transformation with an orthogonal matrix maintains symmetry and therefore preserves the conditioning of the original eigenvalue problem.

Another property of orthogonal matrices that we will find particularly useful is that multiplication by an orthogonal matrix does not change the Euclidean norm of a vector. For, if Q is any orthogonal matrix and $\mathbf{x}$ is any vector, then

$$\sqrt{(Q\mathbf{x})^TQ\mathbf{x}} = \sqrt{\mathbf{x}^TQ^TQ\mathbf{x}} = \sqrt{\mathbf{x}^T\mathbf{x}},$$

since $Q^TQ = I$.

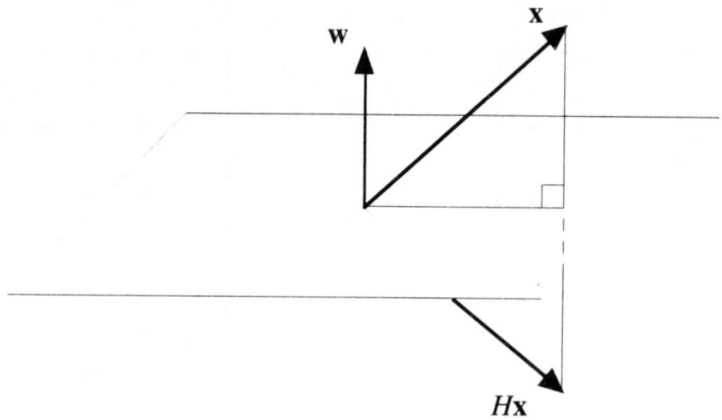

Figure 4.7 Reflection of a vector $\mathbf{x}$ across the hyperplane whose normal vector is $\mathbf{w}$.

Reducing a Symmetric Matrix to Tridiagonal Form

There are several different algorithms available for reducing a symmetric matrix to tridiagonal form. Most work in a sequential manner, applying a succession of similarity transformations which gradually produce the desired form. These techniques differ only in the family of orthogonal matrices used to generate the similarity transformations. There are also direct "tridiagonalization" techniques that determine the single orthogonal matrix needed to reduce the original matrix to tridiagonal form. These techniques work in much the same way that the direct factorization techniques we developed in Section 3.5 compute the LU decomposition of a matrix. One such technique will be treated in the exercises.

Here, we will restrict our attention to a reduction algorithm based on the use of Householder matrices.

Definition. A HOUSEHOLDER MATRIX is any matrix of the form

$$H = I - 2\mathbf{w}\mathbf{w}^T,$$

where $\mathbf{w}$ is a column vector with $\mathbf{w}^T\mathbf{w} = 1$.

It is quite easy to show that Householder matrices are both symmetric and orthogonal; that is, $H^{-1} = H$ (Exercise 2). Geometrically, multiplication of a vector $\mathbf{x}$ by the Householder matrix H results in the reflection of $\mathbf{x}$ across the hyperplane whose normal vector is $\mathbf{w}$ (see Figure 4.7).

In practice, the Householder matrices are not computed explicitly, only the vector $\mathbf{w}$ is computed. For, once the vector $\mathbf{w}$ is known, the similarity transformation HAH is given by

$$(I - 2\mathbf{w}\mathbf{w}^T)A(I - 2\mathbf{w}\mathbf{w}^T) = A - 2\mathbf{w}\mathbf{w}^TA - 2A\mathbf{w}\mathbf{w}^T + 4\mathbf{w}\mathbf{w}^TA\mathbf{w}\mathbf{w}^T,$$

$$\begin{bmatrix} \times & \times & \times & \times & \times \\ \times & \times & \times & \times & \times \\ \times & \times & \times & \times & \times \\ \times & \times & \times & \times & \times \\ \times & \times & \times & \times & \times \end{bmatrix} \xrightarrow{H_1AH_1} \begin{bmatrix} \times & \times & \times & \times & 0 \\ \times & \times & \times & \times & 0 \\ \times & \times & \times & \times & 0 \\ \times & \times & \times & \times & \times \\ 0 & 0 & 0 & \times & \times \end{bmatrix}$$

$$\xrightarrow{H_2H_1AH_1H_2} \begin{bmatrix} \times & \times & \times & 0 & 0 \\ \times & \times & \times & 0 & 0 \\ \times & \times & \times & \times & 0 \\ 0 & 0 & \times & \times & \times \\ 0 & 0 & 0 & \times & \times \end{bmatrix}$$

$$\xrightarrow{H_3H_2H_1AH_1H_2H_3} \begin{bmatrix} \times & \times & 0 & 0 & 0 \\ \times & \times & \times & 0 & 0 \\ 0 & \times & \times & \times & 0 \\ 0 & 0 & \times & \times & \times \\ 0 & 0 & 0 & \times & \times \end{bmatrix}$$

Figure 4.8 Illustration of Householder reduction to symmetric tridiagonal form for a 5×5 matrix. Each $\times$ denotes an element that is not necessarily zero.

which is completely determined by $\mathbf{w}$. The computation of HAH can be simplified tremendously if we define $\mathbf{u} = A\mathbf{w}$ and $K = \mathbf{w}^T\mathbf{u} = \mathbf{w}^T A\mathbf{w}$. Then

$$\begin{aligned} HAH &= A - 2\mathbf{w}\mathbf{w}^T A - 2A\mathbf{w}\mathbf{w}^T + 4\mathbf{w}\mathbf{w}^T A\mathbf{w}\mathbf{w}^T \\ &= A - 2\mathbf{w}\mathbf{u}^T - 2\mathbf{u}\mathbf{w}^T + 4K\mathbf{w}\mathbf{w}^T \\ &= A - 2\mathbf{w}(\mathbf{u}^T - K\mathbf{w}^T) - 2(\mathbf{u} - K\mathbf{w})\mathbf{w}^T. \end{aligned}$$

If we now let $\mathbf{q} = \mathbf{u} - K\mathbf{w}$, then $HAH = A - 2\mathbf{w}\mathbf{q}^T - 2\mathbf{q}\mathbf{w}^T$.

The algorithm to reduce a symmetric matrix to tridiagonal form using Householder matrices involves a sequence of $n-2$ similarity transformations, as illustrated in Figure 4.8 for the case $n = 5$. The first Householder matrix, H_1, is selected so that H_1A will have zeros in the first $n-2$ rows of the nth column and the nth row of A will not be affected. By symmetry, when H_1AH_1 is computed to complete the transformation, the zeros in the nth column will not be changed, but zeros will appear in the first $n-2$ columns of the nth row. Each subsequent Householder matrix, H_i $(i = 2, 3, 4, \ldots, n-2)$, is then selected so that

$$H_iH_{i-1}\cdots H_2H_1AH_1H_2\cdots H_{i-1}$$

will have zeros in the first $n-i-1$ rows of the $(n-i+1)$st column but will not affect the bottom i rows. Completing the ith transformation will place zeros in the first $n-i-1$ columns of the $(n-i+1)$st row.

Determining the appropriate Householder matrix for use in each step of the above algorithm requires the solution of the following fundamental problem:

> Given an integer k and an n-dimensional column vector $\mathbf{x}$, select $\mathbf{w}$ so that $H\mathbf{x} = (I - 2\mathbf{w}\mathbf{w}^T)\mathbf{x}$ has zeros in the first $n-k-1$ rows but leaves the last k elements in $\mathbf{x}$ unchanged.

Note that this problem specification contains only $n-1$ conditions on the vector $\mathbf{w}$. The last condition comes from the requirement that $\mathbf{w}^T\mathbf{w} = 1$, or, equivalently, that the vector $(I - 2\mathbf{w}\mathbf{w}^T)\mathbf{x}$ have the same Euclidean norm as the vector $\mathbf{x}$.

To solve this problem, first note that in order for the last k elements in $\mathbf{x}$ to be unchanged, the last k elements in $\mathbf{w}$ must be zero. This guarantees that the last k rows and columns of H are identical to the identity matrix. Thus $\mathbf{w}$ must be of the form

$$\mathbf{w} = \begin{bmatrix} w_1 & w_2 & w_3 & \cdots & w_{n-k} & 0 & \cdots & 0 \end{bmatrix}^T.$$

Let $\mathbf{b} = (I - 2\mathbf{w}\mathbf{w}^T)\mathbf{x}$, where by construction $\mathbf{b}$ will have the form

$$\mathbf{b} = \begin{bmatrix} 0 & \cdots & 0 & \alpha & x_{n-k+1} & \cdots & x_n \end{bmatrix}^T,$$

with $n-k-1$ zeros at the beginning of the vector. Since multiplication by the Householder matrix must preserve the Euclidean norm, we must have $\mathbf{b}^T\mathbf{b} = \mathbf{x}^T\mathbf{x}$, which implies that

$$\alpha^2 = x_1^2 + x_2^2 + x_3^2 + \cdots + x_{n-k}^2.$$

To proceed further, let's rearrange the equation defining the vector $\mathbf{b}$ as

$$\mathbf{x} - 2\mathbf{w}\mathbf{w}^T\mathbf{x} = \mathbf{b}. \tag{1}$$

Premultiplying equation (1) by $\mathbf{w}^T$ yields

$$\mathbf{w}^T\mathbf{x} - 2\mathbf{w}^T\mathbf{w}\mathbf{w}^T\mathbf{x} = \mathbf{w}^T\mathbf{b},$$

which simplifies to

$$-\mathbf{w}^T\mathbf{x} = \alpha w_{n-k} \tag{2}$$

upon taking into account the form of both $\mathbf{w}$ and $\mathbf{b}$ and using the fact that $\mathbf{w}^T\mathbf{w} = 1$. Substituting equation (2) into equation (1) produces

$$\mathbf{x} + 2\alpha w_{n-k}\mathbf{w} = \mathbf{b},$$

or, in component form,

$$\begin{aligned} x_i + 2\alpha w_{n-k} w_i &= 0 \qquad (i = 1, 2, 3, \ldots, n-k-1) \\ x_{n-k} + 2\alpha w_{n-k}^2 &= \alpha. \end{aligned}$$

From the last of these equations we see that

$$w_{n-k} = \sqrt{\frac{1}{2}\left(1 - \frac{x_{n-k}}{\alpha}\right)}.$$

To avoid cancellation error, we will choose $\text{sgn}(\alpha) = -\text{sgn}(x_{n-k})$. With w_{n-k} determined, the remaining nonzero entries in $\mathbf{w}$ are given by

$$w_i = -\frac{1}{2}\frac{x_i}{\alpha w_{n-k}} \quad (i = 1, 2, 3, \ldots, n-k-1).$$

EXAMPLE 4.8 Reduction to Tridiagonal Form

Consider the symmetric 4×4 matrix

$$A = \begin{bmatrix} -1 & -2 & 1 & 2 \\ -2 & 3 & 0 & -2 \\ 1 & 0 & 2 & 1 \\ 2 & -2 & 1 & 4 \end{bmatrix}.$$

For the first step of the reduction to tridiagonal form, we want to produce zeros in the first two rows of the last column of A and leave the last element in that column alone. Therefore, we are working with $k = 1$ and the vector $\mathbf{x} = \begin{bmatrix} 2 & -2 & 1 & 4 \end{bmatrix}^T$. With this vector, we compute $\alpha^2 = 2^2 + (-2)^2 + 1^2 = 9$ and since $\text{sgn}(x_3) = +1$, we choose $\alpha = -3$. It then follows that

$$w_3 = \sqrt{\frac{1}{2}\left(1 - \frac{1}{-3}\right)} = \frac{\sqrt{6}}{3};$$

$$w_2 = -\frac{1}{2}\frac{-2}{-3(\sqrt{6}/3)} = -\frac{\sqrt{6}}{6}; \text{ and}$$

$$w_1 = -\frac{1}{2}\frac{2}{-3(\sqrt{6}/3)} = \frac{\sqrt{6}}{6}.$$

Hence, $\mathbf{w} = (\sqrt{6}/6)\begin{bmatrix} 1 & -1 & 2 & 0 \end{bmatrix}^T$. Next, we compute

$$\mathbf{u} = A\mathbf{w} = (\sqrt{6}/6)\begin{bmatrix} 3 & -5 & 5 & 6 \end{bmatrix}^T;$$

$$K = \mathbf{w}^T\mathbf{u} = 3; \text{ and}$$

$$\mathbf{q} = \mathbf{u} - K\mathbf{w} = (\sqrt{6}/6)\begin{bmatrix} 0 & -2 & -1 & 6 \end{bmatrix}^T.$$

Therefore,

$$H_1AH_1 = A - \frac{1}{3}\begin{bmatrix} 1 \\ -1 \\ 2 \\ 0 \end{bmatrix}\begin{bmatrix} 0 & -2 & -1 & 6 \end{bmatrix} - \frac{1}{3}\begin{bmatrix} 0 \\ -2 \\ -1 \\ 6 \end{bmatrix}\begin{bmatrix} 1 & -1 & 2 & 0 \end{bmatrix}$$

$$= \begin{bmatrix} -1 & -4/3 & 4/3 & 0 \\ -4/3 & 5/3 & 1 & 0 \\ 4/3 & 1 & 10/3 & -3 \\ 0 & 0 & -3 & 4 \end{bmatrix}.$$

For the second (and final) step of the reduction, we want to produce a zero in the first row of the third column of H_1AH_1 and leave the last two elements in that column alone. Therefore, we are working with $k = 2$ and the vector $\mathbf{x} = \begin{bmatrix} 4/3 & 1 & 10/3 & -3 \end{bmatrix}^T$. With this vector, we compute $\alpha^2 = 25/9$ and since $\operatorname{sgn}(x_2) = +1$, we choose $\alpha = -5/3$. It then follows that

$$w_2 = \sqrt{\frac{1}{2}\left(1 - \frac{1}{-5/3}\right)} = \frac{2\sqrt{5}}{5} \quad \text{and}$$

$$w_1 = -\frac{1}{2}\frac{4/3}{(-5/3)(2\sqrt{5}/5)} = \frac{\sqrt{5}}{5}.$$

Hence, $\mathbf{w} = (\sqrt{5}/5)\begin{bmatrix} 1 & 2 & 0 & 0 \end{bmatrix}^T$. Next, we compute

$$\mathbf{u} = A\mathbf{w} = (\sqrt{5}/5)\begin{bmatrix} -11/3 & 2 & 10/3 & 0 \end{bmatrix}^T;$$
$$K = \mathbf{w}^T\mathbf{u} = 1/15; \quad \text{and}$$
$$\mathbf{q} = \mathbf{u} - K\mathbf{w} = (\sqrt{5}/5)\begin{bmatrix} -56/15 & 28/15 & 10/3 & 0 \end{bmatrix}^T.$$

Therefore,

$$H_2H_1AH_1H_2 = H_1AH_1 - \frac{2}{5}\begin{bmatrix} 1 \\ 2 \\ 0 \\ 0 \end{bmatrix}\begin{bmatrix} -\frac{56}{15} & \frac{28}{15} & \frac{10}{3} & 0 \end{bmatrix}$$
$$- \frac{2}{5}\begin{bmatrix} -56/15 \\ 28/15 \\ 10/3 \\ 0 \end{bmatrix}\begin{bmatrix} 1 & 2 & 0 & 0 \end{bmatrix}$$
$$= \begin{bmatrix} 149/75 & 68/75 & 0 & 0 \\ 68/75 & -33/25 & -5/3 & 0 \\ 0 & -5/3 & 10/3 & -3 \\ 0 & 0 & -3 & 4 \end{bmatrix}.$$

Obtaining the Eigenvectors of a Symmetric Matrix

While computing all of the eigenvalues of a symmetric matrix, it is possible to simultaneously compute the corresponding eigenvectors. The key observation that makes this possible is that when starting from the symmetric matrix A, the end result of all transformations (including those from both the first and second stage) will be the diagonal matrix D. Let the diagonal entries of D be denoted, in order, by $\lambda_1, \lambda_2, \lambda_3, \ldots, \lambda_n$. Not only is each λ_i an eigenvalue of D (and also of A), but the eigenvector associated with λ_i is $\mathbf{e}_i$, the i-th column of the identity matrix. In other words, $D\mathbf{e}_i = \lambda_i \mathbf{e}_i$. But

$$D = M^{-1}H_{n-2}H_{n-3}\cdots H_1AH_1\cdots H_{n-3}H_{n-2}M,$$

where H_i is the ith Householder matrix and M accounts for all of the similarity transformations performed during stage two—the reduction from tridiagonal to diagonal form. Therefore,

$$M^{-1}H_{n-2}H_{n-3}\cdots H_1AH_1\cdots H_{n-3}H_{n-2}M\mathbf{e}_i = \lambda_i \mathbf{e}_i,$$

or

$$AH_1\cdots H_{n-3}H_{n-2}M\mathbf{e}_i = \lambda_i H_1\cdots H_{n-3}H_{n-2}M\mathbf{e}_i.$$

Hence the eigenvector of the original matrix A associated with the eigenvalue λ_i is

$$H_1\cdots H_{n-3}H_{n-2}M\mathbf{e}_i.$$

If we were now to group the eigenvectors of A into a matrix, placing the eigenvector associated with the eigenvalue λ_i in the ith column, the resulting matrix of eigenvectors would be

$$H_1\cdots H_{n-3}H_{n-2}M.$$

Based on this analysis, we see that to be in a position to determine the eigenvalues and eigenvectors of a symmetric matrix simultaneously, the product of all of the Householder matrices used to reduce the original matrix to tridiagonal form must be accumulated so that the computation of the eigenvectors can be continued during the reduction to diagonal form. To accomplish this task, let V denote the matrix of eigenvectors of A, and initialize V to be the identity matrix. As each step of the reduction of A to tridiagonal form is carried out, postmultiply the current V matrix by the corresponding Householder matrix; *i.e.*, replace V by VH_i for each $i = 1, 2, 3, \ldots, n-2$. After the final similarity transformation has been applied, V will be equal to the product $H_1H_2H_3\cdots H_{n-3}H_{n-2}$ and will contain all of the information needed to complete the calculation of the eigenvectors.

EXAMPLE 4.9 Preparing to Find the Eigenvectors from the Previous Example

Suppose that we wish to obtain all of the eigenvectors of the matrix

$$A = \begin{bmatrix} -1 & -2 & 1 & 2 \\ -2 & 3 & 0 & -2 \\ 1 & 0 & 2 & 1 \\ 2 & -2 & 1 & 4 \end{bmatrix},$$

in addition to all of the eigenvalues. In the previous example, we found that A was similar to the tridiagonal matrix

$$\begin{bmatrix} 149/75 & 68/75 & 0 & 0 \\ 68/75 & -33/25 & -5/3 & 0 \\ 0 & -5/3 & 10/3 & -3 \\ 0 & 0 & -3 & 4 \end{bmatrix}.$$

We arrived at this tridiagonal matrix by applying two similarity transformations to A. The first was based on the Householder matrix $H_1 = I - 2\mathbf{w}_1\mathbf{w}_1^T$, where $\mathbf{w}_1 = (\sqrt{6}/6)\begin{bmatrix} 1 & -1 & 2 & 0 \end{bmatrix}^T$, and the second was based on the Householder matrix $H_2 = I - 2\mathbf{w}_2\mathbf{w}_2^T$, where $\mathbf{w}_2 = (\sqrt{5}/5)\begin{bmatrix} 1 & 2 & 0 & 0 \end{bmatrix}^T$.

To be in position to compute the eigenvectors when we reduce the above tridiagonal matrix to diagonal form, we must compute the matrix $V = H_1H_2$. The first step is to initialize the matrix V to the 4×4 identity matrix. Following the first similarity transformation, we replace V by

$$\begin{aligned} VH_1 = H_1 = I - \frac{1}{3}\begin{bmatrix} 1 \\ -1 \\ 2 \\ 0 \end{bmatrix}\begin{bmatrix} 1 & -1 & 2 & 0 \end{bmatrix} \\ = \begin{bmatrix} 2/3 & 1/3 & -2/3 & 0 \\ 1/3 & 2/3 & 2/3 & 0 \\ -2/3 & 2/3 & -1/3 & 0 \\ 0 & 0 & 0 & 1 \end{bmatrix}, \end{aligned}$$

and then, following the second similarity transformation, we replace the current V by

$$\begin{aligned} VH_2 &= \begin{bmatrix} 2/3 & 1/3 & -2/3 & 0 \\ 1/3 & 2/3 & 2/3 & 0 \\ -2/3 & 2/3 & -1/3 & 0 \\ 0 & 0 & 0 & 1 \end{bmatrix}\left(I - \frac{2}{5}\begin{bmatrix} 1 \\ 2 \\ 0 \\ 0 \end{bmatrix}\begin{bmatrix} 1 & 2 & 0 & 0 \end{bmatrix}\right) \\ &= \begin{bmatrix} 2/15 & -11/15 & -2/3 & 0 \\ -1/3 & -2/3 & 2/3 & 0 \\ -14/15 & 2/15 & -1/3 & 0 \\ 0 & 0 & 0 & 1 \end{bmatrix}. \end{aligned}$$

This matrix contains all of the information needed to complete the calculation of the eigenvectors as the above tridiagonal matrix is transformed into a diagonal matrix.

References

1. J. Wilkinson, *The Algebraic Eigenvalue Problem*, Oxford University Press, Oxford, 1965.
2. J. Wilkinson and C. Reinsch, *Linear Algebra: Handbook for Automatic Computation*, Volume 2, Springer-Verlag, New York, 1971.
3. G. H. Golub and C. F. van Loan, *Matrix Computations*, 2nd edition, The Johns Hopkins University Press, Baltimore, 1989.
4. Y. Saad, *Numerical Methods for Large Eigenvalue Problems*, Halsted Press, New York, 1992.

EXERCISES

1. Let Q be an $n \times n$ orthogonal matrix. Show that
   (a) $(Q\mathbf{x})^T Q\mathbf{y} = \mathbf{x}^T\mathbf{y}$ for all n-vectors $\mathbf{x}$ and $\mathbf{y}$.
   (b) if λ is an eigenvalue of Q, then $|\lambda| = 1$.
   (c) $\rho(Q) = 1$.
   (d) $\|Q\|_2 = 1$.
   (e) $\kappa_2(Q) = 1$.
2. Let H be a Householder matrix. Show that
   (a) H is symmetric.
   (b) $H^T H = HH^T = I$.
3. (a) Let Q_1 and Q_2 be orthogonal matrices. Show that the matrices Q_1Q_2 and Q_2Q_1 are orthogonal.
   (b) Let H_1 and H_2 be Householder matrices. Is the matrix H_1H_2 necessarily a Householder matrix?
4. Let $\mathbf{w}_1$ and $\mathbf{w}_2$ be vectors satisfying $\|\mathbf{w}_1\|_2 = \|\mathbf{w}_2\|_2 = 1$ and $\mathbf{w}_1^T\mathbf{w}_2 = 0$, and let H_1 and H_2 denote the Householder matrices associated with $\mathbf{w}_1$ and $\mathbf{w}_2$, respectively.
   (a) Show that H_1H_2 is a Householder matrix. What is the corresponding $\mathbf{w}$ vector?
   (b) Show that H_2H_1 is a Householder matrix. What is the corresponding $\mathbf{w}$ vector?
5. For each vector $\mathbf{w}$, construct the corresponding Householder matrix H. Show that H is symmetric and orthogonal.
   (a) $\mathbf{w} = \begin{bmatrix} \frac{2}{3} & \frac{1}{3} & \frac{2}{3} \end{bmatrix}^T$
   (b) $\mathbf{w} = \begin{bmatrix} \frac{1}{2} & \frac{1}{2} & -\frac{1}{2} & \frac{1}{2} \end{bmatrix}^T$
   (c) $\mathbf{w} = \frac{1}{\sqrt{30}} \begin{bmatrix} 2 & 5 & 1 \end{bmatrix}^T$
   (d) $\mathbf{w} = \frac{1}{\sqrt{22}} \begin{bmatrix} 4 & -1 & 2 & -1 \end{bmatrix}^T$
6. For each vector $\mathbf{x}$ and integer i, determine the Householder matrix H so that $H\mathbf{x}$ has zeros in the first $n - i - 1$ entries.
   (a) $\mathbf{x} = \begin{bmatrix} 4 & 3 & 2 & 5 \end{bmatrix}^T \quad i = 1$
   (b) $\mathbf{x} = \begin{bmatrix} 3 & 2 & 20 & 2 & -6 \end{bmatrix}^T \quad i = 2$
   (c) $\mathbf{x} = \begin{bmatrix} 12 & -1 & 9 & 5 & 2 \end{bmatrix}^T \quad i = 1$
   (d) $\mathbf{x} = \begin{bmatrix} 20 & 28 & 12 & 18 & 12 & 7 \end{bmatrix}^T \quad i = 2$
   (e) $\mathbf{x} = \begin{bmatrix} 1 & -1 & 2 & 3 & -1 & 2 \end{bmatrix}^T \quad i = 3$

For Exercises 7–13, reduce the given symmetric matrix to symmetric tridiagonal form and compute the matrix V which has the information needed to complete the calculation of the eigenvectors.

7. $A = \begin{bmatrix} 5.5 & -2.5 & -2.5 & -1.5 \\ -2.5 & 5.5 & 1.5 & 2.5 \\ -2.5 & 1.5 & 5.5 & 2.5 \\ -1.5 & 2.5 & 2.5 & 5.5 \end{bmatrix}$

8. $A = \begin{bmatrix} 4 & 1 & -2 & 1 \\ 1 & 3 & 1 & -1 \\ -2 & 1 & 2 & 0 \\ 1 & -1 & 0 & 5 \end{bmatrix}$

9. $A = \begin{bmatrix} 12 & -1 & 1 & 0 & 3 \\ -1 & 3 & 0 & 1 & 0 \\ 1 & 0 & -6 & 2 & 1 \\ 0 & 1 & 2 & 9 & 0 \\ 3 & 0 & 1 & 0 & -2 \end{bmatrix}$

10. $A = \begin{bmatrix} 7 & 2 & 1 & 0 & -1 \\ 2 & -8 & -1 & 2 & -1 \\ 1 & -1 & 0 & -1 & 0 \\ 0 & 2 & -1 & 9 & 1 \\ -1 & -1 & 0 & 1 & 12 \end{bmatrix}$

11. $A = \begin{bmatrix} 1 & 0.25 & 0 & 0 & 0 & 0 \\ 0.25 & 0.25 & 0 & 1 & 0.25 & 0 \\ 0 & 0 & 0 & 0 & 1 & 0 \\ 0 & 1 & 0 & 0 & 0 & 0 \\ 0 & 0.25 & 1 & 0 & 0.25 & 0.25 \\ 0 & 0 & 0 & 0 & 0.25 & 1 \end{bmatrix}$

12. $A = \begin{bmatrix} 6 & 2 & -1 & -3 & -5 \\ 2 & 18 & -2 & 5 & 4 \\ -1 & -2 & 20 & -5 & -4 \\ -3 & 5 & -5 & 28 & 1 \\ -5 & 4 & -4 & 1 & 12 \end{bmatrix}$

13. $A = \begin{bmatrix} 3 & -1 & 0 & -1 & 0 & -1 \\ -1 & 2 & -1 & 0 & 0 & 0 \\ 0 & -1 & 2 & -1 & 0 & 0 \\ -1 & 0 & -1 & 3 & -1 & 0 \\ 0 & 0 & 0 & -1 & 2 & -1 \\ -1 & 0 & 0 & 0 & -1 & 2 \end{bmatrix}$

Exercises 14–16, consider the Lanczos method for transforming a symmetric matrix to symmetric tridiagonal form. Let A be an $n \times n$ symmetric matrix and suppose that Q is an orthogonal matrix for which $Q^T AQ = T$, where T is the tridiagonal matrix

$$\begin{bmatrix} a_1 & b_2 & & & & \\ b_2 & a_2 & b_3 & & & \\ & \cdot & \cdot & \cdot & & \\ & & \cdot & \cdot & \cdot & \\ & & & b_{n-1} & a_{n-1} & b_n \\ & & & & b_n & a_n \end{bmatrix}.$$

Further, let $\mathbf{q}_i$ denote the ith column of Q and take $b_1 = b_{n+1} = 0$ and $\mathbf{q}_0 = \mathbf{0}$.

14. (a) Show that $\mathbf{q}_i^T \mathbf{q}_i = 1$ for each $i = 1, 2, 3, \ldots, n$, but $\mathbf{q}_i^T \mathbf{q}_j = 0$ whenever $i \neq j$.

(b) Show that $A\mathbf{q}_i = b_i\mathbf{q}_{i-1} + a_i\mathbf{q}_i + b_{i+1}\mathbf{q}_{i+1}$ for each $i = 1, 2, 3, \ldots, n$. (*Hint*: $Q^T AQ = T$ is equivalent to $AQ = QT$.)

(c) Show that $a_i = \mathbf{q}_i^T A\mathbf{q}_i$.

(d) Show that $b_{i+1} = \pm\|A\mathbf{q}_i - b_i\mathbf{q}_{i-1} - a_i\mathbf{q}_i\|_2$.

15. Use Exercise 14 to construct an algorithm to determine the elements of the tridiagonal matrix T and the columns of the matrix Q given an arbitrary vector $\mathbf{q}_1$ with $\|\mathbf{q}_1\|_2 = 1$.

16. Use the algorithm of Exercise 15 to transform the symmetric matrices of Exercises 7–13 to symmetric tridiagonal form.

In Exercises 17–19, we will examine a deflation technique based on Householder matrices and similarity transformations.

17. Suppose A is an $n \times n$ matrix with eigenvalue λ_1 and associated eigenvector $\mathbf{v}_1$.

(a) Construct a Householder matrix H such that $H\mathbf{v}_1 = \alpha\mathbf{e}_1$, where α is a nonzero constant and $\mathbf{e}_1$ is the first column of the identity matrix.

(b) Let $B = H^T AH$, where H is the Householder matrix constructed in part (a). Show that $B\mathbf{e}_1 = \lambda_1\mathbf{e}_1$. From this we may conclude that

$$B = \begin{bmatrix} \lambda_1 & \times & \cdots & \times \\ 0 & & & \\ \vdots & & \hat{B} & \\ 0 & & & \end{bmatrix},$$

where $\hat{B}$ is an $(n-1) \times (n-1)$ matrix and $\times$ denotes an element that is not necessarily zero.

(c) How are the eigenvalues of $\hat{B}$ related to those of A?

(d) How are the eigenvectors of $\hat{B}$ related to those of A?

18. Use Exercise 17 to construct an algorithm to determine the two dominant eigenvalues and associated eigenvectors for a matrix A.

19. Use the algorithm of Exercise 18 to determine the two dominant eigenvalues and associated eigenvectors for the specified matrix.

(a) $\begin{bmatrix} -101 & 51 & -12 & 0 & 0 \\ -174 & 88 & -20 & 0 & 0 \\ 136 & -68 & 19 & 0 & 0 \\ 840 & -420 & 105 & -32 & 18 \\ 2106 & -1008 & 252 & -84 & 46 \end{bmatrix}$

(b) $\begin{bmatrix} 7 & 2 & 1 & 0 & 1 \\ 1 & -8 & 1 & 2 & -1 \\ 0 & -1 & 0 & -1 & 0 \\ 2 & 0 & 0 & 9 & 1 \\ -1 & 0 & -1 & 0 & 12 \end{bmatrix}$

(c) $\begin{bmatrix} -10 & -4 & 0 & -4 \\ -4 & 5 & 0 & 1 \\ 0 & 0 & 2 & 0 \\ -4 & 1 & 0 & 5 \end{bmatrix}$

(d) $\begin{bmatrix} 52 & 6 & 15 & 1 & 5 & 3 & 5 \\ 6 & 16 & 0 & -13 & 4 & -9 & -2 \\ 15 & 0 & 58 & -3 & 8 & 1 & 6 \\ 1 & -13 & -3 & 30 & 1 & -7 & -3 \\ 5 & 4 & 8 & 1 & 42 & 3 & 5 \\ 3 & -9 & 1 & -7 & 3 & 28 & -1 \\ 5 & -2 & 6 & -3 & 5 & -1 & 44 \end{bmatrix}$

(e) $\begin{bmatrix} 12 & 1 & 1 & 0 & 3 \\ -1 & 3 & 0 & 1 & 0 \\ 1 & 0 & -6 & 2 & 1 \\ 0 & 2 & 1 & 9 & 0 \\ 1 & 0 & 1 & 0 & -2 \end{bmatrix}$

(f) $\begin{bmatrix} 1.00 & 0.01 & 0.97 & 0.44 & 0.02 \\ 0.01 & 1.00 & 0.15 & 0.69 & 0.86 \\ 0.97 & 0.15 & 1.00 & 0.51 & 0.12 \\ 0.44 & 0.69 & 0.51 & 1.00 & 0.78 \\ 0.02 & 0.86 & 0.12 & 0.78 & 1.00 \end{bmatrix}$

4.5 EIGENVALUES OF SYMMETRIC TRIDIAGONAL MATRICES

We will now complete our study of the matrix eigenvalue problem by developing an algorithm to compute all of the eigenvalues of a symmetric tridiagonal matrix. The technique which we will develop in this section is called the QR algorithm. Unlike the technique developed in the previous section, the QR algorithm is iterative in nature. The sequence of matrices generated by the algorithm converges to a diagonal matrix, so the eigenvalues of the "final" matrix in the sequence are just the elements along main diagonal. As we will continue to use similarity transformations, these diagonal elements are also the eigenvalues of the original matrix.

The Very Basics of the QR Algorithm

We will start with a basic description of the QR algorithm and gradually develop the details. Let $A = A^{(0)}$ be a given matrix. The QR algorithm constructs the sequence of matrices $\{A^{(i)}\}$ as follows: for $i = 0, 1, 2, ...,$

- factor $A^{(i)}$ into the product $Q^{(i)}R^{(i)}$, where $Q^{(i)}$ is an orthogonal matrix (i.e., $\left[Q^{(i)}\right]^{-1} = \left[Q^{(i)}\right]^{T}$) and $R^{(i)}$ is an upper triangular matrix; and
- compute $A^{(i+1)} = R^{(i)}Q^{(i)}$.

From the relation $A^{(i)} = Q^{(i)}R^{(i)}$, it follows that $Q^{(i)^T}A^{(i)} = R^{(i)}$, since $Q^{(i)}$ is an orthogonal matrix. The calculation in the second step is then equivalent to $A^{(i+1)} = R^{(i)}Q^{(i)} = Q^{(i)^T}A^{(i)}Q^{(i)}$. Hence, each iteration performs a similarity transformation with an orthogonal matrix, which implies that the eigenvalues of $A^{(i+1)}$ are identical to those of $A^{(i)}$.

As just described, the QR algorithm can be applied to any matrix. We will, however, discuss the implementation of the QR algorithm for symmetric tridiagonal matrices only. For details of the algorithm applied to more general matrices, consult Wilkinson [1], Golub and van Loan [2], or Press, et. al. [3].

What is the effect of performing the iterations of the QR algorithm? Consider the symmetric tridiagonal matrix

$$A^{(0)} = \begin{bmatrix} 4 & 3 & 0 \\ 3 & 1 & -1 \\ 0 & -1 & 3 \end{bmatrix}$$

A portion of the sequence $\{A^{(i)}\}$ is

$$A^{(2)} = \begin{bmatrix} 5.923 & -0.276 & 0 \\ -0.276 & 2.227 & -1.692 \\ 0 & -1.692 & -0.155 \end{bmatrix}$$

$$A^{(4)} = \begin{bmatrix} 5.950 & -0.0664 & 0 \\ -0.0664 & 3.071 & -0.241 \\ 0 & -0.241 & -1.021 \end{bmatrix}$$

$$A^{(6)} = \begin{bmatrix} 5.951 & -0.0178 & 0 \\ -0.0178 & 3.084 & -0.0272 \\ 0 & -0.0272 & -1.035 \end{bmatrix}$$

$$A^{(8)} = \begin{bmatrix} 5.951 & -0.00478 & 0 \\ -0.00478 & 3.084 & -0.00306 \\ 0 & -0.00306 & -1.035 \end{bmatrix}$$

$$A^{(10)} = \begin{bmatrix} 5.951 & -0.00128 & 0 \\ -0.00128 & 3.084 & -0.000345 \\ 0 & -0.000345 & -1.035 \end{bmatrix}.$$

Observe that the off-diagonal elements are converging toward zero, while the diagonal elements are converging toward the eigenvalues of $A^{(0)}$, which, to three decimal places, are 5.951, 3.084 and -1.035. Further, the eigenvalues appear along the diagonal of $A^{(i)}$ in decreasing order of magnitude. Francis [4] has shown that this example demonstrates the general performance of the QR algorithm and that the off-diagonal entries $\left|a_{j,j-1}^{(i+1)}\right|$ converge toward zero with rate of convergence $O(|\lambda_j/\lambda_{j-1}|)$.

Now let's start filling in some of the details.

QR Factorization

The heart of the QR algorithm is the calculation of the QR factorization of the matrix $A^{(i)}$. Given the simple structure of the matrices with which we are going to work (symmetric tridiagonal), we will perform the QR factorization using what are known as *rotation matrices*.

Definition. Let $i < j$. The orthogonal matrix, $P_{(i,j)}$, which is identical to the identity matrix with the exception that

$$p_{i,i} = p_{j,j} = \cos\theta \quad \text{and} \quad p_{i,j} = -p_{j,i} = \sin\theta,$$

for some angle θ, is called a ROTATION MATRIX.

The name *rotation matrix* arises from the geometric fact that $P_{(i,j)}$ represents the rotation of the ith and jth axes about the origin of the coordinate system by an angle of θ. For later use, it is important to note that premultiplication of an arbitrary matrix, M, by $P_{(i,j)}$ affects only the ith and jth rows. In particular,

$$\begin{matrix} i\text{th row of} \\ P_{(i,j)}M \end{matrix} = \cos\theta \cdot \begin{matrix} i\text{th row} \\ \text{of } M \end{matrix} + \sin\theta \cdot \begin{matrix} j\text{th row} \\ \text{of } M \end{matrix}$$

and

$$\begin{matrix} j\text{th row of} \\ P_{(i,j)}M \end{matrix} = -\sin\theta \cdot \begin{matrix} i\text{th row} \\ \text{of } M \end{matrix} + \cos\theta \cdot \begin{matrix} j\text{th row} \\ \text{of } M \end{matrix}$$

The factorization of the symmetric tridiagonal matrix $A^{(i)}$ now proceeds in exactly the same manner as the matrix factorization algorithms we developed in Chapter 3. For an $n \times n$ matrix, we make $n-1$ passes through the matrix, with each pass "zeroing" out a specific element below the main diagonal. Thus, in the first pass, $P_{(1,2)}$ is chosen so that $P_{(1,2)}A^{(i)}$ has a zero in row 2, column 1. Next, $P_{(2,3)}$ is chosen so that $P_{(2,3)}P_{(1,2)}A^{(i)}$ has a zero in the third row of the second column, $P_{(3,4)}$ is chosen so that $P_{(3,4)}P_{(2,3)}P_{(1,2)}A^{(i)}$ has a zero in the fourth row of the third column, and so on. Finally, $P_{(n-1,n)}$ is chosen so that $P_{(n-1,n)} \cdots P_{(3,4)}P_{(2,3)}P_{(1,2)}A^{(i)}$ is an upper triangular matrix. Hence, $R^{(i)} = P_{(n-1,n)} \cdots P_{(3,4)}P_{(2,3)}P_{(1,2)}A^{(i)}$.

To examine the details of this factorization scheme more closely, let

$$A^{(i)} = \begin{bmatrix} a_1 & b_1 & & & & \\ b_1 & a_2 & b_2 & & & \\ & b_2 & a_3 & b_3 & & \\ & & \ddots & \ddots & \ddots & \\ & & & b_{n-2} & a_{n-1} & b_{n-1} \\ & & & & b_{n-1} & a_n \end{bmatrix}.$$

For notational convenience, let the cosine and sine values associated with the rotation matrix $P_{(j,j+1)}$ be denoted by c_j and s_j, respectively. Carrying out the multiplication $P_{(1,2)}A^{(i)}$, we find

$$\begin{bmatrix} c_1 & s_1 & & \\ -s_1 & c_1 & & \\ & & 1 & \\ & & & \ddots \end{bmatrix} \begin{bmatrix} a_1 & b_1 & & \\ b_1 & a_2 & b_2 & \\ & b_2 & a_3 & b_3 \\ & \ddots & \ddots & \ddots \end{bmatrix}$$

$$= \begin{bmatrix} a_1c_1 + b_1s_1 & b_1c_1 + a_2s_1 & b_2s_1 & \\ -a_1s_1 + b_1c_1 & -b_1s_1 + a_2c_1 & b_2c_1 & \\ & b_2 & a_3 & b_3 \\ & \ddots & \ddots & \ddots \end{bmatrix}. \quad (1)$$

We now want to choose c_1 and s_1 so that $-a_1s_1 + b_1c_1 = 0$. One solution of this equation, which also satisfies the fundamental trig identity $c_1^2 + s_1^2 = 1$, is

$$c_1 = \frac{a_1}{\sqrt{a_1^2 + b_1^2}} \quad \text{and} \quad s_1 = \frac{b_1}{\sqrt{a_1^2 + b_1^2}}.$$

With c_1 and s_1 selected so that $-a_1s_1 + b_1c_1 = 0$, note that a_1 appears on the right-hand side of (1) in the first row, first column only—the precise location of a_1 in the matrix $A^{(i)}$. We may therefore overwrite a_1 with the expression

$$a_1c_1 + b_1s_1 = a_1\frac{a_1}{\sqrt{a_1^2 + b_1^2}} + b_1\frac{b_1}{\sqrt{a_1^2 + b_1^2}} = \sqrt{a_1^2 + b_1^2}.$$

In a similar manner, we would like to save the first two elements in the second column of $P_{(1,2)}A^{(i)}$ in place of b_1 and a_2; unfortunately, to calculate these elements, both b_1 and a_2 are required. However, if we save the current value of b_1 in a temporary variable, say t, we may then overwrite b_1 with the expression $tc_1 + a_2s_1$ and a_2 with $-ts_1 + a_2c_1$. Finally, we save the value of b_2 in the variable t and then overwrite b_2 with the quantity $b_2c_1 = tc_1$. We need to save b_2 in order to calculate the sine and cosine values associated with the next rotation matrix.

It turns out that the element in the third column of the first row of $P_{(1,2)}A^{(i)}$, b_2s_1, does not need to be saved. The remaining passes in the factorization step do not involve the first row, so the indicated element will not be needed for any later calculations. Further, as we will see shortly, the calculation of the product $R^{(i)}Q^{(i)}$ can be carried out without knowing this element. Technically, by not saving the value b_2s_1, we are not obtaining the true QR factorization of the matrix $A^{(i)}$. We are, however, maintaining all the information we will need to calculate $R^{(i)}Q^{(i)} = A^{(i+1)}$, which, in our present circumstances, is the real objective.

The calculations required by all subsequent passes in the factorization step are identical to those indicated for the first pass, with two exceptions. First, we of course need to increment the subscripts for each new pass. Second, for the jth pass, with $j = 2, 3, 4, ..., n-1$, c_j and s_j are given by

$$c_j = \frac{a_j}{\sqrt{a_j^2 + t^2}} \quad \text{and} \quad s_j = \frac{t}{\sqrt{a_j^2 + t^2}}$$

since we've used t to save the old value of b_j. We can therefore implement the entire factorization process as follows.

> save b_1 in the temporary variable t
> for $j = 1, 2, 3, ..., n-1$
>   let $r = \sqrt{a_j^2 + t^2}$
>   compute $c_j = a_j/r$ and $s_j = t/r$
>   overwrite a_j with r
>   save b_j in t
>   overwrite b_j with $tc_j + a_{j+1}s_j$

```
        overwrite a_{j+1} with -ts_j + a_{j+1}c_j
        if ( j ≠ n - 1 )
            save b_{j+1} in t
            overwrite b_{j+1} with tc_j
        end
end
```

The first line in this pseudocode has been included so that the first pass can be handled in the same manner as all of the later passes. The final two statements have been placed inside a conditional statement since, during the last pass through the matrix, there is no element $b_{j+1} = b_n$ to overwrite.

EXAMPLE 4.10 The QR Factorization of a Symmetric Tridiagonal Matrix

Consider again the symmetric tridiagonal matrix

$$A = \begin{bmatrix} 4 & 3 & 0 \\ 3 & 1 & -1 \\ 0 & -1 & 3 \end{bmatrix}.$$

For this example, we have

$$a_1 = 4, \quad a_2 = 1, \quad a_3 = 3, \quad b_1 = 3, \quad \text{and} \quad b_2 = -1.$$

To prepare for the first pass, we set $t = b_1 = 3$. We then calculate

$$r = \sqrt{a_1^2 + t^2} = 5, \quad c_1 = \frac{a_1}{r} = \frac{4}{5} \quad \text{and} \quad s_1 = \frac{t}{r} = \frac{3}{5},$$

and set $a_1 = r = 5$. Next, we set $t = b_1 = 3$ and then calculate

$$\begin{aligned} b_1 &= tc_1 + a_2 s_1 = 3; \quad \text{and} \\ a_2 &= -ts_1 + a_2 c_1 = -1. \end{aligned}$$

Finally, set $t = b_2 = -1$ and calculate $b_2 = tc_1 = -\frac{4}{5}$.

The second pass starts with the calculations

$$r = \sqrt{a_1^2 + t^2} = \sqrt{2}, \quad c_2 = \frac{a_2}{r} = -\frac{1}{\sqrt{2}}, \quad \text{and} \quad s_2 = \frac{t}{r} = -\frac{1}{\sqrt{2}}.$$

After setting $a_2 = r = \sqrt{2}$ and $t = b_2 = -\frac{4}{5}$, we then calculate

$$\begin{aligned} b_2 &= tc_2 + a_3 s_2 = -\tfrac{11}{5\sqrt{2}}; \quad \text{and} \\ a_3 &= -ts_2 + a_3 c_2 = -\tfrac{19}{5\sqrt{2}}. \end{aligned}$$

The results of our factorization of A are therefore

$$\begin{array}{lll} a_1 = 5, & a_2 = \sqrt{2}, & a_3 = -\frac{19}{5\sqrt{2}}, \\ b_1 = 3, & b_2 = -\frac{11}{5\sqrt{2}}, & \\ c_1 = \frac{4}{5}, & s_1 = \frac{3}{5}, & \\ c_2 = -\frac{1}{\sqrt{2}}, & s_2 = -\frac{1}{\sqrt{2}}. & \end{array}$$

We will now examine how to use these values to compute the product $R^{(0)}Q^{(0)}$.

The Product $R^{(i)}Q^{(i)}$

Earlier, we established that the upper triangular matrix in the QR factorization of the matrix $A^{(i)}$ is given by $R^{(i)} = P_{(n-1,n)} \cdots P_{(3,4)}P_{(2,3)}P_{(1,2)}A^{(i)}$. Combining this expression with the equation ${Q^{(i)}}^T A^{(i)} = R^{(i)}$, we see that ${Q^{(i)}}^T = P_{(n-1,n)} \cdots P_{(3,4)}P_{(2,3)}P_{(1,2)}$. This, in turn, implies that

$$Q^{(i)} = P^T_{(1,2)}P^T_{(2,3)}P^T_{(3,4)} \cdots P^T_{(n-1,n)}.$$

To form the product $R^{(i)}Q^{(i)}$, however, there is no need to compute the matrix $Q^{(i)}$ explicitly. Instead, we can save the s_j and c_j values associated with each rotation matrix, $P_{(j,j+1)}$, and then postmultiply $R^{(i)}$ by the transpose of each rotation matrix in succession. To carry out each multiplication we make use of the relations

$$\begin{matrix} i\text{th column of} \\ MP^T_{(i,j)} \end{matrix} = \cos\theta \cdot \begin{matrix} i\text{th column} \\ \text{of } M \end{matrix} + \sin\theta \cdot \begin{matrix} j\text{th column} \\ \text{of } M \end{matrix} \tag{2}$$

and

$$\begin{matrix} j\text{th column of} \\ MP^T_{(i,j)} \end{matrix} = -\sin\theta \cdot \begin{matrix} i\text{th column} \\ \text{of } M \end{matrix} + \cos\theta \cdot \begin{matrix} j\text{th column} \\ \text{of } M \end{matrix}. \tag{3}$$

As with the factorization process, we can deduce the complete sequence of calculations for obtaining the product $R^{(i)}Q^{(i)}$ by examining just the first multiplication, $R^{(i)}P^T_{(1,2)}$. We find

$$\begin{bmatrix} a_1 & b_1 & e_1 & & & \\ & a_2 & b_2 & e_2 & & \\ & & a_3 & b_3 & e_3 & \\ & & & \ddots & \ddots & \ddots \end{bmatrix} \begin{bmatrix} c_1 & -s_1 & & \\ s_1 & c_1 & & \\ & & 1 & \\ & & & \ddots \end{bmatrix}$$

$$= \begin{bmatrix} a_1c_1 + b_1s_1 & -a_1s_1 + b_1c_1 & e_1 & & & \\ a_2s_1 & a_2c_1 & b_2 & e_2 & & \\ & & a_3 & b_3 & e_3 & \\ & & & \ddots & \ddots & \ddots \end{bmatrix}. \tag{4}$$

Here, the e_j denote the values that we know are present in $R^{(i)}$ but that we did not save during the factorization step.

We now make two very important observations. First, based on equations (2) and (3), we know that postmultiplication by $P^T_{(2,3)}, P^T_{(3,4)}, \ldots, P^T_{(n-1,n)}$ will have no effect on the first column of $R^{(i)}P^T_{(1,2)}$. Therefore, the first column of $R^{(i)}P^T_{(1,2)}$, as shown on the right-hand side of (4), is the first column of $R^{(i)}Q^{(i)}$. Second, since $A^{(0)}$ is symmetric and each $Q^{(i)}$ is orthogonal, it follows that $A^{(i+1)} = R^{(i)}Q^{(i)}$ must also be symmetric (see Exercise 1). Consequently, not only do we know the first column of $R^{(i)}Q^{(i)}$ after this first multiplication, we know the first row as well. Thus, calculation of the values along the main diagonal and below must be carried out, but calculations above the main diagonal are unnecessary.

Bringing all this information together, it follows that to obtain the product $R^{(i)}Q^{(i)}$, we need to perform the operations

overwrite a_j with $a_jc_j + b_js_j$;
overwrite b_j with $a_{j+1}s_j$; and
overwrite a_{j+1} with $a_{j+1}c_j$,

for $j = 1, 2, 3, ..., n-1$. Note that the e_j do not play a role in any of these computations.

EXAMPLE 4.11 The Product $R^{(0)}Q^{(0)}$ from the Previous Example

The results of our factorization of

$$A = \begin{bmatrix} 3 & 2 & 0 \\ 2 & -5 & -1 \\ 0 & -1 & 4 \end{bmatrix}$$

were

$$\begin{array}{lll} a_1 = 5, & a_2 = \sqrt{2}, & a_3 = -\frac{19}{5\sqrt{2}}, \\ b_1 = 3, & b_2 = -\frac{11}{5\sqrt{2}}, & \\ c_1 = \frac{4}{5}, & s_1 = \frac{3}{5}, & \\ c_2 = -\frac{1}{\sqrt{2}}, & s_2 = -\frac{1}{\sqrt{2}}. & \end{array}$$

The first set of calculations leading to the product $R^{(0)}Q^{(0)}$ yields

$$\begin{aligned} a_1 &= a_1c_1 + b_1s_1 = \frac{29}{5} = 5.8; \\ b_1 &= a_2s_1 = \frac{3\sqrt{2}}{5}; \quad \text{and} \\ a_2 &= a_2c_1 = \frac{4\sqrt{2}}{5}. \end{aligned}$$

The second set of calculations then gives

$$\begin{aligned} a_2 &= a_2c_2 + b_2s_2 = -\frac{4}{5} + \frac{11}{10} = 0.3; \\ b_2 &= a_3s_2 = 1.9; \quad \text{and} \\ a_3 &= a_3c_2 = 1.9. \end{aligned}$$

Hence,

$$A^{(1)} = R^{(0)}Q^{(0)} = \begin{bmatrix} 5.8 & 0.848528 & 0 \\ 0.848528 & 0.3 & 1.9 \\ 0 & 1.9 & 1.9 \end{bmatrix}.$$

Accelerating Convergence

Since the rate of convergence of the off-diagonal elements to zero is $O(|\lambda_j/\lambda_{j-1}|)$, convergence of the sequence $\{A^{(i)}\}$ to diagonal form will be slow whenever the eigenvalues of A are closely spaced in magnitude. To accelerate convergence, we can shift the eigenvalues of $A^{(i)}$ by subtracting a multiple, σ_i, of the identity matrix, much as is done in the inverse power method. This changes the first step of each iteration to

- factor $A^{(i)} - \sigma_i I$ into the product $Q^{(i)}R^{(i)}$.

Technically, the computation of $A^{(i+1)}$ should then be $A^{(i+1)} = R^{(i)}Q^{(i)} + \sigma_i I$ so that $A^{(i+1)}$ will be similar to $A^{(i)}$. The addition of $\sigma_i I$ is typically not done, however. Rather, the shifts, σ_i, from all of the iterations are accumulated, and the accumulated value is added to the diagonal entries once convergence has been obtained.

The two most common choices for σ_i are

(1) $\sigma_i = a_n^{(i)}$, or

(2) $\sigma_i =$ the eigenvalue of $\begin{bmatrix} a_{n-1}^{(i)} & b_{n-1}^{(i)} \\ b_{n-1}^{(i)} & a_n^{(i)} \end{bmatrix}$ that is closest to $a_n^{(i)}$.

Here, and below, $a_j^{(i)}$ denotes the jth diagonal element and $b_j^{(i)}$ the jth off-diagonal element of $A^{(i)}$. The QR algorithm converges much faster with one of these shifts than with no shifting at all (see Wilkinson [1]). The second shift, often called the Wilkinson shift, is generally preferred and usually produces cubic convergence (see Wilkinson and Reinsch [5]).

With either choice of the shift parameter, we are essentially trying to force $b_{n-1}^{(i+1)}$ to be the off-diagonal element which converges to zero fastest. Accordingly, after each of the initial iterations of the algorithm, we check the size of $\left|b_{n-1}^{(i+1)}\right|$, and when this value falls below a specified convergence tolerance, the value $a_n^{(i+1)} + \Sigma$ is accepted as an eigenvalue of $A^{(0)}$. Here, Σ denotes the sum of all shifts that have been carried out, starting from the first iteration up to the current iteration. At this point, we no longer need to include the nth row or the nth column in the calculations. For subsequent iterations, the shift is therefore chosen as either $a_{n-1}^{(i)}$ or the eigenvalue of $\begin{bmatrix} a_{n-2}^{(i)} & b_{n-2}^{(i)} \\ b_{n-2}^{(i)} & a_{n-1}^{(i)} \end{bmatrix}$ that is closest to $a_{n-1}^{(i)}$, and the size of $\left|b_{n-2}^{(i+1)}\right|$ is monitored. When this value has converged to zero, $a_{n-1}^{(i+1)} + \Sigma$ is accepted as another eigenvalue of $A^{(0)}$, and calculations proceed on the first $n-2$ rows and columns. This process,

checking the magnitude of a specific off-diagonal element,
accepting an eigenvalue upon convergence of the off-diagonal element, and
proceeding with calculations on one fewer row and column

continues until $\left|b_1^{(i+1)}\right|$ converges to zero. At this point, $a_2^{(i+1)} + \Sigma$ and $a_1^{(i+1)} + \Sigma$ are the final two eigenvalues of $A^{(0)}$.

The following pseudocode summarizes the QR algorithm. In this code, the variable Σ accumulates the shifts, the variable *last* indicates the portion of the matrix to be included in calculations and the parameter *TOL* is the specified convergence tolerance. Note that n – *last* is the number of eigenvalues that have already been determined.

```
initialize Σ = 0 and last = n
for i = 1, 2, 3, ..., repeat until last = 1
        compute the shift, σ                  \
        Σ = Σ + σ                              |  work only with rows 1, 2, 3, ..., last
        A^(i) = A^(i) − σI                     |  and columns 1, 2, 3, ..., last
        factor A^(i) into Q^(i) R^(i)          |
        compute A^(i+1) = R^(i) Q^(i)         /
    if |b_{last−1}| < TOL
            report a_last + Σ as an eigenvalue
            last = last − 1
    end
end
report a_1 + Σ as an eigenvalue
```

The Algorithm in Action

Now let's put the entire QR algorithm into action. Take

$$A^{(0)} = \begin{bmatrix} 4 & 3 & 0 \\ 3 & 1 & -1 \\ 0 & -1 & 3 \end{bmatrix}.$$

We will let $TOL = 5 \times 10^{-4}$ be the convergence tolerance, and we will implement the Wilkinson shift. Starting with $last = 3$, we compute the eigenvalues of

$$\begin{bmatrix} a_2^{(0)} & b_2^{(0)} \\ b_2^{(0)} & a_3^{(0)} \end{bmatrix} = \begin{bmatrix} 1 & -1 \\ -1 & 3 \end{bmatrix},$$

which are $2 \pm \sqrt{2}$. The eigenvalue that is closest to $a_3^{(0)} = 3$ is $2 + \sqrt{2} \approx 3.4142$, so we take this value for the first shift. Thus, $\sigma = 3.4142$ and $\Sigma = 3.4142$. Factoring the matrix $A^{(0)} - \sigma I$ and computing RQ yields

$$A^{(1)} = \begin{bmatrix} -1.1755 & 3.4850 & 0 \\ 3.4850 & -0.7376 & -0.09673 \\ 0 & -0.09673 & -0.3296 \end{bmatrix}.$$

Since $\left|b_2^{(1)}\right| > TOL$, we continue to work with the entire matrix.

For the second iteration, $\sigma = -0.3078$. Thus $\Sigma = 3.1064$. Factoring $A^{(1)} - \sigma I$ and computing RQ yields

$$A^{(2)} = \begin{bmatrix} -2.0894 & 3.1822 & 0 \\ 3.1822 & 0.7926 & -0.0006623 \\ 0 & -0.0006623 & -0.02244 \end{bmatrix}.$$

Since $|b_2^{(2)}| > TOL$, we work with the entire matrix for yet another iteration.

In the third iteration, we find $\sigma = -0.02244$. Thus $\Sigma = 3.08397$. Factoring $A^{(2)} - \sigma I$ and computing RQ now yields

$$A^{(3)} = \begin{bmatrix} -2.9474 & 2.6102 & 0 \\ 2.6102 & 1.6955 & 9.817 \times 10^{-11} \\ 0 & 9.817 \times 10^{-11} & 4.6138 \times 10^{-7} \end{bmatrix}.$$

At this point, note that $|b_2^{(3)}| < TOL$, so we accept 4.6138×10^{-7} as an eigenvalue of $A^{(3)}$. This implies that $\lambda_3 = 3.08397 + 4.6138 \times 10^{-7} = 3.08397$ is an eigenvalue of $A^{(0)}$.

Having determined one eigenvalue, the fourth iteration works on only the first two rows and columns of $A^{(3)}$. Therefore, the shift is chosen as the eigenvalue of

$$\begin{bmatrix} a_1^{(3)} & b_1^{(3)} \\ b_1^{(3)} & a_2^{(3)} \end{bmatrix} = \begin{bmatrix} -2.9474 & 2.6102 \\ 2.6102 & 1.6955 \end{bmatrix}$$

that is closest in value to $a_2^{(3)} = 1.6955$. This gives $\sigma = 2.8673$, and then $\Sigma = 5.9513$. The matrix $A^{(4)}$ is found to be

$$A^{(4)} = \begin{bmatrix} -6.9865 & 0 & 0 \\ 0 & 0 & 9.817 \times 10^{-11} \\ 0 & 9.817 \times 10^{-11} & 4.6138 \times 10^{-7} \end{bmatrix}.$$

Since $\left|b_1^{(4)}\right| < TOL$, we accept $a_2^{(4)} = 0$ as an eigenvalue of $A^{(4)}$, which produces the second eigenvalue of $A^{(0)}$: $\lambda_2 = 0 + 5.9513 = 5.9513$. With only one eigenvalue left to be found, it follows that $\lambda_1 = -6.9865 + 5.9513 = -1.0352$.

It is interesting to note that had we required the more restrictive convergence tolerance of $TOL = 5 \times 10^{-14}$, just one more iteration of the QR algorithm would have been needed.

Determining Eigenvectors

With the QR algorithm, we can compute the eigenvalues and eigenvectors of a symmetric tridiagonal matrix simultaneously. Let V be the matrix of eigenvectors, with the jth column being the eigenvector associated with λ_j. If $A^{(0)}$ was obtained as the result of reducing a symmetric matrix to symmetric tridiagonal form, then V should be initialized to the matrix of eigenvector information produced by the reduction algorithm; otherwise, V should be initialized to the identity matrix. Within the pseudocode for the QR algorithm, after factoring $A^{(i)}$ but before performing the convergence check, replace V by

$$VP^T_{(1,2)}P^T_{(2,3)}P^T_{(3,4)}\cdots P^T_{(last-1,last)},$$

repeatedly making use of equations (2) and (3) with the appropriate sine and cosine values. This will accumulate the effect of each of the orthogonal matrices $Q^{(i)}$. The justification for this equation follows the same argument as was presented in Section 4.4 regarding the eigenvectors of two matrices related via a similarity transformation.

EXAMPLE 4.12 Determining Eigenvectors

Since the matrix

$$A^{(0)} = \begin{bmatrix} 4 & 3 & 0 \\ 3 & 1 & -1 \\ 0 & -1 & 3 \end{bmatrix}$$

is originally in symmetric tridiagonal form, we initialize V to the identity matrix. During the first iteration of the QR algorithm, the sine and cosine values associated with the rotation matrices $P_{(1,2)}$ and $P_{(2,3)}$ are

$$c_1 = 0.191643, \quad c_2 = -0.959524, \quad s_1 = 0.981465, \quad \text{and} \quad s_2 = -0.281628.$$

Hence, we replace V by

$$\begin{aligned} VP^T_{(1,2)}P^T_{(2,3)} &= I\begin{bmatrix} 0.191643 & -0.981465 & 0 \\ 0.981465 & 0.191643 & 0 \\ 0 & 0 & 1 \end{bmatrix}\begin{bmatrix} 1 & 0 & 0 \\ 0 & -0.959524 & 0.281628 \\ 0 & -0.281628 & -0.959524 \end{bmatrix} \\ &= \begin{bmatrix} 0.191643 & 0.941739 & -0.276407 \\ 0.981465 & -0.183886 & 0.053972 \\ 0 & -0.281628 & -0.959524 \end{bmatrix}. \end{aligned}$$

During the second iteration, the sine and cosine values associated with the rotation matrices $P_{(1,2)}$ and $P_{(2,3)}$ are

$$c_1 = -0.241604, \quad c_2 = -0.999565, \quad s_1 = 0.970375, \quad \text{and} \quad s_2 = -0.029498.$$

The eigenvector matrix V is therefore replaced by

$$\begin{aligned} VP^T_{(1,2)}P^T_{(2,3)} &= V\begin{bmatrix} -0.241604 & -0.970375 & 0 \\ 0.970375 & -0.241604 & 0 \\ 0 & 0 & 1 \end{bmatrix}\begin{bmatrix} 1 & 0 & 0 \\ 0 & -0.999565 & 0.029498 \\ 0 & -0.029498 & -0.999565 \end{bmatrix} \\ &= \begin{bmatrix} 0.867538 & 0.421467 & 0.264091 \\ -0.415564 & 0.905974 & -0.080731 \\ -0.273285 & -0.039709 & 0.961113 \end{bmatrix}. \end{aligned}$$

Following the third iteration

$$V = \begin{bmatrix} -0.119119 & 0.957057 & -0.264294 \\ 0.986132 & 0.145023 & 0.080700 \\ 0.115564 & -0.251016 & -0.961060 \end{bmatrix},$$

and after the final iteration

$$V = \begin{bmatrix} 0.500622 & -0.824334 & -0.264294 \\ -0.840249 & -0.536162 & 0.080700 \\ -0.208228 & 0.181672 & -0.961060 \end{bmatrix}.$$

Thus, the matrix

$$\begin{bmatrix} 4 & 3 & 0 \\ 3 & 1 & -1 \\ 0 & -1 & 3 \end{bmatrix}$$

has eigenpairs

$$(-1.0352, \begin{bmatrix} 0.500622 & -0.840249 & -0.208228 \end{bmatrix}^T),$$

$$(5.9513, \begin{bmatrix} -0.824334 & -0.536162 & 0.181672 \end{bmatrix}^T),$$

$$(3.0840, \begin{bmatrix} -0.264294 & 0.080700 & -0.961060 \end{bmatrix}^T),$$

where the eigenvectors are normalized to unit length in the l_2-norm.

References

1. J. H. Wilkinson, *The Algebraic Eigenvalue Problem*, Oxford University Press, Oxford, 1965.
2. G. H. Golub and C. F. van Loan, *Matrix Computations*, 2nd edition, The Johns Hopkins University Press, Baltimore, 1989.
3. W. H. Press, B. P. Flannery, S. A. Teukolsky, and W. T. Vetterling, *Numerical Recipes: The Art of Scientific Computing*, Cambridge University Press, Cambridge, 1992.
4. J. G. F. Francis, "The QR Transformation I, II," *Computer Journal*, **4**, 265–271, 1961–2.
5. J. H. Wilkinson and C. Reinsch, *Handbook for Automatic Computation. Volume 2: Linear Algebra*, Springer-Verlag, Berlin, 1971.
6. D. Cvetkovic, M. Doob, and H. Sachs, *Spectra of Graphs: Theory and Application*, Academic Press, New York, 1979.
7. H. Longuet-Higgins, "Some Studies in Molecular Orbital Theory, I. Resonance Structures and Molecular Orbitals in Unsaturated Hydrocarbons," *The Journal of Chemical Physics*, **18** (3), 265–275, 1950.

EXERCISES

1. Let $A^{(0)}$ be a symmetric matrix. Prove that the matrices $A^{(i)}$ produced by the QR algorithm are symmetric for all i.

In Exercises 2–5, perform one iteration of the QR algorithm with Wilkinson shift on the indicated matrix.

2. $A = \begin{bmatrix} 3 & 2 & 0 \\ 2 & -5 & -1 \\ 0 & -1 & 4 \end{bmatrix}$

3. $A = \begin{bmatrix} 4 & -2 & 0 \\ -2 & 6 & 1 \\ 0 & 1 & -3 \end{bmatrix}$

4. $A = \begin{bmatrix} 12 & 1 & 0 \\ 1 & 3 & -2 \\ 0 & -2 & 5 \end{bmatrix}$

5. $A = \begin{bmatrix} 7 & 2 & 0 \\ 2 & -8 & 1 \\ 0 & 1 & 12 \end{bmatrix}$

In Exercises 6–11, determine all of the eigenvalues of the indicated matrix using the QR algorithm with Wilkinson shift. Use a convergence tolerance of 5×10^{-6} if working in single precision and 5×10^{-14} if working in double precision. Record the number of iterations needed to obtain the first eigenvalue and the total number of iterations needed to find all of the eigenvalues.

6. $A = \begin{bmatrix} 7 & -2 & & & \\ -2 & 3 & 5 & & \\ & 5 & -6 & 1 & \\ & & 1 & 2 & 2 \\ & & & 2 & -4 \end{bmatrix}$

7. $A = \begin{bmatrix} 3 & 1 & & & & & \\ 1 & 5 & 1 & & & & \\ & 1 & 6 & -1 & & & \\ & & -1 & 4 & 2 & & \\ & & & 2 & 7 & -1 & \\ & & & & -1 & 6 & 3 \\ & & & & & 3 & 8 \end{bmatrix}$

8. $A = \begin{bmatrix} -4 & 1 & & & & \\ 1 & 9 & 3 & & & \\ & 3 & 7 & -4 & & \\ & & -4 & 5 & 2 & \\ & & & 2 & -8 & -1 \\ & & & & -1 & 6 \end{bmatrix}$

9. $A = \begin{bmatrix} 9 & -1 & & & & \\ -1 & 3 & 1 & & & \\ & 1 & -3 & -2 & & \\ & & -2 & -9 & 4 & \\ & & & 4 & 7 & 2 \\ & & & & 2 & -9 \end{bmatrix}$

10. $A = \begin{bmatrix} 5 & 2 & & & & & \\ 2 & 6 & 1 & & & & \\ & 1 & 5 & -3 & & & \\ & & -3 & 8 & 6 & & \\ & & & 6 & 3 & -2 & \\ & & & & -2 & 3 & 1 \\ & & & & & 1 & 4 \end{bmatrix}$

11. $A = \begin{bmatrix} 4 & -1 & & & \\ -1 & 4 & -3 & & \\ & -3 & 8 & 2 & \\ & & 2 & 7 & -1 \\ & & & -1 & 9 \end{bmatrix}$

In Exercises 12–18, compute all of the eigenvalues and corresponding eigenvectors for each matrix.

12. $A = \begin{bmatrix} 12 & -1 & 1 & 0 & 3 \\ -1 & 3 & 0 & 2 & 0 \\ 1 & 0 & -6 & 2 & 1 \\ 0 & 2 & 2 & 9 & 0 \\ 3 & 0 & 1 & 0 & -2 \end{bmatrix}$

13. $A = \begin{bmatrix} 7 & 1 & & & \\ 1 & -8 & -1 & & \\ & -1 & 2 & -1 & \\ & & -1 & 9 & 1 \\ & & & 1 & 12 \end{bmatrix}$

14. $A = \begin{bmatrix} 5 & 1 & & & \\ 1 & 12 & -2 & & \\ & -2 & 8 & 4 & \\ & & 4 & 12 & -3 \\ & & & -3 & 18 \end{bmatrix}$

15. $A = \begin{bmatrix} 6 & 2 & -1 & -3 & -5 \\ 2 & 18 & -2 & 5 & 4 \\ -1 & -2 & 20 & -5 & -4 \\ -3 & 5 & -5 & 28 & 1 \\ -5 & 4 & -4 & 1 & 12 \end{bmatrix}$

16. $A = \begin{bmatrix} 3 & -1 & 0 & -1 & 0 & -1 \\ -1 & 2 & -1 & 0 & 0 & 0 \\ 0 & -1 & 2 & -1 & 0 & 0 \\ -1 & 0 & -1 & 3 & -1 & 0 \\ 0 & 0 & 0 & -1 & 2 & -1 \\ -1 & 0 & 0 & 0 & -1 & 2 \end{bmatrix}$

17. $A = \begin{bmatrix} 1 & 2 & & & & & \\ 2 & 3 & 4 & & & & \\ & 4 & 5 & 6 & & & \\ & & 6 & 7 & -1 & & \\ & & & -1 & 8 & -2 & \\ & & & & -2 & 9 & \end{bmatrix}$

18. $A = \begin{bmatrix} 52 & 6 & 15 & 1 & 5 & 3 & 5 \\ 6 & 16 & 0 & -13 & 4 & -9 & -2 \\ 15 & 0 & 58 & -3 & 8 & 1 & 6 \\ 1 & -13 & -3 & 30 & 1 & -7 & -3 \\ 5 & 4 & 8 & 1 & 42 & 3 & 5 \\ 3 & -9 & 1 & -7 & 3 & 28 & -1 \\ 5 & -2 & 6 & -3 & 5 & -1 & 44 \end{bmatrix}$

19. According to Huckel theory (see Cvetkovic, Doob, and Sachs [6, Chapter 8] or Longuet-Higgins [7]), the energy levels and wave functions of hydrocarbon molecules are related to the eigenvalues and eigenvectors of the adjacency matrix of the graph that represents the carbon skeleton of the molecule. The carbon skeletons for three hydrocarbons are shown below.

(a) Compute all of the eigenvalues and eigenvectors of the adjacency matrix associated with isopropylbenzene.

(b) Repeat part (a) for 4-ethyl, 2-methylhexane.

(c) Repeat part (a) for isobutylcyclopentane.

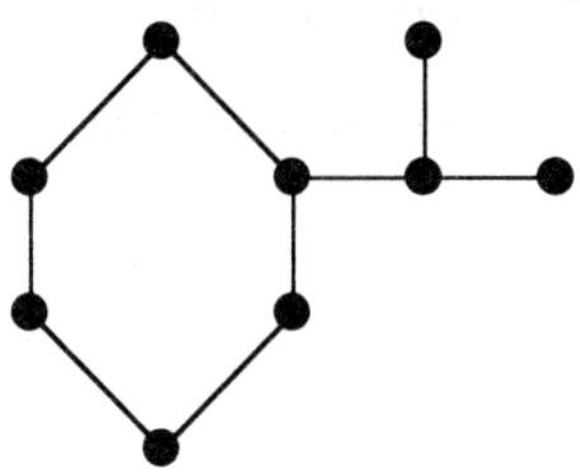

Isopropylbenzene

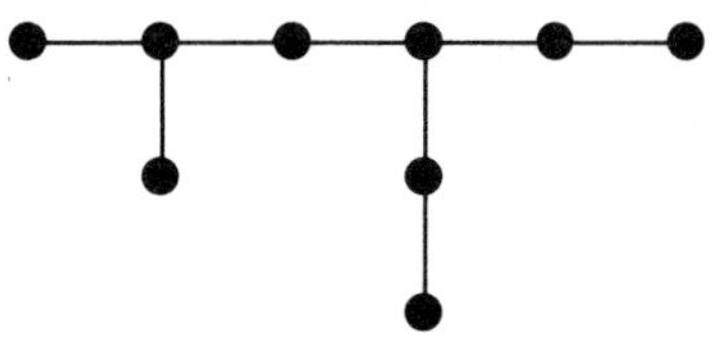

4-Ethyl, 2-methylhexane

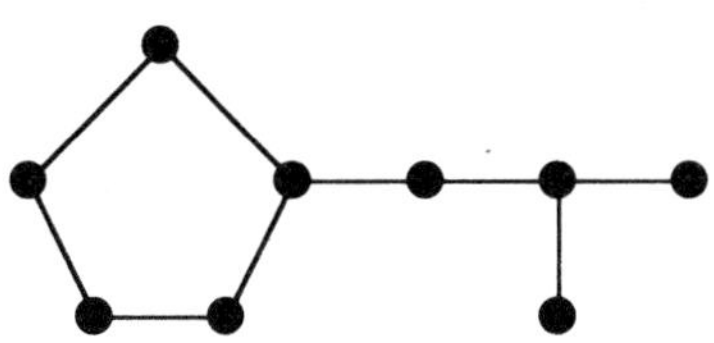

Isobutylcyclopentane

20. In simulating the nuclear magnetic resonance (NMR) spectra of three magnetically inequivalent protons, the transition frequencies and intensities are related to the eigenvalues and eigenvectors of the spin Hamiltonian matrix

$$\begin{bmatrix} s_{11} & 0 & 0 & 0 & 0 & 0 & 0 & 0 \\ 0 & s_{22} & J_{23}/2 & J_{13}/2 & 0 & 0 & 0 & 0 \\ 0 & J_{23}/2 & s_{33} & J_{12}/2 & 0 & 0 & 0 & 0 \\ 0 & J_{13}/2 & J_{12}/2 & s_{44} & 0 & 0 & 0 & 0 \\ 0 & 0 & 0 & 0 & s_{55} & J_{12}/2 & J_{13}/2 & 0 \\ 0 & 0 & 0 & 0 & J_{12}/2 & s_{66} & J_{23}/2 & 0 \\ 0 & 0 & 0 & 0 & J_{13}/2 & J_{23}/2 & s_{77} & 0 \\ 0 & 0 & 0 & 0 & 0 & 0 & 0 & s_{88} \end{bmatrix}$$

The diagonal entries are given by the formulas

$$
\begin{aligned}
s_{11} &= (v_1 + v_2 + v_3)/2 + (J_{12} + J_{13} + J_{23})/4 \\
s_{22} &= (v_1 + v_2 - v_3)/2 + (J_{12} - J_{13} - J_{23})/4 \\
s_{33} &= (v_1 - v_2 + v_3)/2 + (-J_{12} + J_{13} - J_{23})/4 \\
s_{44} &= (-v_1 + v_2 + v_3)/2 + (-J_{12} - J_{13} + J_{23})/4 \\
s_{55} &= (v_1 - v_2 - v_3)/2 + (-J_{12} - J_{13} + J_{23})/4 \\
s_{66} &= (-v_1 + v_2 - v_3)/2 + (-J_{12} + J_{13} - J_{23})/4 \\
s_{77} &= (-v_1 - v_2 + v_3)/2 + (J_{12} - J_{13} - J_{23})/4 \\
s_{88} &= (-v_1 - v_2 - v_3)/2 + (J_{12} + J_{13} + J_{23})/4.
\end{aligned}
$$

The parameters in this system are three chemical shifts: v_1, v_2, and v_3; and three coupling constants: J_{12}, J_{13}, and J_{23}. Take

$$
\begin{aligned}
v_1 &= 342.0 \text{ Hz}, \quad v_2 = 364.6 \text{ Hz}, \quad v_3 = 372.2 \text{ Hz}, \\
J_{12} &= 11.75 \text{ Hz}, \quad J_{13} = 17.90 \text{ Hz}, \quad J_{23} = 0.91 \text{ Hz}
\end{aligned}
$$

and determine all of the eigenvalues of the spin Hamiltonian matrix and their associated eigenvectors.

CHAPTER 5

Interpolation (and Curve Fitting)

AN OVERVIEW

Building a Table of Logarithms

A publisher of mathematics textbooks needs a table of values for the common logarithm function (i.e., the base 10 logarithm) for one of its new precalculus textbooks. Each entry in the table is to be accurate to six (6) decimal places. The table must include entries for uniformly spaced values of x ranging from 1.0 to 10.0, and the increment between x values must be small enough so that linear interpolation between any two entries in the table introduces an error of less than 10^{-6}. What is the maximum possible increment that can be used in the construction of this table?

Properties of Water

Table A.5 in Frank White, *Fluid Mechanics*, lists the following values for the surface tension, Υ, vapor pressure, p_v, and sound speed, a, for water as a function of temperature. Based on these values, what are the surface tension, vapor pressure and sound speed for water when $T = 34°$ C, 68° C, 86° C, and 91° C?

T (°C)	Υ (N/m)	p_v (kPa)	a (m/s)
0	0.0756	0.611	1402
10	0.0742	1.227	1447
20	0.0728	2.337	1482
30	0.0712	4.242	1509
40	0.0696	7.375	1529
50	0.0679	12.34	1542
60	0.0662	19.92	1551
70	0.0644	31.16	1553
80	0.0626	47.35	1554
90	0.0608	70.11	1550
100	0.0589	101.3	1543

Probability of a Shutout in Racquetball

The following table (drawn from Joseph Keller, "Probability of a Shutout in Racquetball" *SIAM Review*, **26**, 267–8, 1984) gives the probability, P, that a given player will shutout an opponent as a function of the probability that the player will win any particular rally, p, regardless of who serves.

p	1.0	0.9	0.85	0.842	0.84	0.5
P	1.0	0.753	0.534	0.500	0.490	0.0001504

Suppose that a player estimates a 60% chance of winning any particular rally against a given opponent, regardless of who serves. What is the probability that this player will shutout the given opponent? With what probability must a player win any particular rally in order to have a 25% chance of shutting out an opponent?

Data Analysis for the Spread of an Epidemic

Suppose that a mathematical model for the spread of an epidemic produces the following estimates for the number of people who have died as a result of the epidemic, $D(t)$, and the rate at which people are dying, $D'(t)$. Here, time is measured in weeks. Using this data, we wish to generate a table which shows the number of dead at half-week increments.

t	$D(t)$	$D'(t)$
0.000000	0.000000	600.000000
0.750000	445.903683	573.579644
1.500000	842.695315	477.074216
2.085600	1095.211197	384.947629
2.676193	1295.955674	296.576145
3.219694	1437.602773	226.796410
3.748513	1542.363644	171.475176
4.279179	1621.280769	127.808738
4.821254	1680.890649	93.728061
5.000000	1696.803710	84.473801

Fundamental Mathematical Problem

Each of the examples just presented illustrates the fundamental mathematical problem to be treated in this chapter:

> Given a set of points (x_i, f_i) for $i = 0, 1, 2, \dots, n$, where the x_i are distinct values of the independent variable and the f_i are corresponding values of some function f, either
>
> approximate the value of f at some value of x not listed among the x_i
>
> or
>
> determine a function g that in some sense approximates the data

Problem data can also include derivative values in addition to function values. In some instances, the problem data will be specified as the function f itself, rather than as a discrete set of points from the graph of f. In these cases, a function g that is less expensive to evaluate and/or easier to manipulate is sought.

The mathematical problem stated above actually gives rise to two different areas of study: interpolation and approximation. In interpolation, the function g is determined by requiring the error to be zero at each of the x_i; that is, by enforcing the conditions $g(x_i) = f_i$ for each $i = 0, 1, 2, \ldots, n$. Hence, interpolation treats error in a somewhat local fashion. Approximation, on the other hand, treats error in a more global manner, requiring that some measure of error, such as the sum of the square of the difference between $g(x_i)$ and f_i, be minimized. Although a final section on least squares regression (one possible approximation technique) is provided for completeness, the objective of this chapter will be to discuss interpolation.

In addition to the variety of applications, such as those given above, for which interpolation is useful, interpolation is also a major tool for the development of other numerical techniques. Most of the algorithms developed in Chapter 6 (Differentiation and Integration) have their basis in interpolation. The same is true for many of the techniques which will be developed in Chapter 7 (Solution of Initial Value Problems).

There are many different types of interpolation, depending upon the class of functions from which g is selected. The most common forms of interpolation are

1. polynomial interpolation
2. piecewise polynomial (spline) interpolation
3. rational interpolation
4. trigonometric interpolation
5. exponential interpolation

The focus of this chapter will be on polynomial and piecewise polynomial interpolation.

There are three major reasons for focusing on interpolation by polynomials and piecewise polynomials. First, as we've seen in Section 2.7 (Polynomial Rootfinding), the polynomial

$$P_n(x) = a_0 + a_1 x + a_2 x^2 + \cdots + a_n x^n$$

can be evaluated very efficiently using the synthetic division algorithm

```
value := a_n
for i from n − 1 downto 0 do
    value := a_i + x · value
```

Second, derivatives and integrals of polynomials are easy to compute and are still polynomials. Third, polynomials satisfy what is known as the uniform approximation property. This is embodied by the following theorem due to Weierstrass, a proof of which may be found in Rivlin [1].

Theorem (Weierstrass Approximation Theorem). Let f be continuous on the closed interval $[a, b]$. Given any $\epsilon > 0$, there exists a polynomial P such that

$$\|f - p\|_\infty \equiv \max_{x\in[a,b]} |f(x) - p(x)| < \epsilon.$$

Everyone who has taken a two-semester sequence in calculus is already familiar with one interpolating polynomial, though it was not referred to as such in calculus. This interpolating polynomial in disguise is none other than the Taylor polynomial:

$$f(x) \approx p_n(x) = \sum_{k=0}^{n} \frac{f^{(k)}(x_0)}{k!}(x - x_0)^k.$$

This is an interpolating polynomial in the sense that it matches the function and the first n derivative values at the location $x = x_0$. Since this polynomial uses a lot of local information about the function f, it is good for making local approximations. The error associated with the Taylor polynomial is given by

$$R_n(x) = \frac{f^{(n+1)}(c)}{(n+1)!}(x - x_0)^{n+1},$$

where c is between x and x_0. The error is therefore bounded by

$$\max |f^{(n+1)}(x)| \frac{|x - x_0|^{n+1}}{(n+1)!}.$$

Note that this bound can be large in two different ways: If $\max |f^{(n+1)}(x)|$ is large or if x is far from x_0. Nothing can really be done to improve results in the former case, but in the latter case, a different polynomial, one that uses information about f at many different locations, can be used.

Remainder of the Chapter

The remainder of this chapter is organized as follows. Section 1 introduces the Lagrange form of the interpolating polynomial and presents the error bound associated with polynomial interpolation. The next two sections provide answers for two important questions associated with polynomial interpolation. Neville's algorithm, presented in Section 2, provides the most efficient means of determining the value of the interpolating polynomial at a single value of the independent variable. Divided differences and the Newton form are covered in Section 3. These provide the most efficient way to explicitly obtain the interpolating polynomial and evaluate it for many x-values. The optimal choice of the interpolating points, x_i, is discussed in Section 4, and the next two sections introduce piecewise polynomial interpolation. Piecewise linear interpolation is treated in Section 5, while cubic spline interpolation is the subject of Section 6. Section 7 discusses Hermite interpolation, which arises when both the value of the function and its first derivative are known at each x_i. The chapter concludes with a section on least squares regression.

References

1. T. J. Rivlin, *An Introduction to the Approximation of Functions*, Dover Publications, Inc. New York, 1981.

5.1 LAGRANGE FORM OF THE INTERPOLATING POLYNOMIAL

Let x_0, x_1, x_2, ..., x_n be $n+1$ distinct, though not necessarily uniformly spaced, points along the real line, and let f_i $(i = 0, 1, 2, \ldots, n)$ denote the function value associated with the point x_i. The x_i may be referred to as abscissas, nodes or interpolating points. At this time, problem data will consist of function values only; derivative values will be introduced in Section 5.7. A polynomial, P_n, of degree at most n that satisfies $P_n(x_i) = f_i$ for each $i = 0, 1, 2, \ldots, n$ is sought. The fundamental concepts of this basic polynomial interpolation problem will be discussed in this section.

Linear Interpolation

Let's start with the simplest case, that of linear interpolation. The data in this case consists of two abscissas, x_0 and x_1, and two corresponding function values, f_0 and f_1. The objective is to find a linear polynomial, $P_1(x) = a_0 + a_1 x$, such that

$$P_1(x_0) = a_0 + a_1 x_0 = f_0 \quad \text{and} \quad P_1(x_1) = a_0 + a_1 x_1 = f_1.$$

The solution of these interpolating conditions is easily found to be

$$a_1 = \frac{f_1 - f_0}{x_1 - x_0} \quad \text{and} \quad a_0 = \frac{x_1 f_0 - x_0 f_1}{x_1 - x_0},$$

so that

$$P_1(x) = \frac{x_1 f_0 - x_0 f_1}{x_1 - x_0} + \frac{f_1 - f_0}{x_1 - x_0} x.$$

The formula for P_1 can be rearranged into the form

$$P_1(x) = \frac{x - x_1}{x_0 - x_1} f_0 + \frac{x - x_0}{x_1 - x_0} f_1.$$

This formula not only clearly identifies and distinguishes the dependence of the interpolating polynomial on the function values from the dependence on the interpolating points, it will also make the generalization to higher-degree interpolating polynomials, to be undertaken momentarily, much easier.

EXAMPLE 5.1 Linear Interpolation from Thermodynamic Tables

A thermodynamics student needs to determine whether Freon-12 under a pressure of $P = 400$ kilopascals (kPa) and with a specific volume (volume per unit mass) of $v = 0.042$ m^3/kg is in a saturated or superheated state. The answer to this question depends upon how the specific volume of $v = 0.042$ m^3/kg compares with the specific volume of saturated Freon-12 vapor, v_g, at a pressure of 400 kPa. If the given vapor pressure is below v_g then the Freon-12 is in a saturated state; otherwise it is in a superheated state.

The available thermodynamic tables (Table A.2.3 of *Fundamentals of Classical Thermodynamics* by Van Wylen and Sonntag) provide the following values for the specific volume of saturated Freon-12 vapor as a function of pressure.

Pressure (kPa)	v_g (m^3/kg)
362.6	0.047485
423.3	0.040914

Linear interpolation can be used to approximate the needed specific volume. The linear interpolating polynomial based on the values given in the table is

$$v_g(P) = \frac{P - 423.3}{362.6 - 423.3} \cdot 0.047485 + \frac{P - 362.6}{423.3 - 362.6} \cdot 0.040914.$$

Evaluating this polynomial at $P = 400$ kPa gives

$$v_g(400) = \frac{-23.3}{362.6 - 423.3} \cdot 0.047485 + \frac{37.4}{423.3 - 362.6} \cdot 0.040914 = 0.043466,$$

which is larger than the specified value for the specific volume. Hence, Freon-12 under a pressure of $P = 400$ kPa and with a specific volume of $v = 0.042$ m^3/kg is in a saturated state.

Interpolation by Higher-Degree Polynomials

If more than two data points are available, a higher-degree interpolating polynomial can be computed. Suppose that $n + 1$ data points are available. Each data point translates into a single interpolation condition; therefore, with $n + 1$ points, there will be $n + 1$ interpolation conditions, which will allow for the determination of $n+1$ polynomial coefficients. Since an nth-degree polynomial has $n+1$ coefficients (one for each power of the independent variable, plus the constant term), it follows that $n + 1$ data points can determine a polynomial of degree at most n.

To determine this polynomial of degree at most n, one could apply the interpolating conditions to produce a system of linear equations for the coefficients of the polynomial. Though a natural process to attempt, this approach is very cumbersome and time consuming. A more efficient scheme for obtaining the interpolating polynomial can be uncovered by making a close examination of the final formula given above for the linear interpolating polynomial. In particular, pay close attention to the coefficients of the function values.

$$P_1(x) = \underbrace{\frac{x - x_1}{x_0 - x_1}}_{\substack{\text{polynomial}\\ \text{degree one}\\ @\, x = x_0,\ \text{value} = 1\\ @\, x = x_1,\ \text{value} = 0}} f_0 + \underbrace{\frac{x - x_0}{x_1 - x_0}}_{\substack{\text{polynomial}\\ \text{degree one}\\ @\, x = x_0,\ \text{value} = 0\\ @\, x = x_1,\ \text{value} = 1}} f_1$$

Note that these coefficients are polynomials of the same degree as the overall interpolating polynomial. Furthermore, the coefficient of f_0 evaluates to 1 at $x = x_0$, the abscissa associated with the function value f_0, and evaluates to zero at the other abscissa. A similar result holds for the coefficient of f_1: The value is 1 at the abscissa associated with f_1, but is zero at the other abscissa. These coefficient polynomials are called *Lagrange polynomials* and are denoted by

$$L_{1,0}(x) = \frac{x - x_1}{x_0 - x_1} \quad \text{and} \quad L_{1,1}(x) = \frac{x - x_0}{x_1 - x_0}.$$

The first subscript indicates the degree of the polynomial, while the second indicates the associated interpolating point. With this notation, P_1 can be expressed very compactly as

$$P_1(x) = L_{1,0}(x)f_0 + L_{1,1}(x)f_1 = \sum_{i=0}^{1} L_{1,i}(x)f_i.$$

The simplicity of this representation suggests obtaining higher degree interpolating polynomials by generalizing the notion of a Lagrange polynomial.

Definition. The LAGRANGE POLYNOMIAL $L_{n,j}(x)$ has degree n and is associated with the interpolating point x_j in the sense

$$L_{n,j}(x_i) = \begin{cases} 1, & i = j \\ 0, & i \neq j \end{cases}.$$

With this family of functions, it is straightforward to demonstrate that

$$P_n(x) = \sum_{i=0}^{n} L_{n,i}(x)f_i$$

interpolates the data (x_j, f_j) for $j = 0, 1, 2, \ldots, n$. For each x_j

$$\begin{aligned} P_n(x_j) &= \sum_{i=0}^{n} \underbrace{L_{n,i}(x_j)}_{\substack{1 \text{ if } i=j, \\ 0 \text{ otherwise}}} f_i \\ &= 0 \cdot f_0 + \cdots + 0 \cdot f_{j-1} + 1 \cdot f_j + 0 \cdot f_{j+1} + \cdots + 0 \cdot f_n \\ &= f_j. \end{aligned}$$

Since $P_n(x) = \sum_{i=0}^{n} L_{n,i}(x)f_i$ is based on Lagrange polynomials, P_n is referred to as the Lagrange Form of the Interpolating Polynomial.

The final piece needed to construct the Lagrange form of the interpolating polynomial is to obtain explicit formulas for the $L_{n,j}$. Fortunately, these can be determined by directly applying the conditions stated in the definition. Since $L_{n,j}$ is an nth-degree polynomial with n roots located at $x = x_i$ $(i \neq j)$, it follows that $L_{n,j}$ must be of the form

$$c(x - x_0)(x - x_1) \cdots (x - x_{j-1})(x - x_{j+1}) \cdots (x - x_n)$$

for some constant c. The final condition of the definition, $L_{n,j}(x_j) = 1$, determines the value of c:

$$c = \frac{1}{(x_j - x_0)(x_j - x_1)\cdots(x_j - x_{j-1})(x_j - x_{j+1})\cdots(x_j - x_n)}.$$

Therefore,

$$\begin{aligned} L_{n,j}(x) &= \frac{(x - x_0)(x - x_1)\cdots(x - x_{j-1})(x - x_{j+1})\cdots(x - x_n)}{(x_j - x_0)(x_j - x_1)\cdots(x_j - x_{j-1})(x_j - x_{j+1})\cdots(x_j - x_n)} \\ &= \prod_{i=0,i\neq j}^{n} \frac{x - x_i}{x_j - x_i}. \end{aligned}$$

EXAMPLE 5.2 Lagrange Polynomials

Consider the following seven interpolating points:

$$x_0 = 0.0, \quad x_1 = 1.6, \quad x_2 = 3.8, \quad x_3 = 4.5, \quad x_4 = 6.3, \quad x_5 = 9.2, \quad x_6 = 10.0.$$

Based on these points, two of the Lagrange polynomials are

$$L_{6,1}(x) = \frac{x(x - 3.8)(x - 4.5)(x - 6.3)(x - 9.2)(x - 10.0)}{1.6(1.6 - 3.8)(1.6 - 4.5)(1.6 - 6.3)(1.6 - 9.2)(1.6 - 10.0)}$$

and

$$L_{6,3}(x) = \frac{x(x - 1.6)(x - 3.8)(x - 6.3)(x - 9.2)(x - 10.0)}{4.5(4.5 - 1.6)(4.5 - 3.8)(4.5 - 6.3)(4.5 - 9.2)(4.5 - 10.0)}.$$

These two polynomials are plotted in Figure 5.1. The circles are located along the x-axis at the positions of the seven interpolating points.

Note the large amplitude oscillations present in Figure 5.1 in $L_{6,3}$. This type of behavior is typical with high degree polynomials and tends to get worse as the degree of the polynomial is increased. Because of this behavior, whenever high degree polynomials are used for interpolation, some sort of consistency check needs to be performed. This could involve simply plotting the data values and the interpolating polynomial on the same graph. This would allow a visual verification of the extent to which the behavior of the polynomial matches that of the underlying data. As an alternative approach, several of the data points could be held in reserve, that is, not used to compute the interpolating polynomial. The differences between the reserved function values and the values of the interpolating polynomial at the abscissas of the reserved points would then serve as a measure of interpolation accuracy.

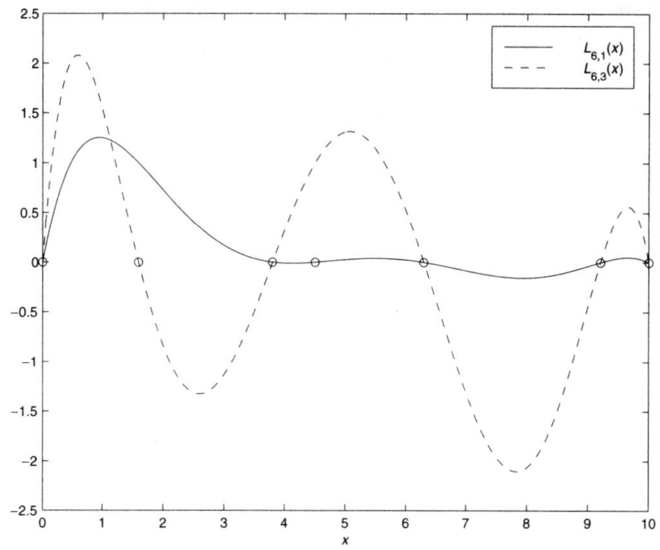

Figure 5.1 Two of the Lagrange polynomials defined by the sequence of seven interpolating points $x_0 = 0.0$, $x_1 = 1.6$, $x_2 = 3.8$, $x_3 = 4.5$, $x_4 = 6.3$, $x_5 = 9.2$, $x_6 = 10.0$. The circles are located along the x-axis at the locations of the interpolating points.

EXAMPLE 5.3 Interpolation from Thermodynamic Tables, Revisited

Reconsider the problem of determining the specific volume of Freon-12 vapor under a pressure of 400 kPa, this time using a higher degree interpolating polynomial. The third-degree polynomial based on the four points

Pressure (kPa)	308.6	362.6	423.3	491.4
v_g (m^3/kg)	0.055389	0.047485	0.040914	0.035413

is given by

$$\begin{aligned}
v_g(P) = & \frac{(P-362.6)(P-423.3)(P-491.4)}{(-54)(-114.7)(-182.8)} \cdot 0.055389 \\
& + \frac{(P-308.6)(P-423.3)(P-491.4)}{(54)(-60.7)(-128.8)} \cdot 0.047485 \\
& + \frac{(P-308.6)(P-362.6)(P-491.4)}{(114.7)(60.7)(-68.1)} \cdot 0.040914 \\
& + \frac{(P-308.6)(P-362.6)(P-423.3)}{(182.8)(128.8)(68.1)} \cdot 0.035413.
\end{aligned}$$

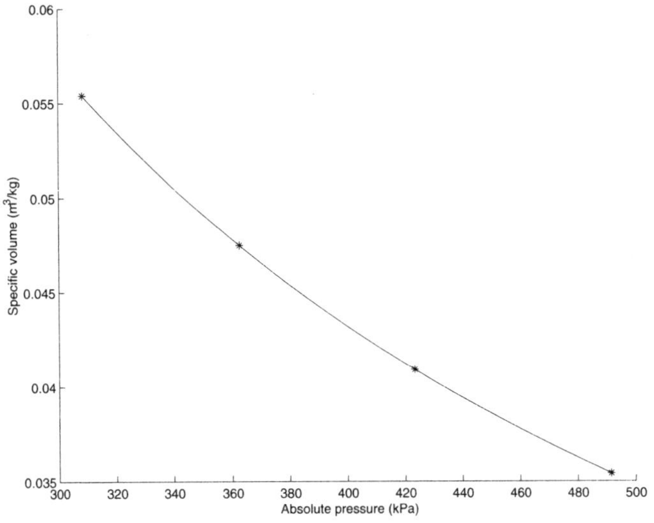

Figure 5.2 Third-degree interpolating polynomial for specific volume as a function of absolute pressure. Each data point used to construct the polynomial is denoted by *.

This polynomial, which is plotted in Figure 5.2, clearly provides a plausible representation for the data. Using this polynomial, the specific volume of saturated Freon-12 vapor at a pressure of 400 kPa is 0.043199 m^3/kg.

EXAMPLE 5.4 Emittance of Tungsten as Function of Temperature

The table below gives experimental values for the emittance of tungsten as a function of temperature.

Temperature (K)	Emittance	Temperature (K)	Emittance
300	0.024	800	0.083
400	0.035	900	0.097
500	0.046	1000	0.111
600	0.058	1100	0.125
700	0.067		

The eighth-degree polynomial which interpolates this data is plotted in Figure 5.3. Note that the behavior of the polynomial, from around 300–500 K and 900–1100 K, does not appear to be consistent with the underlying data. The use of a spline interpolating function (Section 5.6) would be advisable for this problem.

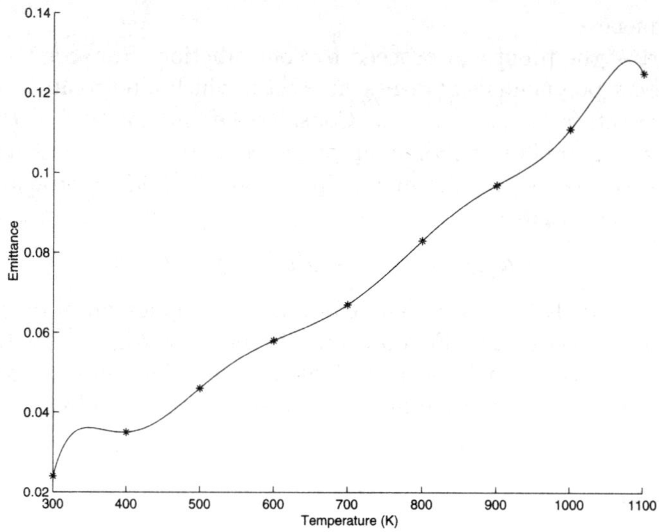

Figure 5.3 Eighth-degree interpolating polynomial for emittance of tungsten as a function of temperature. Each data point used to construct the polynomial is denoted by *.

Uniqueness of the Interpolating Polynomial

In subsequent sections of this chapter, other forms for the interpolating polynomial will be considered. The following theorem shows that given $n+1$ distinct abscissas and the corresponding function values, there is only one polynomial of degree at most n which interpolates the data. This polynomial can, of course, be written in different ways, each having its own advantages and disadvantages, but they all represent exactly the same function.

Theorem. If x_0, x_1, x_2, $\ldots$, x_n are $n+1$ distinct points and f is defined at x_0, x_1, x_2, $\ldots$, x_n, then there exists a unique polynomial, P, of degree at most n such that P interpolates f; that is,

$$P(x_i) = f(x_i)$$

for each $i = 0, 1, 2, \ldots, n$. P is called the INTERPOLATING POLYNOMIAL.

Proof. (1) Existence
This part of the proof is easy. Since the points x_0, x_1, x_2, $\ldots$, x_n are distinct, the polynomial

$$P_n(x) = \sum_{i=0}^{n} L_{n,i}(x) f_i$$

interpolates the data—it is precisely the Lagrange form of the interpolating polynomial. Existence has been established.

(2) Uniqueness
This part of the proof will proceed by contradiction. Suppose that P and Q are different polynomials of degree at most n which interpolate f at the $n+1$ distinct points $x_0, x_1, x_2, \ldots, x_n$. Consider the function $h(x) = P(x) - Q(x)$. Since P and Q are both polynomials of degree at most n, h is also a polynomial of degree at most n. Furthermore, since P and Q both interpolate the same data, it follows that

$$h(x_i) = P(x_i) - Q(x_i) = f_i - f_i = 0$$

for each $i = 0, 1, 2, \ldots, n$. Therefore, h is a polynomial of degree at most n with $n+1$ roots. The Fundamental Theorem of Algebra guarantees that the only way this can happen is if $h(x) \equiv 0$. This implies that $P = Q$, which contradicts our assumption. Hence, the interpolating polynomial is unique. □

Interpolation Error

One of the major benefits of the uniqueness theorem is that it allows for a general discussion of interpolation error. Since the various forms of the interpolating polynomial are just different ways of writing the same function, interpolation error does not depend on the form selected for the interpolating polynomial, and there is no need to treat each form separately.

Theorem. If $x_0, x_1, x_2, \ldots, x_n$ are $n+1$ distinct points in $[a,b]$ and f is continuous on $[a,b]$ and has $n+1$ continuous derivatives on (a,b), then for each $x \in [a,b]$ there exists a $\xi(x) \in [a,b]$ such that

$$f(x) = P(x) + \frac{f^{(n+1)}(\xi)}{(n+1)!}(x-x_0)(x-x_1)(x-x_2)\cdots(x-x_n),$$

where P is the interpolating polynomial.

Proof. First note that since $P(x_i) = f(x_i)$ by the interpolation conditions and since the term involving $f^{(n+1)}$ contains the factor $(x - x_i)$, the error formula holds for each abscissa, $x = x_i$. For all other $x \in [a,b]$, consider the auxiliary function

$$g(t) = f(t) - P(t) - [f(x) - P(x)]\prod_{i=0}^{n} \frac{t-x_i}{x-x_i}.$$

By hypothesis, f has $n+1$ continuous derivatives on (a,b). Since P and $\prod_{i=0}^{n} \frac{t-x_i}{x-x_i}$ are polynomials, they possess infinitely many continuous derivatives on (a,b). By construction, then, g has $n+1$ continuous derivatives on (a,b). Furthermore,

$$\begin{aligned} g(x_j) &= f(x_j) - P(x_j) - [f(x) - P(x)]\prod_{i=0}^{n} \frac{x_j - x_i}{x - x_i} \\ &= f(x_j) - P(x_j) - 0 = 0 \end{aligned}$$

for each $j = 0, 1, 2, \ldots, n$, and

$$\begin{aligned} g(x) &= f(x) - P(x) - [f(x) - P(x)] \prod_{i=0}^{n} \frac{x - x_i}{x - x_i} \\ &= f(x) - P(x) - [f(x) - P(x)] \cdot 1 = 0. \end{aligned}$$

g therefore has $n + 2$ roots on $[a, b]$. Applying the generalized Rolle's theorem (see Appendix A), it follows that there exists $\xi(x) \in [a, b]$ such that $g^{(n+1)}(\xi) = 0$.

Now, P is a polynomial of degree at most n, so $P^{(n+1)}(t) \equiv 0$. On the other hand, $\prod_{i=0}^{n} \frac{t-x_i}{x-x_i}$ is a polynomial of degree $n+1$ with leading coefficient $\left[\prod_{i=0}^{n}(x - x_i)\right]^{-1}$, so

$$\frac{d^{n+1}}{dt^{n+1}} \left[\prod_{i=0}^{n} \frac{t - x_i}{x - x_i} \right] = (n+1)! \cdot \left[\prod_{i=0}^{n} (x - x_i) \right]^{-1}.$$

Differentiating g $n + 1$ times and evaluating at ξ then gives

$$0 = g^{(n+1)}(\xi) = f^{(n+1)}(\xi) - 0 - [f(x) - P(x)](n+1)! \cdot \left[\prod_{i=0}^{n} (x - x_i) \right]^{-1}.$$

Solving this equation for $f(x)$ yields the desired error formula:

$$f(x) = P(x) + \frac{f^{(n+1)}(\xi)}{(n+1)!}(x - x_0)(x - x_1)(x - x_2) \cdots (x - x_n). \qquad \square$$

Advantages and Disadvantages of the Lagrange Form

Each form of the interpolating polynomial that will be studied will have its own set of advantages and disadvantages. One of the advantages of the Lagrange form of the interpolating polynomial is the simplicity of its derivation. Since the Lagrange form isolates the dependence of the interpolating polynomial on the function values, this form is useful when the abscissas are fixed, but the corresponding function values are changed often. The greatest advantage of the Lagrange form, however, is its theoretical value. Almost all of the techniques developed in later chapters which are based on interpolation start from the Lagrange form of the interpolating polynomial. On the negative side, if more data become available, the work performed to generate the original Lagrange form cannot be reused to compute a higher-degree polynomial. Work must begin from scratch. Finally, the Lagrange form of the interpolating polynomial is very cumbersome for common polynomial operations such as evaluation, differentiation and integration.

EXERCISES

1. Let $x_0 = -1$, $x_1 = 1$ and $x_2 = 2$.
   (a) Determine formulas for the Lagrange polynomials $L_{2,0}(x)$, $L_{2,1}(x)$, and $L_{2,2}(x)$ associated with the given interpolating points.

(b) Plot $L_{2,0}(x)$, $L_{2,1}(x)$, and $L_{2,2}(x)$ on the same set of axes over the range $[-1, 2]$.

2. Let $x_0 = -3$, $x_1 = 0$, $x_2 = e$ and $x_3 = \pi$.
   (a) Determine formulas for the Lagrange polynomials $L_{3,0}(x)$, $L_{3,1}(x)$, $L_{3,2}(x)$ and $L_{3,3}(x)$ associated with the given interpolating points.
   (b) Plot $L_{3,0}(x)$, $L_{3,1}(x)$, $L_{3,2}(x)$ and $L_{3,3}(x)$ on the same set of axes over the range $[-3, \pi]$.
3. Let $x_0 = 0.0$, $x_1 = 1.6$, $x_2 = 3.8$, $x_3 = 4.5$, $x_4 = 6.3$, $x_5 = 9.2$, and $x_6 = 10.0$.
   (a) Determine formulas for the Lagrange polynomials $L_{6,0}(x)$, $L_{6,2}(x)$, and $L_{6,5}(x)$ associated with the given interpolating points.
   (b) Plot $L_{6,0}(x)$, $L_{6,2}(x)$, and $L_{6,5}(x)$ on the same set of axes over the range $[0, 10]$.
4. Consider the function $f(x) = \ln x$.
   (a) Construct the Lagrange form of the interpolating polynomial for f passing through the points $(1, \ln 1)$, $(2, \ln 2)$, and $(3, \ln 3)$.
   (b) Plot the polynomial obtained in part (a) on the same set of axes as $f(x) = \ln x$. Use an x range of $[1, 3]$. Next, generate a plot of the difference between the polynomial obtained in part (a) and $f(x) = \ln x$.
   (c) Use the polynomial obtained in part (a) to estimate both $\ln(1.5)$ and $\ln(2.4)$. What is the error in each approximation?
   (d) Establish the theoretical error bound for using the polynomial found in part (a) to approximate $\ln(1.5)$. Compare the theoretical error bound to the error found in part (c).
5. Consider the function $f(x) = \sin x$.
   (a) Construct the Lagrange form of the interpolating polynomial for f passing through the points $(0, \sin 0)$, $(\pi/4, \sin \pi/4)$, and $(\pi/2, \sin \pi/2)$.
   (b) Plot the polynomial obtained in part (a) on the same set of axes as $f(x) = \sin x$. Use an x range of $[0, \pi/2]$. Next, generate a plot of the difference between the polynomial obtained in part (a) and $f(x) = \sin x$.
   (c) Use the polynomial obtained in part (a) to estimate both $\sin(\pi/3)$ and $\sin(\pi/6)$. What is the error in each approximation?
   (d) Establish the theoretical error bound for using the polynomial found in part (a) to approximate $\sin(\pi/3)$. Compare the theoretical error bound to the error found in part (c).
6. Consider the function $f(x) = e^x$.
   (a) Construct the Lagrange form of the interpolating polynomial for f passing through the points $(-1, e^{-1})$, $(0, e^0)$, and $(1, e^1)$.
   (b) Plot the polynomial obtained in part (a) on the same set of axes as $f(x) = e^x$. Use an x range of $[-1, 1]$. Next, generate a plot of the difference between the polynomial obtained in part (a) and $f(x) = e^x$.
   (c) Use the polynomial obtained in part (a) to estimate both $\sqrt{e}$ and $e^{-1/3}$. What is the error in each approximation?
   (d) Establish the theoretical error bound for using the polynomial found in part (a) to approximate $\sqrt{e}$. Compare the theoretical error bound to the error found in part (c).

7. Consider the data set

x	-1	0	1	2
y	5	1	1	11

(a) Show that the polynomials $f(x) = x^3 + 2x^2 - 3x + 1$ and $g(x) = \frac{1}{8}x^4 + \frac{3}{4}x^3 + \frac{15}{8}x^2 - \frac{11}{4}x + 1$ both interpolate all of the data.

(b) Why does this not contradict the uniqueness part of the theorem on existence and uniqueness of polynomial interpolation?

8. Consider the data set

x	-3	1	2	5
y	-23	-11	-23	1

(a) Show that the polynomials $f(x) = x^3 - 3x^2 - 10x + 1$ and $g(x) = -23 + 3(x-3) - 3(x+3)(x-1) + (x+3)(x-1)(x-2)$ both interpolate all of the data.

(b) Why does this not contradict the uniqueness part of the theorem on existence and uniqueness of polynomial interpolation?

9. Suppose that f is continuous and has continuous first and second derivatives on the interval $[x_0, x_1]$. Derive the following bound on the error due to linear interpolation of f:

$$|f(x) - P_1(x)| \le \frac{1}{8}h^2 \max_{x\in[x_0,x_1]} |f''(x)|,$$

where $h = x_1 - x_0$.

10. The interpolation points influence interpolation error through the polynomial $\Pi_{i=0}^{n}(x - x_i)$. Suppose we are interpolating the function f over the interval $[-1, 1]$ using linear interpolation.

(a) If $x_0 = -1$ and $x_1 = 1$, determine the maximum value of the expression $|(x - x_0)(x - x_1)|$ for $-1 \le x \le 1$.

(b) If $x_0 = -\sqrt{2}/2$ and $x_1 = \sqrt{2}/2$, determine the maximum value of the expression $|(x - x_0)(x - x_1)|$ for $-1 \le x \le 1$. How does this compare to the maximum found in part (a)?

(c) Select any two numbers from the interval $[-1, 1]$ to serve as the interpolation points x_0 and x_1. Determine the maximum value of the expression $|(x - x_0)(x - x_1)|$ for $-1 \le x \le 1$, and compare to the maxima found in parts (a) and (b).

11. The interpolation points influence interpolation error through the polynomial $\Pi_{i=0}^{n}(x - x_i)$. Suppose we are interpolating the function f over the interval $[-1, 1]$ using quadratic interpolation.

(a) If $x_0 = -1$, $x_1 = 0$ and $x_2 = 1$, determine the maximum value of the expression $|(x - x_0)(x - x_1)(x - x_2)|$ for $-1 \le x \le 1$.

(b) If $x_0 = -\sqrt{3}/2$, $x_1 = 0$ and $x_2 = \sqrt{3}/2$, determine the maximum value of the expression $|(x - x_0)(x - x_1)(x - x_2)|$ for $-1 \le x \le 1$. How does this compare to the maximum found in part (a)?

(c) Select any three numbers from the interval $[-1, 1]$ to serve as the interpolation points x_0, x_1 and x_2. Determine the maximum value of the expression $|(x - x_0)(x - x_1)(x - x_2)|$ for $-1 \le x \le 1$, and compare to the maxima found in parts (a) and (b).

12. The following data set was taken from a polynomial of degree at most five. Find the polynomial.

x	-2	-1	0	1	2	3
y	39	3	-1	-3	-9	-1

13. Consider the data set

x	0	1.25	1.85	2.40	3.05	3.64	4.25	4.85	5.45
y	0	4	6	8	10	12	14	16	18

Determine the polynomial of degree at most eight (8) which interpolates this data. Over what range of x values would you feel comfortable using the interpolating polynomial to approximate values of y? Explain.

14. A thermodynamics student needs the temperature of saturated steam under a pressure of 6.3 mega-Pascals (MPa).

(a) Estimate the temperature using linear interpolation from the data

Pressure (MPa)	Temperature (°C)
6.0	275.64
7.0	285.88

(b) Estimate the temperature using polynomial interpolation from the data

Pressure (MPa)	4.0	5.0	6.0	7.0	8.0	9.0
Temperature (°C)	250.40	263.99	275.64	285.88	295.06	303.40

(c) Which approximation do you think is more accurate and why?

15. *Perry's Chemical Engineer's Handbook* gives the following values for the heat capacity at constant pressure, c_p, of an aqueous solution of methyl alcohol as a function of the alcohol mole percentage, ϕ:

ϕ (%)	5.88	12.3	27.3	45.8	69.6	100.0
c_p (cal/g °C)	0.995	0.98	0.92	0.83	0.726	0.617

All data are provided at $T = 40°$ C and atmospheric pressure. A table that lists the heat capacity at constant pressure for $\phi = 5, 10, 15, \ldots, 100\%$ is desired.

16. The table below lists the linewidth of a printed feature on a semiconductor device as a function of the dissolution time (the amount of time the silicon wafer is placed in the developer solution).

Dissolution Time (sec)	10	12	14	16	18	20
Linewidth (μm)	0.25	0.36	0.45	0.50	0.53	0.55

(a) Approximate the linewidth of the feature after a dissolution time of 15 seconds.

(b) Plot the values in the table, together with the value obtained in part (a). Does the result from part (a) seem reasonable? Explain.

17. The following table gives the viscosity, in milli-Pascal-seconds (centipoises) of sulfuric acid as a function of concentration, in mass percent.

Concentration	0	20	40	60	80	100
Viscosity	0.89	1.40	2.51	5.37	17.4	24.2

Determine the polynomial of degree at most five which interpolates this data. The viscosity of sulfuric acid with a 5% concentration is 1.01 and with a 10% concentration is 1.12. Use these values to assess the accuracy of the interpolating polynomial.

5.2 NEVILLE'S ALGORITHM

Having established the fundamental concepts behind polynomial interpolation, are two important questions need to be addressed. First, what is the best way to compute the value of the interpolating polynomial for just one value of the independent variable? The answer to this question, which is to be studied in this section, gives rise to what is known as Neville's algorithm. Second, what is the best way to obtain the interpolating polynomial explicitly, whether this is for evaluation or for manipulation? This question leads to the development of divided differences and the Newton form of the interpolating polynomial, topics which will be considered in Section 5.3.

Notation

The following notation will greatly simplify the discussion of Neville's algorithm. Let m_1, m_2, m_3, …, m_k be k distinct integers between 0 and n, inclusive. Denote the unique polynomial of degree at most $k-1$ which interpolates data at the points x_{m_1}, x_{m_2}, x_{m_3}, …, x_{m_k} by $P_{m_1,m_2,m_3,\ldots,m_k}(x)$. The m's are usually listed in ascending order, though this is not necessary. Figure 5.4 provides an illustration of this notational scheme. Arrows indicate the points at which data is interpolated by each polynomial. For example, $P_{1,2,4}(x)$ interpolates the data at the abscissas x_1, x_2, and x_4.

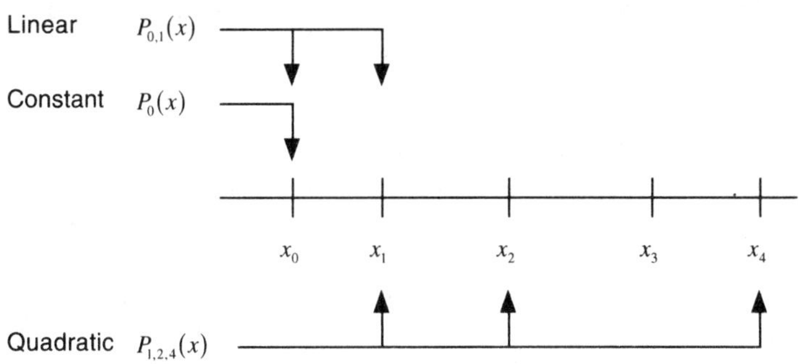

Figure 5.4 Illustration of notation used for the development of Neville's algorithm.

Constructing Higher-Degree Polynomials from Lower-Degree Polynomials

Neville's algorithm is a procedure designed to efficiently determine the value of the interpolating polynomial at a single value of the independent variable. In the process, an explicit formula for the interpolating polynomial is not generated, just the value of the polynomial. The key element in the algorithm is the fact that the interpolating polynomial through a given set of n points can be obtained by combining two polynomials that interpolate different sets of $n-1$ of those points.

For instance, the interpolating polynomial $P_{0,1,2,3}$ can be constructed by combining the polynomials $P_{0,1,3}$ and $P_{1,2,3}$ in a special way.

To determine this special way of combining lower degree interpolating polynomials, consider the case of linear interpolation. From the previous section, it is known that the polynomial which interpolates data at $x = x_0$ and $x = x_1$, $P_{0,1}$ in the current notation, is given by

$$P_{0,1}(x) = \frac{x - x_1}{x_0 - x_1} f_0 + \frac{x - x_0}{x_1 - x_0} f_1.$$

First note that the function values f_0 and f_1 are the values of two constant interpolating polynomials. In particular, $f_0 = P_0(x)$ and $f_1 = P_1(x)$. Making these substitutions into the equation for $P_{0,1}$ and rearranging terms, the linear polynomial $P_{0,1}$ can be written in the form

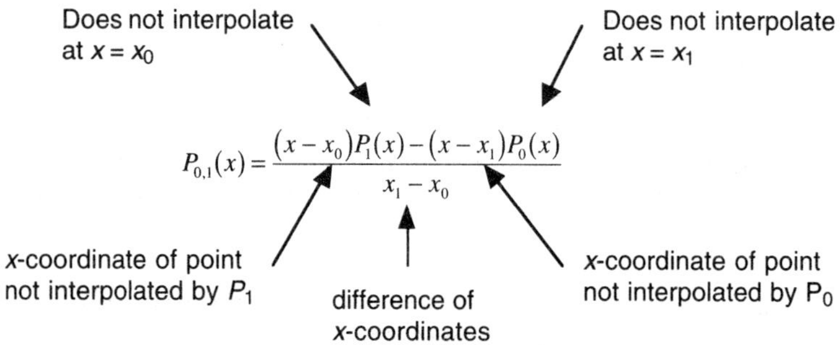

Focus on the manner in which the two lower-degree polynomials have been combined. Note that the polynomial P_1 does not interpolate at $x = x_0$ and it is multiplied by $x - x_0$. Similarly, the polynomial P_0 does not interpolate at $x = x_1$ and it is multiplied by $x - x_1$. Hence, each of the lower-degree polynomials is multiplied by a monomial of the form x minus the x-coordinate of the point not interpolated by that particular polynomial. These two terms are then subtracted, and the construction of the higher-degree interpolating polynomial is completed by dividing by the difference between the abscissas which appear in the coefficient monomials. The order in which the abscissas appear in the denominator is the reverse of the order in which they appear in the numerator.

This result with the linear interpolating polynomial suggests that the previously stated problem of constructing $P_{0,1,2,3}$ from the polynomials $P_{0,1,3}$ and $P_{1,2,3}$ can be carried out as follows. Since $P_{0,1,3}$ does not interpolate at $x = x_2$, multiply this polynomial by $(x - x_2)$. Further, since $P_{1,2,3}$ does not interpolate at $x = x_0$, multiply this polynomial by $(x - x_0)$. Deciding to subtract the term involving $P_{1,2,3}$ from the term involving $P_{0,1,3}$ fixes the order of the abscissas in the denominator as $x_0 - x_2$. Therefore,

$$P_{0,1,2,3}(x) = \frac{(x - x_2)P_{0,1,3}(x) - (x - x_0)P_{1,2,3}(x)}{x_0 - x_2}.$$

The next theorem establishes that this scheme for combining lower-degree interpolating polynomials, which was pieced together from an examination of the linear interpolant, does, in fact, hold in general.

Theorem. Let x_0, x_1, x_2, …, x_n be $n+1$ distinct abscissas; let f_0, f_1, f_2, …, f_n denote the corresponding function values being interpolated; and let m_1, m_2, m_3, …, m_k be k distinct integers between 0 and n, inclusive. Then, for $k = 1$, $P_{m_1}(x) = f_{m_1}$, and for $k > 1$,

$$P_{m_1,m_2,\ldots,m_k}(x) = \frac{(x - x_{m_1})P_{m_2,m_3,\ldots,m_k}(x) - (x - x_{m_k})P_{m_1,m_2,\ldots,m_{k-1}}(x)}{x_{m_k} - x_{m_1}}.$$

Proof. For $k = 1$, $P_{m_1}(x) = f_{m_1}$ follows from the definition of $P_{m_1}(x)$ as the unique polynomial of degree 0 that interpolates at $x = x_{m_1}$. For $k > 1$, consider the polynomial

$$P(x) = \frac{(x - x_{m_1})P_{m_2,m_3,m_4,\ldots,m_k}(x) - (x - x_{m_k})P_{m_1,m_2,m_3,\ldots,m_{k-1}}(x)}{x_{m_k} - x_{m_1}}.$$

First note that $P_{m_2,m_3,m_4,\ldots,m_k}$ and $P_{m_1,m_2,m_3,\ldots,m_{k-1}}$ are polynomials of degree at most $k-2$ (each interpolates data at $k-1$ points), so P is a polynomial of degree at most $k-1$. Since $P_{m_2,m_3,m_4,\ldots,m_k}$ and $P_{m_1,m_2,m_3,\ldots,m_{k-1}}$ both interpolate data at the set of abscissas x_{m_2}, x_{m_3}, x_{m_4}, …, $x_{m_{k-1}}$, it follows that for each $i = 2, 3, 4, \ldots, k-1$

$$\begin{aligned} P(x_{m_i}) &= \frac{(x_{m_i} - x_{m_1})P_{m_2,m_3,\ldots,m_k}(x_{m_i}) - (x_{m_i} - x_{m_k})P_{m_1,m_2,\ldots,m_{k-1}}(x_{m_i})}{x_{m_k} - x_{m_1}} \\ &= \frac{(x_{m_i} - x_{m_1})f_{m_i} - (x_{m_i} - x_{m_k})f_{m_i}}{x_{m_k} - x_{m_1}} \\ &= f_{m_i}. \end{aligned}$$

Furthermore,

$$\begin{aligned} P(x_{m_1}) &= \frac{(x_{m_1} - x_{m_1})P_{m_2,m_3,\ldots,m_k}(x_{m_1}) - (x_{m_1} - x_{m_k})P_{m_1,m_2,\ldots,m_{k-1}}(x_{m_1})}{x_{m_k} - x_{m_1}} \\ &= \frac{(0)P_{m_2,m_3,\ldots,m_k}(x_{m_1}) - (x_{m_1} - x_{m_k})f_{m_1}}{x_{m_k} - x_{m_1}} \\ &= f_{m_1} \end{aligned}$$

and

$$\begin{aligned} P(x_{m_k}) &= \frac{(x_{m_k} - x_{m_1})P_{m_2,m_3,\ldots,m_k}(x_{m_k}) - (x_{m_k} - x_{m_k})P_{m_1,m_2,\ldots,m_{k-1}}(x_{m_k})}{x_{m_k} - x_{m_1}} \\ &= \frac{(x_{m_k} - x_{m_1})f_{m_k} - (0)P_{m_1,m_2,\ldots,m_{k-1}}(x_{m_k})}{x_{m_k} - x_{m_1}} \\ &= f_{m_k}. \end{aligned}$$

Hence, $P(x_{m_i}) = f_{m_i}$ for each $i = 1, 2, 3, \ldots, k$. But, $P_{m_1,m_2,m_3,\ldots,m_k}$ is the unique polynomial of degree at most $k-1$ which interpolates data at the set of abscissas $x_{m_1}, x_{m_2}, x_{m_3}, \ldots, x_{m_k}$. Therefore,

$$P_{m_1,m_2,m_3,\ldots,m_k}(x) = \frac{(x - x_{m_1})P_{m_2,m_3,m_4,\ldots,m_k}(x) - (x - x_{m_k})P_{m_1,m_2,m_3,\ldots,m_{k-1}}(x)}{x_{m_k} - x_{m_1}}. \qquad \square$$

The Algorithm

The problem at hand is that of evaluating the interpolating polynomial at a single point. Problem data consist of $n+1$ distinct abscissas $x_0, x_1, x_2, \ldots, x_n$, a corresponding set of function values $f_0, f_1, f_2, \ldots, f_n$, and one value of the independent variable, $\bar{x}$, at which the interpolating polynomial is to be evaluated. Recall that the function values can be interpreted as a collection of constant (zeroth degree) interpolating polynomials. Starting from these values, the previous theorem, with x replaced by $\bar{x}$, can be used to compute the value of a set of linear polynomials, from which the value of a set of quadratic polynomials can be computed, and so on until the value of the polynomial with highest possible degree has been determined. All calculations for Neville's algorithm can be conveniently organized into a table:

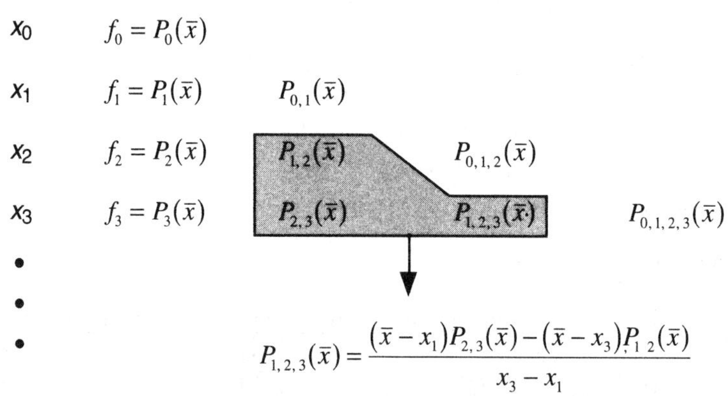

Here is an example to demonstrate the construction of this Neville's algorithm table.

EXAMPLE 5.5 An Arbitrary Set of Data

Based on the following data, approximate the value of y when $x = 1.5$.

x	-1	0	1	2
y	5	1	1	11

The Neville's algorithm table based on this data is given below. From this table, we see that when $x = 1.5$, the value of the interpolating polynomial that uses all four data points is 4.375.

$$\begin{array}{llll}
x_0 = -1 & P_0(1.5) = 5 & & \\
x_1 = 0 & P_1(1.5) = 1 & P_{0,1}(1.5) = -5 & \\
x_2 = 1 & P_2(1.5) = 1 & P_{1,2}(1.5) = 1 & P_{0,1,2}(1.5) = 2.5 \\
x_3 = 2 & P_3(1.5) = 11 & P_{2,3}(1.5) = 6 & P_{1,2,3}(1.5) = 4.75 \quad P_{0,1,2,3}(1.5) = 4.375
\end{array}$$

So where did all of these numbers come from? The first two columns are just the given abscissas and function values. The values in the remaining three columns were computed as follows. Each number gives the value of a different interpolating polynomial when $x = 1.5$.

$$\begin{aligned}
P_{0,1}(1.5) &= \frac{(1.5 - x_0)P_1(1.5) - (1.5 - x_1)P_0(1.5)}{x_1 - x_0} \\
&= \frac{2.5 \times 1 - 1.5 \times 5}{1} = -5 \\
P_{1,2}(1.5) &= \frac{(1.5 - x_1)P_2(1.5) - (1.5 - x_2)P_1(1.5)}{x_2 - x_1} \\
&= \frac{1.5 \times 1 - 0.5 \times 1}{1} = 1 \\
P_{2,3}(1.5) &= \frac{(1.5 - x_2)P_3(1.5) - (1.5 - x_3)P_2(1.5)}{x_3 - x_2} \\
&= \frac{0.5 \times 11 - (-0.5) \times 1}{1} = 6 \\
P_{0,1,2}(1.5) &= \frac{(1.5 - x_0)P_{1,2}(1.5) - (1.5 - x_2)P_{0,1}(1.5)}{x_2 - x_0} \\
&= \frac{2.5 \times 1 - 0.5 \times -5}{2} = 2.5 \\
P_{1,2,3}(1.5) &= \frac{(1.5 - x_1)P_{2,3}(1.5) - (1.5 - x_3)P_{1,2}(1.5)}{x_3 - x_1} \\
&= \frac{1.5 \times 6 - (-0.5) \times 1}{2} = 4.75 \\
P_{0,1,2,3}(1.5) &= \frac{(1.5 - x_0)P_{1,2,3}(1.5) - (1.5 - x_3)P_{0,1,2}(1.5)}{x_3 - x_0} \\
&= \frac{2.5 \times 4.75 - (-0.5) \times 2.5}{3} = 4.375
\end{aligned}$$

To assess the reasonableness of this approximation, Figure 5.5 plots the given data points (represented as circles), the interpolated point (the asterisk), and the interpolating polynomial. The behavior of the polynomial is plausible, giving confidence that the approximation is reasonably accurate.

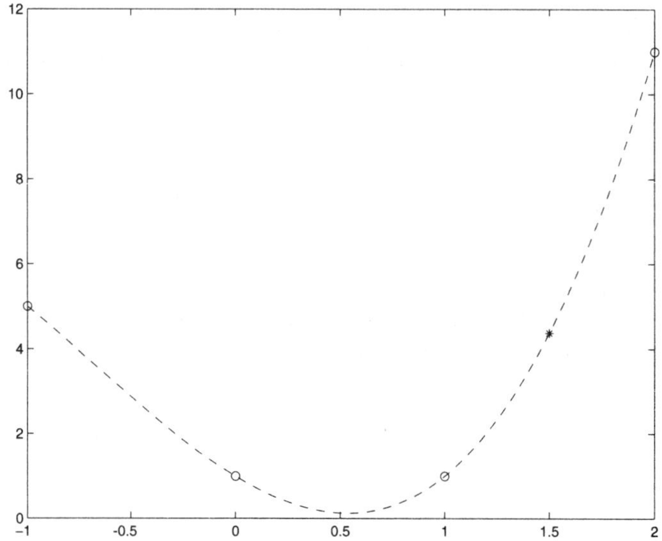

Figure 5.5 Data points (circles), interpolated point (asterisk), and interpolating polynomial for the data in the "An Arbitrary Set of Data" Example.

Although the algorithm just described constructs the Neville's table one column at a time, the table can also be constructed one row at a time. One advantage of working in a row-wise manner is that if additional data becomes available, then there is no need to start from scratch to compute the value of the interpolating polynomial which incorporates the new data as well as the old. As long as the last row from the Neville's table has been saved, all that is needed is to compute one new row for each of the new data points.

Some More Notation

An alternative notation, one that more closely reflects the matrix-like structure of the Neville's table, is typically introduced. For $i \geq j$, let $Q_{i,j} = P_{i-j,i-j+1,i-j+2,\ldots,i}$. Note that the first subscript on $Q_{i,j}$ indicates the index of the last point interpolated, the second subscript denotes the degree of the polynomial and $Q_{i,j}$ interpolates data from $x = x_{i-j}$ through $x = x_i$. In terms of this new notation, the constant (zeroth degree) interpolating polynomials, $Q_{i,0}$, are given by the function values f_i ($i = 0, 1, 2, \ldots, n$); the formula for constructing higher-degree polynomials from lower-degree polynomials takes the form

$$Q_{i,j}(x) = \frac{(x - x_{i-j})Q_{i,j-1}(x) - (x - x_i)Q_{i-1,j-1}(x)}{x_i - x_{i-j}},$$

and the Neville's table appears in Figure 5.6.

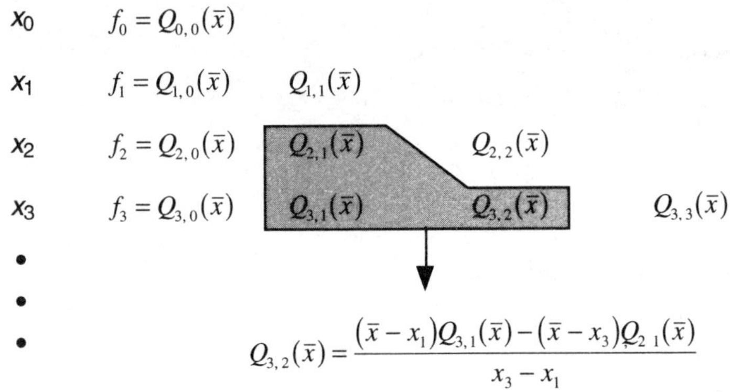

Figure 5.6 Organization of a Neville's Table using $Q_{i,j}$-notation.

EXAMPLE 5.6 Relative Viscosity of Ethanol as a Function of Anhydrous Solute Weight

The table below lists the relative viscosity, V, of ethanol as a function of the percent of anhydrous solute weight, w.

w	10	20	40	60	80	100
V	1.498	2.138	2.840	2.542	1.877	1.201

An estimate of the relative viscosity when $w = 50$ is needed. The Neville's table generated from this data is provided below. All values have been rounded to four decimal places.

10	1.498					
20	2.138	4.058				
40	2.840	3.191	2.902			
60	2.542	2.691	2.816	2.8332		
80	1.877	2.8745	2.7369	2.7764	2.8008	
100	1.201	2.891	2.8704	2.7591	2.7699	2.7871

The relative viscosity of ethanol with a 50% anhydrous solute weight is approximately 2.7871. Figure 5.7 is used to assess the accuracy of this result. Based on this graph, the approximate value for the relative viscosity seems perfectly reasonable.

EXAMPLE 5.7 Probability of a Shutout in Racquetball

The following table (drawn from Joseph Keller, "Probability of a Shutout in Racquetball," *SIAM Review,* **26**, 267–268, 1984) gives the probability, P, that a given player will shutout an opponent as a function of the probability, p, that the player will win any particular rally regardless of who serves.

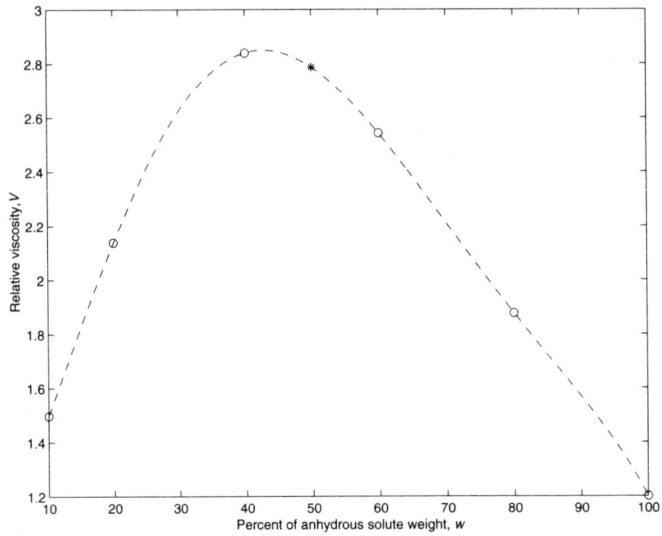

Figure 5.7 Data points (circles), interpolated point (asterisk), and interpolating polynomial for the relative viscosity of ethanol as a function of percent of anhydrous solute weight.

0.5	0.0001504					
0.84	0.490	0.1442				
0.842	0.500	−0.71	−0.1055			
0.85	0.534	−0.5285	−5.066	−1.5228		
0.9	0.753	−0.561	−0.3929	−23.7584	−7.0817	
1.0	1.0	0.0120	−1.516	1.3273	−61.3870	−17.9428

TABLE 5.1: Neville's table for Example 5.7.

p	P
1.0	1.0
0.9	0.753
0.85	0.534
0.842	0.500
0.84	0.490
0.5	0.0001504

Suppose that a player estimates a 60% chance of winning any particular rally against a given opponent, regardless of who serves. What is the probability that this player will shutout that opponent?

The Neville's table generated by this data, with entries rounded to four decimal places is given in Table 5.1. The final value, −17.9428, is clearly inappropriate. Valid probability scores must be between zero and one, inclusive. Figure 5.8 offers

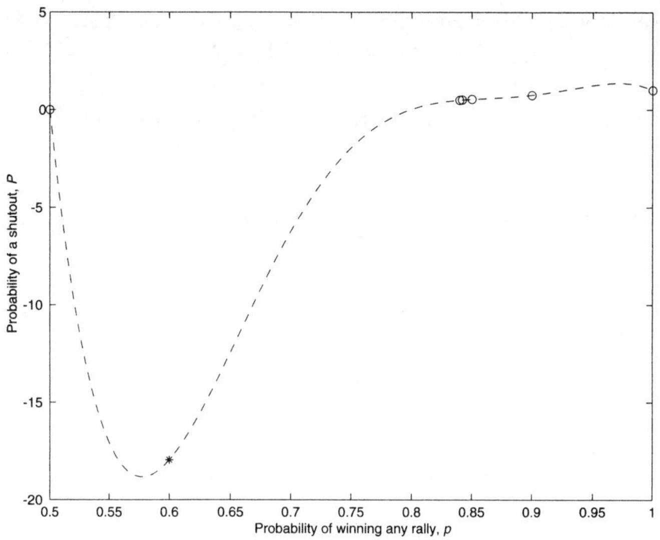

Figure 5.8 Data points (circles), interpolated point (asterisk) and interpolating polynomial for the probability of a shutout in racquetball as a function of the probability of winning any particular rally.

some insight into what has gone wrong: the interpolating polynomial provides a poor representation for the data. Over roughly half of the problem domain, the interpolating polynomial predicts invalid probabilities: negative values on the left portion of the domain, values larger than one on the right.

EXERCISES

1. Indicate how to construct each of the following interpolating polynomials.
   (a) $P_{0,1,2,3}(x)$ from $P_{0,1,2}(x)$ and $P_{1,2,3}(x)$
   (b) $P_{0,1,2,3}(x)$ from $P_{0,2,3}(x)$ and $P_{0,1,3}(x)$
   (c) $P_{0,1,2,3}(x)$ from $P_{1,2,3}(x)$ and $P_{0,2,3}(x)$
   (d) $P_{0,1,2,3}(x)$ from $P_{0,1,3}(x)$ and $P_{0,1,2}(x)$

2. Indicate how to construct each of the following interpolating polynomials.
   (a) $P_{0,1,2}(x)$ from $P_{1,2}(x)$ and $P_{0,2}(x)$
   (b) $P_{1,3,4,6}(x)$ from $P_{1,4,6}(x)$ and $P_{1,3,6}(x)$
   (c) $P_{0,2,3,4,7}(x)$ from $P_{0,2,4,7}(x)$ and $P_{2,3,4,7}(x)$
   (d) $P_{1,2,3,4,5,6}(x)$ from $P_{1,2,3,5,6}(x)$ and $P_{1,3,4,5,6}(x)$

3. Construct the Neville's table for the following data set. Take $\bar{x} = 3.7$.

x	2	4	5
y	-1	4	8

4. Construct the Neville's table for the following data set. Take $\bar{x} = 1.3$.

x	0	1	2
y	2	−1	4

5. Construct the Neville's table for the following data set. Take $\bar{x} = -0.5$.

x	−1	0	1	2
y	3	−1	−3	1

6. Construct the Neville's table for the following data set. Take $\bar{x} = -3$.

x	−7	−5	−4	−1
y	10	5	2	10

7. Given $x_0 = 0$, $x_1 = 1$, $x_2 = 2$, $P_{0,1}(x) = 2x + 2$ and $P_{0,2}(x) = 3x + 2$, what is $P_{0,1,2}(x)$?

8. Given $x_0 = -1$, $x_1 = 0$, $x_2 = 1$, $x_3 = 2$, $P_{0,2}(x) = -3x$, $P_{2,3}(x) = 4x - 7$ and $P_{1,2,3}(1.7) = -0.83$, calculate $P_{0,2,3}(x)$ and $P_{0,1,2,3}(1.7)$.

9. Determine the missing values in the Neville's table provided below.

$x_0 = 0$	$P_0(1.3) = -1$		
$x_1 = 1$	$P_1(1.3) = ?$	$P_{0,1}(1.3) = 5.5$	
$x_2 = 2$	$P_2(1.3) = ?$	$P_{1,2}(1.3) = ?$	$P_{0,1,2}(1.3) = 4.915$

10. Determine the missing values in the Neville's table provided below. For some of the values you will need to work backwards.

$x_0 = 0$	$P_0(2.5) = 1$			
$x_1 = 1$	$P_1(2.5) = 3$	$P_{0,1}(2.5) = 6$		
$x_2 = 2$	$P_2(2.5) = 3$	$P_{1,2}(2.5) = ?$	$P_{0,1,2}(2.5) = ?$	
$x_3 = 3$	$P_3(2.5) = ?$	$P_{2,3}(2.5) = 3$	$P_{1,2,3}(2.5) = 3$	$P_{0,1,2,3}(2.5) = ?$

11. Use Neville's algorithm to evaluate the interpolating polynomial for $f(x) = \ln x$ that passes through the points $(1, \ln 1)$, $(2, \ln 2)$, and $(3, \ln 3)$ at $x = 1.5$.

12. Use Neville's algorithm to evaluate the interpolating polynomial for $f(x) = \sin x$ that passes through the points $(0, \sin 0)$, $(\pi/4, \sin \pi/4)$, and $(\pi/2, \sin \pi/2)$ at $x = \pi/6$.

13. Use Neville's algorithm to evaluate the interpolating polynomial for $f(x) = e^x$ that passes through the points $(-1, e^{-1})$, $(0, e^0)$, and $(1, e^1)$ at $x = 0.5$.

For Exercises 14–17, use Neville's algorithm to estimate the requested value(s). Assess the accuracy of each estimate by plotting the data points and the estimated point(s) on the same set of coordinate axes.

14. The mean activity coefficient at 25°C for silver nitrate, as a function of molality, is given in the table below. Estimate the mean activity coefficient for a molality of 0.032 and for a molality of 1.682.

Molality	0.005	0.010	0.020	0.050	0.100	0.200	0.500	1.000	2.000
Coefficient	0.924	0.896	0.859	0.794	0.732	0.656	0.536	0.430	0.316

15. The values listed in the table provide the surface tension of mercury as a function of temperature. Estimate the surface tension of mercury at 20° C and at 60° C.

Temperature (°C)	10	25	50	75	100
Surface Tension (dyn/cm)	488.55	485.48	480.36	475.23	470.11

16. The thermal conductivity of air as a function of temperature is given in the table below. Estimate the thermal conductivity of air when $T = 240\,\text{K}$ and when $T = 485\,\text{K}$.

Temperature (K)	100	200	300	400	500	600
Thermal Conductivity (mW/m · K)	9.4	18.4	26.2	33.3	39.7	45.7

17. Estimate the viscosity of sulfuric acid with a concentration (in mass percent) of 7.5% and a concentration of 92% given the following values.

Concentration (mass %)	0	5	10	20	40	60	80	100
Viscosity (centipoise)	0.89	1.01	1.12	1.40	2.51	5.37	17.4	24.2

5.3 THE NEWTON FORM OF THE INTERPOLATING POLYNOMIAL

At the beginning of Section 5.2, two important questions were raised regarding polynomial interpolation. The first considered the best way to obtain the value of the interpolating polynomial at a single value of the independent variable. Neville's algorithm provided the answer to this question. The algorithm did not actually construct the interpolating polynomial but rather generated the polynomial value in an incremental fashion. The second question considered the best way to proceed when the value of the interpolating polynomial was needed at several values of the independent variable. In this case, Neville's algorithm is no longer efficient. It is, instead, better to obtain the interpolating polynomial explicitly in a form efficient for evaluation.

Newton Form of the Interpolating Polynomial

The objective of this section is to write $P_{0,1,2,\ldots,n}(x)$, the unique polynomial of degree at most n which interpolates data at the $n+1$ distinct abscissas x_0, x_1, x_2, $\ldots$, x_n, in Newton Form:

$$\begin{aligned} P_{0,1,2,\ldots,n}(x) &= a_0 + a_1(x - x_0) + a_2(x - x_0)(x - x_1) + \cdots \\ &\quad + a_n(x - x_0)(x - x_1)\cdots(x - x_{n-1}) \\ &= \sum_{k=0}^{n} a_k \left(\prod_{i=0}^{k-1} (x - x_i) \right). \end{aligned}$$

Like the more standard representation of an nth-degree polynomial

$$a_0 + a_1 x + a_2 x^2 + \cdots + a_n x^n,$$

the Newton form can be written as a "nested iteration,"

$$\begin{aligned} P_{0,1,2,\ldots,n}(x) = a_0 + (x - x_0)(a_1 + (x - x_1)(a_2 + (x - x_2)(\cdots \\ (x - x_{n-2})(a_{n-1} + a_n(x - x_{n-1}))))) \end{aligned}$$

and evaluated by a modified version of the synthetic division algorithm:

$value := a_n$
for i from $n-1$ downto 0 do
  $value := a_i + (x - x_i) \cdot value$

The primary advantage of the Newton form over the standard representation is in the computation of the polynomial coefficients $a_0, a_1, a_2, \ldots, a_n$. Determining the coefficients for the standard representation requires the solution of a system of $n+1$ linear equations. The number of operations needed to solve this system is $O(n^3)$; that is, roughly a constant times n^3. The algorithm to be developed presently for determining the coefficients of the Newton form requires only $O(n^2)$ operations. For even moderate values of n, this represents a significant computational savings.

Divided Differences

To determine the coefficients of the Newton form of the interpolating polynomial, each of the interpolating conditions

$$P_{0,1,2,\ldots,n}(x_k) = f(x_k)$$

$(k = 0, 1, 2, \ldots, n)$ must be applied. The special structure of the Newton form leads to a system of equations of the form

$$\begin{aligned}
&a_0 &&= f(x_0)\\
&a_0 + a_1(x_1 - x_0) &&= f(x_1)\\
&a_0 + a_1(x_2 - x_0) + a_2(x_2 - x_0)(x_2 - x_1) &&= f(x_2)\\
& &&\vdots\\
&a_0 + a_1(x_n - x_0) + a_2(x_n - x_0)(x_n - x_1) + \cdots + a_n(x_n - x_0)\cdots(x_n - x_{n-1}) &&= f(x_n).
\end{aligned} \tag{1}$$

The solution to this system of equations can be obtained rather easily by forward substitution (solve the first equation for the single unknown a_0, substitute that value into the second equation and solve for the only remaining unknown a_1, substitute the values of a_0 and a_1 into the third equation and solve for the only remaining unknown a_2, etc.) and can be expressed compactly in terms of *divided differences*.

Definition. Let f be a function defined at the distinct points $x_0, x_1, x_2, \ldots, x_n$.

The ZEROTH DIVIDED DIFFERENCE of f with respect to the point x_i is $f[x_i] \equiv f(x_i)$.

For $0 < k \le n$, the kTH DIVIDED DIFFERENCE of f with respect to the points $x_i, x_{i+1}, x_{i+2}, \ldots, x_{i+k}$ is

$$f[x_i, x_{i+1}, x_{i+2}, \ldots, x_{i+k}] = (f[x_{i+1}, x_{i+2}, x_{i+3}, \ldots, x_{i+k}] - f[x_i, x_{i+1}, x_{i+2}, \ldots, x_{i+k-1}])/(x_{i+k} - x_i).$$

Note the way that each kth divided difference with respect to a set of $k+1$ points is constructed from two $(k-1)$-st divided differences with respect to different subsets of k points. This is very similar to the procedure used in Neville's algorithm to construct higher-degree polynomials from lower-degree polynomials.

Now, on to the solution for the coefficients in the Newton form of the interpolating polynomial. The first equation in (1) gives $a_0 = f(x_0) = f[x_0]$, the zeroth divided difference with respect to the first interpolating point $x = x_0$. Substituting this value into the second equation and solving for a_1 gives

$$a_1 = \frac{f(x_1) - f(x_0)}{x_1 - x_0} = \frac{f[x_1] - f[x_0]}{x_1 - x_0} = f[x_0, x_1],$$

the first divided difference with respect to the first two interpolating points. Substituting the values for a_0 and a_1 into the third equation and solving for a_2 produces, after some tedious algebraic manipulation,

$$a_2 = \frac{f[x_1, x_2] - f[x_0, x_1]}{x_2 - x_0} = f[x_0, x_1, x_2].$$

Continuing in this fashion, it is found that

$$a_k = f[x_0, x_1, x_2, \ldots, x_k],$$

and, hence, the Newton form of the interpolating polynomial is

$$P_{0,1,2,\ldots,n}(x) = \sum_{k=0}^{n} f[x_0, x_1, x_2, \ldots, x_k] \left(\prod_{i=0}^{k-1} (x - x_i) \right).$$

The strong similarity between the formula in the definition of the k-th divided difference and the formula used to construct the Neville's algorithm table suggests organizing the calculation of divided differences into a table. Such a divided difference table, based on four interpolating points, is shown in Figure 5.9. Note that the coefficients of the Newton form of the interpolating polynomial are the values at the top of each column.

With $n+1$ interpolating points, n columns need to be computed to form the complete divided difference table. For $i = 1, 2, 3, \ldots, n$, the ith column of the table contains $n+1-i$ values that need to be computed, each of which requires two subtractions and one division. The number of operations to compute the entire table is therefore

$$3\sum_{i=1}^{n}(n+1-i) = 3\frac{n(n+1)}{2},$$

or $O(n^2)$.

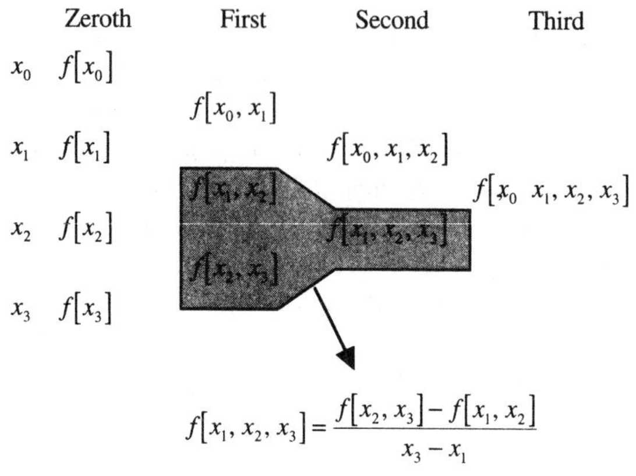

Figure 5.9 Organization of a divided difference table.

EXAMPLE 5.8 An Arbitrary Set of Data

Determine the Newton form of the interpolating polynomial for the following data set. Then use this polynomial to estimate the value of y when $x = 1.5$.

x	-1	0	1	2
y	5	1	1	11

Using the data in the order given produces the following divided difference table.

$x_0 = -1$	$f[x_0] = 5$			
		$f[x_0, x_1] = -4$		
$x_1 = 0$	$f[x_1] = 1$		$f[x_0, x_1, x_2] = 2$	
		$f[x_1, x_2] = 0$		$f[x_0, x_1, x_2, x_3] = 1$
$x_2 = 1$	$f[x_2] = 1$		$f[x_1, x_2, x_3] = 5$	
		$f[x_2, x_3] = 10$		
$x_3 = 2$	$f[x_3] = 11$			

The Newton form of the interpolating polynomial is then

$$\begin{aligned} P_{0,1,2,3}(x) &= f[x_0] + f[x_0, x_1](x - x_0) + f[x_0, x_1, x_2](x - x_0)(x - x_1) + \\ &\quad f[x_0, x_1, x_2, x_3](x - x_0)(x - x_1)(x - x_2) \\ &= 5 - 4(x + 1) + 2(x + 1)x + (x + 1)x(x - 1). \end{aligned}$$

Evaluating this polynomial at $x = 1.5$ produces the value

$$\begin{aligned} P_{0,1,2,3}(1.5) &= 5 - 4(1.5 + 1) + 2(1.5 + 1)(1.5) + (1.5 + 1)(1.5)(1.5 - 1) \\ &= 5 - 10 + 7.5 + 1.875 = 4.375, \end{aligned}$$

precisely the same value as produced by Neville's algorithm in the previous section.

For completeness, each of the values in the divided difference table was computed as follows. For the first divided differences,

$$f[x_0, x_1] = \frac{f[x_1] - f[x_0]}{x_1 - x_0} = \frac{1 - 5}{0 - (-1)} = -4;$$

$$f[x_1, x_2] = \frac{f[x_2] - f[x_1]}{x_2 - x_1} = \frac{1 - 1}{1 - 0} = 0; \quad \text{and}$$

$$f[x_2, x_3] = \frac{f[x_3] - f[x_2]}{x_3 - x_2} = \frac{11 - 1}{2 - 1} = 10.$$

For the second divided differences,

$$f[x_0, x_1, x_2] = \frac{f[x_1, x_2] - f[x_0, x_1]}{x_2 - x_0} = \frac{0 - (-4)}{1 - (-1)} = 2; \quad \text{and}$$

$$f[x_1, x_2, x_3] = \frac{f[x_2, x_3] - f[x_1, x_2]}{x_3 - x_1} = \frac{10 - 0}{2 - 0} = 5.$$

Finally, for the third divided difference,

$$f[x_0, x_1, x_2, x_3] = \frac{f[x_1, x_2, x_3] - f[x_0, x_1, x_2]}{x_3 - x_0} = \frac{5 - 2}{2 - (-1)} = 1.$$

EXAMPLE 5.9 Relative Viscosity of Ethanol as a Function of Anhydrous Solute Weight

The following table lists the relative viscosity, V, of ethanol as a function of the percent of anhydrous solute weight, w.

w	10	20	40	60	80	100
V	1.498	2.138	2.840	2.542	1.877	1.201

A table of relative viscosity values, starting at 5% solute weight and proceeding to 100% solute weight in increments of 5%, is needed. Using the data in the order provided, the divided difference table, with values rounded to five decimal places, is

10	1.498					
		0.064				
20	2.138		−9.6333 e−4			
		0.0351		−5.7334 e−6		
40	2.840		−1.25 e − 3		2.7031 e−7	
		−0.0149		1.3188 e−5		−3.8050 e−9
60	2.542		−4.5875 e−4		−7.2141 e−8	
		−0.03325		7.4167 e−6		
80	1.877		−1.375 e − 5			
		−0.0338				
100	1.201					

The Newton form of the interpolating polynomial relating solute weight and relative viscosity is then

$$\begin{aligned}V = 1.498 + 0.064(w-10) - 0.00096333(w-10)(w-20) - \\ 0.0000057334(w-10)(w-20)(w-40) + \\ 0.00000027031(w-10)(w-20)(w-40)(w-60) - \\ 0.0000000038050(w-10)(w-20)(w-40)(w-60)(w-80).\end{aligned}$$

Evaluating this polynomial at $w = 5\%, 10\%, 15\%, \ldots, 100\%$ produces the desired table:

w	V	w	V
5	1.201	55	2.682
10	1.498	60	2.542
15	1.824	65	2.380
20	2.138	70	2.210
25	2.411	75	2.040
30	2.624	80	1.877
35	2.768	85	1.722
40	2.840	90	1.569
45	2.844	95	1.403
50	2.787	100	1.201

Assessing the accuracy of the values in this table will be left as an exercise.

In addition to being more efficient from the point of view of evaluation and requiring fewer operations to determine its coefficients, the Newton form of the interpolating polynomial has another major advantage over the Lagrange form. Whereas the Lagrange form must be recomputed from scratch when new data become available, the Newton form allows for new data points to be easily incorporated. For each new data point, an additional diagonal must be computed in the divided difference table, and one new term must be added to the interpolating polynomial. The next example demonstrates this process.

EXAMPLE 5.10 An Arbitrary Set of Data Reconsidered

Suppose that in addition to the four data values provided earlier:

x	-1	0	1	2
y	5	1	1	11

it is also known that $y = 5$ when $x = -2$ and $y = 35$ when $x = 3$. Estimate the value of y when $x = 1.5$ using all six data points.

We could arrange the six points in ascending order and generate a whole new divided difference table, but that would be doing more work than is really necessary. Instead, we can simply add the two new points to the bottom of the table we had already computed, and just complete those two new diagonals. The resulting, augmented divided difference table is shown below. New values are displayed in boldface.

$$
\begin{array}{ccccccc}
-1 & 5 & & & & & \\
 & & -4 & & & & \\
0 & 1 & & 2 & & & \\
 & & 0 & & 1 & & \\
1 & 1 & & 5 & & \mathbf{-1/12} & \\
 & & 10 & & \mathbf{13/12} & & \mathbf{0} \\
2 & 11 & & \mathbf{17/6} & & \mathbf{-1/12} & \\
 & & \mathbf{3/2} & & \mathbf{5/6} & & \\
\mathbf{-2} & \mathbf{5} & & \mathbf{9/2} & & & \\
 & & \mathbf{6} & & & & \\
\mathbf{3} & \mathbf{35} & & & & &
\end{array}
$$

The Newton form of the interpolating polynomial is then

$$P_{0,1,2,3,4,5}(x) = 5-4(x+1)+2(x+1)x+(x+1)x(x-1)-\frac{1}{12}(x+1)x(x-1)(x-2).$$

Evaluating this polynomial at $x = 1.5$ produces the value

$$\begin{aligned} P_{0,1,2,3,4,5}(1.5) &= 5-4(1.5+1)+2(1.5+1)(1.5)+(1.5+1)(1.5)(1.5-1)- \\ &\quad \frac{1}{12}(1.5+1)(1.5)(1.5-1)(1.5-2) \\ &= 5-10+7.5+1.875+0.078125 = 4.453125. \end{aligned}$$

Some texts define the divided difference $f[x_0, x_1, x_2, \ldots, x_k]$ to be the leading coefficient in the unique polynomial that interpolates f at the points x_0, x_1, x_2, $\ldots$, x_k. We've already seen that our definition of divided differences leads to this same interpretation. It can also be shown that if $f[x_0, x_1, x_2, \ldots, x_k]$ is the leading coefficient in the unique polynomial that interpolates f at the points x_0, x_1, x_2, $\ldots$, x_k, then $f[x_0, x_1, x_2, \ldots, x_k]$ must satisfy the recursive formula provided in our definition (see Exercise 12). Hence, the two definitions are equivalent.

It is important to note that the value of $f[x_0, x_1, x_2, \ldots, x_k]$ is independent of the order of the interpolating points x_0, x_1, x_2, $\ldots$, x_k. Thus, for example, $f[x_0, x_1, x_2, x_3] = f[x_2, x_1, x_0, x_3] = f[x_3, x_0, x_1, x_2]$. The reason for this independence is tied to the fact that $f[x_0, x_1, x_2, \ldots, x_k]$ is the coefficient of x^k in the polynomial of degree k that interpolates f at the points x_0, x_1, x_2, $\ldots$, x_k. If we let m_0, m_1, m_2, $\ldots$, m_k denote any permutation of the integers 0 through k, inclusive, then $f[x_{m_0}, x_{m_1}, x_{m_2}, \ldots, x_{m_k}]$ is the coefficient of x^k in the polynomial of degree k which interpolates f at the points x_{m_0}, x_{m_1}, x_{m_2}, $\ldots$, x_{m_k}. Since the set of interpolating points is the same in each case, the two interpolating polynomials must be the same, so the coefficients of x^k must be equal.

Interpolation Error Revisited

To finish out this section, we will show that the error in polynomial interpolation can be expressed in terms of divided differences. For, suppose that P is the polynomial that interpolates f at the points x_0, x_1, x_2, $\ldots$, x_n. Further suppose that P^* is the polynomial which interpolates f at the points x_0, x_1, x_2, $\ldots$, x_n, t. It follows that

$$P^*(x) = P(x) + f[x_0, x_1, x_2, \ldots, x_n, t] \prod_{i=0}^{n} (x - x_i).$$

Evaluating this expression at $x = t$ and using the fact that $P^*(t) = f(t)$ yields

$$f(t) = P(t) + f[x_0, x_1, x_2, \ldots, x_n, t] \prod_{i=0}^{n} (t - x_i).$$

Hence, upon rearranging terms and replacing the variable t by x, we arrive at the error formula

$$f(x) - P(x) = f[x_0, x_1, x_2, \ldots, x_n, x] \prod_{i=0}^{n} (x - x_i). \tag{2}$$

Equation (2) may look different from the error term established in Section 5.1, but we will now show that they are the same.

Recall that the first divided difference of a function f with respect to the points x_0 and x_1 is given by

$$f[x_0, x_1] = \frac{f(x_1) - f(x_0)}{x_1 - x_0}.$$

If f is continuous on the closed interval $[x_0, x_1]$ and differentiable on the open interval (x_0, x_1), then the Mean Value Theorem guarantees the existence of $\xi \in (x_0, x_1)$ such that

$$\frac{f(x_1) - f(x_0)}{x_1 - x_0} = f'(\xi).$$

Combining the last two equations gives

$$f[x_0, x_1] = f'(\xi)$$

for some $\xi \in (x_0, x_1)$, provided that f is sufficiently differentiable.

A similar relationship holds between higher divided differences and higher-order derivatives. The specific nature of this relationship is established in the next theorem.

Theorem. Let x_0, x_1, x_2, $\ldots$, x_n be $n+1$ distinct points from the closed interval $[a, b]$. If f is continuous on $[a, b]$ and has n continuous derivatives on the open interval (a, b), then there exists a $\xi \in (a, b)$ such that

$$f[x_0, x_1, x_2, \ldots, x_n] = \frac{f^{(n)}(\xi)}{n!}.$$

Proof. Let x_0, x_1, x_2, ..., x_n be $n+1$ distinct points from the closed interval $[a, b]$, and suppose that f is continuous on $[a, b]$ and has n continuous derivatives on the open interval (a, b). Consider the auxiliary function

$$g(x) = f(x) - P_{0,1,2,\ldots,n}(x).$$

Since $P_{0,1,2,\ldots,n}$ is continuous and infinitely differentiable everywhere, it follows that g is continuous on $[a, b]$ and has n continuous derivatives on the open interval (a, b). Furthermore, since $P_{0,1,2,\ldots,n}$ interpolates f at each of the points x_0, x_1, x_2, ..., x_n,

$$g(x_i) = 0 \qquad \text{for each } i = 0, 1, 2, \ldots, n.$$

Applying the Generalized Rolle's Theorem, there exists $\xi \in (a, b)$ such that $g^{(n)}(\xi) = 0$. Hence

$$0 = g^{(n)}(\xi) = f^{(n)}(\xi) - P^{(n)}_{0,1,2,\ldots,n}(\xi).$$

To complete the proof, recognize that $P_{0,1,2,\ldots,n}$ is a polynomial of degree n with leading coefficient $f[x_0, x_1, x_2, \ldots, x_n]$. Therefore,

$$P^{(n)}_{0,1,2,\ldots,n}(\xi) = n! f[x_0, x_1, x_2, \ldots, x_n].$$

Substituting this result into the previous equation and solving for the divided difference yields

$$f[x_0, x_1, x_2, \ldots, x_n] = \frac{f^{(n)}(\xi)}{n!}. \qquad \square$$

Using this theorem, it follows that for each $x \in (a, b)$, there exists $\xi \in (a, b)$ such that

$$f[x_0, x_1, x_2, \ldots, x_n, x] = \frac{f^{(n+1)}(\xi)}{(n+1)!}. \tag{3}$$

Substituting (3) into (2), we obtain the error formula from Section 5.1.

EXERCISES

1. Assess the accuracy of the values in the relative viscosity table developed earlier in this section by plotting the values from the table and the six given data values on the same set of axes.
2. A more extensive table lists the viscosity of ethanol as 2.209 when the anhydrous solute weight is 70%. Add this value to the bottom of the divided difference table provided in the example in the text and compute the new values at the bottom of each column. What is the interpolating polynomial using seven data points rather than the original six?
3. Construct the divided difference table for the following data set, and then write out the Newton form of the interpolating polynomial.

x	2	4	5
y	−1	4	8

4. Construct the divided difference table for the following data set, and then write out the Newton form of the interpolating polynomial.

x	0	1	2
y	2	-1	4

5. Construct the divided difference table for the following data set, and then write out the Newton form of the interpolating polynomial.

x	-1	0	1	2
y	3	-1	-3	1

6. Construct the divided difference table for the following data set, and then write out the Newton form of the interpolating polynomial.

x	-7	-5	-4	-1
y	10	5	2	10

7. Write out the Newton form of the interpolating polynomial for $f(x) = \ln x$ that passes through the points $(1, \ln 1)$, $(2, \ln 2)$, and $(3, \ln 3)$.

8. Write out the Newton form of the interpolating polynomial for $f(x) = \sin x$ that passes through the points $(0, \sin 0)$, $(\pi/4, \sin \pi/4)$, and $(\pi/2, \sin \pi/2)$.

9. Write out the Newton form of the interpolating polynomial for $f(x) = e^x$ that passes through the points $(-1, e^{-1})$, $(0, e^0)$, and $(1, e^1)$.

10. Determine the missing values in the divided difference table provided below.

$x_0 = 0$	$f[x_0] = -1$		
		$f[x_0, x_1] = 5$	
$x_1 = 1$	$f[x_1] =?$		$f[x_0, x_1, x_2] = -\frac{3}{2}$
		$f[x_1, x_2] =?$	
$x_2 = 2$	$f[x_2] =?$		

11. Determine the missing values in the divided difference table provided below.

$x_0 = 0$	$f[x_0] = 1$			
		$f[x_0, x_1] = 2$		
$x_1 = 1$	$f[x_1] = 3$		$f[x_0, x_1, x_2] =?$	
		$f[x_1, x_2] =?$		$f[x_0, x_1, x_2, x_3] =?$
$x_2 = 2$	$f[x_2] = 3$		$f[x_1, x_2, x_3] = 0$	
		$f[x_2, x_3] = 0$		
$x_3 = 3$	$f[x_3] =?$			

12. Let $f[x_0, x_1, x_2, \ldots, x_k]$ be defined as the leading coefficient in the unique polynomial which interpolates f at the points x_0, x_1, x_2, $\ldots$, x_k. Show that

$$f[x_0, x_1, x_2, \ldots, x_k] = \frac{f[x_1, x_2, \ldots, x_k] - f[x_0, x_1, \ldots, x_{k-1}]}{x_k - x_0}.$$

13. The values listed in the table provide the surface tension of mercury as a function of temperature.

Temperature (°C)	10	25	50	75	100
Surface Tension (dyn/cm)	488.55	485.48	480.36	475.23	470.11

Use these values to determine the Newton form of the interpolating polynomial, and then use the polynomial to produce a table of surface tension values for temperatures ranging from 5°C through 100°C in increments of 5°C. Assess the accuracy of the table by plotting the values from the table and the five given data values on the same set of axes.

14. The thermal conductivity of air as a function of temperature is given in the table below. Estimate the thermal conductivity of air when $T = 240\,\text{K}$ and when $T = 485\,\text{K}$, using the Newton form of the interpolating polynomial.

Temperature (K)	100	200	300	400	500	600
Thermal Conductivity (mW/m · K)	9.4	18.4	26.2	33.3	39.7	45.7

15. Experimentally determined values for the partial pressure of water vapor, p_A, as a function of distance, y, from the surface of a pan of water are given below. Estimate the partial pressure at distances of 0.5 mm, 2.1 mm, and 3.7 mm from the surface of the water.

y (mm)	0	1	2	3	4	5
p_A (atm)	0.100	0.065	0.042	0.029	0.022	0.020

16. Ammonia vapor is compressed inside a cylinder by an external force acting on the piston. The ammonia is initially at 30° C, 500 kPa and the final pressure is 1400 kPa. The following data have been experimentally determined during the process. Use the Newton form of the interpolating polynomial to determine a table of volume as a function of pressure, with pressure ranging from 500 kPa through 1400 kPa in increments of 50 kPa.

Pressure (kPa)	500	653	802	945	1100	1248	1400
Volume (l)	1.25	1.08	0.96	0.84	0.72	0.60	0.50

Table A.5 in Frank White, *Fluid Mechanics*, lists the following values for the surface tension, Υ, vapor pressure, p_v, and sound speed, a, for water as a function of temperature. Use this data for Exercises 17–19.

T (°C)	Υ (N/m)	p_v (kPa)	a (m/s)
0	0.0756	0.611	1402
10	0.0742	1.227	1447
20	0.0728	2.337	1482
30	0.0712	4.242	1509
40	0.0696	7.375	1529
50	0.0679	12.34	1542
60	0.0662	19.92	1551
70	0.0644	31.16	1553
80	0.0626	47.35	1554
90	0.0608	70.11	1550
100	0.0589	101.3	1543

17. Use the Newton form of the interpolating polynomial to determine the surface tension of water when $T = 34°$ C, 68° C, 86° C, and 91° C.

18. Use the Newton form of the interpolating polynomial to determine the vapor pressure of water when $T = 34°$ C, $68°$ C, $86°$ C, and $91°$ C.

19. Use the Newton form of the interpolating polynomial to determine the sound speed of water when $T = 34°$ C, $68°$ C, $86°$ C, and $91°$ C.

5.4 OPTIMAL POINTS FOR INTERPOLATION

In many practical situations, we are not free to choose the interpolating points. For example, the function we wish to interpolate may be known only in terms of experimental data or from some standard engineering table. When we do have the freedom to choose the interpolating points, a natural question arises. Can we select the interpolation points so as to minimize the interpolation error?

It turns out that we can readily minimize a portion of the error, but the details depend upon the norm being used to measure the error. Here, we will consider only the l_∞ and l_2 norms. For a function f, continuous over the interval $[a, b]$, these norms are defined by

$$\|f\|_\infty = \max_{x\in[a,b]} |f(x)|$$

and

$$\|f\|_2 = \left(\int_a^b [f(x)]^2 dx\right)^{1/2}.$$

Remember that when we interpolate the function f over the interval $[a, b]$ at the $n + 1$ points $x_0, x_1, x_2, \ldots, x_n$, the interpolation error is given by

$$f(x) - p_n(x) = \frac{f^{(n+1)}(\xi)}{(n+1)!}(x - x_0)(x - x_1)(x - x_2)\cdots(x - x_n).$$

Here, p_n is the interpolating polynomial of degree at most n based on the $n + 1$ interpolating points, and $\min(x_0, x_1, x_2, \ldots, x_n, x) \le \xi \le \max(x_0, x_1, x_2, \ldots, x_n, x)$. Since f is given and n has been fixed, the only portion of the error formula that we can control is the polynomial

$$\omega(x) = (x - x_0)(x - x_1)(x - x_2)\cdots(x - x_n).$$

In this section, we will determine the interpolation points which make the l_∞ and the l_2 norm of this polynomial as small as possible.

Chebyshev Polynomials

When working with the l_∞-norm, the optimal points for interpolation are related to a special family of functions known as the Chebyshev polynomials.

Definition. For each nonnegative integer n, the CHEBYSHEV POLYNOMIAL T_n is defined for $x \in [-1, 1]$ by

$$T_n(x) = \cos(n \cos^{-1} x).$$

Granted, these functions don't look much like polynomials, but we will shortly see that each T_n is, in fact, an nth-degree polynomial.

The first order of business is to derive a recurrence relation for the Chebyshev polynomials. This is a formula that indicates how to construct one Chebyshev polynomial from others in the family. For notational convenience, let $\theta = \cos^{-1} x$. Then $T_n(\cos\theta) = \cos n\theta$. Using the standard trigonometric identities for the cosine of a sum and of a difference of two angles

$$T_{n+1}(\cos\theta) = \cos[(n+1)\theta] = \cos n\theta \cos\theta - \sin n\theta \sin\theta$$

and

$$T_{n-1}(\cos\theta) = \cos[(n-1)\theta] = \cos n\theta \cos\theta + \sin n\theta \sin\theta.$$

Adding these two expressions gives

$$T_{n+1}(\cos\theta) + T_{n-1}(\cos\theta) = 2\cos\theta \cos n\theta = 2\cos\theta\, T_n(\cos\theta).$$

Finally, solving for T_{n+1} and returning to the variable x yields

$$T_{n+1}(x) = 2xT_n(x) - T_{n-1}(x),$$

for any $n \geq 1$.

Before the recurrence relation can be used, simple expressions for both $T_0(x)$ and $T_1(x)$ are needed. Substituting $n = 0$ into the definition gives

$$T_0(x) = \cos(0 \cdot \cos^{-1} x) = \cos 0 = 1,$$

while, for $n = 1$,

$$T_1(x) = \cos(1 \cdot \cos^{-1} x) = x.$$

Now, applying the recurrence relation, the next three Chebyshev polynomials are found to be

$$T_2(x) = 2xT_1(x) - T_0(x) = 2x^2 - 1,$$

$$\begin{aligned} T_3(x) &= 2xT_2(x) - T_1(x) \\ &= 2x(2x^2 - 1) - x = 4x^3 - 3x, \end{aligned}$$

and

$$\begin{aligned} T_4(x) &= 2xT_3(x) - T_2(x) \\ &= 2x(4x^3 - 3x) - (2x^2 - 1) = 8x^4 - 8x^2 + 1. \end{aligned}$$

These polynomials suggest that

1. for each n, $T_n(x)$ is an nth-degree polynomial;
2. for each $n \geq 1$, the leading coefficient in T_n is 2^{n-1}; and
3. T_n is even (odd) when n is even (odd).

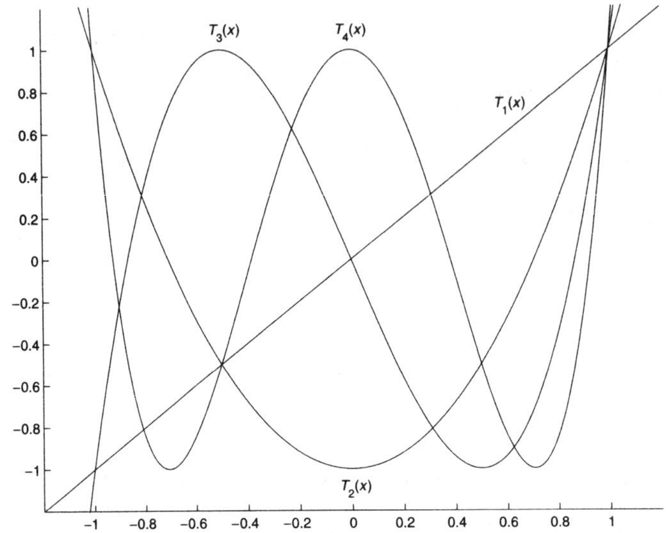

Figure 5.10 Graphs of the Chebyshev polynomials of degree one through four.

A simple inductive argument can be used to show that each of these statements is true in general.

Graphs of the Chebyshev polynomials of degree one through four are shown in Figure 5.10. Two important observations can be made from these graphs. First, each Chebyshev polynomial appears to have all of its roots on the interval $[-1,1]$, with each root being of multiplicity one. Second, on the interval $[-1,1]$, each polynomial appears to oscillate between a maximum value of $+1$ and a minimum value of -1, with the number of extreme values being one more than the number of roots. The following theorem makes these observations more precise.

Theorem. The Chebyshev polynomial $T_n(x)$, of degree $n \geq 1$, has n simple roots on the interval $[-1,1]$ at

$$x_j = \cos\left(\frac{2j-1}{2n}\pi\right)$$

for $j = 1, 2, 3, \ldots, n$. Moreover, $T_n(x)$ has absolute extreme values on $[-1,1]$ at

$$z_j = \cos\left(\frac{j\pi}{n}\right)$$

for $j = 0, 1, 2, \ldots, n$, with $T_n(z_j) = (-1)^j$.

Proof. Let $n \geq 1$, and let

$$x_j = \cos\left(\frac{2j-1}{2n}\pi\right)$$

for $j = 1, 2, 3, \ldots, n$. Then

$$T_n(x_j) = \cos\left\{n\cos^{-1}\left[\cos\left(\frac{2j-1}{2n}\pi\right)\right]\right\}$$
$$= \cos\left(\frac{2j-1}{2}\pi\right) = 0,$$

and each x_j is a root of T_n. Since T_n is a polynomial of degree n, the x_j must be the only zeros of T_n, and since they are distinct, they are all simple roots.

Recall that for a function defined over a closed interval, extreme values can occur only at critical points or at the endpoints of the interval. The critical points of T_n are the roots of

$$T_n'(x) = \frac{n\sin(n\cos^{-1}x)}{\sqrt{1-x^2}}.$$

Now, let

$$z_j = \cos\left(\frac{j\pi}{n}\right)$$

for $j = 0, 1, 2, \ldots, n$. For $j = 1, 2, 3, \ldots, n-1$,

$$T_n'(z_j) = \frac{n\sin j\pi}{\sin(j\pi/n)} = 0.$$

Since T_n is a polynomial of degree n, T_n' will be a polynomial of degree $n-1$; therefore, the points $z_1, z_2, z_3, \ldots, z_{n-1}$ account for all of the roots of T_n' and, hence, all of the critical points of T_n. Note that $z_0 = 1$ and $z_n = -1$ are the endpoints of the interval under consideration, so the z_j are the only possible candidates for the extrema of T_n. At each of the z_j, a direct calculation shows that

$$T_n(z_j) = \cos\left\{n\cos^{-1}\left[\cos\left(\frac{j\pi}{n}\right)\right]\right\}$$
$$= \cos(j\pi) = (-1)^j.$$

For j even, T_n therefore achieves its absolute maximum value on $[-1,1]$ of $+1$, while for j odd, T_n achieves its absolute minimum value on $[-1,1]$ of -1. □

For our current objective, that of determining the interpolating points which will minimize the maximum norm of the interpolation error, the most important property of the Chebyshev polynomials is the *minimax property*. One final preliminary item is needed before this property can be discussed.

Definition. A polynomial is MONIC if its leading coefficient is $+1$. Let $\tilde{\Pi}_n$ denote the set of all monic polynomials of degree n.

With the exception of T_0 and T_1, the Chebyshev polynomials are clearly not monic, as the leading coefficient is of the form 2^{n-1}. However, the polynomials

$$\tilde{T}_n(x) = \begin{cases} T_0(x), & n = 0 \\ 2^{1-n}T_n(x), & n \geq 1 \end{cases}$$

are monic. These are known as the *monic Chebyshev polynomials.* Since $\tilde{T}_n$ is just a multiple of T_n, the roots and extrema of $\tilde{T}_n$ are located at exactly the same points as the roots and extrema of T_n. The extreme values of $\tilde{T}_n$, though, are reduced to $\pm 2^{1-n}$.

Theorem. The monic Chebyshev polynomial $\tilde{T}_n$ $(n \geq 1)$ satisfies

$$\frac{1}{2^{n-1}} = \max_{x\in[-1,1]} |\tilde{T}_n(x)| \leq \max_{x\in[-1,1]} |p_n(x)|$$

for any $p_n \in \tilde{\Pi}_n$, with equality if and only if $p_n = \tilde{T}_n$.

Proof. Suppose $p_n \in \tilde{\Pi}_n$ with

$$\max_{x\in[-1,1]} |p_n(x)| \leq \frac{1}{2^{n-1}} = \max_{x\in[-1,1]} |\tilde{T}_n(x)|.$$

Let $q(x) = \tilde{T}_n(x) - p_n(x)$. Since $\tilde{T}_n$ and p_n are both monic polynomials of degree n, it follows that q is a polynomial of degree at most $n - 1$. At the extrema of $\tilde{T}_n$,

$$\begin{aligned} q(z_j) &= \tilde{T}_n(z_j) - p_n(z_j) \\ &= 2^{1-n}(-1)^j - p_n(z_j). \end{aligned}$$

By supposition, $|p(z_j)| \leq 2^{1-n}$, so

$$\begin{aligned} q(z_j) &\geq 0 \qquad \text{whenever } j \text{ is even} \\ q(z_j) &\leq 0 \qquad \text{whenever } j \text{ is odd.} \end{aligned}$$

Therefore, q is guaranteed to have at least one root between z_j and z_{j+1} for each j =0, 1, 2, $\ldots n-1$. But this implies that q is a polynomial of degree at most $n - 1$ with at least n roots, which is impossible unless $q \equiv 0$. Thus $p_n = \tilde{T}_n$. □

Chebyshev polynomials have many other interesting and important properties, a few of which will be explored in the exercises. For a more complete discussion of Chebyshev polynomials, encompassing their role in both applied mathematics and numerical analysis, consult Rivlin [1] or Fox and Parker [2].

Optimal Interpolating Points Relative to Maximum Norm

Suppose we wish to interpolate the function f over the interval $[-1,1]$ at the $n+1$ points x_0, x_1, x_2, $\dots$, x_n, choosing these points so as to minimize the l_∞ norm of the polynomial

$$\omega(x) = (x - x_0)(x - x_1)(x - x_2)\cdots(x - x_n).$$

Note that $\omega(x)$ happens to be a monic polynomial of degree $n+1$ whose l_∞ norm we desire to make as small as possible. This can only be accomplished by taking $\omega(x) = \tilde{T}_{n+1}(x)$, which can only happen if the interpolating points are chosen to be the roots of $\tilde{T}_{n+1}(x)$. Thus to minimize the l_∞ norm of ω, the interpolating points must be chosen as

$$x_i = \cos\left(\frac{2i+1}{2(n+1)}\pi\right), \quad i = 0, 1, 2, \dots, n.$$

With this choice of interpolating points, it follows that $\|\omega\|_\infty = 2^{-n}$; hence, over the interval $[-1,1]$,

$$\|f - p_n\|_\infty \le \frac{\|f^{(n+1)}\|_\infty}{2^n(n+1)!}.$$

EXAMPLE 5.11 The Optimal Interpolating Points - Maximum Norm

Let's interpolate the function $f(x) = \sin \pi x$ over the interval $[-1,1]$ with a polynomial of degree at most 4. The interpolating points which will minimize the l_∞ norm of ω are therefore the roots of $\tilde{T}_5$:

$$x_0 = \cos\frac{\pi}{10}, \quad x_1 = \cos\frac{3\pi}{10}, \quad x_2 = \cos\frac{\pi}{2} = 0, \quad x_3 = \cos\frac{7\pi}{10}, \quad x_4 = \cos\frac{9\pi}{10}.$$

For comparison, we will also interpolate f at the uniformly spaced points:

$$x_0 = -1, \quad x_1 = -0.5, \quad x_2 = 0, \quad x_3 = 0.5, \quad x_4 = 1,$$

and a third set of interpolating points—an explanation for which will be provided later in the section:

$$x_0 = -0.90618, \quad x_1 = -0.53847, \quad x_2 = 0, \quad x_3 = 0.53847, \quad x_4 = 0.90618.$$

Figure 5.11 displays the absolute value of the interpolation error for the interpolating polynomial of degree at most four based on each of these three sets of points. Although the uniformly spaced points and the third set of points produce smaller errors in the middle section of the domain, the Chebyshev points produce substantially smaller errors at the edges of the domain and a substantially smaller maximum error. In particular, the l_∞ norms are roughly 0.11556, 0.18076, and 0.19216 for the Chebyshev points, the uniformly spaced points, and the third set of points, respectively.

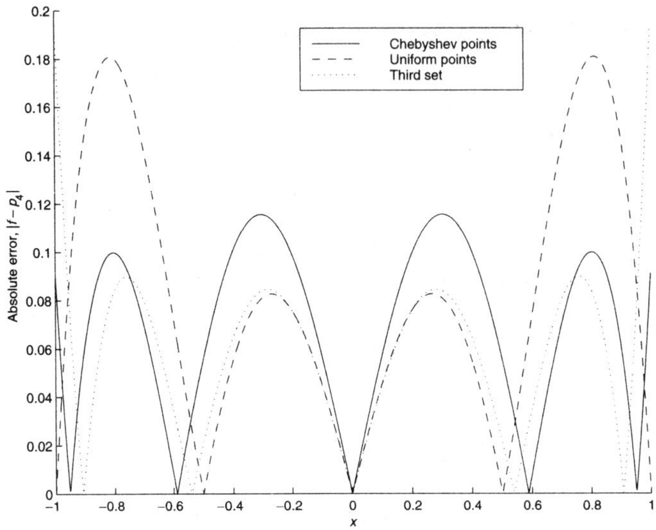

Figure 5.11 Comparison of interpolation error in the interpolating polynomial of degree at most four based on Chebyshev interpolating points, uniformly spaced interpolating points and a third set of interpolating points.

What if the interval under consideration is not $[-1, 1]$? Note that the linear function

$$x = \frac{b-a}{2}t + \frac{a+b}{2}$$

translates and scales the interval $-1 \leq t \leq 1$ onto the interval $a \leq x \leq b$. To minimize the l_∞ norm of ω over the general interval $[a, b]$, we should therefore choose

$$x_i = \frac{b-a}{2}\cos\left(\frac{2i+1}{2(n+1)}\pi\right) + \frac{a+b}{2}, \qquad i = 0, 1, 2, \ldots, n$$

as the interpolating points; that is, the scaled and translated roots of the monic Chebyshev polynomial $\tilde{T}_{n+1}$.

EXAMPLE 5.12 An Interval Other Than $[-1, 1]$

Let's interpolate the function $f(x) = xe^{-x}$ over $[-1, 3]$ using a polynomial of degree at most four. The properly scaled and translated roots of $\tilde{T}_5$ are

$$x_0 = 1 + 2\cos\frac{\pi}{10}, \quad x_1 = 1 + 2\cos\frac{3\pi}{10}, \quad x_2 = 1 + 2\cos\frac{\pi}{2} = 1,$$

$$x_3 = 1 + 2\cos\frac{7\pi}{10}, \quad x_4 = 1 + 2\cos\frac{9\pi}{10}.$$

The interpolating polynomial generated by these points is

$$p_4(x) = -0.06011x^4 + 0.43376x^3 - 1.11011x^2 + 1.08627x + 0.01807,$$

and $\|f - p_4\|_\infty = 0.04191$. It is left as an exercise to verify that interpolation at uniformly spaced points and at the third set of points introduced earlier, translated and scaled to $[-1, 3]$ of course, both produce larger maximum errors.

Legendre Polynomials

Another special family of functions that arises in applied mathematics is the Legendre polynomials.

> **Definition.** For each nonnegative integer n, the LEGENDRE POLYNOMIAL $P_n(x)$ satisfies the recurrence relation
>
> $$P_n(x) = \frac{2n-1}{n}xP_{n-1}(x) - \frac{n-1}{n}P_{n-2}(x),$$
>
> with $P_0(x) = 1$ and $P_1(x) = x$.

Since $P_0(x) = 1$ and $P_1(x) = x$, the recurrence relation implies, through mathematical induction on n, that each P_n is a polynomial of degree n. In particular,

$$P_2(x) = \frac{3}{2}xP_1(x) - \frac{1}{2}P_0(x) = \frac{3}{2}x^2 - \frac{1}{2};$$

$$P_3(x) = \frac{5}{3}xP_2(x) - \frac{2}{3}P_1(x) = \frac{5}{2}x^3 - \frac{3}{2}x; \text{ and}$$

$$P_4(x) = \frac{7}{4}xP_3(x) - \frac{3}{4}P_1(x) = \frac{35}{8}x^4 - \frac{15}{4}x^2 + \frac{3}{8}.$$

We now turn to a discussion of some of the relevant properties of the Legendre polynomials.

First, the Legendre polynomials satisfy the following integral relationship (Hildebrand [3]):

$$\int_{-1}^{1} P_j(x)P_k(x)\,dx = \begin{cases} 0, & j \neq k \\ \frac{2}{2j+1}, & j = k \end{cases}. \tag{1}$$

In general, two functions f and g for which

$$\int_a^b f(x)g(x)w(x)\,dx = 0$$

are said to be *orthogonal on [a,b] with respect to w(x)*. The function w, called a *weight function*, must be nonnegative and integrable on $[a, b]$ and can evaluate to zero only at isolated points along $[a, b]$. When any pair of different functions chosen from a set of functions is orthogonal, those functions are said to form an *orthogonal set*. In light of these definitions, equation (1) tells us that the Legendre polynomials form an orthogonal set on $[-1,1]$ with respect to the weight function $w(x)=1$. In Exercise 3, we will see that the Chebyshev polynomials also form an orthogonal set on the interval $[-1,1]$, but with respect to the weight function $w(x)=(1-x^2)^{-1/2}$.

Next, let k be a nonnegative integer, and consider the set $\{P_0, P_1, P_2, \ldots, P_k\}$. Suppose there exist constants $c_0, c_1, c_2, \ldots, c_k$ such that

$$c_0P_0(x) + c_1P_1(x) + c_2P_2(x) + \cdots + c_kP_k(x) = 0$$

for all x. Multiply this equation by $P_j(x)$ and then integrate from $x=-1$ to $x=+1$. This yields

$$\int_{-1}^{1} \sum_{i=0}^{k} c_iP_i(x)P_j(x)\,dx = \sum_{i=0}^{k} c_i \int_{-1}^{1} P_i(x)P_j(x)\,dx = 0.$$

Orthogonality guarantees that each integral, except the one containing $[P_j(x)]^2$, vanishes. This leaves

$$c_j \int_{-1}^{1} [P_j(x)]^2 dx = \frac{2}{2j+1}c_j = 0,$$

or $c_j = 0$, for each $j = 0, 1, 2, \ldots, k$. Hence, the set $\{P_0, P_1, P_2, \ldots, P_k\}$ is linearly independent. Combining linear independence with the fact that each $P_j(x)$ is a polynomial of degree j, it follows that $\{P_0, P_1, P_2, \ldots, P_k\}$ spans the space of all polynomials of degree at most k.

Finally, let k and n be nonnegative integers with $k < n$. Further, let q be an arbitrary polynomial of degree k. Since $\{P_0, P_1, P_2, \ldots, P_k\}$ spans the space of all polynomials of degree at most k, there exist constants $c_0, c_1, c_2, \ldots, c_k$ such that

$$q(x) = c_0P_0(x) + c_1P_1(x) + c_2P_2(x) + \cdots + c_kP_k(x) = \sum_{j=0}^{k} c_jP_j(x).$$

Then

$$\begin{aligned} \int_{-1}^{1} q(x)P_n(x)\,dx &= \int_{-1}^{1} \left(\sum_{j=0}^{k} c_jP_j(x) \right) P_n(x)\,dx \\ &= \sum_{j=0}^{k} c_j \int_{-1}^{1} P_j(x)P_n(x)\,dx \\ &= 0, \end{aligned}$$

by the orthogonality of the Legendre polynomials. Hence, each Legendre polynomial P_n is orthogonal to all polynomials of degree less than n over the interval $[-1,1]$ with respect to the weight function $w(x)=1$.

If we take each Legendre polynomial and divide through by the leading coefficient, the set of monic Legendre polynomials, $\{\tilde{P}_0, \tilde{P}_1, \tilde{P}_2, \dots\}$, results. Since each $\tilde{P}_j$ is just a constant multiple of P_j, the monic Legendre polynomials inherit all of the linear independence and orthogonality properties of the original Legendre polynomials.

Optimal Interpolating Points Relative to Euclidean Norm

Suppose we wish to interpolate the function f over the interval $[-1, 1]$ at the $n+1$ points x_0, x_1, x_2, ..., x_n, choosing these points so as to minimize the l_2 norm of the polynomial

$$\omega(x) = (x - x_0)(x - x_1)(x - x_2) \cdots (x - x_n).$$

Consider the quantity $\omega(x) - \tilde{P}_{n+1}(x)$. Since both polynomials are monic of degree $n+1$, their difference must be a polynomial of degree at most n, call it $q(x)$. Keep in mind that since q is a polynomial of degree at most n, the integral

$$\int_{-1}^{1} \tilde{P}_{n+1}(x) q(x)\, dx$$

is guaranteed to be equal to 0. Now, with $\omega(x) = \tilde{P}_{n+1}(x) + q(x)$, it follows that

$$\begin{aligned}
\|\omega\|_2^2 &= \|\tilde{P}_{n+1} + q\|_2^2 \\
&= \int_{-1}^{1} \left(\tilde{P}_{n+1}(x) + q(x)\right)^2 dx \\
&= \|\tilde{P}_{n+1}\|_2^2 + 2\int_{-1}^{1} \tilde{P}_{n+1}(x) q(x) dx + \|q\|_2^2 \\
&= \|\tilde{P}_{n+1}\|_2^2 + \|q\|_2^2.
\end{aligned}$$

The l_2 norm of ω will therefore be minimized when $q(x) \equiv 0$; that is, when $\omega(x) = \tilde{P}_{n+1}(x)$. Hence, to minimize the l_2 norm of ω when working on the interval $[-1, 1]$, the interpolation points must be chosen as the roots of the monic Legendre polynomial $\tilde{P}_{n+1}(x)$. In Section 6.6 we will prove that these roots are all simple and all lie inside the interval $[-1, 1]$.

Interpolation over the more general interval $[a, b]$ is handled in precisely the same manner as noted earlier for the l_∞ norm. The roots of the appropriate monic Legendre polynomial are scaled by the factor $(b-a)/2$ and translated by $(a+b)/2$. Interpolation at the resulting points will then minimize the l_2 norm of ω.

EXAMPLE 5.13 The Optimal Interpolating Points—Euclidean Norm

Let's interpolate the function $f(x) = \sin \pi x$ over the interval $[-1, 1]$ with a polynomial of degree at most 4. The interpolating points that will minimize the l_2 norm of ω are the roots of

$$\tilde{P}_5(x) = x^5 - \frac{10}{9}x^3 + \frac{5}{21}x.$$

To five decimal places, the roots of this polynomial are

$$x_0 = -0.90618, \quad x_1 = -0.53847, \quad x_2 = 0, \quad x_3 = 0.53847, \quad x_4 = 0.90618.$$

Note that these points form the third set of interpolating points which were used in the first example of this section. For comparison, we will also interpolate f at the uniformly spaced points

$$x_0 = -1, \quad x_1 = -0.5, \quad x_2 = 0, \quad x_3 = 0.5, \quad x_4 = 1,$$

and the Chebyshev points

$$x_0 = \cos\frac{\pi}{10}, \quad x_1 = \cos\frac{3\pi}{10}, \quad x_2 = \cos\frac{\pi}{2} = 0, \quad x_3 = \cos\frac{7\pi}{10}, \quad x_4 = \cos\frac{9\pi}{10}.$$

Recall that Figure 5.11 displays the absolute value of the interpolation error for the interpolating polynomial of degree at most four based on each of these three sets of points. The l_2 norms are roughly 0.09391, 0.10876, and 0.13964 for the Legendre points, the Chebyshev points and the uniformly spaced points, respectively.

Let's also interpolate the function $f(x) = xe^{-x}$ over $[-1, 3]$ using a polynomial of degree at most four. The properly scaled and translated roots of $\tilde{P}_5$ are

$$x_0 = 1 + 2(-0.90618) = -0.81236, \quad x_1 = 1 + 2(-0.53847) = -0.07694,$$

$$x_2 = 1 + 2(0) = 1, \quad x_3 = 1 + 2(0.53847) = 2.07694,$$

$$x_4 = 1 + 2(0.90618) = 2.81236.$$

The interpolating polynomial generated by these points is

$$p_4(x) = -0.05841x^4 + 0.41820x^3 - 1.07721x^2 + 1.07882x + 0.00648,$$

and $\|f - p_4\|_2 = 0.03916$. It is left as an exercise to verify that interpolation at uniformly spaced points and at the translated and scaled Chebyshev points both produce larger errors in the l_2 norm.

References

1. T. J. Rivlin, *The Chebyshev Polynomials*, John Wiley & Sons, New York, 1974.
2. L. Fox and I. Parker, *Chebyshev Polynomials in Numerical Analysis*, Oxford University Press, Oxford, England, 1968.
3. F. B. Hildebrand, *Advanced Calculus for Applications, 2nd edition*, Prentice-Hall, Englewood Cliffs, NJ, 1976.

EXERCISES

1. Prove each of the following properties of the Chebyshev polynomials:
   (a) for each n, $T_n(1) = 1$.
   (b) for each n, $T_n(-1) = (-1)^n$.

(c) for all $j > k \geq 0$, $T_j(x)T_k(x) = \frac{1}{2}\left[T_{j+k}(x) + T_{j-k}(x)\right]$.

2. Show that the Chebyshev polynomial $T_n(x)$ is a solution to the differential equation

$$(1-x^2)\frac{d^2y}{dx^2} - x\frac{dy}{dx} + n^2y = 0.$$

3. Show that

$$\int_{-1}^{1} \frac{T_n(x)T_m(x)}{\sqrt{1-x^2}}dx = \begin{cases} 0, & m \neq n \\ c_n\frac{\pi}{2}, & m = n \end{cases},$$

where $c_0 = 2$ and $c_n = 1$ $(n \geq 1)$. This implies that the Chebyshev polynomials form an orthogonal set on $[-1, 1]$ with respect to the weight function $w(x) = (1-x^2)^{-1/2}$. (*Hint:* Make the substitution $\theta = \cos^{-1} x$.)

4. Show that the Legendre polynomial $P_n(x)$ is a solution to the differential equation

$$(1-x^2)\frac{d^2y}{dx^2} - 2x\frac{dy}{dx} + n(n+1)y = 0.$$

5. Consider interpolating $f(x) = xe^{-x}$ over $[-1, 3]$ with a polynomial of degree at most four.

(a) Interpolate at uniformly spaced points and at the scaled and translated Legendre points. Determine the l_∞ norm of the interpolation error for both interpolating polynomials and compare with the l_∞ norm associated with the scaled and translated Chebyshev points.

(b) Interpolate at uniformly spaced points and at the scaled and translated Chebyshev points. Determine the l_2 norm of the interpolation error for both interpolating polynomials and compare with the l_2 norm associated with the scaled and translated Legendre points.

6. For each of the following intervals, identify the interpolating points that minimize the l_∞ and the l_2 norm of ω for linear interpolation.

(a) $[-1, 1]$ (b) $[0, 3.5]$ (c) $[-\pi, 0]$ (d) $[-\sqrt{2}, 3]$ (e) $[-2.5, 3.5]$

7. Repeat Exercise 6 for cubic interpolation.

8. Repeat Exercise 6 for interpolation by polynomials of degree at most 5.

For Exercises 9–13, interpolate the given function over the specified interval by a polynomial of the indicated degree. Interpolate at uniformly spaced points, the Chebyshev points and the Legendre points, and compare the errors in the resulting polynomials in both the l_∞ and the l_2 norm.

9. $f(x) = e^x$, $[-1, 1]$, $n = 3$

10. $f(x) = e^{-x}$, $[-1, 2]$, $n = 3$

11. $f(x) = x \ln x$, $[1, 3]$, $n = 4$

12. $f(x) = \ln(x+2)$, $[-1, 1]$, $n = 5$

13. $f(x) = 1/x$, $[1, 4]$, $n = 5$

5.5 PIECEWISE LINEAR INTERPOLATION

Consider approximating the function f by an interpolating polynomial using data from $n+1$ points, and suppose that the resulting approximation is not accurate enough. We could introduce data from more points and use a higher-degree polynomial; however, using a higher-degree polynomial could introduce undesired oscillations, as seen in the examples and exercises of Sections 5.1 through 5.3.

Furthermore, interpolation at equally spaced abscissas can lead to poor results in the sense that there are smooth (infinitely differentiable) functions for which the interpolation error goes off to infinity as the number of data points increases. As an example, consider the function

$$f(x) = \frac{1}{1+25x^2}$$

on the interval $[-1,1]$. If we interpolate this function at the points $x_i = -1 + 2i/n \ (i = 0,1,2,\ldots,n)$, then the maximum norm of the difference between f and the interpolating polynomial as a function of n is summarized in the following table.

n	Maximum Norm
4	0.428
8	1.045
16	14.394
32	5059.033
64	1.078×10^9

Figure 5.12 shows f and the interpolating polynomial with $n = 8$. Note the large oscillations in the interpolating polynomial at the edges of the domain.

By using just one polynomial to approximate a function, we also run the risk that singular behavior at one point could lead to slow convergence over the entire domain. Consider $f(x) = \sqrt{|x|}$ on $[-1,1]$, which has singular derivatives to all orders at $x = 0$. We interpolate this function at the Chebyshev points

$$x_i = \cos\frac{(2i+1)\pi}{2n+2}$$

for $i = 0,\ 1,\ 2,\ \ldots,\ n$. The table below summarizes both the maximum norm and the Euclidean norm of the interpolation error as a function of n.

n	Maximum Norm	Euclidean Norm
4	0.3335	0.2794
8	0.2466	0.1602
16	0.1789	0.08645
32	0.1283	0.04503
64	0.09142	0.02299
128	0.06489	0.01162

Note the maximum norm decreases like $1/\sqrt{n}$, while the Euclidean norm decreases like $1/n$. Hence, to achieve an interpolation error on the order of 10^{-3} measured

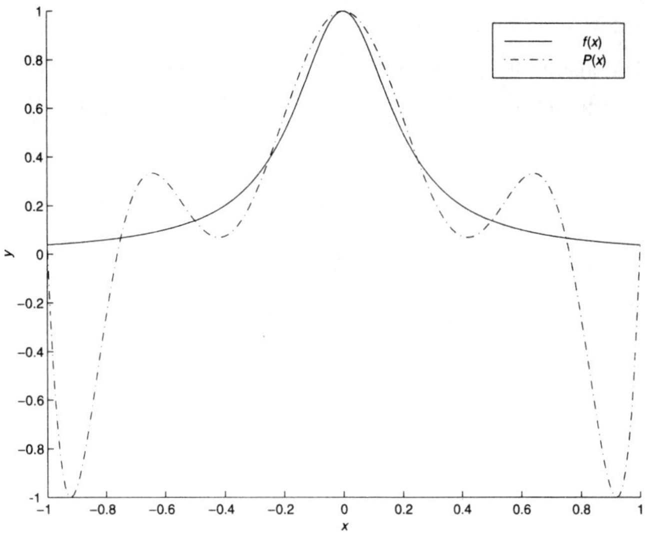

Figure 5.12 Plot of $f(x) = 1/(1+25x^2)$ and the interpolating polynomial, $P(x)$, generated using data from the equally spaced points $x_i = -1 + i/4$ for $i = 0, 1, 2, \ldots, 8$.

in the maximum norm, we would need to select $n \approx 10^6$. For an error on the order of 10^{-3} measured in the Euclidean norm, we would need $n \approx 10^3$.

These phenomena suggest that we should develop an alternative procedure for improving interpolation accuracy. In particular, we will introduce more points, but use a lower-order polynomial on each of the subintervals defined by the interpolation points. This idea gives rise to the notion of piecewise polynomial interpolation. This section is devoted to the simplest case of piecewise polynomial interpolation, that of piecewise linear interpolation. One type of piecewise cubic interpolation, known as cubic spline interpolation, will be developed in the next section. Hermite cubic interpolation, a second type of piecewise cubic interpolation that uses both function and derivative data, will be treated in Section 5.7.

The Piecewise Linear Interpolant

Let f be a function defined on the interval $[a, b]$, and let

$$a = x_0 < x_1 < x_2 < \cdots < x_{n-1} < x_n = b$$

be the $n+1$ distinct points at which f is to be interpolated. Note the collection of n subintervals into which the x_i divide $[a, b]$ is called a partition of $[a, b]$.

Definition. The PIECEWISE LINEAR INTERPOLANT of f relative to the partition

$$a = x_0 < x_1 < x_2 < \cdots < x_{n-1} < x_n = b$$

is a function s that satisfies

(1) s is continuous on $[a, b]$;

(2) on each subinterval $[x_i, x_{i+1}]$, $i = 0, 1, 2, \ldots, n-1$, s coincides with the linear polynomial

$$s(x) = s_i(x) = a_i + b_i(x - x_i);$$

(3) s interpolates f at $x_0, x_1, x_2, \ldots, x_n$.

At $x = x_i$, the interpolation condition gives

$$f(x_i) = s(x_i) = s_i(x_i) = a_i.$$

Continuity of the piecewise linear interpolant requires $s_i(x_{i+1}) = s_{i+1}(x_{i+1})$, which yields the equation

$$a_i + b_i(x_{i+1} - x_i) = a_{i+1},$$

whose solution is

$$b_i = \frac{a_{i+1} - a_i}{x_{i+1} - x_i} = \frac{f(x_{i+1}) - f(x_i)}{x_{i+1} - x_i}.$$

The simplicity of the formulas for the a_i and the b_i makes piecewise linear interpolation ideal for hand calculations from tabulated data. In fact, it is common practice to use piecewise linear interpolation from engineering and thermodynamic tables.

EXAMPLE 5.14 Viscosity of Sulfuric Acid

The following table gives the viscosity of sulfuric acid, in millipascal-seconds (centipoises), as a function of concentration, in mass percent. From these data, we would like to estimate the viscosity when the concentration is 5%, 63%, and 85%. We could attempt to use a fifth-degree interpolating polynomial, but here we will use a piecewise linear interpolant.

Concentration	0	20	40	60	80	100
Viscosity	0.89	1.40	2.51	5.37	17.4	24.2

Using the formulas derived above for the coefficients of a piecewise linear interpolant, we find for this data that

$$a_0 = 0.89, a_1 = 1.40, a_2 = 2.51, a_3 = 5.37, a_4 = 17.4$$

and

$$b_0 = \frac{1.40 - 0.89}{20} = 0.0255, b_1 = \frac{2.51 - 1.40}{20} = 0.0555, b_2 = \frac{5.37 - 2.51}{20} = 0.143$$

$$b_3 = \frac{17.4 - 5.37}{20} = 0.6015, b_4 = \frac{24.2 - 17.4}{20} = 0.34.$$

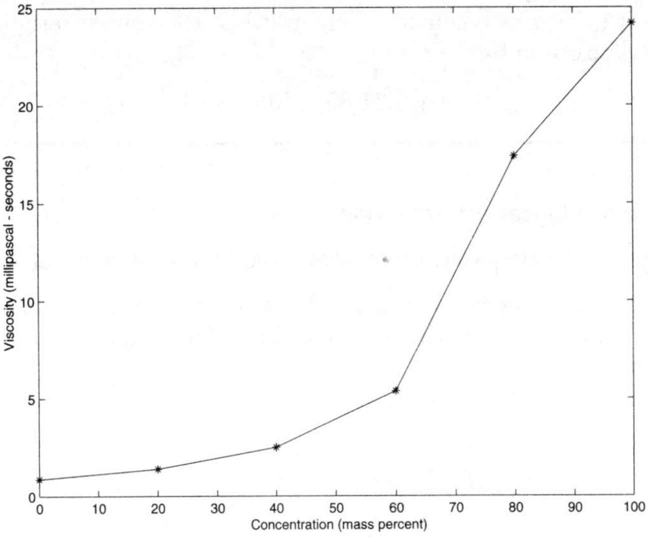

Figure 5.13 Piecewise linear interpolant for viscosity of sulfuric acid as a function of concentration. Data points are indicated by asterisks.

Hence, the interpolating function relating viscosity to concentration is

$$\text{viscosity} = \begin{cases} 0.89 + 0.0255C, & 0 \le C < 20 \\ 1.40 + 0.0555(C - 20), & 20 \le C < 40 \\ 2.51 + 0.143(C - 40), & 40 \le C < 60 \\ 5.37 + 0.6015(C - 60), & 60 \le C < 80 \\ 17.4 + 0.34(C - 80), & 80 \le C \le 100 \end{cases}.$$

A graph of this function is shown in Figure 5.13. Though the function is continuous, it clearly is not differentiable at any of the interpolating points. Regardless, the piecewise linear interpolant provides a reasonable representation for the data.

As with any piecewise function, evaluation of a piecewise linear interpolant is a two-step process. We must first determine which polynomial piece needs to be evaluated, based on the value of the independent variable. Once the polynomial piece has been selected, we can then evaluate at the given value of the independent variable. For example, to estimate the viscosity when the concentration is 5%, the first polynomial piece, $0.89 + 0.0255C$, is selected. We then estimate that the viscosity is

$$0.89 + 0.0255(5) = 1.0175$$

when the concentration is 5%. For the viscosity when the concentration is 63%, we find

$$5.37 + 0.6015(63 - 60) = 7.1745,$$

using the fourth linear polynomial. Finally, when the concentration is 85%, we estimate the viscosity to be

$$17.4 + 0.34(85 - 80) = 19.1.$$

Error in Piecewise Linear Interpolation

The final issue to consider is the error introduced by piecewise linear interpolation.

Theorem. Let f be continuous, with two continuous derivatives, on the interval $[a, b]$, and let s be the piecewise linear interpolant of f relative to the partition

$$a = x_0 < x_1 < x_2 < \cdots < x_{n-1} < x_n = b.$$

Then

$$\max_{x\in[a,b]} |f(x) - s(x)| \le \frac{1}{8}h^2 \max_{x\in[a,b]} |f''(x)|,$$

where $h = \max_{0\le i\le n-1}(x_{i+1} - x_i)$.

Proof. The key to establishing this result is recognizing that on each subinterval, $[x_i, x_{i+1}]$, standard linear interpolation is being performed, so the standard linear interpolation error formula (see Section 5.1),

$$|f(x) - s_i(x)| = \frac{1}{2}|f''(\xi)||(x - x_i)(x - x_{i+1})|$$

holds, where $x_i < \xi < x_{i+1}$. Therefore,

$$\max_{x\in[x_i,x_{i+1}]} |f(x) - s_i(x)| \le \frac{1}{2} \max_{x\in[x_i,x_{i+1}]} |f''(x)| \cdot \max_{x\in[x_i,x_{i+1}]} |(x - x_i)(x - x_{i+1})|.$$

Let $g(x) = |(x - x_i)(x - x_{i+1})|$. On $[x_i, x_{i+1}]$, g attains its maximum value of $h_i^2/4$ when $x = (x_i + x_{i+1})/2$, where $h_i = x_{i+1} - x_i$. Substituting this value into the above error bound produces

$$\begin{aligned}\max_{x\in[x_i,x_{i+1}]} |f(x) - s_i(x)| &\le \frac{1}{2} \max_{x\in[x_i,x_{i+1}]} |f''(x)| \cdot \frac{1}{4}h_i^2 \\ &= \frac{1}{8}h_i^2 \max_{x\in[x_i,x_{i+1}]} |f''(x)|.\end{aligned}$$

Since

$$\max_{x\in[a,b]} |f(x) - s(x)| = \max_{0\le i\le n-1} \Big(\max_{x\in[x_i,x_{i+1}]} |f(x) - s_i(x)| \Big),$$

it follows that

$$\begin{aligned}\max_{x\in[a,b]} |f(x) - s(x)| &\le \frac{1}{8} \max_{0\le i\le n-1} h_i^2 \cdot \max_{0\le i\le n-1} \Big(\max_{x\in[x_i,x_{i+1}]} |f''(x)| \Big) \\ &= \frac{1}{8}h^2 \max_{x\in[a,b]} |f''(x)|,\end{aligned}$$

where $h = \max_{0\le i\le n-1} h_i$. □

EXAMPLE 5.15 Constructing a Table of Sine and Cosine Values

A textbook publisher needs a table of values for the sine and cosine functions. Each function is to be tabulated at equal increments running from $\theta = 0°$ through $\theta = 45°$. The values in the table will be given to six decimal places, and the increment in θ must be selected small enough so that linear interpolation between consecutive values will introduce an error less than 10^{-6}.

To accomplish this objective, we must select h so that

$$\frac{h^2}{8} \max_{0 \le \theta \le \pi/4} |f''(\theta)| < 10^{-6},$$

where the argument θ has been given in radians and f can be taken as either the sine or the cosine function. In either case,

$$\max_{0 \le \theta \le \pi/4} |f''(\theta)| \le 1,$$

leading to $h < 2.828 \times 10^{-3}$ radians. Converting to degrees, any increment less than 0.162 degrees will suffice. For convenience, we might therefore suggest an increment of one-tenth of a degree.

EXERCISES

Exercises 1 through 6 are based on the following data for the density, ρ, viscosity, μ, kinematic viscosity, ν, surface tension, Υ, vapor pressure, p_v, and sound speed, a, of water as a function of temperature. These data were drawn from Tables A.1 and A.5 in Frank White, *Fluid Mechanics*:

T (°C)	ρ (kg/m^3)	μ ($\times 10^{-3}$ N·s/m^2)	ν ($\times 10^{-5}$ m^2/s)	Υ (N/m)	p_v (kPa)	a (m/s)
0	1000	1.788	1.788	0.0756	0.611	1402
10	1000	1.307	1.307	0.0742	1.227	1447
20	998	1.003	1.005	0.0728	2.337	1482
30	996	0.799	0.802	0.0712	4.242	1509
40	992	0.657	0.662	0.0696	7.375	1529
50	988	0.548	0.555	0.0679	12.34	1542
60	983	0.467	0.475	0.0662	19.92	1551
70	978	0.405	0.414	0.0644	31.16	1553
80	972	0.355	0.365	0.0626	47.35	1554
90	965	0.316	0.327	0.0608	70.11	1550
100	958	0.283	0.295	0.0589	101.3	1543

1. Estimate, using piecewise linear interpolation, the density of water when $T =$ 34° C, 68° C, 86° C, and 91° C.
2. Estimate, using piecewise linear interpolation, the viscosity of water when $T =$ 34° C, 68° C, 86° C, and 91° C. At what temperature is the viscosity 1.000×10^{-3} N·s/m^2?

3. Estimate, using piecewise linear interpolation, the kinematic viscosity of water when $T = 34°$ C, $68°$ C, $86°$ C, and $91°$ C. At what temperature is the kinematic viscosity 1.000×10^{-5} m^2/s?
4. Estimate, using piecewise linear interpolation, the surface tension of water when $T = 34°$ C, $68°$ C, $86°$ C, and $91°$ C. At what temperature is the surface tension 0.0650 N/m?
5. Estimate, using piecewise linear interpolation, the vapor pressure of water when $T = 34°$ C, $68°$ C, $86°$ C, and $91°$ C.
6. Estimate, using piecewise linear interpolation, the sound speed of water when $T = 34°$ C, $68°$ C, $86°$ C, and $91°$ C.

For Exercises 7 through 9, use the values given below for the temperature, T, pressure, p, and density, ρ, of the standard atmosphere as a function of altitude. These data were drawn from Table A.6 in Frank White, *Fluid Mechanics*:

z (m)	0	500	1000	1500	2000	2500	3000
T (K)	288.16	284.91	281.66	278.41	275.16	271.91	268.66
p (Pa)	101,350	95,480	89,889	84,565	79,500	74,684	70,107
ρ (kg/m^3)	1.2255	1.1677	1.1120	1.0583	1.0067	0.9570	0.9092

7. Estimate, using piecewise linear interpolation, the temperature of the standard atmosphere at an altitude of $z = 800$ m, 1600 m, 2350 m, and 2790 m. At what altitude is the temperature of the standard atmosphere 273.1 K?
8. Estimate, using piecewise linear interpolation, the pressure of the standard atmosphere at an altitude of $z = 800$ m, 1600 m, 2350 m, and 2790 m.
9. Estimate, using piecewise linear interpolation, the density of the standard atmosphere at an altitude of $z = 800$ m, 1600 m, 2350 m, and 2790 m. At what altitude is the density of the standard atmosphere 1.1000 kg/m^3?
10. A publisher of mathematics textbooks needs a table of values for the common logarithm function (i.e., the base 10 logarithm) for one of its new precalculus textbooks. Each entry in the table is to be accurate to six (6) decimal places. The table must include entries for uniformly spaced values of x ranging from 1.0 to 10.0, and the increment between x-values must be small enough so that linear interpolation between any two consecutive entries in the table introduces an error of less than 10^{-6}. What is the maximum possible increment that can be used in the construction of this table? What increment would you suggest for the construction of the table?
11. Suppose that a table lists the values of the tangent function for angles ranging from $0.0°$ to $45.0°$ in increments of $0.5°$. What is the largest error that we would introduce by performing linear interpolation between successive values in this table?
12. Suppose that a table lists the values of the inverse sine function for inputs ranging from 0.000 to 0.950 in increments of 0.001. What is the largest error that we would introduce by performing linear interpolation between successive values in this table?

5.6 CUBIC SPLINE INTERPOLATION

Although simple to implement and ideal for hand calculations, a major disadvantage associated with piecewise linear interpolation is that the interpolating function generally will not be differentiable at the interpolating points. In many instances, however, physical considerations will require that the interpolating function be continuously differentiable.

To achieve more smoothness (and greater accuracy) from the interpolating function, higher-degree polynomial pieces must be used. The most common choice is cubic polynomials. These cubic polynomial pieces can be combined in different ways to produce the overall interpolating function. Here, we shall develop the technique of cubic spline interpolation, which obtains the highest degree of smoothness from the piecewise interpolating function. In the next section, we will consider a different approach to piecewise cubic interpolation which utilizes both function and derivative data.

The Cubic Spline Interpolant

Let f be a function defined on the interval $[a, b]$, and let

$$a = x_0 < x_1 < x_2 < \cdots < x_{n-1} < x_n = b$$

be the $n+1$ distinct points at which f is to be interpolated. Recall that the x_i divide $[a, b]$ into n subintervals, referred to as a partition of $[a, b]$.

Definition. A CUBIC SPLINE INTERPOLANT of f relative to the partition

$$a = x_0 < x_1 < x_2 < \cdots < x_{n-1} < x_n = b$$

is a function s that satisfies

(1) on each subinterval $[x_j, x_{j+1}]$, $j = 0, 1, 2, \ldots, n-1$, s coincides with the cubic polynomial

$$s(x) = s_j(x) = a_j + b_j(x - x_j) + c_j(x - x_j)^2 + d_j(x - x_j)^3;$$

(2) s interpolates f at $x_0, x_1, x_2, \ldots, x_n$;

(3) s is continuous on $[a, b]$;

(4) s' is continuous on $[a, b]$;

(5) s'' is continuous on $[a, b]$.

Though this definition clarifies the important characteristics of a cubic spline, it does not provide enough information to completely determine the interpolating function. The function s is composed of n different cubic polynomials, each with four coefficients, so there are a total of $4n$ unknowns. Interpolation provides $n+1$ equations. Continuity of the spline and its first two derivatives contribute an additional $3(n-1) = 3n-3$ equations—remember that continuity applies at the interior points $x_1, x_2, x_3, \ldots, x_{n-1}$ only. The definition of a cubic spline therefore provides

$n + 1 + 3(n - 1) = 4n - 2$ equations. To completely determine the interpolating function, two more equations will have to be specified.

Below, we will discuss two different types of additional constraints, or *boundary conditions*: the *not-a-knot* boundary conditions and the *clamped* (or *complete*) boundary conditions. A third type of boundary condition, the *natural* (of *free*) boundary conditions, will be explored in the exercises. For a review of other possible boundary conditions, see Ueberhuber [1].

Fortunately, even after two more equations have been specified, the full system of $4n$ equations in $4n$ unknowns can be solved very efficiently. To see how this is done, let's start by writing out the equations which follow from the definition.

Interpolation:

$$s_j(x_j) = a_j = f(x_j), \quad j = 0, 1, 2, \ldots, n$$

Continuity of spline:

$$a_{j+1} = a_j + b_j h_j + c_j h_j^2 + d_j h_j^3, \quad j = 0, 1, 2, \ldots, n - 2$$

Continuity of spline derivative:

$$b_{j+1} = b_j + 2c_j h_j + 3d_j h_j^2, \quad j = 0, 1, 2, \ldots, n - 2$$

Continuity of spline second derivative:

$$c_{j+1} = c_j + 3d_j h_j, \quad j = 0, 1, 2, \ldots, n - 2$$

To simplify the equations, we have defined $h_j = x_{j+1} - x_j$. Note that we are using $a_n = f(x_n) = f(b)$, which is a slight extension to the notation introduced in the definition of a cubic spline interpolant. This extension will make it easier to express the equations we are about to develop.

The interpolation conditions directly provide the values for the a_j, thereby removing one-quarter of the unknowns. Next, solve the equation for the continuity of the spline second derivative for d_j:

$$d_j = \frac{c_{j+1} - c_j}{3h_j}. \tag{1}$$

Substituting this expression into the equations for the continuity of the spline and its first derivative gives

$$\begin{aligned} a_{j+1} &= a_j + b_j h_j + c_j h_j^2 + \frac{c_{j+1} - c_j}{3} h_j^2 \\ &= a_j + b_j h_j + \frac{c_{j+1} + 2c_j}{3} h_j^2 \end{aligned} \tag{2}$$

and

$$\begin{aligned} b_{j+1} &= b_j + 2c_j h_j + (c_{j+1} - c_j) h_j \\ &= b_j + (c_{j+1} + c_j) h_j. \end{aligned} \tag{3}$$

Finally, solve equation (2) for b_j:

$$b_j = \frac{a_{j+1} - a_j}{h_j} - \frac{2c_j + c_{j+1}}{3}h_j, \tag{4}$$

and substitute the result into equation (3). After performing some algebraic manipulation and shifting the subscripts down by one, we arrive at

$$h_{j-1}c_{j-1} + 2(h_{j-1} + h_j)c_j + h_jc_{j+1} = \frac{3}{h_j}(a_{j+1} - a_j) - \frac{3}{h_{j-1}}(a_j - a_{j-1}). \tag{5}$$

This equation holds for $j = 1, 2, 3, \ldots, n-1$ and forms the basis for a tridiagonal system of equations for determining the c_j. The equations for $j = 0$ and $j = n$ depend on the type of boundary conditions which are being applied.

Regardless of the choice of boundary conditions, computing the coefficients of a cubic spline interpolant is a two-step process. First, the linear system for the c_j must be solved. Recall that an efficient algorithm for solving tridiagonal systems was developed in Chapter 3 and is listed in Appendix B for convenience. Once this has been done, equation (1) is used to compute the d_j and equation (4) is used to compute the b_j. Remember that the a_j are given by the function values, $f(x_j)$. Evaluation of a cubic spline interpolant, as with any piecewise function, is also a two-step process. Based on the value of the independent variable, the polynomial piece which needs to be evaluated must first be determined. Once the polynomial piece has been selected, the value of the interpolant at the given value of the independent variable can then be calculated.

Not-a-Knot Boundary Conditions

When no information other than the value of f at each interpolating point is available, it is recommended that the *not-a-knot* boundary conditions be applied. These conditions require that s''' be continuous at $x = x_1$ and $x = x_{n-1}$. In terms of the spline coefficients, this translates to

$$d_0 = d_1 \quad \text{and} \quad d_{n-2} = d_{n-1}.$$

Using (1), and rearranging terms, these equations can be expressed in terms of the c_j as

$$h_1c_0 - (h_0 + h_1)c_1 + h_0c_2 = 0 \tag{6}$$

$$h_{n-1}c_{n-2} - (h_{n-2} + h_{n-1})c_{n-1} + h_{n-2}c_n = 0. \tag{7}$$

Unfortunately, (6) and (7) do not preserve the tridiagonal structure of (5). This situation, however, can be remedied as follows. Solve equation (6) for c_0 and equation (7) for c_n. This gives

$$c_0 = \left(1 + \frac{h_0}{h_1}\right)c_1 - \frac{h_0}{h_1}c_2 \tag{8}$$

$$c_n = -\frac{h_{n-1}}{h_{n-2}}c_{n-2} + \left(1 + \frac{h_{n-1}}{h_{n-2}}\right)c_{n-1}. \tag{9}$$

Now, substitute c_0 from (8) into (5), for $j = 1$, and group terms to obtain

$$\left(3h_0 + 2h_1 + \frac{h_0^2}{h_1}\right) c_1 + \left(h_1 - \frac{h_0^2}{h_1}\right) c_2 = \frac{3}{h_1}(a_2 - a_1) - \frac{3}{h_0}(a_1 - a_0). \tag{10}$$

Proceed in a similar manner with the expression for c_n from (9) substituted into (5) for $j = n - 1$ to produce

$$\left(h_{n-2} - \frac{h_{n-1}^2}{h_{n-2}}\right) c_{n-2} + \left(3h_{n-1} + 2h_{n-2} + \frac{h_{n-1}^2}{h_{n-2}}\right) c_{n-1}$$
$$= \frac{3}{h_{n-1}}(a_n - a_{n-1}) - \frac{3}{h_{n-2}}(a_{n-1} - a_{n-2}). \tag{11}$$

Equation (5), for $j = 2, 3, 4, \ldots, n-2$, together with equations (10) and (11) constitute a complete tridiagonal system for the coefficients c_1, c_2, c_3, ..., c_{n-1}. Since each h_j is positive by definition, the coefficient matrix for this linear system is strictly diagonally dominant. Hence, there is always a unique solution for the c_j.

EXAMPLE 5.16 Emittance of Tungsten as a Function of Temperature

The table below gives experimental values for the emittance of tungsten as a function of temperature.

Temperature (K)	Emittance	Temperature (K)	Emittance
300	0.024	800	0.083
400	0.035	900	0.097
500	0.046	1000	0.111
600	0.058	1100	0.125
700	0.067		

This problem was previously considered in Section 5.1, where it was shown that the behavior of the eighth-degree interpolating polynomial was not consistent with the data, in particular at the ends of the domain. Here, we develop the not-a-knot cubic spline interpolant (the cubic spline with not-a-knot boundary conditions) for the data. With $h_j = 100$ for each j, the system of equations for the c_j $(1 \le j \le 7)$ takes the form

$$\begin{bmatrix} 600 & 0 & & & & & \\ 100 & 400 & 100 & & & & \\ & 100 & 400 & 100 & & & \\ & & 100 & 400 & 100 & & \\ & & & 100 & 400 & 100 & \\ & & & & 100 & 400 & 100 \\ & & & & & 0 & 600 \end{bmatrix} \begin{bmatrix} c_1 \\ c_2 \\ c_3 \\ c_4 \\ c_5 \\ c_6 \\ c_7 \end{bmatrix} = 0.03 \begin{bmatrix} 0 \\ 0.001 \\ -0.003 \\ 0.007 \\ -0.002 \\ 0 \\ 0 \end{bmatrix}.$$

Solving this linear system and then applying equations (8) and (9) to compute c_0 and c_8, followed by the application of equations (1) and (4), we obtain the complete set of spline coefficients:

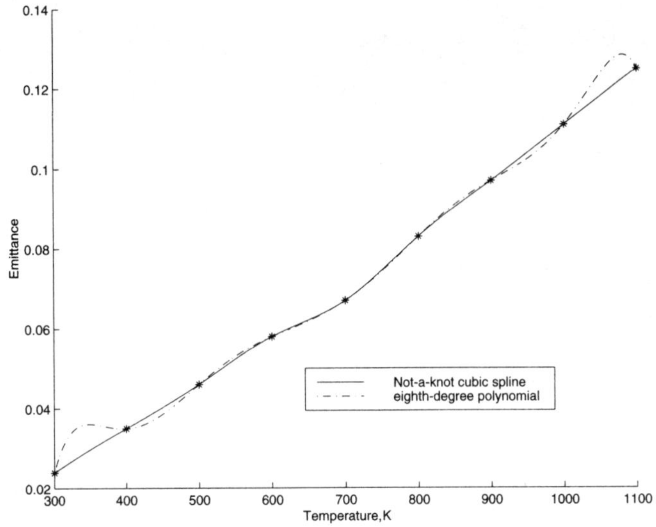

Figure 5.14 Comparison of not-a-knot cubic spline and eight-degree interpolating polynomial for emittance of tungsten as a function of temperature. Data values are indicated by asterisks.

a_j	b_j	c_j	d_j
0.024	0.00012256410	-0.00000018846	0.00000000063
0.035	0.00010371795	0	0.00000000063
0.046	0.00012256410	0.00000018846	-0.00000000214
0.058	0.00009602564	-0.00000045385	0.00000000394
0.067	0.00012333333	0.00000072692	-0.00000000360
0.083	0.00016064103	-0.00000035385	0.00000000147
0.097	0.00013410256	0.00000008846	-0.00000000029
0.111	0.00014294872	0	-0.00000000029

The graph of the not-a-knot cubic spline is given in Figure 5.14. The graph of the eighth-degree interpolating polynomial is also shown for comparison. Clearly, the behavior of the not-a-knot cubic spline is more consistent with the underlying data.

EXAMPLE 5.17 Mean Activity Coefficient of Silver Nitrate

The mean activity coefficient at 25°C for silver nitrate, as a function of molality, is given in the table below. Estimate the mean activity coefficient for a molality of 0.032 and for a molality of 1.682.

Molality	0.005	0.010	0.020	0.050	0.100	0.200	0.500	1.000	2.000
Coefficient	0.924	0.896	0.859	0.794	0.732	0.656	0.536	0.430	0.316

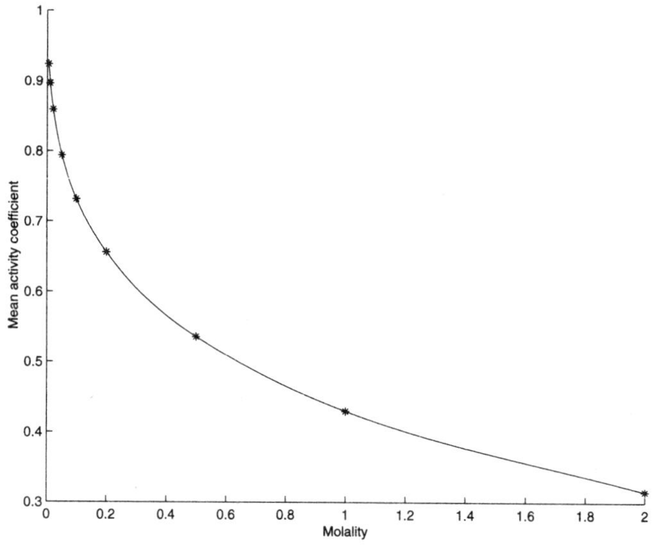

Figure 5.15 Not-a-knot cubic spline for mean activity coefficient of silver nitrate as a function of molality. Data points are indicated by asterisks.

Using an eight-degree interpolating polynomial, the estimates for the mean activity coefficient are 0.831 for a molality of 0.032 and -216711.827 for a molality of 1.682. The former value is reasonable, while the latter clearly is not.

The not-a-knot cubic spline computed from this data is shown in Figure 5.15. Evaluating the spline at a molality of 0.032 gives an estimate for the mean activity coefficient of 0.828. For a molality of 1.682, the estimate for the mean activity coefficient is 0.349.

Clamped Boundary Conditions

If the values $f'(a)$ and $f'(b)$ are known, then it is better to apply the *clamped* (or *complete*) boundary conditions $s'(a) = f'(a)$ and $s'(b) = f'(b)$. Starting with $x = a$, we find $f'(a) = s'(a) = s_0'(a) = b_0$. Equation (4) with $j = 0$ allows us to write this condition in terms of the c_j:

$$f'(a) = \frac{a_1 - a_0}{h_0} - \frac{2c_0 + c_1}{3}h_0,$$

or

$$2h_0c_0 + h_0c_1 = \frac{3}{h_0}(a_1 - a_0) - 3f'(a). \tag{12}$$

At $x = b$, $f'(b) = s'(b) = s_n'(b) = b_n$. Using equation (3) to express b_n in terms of b_{n-1}, c_{n-1} and c_n, followed by equation (4) to rewrite b_{n-1} in terms of a_{n-1}, a_n,

c_{n-1} and c_n, we obtain the equation

$$h_{n-1}c_{n-1} + 2h_{n-1}c_n = 3f'(b) - \frac{3}{h_{n-1}}(a_n - a_{n-1}). \tag{13}$$

Combining equation (5) for $j = 1, 2, 3, \ldots, n-1$ with equations (12) and (13) produces a complete tridiagonal linear system for determining the c_j. As with not-a-knot boundary conditions, the coefficient matrix associated with clamped boundary conditions is strictly diagonally dominant; hence, there is again always a unique solution for the c_j.

EXAMPLE 5.18 A Clamped Cubic Spline

Let's determine the clamped cubic spline for the following data:

x	$f(x)$	$f'(x)$
−1.0	0.00000	2.71828
−0.5	0.82436	
0.0	1.00000	
0.5	0.90980	
1.0	0.73576	−0.36788

Note that this data is taken from the function $f(x) = (x+1)e^{-x}$. Since $h_j = 0.5$ for each j, the tridiagonal system for the c_j takes the form

$$\begin{bmatrix} 1 & 0.5 & & & \\ 0.5 & 2 & 0.5 & & \\ & 0.5 & 2 & 0.5 & \\ & & 0.5 & 2 & 0.5 \\ & & & 0.5 & 1 \end{bmatrix} \begin{bmatrix} c_0 \\ c_1 \\ c_2 \\ c_3 \\ c_4 \end{bmatrix} = \begin{bmatrix} -3.20868 \\ -3.89232 \\ -1.59504 \\ -0.50304 \\ -0.05940 \end{bmatrix}.$$

Solving this system and then applying equation (1) to compute the d_j and equation (4) to compute the b_j, we obtain the complete set of spline coefficients:

a_j	b_j	c_j	d_j
0.00000	2.71828000000	-2.62214571429	0.96605142857
0.82436	0.82067285714	-1.17306857143	0.46856571429
1.00000	-0.00097142857	-0.47022000000	0.22272571429
0.90980	-0.30414714286	-0.13613142857	0.09653142857

The graph of the clamped cubic spline is shown at the top of Figure 5.16. To the resolution of the plotting device, the graph of the cubic spline is indistinguishable from the graph of f. In fact, $\|f - s\|_\infty \approx 0.0015$.

The error in the clamped cubic spline, as a function of x, is shown in the bottom graph of Figure 5.16. For comparison, the error in the not-a-knot cubic spline is also displayed. Clearly, when the derivative values at the ends of the domain are available, the clamped cubic spline is superior to the not-a-knot cubic spline.

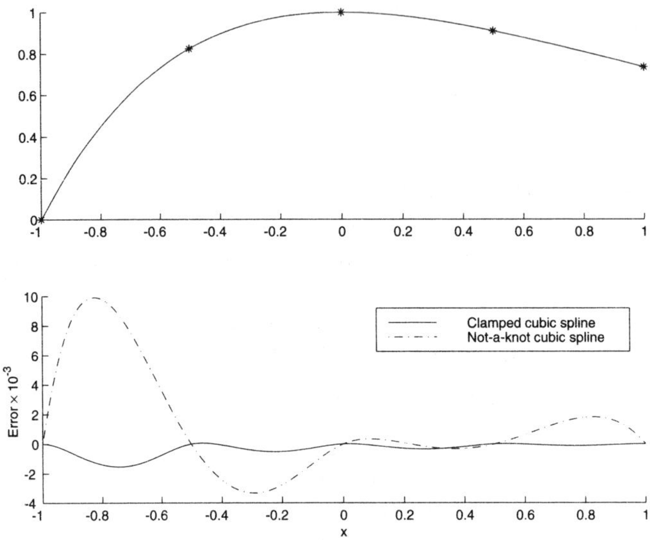

Figure 5.16 (Top graph) Clamped cubic spline for data taken from $f(x) = (x+1)e^{-x}$. Data points are indicated by asterisks. (Bottom graph) Comparison of error in clamped cubic spline and not-a-knot cubic spline.

The clamped cubic spline satisfies an interesting property related to the curvature of a function. For a general function, the curvature at a point is defined by

$$\kappa(x) = \frac{|f''(x)|}{(1+[f'(x)]^2)^{3/2}},$$

which is commonly linearized to $\kappa(x) \approx |f''(x)|$. The quantity $\int_a^b [f''(x)]^2 dx$ can therefore be viewed as a crude measure of the total curvature over an interval. We will now prove that, in this measure, any smooth interpolating function which satisfies clamped boundary conditions must have a total curvature at least as large as that of the clamped cubic spline. This is sometimes referred to as the *minimum curvature property* of the clamped cubic spline.

Theorem. Let g be any function, continuous and twice continuously differentiable on the interval $[a, b]$, that interpolates f over the partition

$$a = x_0 < x_1 < x_2 < \cdots < x_{n-1} < x_n = b$$

and satisfies the clamped boundary conditions $g'(a) = f'(a)$ and $g'(b) = f'(b)$. Then

$$\int_a^b [s''(x)]^2\, dx \le \int_a^b [g''(x)]^2\, dx,$$

where s is the clamped cubic spline.

Proof. First, observe that

$$\begin{aligned}
\int_a^b [g''(x)]^2\,dx &= \int_a^b [g''(x) - s''(x) + s''(x)]^2\,dx \\
&= \int_a^b [g''(x) - s''(x)]^2\,dx + 2\int_a^b s''(x)[g''(x) - s''(x)]\,dx \\
&\quad + \int_a^b [s''(x)]^2\,dx.
\end{aligned}$$

Next, focus on the term

$$\int_a^b s''(x)[g''(x) - s''(x)]\,dx = \sum_{i=0}^{n-1} \int_{x_i}^{x_{i+1}} s''(x)[g''(x) - s''(x)]\,dx.$$

After integrating by parts twice, it follows that

$$\begin{aligned}
\int_{x_i}^{x_{i+1}} s''(x)[g''(x) - s''(x)]\,dx &= \{s''(x)[g'(x) - s'(x)] - s'''(x)[g(x) - s(x)]\}\big|_{x_i}^{x_{i+1}} \\
&\quad + \int_{x_i}^{x_{i+1}} s^{(4)}(x)[g(x) - s(x)]\,dx.
\end{aligned}$$

Since s is a cubic polynomial on $[x_i, x_{i+1}]$, $s^{(4)}(x) \equiv 0$. Furthermore, since both s and g interpolate f at each x_i,

$$\{s'''(x)[g(x) - s(x)]\}\big|_{x_i}^{x_{i+1}} = 0.$$

Therefore,

$$\begin{aligned}
\int_a^b s''(x)[g''(x) - s''(x)]\,dx &= \sum_{i=0}^{n-1} s''(x)[g'(x) - s'(x)]\big|_{x_i}^{x_{i+1}} \\
&= s''(x)[g'(x) - s'(x)]\big|_b - s''(x)[g'(x) - s'(x)]\big|_a \\
&= 0,
\end{aligned}$$

due to the clamped boundary conditions satisfied by both s and g. Thus

$$\int_a^b [g''(x)]^2 dx = \int_a^b [g''(x) - s''(x)]^2 dx + \int_a^b [s''(x)]^2 dx \ge \int_a^b [s''(x)]^2 dx,$$

since the integral of a non-negative function is always nonnegative. □

Error in Cubic Spline Interpolation

We conclude the discussion of cubic spline interpolation with a theorem on the error associated with the clamped cubic spline. For a proof of this result, see de Boor [2], Hall and Meyer [3], or Schultz [4]. An error bound for the not-a-knot cubic spline, also of fourth order, can be found in de Boor [5] or Beatson [6].

Theorem. Let f be continuous, with four continuous derivatives, on the interval $[a, b]$, and let s be the clamped cubic spline interpolant of f relative to the partition

$$a = x_0 < x_1 < x_2 < \cdots < x_{n-1} < x_n = b.$$

Then

$$\max_{x\in[a,b]} |f(x) - s(x)| \le \frac{5}{384} h^4 \max_{x\in[a,b]} |f^{(4)}(x)|,$$

where $h = \max_{0\le i\le n-1}(x_{i+1} - x_i)$.

References

1. C. Ueberhuber, *Numerical Computation 1: Methods, Software and Analysis*, Springer-Verlag, Berlin, 1997.
2. C. de Boor, *A Practical Guide to Splines*, Springer-Verlag, New York, 1978.
3. C. Hall and W. Meyer, "Optimal Error Bounds for Cubic Spline Interpolation," *Journal of Approximation Theory,* **16**, pp. 105–122, 1976.
4. M. H. Schultz, *Spline Analysis*, Prentice-Hall, Englewood Cliffs, NJ, 1973.
5. C. de Boor, "Convergence of Cubic Spline Interpolation with the Not-a-Knot Condition," Mathematics Research Center Preprint, University of Wisconsin, Madison, 1984.
6. R. K. Beatson, "On the Convergence of Some Cubic Spline Interpolation Schemes," *SIAM Journal on Numerical Analysis,* **23**, 903–912, 1986.
7. G. Birkhoff and C. de Boor, "Error Bounds for Spline Interpolation," *Journal of Mathematics and Mechanics,* **13**, 827–836, 1964.

EXERCISES

For Exercises 1 through 3, use the values given below for the temperature, T, pressure, p, and density, ρ, of the standard atmosphere as a function of altitude. This data was drawn from Table A.6 in Frank White, *Fluid Mechanics*:

z (m)	0	500	1000	1500	2000	2500	3000
T (K)	288.16	284.91	281.66	278.41	275.16	271.91	268.66
p (Pa)	101,350	95,480	89,889	84,565	79,500	74,684	70,107
ρ (kg/m^3)	1.2255	1.1677	1.1120	1.0583	1.0067	0.9570	0.9092

1. Using the not-a-knot cubic spline interpolant, estimate the temperature of the standard atmosphere at an altitude of z = 800 m, 1600 m, 2350 m, and 2790 m. At what altitude is the temperature of the standard atmosphere 273.1 K?

2. Using the not-a-knot cubic spline interpolant, estimate the pressure of the standard atmosphere at an altitude of z = 800 m, 1600 m, 2350 m, and 2790 m.

3. Using the not-a-knot cubic spline interpolant, estimate the density of the standard atmosphere at an altitude of z = 800 m, 1600 m, 2350 m, and 2790 m. At what altitude is the density of the standard atmosphere 1.1000 kg/m^3?

Exercises 4 through 9 are based on the following data for the density, ρ, viscosity, μ, kinematic viscosity, ν, surface tension, Υ, vapor pressure, p_v, and sound speed, a, of water as a function of temperature. This data was drawn from Tables A.1 and A.5 in Frank White, *Fluid Mechanics*:

T (°C)	ρ (kg/m^3)	μ ($\times 10^{-3}$ N·s/m^2)	ν ($\times 10^{-5}$ m^2/s)	Υ (N/m)	p_v (kPa)	a (m/s)
0	1000	1.788	1.788	0.0756	0.611	1402
10	1000	1.307	1.307	0.0742	1.227	1447
20	998	1.003	1.005	0.0728	2.337	1482
30	996	0.799	0.802	0.0712	4.242	1509
40	992	0.657	0.662	0.0696	7.375	1529
50	988	0.548	0.555	0.0679	12.34	1542
60	983	0.467	0.475	0.0662	19.92	1551
70	978	0.405	0.414	0.0644	31.16	1553
80	972	0.355	0.365	0.0626	47.35	1554
90	965	0.316	0.327	0.0608	70.11	1550
100	958	0.283	0.295	0.0589	101.3	1543

4. Using the not-a-knot cubic spline interpolant, estimate the density of water when $T = 34°$ C, $68°$ C, $86°$ C, and $91°$ C.

5. Using the not-a-knot cubic spline interpolant, estimate the viscosity of water when $T = 34°$ C, $68°$ C, $86°$ C, and $91°$ C. At what temperature is the viscosity 1.000×10^{-3} N·s/m^2?

6. Using the not-a-knot cubic spline interpolant, estimate the kinematic viscosity of water when $T = 34°$ C, $68°$ C, $86°$ C, and $91°$ C. At what temperature is the kinematic viscosity 1.000×10^{-5} m^2/s?

7. Using the not-a-knot cubic spline interpolant, estimate the surface tension of water when $T = 34°$ C, $68°$ C, $86°$ C, and $91°$ C. At what temperature is the surface tension 0.0650 N/m?

8. Using the not-a-knot cubic spline interpolant, estimate the vapor pressure of water when $T = 34°$ C, $68°$ C, $86°$ C, and $91°$ C.

9. Using the not-a-knot cubic spline interpolant, estimate the sound speed of water when $T = 34°$ C, $68°$ C, $86°$ C, and $91°$ C.

10. Consider the following data set:

x	0.0	0.5	1.0	1.5	2.0
y	0.500000	1.425639	2.640859	4.009155	5.305472
y'	1.500000				2.305472

(a) Construct the not-a-knot cubic spline for this data set.

(b) Construct the clamped cubic spline for this data set.

(c) The data for this problem is taken from the function $y = (x+1)^2 - 0.5e^x$. Plot the error in each of the splines from parts (a) and (b) as a function of x. Which spline produced the better results?

11. Repeat Exercise 10 for the data set

x	1.0	1.5	2.0	2.5	3.0
y	0.000000	0.608198	1.386294	2.290727	3.295837
y'	1.000000				2.098612

which is taken from the function $f(x) = x \ln x$.

12. Repeat Exercise 10 for the data set

x	0.00	0.25	0.50	0.75	1.00
y	0.000000	0.176777	0.500000	0.530330	0.000000
y'	0.000000				−3.141593

which is taken from the function $f(x) = x \sin(\pi x)$.

13. Experimentally determined values for the partial pressure of water vapor, p_A, as a function of distance, y, from the surface of a pan of water are given below. The derivative of the partial pressure with respect to distance is estimated to be −0.0455 atm/mm when $y = 0$ and 0 atm/mm when $y = 5$. Estimate the partial pressure at distances of 0.5 mm, 2.1 mm and 3.7 mm from the surface of the water using a clamped cubic spline.

y (mm)	0	1	2	3	4	5
p_A (atm)	0.100	0.065	0.042	0.029	0.022	0.020

Natural Boundary Conditions

Another set of boundary conditions that can be used when no other information is available about f is the *natural* (or *free*) boundary conditions $s''(a) = s''(b) = 0$. Since $s''(a) = s_0''(a) = c_0$ and $s''(b) = s_n''(b) = c_n$, the natural boundary conditions immediately translate to

$$c_0 = 0 \quad \text{and} \quad c_n = 0.$$

Combining these two equations with equation (5) for $j = 1, 2, 3, \ldots, n-1$ provides a complete linear system for determining the c_j. The coefficient matrix for this system is tridiagonal and strictly diagonally dominant. If $f''(a) = f''(b) = 0$, the natural cubic spline has a fourth-order error bound (see Birkhoff and de Boor [7]); otherwise, the natural cubic spline produces errors that are only second-order near the boundaries (see de Boor [2]). Exercises 14–19 deal with the natural cubic spline.

14. Determine the natural cubic spline for the data in the example "A Clamped Cubic Spline." Compare the error in the natural cubic spline to that of the not-a-knot cubic spline.

15. Determine the natural cubic spline for the data in Exercise 10. Compare the error in the natural cubic spline to that of the not-a-knot cubic spline.

16. Determine the natural cubic spline for the data in Exercise 11. Compare the error in the natural cubic spline to that of the not-a-knot cubic spline.

17. Determine the natural cubic spline for the data in Exercise 12. Compare the error in the natural cubic spline to that of the not-a-knot cubic spline.

18. Determine the natural cubic spline for the following data sets. In each case, compare the natural cubic spline with the not-a-knot cubic spline.
(a) viscosity of water (Exercise 5)
(b) vapor pressure of water (Exercise 8)
(c) sound speed of water (Exercise 9)
(d) pressure of the standard atmosphere (Exercise 2)
(e) density of the standard atmosphere (Exercise 3)

19. Show that the natural cubic spline satisfies the following minimum curvature property: Let g be any function, continuous and twice continuously differentiable on the interval $[a, b]$, which interpolates f over the partition

$$a = x_0 < x_1 < x_2 < \cdots < x_{n-1} < x_n = b.$$

Then

$$\int_a^b [s''(x)]^2\,dx \le \int_a^b [g''(x)]^2\,dx,$$

where s is the natural cubic spline.

5.7 HERMITE AND HERMITE CUBIC INTERPOLATION

To this point, all interpolating data has consisted of function values only. In this section, derivative information will be incorporated into the interpolating polynomial. When derivative values are included in the construction of the interpolating polynomial, the graph of the polynomial will not just intersect the graph of the function being interpolated, but will touch, or "kiss," it. Because the word osculate is a synonym for kiss, interpolation with derivative values is known as *osculatory interpolation.*

Rather than develop general osculatory interpolation, which is of limited use in practice, we will focus on the most important special case, that of Hermite interpolation. In the first half of this section, the Hermite interpolant will be defined, the computation of the Newton form will be described and the basic existence and uniqueness and error theories will be developed. Hermite cubic interpolation, which is another form of piecewise cubic interpolation, will then be presented in the latter half of the section.

Hermite Interpolation

Let $x_0, x_1, x_2, \ldots, x_n$ be $n+1$ distinct points at which the function f and its first derivative are defined. In Hermite interpolation, the function value, $f(x_i)$, and the value of the first derivative, $f'(x_i)$, are known at each interpolating point. Since there are a total of $2n+2$ data values, the objective is to determine a polynomial, P, of degree at most $2n+1$, that satisfies

$$P(x_i) = f(x_i) \quad \text{and} \quad P'(x_i) = f'(x_i)$$

for each $i = 0, 1, 2, \ldots, n$. The next theorem justifies referring to this function as the Hermite interpolating polynomial.

Theorem. Let x_0, x_1, x_2, …, x_n be $n+1$ distinct points on the interval $[a, b]$ and let the function f and its first derivative be defined at each of these points. Then there exists a unique polynomial, P, of degree at most $2n+1$ such that

$$P(x_i) = f(x_i) \quad \text{and} \quad P'(x_i) = f'(x_i)$$

for each $i = 0, 1, 2, \ldots, n$.

Proof. To establish existence, we will construct the Lagrange form of the Hermite interpolating polynomial. Let $L_{n,i}$ denote the Lagrange polynomials developed in Section 5.1, and define

$$H_i(x) = [1 - 2L'_{n,i}(x_i)(x - x_i)]L^2_{n,i}(x)$$
$$\hat{H}_i(x) = (x - x_i)L^2_{n,i}(x).$$

Note that each H_i and $\hat{H}_i$ is a polynomial of degree $2n+1$. Furthermore, it is straightforward to show that (see Exercise 1)

$$H_i(x_j) = \begin{cases} 1, & i = j \\ 0, & \text{otherwise} \end{cases} \qquad \hat{H}_i(x_j) = 0$$

$$H'_i(x_j) = 0 \qquad \hat{H}'_i(x_j) = \begin{cases} 1, & i = j \\ 0, & \text{otherwise} \end{cases}.$$

Hence, H_i is associated with the function value at $x = x_i$, and $\hat{H}_i$ is associated with the derivative value. Now consider the polynomial

$$P(x) = \sum_{i=0}^{n} H_i(x)f(x_i) + \sum_{i=0}^{n} \hat{H}_i(x)f'(x_i).$$

Using the properties of H_i and $\hat{H}_i$, it follows that

$$P(x_j) = \sum_{i=0}^{n} H_i(x_j)f(x_i) + \sum_{i=0}^{n} \hat{H}_i(x_j)f'(x_i) = f(x_j)$$

and

$$P'(x_j) = \sum_{i=0}^{n} H'_i(x_j)f(x_i) + \sum_{i=0}^{n} \hat{H}'_i(x_j)f'(x_i) = f'(x_j)$$

for each $j = 0, 1, 2, \ldots, n$. P therefore interpolates all of the function and all of the derivative values.

To establish uniqueness of the Hermite interpolating polynomial, suppose that Q is a polynomial of degree at most $2n+1$ that interpolates the function and derivative values of f at each x_i with $P \neq Q$. Let $R = P - Q$. Then

$$R(x_i) = P(x_i) - Q(x_i) = f(x_i) - f(x_i) = 0,$$
$$R'(x_i) = P'(x_i) - Q'(x_i) = f'(x_i) - f'(x_i) = 0$$

for each $i = 0, 1, 2, \ldots, n$. This implies that each x_i is a root of R of multiplicity at least 2. Therefore, R is a polynomial of degree at most $2n+1$ with at least $2n+2$ roots. The Fundamental Theorem of Algebra then guarantees that $R \equiv 0$, or $P = Q$. Hence, the Hermite interpolating polynomial is unique. □

Though the Lagrange form of the Hermite interpolating polynomial,

$$P(x) = \sum_{i=0}^{n} H_i(x) f(x_i) + \sum_{i=0}^{n} \hat{H}_i(x) f'(x_i),$$

is useful for theoretical purposes, for practical computations, it is better to develop the Newton form. To do this, first construct the sequence, z_j, of length $2n+2$ according to the rule

$$z_j = x_{(j \text{ div } 2)},$$

where **div** denotes integer division. Next, compute a divided difference table based on the sequence of z values and the corresponding values of the function f. Every other entry in the column of first divided differences will be of the form $f[x_i, x_i]$ for some i. To handle these entries, recall from Section 5.3 that when f is differentiable on $[a, b]$ and $x_i, x_j \in [a, b]$ with $x_i \neq x_j$, there exists ξ between x_i and x_j such that

$$f[x_i, x_j] = f'(\xi).$$

Letting $x_j \to x_i$, it follows from the squeeze theorem that $\xi \to x_i$. We will therefore define $f[x_i, x_i] = f'(x_i)$. This is where the given values of the first derivative enter into the divided difference table. The remaining entries in the table are computed as usual. Once the table has been completed, the Newton form of the Hermite interpolating polynomial is given by

$$P(x) = \sum_{k=0}^{2n+1} f[z_0, z_1, z_2, \ldots, z_k] \left(\prod_{i=0}^{k-1} (x - z_i) \right).$$

For a proof of this result, see Powell [1].

EXAMPLE 5.19 Constructing the Newton Form of the Hermite Interpolating Polynomial

Consider the function $f(x) = xe^{-x}$. We will construct the Newton form of the Hermite interpolating polynomial for this function using the data in the table below.

x_i	$f(x_i)$	$f'(x_i)$
0	0	1
2	$2e^{-2}$	$-e^{-2}$
4	$4e^{-4}$	$-3e^{-4}$

The divided difference table for this data is shown in Figure 5.17(a). Note the repetition of the interpolating points in the first column, the corresponding repetition of the function values in the second column and the placement of the derivative values in the third column. All other entries in the table were computed using the standard divided difference formula. Using the values from the top of each column in the divided difference table, we arrive at the Hermite interpolating polynomial

$$P(x) = x + \frac{e^{-2}-1}{2}x^2 - \frac{3e^{-2}-1}{4}x^2(x-2) + \frac{e^{-4}+4e^{-2}-1}{16}x^2(x-2)^2 - \frac{9e^{-4}+4e^{-2}-1}{64}x^2(x-2)^2(x-4).$$

A plot of this polynomial and the function $f(x) = xe^{-x}$ is shown in Figure 5.17(b). The locations of the interpolating points are indicated by the circles. Note the way the Hermite interpolating polynomial matches both the height and the slope of the interpolated function at each circle.

EXAMPLE 5.20 Data Analysis for the Spread of an Epidemic

Suppose that a mathematical model for the spread of an epidemic produces the following estimates for the number of people who have died as a result of the epidemic, $D(t)$, and the rate at which people are dying, $D'(t)$. Here, time is measured in weeks. Using this data, we wish to generate a table which shows the number of dead at half-week increments.

t	$D(t)$	$D'(t)$
0.000000	0.000000	600.000000
0.750000	445.903683	573.579644
1.500000	842.695315	477.074216
2.085600	1095.211197	384.947629
2.676193	1295.955674	296.576145
3.219694	1437.602773	226.796410
3.748513	1542.363644	171.475176
4.279179	1621.280769	127.808738
4.821254	1680.890649	93.728061
5.000000	1696.803710	84.473801

The graph of the Hermite interpolating polynomial constructed from this data is shown in Figure 5.18. Evaluating the polynomial at t = 0, 0.5, 1.0, 1.5, 2.0, 2.5, 3.0, 3.5, 4.0, 4.5, and 5.0 produces the desired values of $D(t)$:

(a)

z	First	Second	Third	Fourth	Fifth	Sixth
0	$f(0)=0$					
		$f'(0)=1$				
0	$f(0)=0$		$(e^{-2}-1)/2$			
		e^{-2}		$(-3e^{-2}+1)/4$		
2	$f(2)=2e^{-2}$		$-e^{-2}$		$(e^{-4}+4e^{-2}-1)/16$	
		$f'(2)=-e^{-2}$		$(e^{-4}+e^{-2})/4$		$(-9e^{-4}-4e^{-2}+1)/64$
2	$f(2)=2e^{-2}$		e^{-4}		$-e^{-4}/2$	
		$2e^{-4}-e^{-2}$		$(-7e^{-4}+e^{-2})/4$		
4	$f(4)=4e^{-4}$		$(-5e^{-4}+e^{-2})/2$			
		$f'(4)=-3e^{-4}$				
4	$f(4)=4e^{-4}$					

(b)

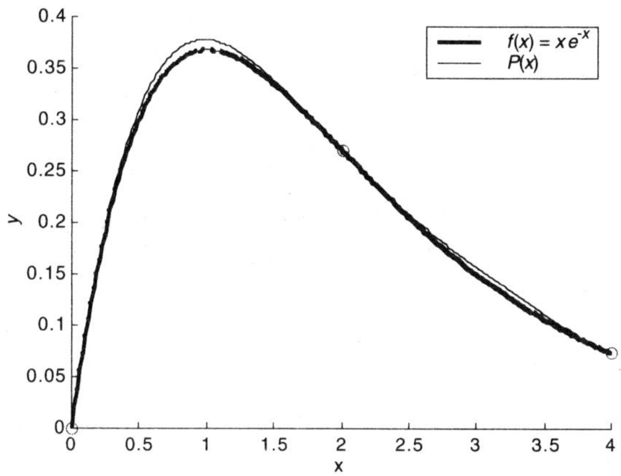

Figure 5.17 Hermite interpolation for the function $f(x) = xe^{-x}$ at $x = 0$, $x = 2$, and $x = 4$. (a) Divided difference table; (b) plot of interpolating polynomial, $P(x)$, and $f(x)$.

t_i	$D(t_i)$
0.0	0.000000
0.5	299.782776
1.0	586.071099
1.5	842.695315
2.0	1061.682772
2.5	1241.495466
3.0	1384.888210
3.5	1496.766805
4.0	1582.655893
4.5	1647.810593
5.0	1696.803710

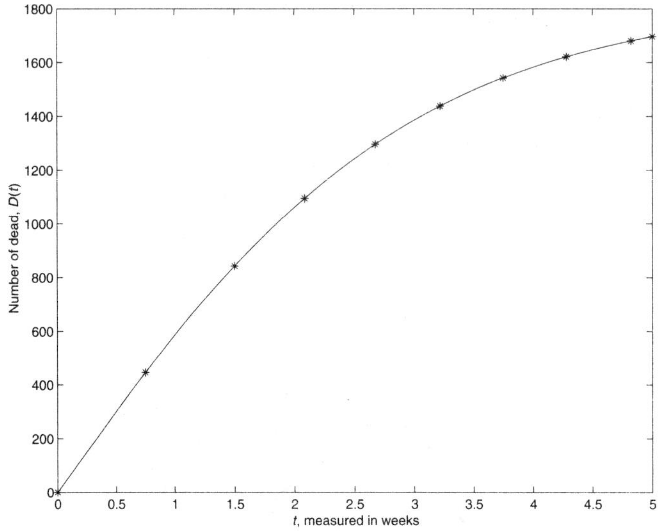

Figure 5.18 Graph of Hermite interpolating polynomial generated from the data for the spread of an epidemic. Data values are indicated by the asterisks.

The last issue to address with regard to Hermite interpolation is interpolation error. If we replace the auxiliary function, g, in the proof of the error theorem in Section 5.1 by

$$g(t) = f(t) - P(t) - [f(x) - P(x)] \prod_{i=0}^{n} \frac{(t - x_i)^2}{(x - x_i)^2},$$

where P is interpreted as the Hermite interpolating polynomial, then we find

Theorem. Let f be continuously differentiable $2n+2$ times on $[a, b]$, and let $x_0, x_1, x_2, \ldots, x_n$ be $n+1$ distinct points from $[a, b]$. Then for each $x \in [a, b]$, there exists a $\xi \in [a, b]$ such that

$$f(x) = P(x) + \frac{f^{(2n+2)}(\xi)}{(2n+2)!} \prod_{i=0}^{n} (x - x_i)^2,$$

where P is the Hermite interpolating polynomial.

Hermite Cubic Interpolation

An alternative scheme for combining cubic polynomial pieces into an interpolating function produces the Hermite cubic interpolant.

Definition. The HERMITE CUBIC INTERPOLANT of f relative to the partition

$$a = x_0 < x_1 < x_2 < \cdots < x_{n-1} < x_n = b$$

is a function s that satisfies

(1) on each subinterval $[x_j, x_{j+1}]$, $j = 0, 1, 2, \ldots, n-1$, s coincides with a cubic polynomial $s_j(x)$;
(2) s interpolates f and f' at $x_0, x_1, x_2, \ldots, x_n$;
(3) s is continuous on $[a, b]$;
(4) s' is continuous on $[a, b]$.

Although the Hermite cubic interpolant has less smoothness than the cubic spline, determining the polynomial pieces for the Hermite cubic requires less work. Focus on the function $s_j(x)$. Interpolation of f and its first derivative at $x = x_j$ requires

$$s_j(x_j) = f(x_j) \quad \text{and} \quad s_j'(x_j) = f'(x_j).$$

Combining the continuity of s and s' at $x = x_{j+1}$ with the interpolation of f and its first derivative at $x = x_{j+1}$ then requires

$$s_j(x_{j+1}) = s_{j+1}(x_{j+1}) = f(x_{j+1})$$

$$s_j'(x_{j+1}) = s_{j+1}'(x_{j+1}) = f'(x_{j+1}).$$

Hence, $s_j(x)$ is a third-degree polynomial that interpolates both f and f' at $x = x_j$ and at $x = x_{j+1}$. However, earlier in this section we established that the Hermite interpolating polynomial is the unique polynomial of degree at most three that interpolates f and f' at $x = x_j$ and at $x = x_{j+1}$. $s_j(x)$ must therefore be the Hermite interpolating polynomial.

Having established that $s_j(x)$ is the Hermite interpolating polynomial, we can immediately write down that

$$s_j(x) = H_{1,j}(x)f(x_j) + H_{1,j+1}(x)f(x_{j+1}) + \hat{H}_{1,j}(x)f'(x_j) + \hat{H}_{1,j+1}(x)f'(x_{j+1}), \quad (1)$$

where

$$H_{1,j}(x) = \left[1 - 2\frac{x - x_j}{x_j - x_{j+1}}\right]\left(\frac{x - x_{j+1}}{x_j - x_{j+1}}\right)^2,$$

$$H_{1,j+1}(x) = \left[1 - 2\frac{x - x_{j+1}}{x_{j+1} - x_j}\right]\left(\frac{x - x_j}{x_{j+1} - x_j}\right)^2,$$

$$\hat{H}_{1,j}(x) = (x - x_j)\left(\frac{x - x_{j+1}}{x_j - x_{j+1}}\right)^2$$

and

$$\hat{H}_{1,j+1}(x) = (x - x_{j+1})\left(\frac{x - x_j}{x_{j+1} - x_j}\right)^2.$$

Note that this is just the Lagrange form of the Hermite interpolating polynomial over the interval $[x_j, x_{j+1}]$. We can simplify equation (1) slightly by introducing the translated and scaled variable $\xi = (x - x_j)/h_j$, where $h_j = x_{j+1} - x_j$, and the functions $\phi(\xi) = (1 + 2\xi)(1 - \xi)^2$ and $\psi(\xi) = \xi(1 - \xi)^2$. The functions ϕ and ψ are usually called *shape functions.*

After some minor algebraic manipulation, the details of which are left as an exercise, it can be shown that

$$H_{1,j}(x) = \phi(\xi),\ H_{1,j+1}(x) = 1 - \phi(\xi),\ \hat{H}_{1,j}(x) = h_j\psi(\xi),\ \text{and}$$
$$\hat{H}_{1,j+1}(x) = -h_j\psi(1 - \xi).$$

Substituting these expressions into equation (1) yields

$$s_j(x) = f(x_{j+1}) + \phi(\xi)\left[f(x_j) - f(x_{j+1})\right] + h_j\left[\psi(\xi)f'(x_j) - \psi(1 - \xi)f'(x_{j+1})\right].$$

Note that in this form, only three function evaluations – $\phi(\xi)$, $\psi(\xi)$ and $\psi(1 - \xi)$ – are needed to evaluate s_j, as opposed to the four function evaluations—$H_{1,j}(x)$, $H_{1,j+1}(x)$, $\hat{H}_{1,j}(x)$, and $\hat{H}_{1,j+1}(x)$—required by equation (1).

EXAMPLE 5.21 Data Analysis for the Spread of an Epidemic—Revisited

Above, a single Hermite interpolating polynomial was used to take the following data and produce a table of values for $D(t)$ at evenly spaced values of t in increments of 0.5.

t	$D(t)$	$D'(t)$
0.000000	0.000000	600.000000
0.750000	445.903683	573.579644
1.500000	842.695315	477.074216
2.085600	1095.211197	384.947629
2.676193	1295.955674	296.576145
3.219694	1437.602773	226.796410
3.748513	1542.363644	171.475176
4.279179	1621.280769	127.808738
4.821254	1680.890649	93.728061
5.000000	1696.803710	84.473801

Here, we will use a Hermite cubic interpolating polynomial to produce the desired table.

For $t = 0$, the value of D is already known: 0.0. For $t = 0.5$, we need to evaluate the polynomial s_0 – since 0.5 is between the first and second values in the first column above. By direct calculation, we find

$$h_0 = 0.750000$$
$$\xi = 0.666667$$
$$\phi(0.666667) = 0.259259$$
$$\psi(0.666667) = 0.074074$$
$$\psi(0.333333) = 0.148148$$

and

$$\begin{aligned} s_0(0.5) &= 445.903683 + 0.259259(0.000000 - 445.903683) + \\ &\quad 0.750000(0.074074 \cdot 600.000000 - 0.148148 \cdot 573.579644) \\ &= 299.901286. \end{aligned}$$

Next, for $t = 1.0$, we need to evaluate s_1. In this case, we find

$$\begin{aligned} &h_1 = 0.750000 \\ &\xi = 0.333333 \\ &\phi(0.333333) = 0.740741 \\ &\psi(0.333333) = 0.148148 \\ &\psi(0.666667) = 0.074074 \end{aligned}$$

and

$$\begin{aligned} s_1(1.0) &= 842.695315 + 0.740741(445.903683 - 842.695315) + \\ &\quad 0.750000(0.148148 \cdot 573.579644 - 0.074074 \cdot 477.074216) \\ &= 586.002536. \end{aligned}$$

Continuing in this manner, we complete the table below. Compare these values with those obtained previously.

t_i	$D(t_i)$
0.0	0.000000
0.5	299.901286
1.0	586.002536
1.5	842.695315
2.0	1061.676605
2.5	1241.486922
3.0	1384.886145
3.5	1496.767832
4.0	1582.658182
4.5	1647.813063
5.0	1696.803710

This last theorem provides an error bound for Hermite cubic interpolation. The proof of this theorem is similar to that of the error bound for piecewise linear interpolation, so the details have been left as an exercise. It may seem odd that the error bound for the Hermite cubic interpolant is smaller than the bound for the clamped cubic spline—a coefficient of $1/384$ versus $5/384$—given that the cubic spline has more smoothness than the Hermite cubic. However, remember that the Hermite cubic was constructed using $2n + 2$ data items, as compared to the $n + 3$ data values used to construct the clamped cubic spline. To level the playing field, we should allow the clamped cubic spline the same number of data values. If we were to use uniformly spaced interpolating points, this would imply that the mesh size for the clamped cubic spline would be roughly half that for the Hermite cubic. We would then find that the error bound for the cubic spline would be superior.

Theorem. Let f be continuous, with four continuous derivatives, on the interval $[a,b]$, and let s be the Hermite cubic interpolant of f relative to the partition

$$a = x_0 < x_1 < x_2 < \cdots < x_{n-1} < x_n = b.$$

Then

$$\max_{x\in[a,b]} |f(x) - s(x)| \le \frac{1}{384} h^4 \max_{x\in[a,b]} |f^{(4)}(x)|,$$

where $h = \max_{0\le i\le n-1}(x_{i+1} - x_i)$.

References

1. M. J. D. Powell, *Approximation Theory and Methods*, Cambridge University Press, Cambridge, 1981.

EXERCISES

1. Show that the polynomials H_i and $\hat{H}_i$ defined by

$$\begin{aligned} H_i(x) &= [1 - 2L'_{n,i}(x_i)(x - x_i)]L^2_{n,i}(x) \\ \hat{H}_i(x) &= (x - x_i)L^2_{n,i}(x), \end{aligned}$$

where $L_{n,i}$ is the Lagrange polynomial associated with the point $x = x_i$ satisfy the relations

$$H_i(x_j) = \begin{cases} 1, & i = j \\ 0, & \text{otherwise} \end{cases} \qquad \hat{H}_i(x_j) = 0$$

$$H'_i(x_j) = 0 \qquad \hat{H}'_i(x_j) = \begin{cases} 1, & i = j \\ 0, & \text{otherwise} \end{cases}.$$

2. Let f be continuously differentiable $2n+2$ times on $[a,b]$, and let x_0, x_1, x_2, $\ldots$, x_n be $n+1$ distinct points from $[a,b]$. Provide the details of the proof that for each $x \in [a,b]$, there exists a $\xi \in [a,b]$ such that

$$f(x) = P(x) + \frac{f^{(2n+2)}(\xi)}{(2n+2)!}\prod_{i=0}^{n}(x - x_i)^2,$$

where P is the Hermite interpolating polynomial.

3. Let $f(x) = x\ln x$, $x_0 = 1$, and $x_1 = 3$.

(a) Construct the Hermite interpolating polynomial for f at the specified interpolating points.

(b) Approximate $f(1.5)$ using the polynomial from part (a), and confirm that the theoretical error bound holds.

4. Let $f(x) = x \ln x$, $x_0 = 1$, $x_1 = 2$, and $x_2 = 3$.
   (a) Construct the Hermite interpolating polynomial for f at the specified interpolating points.
   (b) Approximate $f(1.5)$ using the polynomial from part (a), and confirm that the theoretical error bound holds.
   (c) Construct the Hermite cubic interpolant for f at the specified interpolating points.
   (d) Approximate $f(1.5)$ using the piecewise polynomial from part (c), and confirm that the theoretical error bound holds.
5. Let $f(x) = xe^{-x}$, $x_0 = 1$, $x_1 = 2$, and $x_2 = 3$.
   (a) Construct the Hermite interpolating polynomial for f at the specified interpolating points.
   (b) Approximate $f(1.5)$ using the polynomial from part (a), and confirm that the theoretical error bound holds.
   (c) Construct the Hermite cubic interpolant for f at the specified interpolating points.
   (d) Approximate $f(1.5)$ using the piecewise polynomial from part (c), and confirm that the theoretical error bound holds.
6. Let $f(x) = \frac{1}{1+25x^2}$, $x_0 = -1$, $x_1 = 0$, and $x_2 = 1$.
   (a) Construct the Hermite interpolating polynomial for f at the specified interpolating points.
   (b) Approximate $f(-0.3)$ using the polynomial from part (a), and confirm that the theoretical error bound holds.
   (c) Construct the Hermite cubic interpolant for f at the specified interpolating points.
   (d) Approximate $f(-0.3)$ using the piecewise polynomial from part (c), and confirm that the theoretical error bound holds.
7. A model for the growth of an insect population predicts the following values for the population, $P(t)$, and the rate of increase in the population, $P'(t)$, as functions of time. Here, time is measured in months.

t	$P(t)$	$P'(t)$
0.000000	5.000000	1.850962
0.500000	6.008286	2.179438
0.950023	7.050280	2.443439
1.447286	8.323016	2.658770
1.947286	9.682456	2.756773
2.447286	11.056543	2.716253
2.947286	12.376723	2.544655
3.447286	13.584544	2.273554
3.947286	14.641031	1.946924
4.430434	15.502227	1.618850
4.848017	16.121126	1.348776
5.000000	16.319048	1.256352

   (a) Use the Hermite interpolating polynomial derived from this data to tabulate the population in half-week increments.

(a) Data for Exercises 8, 9, 10			(b) Data for Exercises 11, 12, 13		
Time (sec)	Height (meters)	Velocity (meters/sec)	Time (sec)	Charge (coulombs)	Current (amperes)
0.00	0.290864	−0.16405	0.00	0.000000	0.000000
0.02	0.284279	−0.32857	0.02	0.003293	0.249906
0.04	0.274400	−0.49403	0.04	0.007381	0.121402
0.06	0.260131	−0.71322	0.06	0.007887	−0.053314
0.08	0.241472	−0.93309	0.08	0.006296	−0.080449
0.10	0.219520	−1.09409	0.10	0.005296	−0.015126
0.12	0.189885	−1.47655	0.12	0.005525	0.028800
0.14	0.160250	−1.47891	0.14	0.006086	0.020787
0.16	0.126224	−1.69994	0.16	0.006255	−0.002842
0.18	0.086711	−1.96997	0.18	0.006085	−0.010721
0.20	0.045002	−2.07747	0.20	0.005927	−0.003931
0.22	0.000000	−2.25010			

TABLE 5.2: (a) Data for Exercises 8, 9, 10. (b) Data for Exercises 11, 12, 13.

- **(b)** Use the Hermite cubic interpolating polynomial derived from this data to tabulate the population in half-week increments.
- **(c)** Use the clamped cubic spline derived from this data to tabulate the population in half-week increments.
- **(d)** Use the not-a-knot cubic spline derived from this data to tabulate the population in half-week increments.
- **(e)** Compare the results from (a), (b), (c), and (d).

8. Table 5.2(a) gives the height and velocity of a free-falling object.
- **(a)** Construct the Hermite cubic interpolant for this data set.
- **(b)** What is the height of the object when t = 0.05 seconds? when t = 0.15 seconds?
- **(c)** At what time is the object 0.20 meters above the ground? 0.10 meters above the ground?

9. Repeat Exercise 8 using the Hermite interpolating polynomial.

10. Repeat Exercise 8 using the clamped cubic spline.

11. Table 5.2(b) gives the charge on the capacitor and the current flowing through an RLC circuit. Recall that current is the rate of change of charge.
- **(a)** Construct the Hermite cubic interpolant for this data set.
- **(b)** What is the charge on the capacitor when $t = 0.05$ seconds? when $t = 0.15$ seconds?
- **(c)** At what time is the charge on the capacitor a maximum?

12. Repeat Exercise 11 using the Hermite interpolating polynomial.

13. Repeat Exercise 11 using the clamped cubic spline.

14. Let $\xi = (x - x_j)/h_j$, where $h_j = x_{j+1} - x_j$. Show that

$$H_{1,j}(x) = \phi(\xi), \quad H_{1,j+1}(x) = 1 - \phi(\xi), \quad \hat{H}_{1,j}(x) = h_j\psi(\xi), \quad \text{and}$$
$$\hat{H}_{1,j+1}(x) = -h_j\psi(1-\xi),$$

where

$$H_{1,j}(x) = \left[1 - 2\frac{x - x_j}{x_j - x_{j+1}}\right]\left(\frac{x - x_{j+1}}{x_j - x_{j+1}}\right)^2,$$

$$H_{1,j+1}(x) = \left[1 - 2\frac{x - x_{j+1}}{x_{j+1} - x_j}\right]\left(\frac{x - x_j}{x_{j+1} - x_j}\right)^2,$$

$$\hat{H}_{1,j}(x) = (x - x_j)\left(\frac{x - x_{j+1}}{x_j - x_{j+1}}\right)^2,$$

$$\hat{H}_{1,j+1}(x) = (x - x_{j+1})\left(\frac{x - x_j}{x_{j+1} - x_j}\right)^2,$$

$$\phi(\xi) = (1 + 2\xi)(1 - \xi)^2;$$

and

$$\psi(\xi) = \xi(1 - \xi)^2.$$

15. Prove the theorem that provides the error bound for the Hermite cubic interpolant. (Use the proof of the error bound for piecewise linear interpolation in Section 5.5 as a model.)

16. **(a)** Suppose that f is twice differentiable. Show that

$$f[x_i, x_i, x_i] = \frac{f''(x_i)}{2}.$$

(b) Suppose that f is n times differentiable. Show that

$$f[\overbrace{x_i \quad x_i \quad x_i \quad \cdots \quad x_i}^{n+1\ x_i's}] = \frac{f^{(n)}(x_i)}{n!}.$$

17. Let f be a function defined on the interval $[a, b]$, and let $x_0, x_1, x_2, \ldots, x_n$ be $n + 1$ distinct points from $[a, b]$. For each $i = 0, 1, 2, \ldots, n$, let m_i be a nonnegative integer. The polynomial, P, of degree at most $d = n + \sum_{i=0}^{n} m_i$, such that

$$P^{(k)}(x_i) = f^{(k)}(x_i)$$

for each $i = 0, 1, 2, \ldots, n$ and each $k = 0, 1, 2, \ldots, m_i$ is called the *osculatory interpolating polynomial.* With the Newton form of the Hermite interpolating polynomial as a guide and using the results of Exercise 16, construct the Newton form of the osculatory interpolating polynomial.

18. Determine the osculatory interpolating polynomial for each of the following functions using the indicated amount of data at the specified points.

(a) $f(x) = x\ln x$, $x_0 = 1$, $x_1 = 2$, $x_2 = 3$, $m_0 = 1$, $m_1 = 0$, $m_2 = 2$

(b) $f(x) = \frac{1}{1+25x^2}$, $x_0 = -1$, $x_1 = -1/2$, $x_2 = 0$, $x_3 = 1/2$, $x_4 = 1$, $m_0 = 1$, $m_1 = m_2 = m_3 = 0$, $m_4 = 1$

(b) $f(x) = e^{-x}$, $x_0 = 0$, $x_1 = 1$, $x_2 = 2$, $m_0 = 0$, $m_1 = 1$, $m_2 = 2$

19. Let f be a function defined on the interval $[a, b]$, and let $x_0, x_1, x_2, \ldots, x_n$ be $n + 1$ distinct points from $[a, b]$. For each $i = 0, 1, 2, \ldots, n$, let m_i be a non-negative integer.

(a) Prove that the osculatory interpolating polynomial is unqiue.

(b) If we suppose that f is sufficiently differentiable, what is the error associated with the osculatory interpolating polynomial? Prove it.

5.8 REGRESSION

Regression is a powerful technique for predicting the value of a dependent variable and for estimating the values of model parameters. This method finds application in business, economics, the physical and biological sciences, engineering, and more. Whereas interpolation is fundamentally a local procedure which forces the error to be zero at specific, isolated locations, regression is a global process. In regression analysis, the function that is being fit to the data is overdetermined, meaning that the number of coefficients is smaller than the number of data points. Values for these coefficients are computed by requiring some measure of the total approximation error be minimized. In this section we will deal exclusively with discrete data.

Linear Regression

A manufacturing firm wishes to estimate the production costs for one of its product lines. Over a four week period they monitor the daily production runs, tabulating run size and corresponding total cost. The results of this study were as follows:

Run Size	Total Cost	Run Size	Total Cost
1550	\$17,224	2175	24,095
852	11,314	1213	13,474
2120	22,186	3050	29,349
1128	15,982	1215	14,459
1518	16,497	2207	23,483
786	10,536	1234	14,444
1505	15,888	1616	18,949
1264	13,055	3089	31,237
1963	22,215	2033	21,384
1414	17,510	1467	18,012

A scatter plot of the data (Figure 5.19) shows a clear, roughly linear trend. For simplicity, we will assume that the relationship between run size and total cost is exactly linear.

There are many lines that would approximate the data reasonably well. Let the line that fits the data "best" be given by $\hat{y} = a + bx$, where the parameters a and b are to be determined. The criterion by which the fit of the line to the data is

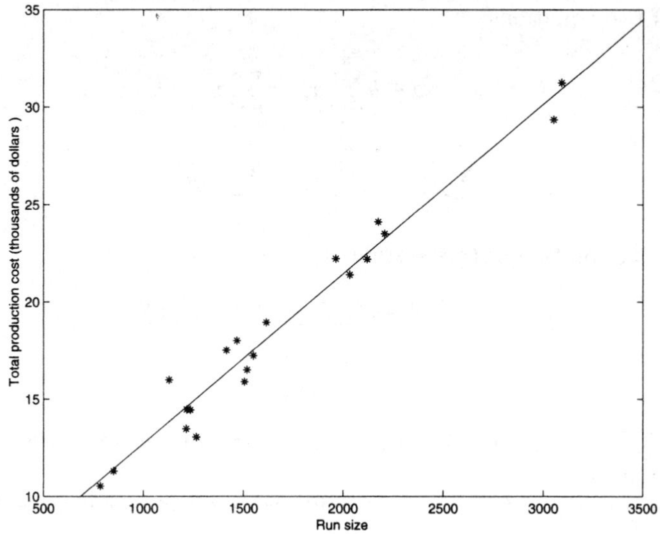

Figure 5.19 Scatter plot of total production cost versus run size data. Least squares regression line is shown.

judged will be given momentarily. Further, let (x_i, y_i) for $i = 1, 2, 3, \ldots, n$ denote the data pairs being examined, and let e_i measure the deviation of the best fit line from the data; that is, for each i

$$e_i = y_i - \hat{y}_i = y_i - (a + bx_i).$$

Define the total error, E, by

$$E = \sum_{i=1}^{n} e_i^2 = \sum_{i=1}^{n} \left[y_i - (a + bx_i)\right]^2.$$

Our objective will be to choose a and b so as to minimize E. This goal is known as the least-squares criterion and produces the least-squares regression line.

To minimize E, we must have

$$\frac{\partial E}{\partial a} = \frac{\partial E}{\partial b} = 0.$$

Since

$$\frac{\partial E}{\partial a} = -2 \sum_{i=1}^{n} \left[y_i - (a + bx_i)\right]$$

and

$$\frac{\partial E}{\partial b} = -2 \sum_{i=1}^{n} \left[y_i - (a + bx_i)\right] x_i,$$

the system of two equations

$$na + b\sum_{i=1}^{n} x_i = \sum_{i=1}^{n} y_i$$
$$a\sum_{i=1}^{n} x_i + b\sum_{i=1}^{n} x_i^2 = \sum_{i=1}^{n} x_i y_i$$

is obtained. The solution of this system is

$$b = \frac{n\sum_{i=1}^{n} x_i y_i - \left(\sum_{i=1}^{n} x_i\right)\left(\sum_{i=1}^{n} y_i\right)}{n\sum_{i=1}^{n} x_i^2 - \left(\sum_{i=1}^{n} x_i\right)^2}$$
$$a = \bar{y} - b\bar{x},$$

where $\bar{x} = \left(\sum_{i=1}^{n} x_i\right)/n$ and $\bar{y} = \left(\sum_{i=1}^{n} y_i\right)/n$ are the mean of the x and the y values, respectively.

Returning to the data from the manufacturing firm problem, we find

$$n = 20, \quad \sum_{i=1}^{n} x_i = 33,399, \quad \sum_{i=1}^{n} y_i = 371,293, \quad \sum_{i=1}^{n} x_i y_i = 685,996,183$$

and

$$\sum_{i=1}^{n} x_i^2 = 63,345,673.$$

Substituting these values into the equations for a and b gives

$$b = \frac{(20)(685,996,183) - (33,399)(371,293)}{(20)(63,345,673) - (33,399)^2} = \frac{1,319,108,753}{151,420,259} \approx \$8.71/\text{unit}$$
$$a = \frac{371,293}{20} - \frac{33,399}{20} \cdot \frac{1,319,108,753}{151,420,259} \approx \$4016.76.$$

Therefore, the regression line for the total production cost as a function of run size is $\hat{y} = 4016.76 + 8.71x$. This line is graphed in Figure 5.19 for comparison. If the standard run size for this product line is 2200 units, we would then predict that the total production cost would be \$4016.76 + (2200 units)(\$8.71/unit), or \$23,178.76. As in most practical regression problems, the parameters a and b have significance beyond being the y-intercept and slope of the regression line, respectively. Here a represents the fixed costs, or overhead, associated with this particular product, whereas b represents the variable costs of production.

Worked Examples

EXAMPLE 5.22 A New Exit on the Jersey Turnpike

The following table gives information about the tolls charged at the main toll booth at the southern end of the New Jersey Turnpike. The distance values are the number of miles from the indicated exit to the main toll booth, and the "toll" values are the toll charge assessed for a vehicle which enters the turnpike at the indicated exit.

Exit	Distance	Toll	Exit	Distance	Toll
2	12	0.45	8	67	1.70
3	25	0.70	8A	73	1.85
4	33	0.95	9	82	2.15
5	43	1.20	10	87	2.20
6	50	1.85	11	90	2.40
7	52	1.45	12	95	2.65
7A	59	1.55			

After examining traffic patterns, the Turnpike Authority has decided to construct a new exit, Exit 5A, a distance of 48 miles from the main toll booth. Based on the information contained in the table, what toll should be charged for a vehicle that enters the turnpike at Exit 5A?

A scatter plot of the toll charge as a function of distance would suggest a roughly linear relationship. We will therefore fit the data to a linear function. Letting x denote the distance and y denote the toll charge, we find

$$n = 13, \quad \sum_{i=1}^{n} x_i = 768, \quad \sum_{i=1}^{n} y_i = 21.1, \quad \sum_{i=1}^{n} x_i y_i = 1449.6 \quad \text{and} \quad \sum_{i=1}^{n} x_i^2 = 53628.$$

Using these values, it follows that

$$b = \frac{(13)(1449.6) - (768)(21.1)}{(13)(53628) - (768)^2} \approx \$0.0246/\text{mile}$$

$$a = \frac{21.1}{13} - \frac{768}{13} \cdot 0.0246 \approx \$0.17.$$

In other words, the toll structure for the listed exits consists of a fixed charge of roughly 17 cents for simply using the turnpike and a variable charge of roughly 2.5 cents per mile. Based on this toll structure, the toll for the new exit 48 miles from the main toll booth should be

$$\hat{y} = 0.17 + (0.0246)(48) = \$1.35.$$

EXAMPLE 5.23 Calibration of a Thermocouple

A group of physics students has constructed a thermocouple for use in a laboratory experiment. To calibrate the device, they have collected the following data.

Temperature (°C)	Reading (mV)
0	0.01
20	0.12
40	0.24
60	0.38
80	0.51
100	0.67
120	0.84
140	1.01
160	1.15
180	1.31

The end result of the calibration process is supposed to be a formula which translates a thermocouple reading, in millivolts, into a temperature reading, in °C. We will therefore let x denote the thermocouple reading, and y denote the corresponding temperature. This assignment leads to the values

$$n = 10, \quad \sum_{i=1}^{n} x_i = 6.24, \quad \sum_{i=1}^{n} y_i = 900, \quad \sum_{i=1}^{n} x_i y_i = 804.6 \quad \text{and} \quad \sum_{i=1}^{n} x_i^2 = 5.6898,$$

which yield the calibration parameters $a \approx 5.574°\text{C}$ and $b \approx 135.298°\text{C/mV}$. The temperature corresponding to a given thermocouple reading, x, is then determined by the equation

$$T = 5.574 + 135.298x.$$

Transformations to Linear

In many circumstances, a linear model is inappropriate. For example, the error sequences generated by fixed point iteration schemes in Chapter 2 satisfied power laws of the form $|e_{n+1}| = C|e_n|^\alpha$, where α represented the order of convergence. Growth and decay phenomena often obey exponential laws of the form $y = ab^x$. Still other phenomena are modeled by logarithmic laws, $y = a + b\log x$; reciprocal laws, $y = 1/(a + bx)$; and higher-degree polynomial laws. Here, we focus on power laws and exponential laws, which can be transformed to linear problems, and logarithmic laws, which are already linear in the model parameters. Other cases will be considered in the exercises.

Let's start with the power law, $y = ax^b$. Taking the logarithm of both sides of the power law (the base of the logarithm is irrelevant) yields

$$\begin{aligned} \log y &= \log(ax^b) \\ &= \log a + b\log x. \end{aligned}$$

Hence, $\log x$ and $\log y$ are related linearly when x and y are related by a power law. Therefore, to fit data to an equation of the form $y = ax^b$, first take the logarithm of the x-values and the y-values, then perform linear regression on the resulting data set. The y-intercept of the regression line is the logarithm of the coefficient in the power law, and the slope of the regression line gives the exponent.

Exponential laws, $y = ab^x$, can be handled in a similar manner. Take the logarithm of both sides of the exponential law to yield

$$\begin{aligned}\log y &= \log(ab^x)\\ &= \log a + x \log b.\end{aligned}$$

Therefore, to fit data to an exponential law, take the logarithm of the y values, then perform linear regression. The y-intercept of the regression line is the logarithm of the coefficient in the exponential law, while the slope gives the logarithm of the base of the exponential. In many instances, the exponential law is written as $y = ae^{bx}$. In these cases, the logarithm of the law becomes

$$\ln y = \ln a + bx,$$

so that the slope of the regression line is the coefficient in the exponent.

Worked Examples

EXAMPLE 5.24 Order of Convergence

In general, a numerical method for approximating the value of a definite integral is said to have order of convergence α if the absolute approximation error, E, is related to the mesh spacing parameter, h, by the power law $E = Ch^\alpha$. Simpson's rule is one general technique for approximating the value of a definite integral. The method involves dividing the integration interval into an even number of equal-sized subintervals and computing a weighted sum of the values of the integrand at the endpoints of the subintervals. The mesh spacing parameter for Simpson's rule is the size of the subintervals into which the integration interval is partitioned.

To determine experimentally the order of convergence of Simpson's rule, we approximate the value of the definite integral

$$\int_0^1 xe^x \, dx$$

for several different values of h and compute the absolute error in the approximation. The results of these experiments are

h	Absolute Error, E
1/2	2.620728×10^{-3}
1/4	1.690471×10^{-4}
1/8	1.065014×10^{-5}
1/16	6.669677×10^{-7}
1/32	4.170636×10^{-8}
1/64	2.606974×10^{-9}
1/128	1.629410×10^{-10}

Next, we fit the data to the power law $E = Ch^\alpha$. Since we are working with a power law, we will perform a linear fit of $\ln E$ versus $\ln h$. Note that we have chosen to use the natural logarithm, but any other logarithm would also work. The resulting regression line is

$$\ln E = -3.159 + 3.992 \ln h.$$

Exponentiating this equation gives $E = 0.0425h^{3.992}$. Therefore, for this problem, it appears that Simpson's rule is roughly of order of convergence 4.

EXAMPLE 5.25 CD Sales versus LP Sales: An Exponential Model

The table below summarizes the sales of compact discs (CDs) and long playing records (LPs) over an eleven year period. Sales are listed in millions of units.

CDs	0	0.8	5.8	23	53	102	150	207	287	333	408
LPs	244	210	205	167	125	107	72	35	12	4.8	2.3

Based on a scatter plot of the data (shown in the upper left panel of Figure 5.20), an economist hypothesizes that the data follows either an exponential law or a power law. To decide between the two, the economist plots the logarithm of the LP sales versus the CD sales (upper right panel in Figure 5.20) and the logarithm of the LP sales versus the logarithm of the CD sales (lower left panel). Since the former plot is more roughly linear, the economist settles on the exponential law.

The resulting regression line is found to be

$$\log L = 2.401 - 0.00479C,$$

where L denotes the level of LP sales, C denotes the level of CD sales and common (base 10) logarithms were used. Exponentiating both sides of the regression equation yields $L = 251.768 \cdot (0.989)^C$. The graph of this equation is shown superimposed on the data in lower right panel of Figure 5.20 to demonstrate the accuracy of the fit.

EXAMPLE 5.26 Break-Even Point for Vitamin A Dosage

To estimate the amount of vitamin A required for maintaining weight, laboratory rats were fed a basic diet devoid of vitamin A, but were given controlled supplementary rations of vitamin A in the form of cod liver oil. The following table summarizes the supplementary dosage of vitamin A and the corresponding weight gain for the test subjects.

Dosage (mg)	0.25	1.00	1.50	2.50	7.50
Weight Gain (grams)	−10.8	13.5	16.4	28.7	51.3

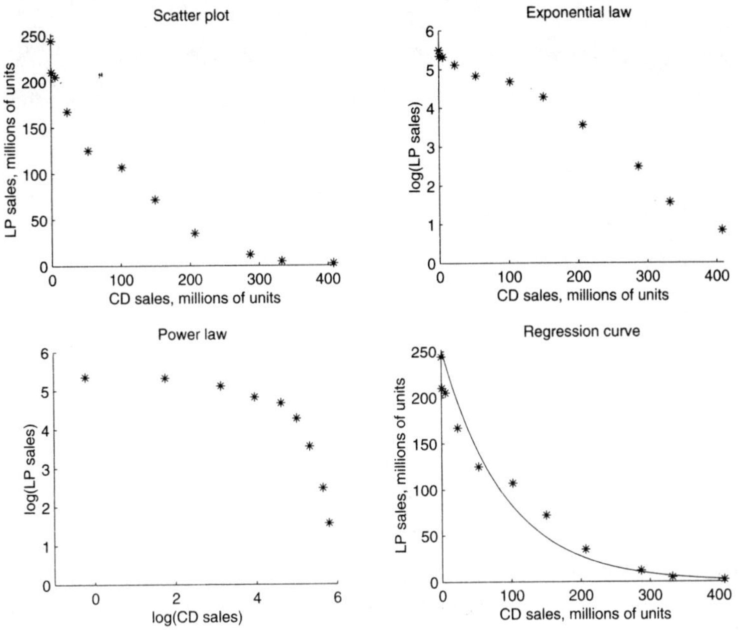

Figure 5.20 (Upper left) Scatter plot of data for CD sales versus LP sales. (Upper right) Logarithm of LP sales versus CD sales. (Lower left) Logarithm of LP sales versus logarithm of CD sales. (Lower right) Exponential law fit to the data points.

A graph of the data (Figure 5.21) suggests fitting the data to a logarithmic law of the form $W = a + b \log D$. Using base 10 logarithms, we find

$$W = 12.762 + 41.661 \log D.$$

Setting $W = 0$, it follows that $D \approx 0.49$ mg of vitamin A is required to maintain weight.

EXERCISES

1. One of the following data sets follows an exponential law and the other follows a power law. Which is which?

x	2.0	2.5	3.0	3.5	4.0	4.5	5.0
y_1	14.79	27.75	47.09	74.07	109.99	156.10	213.69

x	2.0	2.5	3.0	3.5	4.0	4.5	5.0
y_2	12.13	19.58	31.59	50.97	82.21	132.59	213.82

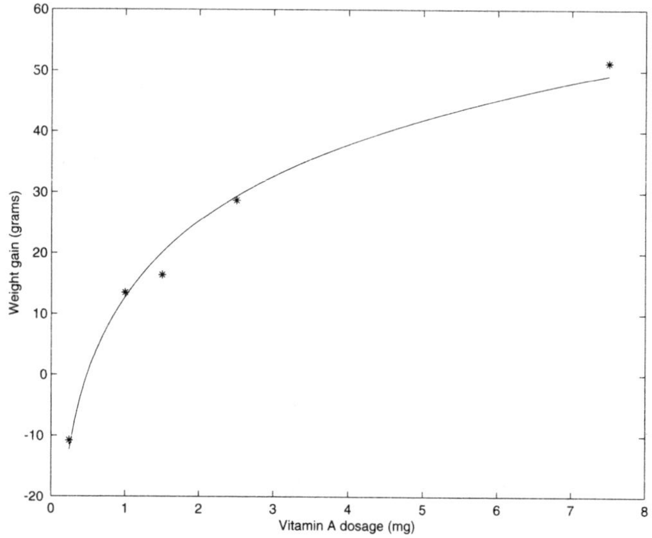

Figure 5.21 Scatter plot of data for weight gain of laboratory rats versus supplementary dosage of vitamin A. Solid curve is the logarithmic law fit to the data points.

2. One of the following data sets follows an exponential law and the other follows a power law. Which is which?

x	2.0	2.5	3.0	3.5	4.0	4.5	5.0
y_1	1.216	1.087	0.972	0.870	0.778	0.696	0.622

x	2.0	2.5	3.0	3.5	4.0	4.5	5.0
y_2	1.108	0.758	0.556	0.427	0.341	0.279	0.233

3. One of the following data sets follows a logarithmic law and the other follows a power law. Which is which?

x	2.0	2.5	3.0	3.5	4.0	4.5	5.0
y_1	16.50	17.77	18.89	19.88	20.79	21.62	22.40

x	2.0	2.5	3.0	3.5	4.0	4.5	5.0
y_2	11.73	14.54	16.84	18.78	20.46	21.95	23.27

4. Experimental data relating the oxide thickness, measured in Angstroms, of a thin film to the baking time of the film, measured in minutes, is given in the table below.

Baking time	20	30	40	60	70	90	100	120	150	180
Oxide thickness	3.5	7.4	7.1	15.6	11.1	14.9	23.5	27.1	22.1	32.9

(a) Construct a scatter plot of this data. What functional form is most appropriate for fitting this data?

(b) Fit the data to the function indicated in part (a). What physical significance do the model parameters have?

(c) Predict the oxide thickness for a film which is baked for 45 minutes.

5. The total production cost as a function of the number of machine hours is provided for a sample of nine production runs. Estimate the fixed costs and the variable costs associated with this process.

Machine hours	22	23	19	12	12	9	7	11	14
Total cost (in 1000s)	23	25	20	20	20	15	14	14	16

6. The resistivity of platinum as a function of temperature is given below. Estimate the parameters in a linear fit to the data and predict the resistivity when the temperature is 365 K.

Temperature (K)	100	200	300	400	500
Resistivity (Ω-cm, $\times 10^6$)	4.1	8.0	12.6	16.3	19.4

7. The table below shows the time (in seconds) required for water to drain through a hole in the bottom of a bottle as a function of the depth (in inches) to which the bottle has been filled.

Depth	0.5	1.0	1.5	2.0	2.5	3.0	3.5	4.0
Time	65.99	120.28	166.69	207.85	245.41	279.95	313.04	344.24

(a) Construct a scatter plot of this data. What functional form is most appropriate for fitting this data?

(b) Fit the data to the function indicated in part (a).

8. The weight, W, of a metallic object decreases over time when exposed to a caustic environment according to the exponential law $W = ae^{-t/\tau}$, where t is the exposure time and τ is known as the decay rate constant. Data for a group of objects made from the same material is given in the following table.

Exposure time (days)	5	10	15	20	25	30	35	40
Weight (grams)	92.7	58.3	59.5	41.7	45.6	31.8	38.3	19.9

Estimate the decay rate constant, τ, for this material.

9. Barometric pressure, P, as a function of elevation above sea level, h, is modeled by the relation $P = \alpha e^{-\beta h}$. Use the data in the table below to estimate the model parameters and to predict the barometric pressure at an elevation of 1200 feet.

Barometric pressure (mm Hg)	29.9	29.4	29.0	28.4	27.7
Elevation above sea level (feet)	0	500	1000	1500	2000

10. When an ideal gas undergoes an isentropic process, the pressure and volume are related by $P = cV^{-\gamma}$, where γ is the ratio of the specific heats of the gas. Estimate the value of γ based on the values in the following table:

Pressure (psi)	16.8	39.7	78.6	115.5	195.0	546.1
Volume (in^3)	50	30	20	15	10	5

11. The results of a tensile strength test for a circular cold-rolled steel specimen are provided in the table below. The specimen had an original diameter of 0.507" and an original length of 2 inches. The normal stress, σ, and the normal strain, ϵ, are given by the equations

$$\sigma = \frac{P}{A} \quad \text{and} \quad \epsilon = \frac{\Delta}{L},$$

where P denotes the load, Δ the elongation, a the original cross-sectional area, and L the original length of the specimen. From the test data, we want to estimate the modulus of elasticity, E, which is defined as the ratio σ/ϵ in the linear portion of the stress-strain curve.

Load (10^3 lb)	Elongation (10^{-4} in)	Load (10^3 lb)	Elongation (10^{-4} in)
0	0	4.85	16
1.25	4	5.45	18
1.85	6	6.05	20
2.4	8	6.7	22
3.05	10	7.25	24
3.64	12	6.9	40
4.25	14	6.95	80

Note: For this problem, you will first need to decide which of the data points correspond to the linear portion of the stress-strain curve.

12. The following table gives the ion concentration, n, as a function of time, t, after an ionization agent has been turned off.

Time (sec)	0	1	2	3	4	5	6	7	8	9	10
n ($\times 10^{-4}$)	5.03	4.71	4.40	3.97	3.88	3.62	3.30	3.15	3.08	2.92	2.70

Theory indicates that ion concentration and time satisfy the reciprocal relationship

$$n = \frac{n_0}{1 + n_0 \alpha t},$$

where n_0 is the initial concentration of ions and α is the coefficient of recombination.

(a) Take the reciprocal of the above equation relating ion concentration and time, and show that n^{-1} and t are related in a linear fashion.

(b) Perform linear regression on n^{-1} versus t to estimate the initial concentration of ions and the coefficient of recombination.

13. Consider the following data relating the amount of varnish additive and the resulting varnish drying time.

Additive (grams)	0.0	1.0	2.0	3.0	4.0	5.0	6.0	7.0	8.0
Drying time (hours)	12.0	10.5	10.0	8.0	7.0	8.0	7.5	8.5	9.0

(a) Produce a scatter plot of the data and show that the data roughly follows the pattern of a quadratic function.

(b) Apply the least squares criterion to the regression equation $\hat{y} = a + bx + cx^2$ to determine formulas for a, b, and c.

(c) Use the results of part (b) to determine the regression parabola for the data given above. What amount of varnish additive will produce the minimum drying time?

CHAPTER 6

Differentiation and Integration

AN OVERVIEW

Fundamental Mathematical Problems

In this chapter we will discuss the concepts and techniques associated with numerical differentiation and integration. In particular, we will address the following three problems.

Problem 1
Approximate the value of a derivative of a function defined by discrete data.

Problem 2
Derive a formula that approximates the derivative of a function in terms of a linear combination of function values.

Problem 3
Approximate the value of the definite integral of a continuous function, defined by a formula or by discrete data, over a specified interval.

The "Estimating a Coefficient of Friction" problem capsule from the Chapter 1 Overview (see page 5) is one example of an application that requires numerical differentiation. Here are two applications that require numerical integration.

Bags of Pine Bark Mulch

Kindly Doc B has an irregularly shaped region in his back yard (see Figure 6.1). He has tried to grow every conceivable plant in that spot, but nothing but weeds ever seem to grow; he has therefore decided to cover the entire plot with pine bark mulch. If Doc B would like to lay down a uniform 3-inch covering of bark mulch over the entire plot and the home improvement store sells bags containing 3 cubic feet of bark mulch, how many bags does Doc B need to buy?

Clearly, the number of bags needed can be obtained by dividing the capacity of each bag (3 cubic feet of mulch) into the total volume of mulch required to cover the plot. In turn, because the plot is to be covered with a layer of mulch 3" thick, the volume of mulch needed for the job is just one-quarter the area of the plot. Consequently,

$$\begin{array}{c}\text{number of}\\ \text{bags needed}\end{array} = \frac{1}{12} \times \text{ area of the plot.}$$

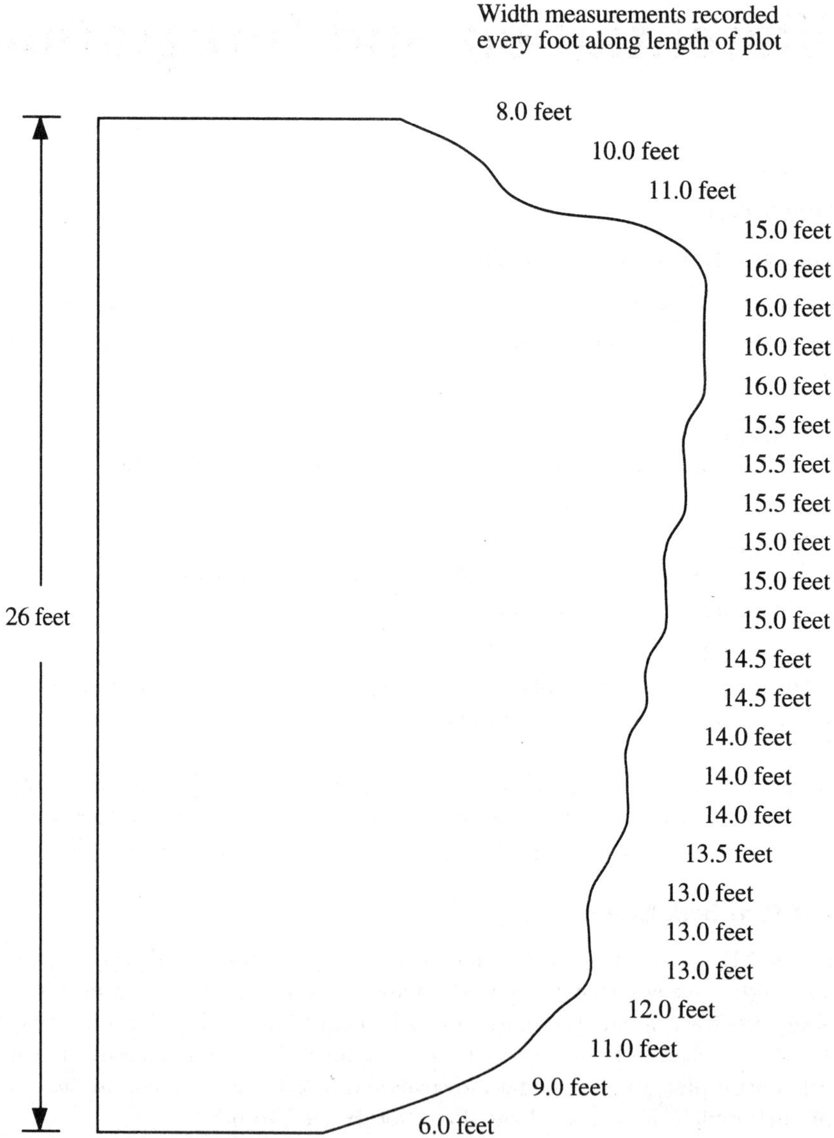

Figure 6.1 Plot of irregularly shaped region for the "Bags of Pine Bark Mulch" problem.

Now, let x denote distance measured along the length of the plot and let $f(x)$ denote the width of the plot at location x, with all distances measured in feet. The area of the plot is then given by

$$\int_0^{26} f(x)\,dx.$$

To determine the number of bags of pine bark mulch Doc B must purchase, we will therefore have to approximate the value of this definite integral using the discrete data from Figure 6.1.

Tabulating the Error Function

There are many special mathematical functions that are defined in terms of definite integrals. One of these is the so-called *error function*, which is given by

$$\text{erf}(x) = \frac{2}{\sqrt{\pi}} \int_0^x e^{-t^2}\,dt.$$

This function has applications in probability, statistics, heat conduction, boundary layer theory, groundwater flow, and so on.

Suppose we want to tabulate values of erf(x) and its derivative,

$$\frac{d}{dx}\text{erf}(x) = \frac{2}{\sqrt{\pi}} e^{-x^2},$$

at equally spaced points from $x = 0$ to $x = 2$. All values are to be accurate to five decimal places, and the increment between x-values, Δx, must be chosen so that Hermite cubic interpolation from the table produces an error less than 5×10^{-6}. Now, the error introduced by Hermite cubic interpolation is bounded above by

$$\frac{1}{384} (\Delta x)^4 \max_{0 \le x \le 2} \left| \frac{d^4}{dx^4}\text{erf}(x) \right|.$$

Since

$$\max_{0 \le x \le 2} \left| \frac{d^4}{dx^4}\text{erf}(x) \right| < 5,$$

the requirement that the interpolation error be less than 5×10^{-6} translates into needing $\Delta x < 0.14$. Let's choose $\Delta x = 0.1$ for convenience.

Evaluating the derivative of the error function at the required x-values is straightforward. To complete the table, however, we must evaluate

$$\frac{2}{\sqrt{\pi}} \int_0^x e^{-t^2}\,dt$$

for x ranging from 0 to 2 in increments of 0.1, guaranteeing that each value is accurate to five decimal places.

The Remainder of the Chapter

The first three sections focus on the techniques for numerical differentiation. In Section 3 the important concept of extrapolation—combining different approximations obtained from the same low-order formula to obtain a higher-order approximation—is introduced. Our study of numerical integration begins in Section 4, where all of the important definitions are provided. This section also includes a discussion of Newton-Cotes quadrature. The following sections consider composite Newton-Cotes quadrature, Gaussian quadrature, Romberg integration, and adaptive quadrature. The chapter concludes with a section dealing with improper integrals and the proper handling of singularities.

6.1 NUMERICAL DIFFERENTIATION, PART I

Numerical differentiation normally arises in one of two contexts. In the first, the objective is to approximate the value of a derivative of a function defined by discrete data. For example, we may want to approximate the endpoint derivative values needed to construct a clamped cubic spline. In the second context, the objective is to derive formulas which approximate the derivatives of a function in terms of a linear combination of function values. These formulas form the basis for finite difference techniques for the solution of boundary value problems and partial differential equations which will be considered in Chapters 8 through 11.

In this section the problem of approximating the value of a derivative from discrete data will be considered. As you read through this section, note that the main concern is assessing the reasonableness of the computed approximation. Deriving formulas for derivatives will be treated in the next section.

Estimating a Coefficient of Friction

Let's begin with the problem of estimating the coefficient of friction between a flexible rope and the post around which it is wrapped from the experimental data given below, which measures the force required to overcome a 5-lb restraining force as a function of the angle through which the rope is wrapped around the post. Recall from the Chapter 1 Overview (see page 5) that the coefficient of friction, μ, is the proportionality constant between the rate of change of the force and the magnitude of the force; that is,

$$\mu = \frac{dF/d\theta}{F(\theta)}.$$

Given the data in Table 6.1, how might we approximate the value of $dF/d\theta$? Considering the material we just covered in Chapter 5, the most appropriate choice might be to pass an interpolating polynomial through the data and then differentiate the polynomial.

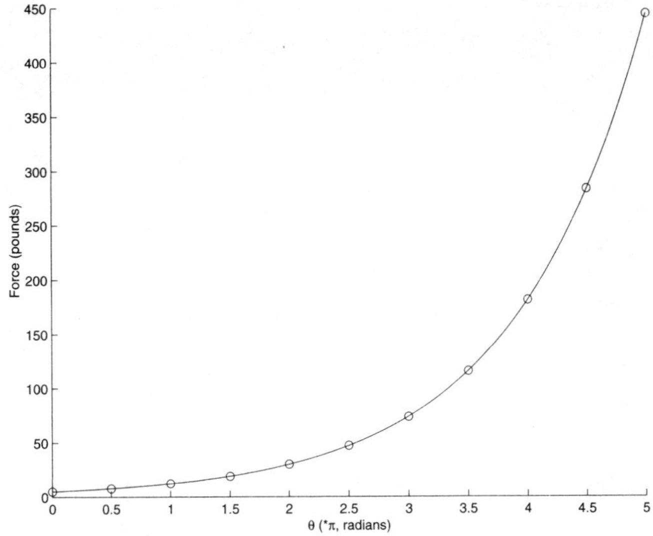

Figure 6.2 Polynomial fit to force data shown in Table 6.1. Experimental data are represented by circles.

θ	0	$\pi/2$	π	$3\pi/2$	2π	$5\pi/2$	3π	$7\pi/2$	4π	$9\pi/2$	5π
$F(\theta)$	5.00	7.83	12.27	19.22	30.10	47.15	73.86	115.70	181.24	283.90	444.71

TABLE 6.1: Force Data

Using all eleven data points, we obtain the interpolating polynomial

$$\begin{aligned} P_{10}(\theta) = 5 &+ 4.634349191\frac{\theta}{\pi} + 1.180101675\left(\frac{\theta}{\pi}\right)^2 + 2.401095921\left(\frac{\theta}{\pi}\right)^3 \\ &- 1.867465364\left(\frac{\theta}{\pi}\right)^4 + 1.331981301\left(\frac{\theta}{\pi}\right)^5 - 0.5264888061\left(\frac{\theta}{\pi}\right)^6 \\ &+ 0.1357248437\left(\frac{\theta}{\pi}\right)^7 - 0.02107936082\left(\frac{\theta}{\pi}\right)^8 \\ &+ 0.001848324094\left(\frac{\theta}{\pi}\right)^9 - 0.00006772484988\left(\frac{\theta}{\pi}\right)^{10}. \end{aligned}$$

Figure 6.2 shows that this polynomial provides a reasonable fit to the data over the entire range of θ values. Evaluating $P_{10}(\theta)$ and its derivative at $\theta = 5\pi/2$ provides the estimate $\mu \approx 0.29$. This result has been rounded to two decimal places, the same precision as the experimental data. Figure 6.3 shows that, with the exception of a small region near $\pi/4$, the same two decimal place estimate would have been obtained throughout the domain.

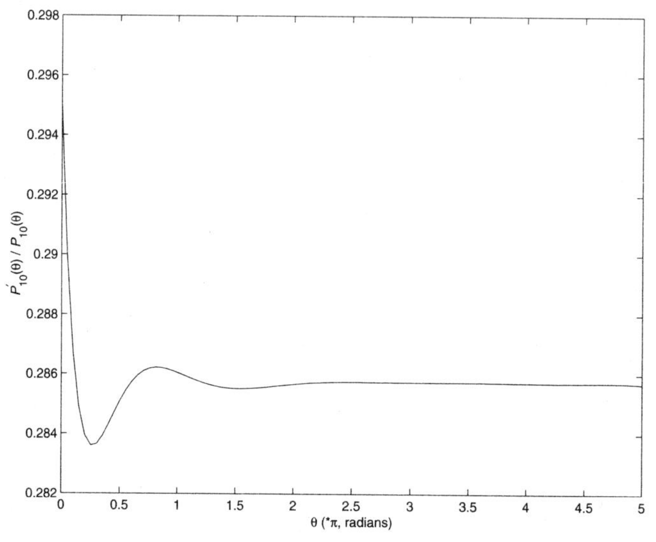

Figure 6.3 Approximate coefficient of friction obtained from $P'_{10}(\theta)/P_{10}(\theta)$.

N	0.0521	0.1028	0.2036	0.4946	0.9863	2.443	5.06
D	1.65	2.10	2.27	2.76	3.12	2.92	2.07

TABLE 6.2: Diffusivity of Copper Compounds

Diffusivity of Copper Compounds

Table 6.2 contains values of the diffusivity, D, of copper compounds from ion-exchange resins for various values of normality, N. A scientist needs to use this data to determine the normality that gives rise to the maximum diffusivity and the corresponding maximum diffusivity value.

Unlike our previous example, using all of the data points to determine an interpolating polynomial in this example provides rather miserable results (see Figure 6.4). An examination of the data suggests that the diffusivity will achieve its maximum somewhere between $N = 0.4946$, $N = 0.9863$, and $N = 2.443$. Interpolating these three values produces the polynomial

$$P(N) = -0.4462381347N^2 + 1.392987806N + 2.180191091.$$

This downward opening parabola achieves its maximum value of 3.26 at $N = 1.5608$. Although this polynomial provides plausible results, note $P(0.0521) = 2.2516$ and $P(5.06) = -2.1966$, which implies that $P(N)$ does not accurately reflect the curvature of the data.

To obtain a more accurate estimate of the maximum diffusivity, we turn to spline interpolation. Figure 6.5 displays the not-a-knot cubic spline interpolant for

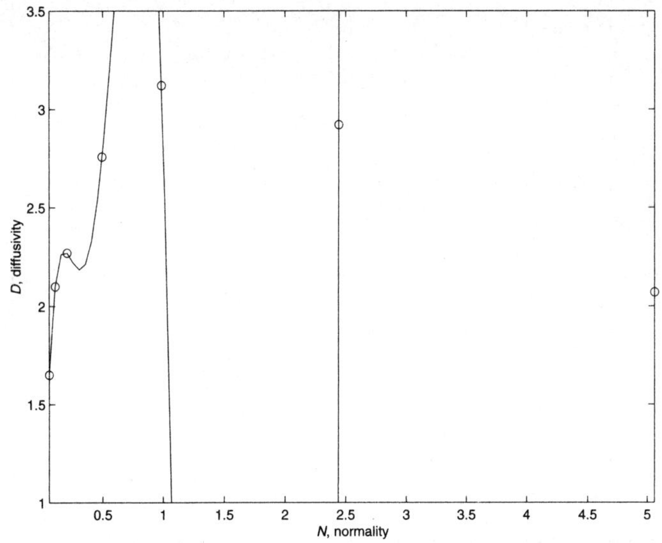

Figure 6.4 Interpolating polynomial for diffusivity data given in Table 6.2. Experimental values are denoted by circles.

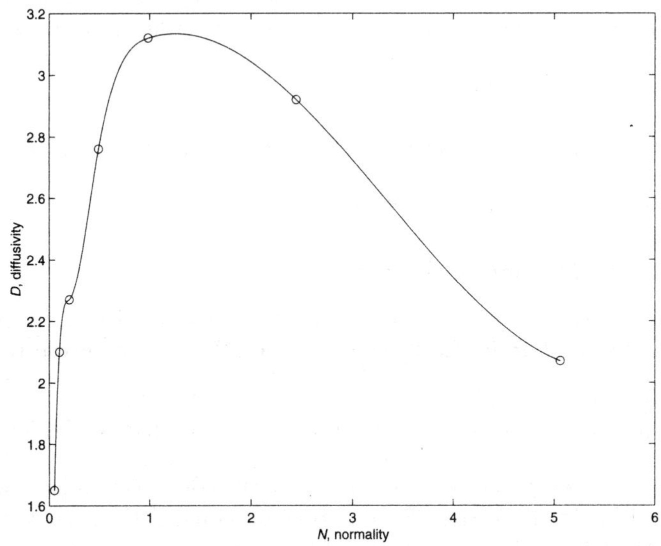

Figure 6.5 Not-a-knot cubic spline interpolant for diffusivity data given in Table 6.2. Experimental data are denoted by circles.

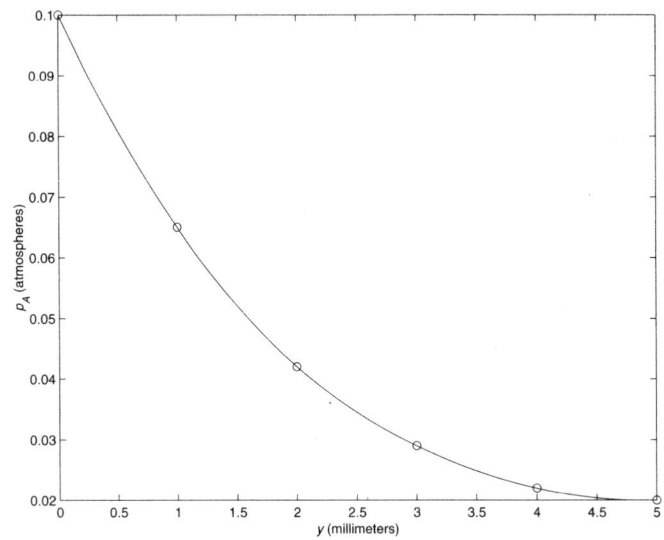

Figure 6.6 Not-a-knot cubic spline interpolant for partial pressure of water data. Experimental data are denoted by circles.

the diffusivity data. There are two observations that we can make from this figure. First, the behavior of the cubic spline is consistent with the experimental data over the entire domain. Second, the maximum diffusivity appears to occur just to the right of $N = 0.9863$. The cubic polynomial associated with that portion of the spline is

$$s(N) = 2.780284551 + 0.6109023877N - 0.2996338780N^2 + 0.02987363472N^3,$$

whose maximum is 3.134 at $N = 1.2549$.

Convection Mass Transfer Coefficient

Experimentally determined values for the partial pressure of water vapor, p_A, as a function of distance, y, from the surface of a pan of water are given below.

y (mm)	0	1	2	3	4	5
p_A (atm)	0.100	0.065	0.042	0.029	0.022	0.020

Approximating the water vapor as an ideal gas, the convection mass transfer coefficient is related to the derivative of the partial pressure at the surface of the water. Estimate the value of this derivative from the experimental data.

Figure 6.6 displays the not-a-knot cubic spline interpolant for the partial pressure data. The portion of the spline that applies when $y = 0$ is

$$0.1 - 0.04142222222y + 0.00663333333y^2 - 0.00021111111y^3,$$

from which it follows that

$$\left.\frac{dp_A}{dy}\right|_{y=0} = -0.04142222222 \text{ atm/mm}.$$

EXERCISES

1. Rework the coefficient of friction problem from the data in Table 6.1 using a not-a-knot cubic spline interpolant rather than a 10th-degree interpolating polynomial.

2. Rework the convection mass transfer coefficient problem using a single interpolating polynomial of degree at most 5.

3. Estimate the temperature, T, at which the sound speed, a, of water is a maximum. What is the corresponding maximum speed of sound in water?

T (°C)	0	10	20	30	40	50	60	70	80	90	100
a (m/s)	1402	1447	1482	1509	1529	1542	1551	1553	1554	1550	1543

4. The following table provides the height of water in a container as a function of time during an experiment dealing with Toricelli's law. Estimate the rate at which the height of water is changing at $t = 90$ seconds.

time (sec)	0.0	13.2	29.4	44.6	61.8	80.1	99.8	121.5	148.3	174.9
height (inches)	5.5	5.0	4.5	4.0	3.5	3.0	2.5	2.0	1.5	1.0

5. The thermal resistance, R, as a function of insulation thickness for a thin-walled copper tube is provided in the table below. Estimate the insulation thickness that corresponds to minimum thermal resistance.

thickness (mm)	0	2	5	10	20	40
thermal resistance R (m·K/W)	6.37	5.52	5.18	5.30	5.93	7.06

6. The specific heat at constant pressure, c_p, is given by

$$c_p = \left(\frac{\partial h}{\partial T}\right)_p,$$

where h denotes enthalpy and T denotes temperature. The parentheses around the partial derivative are used to indicate that the pressure, p, is to be held constant during this calculation.

The enthalpy of superheated nitrogen as a function of temperature is given in the table below. Use these data to estimate the specific heat at constant pressure of superheated nitrogen at a temperature of 200 K. Is the specific heat at constant pressure of superheated nitrogen constant over the range of temperatures 150 K through 250 K? If not, by how much does it vary?

T (K)	100	125	150	175	200
h (kJ/kg)	101.965	128.505	154.779	180.935	207.029

T (K)	225	250	275	300
h (kJ/kg)	233.085	259.122	285.144	311.158

7. In optical microlithography one of the most important performance metrics is the sidewall angle of the photoresist film at the completion of the development phase. Sidewall angle is a function of many different input parameters, including exposure energy, development time, thickness of contrast enhancing film, and numerical aperture. Sensitivity of sidewall to any one of these input parameters is measured by what is known as process latitude. Let θ denote the sidewall angle and u denote one of the input parameters. The process latitude with respect to u is given by

$$\left.\frac{\partial\theta}{\partial u}\right|_{u=u_0},$$

where u_0 is known as the nominal value of the input parameter, and all other input parameters are assumed held fixed in the computation of the derivative.

(a) Sidewall angle as a function of contrast enhancing layer thickness is given in the table below. Estimate the process latitude with respect to film thickness at a nominal value of 0.20 μm.

film thickness (μm)	0.00	0.10	0.20	0.30	0.40
θ (degrees)	80.7	83.8	85.7	86.2	86.3

(b) Sidewall angle as a function of numerical aperture is given in the table below. Estimate the process latitude with respect to numerical aperture at a nominal value of 0.24.

numerical aperture	0.16	0.20	0.24	0.28	0.32
θ (degrees)	76.0	81.1	83.5	85.0	85.7

8. Sidewall angle as a function of resist bleaching rate constant is given below. Estimate the resist bleaching rate constant which gives rise to the maximum sidewall angle.

bleaching rate constant (cm^2/mJ)	0.01	0.02	0.03	0.04	0.06	0.08
θ (degrees)	74.4	76.3	77.5	77.5	76.8	76.1

6.2 NUMERICAL DIFFERENTIATION, PART II

The approach that we used in Section 6.1 to obtain an estimate for the value of the derivative—pass a polynomial through the data and differentiate the resulting interpolating function—can also be applied to obtain generic formulas for approximating the derivatives of a function in terms of a linear combination of function values. Although the formulas developed in this section can be used to solve problems like those posed in Section 6.1, they are used primarily in the solution of differential equations, both ordinary and partial, via the finite difference method. This method will be discussed in Chapters 8 through 11.

Two Formulas for Approximating the First Derivative

Let's begin by developing a formula for approximating the first derivative of an arbitrary function, f, at $x = x_0$. The simplest meaningful approximation that can be obtained for $f'(x_0)$ requires using data from one other point, say $x = x_1$. If f is interpolated through $x = x_0$ and $x = x_1$, interpolation theory guarantees that

$$f(x) = \frac{x - x_1}{x_0 - x_1} f(x_0) + \frac{x - x_0}{x_1 - x_0} f(x_1) + f[x_0, x_1, x](x - x_0)(x - x_1).$$

The Lagrange form of the interpolating polynomial was selected because it clearly isolates the dependence of the polynomial on the function values being interpolated. The divided difference form for the error term has been selected because it is easier to manipulate than the derivative form.

If we now differentiate f with respect to x, we obtain

$$f'(x) = \frac{1}{x_0 - x_1} f(x_0) + \frac{1}{x_1 - x_0} f(x_1) + (x - x_0)(x - x_1)\frac{d}{dx} f[x_0, x_1, x] \\ + f[x_0, x_1, x](2x - x_0 - x_1).$$

Evaluating this expression at $x = x_0$ then yields, after some simplification,

$$f'(x_0) = \frac{f(x_1) - f(x_0)}{x_1 - x_0} + f[x_0, x_1, x_0](x_0 - x_1).$$

Recall that there is a connection between divided differences and derivatives; in particular, if f has two continuous derivatives, then there exists a ξ between x_0 and x_1 such that

$$f[x_0, x_1, x_0] = \frac{f''(\xi)}{2}.$$

Therefore,

$$f'(x_0) = \frac{f(x_1) - f(x_0)}{x_1 - x_0} + \frac{x_0 - x_1}{2} f''(\xi). \tag{1}$$

It is typical to use uniformly, or equally, spaced points when developing numerical differentiation formulas and to denote the spacing between points by the parameter h. Substituting $x_1 = x_0 + h$ into equation (1) produces

$$f'(x_0) = \frac{f(x_0 + h) - f(x_0)}{h} - \frac{h}{2} f''(\xi), \tag{2}$$

where $x_0 < \xi < x_0 + h$, while substituting $x_1 = x_0 - h$ into (1) produces

$$f'(x_0) = \frac{f(x_0) - f(x_0 - h)}{h} + \frac{h}{2} f''(\xi), \tag{3}$$

where $x_0 - h < \xi < x_0$. The first term on the right-hand side of equation (2) and of equation (3) represents our formula for approximating $f'(x_0)$; the second term on the right-hand side of each equation is the error term, which shall be discussed in more detail below. Since the approximation in (2) uses data to the right of x_0, it is referred to as a FORWARD DIFFERENCE APPROXIMATION. For similar reasons, the formula in (3) is referred to as a BACKWARD DIFFERENCE APPROXIMATION.

Alternative Derivation

Difference approximations can also be obtained through the use of Taylor's theorem. For example, suppose that f has two continuous derivatives. Then, by Taylor's theorem, we may expand $f(x_0+h)$ as

$$f(x_0+h) = f(x_0) + hf'(x_0) + \frac{h^2}{2}f''(\xi),$$

where $x_0 < \xi < x_0+h$. Solving this equation for $f'(x_0)$ yields the forward difference approximation in (2). Had we started with an expansion for $f(x_0-h)$, we would have reproduced the backward difference approximation in equation (3).

The Error Term

The error terms in equations (2) and (3) actually provide us with two pieces of information. First, if we assume that f is continuous with two continuous derivatives on $[a,b]$, then the Extreme Value Theorem guarantees that f'' is bounded on either of the closed intervals $[x_0, x_0+h]$ or $[x_0-h, x_0]$. The approximation formulas from (2) and (3) therefore introduce an error which is proportional to the first power of the spacing parameter, or step size, h. Since the error depends on the first power of h, the formulas in (2) and (3) are said to provide *first-order* approximations to the first derivative. From a practical standpoint, when a first-order formula is used and the step size is cut by a factor of 2, say, we expect the error to drop by that same factor. The following example illustrates this point.

EXAMPLE 6.1 Approximating the Derivative of the Natural Logarithm

Consider the function $f(x) = \ln x$. The table below displays the results of approximating the value of the derivative of the natural logarithm at $x_0 = 2$ using the first-order formulas given by (2) and (3).

h	$\frac{f(x_0+h)-f(x_0)}{h}$	error
1.0	$f'(2) \approx \frac{\ln(3.0)-\ln(2.0)}{1.0} = 0.405465$	0.094535
0.1	$f'(2) \approx \frac{\ln(2.1)-\ln(2.0)}{0.1} = 0.487902$	0.012098
0.01	$f'(2) \approx \frac{\ln(2.01)-\ln(2.0)}{0.01} = 0.498754$	1.2458×10^{-3}
0.001	$f'(2) \approx \frac{\ln(2.001)-\ln(2.0)}{0.001} = 0.499875$	1.2496×10^{-4}

h	$\frac{f(x_0)-f(x_0-h)}{h}$	error
1.0	$f'(2) \approx \frac{\ln(2.0)-\ln(1.0)}{1.0} = 0.693147$	0.193147
0.1	$f'(2) \approx \frac{\ln(2.0)-\ln(1.9)}{0.1} = 0.512933$	0.012933
0.01	$f'(2) \approx \frac{\ln(2.0)-\ln(1.99)}{0.01} = 0.501254$	1.2542×10^{-3}
0.001	$f'(2) \approx \frac{\ln(2.0)-\ln(1.999)}{0.001} = 0.500125$	1.2505×10^{-4}

Note how each time the step size, h, is cut by a factor of 10, the corresponding error also drops by a factor of 10. This is characteristic of first-order approximation formulas.

The second piece of information that can be obtained from the error term is related to the order of derivative that appears. In (2) and (3), each error term contains f''. Since every constant and every linear function has a second derivative that is identically zero, we are guaranteed that the formulas in (2) and (3) will provide the exact value of the first derivative of every constant and every linear function, regardless of the step size used. In general, if the error term involves $f^{(n)}$, then the corresponding approximation formula will provide exact results for all polynomial functions of degree up to and including $n-1$.

EXAMPLE 6.2 Formulas in (2) and (3) Are Exact for Constant and Linear Functions

Due to the linearity of both the derivative and our difference approximation formulas, we need only consider the functions $f(x)=1$ and $f(x)=x$.

	$f'(x_0)$	$\frac{f(x_0+h)-f(x_0)}{h}$	$\frac{f(x_0)-f(x_0-h)}{h}$
$f(x)=1$	0	0	0
$f(x)=x$	1	1	1
$f(x)=x^2$	$2x_0$	$2x_0+h$	$2x_0-h$

A Higher-Order Approximation for the First Derivative

Using the function values $f(x_0)$ and $f(x_1)$, the best we can do is generate a first-order approximation to $f'(x_0)$. To obtain a second-order approximation to the first derivative at $x=x_0$—that is, one for which the error term involves h^2—the function value at another point, say x_2, must be included. There are two choices for the placement of x_2 relative to x_0 and x_1: place x_2 on the same side of x_0 as x_1 or place x_2 on the opposite side.

First, suppose we have been using x_0 and $x_1=x_0+h$ and place x_2 at x_0+2h. From interpolation theory, we know

$$f(x)=\frac{(x-x_1)(x-x_2)}{2h^2}f(x_0)-\frac{(x-x_0)(x-x_2)}{h^2}f(x_1)+\frac{(x-x_0)(x-x_1)}{2h^2}f(x_2)+f[x_0,x_1,x_2,x](x-x_0)(x-x_1)(x-x_2).$$

Differentiating this expression with respect to x and evaluating at $x=x_0$ produces

$$f'(x_0)=\frac{-3f(x_0)+4f(x_0+h)-f(x_0+2h)}{2h}+\frac{h^2}{3}f'''(\xi), \tag{4}$$

for some ξ between x_0 and $x_0 + 2h$. The details of this derivation are left as an exercise. Since values of f to the right of x_0 only have been used and an error term that depends on the second power of h has been obtained, this is a SECOND-ORDER FORWARD DIFFERENCE APPROXIMATION to the first derivative. If h is replaced by $-h$ in (4), the SECOND-ORDER BACKWARD DIFFERENCE APPROXIMATION

$$f'(x_0) = \frac{3f(x_0) - 4f(x_0 - h) + f(x_0 - 2h)}{2h} + \frac{h^2}{3}f'''(\xi) \tag{5}$$

is obtained.

Now suppose that x_2 is placed on the opposite side of x_0, and hence data from $x_0 - h$, x_0 and $x_0 + h$ are used to develop an approximation. If the standard procedure is followed (interpolate, differentiate with respect to x, and then evaluate at $x = x_0$), we obtain

$$f'(x_0) = \frac{f(x_0 + h) - f(x_0 - h)}{2h} - \frac{h^2}{6}f'''(\xi). \tag{6}$$

Since this formula uses data from either side of x_0, it is referred to as a *central* difference approximation. In particular, it is the SECOND-ORDER CENTRAL DIFFERENCE APPROXIMATION to the first derivative.

EXAMPLE 6.3 Verifying Second-Order Approximation for Formulas (4), (5), and (6)

Consider the function $f(x) = \ln x$. The table below displays the results of approximating the value of the derivative of the natural logarithm at $x_0 = 2$ using the second-order formulas given by (4), (5), and (6).

h	Equation (4)	Error	Equation (5)	Error
0.1	$f'(2) \approx 0.499252$	7.4762×10^{-4}	$f'(2) \approx 0.499063$	9.3669×10^{-4}
0.01	$f'(2) \approx 0.499992$	8.2405×10^{-6}	$f'(2) \approx 0.499992$	8.4280×10^{-6}
0.001	$f'(2) \approx 0.500000$	8.3245×10^{-8}	$f'(2) \approx 0.500000$	8.3430×10^{-8}

h	Equation (6)	Error
0.1	$f'(2) \approx 0.500417$	4.1729×10^{-4}
0.01	$f'(2) \approx 0.500004$	4.1667×10^{-6}
0.001	$f'(2) \approx 0.500000$	4.1666×10^{-8}

Note how each time the step size, h, is cut by a factor of 10, the corresponding error drops by a factor of 100. This is characteristic of second order approximation formulas.

Although the formulas in equations (4), (5), and (6) are all second order accurate, note that for each fixed value of h, the central difference formula produces an error which is roughly half that of the other formulas. This should not be surprising upon closer examination of the error terms: the coefficient in (6) is half the coefficient in the other equations. Furthermore, the central difference formula required only two function evaluations, as opposed to three function evaluations for the forward and backward difference formulas. As seen in equations (2) and (3), using two function evaluations with a forward or backward difference formula achieves only first order accuracy. In general, central difference formulas require fewer function evaluations to achieve a given order of accuracy. For this reason, central difference formulas are used most often in practice.

A Formula for the Second Derivative

The simplest meaningful approximation to the second derivative requires the use of three points. The central difference formula will be developed here. Derivations of the forward and backward difference formulas will be left for exercises. If f is interpolated at $x_0 - h$, x_0 and $x_0 + h$ and the resulting polynomial is differentiated twice with respect to x and then evaluated at $x = x_0$, we obtain

$$f''(x_0) = \frac{f(x_0+h) - 2f(x_0) + f(x_0-h)}{h^2} - 2h^2 \frac{d}{dx} f[x_0-h, x_0, x_0+h, x]\bigg|_{x=x_0}. \tag{7}$$

Next, using the definition of the derivative and the definition of divided differences, we find that

$$\begin{aligned}
\frac{d}{dx} f[x_0-h, x_0, x_0+h, x]\bigg|_{x=x_0} \\
&= \lim_{\Delta\to 0} \frac{f[x_0-h, x_0, x_0+h, x_0+\Delta] - f[x_0-h, x_0, x_0+h, x_0]}{\Delta} \\
&= \lim_{\Delta\to 0} f[x_0-h, x_0, x_0+h, x_0+\Delta, x_0] \\
&= f[x_0-h, x_0, x_0+h, x_0, x_0].
\end{aligned}$$

Finally, substituting this result into (7) and replacing the divided difference by $\frac{1}{24} f^{(4)}(\xi)$, where $x_0 - h < \xi < x_0 + h$, yields the SECOND-ORDER CENTRAL DIFFERENCE APPROXIMATION for the second derivative:

$$f''(x_0) = \frac{f(x_0+h) - 2f(x_0) + f(x_0-h)}{h^2} - \frac{h^2}{12} f^{(4)}(\xi). \tag{8}$$

The Effect of Roundoff Errors

Considering our discussion of the propagation of roundoff error in Chapter 1, it should not be surprising that numerical differentiation is unstable with respect to roundoff errors. After all, the formulas we just finished deriving involve the subtraction of nearly equal numbers followed by a division by a small number.

To demonstrate this idea, consider the $O(h^2)$ central difference approximation for f':

$$f'(x_0) = \frac{f(x_0+h) - f(x_0-h)}{2h} - \frac{h^2}{6} f'''(\xi).$$

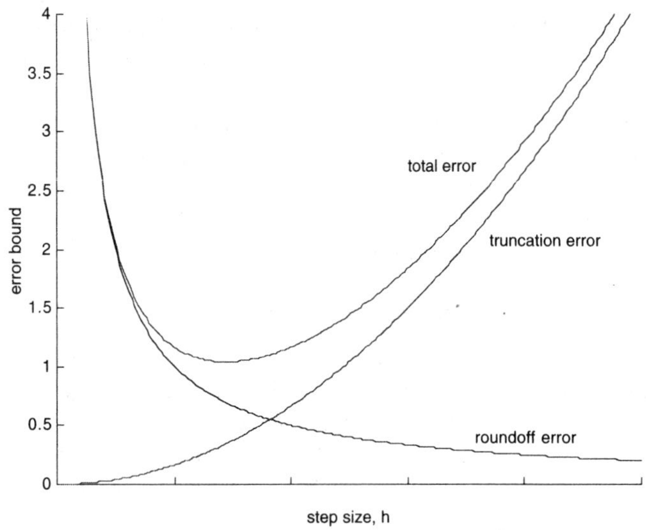

Figure 6.7 Error bound for second-order central difference approximation to the first derivative as a function of step size, when roundoff error is taken into account.

When the values of f are entered into the computer, a roundoff error will be introduced, so that our calculations will actually be made with $\tilde{f}(x_0+h)$ and $\tilde{f}(x_0-h)$, where

$$f(x_0+h) = \tilde{f}(x_0+h) + e(x_0+h),$$
$$f(x_0-h) = \tilde{f}(x_0-h) + e(x_0-h)$$

and $e(x_0+h)$ and $e(x_0-h)$ are the respective roundoff errors. Substituting these values into the central difference formula leads to

$$f'(x_0) = \frac{\tilde{f}(x_0+h) - \tilde{f}(x_0-h)}{2h} + \frac{e(x_0+h) - e(x_0-h)}{2h} - \frac{h^2}{6}f'''(\xi).$$

If we assume that $|e(x_0 \pm h)| \le \epsilon$, then

$$\left| f'(x_0) - \frac{\tilde{f}(x_0+h) - \tilde{f}(x_0-h)}{2h} \right| \le \frac{\epsilon}{h} + \frac{h^2 M}{6},$$

where $M = \max_{a \le x \le b} |f'''(x)| \equiv \|f'''\|_\infty$. Note that as $h \to 0$, the truncation error term $h^2M/6 \to 0$, but the roundoff error term $\epsilon/h \to \infty$ (see Figure 6.7).

EXAMPLE 6.4 Instability of Numerical Differentiation

Consider approximating $f'(0)$ for $f(x) = e^x$ using the second-order central difference formula

$$f'(x_0) \approx \frac{f(x_0+h) - f(x_0-h)}{2h}.$$

Working in double precision, we find that for step sizes from 10^{-1} down to 10^{-5}, the second-order character of the approximation formula is evident. With each reduction by a factor of 10 in the step size, the total error drops by a factor of 100. As the step size is reduced further, the total error transitions from the truncation error dominated domain into the roundoff error dominated domain, and the accuracy of the approximation deteriorates rapidly.

Step Size	Error	Step Size	Error
10^{-1}	1.67×10^{-3}	10^{-9}	2.72×10^{-8}
10^{-2}	1.67×10^{-5}	10^{-10}	8.27×10^{-8}
10^{-3}	1.67×10^{-7}	10^{-11}	8.27×10^{-8}
10^{-4}	1.67×10^{-9}	10^{-12}	3.34×10^{-5}
10^{-5}	1.21×10^{-11}	10^{-13}	2.44×10^{-4}
10^{-6}	2.68×10^{-11}	10^{-14}	8.00×10^{-4}
10^{-7}	5.26×10^{-10}	10^{-15}	5.47×10^{-2}
10^{-8}	6.08×10^{-9}	10^{-16}	0.445

EXERCISES

1. Derive the second-order central difference approximation for the first derivative, including error term:

$$f'(x_0) = \frac{f(x_0+h) - f(x_0-h)}{2h} - \frac{h^2}{6}f'''(\xi).$$

2. Derive equation (4).

3. Derive equation (7).

4. **(a)** Derive the following difference approximation for the first derivative:

$$f'(x_0) \approx \frac{f(x_0+2h) - f(x_0-h)}{3h}.$$

(b) What is the error term associated with this formula?

(c) Numerically verify the order of approximation using $f(x) = \ln x$ and $x_0 = 2$.

5. **(a)** Derive the following forward difference approximation for the second derivative:

$$f''(x_0) \approx \frac{f(x_0) - 2f(x_0+h) + f(x_0+2h)}{h^2}.$$

(b) What is the error term associated with this formula?

(c) Numerically verify the order of approximation using $f(x) = e^x$ and $x_0 = 0$.

6. **(a)** Derive the following backward difference approximation for the second derivative:

$$f''(x_0) \approx \frac{f(x_0-2h) - 2f(x_0-h) + f(x_0)}{h^2}.$$

(b) What is the error term associated with this formula?

(c) Numerically verify the order of approximation using $f(x) = \ln x$ and $x_0 = 2$.

7. (a) Derive a formula for approximating the first derivative of an arbitrary function at $x = x_0$ using four equally spaced points, with two (2) of those points to the left and one (1) to the right of $x = x_0$.
   (b) What is the order of approximation for the formula obtained in part (a)? Completely justify your response.

8. (a) Derive a formula for approximating the first derivative of an arbitrary function at $x = x_0$ by interpolating at $x = x_0 + h$ and $x = x_0 - \alpha h$ for $\alpha > 0$.
   (b) Show, analytically, that the formula from part (a) is second order when $\alpha = 1$, but only first order for $\alpha \neq 1$.

9. (a) Derive a formula for approximating the second derivative of an arbitrary function at $x = x_0$ by interpolating at $x = x_0 + h$, $x = x_0$ and $x = x_0 - \alpha h$ for $\alpha > 0$.
   (b) Show, analytically, that the formula from part (a) is second order when $\alpha = 1$, but only first order for $\alpha \neq 1$.

10. (a) Using $f(x) = \ln x$ and $x_0 = 2$, demonstrate numerically that the central difference approximation for the second derivative given by

$$f''(x_0) \approx \frac{f(x_0 - h) - 2f(x_0) + f(x_0 + h)}{h^2},$$

    is second order accurate.
    (b) Repeat part (a) using $f(x) = e^x$ and $x_0 = 0$.

11. Verify that each of the following difference approximations for the first derivative provides the exact value of the derivative, regardless of h, for the functions $f(x) = 1$, $f(x) = x$ and $f(x) = x^2$, but not for the function $f(x) = x^3$.
    (a) $f'(x_0) \approx \frac{-3f(x_0)+4f(x_0+h)-f(x_0+2h)}{2h}$
    (b) $f'(x_0) \approx \frac{3f(x_0)-4f(x_0-h)+f(x_0-2h)}{2h}$
    (c) $f'(x_0) \approx \frac{f(x_0+h)-f(x_0-h)}{2h}$

12. Verify that the second-order central difference approximation for the second derivative provides the exact value of the second derivative, regardless of the value of h, for the functions $f(x) = 1$, $f(x) = x$, $f(x) = x^2$, and $f(x) = x^3$, but not for the function $f(x) = x^4$.

13. (a) Use the formula

$$f'(x_0) \approx \frac{f(x_0 + h) - f(x_0)}{h}$$

    to approximate the derivative of $f(x) = 1 + x + x^3$ at $x_0 = 1$, taking $h = 1, 0.1, 0.01$, and 0.001. What is the order of approximation?
    (b) Repeat part (a) for $x_0 = 0$.
    (c) Explain any difference between the results from part (a) and those from part (b).

14. (a) Use the formula

$$f'(x_0) \approx \frac{f(x_0) - f(x_0 - h)}{h}$$

to approximate the derivative of $f(x) = \sin x$ at $x_0 = \pi$, taking $h = 1, 0.1, 0.01$, and 0.001. What is the order of approximation?

(b) Repeat part (a) for $x_0 = \pi/2$.

(c) Explain any difference between the results from part (a) and those from part (b).

15. Consider the following formula for approximating the first derivative of an arbitrary function:

$$f'(x_0) = \frac{-2f(x_0 - 3h) + 9f(x_0 - 2h) - 18f(x_0 - h) + 11f(x_0)}{6h} + \frac{1}{4}h^3 f^{(4)}(\xi),$$

where $x_0 - 3h < \xi < x_0$.

(a) Suppose that the function values used in the above formula contain round-off/data errors that are bounded in absolute value by ϵ and that the absolute value of the fourth derivative is bounded by M. Derive a bound for the approximation error associated with the above formula as a function of ϵ, M, and h.

(b) Suppose $\epsilon = 5.96 \times 10^{-8}$ (machine precision in IEEE standard single precision). Determine the value for the step size h that minimizes the bound on the error when approximating the value of the derivative of $f(x) = e^x$ at $x_0 = 1$.

16. Consider the second-order forward difference formula for approximating the first derivative of an arbitrary function:

$$f'(x_0) = \frac{-3f(x_0) + 4f(x_0 + h) - f(x_0 + 2h)}{2h} + \frac{1}{3}h^2 f'''(\xi),$$

where $x_0 < \xi < x_0 + 2h$.

(a) Suppose that the function values used in the above formula contain round-off/data errors that are bounded in absolute value by ϵ and that the absolute value of the third derivative is bounded by M. Derive a bound for the approximation error associated with the above formula as a function of ϵ, M, and h.

(b) Suppose $\epsilon = 1.11 \times 10^{-16}$ (machine precision in IEEE standard double precision). Determine the value for the step size h that minimizes the bound on the error when approximating the value of the derivative of $f(x) = \ln x$ at $x_0 = 2$.

6.3 RICHARDSON EXTRAPOLATION

In the previous section several first- and second-order finite difference approximation formulas for first and second derivatives were obtained. Higher-order formulas can of course be derived by interpolating more data points, but an alternative for obtaining higher-order approximations is to use a procedure known as extrapolation. The basic idea behind extrapolation is that whenever the leading term in the error for an approximation formula is known, we can combine two approximations obtained from that formula using different values of the parameter h to obtain a higher-order approximation. This process will be illustrated in this section for finite difference approximations to derivatives—the technique is known as Richardson extrapolation. In a later section, we will apply extrapolation to numerical integration formulas, and in the next chapter, we will apply extrapolation to numerical solutions of initial value problems.

Small-O Notation

Recall that a function $g(h)$ is said to be big-O of h^k as $h \to 0$, written $g(h) = O(h^k)$, provided there exists a positive constant L such that

$$\left|\frac{g(h)}{h^k}\right| \le L$$

for all sufficiently small h. In other words, $g(h) \to 0$ as $h \to 0$ at least as fast as h^k. If it happens that

$$\lim_{h\to 0}\left|\frac{g(h)}{h^k}\right| = 0,$$

so that $g(h) \to 0$ as $h \to 0$ faster than h^k, then $g(h)$ is said to be small-O of h^k, written $g(h) = o(h^k)$.

Richardson Extrapolation

Let's consider the second-order central difference formula for f':

$$f'(x_0) = \frac{f(x_0+h) - f(x_0-h)}{2h} - \frac{h^2}{6}f'''(\xi).$$

For notational convenience, let D denote the true value of the derivative, and let D_h denote the approximation obtained using a step size of h. From our derivation in the previous section, we know that $x_0 - h < \xi < x_0 + h$; hence, the squeeze theorem guarantees

$$\xi \to x_0 \text{ as } h \to 0.$$

Therefore, as $h \to 0$

$$\left|\frac{D - D_h}{h^2}\right| = \left|\frac{1}{6}f'''(\xi)\right| \to \left|\frac{1}{6}f'''(x_0)\right|$$

$$\Rightarrow D = D_h + O(h^2).$$

Let's now look at the error term more precisely. Since

$$\frac{h^2}{6}f'''(\xi) = \frac{h^2}{6}f'''(x_0) + \frac{h^2}{6}\left[f'''(\xi) - f'''(x_0)\right],$$

we see that

$$\lim_{h\to 0}\left|\frac{D - D_h - (h^2/6)f'''(x_0)}{h^2}\right| = \lim_{h\to 0}\left|\frac{1}{6}\left[f'''(\xi) - f'''(x_0)\right]\right| = 0.$$

Therefore, $D = D_h + K_1 h^2 + o(h^2)$, where $K_1 = \frac{1}{6}f'''(x_0)$.

As stated earlier, the process of extrapolation uses two approximations computed from the same formula, but with different values of h, to obtain a higher-order approximation. For the second-order central difference approximation to f', we have just established

$$D = D_h + K_1 h^2 + o(h^2);$$

hence,

$$D = D_{h/2} + K_1 \left(\frac{h}{2}\right)^2 + o(h^2).$$

Upon subtracting these two expressions, we obtain

$$0 = D_h - D_{h/2} + \frac{3}{4}K_1 h^2 + o(h^2),$$

which can easily be solved for $K_1 h^2$—the leading term in the error of our original approximation:

$$K_1 h^2 = \frac{4}{3}[D_{h/2} - D_h] + o(h^2).$$

We now substitute for $K_1 h^2$ in the original approximation to obtain

$$\begin{aligned} D &= D_h + \frac{4}{3}[D_{h/2} - D_h] + o(h^2) \\ &= \frac{4D_{h/2} - D_h}{3} + o(h^2). \end{aligned}$$

The formula $\frac{4D_{h/2}-D_h}{3}$ is a higher-order approximation for the first derivative in the sense that it converges faster than h^2, as opposed to at the same rate as h^2.

EXAMPLE 6.5 Extrapolating the Derivative of the Natural Logarithm

Let's reconsider the approximation of the derivative of the function $f(x) = \ln x$ using the second-order central difference approximation formula. With $x_0 = 2$, we find

h	$D_h = \frac{f(x_0+h)-f(x_0-h)}{2h}$	Error
0.1	0.500417292	4.17×10^{-4}
0.05	0.500104205	1.04×10^{-4}
0.025	0.500026043	2.60×10^{-5}

Applying the extrapolation formula to these results produces

h	$\frac{4D_{h/2}-D_h}{3}$	Error
0.1	0.499999842	1.58×10^{-7}
0.05	0.499999989	1.10×10^{-8}

Not only has the approximation error been significantly reduced as a result of extrapolation, the cut in error due to the cut in the step size is also larger. In particular, it appears as if the extrapolated values are fourth-order approximations—having cut h by a factor of 2, the error has dropped by a factor of 16!

Seeing the improvement generated in the previous example, the natural question to ask is, Can we extrapolate again? If we know the power of h in the new leading term in the error, then the answer is yes.

Repeated Extrapolation

In order to keep track of the various approximations that will be generated, let's modify our notation slightly. Let $D_h^{(1)}$ denote the original approximation, $D_h^{(2)}$ denote the first extrapolation, $D_h^{(3)}$ denote the next extrapolation, and so on. Furthermore, we will adopt the convention that the step size associated with an extrapolated value is the larger of the two step sizes used to calculate the extrapolated value. For example, for the second-order central difference approximation to f', we would write

$$D_h^{(2)} = \frac{4D_{h/2}^{(1)} - D_h^{(1)}}{3}.$$

Provided that f has five continuous derivatives, it can be shown that

$$D = D_h^{(1)} + K_1 h^2 + K_2 h^4 + o(h^4),$$

where $D_h^{(1)}$ still refers to the second-order central difference approximation for f' and K_1 and K_2 are constants independent of h. To establish this formula, first expand $f(x_0 + h)$ and $f(x_0 - h)$ in Taylor series about the point $x = x_0$:

$$f(x_0+h) = f(x_0) + hf'(x_0) + \frac{h^2}{2}f''(x_0) + \frac{h^3}{6}f'''(x_0) + \frac{h^4}{24}f^{(4)}(x_0) + \frac{h^5}{120}f^{(5)}(\xi_+)$$

$$f(x_0-h) = f(x_0) - hf'(x_0) + \frac{h^2}{2}f''(x_0) - \frac{h^3}{6}f'''(x_0) + \frac{h^4}{24}f^{(4)}(x_0) - \frac{h^5}{120}f^{(5)}(\xi_-),$$

where $x_0 < \xi_+ < x_0 + h$ and $x_0 - h < \xi_- < x_0$. Subtracting the bottom expansion from the top and solving for $f'(x_0)$ gives

$$\begin{aligned} f'(x_0) &= D_h^{(1)} + \frac{h^2}{6}f'''(x_0) + \frac{h^4}{240}\left[f^{(5)}(\xi_+) + f^{(5)}(\xi_-)\right] \\ &= D_h^{(1)} + \frac{h^2}{6}f'''(x_0) + \frac{h^4}{120}f^{(5)}(\xi) \\ &= D_h^{(1)} + \frac{h^2}{6}f'''(x_0) + \frac{h^4}{120}f^{(5)}(x_0) + \frac{h^4}{120}\left[f^{(5)}(\xi) - f^{(5)}(x_0)\right], \end{aligned}$$

where $x_0 - h < \xi < x_0 + h$ and the Intermediate Value Theorem has been applied to the term involving the fifth derivative in going from the first line to the second. Since $\xi \to x_0$ as $h \to 0$, the term in square brackets in the final line is $o(h^4)$. Setting $K_1 = f'''(x_0)/6$ and $K_2 = f^{(5)}(x_0)/120$, we have the desired formula:

$$D = D_h^{(1)} + K_1 h^2 + K_2 h^4 + o(h^4).$$

After one extrapolation, this becomes

$$D = D_h^{(2)} + K_2' h^4 + o(h^4),$$

which verifies our observation at the end of the last example—the extrapolated values are fourth-order approximations. The next extrapolation proceeds in exactly the same manner as the first. We will subtract

$$D = D_{h/2}^{(2)} + K_2' \left(\frac{h}{2}\right)^4 + o(h^4)$$

from

$$D = D_h^{(2)} + K_2' h^4 + o(h^4),$$

leaving

$$0 = D_h^{(2)} - D_{h/2}^{(2)} + \frac{15}{16} K_2' h^4 + o(h^4).$$

This last expression gives

$$K_2' h^4 = \frac{16}{15}\left[D_{h/2}^{(2)} - D_h^{(2)}\right] + o(h^4),$$

so that

$$D = D_h^{(2)} + \frac{16}{15}\left[D_{h/2}^{(2)} - D_h^{(2)}\right] + o(h^4),$$

or

$$D_h^{(3)} = \frac{16 D_{h/2}^{(2)} - D_h^{(2)}}{15}.$$

EXAMPLE 6.6 A Second Extrapolation

In the last example, we found

h	$D_h^{(2)} = \frac{4D_{h/2} - D_h}{3}$	Error
0.1	0.499999842	1.58×10^{-7}
0.05	0.499999989	1.10×10^{-8}

Extrapolating again, we find

$$D_{0.1}^{(3)} = \frac{16 D_{0.05}^{(2)} - D_{0.1}^{(2)}}{15} = 0.499999998.$$

The error in this final approximation is 1.20×10^{-9}.

Note that the formula for $D_h^{(3)}$ has a structure similar to the formula for $D_h^{(2)}$. In particular, the lower order approximation with the larger step size is subtracted from a multiple of the lower order approximation with the smaller step size. This result is then divided by the sum of the coefficients in the numerator. The coefficient of 16 arises from the fact that the step size was cut by a factor of 2 and the lower order approximations were fourth-order (i.e., $16 = 2^4$). In general, if a pth-order formula is extrapolated by cutting the step size by a factor of b, the extrapolation formula will take the form

$$\frac{b^p D_{h/b} - D_h}{b^p - 1}.$$

When performing repeated extrapolations, it is convenient to organize the calculations into an extrapolation table like the following one. Listing the order of approximation associated with each column helps to keep track of the weights needed to compute each successive column.

Step Size	$O(h^{p_1})$	$O(h^{p_2})$	$O(h^{p_3})$	
h	$D_h^{(1)}$			
$h/2$	$D_{h/2}^{(1)}$	$D_h^{(2)}$		
$h/4$	$D_{h/4}^{(1)}$	$D_{h/2}^{(2)}$	$D_h^{(3)}$	
$h/8$	$D_{h/8}^{(1)}$	$D_{h/4}^{(2)}$	$D_{h/2}^{(3)}$	$D_h^{(4)}$

EXAMPLE 6.7 Using a Different Low-Order Approximation Formula

The extrapolation table below displays approximations to the derivative of $f(x) = \tan^{-1} x$ at $x_0 = 2$ starting from the first-order forward difference approximation

$$D_h^{(1)} = \frac{f(x_0 + h) - f(x_0)}{h}.$$

It can be shown (see Exercise 9) that

$$D = D_h^{(1)} + k_1 h + k_2 h^2 + k_3 h^3 + o(h^3),$$

where k_1, k_2, and k_3 are constants independent of h. Since each row of the extrapolation table uses a step size one-half that from the previous row, the weights used to produce the second, third, and fourth columns of the table are 2 and -1, $4/3$, and $-1/3$ and $8/7$, and $-1/7$, respectively.

h	$O(h)$	$O(h^2)$	$O(h^3)$	
1	0.141897054			
0.5	0.166282463	0.190667872		
0.25	0.181693117	0.197103771	0.199249070	
0.125	0.190440208	0.199187299	0.199881808	0.199972199

The error in the final approximation is 2.78×10^{-5}. How small would the step size have to be to obtain the same accuracy using the original approximation formula?

EXERCISES

1. In the last example, extrapolation was used to obtain an approximation to the first derivative of $f(x) = \tan^{-1} x$ at $x_0 = 2$ with an error of 2.78×10^{-5}. The smallest step size used in the construction of the extrapolation table was $h = 0.125$. Starting approximations for the extrapolation table were obtained from the first-order forward difference formula

$$D_h^{(1)} = \frac{f(x_0 + h) - f(x_0)}{h}.$$

What step size would be needed in the first-order forward difference formula to obtain the same accuracy, 2.78×10^{-5}, as the final extrapolated value?

2. In the first example, extrapolation was used to obtain an approximation to the first derivative of $f(x) = \ln x$ at $x_0 = 2$ with an error of 1.20×10^{-9}. The smallest step size used in the construction of the extrapolation table was $h = 0.025$. Starting approximations for the extrapolation table were obtained from the second-order central difference formula

$$D_h^{(1)} = \frac{f(x_0 + h) - f(x_0 - h)}{2h}.$$

What step size would be needed in the second-order central difference formula to obtain the same accuracy, 1.20×10^{-9}, as the final extrapolated value?

In Exercises 3–7, fill in the missing values from the given extrapolation table. The order of approximation associated with each column is indicated above the column, and with each new row, h is reduced by a factor of two.

3.

$O(h^2)$	$O(h^3)$	
0.7398169125		
0.7187845413	?	
0.7104251526	?	?

4.

$O(h)$	$O(h^2)$	
−0.9397248595		
−0.8555953748	?	
−0.7887202658	?	?

5.

$O(h^2)$	$O(h^4)$	$O(h^6)$	
0.7500000000			
0.7083333333	?		
0.6970238095	?	0.6931746034	
?	0.6931545307	?	0.6931474775

6.

$O(h^2)$	$O(h^3)$	$O(h^4)$	
1.0471975512			
?	1.1444682995		
?	?	1.1523449594	
1.1514137785	1.1540323927	1.1544141092	?

7.

$O(h)$	$O(h^2)$	$O(h^3)$	
?			
0.6065306597	0.8451818783		
0.7788007831	0.9510709063	?	
?	?	0.9979003940	0.9995479864

8. Let D denote the true derivative of a function, and let D_h denote the first-order backward difference approximation to the derivative; that is,

$$D_h = \frac{f(x_0) - f(x_0 - h)}{h}.$$

It can be shown that

$$D = D_h + k_1 h + k_2 h^2 + k_3 h^3 + o(h^3),$$

where k_1, k_2, and k_3 are constants independent of h. Let $f(x) = \ln(x^2 + 1)$ and $x_0 = 1$.

(a) Starting from $h = 1$, approximate the value of the first derivative of f at x_0 by applying extrapolation to D_h. Use four rows in your extrapolation table.

(b) What is the error in the final approximation?

(c) What step size would be needed in the first-order backward difference formula to obtain the same accuracy as the final extrapolated value?

9. Assuming that f has four continuous derivatives, show that

$$D = D_h^{(1)} + k_1 h + k_2 h^2 + k_3 h^3 + o(h^3),$$

where D denotes the true derivative of a function, $D_h^{(1)}$ denotes the first-order forward difference approximation to the derivative and k_1, k_2 and k_3 are constants independent of h. [*Hint*: Use Taylor's theorem to expand $f(x_0+h)$ about the point $x = x_0$.]

10. **(a)** Show that

$$D = D_h + k_1 h^2 + k_2 h^3 + k_3 h^4 + o(h^4),$$

where D denotes the true derivative of a function, D_h denotes the second-order forward difference approximation to the derivative

$$D_h = \frac{-3f(x_0) + 4f(x_0 + h) - f(x_0 + 2h)}{2h}$$

and k_1, k_2, and k_3 are constants independent of h. [*Hint*: Use Taylor's theorem to expand $f(x_0 + h)$ and $f(x_0 + 2h)$ about the point $x = x_0$.]

(b) Let $f(x) = x/\sqrt[3]{x^2 + 4}$ and $x_0 = -1$. Starting from $h = 1$, approximate the value of the derivative of f at x_0 by applying extrapolation to D_h. Use four rows in your extrapolation table. What is the error in the final extrapolated value?

11. **(a)** Show that

$$D = D_h + k_1 h^2 + k_2 h^4 + k_3 h^6 + o(h^6),$$

where D denotes the true second derivative of a function, D_h denotes the second-order central difference approximation to the second derivative and

k_1, k_2, and k_3 are constants independent of h. [*Hint*: Use Taylor's theorem to expand $f(x_0+h)$ and $f(x_0-h)$ about the point $x = x_0$.]

(b) Let $f(x) = x^2e^x$ and $x_0 = 0$. Starting from $h = 0.5$, approximate the value of the second derivative of f at x_0 by applying extrapolation to D_h. Use three rows in your extrapolation table. What is the error in the final extrapolated value?

12. (a) Approximate the derivative of $f(x) = 1 + x + x^3$ at $x_0 = 0$ using the first-order forward difference formula. Take $h = 1/4$ and $h = 1/8$, and then extrapolate from these two values.

(b) What is the error associated with each of the approximations computed in part (a)? Explain any unusual behavior in the errors.

13. (a) Approximate the derivative of $f(x) = \sin x$ at $x_0 = \pi$ using the first-order forward difference formula. Take $h = 1/4$ and $h = 1/8$, and then extrapolate from these two values.

(b) What is the error associated with each of the approximations computed in part (a)? Explain any unusual behavior in the errors.

6.4 NUMERICAL INTEGRATION—THE BASICS AND NEWTON-COTES QUADRATURE

The fundamental problem of numerical integration (which is also called numerical quadrature) can be stated as follows:

Given the function, f, continuous on $[a, b]$, approximate $I(f) = \int_a^b f(x)\,dx$.

If an antiderivative of f—that is, a function $F(x)$ such that $F'(x) = f(x)$—can be found, then the Fundamental Theorem of Calculus guarantees that $I(f) = F(b) - F(a)$. There are problems with putting this rule into practice, however. First, there are functions f for which the antiderivative cannot be expressed in terms of standard functions. A simple example is $f(x) = e^{-x^2}$. Second, for other functions, the antiderivative may be so complicated that it would be easier to work with the integrand directly. Third, functions encountered in practice are often defined in terms of discrete data, not a formula. These are the primary motivations for studying the techniques of numerical integration/quadrature.

Most quadrature formulas take the form

$$I(f) \approx I_n(f) = \sum_{i=0}^{n} w_i f(x_i),$$

where the x_i are known as the quadrature points, or abscissas, and the w_i are called the quadrature weights. Note that $I_n(f)$ is a linear operator, just like $I(f)$: that is,

$$\begin{aligned} I_n(f+g) &= I_n(f) + I_n(g) \\ I_n(cf) &= cI_n(f). \end{aligned}$$

Generally, there are two approaches which can be taken to develop numerical quadrature formulas. In the first approach, the x_i are fixed, typically as equally spaced points from within the interval $[a, b]$, and the w_i are then computed by fitting a function to the $f(x_i)$ data and integrating the resulting function exactly. When the chosen interpolating function is a polynomial, a NEWTON-COTES QUADRATURE rule is obtained. In the second approach, given the number of data points n, the weights and abscissas of the quadrature rule are selected to achieve maximum possible accuracy. Formulas developed in this manner are known as GAUSSIAN QUADRATURE rules. The basic theory of Newton-Cotes quadrature is developed in this section and the next; Gaussian quadrature is considered in Section 6.6.

General Newton-Cotes Formulas

The basic procedure for developing Newton-Cotes quadrature rules is to first fix the abscissas $x_0, x_1, x_2, \ldots, x_n \in [a, b]$. Next, interpolate the integrand, f, at the abscissas by the polynomial $P_n(x)$. Finally, integrate the interpolating polynomial and set

$$\underbrace{I(f)}_{\substack{\text{exact value}\\ \text{of integral of}\\ \text{original integrand}}} \approx \underbrace{I_n(f)}_{\substack{\text{Newton-Cotes}\\ \text{formula}}} \equiv \underbrace{I(P_n).}_{\substack{\text{exact value}\\ \text{of integral of}\\ \text{interpolating}\\ \text{polynomial}}}$$

Because we want the final quadrature rule to exhibit a clear dependence on the data values, $f(x_i)$, the Lagrange form of the interpolating polynomial will be used:

$$P_n(x) = \sum_{i=0}^{n} L_{n,i}(x) f(x_i).$$

Hence, Newton-Cotes quadrature rules will take the form

$$\begin{aligned} I_n(f) &\equiv \int_a^b \sum_{i=0}^{n} L_{n,i}(x) f(x_i)\, dx \\ &= \sum_{i=0}^{n} \left(\int_a^b L_{n,i}(x) dx \right) f(x_i) \\ &= \sum_{i=0}^{n} w_i f(x_i), \end{aligned}$$

where

$$w_i = \int_a^b L_{n,i}(x)\, dx.$$

Within this framework, there are two distinct varieties of Newton-Cotes formulas, which differ in their choice of the abscissas within the interval $[a, b]$. The so-called *closed* Newton-Cotes formulas include the endpoints of the integration interval, $x = a$ and $x = b$, among the abscissas. For a given n, we take $\Delta x = (b-a)/n$, and the abscissas are $x_i = a + i\Delta x$ for each $i = 0, 1, 2, \ldots, n$. On the other hand, the *open* Newton-Cotes formulas do not include the endpoints of the integration interval among the abscissas. For these formulas, $\Delta x = (b-a)/(n+2)$, and the abscissas are $x_i = a + (i+1)\Delta x$ for each $i = 0, 1, 2, \ldots, n$.

Some Closed Newton-Cotes Formulas

Let's start with the simplest closed Newton-Cotes formula, which corresponds to $n = 1$. The spacing between abscissas is $\Delta x = b - a$, and the two abscissas are $x_0 = a$ and $x_1 = b$. Since the Lagrange polynomials associated with these points are

$$L_{1,0}(x) = \frac{b-x}{b-a} \quad \text{and} \quad L_{1,1}(x) = \frac{x-a}{b-a},$$

it follows that the quadrature weights are given by

$$w_0 = \int_a^b \frac{b-x}{b-a}\,dx \quad \text{and} \quad w_1 = \int_a^b \frac{x-a}{b-a}\,dx.$$

The substitution $x = a + t\Delta x$ in the integrals defining w_0 and w_1 simplifies the final calculations:

$$w_0 = \Delta x \int_0^1 (1-t)\,dt = \frac{\Delta x}{2}$$
$$w_1 = \Delta x \int_0^1 t\,dt = \frac{\Delta x}{2}.$$

Therefore, the closed Newton-Cotes quadrature formula corresponding to $n = 1$ is

$$I(f) \approx I_{1,\text{closed}}(f) = \frac{\Delta x}{2}[f(a) + f(b)] = \frac{b-a}{2}[f(a) + f(b)].$$

Geometrically, this quadrature rule approximates the value of the definite integral as the area of a trapezoid (see Figure 6.8); hence, this rule is known as the trapezoidal rule.

The case $n = 2$ also produces a well-known formula. Here, $\Delta x = (b-a)/2$, and the three abscissas are $x_0 = a$, $x_1 = a + \Delta x = (a+b)/2$ and $x_2 = a + 2\Delta x = b$. The quadrature weights are found to be

$$w_0 = \int_a^b L_{2,0}(x)\,dx = \frac{\Delta x}{2}\int_0^2 (t-1)(t-2)\,dt = \frac{\Delta x}{3}$$
$$w_1 = \int_a^b L_{2,1}(x)\,dx = -\Delta x\int_0^2 t(t-2)\,dt = \frac{4\Delta x}{3}$$
$$w_2 = \int_a^b L_{2,2}(x)\,dx = \frac{\Delta x}{2}\int_0^2 t(t-1)\,dt = \frac{\Delta x}{3}.$$

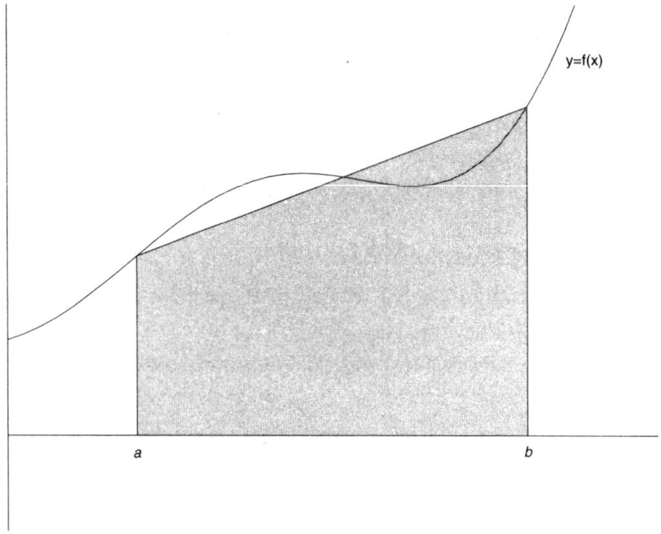

Figure 6.8 The trapezoidal rule.

The substitution $x = a + t\Delta x$ was once again used in the integrals defining w_0, w_1 and w_2 to simplify the final calculation. Therefore,

$$I(f) \approx I_{2,\text{closed}}(f) = \frac{\Delta x}{3}\left[f(a) + 4f\left(\frac{a+b}{2}\right) + f(b)\right]$$
$$= \frac{b-a}{6}\left[f(a) + 4f\left(\frac{a+b}{2}\right) + f(b)\right].$$

This formula may be recognized from calculus as Simpson's rule.

The closed Newton-Cotes formulas with $n = 3$ and $n = 4$ are

$$I(f) \approx I_{3,\text{closed}}(f) = \frac{b-a}{8}\left[f(a) + 3f(a+\Delta x) + 3f(a+2\Delta x) + f(b)\right],$$

where $\Delta x = (b-a)/3$, and

$$I(f) \approx I_{4,\text{closed}}(f) = \frac{b-a}{90}\left[7f(a) + 32f(a+\Delta x) + 12f(a+2\Delta x) + 32f(a+3\Delta x) + 7f(b)\right],$$

where $\Delta x = (b-a)/4$. The $n = 3$ formula is also known as the three-eighths rule, while the $n = 4$ formula is known as Boole's rule. The derivation of these formulas is left as an exercise.

Some Open Newton-Cotes Formulas

The simplest open Newton-Cotes formula corresponds to $n = 0$. With $n = 0$, $\Delta x = (b-a)/2$, and the only abscissa is $x_0 = (a+b)/2$. The quadrature weight is

$$w_0 = \int_a^b L_{0,0}(x)\,dx = \int_a^b 1 \cdot dx = b - a.$$

Therefore, the $n = 0$ open Newton-Cotes quadrature formula is

$$I(f) \approx I_{0,\text{open}}(f) = (b-a) f\left(\frac{a+b}{2}\right).$$

This formula is known as the midpoint rule.

For $n = 1$, $\Delta x = (b-a)/3$, and the abscissas are $x_0 = a+\Delta x$ and $x_1 = a+2\Delta x$. The Lagrange polynomials associated with these points are

$$L_{1,0}(x) = \frac{a + 2\Delta x - x}{\Delta x} \quad \text{and} \quad L_{1,1}(x) = \frac{x - (a + \Delta x)}{\Delta x}.$$

With these functions, the corresponding quadrature weights are

$$w_0 = \int_a^b \frac{a + 2\Delta x - x}{\Delta x}\,dx = \Delta x \int_0^3 (2-t)\,dt = \Delta x \left(2t - \frac{1}{2}t^2\right)\Big|_0^3 = \frac{3\Delta x}{2}$$

and

$$w_1 = \int_a^b \frac{x - (a + \Delta x)}{\Delta x}\,dx = \Delta x \int_0^3 (t-1)\,dt = \Delta x \left(\frac{1}{2}t^2 - t\right)\Big|_0^3 = \frac{3\Delta x}{2},$$

where the change of variable $x = a + t\Delta x$ has been made in each integral. Putting everything together, the open Newton-Cotes formula with $n = 1$ is

$$I(f) \approx I_{1,\text{open}}(f) = \frac{b-a}{2}[f(a+\Delta x) + f(a+2\Delta x)].$$

The open Newton-Cotes formulas with $n = 2$ and $n = 3$ are

$$I(f) \approx I_{2,\text{open}}(f) = \frac{b-a}{3}\left[2f(a+\Delta x) - f(a+2\Delta x) + 2f(a+3\Delta x)\right],$$

where $\Delta x = (b-a)/4$, and

$$I(f) \approx I_{3,\text{open}}(f) = \frac{b-a}{24}\left[11f(a+\Delta x) + f(a+2\Delta x) + f(a+3\Delta x) + 11f(a+4\Delta x)\right],$$

where $\Delta x = (b-a)/5$. The derivation of these formulas is left as an exercise.

Error Analysis

Like the error term associated with a difference approximation, the error term associated with a quadrature rule provides two pieces of information. First, the error term indicates precisely how the error depends on the length of the integration interval. This information will prove extremely useful in later sections. Second, the error term allows us to determine the *degree of precision*, which characterizes the class of polynomials for which the quadrature formula produces exact results.

> **Definition.** The DEGREE OF PRECISION (or ACCURACY) of a quadrature rule $I_n(f)$ is the positive integer m such that
> $I(p) = I_n(p)$ for every polynomial p of degree $\leq m$
> $I(p) \neq I_n(p)$ for some polynomial p of degree $m+1$.

Based on this definition, a quadrature rule that integrates every constant polynomial, every linear polynomial, and every quadratic polynomial exactly but fails to integrate at least one cubic polynomial exactly would be said to have degree of precision equal to 2. Though this may seem to be a cumbersome quantity to determine, since polynomials are just linear combinations of powers of x and both I and I_n are linear operators, we only need to consider whether the rule integrates the powers of x exactly. Therefore, if a rule integrates 1, x, and x^2 exactly but fails to integrate x^3 exactly, the degree of precision is 2.

A powerful tool for deriving error terms associated with quadrature formulas is the following theorem, known as the Weighted Mean-Value Theorem for Integrals.

> **Theorem.** If f is continuous on $[a, b]$, g is integrable on $[a, b]$ and $g(x)$ does not change sign on $[a, b]$, then there exists a number $\xi \in [a, b]$ such that
>
> $$\int_a^b f(x)g(x)\,dx = f(\xi)\int_a^b g(x)\,dx.$$
>
> ***Proof.*** Suppose that $g(x) \geq 0$ on $[a, b]$. The proof for $g(x) \leq 0$ is similar, and the details are left as an exercise. Let m and M denote the minimum and maximum value, respectively, achieved by f on $[a, b]$. Since $g(x) \geq 0$, it follows that
>
> $$mg(x) \leq f(x)g(x) \leq Mg(x)$$
>
> for all $x \in [a, b]$. Consequently,
>
> $$m\int_a^b g(x)\,dx \leq \int_a^b f(x)g(x)\,dx \leq M\int_a^b g(x)\,dx.$$
>
> If $\int_a^b g(x)\,dx = 0$, then $\int_a^b f(x)g(x)\,dx$ must also equal 0, so any $\xi \in [a, b]$ can be chosen to satisfy the requirements of the theorem. Otherwise, $\int_a^b g(x)\,dx > 0$. Therefore,
>
> $$m \leq \frac{\int_a^b f(x)g(x)\,dx}{\int_a^b g(x)\,dx} \leq M.$$

Applying the Intermediate Value Theorem, there exists a $\xi \in [a,b]$ such that

$$f(\xi) = \frac{\int_a^b f(x)g(x)\,dx}{\int_a^b g(x)\,dx},$$

from which the conclusion of the theorem follows. □

As a first example to demonstrate the process for determining the error associated with a Newton-Cotes formula, let's consider the trapezoidal rule. Starting from interpolation theory, we know that

$$f(x) = P_1(x) + f[a,b,x](x-a)(x-b),$$

where $P_1(x)$ is the unique linear polynomial that interpolates the integrand at $x = a$ and $x = b$. Upon integrating both sides of this last expression and applying the definition of $I_{1,\text{closed}}(f)$, we find

$$I(f) = I_{1,\text{closed}}(f) + \int_a^b f[a,b,x](x-a)(x-b)\,dx.$$

Observe that the function $(x-a)(x-b) \le 0$ for all $x \in [a,b]$. Thus, we can apply the Weighted Mean-Value Theorem for Integrals to simplify the error term. The result is

$$\begin{aligned} I(f) - I_{1,\text{closed}}(f) &= f[a,b,\xi]\int_a^b (x-a)(x-b)\,dx \\ &= -\frac{(b-a)^3}{6} f[a,b,\xi] \\ &= -\frac{(b-a)^3}{12} f''(\hat{\xi}), \end{aligned}$$

where $a < \hat{\xi} < b$.

Since the second derivative of every constant and every linear polynomial is identically zero, we see that the trapezoidal rule will integrate every constant and every linear polynomial exactly; hence, the trapezoidal rule has degree of precision 1. In general, when the error term for a quadrature rule involves the n-th derivative of the integrand, the rule has degree of precision $n - 1$.

EXAMPLE 6.8 Verification of Trapezoidal Rule Degree of Precision

The following table demonstrates explicitly that the trapezoidal rule integrates 1 and x exactly, but fails to integrate x^2 exactly; hence, the degree of precision is 1.

$f(x)$	$\int_a^b f(x)\,dx$	$h[f(a)+f(a+h)]/2$
1	$b-a$	$b-a$
x	$(b^2-a^2)/2$	$(b^2-a^2)/2$
x^2	$(b^3-a^3)/3$	$(b^3-a^3+ba^2-ab^2)/2$

$f(x)$	$\int_a^b f(x)\,dx$	$h[f(a)+4f(a+h)+f(a+2h)]/3$
1	$b-a$	$b-a$
x	$(b^2-a^2)/2$	$(b^2-a^2)/2$
x^2	$(b^3-a^3)/3$	$(b^3-a^3)/3$
x^3	$(b^4-a^4)/4$	$(b^4-a^4)/4$
x^4	$(b^5-a^5)/5$	$(5b^5-b^4a+2b^3a^2-2b^2a^3+ba^4-5a^5)/24$

TABLE 6.3: Degree of Precision of Simpson's Rule

Because the error term associated with a quadratic interpolating polynomial involves a third derivative, one might expect that Simpson's rule would have degree of precision equal to 2. The data contained in Table 6.3, however, suggests that Simpson's rule has degree of precision equal to 3.

The error term for Simpson's rule therefore requires a closer look. From interpolation theory and the derivation of Simpson's rule, we have

$$I(f) = I_{2,\text{closed}}(f) + \int_a^b f[a,x_1,b,x](x-a)(x-x_1)(x-b)\,dx,$$

where $x_1 = (a+b)/2$. Note that the function $(x-a)(x-x_1)(x-b)$ changes sign on $[a,b]$, so we cannot apply the weighted mean-value theorem for integrals. Suppose, instead, we integrate the error term by parts, taking $u = f[a,x_1,b,x]$ and $dv = (x-a)(x-x_1)(x-b)\,dx$. Remember that with integration by parts we may choose any antiderivative of dv. Here, we choose the specific antiderivative

$$v = \int_a^x (t-a)(t-x_1)(t-b)\,dt = \frac{1}{4}(x-a)^2(x-b)^2.$$

Then

$$\begin{aligned} I(f) - I_{2,\text{closed}}(f) &= \frac{1}{4}(x-a)^2(x-b)^2 f[a,x_1,b,x]\Big|_a^b - \\ &\quad \frac{1}{4}\int_a^b \left(\frac{d}{dx} f[a,x_1,b,x]\right)(x-a)^2(x-b)^2\,dx \\ &= -\frac{1}{4}\int_a^b f[a,x_1,b,x,x](x-a)^2(x-b)^2\,dx. \end{aligned}$$

Since $(x-a)^2(x-b)^2 \geq 0$ for all $x \in [a,b]$, the weighted mean-value theorem for integrals can now be applied. The end result for the Simpson's rule error term is

$$\begin{aligned} I(f) - I_{2,\text{closed}}(f) &= -\frac{1}{4} f[a,x_1,b,\xi,\xi] \int_a^b (x-a)^2(x-b)^2\,dx \\ &= -\frac{(b-a)^5}{120} f[a,x_1,b,\xi,\xi] \\ &= -\frac{(b-a)^5}{2880} f^{(4)}(\hat{\xi}), \end{aligned}$$

where $a < \hat{\xi} < b$. From here, we see that Simpson's rule does indeed have degree of precision equal to 3.

The error term associated with any Newton-Cotes formula with an even n, be it an open formula or a closed formula, can be obtained in exactly the same manner as the Simpson's rule error term was obtained. In particular, for the midpoint rule, it can be shown that

$$\int_a^b f(x)\,dx = (b-a)f\left(\frac{a+b}{2}\right) + \frac{(b-a)^3}{24} f''(\xi),$$

where $a < \xi < b$. The midpoint rule therefore has degree of precision equal to 1. The details of this derivation are left as an exercise.

With the exception of the trapezoidal rule, the derivation of the error term for a Newton-Cotes formula with n odd is a bit more involved. As an example, consider the open Newton-Cotes formula with $n = 1$. Let $\Delta x = (b-a)/3$, $x_0 = a + \Delta x$ and $x_1 = a + 2\Delta x = b - \Delta x$. The error in $I_{1,\text{open}}(f)$ is then given by

$$I(f) - I_{1,\text{open}}(f) = \int_a^b f[x_0, x_1, x](x - x_0)(x - x_1)\,dx.$$

As a first step in manipulating the error term, split the integration interval at $x = b - \Delta x$; that is, write the error term as

$$\int_a^{b-\Delta x} f[x_0,x_1,x](x-x_0)(x-x_1)\,dx + \int_{b-\Delta x}^b f[x_0,x_1,x](x-x_0)(x-x_1)\,dx. \quad (1)$$

In the second integral, $(x - x_0)(x - x_1) \ge 0$ for all $x \in [b - \Delta x, b]$. Applying the Weighted Mean-Value Theorem for Integrals leads to

$$\int_{b-\Delta x}^b f[x_0,x_1,x](x-x_0)(x-x_1)\,dx = \frac{5}{324}(b-a)^3 f''(\hat{\xi}_1),$$

where $a < \hat{\xi}_1 < b$.

Next, in the first integral in (1), replace the product $f[x_0, x_1, x](x - x_1)$ by $f[x_0, x] - f[x_0, x_1]$, which follows from the definition of divided differences. A straightforward calculation gives

$$\int_a^{b-\Delta x} f[x_0, x_1](x - x_0)\,dx = 0.$$

For the remaining integral, an integration by parts and application of the Weighted

Mean-Value Theorem for Integrals yields

$$\begin{aligned}\int_a^{b-\Delta x} f[x_0,x](x-x_0)\,dx &= \frac{(x-a)(x-x_1)}{2} f[x_0,x]\Big|_a^{b-\Delta x} \\ &\quad - \int_a^{b-\Delta x} f[x_0,x,x]\frac{(x-a)(x-x_1)}{2}\,dx \\ &= \frac{2}{81}(b-a)^3 f[x_0,\xi_2,\xi_2] \\ &= \frac{1}{81}(b-a)^3 f''(\hat{\xi_2}).\end{aligned}$$

Bringing all of these pieces together, we find

$$I(f) - I_{1,\text{open}}(f) = (b-a)^3\left[\frac{5}{324}f''(\hat{\xi_1}) + \frac{1}{81}f''(\hat{\xi_2})\right].$$

Assuming that f'' is continuous, it can be shown (see Exercise 17) that $\hat{\xi_1}$ and $\hat{\xi_2}$ can be replaced by a common value $\hat{\xi}$. Hence,

$$I(f) - I_{1,\text{open}}(f) = \frac{(b-a)^3}{36} f''(\hat{\xi}).$$

Note that the open Newton-Cotes formula with $n = 1$ has degree of precision equal to 1. This is the same degree of precision as the midpoint rule, but the open Newton-Cotes formula with $n = 1$ requires more work than the midpoint rule (two function evaluations versus one).

Using arguments similar to those given above, we can establish the following general theorem for the error associated with Newton-Cotes formulas.

Theorem. Let $I_n(f)$ denote the Newton-Cotes quadrature rule (open or closed) with $n+1$ abscissas.
(a) If n is even and f has $n+2$ continuous derivatives, then there exists a constant c and a $\xi \in [a,b]$ such that

$$I(f) = I_n(f) - c(b-a)^{n+3} f^{(n+2)}(\xi).$$

The degree of precision of $I_n(f)$ is $n+1$.
(b) If n is odd and f has $n+1$ continuous derivatives, then there exists a constant c' and a $\xi' \in [a,b]$ such that

$$I(f) = I_n(f) - c'(b-a)^{n+2} f^{(n+1)}(\xi').$$

The degree of precision of $I_n(f)$ is n.

Remarks. (1) Note that the formulas with $n = 4$ and $n = 5$ both have degree of precision equal to 5. Therefore, formulas with an even n are generally better—they provide similar accuracy with fewer function evaluations.

(2) The constants c and c' depend on both n and the type of formula (open vs. closed). For given n, the constant for the closed formula is typically smaller than the constant for the open formula, so closed formulas are used most often in practice. Because open formulas do not evaluate the integrand at the endpoints of the integration interval, open formulas can be useful for certain problems with endpoint singularities. Open formulas also find use in the construction of numerical methods for the solution of initial value problems.

EXERCISES

1. Approximate the value of each of the following integrals using the trapezoidal rule. Verify that the theoretical error bound holds in each case.

(a) $\int_1^2 \frac{1}{x}\, dx$

(b) $\int_0^1 e^{-x}\, dx$

(c) $\int_0^1 \frac{1}{1+x^2}\, dx$

(d) $\int_0^1 \tan^{-1} x\, dx$.

2. Repeat Exercise 1 using Simpson's rule rather than the trapezoidal rule.

3. Repeat Exercise 1 using the midpoint rule rather than the trapezoidal rule.

4. Verify directly that the midpoint rule has degree of precision equal to 1.

5. Verify directly that the open Newton-Cotes formula with $n = 1$ has degree of precision equal to 1.

6. **(a)** Determine values for the coefficients A_0, A_1, and A_2 so that the quadrature formula

$$I(f) = \int_{-1}^{1} f(x)\, dx = A_0 f\left(-\frac{1}{2}\right) + A_1 f(0) + A_2 f\left(\frac{1}{2}\right)$$

has degree of precision at least 2.

(b) Once the values of A_0, A_1, and A_2 have been computed, determine the overall degree of precision for the quadrature rule.

7. **(a)** Determine values for the coefficients A_0, A_1, and A_2 so that the quadrature formula

$$I(f) = \int_{-1}^{1} f(x)\, dx = A_0 f\left(-\frac{1}{3}\right) + A_1 f\left(\frac{1}{3}\right) + A_2 f(1)$$

has degree of precision at least 2.

(b) Once the values of A_0, A_1, and A_2 have been computed, determine the overall degree of precision for the quadrature rule.

8. **(a)** Determine values for the coefficients A_0, A_1, and x_1 so that the quadrature formula

$$I(f) = \int_{-1}^{1} f(x)\, dx = A_0 f(-1) + A_1 f(x_1)$$

has degree of precision at least 2.

(b) Once the values of A_0, A_1, and x_1 have been computed, determine the overall degree of precision for the quadrature rule.

9. Consider the quadrature rule

$$\int_{-1}^{1} f(x)\,dx \approx f\left(-\frac{\sqrt{3}}{3}\right) + f\left(\frac{\sqrt{3}}{3}\right).$$

Determine the degree of precision of this formula.

10. Consider the quadrature rule

$$\int_{-1}^{1} f(x)\,dx \approx \frac{5}{9} f\left(-\sqrt{\frac{3}{5}}\right) + \frac{8}{9} f(0) + \frac{5}{9} f\left(\sqrt{\frac{3}{5}}\right).$$

Determine the degree of precision of this formula.

11. Derive the error term for the midpoint rule:

$$\frac{(b-a)^3}{24} f''(\xi),$$

where $a < \xi < b$.

12. (a) Derive the closed Newton-Cotes formula with $n = 3$:

$$I(f) \approx I_{3,\text{closed}}(f) = \frac{b-a}{8}[f(a) + 3f(a+\Delta x) + 3f(a+2\Delta x) + f(b)].$$

(b) Verify that this formula has degree of precision equal to 3.

(c) Derive the error term associated with this quadrature rule.

13. (a) Derive the closed Newton-Cotes formula with $n = 4$:

$$I(f) \approx I_{4,\text{closed}}(f) = \frac{b-a}{90}[7f(a) + 32f(a+\Delta x) + 12f(a+2\Delta x) + 32f(a+3\Delta x) + 7f(b)].$$

(b) Verify that this formula has degree of precision equal to 5.

(c) Derive the error term associated with this quadrature rule.

14. (a) Derive the open Newton-Cotes formula with $n = 2$:

$$I(f) \approx I_{2,\text{open}}(f) = \frac{b-a}{3}[2f(a+\Delta x) - f(a+2\Delta x) + 2f(a+3\Delta x)].$$

(b) Verify that this formula has degree of precision equal to 3.

(c) Derive the error term associated with this quadrature rule.

15. (a) Derive the open Newton-Cotes formula with $n = 3$:

$$I(f) \approx I_{3,\text{open}}(f) = \frac{b-a}{24}[11f(a+\Delta x) + f(a+2\Delta x) + f(a+3\Delta x) + 11f(a+4\Delta x)].$$

(b) Verify that this formula has degree of precision equal to 3.

(c) Derive the error term associated with this quadrature rule.

16. Prove the weighted mean-value theorem for integrals when $g(x) \leq 0$ for all $x \in [a, b]$.

17. **(a)** Let g be a continuous function on $[a, b]$ and let $a_1, a_2, a_3, \ldots, a_n$ be any set of nonnegative numbers such that

$$\sum_{i=1}^{n} a_i = A.$$

Show that for any set of points $x_1, x_2, x_3, \ldots, x_n \in [a, b]$, there exists a $\xi \in [a, b]$ such that

$$\sum_{i=1}^{n} a_i g(x_i) = A g(\xi).$$

(b) Use the result of part (a) to show that, provided f'' is continuous, there exists a $\xi \in [a, b]$ such that

$$\frac{5}{324} f''(\xi_1) + \frac{1}{81} f''(\xi_2) = \frac{1}{36} f''(\xi).$$

6.5 COMPOSITE NEWTON-COTES QUADRATURE

Let $I_j(f)$ denote the (open or closed) Newton-Cotes quadrature rule with $j + 1$ abscissas. Suppose one computes

$$I_1(f), I_2(f), I_3(f), \quad \cdots \quad , I_n(f),$$

adding more and more points to obtain greater accuracy and, hopefully, convergence toward the true value of the integral.

Based on our studies of interpolation, we know that there can be some difficulties with this approach. First, since we are dealing with polynomial interpolation at equally spaced points, the sequence of approximations may not converge. Second, for large values of n, some of the quadrature weights may become negative and introduce sensitivity to roundoff errors.

An alternative approach for improving the accuracy of an approximation is to subdivide the integration interval $[a, b]$ into pieces and then apply a low-order Newton-Cotes formula on each subinterval. Numerical integration performed in this manner is referred to as *Composite Newton-Cotes quadrature.*

Composite Trapezoidal Rule

Recall that

$$\begin{aligned} I(f) &= I_{1,\text{closed}}(f) + \text{error} \\ &= \frac{b-a}{2}[f(a) + f(b)] - \frac{(b-a)^3}{12} f''(\xi). \end{aligned}$$

If the integration interval $[a,b]$ is split into n subintervals by defining $h=(b-a)/n$ and $x_j=a+jh$, $0\le j\le n$, and then the trapezoidal rule formula is applied on each subinterval $[x_{j-1},x_j]$, we obtain

$$\begin{aligned} I(f) &= \sum_{j=1}^{n}\int_{x_{j-1}}^{x_j} f(x)\,dx \\ &= \sum_{j=1}^{n}\frac{x_j-x_{j-1}}{2}[f(x_{j-1})+f(x_j)] - \sum_{j=1}^{n}\frac{(x_j-x_{j-1})^3}{12}f''(\xi_j) \\ &= \underbrace{\frac{h}{2}\left[f(x_0)+2\sum_{j=1}^{n-1}f(x_j)+f(x_n)\right]}_{\text{composite trapezoidal rule}} \underbrace{-\frac{h^3}{12}\sum_{j=1}^{n}f''(\xi_j)}_{\text{error}} \end{aligned}$$

where, for each j, $x_{j-1}<\xi_j<x_j$.

The error term needs to be examined more closely. Suppose f has two continuous derivatives. Then the Extreme Value Theorem guarantees that there exist two constants $c_1, c_2 \in [a,b]$ such that

$$\begin{aligned} f''(c_1) &= \max_{a\le x\le b} f''(x) \\ f''(c_2) &= \min_{a\le x\le b} f''(x). \end{aligned}$$

It then follows that for each j

$$f''(c_2)\le f''(\xi_j)\le f''(c_1).$$

Summing over each subinterval $[x_{j-1},x_j]$, we find that

$$nf''(c_2)\le \sum_{j=1}^{n} f''(\xi_j)\le nf''(c_1),$$

or

$$f''(c_2)\le \frac{1}{n}\sum_{j=1}^{n} f''(\xi_j)\le f''(c_1).$$

We can now conclude, by the Intermediate Value Theorem, that there exists $\xi\in[a,b]$ such that $f''(\xi)=\frac{1}{n}\sum_{j=1}^{n}f''(\xi_j)$. This implies that the error for the composite trapezoidal rule can be written as

$$-\frac{nh^3}{12}f''(\xi) = -\frac{(b-a)h^2}{12}f''(\xi).$$

Hence

$$\int_a^b f(x)\,dx = \frac{h}{2}\left[f(x_0)+2\sum_{j=1}^{n-1}f(x_j)+f(x_n)\right] - \frac{(b-a)h^2}{12}f''(\xi).$$

Note that provided the integrand has two continuous derivatives, the composite trapezoidal rule has rate of convergence $O(h^2)$.

EXAMPLE 6.9 Numerical Verification of Rate of Convergence

Consider the integral

$$I(f) = \int_0^{\pi} \sin x \, dx,$$

whose exact value is 2. The table below lists $T_h(f)$, the composite trapezoidal rule approximation to $I(f)$ computed using a subinterval size of h, for several values of h. Note that with each new row, the number of subintervals, n, is doubled and the subinterval size is reduced by a factor of two. Further note that the error ratio in the last column is approaching

$$4 = \left(\frac{2h}{h}\right)^2.$$

This is exactly what we would expect for a sequence converging with rate of convergence $O(h^2)$.

n	h	$T_h(f)$	$e_h = \lvert I(f) - T_h(f)\rvert$	e_{2h}/e_h
1	π	0.0000000	2.0000000	
2	$\frac{\pi}{2}$	1.5707963	0.4292036	4.659792
4	$\frac{\pi}{4}$	1.8961188	0.1038811	4.131681
8	$\frac{\pi}{8}$	1.9742316	0.0257683	4.031337
16	$\frac{\pi}{16}$	1.9935703	0.0064296	4.007741
32	$\frac{\pi}{32}$	1.9983933	0.0016066	4.001929
64	$\frac{\pi}{64}$	1.9995983	0.0004016	4.000482
128	$\frac{\pi}{128}$	1.9998996	0.0001004	4.000120

Composite Simpson's Rule

Since the basic Simpson's rule formula already divides the interval $[a, b]$ into two pieces, $[a, b]$ must be divided into an even number of subintervals to apply Simpson's rule in a composite manner. Therefore, let $n = 2m$, define

$$h = \frac{b-a}{n} = \frac{b-a}{2m},$$
$$x_i = a + ih \quad (0 \le i \le 2m),$$

and apply the Simpson's rule formula m times, once over each $[x_{2j-2}, x_{2j}]$ for $j = 1, 2, 3, \ldots, m$. This produces

$$\begin{aligned}
I(f) &= \sum_{j=1}^{m} \int_{x_{2j-2}}^{x_{2j}} f(x)\, dx \\
&= \sum_{j=1}^{m} \frac{x_{2j} - x_{2j-2}}{6} \left[f(x_{2j-2}) + 4f(x_{2j-1}) + f(x_{2j}) \right] - \sum_{j=1}^{m} \frac{(x_{2j} - x_{2j-2})^5}{2880} f^{(4)}(\xi_j) \\
&= \frac{h}{3} \left[f(x_0) + 4\sum_{j=1}^{m} f(x_{2j-1}) + 2\sum_{j=1}^{m-1} f(x_{2j}) + f(x_{2m}) \right] - \frac{h^5}{90} \sum_{j=1}^{m} f^{(4)}(\xi_j).
\end{aligned}$$

Provided f has four continuous derivatives, an analysis similar to that applied to the error term for the composite trapezoidal rule can be used to show that there exists a $\xi \in [a, b]$ such that $f^{(4)}(\xi) = \frac{1}{m} \sum_{j=1}^{m} f^{(4)}(\xi_j)$. The error term can therefore be written as

$$-\frac{h^5 m}{90} f^{(4)}(\xi) = -\frac{(b-a)h^4}{180} f^{(4)}(\xi),$$

where we have used the fact that $hm = (b-a)/2$. The end result is the composite Simpson's rule:

$$I(f) = \frac{h}{3} \left[f(x_0) + 4\sum_{j=1}^{m} f(x_{2j-1}) + 2\sum_{j=1}^{m-1} f(x_{2j}) + f(x_{2m}) \right] - \frac{(b-a)h^4}{180} f^{(4)}(\xi).$$

Thus, provided the integrand has four continuous derivatives, the composite Simpson's rule has rate of convergence $O(h^4)$.

EXAMPLE 6.10 Numerical Verification of Rate of Convergence

Reconsider the integral

$$I(f) = \int_0^{\pi} \sin x \, dx,$$

whose exact value is 2. The table below lists $S_h(f)$, the composite Simpson's rule approximation to $I(f)$ computed using a subinterval size of h, for several values of h. Note that with each new row, the number of subintervals, n, is doubled and the subinterval size is reduced by a factor of two. Further note that the error ratio in the last column is approaching

$$16 = \left(\frac{2h}{h} \right)^4.$$

This is exactly what we would expect for a sequence converging with rate of convergence $O(h^4)$.

n	h	$S_h(f)$	$e_h = \|I(f) - S_h(f)\|$	e_{2h}/e_h
2	$\frac{\pi}{2}$	2.09439510239	0.09439510239	
4	$\frac{\pi}{4}$	2.00455975498	0.00455975498	20.701792
8	$\frac{\pi}{8}$	2.00026916995	0.00026916995	16.940059
16	$\frac{\pi}{16}$	2.00001659105	0.00001659105	16.223806
32	$\frac{\pi}{32}$	2.00000103337	0.00000103337	16.055292
64	$\frac{\pi}{64}$	2.00000006453	0.00000006453	16.013782
128	$\frac{\pi}{128}$	2.00000000403	0.00000000403	16.003442

A Comment Regarding Numerical Verification of Rates of Convergence

We just got finished with verifying numerically that the composite trapezoidal rule has rate of convergence $O(h^2)$ and that the composite Simpson's rule has rate of convergence $O(h^4)$. We did this by selecting a definite integral for which the exact value was known, computing a sequence of approximations and checking that the approximation error was reduced by an appropriate factor. Do we always have to work with a problem for which the exact solution is known? Fortunately, the answer is no. We can still numerically verify a rate of convergence when the exact solution is not known, we just have to examine a different ratio.

To illustrate the process, suppose we are approximating the value of some definite integral, $I(f)$, using the composite trapezoidal rule. Let $T_h(f)$, $T_{h/2}(f)$ and $T_{h/4}(f)$ denote the composite trapezoidal rule approximations obtained using subinterval sizes of h, $h/2$ and $h/4$, respectively. Consider the ratio

$$\frac{T_h(f) - T_{h/2}(f)}{T_{h/2}(f) - T_{h/4}(f)}.$$

Let $e_h = T_h(f) - I(f)$; that is, e_h is the error associated with $T_h(f)$. Since the composite trapezoidal rule theoretically has rate of convergence $O(h^2)$, we should expect to find $e_h \approx 4e_{h/2}$ for sufficiently small h. Consequently, we should find

$$\begin{aligned}\frac{T_h(f) - T_{h/2}(f)}{T_{h/2}(f) - T_{h/4}(f)} &= \frac{T_h(f) - I(f) - (T_{h/2}(f) - I(f))}{T_{h/2}(f) - I(f) - (T_{h/4}(f) - I(f))} \\ &= \frac{e_h - e_{h/2}}{e_{h/2} - e_{h/4}} \\ &\approx \frac{4e_{h/2} - e_{h/2}}{e_{h/2} - \frac{1}{4}e_{h/2}} = 4\end{aligned}$$

for sufficiently small h.

EXAMPLE 6.11 Numerical Verification of Rate of Convergence

Consider the definite integral

$$I(f) = \int_0^1 \sqrt{1 + x^3}\, dx.$$

The table below lists composite trapezoidal rule approximations to $I(f)$ for several values of h. Observe that the ratio

$$\frac{T_h(f) - T_{h/2}(f)}{T_{h/2}(f) - T_{h/4}(f)}$$

approaches 4 as h is decreased, thereby providing numerical verification that the rate of convergence is $O(h^2)$.

n	h	$T_h(f)$	$\frac{T_h(f)-T_{h/2}(f)}{T_{h/2}(f)-T_{h/4}(f)}$
1	1	1.207106781186	4.335258
2	$\frac{1}{2}$	1.133883476483	4.057269
4	$\frac{1}{4}$	1.116993293318	4.014294
8	$\frac{1}{8}$	1.112830349496	4.003560
16	$\frac{1}{16}$	1.111793319381	4.000889
32	$\frac{1}{32}$	1.111534292393	4.000222
64	$\frac{1}{64}$	1.111469550038	
128	$\frac{1}{128}$	1.111453365349	

A similar process can be used to verify numerically the rate of convergence for any composite quadrature rule. In particular, suppose that $Q_h(f)$ is an approximation to the definite integral $I(f)$ obtained using a generic composite quadrature formula with a subinterval size of h. If the composite quadrature formula has a theoretical rate of convergence of $O(h^k)$, then it can be shown (see Exercise 3) that the ratio

$$\frac{Q_h(f) - Q_{h/b}(f)}{Q_{h/b}(f) - Q_{h/b^2}(f)}$$

should approach b^k as h is decreased toward zero.

Using the Error Term

The next two examples demonstrate the use of the error term associated with a composite quadrature formula for determining the number of subintervals needed to achieve a given level of accuracy. In Sections 6.7 and 6.8, we will see how to use the error terms to construct algorithms which will control the approximation error automatically.

EXAMPLE 6.12 Approximating the Value of π

Since

$$\int_0^1 \frac{1}{1+x^2}\,dx = \tan^{-1} x\Big|_0^1 = \tan^{-1} 1 = \frac{\pi}{4},$$

it follows that the value of π can be approximated by taking four times an approximation for the integral on the left of the above expression. Suppose we want to approximate π to four decimal places. This means that the absolute error must be less than 5.0×10^{-5}, which in turn implies that the error in approximating the integral must be less than 1.25×10^{-5}.

TRAPEZOIDAL RULE

Since $h = (b-a)/n$, the error term associated with the trapezoidal rule can be written in the form

$$\frac{(b-a)h^2}{12}f''(\xi) = \frac{(b-a)^3}{12n^2}f''(\xi).$$

For $f(x) = 1/(1+x^2)$, it can be shown that

$$\max_{x\in[0,1]} |f''(x)| = 2,$$

so the value of n must be selected to satisfy the inequality

$$\frac{(1-0)^3}{12n^2} \cdot 2 < 1.25 \times 10^{-5}$$

in order to guarantee an error of no more than 1.25×10^{-5}. The solution of this inequality is $n > 115.47$; therefore, we use $n = 116$. With $n = 116$,

$$\int_0^1 \frac{1}{1+x^2}\,dx \approx 0.78539506688536 \quad \Rightarrow \quad \pi \approx 3.14158026754144.$$

The absolute error in this approximation is 1.2386×10^{-5}, well within the required accuracy.

SIMPSON'S RULE

For Simpson's rule, $h = (b-a)/n$, so the error term can be written as

$$\frac{(b-a)h^4}{180}f^{(4)}(\xi) = \frac{(b-a)^5}{180n^4}f^{(4)}(\xi).$$

It can be shown that $\max_{x\in[0,1]} |f^{(4)}(x)| = 24$, so that n must be selected to satisfy the inequality

$$\frac{(1-0)^5}{180n^4} \cdot 24 < 1.25 \times 10^{-5}.$$

The solution of this inequality is $n > 10.16$. Simpson's rule requires an even number of subintervals, so 12 is the smallest number of subintervals that will guarantee an error of no more than 1.25×10^{-5}. With $n = 12$,

$$\int_0^1 \frac{1}{1+x^2}\,dx \approx 0.78539816007634 \quad \Rightarrow \quad \pi \approx 3.14159264030538.$$

The absolute error in this approximation is 1.3284×10^{-8}, so we actually achieved seven decimal places of accuracy.

The power of a fourth-order method over a second-order method is quite apparent here. Simpson's Rule produced a more accurate result than the trapezoidal rule using roughly one-tenth the number of function evaluations.

EXAMPLE 6.13 The Cost of a Warranty

The integral

$$I = \int_0^1 e^{-x^4}\, dx$$

arises in the determination of the cost of a treadware warranty on tires [Kevin Hastings, "Reliability and the Cost of Guarantees," in *Applications of Calculus (Resources for Calculus*, volume 3), Philip Straffin, editor, MAA Notes #29, The Mathematical Association of America, 1993, pp. 152–166]. Suppose the value of this integral is needed with a guaranteed error of no more than 0.00001.

For the composite trapezoidal rule, the number of subintervals must be selected to satisfy the inequality

$$\frac{(1-0)^3 \max_{0\le x\le 1} |f''(\xi)|}{12n_{\text{trap}}^2} < 0.00001;$$

for the composite Simpson's rule, the number of subintervals must satisfy the inequality

$$\frac{(1-0)^4 \max_{0\le x\le 1} |f^{(4)}(\xi)|}{180n_{\text{simp}}^4} < 0.00001.$$

It can be shown that $\max_{0\le x\le 1} |f''(\xi)| < 3.5$ and $\max_{0\le x\le 1} |f^{(4)}(\xi)| < 95$, leading to the values $n_{\text{trap}} = 171$ and $n_{\text{simp}} = 16$. Using these values for n produces the estimates:

Trapezoidal rule:	$\int_0^1 e^{-x^4}\, dx \approx 0.8448344011$	(error $= 4.194 \times 10^{-6}$)
Simpson's rule:	$\int_0^1 e^{-x^4}\, dx \approx 0.8448403780$	(error $= 1.783 \times 10^{-6}$)

In computing the absolute error, an "exact" value for the integral was determined by using Simpson's rule with $n = 3200$. The value obtained in this manner was 0.8448385947.

Application Problem: Bags of Pine Bark Mulch

In the Chapter 6 Overview (see page 429), we found that the number of bags of pine bark mulch, each bag with a capacity of 3 cubic feet, needed to cover the irregularly shaped plot of land shown in Figure 6.1 to a uniform depth of three inches is given by

$$\frac{1}{12}\int_0^{26} f(x)\, dx.$$

The integrand, f, is defined in the following table.

x	0	1	2	3	4	5	6	7	8
$f(x)$	8.0	10.0	11.0	15.0	16.0	16.0	16.0	16.0	15.5

x	9	10	11	12	13	14	15	16	17
$f(x)$	15.5	15.5	15.0	15.0	15.0	14.5	14.5	14.0	14.0

x	18	19	20	21	22	23	24	25	26
$f(x)$	14.0	13.5	13.0	13.0	13.0	12.0	11.0	9.0	6.0

Using the trapezoidal rule, we find

$$\begin{aligned}\int_0^{26} f(x)\,dx &\approx \frac{h}{2}\left[f(0) + 2\sum_{j=1}^{25} f(j) + f(26)\right]\\ &= \frac{1}{2}\left[8 + 2(347) + 6\right]\\ &= 354 \text{ square feet.}\end{aligned}$$

Accordingly, we estimate that $\frac{1}{12} \cdot 354 = 29.5$ bags of mulch are needed. Since we need at least this many bags and we must purchase an integer number of bags, we round our estimate up to 30.

Because there are an even number of subintervals in the data set, we can also use Simpson's rule to approximate the definite integral. We then find

$$\begin{aligned}\int_0^{26} f(x)\,dx &\approx \frac{h}{3}\left[f(0) + 4\sum_{j=1}^{12} f(2j) + 2\sum_{j=1}^{13} f(2j-1) + f(26)\right]\\ &= \frac{1}{3}\left[8 + 4(178.5) + 2(168.5) + 6\right]\\ &= 355 \text{ square feet.}\end{aligned}$$

Hence, we would estimate that $\frac{1}{12} \cdot 355 = 29.58$ bags of mulch are needed. Once again, we round this estimate up to 30.

Periodic Integrands

Although the composite trapezoidal rule is generally a second-order method (convergence toward the true value of the integral goes like the square of the mesh spacing), there is a class of integrals for which the composite trapezoidal rule is extremely accurate. Consider evaluating the integral

$$\int_0^{\pi} \sqrt{1+\cos^2 x}\;dx,$$

which computes the length of the sine curve over one half period. The following table shows the approximation to this integral computed using the composite trapezoidal rule for several values of n. The error in the approximation is also shown. All calculations were performed in *Maple* using 110 digits of precision.

n	Approximation	Error
2	3.7922377959	2.79600×10^{-2}
4	3.8199436432	2.54146×10^{-4}
8	3.8201977154	7.36424×10^{-8}
16	3.8201977890	1.90461×10^{-14}
32	3.8201977890	3.74892×10^{-27}
64	3.8201977890	4.18779×10^{-52}
128	3.8201977890	1.49209×10^{-101}

Since the value of n is doubled for each new approximation, which implies that the value of h was cut in half, the error with each new approximation should have dropped by a factor of four. The errors in this example definitely seem to be decreasing at a much more rapid pace. In fact, each new error appears to be almost the square of the previous error.

To make the analysis of the error more quantitative, let E_n denote the error associated with the composite trapezoidal rule computed with n subintervals. Theory indicates that E_n should be inversely proportional to the square of n; that is,

$$E_n \approx c\frac{1}{n^2}$$

for some constant c. Taking the logarithm of this expression yields

$$\log E_n \approx \log c - 2\log n.$$

Hence, if E_n is inversely proportional to the square of n, a plot of $\log n$ versus $\log E_n$ should be linear. Using the values in the above table, the relationship between $\log n$ and $\log E_n$ is clearly not linear (see Figure 6.9). In fact, the rate of decrease in appears to continually accelerate with increasing n.

The plot in Figure 6.9 suggests that the error in this example is exponential in n, not algebraic. To examine this further, suppose

$$E_n \approx cb^n$$

for some constants b and c. Again taking logarithms leads to

$$\log E_n \approx \log c + n\log b,$$

which implies that for an exponential relationship between E_n and n, a plot of n versus $\log E_n$ should be linear. Figure 6.10 shows n versus $\log E_n$ for the current example, verifying an exponential rate of convergence.

The explanation for this phenomenon lies in the Euler-Maclaurin sum formula. Provided f is sufficiently differentiable, it can be shown (see Davis and Rabinowitz [1]) that

$$\begin{aligned} I(f) - T_h(f) = {} & \frac{B_2h^2}{2!}[f'(b) - f'(a)] + \frac{B_4h^4}{4!}[f'''(b) - f'''(a)] + \cdots + \\ & \frac{B_{2k}h^{2k}}{(2k)!}[f^{(2k-1)}(b) - f^{(2k-1)}(a)] + O(h^{2k+1}), \end{aligned}$$

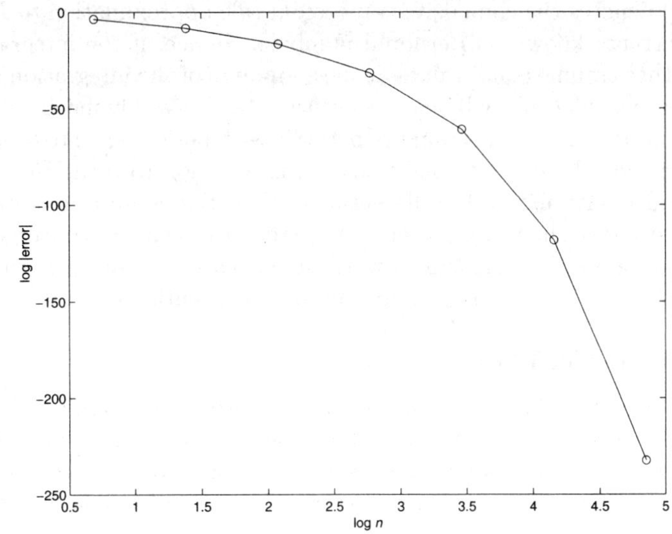

Figure 6.9 Logarithm of the error versus logarithm of number of subintervals for the composite trapezoidal rule approximation to the length of the sine curve over one-half period.

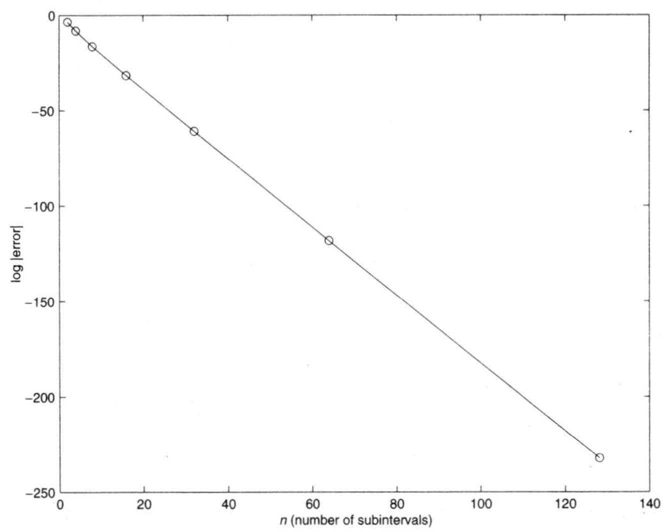

Figure 6.10 Logarithm of the error versus number of subintervals for the composite trapezoidal rule approximation to the length of the sine curve over one-half period.

where $T_h(f)$ denotes the composite trapezoidal rule approximation to $I(f)$ and the B_{2k} are constants known as Bernoulli numbers. Hence, if the integrand has odd derivatives that assume equal values at the endpoints of the integration interval, the composite trapezoidal rule will be more accurate (and possibly much more accurate) than second order. Integrands that are periodic with period equal to $b-a$ will clearly satisfy these special endpoint conditions. If, in addition to being $(b-a)$-periodic, the integrand is also infinitely differentiable, then the composite trapezoidal rule will converge faster than any power of the step size; that is, convergence will be exponentially rapid. It is straightforward in this case to show that the integrand, $f(x) = \sqrt{1+\cos^2 x}$, is π-periodic and infinitely differentiable.

Sensitivity to Roundoff Error

At the end of Section 6.2, the effect of roundoff error on numerical differentiation formulas was investigated. It was found that the roundoff error component of the total error grew without bound as the step size was reduced to zero. The obvious question to ask is whether numerical integration formulas exhibit the same sensitivity to roundoff error.

Consider the composite trapezoidal rule and suppose that

$$f(x_j) = \tilde{f}(x_j) + e_j,$$

for each j, $0 \le j \le n$, where

$f(x_j)$	known function value
$\tilde{f}(x_j)$	floating point representation for $f(x_j)$
e_j	roundoff error associated with $\tilde{f}(x_j)$.

Substituting for $f(x_j)$ in the composite trapezoidal rule, we obtain

$$I(f) = T_h(\tilde{f}) + \underbrace{\frac{h}{2}\left[e_0 + 2\sum_{j=1}^{n-1} e_j + e_n\right]}_{\text{roundoff error}} - \frac{h^2}{12}(b-a)f''(\xi).$$

If $|e_j| \le \epsilon$ for all j, then

$$\left|\begin{array}{c}\text{roundoff}\\ \text{error}\end{array}\right| \le \frac{h}{2}[1 + 2(n-1) + 1]\epsilon = nh\epsilon$$
$$= (b-a)\epsilon.$$

Hence, the amount of roundoff error introduced by the composite trapezoidal rule is bounded independent of the step size.

References

1. P. J. Davis and P. Rabinowitz, *Methods of Numerical Integration*, 2nd edition, Academic Press, New York, 1984.

EXERCISES

1. Provide the details of the transformation of the error term associated with the composite Simpson's rule from

$$\frac{h^5}{90}\sum_{j=1}^{m} f^{(4)}(\xi_j) \qquad \text{to} \qquad \frac{(b-a)h^4}{180}f^{(4)}(\xi).$$

2. Derive the composite midpoint rule with error:

$$\int_a^b f(x)\,dx = 2h\sum_{j=1}^{n} f(x_j) + \frac{(b-a)h^2}{6}f''(\xi),$$

where $h = (b-a)/2n$, $x_j = a + (2j-1)h$ and $\xi \in [a, b]$.

3. (a) Let $Q_h(f)$ be an approximation to the definite integral $I(f)$ obtained using a generic composite quadrature formula with a subinterval size of h. If the composite quadrature formula has a theoretical rate of convergence of $O(h^k)$, show that

$$\frac{Q_h(f) - Q_{h/b}(f)}{Q_{h/b}(f) - Q_{h/b^2}(f)} \approx b^k.$$

   (b) What value do we expect from the ratio

$$\frac{S_h(f) - S_{h/2}(f)}{S_{h/2}(f) - S_{h/4}(f)},$$

   where $S_h(f)$ denotes the composite Simpson's rule approximation to the definite integral $I(f)$ obtained with a subinterval size of h?

4. Verify that the composite Simpson's rule has rate of convergence $O(h^4)$ by approximating the value of $\int_0^1 \sqrt{1+x^3}\,dx$.

5. (a) Verify that the composite midpoint rule has rate of convergence $O(h^2)$ by approximating the value of $\int_0^1 \sqrt{1+x^3}\,dx$.

   (b) Repeat part (a) by approximating the value of $\int_0^\pi \sin x\,dx$.

In Exercises 6–11, verify that the composite trapezoidal rule has rate of convergence $O(h^2)$, the composite midpoint rule has rate of convergence $O(h^2)$, and the composite Simpson's rule has rate of convergence $O(h^4)$, by approximating the value of the indicated definite integral.

6. $\int_1^2 \frac{1}{x}\,dx$
7. $\int_0^1 e^{-x}\,dx$
8. $\int_0^1 \tan^{-1} x\,dx$
9. $\int_1^2 \frac{\sin x}{x}\,dx$
10. $\int_0^1 \frac{1}{\sqrt{1+x^4}}\,dx$

11. $\int_0^4 x\sqrt{x^2+9}\,dx$

12. Suppose that there exists a composite quadrature rule, $Q(f)$, with the property

$$\int_a^b f(x)\,dx = Q(f) - \frac{(b-a)h^4}{240} f^{(5)}(\xi),$$

where $a < \xi < b$ and $h = (b-a)/n$.

(a) What is the rate of convergence associated with this quadrature rule? What conditions must the integrand satisfy to achieve this rate of convergence? Explain how you would numerically verify the rate of convergence.

(b) What is the degree of precision of this quadrature rule? Explain how to verify the degree of precision.

(c) What is the smallest value of n needed to guarantee an approximation to the value of $\int_1^2 \frac{1}{x}\,dx$ to within 10^{-5}? Justify your response.

13. **(a)** Determine the smallest value of n that guarantees that the composite midpoint rule approximates the value of $\int_0^1 \frac{1}{1+x^2}\,dx$ to within 1.25×10^{-5}.

(b) Determine the smallest value of n which guarantees that the composite midpoint rule approximates the value of $\int_0^1 e^{-x^4}\,dx$ to within 10^{-5}.

In Exercises 14–20, approximate the value of the indicated definite integral using the composite trapezoidal rule, the composite midpoint rule and the composite Simpson's rule. For each method, use the smallest value of n that will guarantee an absolute error not greater than 5×10^{-5}.

14. $\int_1^2 \frac{1}{x}\,dx$

15. $\int_0^1 e^{-x}\,dx$

16. $\int_0^1 \tan^{-1} x\,dx$

17. $\int_1^2 \frac{\sin x}{x}\,dx$

18. $\int_0^1 \frac{1}{\sqrt{1+x^4}}\,dx$

19. $\int_0^4 x\sqrt{x^2+9}\,dx$

20. $\int_0^1 \sqrt{1+x^3}\,dx$

21. **(a)** Show that the error associated with the composite Simpson's rule can be approximated by

$$-\frac{h^4}{180}\left[f'''(b) - f'''(a)\right].$$

[*Hint*: Recognize that $2h\sum_{j=1}^{m} f^{(4)}(\xi_j)$ is a Riemann sum for $\int_a^b f^{(4)}(x)\,dx$.]

(b) Show that the error associated with the composite midpoint rule can be approximated by

$$\frac{h^2}{6}\left[f'(b) - f'(a)\right].$$

22. Consider the definite integral $\int_a^b \sin(\sqrt{\pi x})\,dx$. Numerically determine the rate of convergence of the composite trapezoidal rule for each of the following integration intervals.

(a) $[a,b]=[0,1]$

(b) $[a,b]=[\pi/4, 9\pi/4]$

(c) $[a,b]=[\pi, 2\pi]$

(d) Explain any variation among the rates of convergence obtained in parts (a), (b), and (c).

23. Repeat Exercise 22 for the composite midpoint rule.

24. Consider the definite integral $\int_{-1}^{5/4}(x^4+x^3-3x^2-4x-1)\,dx$.

(a) Numerically determine the rate of convergence of the composite trapezoidal rule when applied to the given integral.

(b) Numerically determine the rate of convergence of the composite midpoint rule when applied to the given integral.

(c) Provide an explanation for the results obtained in parts (a) and (b).

25. With an optimal tilting strategy, the theoretical lower bound for the time needed to pour milk from a plastic pouch into a pitcher (N. Curle, "Liquid Flowing from a Container," in *Mathematical Modeling*, Andrews and McLone, eds., Butterworths, 1976, pp. 39–55) requires the calculation of the integrals

$$\int_{0.1763}^{0.8355}(1+x^2)^{1/4}\,dx \quad \text{and} \quad \int_{0.8355}^{1}\frac{2+x^2}{x^3(1+x^2)^{1/4}}\,dx.$$

Approximate the value of each integral with an absolute error no greater than 10^{-4}.

26. Using Newton's Second Law, it can be shown that the period, T (the time for one complete swing), of a pendulum with length L and maximum angle of deflection θ_0 is given by

$$T = 4\sqrt{\frac{L}{g}}\int_0^{\pi/2}\frac{1}{\sqrt{1-k^2\sin^2 x}}\,dx,$$

where $k=\sin(\theta_0)$ and g is the acceleration due to gravity. To calibrate the timing mechanism in their top-of-the-line model, a grandfather clock manufacturer needs to know the period of a pendulum with $L = 1$ meter and $\theta_0 = 12°$ to within 10^{-6} seconds. Calculate the period to the required accuracy.

27. Ammonia vapor is compressed inside a cylinder by an external force acting on the piston. The following data give the volume, v, measured in liters, and the pressure, p, measured in kilopascals.

v	0.50	0.60	0.72	0.84	0.96	1.08	1.25
p	1400	1248	1100	945	802	653	500

The work for the process is given by the integral

$$\int_{0.5}^{1.25} p\,dv.$$

Estimate the work done in the following ways:

(a) using the trapezoidal rule;

(b) by passing a cubic spline through the data and then integrating the spline.

28. Values of the volume (v, measured in cubic inches) and the pressure (p, measured in pounds per square inch) of a gas as it expands from a volume of 1 cubic inch to a volume of 2.5 cubic inches are presented in the table below.

v	1.00	1.25	1.50	1.75	2.00	2.25	2.50
p	68.7	55.0	45.8	39.3	34.4	30.5	27.5

The work done by the gas as it expands is given by

$$W = \int_{1.00}^{2.50} p\, dv.$$

Estimate the value of this integral.

29. Approximate the value of the integral

$$\int_0^1 2x f(x)\, dx,$$

where f is given by

x	0.0	0.1	0.2	0.3	0.4	0.5	0.6	0.7	0.8	0.9	1.0
$f(x)$	0.667	0.671	0.689	0.711	0.742	0.790	0.841	0.910	0.975	1.052	1.130

This integral arises in computing the mean flight distance of birds, randomly dispersed throughout a circular region, to all other points of the region (see J. F. Wittenberger and M. B. Dollinger, "The Effect of Acentric Colony Location on the Energetics of Avian Coloniality," *American Naturalist*, **124**, 189–204, 1984).

6.6 GAUSSIAN QUADRATURE

In the previous two sections, the concept of Newton-Cotes quadrature was developed. In this approach to numerical integration, the value of the definite integral

$$I(f) = \int_a^b f(x)\, dx$$

is approximated by the quadrature rule

$$I_n(f) = \sum_{i=0}^{n} w_i f(x_i).$$

The x_i are called abscissas and are selected as equally spaced points within the integration interval $[a, b]$. The weights, w_i, are then found by integrating the polynomial that interpolates the integrand at the abscissas. The degree of precision of the resulting rule is n when n is odd and $n+1$ when n is even.

In this section, we will develop the concept of Gaussian quadrature. In this approach to numerical integration, the abscissas and weights are selected so as to achieve the highest possible degree of precision.

Method of Undetermined Coefficients

The method of undetermined coefficients is essentially the brute force method for developing Gaussian quadrature rules. It involves a straightforward application of the definition of degree of precision, and proceeds as follows. Given a positive integer n, we wish to determine $2n$ numbers—the abscissas $x_1, x_2, x_3, \ldots, x_n$ and the weights $w_1, w_2, w_3, \ldots, w_n$—so that the summation

$$w_1 f(x_1) + w_2 f(x_2) + w_3 f(x_3) + \cdots + w_n f(x_n)$$

provides the exact value of $\int_a^b f(x)\,dx$ for $f(x) = 1, x, x^2, \ldots, x^{2n-1}$. In other words, the quadrature rule will have degree of precision equal to $2n-1$. Applying each of these conditions produces a system of $2n$ equations, which will be nonlinear for all $n > 1$.

To demonstrate the method of undetermined coefficients process, let's develop a Gaussian quadrature rule with $n = 1$. We want the approximation formula

$$\int_a^b f(x)\,dx = w_1 f(x_1)$$

to have degree of precision 1; that is, this formula should obtain exact results for all constant and for all linear functions. Applying these two conditions produces the following system of equations:

$$\begin{array}{ll} f(x) = 1: & w_1 = \int_a^b dx = b - a \\ f(x) = x: & w_1 x_1 = \int_a^b x\,dx = \frac{1}{2}(b^2 - a^2). \end{array}$$

The solution of this system is $w_1 = b - a$ and $x_1 = (a+b)/2$, so the resulting Gaussian quadrature rule is

$$\int_a^b f(x)\,dx \approx (b-a) f\left(\frac{a+b}{2}\right),$$

which we should recognize as the midpoint rule. The error associated with this quadrature rule is therefore

$$\frac{(b-a)^3}{24} f''(\xi),$$

where $a < \xi < b$.

For determining Gaussian quadrature rules with $n > 1$, it is to our advantage to replace the general integration interval of $[a, b]$ with a standardized interval, the most common choice for which is $[-1, 1]$. With such an interval we can exploit the symmetries in the problem to simplify the solution of the nonlinear system of equations for the abscissas and weights. The conversion from the integral

$$\int_a^b f(x)\,dx$$

to an integral of the form

$$\int_{-1}^{1} \tilde{f}(t)\, dt$$

is most easily accomplished by the change of variable

$$x = \frac{b-a}{2}t + \frac{a+b}{2}.$$

This formula comes from the equation of the line passing through the points $(-1, a)$ and $(1, b)$, where the first coordinate is the t-coordinate and the second is the x-coordinate. The resulting relationship between the two integrals is then

$$\int_a^b f(x)\, dx = \frac{b-a}{2}\int_{-1}^{1} f\left(\frac{b-a}{2}t + \frac{a+b}{2}\right) dt.$$

Let's now construct the two-point Gaussian quadrature rule

$$\int_{-1}^{1} f(x)\, dx \approx w_1 f(x_1) + w_2 f(x_2).$$

Since this formula is to have degree of precision equal to $2(2) - 1 = 3$, the weights and abscissas must satisfy

$$\begin{array}{ll} f(x) = 1: & w_1 + w_2 = \int_{-1}^{1} dx = 2 \\ f(x) = x: & w_1x_1 + w_2x_2 = \int_{-1}^{1} x\, dx = 0 \\ f(x) = x^2: & w_1x_1^2 + w_2x_2^2 = \int_{-1}^{1} x^2\, dx = \frac{2}{3} \\ f(x) = x^3: & w_1x_1^3 + w_2x_2^3 = \int_{-1}^{1} x^3\, dx = 0. \end{array}$$

The symmetry of the integration interval about zero suggests $x_2 = -x_1$ and $w_1 = w_2$. Substituting these relations into the system, the equations for $f(x) = x$ and $f(x) = x^3$ are satisfied exactly, and the remaining equations take the form $2w_1 = 2$ and $2w_1x_1^2 = 2/3$. The solution of the system is then $w_1 = w_2 = 1$, $x_1 = -\sqrt{1/3}$ and $x_2 = \sqrt{1/3}$, giving the quadrature rule

$$\int_{-1}^{1} f(t)\, dt \approx f\left(-\sqrt{\frac{1}{3}}\right) + f\left(\sqrt{\frac{1}{3}}\right).$$

Unlike our previous example, this quadrature rule is not one that was developed in an earlier section, so we will have to derive the error term from scratch. To accomplish this, first note that if we interpolate the integrand, f, at $x_1 = -\sqrt{1/3}$ and $x_2 = \sqrt{1/3}$ and then integrate the resulting interpolating polynomial, we reproduce the two-point Gaussian quadrature rule. This implies that the error term associated with the two-point Gaussian quadrature rule is

$$\int_{-1}^{1} f[x_1, x_2, x](x - x_1)(x - x_2)\, dx.$$

From this starting point, we proceed as follows. Since

$$\frac{f[x_1,x_2,x]-f[x_1,x_2,x_1]}{x-x_1}=f[x_1,x_2,x_1,x],$$

we may replace $f[x_1,x_2,x]$ by $f[x_1,x_2,x_1]+f[x_1,x_2,x_1,x](x-x_1)$. This replacement transforms the error term to

$$\int_{-1}^{1} f[x_1,x_2,x_1](x-x_1)(x-x_2)\,dx+\int_{-1}^{1} f[x_1,x_2,x_1,x](x-x_1)^2(x-x_2)\,dx.$$

The first of these integrals is equal to zero. In the second integral, we use the equation

$$\frac{f[x_1,x_2,x_1,x]-f[x_1,x_2,x_1,x_2]}{x-x_2}=f[x_1,x_2,x_1,x_2,x]$$

to replace $f[x_1,x_2,x_1,x]$ by $f[x_1,x_2,x_1,x_2]+f[x_1,x_2,x_1,x_2,x](x-x_2)$. Now the error term takes the form

$$\int_{-1}^{1} f[x_1,x_2,x_1,x_2](x-x_1)^2(x-x_2)\,dx+\int_{-1}^{1} f[x_1,x_2,x_1,x_2,x](x-x_1)^2(x-x_2)^2\,dx.$$

The first of these integrals is again equal to zero. Finally, an application of the Weighted Mean-Value Theorem for Integrals to the second integral leads to

$$\int_{-1}^{1} f(t)\,dt=f\left(-\sqrt{\frac{1}{3}}\right)+f\left(\sqrt{\frac{1}{3}}\right)+\frac{1}{135}f^{(4)}(\xi),$$

where $-1<\xi<1$.

Converting this rule back to the more general integration interval $[a,b]$ produces

$$\begin{aligned}
&\int_a^b f(x)\,dx\\
&=\frac{b-a}{2}\int_{-1}^{1} f\left(\frac{b-a}{2}t+\frac{a+b}{2}\right)dt\\
&=\frac{b-a}{2}\left[f\left(\frac{a+b}{2}-\sqrt{\frac{1}{3}}\frac{b-a}{2}\right)+f\left(\frac{a+b}{2}+\sqrt{\frac{1}{3}}\frac{b-a}{2}\right)+\frac{1}{135}\frac{d^4f}{dt^4}(\xi)\right]\\
&=\frac{b-a}{2}\left[f\left(\frac{a+b}{2}-\sqrt{\frac{1}{3}}\frac{b-a}{2}\right)+f\left(\frac{a+b}{2}+\sqrt{\frac{1}{3}}\frac{b-a}{2}\right)\right]+\frac{(b-a)^5}{4320}\frac{d^4f}{dx^4}(\hat{\xi})
\end{aligned}$$

where $a<\hat{\xi}<b$ and, in the last line, the chain rule has been used to convert derivatives with respect to t in the error term to derivatives with respect to x:

$$\frac{d}{dt}=\frac{d}{dx}\frac{dx}{dt}=\frac{b-a}{2}\frac{d}{dx}\quad\Rightarrow\quad\frac{d^4}{dt^4}=\left(\frac{b-a}{2}\right)^4\frac{d^4}{dx^4}.$$

EXAMPLE 6.14 Approximating ln(2)

One way to approximate the value of ln(2) is to approximate the value of the integral

$$\int_1^2 \frac{1}{x}\,dx.$$

Using the two-point Gaussian quadrature rule and noting that for this problem $a = 1$, $b = 2$, and $f(x) = 1/x$, we obtain the approximation

$$\begin{aligned}\int_1^2 \frac{1}{x}\,dx &\approx \frac{2-1}{2}\left[\left(\frac{2+1}{2} - \sqrt{\frac{1}{3}}\frac{2-1}{2}\right)^{-1} + \left(\frac{2+1}{2} + \sqrt{\frac{1}{3}}\frac{2-1}{2}\right)^{-1}\right]\\ &= \frac{1}{2}\left[\left(\frac{3}{2} - \frac{\sqrt{3}}{6}\right)^{-1} + \left(\frac{3}{2} + \frac{\sqrt{3}}{6}\right)^{-1}\right]\\ &= 0.6923076923.\end{aligned}$$

Even with only two function evaluations, the absolute error in this approximation is 8.394×10^{-4}.

If greater accuracy is required, we can develop a formula with a higher degree of precision, or we can work in a composite manner, dividing the interval $[a, b]$ into subintervals and applying the lower accuracy formula on each subinterval. In Exercise 2 of Section 6.5, the composite midpoint rule,

$$\int_a^b f(x)\,dx = 2h\sum_{j=1}^{n} f(x_j) + \frac{(b-a)h^2}{6} f''(\xi),$$

where $h = (b-a)/2n$, $x_j = a + (2j-1)h$ and $a < \xi < b$, was considered. For the two-point Gaussian quadrature rule, let $h = (b-a)/n$ and $x_j = a + jh$ for $j = 1, 2, 3, \ldots, n$. Applying the basic quadrature rule on each subinterval $[x_{j-1}, x_j]$ yields

$$\begin{aligned}&\int_a^b f(x)\,dx\\ &= \frac{h}{2}\sum_{j=1}^{n}\left[f\left(x_j - \frac{h}{2} - \sqrt{\frac{1}{3}}\frac{h}{2}\right) + f\left(x_j - \frac{h}{2} + \sqrt{\frac{1}{3}}\frac{h}{2}\right)\right] + \frac{(b-a)h^4}{4320} f^{(4)}(\xi),\end{aligned}$$

where $a < \xi < b$. Details of this derivation are left as an exercise.

EXAMPLE 6.15 Approximating the Value of π

In the previous section, the composite Simpson's rule was used to approximate the value of π to four decimal places by estimating the value of the integral

$$\int_0^1 \frac{1}{1+x^2}\,dx$$

to within 0.0000125. $n = 12$ subintervals were required to guarantee the desired approximation error, and the resulting approximations for the value of the integral and for π were

$$\int_0^1 \frac{1}{1+x^2}\,dx \approx 0.78539816007634$$

and

$$\pi \approx 3.14159264030538.$$

The absolute error in this approximation to π is 1.3284×10^{-8}. For future reference, note that 12 subintervals for Simpson's rule corresponds to 13 function evaluations.

For the composite two-point Gaussian quadrature rule, $h = (b-a)/n$, so the error term can be written as

$$\frac{(b-a)h^4}{4320} f^{(4)}(\xi) = \frac{(b-a)^5}{4320n^4} f^{(4)}(\xi).$$

Since $\max_{x\in[0,1]} |f^{(4)}(x)| = 24$, it follows that n must be selected to satisfy the inequality

$$\frac{(1-0)^5}{4320n^4} \cdot 24 < 1.25 \times 10^{-5}$$

in order to guarantee an estimate for the integral with an absolute error no larger than 0.0000125. The solution of this inequality is $n > 4.59$; therefore, we will use $n = 5$. With $n = 5$,

$$\int_0^1 \frac{1}{1+x^2}\,dx \approx 0.78539817044636$$

$$\Rightarrow \pi \approx 3.14159268178543.$$

The absolute error in this approximation to π is 2.8196×10^{-8}, which is slightly larger than that obtained using Simpson's rule. However, with five subintervals, the composite two-point Gaussian quadrature rule uses only 10 function evaluations (two per subinterval), compared to the 13 used by Simpson's rule.

(The remainder of this section may be omitted without loss of continuity.)

Theoretical Development

Although the method of undetermined coefficients is conceptually straightforward, it tends to give the impression that there is no relationship among Gaussian quadrature rules. This, however, is not the case. There is an underlying theory which connects all Gaussian quadrature rules.

To develop this unifying theory, we will consider the more general class of integrals of the form

$$\int_a^b f(x)w(x)\,dx.$$

The function w is known as a weight function.

Definition. The function w is called a WEIGHT FUNCTION on the interval $[a, b]$ if it satisfies the three properties:

(1) w is integrable on $[a, b]$; *i.e.*, $\int_a^b w(x)\,dx$ exists;
(2) $w(x) \geq 0$ for all $x \in [a, b]$; and
(3) $w(x) = 0$ at isolated points only; that is, there is no open interval $(x_1, x_2) \subset [a, b]$ such that $w(x) = 0$ for all $x \in (x_1, x_2)$.

The weight functions (and corresponding intervals) most commonly encountered in practice are $w(x) = 1$ on $[-1, 1]$, $w(x) = e^{-x}$ on $[0, \infty)$, $w(x) = e^{-x^2}$ on $(-\infty, \infty)$, and $w(x) = (1 - x^2)^{-1/2}$ on $[-1, 1]$.

Associated with each weight function/integration interval pair is a special family of polynomials, unique up to a normalization factor. These functions lie at the heart of Gaussian quadrature. For a given n, the family consists of $n + 1$ polynomials ϕ_0, ϕ_1, $\phi_2, \ldots, \phi_n$, where the degree of each ϕ_k is equal to k. Hence, the family consists of one constant function, one linear, one quadratic and so on. The most important property of these functions is ORTHOGONALITY:

$$\int_a^b \phi_i(x)\phi_j(x)w(x)\,dx = 0$$

whenever $i \neq j$. The families of orthogonal polynomials which correspond to the specific weight functions mentioned above are summarized in Table 6.4. For an arbitrary weight function, the corresponding family of orthogonal polynomials can be generated by applying the Gram-Schmidt process (see any text on linear algebra) to the polynomials 1, x, $x^2, \ldots, x^n$.

We are now in a position to establish the main result regarding Gaussian quadrature. To simplify notation, let Π_n = {all polynomials of degree $\leq n$}. In the proof of the following theorem, we will need this basic fact from linear algebra: if $B = \{\phi_0, \phi_1, \phi_2, \ldots, \phi_n\} \subset \Pi_n$ is an orthogonal family, then B forms a basis for Π_n. In other words, for any polynomial $p \in \Pi_n$, there exist unique constants c_0, c_1, $c_2, \ldots, c_n$ such that p can be expressed as

$$p = c_0\phi_0 + c_1\phi_1 + c_2\phi_2 + \cdots + c_n\phi_n = \sum_{i=0}^{n} c_i\phi_i.$$

Weight	Family	First polynomials
$w(x) = 1$ on $[-1, 1]$	Legendre polynomials $P_n(x)$	$P_0(x) = 1,\ P_1(x) = x$ $P_2(x) = (3x^2 - 1)/2$ $P_3(x) = (5x^3 - 3x)/2$
Recurrence relation: $P_{n+1}(x) = \frac{2n+1}{n+1} x P_n(x) - \frac{n}{n+1} P_{n-1}(x)$		
$w(x) = e^{-x}$ on $[0, \infty)$	Laguerre polynomials $L_n(x)$	$L_0(x) = 1,\ L_1(x) = 1 - x$ $L_2(x) = x^2 - 4x + 2$ $L_3(x) = -x^3 + 9x^2 - 18x + 6$
Recurrence relation: $L_{n+1}(x) = (1 + 2n - x)L_n(x) - n^2 L_{n-1}(x)$		
$w(x) = e^{-x^2}$ on $(-\infty, \infty)$	Hermite polynomials $H_n(x)$	$H_0(x) = 1,\ H_1(x) = 2x$ $H_2(x) = 4x^2 - 2$ $H_3(x) = 8x^3 - 12x$
Recurrence relation: $H_{n+1}(x) = 2xH_n(x) - 2nH_{n-1}(x)$		
$w(x) = (1 - x^2)^{-1/2}$ on $[-1, 1]$	Chebyshev polynomials $T_n(x)$	$T_0(x) = 1,\ T_1(x) = x$ $T_2(x) = 2x^2 - 1$ $T_3(x) = 4x^3 - 3x$
Recurrence relation: $T_{n+1}(x) = 2xT_n(x) - T_{n-1}(x)$		

TABLE 6.4: Common Families of Orthogonal Polynomials

Theorem (Gaussian Quadrature). Let w be a weight function on $[a, b]$, let n be a positive integer and let $\{\phi_0, \phi_1, \phi_2, \ldots, \phi_n\} \subset \Pi_n$ be an orthogonal family with degree of $\phi_k = k$ for each k. Let $x_1, x_2, x_3, \ldots, x_n$ be the roots of $\phi_n(x)$ and define

$$L_i(x) = \prod_{j=1, j\neq i}^{n} \frac{(x - x_j)}{(x_i - x_j)} \quad \text{for each } i.$$

Then the corresponding Gaussian quadrature formula is given by

$$I(f) = \int_a^b f(x)w(x)\,dx \approx I_n(f) \equiv \sum_{i=1}^{n} w_i f(x_i),$$

where $w_i = \int_a^b L_i(x)w(x)\,dx$. The formula $I_n(f)$ has degree of precision $2n-1$.

Proof. The only thing that needs to be proven is that the quadrature rule

$$I_n(f) = \sum_{i=1}^{n} w_i f(x_i),$$

where $w_i = \int_a^b L_i(x)w(x)\,dx$, has degree of precision equal to $2n-1$. We will establish this result in three steps. First, we will show that the quadrature rule has degree of precision at least $n-1$, then that it has degree of precision at least $2n-1$, and finally that the degree of precision is exactly $2n-1$.

(a) Degree of Precision $\geq n-1$
Let $g \in \Pi_{n-1}$. Note that the functions $L_i(x)$ defined in the statement of the theorem are the Lagrange polynomials associated with the roots of the polynomial $\phi_n(x)$. The function $\sum_{i=1}^{n} g(x_i)L_i(x)$ is therefore the Lagrange form of the interpolating polynomial that interpolates g at the roots of ϕ_n. Since g is itself a polynomial, it follows from the uniqueness of the interpolating polynomial that $g(x) = \sum_{i=1}^{n} g(x_i)L_i(x)$ for all x.

Then

$$I(g) = \int_a^b g(x)w(x)\,dx = \sum_{i=1}^{n} g(x_i) \int_a^b L_i(x)w(x)\,dx = \sum_{i=1}^{n} w_i g(x_i) = I_n(g).$$

I_n therefore produces the exact integral for any polynomial of degree up to and including $n-1$, so, by definition, the quadrature rule has degree of precision at least equal to $n-1$.

(b) Degree of Precision $\geq 2n-1$
Let $p \in \Pi_{2n-1}$ and divide p by ϕ_n to obtain

$$p(x) = q(x)\phi_n(x) + r(x),$$

where $q, r \in \Pi_{n-1}$. Since the polynomials $\{\phi_0, \phi_1, \phi_2, \ldots, \phi_{n-1}\}$ form a basis for Π_{n-1} and $q \in \Pi_{n-1}$, there exist constants c_0, c_1, $c_2, \ldots, c_{n-1}$ such that $q = \sum_{i=0}^{n-1} c_i\phi_i$. Then

$$\begin{aligned} I(p) &= I(q\phi_n + r) \\ &= \sum_{i=0}^{n-1} c_i \int_a^b \phi_i(x)\phi_n(x)w(x)\,dx + I(r) \\ &= 0 + I(r) = I(r), \end{aligned}$$

where the summation term vanishes due to orthogonality. Now, $r \in \Pi_{n-1}$ so that $I(r) = I_n(r)$ by part (a). Since the x_i are roots of ϕ_n, it follows that

$$p(x_i) = q(x_i)\phi_n(x_i) + r(x_i) = 0 + r(x_i) = r(x_i),$$

which implies $I_n(r) = I_n(p)$. Combining these results yields $I(p) = I_n(p)$.

(c) Degree of Precision $= 2n-1$
To establish that the degree of precision is exactly $2n-1$, it is only necessary to find one polynomial of degree $2n$ for which $I_n(p) \neq I(p)$. Consider the polynomial $\phi_n^2(x)$, which has degree $2n$. Note that

$$I(\phi_n^2(x)) = \int_a^b \phi_n^2(x)w(x)\,dx > 0,$$

but

$$I_n(\phi_n^2(x)) = \sum_{i=1}^{n} w_i \phi_n^2(x_i) = 0. \qquad \square$$

Remarks. (1) For a given number of quadrature points n, weight function w, and integration interval $[a,b]$, the quadrature rule $I_n(f) = \sum_{i=1}^{n} w_i f(x_i)$ is unique in the sense that it is the only quadrature rule based on the weight function $w(x)$ with degree of precision equal to $2n-1$.
(2) If f has $2n$ continuous derivatives, then there exists $\xi \in [a,b]$ such that

$$I(f) = I_n(f) + \frac{\alpha_n}{a_n^2(2n)!} f^{(2n)}(\xi),$$

where $\alpha_n = \int_a^b \phi_n^2(x)w(x)\,dx$ and a_n is the leading coefficient of $\phi_n(x)$ (see Exercises 31 and 32).

In establishing the previous theorem, we have overlooked one important item. The proof implicitly assumes that ϕ_n has n simple roots, all of which lie inside the interval (a,b). The next theorem establishes that this is true.

Theorem. If $\{\phi_0, \phi_1, \phi_2, \ldots, \phi_n\}$ is a set of polynomials with the following properties:

1. ϕ_k has degree k for each k, and
2. the set is orthogonal with respect to the weight function $w(x)$ on $[a,b]$,

then for each k, ϕ_k has precisely k real roots, which are all simple and all lie in (a,b).

Proof. The proof of this theorem will proceed in two steps. First we will establish the existence of real roots for ϕ_k, and then we will count those roots.

(a) Existence of Real Roots
Since the degree of ϕ_0 is zero, it follows that $\phi_0 = \alpha$ for some nonzero constant α. Then, for $k \geq 0$,

$$0 = \int_a^b \phi_0(x)\phi_k(x)w(x)\,dx = \alpha \int_a^b \phi_k(x)w(x)\,dx,$$

where the first equality holds due to orthogonality. However,

$$\left.\begin{array}{r} w(x) \geq 0 \text{ on } [a,b] \\ \text{integral } = 0 \end{array}\right\} \Rightarrow \phi_k \text{ must change sign in } (a,b).$$

Since ϕ_k is a continuous function on (a,b), the Intermediate Value Theorem guarantees the existence of a real root somewhere on (a,b).

(b) Count Roots

Suppose ϕ_k changes sign at exactly j points $r_1, r_2, r_3, \ldots, r_j$ in (a,b) such that

$$a < r_1 < r_2 < r_3 < \cdots < r_j < b.$$

Without loss of generality, assume that $\phi_k(x) > 0$ on (a, r_1). Then ϕ_k alternates sign on (r_1, r_2), (r_2, r_3), $(r_3, r_4), \ldots, (r_j, b)$. Define the auxiliary function

$$p(x) = (-1)^j \prod_{i=1}^{j} (x - r_i) \in \Pi_j.$$

By construction, $p(x)$ and $\phi_k(x)$ have the same sign for all $x \in [a,b]$. From this it follows that

$$\int_a^b p(x)\phi_k(x)w(x)\,dx > 0. \tag{1}$$

Suppose that $j < k$. Since $\{\phi_0, \phi_1, \phi_2, \ldots, \phi_j\}$ forms a basis for Π_j, there exist constants $c_0, c_1, c_2, \ldots, c_j$ such that $p(x) = \sum_{i=0}^{j} c_i\phi_i(x)$. Substituting this expression into (1) yields

$$\begin{aligned} \int_a^b p(x)\phi_k(x)w(x)\,dx &= \int_a^b \sum_{i=0}^{j} c_i\phi_i(x)\phi_k(x)w(x)\,dx \\ &= \sum_{i=0}^{j} c_i \int_a^b \phi_i(x)\phi_k(x)w(x)\,dx = 0, \end{aligned}$$

where the final equality follows from the assumption that $j < k$ and from orthogonality. We have therefore arrived at a contradiction, which implies that $j \geq k$.

However, each $r_1, r_2, r_3, \ldots, r_j$ is a root of ϕ_k, and ϕ_k has degree k, which means ϕ_k can have at most k roots. Therefore, we must have $j = k$, so that ϕ_k has k simple real roots in (a,b). □

EXERCISES

1. Approximate the value of each of the following integrals using the two-point Gaussian quadrature rule (the basic formula, not the composite rule). Verify that the theoretical error bound holds in each case.

   (a) $\int_{-1}^{1} e^{-x}\,dx$ **(b)** $\int_{-1}^{1} \frac{1}{1+x^2}\,dx$

   (c) $\int_0^{\pi} \sin x\,dx$ **(d)** $\int_0^1 \tan^{-1} x\,dx$

2. Derive the composite two-point Gaussian quadrature rule:

$$\int_a^b f(x)\,dx = \frac{h}{2}\sum_{j=1}^{n}\left[f\left(x_j - \frac{h}{2} - \sqrt{\frac{1}{3}}\frac{h}{2}\right) + f\left(x_j - \frac{h}{2} + \sqrt{\frac{1}{3}}\frac{h}{2}\right)\right] + \frac{(b-a)h^4}{4320}f^{(4)}(\xi),$$

where $h = (b-a)/n$, $x_j = a + jh$ and $a < \xi < b$.

3. Approximate the value of each of the following integrals using the composite two-point Gaussian quadrature rule with the specified number of subintervals. Verify that the theoretical error bound holds in each case.

(a) $\int_{-1}^{1} e^{-x}\,dx, \quad n = 2$ **(b)** $\int_{-1}^{1} \frac{1}{1+x^2}\,dx, \quad n = 2$

(c) $\int_0^{\pi} \sin x\,dx, \quad n = 3$ **(d)** $\int_0^1 \tan^{-1} x\,dx, \quad n = 3$

4. Let $x_1 = -\sqrt{1/3}$ and $x_2 = \sqrt{1/3}$. Show that

(a) $\int_{-1}^{1} f[x_1, x_2, x_1](x - x_1)(x - x_2)\,dx = 0$;

(b) $\int_{-1}^{1} f[x_1, x_2, x_1, x_2](x - x_1)^2(x - x_2)\,dx = 0$; and

(c) $\int_{-1}^{1} f[x_1, x_2, x_1, x_2, x](x - x_1)^2(x - x_2)^2\,dx = \frac{1}{135}f^{(4)}(\xi)$, where $a < \xi < b$.

5. (a) Derive the three-point Gaussian quadrature rule

$$\int_{-1}^{1} f(x)\,dx = \frac{5}{9}f\left(-\sqrt{\frac{3}{5}}\right) + \frac{8}{9}f(0) + \frac{5}{9}f\left(\sqrt{\frac{3}{5}}\right) + \frac{1}{15750}f^{(6)}(\xi),$$

where $-1 < \xi < 1$.

(b) Convert the quadrature rule from part (a) to the general integration interval $[a, b]$.

(c) Derive the composite three-point Gaussian quadrature rule. [*Note*: The rate of convergence should be $O(h^6)$.]

6. Use the three-point Gaussian quadrature rule to approximate the value of the definite integral $\int_1^2 \frac{1}{x}\,dx$. What is the absolute error in this approximation?

7. Repeat Exercise 1 using the three-point Gaussian quadrature rule.

8. Repeat Exercise 3 using the composite three-point Gaussian quadrature rule.

In Exercises 9–16, verify that the composite two-point Gaussian quadrature rule has rate of convergence $O(h^4)$ and the composite three-point Gaussian quadrature rule has rate of convergence $O(h^6)$ by approximating the value of the indicated definite integral.

9. $\int_0^1 \sqrt{1 + x^3}\,dx$

10. $\int_0^{\pi} \sin x\,dx$

11. $\int_1^2 \frac{1}{x}\,dx$

12. $\int_0^1 e^{-x}\,dx$

13. $\int_0^1 \tan^{-1} x\,dx$

14. $\int_1^2 \frac{\sin x}{x}\,dx$

15. $\int_0^1 \frac{1}{\sqrt{1+x^4}}\,dx$

16. $\int_0^4 x\sqrt{x^2+9}\,dx$

In Exercises 17–24, approximate the value of the indicated definite integral using the composite two-point Gaussian quadrature rule and the composite three-point Gaussian quadrature rule. For each method, use the smallest value of n which will guarantee an absolute error of no greater than 5×10^{-5}.

17. $\int_1^2 \frac{1}{x}\,dx$

18. $\int_0^1 e^{-x}\,dx$

19. $\int_0^1 \tan^{-1} x\,dx$

20. $\int_1^2 \frac{\sin x}{x}\,dx$

21. $\int_0^1 \frac{1}{\sqrt{1+x^4}}\,dx$

22. $\int_0^4 x\sqrt{x^2+9}\,dx$

23. $\int_0^1 \sqrt{1+x^3}\,dx$

24. $\int_0^1 e^{-x^4}\,dx$

25. Consider the definite integral $\int_a^b \sin(\sqrt{\pi x})\,dx$. Numerically determine the rate of convergence of the composite two-point Gaussian quadrature rule for each of the following integration intervals.
 (a) $[a,b] = [0,1]$
 (b) $[a,b] = [\pi/4, 9\pi/4]$
 (c) $[a,b] = [\pi, 2\pi]$
 (d) Explain any variation among the rates of convergence obtained in parts (a), (b), and (c).

26. Repeat Exercise 25 for the composite three-point Gaussian quadrature rule.

27. Consider the definite integral $\int_a^b x^2 e^{-x}\,dx$. Numerically determine the rate of convergence of the composite two-point Gaussian quadrature rule for each of the following integration intervals.
 (a) $[a,b] = [0,2]$
 (b) $[a,b] = [3-\sqrt{3}, 3+\sqrt{3}]$
 (c) $[a,b] = [-1,1]$
 (d) Explain any variation among the rates of convergence obtained in parts (a), (b), and (c).

Optional Material

28. **(a)** Find the abscissas, x_i, and the weights, w_i, of the three-point Gauss-Hermite quadrature formula

$$\int_{-\infty}^{\infty} e^{-x^2} f(x)\,dx \approx w_1 f(x_1) + w_2 f(x_2) + w_3 f(x_3).$$

Use the fact that the Hermite polynomials, $H_n(x)$, are orthogonal in the corresponding inner product

$$(f,g) = \int_{-\infty}^{\infty} e^{-x^2} f(x)g(x)\,dx$$

and that $H_3(x) = 8x^3 - 12x$. Find the weights by undetermined coefficients using the values

$$\int_{-\infty}^{\infty} e^{-x^2}\,dx = \sqrt{\pi} \qquad \int_{-\infty}^{\infty} xe^{-x^2}\,dx = 0 \qquad \int_{-\infty}^{\infty} x^2e^{-x^2}\,dx = \frac{\sqrt{\pi}}{2}.$$

(b) Use your results from part (a) to evaluate both

$$\int_{-\infty}^{\infty} \frac{e^{-x^2}}{1+x^2}\,dx \qquad \text{and} \qquad \int_{-\infty}^{\infty} \frac{1}{1+x^2}\,dx.$$

29. (a) Find the abscissas and the weights of the three-point Gauss-Chebyshev quadrature formula

$$\int_{-1}^{1} \frac{f(x)}{\sqrt{1-x^2}}\,dx \approx w_1 f(x_1) + w_2 f(x_2) + w_3 f(x_3).$$

Use the fact that the Chebyshev polynomials, $T_n(x)$, are orthogonal in the corresponding inner product

$$(f,g) = \int_{-1}^{1} \frac{f(x)g(x)}{\sqrt{1-x^2}}\,dx$$

and that $T_3(x) = 4x^3 - 3x$. Find the weights by undetermined coefficients using the values

$$\int_{-1}^{1} \frac{1}{\sqrt{1-x^2}}\,dx = \pi \qquad \int_{-1}^{1} \frac{x}{\sqrt{1-x^2}}\,dx = 0 \qquad \int_{-1}^{1} \frac{x^2}{\sqrt{1-x^2}}\,dx = \frac{\pi}{2}.$$

(b) Use your results from part (a) to evaluate

$$\int_{-1}^{1} \frac{\cos x}{\sqrt{1-x^2}}\,dx.$$

30. (a) Find the abscissas and the weights of the three-point Gauss-Laguerre quadrature formula

$$\int_{0}^{\infty} e^{-x} f(x)\,dx \approx w_1 f(x_1) + w_2 f(x_2) + w_3 f(x_3).$$

Use the fact that the Laguerre polynomials, $L_n(x)$, are orthogonal in the corresponding inner product

$$(f,g) = \int_{0}^{\infty} e^{-x} f(x)g(x)\,dx$$

and that $L_3(x) = -x^3 + 9x^2 - 18x + 6$. Find the weights by undetermined coefficients using the values

$$\int_0^\infty e^{-x} = 1 \quad \int_0^\infty xe^{-x} = 1 \quad \int_0^\infty x^2e^{-x} = 2.$$

(b) Use your results from part (a) to evaluate both

$$\int_0^\infty \frac{e^{-x}}{1+x^2}\,dx \quad \text{and} \quad \int_0^\infty \frac{1}{1+x^2}\,dx.$$

31. Let w be a weight function on $[a, b]$, let $\{\phi_0, \phi_1, \phi_2, \ldots \phi_n\} \subset \Pi_n$ be an orthogonal family with respect to w with degree of $\phi_k = k$ for each k and let x_1, x_2, $x_3, \ldots, x_n$ be the roots of $\phi_n(x)$. Show that

$$\int_a^b w(x)\prod_{i=1}^{n}(x - x_i)\,dx = 0$$

and

$$\int_a^b w(x)\prod_{i=1}^{k}(x - x_i)^2 \prod_{j=k+1}^{n}(x - x_j)\,dx = 0$$

for $k = 1, 2, 3, \ldots, n-1$.

32. Let w be a weight function on $[a, b]$, let n be a positive integer, let $\{\phi_0, \phi_1, \phi_2, \ldots, \phi_n\} \subset \Pi_n$ be an orthogonal family with respect to w with degree of $\phi_k = k$ for each k, and let $I_n(f)$ denote the corresponding Gaussian quadrature rule for approximating

$$I(f) = \int_a^b f(x)w(x)\,dx.$$

Suppose f has $2n$ continuous derivatives. Show there exists $\xi \in [a, b]$ such that

$$I(f) = I_n(f) + \frac{\alpha_n}{a_n^2(2n)!}f^{(2n)}(\xi),$$

where $\alpha_n = \int_a^b \phi_n^2(x)w(x)\,dx$ and a_n is the leading coefficient of $\phi_n(x)$.

6.7 ROMBERG INTEGRATION

Though composite Newton-Cotes formulas and composite Gaussian quadrature formulas come equipped with error terms that can, in principle, be used to determine the number of subintervals needed to guarantee a given accuracy, the amount of work required to determine the number of subintervals can be greater than the amount of work required to compute the final approximation. The error terms also tend to provide pessimistic bounds, thereby requiring more work than is actually necessary.

In practice, we would prefer to have a quadrature scheme that automatically determines the amount of work (measured in terms of the number of function evaluations of the integrand) needed to achieve a desired level of accuracy. The scheme should also be able to report an estimate of the error in the computed approximation. In this section, we study one such scheme known as Romberg integration.

The Basic Process

Romberg integration is extrapolation applied to the composite trapezoidal rule. Recall from Section 6.3 that extrapolation is a general procedure which takes two approximations, computed using different values of some parameter h, and combines them in such a way that the result is a higher-order approximation than the original values. To apply this process, the power of h which appears in the leading term of the error of the original approximations must be known. Extrapolation works well with the composite trapezoidal rule because, by the Euler-Maclaurin summation formula (Section 6.5),

$$\int_a^b f(x)\,dx = T_h(f) + K_1 h^2 + K_2 h^4 + \cdots + K_n h^{2n} + o(h^{2n})$$

provided the integrand is sufficiently differentiable; that is, provided f is continuously differentiable $2n$ times on $[a,b]$. Here, the K_i's are constants which are independent of h. Hence, the original approximations are second order, the first extrapolated values will be fourth order, the next extrapolated values will be sixth order, and so on.

The conventional notation used for Romberg approximations is $R_{k,j}$, where the first subscript controls the step size, h, the second subscript indicates the level of extrapolation, and calculations are organized into the usual extrapolation table as follows:

$$\begin{array}{llll}
R_{1,1} & & & \\
R_{2,1} & R_{2,2} & & \\
R_{3,1} & R_{3,2} & R_{3,3} & \\
R_{4,1} & R_{4,2} & R_{4,3} & R_{4,4} \\
\vdots & & & \ddots
\end{array}$$

The first column of the table contains the composite trapezoidal rule approximations; that is,

$$R_{k,1} = T_h(f), \qquad \text{with } h = \frac{b-a}{2^{k-1}}.$$

Since $R_{k,j}$ is an $O(h^{2j})$ approximation for each j and the step size is cut in half with each new row, the remaining entries in the table are computed according to the formula

$$R_{k,j} = \frac{4^{j-1}R_{k,j-1} - R_{k-1,j-1}}{4^{j-1} - 1}.$$

EXAMPLE 6.16 Approximating $\ln(2)$

A five row Romberg Integration table for the value of

$$\int_1^2 \frac{1}{x}\, dx,$$

whose exact value is $\ln(2)$, is shown below. A single subinterval was used to compute the first entry in the first column. The error corresponding to each value in the integration table appears in the second table. Although the best trapezoidal rule approximation is correct to just 3 decimal places, the final tabulated value is accurate to 8 decimal places. Further, note that the errors in the first three columns drop by factors of roughly 4, 16, and 64 with each new row. This is exactly what we would expect from second-, fourth-, and sixth-order approximations.

Romberg Integration Extrapolation Table

0.7500000000				
0.7083333333	0.6944444444			
0.6970238095	0.6932539683	0.6931746032		
0.6941218504	0.6931545307	0.6931479015	0.6931474776	
0.6933912022	0.6931476528	0.6931471943	0.6931471831	0.6931471819

Corresponding Errors

5.6853×10^{-2}				
1.5186×10^{-2}	1.2973×10^{-3}			
3.8766×10^{-3}	1.0679×10^{-4}	2.7423×10^{-5}		
9.7467×10^{-4}	7.3501×10^{-6}	7.2092×10^{-7}	2.9708×10^{-7}	
2.4402×10^{-4}	4.7226×10^{-7}	1.3737×10^{-8}	2.5120×10^{-9}	1.3568×10^{-9}

Efficient Calculation of First Column

When computing the first column of the Romberg integration table, it is not necessary to start each trapezoidal rule calculation from scratch. In fact, all of the information used to compute one trapezoidal rule approximation can be reused in the calculation of the approximation with half the step size. To see how this works, suppose that $T_h(f)$ has been computed, where $h = (b-a)/n$ for some n. Then

$$\begin{aligned} T_{h/2}(f) &= \frac{h/2}{2}\left[f(a) + 2\sum_{j=1}^{2n-1} f\left(a + j\frac{h}{2}\right) + f(b)\right] \\ &= \frac{h/2}{2}\left[f(a) + 2\sum_{j=2,\text{even}}^{2n-2} f\left(a + j\frac{h}{2}\right) + 2\sum_{j=1,\text{odd}}^{2n-1} f\left(a + j\frac{h}{2}\right) + f(b)\right] \end{aligned}$$

$$= \frac{h/2}{2}\left[f(a) + 2\sum_{k=1}^{n-1} f(a+kh) + f(b)\right] + \frac{h}{2}\sum_{j=1,\text{odd}}^{2n-1} f\left(a + j\frac{h}{2}\right)$$

$$= \frac{1}{2}T_h(f) + \frac{h}{2}\sum_{j=1,\text{odd}}^{2n-1} f\left(a + j\frac{h}{2}\right).$$

The second term in this last expression consists only of those function evaluations which were not used in the computation of $T_h(f)$. Working in this fashion, we can calculate the entire first column of the Romberg integration table for the same number of function evaluations needed just for the very last entry.

Error Estimate

The Romberg integration table can be computed either in a column-by-column or in a row-by-row manner. The principal advantage of computing the table one row at a time is that once a row has been completed, the error in the final value can be estimated and used to decide whether a new row should be computed. To establish an error estimate, recall from Section 6.3 that the formula

$$R_{n,n} = \frac{4^{n-1}R_{n,n-1} - R_{n-1,n-1}}{4^{n-1} - 1}$$

arises from

$$R_{n,n} = R_{n-1,n-1} + \frac{4^{n-1}}{4^{n-1}-1}(R_{n,n-1} - R_{n-1,n-1}),$$

where the second term on the right is the extrapolation estimate for the leading term in the error of $R_{n-1,n-1}$. Since $R_{n,n}$ should be a better approximation than $R_{n-1,n-1}$, we see that the difference $|R_{n,n} - R_{n-1,n-1}|$ can serve as an (albeit crude) estimate for the error in $R_{n,n}$.

The formula

$$R_{n,n} = \frac{4^{n-1}R_{n,n-1} - R_{n-1,n-1}}{4^{n-1} - 1}$$

can also be rearranged as

$$R_{n,n} = R_{n,n-1} + \frac{1}{4^{n-1}-1}(R_{n,n-1} - R_{n-1,n-1}),$$

where the second term on the right is the extrapolation estimate for the leading term in the error of $R_{n,n-1}$. Therefore, the difference $|R_{n,n} - R_{n,n-1}|$ can also serve as an estimate for the error in $R_{n,n}$.

Since several approximations and assumptions have been made along the way to arrive at these error estimates, it is advisable, in practice, to be conservative and strike some balance between the two estimates. Note that

$$R_{n,n} - R_{n,n-1} = \frac{1}{4^{n-1}-1}(R_{n,n-1} - R_{n-1,n-1}) = \frac{R_{n,n} - R_{n-1,n-1}}{4^{n-1}},$$

or one of the error estimates is equal to the other divided by 4^{n-1}. As a compromise, we will "split the difference" in the denominator and use $(R_{n,n} - R_{n-1,n-1})/2^{n-1}$ as an error estimate in the stopping criterion for Romberg integration. In other words, we will construct the Romberg integration table one row at a time. After each row has been completed, the condition

$$(R_{n,n} - R_{n-1,n-1})/2^{n-1} < \epsilon,$$

where ϵ is a specified convergence tolerance, will be tested. If this test passes, the algorithm terminates, returning $R_{n,n}$ as the estimate for the value of the integral. If this test fails, another row in the table is calculated.

EXAMPLE 6.17 Approximating $\pi/4$

Suppose we wish to use Romberg integration to approximate the value of the definite integral

$$\int_0^1 \frac{1}{1+x^2},$$

whose exact value is $\pi/4$, to within an absolute error of 5×10^{-5}. We start out computing the first two rows of the Romberg integration table, which are

$$\begin{array}{ll} R_{1,1} = T_1(f) = 0.7500000000 & \\ R_{2,1} = T_{1/2}(f) = 0.7750000000 & R_{2,2} = 0.7833333333 \end{array}$$

At this point, we estimate the error in $R_{2,2}$ to be

$$\frac{|R_{2,2} - R_{1,1}|}{2} = 0.0166666666.$$

Since this is larger than $\epsilon = 5 \times 10^{-5}$, we compute the next row of the table.

The third row of the Romberg integration table contains

$$R_{3,1} = T_{1/4}(f) = 0.7827941176, \quad R_{3,2} = 0.7853921569$$

and $R_{3,3} = 0.7855294118$. The error estimate associated with $R_{3,3}$ is

$$\frac{|R_{3,3} - R_{2,2}|}{4} = 5.4902 \times 10^{-4},$$

which is still too large. We therefore move on to compute the fourth row of the integration table.

The fourth row of the Romberg integration table contains

$$R_{4,1} = T_{1/8}(f) = 0.7847471236, \quad R_{4,2} = 0.7853981256, \quad R_{4,3} = 0.7853985235$$

and $R_{4,4} = 0.7853964459$. The corresponding estimate for the error in $R_{4,4}$ is

$$\frac{|R_{4,4} - R_{3,3}|}{8} = 1.6621 \times 10^{-5}.$$

Since this error estimate is smaller than $\epsilon = 5 \times 10^{-5}$, we terminate computations. Our final results are summarized below:

Approximate value of integral:	0.7853964459
Error estimate:	1.6621e-5
Number of function evaluations:	9

Note that the actual error is 1.7175×10^{-6}.

EXAMPLE 6.18 The Cost of a Warranty

In Section 6.5, the value of the integral

$$\int_0^1 e^{-x^4}\, dx,$$

which arises in the determination of the cost of a treadware warranty on tires, was approximated using the composite trapezoidal rule and the composite Simpson's rule, with a requirement that the absolute error be no more than 10^{-5}. The results obtained were

Trapezoidal rule:	$\int_0^1 e^{-x^4}\, dx \approx 0.8448344011$	(error $= 4.194 \times 10^{-6}$)
Simpson's rule:	$\int_0^1 e^{-x^4}\, dx \approx 0.8448403780$	(error $= 1.783 \times 10^{-6}$)

To compute the trapezoidal rule approximation, 172 function evaluations were needed; 17 function evaluations were needed to obtain the Simpson's rule approximation.

Using Romberg integration with a convergence tolerance of 10^{-5} produces the results:

Approximate value of integral:	0.844838710518
Error estimate:	8.8728e-07
Number of function evaluations:	17

The actual error in this approximation is 1.1576×10^{-7}. Hence, with the same number of function evaluations as the composite Simpson's rule, Romberg integration produces an error that is one order of magnitude smaller.

Application Problem: Tabulating the Error Function

Based on our analysis in the Chapter 6 Overview (see page 431), if we tabulate values of the error function,

$$\operatorname{erf}(x) = \frac{2}{\sqrt{\pi}} \int_0^x e^{-t^2} dt,$$

and its first derivative, $\frac{2}{\sqrt{\pi}} e^{-x^2}$, from $x = 0$ to $x = 2$ in increments of $\Delta x = 0.1$, then we are guaranteed that Hermite cubic interpolation between successive entries in the table will introduce an error smaller than 5×10^{-6}. Using Romberg integration with a tolerance of 5×10^{-6} to evaluate erf(x) for each x, we produce the following table.

x	erf(x)	$\frac{2}{\sqrt{\pi}}e^{-x^2}$	x	erf(x)	$\frac{2}{\sqrt{\pi}}e^{-x^2}$
0.0	0.00000	1.12838			
0.1	0.11246	1.11715	1.1	0.88020	0.33648
0.2	0.22270	1.08413	1.2	0.91031	0.26734
0.3	0.32863	1.03126	1.3	0.93401	0.20821
0.4	0.42839	0.96154	1.4	0.95229	0.15894
0.5	0.52050	0.87878	1.5	0.96611	0.11893
0.6	0.60386	0.78724	1.6	0.97635	0.08723
0.7	0.67780	0.69127	1.7	0.98379	0.06271
0.8	0.74210	0.59499	1.8	0.98909	0.04419
0.9	0.79691	0.50197	1.9	0.99279	0.03052
1.0	0.84270	0.41511	2.0	0.99532	0.02067

EXERCISES

1. Romberg integration approximates the value of the integral

$$\int_0^1 \frac{1}{1+x^2}\, dx$$

with an error of 1.2113×10^{-11} using only 33 function evaluations. How many function evaluations would be needed to achieve the same level of accuracy using the composite trapezoidal rule, the composite midpoint rule, the composite Simpson's rule, and the composite two-point Gaussian quadrature rule?

2. Romberg integration approximates the value of the integral

$$\int_{-1}^1 e^{-x}\, dx$$

with an error of 4.2399×10^{-11} using only 17 function evaluations. How many function evaluations would be needed to achieve the same level of accuracy using the composite trapezoidal rule, the composite midpoint rule, the composite Simpson's rule, and the composite two-point Gaussian quadrature rule?

3. Romberg integration approximates the value of the integral

$$\int_0^\pi \sin x\, dx$$

with an error of 1.3207×10^{-12} using only 33 function evaluations. How many function evaluations would be needed to achieve the same level of accuracy using the composite trapezoidal rule, the composite midpoint rule, the composite Simpson's rule and the composite two-point Gaussian quadrature rule?

In Exercises 4–7, the first column of the Romberg integration table for the specified definite integral is provided. Complete the table and determine the absolute error in the final approximation.

4. $\int_0^{3\pi/2} \cos x \, dx$

2.3561944902
−0.4879838567
−0.8815735630
−0.9709165361

5. $\int_0^2 e^x \, dx$

8.3890560989
6.9128098779
6.5216101095
6.4222978214

6. $\int_0^4 x\sqrt{x^2+9} \, dx$

40.0000000000
34.4222051019
33.1013022725
32.7750803748

7. $\int_1^3 \frac{1}{x} \, dx$

1.3333333333
1.1666666667
1.1166666667
1.1032106782

In Exercises 8–13,

(a) Starting with only one subinterval, construct the four row Romberg integration table for the indicated integral.

(b) What is the error estimate for the final approximation? How does this compare with the actual error?

(c) How many subintervals would have been necessary to achieve the same accuracy using the composite trapezoidal rule without extrapolation?

8. $\int_3^{3.5} \frac{x}{\sqrt{x^2-4}} \, dx$

9. $\int_0^1 x^2 e^{-x} \, dx$

10. $\int_0^1 \sqrt{1+x^2}x \, dx$

11. $\int_0^1 \tan^{-1} x \, dx$

12. $\int_0^2 \frac{1}{\sqrt{1+x}} \, dx$

13. $\int_0^{\pi/2} \frac{\sin x}{1+\cos x} \, dx$

In Exercises 14–19, approximate the value of the indicated definite integral to within an absolute error tolerance of 5×10^{-7} using Romberg integration. How many function evaluations are needed?

14. $\int_1^2 \frac{\sin x}{x} \, dx$

15. $\int_0^1 \frac{1}{\sqrt{1+x^4}} \, dx$

16. $\int_0^1 \sqrt{1+x^3} \, dx$

17. $\int_0^1 \sin(x^2)\, dx$

18. $\int_0^1 \frac{1}{1+x^6}\, dx$

19. $\int_0^1 x^2 \tan^{-1}(x^4)\, dx$

20. Use the table generated in the "Tabulating the Error Function" application problem and Hermite cubic interpolation to approximate the value of the error function at the indicated value of x. How well does the value obtained in this manner compare to the actual value of the error function?

(a) $x = 0.799$ **(b)** $x = 1.265$

(c) $x = 0.156$ **(d)** $x = 1.771$

(e) $x = 0.301$ **(f)** $x = 1.545$

21. Show that, for any k, $R_{k,2}$ is the composite Simpson's rule with $h = (b-a)/2^{k-1}$.

22. The table below gives the volume v (measured in cubic inches) and the pressure p (measured in pounds per square inch) of a gas as it expands.

v	0.75	1.00	1.25	1.50	1.75	2.00	2.25	2.50	2.75
p	89.8	68.7	55.0	45.8	39.3	34.4	30.5	27.5	26.0

Estimate the work done by the gas,

$$W = \int_{0.75}^{2.75} p\, dv,$$

as follows: Use the trapezoidal rule with $h = 2.0$, $h = 1.0$, $h = 0.5$, and $h = 0.25$, and then extrapolate.

23. Consider the integral

$$\int_{\exp(-ICt)}^{M} \frac{dy}{y[A(1-y) - B\ln y]},$$

which arises in the projection printing of a photoresist film. Here, M denotes the normalized photoactive compound concentration present in the resist film after exposure to light; A, B, and C are material properties of the resist film; and the product It is the exposure energy of the light source used during the printing phase. For the resist material AZ2400, $A = 0.162/\mu\text{m}$, $B = 0.184/\mu\text{m}$, and $C = 0.0128\ \text{cm}^2/\text{mJ}$. Suppose the exposure energy is $110\ \text{mJ/cm}^2$.

(a) For the resist material AZ2400, evaluate the above integral for $M = 0.32$ to five decimal places.

(b) Determine the value of M, correct to four decimal places, so that

$$\int_{\exp(-ICt)}^{M} \frac{dy}{y[A(1-y) - B\ln y]} = 1.$$

6.8 ADAPTIVE QUADRATURE

Consider approximating the value of

$$\int_0^1 x^{16}\cos(x^{16})\,dx$$

using Simpson's rule. An accuracy of six decimal places (i.e., a maximum absolute error of 5×10^{-7}) is desired. Since for $f(x) = x^{16}\cos(x^{16})$

$$\max_{0\le x\le 1} |f^{(4)}(x)| = 906753.5482,$$

$n = 318$ (that amounts to 319 function evaluations) would be required to guarantee the desired accuracy. An examination of the graph of the integrand (Figure 6.11) suggests partitioning the integration interval into say $[0, 0.7] \cup [0.7, 1]$ to isolate the nearly constant left-hand portion of the integrand from the rapidly varying right-hand portion. Dividing the allowable error equally between the two pieces, the number of function evaluations required to guarantee a total absolute error of less than 5×10^{-7} from Simpson's rule would be reduced to 124: 40 evaluations for the subinterval $[0, 0.7]$ and 84 more evaluations for the subinterval $[0.7, 1]$.

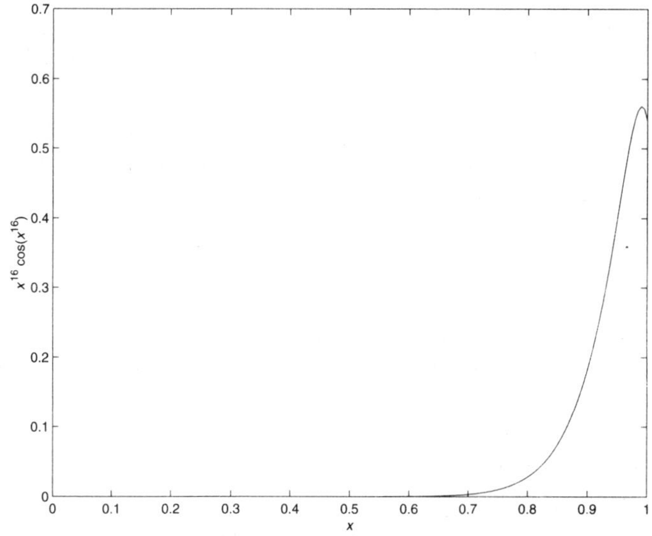

Figure 6.11 Graph of $x^{16}\cos(x^{16})$ over the interval $[0, 1]$.

This example illustrates a general problem associated with composite Newton-Cotes formulas. If the step size is chosen small enough to resolve the most rapidly varying portion(s) of the integrand, then the use of equally spaced points implies that the step size will be smaller (possibly much smaller) than is necessary to resolve the slowly varying portion(s) of the integrand. We would therefore like to use nonuniformly spaced abscissas. These abscissas could be fixed ahead of time, as done above, by using a priori knowledge of the integrand. It would, however, be preferable to allow the points to be chosen automatically as the approximation evolves, making for a more robust algorithm that could be used on a wide variety of problems. Such algorithms are referred to as *adaptive quadrature* routines. As we will see below, the key to producing such schemes is the ability to estimate the error in the approximation as it is being computed.

A Simple Adaptive Strategy

Given a function f, an interval $[a, b]$ and an error tolerance ϵ, the objective of an adaptive quadrature scheme is to automatically select the abscissas in such a way that the estimated error in the final approximation will be less than ϵ. A simple adaptive strategy is outlined in the following four-step algorithm.

1. Compute an approximation, I_0, to $\int_a^b f(x)\,dx$;
2. Split the interval $[a, b]$ into two pieces, $[a, c]$ and $[c, b]$, where $c = (a + b)/2$, and then compute $I_1 \approx \int_a^c f(x)\,dx$ and $I_2 \approx \int_c^b f(x)\,dx$;
3. Compare $I_1 + I_2$ with I_0 to estimate the error in $I_1 + I_2$; and
4. If |estimated error| $\le \epsilon$, then accept $I_1 + I_2$ as an approximation to $\int_a^b f(x)\,dx$
   ... otherwise ...
   apply the same procedure to $[a, c]$ and $[c, b]$, allowing each piece a tolerance of $\epsilon/2$.

In general, there is no requirement that the same basic quadrature scheme be used to compute all three approximations I_0, I_1, and I_2. The key to implementing this algorithm, however, is step 3—the ability to combine I_0, I_1, and I_2 into an estimate of the error in the overall approximation as it is evolving.

Estimating Error on the Fly

The specific mechanism by which the error is estimated in step 3 depends on the basic quadrature scheme used to generate the approximations I_0, I_1 and I_2. Let's suppose that each of these values was computed using Simpson's rule and that $h = (b - a)/2$. Since Simpson's rule is a fourth-order method, provided f is sufficiently differentiable, we know that

$$\int_a^b f(x)\,dx = S(a, b) + k_1 h^4 + o(h^4),$$

where k_1 is a constant independent of h. Here, we have used $S(a, b)$ to denote the Simpson's rule approximation computed over the interval $[a, b]$. Using similar notation for the Simpson's rule approximations over $[a, c]$ and $[c, b]$, and having selected c to be the midpoint between a and b, we also know that

$$\int_a^b f(x)\,dx = S(a, c) + S(c, b) + k_1 \left(\frac{h}{2}\right)^4 + o(h^4).$$

Subtracting these last two expressions and solving for the leading term in the error of $S(a,c) + S(c,b)$, we find

$$k_1 \left(\frac{h}{2}\right)^4 = \frac{1}{15}[S(a,c) + S(c,b) - S(a,b)] + o(h^4).$$

If we assume that the $o(h^4)$ term is in fact negligible, then we will accept

$$\frac{1}{15}[S(a,c) + S(c,b) - S(a,b)]$$

as an estimate of the error in $S(a,c) + S(c,b)$. Therefore, we can test

$$|S(a,c) + S(c,b) - S(a,b)| \leq 15\epsilon$$

in step 3 of our basic algorithm. Since we have made several approximations and assumptions along the way, in practice, we would be conservative and use a number smaller than 15 in step 3—say 10.

It is important to note that the only characteristic of Simpson's rule that is used in obtaining the above error estimate is the fact that the rule is fourth order. It therefore follows that if $Q(a,b)$ denotes any fourth order quadrature rule (such as the two-point Gaussian quadrature rule or the open Newton-Cotes formula with $n = 2$) applied over the integration interval $[a,b]$ and c denotes the midpoint of that interval, then

$$\frac{1}{15}[Q(a,c) + Q(c,b) - Q(a,b)]$$

represents an estimate for the error in the approximation $Q(a,c) + Q(c,b)$.

This technique for obtaining error estimates is easily extended to methods of other orders. In general, if $Q(a,b)$ is a method of order p—that is, the leading term in the error is $O(h^p)$—then

$$\frac{1}{2^p - 1}[Q(a,c) + Q(c,b) - Q(a,b)]$$

is an estimate for the error in the approximation $Q(a,c) + Q(c,b)$. As discussed above, when putting this error estimate into practice, a number smaller than $2^p - 1$ should be used to be conservative.

Adaptive Quadrature in Action

Recall the problem posed at the beginning of this section: Approximate the value of the integral

$$\int_0^1 x^{16} \cos(x^{16})\, dx$$

to six decimal places of accuracy. Let's use Simpson's rule as the underlying quadrature rule and continue to use $S(a,b)$ to denote the Simpson's rule approximation computed over the interval $[a,b]$.

Our first step is to calculate

$$S(0,1) = \frac{1}{6}\left[f(0) + 4f\left(\frac{1}{2}\right) + f(1)\right] = 0.09006055683740.$$

Next, we cut the integration interval in half and calculate

$$S\left(0,\frac{1}{2}\right) = \frac{1}{12}\left[f(0) + 4f\left(\frac{1}{4}\right) + f\left(\frac{1}{2}\right)\right] = 0.00000127164336; \quad \text{and}$$

$$S\left(\frac{1}{2},1\right) = \frac{1}{12}\left[f\left(\frac{1}{2}\right) + 4f\left(\frac{3}{4}\right) + f(1)\right] = 0.04836716117637.$$

Our estimate for the error in $S(0,\frac{1}{2}) + S(\frac{1}{2},1)$ is therefore

$$\frac{1}{10}\left|S\left(0,\frac{1}{2}\right) + S\left(\frac{1}{2},1\right) - S(0,1)\right| = 4.169 \times 10^{-3}.$$

Since this error estimate is larger than the specified tolerance of 5×10^{-7}, we repeat the entire process on the subintervals $[0,\frac{1}{2}]$ and $[\frac{1}{2},1]$, allowing each subinterval an error of 2.5×10^{-7}.

On $[0,\frac{1}{2}]$, we already know that $S(0,\frac{1}{2}) = 0.00000127164336$. Cutting the interval in half, we then calculate

$$S\left(0,\frac{1}{4}\right) = \frac{1}{24}\left[f(0) + 4f\left(\frac{1}{8}\right) + f\left(\frac{1}{4}\right)\right] = 0.00000000000970; \quad \text{and}$$

$$S\left(\frac{1}{4},\frac{1}{2}\right) = \frac{1}{24}\left[f\left(\frac{1}{4}\right) + 4f\left(\frac{3}{8}\right) + f\left(\frac{1}{2}\right)\right] = 0.00000066128135.$$

Accordingly, we estimate that the error in $S(0,\frac{1}{4}) + S(\frac{1}{4},\frac{1}{2})$ is

$$\frac{1}{10}\left|S\left(0,\frac{1}{4}\right) + S\left(\frac{1}{4},\frac{1}{2}\right) - S(0,\frac{1}{2})\right| = 6.103 \times 10^{-8},$$

which is smaller than 2.5×10^{-7}. Hence, we are finished with the subinterval $[0,\frac{1}{2}]$ and have found that

$$\int_0^{1/2} x^{16}\cos(x^{16})\,dx \approx 0.00000000000970 + 0.00000066128135$$
$$= 0.00000066129105$$

with an estimated error of 6.103×10^{-8}.

Moving on to $[\frac{1}{2},1]$, we already know that $S(\frac{1}{2},1) = 0.04836716117637$. Next, we calculate

$$S\left(\frac{1}{2},\frac{3}{4}\right) = \frac{1}{24}\left[f\left(\frac{1}{2}\right) + 4f\left(\frac{5}{8}\right) + f\left(\frac{3}{4}\right)\right] = 0.00050857313250,$$

$$S\left(\frac{3}{4},1\right) = \frac{1}{24}\left[f\left(\frac{3}{4}\right) + 4f\left(\frac{7}{8}\right) + f(1)\right] = 0.04247103735072,$$

and the error estimate

$$\frac{1}{10}\left|S\left(\frac{1}{2},\frac{3}{4}\right)+S\left(\frac{3}{4},1\right)-S(\frac{1}{2},1)\right|=5.388\times 10^{-4}.$$

Since this error estimate is larger than 2.5×10^{-7}, we must further subdivide $[\frac{1}{2}, 1]$ into $[\frac{1}{2}, \frac{3}{4}]$ and $[\frac{3}{4}, 1]$. Each of these new subintervals is allotted an error of 1.25×10^{-7}.

If we continue to work in this fashion, we eventually find

$$\int_0^1 x^{16}\cos(x^{16})\,dx \approx 0.049121999503,$$

with an error estimate of 1.6873×10^{-7}. To thirteen decimal places, the value of this integral is 0.0491217295177. The actual error in the approximate value is therefore 2.6999×10^{-7}. Thus, the reported error estimate is slightly low, but of the correct order of magnitude, and the overall error is well within the requested accuracy.

Figure 6.12 displays the 73 non-uniformly spaced points, represented by circles, at which the integrand was sampled to arrive at the above approximation. Compare this with the 319 uniformly spaced abscissas which would have been needed to guarantee similar accuracy. The asterisks along the top of the figure display the locations of the abscissas and have been included to more clearly demonstrate the varying density of the abscissas within the integration interval. At the left end of the interval, where the integrand is relatively flat, few quadrature points are selected. As we move to the right, the integrand varies more widely, and the density of quadrature points increases accordingly.

An Efficient Implementation

To achieve the efficiency indicated by Figure 6.12 (i.e., using only 73 function evaluations to obtain the final approximation), we need to be careful how we construct our algorithm. A naive translation of the basic four-step strategy produces

```
function adapt ( f, a, b, ε )
set sab = S(a, b)
set c = (a + b)/2
set sac = S(a, c)
set scb = S(c, b)
if ( |sac + scb − sab|/10 < ε )
          return ( sac + scb )
else
          return ( adapt ( f, a, c, ε/2 ) + adapt ( f, c, b, ε/2 ) )
```

This algorithm, however, is very wasteful of computational effort. For instance, suppose we approximate $\int_0^1 x^{16}\cos(x^{16})\,dx$ to an accuracy of six decimal places. Though the integrand is evaluated at only those 73 points shown in Figure 6.12, the integrand is evaluated a total of 315 times.

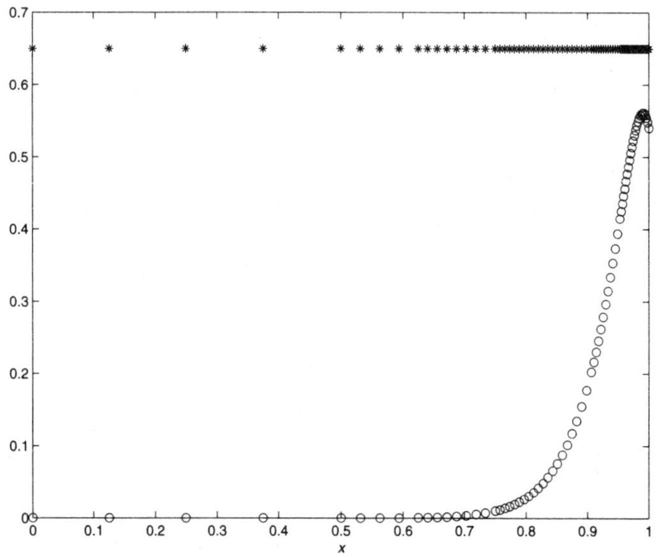

Figure 6.12 Circles represent the 73 function values sampled from the integrand $x^{16}\cos(x^{16})$ in calculating an estimate for the integral accurate to six decimal places. Asterisks denote the locations of the quadrature abscissas to demonstrate the variable density within the integration interval.

Why is this algorithm so inefficient? First, with each recursive call to *adapt*, one Simpson's rule approximation is entirely recalculated. That's three unnecessary function evaluations per recursive call. Second, the algorithm continually recalculates function values used to construct different Simpson's rule approximations. For example, $f(0)$ is calculated as part of $S(0,1)$, $S(0,\frac{1}{2})$, and $S(0,\frac{1}{4})$, while $f(\frac{1}{2})$ is calculated as part of

$$S(0,1), S(0,\tfrac{1}{2}), S(\tfrac{1}{2},1), S(\tfrac{1}{4},\tfrac{1}{2}), S(\tfrac{1}{2},\tfrac{3}{4}), S(\tfrac{1}{2},\tfrac{5}{8}), \text{ and, } S(\tfrac{1}{2},\tfrac{9}{16}).$$

To improve efficiency, we have to find some way to save Simpson's rule approximations and function values for later use.

An algorithm which achieves this objective is given below. Note that this algorithm is separated into two parts: an initialization component, *adapt*, and a recursive component, *adapt1*. The initialization component simply calculates the first three function values and the first Simpson's rule approximation. This information is then passed along to the recursive component. Each call to the recursive component carries out the second, third, and fourth steps of the basic strategy for a particular portion of the integration interval. By passing the Simpson's rule approximation and the function values needed to process each subinterval through the argument list, this algorithm calculates each Simpson's rule approximation and each function value only once.

```
function adapt ( f, a, b, ε )
set fa = f(a)
set fc = f((a + b)/2)
set fb = f(b)
set sab = (b − a) * (fa + 4 * fc + fb)/6
return ( adapt1 ( sab, fa, fc, fb, f, a, b, ε ) )

function adapt1 ( sab, fa, fc, fb, f, a, b, ε )
set c = (a + b)/2
set fd = f((a + c)/2)
set fe = f((c + b)/2)
set sac = (c − a) * (fa + 4 * fd + fc)/6
set scb = (b − c) * (fc + 4 * fe + fb)/6
if ( |sac + scb − sab|/10 < ε )
          return ( sac + scb )
else
          return ( adapt1 ( sac, fa, fd, fc, f, a, c, ε/2 ) +
                   adapt1 ( scb, fc, fe, fb, f, c, b, ε/2 ) )
```

A Second Strategy

An alternative to the adaptive strategy presented above is the four-step algorithm:

1. Compute an approximation, I_0, to $\int_a^b f(x)\,dx$;
2. Compute a second approximation, I_1, using a method of higher order than the one used to compute I_0;
3. Use $I_1 - I_0$ as an estimate of the error in I_1; and
4. If $|\text{estimated error}| \leq \epsilon$, then accept I_1 as an approximation to $\int_a^b f(x)\,dx$
   ... otherwise ...
   apply the same procedure to $[a, c]$ and $[c, b]$, allowing each piece a tolerance of $\epsilon/2$, where $c = (a + b)/2$.

The error estimate in step 3 is justified as follows. Suppose the approximation I_0 is computed using a quadrature rule that is $O(h^p)$, while I_1 is computed using a rule that is $O(h^q)$, with $q > p$. Then

$$\int_a^b f(x)\,dx = I_0 + c_0 h^p + o(h^p)$$

and

$$\int_a^b f(x)\,dx = I_1 + c_1 h^q + o(h^q),$$

where c_0 and c_1 are constants indepedent of h. Subtracting these two equations gives

$$0 = I_0 - I_1 + c_0 h^p - c_1 h^q + o(h^p) - o(h^q).$$

Assuming all terms $o(h^p)$ and smaller can be neglected, it follows that

$$c_0 h^p \approx I_1 - I_0;$$

hence, the leading term in the error in I_0 is approximately $I_1 - I_0$. Since I_1 is a more accurate approximation than I_0, it follows that $I_1 - I_0$ is a conservative estimate for the error in I_1.

Applying this second adaptive strategy, with Simpson's rule as the lower-order method, Boole's rule as the higher-order method and $\epsilon = 5 \times 10^{-7}$, we find

$$\int_0^1 x^{16} \cos(x^{16})\, dx \approx 0.049121730001,$$

with an error estimate of 1.0920×10^{-7}. The actual error in this approximation is 4.8378×10^{-10}, which is quite a bit smaller than the reported error estimate.

By constructing the algorithm to reuse function values whenever possible, the above approximation was obtained with 141 function evaluations. Though this second approach needed nearly twice as many function evaluations as the first adaptive scheme to achieve six decimal places of accuracy, this scheme still used fewer than half the number of function evaluations required by the composite Simpson's rule.

Two Comparison Problems

The following examples compare the performance of the adaptive strategies described in this section applied to several different basic quadrature rules. Each scheme based upon a fourth-order method (the adaptive Simpson's rule and the adaptive two-point Gaussian quadrature rule) uses

$$[Q(a, c) + Q(c, b) - Q(a, b)]/10$$

as a conservative estimate of the error in $Q(a, c) + Q(c, b)$. The schemes based upon sixth-order methods (the adaptive Boole's rule and the adaptive three-point Gaussian quadrature rule) use

$$[Q(a, c) + Q(c, b) - Q(a, b)]/42$$

as an error estimate. A brief explanation is warranted for the choice of constant in this error estimate. Theory indicates that division by 15 should be used in the error estimate for fourth-order methods. We cut one-third off of this value and used 10 to be conservative. For sixth-order methods, theory indicates that the error estimate should involve division by 63—cutting one-third off this value leaves 42.

EXAMPLE 6.19 Comparison of Adaptive Quadrature Rules 1

Consider the integral

$$I = \int_0^2 e^x \sin(x^2 \cos e^x)\, dx \approx -1.11595799093275.$$

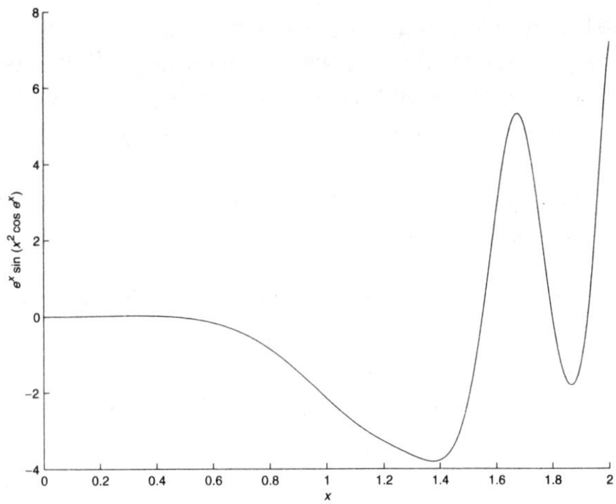

Figure 6.13 Graph of $e^x \sin(x2 \cos e^x)$ on $[0, 2]$.

The graph of the integrand is shown in Figure 6.13. This integrand is similar to the one considered earlier in this section in that the left portion is quite flat and the amount of variation increases as we move to the right. Adaptive quadrature should therefore prove very helpful in efficiently approximating the value of this integral.

The following table lists the number of function evaluations needed to approximate I to five decimal places of accuracy ($\epsilon = 5 \times 10^{-6}$) and to ten decimal places of accuracy ($\epsilon = 5 \times 10^{-11}$). To save space, we have abbreviated "two-point Gaussian quadrature" as GQ2 and "three-point Gaussian quadrature" as GQ3.

	Accuracy	
	5 decimal places	10 decimal places
Adaptive Simpson	217	3697
Adaptive GQ2	414	7054
Adaptive Simpson/Boole	401	7029
Adaptive Boole	113	689
Adaptive GQ3	137	857
Romberg Integration	129	1025

Some of the function evaluation counts given above (especially those associated with the fourth-order basic quadrature rules) may seem rather high. To place these counts into perspective, let's determine the number of function evaluations which would be required by the corresponding composite quadrature rules to achieve each level of accuracy. For $f(x) = e^x \sin(x^2 \cos e^x)$,

$$\max_{0 \le x \le 2} |f^{(4)}| = 2.5823 \times 10^7.$$

Substituting this value into the error term for the composite Simpson's rule leads to an estimate of 981 function evaluations to achieve five decimal places of accuracy and 17408 to achieve ten decimal places of accuracy. The function evaluation counts for the composite two-point Gaussian quadrature rule to achieve five and ten decimal places of accuracy are 886 and 15730, respectively. Comparing these counts to those listed above, we see that the adaptive Simpson's rule reduces the work effort by better than a factor of 4, while the adaptive two-point Gaussian quadrature rule and the adaptive Simpson/Boole rule reduce the work effort by better than a factor of 2.

EXAMPLE 6.20 Comparison of Adaptive Quadrature Rules 2

As another test problem, consider the integral

$$I = \int_0^5 \frac{50}{\pi(1+2500x^2)}\,dx = \tan^{-1}\frac{250}{\pi} \approx 0.49872676724581.$$

The table below lists the number of function evaluations needed to approximate I to five decimal places of accuracy ($\epsilon = 5\times10^{-6}$) and to ten decimal places of accuracy ($\epsilon = 5\times10^{-11}$).

	Accuracy	
	5 decimal places	10 decimal places
Adaptive Simpson	177	3161
Adaptive GQ2	326	5630
Adaptive Simpson/Boole	309	5629
Adaptive Boole	105	633
Adaptive GQ3	127	787
Romberg Integration	1025	8193

Comparing the function evaluation counts for the adaptive schemes with those required by the corresponding composite routines is left as an exercise.

It is interesting to note that the adaptive Gaussian quadrature routines (both two-point and three-point) underperform the routines based on Newton-Cotes quadrature rules of similar order. In Section 6.6, we found that Gaussian quadrature rules have smaller coefficients on their error terms, and therefore, generally require fewer function evaluations than Newton-Cotes rules when used in a composite manner. When applied in an adaptive manner, however, Newton-Cotes rules have the advantage of being able to reuse function values. With Gaussian quadrature rules, the only time that function values can be reused is when we implement the second adaptive strategy and choose rules that both evaluate the integrand at the midpoint of the integration interval. Even then, the only function value that can be reused is that value at the midpoint of the integration interval.

The underperformance of the adaptive two-point and three-point Gaussian quadrature rules should not be taken to imply that all adaptive Gaussian quadrature rules underperform their Newton-Cotes counterparts. For, note that while the adaptive Simpson's rule uses fewer than 60% the number of function evaluations of the two-point Gaussian quadrature scheme, the adaptive Boole's rule uses more than 80% the number of evaluations of the three-point scheme. Eventually, adaptive Gaussian quadrature surpasses adaptive Newton-Cotes quadrature.

We make one final comment regarding adaptive Gaussian quadrature. Though different Gaussian quadrature rules do not share abscissa (with the possible exception of the integration interval midpoint), it is possible to construct higher-order quadrature rules which use the abscissas of a given Gaussian quadrature rule. These rules can be paired with the Gaussian quadrature rule to form efficient adaptive schemes based on our second basic strategy. For example, Kronrod [1] showed how to supplement the abscissas of any n-point Gaussian quadrature rule with $n+1$ new abscissas so as to construct a formula with maximum degree of precision. The resulting $(2n+1)$-point Kronrod rule has degree of precision $3n+1$ when n is even and $3n+2$ when n is odd. Patterson [2] built upon Kronrod's idea by adding an additional $2n+2$ abscissas to obtain a quadrature rule with degree of precision $6n+4$.

Application Problem: Flow between Parallel Plates

A classic problem in fluid mechanics is determining the flow induced in a viscous fluid filling the gap between two large parallel plates, as shown in Figure 6.14(a). The lower plate is fixed and the upper plate moves steadily with velocity U_0. The distance between the plates is h, and no-slip conditions are assumed at both plates. This means that the fluid velocity is zero at the lower plate and U_0 at the upper plate.

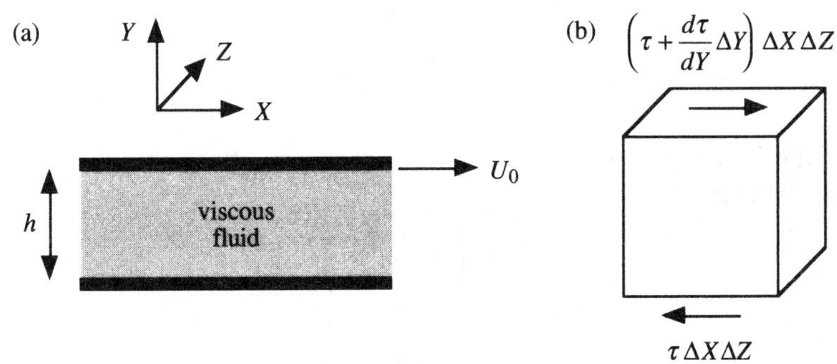

Figure 6.14 (a) Viscous fluid filling the gap between two parallel plates, one fixed, the other moving with constant velocity. (b) Force balance on a representative fluid element.

Let $U(Y)$ denote the steady-state velocity distribution established in the fluid as a result of the shearing motion of the upper plate. To determine $U(Y)$, we first consider the representative fluid element shown in Figure 6.14(b). Assuming no pressure variation in the flow direction and that gravitational forces may be neglected, the only force acting on the fluid element is the shear stress, τ, acting along the top and bottom surfaces. Summing forces in the X-direction yields the equation

$$\left(\tau + \frac{d\tau}{dY}\Delta Y\right)\Delta X\ \Delta Z - \tau \Delta X\ \Delta Z = 0,$$

or, after simplification,

$$\frac{d\tau}{dY} = 0. \qquad (1)$$

If the fluid is newtonian, then the shear stress is related to the velocity by the law

$$\tau = \mu\frac{dU}{dY}, \qquad (2)$$

where μ is a property of the fluid called the coefficient of viscosity. In the simplest case, we would assume that μ was constant; however, to add a little twist, let's suppose that the two plates are maintained at different temperatures and a linear temperature gradient exists within the fluid. Now, temperature has a strong effect on viscosity. Hence, assuming that temperature is a function of Y implies that μ would also be a function of Y. Using this assumption in (2) and then substituting into (1), we find that $U(Y)$ satisfies the differential equation

$$\frac{d}{dY}\left(\mu(Y)\frac{dU}{dY}\right) = 0. \qquad (3)$$

Additionally, we have the boundary conditions $U(0) = 0$ and $U(h) = U_0$.

Integrating (3) twice with respect to Y, it follows that

$$U(Y) = c_1 \int_0^Y \frac{d\xi}{\mu(\xi)} + c_2.$$

The boundary condition $U(0) = 0$ requires that $c_2 = 0$. Next, the condition at $Y = h$ requires that

$$c_1 = U_0\left(\int_0^h \frac{d\xi}{\mu(\xi)}\right)^{-1}.$$

Therefore,

$$U(Y) = U_0\frac{\int_0^Y d\xi/\mu(\xi)}{\int_0^h d\xi/\mu(\xi)}. \qquad (4)$$

Let's suppose the fluid is water, the bottom plate is maintained at a temperature of 20° C and the top plate is maintained at 100° C. The following table lists the viscosity of water as a function of temperature.

$T\,(^\circ\text{C})$	20	30	40	50	60	70	80	90	100
$\mu(\times 10^{-3}\ \text{N}\cdot\text{s/m}^2)$	1.003	0.799	0.657	0.548	0.467	0.405	0.355	0.316	0.283

According to White [3], the best fit to this data is the empirical result that $\ln\mu$ is quadratic in $1/T$, when T is measured in the Kelvin scale. Temperature in the Kelvin scale is obtained from temperature in the Celsius scale by adding 273.16. After performing the indicated data transformations, regression yields

$$\mu(T) = \exp\left(-8.944 - \frac{839.456}{T} + \frac{421194.298}{T^2}\right). \tag{5}$$

With the assumption of a linear temperature gradient between the two plates, it follows that

$$T(Y) = \left(20 + 80\frac{Y}{h}\right)\ ^\circ\text{C} = \left(293.16 + 80\frac{Y}{h}\right)\ \text{K}. \tag{6}$$

Substituting (5) and (6) into (4) and then introducing the dimensionless variables $u = U/U_0$ and $y = Y/h$, the expression for the velocity becomes

$$u(y) = \frac{\int_0^y \exp\left(8.944 + \frac{839.456}{293.16+80\xi} - \frac{421194.298}{(293.16+80\xi)^2}\right) d\xi}{\int_0^1 \exp\left(8.944 + \frac{839.456}{293.16+80\xi} - \frac{421194.298}{(293.16+80\xi)^2}\right) d\xi}. \tag{7}$$

Figure 6.15 displays the nondimensional velocity distribution obtained from equation (7). To produce this graph, values of u were calculated at $y_i = 0.01i$ for $i = 0, 1, 2, \ldots, 100$. All integrals were evaluated using adaptive three-point Gaussian quadrature with $\epsilon = 5 \times 10^{-7}$. The independent variable has been plotted along the vertical axis to match the geometry depicted in Figure 6.14(a).

References

1. A. S. Kronrod, *Nodes and Weights of Quadrature Formulas*, Consultants Bureau, New York, 1965.
2. T. N. L. Patterson, "The Optimum Addition of Points to Quadrature Formulae," Mathematics of Computation, **22**, 847–856, 1968.
3. F. M. White, *Fluid Mechanics*, 2nd edition, McGraw-Hill, New York, 1986.

EXERCISES

1. For each of the following integrals, compute $S(a,b)$, $S(a,c)$, and $S(c,b)$, where $c = (a+b)/2$. Compute the estimate for the error in $S(a,c)+S(c,b)$ and compare this to the actual error is $S(a,c) + S(c,b)$.

(a) $\int_0^1 e^{-x}\,dx$ **(b)** $\int_1^2 \frac{1}{x}\,dx$

(c) $\int_0^4 x\sqrt{x^2+9}\,dx$ **(d)** $\int_0^1 \tan^{-1} x\,dx$

2. Repeat Exercise 1 using Boole's rule (the closed Newton-Cotes formula with $n = 4$).

3. Repeat Exercise 1 using the two-point Gaussian quadrature rule.

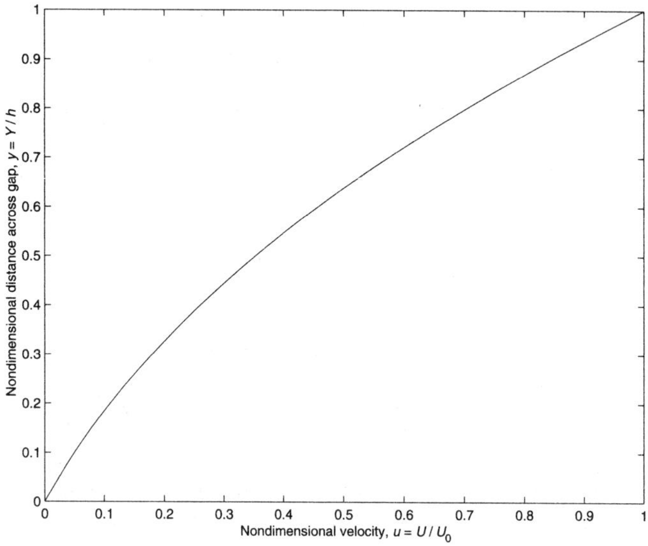

Figure 6.15 Nondimensional velocity distribution in viscous fluid filling gap between two large parallel plates.

4. Repeat Exercise 1 using the three-point Gaussian quadrature rule.

5. For each of the integrals in Exercise 1, compute the Simpson's rule approximation and the Boole's rule approximation. Confirm that the difference between these two values approximates the error in the Simpson's rule value.

6. For each of the integrals in Exercise 1, compute the two-point Gausssian quadrature rule approximation and the three-point Gaussian quadrature rule approximation. Confirm that the difference between these two values approximates the error in the two-point Gaussian quadrature rule value.

7. Determine the number of function evaluations which would be needed to guarantee an accuracy of 10 decimal places in the approximation to the value of

$$I = \int_0^5 \frac{50}{\pi(1+2500x^2)}\,dx$$

using the composite Simpson's rule and the composite two-point Gaussian quadrature rule. Compare with the number of function evaluations required by the corresponding adaptive routines listed in the second example above.

In Exercises 8–16, approximate the value of the given integral to six (6) and to ten (10) decimal places using the adaptive quadrature scheme of your choice. Compare the number of function evaluations used to the number that would be required by the corresponding composite quadrature rule to achieve the same accuracy.

8. $\int_0^1 e^{-x^4}\,dx$

9. $\int_0^5 \frac{1}{\sqrt{1+x^3}}\, dx$

10. $\int_1^2 \frac{\sin x}{x}\, dx$

11. $\int_0^2 e^{-x} \sin(x^2 \cos e^{-x})\, dx$

12. $\int_0^1 \sqrt{1+x^4}\, dx$

13. $\int_0^1 \frac{u^7}{1+u^{14}} du$

14. $\int_0^{10} 25e^{-25x}\, dx$

15. $\int_0^1 \frac{1}{1+e^x}\, dx$

16. $\int_0^\pi \cos(\cos x + 3\sin x + 2\cos(2x) + 3\cos(3x) + 3\sin(2x))\, dx$

17. (a) Evaluate the integral

$$\int_0^1 \sin(\sqrt{\pi x})\, dx$$

to six decimal places of accuracy using the adaptive Simpson's rule. How many function evaluations were needed?

(b) Make the change of variable $u^2 = \pi x$ in the integral from part (a) and reevaluate using the adaptive Simpson's rule. How does the number of function evaluations compare with the number from part (a)?

18. (a) Evaluate the integral

$$\int_0^1 \frac{2}{2+\sin(10\pi x)}\, dx$$

to ten decimal places of accuracy using the adaptive Simpson's rule. How many function evaluations were needed?

(b) Recognizing that the integrand in part (a) is periodic with period 1/5, recompute the value from part (a) as

$$5\int_0^{0.2} \frac{2}{2+\sin(10\pi x)}\, dx$$

using the adaptive Simpson's rule. How does the number of function evaluations compare with the number from part (a)?

19. The Fresnel integrals

$$c(x) = \int_0^x \cos\left(\frac{\pi}{2}t^2\right) dt \qquad s(x) = \int_0^x \sin\left(\frac{\pi}{2}t^2\right) dt$$

arise in the study of light diffraction at a rectangular aperture.

(a) Construct a table of values for $c(x)$ and $s(x)$ for x ranging from 0 through 2 in increments of 0.2. Each entry in the table should be accurate to five decimal places.

(b) Determine the two smallest positive values for x such that $c(x) = 0.5$, accurate to four decimal places. Repeat for the equation $s(x) = 0.5$.

20. Consider the integral

$$\int_0^x \frac{\sin t}{t}\,dt.$$

(a) Use the adaptive two-point Gaussian quadrature scheme to tabulate the value of this integral for x ranging from 0 through 10 in increments of 0.5. Each tabulated value should be accurate to six decimal places.

(b) What happens if you try to use the adaptive Simpson's rule to tabulate the values of this integral? Can you think of a way to alleviate this problem?

21. Evaluate

$$\frac{\int_0^{1/2} e^{2(1-x)^3/3}\,dx}{\int_0^1 e^{2(1-x)^3/3}\,dx}$$

and

$$\frac{\int_0^{1/2} e^{-x^2+2x^3/3}\,dx}{\int_0^1 e^{-x^2+2x^3/3}\,dx}.$$

These expressions arise in determining the probability that an allele with a selective advantage over its competitors will become fixed in a population (see P. D. Taylor and A. Sauer, "The Selective Advantage of Sex-Ratio Homeostasis," *American Naturalist*, **116**, 305–310, 1980).

22. Rework the "Flow between Parallel Plates" problem assuming that the lower plate is maintained at 100°C and the upper plate is maintained at 20°C.

6.9 IMPROPER INTEGRALS AND OTHER DISCONTINUITIES

The definite integral

$$\int_a^b f(x)\,dx$$

is called *improper* if

- one or both of the integration limits is infinite; or
- the integrand, f, is discontinuous somewhere on the integration interval $[a, b]$.

Provided the integral exists, the quadrature rules that have been developed in this chapter can be used to approximate the value, though some change of variable generally must be made to achieve theoretical order of convergence. The objective of this section is to demonstrate the substitutions appropriate for a variety of different situations.

n	Trapezoidal Rule	Simpson's Rule	Midpoint Rule	Two-Point Gaussian
2	3.502×10^{-2}	1.336×10^{-2}	1.117×10^{-2}	1.338×10^{-3}
4	1.193×10^{-2}	4.229×10^{-3}	3.963×10^{-3}	4.225×10^{-4}
8	3.982×10^{-3}	1.334×10^{-3}	1.361×10^{-3}	1.331×10^{-4}
16	1.311×10^{-3}	4.202×10^{-4}	4.568×10^{-4}	4.194×10^{-5}
32	4.269×10^{-4}	1.324×10^{-4}	1.509×10^{-4}	1.321×10^{-5}
64	1.380×10^{-4}	4.169×10^{-5}	4.930×10^{-5}	4.161×10^{-6}
Order of Convergence:				
Theoretical	2	4	2	4
Experimental	1.5986	1.6651	1.5667	1.6659

TABLE 6.5: Error in Approximate Value of $\int_0^1 x^{2/3}\,dx$ Computed Using Four Basic Quadrature Rules

Discontinuous Derivatives

Consider the definite integral

$$\int_0^1 x^{2/3}\,dx.$$

Technically, this integral does not fit into this section because it is not improper. However, the development of efficient procedures for evaluating certain improper integrals will depend upon the information we obtain from studying this problem. Table 6.5 lists the error in the approximate value of the above integral computed using four basic quadrature rules: the trapezoidal rule, Simpson's rule, the midpoint rule, and the two-point Gaussian quadrature rule. Each method works in the sense that as n is increased, the approximation error converges to zero; however, none of the methods performs at its theoretical order of convergence.

What caused this reduction in the order of convergence? Note that though the integrand in this problem is defined throughout the integration interval, all of its derivatives are unbounded at $x = 0$. Every quadrature rule that has been developed in this chapter has an error term which contains some order derivative of the integrand. For example, the error term for the trapezoidal rule contains the second derivative of the integrand, while the error term for the two-point Gaussian quadrature rule contains the fourth derivative. The theoretical order of convergence of each method is therefore predicated on sufficient smoothness in the integrand. As this example demonstrates, when the integrand is not sufficiently differentiable, the methods may still work, but if they do, they may perform below theoretical levels of efficiency.

The derivatives of $x^{2/3}$ become unbounded at $x = 0$ because of the presence of a fractional exponent. We can easily clear the fractional exponent by making the

n	Trapezoidal Rule	Simpson's Rule	Midpoint Rule	Two-Point Gaussian
2	2.438×10^{-1}	2.500×10^{-2}	1.195×10^{-1}	1.042×10^{-3}
4	6.211×10^{-2}	1.563×10^{-3}	3.091×10^{-2}	6.510×10^{-5}
8	1.560×10^{-2}	9.766×10^{-5}	7.791×10^{-3}	4.069×10^{-6}
16	3.905×10^{-3}	6.104×10^{-6}	1.952×10^{-3}	2.543×10^{-7}
32	9.765×10^{-4}	3.815×10^{-7}	4.882×10^{-4}	1.589×10^{-8}
64	2.441×10^{-4}	2.384×10^{-8}	1.221×10^{-4}	9.934×10^{-10}
Order of Convergence:				
Theoretical	2	4	2	4
Experimental	1.9939	4.0000	1.9894	4.0000

TABLE 6.6: Error in Approximate Value of $\int_0^1 3u^4\,du$ Computed Using Four Basic Quadrature Rules

change of variables $x = u^3$. With this substitution, the original integral becomes

$$\int_0^1 3u^4\,du.$$

The new integrand, $\tilde{f}(u) = 3u^4$, is now infinitely differentiable. The data in Table 6.6 confirm that when the trapezoidal rule, Simpson's rule, the midpoint rule, and the two-point Gaussian quadrature rule are applied to this transformed integral, each method performs at its theoretical order of convergence.

Removable Discontinuities

Recall that when a function f has a discontinuity at $x = a$, but $\lim_{x\to a} f(x)$ exists, that discontinuity is said to be *removable.* The proper handling of a removable discontinuity depends on whether or not the derivatives of f are bounded as x approaches a. This issue can be determined by expanding f in an infinite series about $x = a$.

Consider the function $f(x) = (e^x - 1)/x$, which is discontinuous at $x = 0$. Expanding the exponential function into its MacLaurin series and simplifying, we find

$$f(x) = \frac{e^x - 1}{x} = \frac{\left(1 + x + \frac{1}{2}x^2 + \frac{1}{6}x^3 + \cdots\right) - 1}{x} = 1 + \frac{1}{2}x + \frac{1}{6}x^2 + \cdots.$$

Two observations can be made from this expansion. First, $\lim_{x\to 0} f(x) = 1$, so the discontinuity is removable. Second, since the expansion contains only positive integer powers of x, all derivatives of f are bounded as x approaches 0.

What do these observations imply regarding the evaluation of the integral

$$\int_{-1}^1 \frac{e^x - 1}{x}\,dx?$$

Since the derivatives of the integrand are bounded throughout the integration interval, there is no need to make a change of variables, and we can expect that any method will perform at its theoretical order of convergence. All we have to do is avoid evaluating $(e^x - 1)/x$ at $x = 0$. This can be accomplished by carefully selecting the method to guarantee that $x = 0$ is not used as an abscissa. A better approach is to program the integrand as the piecewise function

$$\tilde{f}(x) = \begin{cases} (e^x - 1)/x, & x \neq 0 \\ 1 & x = 0, \end{cases}$$

explicitly taking into account the limit as x approaches 0, and then we can use any method. Following the latter approach and using the adaptive three-point Gaussian quadrature rule with an error tolerance of 5×10^{-11}, the value of the integral is found to be 2.1145017507.

Next consider the integral

$$\int_0^1 \frac{\sin x}{\sqrt{x}}\, dx.$$

Expanding $\sin x$ into its MacLaurin series produces

$$\frac{\sin x}{\sqrt{x}} = \frac{x - \frac{1}{6}x^3 + \frac{1}{120}x^5 - + \cdots}{\sqrt{x}} = x^{1/2} - \frac{1}{6}x^{5/2} + \frac{1}{120}x^{9/2} - + \cdots.$$

Hence, $\lim_{x\to 0} \sin x/\sqrt{x} = 0$, and the discontinuity at $x = 0$ is removable. This expansion also establishes that the derivatives of the integrand are discontinuous at $x = 0$. Since the powers of x are all multiples of $1/2$, the appropriate change of variables is $x = u^2$. With this substitution, the original integral becomes

$$\int_0^1 2\sin(u^2)\, du.$$

The adaptive three-point Gaussian quadrature rule, with a tolerance of 5×10^{-11}, produces the results

Approximate value of integral:	0.620536603446
Error estimate:	1.0248e-11
Number of function evaluations:	87

For comparison, had the change of variable not been carried out, the same quadrature rule with the same tolerance parameter would have yielded

Approximate value of integral:	0.620536603453
Error estimate:	8.8688e-12
Number of function evaluations:	677

Thus, nearly eight times as many function evaluations are needed when the discontinuous nature of the derivatives is not taken into account.

Nonremovable Algebraic Discontinuities

Nonremovable discontinuities—those for which $\lim_{x\to a} f(x)$ does not exist—can be subdivided into two categories: algebraic and logarithmic. Algebraic discontinuities are those for which the integrand behaves like $(x-a)^{-\alpha}$ for some $\alpha > 0$ as x approaches a. On the other hand, with a logarithmic discontinuity, the integrand tends to infinity like $\ln(x-a)$ as x approaches a. We will start by examining the proper way to deal with an algebraic discontinuity.

When working with a nonremovable algebraic discontinuity, some change of variables will have to be made. To determine the appropriate substitution, a series expansion of the integrand is an extremely valuable tool. As an example, take the definite integral

$$\int_0^1 \frac{(\cos(\sqrt[5]{x})-1)\,e^{\sqrt{x}}}{\sqrt[3]{x^2}(1+\sqrt[3]{x})}\,dx.$$

The integrand is clearly undefined at the lower limit of integration. Expanding the cosine, the exponential and the reciprocal of $1+\sqrt[3]{x}$ into their respective MacLaurin series and simplifying yields

$$\begin{aligned}\frac{(\cos(\sqrt[5]{x})-1)\,e^{\sqrt{x}}}{\sqrt[3]{x^2}(1+\sqrt[3]{x})} &= x^{-2/3}\left(\sum_{k=1}^{\infty}(-1)^k\frac{x^{2k/5}}{(2k)!}\right)\left(\sum_{k=0}^{\infty}\frac{x^{k/2}}{k!}\right)\left(\sum_{k=0}^{\infty}(-1)^k x^{k/3}\right)\\ &= -\frac{1}{2}x^{-4/15}+\frac{1}{2}x^{1/15}+\frac{1}{24}x^{2/15}-\frac{1}{2}x^{7/30}-\cdots.\end{aligned}$$

There are three observations to make from this expansion. First, since the leading power in the expansion is larger than -1, the discontinuity at $x=0$ is integrable. Second,

$$\lim_{x\to 0+}\frac{(\cos(\sqrt[5]{x})-1)\,e^{\sqrt{x}}}{\sqrt[3]{x^2}(1+\sqrt[3]{x})}\to -\infty,$$

so the discontinuity is not removable. Finally, the least common denominator of the rational exponents in the expansion is 30. This indicates that the appropriate change of variable for this problem is $x=u^{30}$. In terms of the variable u, the integral reads

$$\int_0^1 30u^9\frac{(\cos u^6-1)e^{u^{15}}}{1+u^{10}}\,du.$$

The value of this integral is found to be -0.7043797072 using the adaptive Boole's rule with an error tolerance of 5×10^{-11}.

Logarithmic Discontinuities

Whereas algebraic discontinuities are typically handled by making a change of variable, logarithmic discontinuities are typically treated by subtracting out the leading term in the behavior near the discontinuity. Consider the integral

$$\int_0^1 \frac{\ln x}{1+x^2}\,dx.$$

As x approaches 0, the value of the natural logarithm tends toward negative infinity, and thus the integrand has a logarithmic discontinuity. Expanding $(1+x^2)^{-1}$ in series, we find

$$\begin{aligned}\frac{\ln x}{1+x^2} &= (1 - x^2 + x^4 - x^6 + - \cdots)\ln x \\ &= \ln x - x^2 \ln x + x^4 \ln x - x^6 \ln x + - \cdots,\end{aligned}$$

or the discontinuous behavior of the integrand is controlled by $\ln x$. Note that each of the terms following the first is well behaved since $\lim_{x\to 0^+} x^\alpha \ln x = 0$ for all $\alpha > 0$. Subtracting away the discontinuous behavior of the integrand, we rewrite the original problem as

$$\begin{aligned}\int_0^1 \frac{\ln x}{1+x^2}\,dx &= \int_0^1 \left[\frac{\ln x}{1+x^2} - \ln x\right] dx + \int_0^1 \ln x\,dx \\ &= -\int_0^1 \frac{x^2 \ln x}{1+x^2}\,dx + \int_0^1 \ln x\,dx.\end{aligned}$$

The second integral can be evaluated analytically:

$$\int_0^1 \ln x\,dx = (x\ln x - x)\big|_0^1 = -1.$$

In the first integral, the logarithmic discontinuity has been replaced by a removable discontinuity. Programming the integrand as the piecewise function

$$\tilde{f}(x) = \begin{cases} -\frac{x^2 \ln x}{1+x^2}, & x \neq 0 \\ 0, & x = 0 \end{cases}$$

and using the adaptive Boole's rule, the value of the first integral is found to be

$$-\int_0^1 \frac{x^2 \ln x}{1+x^2}\,dx \approx 0.0840344058.$$

Therefore,

$$\int_0^1 \frac{\ln x}{1+x^2}\,dx \approx 0.0840344058 + (-1) = -0.9159655942.$$

EXAMPLE 6.21 Another Logarithmic Discontinuity

As a second problem with a logarithmic discontinuity, consider the definite integral

$$\int_0^1 \ln(x - \sin x)\, dx.$$

As in the previous problem, this integral has a logarithmic discontinuity at $x = 0$. To determine the leading behavior in the discontinuity, we expand the sine function into its MacLaurin series and simplify:

$$\begin{aligned}\ln(x - \sin x) &= \ln\left[x - \left(x - \frac{1}{6}x^3 + \frac{1}{120}x^5 - + \cdots\right)\right]\\ &= \ln\left[x^3\left(\frac{1}{6} - \frac{1}{120}x^2 + - \cdots\right)\right]\\ &= \ln x^3 + \ln\left(\frac{1}{6} - \frac{1}{120}x^2 + - \cdots\right).\end{aligned}$$

Thus, we want to subtract $\ln x^3$ from the integrand. Balancing this with the addition of $\ln x^3$, followed by the application of the properties of logarithms, we rewrite the original integral as

$$\begin{aligned}\int_0^1 \ln(x - \sin x)\, dx &= \int_0^1 \left[\ln(x - \sin x) - \ln x^3\right] dx + \int_0^1 \ln x^3\, dx\\ &= \int_0^1 \ln\left(\frac{x - \sin x}{x^3}\right) dx + 3\int_0^1 \ln x\, dx.\end{aligned}$$

The discontinuity in the first integral on the right-hand side is now removable and is handled by programming the integrand as the piecewise function

$$\tilde{f}(x) = \begin{cases} \ln\left(\frac{x - \sin x}{x^3}\right), & x \neq 0\\ \ln\left(\frac{1}{6}\right), & x = 0.\end{cases}$$

The adaptive Boole's rule provides the estimate

$$\int_0^1 \ln\left(\frac{x - \sin x}{x^3}\right) dx \approx -1.8084378485;$$

therefore,

$$\begin{aligned}\int_0^1 \ln(x - \sin x)\, dx &\approx -1.8084378485 + 3(-1)\\ &= -4.8084378485.\end{aligned}$$

Infinite Limits of Integration

A variety of different change of variables can be used to transform infinite integration intervals into finite intervals. Here, we will focus on the two substitutions $x = 1/u$ and $x = \tan\theta$. For a review of other substitutions used to treat infinite limits of integration, see Ueberhuber [1].

As a first example with an infinite limit of integration, consider

$$\int_1^\infty \frac{e^{-x}}{x}\,dx.$$

We will make the substitution $x = 1/u$. This particular change of variable can be applied to transform any infinite integration interval into a finite length interval, provided zero is not contained in the original interval. For this problem, $x = 1/u$ transforms the interval $[1, \infty)$ into $[0, 1]$. In terms of the new variable, u, the integration problem becomes

$$\int_1^\infty \frac{e^{-x}}{x}\,dx = \int_1^0 \frac{e^{-1/u}}{1/u}\left(-\frac{du}{u^2}\right) = \int_0^1 \frac{e^{-1/u}}{u}\,du.$$

Note that we have traded an infinite limit of integration for a discontinuity at $u = 0$. Since

$$\lim_{u\to 0+} \frac{e^{-1/u}}{u} = 0,$$

the discontinuity at $u = 0$ is removable. Proceeding as we have done with other removable discontinuities (programming the integrand as a piecewise function), we find the approximate value of the original integral to be 0.2193839344.

For the integral

$$\int_0^\infty t^2 e^{-t^2}\,dt,$$

the substitution $t = 1/u$ will not have the desired effect since zero is contained in the original integration interval: After the change of variable, we would still have an infinite limit of integration. To circumvent this problem, we simply split the integration interval into two pieces, $[0, a]$ and $[a, \infty)$. Any nonzero value for a will work. For this problem, let's choose $a = 1$ and break the integral into

$$\int_0^\infty t^2 e^{-t^2}\,dt = \int_0^1 t^2 e^{-t^2}\,dt + \int_1^\infty t^2 e^{-t^2}\,dt.$$

The first integral on the right-hand side is not improper and can be approximated directly. We find

$$\int_0^1 t^2 e^{-t^2}\,dt \approx 0.189472345819$$

using the adaptive Boole's rule with an error tolerance of 2.5×10^{-11}. In the second integral, we make the substitution $t = 1/u$, producing

$$\int_1^\infty t^2 e^{-t^2}\,dt = \int_1^0 u^{-2} e^{-1/u^2}\,\frac{du}{-u^2} = \int_0^1 \frac{e^{-1/u^2}}{u^4}\,du.$$

The discontinuity at $t = 0$ is removable, with $\lim_{u \to 0^+} u^{-4} e^{-1/u^2} = 0$. Using the adaptive Boole's rule with the same error tolerance as specified above, we find

$$\int_0^1 \frac{e^{-1/u^2}}{u^4} \, du \approx 0.253641116905.$$

Therefore,

$$\int_0^\infty t^2 e^{-t^2} \, dt \approx 0.189472345819 + 0.253641116905$$
$$= 0.443113462724.$$

The exact value for this integral is $\sqrt{\pi}/4$, so the error in the approximate value is roughly 2×10^{-12}.

As a final general example, consider

$$\int_{-\infty}^\infty \frac{1}{1 + x + x^2} e^{-x^2} \, dx.$$

We could split the integration interval into 3 pieces, say $(-\infty, -1] \cup [-1, 1] \cup [1, \infty)$, and then apply the substitution $x = 1/u$ to the first and last subintervals. This approach is a bit cumbersome, though, so we choose instead to let $x = \tan\theta$, which allows the interval $-\infty < x < \infty$ to be treated as a single piece. With $x = \tan\theta$, the integral is transformed to

$$\begin{aligned} \int_{-\infty}^\infty \frac{1}{1 + x + x^2} e^{-x^2} \, dx &= \int_{-\pi/2}^{\pi/2} \frac{1}{\sec^2\theta + \tan\theta} e^{-\tan^2\theta} \sec^2\theta \, d\theta \\ &= \int_{-\pi/2}^{\pi/2} \frac{1}{1 + \sin\theta\cos\theta} e^{-\tan^2\theta} \, d\theta. \end{aligned}$$

The integral in θ has two removable discontinuities—one at each endpoint. Approximating the value of this integral, we find

$$\int_{-\infty}^\infty \frac{1}{1 + x + x^2} e^{-x^2} \, dx \approx 1.5327082101.$$

Miscellaneous Examples

EXAMPLE 6.22 An Integral from Electron Gas Theory

The integral

$$\int_0^\infty e^{-x^2} \tan^{-1}(1/x) \, dx$$

arises in calculating the exchange-correlation energy of an electron gas in a strong magnetic field (M. L. Glasser, "The Electron Gas in a Magnetic Field: Nonrelativistic Ground State Properties," in *Theoretical Chemistry: Advances and Perspectives,*

Volume 2, H. Eyring and D. Henderson, eds., Academic Press, New York, 1976, pp. 67–129). To isolate the handling of the discontinuity at the lower limit of integration from the handling of the infinite upper limit, we break the problem into two parts as follows:

$$\int_0^\infty e^{-x^2}\tan^{-1}(1/x)\,dx = \int_0^1 e^{-x^2}\tan^{-1}(1/x)\,dx + \int_1^\infty e^{-x^2}\tan^{-1}(1/x)\,dx.$$

The integration interval could have been split at any value of x; $x = 1$ is just convenient. On $[0,1]$, the discontinuity at $x = 0$ is removable since

$$\lim_{x\to 0+} e^{-x^2}\tan^{-1}(1/x) = \pi/2.$$

Taking this limit into account, we find

$$\int_0^1 e^{-x^2}\tan^{-1}(1/x)\,dx \approx 0.8901979553.$$

On $[1,\infty)$, we introduce the change of variable $x = 1/u$, which yields

$$\int_1^\infty e^{-x^2}\tan^{-1}(1/x)\,dx = \int_0^1 \frac{e^{-1/u^2}\tan^{-1} u}{u^2}\,du.$$

With

$$\lim_{u\to 0+} \frac{e^{-1/u^2}\tan^{-1} u}{u^2} = 0,$$

we then find

$$\int_0^1 \frac{e^{-1/u^2}\tan^{-1} u}{u^2}\,du \approx 0.0921039610.$$

Recombining the two pieces, we have

$$\int_0^\infty e^{-x^2}\tan^{-1}(1/x)\,dx \approx 0.9823019163.$$

EXAMPLE 6.23 The Period of Shock Wave Oscillations

In obtaining an asymptotic estimate for the period of shock wave oscillations produced during lithotripsy [see L. Howle, D. Schaeffer, M. Shearer, and P. Zhong, "Lithotripsy: The Treatment of Kidney Stones with Shock Waves," *SIAM Review*, **40** (2), pp. 356–371, 1998], the integral

$$\sqrt{\frac{3}{2}}\int_0^1 \sqrt{\frac{y^3}{1-y^3}}\,dy$$

must be evaluated. The integrand is discontinuous at $y = 1$, and derivatives beyond the first order are discontinuous at $y = 0$. To handle these discontinuities separately, we will split the integration interval at $y = 1/2$. On $[0, 1/2]$, we let $y = x^2$ and find

$$\int_0^{1/2} \sqrt{\frac{y^3}{1-y^3}}\,dy = \int_0^{\sqrt{2}/2} 2x^4\sqrt{\frac{1}{1-x^6}}\,dx \approx 0.0728516434.$$

Next, on $[1/2, 1]$, we let $1 - y = x^2$ and find

$$\int_{1/2}^{1} \sqrt{\frac{y^3}{1-y^3}}\,dy = \int_0^{\sqrt{2}/2} 2\sqrt{\frac{(1-x^2)^3}{3-3x^2+x^4}}\,dx \approx 0.6739825568.$$

Therefore,

$$\sqrt{\frac{3}{2}}\int_0^1 \sqrt{\frac{y^3}{1-y^3}}\,dy \approx 0.9146813565.$$

EXAMPLE 6.24 Light Transmission Through a Crystal

Suppose we need to evaluate

$$\int_0^{\infty} \frac{\cosh\left[(1+t^2)^{-1/2}\right]}{1+t^2}\,dt,$$

which is a particular member of a family of integrals that arise in the study of light transmission through a crystal (M. L. Glasser, "Two Definite Integrals Arising in Light Transmission Through a Crystal: Problem 93-4," *SIAM Review*, **35** (1), p. 136, 1993). We could proceed with this problem as we have done on previous problems of the same type, splitting the integration interval into $[0, 1] \cup [1, \infty)$ and using the change of variable $t = 1/u$ on the latter subinterval. Given the structure of this integrand, however, it will be more efficient to let $t = \tan\theta$. This trigonometric substitution transforms the integral into

$$\int_0^{\pi/2} \cosh(\cos\theta)\,d\theta,$$

which has no discontinuities. Evaluating this integral, we find

$$\int_0^{\infty} \frac{\cosh\left[(1+t^2)^{-1/2}\right]}{1+t^2}\,dt \approx 1.9887316303.$$

References

1. C. Ueberhuber, *Numerical Computation 2: Methods, Software and Analysis*, Springer-Verlag, Berlin, 1997.

EXERCISES

In Exercises 1–3,

(a) Compute the value of the indicated definite integral using the trapezoidal rule, Simpson's rule, the midpoint rule, and the two-point Gaussian quadrature rule, programming the integrand as given. Use $n = 2, 4, 8, 16, 32$, and 64 for each method. Compare the observed order of convergence with the theoretical value.

(b) Repeat part (a) after making an appropriate change of variable in the integrand.

1. $\int_0^1 e^{\sqrt{x}}\,dx$

2. $\int_0^1 x^{5/2}\,dx$

3. $\int_0^1 \sin(\sqrt{x})\,dx$

For the integrals given in Exercises 4–13, identify each discontinuity/limit of integration which must be handled, then take appropriate action, and compute the value of the integral, accurate to at least ten decimal places.

4. $\int_0^1 \frac{\sin x}{x}\,dx$

5. $\int_0^1 \frac{x^{1/7}}{1+x^2}\,dx$

6. $\int_0^1 \frac{\ln(1-x)}{\sqrt{x}}\,dx$

7. $\int_0^\infty e^{-x^4}\,dx$

8. $\int_0^1 \frac{e^x}{\sqrt{1-x}}\,dx$

9. $\int_0^\infty \frac{dx}{\sqrt{x}(x+1)}$

10. $\int_0^\infty \frac{dx}{1+x^3} x\,dx$

11. $\int_0^1 \left[\frac{e^{-x^2}\ln(1+x)}{x^2} - \frac{1}{x}\right] dx$

12. $\int_1^\infty \frac{e^{-x^2}\ln(1+x)}{x^2}\,dx$

13. $\int_{-\infty}^\infty \frac{x^2}{(x^2+1)(x^2-x+1)}\,dx$

14. Compute the value of the integral

$$\int_1^\infty \frac{\ln x}{1+x^2}\,dx,$$

accurate to at least ten decimal places in two ways:

(a) making the substitution $x = 1/u$; and

(b) making the substitution $x = \tan\theta$.

15. An integral of the form

$$\int_{-1}^1 \frac{f(x)}{\sqrt{1-x^2}}\,dx$$

has discontinuities at both endpoints of the integration interval. For integrals of this type, the substitution $x = \sin\theta$ transforms the problem to

$$\int_{-1}^{1} \frac{f(x)}{\sqrt{1-x^2}}\,dx = \int_{-\pi/2}^{\pi/2} f(\sin\theta)\,d\theta.$$

Evaluate each of the following integrals using this approach.

(a) $\int_{-1}^{1} \frac{e^x}{\sqrt{1-x^2}}\,dx$

(b) $\int_{-1}^{1} \frac{x^4}{\sqrt{1-x^2}}\,dx$

(c) $\int_{-1}^{1} \frac{\cos(\pi x)}{\sqrt{1-x^2}}\,dx$

16. Repeat Exercise 15, but make the substitution $x = \cos\theta$.

17. The integral

$$G(t) = \int_0^\infty e^{-t/x} e^{-x^2/2}\,dx$$

arises in studies of hopping transport for one-dimensional percolation (see J. Bernasconi, "Hopping Transport in One-Dimensional Percolation Model: A Comment," *Phys. Rev. B*, **25**, 1982, pp. 1394–1395). Evaluate $G(1)$ and $G(5)$.

18. In determining the overlap interaction for the kinetic energy of a free electron gas, the integral

$$K(\alpha) = \int_0^\infty \left[(e^{-x} + e^{x})^\alpha - (e^{-\alpha x} + e^{\alpha x})\right] dx$$

arises (see W. Harrison, "Total Energies in the Tight-Binding Theory," *Phys. Rev. B*, **23**, 1981, pp. 5230–5245). In particular, the value of $K(5/3)$ is needed. Evaluate $K(5/3)$.

19. Evaluate the integrals

$$\int_0^\infty \frac{x^2}{e^x - 1}\,dx \quad \text{and} \quad \int_0^\infty \frac{x^3}{e^x - 1}\,dx,$$

whise arise in determining the photon density and the energy density, respectively, associated with blackbody radiation (see A. Beiser, *Concepts of Modern Physics*, McGraw-Hill, New York, 1981).

CHAPTER 7

Initial Value Problems of Ordinary Differential Equations

AN OVERVIEW

Some Background

An equation in which an unknown function appears inside one or more derivatives is called a *differential equation.* When the unknown is a function of only one independent variable, the equation is referred to as an *ordinary differential equation.* On the other hand, when the unknown is a function of more than one independent variable, we call the equation a *partial differential equation.* Numerical techniques for approximating the solution of a partial differential equation will be treated in Chapters 9, 10, and 11.

In principle, solving a differential equation requires integration. From our knowledge of calculus, we know that integration introduces arbitrary constants; therefore, the solution to a differential equation will contain one or more arbitrary constants. Values can be determined for these constants by imposing extra conditions on the solution. If these extra conditions are specified at more than one value of the independent variable, they are called *boundary conditions.* A differential equation combined with a set of boundary conditions is called a *boundary value problem.* Chapter 8 discusses techniques for approximating the solution of a boundary value problem. When all of the extra conditions that have been imposed on the solution are specified at a single value of the independent variable, they are called *initial conditions*, and the combination of a differential equation and a set of initial conditions is called an *initial value problem.*

Fundamental Mathematical Problem

In this chapter, we will develop and investigate the performance of numerical techniques for approximating the solution of initial value problems of ordinary differential equations. The basic problem we will address can be stated as follows: Find the function $y(t)$ that satisfies

$$\begin{aligned} y'(t) &= f(t, y(t)), \quad a \le t \le b \\ y(a) &= \alpha, \end{aligned}$$

where the right-hand side function, f, and the initial value, α, are given. In the case of a single equation (also known as the scalar case), y and f will be real-valued functions and α a real number. For a system of equations, y and f will be vector-valued functions [i.e., for each t, $y(t), f(t, y(t)) \in R^n$ for some integer n] and α a vector. We will focus our attention at the beginning of the chapter on scalar initial value problems and then discuss how to generalize methods for scalar problems in order to solve systems of equations. Higher-order differential equations will be handled by recasting the equation as a system of first-order equations.

The "Projection Printing" problem capsule that was presented in the overview to Chapter 1 (see page 6), is one application that gives rise to an initial value problem of ordinary differential equations. Here are two more examples.

Population Growth in a Closed System

Consider the growth of a species in a closed system. Let $p(\tilde{t})$ denote the population of the species at any time $\tilde{t}$, and suppose the initial population $p(0) = p_0$ is known. A standard assumption in population modeling is that the instantaneous rate of change in the population is proportional to the population; that is,

$$\frac{dp}{d\tilde{t}} = kp,$$

for some proportionality factor k. Note that k represents the overall growth rate of the population. To determine a specific functional form for k we make three additional assumptions.

- The population exhibits a constant, intrinsic per capita growth rate;
- Due to limited resources (such as food, water and shelter), a "crowding" effect causes a reduction in the overall growth rate which is proportional to the population; and
- The accumulation of waste products in the closed system produces a "toxicity" effect which further reduces the overall growth rate by an amount proportional to $\int_0^{\tilde{t}} p(\tau)\, d\tau$.

Bringing all of these effects together, we arrive at Volterra's model for population growth

$$\frac{dp}{d\tilde{t}} = \left[r - \lambda p - c\int_0^{\tilde{t}} p(\tau)\, d\tau\right] p,$$

where r is the intrinsic growth rate, λ is the crowding coefficient, and c is the toxicity coefficient.

As formulated, Volterra's model is an integro-differential equation. Note that the unknown appears inside both an integral and a derivative. Following the procedure of TeBeest [1], we can reduce this to a second-order differential equation. First, introduce the dimensionless variables

$$t = \frac{\tilde{t}}{\lambda/c} \quad \text{and} \quad u = \frac{p}{r/\lambda}.$$

This produces the nondimensional problem

$$\frac{du}{dt} = \frac{u}{\kappa}\left[1 - u - \int_0^t u(\tau)\, d\tau\right], \qquad u(0) = u_0 \equiv \frac{p_0}{r/\lambda},$$

where $\kappa = c/(r\lambda)$. Now, change the dependent variable to $y = \ln u$ and differentiate the resulting equation with respect to t. This yields

$$\frac{d^2y}{dt^2} = -\frac{e^y}{\kappa}\left(1 + \frac{dy}{dt}\right). \tag{1}$$

The initial conditions associated with (1) are

$$y(0) = \ln u(0) = \ln u_0 \tag{2}$$

and

$$y'(0) = \frac{u'(0)}{u(0)} = \frac{1 - u_0}{\kappa}. \tag{3}$$

Equations (1), (2), and (3) form an initial value problem for determining the time evolution of the population.

A Catalyzed Reaction

Suppose two chemical compounds, A and B, react to form a product, P. When this reaction takes place in isolation, it proceeds slowly; however, when this reaction takes place in the presence of chemical compound C, the rate of reaction is increased. Since its presence causes an increase in reaction rate, the chemical C is called a *catalyst.* It is believed that the catalyzed reaction proceeds according to the following two-step mechanism:

$$\begin{aligned} A + C &\longrightarrow X \\ X + B &\longrightarrow P + C. \end{aligned}$$

The compound X is called a catalyst-reactant complex.

To model the evolution of this reaction, suppose that each step occurs at a rate proportional to the product of the concentrations of the reactants. Taking into account the rates at which each compound is produced and consumed, we arrive at the following system of five equations with associated initial conditions:

$$\begin{aligned} [A]' &= -k_1\,[A]\,[C], & [A]\,(0) &= A_0 \\ [B]' &= -k_2\,[B]\,[X], & [B]\,(0) &= B_0 \\ [C]' &= -k_1\,[A]\,[C] + k_2\,[B]\,[X], & [C]\,(0) &= C_0 \\ [X]' &= k_1\,[A]\,[C] - k_2\,[B]\,[X], & [X]\,(0) &= 0 \\ [P]' &= k_2\,[B]\,[X], & [P]\,(0) &= 0. \end{aligned} \tag{4}$$

Here, $[\cdot]$ denotes the concentration of the indicated compound, primes indicate differentiation with respect to time, k_1 and k_2 are the rate constants for the two steps of the reaction, and A_0, B_0, and C_0 are the initial concentrations of A, B, and C, respectively.

Given values for the parameters k_1, k_2, A_0, B_0, and C_0, a numerical solution to the initial value problem in (4) can be obtained. With a little extra analysis, however, the size of the problem can be reduced to just two equations. First, add the second and fifth differential equations in (4) to obtain

$$([B]+[P])' = 0 \qquad \Rightarrow \qquad [B]+[P] = \text{constant}.$$

Evaluating this last expression at $t = 0$ determines the value of the constant to be B_0. Solving for $[P]$, we then find

$$[P] = B_0 - [B]\,. \tag{5}$$

Proceeding in a similar manner with the third and fourth differential equations in (4), we find

$$[C] = C_0 - [X]\,. \tag{6}$$

A third relationship, $[A]-[B]+[X] = A_0 - B_0$, is obtained after adding the fourth differential equation to the first and then subtracting the second. To simplify matters, from this point forward we will assume that $A_0 = B_0$. With this assumption, it follows that

$$[X] = [B] - [A]\,. \tag{7}$$

Now, substitute (6) and (7) into the first two equations in (4) to obtain

$$\begin{aligned} [A]' &= -k_1\,[A]\,(C_0 - [B] + [A])\,, \qquad & [A]\,(0) = A_0 \\ [B]' &= -k_2\,[B]\,([B]-[A])\,, & [B]\,(0) = A_0. \end{aligned} \tag{8}$$

Once the system (8) has been solved for $[A]$ and $[B]$, equations (5), (6), and (7) can be used to calculate $[P]$, $[C]$, and $[X]$.

As a final preliminary step, let's introduce the dimensionless variables

$$\alpha = \frac{[A]}{A_0}, \quad \beta = \frac{[B]}{A_0} \quad \text{and} \quad \tau = k_1 A_0 t$$

and define the dimensionless parameters

$$\kappa = \frac{k_2}{k_1} \quad \text{and} \quad \lambda = \frac{C_0}{A_0}.$$

This transforms the system (8) into

$$\begin{aligned} \alpha' &= -\alpha(\lambda - \beta + \alpha), \qquad & \alpha(0) = 1 \\ \beta' &= -\kappa\beta(\beta - \alpha), & \beta(0) = 1. \end{aligned}$$

We would like to study the dynamics of this last system for various values of λ and κ.

Remainder of the Chapter

Section 1 provides basic information related to the theory of initial value problems for ordinary differential equations and introduces some of the key numerical concepts that will be addressed throughout the chapter. Euler's method, the simplest numerical technique for approximating the solution of an initial value problem, is then presented in Section 2. More advanced techniques, including Taylor methods, Runge-Kutta methods, Adams-Bashforth methods, and Adams-Moulton methods, are developed in the next three sections. Convergence analysis for these techniques is presented in Section 6. Section 7 explores the issue of error control and the construction of variable step size algorithms, and Section 8 discusses the solution of higher-order differential equations and systems of differential equations. The chapter concludes with a discussion of absolute stability and stiff equations.

References

1. K. G. TeBeest, "Numerical and Analytical Solutions of Volterra's Population Model," *SIAM Review*, **39**, 484–493, 1997.

7.1 KEY NUMERICAL CONCEPTS AND ELEMENTS OF CONTINUOUS THEORY

For the time being, we will restrict our attention to the scalar first-order initial value problem

$$\begin{aligned} y'(t) &= f(t, y(t)), \quad a \leq t \leq b \\ y(a) &= \alpha. \end{aligned} \tag{1}$$

Although the true solution of (1), $y(t)$, will be a continuous function defined for all $a \leq t \leq b$, the approximate solution we obtain will consist of individual approximations to the values of y at only a discrete set of times, say

$$a = t_0 < t_1 < t_2 < \cdots < t_{N-1} < t_N = b.$$

Furthermore, we will adopt the following notation: let y_i denote the value of the true solution at $t = t_i$ and let w_i denote the approximation to y_i. In other words, throughout our discussions, we will have

$$w_i \approx y_i = y(t_i).$$

The techniques developed in this chapter are all based on the replacement of the differential equation in (1) by a difference equation. Once this replacement has been made, the approximate solution is computed via a "time marching," or "time stepping," procedure. Starting from $w_0 = \alpha$, the initial value of the true solution, the difference equation is used to determine w_1—this essentially marches the solution from $t = t_0$ forward to $t = t_1$. Next, w_2 is determined from w_1 (and possibly also w_0), w_3 is determined from w_2 (and possibly also w_1 and w_0), and so on. The procedure terminates when w_N has been computed—time has been marched out to $t_N = b$.

One-Step versus Multistep Methods

We will consider two different types of initial value problem solvers in this text: one-step methods and multistep methods. The general form for a one-step method is

$$\frac{w_{i+1} - w_i}{h_i} = \phi(f, t_i, w_i, w_{i+1}, h_i), \tag{2}$$

where $h_i = t_{i+1} - t_i$. The rationale behind the nomenclature of a "one-step method" is clear from the form of the difference equation in (2): The computation of w_{i+1} requires knowledge of w_i only, where w_i is the approximation made one step prior to the present value.

If the function ϕ is independent of w_{i+1}, then the difference equation can be solved explicitly for w_{i+1}, so the method is said to be *explicit*. For an explicit method, knowing the right-hand side function f and given values for t_i, h_i, and w_i, we can immediately calculate the value of w_{i+1}. When ϕ does depend of w_{i+1}, the difference equation defines the value of w_{i+1} only implicitly, so the method is said to be *implicit*. For an implicit method, some sort of rootfinding technique, such as fixed-point iteration or Newton's method, must be applied during each time step to determine the value of w_{i+1}.

EXAMPLE 7.1 Explicit versus Implicit

Consider the difference equation

$$\frac{w_{i+1} - w_i}{h_i} = f(t_i, w_i).$$

Here, $\phi(f, t_i, h_i, w_i, w_{i+1}) = f(t_i, w_i)$, so this equation fits the pattern of an explicit one-step method. In particular, this difference equation is known as Euler's method, which we will examine in depth in Section 7.2. As a second example consider the difference equation

$$\frac{w_{i+1} - w_i}{h_i} = \frac{1}{2}[f(t_i, w_i) + f(t_i + h, w_{i+1})],$$

which has

$$\phi(f, t_i, h_i, w_i, w_{i+1}) = \frac{1}{2}[f(t_i, w_i) + f(t_i + h_i, w_{i+1})].$$

This equation, which is known as the trapezoidal method, therefore fits the pattern of an implicit one-step method.

To illustrate more clearly the practical difference between an explicit method and an implicit method, suppose we were to apply Euler's method and the trapezoidal method to the differential equation

$$\frac{dy}{dt} = t \sin y.$$

Comparing this specific equation to our model problem, we identify $f(t,y) = t\sin y$. The difference equation for Euler's method then becomes

$$\frac{w_{i+1} - w_i}{h_i} = t_i \sin w_i,$$

from which we obtain the explicit expression $w_{i+1} = w_i + h_i t_i \sin w_i$. To calculate w_{i+1}, we only need to have values for w_i, h_i, and t_i and then perform some basic arithmetic operations. On the other hand, substituting $f(t,y) = t\sin y$ into the difference equation for the trapezoidal method yields

$$\frac{w_{i+1} - w_i}{h_i} = \frac{1}{2}\left[t_i \sin w_i + (t_i + h_i)\sin w_{i+1}\right].$$

There is no way to explicitly solve this equation for w_{i+1}. Hence, even given values for w_i, h_i, and t_i, we will have to solve a nonlinear equation to determine w_{i+1}.

If one-step methods are thought of as using information regarding where we currently are to predict where we should be at the next time step, then multistep methods use information regarding where we are and where we have been to make that prediction. For example, a two-step method would use both w_i and w_{i-1} to compute w_{i+1}, while a four-step method would use w_i, w_{i-1}, w_{i-2}, and w_{i-3}. The general linear m-step method can be written in the form

$$\frac{w_{i+1} - a_1 w_i - a_2 w_{i-1} - \cdots - a_m w_{i+1-m}}{h_i} =$$
$$b_0 f(t_{i+1}, w_{i+1}) + b_1 f(t_i, w_i) + b_2 f(t_{i-1}, w_{i-1}) + \cdots + b_m f(t_{i+1-m}, w_{i+1-m}). \quad (3)$$

The method is explicit when $b_0 = 0$, and implicit otherwise.

EXAMPLE 7.2 A Couple of Multistep Methods

Consider the difference equation

$$\frac{w_{i+1} - w_i}{h_i} = \frac{23}{12} f(t_i, w_i) - \frac{4}{3} f(t_{i-1}, w_{i-1}) + \frac{5}{12} f(t_{i-2}, w_{i-2}).$$

Since the smallest subscript that appears in this equation is $i - 2$, which is three less than $i + 1$, this is a three-step method. Furthermore, because w_{i+1} does not appear inside the function f, this is an explicit method. Matching the difference equation to equation (3), we see that the values of the various coefficients are

$$a_1 = 1, \quad a_2 = a_3 = 0, \quad b_0 = 0, \quad b_1 = \frac{23}{12}, \quad b_2 = -\frac{4}{3}, \quad b_3 = \frac{5}{12}.$$

As a second example, consider the difference equation

$$\frac{w_{i+1} - \frac{4}{3} w_i + \frac{1}{3} w_{i-1}}{h_i} = \frac{2}{3} f(t_{i+1}, w_{i+1}).$$

This is an implicit two-step method because w_{i+1} does appear inside the function f and the smallest subscript in the equation is $i-1$. Matching the difference equation to equation (3), we see that the values of the various coefficients are

$$a_1 = \frac{4}{3}, \quad a_2 = -\frac{1}{3}, \quad b_0 = \frac{2}{3}, \quad b_1 = b_2 = 0.$$

Fundamental Numerical Concepts

The analysis of numerical methods for initial value problems, be they one-step or multistep methods, involves two different types of errors. One measures how well the difference equation that defines the method approximates the differential equation; the other measures how well the solution of the difference equation approximates the solution of the differential equation. The sensitivity of the solution of the difference equation to changes in the initial condition is also considered.

The error associated with the approximation of the differential equation is called the *local truncation error*, which we shall denote by the symbol τ_i. We refer to this error as "local" because it measures the error generated by one step of the method, assuming the solution at previous steps was exact. As such, truncation error is determined by substituting the solution of the differential equation into the method's difference equation. Recalling equations (2) and (3), it follows that for a one-step method

$$\tau_i = \frac{y_{i+1} - y_i}{h_i} - \phi(f, t_i, y_i, y_{i+1}, h_i)$$

and for a linear multistep method

$$\tau_i = \frac{y_{i+1} - \sum_{j=1}^{m} a_j y_{i+1-j}}{h_i} - \sum_{j=0}^{m} b_j f(t_{i+1-j}, y_{i+1-j}).$$

If $\tau_i \to 0$ as $h_i \to 0$, then the difference equation is said to be consistent with the differential equation $y' = f(t, y)$ and the numerical method is called *consistent*. Additionally, if $\tau_i = O(h_i^p)$ for some constant p, the method is said to be of *order* p.

Of course, the error in which we are most interested is the difference between the solution of the differential equation and the solution of the difference equation, $y_i - w_i$. This is known as the *global discretization error* and measures the cumulative effect of the errors introduced by all of the time steps taken. One might expect that global error would be a monotonically increasing function of the number of steps taken, but this is not necessarily the case. We will encounter examples in subsequent sections for which global error oscillates, as well as examples for which global error monotonically decreases.

Our true interest, however, is not in the behavior of global error for a fixed partition of the interval $a \le t \le b$. Rather, our concern is with the behavior of global error as the partition is refined. In particular, we want to increase the number of time steps, N, in such a way that $h = \max(t_{i+1} - t_i)$ tends to zero, but fix $t_N = b$ regardless of the value of N. If

$$\lim_{h\to 0} \max_{1\le i\le N} |y_i - w_i| = 0,$$

we say the method is *convergent*.

Another desirable property for a numerical method to have is stability. For a given numerical method and a fixed partition of $a \le t \le b$, let $\{w_i\}$ denote the sequence of values generated from an initial condition of α and $\{\tilde{w}_i\}$ denote the sequence generated from an initial condition of $\tilde{\alpha}$. The numerical method is *stable* if and only if there exists a function $k(t) > 0$ such that

$$|\tilde{w}_i - w_i| \le k(t_i)|\tilde{\alpha} - \alpha|$$

for all i. The key point is that the function $k(t_i)$ must be independent of h. The method is *unstable* when no such bound exists.

We've just introduced three important properties for numerical methods for initial value problems—consistency, convergence, and stability. The most important of these properties is convergence. If the solution of the difference equation does not tend toward the solution of the differential equation as the computational partition is refined, the method is useless.

Elements of Continuous Theory

Before we start developing and investigating specific numerical methods for initial value problems, it is worthwhile taking a few moments to examine certain aspects of the theory of initial value problems. We will focus on three areas: existence, uniqueness, and stability. Existence deals with the question of whether a given initial value problem has a solution. Uniqueness takes this issue one step further. If an initial value problem has a solution, does it have only one? The concept of stability for an initial value problem is similar to that of stability of a numerical method and deals with sensitivity of the solution to changes in initial conditions. We will make this statement more precise below.

We are interested in the theory of initial value problems for two reasons. On the one hand, the conditions under which an initial value problem will have a unique solution and be stable will place the conditions under which a numerical method is consistent, convergent, and stable into perspective. On the other hand, the theory of initial value problems has certain practical implications. For example, suppose that a particular initial value problem does not have a solution. In this case, it will not matter how sophisticated a numerical method we have at our disposal. Any calculated values will be meaningless. The same is true if an initial value problem does not have a unique solution.

What practical implications are associated with sensitivity to changes in initial data? Due to the presence of roundoff error, when we implement our numerical methods, we will never actually be solving the original initial value problem

$$y' = f(t, y), \quad y(a) = \alpha.$$

Instead, we will be solving the *perturbed problem*

$$z' = f(t, z) + \delta(t), \quad z(a) = \alpha + \epsilon_0.$$

Our hope is that the solutions of these problems are "close"; that is, that $|z(t)-y(t)|$ is bounded on some interval $a \le t \le b$. More precisely, we are counting on the existence of positive constants ϵ and k such that, whenever

$$|\epsilon_0| < \epsilon \quad \text{and} \quad |\delta(t)| < \epsilon \quad \text{for all } t \in [a, b],$$

the perturbed problem has a unique solution that satisfies

$$|z(t) - y(t)| < k\epsilon$$

for all $t \in [a, b]$. If these conditions hold, then the original initial value problem is said to be *stable.* Thus, only when the original problem is stable can we expect the numerical solution to provide a reasonable approximation to the solution of that original problem.

An initial value problem that has a unique solution and is stable is said to be *well-posed.* An important tool for establishing that an initial value problem is well posed is the *Lipschitz condition.*

> **Definition.** A function $f(t, y)$ satisfies a LIPSCHITZ CONDITION in y on the set $D \subset R^2$ if there exists a constant $L > 0$ such that
>
> $$|f(t, y_1) - f(t, y_2)| \le L|y_1 - y_2|$$
>
> for all $(t, y_1), (t, y_2) \in D$. The constant L is called the LIPSCHITZ CONSTANT for f.

Note the central role that the set D plays in this definition. A given function may satisfy a Lipschitz condition on one set, but not on another. Furthermore, the value of L may depend on D.

EXAMPLE 7.3 Determining Lipschitz Conditions

Let $f(t, y) = y \sin t$ and take $D = R^2$. Since

$$f(t, y_1) - f(t, y_2) = y_1 \sin t - y_2 \sin t = (y_1 - y_2) \sin t$$

for all $(t, y_1), (t, y_2) \in D$, it follows that

$$|f(t, y_1) - f(t, y_2)| = |\sin t||y_1 - y_2| \le |y_1 - y_2|$$

for all $(t, y_1), (t, y_2) \in D$. Hence, f satisfies a Lipschitz condition in y on all of R^2, with Lipschitz constant $L = 1$.

As a second example, consider the function

$$f(t, y) = \frac{t^2 y^2}{1 + t^2}.$$

Let $M > 0$, and take $D = \{(t,y) | t \in R, -M \leq y \leq M\}$. Here, we have

$$\begin{aligned} f(t,y_1) - f(t,y_2) &= \frac{t^2 y_1^2}{1+t^2} - \frac{t^2 y_2^2}{1+t^2} \\ &= \frac{t^2}{1+t^2}(y_1^2 - y_2^2) = \frac{t^2}{1+t^2}(y_1+y_2)(y_1-y_2), \end{aligned}$$

from which it follows that

$$|f(t,y_1) - f(t,y_2)| = \left|\frac{t^2}{1+t^2}\right| |y_1+y_2||y_1-y_2|.$$

Now suppose (t,y_1) and (t,y_2) are both in D. For any $t \in R$, $0 \leq t^2/(1+t^2) \leq 1$. Further, since $-M \leq y_1, y_2 \leq M$, we have $|y_1+y_2| \leq |y_1| + |y_2| \leq 2M$. Therefore, for all $(t,y_1), (t,y_2) \in D$,

$$|f(t,y_1) - f(t,y_2)| \leq 2M|y_1-y_2|,$$

and f satisfies a Lipschitz condition in y on D with Lipschitz constant $L = 2M$. Note the dependence of the Lipschitz constant on the set D. Also note that, although f satisfies a Lipschitz condition on D for any *fixed* M, we cannot extend this result to all of R^2. If we try to take $M \to \infty$, we will no longer be able to obtain an upper bound for the quantity $|y_1+y_2|$.

It can be difficult using the definition to establish that a function satisfies a Lipschitz condition. The following result can be useful in certain instances.

Theorem. If f is defined on $D = \{(t,y) | a \leq t \leq b, c \leq y \leq d\}$ and there exists a constant $L > 0$ such that

$$\left|\frac{\partial f}{\partial y}(t,y)\right| \leq L$$

for all $(t,y) \in D$, then f satisfies a Lipschitz condition in y on the set D with Lipschitz constant L.

Proof. This is a generalization of the Mean Value Theorem. Fix $t \in [a,b]$, and let $(t,y_1), (t,y_2) \in D$. Apply the Mean Value Theorem to f, with t fixed, to obtain

$$f(t,y_1) - f(t,y_2) = \frac{\partial f}{\partial y}(t,\xi)(y_1-y_2)$$

for some ξ between y_1 and y_2. The result follows upon taking the absolute value of this last expression and applying the stated bound on $|\partial f/\partial y|$. □

EXAMPLE 7.4 Another Lipschitz Condition

Consider the set $D = \{(t,y) | 0 \le t \le 1, 0 \le y \le \pi/4\}$, and let

$$f(t,y) = \frac{t \sec y}{t-2}.$$

Here, we have

$$\frac{\partial f}{\partial y} = \frac{t}{t-2} \sec y \tan y.$$

For $0 \le t \le 1$, $|t/(t-2)| \le 1$, and for $0 \le y \le \pi/4$, $|\sec y \tan y| \le \sqrt{2}$. Therefore,

$$\left| \frac{\partial f}{\partial y} \right| \le \sqrt{2} \quad \text{for all } (t,y) \in D,$$

so f satisfies a Lipschitz condition in y on D with $L = \sqrt{2}$.

The following theorem links the Lipschitz condition with the concept of well posedness. More general statements can be made, but this will be sufficient for our purposes. For a proof of this theorem, see Coddington [1].

Theorem. Let $D = \{(t,y) | a \le t \le T, |y - \alpha| \le \beta\}$ for some $\beta > 0$. Suppose that the function $f(t,y)$ is continuous on D and satisfies a Lipschitz condition in y on D. Then, there exists an interval $[a,b] \subset [a,T]$ such that the initial value problem

$$y' = f(t,y), \quad y(a) = \alpha$$

is well posed on $a \le t \le b$.

References

1. E. A. Coddington, *An Introduction to Ordinary Differential Equations*, Dover Publications, New York, 1989.

EXERCISES

1. Identify each of the following difference equations as representing a one-step method or a multistep method and as being implicit or explicit.

(a) $\dfrac{w_{i+1} - w_i}{h} = \dfrac{3}{2} f(t_i, w_i) - \dfrac{1}{2} f(t_{i-1}, w_{i-1})$

(b) $\dfrac{w_{i+1} - w_i}{h} = f\left(t_i + \dfrac{h}{2}, w_i + \dfrac{h}{2} f(t_i, w_i)\right)$

(c) $\dfrac{w_{i+1} - w_i}{h} = \dfrac{5}{12} f(t_{i+1}, w_{i+1}) + \dfrac{2}{3} f(t_i, w_i) - \dfrac{1}{12} f(t_{i-1}, w_{i-1})$

(d) $\dfrac{w_{i+1} - w_i}{h} = f(t_{i+1}, w_{i+1})$

(e) $\dfrac{w_{i+1} - 4w_i + 3w_{i-1}}{h} = -2f(t_{i-1}, w_{i-1})$

2. Each of the following difference equations represents a linear multistep method. Identify the number of steps, m, and the values of the coefficients a_j and b_j.

(a) $\dfrac{w_{i+1} - w_{i-1}}{h} = \dfrac{1}{3}\left[f(t_{i+1}, w_{i+1}) + 4f(t_i, w_i) + f(t_{i-1}, w_{i-1})\right]$

(b) $\dfrac{w_{i+1} - w_{i-3}}{h} = \dfrac{4}{3}\left[2f(t_i, w_i) - f(t_{i-1}, w_{i-1}) + 2f(t_{i-2}, w_{i-2})\right]$

(c) $\dfrac{w_{i+1} - w_{i-1}}{h} = 2f(t_i, w_i)$

(d) $\dfrac{w_{i+1} - w_{i-2}}{h} =$
$\dfrac{3}{8}\left[f(t_{i+1}, w_{i+1}) + 3f(t_i, w_i) + 3f(t_{i-1}, w_{i-1}) + f(t_{i-2}, w_{i-2})\right]$

(e) $\dfrac{w_{i+1} - w_i}{h} =$
$\dfrac{9}{24}f(t_{i+1}, w_{i+1}) + \dfrac{19}{24}f(t_i, w_i) - \dfrac{5}{24}f(t_{i-1}, w_{i-1}) + \dfrac{1}{24}f(t_{i-2}, w_{i-2})$

(f) $\dfrac{w_{i+1} - \frac{1}{2}w_i - \frac{1}{2}w_{i-1}}{h} = f(t_{i+1}, w_{i+1}) - \dfrac{1}{4}f(t_i, w_i) + \dfrac{3}{4}f(t_{i-1}, w_{i-1})$

3. Show that each of the following functions satisfies a Lipschitz condition in y on the indicated set D.

(a) $f(t, y) = ty^3, \quad D = \{(t, y) \mid -1 \le t \le 1, 0 \le y \le 10\}$

(b) $f(t, y) = t\sqrt{y}/(1 + t^2), \quad D = \{(t, y) \mid t \in R, y \ge 1\}$

(c) $f(t, y) = e^t/y, \quad D = \{(t, y) \mid 0 \le t \le 2, y \ge 1\}$

(d) $f(t, y) = 1 - y + e^{2t}y^2, \quad D = \{(t, y) \mid 0 \le t \le 1, -5 \le y \le 5\}$

(e) $f(t, y) = y^2 - t/y, \quad D = \{(t, y) \mid -2 \le t \le 2, 1 \le y \le 10\}$

4. Let $M > 0$. Show that the function $f(t, y) = t^4 - 4y/t$ satisfies a Lipschitz condition in y on any set D of the form $\{(t, y) \mid t \ge M, y \in R\}$. Does f satisfy a Lipschitz condition in y on the set $D = \{(t, y) \mid t > 0, y \in R\}$?

5. Let $M > 0$. Show that the function $f(t, y) = t(y^2 - y)$ satisfies a Lipschitz condition in y on any set D of the form $\{(t, y) \mid -1 \le t \le 1, -M \le y \le M\}$. Does f satisfy a Lipschitz condition in y on the set $D = \{(t, y) \mid -1 \le t \le 1, y \in R\}$?

6. Let $M > -1$. Show that the function $f(t, y) = 2ty/(y + 1)$ satisfies a Lipschitz condition in y on any set D of the form $\{(t, y) \mid 0 \le t \le 1, y \ge M\}$. Does f satisfy a Lipschitz condition in y on the set $D = \{(t, y) \mid 0 \le t \le 1, y > -1\}$?

Consider the initial value problem $y' = f(t, y), \; y(a) = \alpha$. If we integrate both sides of the differential equation between a and t, we obtain

$$y(t) - y(a) = \int_a^t f(z, y(z))\, dz.$$

Using the initial condition and some algebra, this becomes

$$y(t) = \alpha + \int_a^t f(z, y(z))\, dz.$$

Now, let $y_0(t) = \alpha$ and define the sequence of functions $\{y_k(t)\}$ by

$$y_k(t) = \alpha + \int_a^t f(z, y_{k-1}(z))\, dz$$

for $k = 1, 2, 3, \ldots$. This technique is called *Picard iteration.* Under certain conditions, this sequence can be shown to converge to the solution of the initial value problem (see Coddington [1]).

In Exercises 7–10, perform Picard iteration to determine the indicated function in the sequence $\{y_k(t)\}$.

7. $y' = t^2 - 2y^2 - 1, \quad y(0) = 0, \quad y_2(t)$

8. $y' = 2e^t - y, \quad y(0) = 1, \quad y_4(t)$

9. $y' = t - y, \quad y(0) = 1, \quad y_4(t)$

10. $y' = t^2 + y^2, \quad y(0) = 1, \quad y_2(t)$

Recall that the Taylor series for the function $y(t)$ about the point $t = a$ is given by

$$y(a) + y'(a)(t-a) + \frac{y''(a)}{2}(t-a)^2 + \cdots + \frac{y^{(k)}(a)}{k!}(t-a)^k + \cdots.$$

If we are trying to solve the initial value problem $y' = f(t, y), \; y(a) = \alpha$, notice that the initial condition provides the value of $y(a)$ and, if we substitute $t = a$ into the differential equation, we can calculate $y'(a)$. Differentiating the differential equation with respect to t and substituting $t = a$ will then give the value for $y''(a)$. Continuing in this fashion, we can, in principle, obtain as many terms in the Taylor series for $y(t)$ as we desire.

In Exercises 11–14, use this approach to determine the first five terms in the Taylor series expansion of the solution of the indicated initial value problem.

11. $y' = y^2 - t^2, \quad y(0) = 1$

12. $y' = y - t, \quad y(0) = 2$

13. $y' = e^t/y, \quad y(0) = 1$

14. $y' = \sin t - y, \quad y(0) = 1$

7.2 EULER'S METHOD

Consider the scalar, first-order initial value problem:

$$\begin{aligned} y'(t) &= f(t, y(t)), \quad a \le t \le b \\ y(a) &= \alpha. \end{aligned} \tag{1}$$

Our objective is to determine a numerical approximation $w \approx y$, where $y(t)$ is the exact solution of (1). As indicated in Section 7.1, we will determine values of w at the discrete set of points

$$a = t_0 < t_1 < t_2 < \cdots < t_{N-1} < t_N = b,$$

only, and we will adopt the notational convention that w_i represents the approximation to $y_i = y(t_i)$. For simplicity, the approximate solution will be sought at equally spaced points; that is, for some positive integer N, we will define the step size

$$h = (b - a)/N,$$

and then the t_i will be given by

$$t_i = a + ih \quad (i = 0, 1, 2, \ldots, N).$$

Derivation of Method

Euler's method is the simplest of the one-step methods for approximating the solution to the initial value problem (1). The derivation of the method begins by assuming that the true solution of (1), $y(t)$, has two continuous derivatives. Expanding this true solution in a Taylor series about the point $t = t_i$ produces

$$y(t) = y_i + (t - t_i)y_i' + \frac{1}{2}(t - t_i)^2 y''(\xi),$$

where ξ is guaranteed to lie between t and t_i. Evaluating the above Taylor expansion at $t = t_{i+1}$ and substituting for y_i' from the right-hand side of the differential equation, we obtain

$$y_{i+1} = y_i + hf(t_i, y_i) + \frac{1}{2}h^2 y''(\xi).$$

Euler's method arises by dropping the error term and replacing y_i (exact solution) by w_i (approximate solution):

$$\begin{aligned} w_0 &= \alpha \\ w_{i+1} &= w_i + hf(t_i, w_i) \qquad i = 0, 1, 2, \ldots, N-1. \end{aligned} \tag{2}$$

It is worth noting that Euler's method can also be derived in other ways. For example, if we were to first evaluate the differential equation in (1) at $t = t_i$ and then replace the derivative term by the first-order forward difference approximation

$$y'(t_i) = \frac{y_{i+1} - y_i}{h} - \frac{h}{2}y''(\xi),$$

we would obtain

$$\frac{y_{i+1} - y_i}{h} - \frac{h}{2}y''(\xi) = f(t_i, y_i).$$

Upon dropping the error term, replacing y_i by w_i, and solving for w_{i+1}, we would reproduce (2). Yet another derivation begins by integrating both sides of the differential equation in (1) from $t = t_i$ to $t = t_{i+1}$ to give

$$y_{i+1} - y_i = \int_{t_i}^{t_{i+1}} f(t, y)\, dt.$$

Approximating the integral of $f(t, y)$ using a left-endpoint approximation again reproduces (2).

EXAMPLE 7.5 Euler's Method in Action, Problem 1

The initial value problem

$$\frac{dx}{dt} = 1 + \frac{x}{t}, \quad 1 \le t \le 6$$
$$x(1) = 1$$

has as its exact solution $x(t) = t(1 + \ln t)$. For this problem, $f(t, x)$ is given by $f(t, x) = 1 + x/t$, so that the Euler's method difference equation takes the form

$$w_0 = 1$$
$$w_{i+1} = w_i + h\left(1 + \frac{w_i}{t_i}\right).$$

Let's use a step size of $h = 0.5$, which will require ten steps to advance from $t = 1$ to $t = 6$. With $t_0 = 1$ and $w_0 = 1$, we calculate

$$w_1 = w_0 + h\left(1 + \frac{w_0}{t_0}\right) = 1 + 0.5\left(1 + \frac{1}{1}\right) = 2.$$

Advancing the value of the independent variable from t_0 to $t_1 = t_0 + h = 1.5$, we then calculate

$$w_2 = w_1 + h\left(1 + \frac{w_1}{t_1}\right) = 2 + 0.5\left(1 + \frac{2}{1.5}\right) = 3.16666667.$$

Continuing in this fashion, we obtain the following results, with the value of the exact solution listed for comparison.

t	Approximate Solution	Exact Solution	$\lvert y(t_i) - w_i\rvert$
1.0	1.00000000	1.00000000	
1.5	2.00000000	2.10819766	0.108198
2.0	3.16666667	3.38629436	0.219628
2.5	4.45833333	4.79072683	0.332393
3.0	5.85000000	6.29583687	0.445837
3.5	7.32500000	7.88467039	0.559670
4.0	8.87142857	9.54517744	0.673749
4.5	10.48035714	11.26834829	0.787991
5.0	12.14484127	13.04718956	0.902348
5.5	13.85932540	14.87611451	1.016789
6.0	15.61926407	16.75055682	1.131293

Observe the slow but steady growth in the global error as t increases. Since each step introduces new error into the computed approximate solution, we might expect this type of behavior in every problem; however, the actual accumulation of global error is very problem dependent. Essentially, the error introduced by each step of the time marching process moves us from one solution of the differential equation onto a different solution. If, as in the previous example, nearby solutions separate from one another as t increases, we can expect to see a steady increase in the global error. On the other hand, if nearby solutions move closer together as t increases, we could expect to observe a steady decline in the global error. The next example demonstrates this situation.

EXAMPLE 7.6 Euler's Method in Action, Problem 2

The initial value problem

$$\frac{dx}{dt} = \frac{t}{x}, \quad 0 \le t \le 5$$
$$x(0) = 1$$

has as its exact solution $x(t) = \sqrt{t^2+1}$. Here the Euler's method difference equation takes the form

$$w_0 = 1$$
$$w_{i+1} = w_i + h\frac{t_i}{w_i}.$$

Let's once again use a step size of $h = 0.5$. In the first time step we calculate

$$w_1 = w_0 + h\frac{t_0}{w_0} = 1 + 0.5\frac{0}{1} = 1.$$

Advancing the value of the independent variable from t_0 to $t_1 = t_0 + h = 0.5$, we then calculate

$$w_2 = w_1 + h\frac{t_1}{w_1} = 1 + 0.5\frac{0.5}{1} = 1.25.$$

Continuing to time step in this fashion until we reach $t = 5$, we obtain the following results, with the value of the exact solution listed for comparison. After an initial increase in the global error, note the steady decline as t advances.

t	Approximate Solution	Exact Solution	$\lvert y(t_i) - w_i\rvert$
0.0	1.00000000	1.00000000	
0.5	1.00000000	1.11803399	0.118034
1.0	1.25000000	1.41421356	0.164214
1.5	1.65000000	1.80277564	0.152776
2.0	2.10454545	2.23606798	0.131523
2.5	2.57970744	2.69258240	0.112875
3.0	3.06425851	3.16227766	0.098019
3.5	3.55377335	3.64005494	0.086282
4.0	4.04620768	4.12310563	0.076898
4.5	4.54049768	4.60977223	0.069275
5.0	5.03603807	5.09901951	0.062981

Let's perform one more numerical experiment before examining the theoretical aspects of Euler's method. We expect that the accuracy of the approximate solution generated by Euler's method will improve if we decrease the step size h, but how much improvement will we obtain? Is the global error $O(h)$? Is it $O(h^2)$? Is it $O(h^3)$?

EXAMPLE 7.7 Numerical Investigation of the Rate of Convergence of Euler's Method

Reconsider the initial value problem

$$\frac{dx}{dt} = 1 + \frac{x}{t}, \quad 1 \le t \le 6, \quad x(1) = 1$$

whose exact solution is $x(t) = t(1 + \ln t)$. The following table displays the absolute error in the approximate solution at $t = 6$, obtained from Euler's method using successively smaller step sizes. Note each time the step size is cut by a factor of 2, the absolute error shrinks by roughly the same factor. This suggests that the global error associated with Euler's method is $O(h)$.

h	Approximate Solution	Absolute Error	Error Ratio
1/2	15.61926407	1.131293	
1/4	16.15574907	0.594808	1.90195
1/8	16.44564019	0.304917	1.95072
1/16	16.59620493	0.154352	1.97546
1/32	16.67290649	0.077650	1.98778
1/64	16.71161299	0.038944	1.99391
1/128	16.73105524	0.019502	1.99696
1/256	16.74079861	0.009758	1.99848

We arrive at the same conclusion when we reconsider the initial value problem

$$\frac{dx}{dt} = \frac{t}{x}, \quad 0 \le t \le 5, \quad x(0) = 1.$$

The table below displays the absolute error in the approximate solution at $t = 5$. Once again, each time the step size is cut by a factor of 2, the absolute error shrinks by roughly the same factor.

h	Approximate Solution	Absolute Error	Error Ratio
1/2	5.03603807	0.062981	
1/4	5.06611825	0.032901	1.91426
1/8	5.08234203	0.016677	1.97280
1/16	5.09063747	0.008382	1.98967
1/32	5.09481923	0.004200	1.99559
1/64	5.09691725	0.002102	1.99798
1/128	5.09796787	0.001052	1.99903
1/256	5.09849357	0.000526	1.99953

Analysis of Euler's Method

We will begin by examining the local truncation error for Euler's method. Recall that truncation error measures how well the continuous differential equation has been approximated by the discrete difference equation and is determined by substituting the true solution into the difference equation. Given the first derivation of Euler's method, it is clear that

$$\tau_i = \frac{y_{i+1} - y_i}{h} - f(t_i, y_i) = \frac{h}{2}y''(\xi_i).$$

Assuming that y has two continuous derivatives, we have

$$|\tau_i| \le \frac{h}{2} \max_{t \in [a,b]} |y''|,$$

which implies that $\tau_i = O(h)$.

From this last expression we can make two important observations. First, since $\tau_i = O(h)$ (i.e., the local truncation error goes like the first power of the step size), it follows that Euler's method is a first-order scheme for approximating the solution of an initial value problem. Second, since $\tau_i \to 0$ as $h \to 0$, we see that Euler's method is consistent.

Next, consider the global discretization error associated with Euler's method. Recall that this measures how well the true solution of the differential equation has been approximated. In particular, the global error at $t = t_i$ is given by $y_i - w_i$. Note that this a kind of accumulated error, since w_i depends on the previous approximations $w_{i-1}, w_{i-2}, w_{i-3}, \ldots, w_1$.

Theorem (Global Error for Euler's method). Let $y(t)$ be the unique solution to the initial value problem

$$\begin{aligned} y'(t) &= f(t, y(t)), \quad a \le t \le b \\ y(a) &= \alpha. \end{aligned}$$

Let $w_0, w_1, w_2, \ldots, w_N$ be generated by Euler's method:

$$\begin{aligned} w_0 &= \alpha \\ w_{i+1} &= w_i + hf(t_i, w_i), \end{aligned}$$

where $h = (b - a)/N$ and $t_i = a + ih$. If f satisfies a Lipschitz condition in y on $D = \{(t, y) | \, a \le t \le b, y \in R\}$ with constant L and there exists a constant M such that

$$\max_{t \in [a,b]} |y''(t)| \le M,$$

then

$$|y_i - w_i| \le \frac{hM}{2L}\left[e^{L(t_i - a)} - 1\right].$$

Remarks. (1) Note that $|y_i - w_i| = O(h)$, which confirms our earlier numerical evidence.
(2) It is clear that $|y_i - w_i| \to 0$ as $h \to 0$, so Euler's method is convergent.
(3) One weakness of this theorem is that we don't know M, the bound on the second derivative of the true solution. If $\partial f/\partial y$ and $\partial f/\partial t$ exist, then

$$y'' = \frac{\partial f}{\partial t} + \frac{\partial f}{\partial y}y' = \frac{\partial f}{\partial t} + \frac{\partial f}{\partial y}f,$$

which can be used to obtain M.

The proof of this theorem requires two lemmas. We will state and prove both lemmas now and then present the proof of the theorem.

Lemma 1. For all $x \geq -1$ and any positive m,

$$0 \leq (1+x)^m \leq e^{mx}.$$

Proof. Since m must be positive, it is sufficient to establish that $0 \leq 1+x \leq e^x$. Expanding e^x in a Taylor series about $x_0 = 0$ with $n = 1$ gives

$$e^x = 1 + x + \frac{1}{2}x^2 e^{\xi},$$

where ξ is between 0 and x. Then, for all $x \geq -1$, it follows that

$$0 \leq 1 + x \leq 1 + x + \frac{1}{2}x^2 e^{\xi} = e^x.$$

□

Lemma 2. If s and $t \in R^+$ and $\{a_i\}$ is a sequence satisfying $a_0 \geq -t/s$ and

$$a_{i+1} \leq (1+s)a_i + t,$$

then

$$a_{i+1} \leq e^{(i+1)s}\left(\frac{t}{s} + a_0\right) - \frac{t}{s}.$$

Proof. For each fixed t,

$$\begin{aligned}
a_{i+1} &\leq (1+s)a_i + t \\
&\leq (1+s)[(1+s)a_{i-1} + t] + t = (1+s)^2 a_{i-1} + t[1 + (1+s)] \\
&\leq (1+s)^2[(1+s)a_{i-2} + t] + t[1 + (1+s)] \\
&= (1+s)^3 a_{i-2} + t[1 + (1+s) + (1+s)^2] \\
&\ \ \vdots \\
&\leq (1+s)^{i+1} a_0 + [1 + (1+s) + (1+s)^2 + \cdots + (1+s)^i]t.
\end{aligned}$$

The factor multiplying t in the last line is a geometric series whose sum is $[(1+s)^{i+1}-1]/s$. Therefore,

$$a_{i+1} \le (1+s)^{i+1} + \frac{t}{s}\left[(1+s)^{i+1}-1\right]$$
$$= (1+s)^{i+1}\left(a_0 + \frac{t}{s}\right) - \frac{t}{s}.$$

Applying Lemma 1, we arrive at the required formula:

$$a_{i+1} \le e^{(i+1)s}\left(a_0 + \frac{t}{s}\right) - \frac{t}{s}.$$ □

We are now in a position to prove the theorem on the global error for Euler's method.

Proof. The error formula is clearly true for $i=0$ since $y_0 = w_0 = \alpha$. For $i = 0, 1, 2, \ldots, N-1$,

$$y(t_{i+1}) = y(t_i) + hf(t_i, y(t_i)) + \frac{h^2}{2}y''(\xi_i)$$

by Taylor's Theorem, and

$$w_{i+1} = w_i + hf(t_i, w_i)$$

by the definition of Euler's method. Upon subtracting these two equations, we have

$$y_{i+1} - w_{i+1} = y_i - w_i + h[f(t_i, y_i) - f(t_i, w_i)] + \frac{h^2}{2}y''(\xi_i)$$

and

$$|y_{i+1} - w_{i+1}| \le |y_i - w_i| + h\,|[f(t_i, y_i) - f(t_i, w_i)]| + \frac{h^2}{2}|y''(\xi_i)|.$$

Using the Lipschitz condition on f and the bound on y'' gives

$$|y_{i+1} - w_{i+1}| \le |y_i - w_i|(1+hL) + \frac{h^2}{2}M.$$

Identifying $a_i = |y_i - w_i|$, $s = hL$ and $t = h^2M/2$ in Lemma 2 gives

$$|y_{i+1} - w_{i+1}| \le e^{(i+1)hL}\left(|y_0 - w_0| + \frac{hM}{2L}\right) - \frac{hM}{2L}$$
$$= \left[e^{(i+1)hL} - 1\right]\frac{hM}{2L}.$$

But $(i+1)h = t_{i+1} - t_0 = t_{i+1} - a$, so

$$|y_{i+1} - w_{i+1}| = \frac{hM}{2L}\left[e^{(t_{i+1}-a)L} - 1\right].$$ □

EXAMPLE 7.8 Confirming the Global Error Bound for Euler's Method

Let's once again consider the initial value problem

$$\frac{dx}{dt} = 1 + \frac{x}{t}, \quad 1 \le t \le 6$$
$$x(1) = 1.$$

Since $f(t, x) = 1 + x/t$, it follows that

$$\begin{aligned} |f(t, x_1) - f(t, x_2)| &= \frac{|x_1 - x_2|}{t} \\ &\le |x_1 - x_2| \qquad \text{for } 1 \le t \le 6. \end{aligned}$$

Hence, $L = 1$. The exact solution of the initial value problem is $x(t) = t(1 + \ln t)$, so $x''(t) = 1/t$ and

$$M = \max_{1 \le t \le 6} |x''(t)| = 1.$$

With $h = 0.5$ and $a = 1$, the Global Error Theorem guarantees that

$$|y_i - w_i| \le \frac{0.5 \cdot 1}{2 \cdot 1}\left[e^{(t_i - 1)\cdot 1} - 1\right] = \frac{1}{4}\left[e^{(t_i - 1)} - 1\right].$$

The table below compares the actual error with this error bound for each t_i. Note that for this problem, the error bound is significantly larger than the actual error.

t	$\lvert y_i - w_i\rvert$	Error Bound
1.5	0.108198	0.162180
2.0	0.219628	0.429570
2.5	0.332393	0.870422
3.0	0.445837	1.597264
3.5	0.559670	2.795623
4.0	0.673749	4.771384
4.5	0.787991	8.028863
5.0	0.902348	13.399538
5.5	1.016789	22.254283
6.0	1.131293	36.853290

The global error theorem for Euler's method neglected the effect of roundoff error in the calculation of the approximate solution. To examine this effect, let

$$\begin{aligned} w_0 &= \alpha \\ w_{i+1} &= w_i + hf(t_i, w_i) \end{aligned}$$

denote the Euler's method approximation obtained using exact arithmetic, and let

$$\begin{aligned} u_0 &= \alpha + \delta_0 \\ u_{i+1} &= u_i + hf(t_i, u_i) + \delta_i \end{aligned}$$

denote the approximation obtained with finite precision arithmetic, where each δ_i represents the roundoff error introduced during the ith step of the solution process. We then have the following result:

Theorem. Let $y(t)$ be the unique solution to the initial value problem

$$y'(t) = f(t, y(t)), \quad a \le t \le b$$
$$y(a) = \alpha,$$

and let $u_0, u_1, u_2, \ldots, u_N$ be generated by Euler's method with finite precision arithmetic:

$$u_0 = \alpha + \delta_0$$
$$u_{i+1} = u_i + hf(t_i, u_i) + \delta_i,$$

where $h = (b - a)/N$, $t_i = a + ih$ and $|\delta_i| < \delta$ for each $i = 0, 1, 2, \ldots, N$. If f satisfies a Lipschitz condition in y on $D = \{(t, y) | a \le t \le b, y \in R\}$ with constant L and there exists a constant M such that

$$\max_{t \in [a,b]} |y''(t)| \le M,$$

then

$$|y_i - u_i| \le \frac{1}{L}\left(\frac{hM}{2} + \frac{\delta}{h}\right)\left[e^{L(t_i - a)} - 1\right] + \delta e^{L(t_i - a)}.$$

Proof. The proof of this theorem follows the same steps as the proof of the previous theorem with all w's replaced by u's, except that

$$|y_0 - u_0| = \delta_0,$$

and in the application of the Lemma 2

$$t = \frac{h^2 M}{2} + \delta. \qquad \square$$

Note that when the effect of roundoff error is taken into account, the global error is composed of two competing forces. On one hand, there is the local truncation error, which decreases linearly with h. On the other hand, there is the roundoff error, which is inversely proportional to the step size. Thus, when $hM/2$ is much larger than δ/h, truncation error will dominate, and we can expect to observe $O(h)$ convergence toward the exact solution. However, for sufficiently small values of the step size, roundoff error will dominate, and we can expect global error to grow with decreasing h. This situation is identical to the phenomenon we observed with numerical differentiation in Section 6-2.

The final analysis issue that we will address is stability. Suppose we apply Euler's method to the initial value problems

$$\frac{dy}{dt} = f(t, y), \quad y(a) = \alpha$$

and

$$\frac{dy}{dt} = f(t,y), \qquad y(a) = \tilde{\alpha}.$$

Let w_i denote the approximate solution associated with the initial condition $y(a) = \alpha$ and $\tilde{w}_i$ denote the approximate solution associated with the initial condition $y(a) = \tilde{\alpha}$. Provided f satisfies a Lipschitz condition in y, a procedure similar to the one used to prove the global error theorem can be used here to establish that

$$|w_i - \tilde{w}_i| \le k(t_i)|\alpha - \tilde{\alpha}|, \tag{3}$$

where $k(t_i) = e^{(t_i-a)L}$ and L is the Lipschitz constant associated with f. The details of this derivation are left as an exercise. Since $k(t_i)$ is independent of h, it follows that Euler's method is stable with respect to perturbations of the initial condition.

An Application: Modeling the Spread of an Epidemic

Consider the following simple model, proposed by Kermack and McKendrick [1], for the spread of an epidemic. The population is divided into three categories: healthy individuals, infected individuals, and the dead. It is assumed that the epidemic spreads so rapidly that changes in the population due to birth, death by other causes and migration can be ignored. It is also assumed that the disease is transmitted to healthy individuals at a rate proportional to the product of the healthy and infected populations.

Let H denote the number of healthy individuals, I the number of infected individuals, and D the number of dead as functions of time. Time, here, will be measured in weeks. The basic assumptions stated in the previous paragraph lead to the model

$$\frac{dH}{dt} = -cHI, \qquad \frac{dI}{dt} = cHI - mI, \qquad \frac{dD}{dt} = mI,$$

where c is the transmission rate of the disease to healthy individuals and m is the mortality rate of infected individuals.

The model can be reduced to a single equation as follows. Divide the H equation by the D equation to obtain

$$\frac{dH}{dD} = -\frac{c}{m}H.$$

The solution of this equation is

$$H = H_0 \exp(-cD/m), \tag{4}$$

where H_0 is the initial number of healthy individuals. Next, sum the three original model equations to obtain

$$\frac{d(H+I+D)}{dt} = 0 \quad \Rightarrow \quad H + I + D = N$$

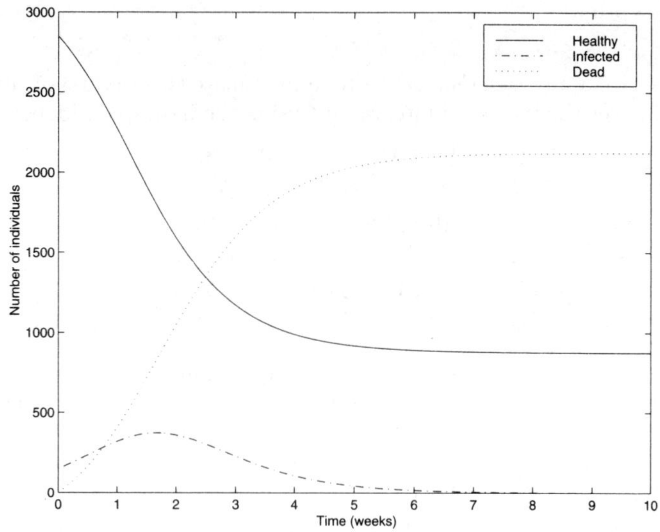

Figure 7.1 Time evolution of the number of healthy, the number of infected and the number of dead due to the spread of an epidemic.

for some (constant) total population N. Solving this last expression for I, we find

$$I = N - H - D. \tag{5}$$

Finally, substitute (4) into (5) and then substitute the resulting expression for I into the original model equation for D. This yields

$$\frac{dD}{dt} = m[N - D - H_0 \exp(-cD/m)]. \tag{6}$$

Once the number of dead has been computed from (6), H and I can be obtained from the algebraic equations (4) and (5).

Suppose that in a remote village of 3000 people, 150 are initially infected with the disease and the other 2850 people are healthy. How many people will eventually die due to the disease? How long will it take for the disease to run its course? With $m = 1.8$ week^{-1} and $c = 0.001$(person·week)$^{-1}$, the time evolution of the three categories (healthy, infected and dead) is shown in Figure 7.1. The initial value problem for D was solved using Euler's method with a step size of $h = 0.1$ weeks. Equilibrium seems to be reached in roughly eight weeks, with 2124 dead at that time.

References

1. W. O. Kermack and A. G. McKendrick, "Contributions to the Mathematical Theory of Epidemics, I.," *Proc. Roy. Soc.*, **115A**, 700, 1927.

EXERCISES

For Exercises 1–6, apply Euler's method to approximate the solution of the given initial value problem over the indicated interval in t using the indicated number of time steps.

1. $x' = tx^3 - x \quad (0 \le t \le 1), \quad x(0) = 1, \quad N = 4$

2. $x' + (4/t)x = t^4 \quad (1 \le t \le 3), \quad x(1) = 1, \quad N = 5$

3. $x' = (\sin x - e^t)/\cos x \quad (0 \le t \le 1), \quad x(0) = 0, \quad N = 3$

4. $x' = (1 + x^2)/t \quad (1 \le t \le 4), \quad x(1) = 0, \quad N = 5$

5. $x' = t^2 - 2x^2 - 1 \quad (0 \le t \le 1), \quad x(0) = 0, \quad N = 4$

6. $x' = 2(1 - x)/(t^2 \sin x) \quad (1 \le t \le 2), \quad x(1) = 2, \quad N = 3$

For Exercises 7–10, apply Euler's method to approximate the solution of the given initial value problem over the indicated interval in t using the indicated number of time steps. Compare the approximate solution with the given exact solution, and compare the actual error with the theoretical error bound. When determining the Lipschitz constant, consider the indicated set D.

7. $x' = e^t/x \quad (0 \le t \le 2), \quad x(0) = 1, \quad N = 4, \quad x(t) = \sqrt{2e^t - 1},$
$D = \{(t, x) |\ 0 \le t \le 2,\ x \ge 1\}$

8. $x' = -t \tan x/(1 + t^2) \quad (0 \le t \le 1), \quad x(0) = \pi/4, \quad N = 4,$
$x(t) = \sin^{-1}\sqrt{(2 + 2t^2)^{-1}}, \quad D = \{(t, x) |\ 0 \le t \le 1,\ 0 \le x \le \pi/4\}$

9. $x' = t - x \quad (0 \le t \le 4), \quad x(0) = 1, \quad N = 4, \quad x(t) = 2e^{-t} + t - 1,$
$D = \{(t, x) |\ 0 \le t \le 4,\ x \in R\}$

10. $x' = x - t \quad (0 \le t \le 2), \quad x(0) = 2, \quad N = 4, \quad x(t) = e^t + t + 1,$
$D = \{(t, x) |\ 0 \le t \le 4,\ x \in R\}$

Two different initial conditions are specified for the initial value problems in Exercises 11–14. Compare the approximate solutions corresponding to each initial condition, and verify that the bound specified in equation (3) holds.

11. $x' = -2tx/(1 + t^2) \quad (2 \le t \le 4), \quad N = 5, \quad x(2) = -5$ versus $x(2) = -5.1$

12. $x' = \sin t - x \quad (0 \le t \le 2), \quad N = 5, \quad x(0) = 1$ versus $x(0) = 1.1$

13. $x' = 2e^t - x \quad (0 \le t \le 5), \quad N = 5, \quad x(0) = 1$ versus $x(0) = 0.9$

14. $x' = 1 + x/t \quad (1 \le t \le 6), \quad N = 5, \quad x(1) = 1$ versus $x(1) = 1.1$

15. Derive equation (3).

16. Suppose Euler's method has been used to obtain approximate values for the solution to the initial value problem $y' = f(t, y), y(a) = \alpha$ at $t = t_1$ and $t = t_2$. Would it be appropriate to approximate the solution for t between t_1 and t_2 using linear interpolation? Explain.

17. Suppose we use Euler's method to approximate the solution of the initial value problem $y' = f(t, y), y(a) = \alpha$ over the interval $a \le t \le b$. Let $w_h(b)$ denote the approximation to $y(b)$ obtained with a step size of h. Since the global error associated with Euler's method is $O(h)$, toward what value do we expect the expression

$$\frac{w_h(b) - w_{h/2}(b)}{w_{h/2}(b) - w_{h/4}(b)}$$

to converge as h is reduced?

In Exercises 18–22, confirm that the global error associated with Euler's method is $O(h)$.

18. $x' = 4t\sqrt{x^2+1}/x \quad (0 \le t \le 5), \quad x(0) = 1, \quad x(t) = \sqrt{(2t^2+\sqrt{2})^2 - 1}$

19. $tx' - 4x = t^5 e^t \quad (1 \le t \le 2), \quad x(1) = 0, \quad x(t) = t^4(e^t - e)$

20. $x' = 1 - x + e^{2t}x^2 \quad (0 \le t \le 0.9), \quad x(0) = 0, \quad x(t) = e^{-t}\tan(e^t - 1)$

21. $x' = 2t(x+1)/x \quad (0 \le t \le 2), \quad x(0) = -2$

22. $x' = (t^2 - tx + x^2)/(tx) \quad (1 \le t \le 2), \quad x(1) = 2$

23. **(a)** Consider the initial value problem

$$\frac{dx}{dt} = 2 - \frac{x}{t} \qquad (1 \le t \le 6), \qquad x(1) = 2.$$

The exact solution of this problem is $x(t) = t + 1/t$. With what rate does Euler's method converge to this exact solution?

(b) Repeat part (a), but change the initial condition to $x(1) = 1$. The exact solution in this case is $x(t) = t$.

(c) Explain any difference in rate of convergence between parts (a) and (b).

24. **(a)** Consider the initial value problem

$$\frac{dx}{dt} = -(1+t+t^2) - (2t+1)x - x^2 \qquad (0 \le t \le 3), \qquad x(0) = -\frac{1}{2}.$$

The exact solution of this problem is $x(t) = -t - 1/(e^t+1)$. With what rate does Euler's method converge to this exact solution?

(b) Repeat part (a), but change the initial condition to $x(0) = -1$. The exact solution in this case is $x(t) = -t - 1$.

(c) Explain any difference in rate of convergence between parts (a) and (b).

25. In the "Modeling the Spread of an Epidemic" problem, vary the parameters c and m. What effect does each parameter apparently have upon the dynamics of the epidemic?

26. The liquid level in an accumulator for a pumped-fluid system satisfies the model

$$\frac{dh}{dt} + 0.002\left(52.1h + \frac{10.3}{10.3+h}\right) - 1.17(1+\sin 3t) = 0.0308$$

$$h(0) = 5.0,$$

where h denotes the liquid level (measured in meters) and t denotes time (measured in minutes). How does the liquid level evolve in time? Estimate the amplitude and period of the oscillation in the liquid level.

27. Consider the population model

$$\frac{dx}{dt} = rx\left(1 - \frac{x}{k}\right) - \frac{x^2}{1+x^2}.$$

The first term on the right-hand side is known as the logistic growth term. This term results from the assumption that for small population levels, the population will grow at a rate proportional to the current level, while, for large population levels, limited resources will cause the growth rate to decrease and eventually become negative. The parameters r and k are called the natural growth rate of the population and the environmental carrying capacity, respectively. The second term on the right-hand side represents harvesting/predation of the species by some other species (e.g., fish being caught by fishermen or insects being eaten by birds).

(a) For $r = 0.4$ and $k = 20$, use Euler's method to determine the eventual population level reached from an initial population of 2.44.

(b) Repeat part (a), but with the initial population changed to 2.40.

7.3 HIGHER-ORDER ONE-STEP METHODS: TAYLOR METHODS

Although Euler's method is both straightforward to develop and to implement, it is not very accurate. In particular, both the local truncation error, τ_i, and the global discretization error, $|y_i - w_i|$, are only $O(h)$. In this section, we will develop several higher-order one-step methods for scalar first-order initial value problems.

Taylor Methods

The most natural approach to develop a more accurate scheme than Euler's method is to follow the derivation of Euler's method, but to assume the true solution has more continuous derivatives and, accordingly, to retain more terms in the Taylor expansion. In particular, assume y has $n+1$ continuous derivatives on $[a, b]$ and expand in a Taylor series about $t = t_i$:

$$y(t) = y_i + (t - t_i)y_i' + \frac{(t - t_i)^2}{2}y_i'' + \cdots + \frac{(t - t_i)^n}{n!}y_i^{(n)} + \frac{(t - t_i)^{n+1}}{(n+1)!}y^{(n+1)}(\xi_i). \quad (1)$$

From the differential equation, we know that we can replace y_i' by $f(t_i, y_i)$. Each higher-order derivative of y that appears in (1) can be replaced by the appropriate order total derivative (d/dt) of f. For instance,

$$\begin{aligned} y_i'' &= \tfrac{d}{dt} f(t,y)\big|_{t=t_i} = \left(\tfrac{\partial f}{\partial t} + \tfrac{\partial f}{\partial y} f\right)\Big|_{t=t_i} \\ y_i''' &= \tfrac{d^2}{dt^2} f(t,y)\Big|_{t=t_i} = \left(\tfrac{\partial^2 f}{\partial t^2} + 2f\tfrac{\partial^2 f}{\partial t \partial y} + \tfrac{\partial^2 f}{\partial y^2} f^2 + \tfrac{\partial f}{\partial y}\tfrac{\partial f}{\partial t} + \left(\tfrac{\partial f}{\partial y}\right)^2 f\right)\Bigg|_{t=t_i} \end{aligned} \quad (2)$$

etc.

Evaluating the resulting expression at $t = t_{i+1}$, dropping the remainder term and replacing all y's by w's produces the one-step method

$$\frac{w_{i+1} - w_i}{h} = f(t_i, w_i) + \frac{h}{2}\frac{d}{dt}f(t,y)\Big|_{(t_i,w_i)} + \cdots + \frac{h^{n-1}}{n!}\frac{d^{n-1}}{dt^{n-1}}f(t,y)\Big|_{(t_i,w_i)}, \quad (3)$$

where $h = t_{i+1} - t_i$. Numerical methods derived in this manner are called *Taylor methods*. Note that Euler's method is the Taylor method corresponding to $n = 1$.

The local truncation error associated with (3) is clearly $O(h^n)$. Equation (3) therefore represents the general nth-order Taylor method. Below, and in the exercises, we will undertake a numerical investigation of the rate of convergence of the global error of Taylor methods. A theoretical analysis of the global error and stability of Taylor methods will be deferred until Section 7.6.

While retaining more terms in the Taylor series provides a more accurate representation of the solution, it also produces numerical methods which require more work per time step. With initial value problems, as with rootfinding problems and numerical integration, the standard measure of work is the number of function evaluations. For instance, the second-order Taylor method,

$$w_{i+1} = w_i + hf(t_i, w_i) + \frac{h^2}{2}\frac{df}{dt}(t_i, w_i),$$

requires two function evaluations per time step, while the fourth-order Taylor method,

$$w_{i+1} = w_i + hf(t_i, w_i) + \frac{h^2}{2}\frac{df}{dt}(t_i, w_i) + \frac{h^3}{6}\frac{d^2f}{dt^2}(t_i, w_i) + \frac{h^4}{24}\frac{d^3f}{dt^3}(t_i, w_i),$$

requires four function evaluations. Compare these counts with the one function evaluation per time step used by Euler's method.

This brings up a very important question. Is the extra expense of a higher-order Taylor method justified in terms of performance? If several methods of different order are applied to the same initial value problem and the same step size is used for each method, then, of course, the higher-order methods will produce more accurate approximations. But what happens if we vary the step size from method to method in such a way that each method uses the same total number of function evaluations? Will the higher-order methods still outperform the lower-order methods? Let's find out.

EXAMPLE 7.9 Euler's Method versus the Second- and Fourth-Order Taylor Methods

Consider the initial value problem

$$\frac{dx}{dt} = 1 + \frac{x}{t} \quad (1 \le t \le 6), \quad x(1) = 1,$$

whose exact solution is $x(t) = t(1 + \ln t)$. Figure 7.2 displays the base 10 logarithm of the absolute error in the approximate solutions obtained from Euler's method, the second-order Taylor method, and the fourth-order Taylor method. Step sizes of $h = 0.125$, $h = 0.25$, and $h = 0.5$ were used for the three methods, respectively. These choices allowed each method the same number of function evaluations (40) to advance from $t = 1$ to $t = 6$. Clearly, even with larger step sizes, higher-order Taylor methods outperform their lower-order counterparts.

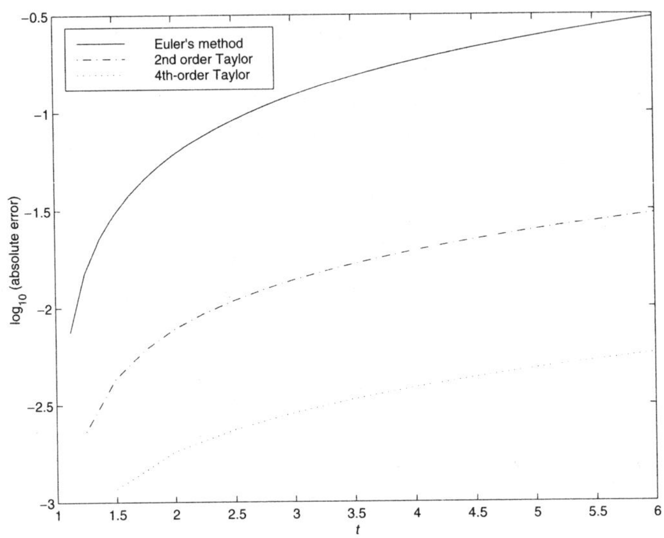

Figure 7.2 Logarithm of error in approximate solution to the initial value problem $x' = 1 + x/t$, $x(1) = 1$ computed using Euler's method (with $h = 0.125$), the second-order Taylor method (with $h = 0.25$), and the fourth-order Taylor method (with $h = 0.5$).

Now, let's take a closer look at the calculations involved in using higher-order Taylor methods. In the next two examples, pay particular attention to the derivatives of the right-hand side function f.

EXAMPLE 7.10 Second-Order Taylor Method in Action

The second-order Taylor method requires the evaluation of both the right-hand side function and its first derivative during each time step. For the initial value problem

$$\frac{dx}{dt} = 1 + x/t, \qquad (1 \le t \le 6), \qquad x(1) = 1,$$

$f(t, x) = 1 + x/t$, so, applying the quotient rule,

$$\frac{df}{dt}(t, x) = \frac{tx' - x}{t^2} = \frac{tf - x}{t^2}.$$

Take $h = 0.25$, the same value used in the previous example. With $t_0 = 1$ and $w_0 = 1$, it follows that

$$f(t_0, w_0) = 1 + \frac{w_0}{t_0} = 1 + \frac{1}{1} = 2; \quad \text{and}$$

$$\frac{df}{dt}(t_0, w_0) = \frac{t_0 f - w_0}{t_0^2} = \frac{1 \cdot 2 - 1}{1^2} = 1.$$

Therefore,

$$\begin{aligned} w_1 &= w_0 + hf(t_0, w_0) + \frac{h^2}{2}\frac{df}{dt}(t_0, w_0) \\ &= 1 + (0.25)(2) + \frac{0.25^2}{2}(1) = 1.53125. \end{aligned}$$

Hence, $y(1.25) \approx 1.53125$, an approximation that is in error by 0.00232. In the next time step, we calculate, with $t_1 = t_0 + h = 1.25$,

$$\begin{aligned} f(t_1, w_1) &= 1 + \frac{w_1}{t_1} = 1 + \frac{1.53125}{1.25} = 2.225; \quad \text{and} \\ \frac{df}{dt}(t_1, w_1) &= \frac{t_1 f - w_1}{t_1^2} = \frac{(1.25)(2.225) - 1.53125}{1.25^2} = 0.8. \end{aligned}$$

Finally,

$$\begin{aligned} w_2 &= w_1 + hf(t_1, w_1) + \frac{h^2}{2}\frac{df}{dt}(t_1, w_1) \\ &= 1.53125 + (0.25)(2.225) + \frac{0.25^2}{2}(0.8) = 2.1125. \end{aligned}$$

This approximation to $y(1.5)$ is in error by roughly 0.00430.

EXAMPLE 7.11 Fourth-Order Taylor Method in Action

The fourth-order Taylor method requires the evaluation of f and its first three derivatives each time step. For our standard test problem, $f(t, x) = 1 + x/t$, and we have already calculated

$$\frac{df}{dt}(t, x) = \frac{tf - x}{t^2}.$$

To obtain the second derivative of f, let's first rewrite the first derivative as $f/t - x/t^2$. Using the quotient rule twice yields

$$\frac{d^2 f}{dt^2}(t, x) = \frac{tf' - f}{t^2} - \frac{t^2 f - 2tx}{t^4} = \frac{f'}{t} - \frac{2f}{t^2} + \frac{2x}{t^3}.$$

The third derivative of f is then given by

$$\frac{d^3 f}{dt^3}(t, x) = \frac{tf'' - f'}{t^2} - 2\frac{t^2 f' - 2tf}{t^4} + 2\frac{t^3 f - 3t^2 x}{t^6} = \frac{f''}{t} - \frac{3f'}{t^2} + \frac{6f}{t^3} - \frac{6x}{t^4}.$$

With $h = 0.5$ and $t_0 = w_0 = 1$, we calculate

$$\begin{aligned} f(t_0, w_0) &= 2; \qquad \frac{df}{dt}(t_0, w_0) = 1; \\ \frac{d^2 f}{dt^2}(t_0, w_0) &= \frac{1}{1} - \frac{2(2)}{1} + \frac{2(1)}{1} = -1; \quad \text{and} \\ \frac{d^3 f}{dt^3}(t_0, w_0) &= \frac{-1}{1} - \frac{3(1)}{1} + \frac{6(2)}{1} - \frac{6(1)}{1} = 2. \end{aligned}$$

Therefore,

$$w_1 = 1 + (0.5)(2) + \frac{0.5^2}{2}(1) + \frac{0.5^3}{6}(-1) + \frac{0.5^4}{24}(2) = 2.109375.$$

This approximation to $y(1.5)$ is in error by only 0.00118, which is roughly one-fourth the error produced by the second-order Taylor method. Calculations for the next time step produce

$$f(t_1, w_1) = 2.40625 \qquad \frac{df}{dt}(t_1, w_1) = 0.666667$$
$$\frac{d^2 f}{dt^2}(t_1, w_1) = -0.444444 \qquad \frac{d^3 f}{dt^3}(t_1, w_1) = 0.592593$$

and $w_2 = 3.388117$.

In the previous section, we found that the global error associated with Euler's method was, like its local truncation error, $O(h)$. Is the same thing true for higher-order Taylor methods? For example, the local truncation error of the second-order Taylor method is $O(h^2)$. Is the global error for this method also $O(h^2)$? Is the global error of the fourth-order Taylor method $O(h^4)$? As noted earlier, we will undertake a numerical investigation of global error here and defer a theoretical analysis to a later section.

EXAMPLE 7.12 Rate of Convergence of Higher-Order Taylor Methods

The following table below lists the absolute error, as a function of step size, in the approximate value for $x(6)$, where $x(t) = t(1 + \ln t)$ is the solution of the initial value problem

$$\frac{dx}{dt} = 1 + \frac{x}{t} \quad (1 \le t \le 6), \quad x(1) = 1.$$

Approximate solutions were computed using both the second-order and the fourth-order Taylor methods. Observe the values in the error ratio columns. With each decrease by a factor of two in the step size, the error in the approximation obtained from the second-order Taylor method drops by roughly a factor of 4, and the error in the approximation obtained from the fourth-order Taylor method drops by roughly a factor of 16. Hence, our numerical evidence suggests that the global error for the second-order method is $O(h^2)$, while the global error for the fourth-order method is $O(h^4)$. Further evidence will be gathered in the exercises.

	Second-Order Taylor		Fourth-Order Taylor	
h	Error	Error Ratio	Error	Error Ratio
1/2	1.187073e-01		5.800613e-03	
1/4	3.019225e-02	3.9317	3.354233e-04	17.2934
1/8	7.583378e-03	3.9814	1.973828e-05	16.9935
1/16	1.898111e-03	3.9952	1.190011e-06	16.5866

(table continued on next page)

1/32	4.746703e-04	3.9988	7.294264e-08	16.3143
1/64	1.186765e-04	3.9997	4.513211e-09	16.1620
1/128	2.966968e-05	3.9999	2.806431e-10	16.0817
1/256	7.417455e-06	4.0000	1.745804e-11	16.0753

Although Taylor methods are straightforward to derive, for $n > 1$ they all suffer the practical disadvantage of requiring the user to compute and supply the appropriate number of derivatives of the function f. As illustrated by equation (2) and the worked examples involving the second- and fourth-order methods, these computations become very cumbersome, very quickly. For this reason one would prefer to develop higher-order methods which involve evaluations of f only. We will derive such methods in the next section.

An Application: Radiative Heat Transfer to a Thin Metal Plate

This problem was adapted from Cutlip and Shacham [1]. A metal plate is to be heat treated by placing it into a high-temperature furnace. The plate will be suspended inside the furnace so that both sides may be rapidly heated by radiation from the furnace walls. We will assume that the interior of the furnace and the surfaces of the plate radiate as black bodies. We will also assume that the plate is sufficiently thin and the metal of sufficiently high thermal conductivity so that temperature variations across the thickness of the plate can be ignored.

To model the time variation of the temperature of the plate, we perform a basic energy balance. This balance will consist of two components: the rate of change of the thermal energy stored in the plate and the rate at which radiation energy is absorbed. The total thermal energy stored within the plate, E_s, is given by

$$E_s = \rho V c_p T,$$

where ρ is the mass density, V the volume, c_p the specific heat, and T the temperature of the plate. Treating the density and volume as constant, but taking into account the temperature dependence of the specific heat, it follows that

$$\begin{aligned}\frac{dE_s}{dt} &= \rho V \frac{d}{dt}(c_p T) \\ &= \rho V \left(c_p \frac{dT}{dt} + T \frac{dc_p}{dT}\frac{dT}{dt} \right) \\ &= \rho V \left(c_p + T\frac{dc_p}{dT} \right) \frac{dT}{dt}.\end{aligned}$$

The rate at which radiation energy is absorbed by the plate, q_{rad}, is governed by the Stefan-Boltzmann law

$$q_{\text{rad}} = \sigma A \left(T_F^4 - T^4 \right),$$

where $\sigma = 5.676 \times 10^{-8}$ W/m$^2\cdot$ K^4 is the Stefan-Boltzmann constant, A is the surface area of both sides of the plate, and T_F is the temperature of the furnace.

Equating dE_s/dt with q_{rad}, solving for dT/dt, and using the fact that $A/V = 2/d$, where d is the thickness of the plate, yields

$$\frac{dT}{dt} = \frac{2\sigma}{\rho d} \frac{T_F^4 - T^4}{c_p + T\frac{dc_p}{dT}}. \tag{4}$$

Suppose a copper plate of thickness $d = 0.002$ m, mass density $\rho = 8933$ kg/m^3, and initial temperature $T(0) = 300$ K is suspended in a furnace of temperature $T_F = 1200$ K. Standard thermodynamic tables (see, for example, Incropera and DeWitt [2]) list the following values for the specific heat of copper:

Temperature (K)	300	400	600	800	1000	1200
Specific Heat (J/kg · K)	385	397	417	433	451	480

We will model this data using the least-squares regression line

$$c_p(T) = 355.2 + 0.1004T.$$

Substituting the given values into (4) leads to

$$\frac{dT}{dt} = 6.354 \times 10^{-9} \frac{T_F^4 - T^4}{355.2 + 0.2008T},$$

together with the initial condition $T(0) = 300$. To improve the scaling of the problem, introduce the nondimensional variable $\theta = T/T_F$. The initial value problem for θ becomes

$$\frac{d\theta}{dt} = 10.980 \frac{1 - \theta^4}{355.2 + 240.96\theta}, \qquad \theta(0) = 0.25.$$

Let's use the second-order Taylor method to approximate the solution of this initial value problem. For this problem

$$f(t, \theta) = 10.980 \frac{1 - \theta^4}{355.2 + 240.96\theta},$$

from which we calculate

$$\frac{df}{dt} = -10.980 \frac{(355.2 + 240.96\theta)(4\theta^3 f) + 240.96(1 - \theta^4)f}{(355.2 + 240.96\theta)^2}.$$

With a step size of $h = 0.25$ seconds, time was advanced from $t = 0$ to $t = 100$ seconds, and the results are displayed in Figure 7.3. Note that it takes roughly 90 seconds for the temperature of the plate to reach equilibrium.

References

1. M. B. Cutlip and M. Shacham, *Problem Solving in Chemical Engineering with Numerical Methods*, Prentice Hall PTR, Upper Saddle River, NJ, 1999.
2. F. P. Incropera and D. P. DeWitt, *Fundamentals of Heat and Mass Transfer*, 2nd edition, John Wiley and Sons, New York, 1985.
3. J. D. Murray, *Mathematical Biology*, Springer-Verlag, Berlin, 1993.

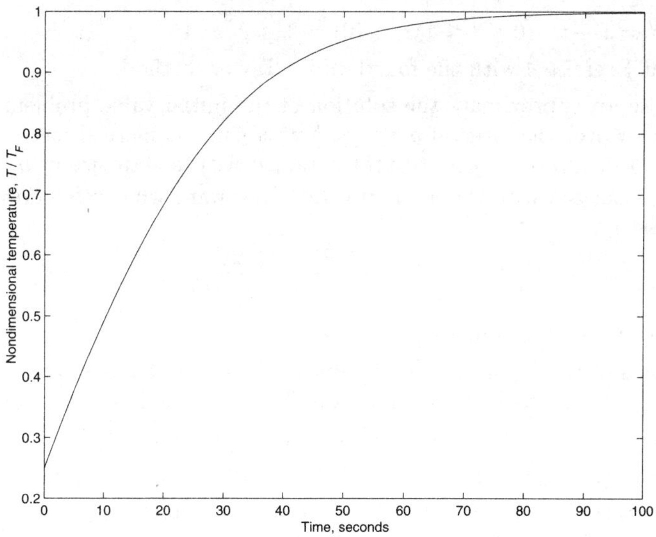

Figure 7.3 Temperature of thin copper plate suspended in a high-temperature furnace.

EXERCISES

1. For each of the following differential equations, identify the function $f(t,x)$ and calculate df/dt.
   (a) $x' = e^t/x$ **(b)** $x' + tx = tx^2$
   (c) $x' = te^{x+t} - 1$ **(d)** $x' = e^{2t} + (1 + \frac{5}{2}e^t)x + x^2$
2. For each of the following differential equations, identify the function $f(t,x)$ and calculate df/dt, d^2f/dt^2 and d^3f/dt^3.
   (a) $x' + 2x^2 = t^2 - 1$ **(b)** $x' = \sin t - x$
   (c) $x' + \frac{4x}{t} = t^4$ **(d)** $x' = x - t$
3. Apply the second-order Taylor method to approximate the solution of the given initial value problem over the indicated interval in t using the indicated number of time steps.
   (a) $x' = e^t/x \quad (0 \leq t \leq 1), \quad x(0) = 1, \quad N = 4$
   (b) $x' + tx = tx^2 \quad (0 \leq t \leq 2), \quad x(0) = 1/2, \quad N = 4$
   (c) $x' = te^{x+t} - 1 \quad (0 \leq t \leq 2), \quad x(0) = -1, \quad N = 3$
   (d) $x' = e^{2t} + (1 + \frac{5}{2}e^t)x + x^2 \quad (0 \leq t \leq 1), \quad x(0) = -1, \quad N = 2$
4. Apply the third-order Taylor method to approximate the solution of the given initial value problem over the indicated interval in t using the indicated number of time steps.
   (a) $x' + 2x^2 = t^2 - 1 \quad (0 \leq t \leq 1), \quad x(0) = 0, \quad N = 2$
   (b) $x' = \sin t - x \quad (\pi \leq t \leq 2\pi), \quad x(\pi) = 1, \quad N = 2$
   (c) $x' + \frac{4x}{t} = t^4 \quad (1 \leq t \leq 2), \quad x(1) = 1, \quad N = 4$

(d) $x' = x - t \quad (0 \le t \le 1), \quad x(0) = 2, \quad N = 4$

5. Repeat Exercise 4 with the fourth-order Taylor method.

6. Suppose we approximate the solution of the initial value problem $y' = f(t, y)$, $y(a) = \alpha$ over the interval $a \le t \le b$ with some numerical method. Let $w_h(b)$ denote the approximation to $y(b)$ obtained with a step size of h. If the global error associated with the method is $O(h^k)$, toward what value do we expect the expression

$$\frac{w_h(b) - w_{h/2}(b)}{w_{h/2}(b) - w_{h/4}(b)}$$

to converge as h is reduced?

7. Use each of the following initial value problems to demonstrate that the global error associated with the second-order Taylor method is $O(h^2)$.
   (a) $x' = e^t/x \quad (0 \le t \le 1), \quad x(0) = 1, \quad x(t) = \sqrt{2e^t - 1}$
   (b) $x' + tx = tx^2 \quad (0 \le t \le 2), \quad x(0) = 1/2, \quad x(t) = (1 + e^{t^2/2})^{-1}$
   (c) $x' = te^{x+t} - 1 \quad (0 \le t \le 2), \quad x(0) = -1, \quad x(t) = -t - \ln(e - t^2/2)$
   (d) $x' = e^{2t} + (1 + \frac{5}{2}e^t)x + x^2 \quad (0 \le t \le 1), \quad x(0) = -1$

8. Use each of the following initial value problems to demonstrate that the global error associated with the third-order Taylor method is $O(h^3)$.
   (a) $x' + 2x^2 = t^2 - 1 \quad (0 \le t \le 1), \quad x(0) = 0$
   (b) $x' = \sin t - x \quad (\pi \le t \le 2\pi), \quad x(\pi) = 1, \quad x(t) = \frac{1}{2}e^{\pi - t} + \frac{1}{2}\sin t - \frac{1}{2}\cos t$
   (c) $x' + \frac{4x}{t} = t^4 \quad (1 \le t \le 2), \quad x(1) = 1, \quad x(t) = \frac{1}{9}t^5 + \frac{8}{9}t^{-4}$
   (d) $x' = x - t \quad (0 \le t \le 1), \quad x(0) = 2, \quad x(t) = e^t + t + 1$

9. Repeat Exercise 8 with the fourth-order Taylor method.

10. Compare the performance of Euler's method, the second-order Taylor method and the fourth-order Taylor method on the initial value problem

$$\frac{dx}{dt} = \frac{t}{x}, \quad (0 \le t \le 5), \quad x(0) = 1.$$

The exact solution for this problem is $x(t) = \sqrt{t^2 + 1}$.

11. Repeat Exercise 10 using the initial value problem

$$\frac{dx}{dt} = \frac{tx^2 - x}{t} \quad (1 \le t \le 5), \quad x(1) = -\frac{1}{\ln 2},$$

whose exact solution is $x(t) = -1/(t \ln(2t))$.

12. (a) Consider the initial value problem

$$\frac{dx}{dt} = 3t - \frac{x}{t} \quad (1 \le t \le 6), \quad x(1) = 2.$$

The exact solution of this problem is $x(t) = t^2 + 1/t$. With what rate does the second-order Taylor method converge to this exact solution?

(b) Repeat part (a), but change the initial condition to $x(1) = 1$. The exact solution in this case is $x(t) = t^2$.

(c) Explain any difference in rate of convergence between parts (a) and (b).

13. Repeat Exercise 12 using the fourth-order Taylor method.

14. (a) Consider the initial value problem

$$\frac{dx}{dt} = -(1+t+t^2) - (2t+1)x - x^2 \quad (0 \le t \le 3), \quad x(0) = -\frac{1}{2}.$$

The exact solution of this problem is $x(t) = -t - 1/(e^t + 1)$. With what rate does the second-order Taylor method converge to this exact solution?

(b) Repeat part (a), but change the initial condition to $x(0) = -1$. The exact solution in this case is $x(t) = -t - 1$.

(c) Explain any difference in rate of convergence between parts (a) and (b).

15. Repeat Exercise 14 using the fourth-order Taylor method.

16. (a) Consider the initial value problem

$$\frac{dx}{dt} = 7t^2 - \frac{4x}{t} \quad (1 \le t \le 6), \quad x(1) = 2.$$

The exact solution of this problem is $x(t) = t^3 + 1/t^4$. With what rate does the fourth-order Taylor method converge to this exact solution?

(b) Repeat part (a), but change the initial condition to $x(1) = 1$. The exact solution in this case is $x(t) = t^3$.

(c) Explain any difference in rate of convergence between parts (a) and (b).

17. Suppose a plate of AISI 304 stainless steel has been suspended in a furnace of temperature $T_F = 1500$ K. The plate has a thickness of $d = 0.002$ m, a mass density of $\rho = 7900$ kg/m^3 and an initial temperature of $T(0) = 300$ K. The specific heat data for AISI 304 is provided below. Determine how long it takes for the temperature of the plate to come into equilibrium with the furnace.

Temperature (K)	300	400	600	800	1000	1200	1500
Specific Heat (J/kg · K)	477	515	557	582	611	640	682

18. Suppose a plate of tungsten has been suspended in a furnace of temperature $T_F = 1500$ K. The plate has a thickness of $d = 0.002$ m, a mass density of $\rho = 19300$ kg/m^3 and an initial temperature of $T(0) = 300$ K. The specific heat data for tungsten is provided below. Determine how long it takes for the temperature of the plate to come into equilibrium with the furnace.

Temperature (K)	300	400	600	800	1000	1200	1500
Specific Heat (J/kg · K)	132	137	142	145	148	152	157

19. Recall the "Modeling the Spread of an Epidemic" problem of Section 7.2. Murray [3] reports the following parameter values for a flu epidemic that struck a boys boarding school in the early part of 1978: $N = 763$, $H_0 = 762$, $D(0) = 0$, $c = 0.00218$, and $m = 0.44036$. Simulate 15 days of the spread of this epidemic, and produce a plot like Figure 7.1. In this situation, D denotes the number of boys confined to bed as a result of the flu.

7.4 RUNGE-KUTTA METHODS

As indicated in the previous section, the biggest disadvantage associated with Taylor methods of order $n > 1$ is the need to compute derivatives of the right-hand side function f. In this section, we will develop a class of higher-order one-step methods that use values of f exclusively. These techniques, collectively, are called *Runge-Kutta methods.*

The fundamental idea behind the development of Runge-Kutta methods is to approximate, to appropriate order, the right-hand side of the Taylor method. In formulating this approximation, only evaluations of f are to be used. Derivatives of f are not to be included. For example, to develop a third-order Runge-Kutta method, we would attempt to determine an $O(h^3)$ approximation to the right-hand side of the third-order Taylor method.

Second-Order Runge-Kutta Methods

We will develop second-order Runge-Kutta methods here to illustrate the process. Consider the explicit one-step method

$$\frac{w_{i+1} - w_i}{h} = \phi(f, t_i, w_i, h) \tag{1}$$

with

$$\phi(f, t, y, h) = a_1 f(t, y) + a_2 f(t + \alpha_2, y + \delta_2 f(t, y)). \tag{2}$$

Below, our objective will be to determine values for the parameters a_1, a_2, α_2, and δ_2 so that $\phi(f, t, y, h)$ provides an $O(h^2)$ approximation to the right-hand side of the $O(h^2)$ Taylor method.

In practice, the numerical method given by equations (1) and (2) would be implemented in two stages. In the first stage, we calculate

$$\tilde{w} = w_i + \delta_2 f(t_i, w_i),$$

which is just Euler's method with $h = \delta_2$. Therefore, $\tilde{w} \approx y(t_i + \delta_2)$. In the second stage, we determine w_{i+1} from the equation

$$\frac{w_{i+1} - w_i}{h} = a_1 f(t_i, w_i) + a_2 f(t_i + \alpha_2, \tilde{w}).$$

To avoid an unnecessary function evaluation, the value of $f(t_i, w_i)$ from the first stage should be saved and reused in the second stage.

Since $\tilde{w} \approx y(t_i + \delta_2)$, the last term in the equation for w_{i+1} suggests that we should select $\alpha_2 = \delta_2$. Furthermore, since the linear combination of f values in the w_{i+1} equation has the appearance of a weighted average, we should also expect to find that $a_1 + a_2 = 1$. To confirm our intuition (and to determine what other relationships must hold between the parameters), let's continue with our analysis.

To proceed with our development, we will need the following simplified version of Taylor's theorem in two variables. For a general version of the theorem and a proof, see, for example, Douglass [1].

Theorem. Let $f(t,y)$ and all of its first and second partial derivatives be continuous on $D = \{(t,y) | a \le t \le b, c \le y \le d\}$ and let $(t,y) \in D$. For any Δt and Δy such that $(t+\Delta t, y+\Delta y) \in D$, there exists ξ between t and $t+\Delta t$ and η between y and $y+\Delta y$ with

$$
\begin{aligned}
f(t+\Delta t, y+\Delta y) = f(t,y) &+ \left[\Delta t \frac{\partial f}{\partial t}(t,y) + \Delta y \frac{\partial f}{\partial y}(t,y)\right] \\
&+ \left[\frac{(\Delta t)^2}{2}\frac{\partial^2 f}{\partial t^2}(\xi,\eta) + \Delta t \Delta y \frac{\partial^2 f}{\partial t\, \partial y}(\xi,\eta) + \frac{(\Delta y)^2}{2}\frac{\partial^2 f}{\partial y^2}(\xi,\eta)\right].
\end{aligned}
$$

The right-hand side of the second-order Taylor method is

$$
f(t,y) + \frac{h}{2} f'(t,y) = f(t,y) + \frac{h}{2}\left[\frac{\partial f}{\partial t} + \frac{\partial f}{\partial y} f\right]
$$

and, using Taylor's theorem in two variables, the right-hand side for the second-order Runge-Kutta method can be written as

$$
\begin{aligned}
\phi(f,t,y,h) &= a_1 f(t,y) + a_2 f(t+\alpha_2, y+\delta_2 f(t,y)) \\
&= a_1 f(t,y) + a_2\left[f(t,y) + \alpha_2 \frac{\partial f}{\partial t} + \delta_2 f(t,y)\frac{\partial f}{\partial y} + R_1\right] \\
&= (a_1+a_2) f(t,y) + a_2\alpha_2 \frac{\partial f}{\partial t} + a_2\delta_2 f(t,y)\frac{\partial f}{\partial y} + a_2 R_1,
\end{aligned}
$$

where

$$
R_1 = \alpha_2^2 \frac{\partial^2 f}{\partial t^2}(\xi,\eta) + \alpha_2\delta_2 f(t,y)\frac{\partial^2 f}{\partial t\, \partial y}(\xi,\eta) + \delta_2^2 f^2(t,y)\frac{\partial^2 f}{\partial y^2}(\xi,\eta).
$$

Notice that every term in R_1 involves a second derivative of f. Equating the coefficients of like terms between the Taylor method and the Runge-Kutta method provides the equations

$$
\begin{gathered}
a_1 + a_2 = 1 \\
\left.\begin{aligned} a_2\alpha_2 &= \tfrac{h}{2} \\ a_2\delta_2 &= \tfrac{h}{2} \end{aligned}\right\} \Rightarrow \alpha_2 = \delta_2 = \frac{h}{2a_2}.
\end{gathered}
$$

There are therefore an infinite number of second-order Runge-Kutta methods of the type considered. The most common second-order Runge-Kutta methods are as follows:

Modified Euler method ($a_1 = 0$, $a_2 = 1$, $\alpha_2 = \delta_2 = h/2$)

$\tilde{w} = w_i + \frac{h}{2} f(t_i, w_i)$	Euler's method with step $h/2$
$w_{i+1} = w_i + hf\left(t_i + \frac{h}{2}, \tilde{w}\right)$	Midpoint integration

Heun method ($a_1 = a_2 = 1/2$, $\alpha_2 = \delta_2 = h$)

$$\begin{aligned} \tilde{w} &= w_i + hf(t_i, w_i) && \text{Euler's method with step } h \\ w_{i+1} &= w_i + \tfrac{h}{2}\left[f(t_i, w_i) + f(t_i + h, \tilde{w})\right] && \text{Trapezoidal integration} \end{aligned}$$

Optimal RK2 method ($a_1 = 1/4$, $a_2 = 3/4$, $\alpha_2 = \delta_2 = 2h/3$)

$$\begin{aligned} \tilde{w} &= w_i + \tfrac{2h}{3} f(t_i, w_i) \\ w_{i+1} &= w_i + \tfrac{h}{4} f(t_i, w_i) + \tfrac{3h}{4} f\left(t_i + \tfrac{2h}{3}, \tilde{w}\right) \end{aligned}$$

The last of these methods is *optimal* in the sense that the choice $a_2 = 3/4$ minimizes the numerical coefficient on the truncation error (see, for example, Ralston [2]).

EXAMPLE 7.13 A Second-Order Runge-Kutta Method in Action

All second-order Runge-Kutta methods operate in fundamentally the same two-stage fashion, so here we will demonstrate only the Heun method. Consider the initial value problem

$$\frac{dx}{dt} = 1 + \frac{x}{t} \quad (1 \le t \le 6), \quad x(1) = 1,$$

and let the step size be $h = 0.5$.

For the first time step, we have $t_0 = w_0 = 1$. Using these values in the first stage of the calculation, we find

$$f(t_0, w_0) = 1 + \frac{1}{1} = 2 \ \Rightarrow \ \tilde{w} = 1 + (0.5)(2) = 2.$$

Now, in the second stage, we obtain

$$f(t_0 + h, \tilde{w}) = 1 + \frac{2}{1.5} = 2.333333$$

and

$$w_1 = 1 + \frac{0.5}{2}\left[2 + 2.333333\right] = 2.083333.$$

To perform the next time step, first set $t_1 = t_0 + h = 1.5$. The first stage of the time step then yields

$$f(t_1, w_1) = 1 + \frac{2.083333}{1.5} = 2.388889 \quad \text{and}$$
$$\tilde{w} = 2.083333 + (0.5)(2.388889) = 3.277778.$$

In the second stage of the time step, we calculate

$$f(t_1 + h, \tilde{w}) = 1 + \frac{3.277778}{2} = 2.638889$$

and

$$w_2 = 2.083333 + \frac{0.5}{2}[2.388889 + 2.638889] = 3.340278.$$

Continuing in this fashion, we obtain the following results, with the value of the exact solution listed for comparison.

t_i	w_i	$x(t_i)$	$\lvert x(t_i) - w_i\rvert$
1.0	1.000000	1.000000	
1.5	2.083333	2.108198	0.024864
2.0	3.340278	3.386294	0.046017
2.5	4.725347	4.790727	0.065380
3.0	6.212083	6.295837	0.083754
3.5	7.783145	7.884670	0.101526
4.0	9.426273	9.545177	0.118905
4.5	11.132335	11.268348	0.136014
5.0	12.894261	13.047190	0.152929
5.5	14.706414	14.876115	0.169701
6.0	16.564194	16.750557	0.186363

We know that, by construction, the local truncation error associated with a second-order Runge-Kutta method is $O(h^2)$. What about the global error? It turns out that, like the second-order Taylor method, the global error associated with second-order Runge-Kutta methods is also $O(h^2)$. The theoretical justification for this statement will be provided in Section 7.6. Numerical investigations of the global error for second-order Runge-Kutta methods will be carried out in the exercises.

There is one more issue we wish to address relating to second-order Runge-Kutta methods. How do these methods compare to one another in terms of performance? Further, how do these methods compare to the Taylor method from which they were derived? Figure 7.4 displays the logarithm of the absolute error in the approximate solutions obtained from the modified Euler method, the Heun method, the optimal RK2 method, and the second-order Taylor method. The top graph corresponds to the initial value problem

$$\frac{dx}{dt} = 1 + \frac{x}{t} \quad (1 \le t \le 6), \quad x(1) = 1,$$

and the bottom graph corresponds to the initial value problem

$$\frac{dx}{dt} = \frac{t}{x} \quad (0 \le t \le 5), \quad x(0) = 1.$$

For both initial value problems and for all numerical methods, a step size of $h = 0.05$ was used.

Observe that for the first initial value problem, there is not much more than half an order of magnitude difference between the most accurate solution (obtained from the modified Euler method) and the least accurate solution (the Heun

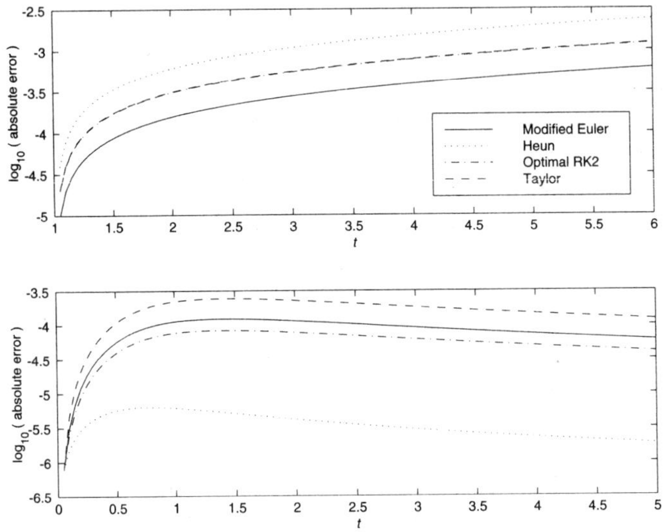

Figure 7.4 Comparison between second-order Runge-Kutta methods and the second-order Taylor method. The top graph corresponds to the initial value problem $x' = 1 + x/t, \quad x(1) = 1$, while the bottom graph corresponds to $x' = t/x$, $x(0) = 1$. A step size of $h = 0.05$ was used for all methods. In the top graph, the errors generated by the Taylor method and the optimal RK2 method are indistinguishable.

method). The errors generated by the Taylor method and the optimal RK2 method are indistinguishable to the resolution of the graph. For the second initial value problem, the Taylor method now produces the least accurate solution, followed by the modified Euler method, the optimal RK2 method, and the Heun method. The errors generated by the first three methods are again within roughly half an order of magnitude of one another; however, the errors generated by the Heun method are more than an order of magnitude smaller than any of the other errors.

So what's the moral of this story? Generally, any of the second-order methods we have considered will produce results that are roughly as accurate as any of the other second-order methods. For a particular problem, one technique may significantly outperform its counterparts, as was the case with the Heun method on the second initial value problem considered above, but there is no theory to indicate when this is going to happen. There is also no theory to indicate which second-order method will produce the most accurate approximation for a given problem.

Classical Fourth-Order Runge-Kutta Method

By far, the most common Runge-Kutta method is the classical fourth-order scheme. (Like the second-order methods we derived earlier in this section, there is more than one fourth-order Runge-Kutta method.) The classical fourth-order scheme updates the approximate solution at each time step according to the formula

$$w_{i+1} = w_i + \frac{1}{6}(k_1 + 2k_2 + 2k_3 + k_4),$$

where

$$\begin{aligned} k_1 &= hf(t_i, w_i) \\ k_2 &= hf\left(t_i + \frac{h}{2}, w_i + \frac{k_1}{2}\right) \\ k_3 &= hf\left(t_i + \frac{h}{2}, w_i + \frac{k_2}{2}\right) \\ k_4 &= hf(t_i + h, w_i + k_3). \end{aligned}$$

If f is a function of t only, this scheme reduces to Simpson's rule of integration. Note that this method requires four function evaluations per time step. Experimental verification of the fourth-order accuracy of this method is left as an exercise.

EXAMPLE 7.14 Fourth-Order Runge Kutta Method in Action

Let's approximate the solution of the initial value problem

$$\frac{dx}{dt} = 1 + \frac{x}{t} \quad (1 \le t \le 6), \quad x(1) = 1$$

using the classical fourth-order Runge-Kutta method with a step size of $h = 1$. For the first time step, with $t_0 = w_0 = 1$, we calculate k_1, k_2, k_3, and k_4 as follows:

$$\begin{aligned} k_1 &= hf(t_0, w_0) = f(1, 1) = 2 \\ k_2 &= hf(t_0 + h/2, w_0 + k_1/2) = f(1.5, 2) = 2.333333 \\ k_3 &= hf(t_0 + h/2, w_0 + k_2/2) = f(1.5, 2.166667) = 2.444444 \\ k_4 &= hf(t_0 + h, w_0 + k_3) = f(2, 3.444444) = 2.722222. \end{aligned}$$

From here, we find that

$$w_1 = 1 + \frac{1}{6}[2 + 2(2.333333) + 2(2.444444) + 2.722222] = 3.379630.$$

Continuing in this fashion, we obtain the results listed below, with the value of the exact solution included for comparison.

t_i	w_i	$x(t_i)$	$\lvert x(t_i) - w_i \rvert$
1.0	1.000000	1.000000	
2.0	3.379630	3.386294	0.006665
3.0	6.285000	6.295837	0.010837
4.0	9.530510	9.545177	0.014667
5.0	13.028776	13.047190	0.018414
6.0	16.728424	16.750557	0.022133

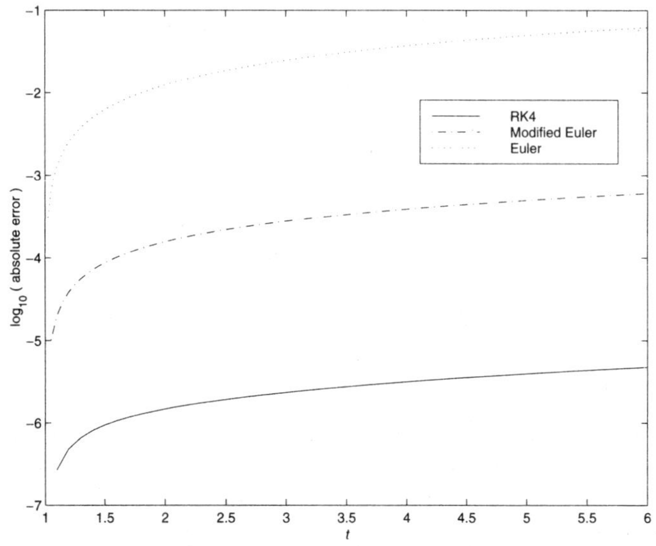

Figure 7.5 Logarithm of error in approximate solution to the initial value problem $x' = 1 + x/t$, $x(1) = 1$ computed using the classical fourth-order Runge-Kutta method (RK4, with $h = 0.1$), the modified Euler method (with $h = 0.05$), and Euler's method (with $h = 0.025$).

Let's perform a comparison between the various methods which have been introduced thus far. In particular, let's apply Euler's method, the modified Euler method, and the classical fourth-order Runge-Kutta method (which we subsequently refer to as RK4) to the standard test problem

$$\frac{dx}{dt} = 1 + \frac{x}{t} \quad (1 \le t \le 6), \qquad x(1) = 1.$$

Step sizes of $h = 0.1$, $h = 0.05$, and $h = 0.025$ were used for the RK4 method, the modified Euler method and Euler's method, respectively, so as to allow each method exactly four function evaluations for each advancement of the solution by $\Delta t = 0.1$. The logarithm of the absolute error in each computed approximation is shown in Figure 7.5. Even with the use of a smaller step size, the errors generated by Euler's method are about two orders of magnitude larger than the errors generated by the modified Euler method, which in turn are about two orders of magnitude larger than the errors generated by the RK4 method. A comparison between the fourth-order Taylor method and the RK4 method is left as an exercise.

General Explicit Runge-Kutta Methods

For completeness, it is worth noting that the Runge-Kutta methods we have discussed are special instances of the general s-stage explicit Runge-Kutta method.

This general scheme can be expressed in the form

$$k_j = hf\left(t_i + c_j h, w_i + \sum_{l=1}^{j-1} a_{j,l} k_l\right) \qquad j = 1, 2, 3, \ldots, s$$

$$w_{i+1} = w_i + \sum_{j=1}^{s} b_j k_j,$$

where $c_1 = 0$. The coefficients $a_{j,l}$ are collectively referred to as the *RK matrix*. The b_j are called the *RK weights*, and the c_j are called the *RK nodes*. The RK matrix, weights, and nodes are often displayed graphically in the *RK tableau*:

$$\begin{array}{c|ccccc}
c_1 & & & & & \\
c_2 & a_{2,1} & & & & \\
c_3 & a_{3,1} & a_{3,2} & & & \\
\vdots & \vdots & & \ddots & & \\
c_s & a_{s,1} & a_{s,2} & \cdots & a_{s,s-1} & \\
\hline
 & b_1 & b_2 & \cdots & b_{s-1} & b_s
\end{array}$$

The tableaus for the methods presented in this section are

$$\begin{array}{c|cc}
0 & & \\
\frac{1}{2} & \frac{1}{2} & \\
\hline
 & 0 & 1
\end{array}
\qquad
\begin{array}{c|cc}
0 & & \\
1 & 1 & \\
\hline
 & \frac{1}{2} & \frac{1}{2}
\end{array}
\qquad
\begin{array}{c|cc}
0 & & \\
\frac{2}{3} & \frac{2}{3} & \\
\hline
 & \frac{1}{4} & \frac{3}{4}
\end{array}$$

Modified Euler Heun Optimal RK2

$$\begin{array}{c|cccc}
0 & & & & \\
\frac{1}{2} & \frac{1}{2} & & & \\
\frac{1}{2} & 0 & \frac{1}{2} & & \\
1 & 0 & 0 & 1 & \\
\hline
 & \frac{1}{6} & \frac{1}{3} & \frac{1}{3} & \frac{1}{6}
\end{array}$$

Classical RK4

For a detailed development of general Runge-Kutta methods requiring a modest theoretical background, consult Lambert [3]. More advanced treatments can be found in Butcher [4] and Hairer, Norsett, and Wanner [5].

An Application: A Model for a Genetic Switch

A genetic switch is a biochemical mechanism that governs whether a particular protein product of a cell (e.g., a pigment) is synthesized or not. The following initial value problem has been proposed as a model for a genetic switch:

$$\frac{dg}{dt} = s - 1.51g + 3.03\frac{g^2}{1+g^2}, \qquad g(0) = 0.$$

The variable g denotes the concentration of the protein product, and the parameter s denotes the concentration of the chemical that activates the gene to produce the protein. The second term on the right-hand side of the differential equation is based on the assumption that, once produced, the protein naturally decays at a rate proportional to its concentration. The final term in the differential equation models a positive feedback effect that the protein exerts on its formation.

To be considered a reasonably accurate model for a genetic switch, this initial value problem must capture two important phenomena. The first is known as the threshold effect. This means that there must exist a critical, or threshold, value of the parameter s such that for values of s below the threshold, the equilibrium concentration of the gene remains near zero, but for values of s above the threshold, the equilibrium gene concentration jumps to a higher level. In other words, there must be a jump discontinuity in the equilibrium value of g as a function of the parameter. When in the higher equilibrium level, the gene is considered to be "on." The second phenomenon that the initial value problem must capture is a hysteresis effect. Once the gene concentration has reached the "on" state, if the parameter value is set to zero, the gene concentration should approach some fixed, nonzero level—the gene stays "on." Here, we will examine the threshold effect. An examination of the hysteresis effect will be left as an exercise.

To determine whether the given model exhibits the threshold effect, we need to establish that the equilibrium gene concentration jumps from a low value to a high value as the parameter s is varied. Figure 7.6 displays the approximate solution to the model initial value problem for four different values of s: $s = 0.1$, $s = 0.2$, $s = 0.3$, and $s = 0.4$. The classical fourth-order Runge-Kutta method was used with a step size of $h = 0.2$. There is a definite jump in equilibrium gene concentration as the parameter is changed from $s = 0.2$ to $s = 0.3$. To be certain that the equilibrium concentration doesn't vary continuously from the lower level to the higher level, we examine other parameter values between 0.2 and 0.3. The approximate solutions for $s = 0.2$, $s = 0.202$, $s = 0.204$, and $s = 0.206$ are shown in Figure 7.7. Again, RK4 with a step size of 0.2 was used. There is no doubt that this initial value problem exhibits the threshold effect. Furthermore, the critical value for s is somewhere between 0.202 and 0.204.

References

1. S. A. Douglass, *Introduction to Mathematical Analysis*, Addison-Wesley, Massachusetts, 1996.

2. A. Ralston, *A First Course in Numerical Analysis*, McGraw-Hill, New York, 1965.

3. J. D. Lambert, *Numerical Methods for Ordinary Differential Systems*, Wiley, London, 1991.

4. J. C. Butcher, *The Numerical Analysis of Ordinary Differential Equations*, John Wiley & Sons, New York, 1987.

5. E. Hairer, S. P. Norsett, and G. Wanner, *Solving Ordinary Differential Equations I: Nonstiff Problems*, Springer-Verlag, Berlin, 1991.

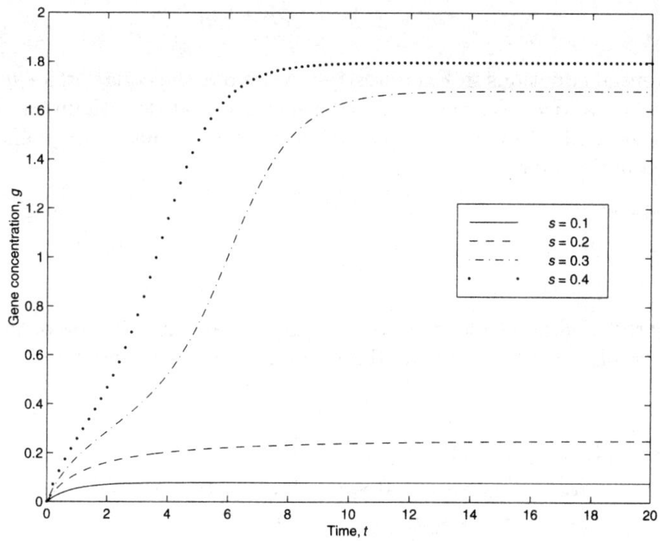

Figure 7.6 Approximate solutions to genetic switch model initial value problem for different values of the control parameter s. Note the jump in the equilibrium value of the gene concentration as s changes from 0.2 to 0.3.

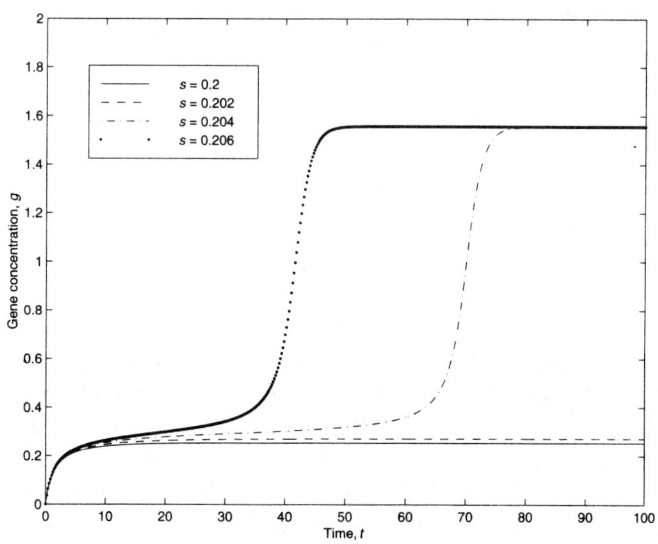

Figure 7.7 Approximate solutions to genetic switch model initial value problem for different values of the control parameter s. Note the jump in the equilibrium value of the gene concentration as s changes from 0.202 to 0.204.

EXERCISES

For the differential equations in Exercises 1–3, write out the equation for w_{i+1} explicitly in terms of h, t_i, and w_i for the modified Euler method, the Heun method, and the optimal RK4 method. Provide an explanation for the peculiar phenomena you observe in the three equations for w_{i+1}.

1. $x' = t - x$

2. $x' = \frac{1}{2}x - 2t$

3. $x' = \frac{1}{3}t + 2 - 4x$

4. Apply the modified Euler method to approximate the solution of the given initial value problem over the indicated interval in t using the indicated number of time steps.

(a) $x' = tx^3 - x \quad (0 \le t \le 1), \quad x(0) = 1, \quad N = 4$

(b) $x' + (4/t)x = t^4 \quad (1 \le t \le 3), \quad x(1) = 1, \quad N = 5$

(c) $x' = (\sin x - e^t)/\cos x \quad (0 \le t \le 1), \quad x(0) = 0, \quad N = 3$

(d) $x' = (1 + x^2)/t \quad (1 \le t \le 4), \quad x(1) = 0, \quad N = 5$

(e) $x' = t^2 - 2x^2 - 1 \quad (0 \le t \le 1), \quad x(0) = 0, \quad N = 4$

(f) $x' = 2(1 - x)/(t^2 \sin x) \quad (1 \le t \le 2), \quad x(1) = 2, \quad N = 3$

5. Repeat Exercise 4 using the Heun method.

6. Repeat Exercise 4 using the optimal RK2 method.

7. Repeat Exercise 4 using the classical fourth-order Runge-Kutta method, but take $N = 2$ in each case.

8. Use each of the following initial value problems to demonstrate that the global error associated with the modified Euler method is $O(h^2)$.

(a) $x' = 4t\sqrt{x^2+1}/x \quad (0 \le t \le 5), \quad x(0) = 1, \quad x(t) = \sqrt{(2t^2 + \sqrt{2})^2 - 1}$

(b) $tx' - 4x = t^5 e^t \quad (1 \le t \le 2), \quad x(1) = 0, \quad x(t) = t^4(e^t - e)$

(c) $x' = 1 - x + e^{2t}x^2 \quad (0 \le t \le 0.9), \quad x(0) = 0, \quad x(t) = e^{-t}\tan(e^t - 1)$

(d) $x' = 2t(x + 1)/x \quad (0 \le t \le 2), \quad x(0) = -2$

(e) $x' = (t^2 - tx + x^2)/(tx) \quad (1 \le t \le 2), \quad x(1) = 2$

9. Repeat Exercise 8 for the Heun method.

10. Repeat Exercise 8 for the optimal RK2 method.

11. Select an arbitrary value for a_2 between 0 and 1 (but not 1/2, 3/4 or 1), and implement the resulting second-order Runge-Kutta scheme. Numerically verify the global error is $O(h^2)$ using the initial value problems from Exercise 8.

12. Numerically demonstrate the global error of the classical fourth-order Runge-Kutta scheme is $O(h^4)$ using the initial value problems from Exercise 8.

13. Nystrom's method is a three-stage Runge-Kutta method whose tableau is

$$\begin{array}{c|ccc} 0 & & & \\ \frac{2}{3} & \frac{2}{3} & & \\ \frac{2}{3} & 0 & \frac{2}{3} & \\ \hline & \frac{1}{4} & \frac{3}{8} & \frac{3}{8} \end{array}$$

Implement this scheme and numerically determine the rate of convergence of the global error using the initial value problems from Exercise 8.

14. Kutta's method is a four-stage Runge-Kutta method whose tableau is

$$\begin{array}{c|cccc} 0 & & & & \\ \frac{1}{3} & \frac{1}{3} & & & \\ \frac{2}{3} & -\frac{1}{3} & 1 & & \\ 1 & 1 & -1 & 1 & \\ \hline & \frac{1}{8} & \frac{3}{8} & \frac{3}{8} & \frac{1}{8} \end{array}$$

Implement this scheme and numerically determine the rate of convergence of the global error using the initial value problems from Exercise 8.

15. Compare the performance of Euler's method, one of the second-order Runge-Kutta methods, and the classical fourth-order Runge-Kutta method on the initial value problem

$$\frac{dx}{dt} = \frac{t}{x} \quad (0 \le t \le 5), \quad x(0) = 1.$$

The exact solution for this problem is $x(t) = \sqrt{t^2+1}$.

16. Repeat Exercise 15 using the initial value problem

$$\frac{dx}{dt} = \frac{tx^2 - x}{t} \quad (1 \le t \le 5), \quad x(1) = -\frac{1}{\ln 2},$$

whose exact solution is $x(t) = -1/(t\ln(2t))$.

17. Compare the performance of the fourth-order Taylor method and the classical fourth-order Runge-Kutta method on each of the following initial value problems.

(a) $x' = 1 + x/t \quad (1 \le t \le 6), \quad x(1) = 1, \quad x(t) = t(1 + \ln t)$

(b) $x' = t/x \quad (0 \le t \le 5), \quad x(0) = 1, \quad x(t) = \sqrt{t^2+1}$

(c) $x' = (tx^2 - x)/t \quad (1 \le t \le 5), \quad x(1) = -1/\ln 2, \quad x(t) = -1/(t\ln(2t))$

18. (a) Consider the initial value problem

$$\frac{dx}{dt} = 3t - \frac{x}{t} \quad (1 \le t \le 6), \quad x(1) = 2.$$

The exact solution of this problem is $x(t) = t^2 + 1/t$. With what rate does the modified Euler method converge to this exact solution?

(b) Repeat part (a), but change the initial condition to $x(1) = 1$. The exact solution in this case is $x(t) = t^2$.

(c) Explain any difference in rate of convergence between parts (a) and (b).

19. Repeat Exercise 18 using the classical fourth-order Runge-Kutta method.

20. (a) Consider the initial value problem

$$\frac{dx}{dt} = -(1+t+t^2) - (2t+1)x - x^2 \quad (0 \le t \le 3), \quad x(0) = -\frac{1}{2}.$$

The exact solution of this problem is $x(t) = -t - 1/(e^t + 1)$. With what rate does the optimal RK2 method converge to this exact solution?

(b) Repeat part (a), but change the initial condition to $x(0) = -1$. The exact solution in this case is $x(t) = -t - 1$.

(c) Explain any difference in rate of convergence between parts (a) and (b).

21. Repeat Exercise 20 using the classical fourth-order Runge-Kutta method.

22. (a) Consider the initial value problem

$$\frac{dx}{dt} = 7t^2 - \frac{4x}{t} \qquad (1 \le t \le 6), \qquad x(1) = 2.$$

The exact solution of this problem is $x(t) = t^3 + 1/t^4$. With what rate does the classical fourth-order Runge-Kutta method converge to this exact solution?

(b) Repeat part (a), but change the initial condition to $x(1) = 1$. The exact solution in this case is $x(t) = t^3$.

(c) Explain any difference in rate of convergence between parts (a) and (b).

23. Reexamine the genetic switch problem

$$\frac{dg}{dt} = s + 3.03\frac{g^2}{1+g^2} - 1.51g,$$

paying attention to the hysteresis effect. Use initial conditions $g(0) = 1.8$, $g(0) = 1.68$, and $g(0) = 1.57$ (roughly the equilibrium values attained in Figures 7.6 and 7.7 and set $s = 0$. What new equilibrium level is reached? Does the gene stay "on" even after the parameter s is set to zero?

24. The following initial value problem has been proposed as a model for a genetic switch:

$$\frac{dg}{dt} = s + 2.43\frac{g^2}{1+g^2} - 1.61g, \qquad g(0) = 0.$$

Does this initial value problem exhibit the threshold effect?

25. Recall the "Radiative Heat Transfer to a Thin Metal Plate" problem from Section 7.3. Suppose a plate of chromium has been suspended in a furnace of temperature $T_F = 1500$ K. The plate has a thickness of $d = 0.002$ m, a mass density of $\rho = 7160$ kg/m^3 and an initial temperature of $T(0) = 300$ K. The specific heat data for chromium is provided below. Determine how long it takes for the temperature of the plate to come into equilibrium with the furnace.

Temperature (K)	300	400	600	800	1000	1200	1500
Specific Heat (J/kg · K)	449	484	542	581	616	682	779

26. Repeat Exercise 25 for a silver plate of thickness $d = 0.002$ m, mass density $\rho = 10500$ kg/m^3 and initial temperature $T(0) = 300$ K suspended in a furnace of temperature $T_F = 1200$ K. The specific heat data for silver is provided below.

Temperature (K)	300	400	600	800	1000	1200
Specific Heat (J/kg · K)	235	239	250	262	277	292

7.5 MULTISTEP METHODS

Recall from the introduction to this chapter that the general form for a linear m-step multistep method is

$$\frac{w_{i+1} - a_1 w_i - a_2 w_{i-1} - \cdots - a_m w_{i+1-m}}{h_i} =$$
$$b_0 f(t_{i+1}, w_{i+1}) + b_1 f(t_i, w_i) + b_2 f(t_{i-1}, w_{i-1}) + \cdots + b_m f(t_{i+1-m}, w_{i+1-m}).$$

When $b_0 = 0$, the method is said to be explicit; otherwise, it is said to be implicit. We will discuss the advantages and disadvantages of explicit versus implicit methods later in this section.

Note that there is one immediate obstacle to the practical implementation of a multistep method—obtaining starting values for the difference equation. Unlike a one-step method, for which knowledge of w_0 is sufficient to begin calculations, multistep methods require multiple starting values. For example, since a four-step method makes use of w_i, w_{i-1}, w_{i-2}, and w_{i-3}, the starting values w_0, w_1, w_2, and w_3 would be needed to begin calculations. Any starting values needed beyond w_0 are best determined using a one-step method of appropriate order. By this we mean that for a fourth-order multistep method, starting values should be determined using a fourth-order one-step method such as the classical fourth-order Runge-Kutta method.

Once the starting values have been obtained, however, multistep methods offer the advantage of requiring just one new function evaluation per time step, regardless of the order of approximation of the method. Compare this with the one-step methods we studied in Sections 7.2, 7.3, and 7.4: The first-order method used one function evaluation, the second-order methods used two function evaluations, and the fourth-order methods used four new function evaluations per time step.

In this section, we will focus on the derivation of two special classes of multistep methods. First we will derive the so-called Adams-Bashforth methods, which are explicit methods. Next, the derivation of the implicit Adams-Moulton methods will be considered. Predictor-Corrector methods, which are a special combination of explicit and implicit methods, will also be discussed.

Adams-Bashforth Methods

The general procedure for deriving an m-step Adams-Bashforth method is rather straightforward. Throughout, we will assume a uniform discretization in the t-domain; that is, we define $t_i = a + ih$, where $h = (b - a)/N$ for some positive integer N. We begin by integrating both sides of the model differential equation

$$y'(t) = f(t, y(t))$$

from $t = t_i$ to $t = t_{i+1}$. This yields the equation

$$y(t_{i+1}) - y(t_i) = \int_{t_i}^{t_{i+1}} f(t, y(t))\, dt.$$

Next, we write

$$f(t, y(t)) = P_{m-1}(t) + R_{m-1}(t),$$

where

$$P_{m-1}(t) = \sum_{j=1}^{m} L_{m-1,j}(t) f(t_{i+1-j}, y(t_{i+1-j}))$$

is the Lagrange form of the polynomial of degree at most $m - 1$ that interpolates f at the m points $t_i, t_{i-1}, t_{i-2}, \ldots, t_{i+1-m}$ and

$$R_{m-1}(t) = \frac{f^{(m)}(\xi, y(\xi))}{m!} \prod_{j=1}^{m} (t - t_{i+1-j})$$

is the corresponding remainder term. As with the interpolation-based differentiation and integration formulas we derived in Chapter 6, the Lagrange form of the interpolating polynomial is chosen here because of its clear and explicit dependence on the function values at the interpolating points. The integrals of the Lagrange polynomials, $L_{m-1,j}(t)$, provide the values of the coefficients, b_j, on the right-hand side of the difference equation, while the integral of the remainder term provides the truncation error term. Upon dropping the truncation error term and replacing all y's with w's, derivation of the method is complete.

Let's demonstrate this procedure by deriving the 2-step Adams-Bashforth method. Since $m = 2$, we write

$$f(t, y(t)) = \frac{t - t_{i-1}}{t_i - t_{i-1}} f(t_i, y_i) + \frac{t - t_i}{t_{i-1} - t_i} f(t_{i-1}, y_{i-1}) + \frac{f''(\xi, y(\xi))}{2}(t - t_i)(t - t_{i-1}).$$

Integrating the Lagrange polynomials yields

$$b_1 = \int_{t_i}^{t_{i+1}} \frac{t - t_{i-1}}{t_i - t_{i-1}}\, dt = h \int_0^1 (s + 1)\, ds = \frac{3h}{2}$$

and

$$b_2 = \int_{t_i}^{t_{i+1}} \frac{t - t_i}{t_{i-1} - t_i}\, dt = -h \int_0^1 s\, ds = -\frac{h}{2}.$$

The change of variable $t = t_i + sh$ was introduced to simplify the calculations. Using this same change of variable in and applying the Weighted Mean-Value Theorem for integrals to the integral of the remainder term produces

$$\int_{t_i}^{t_{i+1}} \frac{f''(\xi, y(\xi))}{2}(t - t_i)(t - t_{i-1})\, dt = \frac{h^3 f''(\hat{\xi}, y(\hat{\xi}))}{2} \int_0^1 s(s + 1)\, ds = \frac{5h^3}{12} y'''(\hat{\xi}).$$

In the final step, the differential equation was used to replace f'' by y'''. Gathering everything together and rearranging into standard form, we have

$$\frac{y_{i+1} - y_i}{h} = \frac{3}{2} f(t_i, y_i) - \frac{1}{2} f(t_{i-1}, y_{i-1}) + \frac{5h^2}{12} y'''(\hat{\xi}).$$

Therefore, the two-step Adams-Bashforth method is

$$\frac{w_{i+1} - w_i}{h} = \frac{3}{2} f(t_i, w_i) - \frac{1}{2} f(t_{i-1}, w_{i-1})$$

with local truncation error

$$\tau_i = \frac{5h^2}{12} y'''(\hat{\xi}) = O(h^2).$$

Note that the two-step method is second-order. To implement this scheme, values are needed for both w_0 and w_1. The initial condition is, of course, assigned to w_0, while w_1 can be obtained using any second-order one-step method.

Proceeding in a similar manner, the three-step Adams-Bashforth method

$$\frac{w_{i+1} - w_i}{h} = \frac{23}{12} f(t_i, w_i) - \frac{4}{3} f(t_{i-1}, w_{i-1}) + \frac{5}{12} f(t_{i-2}, w_{i-2})$$
$$\tau_i = \frac{3h^3}{8} y^{(4)}(\hat{\xi}) = O(h^3)$$

and the four-step Adams-Bashforth method

$$\frac{w_{i+1} - w_i}{h} = \frac{55}{24} f(t_i, w_i) - \frac{59}{24} f(t_{i-1}, w_{i-1}) + \frac{37}{24} f(t_{i-2}, w_{i-2}) - \frac{9}{24} f(t_{i-3}, w_{i-3})$$
$$\tau_i = \frac{251h^4}{720} y^{(5)}(\hat{\xi}) = O(h^4)$$

can be derived. The starting values, w_1 and w_2, for the three-step method should be obtained from a third-order one-step method; the values w_1, w_2, and w_3 for the four-step method should be obtained from a fourth-order one-step method. In general, the m-step Adams-Bashforth method has local truncation error that is $O(h^m)$, and therefore the method is of order m. Numerical verification that the global error of the m-step Adams-Bashforth method is also $O(h^m)$ will be considered in the exercises.

EXAMPLE 7.15 The Two-Step Adams-Bashforth Method in Action

Consider our standard test problem

$$\frac{dx}{dt} = 1 + \frac{x}{t} \quad (1 \le t \le 6), \quad x(1) = 1.$$

Before we can begin calculating with the two-step Adams-Bashforth method, we need a value for w_1. We can use any second-order one-step method to do this, but here we will use the second-order Taylor method. In practice, we would not choose this one-step method since it requires the derivative of the right-hand side function, whereas the Adams-Bashforth method does not. We have chosen to use the Taylor method in this example simply to facilitate comparisons between the Adams-Bashforth and the Taylor methods.

For the first time step, we have $t_0 = w_0 = 1$, from which it follows that $f(t_0, w_0) = 2$. The derivative of f is

$$f'(t,x) = \frac{tf(t,x) - x}{t^2} \quad \Rightarrow \quad f'(t_0, w_0) = \frac{1 \cdot 2 - 1}{1^2} = 1.$$

Using a step size of $h = 0.5$, the second-order Taylor method produces

$$w_1 = 1 + (0.5)(2) + \frac{(0.5)^2}{2}(1) = 2.125.$$

For all remaining time steps we use the difference equation for the two-step Adams-Bashforth method. In order to proceed as efficiently as possible, we save the value of $f(t_0, w_0)$ and calculate

$$f(t_1, w_1) = 1 + \frac{2.125}{1.5} = 2.416667.$$

It then follows that

$$\begin{aligned} w_2 &= w_1 + \frac{h}{2}\left[3f(t_1, w_1) - f(t_0, w_0)\right] \\ &= 2.125 + 0.25\left[3(2.416667) - 2\right] = 3.4375. \end{aligned}$$

At this point, the value of $f(t_0, w_0)$ is no longer needed, but we will need $f(t_1, w_1)$ for the next time step. We therefore save the value of $f(t_1, w_1)$ in place of $f(t_0, w_0)$ and then calculate $f(t_2, w_2) = 2.71875$ and

$$w_3 = 3.4375 + 0.25\left[3(2.71875) - 2.416667\right] = 4.872396.$$

Continuing in this fashion, we obtain the following results.

t_i	Adams-Bashforth	Error	Taylor	Error	Ratio of Errors
1.0	1.000000	0.000000	1.000000	0.000000	
1.5	2.125000	0.016802	2.125000	0.016802	
2.0	3.437500	0.051206	3.416667	0.030372	1.6859
2.5	4.872396	0.081669	4.833333	0.042607	1.9168
3.0	6.404427	0.108590	6.350000	0.054163	2.0049
3.5	8.018294	0.133624	7.950000	0.065330	2.0454
4.0	9.702798	0.157620	9.621429	0.076251	2.0671
4.5	11.449337	0.180989	11.355357	0.087009	2.0801
5.0	13.251135	0.203946	13.144841	0.097652	2.0885
5.5	15.102731	0.226617	14.984325	0.108211	2.0942
6.0	16.999638	0.249081	16.869264	0.118707	2.0983

Observe that the error in the Adams-Bashforth approximation for each $t_i \geq 2$ is roughly twice as large as the error in the corresponding Taylor approximation. If we examine the truncation error terms of the two methods, this result is not very surprising. The coefficient on the truncation error for the two-step Adams-Bashforth method is $\frac{5}{12}$, which is 2.5 times the truncation error coefficient for the second-order Taylor method. The ratio of truncation error coefficients grows even larger with increasing order. For example, the truncation error coefficient for the three-step Adams-Bashforth method is nine times larger than the coefficient for the third-order Taylor method. For fourth-order methods, the ratio of truncation error coefficients is $\frac{251}{6} > 40$. Clearly, to achieve similar accuracy, Adams-Bashforth methods will have to use smaller step sizes than Taylor methods.

The astute reader is commenting at this point that the comparisons we've just made are not quite fair and that using smaller step sizes for Adams-Bashforth methods is not really a problem. After all, a Taylor method uses more function evaluations per time step than the Adams-Bashforth method of similar order. To level the playing field, we have to allow the Adams-Bashforth method to use a smaller step size and take more time steps. In particular, the two-step Adams-Bashforth method should be allowed a step size that is one-half the step size used by the second-order Taylor method. For the same number of function evaluations, we might therefore expect the errors from the Adams-Bashforth method to be roughly $\frac{5}{2}/4 = 5/8$ as large as the errors from the Taylor method. Figure 7.8 shows this to be the case for our two standard initial value problems. In each graph, the vertical shift between the errors is roughly $0.21 \approx |\log_{10}(5/8)|$. When we allow the same number of function evaluations for fourth-order methods, we might expect the errors from the Adams-Bashforth method to be roughly $\frac{251}{6}/256 < 1/6$ as large as the errors from the Taylor method. We will examine this in the exercises.

So, Adams-Bashforth methods appear to outperform Taylor methods when we fix the number of function evaluations. What about Runge-Kutta methods? In Section 7.4, we found that on some problems a Runge-Kutta method will outperform the Taylor method of the same order, while on other problems the Runge-Kutta method will underperform the Taylor method. It is likely that a similar statement can be made regarding Runge-Kutta methods and Adams-Bashforth methods. In particular, Figure 7.9 shows that the two-step Adams-Bashforth method produces more accurate results than the Heun method and the optimal RK2 method but is slightly less accurate than the modified Euler method for the initial value problem $x' = 1 + x/t, x(1) = 1$. On the other hand, for the initial value problem $x' = t/x, x(0) = 1$, the Adams-Bashforth method underperforms all three Runge-Kutta methods. We will compare the fourth-order methods in the exercises.

Adams-Moulton Methods

The derivation of Adams-Moulton methods follows exactly the same procedure as the derivation of the Adams-Bashforth methods just presented, with one exception.

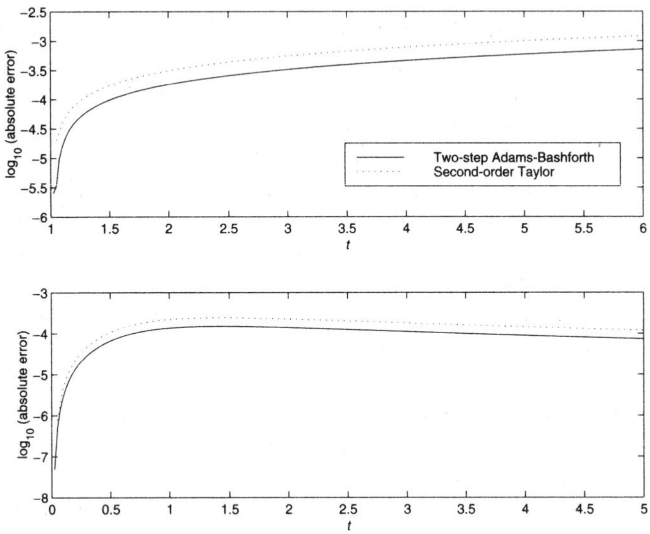

Figure 7.8 Comparison between the two-step Adams-Bashforth method (with $h = 0.025$ and w_1 calculated using the optimal RK2 method) and the second-order Taylor method (with $h = 0.05$). The top graph corresponds to the initial value problem $x' = 1 + x/t$, $x(1) = 1$, while the bottom graph corresponds to $x' = t/x$, $x(1) = 0$.

In addition to interpolating f at $t_i, t_{i-1}, t_{i-2}, \ldots,$ and t_{i+1-m}, we also interpolate at t_{i+1}. Hence, we write

$$f(t, y(t)) = P_m(t) + R_m(t),$$

where

$$P_m(t) = \sum_{j=0}^{m} L_{m,j}(t) f(t_{i+1-j}, y(t_{i+1-j}))$$

and

$$R_m(t) = \frac{f^{(m+1)}(\xi, y(\xi))}{(m+1)!} \prod_{j=0}^{m} (t - t_{i+1-j}).$$

Since $P_m(t)$ contains a term involving $f(t_{i+1}, y(t_{i+1}))$, the resulting method will be implicit. Furthermore, by using an additional point in the interpolating polynomial, the degree of the remainder term is increased by one over the Adams-Bashforth case, which implies that we get one more power of h in the truncation error term. Therefore, in general, an m-step Adams-Moulton method has local truncation error which is $O(h^{m+1})$, so the method is of order $m + 1$.

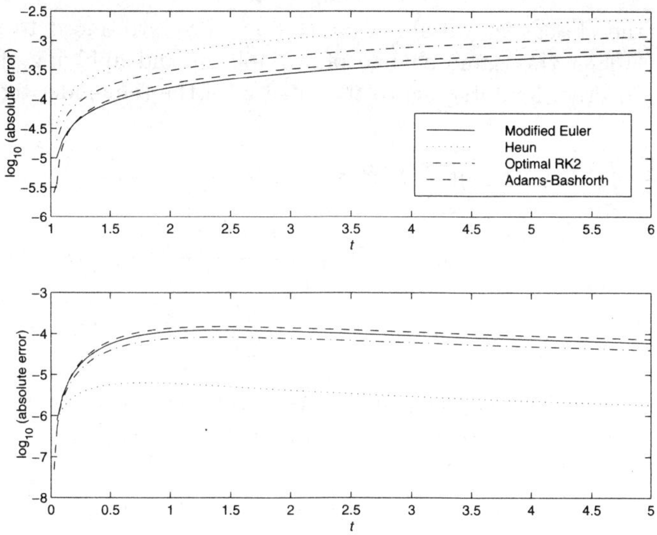

Figure 7.9 Comparison between the two-step Adams-Bashforth method (with $h = 0.025$ and w_1 calculated using the optimal RK2 method) and the second-order Runge-Kutta methods (with $h = 0.05$). The top graph corresponds to the initial value problem $x' = 1 + x/t$, $x(1) = 1$, while the bottom graph corresponds to $x' = t/x$, $x(1) = 0$.

Let us demonstrate this procedure by deriving the 2-step Adams-Moulton method. Since $m = 2$, we write

$$f(t, y(t)) = \frac{(t-t_i)(t-t_{i-1})}{(t_{i+1}-t_i)(t_{i+1}-t_{i-1})} f(t_{i+1}, y_{i+1}) + \frac{(t-t_{i+1})(t-t_{i-1})}{(t_i-t_{i+1})(t_i-t_{i-1})} f(t_i, y_i) + \frac{(t-t_{i+1})(t-t_i)}{(t_{i-1}-t_{i+1})(t_{i-1}-t_i)} f(t_{i-1}, y_{i-1}) + \frac{f'''(\xi, y(\xi))}{6}(t-t_{i+1})(t-t_i)(t-t_{i-1}).$$

Integrating the Lagrange polynomials yields

$$b_0 = \int_{t_i}^{t_{i+1}} \frac{(t-t_i)(t-t_{i-1})}{(t_{i+1}-t_i)(t_{i+1}-t_{i-1})}\,dt = \frac{h}{2}\int_0^1 s(s+1)\,ds = \frac{5h}{12}$$

$$b_1 = \int_{t_i}^{t_{i+1}} \frac{(t-t_{i+1})(t-t_{i-1})}{(t_i-t_{i+1})(t_i-t_{i-1})}\,dt = -h\int_0^1 (s-1)(s+1)\,ds = \frac{2h}{3}$$

and

$$b_2 = \int_{t_i}^{t_{i+1}} \frac{(t-t_{i+1})(t-t_i)}{(t_{i-1}-t_{i+1})(t_{i-1}-t_i)}\,dt = \frac{h}{2}\int_0^1 s(s-1)\,ds = -\frac{h}{12}.$$

Once again, the change of variable $t = t_i + sh$ was introduced to simplify the calculations. Using this same change of variable in and applying the Weighted Mean-Value Theorem for Integrals to the integral of the remainder term produces

$$\int_{t_i}^{t_{i+1}} \frac{f'''(\xi, y(\xi))}{6}(t - t_{i+1})(t - t_i)(t - t_{i-1})\, dt$$

$$= \frac{h^4 f'''(\hat{\xi}, y(\hat{\xi}))}{6} \int_0^1 (s-1)s(s+1)\, ds$$

$$= -\frac{h^4}{24} y^{(4)}(\hat{\xi}).$$

In the final step, the differential equation was used to replace f''' by $y^{(4)}$. Gathering everything together and rearranging into standard form, we have

$$\frac{y_{i+1} - y_i}{h} = \frac{5}{12} f(t_{i+1}, y_{i+1}) + \frac{2}{3} f(t_i, y_i) - \frac{1}{12} f(t_{i-1}, y_{i-1}) - \frac{h^3}{24} y^{(4)}(\hat{\xi}).$$

Therefore, the two-step Adams-Moulton method is

$$\frac{w_{i+1} - w_i}{h} = \frac{5}{12} f(t_{i+1}, w_{i+1}) + \frac{2}{3} f(t_i, w_i) - \frac{1}{12} f(t_{i-1}, w_{i-1})$$

with local truncation error

$$\tau_i = -\frac{h^3}{24} y^{(4)}(\hat{\xi}) = O(h^3).$$

To implement this scheme, values are needed for both w_0 and w_1. The initial condition is, of course, assigned to w_0, while w_1 can be obtained using any third-order one-step method.

The three-step Adams-Moulton method is given by the difference equation

$$\frac{w_{i+1} - w_i}{h} = \frac{9}{24} f(t_{i+1}, w_{i+1}) + \frac{19}{24} f(t_i, w_i) - \frac{5}{24} f(t_{i-1}, w_{i-1}) + \frac{1}{24} f(t_{i-2}, w_{i-2}),$$

with local truncation error

$$\tau_i = -\frac{19h^4}{720} y^{(5)}(\hat{\xi}).$$

Derivation of this method follows the same steps as the derivation of the two-step method. The required starting values for the three-step method should be obtained using a fourth-order one-step method.

A more substantial problem than obtaining the required number of starting values associated with implementing any Adams-Moulton method is the fact that the equation defining w_{i+1} is implicit. We may be able to transform the equation into explicit form, but this happens very rarely and is very problem dependent. More than likely, some sort of rootfinding scheme (such as fixed-point iteration or Newton's method) will need to be implemented. Of course, regardless of which rootfinding scheme we choose, the question of convergence arises. With fixed-point iteration, convergence will generally be obtained provided h is small enough. For those situations when we want to avoid such restrictions on h, which is an issue we will discuss in Section 7.9, Newton's method and the Secant method are common choices. But, as we know from Chapter 2, these methods may not converge.

Before moving on, it is interesting to note that if we were to apply the Adams-Moulton approach to deriving numerical methods with $m = 1$, we would obtain the implicit one-step method

$$\frac{w_{i+1} - w_i}{h} = \frac{1}{2}\left[f(t_{i+1}, w_{i+1}) + f(t_i, w_i)\right].$$

This scheme, which can also be derived by integrating our model initial value problem from $t = t_i$ to $t = t_{i+1}$ and applying the trapezoidal rule to the integral of f, is known as the trapezoidal method. You may recall that we encountered this method in Section 7.4 when discussing the implementation of the Heun method.

Predictor-Corrector Schemes

One common approach to circumventing the need to solve the implicit equation associated with an implicit method, such as an Adams-Moulton method, is to use an explicit method to "predict" an approximate value, $\tilde{w}_{i+1}$, and then to "correct" $\tilde{w}_{i+1}$ to w_{i+1} with the equation of the implicit method. This is the basic idea behind a *predictor-corrector scheme.*

The Heun method which was developed in Section 7.4 is an example of a predictor-corrector scheme involving one-step methods. Recall that the Heun method advances the approximate solution from w_i to w_{i+1} according to the rules

$$\frac{\tilde{w}_{i+1} - w_i}{h} = f(t_i, w_i)$$

and

$$\frac{w_{i+1} - w_i}{h} = \frac{1}{2}\left[f(t_i, w_i) + f(t_{i+1}, \tilde{w}_{i+1})\right].$$

These equations can be interpreted as using the explicit Euler's method to predict w_{i+1}, followed by the implicit trapezoidal method as a corrector.

Perhaps the most popular of the predictor-corrector schemes is the Adams fourth-order predictor-corrector method. This uses the four-step, fourth-order Adams-Bashforth method

$$\frac{\tilde{w}_{i+1} - w_i}{h} = \frac{1}{24}\left[55f(t_i, w_i) - 59f(t_{i-1}, w_{i-1}) + 37f(t_{i-2}, w_{i-2}) - 9f(t_{i-3}, w_{i-3})\right]$$

as a predictor, followed by the three-step, fourth-order Adams-Moulton method

$$\frac{w_{i+1} - w_i}{h} = \frac{1}{24}\left[9f(t_{i+1}, \tilde{w}_{i+1}) + 19f(t_i, w_i) - 5f(t_{i-1}, w_{i-1}) + f(t_{i-2}, w_{i-2})\right]$$

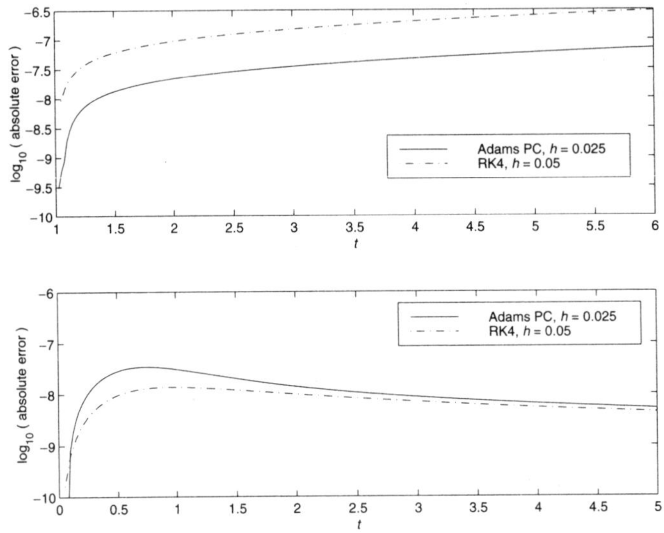

Figure 7.10 Comparison between the Adams fourth-order predictor-corrector (Adams PC) and the classical fourth-order Runge-Kutta method (RK4). The top graph corresponds to the initial value problem $x' = 1 + x/t$, $x(1) = 1$, while the bottom graph corresponds to $x' = t/x$, $x(1) = 0$.

as a corrector. This scheme requires only two new function evaluations, $f(t_i, w_i)$ and $f(t_{i+1}, \tilde{w}_{i+1})$, per time step. The required starting values (w_1, w_2, and w_3) are typically obtained from the classical fourth-order Runge-Kutta method.

When applied to the first of our standard initial value problems

$$\frac{dx}{dt} = 1 + \frac{x}{t}, \qquad (1 \le t \le 6), \qquad x(1) = 1,$$

the top graph in Figure 7.10 shows that the Adams fourth-order predictor-corrector scheme produces errors that are roughly half an order of magnitude smaller than the errors introduced by the classical fourth-order Runge-Kutta method (RK4). In these calculations, the Adams method used a step size of $h = 0.025$ as compared to $h = 0.05$ for the RK4 method so as to allow each method roughly the same number of function evaluations. When applied to the initial value problem

$$\frac{dx}{dt} = t/x, \qquad (0 \le t \le 5), \qquad x(0) = 1,$$

however, we see in the bottom graph of Figure 7.10 that the RK4 method provides better accuracy than the Adams predictor-corrector scheme. Once again, step sizes were selected to allow each method the same number of function evaluations.

The Order of General Linear Multistep Methods

Since not all linear multistep methods are either Adams-Bashforth or Adams-Moulton methods, it is worth demonstrating that the order of a linear multistep method can be determined by checking certain conditions on the coefficients a_j and b_j. Recall that the local truncation error for the general linear m-step multistep method is

$$\tau_i = \frac{y(t_{i+1}) - \sum_{j=1}^{m} a_j y(t_{i+1-j})}{h} - \sum_{j=0}^{m} b_j y'(t_{i+1-j}),$$

where the differential equation $y' = f(t, y)$ has been used to replace each $f(t_{i+1-j}, y_{i+1-j})$ by $y'(t_{i+1-j})$. In order for the method to be of order p, the truncation error must be $O(h^p)$. In light of the above expression for τ_i, this requires

$$y(t_{i+1}) - \sum_{j=1}^{m} a_j y(t_{i+1-j}) - h \sum_{j=0}^{m} b_j y'(t_{i+1-j}) = O(h^{p+1}).$$

If we now expand each y and y' in a Taylor series about $t^* = t_{i+1-m}$, we find

$$\begin{aligned}
y(t_{i+1}) &- \sum_{j=1}^{m} a_j y(t_{i+1-j}) - h \sum_{j=0}^{m} b_j y'(t_{i+1-j}) \\
&= \sum_{k=0}^{\infty} \frac{(mh)^k}{k!} y^{(k)}(t^*) - \sum_{j=1}^{m} a_j \sum_{k=0}^{\infty} \frac{[(m-j)h]^k}{k!} y^{(k)}(t^*) \\
&\quad - h \sum_{j=0}^{m} b_j \sum_{k=0}^{\infty} \frac{[(m-j)h]^k}{k!} y^{(k+1)}(t^*) \\
&= \left(1 - \sum_{j=1}^{m} a_j\right) y(t^*) \\
&\quad + \sum_{k=1}^{\infty} \left(m^k - \sum_{j=1}^{m} a_j (m-j)^k - k \sum_{j=0}^{m} b_j (m-j)^{k-1}\right) \frac{h^k}{k!} y^{(k)}(t^*).
\end{aligned}$$

For this last expression to be $O(h^{p+1})$, the a_j and b_j must satisfy

$$\begin{aligned}
1 &= \sum_{j=1}^{m} a_j \\
m^k &= \sum_{j=1}^{m} a_j (m-j)^k + k \sum_{j=0}^{m} b_j (m-j)^{k-1} \qquad (k = 1, 2, 3, \ldots, p); \\
m^{p+1} &\neq \sum_{j=1}^{m} a_j (m-j)^{p+1} + (p+1) \sum_{j=0}^{m} b_j (m-j)^{p}.
\end{aligned}$$

Thus, by determining the value of p for which the above algebraic conditions are satisfied, we obtain the order of the linear multistep method.

EXAMPLE 7.16 The Two-Step Adams-Bashforth Method

For the two-step Adams-Bashforth method, we have

$$m = 2, \quad a_1 = 1, \quad a_2 = 0, \quad b_0 = 0, \quad b_1 = \frac{3}{2}, \quad \text{and} \quad b_2 = -\frac{1}{2}.$$

The condition $\sum_{j=1}^{2} a_j = 1$ is clearly satisfied. Starting from $k = 1$, we then find

$$\begin{array}{ll} k = 1: & a_1 + b_0 + b_1 + b_2 = 2 = m^1 \\ k = 2: & a_1 + 2(2b_0 + b_1) = 4 = m^2 \\ k = 3: & a_1 + 3(4b_0 + b_1) = \frac{11}{2} \neq m^3. \end{array}$$

Therefore, the two-step Adams-Bashforth method is, as was determined earlier, of order 2.

EXAMPLE 7.17 Milne's Method

The explicit four-step multistep method

$$\frac{w_{i+1} - w_{i-3}}{h} = \frac{4}{3}\left[2f(t_i, w_i) - f(t_{i-1}, w_{i-1}) + 2f(t_{i-2}, w_{i-2})\right]$$

is known as Milne's method. For this method, we have

$$m = 4, \quad a_1 = a_2 = a_3 = 0, \quad a_4 = 1,$$

$$b_0 = 0, \quad b_1 = \frac{8}{3}, \quad b_2 = -\frac{4}{3}, \quad b_3 = \frac{8}{3}, \quad b_4 = 0.$$

As in the previous example, the condition $\sum_{j=1}^{2} a_j = 1$ is clearly satisfied. Starting with $k = 1$, we then calculate

$$\begin{array}{ll} k = 1: & b_1 + b_2 + b_3 = 4 = m^1 \\ k = 2: & 2(3b_1 + 2b_2 + b_3) = 16 = m^2 \\ k = 3: & 3(9b_1 + 4b_2 + b_3) = 64 = m^3 \\ k = 4: & 4(27b_1 + 8b_2 + b_3) = 256 = m^4 \\ k = 5: & 5(81b_1 + 16b_2 + b_3) = \frac{2960}{3} \neq m^5. \end{array}$$

Hence, Milne's method is of order 4.

EXERCISES

1. Derive the difference equation for the three-step Adams-Bashforth method:

$$\frac{w_{i+1} - w_i}{h} = \frac{23}{12} f(t_i, w_i) - \frac{4}{3} f(t_{i-1}, w_{i-1}) + \frac{5}{12} f(t_{i-2}, w_{i-2}).$$

Also derive the associated truncation error:

$$\tau_i = \frac{3h^3}{8} y^{(4)}(\hat{\xi}).$$

2. Derive the difference equation for the four-step Adams-Bashforth method:

$$\frac{w_{i+1} - w_i}{h} = \frac{55}{24} f(t_i, w_i) - \frac{59}{24} f(t_{i-1}, w_{i-1}) + \frac{37}{24} f(t_{i-2}, w_{i-2}) - \frac{9}{24} f(t_{i-3}, w_{i-3}).$$

Also derive the associated truncation error:

$$\tau_i = \frac{251h^4}{720} y^{(5)}(\hat{\xi}).$$

3. Derive the difference equation for the three-step Adams-Moulton method:

$$\frac{w_{i+1} - w_i}{h} = \frac{9}{24} f(t_{i+1}, w_{i+1}) + \frac{19}{24} f(t_i, w_i) - \frac{5}{24} f(t_{i-1}, w_{i-1}) + \frac{1}{24} f(t_{i-2}, w_{i-2}).$$

Also derive the associated truncation error:

$$\tau_i = -\frac{19h^4}{720} y^{(5)}(\hat{\xi}).$$

4. Derive the difference equation for the trapezoidal method (i.e., the Adams-Moulton method with $m = 1$):

$$\frac{w_{i+1} - w_i}{h} = \frac{1}{2} \left[f(t_{i+1}, w_{i+1}) + f(t_i, w_i) \right].$$

What is the truncation error term associated with this method?

5. Apply the two-step Adams-Bashforth method to approximate the solution of the given initial value problem over the indicated interval in t using the indicated number of time steps. Use any second-order one-step method to determine w_1.
   (a) $x' = tx^3 - x \quad (0 \le t \le 1), \quad x(0) = 1, \quad N = 4$
   (b) $x' + (4/t)x = t^4 \quad (1 \le t \le 3), \quad x(1) = 1, \quad N = 5$
   (c) $x' = (\sin x - e^t)/\cos x \quad (0 \le t \le 1), \quad x(0) = 0, \quad N = 3$
   (d) $x' = (1 + x^2)/t \quad (1 \le t \le 4), \quad x(1) = 0, \quad N = 5$
   (e) $x' = t^2 - 2x^2 - 1 \quad (0 \le t \le 1), \quad x(0) = 0, \quad N = 4$
   (f) $x' = 2(1 - x)/(t^2 \sin x) \quad (1 \le t \le 2), \quad x(1) = 2, \quad N = 3$

6. Repeat Exercise 5 using the four-step Adams-Bashforth method, but take $N = 6$ in each case. Use the classical fourth-order Runge-Kutta method to calculate w_1, w_2, and w_3.

7. Repeat Exercise 5 using the Adams fourth-order predictor-corrector method, but take $N = 6$ in each case. Use the classical fourth-order Runge-Kutta method to calculate w_1, w_2, and w_3.

8. In the last worked example of the section, we determined that Milne's method is of order 4. Determine the complete form of the truncation error term for Milne's method.

9. Determine the order and the complete form of the truncation error term for the indicated linear multistep method.
   (a) $w_{i+1} - w_{i-1} = \frac{1}{3}h\left[f(t_{i+1}, w_{i+1}) + 4f(t_i, w_i) + f(t_{i-1}, w_{i-1})\right]$
   (b) $w_{i+1} - 4w_i + 3w_{i-1} = -2hf(t_{i-1}, w_{i-1})$
   (c) $w_{i+1} - w_{i-1} = 2hf(t_i, w_i)$
   (d) $w_{i+1} - w_{i-2}$
   $= \frac{3}{8}h\left[f(t_{i+1}, w_{i+1}) + 3f(t_i, w_i) + 3f(t_{i-1}, w_{i-1}) + f(t_{i-2}, w_{i-2})\right]$
   (e) $w_{i+1} - \frac{4}{3}w_i + \frac{1}{3}w_{i-1} = \frac{2}{3}hf(t_{i+1}, w_{i+1})$
   (f) $w_{i+1} - \frac{1}{2}w_i - \frac{1}{2}w_{i-1} = h\left[f(t_{i+1}, w_{i+1}) - \frac{1}{4}f(t_i, w_i) + \frac{3}{4}f(t_{i-1}, w_{i-1})\right]$

10. Use each of the following initial value problems to demonstrate that the global error associated with the two-step Adams-Bashforth method is $O(h^2)$.
    (a) $x' = 4t\sqrt{x^2+1}/x \quad (0 \le t \le 5), \quad x(0) = 1, \quad x(t) = \sqrt{(2t^2+\sqrt{2})^2 - 1}$
    (b) $x' = 1 + x/t \quad (1 \le t \le 6), \quad x(1) = 1, \quad x(t) = t(1 + \ln t)$
    (c) $x' = t/x \quad (0 \le t \le 5), \quad x(0) = 1, \quad x(t) = \sqrt{t^2+1}$
    (d) $x' = 2t(x+1)/x \quad (0 \le t \le 2), \quad x(0) = -2$
    (e) $x' = (t^2 - tx + x^2)/(tx) \quad (1 \le t \le 2), \quad x(1) = 2$

11. Numerically demonstrate that the global error of the four-step Adams-Bashforth method is $O(h^4)$ using the initial value problems from Exercise 10.

12. Numerically demonstrate the global error of the Adams fourth-order predictor-corrector method is $O(h^4)$ using the initial value problems from Exercise 10.

13. Use each of the following initial value problems to compare the performance of the two-step Adams-Bashforth method with the second-order Taylor method. Remember to adjust step sizes so that each method uses roughly the same number of function evaluations.
    (a) $x' = 1 - x + e^{2t}x^2 \quad (0 \le t \le 0.9), \quad x(0) = 0, \quad x(t) = e^{-t}\tan(e^t - 1)$
    (b) $x' = (tx^2 - x)/t \quad (1 \le t \le 5), \quad x(1) = -1/\ln(2), \quad x(t) = -1/(t\ln(2t))$
    (c) $x' = -(1 + t + t^2) - (2t+1)x - x^2 \quad (0 \le t \le 3), \quad x(0) = -1/2,$
    $x(t) = -t - 1/(e^t + 1)$
    (d) $x' = (4/t)x + t^4e^t \quad (1 \le t \le 2), \quad x(1) = 0, \quad x(t) = t^4(e^t - e)$

14. Compare the performance of the two-step Adams-Bashforth method with the modified Euler method, Heun's method and the optimal RK2 method. Use the initial value problems from Exercise 13, and remember to adjust step sizes so that each method uses roughly the same number of function evaluations.

15. Compare the performance of the four-step Adams-Bashforth method with the classical fourth-order Runge-Kutta method. Use the initial value problems from Exercise 13, and remember to adjust step sizes so that each method uses roughly the same number of function evaluations.

16. Compare the performance of the four-step Adams-Bashforth method with the fourth-order Taylor method. Use the initial value problems from Exercise 13, and remember to adjust step sizes so that each method uses roughly the same number of function evaluations.

17. Compare the performance of the Adams fourth-order predictor-corrector method with the classical fourth-order Runge-Kutta method. Use the initial value problems from Exercise 13, and remember to adjust step sizes so that each method uses roughly the same number of function evaluations.

18. **(a)** Consider the initial value problem

$$\frac{dx}{dt} = -(1+t+t^2) - (2t+1)x - x^2 \quad (0 \le t \le 3), \quad x(0) = -\frac{1}{2}.$$

The exact solution of this problem is $x(t) = -t - 1/(e^t + 1)$. With what rate does the two-step Adams-Bashforth method converge to this exact solution?

(b) Repeat part (a), but change the initial condition to $x(0) = -1$. The exact solution in this case is $x(t) = -t - 1$.

(c) Explain any difference in rate of convergence between parts (a) and (b).

19. Repeat Exercise 18 using the four-step Adams-Bashforth method.

20. Repeat Exercise 18 using the Adams fourth-order predictor-corrector method.

21. **(a)** Consider the initial value problem

$$\frac{dx}{dt} = 7t^2 - \frac{4x}{t} \quad (1 \le t \le 6), \quad x(1) = 2.$$

The exact solution of this problem is $x(t) = t^3 + 1/t^4$. With what rate does the four-step Adams-Bashforth method converge to this exact solution?

(b) Repeat part (a), but change the initial condition to $x(1) = 1$. The exact solution in this case is $x(t) = t^3$.

(c) Explain any difference in rate of convergence between parts (a) and (b).

(d) Repeat parts (a), (b), and (c) using the Adams fourth-order predictor-corrector method.

22. In the "Projection Printing" problem capsule of the Overview to Chapter 1 (see page 6), the following initial value problem for the normalized photoactive compound concentration, M, inside a resist film after exposure was developed:

$$\frac{dM(z, t_{\text{exp}})}{dz} = M(z, t_{\text{exp}})\left[A\left(1 - M(z, t_{\text{exp}})\right) - B \ln M(z, t_{\text{exp}})\right]$$

$$M(0, t_{\text{exp}}) = \exp(-I_0 C t_{\text{exp}}).$$

In this problem z denotes depth into the resist film, A, B, and C are properties of the resist material, and the product It_{exp} is the exposure energy of the light source used during the illumination phase. For the resist material AZ2400, it has been determined that $A = 0.162/\mu\text{m}$, $B = 0.184/\mu\text{m}$, and $C = 0.0128\ \text{cm}^2/\text{mJ}$. If the exposure energy is $110\ \text{mJ/cm}^2$, what is the normalized photoactive compound concentration as a function of depth? Tabulate M for z ranging from 0 μm through 1 μm, in increments of 0.05 μm.

23. A common method for improving the performance of projection printing in optical microlithography is to place a layer of film, known as a contrast enhancing layer (CEL), on top of the photoresist material. The following initial value problem serves as a model for the concentration of photoactive compound within the CEL, M_c:

$$\frac{dM_c}{dz} = M_c\left[\frac{\alpha}{2}(1 - M_c^2) + \beta(1 - M_c) - B_c \ln M_c\right]$$

$$M_c(0) = \exp(-IC_c t).$$

Here, $\alpha = 1.4A_c$; $\beta = -0.06A_c$; A_c, B_c, and C_c are material properties of the CEL material; the product It is the exposure energy of the light source; and z denotes depth into the CEL film. For a particular polysilane CEL, it has been determined that $A_c = 8.93/\mu\text{m}$, $B_c = 0.175/\mu\text{m}$, and $C_c = 0.0376\ \text{cm}^2/\text{mJ}$. If the exposure energy is 110 mJ/cm^2, tabulate M_c for z ranging from 0 μm through 0.2 μm, in increments of 0.005 μm.

7.6 CONVERGENCE AND STABILITY ANALYSIS

Earlier in the chapter it was noted that the analysis of numerical methods for initial value problems focuses on two types of error, local truncation error and global discretization error, and on the three important properties of consistency, convergence and stability. When we investigated Euler's method in Section 7.2, the simplicity of the difference equation allowed us to perform a complete analysis of the method. In subsequent sections, however, the difference equations became more complicated. Rather than perform a complete analysis of each of these methods, we dealt almost exclusively with the local truncation error. Global error was treated only with numerical experiments, and stability was ignored entirely. In this section, we will remedy this situation by developing a general framework for convergence and stability analysis of both one-step and linear multistep methods. For completeness, consistency will also be included in our discussions.

Recalling the Key Concepts

Since we've covered a lot of material since Section 7.1, it is probably worthwhile to review the basic notation we will use and the central concepts we will discuss. We seek to approximate the solution of the initial value problem

$$y' = f(t, y), \quad y(a) = \alpha,$$

over the interval $a \le t \le b$. For simplicity, the approximate solution is obtained at equally spaced points; that is, for some positive integer N, we define the step size $h = (b - a)/N$ and then set $t_i = a + ih$ for each $i = 0, 1, 2, \ldots, N$. The value of the exact solution of the initial value problem at $t = t_i$, $y(t_i)$, is denoted by y_i, and the approximation to y_i is denoted by w_i.

Local truncation error, τ_i, measures the amount by which the solution of the differential equation fails to satisfy the difference equation that defines the numerical method. If $\tau_i \to 0$ as $h \to 0$, then the method is called consistent. Additionally, if $\tau_i = O(h^p)$ for some constant p, the method is said to be of order p.

Global discretization error measures the difference between the solution of the differential equation and the solution of the difference equation, $y_i - w_i$. Suppose we let N tend toward infinity [note that this will force the step size $h = (b-a)/N$ toward zero], keeping $t_N = b$ regardless of the value of N. If it follows that

$$\lim_{h\to 0} \max_{1\le i\le N} |y_i - w_i| = 0,$$

we say the method is convergent.

Stability deals with sensitivity of the solution of the difference equations to changes in initial conditions. For a given numerical method and a fixed N, let $\{w_i\}$ denote the sequence of values generated from an initial condition of α and $\{\tilde{w}_i\}$ denote the sequence generated from an initial condition of $\tilde{\alpha}$. The numerical method is stable if and only if there exists a function $k(t) > 0$ such that

$$|\tilde{w}_i - w_i| \le k(t_i)|\tilde{\alpha} - \alpha|$$

for all i. The key point is that the function $k(t_i)$ must be independent of h.

Of course, the most important property for a numerical method to have is convergence. If a method does not converge (i.e., if the solution of the difference equation does not tend toward the solution of the differential equation), then the method is useless. Unfortunately, direct proofs of convergence can be hard to obtain, whereas checks for consistency and stability tend to be much easier. In what follows, we therefore want to determine whether we can link the consistency and stability of a method with convergence of that method.

One-Step Methods

Analysis for the one-step method

$$\frac{w_{i+1} - w_i}{h} = \phi(f, t_i, w_i, h) \tag{1}$$

is rather straightforward. We start with consistency. The local truncation error associated with (1) is

$$\tau_i = \frac{y_{i+1} - y_i}{h} - \phi(f, t_i, y_i, h).$$

If we take the limit as $h \to 0$ of this last equation and assume that ϕ is continuous in h, we obtain

$$\begin{aligned}\lim_{h\to 0} \tau_i &= y'(t_i) - \phi(f, t_i, y_i, 0) \\ &= f(t_i, y_i) - \phi(f, t_i, y_i, 0).\end{aligned}$$

Therefore, the method is consistent provided

$$f(t_i, y_i) = \phi(f, t_i, y_i, 0),$$

which is a simple condition to verify.

EXAMPLE 7.18 Consistency of Modified Euler Method

Recall that the difference equation for the modified Euler method is

$$\frac{w_{i+1} - w_i}{h} = f\left(t_i + \frac{h}{2}, w_i + \frac{h}{2} f(t_i, w_i)\right);$$

hence,

$$\phi(f, t_i, w_i, h) = f\left(t_i + \frac{h}{2}, w_i + \frac{h}{2} f(t_i, w_i)\right).$$

It then follows that

$$\phi(f, t_i, y_i, 0) = f\left(t_i + \frac{0}{2}, y_i + \frac{0}{2} f(t_i, y_i)\right) = f(t_i, y_i),$$

so the modified Euler method is consistent.

The following theorem provides the conditions for a one-step method to be stable, as well as for a one-step method to be convergent. Note the role of the Lipschitz condition in this theorem, and compare the hypotheses of this theorem with those for a well-posed initial value problem.

Theorem. Suppose the one-step method given in (1) is applied to the initial value problem $y' = f(t, y),\ y(a) = \alpha$, where $h = (b - a)/N$ for some positive integer N and $t_i = a + ih$. Let $D = \{(t, y) | \, a \le t \le b, y \in R\}$, and $D_\phi = \{(f, t, y, h) | \, f(t, y) \text{ is Lipschitz in } y \text{ on } D, (t, y) \in D, 0 \le h \le h_0\}$ for some h_0. If $\phi(f, t, y, h)$ satisfies a Lipschitz condition in y on D_ϕ with Lipschitz contant L_ϕ, then

1. the one-step method is stable; and
2. if there exist positive constants c and p such that $|\tau_i| \le ch^p$ for each $i = 1, 2, 3, \ldots, N$ whenever $0 \le h \le h_0$, then

$$|y_i - w_i| \le \frac{ch^p}{L_\phi}\left[e^{L_\phi(t_i - a)} - 1\right].$$

Proof. (1) Let $\{w_i\}$ and $\{\tilde{w}_i\}$ be numerical solutions generated by (1) from initial conditions of α and $\tilde{\alpha}$, respectively. Then

$$\begin{aligned}\left|\frac{\tilde{w}_{i+1} - \tilde{w}_i}{h} - \frac{w_{i+1} - w_i}{h}\right| &= |\phi(f, t_i, \tilde{w}_i, h) - \phi(f, t_i, w_i, h)| \\ &\le L_\phi |\tilde{w}_i - w_i|.\end{aligned}$$

After some algebraic manipulation, this yields

$$|\tilde{w}_{i+1} - w_{i+1}| \leq (1 + hL_\phi)|\tilde{w}_i - w_i|.$$

Next, apply this last formula recursively to obtain

$$|\tilde{w}_{i+1} - w_{i+1}| \leq (1 + hL_\phi)^{(i+1)}|\tilde{w}_0 - w_0|.$$

But $\tilde{w}_0 = \tilde{\alpha}$, $w_0 = \alpha$, and $0 \leq 1 + hL_\phi \leq e^{hL_\phi}$, so

$$|\tilde{w}_{i+1} - w_{i+1}| \leq e^{(i+1)hL_\phi}|\tilde{\alpha} - \alpha| = e^{L_\phi(t_{i+1}-a)}|\tilde{\alpha} - \alpha|,$$

since $(i + 1)h = t_{i+1} - a$. Thus $k(t) = e^{L_\phi(t-a)}$, which is independent of h, and the method is stable.
(2) The proof of this part is similar to that of part (1) and will be left as an exercise. □

The second part of this theorem tells us two important things. First, if a one-step method is consistent, then it is also convergent. Therefore, we can guarantee that a particular one-step method converges by showing that the method is consistent and that the function ϕ satisfies a Lipschitz condition. Second, if the truncation error is of order p, then the global error will also be of order p. This provides theoretical justification for the conclusions we drew from the results of our numerical investigations with Taylor and Runge-Kutta methods in Sections 7.3 and 7.4.

The connection between consistency and convergence for one-step methods is actually stronger than is suggested by this theorem. Under the same hypothesis on ϕ, it can be shown (see Gear [1] or Henrici [2] for details) that convergence implies consistency. Hence, a one-step method converges if and only if it is consistent.

EXAMPLE 7.19 Convergence and Stability for Modified Euler Method

For the modified Euler method

$$\phi(f, t, y, h) = f\left(t + \frac{h}{2}, y + \frac{h}{2}f(t, y)\right).$$

Suppose that f satisfies a Lipschitz condition in y on the set $D = \{(t, y)|\, a \leq t \leq b, y \in R\}$ with Lipschitz constant L. Then

$$\begin{aligned}
&|\phi(f, t, y_1, h) - \phi(f, t, y_2, h)| \\
&\quad = \left|f\left(t + \frac{h}{2}, y_1 + \frac{h}{2}f(t, y_1)\right) - f\left(t + \frac{h}{2}, y_2 + \frac{h}{2}f(t, y_2)\right)\right| \\
&\quad \leq L\left|y_1 + \frac{h}{2}f(t, y_1) - y_2 - \frac{h}{2}f(t, y_2)\right| \\
&\quad \leq L|y_1 - y_2| + \frac{hL}{2}|f(t, y_1) - f(t, y_2)| \\
&\quad \leq L|y_1 - y_2| + \frac{hL^2}{2}|y_1 - y_2| = \left(L + \frac{hL^2}{2}\right)|y_1 - y_2|.
\end{aligned}$$

Therefore, ϕ satisfies a Lipschitz condition in y on the set

$$\{(f,t,y,h)|\, f(t,y) \text{ is Lipschitz in } y \text{ on } D, (t,y) \in D, 0 \leq h \leq h_0\}$$

for any $h_0 > 0$ with Lipschitz constant

$$L_\phi = L + \frac{h_0 L^2}{2}.$$

Hence, we may conclude that the modified Euler method is stable. Since we have already established that the modified Euler method is consistent, we may also conclude that the method is convergent. Furthermore, because the local truncation error for this method is $O(h^2)$, it follows that the global error is also $O(h^2)$.

Linear Multistep Methods

Consider the general linear m-step multistep method

$$\frac{w_{i+1} - a_1 w_i - a_2 w_{i-1} - \cdots - a_m w_{i+1-m}}{h_i} =$$

$$b_0 f(t_{i+1}, w_{i+1}) + b_1 f(t_i, w_i) + b_2 f(t_{i-1}, w_{i-1}) + \cdots + b_m f(t_{i+1-m}, w_{i+1-m}). \quad (2)$$

Before we turn to a development of the conditions required for (2) to be consistent, stable and convergent, there is a technical detail that must be addressed. Recall that the values for $w_1, w_2, w_3, \ldots, w_{m-1}$ are obtained from a different numerical method. We cannot reasonably expect the values that are then obtained from (2) to converge to the solution of the initial value problem if these starting values don't converge. For each $i = 1, 2, 3, \ldots, m-1$, let ϵ_i denote the error associated with the starting value w_i. Throughout the remainder of our analysis, we will assume that there exists a function $\epsilon(h)$ such that $|\epsilon_i| \leq \epsilon(h)$ for each $i = 1, 2, 3, \ldots m-1$ and such that $\epsilon(h) \to 0$ as $h \to 0$.

For (2) to be consistent, we must have $\tau_i \to 0$ as $h \to 0$. This requires that the local truncation error be at least $O(h)$; that is, the method must be at least first order. From our work at the end of Section 7.5, the coefficients a_j and b_j must satisfy

$$\sum_{j=1}^{m} a_j = 1 \qquad \text{and} \qquad \sum_{j=1}^{m} a_j(m-j) + \sum_{j=0}^{m} b_j = m$$

for the method to be at least first order. By using the first of these relations, the second relation can be simplified to $-\sum_{j=1}^{m} j a_j + \sum_{j=0}^{m} b_j = 0$. Putting all of this together, it follows that the *consistency conditions* for linear multistep methods are given by

$$\sum_{j=1}^{m} a_j = 1 \qquad \text{and} \qquad -\sum_{j=1}^{m} j a_j + \sum_{j=0}^{m} b_j = 0. \quad (3)$$

EXAMPLE 7.20 The Leapfrog Method

The explicit two-step method

$$\frac{w_{i+1} - w_{i-1}}{h} = 2f(t_i, w_i)$$

is known as the leapfrog method, or the midpoint method. Here, we have

$$m = 2, \quad a_1 = 0, \quad a_2 = 1, \quad b_0 = 0, \quad b_1 = 2, \quad \text{and} \quad b_2 = 0.$$

Substituting these values into the consistency conditions (3), we find

$$\begin{array}{l} a_1 + a_2 = 0 + 1 = 1; \quad \text{and} \\ -(a_1 + 2a_2) + b_0 + b_1 + b_2 = -(0 + 2) + 0 + 2 + 0 = 0. \end{array}$$

Therefore, the leapfrog method is consistent.

Let's now turn our attention to stability. For the moment, suppose that $f(t, y) \equiv 0$. With this assumption, equation (2) becomes

$$w_{i+1} - a_1 w_i - a_2 w_{i-1} - \cdots - a_m w_{i+1-m} = 0.$$

Let $\{w_i\}$ and $\{\tilde{w}_i\}$ denote solutions obtained from this simplified difference equation for different initial conditions. Subtract the equation for the w_i from the equation for the $\tilde{w}_i$, and define $v_i = \tilde{w}_i - w_i$. It follows that v_i satisfies the difference equation

$$v_{i+1} - a_1 v_i - a_2 v_{i-1} - \cdots - a_m v_{i+1-m} = 0. \tag{4}$$

Associated with (4) is the characteristic polynomial of the multistep method

$$P(\lambda) = \lambda^m - a_1 \lambda^{m-1} - a_2 \lambda^{m-2} - \cdots - a_m.$$

If λ is a zero of p, then $v_i = \lambda^i$ is a solution of (4) since

$$\lambda^{i+1} - a_1 \lambda^i - a_2 \lambda^{i-1} - \cdots - a_m \lambda^{i+1-m} = \lambda^{i+1-m} P(\lambda) = 0.$$

Now, suppose that $|\lambda| > 1$. This implies that $|v_i| = |\lambda|^i$ is unbounded as $i \to \infty$. In other words, the difference between $\tilde{w}_i$ and w_i is unbounded, and the method is unstable. Alternatively, suppose that $|\lambda| = 1$ but that λ has multiplicity greater than one. The expression $v_i = i\lambda^i$ is then a solution of (4). As $i \to \infty$, $|v_i| = i|\lambda|^i = i$ is unbounded and the method is again unstable. These results motivate the following definition.

Definition. The linear multistep method given by (2) satisfies the ROOT CONDITION if the zeros of the characteristic polynomial associated with the method,

$$P(\lambda) = \lambda^m - a_1 \lambda^{m-1} - a_2 \lambda^{m-2} - \cdots - a_m,$$

satisfy

1. $|\lambda| \le 1$ for all zeros λ; and

2. if $|\lambda| = 1$, then λ is a simple zero.

We have essentially just established that when $f(t, y) \equiv 0$, if a method does not satisfy the root condition, then the method is not stable. Equivalently, if a method is stable, then it must satisfy the root condition. It can be shown (see Isaacson and Keller [3]) that the same result holds when $f(t, y)$ is not identically zero and that the root condition implies stability, whether $f(t, y)$ is identically zero or not. Hence, we have the following result.

Theorem. A linear multistep method is stable if and only if it satisfies the root condition.

Note that $P(1) = 1 - a_1 - a_2 - \cdots - a_m$. Using the first of the relations in (3), we see that for a consistent linear mutlistep method $P(1) = 0$. Hence, a consistent and stable linear multistep method will always have at least one zero with magnitude equal to 1. We subclassify the stability of a method based on whether or not there are other zeros of unit magnitude.

Definition. A stable multistep method is said to be STRONGLY STABLE if $\lambda = 1$ is the only zero of $P(\lambda)$ with $|\lambda| = 1$ and is said to be WEAKLY STABLE otherwise.

We will explore the consequences of weak stability below and in the exercises.

EXAMPLE 7.21 Stability of Adams-Bashforth and Adams-Moulton Methods

Regardless of the number of steps, m, all Adams-Bashforth and Adams-Moulton methods have $a_1 = 1$ and $a_j = 0$ for each $j = 2, 3, 4, \ldots, m$. Thus, for each m, the characteristic polynomial associated with one of these methods is of the form

$$P(\lambda) = \lambda^m - \lambda^{m-1} = \lambda^{m-1}(\lambda - 1).$$

The zeros of this polynomial are $\lambda = 0$ (of multiplicity $m - 1$) and $\lambda = 1$. Consequently, Adams-Bashforth and Adams-Moulton methods of all orders are strongly stable.

EXAMPLE 7.22 Stability of Leapfrog Method

For the leapfrog method, we know that $m = 2$, $a_1 = 0$ and $a_2 = 1$. The characteristic polynomial for this method is then

$$P(\lambda) = \lambda^2 - 1 = (\lambda - 1)(\lambda + 1).$$

The zeros of $P(\lambda)$ are clearly 1 and -1, which implies that the leapfrog method is weakly stable.

Finally, let's tackle the question of convergence for linear multistep methods. Suppose (2) is convergent, and consider the initial value problem $y' = 0$, $y(0) = 1$, whose exact solution is $y(t) = 1$. Applying (2) to this initial value problem, the resulting difference equation is

$$w_{i+1} - \sum_{j=1}^{m} a_j w_{i+1-j} = 0. \tag{5}$$

Since the solution of the initial value problem is constant, we seek a constant solution of (5). Substituting $w_i = c$, for some constant c, into (5) yields

$$c\left(1 - \sum_{j=1}^{m} a_j\right) = 0.$$

If $\sum_{j=1}^{m} a_j \neq 1$, the only constant solution of (5) is $w_i = 0$. This, however, violates the assumption that the method is convergent, so we must have $\sum_{j=1}^{m} a_j = 1$.

Continue to suppose that (2) is convergent. Consider the initial value problem $y' = 1, y(0) = 0$, which we will solve over the interval $0 \leq t \leq T$ for some fixed T. Furthermore, we will always select the step size, h, and the number of time steps, N, so that $Nh = T$. The exact solution in this case is $y(t) = t$, and the resulting difference equation is

$$w_{i+1} - \sum_{j=1}^{m} a_j w_{i+1-j} = h \sum_{j=0}^{m} b_j. \tag{6}$$

Since the solution of the initial value problem is linear in the independent variable, we seek a solution of (6) in the form $w_i = ci$ for some constant c. After some algebraic manipulation, and using our previous result that $\sum_{j=1}^{m} a_j = 1$, we find that

$$c = \frac{h \sum_{j=0}^{m} b_j}{\sum_{j=1}^{m} j a_j}.$$

For a convergent method, it can be shown that $\sum_{j=1}^{m} j a_j \neq 0$ (Exercise 14). Therefore,

$$w_i = ih \frac{\sum_{j=0}^{m} b_j}{\sum_{j=1}^{m} j a_j}, \qquad \text{and, in particular,} \qquad w_N = Nh \frac{\sum_{j=0}^{m} b_j}{\sum_{j=1}^{m} j a_j} = T \frac{\sum_{j=0}^{m} b_j}{\sum_{j=1}^{m} j a_j}.$$

Now, because the method is convergent,

$$T = y(T) = \lim_{N \to \infty} w_N = T \frac{\sum_{j=0}^{m} b_j}{\sum_{j=1}^{m} j a_j},$$

which implies that $\sum_{j=0}^{m} b_j / \sum_{j=1}^{m} j a_j = 1$, or, equivalently, that $-\sum_{j=1}^{m} j a_j + \sum_{j=0}^{m} b_j = 0$.

We have just shown that if a linear multistep method is convergent, then the coefficients a_j and b_j must satisfy

$$\sum_{j=1}^{m} a_j = 1 \qquad \text{and} \qquad -\sum_{j=1}^{m} j a_j + \sum_{j=0}^{m} b_j = 0.$$

These are precisely the consistency conditions we stated in equation (3). Therefore, a convergent linear multistep method must be consistent. By considering the initial value problem $y' = 0, y(0) = 0$, we can show that a convergent linear multistep method must satisfy the root condition (see Exercise 13), and hence must be stable. Therefore, convergence implies both consistency and stability. It can also be shown (see Henrici [2] or Isaacson and Keller [3]) that consistency and stability together imply convergence. Our final result is then

Theorem. A linear multistep method is convergent if and only if it is both consistent and stable.

EXAMPLE 7.23 Convergent but Weakly Stable versus Convergent and Strongly Stable

In previous examples we have established that the leapfrog method is consistent and stable, though only weakly stable. It follows that the leapfrog method is convergent. To demonstrate the consequences of the weak stability of this method, we will compare the leapfrog method with the stongly stable two-step Adams-Bashforth method. We have chosen this strongly stable method because both the leapfrog method and the two-step Adams-Bashforth method are second order.

The test problem is

$$\frac{dx}{dt} + x = \sin t, \qquad x(0) = 1.$$

Figure 7.11 displays the approximate solutions for this problem, leapfrog method in the top graph and Adams-Bashforth method in the bottom graph, over the interval $0 \le t \le 10$ using $N = 100$ time steps. Observe the "saw-tooth" oscillations in the leapfrog method solution that appear around $t = 7$ and grow in amplitude for increasing t. This phenomenon is a result of the weak stability of the numerical method. Reducing the step size will not eliminate this problem, it will only delay its onset (see Exercise 15).

References

1. C. W. Gear, *Numerical Initial-Value Problems in Ordinary Differential Equations*, Prentice Hall, Englewood Cliffs, NJ, 1971.
2. P. Henrici, *Discrete Variable Methods in Ordinary Differential Equations*, John Wiley and Sons, New York, 1962.
3. E. Isaacson and H. B. Keller, *Analysis of Numerical Methods*, John Wiley and Sons, New York, 1966.

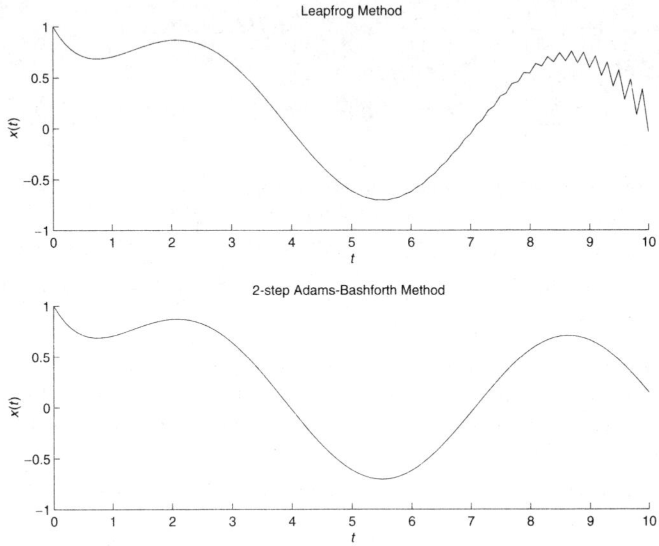

Figure 7.11 Comparison between the leapfrog method (top graph) and the two-step Adams-Bashforth method (bottom graph) for the initial value problem $x' + x = \sin t$, $x(0) = 1$ using a step size of $h = 0.1$.

EXERCISES

1. Prove the second part of the stability and convergence theorem for one-step methods.

2. Comment on the consistency, stability, and convergence of Heun's method, assuming the function $f(t, y)$ satisfies a Lipschitz condition in y on the set $D = \{(t, y) | a \le t \le b, y \in R\}$ with Lipschitz constant L.

3. Repeat Exercise 2 for the optimal RK2 method.

4. Repeat Exercise 2 for the classical fourth-order Runge-Kutta method.

5. Repeat Exercise 2 for the Taylor method of order p. Here, assume that $f, f', f'', \ldots, f^{(p-1)}$ all satisfy a Lipschitz condition in y on the set $D = \{(t, y) | a \le t \le b, y \in R\}$ with common Lipschitz constant L.

In Exercises 6–11, comment on the consistency, stability, and convergence of the indicated linear multistep method.

6. $w_{i+1} - w_{i-1} = \frac{1}{3}h\,[f(t_{i+1}, w_{i+1}) + 4f(t_i, w_i) + f(t_{i-1}, w_{i-1})]$

7. $w_{i+1} - \frac{4}{3}w_i + \frac{1}{3}w_{i-1} = \frac{2}{3}hf(t_{i+1}, w_{i+1})$

8. $w_{i+1} - w_{i-2} = \frac{3}{8}h\,[f(t_{i+1}, w_{i+1}) + 3f(t_i, w_i) + 3f(t_{i-1}, w_{i-1}) + f(t_{i-2}, w_{i-2})]$

9. $w_{i+1} - \frac{1}{2}w_i - \frac{1}{2}w_{i-1} = h\left[f(t_{i+1}, w_{i+1}) - \frac{1}{4}f(t_i, w_i) + \frac{3}{4}f(t_{i-1}, w_{i-1})\right]$

10. $w_{i+1} - \frac{18}{11}w_i + \frac{9}{11}w_{i-1} - \frac{2}{11}w_{i-2} = \frac{6}{11}hf(t_{i+1}, w_{i+1})$

11. $w_{i+1} - 4w_i + 3w_{i-1} = -2hf(t_{i-1}, w_{i-1})$

12. Suppose that $\sum_{j=1}^{m} a_j = 1$. Show that $\sum_{j=1}^{m} a_j(m-j) + \sum_{j=0}^{m} b_j = m$ is equivalent to $-\sum_{j=1}^{m} ja_j = \sum_{j=0}^{m} b_j$.

13. Show that a convergent linear multistep method must satisfy the root condition by considering the initial value problem $y' = 0, y(0) = 0$.

14. Show that for a convergent linear multistep method $\sum_{j=1}^{m} ja_j \neq 0$. [*Hint*: Show that the root condition and $\sum_{j=1}^{m} a_j = 1$ imply that $P'(1) \neq 0$, and then show that this implies $\sum_{j=1}^{m} ja_j \neq 0$.]

15. Reconsider the solution of the initial value problem $x' + x = \sin t, \quad x(0) = 1$ using the leapfrog method. Show that if the step size is reduced to $h = 0.01$, but the interval is extended to $0 \leq t \leq 20$, sawtooth oscillations still appear in the approximate solution.

16. Recall that Milne's method is the linear multistep method given by

$$\frac{w_{i+1} - w_{i-3}}{h} = \frac{4}{3}\left[2f(t_i, w_i) - f(t_{i-1}, w_{i-1}) + 2f(t_{i-2}, w_{i-2})\right].$$

(a) Show that Milne's method is weakly stable.

(b) Compare Milne's method and the four-step Adams-Bashforth method for solving the initial value problem $x' + x = \sin t, \quad x(0) = 1$. Use $h = 0.1$ and compute the approximate solutions over the interval $0 \leq t \leq 10$.

(c) Show that if the step size is reduced to $h = 0.01$, but the interval is extended to $0 \leq t \leq 15$, sawtooth oscillations still appear in the approximate solution obtained from Milne's method.

7.7 ERROR CONTROL AND VARIABLE STEP SIZE ALGORITHMS

The function whose graph is shown in Figure 7.12, $x(t) = t/(1+t^3)$, is the solution of the initial value problem

$$\frac{dx}{dt} = -3tx^2 + \frac{1}{1+t^3}, \qquad x(0) = 0.$$

If we were to use one of the numerical methods developed earlier in the chapter to solve this initial value problem, we would have to select a step size small enough to resolve the rapid variation in the solution that occurs from $t = 0$ through roughly $t = 2$. With that step size, we would then be doing much more work than necessary to resolve the more slowly varying portions of the solution for $t > 2$.

This situation is reminiscent of the one encountered in Chapter 6 with numerical integration. There, it was advantageous to develop algorithms which automatically selected the quadrature points to control the error in the computed approximation. In this section, we will discuss several different approaches for the construction of variable step size initial value problem solvers, which adapt the step size as the approximate solution evolves.

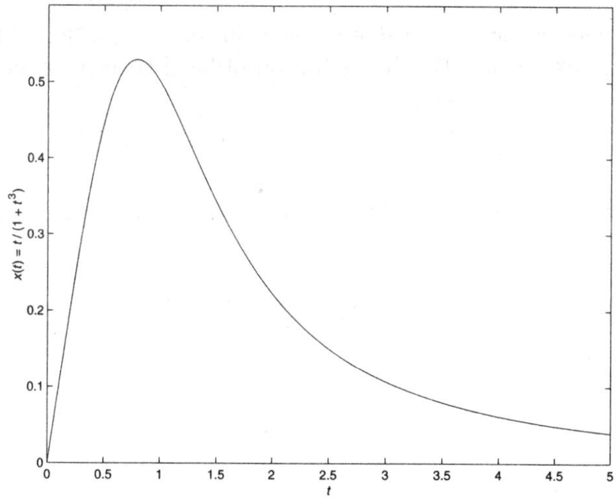

Figure 7.12 Solution of the initial value problem $x' = -3tx^2 + 1/(1+t^3)$, $x(0) = 0$.

One-Step Methods

Let's start with one-step methods. Multistep methods will be considered toward the end of the section. The first component of any variable step size algorithm for the solution of initial value problems is a procedure for estimating the error in the approximate solution. Ideally, we want an estimate for the global error in the approximate solution; however, such estimates are very difficult to come by. We will therefore focus on controlling the local truncation error in the approximate solution.

Let w_i be an approximation to the true solution of the initial value problem

$$y'(t) = f(t, y), \quad y(t_0) = y_0 \tag{1}$$

at $t = t_i$. Suppose that we want to advance the approximation to $t = t_{i+1} = t_i + h$ using the one-step method

$$w_{i+1} = w_i + h\phi(f, t_i, h, w_i), \tag{2}$$

which has local truncation error of order α_1. We can estimate the local truncation error that is introduced by this time step by comparing the value w_{i+1} computed above with the value generated by a different one-step method

$$\tilde{w}_{i+1} = w_i + h\tilde{\phi}(f, t_i, h, w_i),$$

which has local truncation error of order α_2, where $\alpha_2 > \alpha_1$. The details of this process are as follows.

Let $\hat{y}_{i+1}$ denote the exact value of the solution of equation (1) subject to the initial condition $\hat{y}(t_i) = w_i$. By the definition of local truncation error,

$$\frac{\hat{y}_{i+1} - w_i}{h} - \phi(f, t_i, h, w_i) = k_1 h^{\alpha_1} + o(h^{\alpha_1}). \tag{3}$$

Combining the terms on the left-hand side of equation (3) and substituting from equation (2) yields

$$\frac{\hat{y}_{i+1} - w_i}{h} - \phi(f, t_i, h, w_i) = \frac{\hat{y}_{i+1} - w_i - h\phi(f, t_i, h, w_i)}{h} = \frac{\hat{y}_{i+1} - w_{i+1}}{h}.$$

Therefore,

$$\frac{\hat{y}_{i+1} - w_{i+1}}{h} = k_1 h^{\alpha_1} + o(h^{\alpha_1}). \tag{4}$$

In a similar manner, we can establish that

$$\frac{\hat{y}_{i+1} - \tilde{w}_{i+1}}{h} = O(h^{\alpha_2}). \tag{5}$$

Subtracting equation (5) from equation (4) and using the fact that $O(h^{\alpha_2})$ is $o(h^{\alpha_1})$ since $\alpha_2 > \alpha_1$, it follows that

$$\frac{\tilde{w}_{i+1} - w_{i+1}}{h} = k_1 h^{\alpha_1} + o(h^{\alpha_1}).$$

Finally, assuming that the $o(h^{\alpha_1})$ term can be neglected, we see that

$$\epsilon = \frac{\tilde{w}_{i+1} - w_{i+1}}{h}$$

is an estimate for the local truncation error introduced by the time step taken with the lower-order method.

Having a credible estimate for the local truncation error in w_{i+1} available, it seems reasonable to use this estimate to improve the accuracy of w_{i+1}. This process, being similar to the extrapolation we applied to numerical differentiation and integration formulas in Chapter 6, but here being based on a local error estimate, is commonly known as *local extrapolation.* By substituting ϵ for $k_1 h^{\alpha_1}$ in (4) and simplifying, we find $\hat{y}_{i+1} \approx \tilde{w}_{i+1}$; hence, local extrapolation is equivalent to advancing the approximate solution using the higher-order method. Because local extrapolation increases accuracy at no additional computational expense, we will use it in the adaptive one-step methods we implement below.

Let TOL denote the maximum allowable local truncation error. If $|\epsilon| > TOL$, we must reduce the step size and repeat the step. If $|\epsilon| < TOL$, we may want to increase the step size so as to reduce the overall workload. This leads to the second component needed to construct a variable step size algorithm for the solution of an initial value problem: a procedure for automatically adjusting the step size. To determine an appropriate size for the next time step, recall that ϵ is an estimate for the error introduced when the step size is h. Since we are working with an underlying numerical method of order α_1, if the next step is taken to be of size qh, for some real number q, it follows that

$$\begin{array}{c}\text{error with step}\\ \text{of size } qh\end{array} \approx q^{\alpha_1} \begin{array}{c}\text{error with step}\\ \text{of size } h\end{array} = q^{\alpha_1}\epsilon.$$

We can therefore select q to satisfy $q^{\alpha_1}|\epsilon| < TOL$, or

$$q < \left(\frac{TOL}{|\epsilon|}\right)^{1/\alpha_1}.$$

Since several approximations have been made in arriving at this expression, we will be conservative and choose q to satisfy

$$q < \left(\frac{TOL}{2|\epsilon|}\right)^{1/\alpha_1}.$$

To prevent one very accurate time step from producing a large increase in the step size, as well as to prevent one very inaccurate time step from producing a large decrease in the step size, it is common to restrict the allowable range of q values. We will adopt the convention that $0.1 < q < 4.0$.

Any two methods of different order can be used to construct a variable step size algorithm according to the above process. For example, the classical fourth-order Runge-Kutta method can be used to estimate the local truncation error introduced by the third-order Taylor method. Using arbitrary methods, however, leads to a large overhead expense in terms of additional function evaluations. For the two methods listed in the previous sentence, the third-order Taylor method, by itself, requires three function evaluations per time step. Introducing the RK4 method to estimate the truncation error requires three more function evaluations—the value $f(t_i, w_i)$ can be reused. Hence, estimating the truncation error with these two methods doubles the computational effort per time step. To reduce the cost of estimating the truncation error, we need methods that share function values. Runge-Kutta methods are ideal for this situation. Given that there are many different Runge-Kutta methods of each order (recall that there are an infinite number of different second-order Runge-Kutta methods), it is possible to design special pairs of methods, called *embedded pairs*, that share several function values. For details of this design process, see Butcher [1] or Iserles [2].

A popular technique that falls into this category is the Runge-Kutta-Fehlberg fourth-order–fifth-order scheme, which will hereafter be referred to as the RKF45 method. This algorithm uses the fifth-order method

$$\tilde{w}_{i+1} = w_i + \frac{16}{135}k_1 + \frac{6656}{12825}k_3 + \frac{28561}{56430}k_4 - \frac{9}{50}k_5 + \frac{2}{55}k_6$$

to estimate the local truncation error in one time step of the fourth-order method

$$w_{i+1} = w_i + \frac{25}{216}k_1 + \frac{1408}{2565}k_3 + \frac{2197}{4104}k_4 - \frac{1}{5}k_5,$$

where

$$\begin{aligned}
k_1 &= hf(t_i, w_i) \\
k_2 &= hf\left(t_i + \frac{1}{4}h, w_i + \frac{1}{4}k_1\right) \\
k_3 &= hf\left(t_i + \frac{3}{8}h, w_i + \frac{3}{32}k_1 + \frac{9}{32}k_2\right) \\
k_4 &= hf\left(t_i + \frac{12}{13}h, w_i + \frac{1932}{2197}k_1 - \frac{7200}{2197}k_2 + \frac{7296}{2197}k_3\right) \\
k_5 &= hf\left(t_i + h, w_i + \frac{439}{216}k_1 - 8k_2 + \frac{3680}{513}k_3 - \frac{845}{4104}k_4\right) \\
k_6 &= hf\left(t_i + \frac{1}{2}h, w_i - \frac{8}{27}k_1 + 2k_2 - \frac{3544}{2565}k_3 + \frac{1859}{4104}k_4 - \frac{11}{40}k_5\right).
\end{aligned}$$

Note this technique uses just six function evaluations per time step. (An arbitrary combination of fourth- and fifth-order methods would require at least nine function evaluations—four for the fourth-order method plus six for the fifth-order method minus the reused value $f(t_i, w_i)$. See Butcher [1] for a discussion of why at least six values are needed to achieve fifth-order accuracy.) The truncation error estimate for the RKF45 method takes the form

$$\epsilon = \frac{\tilde{w}_{i+1} - w_{i+1}}{h} = \frac{\frac{1}{360}k_1 - \frac{128}{4275}k_3 - \frac{2197}{75240}k_4 + \frac{1}{50}k_5 + \frac{2}{55}k_6}{h}.$$

Since the lower-order method in this pair is fourth order, the step size adjustment factor is given by

$$q = \left(\frac{TOL}{2|\epsilon|}\right)^{1/4} \approx 0.84\left(\frac{TOL}{|\epsilon|}\right)^{1/4}.$$

Let's apply the RKF45 method to the initial value problem posed at the beginning of the section:

$$\frac{dx}{dt} = -3tx^2 + \frac{1}{1+t^3}, \qquad x(0) = 0.$$

We will approximate the solution over the range $0 \le t \le 5$. In addition to the function on the right-hand side of the differential equation, the initial condition and the final time, the algorithm requires three other parameters: the smallest allowable step size, $h_{\min}$, the largest allowable step size, $h_{\max}$, and TOL. For this problem, the values $h_{\min} = 0.01$, $h_{\max} = 0.25$, and $TOL = 5 \times 10^{-7}$ were used. Results are summarized in Figure 7.13. The graph in the top left panel displays the approximate solution, while the graph in the top right panel displays the global error in the approximate solution. The graph in the bottom left panel illustrates the evolution of the step size. Note how the step size automatically adjusts in response to the variation in the solution curve, decreasing to resolve smaller scale features and increasing to reduce computational effort in regions of larger scale features. In

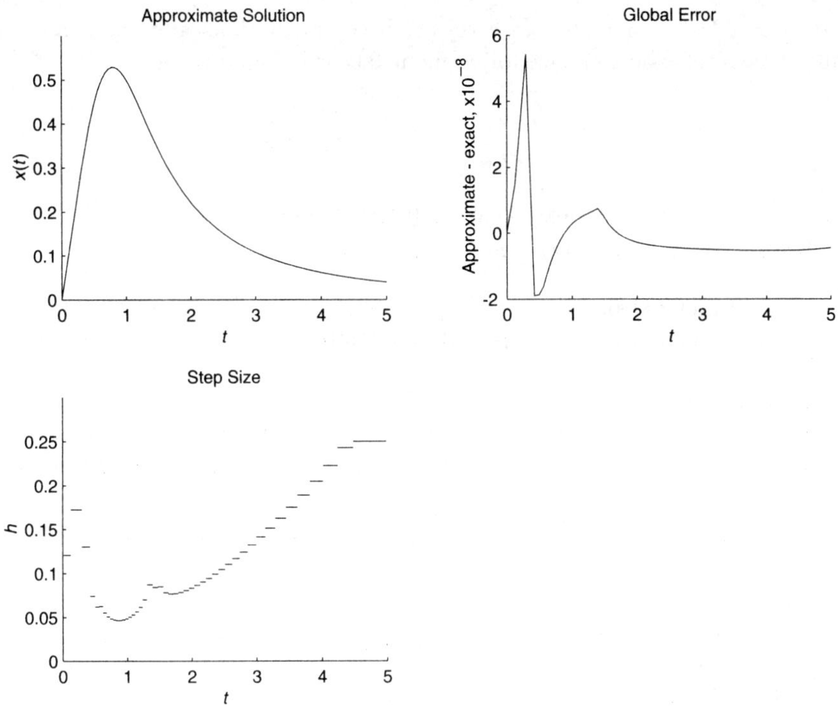

Figure 7.13 (Top left) Approximate solution of $x' = -3tx^2 + 1/(1+t^3)$, $x(0) = 0$ obtained using the RKF45 method. (Top right) Global error in approximate solution. (Bottom left) Step size used by RKF45 method as a function of location.

total, the RKF45 method required 48 time steps to traverse from $t = 0$ to $t = 5$ and used 318 function evaluations.

EXAMPLE 7.24 Some of the RKF45 Calculations

Let's take a look at how some of the calculations that led to Figure 7.13 were carried out. From $t_0 = w_0 = 0$, we attempt a step of length $h = h_{\max} = 0.25$. With these values and $f(t,x) = -3tx^2 + 1/(1+t^3)$, we find

$$k_1 = 0.250000000, \quad k_2 = 0.249755874, \quad k_3 = 0.249177100,$$
$$k_4 = 0.237901556, \quad k_5 = 0.234571481, \quad k_6 = 0.248061592.$$

The resulting estimate for the local truncation error is

$$\epsilon = \frac{\frac{1}{360}k_1 - \frac{128}{4275}k_3 - \frac{2197}{75240}k_4 + \frac{1}{50}k_5 + \frac{2}{55}k_6}{h} = -4.5831 \times 10^{-6}.$$

Since $|\epsilon| > TOL = 5 \times 10^{-7}$, we reject these calculations and try again with a smaller step size. The step size adjustment factor is found to be

$$q = \left(\frac{TOL}{2|\epsilon|}\right)^{1/4} = 0.482759999,$$

so we choose $h = (0.482759999)(0.25) = 0.120690000$.

With $t_0 = w_0 = 0$ and this new value for h, we obtain

$$\begin{array}{lll} k_1 = 0.120690000, & k_2 = 0.120676739, & k_3 = 0.120645252, \\ k_4 = 0.120023663, & k_5 = 0.119842509, & k_6 = 0.120584009 \end{array}$$

and the truncation error estimate $\epsilon = -6.0064 \times 10^{-8}$. Now $|\epsilon| < TOL$, so these calculations are deemed sufficiently accurate. We advance the independent variable to $t_1 = t_0 + h = 0.120690000$ and, taking advantage of local extrapolation, accept

$$\tilde{w}_1 = w_0 + \frac{16}{135}k_1 + \frac{6656}{12825}k_3 + \frac{28561}{56430}k_4 - \frac{9}{50}k_5 + \frac{2}{55}k_6 = 0.120478216$$

as an approximation to $x(t_1)$.

To prepare for the second time step, we compute a step size adjustment factor of $q = 1.426818220$ and therefore select $h = (1.426818220)(0.120690000) = 0.172202690$. The next round of calculations then yields

$$\begin{array}{lll} k_1 = 0.170995491, & k_2 = 0.169196284, & k_3 = 0.167870858, \\ k_4 = 0.157684015, & k_5 = 0.155588578, & k_6 = 0.166208019 \end{array}$$

and $\epsilon = 1.2658 \times 10^{-7}$. This value is, in magnitude, smaller than TOL, so the independent variable is advanced to $t_2 = t_1 + h = 0.292892690$ and $\tilde{w}_2 = 0.285713862$ is accepted as an approximation to $x(t_2)$.

The third time step is attempted with a step size of $h = 0.203923299$. Unfortunately, this leads to a truncation error estimate of $\epsilon = 1.4976 \times 10^{-6}$, which is too large. The third time step is then recomputed with $h = 0.130207704$. Now, $\epsilon = 1.8269 \times 10^{-7} < TOL$. The independent variable is advanced to $t_3 = 0.423100394$ and $\tilde{w}_3 = 0.393310681$ is accepeted as an approximation to $x(t_3)$. This process is continued until the independent variable has been advanced to $t = 5$.

Another common embedded pair is the Runge-Kutta-Verner fifth-order–sixth-order method (RKV56), which uses the sixth-order method

$$\tilde{w}_{i+1} = w_i + \frac{3}{40}k_1 + \frac{875}{2244}k_3 + \frac{23}{72}k_4 + \frac{264}{1955}k_5 + \frac{125}{11592}k_7 + \frac{43}{616}k_8$$

to estimate the local error in the fifth-order method

$$w_{i+1} = w_i + \frac{13}{160}k_1 + \frac{2375}{5984}k_3 + \frac{5}{16}k_4 + \frac{12}{85}k_5 + \frac{3}{44}k_6,$$

where

$$
\begin{aligned}
k_1 &= hf(t_i, w_i)\\
k_2 &= hf\left(t_i + \frac{1}{6}h, w_i + \frac{1}{6}k_1\right)\\
k_3 &= hf\left(t_i + \frac{4}{15}h, w_i + \frac{4}{75}k_1 + \frac{16}{75}k_2\right)\\
k_4 &= hf\left(t_i + \frac{2}{3}h, w_i + \frac{5}{6}k_1 - \frac{8}{3}k_2 + \frac{5}{2}k_3\right)\\
k_5 &= hf\left(t_i + \frac{5}{6}h, w_i - \frac{165}{64}k_1 + \frac{55}{6}k_2 - \frac{425}{64}k_3 + \frac{85}{96}k_4\right)\\
k_6 &= hf\left(t_i + h, w_i + \frac{12}{5}k_1 - 8k_2 + \frac{4015}{612}k_3 - \frac{11}{36}k_4 + \frac{88}{255}k_5\right)\\
k_7 &= hf\left(t_i + \frac{1}{15}h, w_i - \frac{8263}{15000}k_1 + \frac{124}{75}k_2 - \frac{643}{680}k_3 - \frac{81}{250}k_4 + \frac{2484}{10625}k_5\right)\\
k_8 &= hf\left(t_i + h, w_i + \frac{3501}{1720}k_1 - \frac{300}{43}k_2 + \frac{297275}{52632}k_3 - \frac{319}{2322}k_4 + \frac{24068}{84065}k_5 + \frac{3850}{26703}k_7\right).
\end{aligned}
$$

This method uses just eight new function evaluations per time step, as compared to the at least 12 evaluations that would be required of an arbitrary fifth-order, sixth-order combination. The truncation error estimate and the step size adjustment factor for the RKV56 method are, respectively,

$$
\epsilon = \frac{-\frac{1}{160}k_1 - \frac{125}{17952}k_3 + \frac{1}{144}k_4 - \frac{12}{1955}k_5 - \frac{3}{44}k_6 + \frac{125}{11592}k_7 + \frac{43}{616}k_8}{h}
$$

and

$$
q = \left(\frac{TOL}{2|\epsilon|}\right)^{1/5} \approx 0.87\left(\frac{TOL}{|\epsilon|}\right)^{1/5}.
$$

With the same parameter values as listed above ($h_{\min} = 0.01$, $h_{\max} = 0.25$ and $TOL = 5 \times 10^{-7}$), the RKV56 method was used to approximate the solution of

$$
\frac{dx}{dt} = -3tx^2 + \frac{1}{1+t^3}, \qquad x(0) = 0,
$$

and the results are shown in Figure 7.14. This figure is structured similar to Figure 7.13. The top left graph shows the approximate solution, the top right the global error, and the bottom left the evolution of the step size. Higher-order component methods allow the RKV56 method to use larger step sizes than the RKF45 method. In total, the RKV56 method required 33 time steps and 288 function evaluations for this problem.

Multistep Methods

Variable step size algorithms can also be constructed from multistep methods. Local truncation error is still estimated by comparing approximations generated from two different methods; however, when working with multistep methods, the convention is to combine an explicit method with an implicit method of the same order, as

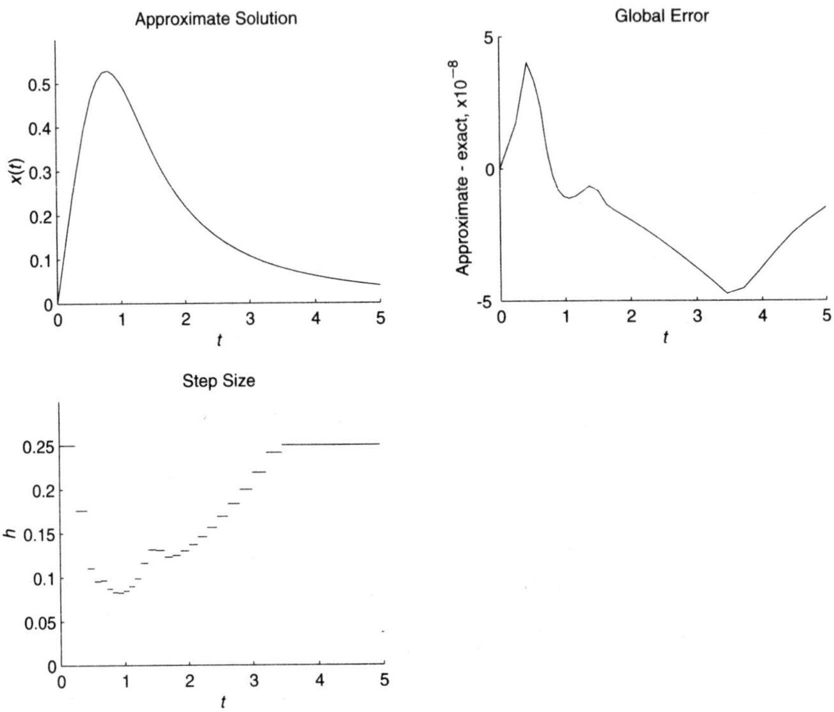

Figure 7.14 (Top left) Approximate solution of $x' = -3tx^2 + 1/(1+t^3)$, $x(0) = 0$ obtained using the RKV56 method. (Top right) Global error in approximate solution. (Bottom left) Step size used by RKV56 method as a function of location.

in predictor-corrector schemes. Let w_{i+1} be the approximation produced by the explicit method and $\tilde{w}_{i+1}$ the approximation from the implicit method. If each method is of order α and $\hat{y}_{i+1}$ is the exact value of the solution of equation (1) subject to the initial condition $\hat{y}(t_i) = w_i$—and we make the assumption that all other values used in the multistep equations are exact—then an analysis similar to that which led to equations (4) and (5) yields

$$\tau_i = \frac{\hat{y}_{i+1} - w_{i+1}}{h} = kh^\alpha + o(h^\alpha) \tag{6}$$

and

$$\tilde{\tau}_i = \frac{\hat{y}_{i+1} - \tilde{w}_{i+1}}{h} = \tilde{k}h^\alpha + o(h^\alpha). \tag{7}$$

Assuming that the $o(h^\alpha)$ terms can be neglected and then subtracting equation (7) from equation (6) gives

$$\frac{\tilde{w}_{i+1} - w_{i+1}}{h} \approx (\tilde{k} - k)h^\alpha.$$

Solving this last expression for h^α and substituting into equation (7), we find that the truncation error for the implicit method is approximately given by

$$\tilde{\tau}_i \approx \epsilon = \frac{\tilde{k}}{\tilde{k}-k}\frac{\tilde{w}_{i+1}-w_{i+1}}{h}.$$

This process for producing a truncation error estimate is known as the *Milne device.* The formula for the step size adjustment factor,

$$q = \left(\frac{TOL}{2|\epsilon|}\right)^{1/\alpha},$$

is determined in the same manner as for one-step methods. The factor of 2 in the denominator is included to be conservative.

As an example, suppose we construct a variable step size algorithm from the four-step, fourth-order Adams-Bashforth method

$$\frac{w_{i+1}-w_i}{h} = \frac{1}{24}\left[55f(t_i,w_i) - 59f(t_{i-1},w_{i-1}) + 37f(t_{i-2},w_{i-2}) - 9f(t_{i-3},w_{i-3})\right]$$

and the three-step, fourth-order Adams-Moulton method

$$\frac{\tilde{w}_{i+1}-w_i}{h} = \frac{1}{24}\left[9f(t_{i+1},w_{i+1}) + 19f(t_i,w_i) - 5f(t_{i-1},w_{i-1}) + f(t_{i-2},w_{i-2})\right].$$

The local truncation error for the Adams-Bashforth method is

$$\tau_i = \frac{251}{720}h^4 y^{(5)}(\xi),$$

where $t_i < \xi < t_{i+1}$. Since we can write

$$y^{(5)}(\xi) = y^{(5)}(t_i) + \left[y^{(5)}(\xi) - y^{(5)}(t_i)\right]$$

and we know that $\xi \to t_i$ as $h \to 0$, it follows that

$$\tau_i = \frac{251}{720}h^4 y^{(5)}(t_i) + o(h^4).$$

In a similar manner, we can establish that the local truncation error for the Adams-Moulton method is

$$\tilde{\tau}_i = -\frac{19}{720}h^4 y^{(5)}(t_i) + o(h^4).$$

Therefore, for this pair of methods, $k = \frac{251}{720}y^{(5)}(t_i)$ and $\tilde{k} = -\frac{19}{720}y^{(5)}(t_i)$, so that

$$\frac{\tilde{k}}{\tilde{k}-k} = \frac{-19/720}{-270/720} = \frac{19}{270}$$

and

$$\epsilon = \frac{19}{720}\frac{\tilde{w}_{i+1}-w_{i+1}}{h}.$$

Furthermore, the step size adjustment factor is

$$q = \left(\frac{TOL}{2|\epsilon|}\right)^{1/4} \approx 0.84 \left(\frac{TOL}{|\epsilon|}\right)^{1/4},$$

since the component methods are both fourth order.

When it comes to the implementation of a variable step size algorithm based on multistep methods, there are a few more details to take into account than there were with one-step methods. These are all related to the fact that multistep methods require starting values at equally spaced points, so that any change in step size necessitates computing new starting values. Therefore, when $|\epsilon| < TOL$ and $q > 1$, we do not automatically increase the step size. We have to balance the expense of the function evaluations needed to obtain new equally spaced starting values with the savings attributed to taking fewer steps. For the fourth-order Adams methods, experience suggests increasing the step size only when $|\epsilon| < 0.1 \cdot TOL$.

A second consideration involves determining the amount of work that needs to be redone when $|\epsilon| > TOL$. If the steps that preceded the one that just failed were used to compute new equally spaced starting values, then those starting values must be rejected along with the current step. We must return to the last accepted values and start over. On the other hand, if the preceding steps were ones computed with the multistep methods and accepted, we simply need to adjust the step size, compute new starting values, and continue. No prior results need to be rejected.

The last issue to address is the final time steps, those that advance the approximate solution to the final solution time, t_f. Generally, it will be necessary to adjust the step size so that the solution time exactly reaches t_f. Unfortunately, this means that several time steps will be needed to finish the solution process—those associated with computing new equally spaced starting values plus one step with the multistep methods. Therefore, the step size needs to be set not to t_f minus the current time, but to some fraction of this value, with the fraction determined by the number of starting values that need to be computed. For the fourth-order Adams methods, three new starting values must be computed, so we would set the step size to one-fourth the difference between t_f and the current time.

With the same parameter values applied earlier ($h_{\min} = 0.01$, $h_{\max} = 0.25$, and $TOL = 5 \times 10^{-7}$), the approximate solution of the equation posed at the beginning of the section was computed using a variable step size Adams fourth-order predictor-corrector (VS_PC4) method, and the results are shown in Figure 7.15. The top left graph shows the approximate solution, the top right the global error, and the bottom left the evolution of the step size. Note that the step sizes used by this method are significantly smaller than those used by either the RKF45 or the RKV56 method. This translates into $2\frac{1}{2}$ times and 4 times as many time steps (133 versus 48 and 33) as the RKF45 method and the RKV56 method, respectively. In terms of the number of function evaluations, which is a better measure of computational effort, the difference between the three methods is less pronounced. The current algorithm required 370 evaluations, just 16% more than RKF45 and 28% more than RKV56.

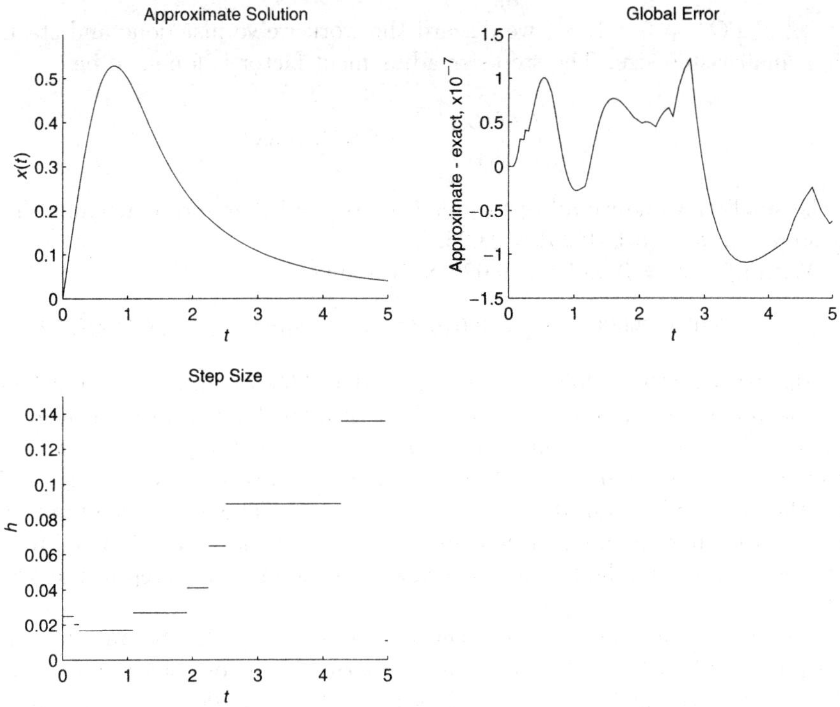

Figure 7.15 (Top left) Approximate solution of $x' = -3tx^2 + 1/(1+t^3)$, $x(0) = 0$ obtained using the VS_PC4 method. (Top right) Global error in approximate solution. (Bottom left) Step size used by VS_PC4 method as a function of location.

EXAMPLE 7.25 Some of the VS_PC4 Calculations

As we did earlier with the RKF45 method, we will now examine some of the calculations which went into the construction of Figure 7.15. From $t_0 = w_0 = 0$, we attempt a step of length $h = h_{\max} = 0.25$. To initialize the multistep methods, we use the classical fourth-order Runge-Kutta (RK4) method to calculate

$$w_1 = 0.246142027, \quad w_2 = 0.444346535, \quad \text{and} \quad w_3 = 0.527057387.$$

The Adams fourth-order predictor-corrector method then yields

$$w_4 = 0.475846091 \qquad \text{and} \qquad \tilde{w}_4 = 0.504652183.$$

The resulting estimate for the local truncation error is

$$\epsilon = \frac{19}{720}\frac{\tilde{w}_4 - w_4}{h} = 0.008108381.$$

Since $|\epsilon| > TOL = 5 \times 10^{-7}$, we discard the work we've just done and start over with a smaller step size. The step size adjustment factor is found to be

$$q = \left(\frac{TOL}{2|\epsilon|}\right)^{1/4} = 0.074516330,$$

but the smallest value we allow for q is 0.1. We will therefore reattempt the first time step with $h = (0.1)(0.25) = 0.025$.

With $t_0 = w_0 = 0$ and $h = 0.025$, we obtain

$$w_1 = 0.024999609, \quad w_2 = 0.049993751, \quad \text{and} \quad w_3 = 0.074968373$$

from the RK4 method, followed by $w_4 = 0.099900078$ and $\tilde{w}_4 = 0.099900103$ from the predictor-corrector method. The estimate for the local truncation error is $\epsilon = 7.0200 \times 10^{-8}$. Since $|\epsilon| < TOL$, we accept w_1, w_2, w_3, and $\tilde{w}_4$ as approximations to $x(0.025)$, $x(0.05)$, $x(0.075)$, and $x(0.1)$, respectively, and advance the independent variable to $t_4 = t_0 + 4h = 0.1$. Though we've obtained four values for the approximate solution, since only one of these was obtained from the predictor-corrector method, this is considered to be one time step of the VS_PC4 method.

For the second time step we continue to use $h = 0.025$ because $|\epsilon|$ was not smaller than $0.1 \cdot TOL$. Working only with the predictor-corrector method, we find $w_5 = 0.124756290$, $\tilde{w}_5 = 0.124756344$, and $\epsilon = 1.5064 \times 10^{-7}$. This local truncation error estimate is below TOL, so we advance the independent variable to $t_5 = 0.125$ and accept $x(t_5) \approx \tilde{w}_5$. The third and fourth time steps, both computed with $h = 0.025$, also produce sufficiently accurate results. In particular, we find $x(0.15) \approx \tilde{w}_6 = 0.149495471$ and $x(0.175) \approx \tilde{w}_7 = 0.174067140$.

Attempting another step of length $h = 0.025$, however, produces $\epsilon = 5.6576 \times 10^{-7}$, which is too large. The step size adjustment factor is $q = 0.815316647$, so we reduce the step size to $h = (0.815316647)(0.025) = 0.020382916$. The RK4 method is used three times to initialize the predictor-corrector method, which is then used to produce a truncation error estimate. With the smaller step size, we find $\epsilon = 4.7438 \times 10^{-7} < TOL$. The accepted approximations are

$$\begin{aligned} x(0.195382916) &\approx 0.193936444; \\ x(0.215765832) &\approx 0.213620059; \\ x(0.236148748) &\approx 0.233079328; \quad \text{and} \\ x(0.256531664) &\approx 0.252272845. \end{aligned}$$

The independent variable is advanced to 0.256531664, and we continue until we reach $t = 5$.

The results of this one test problem should not be taken to imply that the RKV56 method is always more efficient than the RKF45 method, which is always more efficient than the VS_PC4 method. There are problems for which the RKF45 method will be the most efficient, other problems for which the VS_PC4 method will be the most efficient, and still others for which all three algorithms will perform roughly the same. In fact, for the test problem we've been examining in this section, if we were to advance the independent variable to $t = 10$, we would find that the VS_PC4 method was the most efficient in terms of required function evaluations, followed by the RKF45 method and the RKV56 method, in that order.

References

1. J. C. Butcher, "The Non-existence of Ten-Stage Eighth-Order Runge-Kutta Methods," *BIT*, **25**, pp. 521–542, 1985.
2. A. Iserles, *A First Course in the Numerical Analysis of Differential Equations*, Cambridge Texts in Applied Mathematics, Cambridge University Press, Cambridge, 1996.

EXERCISES

1. Perform three steps of the RKF45 method for each of the following initial value problems. Take $h_{\min} = 0.0001$, $h_{\max} = 0.25$ and $TOL = 5 \times 10^{-7}$.
- **(a)** $x' = 1 + x/t, \quad x(1) = 1$
- **(b)** $x' = t/x, \quad x(0) = 1$
- **(c)** $x' = 2x(1 - x), \quad x(0) = 0.25$
- **(d)** $x' = 1.09 + 2x - x^3, \quad x(0) = -1.5$
- **(e)** $x' = -2x^2 + 1/(1 + t^2), \quad x(0) = 0$

2. Repeat Exercise 1 with the RKV56 method.

3. Repeat Exercise 1 with the VS_PC4 method, but change $h_{\max}$ to 0.15.

4. Suppose we were to construct a variable step size algorithm from the following embeddded pair: The third-order method

$$\tilde{w}_{i+1} = w_i + \frac{1}{4}k_1 + \frac{3}{8}k_2 + \frac{3}{8}k_3$$

is used to approximate the local truncation error in the second-order method

$$w_{i+1} = w_1 + \frac{1}{4}k_1 + \frac{3}{4}k_2,$$

where

$$\begin{aligned} k_1 &= hf(t_i, w_i) \\ k_2 &= hf\left(t_i + \frac{2}{3}h, w_i + \frac{2}{3}k_1\right) \\ k_3 &= hf\left(t_i + \frac{2}{3}h, w_i + \frac{2}{3}k_2\right). \end{aligned}$$

- **(a)** In terms of k_1, k_2 and k_3, what is the formula for the local truncation error estimate?
- **(b)** What is the formula for the step size adjustment factor q?

5. Repeat Exercise 1 with the variable step size algorithm described in Exercise 4, but change TOL to 5×10^{-4}.

6. Suppose we were to construct a variable step size algorithm from the two-step Adams-Bashforth method and the one-step Adams-Moulton method (the trapezoidal method).

(a) Using the Milne device, what formula would we use to estimate the local truncation error in the one-step Adams-Moulton method?

(b) What would be the formula for the step size adjustment factor q?

(c) What would be an appropriate choice for the method to initialize the predictor-corrector?

7. Repeat Exercise 1 with the variable step size algorithm described in Exercise 6, but change TOL to 5×10^{-4} and change h_{max} to 0.15.

8. Suppose we were to construct a variable step size algorithm from the three-step Adams-Bashforth method and the two-step Adams-Moulton method.

(a) Using the Milne device, what formula would we use to estimate the local truncation error in the two-step Adams-Moulton method?

(b) What would be the formula for the step size adjustment factor q?

(c) What would be an appropriate choice for the method to initialize the predictor-corrector?

In Exercises 9–11, use the RKF45 method to approximate the solution of the given initial value problem using the indicated parameter values. Plot the approximate solution and the step size selected by the algorithm as a function of the independent variable. Note any unusual behavior in the step size.

9. $x' = 100(1-x)$ $(0 \le t \le 2)$, $x(0) = 0$, $h_{\text{min}} = 0.0001$, $h_{\text{max}} = 0.25$, $TOL = 5 \times 10^{-7}$

10. $x' = -50(x - \cos t)$ $(0 \le t \le 10)$, $x(0) = 1$, $h_{\text{min}} = 0.001$, $h_{\text{max}} = 0.2$, $TOL = 10^{-4}$

11. $x' = 1 + 32(t - x)$ $(0 \le t \le 4)$, $x(0) = 1$, $h_{\text{min}} = 0.0001$, $h_{\text{max}} = 0.25$, $TOL = 5 \times 10^{-7}$

In Exercises 12–18, compare the performance (as measured by the number of function evaluations) of the RKF45 method, the RKV56 method, and the VS_PC4 method for the given initial value problem. Use the specified parameter values for each method.

12. $x' = e^x$ $(0 \le t \le 0.9)$, $x(0) = 0$, $h_{\text{min}} = 0.001$, $h_{\text{max}} = 0.25$, $TOL = 10^{-6}$

13. $x' = x - t^2 + 1$ $(0 \le t \le 2)$, $x(0) = 0.5$, $h_{\text{min}} = 0.01$, $h_{\text{max}} = 0.5$, $TOL = 5 \times 10^{-6}$

14. $x' = -3tx^2 + 1/(1+t^3)$ $(0 \le t \le 10)$, $x(0) = 0$, $h_{\text{min}} = 0.01$, $h_{\text{max}} = 0.25$ $TOL = 5 \times 10^{-7}$

15. $x' = 1.09 + 2x - x^3$ $(0 \le t \le 60)$, $x(0) = -3$, $h_{\text{min}} = 0.0001$, $h_{\text{max}} = 0.5$ $TOL = 5 \times 10^{-5}$

16. The "Spread of an Epidemic" Problem from Section 7.2

$$\frac{dD}{dt} = m\left[N - D - H_0 \exp(-cD/m)\right] \quad (0 \le t \le 15), \quad D(0) = 0$$
$$m = 1.8 \text{ week}^{-1}, \quad c = 0.001 \text{ (person} \cdot \text{week)}^{-1}$$
$$N = 3000 \text{ people}, \quad H_0 = 2850 \text{ people}$$
$$h_{\min} = 0.1 \text{ weeks}, \quad h_{\max} = 2 \text{ weeks}, \quad TOL = 5 \times 10^{-3}$$

17. The "Contrast Enhanced Lithography" Problem from Section 7.5

$$\frac{dM_c}{dz} = M_c\left[\frac{\alpha}{2}(1 - M_c^2) + \beta(1 - M_c) - B_c \ln M_c\right] \quad (0 \le z \le 0.4)$$
$$M_c(0) = \exp(-IC_c t)$$
$$\alpha = 1.4A_c, \quad \beta = -0.06A_c, \quad A_c = 8.93/\mu\text{m}$$
$$B_c = 0.175/\mu\text{m}, \quad C_c = 0.0376 \text{ cm}^2/\text{mJ},$$
$$It = 110 \text{ mJ/cm}^2, \quad h_{\min} = 0.001\mu\text{m}, \quad h_{\max} = 0.05\mu\text{m}, \quad TOL = 10^{-6}$$

18. The "Genetic Switch" Problem from Section 7.4

$$\frac{dg}{dt} = 0.206 + 3.03\frac{g^2}{1 + g^2} - 1.51g \quad (0 \le t \le 100), \quad g(0) = 0$$
$$h_{\min} = 0.1, \quad h_{\max} = 5, \quad TOL = 5 \times 10^{-5}$$

7.8 SYSTEMS OF EQUATIONS AND HIGHER-ORDER EQUATIONS

Systems of Equations

Many of the phenomena that interest scientists give rise not to a scalar initial value problem but rather to a system of initial value problems. To study the dynamics of interacting populations, (at least) one equation is needed for each of the relevant populations. Even in the simplest models of two populations interacting either in a predator-prey or a competitive relationship, a system of equations of the form

$$\begin{aligned}
\frac{dx_1}{dt} &= a_1x_1 - b_1x_1^2 + c_1x_1x_2 \\
\frac{dx_2}{dt} &= a_2x_2 - b_2x_2^2 + c_2x_1x_2
\end{aligned}$$

must be studied. The system of three equations

$$\begin{aligned}
\frac{dL}{dt} &= -3.6L + 1.2\left[P(1 - P^2) - 1\right] \\
\frac{dP}{dt} &= -1.2P + 6\left(L + \frac{2}{1 + Z}\right) \\
\frac{dZ}{dt} &= -0.12Z + 12P
\end{aligned}$$

has been used to model the emotional and inspirational cycle of a fourteenth-century Italian poet (see Exercise 16). Larger systems are also easy to find. For example, the study of combustion dynamics can give rise to extermely large systems. In addition to equations for the relevant thermodynamic properties, there could be hundreds, or even thousands, of species involved in the chemical reactions taking place, and each species requires its own differential equation.

Those who have studied systems of differential equations know that the analysis of systems generally requires more advanced mathematical techniques than the analysis of scalar equations. Systems of equations also exhibit a wider variety of possible behaviors than do scalar equations. Fortunately, however, as we will see below, the numerical analysis of systems requires little more than a notational change from the scalar case. The derivations of the methods will be identical, and the advantages and disadvantages of the various methods will be unchanged.

The most general system of m first-order initial value problems can be written in the form

$$\begin{array}{ll} u_1'(t) = f_1(t, u_1, u_2, u_3, \ldots, u_m) & u_1(a) = \alpha_1 \\ u_2'(t) = f_2(t, u_1, u_2, u_3, \ldots, u_m) & u_2(a) = \alpha_2 \\ u_3'(t) = f_3(t, u_1, u_2, u_3, \ldots, u_m) & u_3(a) = \alpha_3 \\ \vdots & \vdots \\ u_m'(t) = f_m(t, u_1, u_2, u_3, \ldots, u_m) & u_m(a) = \alpha_m, \end{array}$$

where t is the independent variable (we will once again assume that we are interested in the range $a \leq t \leq b$) and $u_1, u_2, u_3, \ldots, u_m$ are the dependent variables. This system is most efficiently represented using vector notation. Toward that end, let

$$\mathbf{u}(t) = \begin{bmatrix} u_1(t) \\ u_2(t) \\ u_3(t) \\ \vdots \\ u_m(t) \end{bmatrix} \qquad \boldsymbol{\alpha} = \begin{bmatrix} \alpha_1 \\ \alpha_2 \\ \alpha_3 \\ \vdots \\ \alpha_m \end{bmatrix}$$

and

$$\mathbf{f}(t, \mathbf{u}) = \begin{bmatrix} f_1(t, u_1, u_2, u_3, \ldots, u_m) \\ f_2(t, u_1, u_2, u_3, \ldots, u_m) \\ f_3(t, u_1, u_2, u_3, \ldots, u_m) \\ \vdots \\ f_m(t, u_1, u_2, u_3, \ldots, u_m) \end{bmatrix}.$$

Then

$$\mathbf{u}'(t) = \begin{bmatrix} u_1'(t) \\ u_2'(t) \\ u_3'(t) \\ \vdots \\ u_m'(t) \end{bmatrix},$$

and the original system of equations can be written very compactly as

$$\mathbf{u}'(t) = \mathbf{f}(t, \mathbf{u}), \quad a \le t \le b$$
$$\mathbf{u}(a) = \boldsymbol{\alpha}.$$

Once we have recognized that a system of first-order differential equations can be written in a form similar to that of a scalar first-order equation, but with scalar notation replaced by vector notation, we see that all of the key concepts and definitions (such as local truncation error and global discretization error) and all of the algorithms we have spent so much time developing carry over directly from the scalar case. We simply have to replace all of the scalar notation with the appropriate vector notation.

For example, Euler's method takes the form

$$w_0 = \alpha$$
$$\frac{w_{i+1} - w_i}{h} = f(t_i, w_i)$$

in the scalar case; whereas, for a system of equations, Euler's method takes the form

$$\mathbf{w}^{(0)} = \boldsymbol{\alpha}$$
$$\frac{\mathbf{w}^{(j+1)} - \mathbf{w}^{(j)}}{h} = \mathbf{f}\left(t_j, \mathbf{w}^{(j)}\right).$$

Note that in the vector notation, we have used a superscript to denote the time step so that subscripts can be used in the conventional manner to denote components within a vector. As a second example, the two-step Adams-Bashforth method for a scalar equation is

$$\frac{w_{i+1} - w_i}{h} = \frac{3}{2} f(t_i, w_i) - \frac{1}{2} f(t_{i-1}, w_{i-1}).$$

For a system of equations, the method equation reads

$$\frac{\mathbf{w}^{(j+1)} - \mathbf{w}^{(j)}}{h} = \frac{3}{2}\mathbf{f}\left(t_j, \mathbf{w}^{(j)}\right) - \frac{1}{2}\mathbf{f}\left(t_{j-1}, \mathbf{w}^{(j-1)}\right);$$

that is, the method retains the exact same functional form but with vector quantities substituted for the appropriate scalar values.

Although the concepts, definitions, and derivations of the various numerical methods are unchanged when dealing with systems of equations, there are certain programming considerations that must be handled. First, the user-supplied function that computes the values of **f** must take an array of input values for the dependent variables, in addition to the value of the independent variable, and return an array of values. Second, the code must contain a loop to update each component of the **w** vector, not just a single value. Next, for those methods that make use of intermediate values, such as the RK4 method, or that save previous function values for efficiency reasons, such as the Adams-Bashforth and Adams-Moulton methods, intermediate storage may be needed. This issue is, of course, heavily dependent on the choice of programming language. For adaptive schemes, like the RKF45 method, a norm must be used to compute the error estimate. Finally, Newton's method for systems of nonlinear equations will be needed for implicit schemes like the trapezoidal method. These are the only changes needed to translate the code for a single equation to handle a system of equations.

Worked Examples: Systems of Equations

EXAMPLE 7.26 Cooling of a Container and its Liquid Contents

Following an experiment, a small container and its liquid contents are at a temperature of 150°F. To cool both the liquid and the container to room temperature (70°F), the container is immersed in a bath held at 32°F. To model the cooling of both the container and the liquid, let L denote the temperature of the liquid and C denote the temperature of the container. Balancing the rate of change of energy storage in the liquid and the container with the rate of convective heat transfer (between the liquid and the container and between the container and the bath) leads to the coupled system of differential equations:

$$\frac{dL}{dt} = \frac{A_1 h}{\rho_1 c_{p,1} v_1}(C - L)$$

$$\frac{dC}{dt} = \frac{A_2 h}{\rho_2 c_{p,2} v_2}(32 - C) + \frac{A_1 h}{\rho_2 c_{p,2} v_2}(L - C).$$

The parameter values for this system are summarized in the table below.

	Liquid	Container
Mass density	$\rho_1 = 62\ \text{lb}_m/\text{ft}^3$	$\rho_2 = 139\ \text{lb}_m/\text{ft}^3$
Specific heat	$c_{p,1} = 1.00\ \text{Btu}/\text{lb}_m \cdot {}^\circ\text{F}$	$c_{p,2} = 0.2\ \text{Btu}/\text{lb}_m \cdot {}^\circ\text{F}$
Volume	$v_1 = 0.03\ \text{ft}^3$	$v_2 = 0.003\ \text{ft}^3$

The convective heat transfer coefficient, h, is assumed to be 8.8 Btu/hr · ft^3 · °F, and the surface contact areas are $A_1 = 0.4\ \text{ft}^2$ and $A_2 = 0.5\ \text{ft}^2$.

Anticipating a rapid initial temperature drop for the container, this model was simulated using the RKF45 method with a minimum step size of 0.0001, a maximum step size of 0.2, an error tolerance of 0.0001 per unit step, and the infinity norm to measure the error estimates. The resulting temperature profiles are shown in the graph in Figure 7.16. The container reaches room temperature in roughly 15 minutes, but it takes slightly more than an hour for the liquid temperature to reach 70°F.

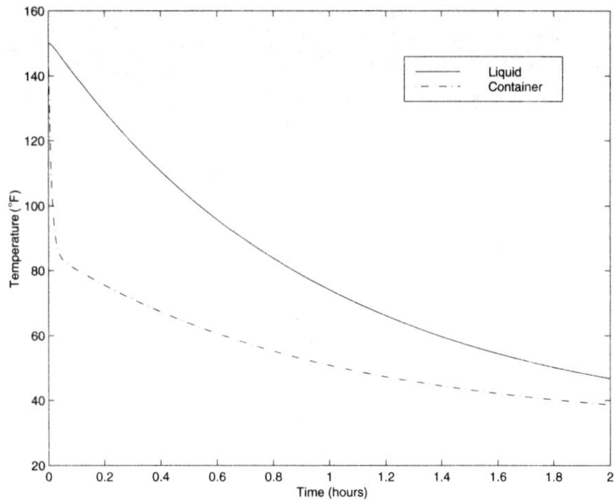

Figure 7.16 Temperature profiles for Example 7.26.

EXAMPLE 7.27 A Catalyzed Reaction

In the Overview to this chapter (see "A Catalyzed Reaction," page 535), we developed a model for a catalyzed chemical reaction between two reactants, which we called A and B. The model originally consisted of five equations, which we reduced to the system

$$\alpha' = -\alpha(\lambda - \beta + \alpha), \qquad \alpha(0) = 1$$
$$\beta' = -\kappa\beta(\beta - \alpha), \qquad \beta(0) = 1.$$

Here, α and β denote the normalized concentration of A and B, respectively, λ is the ratio of the initial concentration of the catalyst to the initial concentration of A, and κ is the ratio of the reaction rates of the two pathways by which the reaction takes place.

Let's fix $\kappa = 1$ and examine the effect of λ on the dynamics of the reaction. We will investigate the effect of κ on reaction dynamics in the exercises. For $\lambda = 10$, 1, 0.1, and 0.01, the model equations were solved using the RKF45 method with parameter values $h_{\min} = 0.001$, $h_{\max} = 0.5$ and $TOL = 5 \times 10^{-5}$. The independent variable was advanced from $t = 0$ to $t = 100$, 150, 200, and 1000 for the respective values of λ. Since the presence of the catalyst is supposed to speed up the reaction, it should not be surprising that larger catalyst concentrations (larger values of λ) cause the reaction to reach equilibrium more rapidly. The results of the calculations are shown below in Figure 7.17.

Apparently, the value of λ influences the relative rates at which the reactants A and B are depleted during the reaction. When λ is "small," the solution trajectories remain near the $\alpha = \beta$ line of the phase plane, indicating that the two reactants

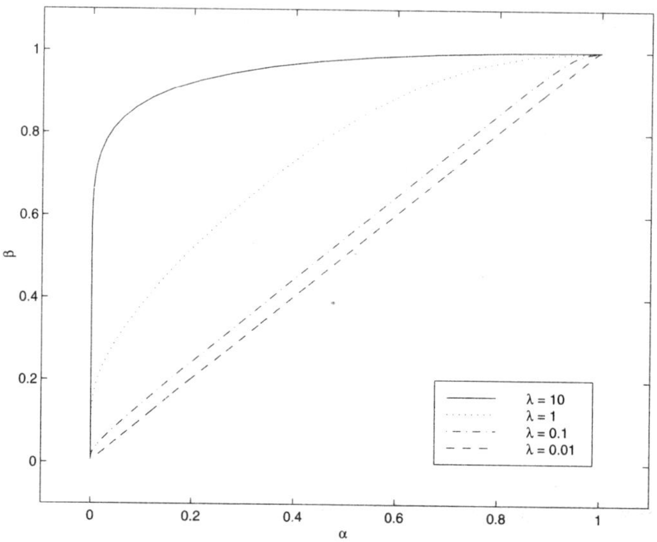

Figure 7.17 Influence of λ on the dynamics of a catalyzed reaction.

are being depleted at roughly the same rate. For larger λ, however, we observe that reactant A is consumed more rapidly than reactant B and that the disparity between the depletion rates grows with increasing λ.

Higher-Order Equations

Consider the mth-order initial value problem

$$y^{(m)}(t) = f(t, y, y', y'', \dots, y^{(m-1)})$$

$$y(a) = \alpha_1, y'(a) = \alpha_2, y''(a) = \alpha_3, \dots, y^{(m-1)}(a) = \alpha_m.$$

Any mth-order initial value problem can always be reduced to a system of m first-order equations by the introduction of an appropriate set of variables. Once this conversion has been performed, any of the numerical techniques that we have discussed in this chapter and have extended to systems of equations earlier in this section can be used.

The reduction to a system of equations begins with the definition of the intermediate variables

$$u_1(t) = y(t)$$
$$u_2(t) = y'(t)$$

$$u_3(t) = y''(t)$$
$$\vdots$$
$$u_m(t) = y^{(m-1)}(t).$$

These definitions, in turn, imply

$$\begin{aligned}
u_1'(t) &= y'(t) = u_2(t)\\
u_2'(t) &= y''(t) = u_3(t)\\
u_3'(t) &= y'''(t) = u_4(t)\\
&\vdots\\
u_m'(t) &= y^{(m)}(t) = f(t, u_1, u_2, u_3, \ldots, u_m),
\end{aligned}$$

together with the associated initial conditions

$$u_1(a) = \alpha_1, u_2(a) = \alpha_2, u_3(a) = \alpha_3, \ldots, u_m(a) = \alpha_m.$$

In vector form, the original mth-order equation can now be written as the system of m first-order equations

$$\mathbf{u}'(t) = \mathbf{f}(t, \mathbf{u}(t)),$$

where

$$\mathbf{u}(t) = \begin{bmatrix} u_1(t) \\ u_2(t) \\ u_3(t) \\ \vdots \\ u_m(t) \end{bmatrix} \quad \text{and} \quad \mathbf{f}(t, \mathbf{u}) = \begin{bmatrix} u_2(t) \\ u_3(t) \\ u_4(t) \\ \vdots \\ f(t, \mathbf{u}) \end{bmatrix}.$$

Worked Examples: Higher-Order Equations

EXAMPLE 7.28 The van der Pol Equation

A famous, nonlinear second-order differential equation is the van der Pol equation:

$$x'' + \epsilon(x^2 - 1)x' + x = 0,$$

where ϵ is a positive constant. One can think of this equation as a model for a mass-spring-damper system with nonlinear damping. Since the damping coefficient, $\epsilon(x^2 - 1)$, is positive for $|x| > 1$, energy is drained from the system when the amplitude of the motion is large, which tends to decrease the amplitude of the motion; however, for $|x| < 1$, the damping term becomes negative, implying that energy is supplied to the system. It seems reasonable, therefore, to expect that the solution will exhibit some sort of nonlinear oscillatory, or limit cycle, behavior.

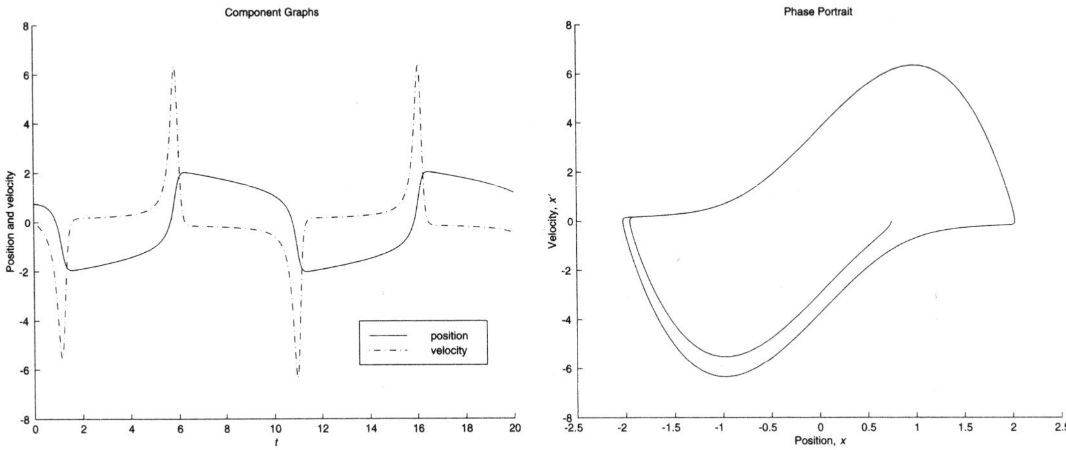

Figure 7.18 Solution to van der Pol equation with $\epsilon = 4$.

We would like to investigate this behavior numerically. To accomplish this, we first convert the second-order differential equation into a system of first-order equations. Define the intermediate variables

$$\begin{aligned} u_1 &= x \\ u_2 &= x'; \end{aligned}$$

the resulting system then takes the form

$$\begin{aligned} u_1' &= u_2 \\ u_2' &= \epsilon(1 - u_1^2)u_2 - u_1. \end{aligned}$$

Let's take $\epsilon = 4$ and apply the initial conditions $x(0) = 0.75, x'(0) = 0$. Using the RKF45 method with a minimum allowed step size of 0.01, a maximum step size of 0.5, an error tolerance of 0.0001 per unit step, and the infinity norm to measure error estimates, the results shown in Figure 7.18 are obtained. The limit cycle behavior of the solution is clear. From the component graphs, the period of the nonlinear oscillation appears to be slightly more than 10 time units. In the phase portrait, motion is clockwise around the limit cycle.

EXAMPLE 7.29 Bubble Dynamics

Lithotripsy is a medical procedure whereby brief, intense ultrasound pulses are used to break kidney stones into small enough pieces that can be passed naturally. The lithotripsy pulses induce cavitation, and the oscillations in the radius of the resulting bubbles play a role in the break down of the kidney stones. Howle, Schaeffer,

Shearer, and Zhong ("Lithotripsy: The Treatment of Kidney Stones with Shock Waves," *SIAM Review*, pp. 356–371, 1998) investigate the free vibrations of these cavitation induced air bubbles. The model they consider is the second-order initial value problem

$$R\ddot{R} + \frac{3}{2}\dot{R}^2 = a^2\left[\left(\frac{R_0}{R}\right)^{3\gamma} - 1\right] \qquad R(0) = AR_0, \quad \dot{R}(0) = 0,$$

where dots denote differentiation with respect to time. R is the time-varying radius of the air bubble, $R_0 = 3 \times 10^{-6}$ meters is the equilibrium radius of the bubble, $\gamma = 1.4$ is the adiabatic exponent, and $a = \sqrt{p_{\text{atm}}/\rho} \approx 10$ m/s.

Note the large difference in order of magnitude between the value of R_0 and that of a. Whenever system parameters vary in size by many orders of magnitude, it is best to nondimensionalize the variables, taking into account the natural scales of the problem, before running simulations. For this model, R_0 is the most obvious choice for the length scale. As for the time scale, note that R_0/a has dimensions of time. We will therefore introduce the nondimensional radius and nondimensional time variables

$$r = R/R_0 \qquad \text{and} \qquad \tau = (a/R_0)t$$

into the original initial value problem, producing

$$rr'' + \frac{3}{2}(r')^2 = \left[r^{-3\gamma} - 1\right] \qquad r(0) = A, \quad r'(0) = 0,$$

where primes denote differentiation with respect to the variable τ.

To solve this initial value problem, we next introduce the intermediate variables

$$y_1 = r \qquad \text{and} \qquad y_2 = dr/d\tau,$$

which transform the second-order differential equation into the system of first-order equations:

$$\begin{aligned} y_1' &= y_2, & y_1(0) &= A \\ y_2' &= (y_1^{-3\gamma} - 1 - 1.5y_2^2)/y_1, & y_2(0) &= 0. \end{aligned}$$

Take $A = 2.5$. Using the RKV56 method with a minimum step size of 0.0001, a maximum step size of 0.05, and a tolerance of 0.00001 per unit step produces the results summarized in Figure 7.19. The period of the oscillation is roughly 4.7 time units, which translates to a dimensional period of approximately 1.4 microseconds: $(4.7)(3 \times 10^{-6}\text{ m})/(10\text{ m/s})$.

EXERCISES

1. Advance the solution of each of the following systems through $N = 4$ time steps using Euler's method and a step size of $h = 0.25$.
   (a) $x' = y, \; y' = -x - 2y, \quad x(0) = 2, \; y(0) = -1$

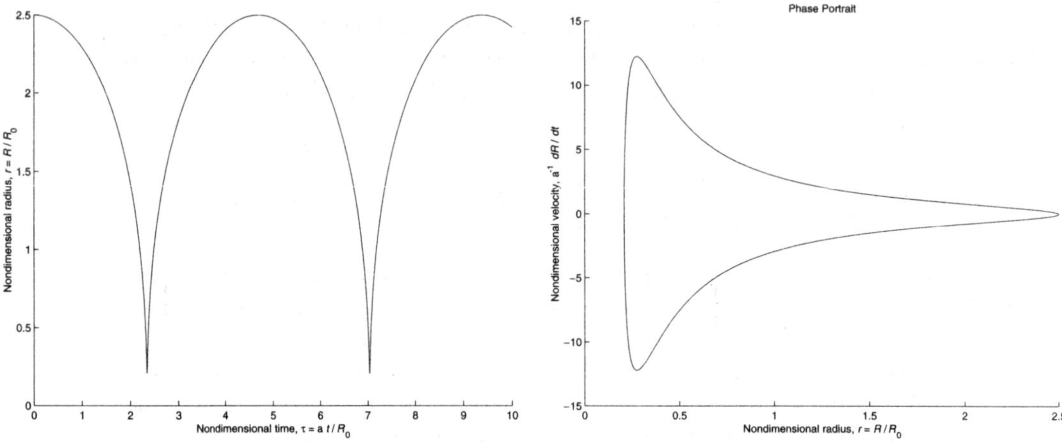

Figure 7.19 Radius of cavitation induced bubbles curing lithotripsy.

(b) $x' = xy - 1, \quad y' = x - y^3, \qquad x(0) = 1, \quad y(0) = 0$
(c) $x' = 2x + z, \quad y' = 2y + z, \quad z' = -z, \qquad x(0) = 1, \quad y(0) = -1, \quad z(0) = 1$
(d) $x' = y - x, \quad y' = \frac{1}{2}x - y - xz, \quad z' = xy - z, \qquad x(0) = y(0) = z(0) = 1$

2. Repeat Exercise 1 using the second-order Runge-Kutta method of your choice.

3. Repeat Exercise 1 using the classical fourth-order Runge-Kutta method.

4. Repeat Exercise 1 using the two-step Adams-Bashforth method.

5. Repeat Exercise 1 using the four-step Adams-Bashforth method and $N = 6$ time steps.

6. Repeat Exercise 1 using the Adams fourth-order predictor-corrector and $N = 6$ time steps.

7. Convert each of the following nth-order differential equations to a system of first-order equations.
(a) $x''' + 4x'' + 5x' = 0$
(b) $x^{(4)} = x$
(c) $x''' + \frac{1}{2}xx'' = 0$
(d) $x''' = x' \ln(x'') + \sin(x)$
(e) $x'' + \sin(xx') = 1$
(f) $x^{(4)} = \cos(x''') - \sqrt{xx''}$

8. Convert each of the following to a system of first-order differential equations.
(a) $x' + \sin(x) = 1 - z', \quad y' + \sin(y) = 1 - z', \quad z'' + z' + z = x' + y'$
(b) $x'' = (x')^2 + x - \sin(t), \quad y'' = \sqrt{y} - ty'$
(c) $x'' = y - x, \quad y'' = x - 2y + z, \quad z'' = y - 2z$

9. Advance the solution of each of the following systems through $N = 4$ time steps using Euler's method and a step size of $h = 0.25$.
(a) $x''' + \frac{1}{2}xx'' = 0, \quad x(0) = 0, x'(0) = 0, x''(0) = 1$

(b) $x''' + 4x'' + 5x' = 0, \quad x(0) = 1, x'(0) = 0, x''(0) = -1$
(c) $x'' + (x^2 - 1)x' + x = 0, \quad x(0) = 0.5, x'(0) = 0.1$
(d) $x'' + 4x' + x = \sin(t), \quad x(0) = 1, x'(0) = 0$

10. Repeat Exercise 9 using the second-order Runge-Kutta method of your choice.

11. Repeat Exercise 9 using the classical fourth-order Runge-Kutta method.

12. Repeat Exercise 9 using the two-step Adams-Bashforth method.

13. Repeat Exercise 9 using the four-step Adams-Bashforth method and $N = 6$ time steps.

14. Repeat Exercise 9 using the Adams fourth-order predictor-corrector and $N = 6$ time steps.

15. To help understand the long term effects of stress on plants and animals, the following food-limited population model has been proposed (Nisbet, McCauley, Gurney, Murdoch, and de Roos, "Simple Representations of Biomass Dynamics in Structured Populations," in *Case Studies in Mathematical Modeling—Ecology, Physiology and Cell Biology*, Othmer, Adler, Lewis and Dallon, eds., Prentice Hall, 1997, pp. 61–79):

$$\begin{aligned}\frac{dF}{dt} &= \Phi - \frac{I_{\max}FC}{F + F_h}\\ \frac{dC}{dt} &= \frac{\epsilon I_{\max}FC}{F + F_h} - (m + b)C.\end{aligned}$$

Here, $F(t)$ denotes the food biomass density, and $C(t)$ denotes the herbivore (consumer) biomass density. Φ represents the net rate of food supply to the model. The various model parameters include

$I_{\max} = 6.5 \text{ day}^{-1}$	maximum specific ingestion rate;
$F_h = 0.98 \text{ mg/L}$	half-saturation constant;
$\epsilon = 0.75$	assimilation efficiency of herbivore;
$m = 0.04 \text{ day}^{-1}$	death rate of herbivore; and
$b = 0.23 \text{ day}^{-1}$	respiration rate of herbivore.

Simulate 30 days of the biomass dynamics for a constant food supply rate of $\Phi = 0.125 \text{ mg/L} \cdot \text{day}$. Take $F(0) = 0$ and $C(0) = 0.1$.

16. S. Rinaldi ("Laura and Petrarch: An Intriguing Case of Cyclical Love Dynamics," *SIAM J. Appl. Math.*, **58**, pp. 1205–1221, 1998) presents the following model for the emotional and inspirational cycle of the fourteenth-century Italian poet Petrarch:

$$\begin{aligned}\frac{dL}{dt} &= -3.6L + 1.2\left[P(1 - P^2) - 1\right]\\ \frac{dP}{dt} &= -1.2P + 6\left(L + \frac{2}{1 + Z}\right)\\ \frac{dZ}{dt} &= -0.12Z + 12P.\end{aligned}$$

Here, L represents the love of Laura (a beautiful woman who was Petrarch's inspiration) for Petrarch, P represents the magnitude of Petrarch's love for Laura, and Z represents the poet's inspiration level. Time is measured in years. Starting from the initial conditions

$$L(0) = P(0) = Z(0) = 0,$$

simulate 21 years of Petrarch's emotional cycle. Display your results as functions of time and in the $P - L$ and $Z - P$ phase planes.

17. Neglecting the effect of air resistance, the motion of a pendulum can be modeled by the second-order initial value problem

$$L\theta'' + g\sin\theta = 0, \qquad \theta(0) = \theta_0, \quad \theta'(0) = 0,$$

where θ denotes the angle which the pendulum rod makes with the vertical, L is the length of the pendulum rod, and g is the acceleration due to gravity. For this problem, take $L = 1$ meter and $g = 9.8$ m/s^2. Estimate the period of the pendulum for values of θ_0 ranging from 0.1 through 2.0 in increments of 0.1.

18. Recall the catalyzed reaction problem considered earlier in this section:

$$\begin{aligned} \alpha' &= -\alpha(\lambda - \beta + \alpha), \qquad & \alpha(0) = 1 \\ \beta' &= -\kappa\beta(\beta - \alpha), \qquad & \beta(0) = 1. \end{aligned}$$

Fix $\lambda = 1$, and investigate the effect of κ on the reaction dynamics. Consider $\kappa = 10$, 1, 0.1, and 0.01.

19. Recall the van der Pol equation considered earlier in this section:

$$x'' + \epsilon(x^2 - 1)x' + x = 0.$$

Using the initial conditions $x(0) = 0.75$ and $x'(0) = 0$, estimate the period of the motion for values of ϵ ranging from 2.0 through 6.0 in increments of 0.25.

20. Howle, Schaeffer, Shearer, and Zhong ("Lithotripsy: The Treatment of Kidney Stones with Shock Waves," *SIAM Review*, pp. 356–371, 1998) also investigate the direct effect of ultrasound pulses on the cavitation induced air bubbles. In this case, the model they consider is the second-order initial value problem

$$\frac{1}{a^2}\left[R\ddot{R} + \frac{3}{2}\dot{R}^2\right] = \left(\frac{R_0}{R}\right)^{3\gamma} - \tilde{p}_\infty(t), \qquad R(0) = R_0, \quad \dot{R}(0) = 0,$$

where dots denote differentiation with respect to time and

$$\tilde{p}_\infty(t) = \begin{cases} 2000e^{-t/\tau_1}\cos(t/\tau_2 + \pi/3) + 1, & 0 < t < (7\pi/6)\tau_2 \\ 1, & \text{otherwise.} \end{cases}$$

Here, R is the time-varying radius of the air bubble, $R_0 = 3 \times 10^{-6}$ meters is the equilibrium radius of the bubble, $\gamma = 1.4$ is the adiabatic exponent, and $a = \sqrt{p_{\text{atm}}/\rho} \approx 10$ m/s. The time constants in $\tilde{p}_\infty(t)$ are $\tau_1 = 1.1 \times 10^{-6}$ seconds and $\tau_2 = 1.9 \times 10^{-6}$ seconds. Estimate the period of the bubble motion. Remember to introduce nondimensional radius and time variables, and don't forget to nondimensionalize the time constants τ_1 and τ_2.

21. In a study of nonlinear spatial developments of two-dimensional wall jets on curved surfaces, Le Cunff and Zebib ("Nonlinear Spatially Developing Görtler Vortices in Curved Wall Jet Flow," *Phys. Fluids*, **8**, pp. 2375–2384, 1996) require the solution of the initial value problem

$$x''' + \frac{1}{4}(xx'' + 2(x')^2) = 0, \quad x(0) = x'(0) = 0, x''(0) = \frac{2^{5/2}}{9}.$$

Approximate the solution of this problem from $t = 0$ to $t = 20$.

22. In the Overview to this chapter (see page 534), we developed the model

$$\frac{d^2y}{dt^2} = -\frac{e^y}{\kappa}\left(1 + \frac{dy}{dt}\right), \qquad y(0) = \ln u_0, \quad y'(0) = \frac{1 - u_0}{\kappa}$$

for the growth of a population in a closed system. Here, y is the natural logarithm of the population, u_0 is the initial population, and κ is a dimensionless parameter. Investigate the evolution of the population from $t = 0$ to $t = 5$ for $\kappa = 0.05$, 0.1, 0.25, and 0.5. Take $u_0 = 0.1$.

23. D. Winter ("On the Stem Curve of a Tall Palm in a Strong Wind," *SIAM Review*, **35**, pp. 567–579, 1993) develops the following model for the stem curve of a palm tree subject to wind loading:

$$\begin{aligned}\frac{d^2\theta}{ds^2} &= \frac{W_s}{EI}\left(1 - \frac{s}{L} + \frac{W_c}{W_s}\right)\sin\theta + \frac{D}{EI}\cos\theta \\ \frac{dx}{ds} &= \sin\theta \\ \frac{dz}{ds} &= \cos\theta.\end{aligned}$$

The variables in the problem are the angle of the stem relative to the vertical position, θ, the arc length measured along the stem, s, the horizontal displacement of the stem, x, and height of a location along the stem, z. Both x and z are treated as functions of s. The parameters are the total stem weight $W_s = 22700$ N, the Young's modulus of the stem $E = 6 \times 10^9$ N/m^2, the moment of inertia of the stem $I = 5.147 \times 10^{-4}$ m^4, the length of the stem $L = 30$ m, the total canopy weight $W_c = 1385.5$ N, and the wind drag force on the canopy $D = 4.135U^2$ N, where U is the wind speed in m/s. With initial conditions of

$$\theta(0) = \theta'(0) = x(0) = z(0) = 0,$$

simulate the stem curve of a palm tree subject to a wind speed of 18 m/s.

7.9 ABSOLUTE STABILITY AND STIFF EQUATIONS

An Example to Motivate the Discussion

A chemostat is a completely mixed, continuously stirred tank reactor used for growing cells—such as yeast or bacteria—in a controlled environment. The chemostat consists of a large vessel with a mechanical stirring paddle. An excess of nutrients are provided for the growth of the cells, with the exception of one controlled medium, known as the substrate. A tube leading into the chemostat provides a controlled flow of substrate, along with chemical compounds needed to maintain an environment favorable to cell growth. A second tube, leading out of the vessel, is used to siphon fluid from the chemostat so that cells can be harvested. The flow rates into and out from the chemostat are held equal so as to maintain constant volume within the vessel.

Boyd and Wang ("Optimizing Cell Production in a Chemostat," in Proceedings of *Mathematical Modeling in the Undergraduate Curriculum*, H. Skala, editor, University of Wisconsin/La Crosse, 1994) present the following model for the dynamics of a chemostat:

$$\frac{ds}{dt} = 1 - s - \frac{msx}{a+s} \tag{1}$$

$$\frac{dx}{dt} = \frac{msx}{a+s} - x \tag{2}$$

Here, s is the nondimensional mass of substrate within the chemostat per unit volume, and x is the nondimensional cell mass per unit volume. The parameters in this system, m and a, denote the maximal cell growth rate (due to nutrient consumption) relative to the volumetric flow rate into and out from the chemostat and the half-saturation level relative to the inflow substrate concentration, respectively. The half-saturation level is the concentration of substrate at which the per-capita cell growth rate achieves half its maximal value.

The chemostat equations, (1) and (2), possess two equilibrium solutions. One equilibrium solution is located at the point (1, 0), and the other is located at

$$(s^*, x^*) = \left(\frac{a}{m-1}, 1 - \frac{a}{m-1}\right).$$

Of course, the second equilibrium solution is physically meaningful only for $m > 1$ and $a < m - 1$. Suppose $m = 16$ and $a = 0.25$. Then

$$(s^*, x^*) = \left(\frac{1}{60}, \frac{59}{60}\right).$$

Linear stability analysis about this point indicates that the solution is asymptotically stable with characteristic exponents $\lambda_1 = -1$ and $\lambda_2 = -55.31$. This means that near the point (s^*, x^*), the solution of (1) and (2) behaves like

$$\begin{bmatrix} s(t) \\ x(t) \end{bmatrix} \approx \begin{bmatrix} s^* \\ x^* \end{bmatrix} + c_1 e^{-t}\mathbf{v}_1 + c_2 e^{-55.31t}\mathbf{v}_2,$$

for some constants c_1 and c_2, where $\mathbf{v}_1$ and $\mathbf{v}_2$ are the eigenvectors associated with the characteristic exponents.

The top graph of Figure 7.20 displays the approximate solution to equations (1) and (2) with $m = 16$ and $a = 0.25$ and initial conditions $s(0) = 0.5$, and $x(0) = 0.02$. The solution was computed using the RKF45 method with

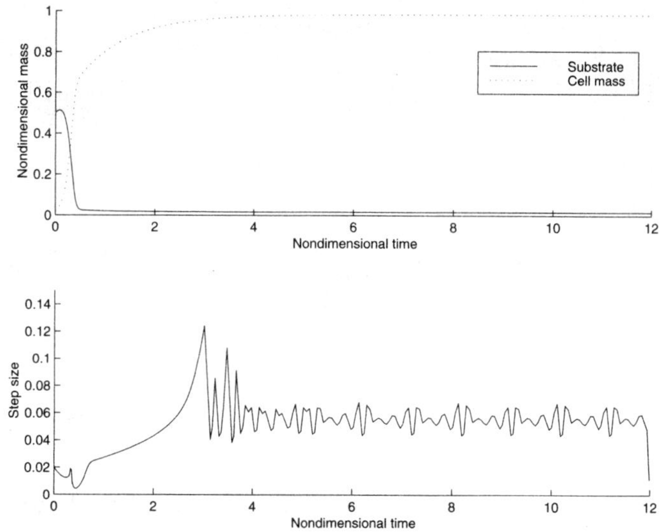

Figure 7.20 Solution of equations (1) and (2) with $m = 16$ and $a = 0.25$ obtained using the RKF45 method. (Top graph) Time evolution of substrate and cell mass concentrations within the chemostat. (Bottom graph) Evolution of step sizes used to compute approximate solution.

$h_{\min} = 0.001$, $h_{\max} = 0.5$, and $TOL = 5 \times 10^{-7}$. As expected, the solution exhibits fairly rapid convergence toward equilibrium. What is not expected, however, is the time evolution of the step sizes used to obtain the approximate solution, as displayed in the bottom graph of Figure 7.20. For $t > 2$, both components of the solution remain essentially constant, with changes in the solution occurring on a time scale on the order of 1—the reciprocal of the absolute value of the less negative characteristic exponent λ_1. Despite the observed behavior in the solution, the time steps selected by the RKF45 method hover around the 0.04–0.06 range, which is roughly on the same order as the time scale associated with the more negative characteristic exponent λ_2 ($\approx 1/55.31 = 0.018$).

The results contained in Figure 7.21 are equally unexpected. The top graph shows the approximate solution of the chemostat equations computed using the classical fourth-order Runge-Kutta (RK4) method with a step size of $h = 0.05$. The parameter values and initial conditions are the same as above: $m = 16$, $a = 0.25$, $s(0) = 0.5$, and $x(0) = 0.02$. The bottom graph shows the solution computed with the RK4 method and a time step of $h = 0.06$. The solution computed with $h = 0.05$ is essentially identical to the solution obtained with the RKF45 method. With $h = 0.06$, however, the approximate solution initially evolves much like the previous solutions, but sometime after $t = 2$ breaks down and settles onto a long-term behavior which is quite different from the asymptotic nature of the exact solution.

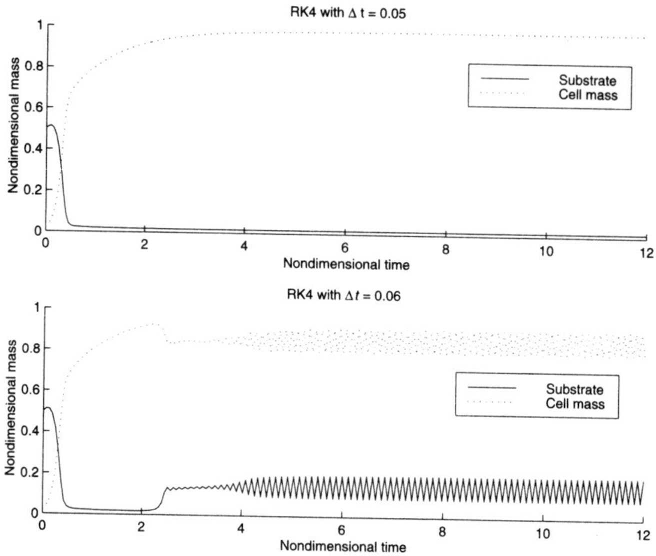

Figure 7.21 Solution of equations (1) and (2) with $m = 16$ and $a = 0.25$ obtained using the RK4 method with $h = 0.05$ (top graph) and $h = 0.06$ (bottom graph).

Figures 7.20 and 7.21 raise several important questions. Why does the RKF45 method select such small step sizes to approximate a nearly constant solution? Furthermore, why is there so much variation in the selected step sizes? With regard to the RK4 method, why did such a small change in the step size cause the approximate solution to fail to maintain the asymptotic character of the exact solution? Finally, why do the step sizes required by these two methods to obtain accurate approximations appear to be determined by the most rapidly decaying component of the solution? As we will now establish, the answers to these questions are related to the ability of the difference equations which define the numerical methods to approximate decaying exponentials.

Test Problem

To investigate the issues raised by Figures 7.20 and 7.21, consider the standard test problem

$$y' = \lambda y,$$

where λ is complex valued with $Re(\lambda) < 0$. This is the simplest differential equation that produces exponential solutions, and the real part of the parameter λ completely controls the time scale of the decay in the solution. We allow λ to be complex valued in order for our analysis to be applicable to systems of differential equations, in which case λ would represent an eigenvalue of the Jacobian associated with the right-hand side of the system. Recall that the Jacobian of a vector-valued function is a matrix of partial derivatives (see Section 3.10) and that the eigenvalues of a real matrix can be complex.

The next step is to apply the numerical methods developed earlier in this chapter to the test problem. The objective is to determine what restrictions, if any, are placed on the step size h to guarantee that the asymptotic character of the approximate solution matches that of the analytical solution. Since the analytical solution of the test problem decays to zero as $t \to \infty$, we will want the approximate solution of the test problem to decay to zero as $n \to \infty$. When the asymptotic character of the approximate solution produced by a given numerical method matches that of the analytical solution, the numerical method is said to be *absolutely stable*.

One-Step Methods

The one-step methods introduced in Sections 7.2, 7.3, and 7.4 fell into two categories: Taylor methods and Runge-Kutta methods. Our analysis of absolute stability will be simplified, however, by the fact that the right-hand side of the test problem is linear in both the independent and the dependent variables. In this case, all of the Runge-Kutta methods reduce to the corresponding Taylor method of the same order. We can therefore restrict attention to the Taylor methods.

The mth-order Taylor method is defined by the difference equation

$$w_n = w_{n-1} + hf(t_{n-1}, w_{n-1}) + \frac{h^2}{2}\frac{df}{dt}(t_{n-1}, w_{n-1}) + \cdots + \frac{h^m}{m!}\frac{d^{m-1}f}{dt^{m-1}}(t_{n-1}, w_{n-1}).$$

For the test problem under consideration, $f(t, y) = \lambda y$, so

$$\frac{d^k f}{dt^k} = \frac{d^k(\lambda y)}{dt^k} = \lambda \frac{d^k y}{dt^k}.$$

Since $y' = \lambda y$, it is straightforward to establish that $y^{(k)} = \lambda^k y$, and therefore

$$\frac{d^k f}{dt^k} = \lambda^{k+1} y.$$

Substituting this last expression into the equation for the Taylor method yields

$$\begin{aligned} w_n &= w_{n-1} + h\lambda w_{n-1} + \frac{h^2}{2}\lambda^2 w_{n-1} + \cdots + \frac{h^m}{m!}\lambda^m w_{n-1} \\ &= \left(1 + h\lambda + \frac{1}{2}(h\lambda)^2 + \cdots + \frac{1}{m!}(h\lambda)^m\right) w_{n-1}. \end{aligned}$$

Now, let

$$Q(z) = 1 + z + \frac{1}{2}z^2 + \cdots + \frac{1}{m!}z^m.$$

Then the Taylor method applied to the test problem becomes

$$w_n = [Q(h\lambda)]\, w_{n-1}. \tag{3}$$

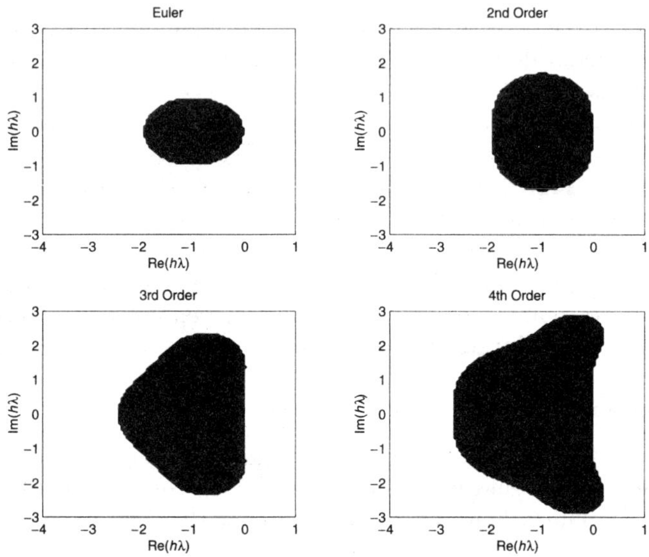

Figure 7.22 Regions of absolute stability for the Taylor/Runge-Kutta methods of order one (Euler's method) through four.

The solution to the elementary difference equation, (3), is given by

$$w_n = [Q(h\lambda)]^n \, w_0;$$

hence, w_n will decay toward zero with increasing n, matching the asymptotic character of the exact solution of the test problem, provided that $|Q(h\lambda)| < 1$. The set of all values of $h\lambda$ for which $|Q(h\lambda)| < 1$ is called the region of absolute stability for the given one-step method.

Definition. The REGION OF ABSOLUTE STABILITY for a one-step method is the set

$$R = \{h\lambda \in \mathbf{C} \mid |Q(h\lambda)| < 1\}.$$

Figure 7.22 displays the region of absolute stability associated with the order one through four Taylor/Runge-Kutta methods. Remember that Euler's method is the first-order Taylor method. Two observations can be made from this figure. First, the regions of absolute stability grow with increasing order of the method. This should not be surprising considering that with increasing order, more terms in the Taylor expansion of the exact solution are retained, so the method provides a better approximation to the exponential function. The second observation is that the regions are of finite size. Thus, for a given λ, absolute stability considerations will place an upper bound on the step size that can be used to compute the approximate solution.

How does this information help to explain the phenomena that were observed in the numerical solution of the chemostat problem at the beginning of the section? Consider a fourth-order Runge-Kutta method, and suppose that λ is real. For a fourth-order Runge-Kutta method,

$$Q(z) = 1 + z + \frac{1}{2}z^2 + \frac{1}{6}z^3 + \frac{1}{24}z^4,$$

so absolute stability requires

$$\left|1 + h\lambda + \frac{1}{2}(h\lambda)^2 + \frac{1}{6}(h\lambda)^3 + \frac{1}{24}(h\lambda)^4\right| < 1.$$

The solution of the above inequality is

$$-2.7853 < h\lambda < 0,$$

meaning that we must choose $h < -2.7853/\lambda$. Recall that there were two different values of λ for the chemostat problem. Since h is inversely proportional to λ, it follows that the upper bound on h will be derived from $\lambda_2 = -55.31$, which is larger in magnitude than $\lambda_1 = -1$. This state of affairs will hold true in general. When a problem possesses multiple time scales, the product $h\lambda$ must lie in the region of absolute stability for all values of λ. The maximum allowable time step based on absolute stability considerations will therefore be dictated by the λ that is largest in magnitude. Returning to the specifics of the chemostat problem, we see that absolute stability requires

$$h < \frac{-2.7853}{\lambda_2} \approx 0.05036 \equiv h_c.$$

Now reexamine Figure 7.21. With $h = 0.05 < h_c$, the RK4 method is absolutely stable, but with $h = 0.06 > h_c$, the method is not absolutely stable. The breakdown in the approximation to $e^{\lambda_2 t}$ for $h = 0.06$ causes the approximate solution to exhibit an asymptotic behavior different from that of the exact solution. As for the time steps selected by the RKF45 method for $t > 2$ (bottom graph of Figure 7.20), these can be explained as follows. When $h < h_c$, the underlying methods are absolutely stable, so estimates of the local truncation error result in the selection of a larger step size. Eventually, h grows larger than h_c, and absolute stability is lost. After a few time steps, the breakdown in the approximation to $e^{\lambda_2 t}$ produces large local truncation error estimates and forces the step size below h_c. The cycle then repeats, generating oscillation about the cutoff value for absolute stability.

Multistep Methods

Recall that the general linear m-step multistep method is of the form

$$\frac{w_{i+1} - a_1 w_i - a_2 w_{i-1} - \cdots - a_m w_{i+1-m}}{h} =$$
$$b_0 f(t_{i+1}, w_{i+1}) + b_1 f(t_i, w_i) + b_2 f(t_{i-1}, w_{i-1}) + \cdots + b_m f(t_{i+1-m}, w_{i+1-m}). \quad (4)$$

Applying equation (4) to the test problem $y' = \lambda y$ and collecting like terms leads to

$$(1-b_0h\lambda)w_{i+1}-(a_1+b_1h\lambda)w_i-(a_2+b_2h\lambda)w_{i-1}-\cdots-(a_m+b_mh\lambda)w_{i+1-m}=0. \quad (5)$$

The solution of this linear difference equation is related to the roots of the corresponding characteristic polynomial

$$Q(z,h\lambda)=(1-b_0h\lambda)z^m-(a_1+b_1h\lambda)z^{m-1}-(a_2+b_2h\lambda)z^{m-2}-\cdots-(a_m+b_mh\lambda).$$

In particular, suppose that $\beta_1(h\lambda)$, $\beta_2(h\lambda)$, $\beta_3(h\lambda), \ldots, \beta_m(h\lambda)$ are the roots of $Q(z,h\lambda)$. If the β_k are all distinct, then the solution of (5) can be written as

$$w_n=\sum_{k=1}^{m} c_k\,[\beta_k(h\lambda)]^n,$$

where the coefficients c_k are determined by the values of w_0, w_1, $w_2, \ldots w_{m-1}$. When the roots are not distinct, the form of the solution changes somewhat but will still depend on powers of each of the roots of $Q(z,h\lambda)$. Thus, it is clear that the general multistep method will be absolutely stable only when $|\beta_k(h\lambda)|<1$ for each k.

Definition. The REGION OF ABSOLUTE STABILITY for a multistep method is the set

$$R=\{h\lambda\in\mathbf{C}\mid |\beta_k(h\lambda)|<1 \text{ for each root of } Q(z,h\lambda)\}.$$

EXAMPLE 7.30 Two Charateristic Polynomials and Their Roots

Recall that the two-step Adams-Bashforth method is defined by the difference equation

$$\frac{w_{n+1}-w_n}{h}=\frac{3}{2}f(t_n,w_n)-\frac{1}{2}f(t_{n-1},w_{n-1}).$$

Applying this method to the test problem $y' = \lambda y$ leads to, after collecting like terms,

$$w_{n+1}-(1+\frac{3}{2}h\lambda)w_n+\frac{1}{2}h\lambda w_{n-1}=0.$$

The characteristic polynomial associated with this difference equation is

$$Q(z,h\lambda)=z^2-(1+\frac{3}{2}h\lambda)z+\frac{1}{2}h\lambda,$$

whose roots can be obtained from the quadratic formula:

$$\beta_1(h\lambda)=\frac{1}{2}\left[\left(1+\frac{3}{2}h\lambda\right)+\sqrt{1+h\lambda+\frac{9}{4}(h\lambda)^2}\right]$$

$$\beta_2(h\lambda)=\frac{1}{2}\left[\left(1+\frac{3}{2}h\lambda\right)-\sqrt{1+h\lambda+\frac{9}{4}(h\lambda)^2}\right].$$

The set of $h\lambda$ values for which both of these roots have magnitude less than unity defines the region of absolute stability for this numerical method—see the upper left panel of Figure 7.23.

As a second illustration, consider the two-step Adams-Moulton method:

$$\frac{w_{n+1} - w_n}{h} = \frac{5}{12}f(t_{n+1}, w_{n+1}) + \frac{2}{3}f(t_n, w_n) - \frac{1}{12}f(t_{n-1}, w_{n-1}).$$

The characteristic polynomial associated with this difference equation is

$$Q(z, h\lambda) = (1 - \frac{5}{12}h\lambda)z^2 - (1 + \frac{2}{3}h\lambda)z + \frac{1}{12}h\lambda,$$

whose roots can again be obtained from the quadratic formula:

$$\beta_1(h\lambda) = \frac{(1 + \frac{2}{3}h\lambda) + \sqrt{1 + h\lambda + \frac{7}{12}(h\lambda)^2}}{2(1 - \frac{5}{12}h\lambda)}$$

$$\beta_2(h\lambda) = \frac{(1 + \frac{2}{3}h\lambda) - \sqrt{1 + h\lambda + \frac{7}{12}(h\lambda)^2}}{2(1 - \frac{5}{12}h\lambda)}.$$

The region of absolute stability determined by these roots is plotted in the upper right panel of Figure 7.23.

Figure 7.23 displays the region of absolute stability for the two- and three-step Adams-Bashforth and Adams-Moulton methods. Two important observations can be made from this figure. First, the regions of absolute stability for the implicit methods (the Adams-Moulton methods) are significantly larger than the regions for their explicit counterparts. Second, unlike the Taylor/Runge-Kutta methods for which the regions of absolute stability increase in size with increasing order of approximation, here we see that the regions decrease in size with increasing order of approximation. Thus higher-order Adams-Bashforth and Adams-Moulton methods will require smaller values of h to maintain absolute stability.

Stiff Equations and A-stable Methods

The chemostat problem used to open this section belongs to a large class of differential equations that are called *stiff*. Loosely speaking, a stiff equation is one for which the step size of certain numerical methods must be drastically decreased, over at least a portion of the domain, to maintain absolute stability. The analysis we have just completed tells us that the mechanism that gives rise to stiffness is the presence of vastly different evolutionary time scales in the solution. Thus, whenever a differential equation is modeling a process with multiple, widely varied time scales, we must be on guard for stiffness. Electronics, weather prediction, chemical kinetics, and mathematical biology are common sources of stiff differential equations.

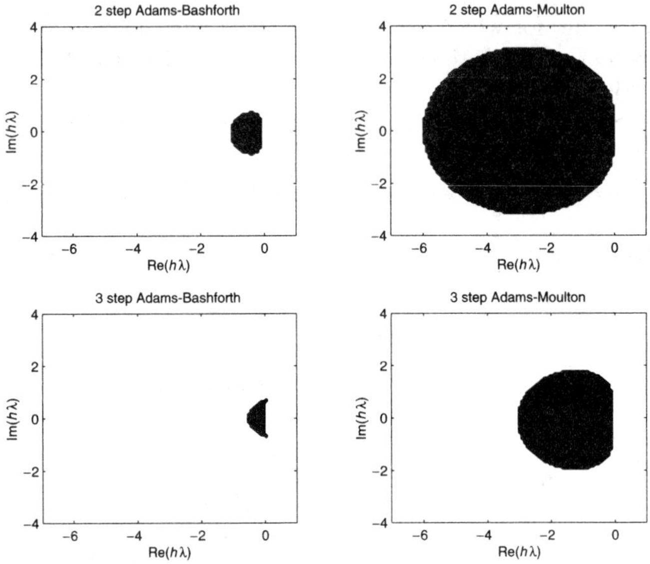

Figure 7.23 Regions of absolute stability for the two- and three-step Adams-Bashforth (left panels) and Adams-Moulton methods (right panels).

Suppose $\{\lambda_k\}$ is the set of characteristic exponents associated with a particular differential equation. The quantity

$$S = \frac{\max_k |\text{Re } \lambda_k|}{\min_k |\text{Re } \lambda_k|}$$

is called the *stiffness ratio* and is often quoted as a measure of the degree of stiffness inherent in a given problem. The stiffness ratio for the chemostat problem is $S = 55.31$. As stiffness ratios go, this value is quite tame. Problems in chemical kinetics routinely involve ratios on the order of 10^{17}, and a model of the Big Bang has a ratio on the order of 10^{31}.

Since the techniques we have examined thus far have such small regions of absolute stability, the application of these techniques to stiff problems is not advised. Even the variable step size algorithms are poor choices. Though the variable step size methods will automatically select an appropriate step size, these step sizes may become so ridiculously small as to make the total computational cost prohibitive. Further, when exceedingly small step sizes are used, the propagation of roundoff error can become a significant problem. The bottom line is that to approximate efficiently the solution of stiff problems, we need methods that have regions of absolute stability that are as large as possible. The ideal situation would be for the region of absolute stability to contain the entire left-hand side of the complex plane, as then the step size could be selected solely on the basis of accuracy, and not absolute stability. Such methods do exist and are called *A-stable*.

Definition. A numerical method is A-STABLE if it is absolutely stable for all $h\lambda$ such that Re $h\lambda < 0$.

The Trapezoidal Method

The only A-stable method among those introduced in the previous sections of this chapter is the trapezoidal method:

$$\frac{w_{n+1} - w_n}{h} = \frac{1}{2}\left[f(t_{n+1}, w_{n+1}) + f(t_n, w_n)\right]. \tag{6}$$

Recall that this is a second-order, implicit one-step method. When applied to the test problem $y' = \lambda y$, equation (6) reduces to

$$w_{n+1} = \left[\frac{2 + h\lambda}{2 - h\lambda}\right] w_n.$$

Thus

$$Q(h\lambda) = \frac{2 + h\lambda}{2 - h\lambda},$$

and the region of absolute stability for the trapezoidal method is given by

$$R = \left\{ h\lambda \in \mathbf{C} \,\middle|\, \left|\frac{2 + h\lambda}{2 - h\lambda}\right| < 1 \right\}.$$

Since $|z| < 1$ if and only if $|z|^2 < 1$, we can continue our analysis by examining $|Q(h\lambda)|^2$. Separating $h\lambda$ into real and imaginary parts, we find

$$\begin{aligned}\left|\frac{2 + h\lambda}{2 - h\lambda}\right|^2 &= \left|\frac{(2 + \text{Re } h\lambda) + i\text{Im } h\lambda}{(2 - \text{Re } h\lambda) - i\text{Im } h\lambda}\right|^2 \\ &= \frac{(2 + \text{Re } h\lambda)^2 + (\text{Im } h\lambda)^2}{(2 - \text{Re } h\lambda)^2 + (\text{Im } h\lambda)^2}.\end{aligned}$$

Next, set the last expression less than one, clear the fraction and expand both terms involving the real part of $h\lambda$ to yield

$$4 + \text{Re } h\lambda + (\text{Re } h\lambda)^2 + (\text{Im } h\lambda)^2 < 4 - \text{Re } h\lambda + (\text{Re } h\lambda)^2 + (\text{Im } h\lambda)^2.$$

Some final simplification produces Re $h\lambda < 0$. Hence, the region of absolute stability for the trapezoidal method is the entire left-hand side of the complex plane, and the method is A-stable.

For a linear differential equation, equation (6) can be solved explicitly for w_{n+1} in terms of w_n, t_n, t_{n+1} and h; however, for nonlinear problems, each time step of the trapezoidal method will require the solution of an implicit equation for w_{n+1}. Solving this implicit equation by functional iteration (e.g., in predictor-corrector fashion) should not be attempted as this would impose a convergence condition on the maximum size of h with the same effect as a stability condition. Instead, a rootfinding technique such as Newton's method or the Secant method should be employed. With Newton's method, it is customary to select the value of w_n as the initial guess for the value of w_{n+1}. The secant method, of course, requires two starting values. w_n can serve as one of those starting values, and the usual practice is to obtain the second starting value from an explicit method. When working with a system of differential equations, Newton's method, Broyden's method, or one of the other quasi-Newton methods mentioned in Section 3.10 can be used.

One major drawback to the trapezoidal method is that although the method is A-stable and hence will correctly predict the long-term behavior of the solution for any value of h, the method will generally fail to produce an accurate approximation to the initial behavior of the solution for "large" step sizes. Within this context, "large" refers to any value of h for which $h\lambda < -2$. When $h\lambda < -2$, $Q(h\lambda) < 0$ and a sawtooth-like oscillation is introduced into the approximate solution. The good news is that since $|Q(h\lambda)| < 1$, the unwanted oscillations will eventually die out. The bad news is that the more negative the value of $h\lambda$, the more negative the value of $Q(h\lambda)$, which translates to an increase in the initial amplitude of the oscillation, and the closer the value of $|Q(h\lambda)|$ becomes to one, which implies that the oscillations will persist longer.

EXAMPLE 7.31 The Trapezoidal Method Applied to a Stiff Equation

Consider the differential equation

$$y' = -1000y - e^{-t}, \quad y(0) = 0.$$

The exact solution for this problem is

$$y(t) = \frac{1}{999}(e^{-1000t} - e^{-t}),$$

which clearly evolves with two markedly different time scales: 1 versus 1/1000. The stiffness ratio for this problem is thus 1000.

The top graph in Figure 7.24 displays the exact solution to the differential equation together with the approximate solution computed with the trapezoidal method. A step size of $h = 1/600$ was used. Since the more negative characteristic exponent associated with the solution is $\lambda = -1000$, it follows that $h\lambda = -5/3 > -2$. For comparison, the approximate solution obtained from the second-order Taylor method, with $h = 1/600$, is also plotted. Note that $h\lambda = -5/3$ falls inside the region of absolute stability for the second-order Taylor method, which, for real λ, is the interval $-2 < h\lambda < 0$. Both methods accurately reproduce the asymptotic behavior of the exact solution. The trapezoidal method, however, does a significantly better job approximating the initial, transient behavior.

The bottom graph of Figure 7.24 demonstrates the drawback to the trapezoidal method. Three approximate solutions, computed with $h = 1/400$, $h = 1/300$, and $h = 1/200$, are plotted. These step sizes correspond to values of $h\lambda$ of -2.5, $-10/3$, and -5, respectively. Since the method is A-stable, the approximate solution continues to track the asymptotic behavior of the exact solution with these

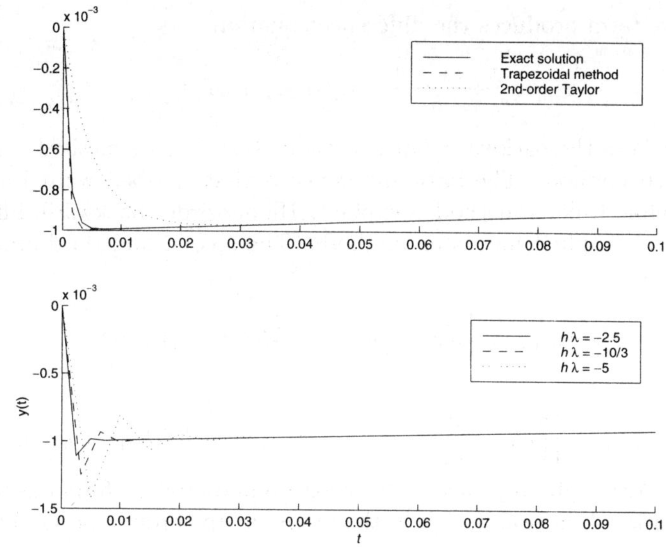

Figure 7.24 Solution of $y' = -1000y - e^{-t}$, $y(0) = 0$ using the trapezoidal method. (Top panel) Approximate solution, computed with $h = 1/600$, versus the exact solution. Solution obtained from second-order Taylor method included for comparison. (Bottom panel) Approximate solutions (trapezoidal method) computed with $h\lambda = -2.5$, $h\lambda = -10/3$, and $h\lambda = -5$.

larger step sizes. (The second-order Taylor method, which is not A-stable, produces exponentially increasing solutions for any step size larger than $h = 1/500$.) On the other hand, as noted above, values of $h\lambda$ smaller than -2 introduce sawtooth oscillations into the approximate solution. Furthermore, with decreasing $h\lambda$, there is an increase in both the amplitude and the duration of the oscillations.

Backward Differentiation Formulas

Among the most widely used numerical techniques for approximating the solution of stiff problems are the backward differentiation formulas popularized by Gear [1]. The process for constructing backward differentiation formulas is quite straightforward. Simply evaluate the differential equation

$$y'(t) = f(t, y(t))$$

at time level $t = t_{n+1}$ and then replace $y'(t_{n+1})$ by a backward difference approximation of desired order. Using a first-order backward difference approximation for the derivative term produces the difference equation

$$\frac{w_{n+1} - w_n}{h} = f(t_{n+1}, w_{n+1}),$$

which is known as the backward Euler method. Like Euler's method, this is a first-order one-step method. The basic difference is that the backward Euler method is implicit, while Euler's method is explicit. Higher-order backward differentiation formulas are all implicit multistep methods. The second and third order schemes, for example, are

$$w_{n+1} = \frac{4}{3}w_n - \frac{1}{3}w_{n-1} + \frac{2}{3}hf(t_{n+1}, w_{n+1})$$

and

$$w_{n+1} = \frac{18}{11}w_n - \frac{9}{11}w_{n-1} + \frac{2}{11}w_{n-1} + \frac{6}{11}hf(t_{n+1}, w_{n+1}),$$

respectively. Although, in principle, backward differentiation formulas to any order can be constructed, in practice, only the methods up through order six are useful. The methods of order seven and beyond are not convergent.

What has made the backward differentiation formulas so popular is their excellent stability properties. For example, the backward Euler method is A-stable. To establish this result, note that for the backward Euler method

$$Q(h\lambda) = \frac{1}{1 - h\lambda}.$$

Therefore,

$$|Q(h\lambda)| = \frac{1}{\sqrt{(1 - \text{Re } h\lambda)^2 + (\text{Im } h\lambda)^2}}.$$

For any $h\lambda$ with Re $h\lambda < 0$, it follows that $1 - \text{Re } h\lambda > 1$. This implies that $(1 - \text{Re } h\lambda)^2 + (\text{Im } h\lambda)^2 > 1$ and finally that $|Q(h\lambda)| < 1$. Hence the backward Euler method is absolutely stable for all $h\lambda$ with Re $h\lambda < 0$, and the method is A-stable. The second-order backward differentiation formula is also A-stable, but the proof is much more involved (see Iserles [2]). The methods of order three through six are not A-stable. In fact, no multistep method of order larger than two can be A-stable. This is known as the Dahlquist barrier [3]. The regions of absolute stability for the backward differentiation formulas of order three through six do, however, contain the entire negative real line and, compared to the other methods we have studied, large portions of the left side of the complex plane.

EXAMPLE 7.32 Backward Differentiation Formulas in Action

Let's reconsider the initial value problem

$$y' = -1000y - e^{-t}, \quad y(0) = 0,$$

whose exact solution is

$$y(t) = \frac{1}{999}(e^{-1000t} - e^{-t}).$$

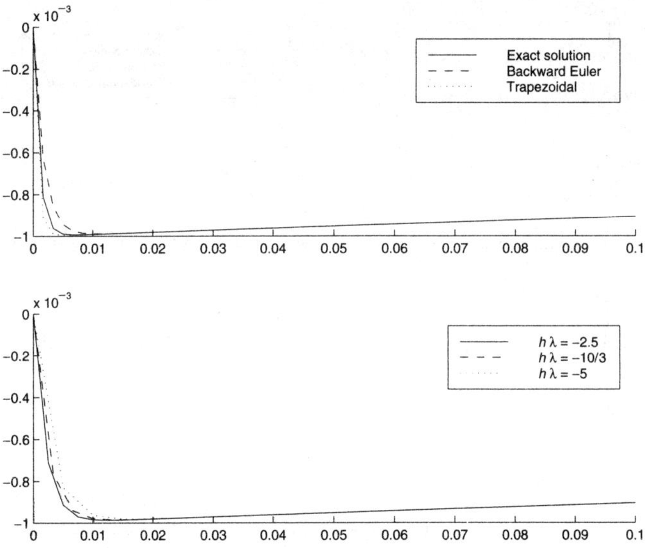

Figure 7.25 Solution of $y' = -1000y - e^{-t}$, $y(0) = 0$ using the backward Euler method. (Top panel) Approximate solution, computed with $h = 1/600$, versus the exact solution. Solution obtained from trapezoidal method included for comparison. (Bottom panel) Approximate solutions (backward Euler method) computed with $h\lambda = -2.5$, $h\lambda = -10/3$, and $h\lambda = -5$.

The top graph of Figure 7.25 displays the exact solution of the initial value problem together with the approximate solutions obtained from the backward Euler method and the trapezoidal method. A step size of $h = 1/600$ was used with each numerical method. For this problem at least, we see that the accuracy of the backward Euler method compares quite favorably with that of the trapezoidal method, even though the backward Euler method is only first-order, whereas the trapezoidal method is second-order. When larger step sizes are used (in particular $h = 1/400$, $h = 1/300$, and $h = 1/200$), the bottom graph of Figure 7.25 shows that the backward Euler method does not introduce spurious oscillations into the approximate solution.

The same is not true for the second-order backward differentiation formula. Observe in the top graph of Figure 7.26 that the second-order backward differentiation formula overshoots the minimum value of the exact solution, though it does more closely follow the initial, decreasing portion than does the trapezoidal method. This overshoot is caused by the characteristic polynomial associated with the formula having complex roots when $h = 1/600$. These complex roots introduce a sinusoidal oscillation into the approximate solution. In fact, with real λ, the roots of the characteristic polynomial associated with the second-order backward differentiation formula will be complex whenever $h\lambda < -1/2$ (see Exercise 7). Since the magnitude of the roots will always be less than 1 (recall that the method is

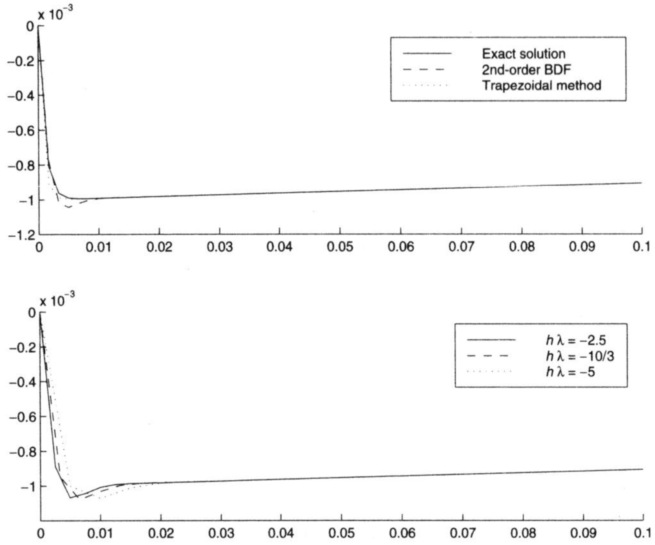

Figure 7.26 Solution of $y' = -1000y - e^{-t}$, $y(0) = 0$ using the second-order backward differentiation formula (BDF). (Top panel) Approximate solution, computed with $h = 1/600$, versus the exact solution. Solution obtained from trapezoidal method included for comparison. (Bottom panel) Approximate solutions (second-order BDF) computed with $h\lambda = -2.5$, $h\lambda = -10/3$, and $h\lambda = -5$.

A-stable), the amplitude of the oscillation will decay as additional time steps are computed. In contrast to the trapezoidal method, however, the initial amplitude of the spurious oscillation decreases and the decay rate increases as h increases, as demonstrated in the bottom graph of Figure 7.26.

One final comment needs to be made before leaving this example. Since the second-order backward differentiation formula is a two-step method, a one-step method must be used to compute w_1. For this problem, w_1 was obtained as follows. First, apply the backward Euler method to advance the solution from t_0 to $t_0 + h$ in a single step. Call this value w_1^h. Next, apply the backward Euler method to advance the solution from t_0 to $t_0 + h$ in two steps, calling the result $w_1^{h/2}$. Finally, extrapolate from these two values to give

$$w_1 = 2w_1^{h/2} - w_1^h.$$

This procedure, in principle, produces a second-order accurate starting value.

Some Final Thoughts

In this section, the concepts of absolute stability and A-stable numerical methods were introduced. Three different A-stable methods were then investigated, and the relative merits of each method noted. All of the results derived above were based on the examination of a simple linear test problem and a single linear initial value problem. Though similar performance can be expected for general linear equations and systems of equations, what happens with nonlinear problems? Unfortunately, there is no simple answer to this question, and stability analysis of discretized nonlinear ordinary differential equations is currently a very active area of research. The major difficulty with nonlinear equations is that the characteristic exponents, λ, change from point to point, and, hence, the performance of a given numerical method can vary significantly as the approximate solution is computed. For example, in the event that λ becomes positive, it is possible for the backward Euler method to introduce spurious oscillations. This does not mean that the work put forth in this section has been wasted. On the contrary, the linear model was a convenient starting point that served well in illuminating the basic processes which give rise to stiff equations.

Though we shall not undertake such a development here, variable step size algorithms can be constructed based on backward differentiation formulas. Extensive effort has been made to develop computationally efficient implementations of these schemes. Substantial research has also gone into the derivation and implementation of implicit Runge-Kutta methods: the k values on which these methods are based (such as the k_1, k_2, k_3, and k_4 values which define the classical fourth-order Runge-Kutta method) appear implicitly. It is known that implicit Runge-Kutta methods are A-stable to all orders (Butcher [4]) and hence are suitable for the solution of stiff problems. Suggested references for further study on stiff problems include Gear [1], Shampine and Gear [5], Aiken [6], Shampine and Gordon [7], Dekker and Verwer [8], Lambert [9], and Hairer and Wanner [10].

References

1. C. W. Gear, *Numerical Initial Value Problems in Ordinary Differential Equations*, Prentice Hall, Englewood Cliffs, NJ, 1971.

2. A. Iserles, *A First Course in the Numerical Analysis of Differential Equations*, Cambridge Texts in Applied Mathematics, Cambridge University Press, Cambridge, 1996.

3. G. Dahlquist, "A Special Stability Property for Linear Multistep Methods," *BIT*, **3**, 27–43, 1963.

4. J. C. Butcher, *The Numerical Analysis of Ordinary Differential Equations*, John Wiley and Sons, New York, 1987.

5. L. Shampine and C. Gear, "A User's View of Solving Stiff Ordinary Differential Equations," *SIAM Review*, **21**, 1–17, 1979.

6. R. Aiken, ed., *Stiff Computation*, Oxford University Press, Oxford, England, 1985.

7. L. Shampine and M. Gordon, *Computer Solution of Ordinary Differential Equations*, W. H. Freeman, San Francisco, 1975.
8. K. Dekker and J. G. Verwer, *Stability of Runge-Kutta Methods for Stiff Nonlinear Differential Equations*, CWI Monographs **2**, North-Holland, Amsterdam, 1984.
9. J. D. Lambert, *Numerical Methods for Ordinary Differential Equations*, John Wiley and Sons, London, 1991.
10. E. Hairer and G. Wanner, *Solving Ordinary Differential Equations II: Stiff Problems and Differential-Algebraic Equations*, Springer-Verlag, Berlin, 1991.

EXERCISES

1. Consider the fifth-order Taylor method.

(a) What is the polynomial $Q(z)$ associated with this method?

(b) For real λ, to what interval must the value of $h\lambda$ be restricted to maintain absolute stability?

(c) Plot the region of absolute stability for the fifth-order Taylor method.

2. Repeat Exercise 1 for the sixth-order Taylor method.

3. Consider the four-step Adams-Bashforth method

$$\frac{w_{i+1}-w_i}{h} = \frac{55}{24}f(t_i,w_i)-\frac{59}{24}f(t_{i-1},w_{i-1})+\frac{37}{24}f(t_{i-2},w_{i-2})-\frac{9}{24}f(t_{i-3},w_{i-3}).$$

(a) What is the characteristic polynomial, $Q(z,h\lambda)$, associated with this method?

(b) Plot the region of absolute stability for the four-step Adams-Bashforth method.

4. Repeat Exercise 3 for Milne's method

$$\frac{w_{i+1}-w_{i-3}}{h} = \frac{4}{3}\left[2f(t_i,w_i)-f(t_{i-1},w_{i-1})+2f(t_{i-2},w_{i-2})\right].$$

5. Repeat Exercise 3 for Simpson's method

$$\frac{w_{i+1}-w_{i-1}}{h} = \frac{1}{3}\left[f(t_{i+1},w_{i+1})+4f(t_i,w_i)+f(t_{i-1},w_{i-1})\right].$$

6. Repeat Exercise 3 for the leapfrog method

$$\frac{w_{i+1}-w_{i-1}}{h} = 2f(t_i,w_i).$$

7. Consider the second-order backward differentiation formula

$$w_{n+1} = \frac{4}{3}w_n - \frac{1}{3}w_{n-1} + \frac{2}{3}hf(t_{n+1},w_{n+1}).$$

(a) What is the characteristic polynomial, $Q(z,h\lambda)$, associated with this method?

(b) What are the roots of $Q(z,h\lambda)$?

(c) Suppose that λ is real. Show that the roots of $Q(z, h\lambda)$ are complex whenever $h\lambda < -1/2$.

8. Apply the backward Euler method to approximate the solution of the given initial value problem over the indicated interval in t and using the indicated number of time steps. Solve any nonlinear algebraic equations using Newton's method.

(a) $x' = tx^3 - x \quad (0 \le t \le 1), \quad x(0) = 1, \quad N = 4$

(b) $x' + (4/t)x = t^4 \quad (1 \le t \le 3), \quad x(1) = 1, \quad N = 5$

(c) $x' = (\sin x - e^t)/\cos x \quad (0 \le t \le 1), \quad x(0) = 0, \quad N = 3$

(d) $x' = (1 + x^2)/t \quad (1 \le t \le 4), \quad x(1) = 0, \quad N = 5$

(e) $x' = t^2 - 2x^2 - 1 \quad (0 \le t \le 1), \quad x(0) = 0, \quad N = 4$

(f) $x' = 2(1 - x)/(t^2 \sin x) \quad (1 \le t \le 2), \quad x(1) = 2, \quad N = 3$

9. Repeat Exercise 8 using the trapezoidal method.

10. Repeat Exercise 8 using the second-order backward differentiation formula. Use the trapezoidal method to determine w_1.

11. Recall that when λ is real and $Q(h\lambda) < 0$, a one-step method will introduce spurious sawtooth-like oscillations into the approximate solution. Consider Euler's method.

(a) What is the polynomial $Q(h\lambda)$ for Euler's method?

(b) Use the initial value problems

$$\text{IVP\#1}: \quad y' = -1000y - e^{-t}, \quad y(0) = 0$$

$$\text{IVP\#2}: \quad \begin{array}{ll} u_1' = 9u_1 + 24u_2 + 5\cos t - \frac{1}{3}\sin t & u_1(0) = 4/3 \\ u_2' = -24u_1 - 51u_2 - 9\cos t + \frac{1}{3}\sin t & u_2(0) = 2/3 \end{array}$$

to demonstrate the presence of spurious oscillations in the approximate solution generated by Euler's method.

The following three initial value problems are for use with Exercises 12–15.

$$y' = -200y + 200\sin t + \cos t, \quad y(0) = 1$$

$$\begin{array}{ll} u_1' = 9u_1 + 24u_2 + 5\cos t - \frac{1}{3}\sin t & u_1(0) = 4/3 \\ u_2' = -24u_1 - 51u_2 - 9\cos t + \frac{1}{3}\sin t & u_2(0) = 2/3 \end{array}$$

$$\begin{array}{ll} u_1' = -20u_1 - 19u_2 & u_1(0) = 2 \\ u_2' = -19u_1 - 20u_2 & u_2(0) = 0 \end{array}$$

12. For the second-order Runge-Kutta method of your choice and for each of the three initial value problems listed above,

(a) determine the maximum allowable time step to maintain absolute stability;

(b) compute the approximate solution using a step size which is roughly 20% smaller than the value found in part (a); and

(c) compute the approximate solution using a step size that is roughly 20% larger than the value found in part (a).

13. Repeat Exercise 12 with the second-order Adams-Bashforth method.

14. Repeat Exercise 12 with the classical fourth-order Runge-Kutta method.

15. On the three problems listed before Exercise 12, compare the performance of the trapezoidal method, the backward Euler method, and the second-order backward differentiation formula. Choose one value for h such that $h\lambda > -2$ and several (increasing) values for h such that $h\lambda < -2$.

16. Compare the approximate solutions of the initial value problem

$$y' = 5e^{5t}(y-t)^2 + 1, \quad y(0) = 1$$

obtained using

(a) the trapezoidal method;

(b) the backward Euler method;

(c) the second-order backward differentiation formula;

(d) the classical fourth-order Runge-Kutta method; and

(e) the RKF45 method.

For the fixed step size methods, use $h = 0.25$, and for the RKF45 method use an error tolerance of 5×10^{-3}. Advance each solution out to $t = 1$. The exact solution of the initial value problem is $y(t) = t - e^{-5t}$.

17. Reconsider the chemostat problem from the beginning of the section:

$$\begin{aligned} \frac{ds}{dt} &= 1 - s - \frac{msx}{a+s} \\ \frac{dx}{dt} &= \frac{msx}{a+s} - x. \end{aligned}$$

Take $m = 16$, $a = 0.25$, $s(0) = 0.5$, and $x(0) = 0.02$. Approximate the solution to this problem, out to $t = 12$, using the trapezoidal method, the backward Euler method and the second-order backward differentiation formula. Compare performance for $h = 0.05$, $h = 0.1$, and $h = 0.12$.

18. Consider the system of differential equations

$$\begin{aligned} \frac{dy_1}{dt} &= -0.013y_1 - 1000y_1y_3 \\ \frac{dy_2}{dt} &= -2500y_2y_3 \\ \frac{dy_3}{dt} &= -0.013y_1 - 1000y_1y_3 - 2500y_2y_3 \end{aligned}$$

subject to the initial conditions $y_1(0) = 1$, $y_2(0) = 1$, and $y_3(0) = 0$.

(a) Approximate the solution using the RKF45 method with an error tolerance of 5×10^{-3}. Advance the solution to $t = 5$. In what range do the majority of time steps taken during the calculation of the solution fall? Assuming this range persists forward in time, roughly how many time steps would be needed to advance the solution to $t = 50$?

(b) Approximate the solution, out to $t = 5$, using the trapezoidal method, the backward Euler method, and the second-order backward differentiation formula with a step size of $h = 0.01$. Compare with the results from part (a).

(c) Approximate the solution, out to $t = 5$, using the trapezoidal method, the backward Euler method, and the second-order backward differentiation formula with a step size of $h = 0.1$. Compare with the results from parts (a) and (b).

Note: This problem was proposed by Gear ("The Automatic Integration of Stiff Ordinary Differential Equations," *Proceedings of the IP68 Conference*, North-Holland, Amsterdam, 1969) as a test problem for software to solve stiff ordinary differential equations (ODEs).

CHAPTER 8

Two-Point Boundary Value Problems

AN OVERVIEW

Fundamental Mathematical Problem

In this chapter, we will develop numerical techniques for approximating the solution of the two-point boundary value problem

$$\begin{aligned} y'' &= f(x, y, y'), \quad a \le x \le b, \\ \alpha_1 y(a) &+ \alpha_2 y'(a) = \alpha_3 \\ \beta_1 y(b) &+ \beta_2 y'(b) = \beta_3, \end{aligned}$$

where f is an arbitrary function of its three arguments. When f is of the form

$$f(x, y, y') = p(x)y' + q(x)y + r(x)$$

for some functions p, q and r, the boundary value problem is called linear; otherwise, it is nonlinear. The boundary conditions given above, in which a value is specified for a linear combination of the unknown function and its first derivative, are called Robin (or mixed) boundary conditions. The special case in which the value of the unknown function is specified [e.g., $y(a) = \alpha$ and/or $y(b) = \beta$] is called a Dirichlet boundary condition, while the special case in which the value of the first derivative is specified [e.g., $y'(a) = \alpha$ and/or $y'(b) = \beta$] is called a Neumann boundary condition. More general boundary conditions, such as periodic conditions, can be specified, but we will not consider them here.

Steady-State Temperature Distribution in a Pin Fin

In heat transfer, the term *extended surface* is used in reference to a solid that experiences energy transfer by conduction within its boundaries and by convection between its boundaries and the surroundings. The most common application of an extended surface is to enhance the heat transfer rate between a solid and an adjoining fluid (such as air or water). Extended surfaces used in this manner are called fins. A fin that has a circular cross section is called a pin fin.

Suppose we have a pin fin of length L and varying radius $r(x)$ attached to a surface, as shown in Figure 8.1(a). We would like to determine both the temperature distribution along the length of the fin, $T(x)$, and the total fin heat transfer rate, q_f.

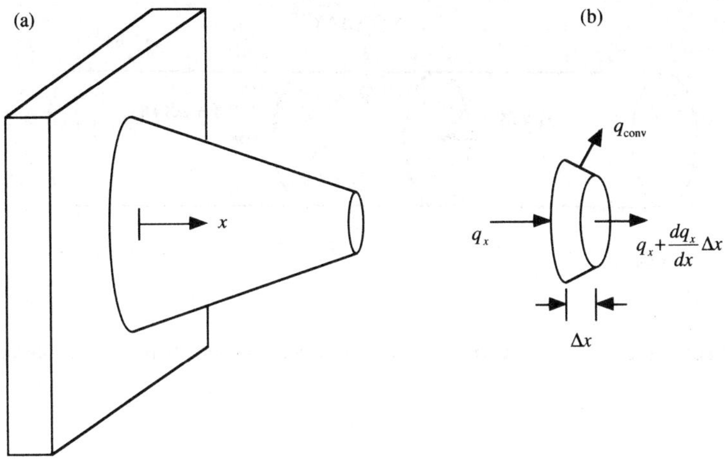

Figure 8.1 (a) A pin fin with nonuniform cross section. (b) Energy balance on an arbitrary slice of a pin fin that experiences convective heat loss from its lateral surface.

Performing an energy balance on the arbitrary slice of the fin shown in Figure 8.1(b) gives

$$q_x = q_x + \frac{dq_x}{dx}\Delta x + q_{\text{conv}}, \tag{1}$$

where q_x is the rate of heat transfer along the length of the fin due to conduction and q_{conv} is the rate of heat transfer from the lateral surface of the fin due to convection. According to Fourier's law

$$q_x = -kA\frac{dT}{dx}, \tag{2}$$

where K is the thermal conductivity of the fin and A is the area through which the heat flows. The minus sign indicates that heat flows from regions of high temperature to regions of low temperature. The convection heat transfer rate is given by

$$q_{\text{conv}} = hA_s(T - T_\infty). \tag{3}$$

Here, h is the convection heat transfer coefficient, A_s is the area of the surface from which convection takes place and T_∞ is the temperature of the adjoining fluid.

Substituting (2) and (3) into (1), taking into account the geometry of the fin that implies that $A = \pi[r(x)]^2$ and $A_s = 2\pi r(x)\Delta x$, and simplifying the resulting equation yields

$$\frac{d}{dx}\left([r(x)]^2\frac{dT}{dx}\right) - \frac{2h}{k}r(x)(T - T_\infty) = 0. \tag{4}$$

At $x = 0$, where the fin meets the solid, the temperature of the fin equals that of

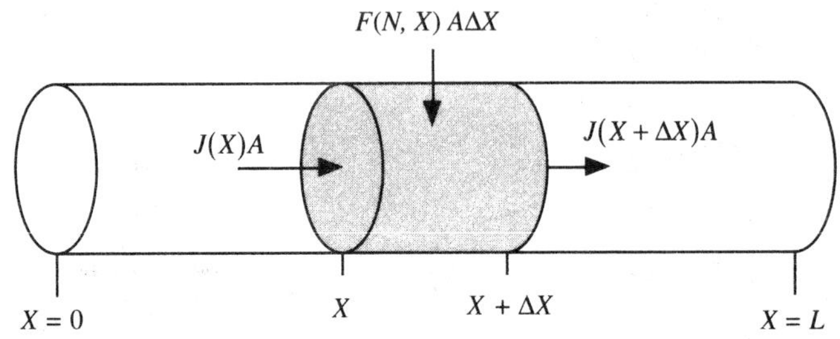

Figure 8.2 Geometry and control volume for insect dispersal problem.

the solid, T_0. Thus, we have the boundary condition

$$T(0) = T_0. \tag{5}$$

At the tip of the fin, the heat conducted to the tip must balance the heat convected away from the tip, so we have the boundary condition

$$-kT'(L) = h(T(L) - T_\infty). \tag{6}$$

The boundary value problem given by (4), (5) and (6) determines $T(x)$. Once $T(x)$ is known, the total fin heat transfer rate is given by

$$q_f = -k\pi[r(0)]^2 T'(0). \tag{7}$$

Spatial Distribution of an Insect Population

Suppose a population of insects is placed into a closed environment in the shape of a circular cylinder of length L (see Figure 8.2). We would like to model the spread of the population throughout the environment, with the objective of determining the eventual steady-state distribution of the insects. For simplicity, we will assume the insects disperse along the axis of the cylinder only. At $X = 0$, such harsh environmental conditions are maintained that the insects cannot survive, while a barrier prevents the insects from migrating beyond $X = L$.

Let $N(X)$ denote the steady-state population density of the insects, measured in insects per unit volume. Consider the arbitrary slice of the cylinder which is highlighted in Figure 8.2. This slice is called a *control volume*. The flux function $J(X)$ measures the number of insects that cross location x in the positive direction per unit area per unit time. The function $F(N, X)$ describes the net birth rate of insects per unit volume. Finally, A denotes the cross-sectional area of the cylinder.

At steady state, the number of insects within the control volume must remain constant. This requires that the conservation law

$$\begin{pmatrix}\text{the rate at which}\\ \text{insects enter}\\ \text{the control volume}\end{pmatrix} - \begin{pmatrix}\text{the rate at which}\\ \text{insects leave}\\ \text{the control volume}\end{pmatrix} + \begin{pmatrix}\text{the rate at which}\\ \text{insects are}\\ \text{generated inside}\\ \text{the control volume}\end{pmatrix} = 0$$

hold. Using the information from Figure 8.2, the conservation law becomes

$$[J(X) - J(X + \Delta X)]\, A + F(N, X) A \Delta X = 0.$$

Dividing this equation by $A\Delta X$ and taking the limit as $\Delta X \to 0$ yields

$$-\frac{dJ}{dX} + F(N, X) = 0. \tag{8}$$

To proceed further, we relate the flux to the population density by assuming Fick's law, which states that j is proportional to the gradient of N. Thus,

$$J = -D\frac{dN}{dX}. \tag{9}$$

The coefficient of diffusion D measures the efficiency with which the insects disperse, and the minus sign indicates that migration takes place from regions of high density to regions of low density. For the source term we will use the logistic growth law; that is,

$$F(N, X) = rN(1 - N/K), \tag{10}$$

where r is the reproductive rate of the insects and K is the environmental carrying capacity. Substituting (9) and (10) into (8) and assuming that D is constant yields

$$D\frac{d^2N}{dX^2} + rN\left(1 - \frac{N}{K}\right) = 0. \tag{11}$$

The boundary conditions associated with (11) are

$$N(0) = 0 \quad \text{and} \quad \frac{dN}{dX}(L) = 0.$$

If we now introduce the nondimensional variables $n = N/K$ and $x = X\sqrt{r/D}$, we arrive at the boundary value problem

$$n'' + n(1 - n) = 0, \quad n(0) = 0, \quad n'(l) = 0, \tag{12}$$

where primes indicate differentiation with respect to x and $l = L\sqrt{r/D}$.

The Remainder of the Chapter

In this chapter, we will develop two techniques for the numerical solution of two-point boundary value problems: the finite difference method and the shooting method. We will start with the finite difference method. In Section 8.1, the linear boundary value problem with Dirichlet boundary conditions is considered. Non-Dirichlet boundary conditions (Neumann and Robin conditions) are considered in Section 8.2. The solution of nonlinear boundary value problems is discussed in Section 8.3. The final two sections will present the shooting method, covering linear problems and nonlinear problems, in that order.

An Artificial Singularity

When a partial differential equation is reduced to an ordinary differential equation using symmetry considerations, an artificial singularity often appears in one or more of the coefficients of the ordinary differential equation. The singularity is artificial in the sense that the exact solution is smooth near the coefficient singularity. For example, the boundary value problem

$$\begin{gathered} u'' + \frac{1}{x}u' = \left(\frac{8}{8-x^2}\right)^2 \\ u'(0) = u(1) = 0 \end{gathered}$$

has a coefficient singularity at $x = 0$, but the exact solution

$$u(x) = 2\ln\left(\frac{7}{8-x^2}\right)$$

is smooth at $x = 0$. The handling of artificial singularities will be treated where appropriate throughout the chapter.

8.1 FINITE DIFFERENCE METHOD, PART I: THE LINEAR PROBLEM WITH DIRICHLET BOUNDARY CONDITIONS

Rather than jumping straight into a treatment of the general second-order one-dimensional two-point boundary value problem

$$\begin{aligned} y'' &= f(x, y, y'), \quad a \le x \le b, \\ \alpha_1 y(a) &+ \alpha_2 y'(a) = \alpha_3 \\ \beta_1 y(b) &+ \beta_2 y'(b) = \beta_3, \end{aligned}$$

we will begin our study of finite difference methods by investigating the linear problem

$$y'' = p(x)y' + q(x)y + r(x), \quad x \in [a, b] \tag{1}$$

subject to the Dirichlet boundary conditions

$$y(a) = \alpha, \quad y(b) = \beta.$$

As with the initial value problem techniques we developed in Chapter 7, the objective of a finite difference method is to approximate the value of the exact solution to the boundary value problem, $y(x)$, at a discrete set of points x_0, x_1, x_2, ..., $x_N \in [a, b]$. Throughout our discussion, we will let y_i denote the value of the exact solution at $x = x_i$, and we will denote our finite difference approximation to y_i by w_i. This finite difference approximation is obtained by replacing each derivative which appears in the boundary value problem with an appropriate finite difference formula (see Chapter 6). This converts the single continuous ordinary differential equation for $y(x)$ into a system of discrete algebraic equations for the values w_0, w_1, w_2, ..., w_N.

The Computational Grid

To begin the approximation process, we must first introduce a partition, or computational grid, over the interval $[a, b]$. Let N be a positive integer that denotes the number of subintervals in the partition and denote the partition itself by

$$a = x_0 < x_1 < x_2 < \cdots < x_{N-1} < x_N = b.$$

The points x_0 and x_N are called boundary grid points, while the points x_1, x_2, x_3, ..., x_{N-1} are called interior grid points. For simplicity, a uniform grid will be assumed. That is, we will assume that $x_i = a + ih$, where $h = (b - a)/N$. The parameter h, known as the step size or the mesh size, is the key parameter governing the accuracy of the finite difference approximation.

The Finite Difference Approximation

Once the computational grid has been established, we evaluate the differential equation, that is, equation (1), at each interior grid point

$$\{y'' = p(x)y' + q(x)y + r(x)\}|_{x=x_i} \qquad (1 \le i \le N-1)$$

and then replace all of the derivatives by second-order central finite difference approximations:

$$\frac{y_{i+1} - 2y_i + y_{i-1}}{h^2} + O(h^2) = p_i \frac{y_{i+1} - y_{i-1}}{2h} + q_i y_i + r_i + O(h^2).$$

To simplify notation, we have used p_i, q_i, and r_i to denote the coefficient function values $p(x_i)$, $q(x_i)$ and $r(x_i)$, respectively. Next, we drop the truncation error terms and replace all of the y's (exact solution) by w's (approximate solution). Thus, for $i = 1, 2, 3, \ldots, N - 1$

$$\frac{w_{i+1} - 2w_i + w_{i-1}}{h^2} = p_i \frac{w_{i+1} - w_{i-1}}{2h} + q_i w_i + r_i. \tag{2}$$

To completely determine the unknowns w_0, w_1, w_2, ..., w_N, two more equations are needed. These will come from the boundary conditions. At the boundary grid points, the values $y(x_0) = \alpha$ and $y(x_N) = \beta$ have been specified; therefore, we will set $w_0 = \alpha$ and $w_N = \beta$. Combining these equations with equation (2) constitutes the finite difference method. Since the truncation error terms that were dropped to form equation (2) were $O(h^2)$, the resulting numerical method will be second order. This point will be demonstrated below.

Second-Order Finite Difference Method

$$w_0 = \alpha$$

$$\frac{w_{i+1} - 2w_i + w_{i-1}}{h^2} = p_i \frac{w_{i+1} - w_{i-1}}{2h} + q_i w_i + r_i \qquad 1 \leq i \leq N-1$$

$$w_N = \beta$$

Matrix Formulation

Since the equations of our numerical method are linear in the unknowns, we will express the system in matrix form. With w_0 and w_N known from the Dirichlet boundary conditions, there are $N-1$ unknowns to be determined. To construct the coefficient matrix and the right-hand-side vector for the system, start by multiplying equation (2) through by $-h^2$ (to avoid division by a small number—the negative sign is included so as to obtain a positive coefficient on w_i) and then collect terms. For $i = 1, 2, 3, \ldots, N-1$, we therefore have

$$\left(-1 - \frac{h}{2}p_i\right) w_{i-1} + (2 + h^2 q_i) w_i + \left(-1 + \frac{h}{2}p_i\right) w_{i+1} = -h^2 r_i. \tag{3}$$

We can think of this equation as a *computational template* or *stencil*, which is to be applied at each grid point where the value of the approximate solution is unknown. From equation (3), we recognize that each row of the coefficient matrix will have only three nonzero entries: the entry along the main diagonal and the entries one position to the right and left of the main diagonal. Hence, the coefficient matrix will be tridiagonal. The first row (corresponding to $i = 1$) and the last row (corresponding to $i = N-1$) of the coefficient matrix and the right-hand-side vector require some care in writing out because w_0 and w_N are not unknown. For $i = 1$, equation (3) gives

$$\left(-1 - \frac{h}{2}p_1\right) w_0 + (2 + h^2 q_1) w_1 + \left(-1 + \frac{h}{2}p_1\right) w_2 = -h^2 r_1.$$

Substituting $w_0 = \alpha$ and transposing the first term to the right-hand side, we arrive at

$$(2 + h^2 q_1) w_1 + \left(-1 + \frac{h}{2}p_1\right) w_2 = -h^2 r_1 + \left(1 + \frac{h}{2}p_1\right) \alpha$$

as the first equation in the system. Working in a similar manner for $i = N-1$, we find the last equation to be

$$\left(-1 - \frac{h}{2}p_{N-1}\right) w_{N-2} + (2 + h^2 q_{N-1}) w_{N-1} = -h^2 r_{N-1} + \left(1 - \frac{h}{2}p_{N-1}\right) \beta.$$

The system of equations for w_1, w_2, w_3, $\ldots$, w_{N-1} can therefore be written in the form $A\mathbf{w} = \mathbf{b}$, where

$$A = \begin{bmatrix} d_1 & u_1 & & & & & \\ l_2 & d_2 & u_2 & & & & \\ & l_3 & d_3 & u_3 & & & \\ & & \cdot & \cdot & \cdot & & \\ & & & \cdot & \cdot & \cdot & \\ & & & & \cdot & \cdot & \cdot \\ & & & & l_{N-3} & d_{N-3} & u_{N-3} \\ & & & & & l_{N-2} & d_{N-2} & u_{N-2} \\ & & & & & & l_{N-1} & d_{N-1} \end{bmatrix},$$

with

$$d_i = 2 + h^2 q_i \tag{4}$$

$$u_i = -1 + \frac{h}{2} p_i \tag{5}$$

$$l_i = -1 - \frac{h}{2} p_i. \tag{6}$$

The vectors $\mathbf{w}$ and $\mathbf{b}$ are given by

$$\mathbf{w} = \begin{bmatrix} w_1 \\ w_2 \\ \cdot \\ \cdot \\ \cdot \\ w_{N-2} \\ w_{N-1} \end{bmatrix} \quad \text{and} \quad \mathbf{b} = \begin{bmatrix} -h^2 r_1 + \left(1 + \frac{h}{2} p_1\right) \alpha \\ -h^2 r_2 \\ \cdot \\ \cdot \\ \cdot \\ -h^2 r_{N-2} \\ -h^2 r_{N-1} + \left(1 - \frac{h}{2} p_{N-1}\right) \beta \end{bmatrix}.$$

Note how the boundary conditions have been incorporated into the first and last entries of the vector $\mathbf{b}$.

Solvability of Discrete Equations

Does this system of algebraic equations have a unique solution? Let's suppose that p is continuous and that $q(x) \geq 0$ on $[a, b]$. By the Extreme Value Theorem, the continuity of p over the closed interval $[a, b]$ guarantees the existence of a positive constant l such that $|p(x)| \leq L$ on $[a, b]$. If the step size h is chosen to be smaller than $2/L$, then for each i, $-1 < hp_i/2 < 1$, which in turn implies that the terms

$$-1 - \frac{h}{2} p_i \quad \text{and} \quad -1 + \frac{h}{2} p_i$$

are always negative. Therefore,

$$\left|-1-\frac{h}{2}p_i\right| = 1+\frac{h}{2}p_i$$
$$\left|-1+\frac{h}{2}p_i\right| = 1-\frac{h}{2}p_i$$

so that along rows 2 through $N-2$ of the matrix A,

$$\left|-1-\frac{h}{2}p_i\right| + \left|-1+\frac{h}{2}p_i\right| = 2 \le |2+h^2q_i|;$$

that is, in rows 2 through $N-2$ the sum of the absolute values of the off-diagonal entries is less than or equal to the absolute value of the diagonal entry. The first and last rows satisfy an even stronger condition: the sum of the absolute values of the off-diagonal entries is strictly less than the absolute value of the diagonal element. Hence, A is diagonally dominant, with its first and last rows being strictly diagonally dominant. These conditions are sufficient to guarantee a unique solution to our finite difference equations (see Isaacson and Keller [1]), provided the step size, h, is selected smaller than $2/L$.

Alternative Matrix Formulation

When we deal with Neumann and Robin boundary conditions in Section 8.2, it will be easier to generalize the matrix formulation of the finite difference method if we include w_0 and w_N in the vector of unknowns; that is, if we formulate the matrix equation based on all $N+1$ equations

$$w_0 = \alpha$$
$$\left(-1-\frac{h}{2}p_i\right)w_{i-1} + (2+h^2q_i)w_i + \left(-1+\frac{h}{2}p_i\right)w_{i+1} = -h^2r_i \qquad 1 \le i \le N-1$$
$$w_N = \beta.$$

Taking this approach, the system of finite difference equations can be written in the form $A\mathbf{w} = \mathbf{b}$, where A is the $(N+1)\times(N+1)$ tridiagonal matrix

$$A = \begin{bmatrix} 1 & 0 & & & & & \\ l_1 & d_1 & u_1 & & & & \\ & l_2 & d_2 & u_2 & & & \\ & & \ddots & \ddots & \ddots & & \\ & & & l_{N-2} & d_{N-2} & u_{N-2} & \\ & & & & l_{N-1} & d_{N-1} & u_{N-1} \\ & & & & & 0 & 1 \end{bmatrix}$$

and the vectors $\mathbf{w}$ and $\mathbf{b}$ are given by

$$\mathbf{w} = \begin{bmatrix} w_0 \\ w_1 \\ w_2 \\ \cdot \\ \cdot \\ \cdot \\ w_{N-2} \\ w_{N-1} \\ w_N \end{bmatrix} \quad \text{and} \quad \mathbf{b} = \begin{bmatrix} \alpha \\ -h^2 r_1 \\ -h^2 r_2 \\ \cdot \\ \cdot \\ \cdot \\ -h^2 r_{N-2} \\ -h^2 r_{N-1} \\ \beta \end{bmatrix}.$$

Recall that d_i, u_i, and l_i were defined in equations (4), (5) and (6). Following the same analysis given above, this newly formulated matrix A is diagonally dominant with strictly diagonally dominant first and last rows and, hence, is guaranteed to be nonsingular when the step size is selected to satisfy $h < 2/L$.

Worked Examples

EXAMPLE 8.1 Demonstration Problem 1

Consider the boundary value problem

$$\begin{aligned} -u'' + \pi^2 u &= 2\pi^2 \sin(\pi x) \\ u(0) &= u(1) = 0. \end{aligned}$$

Comparing this problem with the prototype problem given by equation (1), we see that $p(x) = 0$, $q(x) = \pi^2$ and $r(x) = -2\pi^2 \sin(\pi x)$. Since $p(x) = 0$, we are guaranteed of a unique solution to the finite difference equations for any value of h.

Let's start with a uniform partition of the interval $[0, 1]$ containing $N = 4$ subintervals. Then $h = 1/4$ and the grid points are given by $x_i = a + ih = 0 + i(1/4) = i/4$. This partition is illustrated in the Figure 8.3.

Evaluate the differential equation at the interior grid point $x = x_i$ and then replace the second derivative by its second-order central finite difference approximation. This produces the equation

$$\frac{-u_{i-1} + 2u_i - u_{i+1}}{(1/4)^2} + O(h^2) + \pi^2 u_i = 2\pi^2 \sin(i\pi/4).$$

Next, drop the truncation error term and replace each u_i, which is a value of the exact solution, by w_i, which is a value of the approximate solution. Multiply both sides of the resulting expression by $(1/4)^2$ and group like terms to yield the computational template

$$-w_{i-1} + \left[2 + (\pi/4)^2\right] w_i - w_{i+1} = 2(\pi/4)^2 \sin(i\pi/4).$$

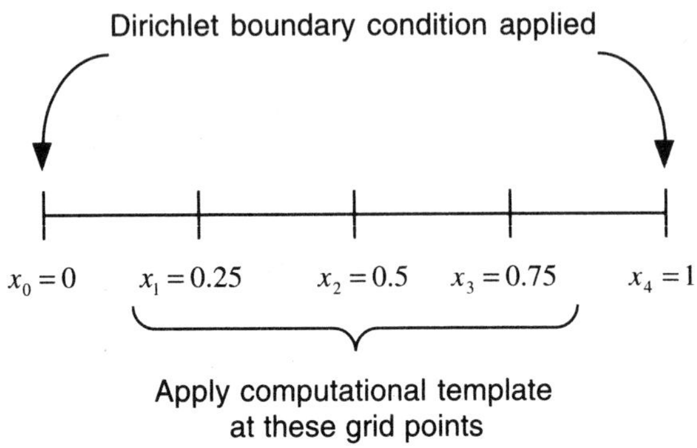

Figure 8.3 Figure for Example 8.1.

Writing out the equations that correspond to $i = 1$, $i = 2$, and $i = 3$, and combining these with the equations $w_0 = 0$ and $w_N = 0$, which are derived from the Dirichlet boundary conditions, the complete system of finite difference equations is given by

$$\begin{bmatrix} 1 & 0 & & & \\ -1 & 2+(\pi/4)^2 & -1 & & \\ & -1 & 2+(\pi/4)^2 & -1 & \\ & & -1 & 2+(\pi/4)^2 & -1 \\ & & & 0 & 1 \end{bmatrix} \begin{bmatrix} w_0 \\ w_1 \\ w_2 \\ w_3 \\ w_4 \end{bmatrix} = \begin{bmatrix} 0 \\ \sqrt{2}(\pi/4)^2 \\ 2(\pi/4)^2 \\ \sqrt{2}(\pi/4)^2 \\ 0 \end{bmatrix}.$$

The solution of this tridiagonal linear system is

$$\mathbf{w} = \begin{bmatrix} 0 & 0.725371 & 1.025830 & 0.725371 & 0 \end{bmatrix}^T.$$

The following table compares this approximate solution with the exact solution: $u(x) = \sin(\pi x)$. The accuracy of the approximate solution is quite reasonable given the crudeness of the computational grid.

x_i	Approximate Solution, w_i	Exact Solution, u_i	Absolute Error
0.00	0.000000	0.000000	
0.25	0.725371	0.707107	0.018264
0.50	1.025830	1.000000	0.025830
0.75	0.725371	0.707107	0.018264
1.00	0.000000	0.000000	

A numerical verification of the second-order accuracy of the scheme is presented in the next table, which displays both the maximum absolute error and the root mean square (rms) error in the approximate solution as a function of the number

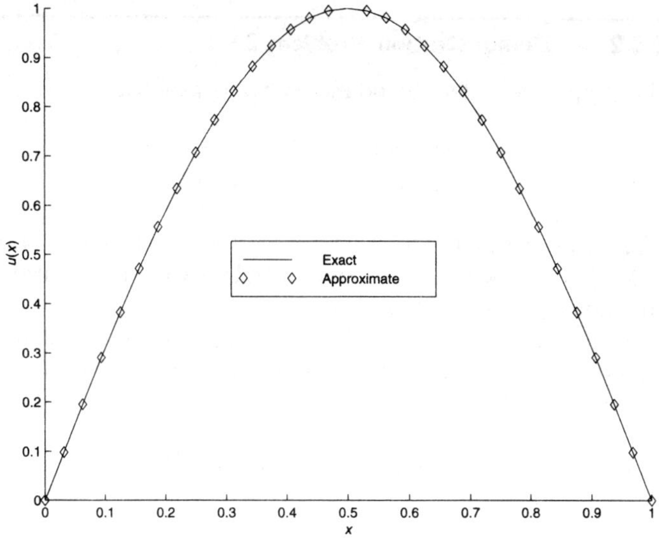

Figure 8.4 Comparison between the exact solution to the boundary value problem $-u'' + \pi^2 u = 2\pi^2 \sin(\pi x)$, $u(0) = u(1) = 0$, and the finite difference approximation generated with a uniform partition containing 32 subintervals.

of subintervals, N. The rms error was computed according to the formula

$$\text{rms error} = \sqrt{\frac{1}{N}\sum_{i=1}^{N-1}(w_i - u_i)^2}.$$

Note that with each doubling of N, the step size is cut in half, and the approximation error is reduced by roughly a factor of 4, which is what one would expect from a second order numerical method. The finite difference approximation obtained using $N = 32$ is shown in Figure 8.4—denoted by the diamonds—superimposed on the exact solution.

N	Maximum Absolute Error	Error Ratio	rms Error	Error Ratio
4	0.0258297765		0.0182644101	
8	0.0064337127	4.014754	0.0045493219	4.014754
16	0.0016068959	4.003814	0.0011362470	4.003814
32	0.0004016275	4.000961	0.0002839935	4.000961
64	0.0001004008	4.000241	0.0000709941	4.000241
128	0.0000250998	4.000060	0.0000177483	4.000060
256	0.0000062749	4.000015	0.0000044370	4.000015
512	0.0000015687	4.000000	0.0000011093	4.000000

EXAMPLE 8.2 Demonstration Problem 2

As a second example, consider the boundary value problem

$$u'' = -(x+1)u' + 2u + (1-x^2)e^{-x}$$
$$u(0) = -1, \quad u(1) = 0.$$

For this problem, we see that $p(x) = -(x+1)$, whose maximum absolute value on the interval $[0,1]$ is 2; hence, we are guaranteed of a unique solution to the finite difference equations for any value of h less than 1.

Let's use the same partition as in the previous example; that is, divide the interval $[0,1]$ into $N = 4$ equal subintervals, so that $h = 1/4$ and $x_i = i/4$. Applying the computational template

$$\left(-1 - \frac{h}{2}p_i\right) w_{i-1} + (2 + h^2 q_i)w_i + \left(-1 + \frac{h}{2}p_i\right) w_{i+1} = -h^2 r_i,$$

for $i = 1$, 2, and 3, where

$$p_i = -(x_i + 1) = -(1 + i/4);$$
$$q_i = 2; \text{ and}$$
$$r_i = (1 - x_i^2)e^{-x_i} = \left[1 - (i/4)^2\right] e^{-i/4},$$

and including the equations $w_0 = -1$ and $w_4 = 0$ obtained from the boundary conditions, we arrive at the system of finite difference equations

$$\begin{bmatrix} 1 & 0 & & & \\ -\frac{27}{32} & \frac{17}{8} & -\frac{37}{32} & & \\ & -\frac{13}{16} & \frac{17}{8} & -\frac{19}{16} & \\ & & -\frac{25}{32} & \frac{17}{8} & -\frac{39}{32} \\ & & & 0 & 1 \end{bmatrix} \begin{bmatrix} w_0 \\ w_1 \\ w_2 \\ w_3 \\ w_4 \end{bmatrix} = \begin{bmatrix} -1 \\ -\frac{15}{256}e^{-1/4} \\ -\frac{3}{64}e^{-1/2} \\ -\frac{7}{256}e^{-3/4} \\ 0 \end{bmatrix}.$$

The solution of these equations is

$$\mathbf{w} = \begin{bmatrix} -1 & -0.582559 & -0.301452 & -0.116906 & 0 \end{bmatrix}^T,$$

which compares favorably with the exact solution $u(x) = (x-1)e^{-x}$, as shown in the following table. A numerical verification of the second-order accuracy of the finite difference scheme is left as an exercise.

x_i	Approximate Solution, w_i	Exact Solution, u_i	Absolute error
0.00	−1.000000	−1.000000	
0.25	−0.582559	−0.584101	0.001542
0.50	−0.301452	−0.303265	0.001813
0.75	−0.116906	−0.118092	0.001186
1.00	0.000000	0.000000	

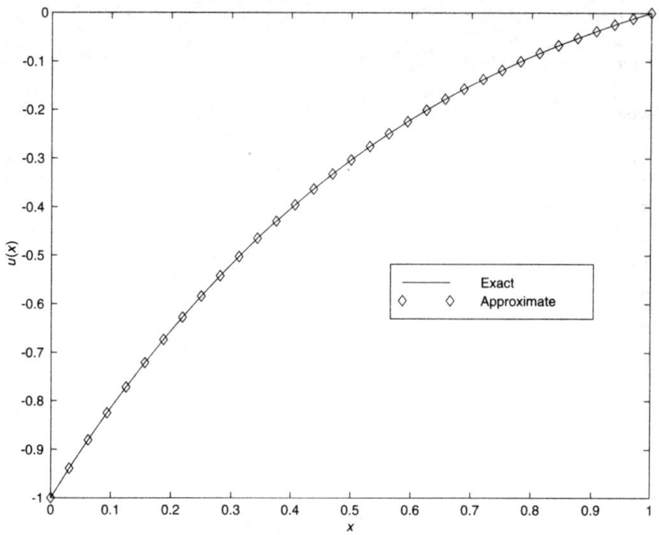

Figure 8.5 Comparison between the exact solution to the boundary value problem $u'' = -(x+1)u' + 2u + (1-x^2)e^{-x}$, $u(0) = -1$, $u(1) = 0$ and the finite difference approximation generated with a uniform partition containing 32 subintervals.

The approximate solution obtained using a uniform partition with $N = 32$ subintervals is shown in Figure 8.5—denoted by the diamonds—superimposed on the exact solution.

An Application Problem: Flow between Parallel Plates

In Chapter 6 (see page 515), we investigated the flow of a viscous fluid filling a gap between two large parallel plates. One plate was stationary, while the other moved with constant velocity, and there was a linear temperature gradient between the plates. The velocity distribution established within the fluid, $U(Y)$, was found to satisfy the boundary value problem

$$\frac{d}{dY}\left(\mu(Y)\frac{dU}{dY}\right) = 0 \tag{7}$$

$$U(0) = 0, \quad U(h) = U_0.$$

Here, $\mu(Y)$ denotes the viscosity of the fluid, h the separation between the plates, and U_0 the velocity of the moving plate. The solution of (7) was expressed in terms of two definite integrals that had to be calculated numerically.

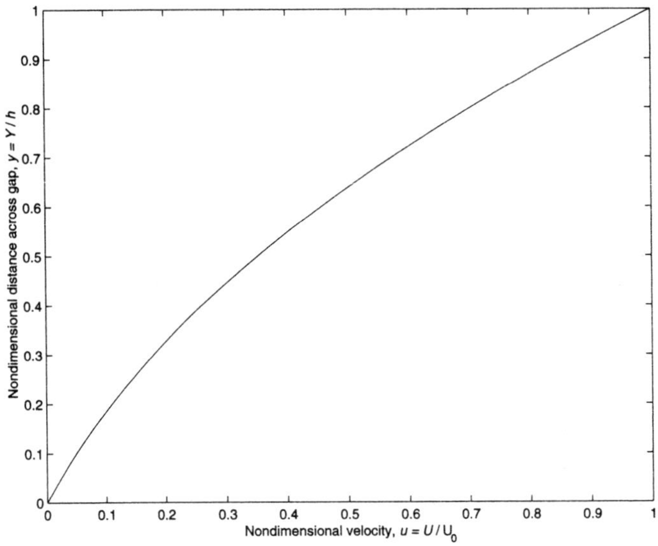

Figure 8.6 Velocity distribution established in viscous fluid filling gap between two large parallel plates.

Here, we will determine the velocity distribution by applying the finite difference method. As in Chapter 6, we take the fluid to be water and use

$$\mu(Y) = \exp\left\{-8.944 - \frac{839.456}{293.16 + 80Y/h} + \frac{421194.298}{(293.16 + 80Y/h)^2}\right\}. \tag{8}$$

Substituting (8) into the differential equation in (7), expanding the derivative of the product and simplifying yields

$$\frac{d^2U}{dY^2} = \left[\frac{160}{h}\frac{421194.298}{(293.16 + 80Y/h)^3} - \frac{80}{h}\frac{839.456}{(293.16 + 80Y/h)^2}\right]\frac{dU}{dY}.$$

Now introduce the nondimensional variables $u = U/U_0$ and $y = Y/h$ to obtain the boundary value problem

$$u'' = 80\left[\frac{842388.596}{(293.16 + 80y)^3} - \frac{839.456}{(293.16 + 80y)^2}\right]u' \tag{9}$$
$$u(0) = 0, \quad u(1) = 1,$$

where primes denote differentiation with respect to y. Finally, taking a uniform partition of $[0, 1]$ with $N = 100$ subintervals, the velocity distribution shown in Figure 8.6 is obtained. Compare the velocity distribution in this figure with the distribution displayed in Figure 6.15 (page 517).

References

1. E. Isaacson and H. B. Keller, *Analysis of Numerical Methods*, John Wiley & Sons, New York, 1966.

EXERCISES

In Exercises 1–5,

(a) identify the interval $[a, b]$ and the functions p, q, and r; and

(b) using a uniform partition of $[a, b]$ with $N = 4$ subintervals, write out the system of finite difference equations.

1. $y'' = -\frac{3}{2+x}y' - \frac{2}{(2+x)^3}, \quad y(0) = y(1) = 0$

2. $y'' = 120y - 2560, \quad y(0) = y(0.15) = 20$

3. $xy'' - (2x+1)y' + (x+1)y = 0, \quad y(1) = 2e, \quad y(3) = 10e^3$

4. $(1 - x\cot x)y'' - xy' + y = 0, \quad y(1) = 1 + \sin(1), \quad y(2) = 2 + \sin(2)$

5. $\frac{1}{x}\left(xy'\right)' + y = 10, \quad y(1) = 20, \quad y(3) = 100$

6. Suppose we use the finite difference method to approximate the solution of the boundary value problem $y'' = p(x)y' + q(x)y + r(x), \; y(a) = \alpha, y(b) = \beta$. Choose any point $x = c$ with $a < c < b$, and select N so that c is one of the interior grid points. Let $w_M(c)$ denote the approximation to $y(c)$ obtained with a uniform partition containing m subintervals, and let j be a nonnegative integer. If the errors associated with the finite difference method are $O(h^2)$, toward what value should the ratio

$$\frac{w_{2^j N}(c) - w_{2^{j+1} N}(c)}{w_{2^{j+1} N}(c) - w_{2^{j+2} N}(c)}$$

converge as j is increased?

In Exercises 7–14, approximate the solution of the indicated boundary value problem using the finite difference method. If the exact solution is given, confirm the second-order accuracy of the numerical method using both the maximum absolute error and the root mean square error. If the exact solution is not given, use the technique outlined in Exercise 6 to confirm the second order accuracy of the numerical method. Explain any unusual behavior.

7. $u'' = -(x+1)u' + 2u + (1-x^2)e^{-x}, \quad u(0) = -1, \quad u(1) = 0, \quad u(x) = (x-1)e^{-x}$

8. $e^x \frac{d}{dx}\left(e^x \frac{dy}{dx}\right) = -1, \quad y(0) = y(1) = 0, \quad y(x) = -\frac{1}{2}e^{-2x} + \frac{1}{2}(1 + e^{-1})e^{-x} - \frac{1}{2}e^{-1}$

9. $y'' + xy' + y = x^2, \quad y(0) = 0, \quad y(1) = 1$

10. $u'' + 3u' = x^2 + \sin x, \quad u(-5) = 10, \quad u(13.2) = 23$

11. $\frac{1}{\rho^2}\frac{d}{d\rho}\left(\rho^2 \frac{du}{d\rho}\right) = -1, \quad u(1) = 0, \quad u(2) = -1/2, \quad u(\rho) = \frac{1}{6}(1 - \rho^2)$

12. $xy'' - (x+5)y' + 4y = x, \quad y(1) = -1, \quad y(2) = 1$

13. $x^2y'' - 2xy' + 2y = 3x^2 + 2\ln x, \quad y(1) = 9/4, \quad y(2) = 13\ln 2,$
$y(x) = \frac{3}{2}\left(1 + \frac{3}{2}x - x^2\right) + (1 + 3x^2)\ln x$

14. $y'' - 2xy' + 2y = -1, \quad y(-1) = 0, \quad y(1) = 1$

15. A wooden beam of square cross section is supported at both ends and is carrying a distributed lateral load of uniform intensity $w = 20$ lb/ft and an axial tension load $T = 100$ lb. The deflection, $u(x)$, of the beam's centerline satisfies the boundary value problem

$$u'' - \frac{T}{EI}u = -\frac{w}{2EI}x(L - x), \quad u(0) = u(L) = 0,$$

where $L = 6$ ft is the length, $E = 1.3 \times 10^6$ lb/in^2 is the modulus of elasticity and $I = s^4$ is the moment of inertia of the beam. The side length of the square cross section is $s = 4$ inches.

(a) Determine the deflection of the beam at 1-inch intervals along its length.

(b) Repeat part (a) assuming that the beam tapers along its length so that $s = (4 - x/2L)$ inches.

16. Repeat Exercise 15 for a metal rod of circular cross section. Use the parameter values

$$w = 200 \text{ lb/ft}, \quad T = 750 \text{ lb}$$
$$L = 10 \text{ ft}, \quad E = 3.0 \times 10^7 \text{ lb/in}^2, \quad \text{and} \quad I = \pi r^4/4.$$

For part (a), take $r = 3$ inches. For part (b), use $r = (3 + 0.25\sin(\pi x/L))$ inches.

17. Rework the "Flow between Parallel Plates" problem, taking

$$\mu(Y) = \exp\left\{-8.944 - \frac{839.456}{373.16 - 80Y/h} + \frac{421194.298}{(373.16 - 80Y/h)^2}\right\}.$$

This models the situation where the lower, stationary plate is maintained at 100° C, the upper, moving plate is maintained at 20° C and there is a linear temperature gradient between the two plates.

18. In the "Flow between Parallel Plates" problem, suppose we introduce the effect of a constant pressure gradient, which we denote by dp/dx, in the direction of the flow. The boundary value problem (9) then becomes

$$u'' = 80\left[\frac{842388.596}{(293.16 + 80y)^3} - \frac{839.456}{(293.16 + 80y)^2}\right]u' + \frac{h^2}{U_0}\frac{dp}{dx}$$
$$u(0) = 0, \quad u(1) = 1.$$

The constant $\frac{h^2}{U_0}\frac{dp}{dx}$ is called the pressure parameter. Calculate the velocity distribution for values of the pressure parameter of -4, -2, 0, 2, and 4.

8.2 FINITE DIFFERENCE METHOD, PART II: THE LINEAR PROBLEM WITH NON-DIRICHLET BOUNDARY CONDITIONS

In Section 8.1, we considered the prototype one dimensional two-point boundary value problem

$$y'' = p(x)y' + q(x)y + r(x), \quad x \in [a,b]$$

subject to the Dirichlet boundary conditions

$$y(a) = \alpha, \quad y(b) = \beta.$$

If our boundary value problem were a model for the one-dimensional steady-state conduction of heat in a metal rod, Dirichlet boundary conditions would correspond to prescribed temperatures at each end of the rod.

In practice, however, the temperature at each end of the rod might not be known. For example, we might only know that the end at $x = b$ was insulated, so that there was no heat flux from that end. This would give rise to a boundary condition of the form

$$y'(b) = 0.$$

A boundary condition of this form, in which the value of the derivative is specified, is known as a Neumann boundary condition. We also might only know that convective heat transfer is taking place at the end $x = a$. This would give rise to the boundary condition

$$-ky'(a) = h\left[T_\infty - y(a)\right],$$

where K is the thermal conductivity of the rod, h is the convective heat transfer coefficient between the rod and its surroundings, and T_∞ is the ambient temperature of the surroundings. A boundary condition of this type, in which a linear combination of the value of the function and the value of the first derivative is specified, is known as a Robin boundary condition.

In this section we will investigate the formulation of finite difference approximations for linear boundary value problems subject to both Neumann and Robin boundary conditions.

Non-Dirichlet Boundary Conditions

Because the general Neumann boundary condition

$$y'(a) = \alpha \quad \text{or} \quad y'(b) = \beta$$

is just a special case of the general Robin boundary condition

$$\alpha_1 y(a) + \alpha_2 y'(a) = \alpha_3 \quad \text{or} \quad \beta_1 y(b) + \beta_2 y'(b) = \beta_3$$

(set $\alpha_1 = 0$ or $\beta_1 = 0$), we will develop the system of algebraic equations for the finite difference approximation to the linear boundary value problem

$$y'' = p(x)y' + q(x)y + r(x), \quad x \in [a,b]$$

subject to the Robin boundary conditions

$$\alpha_1 y(a) + \alpha_2 y'(a) = \alpha_3$$
$$\beta_1 y(b) + \beta_2 y'(b) = \beta_3.$$

We have already discussed the handling of Dirichlet boundary conditions, so in what follows, we will assume that $\alpha_2 \neq 0$ and $\beta_2 \neq 0$.

For the computational grid, let N be a positive integer, and partition the interval $[a, b]$ into

$$a = x_0 < x_1 < x_2 < \cdots < x_{N-1} < x_N = b,$$

where $x_i = a + ih$ and $h = (b-a)/N$. Further, let w_i denote the approximation to the exact solution, $y(x)$, at $x = x_i$.

We need $N+1$ equations to determine the values w_0, w_1, w_2, ..., w_N. $N-1$ of these equations are obtained as in the previous section: Evaluate the differential equation at each interior grid point $x = x_i (1 \leq i \leq N-1)$, replace the derivatives by second-order central difference approximations, drop the truncation error terms, and collect like terms. The resulting computational template is

$$\left(-1 - \frac{h}{2}p_i\right) w_{i-1} + (2 + h^2 q_i) w_i + \left(-1 + \frac{h}{2}p_i\right) w_{i+1} = -h^2 r_i.$$

The only remaining question is what we do at $x_0 = a$ and at $x_N = b$.

Let's focus on the treatment of the boundary condition at $x_0 = a$:

$$\alpha_1 y(a) + \alpha_2 y'(a) = \alpha_3.$$

To maintain the second-order accuracy of the other equations, we could replace the derivative in the boundary condition by the $O(h^2)$ forward difference approximation

$$y_i' \approx \frac{-3y_i + 4y_{i+1} - y_{i+2}}{2h};$$

unfortunately, this would destroy the tridiagonal structure of the coefficient matrix. Using a first-order forward difference formula would maintain the tridiagonal structure of the coefficient matrix but would degrade the accuracy of the overall approximation.

A third possibility, which will maintain both the structure of the coefficient matrix and the second-order accuracy of the approximation, is to introduce a "fictitious node" to the computational grid.

Applying the computational template for the differential equation at $x = x_0$ produces

$$\left(-1 - \frac{h}{2}p_0\right) w_f + (2 + h^2 q_0) w_0 + \left(-1 + \frac{h}{2}p_0\right) w_1 = -h^2 r_0; \tag{1}$$

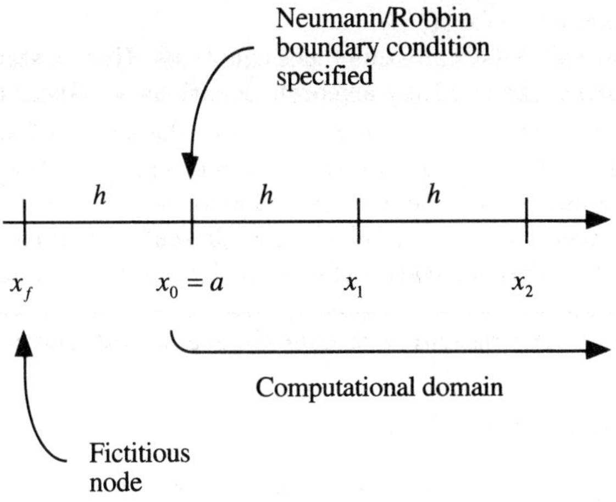

of course, w_f must be eliminated from this equation. This is accomplished by applying the Robin boundary condition:

$$\alpha_1 y(a) + \alpha_2 y'(a) = \alpha_3 \quad \Rightarrow \quad \alpha_1 w_0 + \alpha_2 \frac{w_1 - w_f}{2h} = \alpha_3,$$

where we have replaced the first derivative with its second-order central difference approximation. Solving this last expression for w_f yields

$$w_f = w_1 - \frac{2h}{\alpha_2}(\alpha_3 - \alpha_1 w_0).$$

Substituting this relation into equation (1), we obtain the finite difference equation associated with $x = a$:

$$\left[2 + h^2 q_0 - (2 + hp_0)h\frac{\alpha_1}{\alpha_2}\right] w_0 - 2w_1 = -h^2 r_0 - (2 + hp_0)h\frac{\alpha_3}{\alpha_2}.$$

For a Neumann boundary condition, $\alpha_1 = 0$, so the corresponding finite difference equation would read

$$(2 + h^2 q_0)w_0 - 2w_1 = -h^2 r_0 - (2 + hp_0)h\alpha,$$

where we have written α for the ratio α_3/α_2.

Performing a similar analysis for a Robin boundary condition at $x = b$, we find the corresponding finite difference equation to be

$$-2w_{N-1} + \left[2 + h^2 q_N + (2 - hp_N)h\frac{\beta_1}{\beta_2}\right] w_N = -h^2 r_N + (2 - hp_N)h\frac{\beta_3}{\beta_2}.$$

The derivation of this equation is left as an exercise. For a Neumann boundary condition at $x = b$, the equation would be

$$-2w_{N-1} + (2 + h^2 q_N)w_N = -h^2 r_N + (2 - hp_N)h\beta,$$

where we have written β for β_3/β_2.

We now have all $N+1$ finite difference equations. Having started from a linear differential equation, the resulting algebraic equations are linear in the unknowns. Let $\mathbf{w} = \begin{bmatrix} w_0 & w_1 & w_2 & \cdot & \cdot & \cdot & w_N \end{bmatrix}^T$ denote the vector of unknowns, and let the matrix A and the vector $\mathbf{b}$ have the structure shown in Table 8.1. The finite difference equations can then be written in the form $A\mathbf{w} = \mathbf{b}$. Dirichlet boundary conditions have been included in Table 8.1 to present a complete summary of the second-order finite difference method for linear boundary value problems.

EXAMPLE 8.3 A Problem with One Neumann and One Robin Boundary Condition

Consider the linear ordinary differential equation

$$u'' + u = \sin(3x), \quad x \in [0, \pi/2]$$

subject to a Robin boundary condition at $x = 0$:

$$u(0) + u'(0) = -1$$

and a Neumann boundary condition at $x = \pi/2$:

$$u'(\pi/2) = 1.$$

Let's take a uniform partition of $[0, \pi/2]$ with four subintervals. Then $h = \pi/8$ and $x_i = i\pi/8$ for $i = 0$, 1, 2, 3, and 4. Comparing the given differential equation with the prototype, we see that $p(x) = 0$, $q(x) = -1$ and $r(x) = \sin(3x)$. Therefore, for each i,

$$\begin{aligned} p_i &= 0; \\ q_i &= -1; \qquad \text{and} \\ r_i &= \sin(3i\pi/8). \end{aligned}$$

Furthermore, for the Robin boundary condition at $x = 0$, we have $\alpha_1 = \alpha_2 = 1$ and $\alpha_3 = -1$. Finally, with the Neumann boundary condition at $x = \pi/2$, we have $\beta = 1$.

Using Table 8.1, we then find that the system of finite difference equations is given by

$$\begin{bmatrix} d - \frac{\pi}{4} & -2 & & & \\ -1 & d & -1 & & \\ & -1 & d & -1 & \\ & & -1 & d & -1 \\ & & & -2 & d \end{bmatrix} \begin{bmatrix} w_0 \\ w_1 \\ w_2 \\ w_3 \\ w_4 \end{bmatrix} = \begin{bmatrix} \pi/4 \\ -\left(\frac{\pi}{8}\right)^2 \sin\left(\frac{3\pi}{8}\right) \\ -\left(\frac{\pi}{8}\right)^2 \sin\left(\frac{3\pi}{4}\right) \\ -\left(\frac{\pi}{8}\right)^2 \sin\left(\frac{9\pi}{8}\right) \\ \left(\frac{\pi}{8}\right)^2 + \pi/4 \end{bmatrix},$$

where

$$d = 2 - \left(\frac{\pi}{8}\right)^2.$$

$$A = \begin{bmatrix} a_{11} & a_{12} & & & & & \\ l_1 & d_1 & u_1 & & & & \\ & l_2 & d_2 & u_2 & & & \\ & & \cdot & \cdot & \cdot & & \\ & & & \cdot & \cdot & \cdot & \\ & & & & \cdot & \cdot & \cdot \\ & & & & l_{N-1} & d_{N-1} & u_{N-1} \\ & & & & & a_{N+1,N} & a_{N+1,N+1} \end{bmatrix}$$

$$\mathbf{b} = \begin{bmatrix} b_1 \\ -h^2 r_1 \\ -h^2 r_2 \\ \cdot \\ \cdot \\ \cdot \\ -h^2 r_{N-1} \\ b_{N+1} \end{bmatrix}$$

$$d_i = 2 + h^2 q_i, \qquad u_i = -1 + \frac{h}{2} p_i, \qquad l_i = -1 - \frac{h}{2} p_i$$

$$a_{11} = \begin{cases} 1, & \text{Dirichlet BC at } x = a \\ d_0, & \text{Neumann BC at } x = a \\ d_0 + 2hl_0\alpha_1/\alpha_2, & \text{Robin BC at } x = a \end{cases}$$

$$a_{12} = \begin{cases} 0, & \text{Dirichlet BC at } x = a \\ -2, & \text{otherwise} \end{cases}$$

$$a_{N+1,N+1} = \begin{cases} 1, & \text{Dirichlet BC at } x = b \\ d_N, & \text{Neumann BC at } x = b \\ d_N - 2hu_N\beta_1/\beta_2, & \text{Robin BC at } x = b \end{cases}$$

$$a_{N+1,N} = \begin{cases} 0, & \text{Dirichlet BC at } x = b \\ -2, & \text{otherwise} \end{cases}$$

$$b_1 = \begin{cases} \alpha, & \text{Dirichlet BC at } x = a \\ -h^2 r_0 + 2hl_0\alpha, & \text{Neumann BC at } x = a \\ -h^2 r_0 + 2hl_0\alpha_3/\alpha_2, & \text{Robin BC at } x = a \end{cases}$$

$$b_{N+1} = \begin{cases} \beta, & \text{Dirichlet BC at } x = b \\ -h^2 r_N - 2hu_N\beta, & \text{Neumann BC at } x = b \\ -h^2 r_N - 2hu_N\beta_3/\beta_2, & \text{Robin BC at } x = b \end{cases}$$

BC = boundary conditions

TABLE 8.1: Matrix Formulation of Second-Order Finite Difference Method for the Linear Boundary Value Problem $y'' = p(x)y' + q(x)y + r(x)$, $x \in [a, b]$ Subject to Some Combination of Dirichlet Boundary Conditions—$y(a) = \alpha, y(b) = \beta$—Neumann Boundary Conditions—$y'(a) = \alpha, y'(b) = \beta$—and Robin Boundary Conditions—$\alpha_1 y(a) + \alpha_2 y'(a) = \alpha_3, \beta_1 y(b) + \beta_2 y'(b) = \beta_3$

In the first and last entries of the right-hand-side vector, we have used the fact that $\sin 0 = 0$ and $\sin 3\pi/2 = -1$, respectively. The solution of this tridiagonal linear system is

$$\mathbf{w} = \begin{bmatrix} -1.023672 & -0.935445 & -0.560486 & 0.00995175 & 0.519840 \end{bmatrix}^T.$$

The following table compares this approximate solution with the exact solution: $u(x) = -\cos x + (3/8)\sin x - (1/8)\sin 3x$. The accuracy of the approximate solution is quite reasonable given the crudeness of the computational grid.

x_i	Approximate Solution, w_i	Exact Solution, u_i	Absolute Error
0	-1.023672	-1.000000	0.023672
$\pi/8$	-0.935445	-0.895858	0.039587
$\pi/4$	-0.560486	-0.530330	0.030156
$3\pi/8$	0.00995175	0.0116068	0.001655
$\pi/2$	0.519840	0.500000	0.019840

A numerical verification of the second-order accuracy of the scheme is presented in the next table, which displays both the maximum absolute error and the root mean square (rms) error in the approximate solution as a function of the number of subintervals, N. The rms error was computed according to the formula

$$\text{rms error} = \sqrt{\frac{1}{N+1}\sum_{i=0}^{N}(w_i - u_i)^2}.$$

Note that with each doubling of N, the step size is cut in half, and the approximation error is reduced by roughly a factor of 4—which is what one would expect from a second order numerical method.

N	Maximum Absolute Error	Error Ratio	rms Error	Error Ratio
4	0.0395865088		0.0262038549	
8	0.0094846260	4.173755	0.0064356848	4.071650
16	0.0023587346	4.021065	0.0016105638	3.995920
32	0.0005899465	3.998218	0.0004038284	3.988239
64	0.0001473906	4.002605	0.0001011678	3.991667
128	0.0000368515	3.999585	0.0000253223	3.995214
256	0.0000092125	4.000162	0.0000063346	3.997450
512	0.0000023031	4.000031	0.0000015842	3.998676

Handling an Artificial Singularity

Consider the boundary value problem

$$u'' + \frac{1}{x}u' = \left(\frac{8}{8-x^2}\right)^2$$

$$u'(0) = u(1) = 0.$$

Note the coefficient of the first derivative is singular at the left endpoint of the problem domain, $x = 0$; however, the exact solution to the problem,

$$u(x) = 2\ln\left(\frac{7}{8 - x^2}\right),$$

is not singular at $x = 0$. We could have anticipated this eventuality since the Neumann boundary condition at $x = 0$ implies that

$$\lim_{x\to 0} \underbrace{\frac{u'(x)}{x}}_{0/0 \text{ form}} \overset{\substack{\text{Using} \\ \text{L'Hôpital's} \\ \text{Rule} \\ \downarrow}}{=} \lim_{x\to 0} \frac{u''(x)}{1} = u''(0), \tag{2}$$

so the first derivative term in the differential equation is not singular. Such a situation is referred to as an *artificial singularity.* Artificial singularities occur frequently in problems in polar, cylindrical, and spherical coordinates.

How do we handle the artificial singularity in this problem, within the context of constructing a finite difference approximation? Once we establish a computational grid, at every grid point but $x_0 = 0$ we can employ our standard finite difference procedures (summarized in Table 8.1) to generate N of the $N+1$ finite difference equations. At $x = 0$, we cannot use Table 8.1, since $p(0)$ is undefined. However, making use of equation (2), we find that, in the limit as $x \to 0$, the differential equation reduces to

$$u''(0) + u''(0) = \left(\frac{8}{8 - 0^2}\right)^2 = 1,$$

or $u''(0) = 1/2$. If we replace the second derivative in this expression with its second-order finite difference formula (which will involve the use of a fictitious node), drop the truncation error term, and use the Neumann boundary condition to eliminate the fictitious node, the finite difference equation for $x = 0$ is found to be

$$2w_0 - 2w_1 = -\frac{1}{2}h^2.$$

The details of this derivation are left as an exercise.

It is important to note that we can arrive at the same system of equations, including the correct equation corresponding to $x = 0$, if we consolidate the equations

$$u''(0) = 1/2 \quad \text{and} \quad u'' + \frac{1}{x}u' = \left(\frac{8}{8 - x^2}\right)^2$$

into the single differential equation $u'' = p(x)u' + r(x)$, where

$$p(x) = \begin{cases} 0, & x = 0 \\ -\frac{1}{x}, & x \neq 0 \end{cases} \quad \text{and} \quad r(x) = \begin{cases} \frac{1}{2}, & x = 0 \\ \left(\frac{8}{8-x^2}\right)^2, & x \neq 0 \end{cases}.$$

This procedure will allow us to handle problems with artificial singularities using the material contained in Table 8.1. Proceeding in this manner for the current problem, we obtain the results tabulated below, which demonstrate that the scheme maintains full second-order accuracy.

N	Maximum Absolute Error	Error Ratio	rms Error	Error Ratio
4	0.0016976480		0.0012738555	
8	0.0004369617	3.885119	0.0003270583	3.894889
16	0.0001101553	3.966778	0.0000824331	3.967562
32	0.0000276039	3.990575	0.0000206671	3.988619
64	0.0000069055	3.997364	0.0000051726	3.995466
128	0.0000017267	3.999271	0.0000012938	3.998011
256	0.0000004317	3.999800	0.0000003235	3.999073
512	0.0000001079	3.999946	0.0000000809	3.999554

Application Problem 1: Steady-State Temperature Distribution in a Pin Fin

In the Overview to this chapter (see page 656), we developed a model for the steady-state temperature distribution, $T(x)$, along the length of a pin fin with nonuniform cross section. Recall that the boundary value problem satisfied by $T(x)$ is

$$\frac{d}{dx}\left([r(x)]^2\frac{dT}{dx}\right) - \frac{2h}{k}r(x)(T - T_\infty) = 0$$
$$T(0) = T_0, \qquad -kT'(L) = h(T(L) - T_\infty),$$

where $r(x)$ is the radius of the fin cross section, h is the convection heat transfer coefficient, K is the thermal conductivity of the fin, T_∞ is the ambient temperature of the fluid surrounding the fin, T_0 is the temperature of the solid from which the fin extends and l is the length of the fin. Once the temperature distribution has been determined, the total fin heat transfer rate is given by

$$q_f = -k\pi[r(0)]^2T'(0).$$

Suppose a fin made from AISI stainless steel, with a thermal conductivity of $k = 14$ W/m · K, extends from a solid whose temperature is $T_0 = 100°$ C. The pin has a length of $L = 10$ cm and tapers linearly along its length from a radius of 2 cm where the pin attachs to the solid to a radius of 1 cm at its tip; that is,

$$r(x) = \left(2 - \frac{x}{10}\right) \text{ cm}.$$

The temperature of the surrounding fluid (air in this case) is $T_\infty = 20°$ C, and the convection heat transfer coefficient is $h = 20$ W/m^2 · K.

Substituting the indicated parameter values into the model boundary value problem, converting all distance units to meters and rearranging into standard form

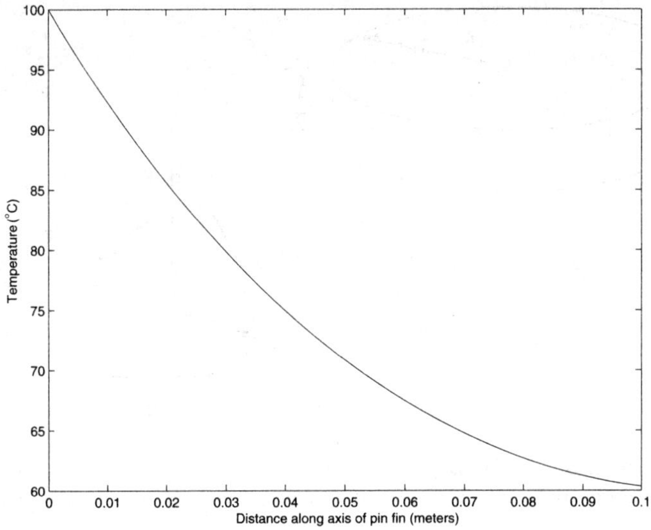

Figure 8.7 Temperature distribution along length of a pin fin with non-uniform cross section and experiencing convective heat loss from its lateral surface.

yields

$$T'' = \frac{2}{20-x}T' + \frac{20000}{140-7x}T - \frac{400000}{140-7x}$$
$$T(0) = 100, \quad 20T(0.1) + 14T'(0.1) = 400.$$

Taking a uniform partition of the interval $[0, 0.1]$ with $N = 100$ subintervals produces the temperature distribution shown in Figure 8.7. The temperature at the tip of the fin ($x = 0.1$) is approximately 60.36° C. To determine the total fin heat transfer rate, we need to know $T'(0)$. Using the formula for the second-order forward difference approximation to the first derivative, we calculate

$$T'(0) \approx \frac{-3T(0) + 4T(0.001) - T(0.002)}{0.002} = -825.69\frac{\text{K}}{\text{m}}.$$

Accordingly,

$$q_f = -14\pi(0.02)^2(-825.69) = 14.53 \text{ W}.$$

Application Problem 2: The Heat Pack

A heat pack is in the shape of a thin circular cylinder with radius r and thickness T, as shown in Figure 8.8(a). When the pack is squeezed, a bubble inside the pack breaks releasing chemicals that initiate an exothermic reaction. We will

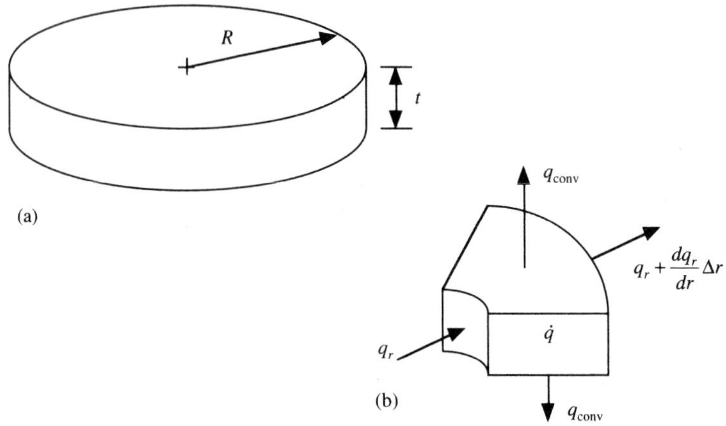

Figure 8.8 (a) Heat pack in the shape of a thin circular cylinder. (b) Energy balance on an arbitrary control volume of the heat pack.

suppose that the generation of heat resulting from this reaction is uniformly distributed throughout the pack. Since the pack is thin, we will further assume that temperature variations across the thickness are negligible.

To model the temperature variation in the radial direction, we perform an energy balance on the control volume shown in Figure 8.8(b). This gives

$$-\frac{dq_r}{dr}\Delta r - 2q_{\text{conv}} + \dot{q}r\,\Delta r\,\Delta\theta t = 0, \tag{3}$$

where q_r is the rate of heat transfer due to conduction, q_{conv} is the rate of heat transfer due to convection, $\dot{q}$ is the rate of heat generation per unit volume, and $\Delta\theta$ is the angle subtended by the control volume. In this geometry, Fourier's law gives

$$q_r = -kr\Delta\theta t\frac{dT}{dr}, \tag{4}$$

where K is the thermal conductivity of the material inside the heat pack. The convection heat transfer rate is

$$q_{\text{conv}} = hr\Delta r\Delta\theta(T - T_\infty), \tag{5}$$

where h is the convection heat transfer coefficient and T_∞ is the temperature of the air surrounding the heat pack.

Substituting (4) and (5) into (3) and dividing by $r\Delta r\Delta\theta t$ yields

$$T'' + \frac{1}{r}T' - \frac{2h}{kt}T + \left(\frac{2hT_\infty}{kt} + \frac{\dot{q}}{k}\right) = 0. \tag{6}$$

Symmetry about $r = 0$ gives rise to the boundary condition $T'(0) = 0$, while convective heat loss at the outer edge of the cylinder leads to the boundary condition

$$-kT'(R) = h(T(R) - T_\infty).$$

Observe that equation (6) has an artificial singularity at $r = 0$. Taking the limit of (6) as r approaches zero and using $T'(0) = 0$, we find that, at $r = 0$, the governing differential equation reduces to

$$T''(0) - \frac{h}{kt}T(0) + \left(\frac{hT_\infty}{kt} + \frac{\dot{q}}{2k}\right) = 0. \tag{7}$$

Thus, we can identify the coefficient functions, for use with Table 8.1, as

$$p(r) = \begin{cases} 0, & r = 0 \\ -1/r, & r \neq 0 \end{cases}, \qquad q(r) = \begin{cases} h/kt, & r = 0 \\ 2h/kt, & r \neq 0 \end{cases}, \qquad \text{and}$$

$$f(r) = \begin{cases} -(hT_\infty/kt + \dot{q}/2k), & r = 0 \\ -(2hT_\infty/kt + \dot{q}/k), & r \neq 0 \end{cases}.$$

Note that we have denoted the nonhomogeneous term by f, rather than r, because r is already serving duty as the independent variable in this problem.

Suppose the heat pack has a radius of $R = 10$ cm and thickness of $t = 0.6$ cm. The convection heat transfer coefficient is $h = 20\ \text{W/m}^2 \cdot \text{K}$, and the temperature of the air surrounding the heat pack is $T_\infty = 20^\circ$ C. Take $k = 0.4\ \text{W/m} \cdot \text{K}$ as the thermal conductivity of the material inside the pack. If the chemical reaction releases 30 W of heat energy, then

$$\dot{q} = \frac{30}{\pi(0.1)^2(0.006)} = \frac{500000}{\pi}.$$

With these parameter values and a uniform partition of $[0, 0.1]$ with $N = 100$ subintervals, the temperature profile shown in Figure 8.9 is obtained.

EXERCISES

1. Derive the finite difference equation

$$-2w_{N-1} + \left[2 + h^2 q_N + (2 - hp_N)h\frac{\beta_1}{\beta_2}\right] w_N = -h^2 r_N + (2 - hp_N)h\frac{\beta_3}{\beta_2}$$

corresponding to the Robin boundary condition $\beta_1 y(b) + \beta_2 y'(b) = \beta_3$, where $x = b$ is the right endpoint of the problem domain.

2. Derive the finite difference equation

$$2w_0 - 2w_1 = -\frac{1}{2}h^2$$

corresponding to $x = 0$ for the boundary value problem

$$u'' + \frac{1}{x}u' = \left(\frac{8}{8 - x^2}\right)^2, \qquad u'(0) = u(1) = 0.$$

Recall that at $x = 0$, the differential equation reduces to $u''(0) = 1/2$.

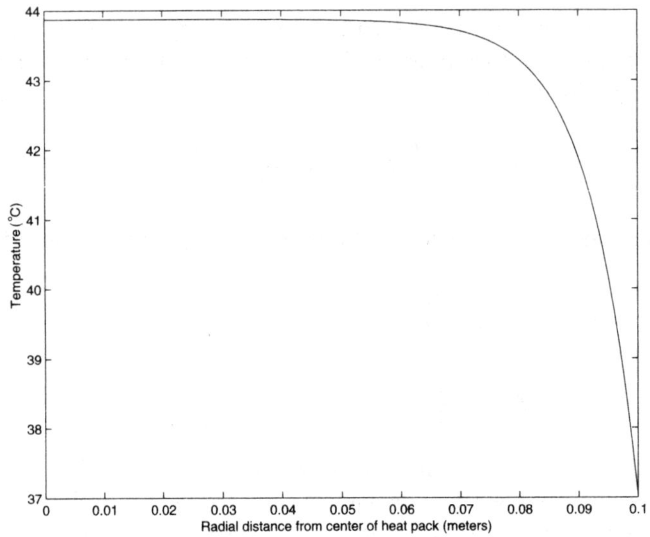

Figure 8.9 Radial temperature distribution in heat pack.

In Exercises 3–10,

(a) identify the interval $[a, b]$ and the functions p, q and r; and

(b) using a uniform partition of $[a.b]$ with $N = 4$ subintervals, write out the system of finite difference equations.

Note that some of these problems have artificial singularities.

3. $y'' = -\dfrac{3}{2+x}y' - \dfrac{2}{(2+x)^3}, \quad y(0) = y'(1) = 0$

4. $\dfrac{1}{x}\left(xy'\right)' = 1, \quad y'(0) = 0, \quad y(1) = 10$

5. $y'' = 120y - 2560, \quad 10y(0) + 35y'(0) = 200, \quad y(0.15) = 20$

6. $\dfrac{1}{x^2}\left(x^2y'\right)' = 1 + x^2, \quad y'(0) = 0, \quad 4y(1) + y'(1) = 40$

7. $xy'' - (2x+1)y' + (x+1)y = 0, \quad y(1) - 2y'(1) = -6e, \quad y(3) + y'(3) = 26e^3$

8. $\dfrac{1}{x}\left(xy'\right)' + y = 10, \quad y(1) = 20, \quad y(3) + y'(3) = 10$

9. $y'' + \dfrac{1}{x}y' + y = 1, \quad y'(0) = 0, \quad y(1) = 1$

10. $(1 - x\cot x)y'' - xy' + y = 0, \quad y'(1) = 1 + \cos(1), \quad y(2) + y'(2) = 3 + \sin(2) + \cos(2)$

In Exercises 11–16, approximate the solution of the indicated boundary value problem using the finite difference method and confirm the second order accuracy of the numerical method. Explain any unusual behavior.

11. $y'' - y = 1, \quad y(0) = 0, \quad y(1) + y'(1) = 1, \quad y(x) = e^{x-1} + \left(1 - \frac{1}{e}\right) e^{-x} - 1$

12. $\frac{1}{\rho^2}\left(\rho^2 u'\right)' = -1, \quad u'(0) = 0, \quad u(1) = 1, \quad u(\rho) = \frac{1}{6}(7 - \rho^2)$

13. $xy'' - (2x+1)y' + (x+1)y = 0, \quad y(1) - 2y'(1) = -6e, \quad y(3) + y'(3) = 26e^3,$
$y(x) = (1 + x^2)e^x$

14. $y'' - 2xy' + 2y = -1, \quad y'(-1) = 0, \quad y(1) = 1$

15. $y'' + xy' + y = x^2, \quad y(0) + y'(0) = 0, \quad y(1) = 1$

16. $y'' + \frac{1}{x}y' + y = 1, \quad y'(0) = 0, \quad y'(1) = 1$

17. One component of a model for a styrene monomer tubular reactor is the steady-state temperature profile of the solid phase catalyst. The governing boundary value problem is

$$\frac{d^2\tau}{dx^2} + \frac{1}{x}\frac{d\tau}{dx} - \beta^2\tau = 0$$

$$\left.\frac{d\tau}{dx}\right|_{x=0} = 0, \quad \tau(1) = 1.$$

Here, $\tau = (T_C - T)/(T_w - T)$ is the nondimensional catalyst temperature, x is the nondimensional radial position, T is the constant temperature of the fluid in the reactor, and T_w is the temperature at the wall of the reactor. The parameter β^2 is given by

$$\beta^2 = \frac{R^2 hA}{k(1-\epsilon)},$$

where $R = 1.3$ cm is the radius of the reactor, $h = 10^{-3}$ cal/cm$^2 \cdot$ s $\cdot\ ^\circ$C is the heat transfer coefficient, $\epsilon = 0.36$ is the porosity of the packed bed reactor, $A = 15$ cm^{-1} is the surface area of the catalyst per unit volume, and $k = 0.0034$ cal/cm·s·$^\circ$ C is the thermal diffusivity of the catalyst. Approximate $\tau(x)$ using $\Delta x = 0.0025$.

18. A thin cylindrical fiber, ten inches in length, has its left end maintained at a constant temperature T_0 and experiences convective heat loss along its lateral surface and from its right end. The temperature within the fiber is governed by the boundary value problem

$$-\frac{d}{dx}\left(kr^2\frac{dT}{dx}\right) + 2hrT = 2hrT_\infty,$$

$$T(-5) = T_0, \qquad -k\left.\frac{dT}{dx}\right|_{x=5} = h[T(5) - T_\infty].$$

The parameters in this problem are the thermal conductivity of the fiber $k =$ 2 BTU/sec $\cdot$ in $\cdot\ ^\circ$ F, the convective heat transfer coefficient $h = 10^{-5}$ BTU/sec $\cdot$ in$^2 \cdot\ ^\circ$ F, the radius of the fiber

$$r = 0.002\left(1 + \frac{0.1}{1 + (x-5)^2}\right) \text{ inches,}$$

the ambient temperature of the surroundings $T_\infty = 50^\circ$ F and the constant temperature maintained at the left end of the fiber $T_0 = 200^\circ$ F. Determine the temperature within the fiber at increments of 0.1 inches.

19. Rework the first application problem from this section, "Steady-State Temperature Distribution in a Pin Fin," for an iron fin with thermal conductivity $k = 80$ W/m · K. Use the values given in the text for all other parameters.

20. Rework the first application problem from this section, "Steady-State Temperature Distribution in a Pin Fin," for a copper fin with thermal conductivity $k = 401$ W/m · K. Take

$$r(x) = 2 - 2\left(\frac{x}{L}\right) + \left(\frac{x}{L}\right)^2 \text{ cm},$$

and use the values given in the text for all other parameters.

21. Reconsider the second application problem from this section, "The Heat Pack." Calculate the radial temperature profile for convection heat transfer coefficients of $h = 5$, 10, 15, and 25 W/m^2 · K. Use the values given in the text for all other parameters. What effect does changing the value of h seem to have on the resulting temperature profile? Examine the temperature at the center of the pack, the length of the nearly constant portion of the profile, and the temperature at the outer edge of the pack.

22. Suppose we had chosen to handle Neumann and Robin boundary conditions by replacing the derivative that appears in the conditions with a first-order finite difference formula. This would mean, for example, that the condition $y'(a) = \alpha$ would translate into the finite difference equation $w_0 - w_1 = -h\alpha$, assuming that $x = a$ were the left endpoint of the domain. As stated earlier, following this approach maintains the tridiagonal structure of the coefficient matrix but introduces $O(h)$ errors where all other errors were $O(h^2)$. In this exercise we will investigate the effect of these lower-order errors on the overall accuracy of the approximation.

Consider the boundary value problem

$$\begin{gathered} u'' + u = \sin 3x, \quad x \in [0, \pi/2] \\ u(0) + u'(0) = -1, \quad u'(\pi/2) = 1, \end{gathered}$$

whose exact solution is $u(x) = -\cos x + (3/8)\sin x - (1/8)\sin 3x$. Use Table 8.1 to construct the system of finite difference equations for this BVP, but replace the first equation by $(1 - h)w_0 - w_1 = h$ and replace the last equation by $w_N - w_{N-1} = h$. By computing the approximate solution for various values of N and comparing with the exact solution, numerically estimate the order of convergence of this modified finite difference method.

8.3 FINITE DIFFERENCE METHOD, PART III: NONLINEAR PROBLEMS

Having treated linear boundary value problems in detail, we now turn our attention to the nonlinear boundary value problem

$$
\begin{aligned}
y'' &= f(x, y, y'), \quad x \in [a, b] \\
\alpha_1 y(a) + \alpha_2 y'(a) &= \alpha_3 \\
\beta_1 y(b) + \beta_2 y'(b) &= \beta_3.
\end{aligned}
$$

For completeness, we note that when

- f, $\partial f/\partial y$ and $\partial f/\partial y'$ are continuous on the set

$$D = \{(x, y) : a \le x \le b, y, y' \in \mathbf{R}\};$$

- $\partial f/\partial y > 0$ for all $(x, y, y') \in D$; and
- there exists a constant l such that $|\partial f/\partial y'| \le L$ for all $(x, y, y') \in D$

the given boundary value problem is guaranteed to have a unique solution. For a proof of this result, see Keller [1].

The techniques that were introduced in the previous two sections for developing finite difference equations can still be applied to a nonlinear differential equation; however, the resulting system of algebraic equations will be nonlinear. In this section, we will derive the second-order finite difference equations associated with the general nonlinear boundary value problem cited above. The solution of the nonlinear algebraic equations will also be discussed.

Dirichlet Boundary Conditions

Let's begin our investigation of finite difference methods for nonlinear boundary value problems by considering the case of Dirichlet boundary conditions:

$$
\begin{aligned}
y'' &= f(x, y, y'), \quad x \in [a, b] \\
y(a) &= \alpha, \quad y(b) = \beta.
\end{aligned}
$$

First, we introduce our standard computational grid, which is defined by

$$x_i = a + ih \quad (i = 0, 1, 2, \dots, N), \qquad h = (b - a)/N.$$

Next, we evaluate the governing differential equation at an arbitrary interior grid point $x = x_i$,

$$[y'' = f(x, y, y')]|_{x=x_i},$$

and substitute the second-order finite difference formulas

$$
\begin{aligned}
y'' &= \frac{y_{i-1} - 2y_i + y_{i+1}}{h^2} + O(h^2) \\
y' &= \frac{y_{i+1} - y_{i-1}}{2h} + O(h^2)
\end{aligned}
$$

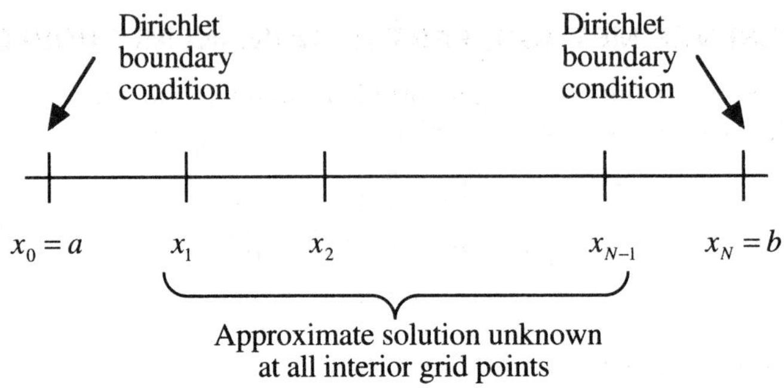

for the derivatives. The truncation error terms are then dropped, and the y's (exact solution) are replaced by w's (approximate solution). The resulting equation is

$$\frac{w_{i-1} - 2w_i + w_{i+1}}{h^2} = f\left(x_i, w_i, \frac{w_{i+1} - w_{i-1}}{2h}\right). \tag{1}$$

With the assumed bound on $\partial f/\partial y'$, the overall truncation error for this approximation is $O(h^2)$. Rearranging equation (1) produces the computational template

$$-w_{i-1} + 2w_i - w_{i+1} + h^2 f\left(x_i, w_i, \frac{w_{i+1} - w_{i-1}}{2h}\right) = 0, \tag{2}$$

which holds for $i = 1, 2, 3, \ldots, N-1$. To these equations, we add

$$w_0 = \alpha \quad \text{and} \quad w_N = \beta,$$

both of which are obtained from the boundary conditions.

If we let

$$\mathbf{w} = \begin{bmatrix} w_0 & w_1 & w_2 & \cdots & w_N \end{bmatrix}^T$$

and

$$\mathbf{G}(\mathbf{w}) = \begin{bmatrix} g_0(\mathbf{w}) & g_1(\mathbf{w}) & g_2(\mathbf{w}) & \cdots & g_N(\mathbf{w}) \end{bmatrix}^T,$$

where

$$g_0(\mathbf{w}) = w_0 - \alpha$$

$$g_i(\mathbf{w}) = -w_{i-1} + 2w_i - w_{i+1} + h^2 f\left(x_i, w_i, \frac{w_{i+1} - w_{i-1}}{2h}\right) \quad (i = 1, 2, 3, \ldots, N-1)$$

$$g_N(\mathbf{w}) = w_N - \beta,$$

then the system of finite difference equations can be written in vector form as

$$\mathbf{G}(\mathbf{w}) = \mathbf{0}.$$

This is a nonlinear system of equations that we will solve with Newton's method.

Recall from Section 3.10 that Newton's method for the general system of m equations, $\mathbf{F}(\mathbf{x}) = \mathbf{0}$, generates the sequence $\{\mathbf{x}^{(n+1)}\}$ according to the rule

$$\mathbf{x}^{(n+1)} = \mathbf{x}^{(n)} + \mathbf{v}^{(n)}.$$

The update vector, $\mathbf{v}^{(n)}$, is the solution of the linear system

$$\left[J(\mathbf{x}^{(n)})\right] \mathbf{v}^{(n)} = -\mathbf{F}(\mathbf{x}^{(n)}),$$

where J is the Jacobian matrix for the system. The Jacobian is given by

$$J(\mathbf{x}) = \begin{bmatrix} \partial f_1/\partial x_1 & \partial f_1/\partial x_2 & \partial f_1/\partial x_3 & \cdots & \partial f_1/\partial x_m \\ \partial f_2/\partial x_1 & \partial f_2/\partial x_2 & \partial f_2/\partial x_3 & \cdots & \partial f_2/\partial x_m \\ \partial f_3/\partial x_1 & \partial f_3/\partial x_2 & \partial f_3/\partial x_3 & \cdots & \partial f_3/\partial x_m \\ & \vdots & & & \vdots \\ \partial f_m/\partial x_1 & \partial f_m/\partial x_2 & \partial f_m/\partial x_3 & \cdots & \partial f_m/\partial x_m \end{bmatrix}.$$

Since each iteration requires the solution of a linear system of equations, if the Jacobian matrix for the nonlinear equations is a full matrix, Newton's method will be very expensive.

Fortunately, the Jacobian matrix for the system of finite difference equations $\mathbf{G}(\mathbf{w}) = \mathbf{0}$ has a special structure. With the exception of the first and last equations, the finite difference equations take the form

$$g_i(\mathbf{w}) = -w_{i-1} + 2w_i - w_{i+1} + h^2 f\left(x_i, w_i, \frac{w_{i+1} - w_{i-1}}{2h}\right).$$

Since $g_i(\mathbf{w})$ depends only on the unknowns w_{i-1}, w_i and w_{i+1}, the only nonzero entries along the ith row of the Jacobian will be

$$\frac{\partial g_i}{\partial w_{i-1}} = -1 - \frac{h}{2}\frac{\partial f}{\partial y'}\left(x_i, w_i, \frac{w_{i+1} - w_{i-1}}{2h}\right),$$

$$\frac{\partial g_i}{\partial w_i} = 2 + h^2 \frac{\partial f}{\partial y}\left(x_i, w_i, \frac{w_{i+1} - w_{i-1}}{2h}\right), \quad \text{and}$$

$$\frac{\partial g_i}{\partial w_{i+1}} = -1 + \frac{h}{2}\frac{\partial f}{\partial y'}\left(x_i, w_i, \frac{w_{i+1} - w_{i-1}}{2h}\right).$$

The first and last rows of the Jacobian will contain a 1 along the main diagonal and zeros everywhere else.

Therefore, the overall structure of the Jacobian will be

$$J(\mathbf{w}) = \begin{bmatrix} 1 & 0 & & & & & & \\ l_1 & d_1 & u_1 & & & & & \\ & l_2 & d_2 & u_2 & & & & \\ & & \cdot & \cdot & \cdot & & & \\ & & & \cdot & \cdot & \cdot & & \\ & & & & \cdot & \cdot & \cdot & \\ & & & & l_{N-2} & d_{N-2} & u_{N-2} & \\ & & & & & l_{N-1} & d_{N-1} & u_{N-1} \\ & & & & & & 0 & 1 \end{bmatrix},$$

where

$$d_i = 2 + h^2 \frac{\partial f}{\partial y}\left(x_i, w_i, \frac{w_{i+1} - w_{i-1}}{2h}\right)$$

$$u_i = -1 + \frac{h}{2}\frac{\partial f}{\partial y'}\left(x_i, w_i, \frac{w_{i+1} - w_{i-1}}{2h}\right)$$

$$l_i = -1 - \frac{h}{2}\frac{\partial f}{\partial y'}\left(x_i, w_i, \frac{w_{i+1} - w_{i-1}}{2h}\right).$$

Note that this is a tridiagonal matrix. The number of operations needed to solve a linear system with a tridiagonal coefficient matrix is only $O(n)$—a significant savings over the general case.

The final issue to discuss regarding the use of Newton's method to solve the finite difference equations is the choice of the initial vector $\mathbf{w}^{(0)}$. Unless a previous approximate solution is available, whenever Dirichlet boundary conditions have been specified, the simplest scheme for obtaining an initial vector is to pass a line through the points (a, α) and (b, β) and evaluate that function at each $x = x_i$. This procedure yields

$$w_i^{(0)} = \frac{\beta - \alpha}{b - a}(x_i - a) + \alpha,$$

or, after substituting $x_i = a + ih$ and $h = (b - a)/N$,

$$w_i^{(0)} = i\frac{\beta - \alpha}{N} + \alpha \quad (i = 0, 1, 2, \ldots, N).$$

EXAMPLE 8.4 A Sample Nonlinear Boundary Value Problem

Let's use the finite difference method to approximate the solution of the nonlinear boundary value problem with Dirichlet boundary conditions

$$yy'' + (y')^2 + 1 = 0$$
$$y(1) = 1, \quad y(2) = 2.$$

Solving the differential equation for the second derivative, we find

$$y'' = -\frac{1 + (y')^2}{y},$$

so that

$$f(x, y, y') = -\frac{1 + (y')^2}{y}.$$

Applying the basic rules of differential calculus, we compute the partial derivatives

$$\frac{\partial f}{\partial y}(x, y, y') = \frac{1 + (y')^2}{y^2} \quad \text{and} \quad \frac{\partial f}{\partial y'}(x, y, y') = -\frac{2y'}{y}.$$

Thus, the elements along the diagonal of the Jacobian are given by

$$d_i = 2 + h^2 \frac{1 + \left(\frac{w_{i+1} - w_{i-1}}{2h}\right)^2}{w_i^2} = 2 + \frac{4h^2 + (w_{i+1} - w_{i-1})^2}{4w_i^2},$$

while the off-diagonal elements are given by

$$u_i = -1 + \frac{h}{2}\left(-\frac{2\frac{w_{i+1} - w_{i-1}}{2h}}{w_i}\right) = -1 - \frac{w_{i+1} - w_{i-1}}{2w_i}$$

and

$$l_i = -1 - \frac{h}{2}\left(-\frac{2\frac{w_{i+1} - w_{i-1}}{2h}}{w_i}\right) = -1 + \frac{w_{i+1} - w_{i-1}}{2w_i}.$$

Using a uniform grid with $N = 8$ subintervals, we obtain the results listed in the second column of the following table. A convergence tolerance of $TOL = 5 \times 10^{-14}$ was used to terminate the Newton iterations. The values in the third column were obtained by evaluating the exact solution to the boundary value problem, $y(x) = \sqrt{6x - 4 - x^2}$, at each grid point.

x_i	Approximate Solution, w_i	Exact Solution, y_i	Absolute Error
1.000	1.000000	1.000000	0.000000
1.125	1.217747	1.218349	0.000602
1.250	1.391239	1.391941	0.000702
1.375	1.535371	1.536026	0.000655
1.500	1.657760	1.658312	0.000552
1.625	1.762916	1.763342	0.000426
1.750	1.853761	1.854050	0.000289
1.875	1.932307	1.932453	0.000146
2.000	2.000000	2.000000	0.000000

We complete this example by demonstrating the second-order accuracy of the numerical method. The next table lists the maximum absolute error and the root mean square (rms) error in the approximate solution as a function of the number of subintervals in the computational grid. With each doubling of the number of subintervals, the step size is cut in half. For a second-order method we would expect the error to decrease by a factor of $2^2 = 4$. The error ratio values listed in the third and fifth columns are clearly approaching the expected value.

N	Maximum Absolute Error	Error Ratio	rms Error	Error Ratio
4	0.0025761053		0.0015533493	
8	0.0007021651	3.668803	0.0004564927	3.402791
16	0.0001804347	3.891520	0.0001209483	3.774280
32	0.0000454487	3.970073	0.0000309207	3.911566
64	0.0000113841	3.992300	0.0000078039	3.962232
128	0.0000028474	3.998060	0.0000019594	3.982742
256	0.0000007120	3.999337	0.0000004909	3.991779
512	0.0000001780	3.999875	0.0000001228	3.995992

Neumann and Robin Boundary Conditions

To modify our finite difference method to handle non-Dirichlet boundary conditions, we will have to derive a new function $g_0(\mathbf{w})$ and/or a new function $g_N(\mathbf{w})$, and, accordingly, determine a new first row and/or a new last row for the Jacobian. To maintain the second-order accuracy of our scheme and to maintain the tridiagonal structure of the Jacobian matrix, we will once again make use of fictitious nodes. Furthermore, since a Neumann boundary condition is just a special case of a Robin boundary condition, we will focus our attention on the equations corresponding to Robin conditions.

Let's start at $x = x_0$. Applying the computational template given by equation (2) with $i = 0$ yields

$$-w_f + 2w_0 - w_1 + h^2 f\left(x_0, w_0, \frac{w_1 - w_f}{2h}\right) = 0. \tag{3}$$

We must, of course, eliminate w_f from this equation. The Robin boundary condition $\alpha_1 y(a) + \alpha_2 y'(a) = \alpha_3$ leads to the finite difference equation

$$\alpha_1 w_0 + \alpha_2 \frac{w_1 - w_f}{2h} = \alpha_3.$$

An intermediate result arrived at during the solution of this equation for w_f is

$$\frac{w_1 - w_f}{2h} = \frac{\alpha_3 - \alpha_1 w_0}{\alpha_2}, \tag{4}$$

which can be recognized as the third argument to the function f in equation (3). Completing the solution for w_f yields

$$w_f = w_1 - 2h\frac{\alpha_3}{\alpha_2} + 2h\frac{\alpha_1}{\alpha_2} w_0. \tag{5}$$

Substituting equations (4) and (5) into equation (3) gives

$$g_0(\mathbf{w}) = 2\left(1 - h\frac{\alpha_1}{\alpha_2}\right)w_0 - 2w_1 + h^2 f\left(x_0, w_0, \frac{\alpha_3 - \alpha_1 w_0}{\alpha_2}\right) + 2h\frac{\alpha_3}{\alpha_2}. \tag{6}$$

For a Neumann boundary condition at $x = x_0$, $\alpha_1 = 0$ and equation (6) reduces to

$$g_0(\mathbf{w}) = 2w_0 - 2w_1 + h^2 f(x_0, w_0, \alpha) + 2h\alpha, \tag{7}$$

where we have written α for the ratio α_3/α_2.

Proceeding in an analogous manner, the Robin boundary condition $\beta_1 y(b) + \beta_2 y'(b) = \beta_3$ leads to

$$g_N(\mathbf{w}) = -2w_{N-1} + 2\left(1 + h\frac{\beta_1}{\beta_2}\right)w_N + h^2 f\left(x_N, w_N, \frac{\beta_3 - \beta_1 w_N}{\beta_2}\right) - 2h\frac{\beta_3}{\beta_2}, \tag{8}$$

while the Neumann boundary condition $y'(b) = \beta$ leads to

$$g_N(\mathbf{w}) = -2w_{N-1} + 2w_N + h^2 f(x_N, w_N, \beta) - 2h\beta. \tag{9}$$

The derivation of equation (8) is left as an exercise.

Having determined the functions $g_0(\mathbf{w})$ and $g_N(\mathbf{w})$ for Robin and Neumann boundary conditions, we can now compute the corresponding entries along the first and last rows of the Jacobian matrix. By design, the only nonzero entries along the first row are

$$J_{1,1} = \frac{\partial g_0}{\partial w_0} \quad \text{and} \quad J_{1,2} = \frac{\partial g_0}{\partial w_1},$$

and the only nonzero entries along the last row are

$$J_{N+1,N} = \frac{\partial g_N}{\partial w_{N-1}} \quad \text{and} \quad J_{N+1,N+1} = \frac{\partial g_N}{\partial w_N}.$$

From equations (6) and (7)

$$J_{1,1} = \begin{cases} 2\left(1 - h\frac{\alpha_1}{\alpha_2}\right) + h^2\frac{\partial f}{\partial y}(x_0, w_0, \bar{\alpha}) - h^2\frac{\alpha_1}{\alpha_2}\frac{\partial f}{\partial y'}(x_0, w_0, \bar{\alpha}) & \text{Robin} \\ 2 + h^2\frac{\partial f}{\partial y}(x_0, w_0, \alpha) & \text{Neumann,} \end{cases}$$

where $\bar{\alpha} = (\alpha_3 - \alpha_1 w_0)/\alpha_2$, and $J_{1,2} = -2$. From equations (8) and (9)

$$J_{N+1,N+1} = \begin{cases} 2\left(1 + h\frac{\beta_1}{\beta_2}\right) + h^2\frac{\partial f}{\partial y}(x_N, w_N, \bar{\beta}) - h^2\frac{\beta_1}{\beta_2}\frac{\partial f}{\partial y'}(x_N, w_N, \bar{\beta}) & \text{Robin} \\ 2 + h^2\frac{\partial f}{\partial y}(x_N, w_N, \beta) & \text{Neumann,} \end{cases}$$

where $\bar{\beta} = (\beta_3 - \beta_1 w_N)/\beta_2$, and $J_{N+1,N} = -2$.

The selection of the initial Newton iterate, $\mathbf{w}^{(0)}$, is not as straightforward for non-Dirichlet boundary conditions. In most cases, we can still use the boundary conditions to determine a linear function that can then be evaluated at the grid points to produce an initial vector. For instance, the boundary conditions

$$y'(a) = \alpha \quad \text{and} \quad \beta_1 y(b) + \beta_2 y'(b) = \beta_3$$

determine the linear function

$$\alpha x + \left(\frac{\beta_3 - \beta_2 \alpha}{\beta_1} - \alpha b \right).$$

There are situations, however, when a linear fit to the boundary conditions is not possible, such as when Neumann conditions are specified at both endpoints, with different values for the derivative. After all, a linear function can have only one slope. Furthermore, even when a linear function can be fit to the boundary conditions, the resulting function may not produce an appropriate initial vector. Consider the boundary value problem

$$\begin{gathered} y'' = -(1 + (y')^2)/y \\ y(1) - y'(1) = -1, \quad y'(2) = 1/2. \end{gathered}$$

The function $(1/2)x - 1$ satisfies the boundary conditions, but unfortunately evaluates to zero at $x = 2$, which generates division by zero in the differential equation. In these cases (the linear fit fails or does not provide an appropriate initial vector), it may be best to just use an arbitrary constant vector as $\mathbf{w}^{(0)}$. The bottom line is that for nonlinear boundary value problems with non-Dirichlet boundary conditions, some trial and error may be necessary to find a good starting vector for Newton's method.

EXAMPLE 8.5 A Problem with One Robin and One Neumann Boundary Condition

Consider the nonlinear boundary value problem

$$\begin{gathered} y'' - 2y^3 = 0, \quad x \in [0, 1] \\ 3y(0) - 9y'(0) = 2, \quad y'(1) = -1/16. \end{gathered}$$

The exact solution for this problem is $y(x) = 1/(x + 3)$.

Using a uniform partition with $N = 8$ subintervals, the results shown in the first table were obtained. The initial vector for Newton's method was taken as

$$w_i^{(0)} = -\frac{1}{16} x_i + \frac{23}{48}.$$

These values were determined by evaluating the linear function that satisfies the boundary conditions at each grid point. A convergence tolerance of $TOL = 5 \times 10^{-14}$ was used to terminate the iteration.

The second table demonstrates the second-order accuracy of the numerical method. Each time the number of subintervals in the uniform partition was doubled, the maximum absolute error and the rms error were reduced roughly by a factor of four. For each value of N, the initial vector and convergence tolerance cited above were used.

x_i	Approximate Solution, w_i	Exact Solution, y_i	Absolute Error
0.000	0.333224	0.333333	0.000109
0.125	0.319909	0.320000	0.000091
0.250	0.307617	0.307692	0.000076
0.375	0.296234	0.296296	0.000062
0.500	0.285664	0.285714	0.000050
0.625	0.275822	0.275862	0.000040
0.750	0.266636	0.266667	0.000030
0.875	0.258043	0.258065	0.000022
1.000	0.249986	0.250000	0.000014

N	Maximum Absolute Error	Error Ratio	rms Error	Error Ratio
4	0.0004333909		0.0002590402	
8	0.0001093182	3.964488	0.0000629312	4.116244
16	0.0000273913	3.990991	0.0000154601	4.070552
32	0.0000068517	3.997739	0.0000038283	4.038336
64	0.0000017132	3.999434	0.0000009523	4.019922
128	0.0000004283	3.999859	0.0000002375	4.010148
256	0.0000001071	3.999965	0.0000000593	4.005121
512	0.0000000268	3.999989	0.0000000148	4.002569

Application Problem: Spatial Distribution of an Insect Population

In the Overview to this chapter (see page 658), we showed that under certain assumptions (one-dimensional diffusion, Fick's law, logistic growth, etc.) the nondimensional steady-state population density, $n(x)$, of an insect population that has been released into a cylindrical environment satisfies the boundary value problem

$$n'' + n(1-n) = 0, \quad n(0) = 0, \quad n'(l) = 0. \tag{10}$$

Here, primes denote differentiation with respect to x, x measures nondimensional distance along the axis of the cylinder, and l is the nondimensional length of the cylinder. Clearly, the so-called *trivial solution*, $n(x) \equiv 0$, is a solution of (10). It can be shown (see, for example, Ludwig, Aronson and Weinberger [2]) that the trivial solution is the only solution to (10) whenever $l \le \pi/2$; but, when $l > \pi/2$, there exists a unique nonnegative nontrivial solution.

The nonnegative nontrivial solution corresponding to $l = 3$ is displayed in the top graph of Figure 8.10. This solution was determined using a uniform partition of $[0, 3]$ with $N = 300$ subintervals. The initial vector for Newton's method was

$$w_i^{(0)} = 1 - \frac{(x_i - 3)^2}{9},$$

which was obtained by evaluating, at each of the grid points, the quadratic polynomial satisfying both boundary conditions and taking the value 1 at $x = 3$. (Why

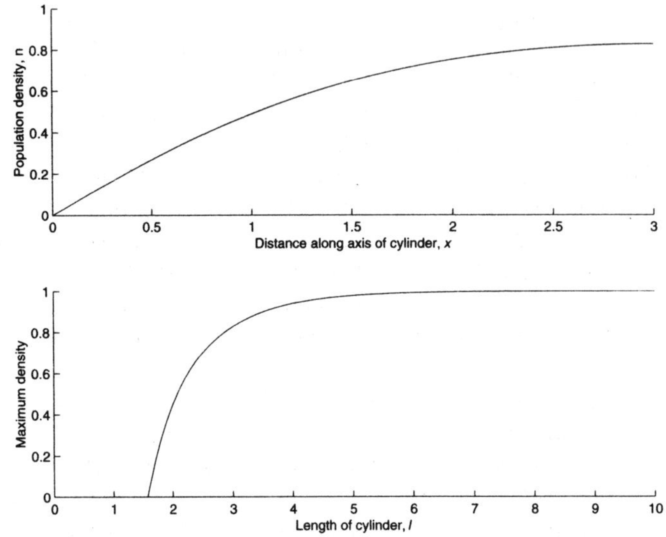

Figure 8.10 Numerical results for the "Spatial Distribution of an Insect Population" problem. (Top graph) Nonzero steady-state population density distribution for $l = 3$. (Bottom graph) Maximum population density as a function of cylinder length, l.

would the linear function that satisfied the boundary conditions not have provided an appropriate choice for the initial vector?) A convergence tolerance of 5×10^{-10} was used to terminate iteration.

The bottom graph in Figure 8.10 displays the maximum value of the population density as a function of the cylinder length, l. For $l \leq \pi/2$, we know that the density is zero everywhere; hence, the maximum density is zero. The remainder of the graph was obtained by solving (10) for values of l ranging from 1.6 to 10.0 in increments of 0.1 and recording the maximum value of the resulting density profile. For each l, calculations were performed with

$$N = 100l, \qquad w_i^{(0)} = 1 - \frac{(x_i - l)^2}{l^2}$$

and a convergence tolerance of 5×10^{-10}. Note that for $l \geq 5.8$ the maximum density is within one percent of the carrying capacity.

References

1. H. B. Keller, *Numerical Methods for Two-Point Boundary Value Problems*, Blaisdell, Waltham, 1968.

2. D. Ludwig, D. G. Aronson and H. F. Weinberger, "Spatial Patterning of the Spruce Budworm," *Journal of Mathematical Biology*, **8**, 217–258, 1979.

EXERCISES

1. Determine the conditions under which a linear function cannot be fit to the Dirichlet boundary condition $y(a) = \alpha$ at the left endpoint and the Robin boundary condition $\beta_1 y(b) + \beta_2 y'(b) = \beta_3$ at the right endpoint.
2. Derive Equation (8) – *i.e.*, the function $g_N(\mathbf{w})$ corresponding to a Robin boundary condition at the right endpoint of the problem domain.
3. If the function $f(x, y, y')$ is of the form

$$f(x, y, y') = p(x)y' + q(x)y + r(x)$$

for some functions p, q and r, show that the formulas for d_i, u_i, and l_i reduce to formulas (4), (5), and (6) from Section 8.1.

For the boundary value problems in Exercises 4–9,

(a) identify the function $f(x, y, y')$ and compute the partial derivatives $\partial f/\partial y$ and $\partial f/\partial y'$;

(b) for $N = 4$, write out the system of finite difference equations.

4. $y'' + (y')^2 = 1, \quad y(0) = 1, \quad y(1) = 2$
5. $y'' + yy' + (y')^2 = 0, \quad y(0) = 0, \quad y(1) + y'(1) = 1$
6. $y'' = -2y^2 - 4xyy', \quad y(0) = 1, \quad y(1) = 1/2$
7. $y'' + 4yy' = -2y/(1 + x^2), \quad y(0) + y'(0) = 1, \quad y'(1) = 0$
8. $y'' + 2y^2 - 8x^2y^3 = 0, \quad y(0) = 1, \quad y'(1) = -1/2$
9. $y'' = 6y^2, \quad y(0) - y'(0) = 0, \quad y(1) + y'(1) = 3$

In Exercises 10–16, approximate the solution of the indicated boundary value problem using the finite difference method and confirm the second-order accuracy of the numerical method.

10. $y'' = -(1 + (y')^2)/y, \quad y(1) - y'(1) = -1, \quad y'(2) = 1/2, \quad y(x) = \sqrt{6x - 4 - x^2}$
11. $y'' = -2y^2 + 8x^2y^3, \quad y(0) = 1, \quad y(1) = 1/2, \quad y(x) = 1/(1 + x^2)$
12. $y'' + 4yy' = -2y/(1 + x^2), \quad y(0) = 0, \quad y'(0) = 1/2, \quad y(x) = x/(1 + x^2)$
13. $y'' + (y')^2 = 1, \quad y'(0) = 0, \quad y(1) = 2$
14. $y'' + yy' + (y')^2 = 0, \quad y(0) = 0, \quad y(1) + y'(1) = 1, \quad y(x) = 1 - e^{-x}$
15. $y'' + 4xyy' = -2y^2, \quad y(0) + y'(0) = 1, \quad y(1) + y'(1) = 0, \quad y(x) = 1/(1 + x^2)$
16. $yy'' = y' - 1, \quad y(1) = 0, \quad y'(2) = 1 + \ln 2$
17. Consider the nonlinear differential equation

$$2xy'' + (y')^2 - 4y = 4x.$$

Use the finite difference method to solve this differential equation subject to each of the following sets of boundary conditions. In each case, the exact solution is $y(x) = (x + 1)^2$. How rapidly does the approximate solution converge toward the exact solution as a function of the number of subintervals? Provide an explanation for your observation.

(a) $y(1) = 4, \ y(2) = 9$ (b) $y(1) = 4, \ y'(2) = 6$

(c) $y'(1) = 4, \ y'(2) = 6$ (d) $y(1) - y'(1) = 0, \quad y'(2) = 6$

18. Ramirez (*Computational Methods for Process Simulation*, Butterworths, Boston, 1989) develops the following nonlinear boundary value problem for the concentration, $y(x)$, of the reactant in a second-order chemical reaction taking place within a tubular reactor with dispersion:

$$\frac{d^2y}{dx^2} - 10\frac{dy}{dx} - 5y^2 = 0$$

$$y(0) - \frac{1}{10}\left.\frac{dy}{dx}\right|_{x=0} = 5, \qquad \left.\frac{dy}{dx}\right|_{x=1} = 0.$$

Determine the concentration in increments of 0.01 along the length of the reactor.

19. C. Philipsen, S. Markvorsen, and W. Kleim ["Modelling the Stem Curve of a Palm in a Strong Wind," *SIAM Review,* **38** (3), pp. 483–484, 1996] present the following model for the angle of the stem of a tall palm tree, relative to its vertical position, when the tree is subjected to wind loading:

$$EI\frac{d^2\theta}{ds^2} = -W_s\left(1 - \frac{s}{L}\right)\sin\theta - W_c\sin\theta - D\cos\theta.$$

Here, θ is the angle of the stem relative to the vertical position, and s is arc length measured along the stem. The parameters in the model are the total stem weight $W_s = 22700$ N, the Young's modulus of the stem $E = 0.9 \times 10^9$ N/m^2, the moment of inertia of the stem $I = 5.147 \times 10^{-4}$ m^4, the length of the stem $L = 30$ m, the total canopy weight $W_c = 1385.5$ N, and the wind drag force on the canopy $D = 1.2405U^2$ N, where U is the wind speed in m/s. The boundary conditions imposed on the stem angle are

$$\theta(0) = 0 \quad \text{and} \quad \theta'(L) = 0.$$

Determine the function $\theta(s)$ when the wind speed is 8 meters/second.

20. Subramanian and Balakotaiah ["Convective Instabilities Induced by Exothermic Reactions Occurring in a Porous Medium," *Phys. Fluids,* **6** (9), pp. 2907–2922, 1994] develop the boundary value problem

$$\frac{d^2\theta}{dz^2} + B\phi^2\left(1 - \frac{\theta}{B}\right)\exp\left(\frac{\gamma\theta}{\gamma+\theta}\right) = 0$$

$$\theta'(0) = 0, \quad \theta(1) = 0$$

for the steady-state temperature profile, $\theta(z)$, in a porous medium undergoing an exothermic reaction. The parameter B is the maximum possible temperature in the absence of natural convection, ϕ^2 is the ratio of the characteristic time for conduction to that for heat generation and γ is the dimensionless activation energy. For $B = 6.0$, $\phi^2 = 0.25$, and $\gamma = 30.0$, determine $\theta(z)$.

21. In the "Spatial Distribution of an Insect Population" problem, suppose that rather than a barrier at $x = l$, a steady influx of insects is maintained. This changes the boundary condition from $n'(l) = 0$ to $n'(l) = j$, where j is a nondimensional flux parameter. For $l = 3$, determine the density distribution for values of j ranging from 0.05 to 0.25 in increments of 0.05. Approximate the value of j for which the maximum density is equal the carrying capacity.

8.4 THE SHOOTING METHOD, PART I: LINEAR BOUNDARY VALUE PROBLEMS

The shooting method is an alternative numerical method for solving boundary value problems. The basic idea is to convert the boundary value problem into two or more initial value problems that can be solved using the techniques developed in Chapter 7. For linear boundary value problems, it is a simple matter to combine the solutions of the initial value problems to generate the solution to the original boundary value problem.

Dirichlet Boundary Conditions

Let's start simple and demonstrate the technique for a boundary value problem with Dirichlet boundary conditions:

$$y'' = p(x)y' + q(x)y + r(x)$$
$$y(a) = \alpha, \quad y(b) = \beta.$$

The basis for the shooting method is as follows. The above boundary value problem is almost an initial value problem. We have the differential equation and the value of the solution at $x = a$. The only piece of information that is missing is the value of the first derivative at $x = a$. Why not then guess a value for $y'(a)$, use any available initial value problem solver to march the solution out to $x = b$ and check whether the boundary condition at $x = b$ has been satisfied? If it has, then we have found the solution to the boundary value problem; if the boundary condition at $x = b$ has not been satisfied, then we make a "better" guess for the value of $y'(a)$ and repeat the process. This approach essentially transforms the boundary value problem into a rootfinding problem. When we work with nonlinear problems in the next section, we will use precisely this approach.

For linear boundary value problems, however, a slightly different plan of attack will produce the approximate solution in a much more direct manner. The key observation is that every solution to a linear, nonhomogeneous differential equation can be written as a particular solution plus a constant times a solution to the corresponding homogeneous problem. This suggests working with not one initial value problem, but two. The first of these has the original nonhomogeneous differential equation, with the function value at $x = a$ given by the boundary condition $y(a) = \alpha$ and with an arbitrary value specified for the first derivative. The solution of this problem is not expected to match the boundary condition at $x = b$. Therefore, the second initial value problem, which has the corresponding homogeneous differential equation subject to $y(a) = 0$ and an arbitrary, nonzero value for the first derivative, is also solved. Multiplying the solution of this second initial value problem by an appropriate constant and adding the result to the solution of the first initial value problem will allow the boundary condition at $x = b$ to be satisfied.

Let's examine this computational scheme in detail. Consider the two initial value problems

$$\text{IVP1} \quad \begin{cases} y'' = p(x)y' + q(x)y + r(x) \\ y(a) = \alpha, \quad y'(a) = 0 \end{cases}$$

$$\text{IVP2} \quad \begin{cases} y'' = p(x)y' + q(x)y \\ y(a) = 0, \quad y'(a) = 1 \end{cases}.$$

The initial values shown for the first derivative are the standard choices. Let $y_1(x)$ denote the solution of IVP1 and $y_2(x)$ denote the solution of IVP2. Due to the linearity of the differential equation, it follows that

$$y(x) = y_1(x) + cy_2(x) \tag{1}$$

is a solution of $y'' = p(x)y' + q(x)y + r(x)$ for any value of the constant c. Furthermore,

$$\begin{aligned} y(a) &= y_1(a) + cy_2(a) \\ &= \alpha + c \cdot 0 = \alpha. \end{aligned}$$

At $x = b$ we have $y(b) = y_1(b) + cy_2(b)$. Equating this value to β and solving for c, we find that with

$$c = \frac{\beta - y_1(b)}{y_2(b)} \tag{2}$$

the function $y(x) = y_1(x) + cy_2(x)$ will satisfy the original boundary value problem.

To summarize, the shooting method for approximating the solution of the linear boundary value problem with Dirichlet boundary conditions,

$$\begin{aligned} y'' &= p(x)y' + q(x)y + r(x) \\ y(a) &= \alpha, \quad y(b) = \beta, \end{aligned}$$

consists of three steps. First, solve the initial value problems IVP1 and IVP2 using any initial value problem solver. It is not required that both problems be solved with the same numerical method. Second, using the values of $y_1(b)$ and $y_2(b)$ obtained in the first step and the value of β given in the boundary condition at $x = b$, compute c from equation (2). Finally, pointwise combine the solutions $y_1(x)$ and $y_2(x)$ according to equation (1).

EXAMPLE 8.6 Demonstration of the Shooting Method for a Linear Boundary Value Problem

Consider the linear boundary value problem with Dirichlet boundary conditions

$$\begin{aligned} -u'' + \pi^2 u &= 2\pi^2 \sin(\pi x) \\ u(0) = u(1) &= 0. \end{aligned}$$

To approximate the solution of this problem using the shooting method, we first convert the boundary value problem into the two initial value problems

$$\text{IVP1} \quad \begin{cases} u'' = \pi^2 u - 2\pi^2 \sin(\pi x) \\ u(0) = 0, \quad u'(0) = 0 \end{cases}$$

and

$$\text{IVP2} \quad \begin{cases} u'' = \pi^2 u \\ u(0) = 0, \quad u'(0) = 1 \end{cases}.$$

Let u_1 denote the approximate solution of IVP1 and u_2 the approximate solution of IVP2. Regardless of which initial value problem solver we choose, each initial value problem must be converted to a system. The systems corresponding to IVP1 and IVP2 are, respectively,

$$\begin{aligned} u_{1,1}' &= u_{1,2} \\ u_{1,2}' &= \pi^2 u_{1,1} - 2\pi^2 \sin(\pi x) \end{aligned}$$

and

$$\begin{aligned} u_{2,1}' &= u_{2,2} \\ u_{2,2}' &= \pi^2 u_{2,1}. \end{aligned}$$

Using the classical fourth-order Runge-Kutta method (RK4) with $N = 4$ steps to march from $x = 0$ to $x = 1$, the results given below are obtained.

x_i	$u_1(x_i)$	$u_2(x_i)$
0.00	0.000000	0.000000
0.25	-0.157372	0.275702
0.50	-1.290357	0.730213
0.75	-4.490694	1.657343
1.00	-11.466375	3.656793

Note that neither solution satisfies the boundary condition at $x = 1$.

Now, using equation (2), we compute

$$c = \frac{0 - (-11.466375)}{3.656793} = 3.135637.$$

The function $w(x) = u_1(x) + 3.135637u_2(x)$ is then guaranteed to satisfy both boundary conditions. The value of w at each x_i, w_i, is given in the second column of the following table and is compared to the corresponding value of the exact solution, $u(x) = \sin(\pi x)$.

x_i	Approximate Solution, w_i	Exact Solution, u_i	Absolute Error
0.00	0.000000	0.000000	
0.25	0.707129	0.707107	0.000022
0.50	0.999327	1.000000	0.000673
0.75	0.706132	0.707107	0.000975
1.00	0.000000	0.000000	

The accuracy of this solution is excellent, especially considering the crudeness of the computational grid. Since we have used RK4 to obtain our approximate solution, we expect fourth-order convergence toward the exact solution. Numerical verification of the order of convergence is left as an exercise.

Other Types of Boundary Conditions

The shooting method for linear boundary value problems that has just been described can also be applied when boundary conditions other than Dirichlet are specified. Suppose, for example, that we have the same Dirichlet condition at $x = a$, $y(a) = \alpha$, but we replace the condition at $x = b$ with the Robin condition $\beta_1 y(b) + \beta_2 y'(b) = \beta_3$. Since the condition at $x = a$ has not changed, the basic structure of the problem remains the same: The initial value of the function is known, but the initial value for the first derivative must be "guessed." Therefore, IVP1 and IVP2 need not change, and the solutions of these two problems will still be combined according to equation (1). Only the equation for computing the constant c must change.

From equation (1), at $x = b$ we have

$$y(b) = y_1(b) + cy_2(b)$$

and

$$y'(b) = y_1'(b) + cy_2'(b).$$

Therefore, to satisfy the boundary condition at $x = b$, c must be selected to satisfy

$$\beta_1[y_1(b) + cy_2(b)] + \beta_2[y_1'(b) + cy_2'(b)] = \beta_3.$$

Solving for c, we obtain

$$c = \frac{\beta_3 - \beta_1 y_1(b) - \beta_2 y_1'(b)}{\beta_1 y_2(b) + \beta_2 y_2'(b)}. \tag{3}$$

For the Neumann condition $y'(b) = \beta$ at $x = b$, the equation for c becomes

$$c = \frac{\beta - y_1'(b)}{y_2'(b)}. \tag{4}$$

Next, suppose we have the Neumann condition $y'(a) = \alpha$ at $x = a$. This changes the basic nature of the problem. Instead of knowing the initial function value and needing to guess the initial value for the derivative, now we know the initial value for the derivative and need to guess the initial function value. This suggests converting the boundary value problem into the two initial value problems:

$$\text{IVP1} \quad \begin{cases} y'' = p(x)y' + q(x)y + r(x) \\ y(a) = 0, \quad y'(a) = \alpha \end{cases}$$

and

$$\text{IVP2} \quad \begin{cases} y'' = p(x)y' + q(x)y \\ y(a) = 1, \quad y'(a) = 0 \end{cases}.$$

Once again, let $y_1(x)$ denote the solution of IVP1 and $y_2(x)$ denote the solution of IVP2, and let $y(x) = y_1(x) + cy_2(x)$ for some constant c. The value of c is given by equation (2), (3) or (4), depending on whether the boundary condition at $x = b$ is a Dirichlet condition, a Robin condition, or a Neumann condition, respectively.

Finally, suppose that the Robin condition $\alpha_1 y(a) + \alpha_2 y'(a) = \alpha_3$ is specified at $x = a$. Now we know neither the initial function value nor the initial value for the first derivative, so both will have to be guessed. This requires that the boundary value problem be replaced by three initial value problems:

$$\text{IVP1} \quad \begin{cases} y'' = p(x)y' + q(x)y + r(x) \\ y(a) = 0, \quad y'(a) = 0 \end{cases},$$

$$\text{IVP2} \quad \begin{cases} y'' = p(x)y' + q(x)y \\ y(a) = 1, \quad y'(a) = 0 \end{cases},$$

and

$$\text{IVP3} \quad \begin{cases} y'' = p(x)y' + q(x)y \\ y(a) = 0, \quad y'(a) = 1 \end{cases}.$$

The solutions to these problems will be combined according to the rule

$$y(x) = y_1(x) + c_1 y_2(x) + c_2 y_3(x), \tag{5}$$

where $y_1(x)$, $y_2(x)$, and $y_3(x)$ are the solutions to IVP1, IVP2, and IVP3, respectively. To determine the appropriate values for the constants c_1 and c_2, we have to satisfy both boundary conditions. From the boundary condition at $x = a$ and equation (5), we obtain the equation

$$[\alpha_1 y_2(a) + \alpha_2 y_2'(a)]c_1 + [\alpha_1 y_3(a) + \alpha_2 y_3'(a)]c_2 = \alpha_3 - \alpha_1 y_1(a) - \alpha_2 y_1'(a),$$

which simplifies to

$$\alpha_1 c_1 + \alpha_2 c_2 = \alpha_3 \tag{6}$$

upon substituting the initial values from IVP1, IVP2, and IVP3. The second equation for the constants c_1 and c_2 depends on the type of boundary condition specified at $x = b$. The three possibilities are

$$y_2(b)c_1 + y_3(b)c_2 = \beta - y_1(b) \tag{7}$$

$$y_2'(b)c_1 + y_3'(b)c_2 = \beta - y_1'(b) \tag{8}$$

$$[\beta_1 y_2(b) + \beta_2 y_2'(b)]\, c_1 + [\beta_1 y_3(b) + \beta_2 y_3'(b)]c_2 = \beta_3 - \beta_1 y_1(b) - \beta_2 y_1'(b) \tag{9}$$

for a Dirichlet condition, a Neumann condition, and a Robin condition, respectively.

EXAMPLE 8.7 A Problem with One Neumann and One Robin Boundary Condition

Consider the ordinary differential equation

$$u'' + u = \sin(3x), \quad 0 \le x \le \pi/2$$

subject to a Neumann boundary condition at $x = 0$:

$$u'(0) = 1$$

and a Robin boundary condition at $x = 1$:

$$u(\pi/2) + u'(\pi/2) = -1.$$

With the Neumann boundary condition at $x = 0$, we first convert the boundary value problem into the two initial value problems

$$\text{IVP1} \quad \begin{cases} u'' = -u + \sin(3x) \\ u(0) = 0, \quad u'(0) = 1 \end{cases}$$

and

$$\text{IVP2} \quad \begin{cases} u'' = -u \\ u(0) = 1, \quad u'(0) = 0 \end{cases}.$$

Let u_1 denote the approximate solution of IVP1 and u_2 the approximate solution of IVP2. The systems corresponding to IVP1 and IVP2 are, respectively,

$$\begin{aligned} u'_{1,1} &= u_{1,2} \\ u'_{1,2} &= -u_{1,1} + \sin(3x) \end{aligned}$$

and

$$\begin{aligned} u'_{2,1} &= u_{2,2} \\ u'_{2,2} &= -u_{2,1}. \end{aligned}$$

The RK4 method, with $N = 4$ steps to march from $x = 0$ to $x = \pi/2$, produces the results

x_i	$u_1(x_i)$	$u'_1(x_i)$	$u_2(x_i)$	$u'_2(x_i)$
0	0.000000	1.000000	1.000000	0.000000
$\pi/8$	0.411165	1.126997	0.923885	-0.382606
$\pi/4$	0.884311	1.237806	0.707176	-0.706967
$3\pi/8$	1.318095	0.873002	0.382859	-0.923726
$\pi/2$	1.499586	0.000195	0.000294	-0.999900

With a Robin condition at $x = \pi/2$, we use equation (3) to compute

$$c = \frac{-1 - 1.499586 - 0.000195}{0.000294 - 0.999900} = 2.500766.$$

The function $w(x) = u_1(x) + 2.500766 u_2(x)$ is then guaranteed to satisfy both boundary conditions. The value of w at each x_i, w_i, is given in the second column of the following table and is compared to the corresponding value of the exact solution, $u(x) = (11/8)\sin x + (5/2)\cos x - (1/8)\sin(3x)$.

x_i	Approximate Solution, w_i	Exact Solution, u_i	Absolute Error
0	2.500766	2.500000	0.000766
$\pi/8$	2.721585	2.720404	0.001181
$\pi/4$	2.652793	2.651650	0.001143
$3\pi/8$	2.275536	2.274878	0.000658
$\pi/2$	1.500321	1.500000	0.000321

Once again, numerical verification of the order of convergence is left as an exercise.

EXAMPLE 8.8 A Problem with a Robin Boundary Condition at $x = a$

Consider the ordinary differential equation

$$u'' + u = \sin(3x), \quad 0 \le x \le \pi/2$$

subject to a Robin boundary condition at $x = 0$:

$$u(0) + u'(0) = -1$$

and a Neumann boundary condition at $x = \pi/2$:

$$u'(\pi/2) = 1.$$

With the Robin boundary condition at $x = 0$, we replace the boundary value problem with the three initial value problems

$$\text{IVP1} \quad \begin{cases} u'' = -u + \sin(3x) \\ u(0) = 0, \quad u'(0) = 0 \end{cases},$$

$$\text{IVP2} \quad \begin{cases} u'' = -u \\ u(0) = 1, \quad u'(0) = 0 \end{cases},$$

and

$$\text{IVP3} \quad \begin{cases} u'' = -u \\ u(0) = 0, \quad u'(0) = 1 \end{cases}.$$

Let u_1, u_2, and u_3 denote the approximate solution of IVP1, IVP2, and IVP3, respectively. Converting each initial value problem into a system and then using the RK4 method, with $N = 4$ steps to march from $x = 0$ to $x = \pi/2$, the results given below are obtained.

x_i	$u_1(x_i)$	$u_2(x_i)$	$u_3(x_i)$
0	0.000000	1.000000	0.000000
$\pi/8$	0.028559	0.923885	0.382606
$\pi/4$	0.177343	0.707176	0.706967
$3\pi/8$	0.394369	0.382859	0.923726
$\pi/2$	0.499685	0.000294	0.999900

We want to combine these three solutions into a single function: $w = u_1 + c_1u_2 + c_2u_3$. To determine the appropriate values for c_1 and c_2, we must solve the system of algebraic equations composed of equation (5) with $\alpha_1 = \alpha_2 = 1$ and $\alpha_3 = -1$ and equation (8) with $\beta = 1$; that is,

$$\begin{aligned} c_1 + c_2 &= -1 \\ u_2'(\pi/2)c_1 + u_3'(\pi/2)c_2 &= 1 - u_1'(\pi/2). \end{aligned}$$

The derivative values at $x = \pi/2$ are

$$\begin{aligned} u_1'(\pi/2) &= -0.00009934793668 \\ u_2'(\pi/2) &= -0.99990005047118 \\ u_3'(\pi/2) &= -0.00029430281825. \end{aligned}$$

With these values, the constants are found to be $c_1 = -1.000199$ and $c_2 = 0.000199259$. Therefore, $w(x) = u_1(x) - 1.000199u_2(x) + 0.000199259u_3(x)$. The value of w at each x_i, w_i, is given in the second column of the following table and is compared to the corresponding value of the exact solution, $u(x) = (3/8)\sin x - \cos x - (1/8)\sin(3x)$.

x_i	Approximate Solution, w_i	Exact Solution, u_i	Absolute Error
0	−1.000199	−1.000000	0.000199
$\pi/8$	−0.895434	−0.895858	0.000424
$\pi/4$	−0.529833	−0.530330	0.000497
$3\pi/8$	0.0116179	0.0116068	0.000011
$\pi/2$	0.499590	0.500000	0.000410

EXAMPLE 8.9 A Problem with an Artificial Singularity

The boundary value problem

$$\begin{gathered} y'' + \frac{2}{x}y' = -1 \\ y'(0) = 0, \quad y(1) = 1 \end{gathered}$$

has an artificial singularity at $x = 0$. We handle the artificial singularity here in exactly the same manner as we did with the finite difference method (see Section 8.2); that is, we determine the equation that applies at $x = 0$ and then define the coefficient functions p, q and r in a piecewise manner. Taking the limit as $x \to 0$, the differential equation becomes

$$y''(0) = -1/3.$$

The appropriate coefficient functions for the numerical solution of the boundary value problem are therefore

$$p(x) = \begin{cases} 0, & x = 0 \\ -2/x, & x \neq 0 \end{cases} \quad q(x) = 0, \text{ and } r(x) = \begin{cases} -1/3, & x = 0 \\ -1, & x \neq 0 \end{cases}.$$

With these functions, the two initial value problems which must be solved are

$$\text{IVP1} \quad \begin{cases} y'' = p(x)y' + r(x) \\ y(0) = 0, \quad y'(0) = 0 \end{cases}.$$

$$\text{IVP2} \quad \begin{cases} y'' = p(x)y' \\ y(0) = 1, \quad y'(0) = 0 \end{cases}$$

Converting these problems into systems and using the RK4 method with $N = 4$, we obtain the approximate solutions

x_i	$y_1(x_i)$	$y_2(x_i)$
0.00	0.000000	1.000000
0.25	−0.010417	1.000000
0.50	−0.041667	1.000000
0.75	−0.093750	1.000000
1.00	−0.166667	1.000000

With a Dirichlet boundary condition at the right endpoint, equation (2) is used to compute

$$c = \frac{1 - (-0.166667)}{1} = 1.166667.$$

The value of $w(x) = y_1(x) + 1.166667y_2(x)$ at each x_i is given in the second column of the following table and is compared to the corresponding value of the exact solution, $y(x) = (7 - x^2)/6$.

x_i	Approximate Solution, w_i	Exact Solution, y_i	Absolute Error
0	1.166667	1.166667	0
0.25	1.156250	1.156250	0
0.50	1.125000	1.125000	0
0.75	1.072917	1.072917	0
1	1.000000	1.000000	0

Note that, even with $N = 4$, we have obtained the exact solution. This happens because the error term associated with the RK4 method contains the fifth derivative of the solution. In this case, the exact solution is a second degree polynomial, for which the fifth derivative is identically zero. The RK4 method will therefore produce the exact solution with any value of N.

Throughout our development of the shooting method, we have assumed that we would march "forward" from $x = a$ to $x = b$. There is no reason, however, why we couldn't march "backward" from $x = b$ to $x = a$. This approach can be particularly useful when a Robin condition is specified at $x = a$, but not at $x = b$. In this case, starting from $x = b$ reduces the number of initial value problems that must be solved from three to two.

There are two other circumstances under which reversing direction in the shooting method would be an appropriate course of action. First, the shooting method, working in either direction, can be prone to roundoff error. If the solution obtained marching in one direction is found to be overly contaminated by roundoff error, marching in the opposite direction may improve performance. Second, suppose the solutions to the initial value problems grow exponentially in one direction. The solution for the shooting parameter(s)–c or c_1 and c_2—may then be extremely ill conditioned. Marching in the other direction should alleviate this problem.

Application Problem: Cooling Fin on the Cylinder Barrel of a Motorcycle

This problem is adapted from Incropera and DeWitt [1]. The cylinder barrel of a motorcycle is made from an aluminum alloy with a thermal conductivity of $k = 186$ W/m · K. The outside diameter of the barrel is 50 mm. Under normal operating conditions, the outer surface of the barrel has a temperature of 220° C and experiences heat loss to air at $T_\infty = 20°$ C with a convection coefficient of $h = 50$ W/m^2 · K.

To improve heat transfer from the barrel to the surroundings, annular fins of length 20 mm are added, as shown in Figure 8.11. Note from the cross-sectional view that the fins are not of uniform thickness; rather, they taper linearly from a thickness of 6 mm at the point of contact with the barrel to a thickness of 4 mm at the outer edge. Thus, if r denotes distance from the center of the barrel, measured in meters, and $t(r)$ denotes the thickness of the fin, also measured in meters, then

$$t(r) = 0.0085 - \frac{r}{10}.$$

We would like to assess the improvement in heat transfer rate due to the presence of the cooling fins. To do this, we must first determine the temperature distribution, $T(r)$, within the fin. Let's assume that temperature variations across the thickness of the fin are negligible. Following the same procedure which led to equation (6) of Section 8.2, but omitting the internal heat generation term and taking into account the nonuniform thickness of the fin, yields the differential equation

$$\frac{d^2T}{dr^2} + \left(\frac{t'(r)}{t(r)} + \frac{1}{r}\right)\frac{dT}{dr} - \frac{2h}{kt(r)}(T - T_\infty) = 0.$$

The boundary conditions associated with this problem are

$$T(0.025) = 220 \quad \text{and} \quad hT(0.045) + kT'(0.045) = hT_\infty.$$

Applying the shooting method with the RK4 method as the underlying initial value problem solver and taking $N = 100$ steps to march from $r = 0.025$ meters to $r = 0.045$ meters, we obtain the temperature profile shown in Figure 8.12. The total fin heat transfer rate is given by

$$q_f = -k(2\pi r_o t(0.025))T'(0.025),$$

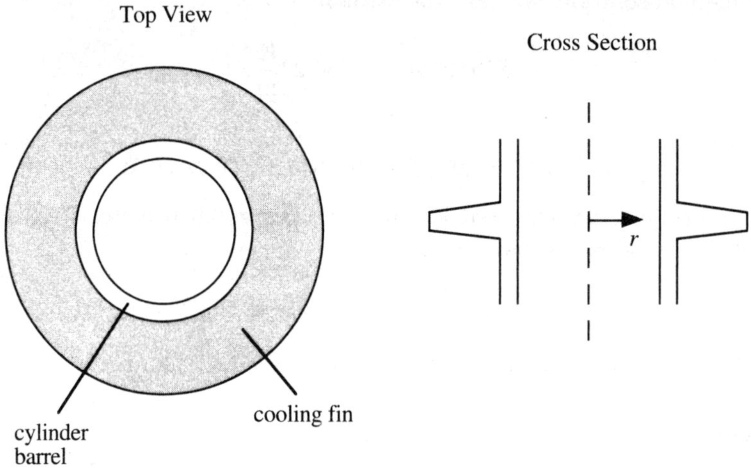

Figure 8.11 Top view and cross-sectional view of annular cooling fins attached to the cylinder barrel of a motorcycle.

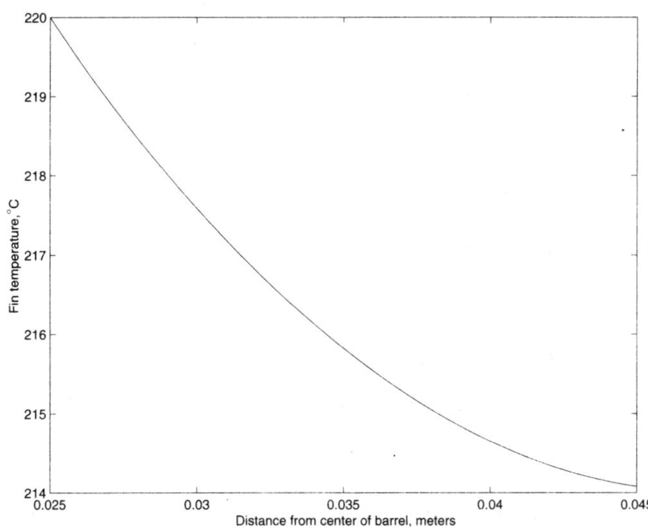

Figure 8.12 Temperature distribution in an annular fin of nonuniform thickness attached to the cylinder barrel of a motorcycle.

where $r_o = 0.025$ meters is the radius of the outer surface of the barrel. From the shooting method solution, we have the estimate

$$T'(0.025) = -554.22\frac{\text{K}}{\text{m}}.$$

Accordingly,

$$q_f = -372\pi(0.025)(0.006)(-554.22) = 97.16 \text{ W}.$$

Without the cooling fin, the heat loss from the same 0.006 meter strip along the surface of the barrel would have been

$$q = 2\pi h r_o t(0.025)(220 - 20) = 9.42 \text{ W}.$$

The cooling fin therefore increases the heat loss by more than tenfold.

References

1. F. P. Incropera and D. P. DeWitt, *Fundamentals of Heat and Mass Transfer*, John Wiley and Sons, New York, 1985.

EXERCISES

In Exercises 1–10, suppose we use the shooting method to approximate the solution of the indicated boundary value problem. Write out the initial value problems that must be solved and the equation(s) to determine the shooting parameter(s).

1. $u'' = -(x+1)u' + 2u + (1-x^2)e^{-x}, \quad u(0) = -1, \quad u(1) = 0$

2. $xy'' - (x+5)y' + 4y = x, \quad y(1) = -1, \quad y(2) = 1$

3. $y'' = -\dfrac{3}{2+x}y' - \dfrac{2}{(2+x)^3}, \quad y(0) = y'(1) = 0$

4. $\dfrac{1}{x}\left(xy'\right)' = 1, \quad y'(0) = 0, \quad y'(1) = 10$

5. $y'' = 120y - 2560, \quad 10y(0) + 35y'(0) = 200, \quad y(0.15) = 20$

6. $\dfrac{1}{x^2}\left(x^2y'\right)' = 1 + x^2, \quad y'(0) = 0, \quad 4y(1) + y'(1) = 40$

7. $xy'' - (2x+1)y' + (x+1)y = 0, \quad y(1) - 2y'(1) = -6e, \quad y(3) + y'(3) = 26e^3$

8. $\dfrac{1}{x}\left(xy'\right)' + y = 10, \quad y(1) = 20, \quad y(3) + y'(3) = 10$

9. $y'' + \dfrac{1}{x}y' + y = 1, \quad y'(0) = 0, \quad y(1) = 1$

10. $(1 - x\cot x)y'' - xy' + y = 0, \quad y(1) + y'(1) = 1 + \sin(1) + \cos(1), \quad y'(2) = 1 + \cos(2)$

In Exercises 11–17, numerically verify that the shooting method, with the RK4 method as the underlying initial value problem solver, produces results that converge toward the exact solution with rate of convergence $O(h^4)$.

11. $-u'' + \pi^2 u = 2\pi^2\sin(\pi x), \quad u(0) = u(1) = 0, \quad u(x) = \sin(\pi x)$

12. $u'' + u = \sin(3x)$, $u'(0) = 1$, $u(\pi/2) + u'(\pi/2) = -1$, $u(x) = (11/8)\sin x + (5/2)\cos x - (1/8)\sin(3x)$

13. $u'' + u = \sin(3x)$, $u(0) + u'(0) = -1$, $u'(\pi/2) = 1$, $u(x) = (3/8)\sin x - \cos x - (1/8)\sin(3x)$

14. $u'' + \frac{1}{x}u' = \left(\frac{8}{8-x^2}\right)^2$, $u'(0) = u(1) = 0$, $u(x) = 2\ln\left(\frac{7}{8-x^2}\right)$

15. $x^2y'' - 2xy' + 2y = 3x^2 + 2\ln x$, $y(1) = 9/4$, $y(2) = 13\ln 2$, $y(x) = \frac{3}{2}\left(1 + \frac{3}{2}x - x^2\right) + (1 + 3x^2)\ln x$

16. $y'' + xy' + y = x^2$, $y(0) = 0$, $y(1) = 1$

17. $y'' + \frac{1}{x}y' + y = 1$, $y'(0) = 0$, $y'(1) = 1$

18. One component of a model for a styrene monomer tubular reactor is the steady-state temperature profile of the solid phase catalyst. The governing boundary value problem is

$$\frac{d^2\tau}{dx^2} + \frac{1}{x}\frac{d\tau}{dx} - \beta^2\tau = 0$$

$$\left.\frac{d\tau}{dx}\right|_{x=0} = 0, \quad \tau(1) = 1.$$

Here, $\tau = (T_c - T)/(T_w - T)$ is the nondimensional catalyst temperature, x is the nondimensional radial position, T is the constant temperature of the fluid in the reactor, and T_w is the temperature at the wall of the reactor. The parameter β^2 is given by

$$\beta^2 = \frac{R^2hA}{k(1-\epsilon)},$$

where $R = 1.3$ cm is the radius of the reactor, $h = 10^{-3}$ cal/cm$^2 \cdot$ s $\cdot\,^\circ$C is the heat transfer coefficient, $\epsilon = 0.36$ is the porosity of the packed bed reactor, $A = 15$ cm^{-1} is the surface area of the catalyst per unit volume and $k = 0.0034$ cal/cm$\cdot$s$\cdot\,^\circ$C is the thermal diffusivity of the catalyst. Approximate $\tau(x)$ using $\Delta x = 0.0025$.

19. A thin cylindrical fiber, ten inches in length, has its left end maintained at a constant temperature T_0 and experiences convective heat loss along its lateral surface and from its right end. The temperature within the fiber is governed by the differential equation

$$-\frac{d}{dx}\left(kr^2\frac{dT}{dx}\right) + 2hrT = 2hrT_\infty,$$

subject to the boundary conditions

$$T(-5) = T_0$$

$$-k\left.\frac{dT}{dx}\right|_{x=5} = h[T(5) - T_\infty].$$

The parameters in this problem are the thermal conductivity of the fiber $k = 2\text{BTU/sec}\cdot\text{in}\cdot{}^\circ\text{F}$, the convective heat transfer coefficient $h = 10^{-5}\text{BTU/sec}\cdot\text{in}^{2\circ}\text{F}$, the radius of the fiber

$$r = 0.002\left(1 + \frac{0.1}{1+(x-5)^2}\right) \text{ inches,}$$

the ambient temperature of the surroundings $T_\infty = 50^\circ$ F and the constant temperature maintained at the left end of the fiber $T_0 = 200^\circ$ F. Determine the temperature within the fiber at increments of 0.1 inches.

20. A wooden beam of square cross section is supported at both ends and is carrying a distributed lateral load of uniform intensity $w = 20$ lb/ft and an axial tension load $T = 100$ lb. The deflection, $u(x)$, of the beam's centerline satisfies the boundary value problem

$$u'' - \frac{T}{EI}u = -\frac{w}{2EI}x(L-x)$$
$$u(0) = u(L) = 0,$$

where $L = 6$ ft is the length, $E = 1.3 \times 10^6$ lb/in^2 is the modulus of elasticity and $I = s^4$ is the moment of inertia of the beam. The side length of the square cross section is $s = 4$ inches.

(a) Determine the deflection of the beam at 1 inch intervals along its length.

(b) Repeat part (a) assuming that the beam tapers along its length so that $s = (4 - x/2L)$ inches.

21. Repeat Exercise 20 for a metal rod of circular cross section. Use the parameter values

$$w = 200 \text{ lb/ft}, \quad T = 750 \text{ lb}$$
$$L = 10 \text{ ft}, \quad E = 3.0 \times 10^7 \text{ lb/in}^2, \quad \text{and} \quad I = \pi r^4/4.$$

For part (a), take $r = 3$ inches. For part (b), use $r = (3 + 0.25\sin(\pi x/L))$ inches.

22. In the "Flow between Parallel Plates" problem of Section 8.1, suppose we introduce the effect of a constant pressure gradient, which we denote by dp/dx, in the direction of the flow. The boundary value problem for determining the flow velocity then becomes

$$u'' = 80\left[\frac{842388.596}{(293.16+80y)^3} - \frac{839.456}{(293.16+80y)^2}\right]u' + \frac{h^2}{U_0}\frac{dp}{dx}$$

$$u(0) = 0, \quad u(1) = 1.$$

The constant $\frac{h^2}{U_0}\frac{dp}{dx}$ is called the pressure parameter. Calculate the velocity distribution for values of the pressure parameter of -4, -2, 0, 2, and 4.

23. Rework the first application problem from Section 8.2, "Steady-State Temperature Distribution in a Pin Fin," using the shooting method.

24. Rework the second application problem from Section 8.2, "The Heat Pack," using the shooting method.

25. Rework the application problem from this section, "Cooling Fin on the Cylinder Barrel of a Motorcycle," changing the temperature of the outer surface of the barrel to 250° C and the convection coefficient to $h = 100$ W/m · K. Use the values given in the text for all other parameters.

26. Rework the application problem from this section, "Cooling Fin on the Cylinder Barrel of a Motorcycle," but assume the fin has a uniform thickness of 6 mm. Use the values given in the text for all other parameters.

8.5 THE SHOOTING METHOD, PART II: NONLINEAR BOUNDARY VALUE PROBLEMS

We now turn our attention to the development of the shooting method for the general nonlinear boundary value problem

$$y'' = f(x, y, y'), \quad \alpha_1 y(a) + \alpha_2 y'(a) = \alpha_3, \quad \beta_1 y(b) + \beta_2 y'(b) = \beta_3.$$

As indicated in the previous section, the fundamental strategy behind the shooting method is the transformation of a boundary value problem into a rootfinding problem that is, itself, based on the solution of an initial value problem.

A Specific Boundary Value Problem

Let's develop the details of this approach within the context of a specific example. Suppose we wish to approximate the solution of the nonlinear differential equation

$$yy'' + (y')^2 + 1 = 0$$

subject to Dirichlet boundary conditions

$$y(1) = 1, \quad y(2) = 2.$$

Since the value of the solution is known at $x = 1$, only a value for the first derivative needs to be supplied to complete the specification of an initial value problem. Let $y(x;p)$ denote the solution of the initial value problem that results upon setting $y'(1) = p$. To obtain an approximation to the solution of the original boundary value problem, we must determine the value $p = p^*$ for which $y(2;p^*) = 2$. If we define the *objective function* $F(p) = 2 - y(2;p)$, then solving the boundary value problem becomes equivalent to locating the root of F. This is the aforementioned rootfinding problem.

To approximate p^* we can use any of the rootfinding techniques that were developed in Chapter 2. In the present framework, each "evaluation" of the function F requires the solution of an initial value problem. Therefore, we will avoid the use of the bisection method: The linear convergence of the method is much too slow. We will also bypass Newton's method. Although Newton's method would provide quadratic convergence, it requires the calculation of

$$F'(p) = \frac{d}{dp} y(2;p).$$

This quantity can be obtained by solving a second initial value problem (see Asaithambi [1] or Burden and Faires [2] for details), but that doubles the computational cost of each iteration. Hence, we will implement the secant method. This may require a few more iterations than Newton's method but generally will result in the numerical solution of fewer initial value problems. As a reminder, recall that starting from the initial iterates p_0 and p_1, the secant method generates a sequence of approximations according to the rule

$$p_n = p_{n-1} - F(p_{n-1})\frac{p_{n-1} - p_{n-2}}{F(p_{n-1}) - F(p_{n-2})}.$$

Returning to the problem at hand, we will use the classical fourth-order Runge-Kutta (RK4) method to solve all initial value problems, taking ten steps to integrate from $x = 1$ to $x = 2$. With $p_0 = 0$ [i.e., setting $y'(1) = 0$], we find $y(2;0) \approx 0.104101$, so that $F(0) = 2 - y(2;0) = 1.895899$. If we next choose $p_1 = 1$ the RK4 method produces the value $y(2;1) \approx 1.414197$, from which we compute $F(1) = 0.585803$. Applying the secant method then yields

$$p_2 = 1 - 0.585803\frac{1-0}{0.585803 - 1.895899} = 1.447145.$$

Assigning this value to $y'(1)$ leads to $y(2;p_2) \approx 1.701210$ and $F(p_2) = 0.298790$. Thus,

$$p_3 = 1.447145 - 0.298790\frac{1.447145 - 1}{0.298790 - 0.585803} = 1.912638.$$

Continuing in this fashion, the next four iterations produce the results

$$\begin{array}{lll} p_3 = 1.912638, & y(2;p_3) \approx 1.955692, & F(p_3) = 0.044308 \\ p_4 = 1.993685, & y(2;p_4) \approx 1.996677, & F(p_4) = 0.003322 \\ p_5 = 2.000256, & y(2;p_5) \approx 1.999963, & F(p_5) = 3.698 \times 10^{-5} \\ p_6 = 2.000330, & y(2;p_6) \approx 1.999999969, & F(p_6) = 3.088 \times 10^{-8}. \end{array}$$

With $|F(p_6)| < 5 \times 10^{-7}$, we will terminate the iteration and accept $y(x;p_6)$ as the approximate solution to the boundary value problem

$$\begin{gathered} yy'' + (y')^2 + 1 = 0 \\ y(1) = 1, \quad y(2) = 2. \end{gathered}$$

The convergence tolerance has been applied to $|F(p_n)|$, and not to the difference between successive p values, since the true objective is to match the second boundary condition, not to approximate the necessary initial condition for the first derivative.

The values of the approximate solution, $y(x_i;p_6)$, are tabulated below. For comparison, the values of the exact solution,

$$y(x) = \sqrt{6x - 4 - x^2},$$

are also displayed. The final column of the table lists the absolute error in the approximate solution.

x_i	$y(x_i; p_6)$	$y(x_i)$	Absolute Error
1.00	1.000000	1.000000	
1.10	1.178956	1.178983	0.000027
1.20	1.326623	1.326650	0.000027
1.30	1.452560	1.452584	0.000024
1.40	1.562030	1.562050	0.000020
1.50	1.658296	1.658312	0.000016
1.60	1.743547	1.743560	0.000013
1.70	1.819332	1.819340	0.000008
1.80	1.886790	1.886796	0.000006
1.90	1.946789	1.946792	0.000003
2.00	2.000000	2.000000	

Finally, we investigate the order of convergence. Since the RK4 method has been used to compute the solution of all of the initial value problems, we expect that the approximate solution of the boundary value problem should converge toward the exact solution with rate of convergence $O(h^4)$. The following table lists the maximum absolute error and the root mean square (rms) error in the approximate solution as a function of N, the number of steps taken to integrate from $x = 1$ to $x = 2$. The convergence tolerance for the secant method was set at 5×10^{-14} for all cases. Note that with each doubling of N, the error drops by roughly a factor of $16 = 2^4$, exactly what one would expect from a fourth-order method.

N	Maximum Absolute Error	Error Ratio	rms Error	Error Ratio
5	3.048872×10^{-4}		1.851011×10^{-4}	
10	2.726369×10^{-5}	11.182906	1.710854×10^{-5}	10.819221
20	1.812206×10^{-6}	15.044472	1.116743×10^{-6}	15.320038
40	1.116332×10^{-7}	16.233579	6.885077×10^{-8}	16.219755
80	6.870329×10^{-9}	16.248596	4.239422×10^{-9}	16.240602
160	4.252689×10^{-10}	16.155256	2.625002×10^{-10}	16.150171
320	2.643930×10^{-11}	16.084730	1.632357×10^{-11}	16.081050
640	1.648681×10^{-12}	16.036633	1.017769×10^{-12}	16.038575
1280	1.023626×10^{-13}	16.106291	6.221091×10^{-14}	16.359982

Other Boundary Conditions

The shooting method can also be used to approximate the solution of a boundary value problem which has boundary conditions which are not Dirichlet. We simply have to adjust the initial conditions and/or the objective function to reflect the specific set of boundary conditions given in the problem statement. For instance, if the Neumann boundary condition $y'(a) = \alpha$ is specified at $x = a$, then the initial value of the solution becomes the variable in the rootfinding problem, and the initial conditions applied to the differential equation must take the form

$$y(a) = p, \quad y'(a) = \alpha.$$

The case of a Robin boundary condition at $x = a$ will be discussed momentarily. As for the objective function, this is derived from the boundary condition at the right endpoint. In the case of the Neumann condition $y'(b) = \beta$, the objective function would be $F(p) = \beta - y'(b;p)$; whereas, in the case of the Robin condition $\beta_1 y(b) + \beta_2 y'(b) = \beta_3$, the objective function would be $F(p) = \beta_3 - \beta_1 y(b;p) - \beta_2 y'(b;p)$.

EXAMPLE 8.10 A Problem with One Neumann and One Robin Boundary Condition

The exact solution of the nonlinear boundary value problem

$$\begin{gathered} 2xy'' + (y')^2 - 4y = 4x \\ y'(1) = 4, \quad y(3) + 2y'(3) = 32 \end{gathered}$$

is $y(x) = (x+1)^2$. To approximate the solution of this problem using the shooting method, we convert the boundary value problem into the rootfinding problem

> determine the value of p for which $F(p) = 32 - y(3;p) - 2y'(3;p)$ is equal to zero.

Here, $y(x;p)$ denotes the solution of the initial value problem

$$\begin{gathered} 2xy'' + (y')^2 - 4y = 4x \\ y(1) = p, \quad y'(1) = 4. \end{gathered}$$

Note how the boundary condition at $x = 1$ (the left endpoint) dictates the initial conditions on the initial value problem, while the boundary condition at $x = 3$ (the right endpoint) dictates the objective function for the rootfinding problem.

The RK4 method is used to compute $y(x;p)$ for each $p = p_n$, with a step size of $\Delta x = 0.2$. The secant method is initialized with $p_0 = 0$ and $p_1 = 1$, and iterations are terminated when $|F(p)|$ falls below a tolerance of 5×10^{-7}. Five iterations are needed to achieve convergence. The results of each iteration are listed below, where the values of $y(3;p)$ and $y'(3;p)$ have been reported to six decimal places for display purposes.

p	$y(3;p)$	$y'(3;p)$	$F(p)$
0	8.303482	5.781270	12.133978
1	10.361601	6.445059	8.748282
3.583894	15.254059	7.811455	1.123031
3.964445	15.936660	7.984070	0.0952008
3.999693	15.999486	7.999773	9.677×10^{-4}
4.000054	16.000131	7.999934	8.265×10^{-7}
4.000055	16.000132	7.999934	7.168×10^{-12}

The values of $y(x_i;p_6)$ are shown in the following table and compared with the values of the exact solution. The absolute errors are all on the order of 10^{-4}, even with a step size as large as $\Delta x = 0.2$.

x_i	$y(x_i;p_6)$	$y(x_i)$	Absolute Error
1.00	4.000055	4.00	0.000055
1.20	4.840102	4.84	0.000102
1.40	5.760118	5.76	0.000118
1.60	6.760124	6.76	0.000124
1.80	7.840125	7.84	0.000125
2.00	9.000125	9.00	0.000125
2.20	10.240126	10.24	0.000126
2.40	11.560127	11.56	0.000127
2.60	12.960128	12.96	0.000128
2.80	14.440130	14.44	0.000130
3.00	16.000132	16.00	0.000132

The final case to consider is that of a Robin boundary condition specified at the left endpoint, $x = a$. Theoretically, we can let either the initial value of the solution or the initial value of the first derivative serve as the variable for the rootfinding problem and compute the other initial value from the Robin boundary condition. That is, if the boundary condition at $x = a$ is $\alpha_1 y(a) + \alpha_2 y'(a) = \alpha_3$, then we can either apply the initial conditions

$$y(a) = p, \quad y'(a) = (\alpha_3 - \alpha_1 p)/\alpha_2,$$

or the initial conditions

$$y(a) = (\alpha_3 - \alpha_2 p)/\alpha_1, \quad y'(a) = p.$$

Unfortunately, the solution of an initial value problem with initial conditions of this type is significantly more sensitive to the choice of p than is the solution of an initial value problem for which one of the initial values is fixed.

To illustrate this point, consider the differential equation $y'' = 2y^3$. Whether the boundary conditions are

$$y(0) = 1/3, \quad y(1) = 1/4,$$
$$y'(0) = -1/9, \quad y(1) = 1/4$$

or

$$3y(0) - 9y'(0) = 2, \quad y(1) = 1/4,$$

the exact solution to the resulting boundary value problem is $y(x) = 1/(x+3)$. If we were to use the shooting method to solve each of these boundary value problems, the first set of boundary conditions would give rise to the initial conditions

$$y(0) = 1/3, \quad y'(0) = p, \tag{1}$$

the second set would lead to the initial conditions

$$y(0) = p, \quad y'(0) = -1/9, \tag{2}$$

and the third set would give either

$$y(0) = (2/3) + 3p, \quad y'(0) = p \text{ or} \tag{3}$$
$$y(0) = p, \quad y'(0) = -(2/9) + (1/3)p. \tag{4}$$

Figure 8.13 displays the solutions of $y'' = 2y^3$ subject to initial condition (1)—top graph—and initial condition (3)—bottom graph—for $p = -1, -1/2, 0, 1/2$ and 1. The solution is clearly much more sensitive to the choice of p with initial condition (3). This sensitivity can be expected to play havoc with the convergence of the shooting method.

Next, the solutions of $y'' = 2y^3$ subject to initial condition (2)—top graph—and initial condition (4)—bottom graph—for $p = -1, -1/2, 0, 1/2$ and 1 are displayed in Figure 8.14. Though not as dramatic as the situation depicted in Figure 8.13, the solution of the initial value problem is still clearly more sensitive to the choice of p with initial condition (4).

The moral of this story is simple. When a boundary value problem has a Robin boundary condition at the left endpoint of the domain, but not at the right endpoint, it is advisable to reverse direction and integrate from $x = b$ back to $x = a$. This can be accomplished by introducing the change of independent variable

$$z = \frac{1}{b-a}(b-x).$$

Note that $x = b \to a$ corresponds to $z = 0 \to 1$. To replace the derivatives in the differential equation and in the boundary conditions, we make use of the chain rule to determine

$$\frac{d}{dx} = \frac{d}{dz}\frac{dz}{dx} = -\frac{1}{b-a}\frac{d}{dz}$$

and

$$\frac{d^2}{dx^2} = \frac{d}{dx}\left(\frac{d}{dx}\right) = -\frac{1}{b-a}\frac{d}{dz}\left(-\frac{1}{b-a}\frac{d}{dz}\right) = \frac{1}{(b-a)^2}\frac{d^2}{dz^2}.$$

EXAMPLE 8.11 Reversing the Direction of Integration

Let's approximate the solution of the nonlinear boundary value problem

$$y'' = 2y^3$$
$$3y(0) - 9y'(0) = 2, \quad y(1) = 1/4$$

using the shooting method. Since this problem has a Robin condition specified at the left end of the domain, we first introduce the new independent variable $z = 1-x$.

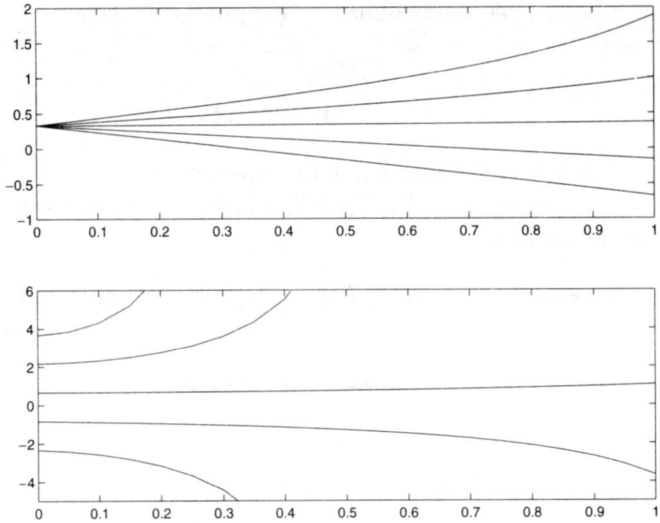

Figure 8.13 (Top) Solution of $y'' = 2y^3$ subject to the initial conditions $y(0) = 1/3$ and $y'(0) = -1, -1/2, 0, 1/2, 1$. (Bottom) Solution of $y'' = 2y^3$ subject to the initial conditions $y(0) = (2/3) + 3y'(0)$ and $y'(0) = -1, -1/2, 0, 1/2, 1$.

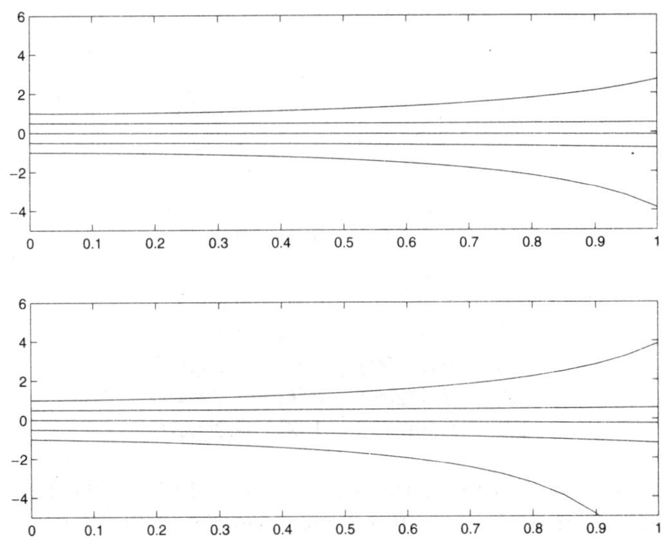

Figure 8.14 (Top) Solution of $y'' = 2y^3$ subject to the initial conditions $y(0) = -1, -1/2, 0, 1/2, 1$ and $y'(0) = -1/9$ (Bottom) Solution of $y'' = 2y^3$ subject to the initial conditions $y(0) = -1, -1/2, 0, 1/2, 1$ and $y'(0) = -(2/9) + (1/3)y(0)$.

In terms of this new variable the boundary value problem becomes

$$y'' = 2y^3$$
$$y(0) = 1/4, \quad 3y(1) + 9y'(1) = 2,$$

where primes now denote differentiation with respect to z.

To apply the shooting method to the latter boundary value problem, we need to solve the rootfinding problem

> determine the value of p for which $F(p) = 2 - 3y(1;p) - 9y'(1;p)$ is equal to zero,

where $y(z;p)$ is the solution to the initial value problem

$$y'' = 2y^3$$
$$y(0) = 1/4, \quad y'(0) = p.$$

Let's solve the initial value problems using the RK4 method with a step size of $\Delta z = 0.1$. For the rootfinding problem, take $p_0 = 0$ and $p_1 = 1$ and use a convergence tolerance of 5×10^{-14}. With six iterations of the secant method, the solution converges to the values listed below. Since $z = 1 - x$, the values in the last column were obtained via the formula

$$y(x_i;p_7) = y(1 - z_i;p_7).$$

z_1	$y(z_i;p_7)$	x_i	$y(x_i;p_7)$
0.00	0.250000	0.00	0.333333
0.10	0.256410	0.10	0.322581
0.20	0.263158	0.20	0.312500
0.30	0.270270	0.30	0.303030
0.40	0.277777	0.40	0.294118
0.50	0.285714	0.50	0.285714
0.60	0.294118	0.60	0.277777
0.70	0.303030	0.70	0.270270
0.80	0.312500	0.80	0.263158
0.90	0.322581	0.90	0.256410
1.00	0.333333	1.00	0.250000

All solution values are correct to the digits shown. In fact, the largest absolute error occurs in the value of $y(x_0 = 0;p_7)$ and is roughly 2.13×10^{-8}.

Application Problem: Density-Dependent Dispersal of an Insect Population

When we investigated the spread of an insect population in the Chapter 8 Overview and in Section 8.3, we assumed the coefficient of diffusion, D, was constant. Suppose we now make the more realistic assumption that D is an increasing function of the population density, N. In particular, let's take

$$D = D_0 \left(\frac{N}{K}\right)^m,$$

where D_0 is a positive constant, K is the environmental carrying capacity and $m > 0$. Using this density-dependent diffusion coefficient, equation (11) from the Chapter 8 Overview becomes

$$\frac{d}{dX}\left[D_0 \left(\frac{N}{K}\right)^m \frac{dN}{dX}\right] + rN\left(1 - \frac{N}{K}\right) = 0,$$

or, equivalently,

$$\frac{D_0}{r}\frac{d^2(N/K)^{m+1}}{dX^2} + (m+1)\frac{N}{K}\left(1 - \frac{N}{K}\right) = 0.$$

Introducing the nondimensional variables $n = (N/K)^{m+1}$ and $x = X\sqrt{r/D_0}$ yields the differential equation

$$n'' + (m+1)n^{1/(m+1)}\left[1 - n^{1/(m+1)}\right] = 0, \qquad (5)$$

where primes denote differentiation with respect to x.

To investigate the effect of density-dependent dispersal on the steady-state density profile, (5) was solved, subject to the boundary conditions

$$n(0) = 0 \quad \text{and} \quad n'(3) = 0,$$

for $m = 1$, 2, and 3. When applying the shooting method, all initial value problems were handled using the RK4 method with a step size of $\Delta x = 0.01$. The secant method was initialized with $p_0 = 1$ and $p_1 = 2$, and iterations were terminated when

$$|F(p)| = |n'(3;p)| < 5 \times 10^{-10}.$$

Seven, four, and five iterations were needed to achieve convergence for the case $m = 1$, 2 and 3, respectively.

Results of these experiments are summarized in Figure 8.15. For each value of m, the quantity

$$[n(x)]^{1/(m+1)} = \frac{N(x)}{K}$$

is plotted. The density profile corresponding to a constant coefficient of diffusion has been included for comparison. To obtain the constant D profile, the secant method was initialized with $p_0 = 1$ and $p_1 = 0.5$, and seven iterations were needed to achieve convergence. Observe that increasing the value of m leads to an overall increase in the density profile.

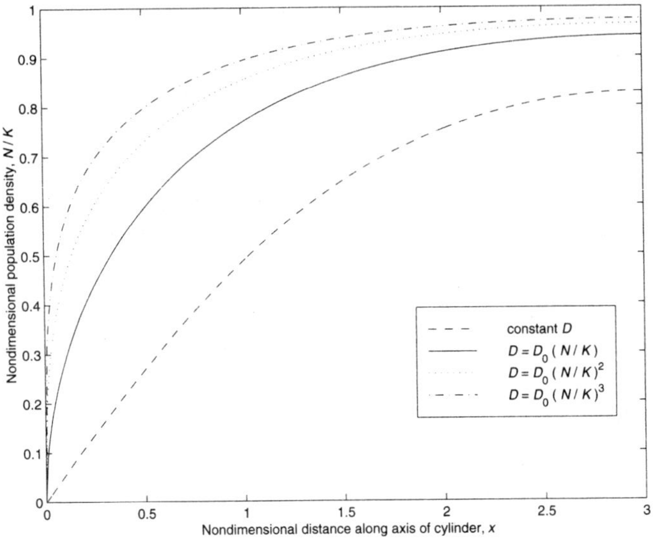

Figure 8.15 Steady-state population density profiles for an insect population which disperses with a density-dependent dispersal rate.

References

1. N. S. Asaithambi, *Numerical Analysis: Theory and Practice*, Saunders College Publishing, Fort Worth, 1995.
2. R. Burden and J. D. Faires, *Numerical Analysis, 5th edition*, PWS Publishing Company, Boston, 1993.

EXERCISES

In Exercises 1–6, suppose we use the shooting method to approximate the solution of the indicated boundary value problem. Write out the initial value problem that must be solved and the objective function for the corresponding rootfinding problem.

1. $y'' + (y')^2 = 1, \quad y(0) = 1, \quad y(1) = 2$

2. $y'' + yy' + (y')^2 = 0, \quad y(0) = 0, \quad y(1) + y'(1) = 1$

3. $y'' = -2y^2 - 4xyy', \quad y'(0) = 0, \quad y(1) = 1/2$

4. $y'' + 4yy' = -2y/(1 + x^2), \quad y'(0) = 1, \quad y'(1) = 0$

5. $y'' + 2y^2 - 8x^2y^3 = 0, \quad y(0) = 1, \quad y'(1) = -1/2$

6. $y'' = 6y^2, \quad y'(0) = 1/4, \quad y(1) + y'(1) = 3$

In Exercises 7–10, make a change of independent variable to reverse the direction of integration, and write out the resulting boundary value problem.

7. $y'' + yy' + (y')^2 = 0, \quad y(0) + y'(0) = 0, \quad y'(1) = 1/e$

8. $y'' = -2y^2 - 4xyy', \quad y(0) + y'(0) = 1, \quad y(1) = 1/2$

9. $y'' + 4yy' = -2y/(1 + x^2), \quad y(0) + y'(0) = 1, \quad y'(1) = 0$

10. $y'' + 2y^2 - 8x^2y^3 = 0, \quad y(0) + y'(0) = 1, \quad y(1) = 1/2$

In Exercises 11–16, numerically verify that the shooting method, with the RK4 method as the underlying initial value problem solver, produces results that converge toward the exact solution with rate of convergence $O(h^4)$.

11. $y'' = -2y^2 + 8x^2y^3, \quad y(0) = 1, \quad y(1) = 1/2, \quad y(x) = 1/(1 + x^2)$

12. $y'' + 4yy' = -2y/(1 + x^2), \quad y(0) = 0, \quad y'(0) = 1/2, \quad y(x) = x/(1 + x^2)$

13. $y'' + (y')^2 = 1, \quad y'(0) = 0, \quad y(1) = 2$

14. $y'' + yy' + (y')^2 = 0, \quad y(0) = 0, \quad y(1) + y'(1) = 1, \quad y(x) = 1 - e^{-x}$

15. $y'' + 4xyy' = -2y^2, \quad y(0) + y'(0) = 1, \quad y(1) = 1/2, \quad y(x) = 1/(1 + x^2)$

16. $yy'' = y' - 1, \quad y(1) = 0, \quad y'(2) = 1 + \ln 2$

17. Consider the nonlinear differential equation

$$2xy'' + (y')^2 - 4y = 4x.$$

Use the shooting method to solve this differential equation subject to each of the following sets of boundary conditions. In each case, the exact solution is $y(x) = (x + 1)^2$. How rapidly does the approximate solution converge toward the exact solution as a function of the number of subintervals?

(a) $y(1) = 4, \quad y(2) = 9$ **(b)** $y(1) = 4, \quad y'(2) = 6$

(c) $y'(1) = 4, \quad y'(2) = 6$ **(d)** $y(1) - y'(1) = 0, \quad y'(2) = 6$

18. Ramirez (*Computational Methods for Process Simulation*, Butterworths, Boston, 1989) develops the following nonlinear boundary value problem for the concentration, $y(x)$, of the reactant in a second-order chemical reaction taking place within a tubular reactor with dispersion:

$$\frac{d^2y}{dx^2} - 10\frac{dy}{dx} - 5y^2 = 0$$

$$y(0) - \frac{1}{10}\left.\frac{dy}{dx}\right|_{x=0} = 5, \qquad \left.\frac{dy}{dx}\right|_{x=1} = 0.$$

Determine the concentration in increments of 0.01 along the length of the reactor.

19. C. Philipsen, S. Markvorsen and W. Kleim ["Modelling the Stem Curve of a Palm in a Strong Wind," *SIAM Review,* **38** (3), pp. 483–484, 1996] present the following model for the angle of the stem of a tall palm tree, relative to its vertical position, when the tree is subjected to wind loading:

$$EI\frac{d^2\theta}{ds^2} = -W_s\left(1 - \frac{s}{L}\right)\sin\theta - W_c\sin\theta - D\cos\theta.$$

Here, θ is the angle of the stem relative to the vertical position, and s is arc length measured along the stem. The parameters in the model are the total stem weight $W_s = 22700$ N, the Young's modulus of the stem $E = 0.9 \times 10^9$ N/m^2, the moment of inertia of the stem $I = 5.147 \times 10^{-4}$ m^4, the length of the stem $L = 30$ m, the total canopy weight $W_c = 1385.5$ N, and the wind drag force on the canopy $D = 1.2405U^2$ N, where U is the wind speed in m/s. The boundary conditions imposed on the stem angle are

$$\theta(0) = 0 \quad \text{and} \quad \theta'(L) = 0.$$

Determine the function $\theta(s)$ when the wind speed is 8 meters/second.

20. Rework the "Density-Dependent Dispersal of an Insect Population" problem for an environment with a nondimensional length of 5; that is, replace the boundary condition $n'(3) = 0$ with the condition $n'(5) = 0$. Produce a plot similar to Figure 8.15 to display the results.

21. Subramanian and Balakotaiah ["Convective Instabilities Induced by Exothermic Reactions Occurring in a Porous Medium," *Phys. Fluids,* **6** (9), pp. 2907–2922, 1994] develop the boundary value problem

$$\frac{d^2\theta}{dz^2} + B\phi^2\left(1 - \frac{\theta}{B}\right)\exp\left(\frac{\gamma\theta}{\gamma + \theta}\right) = 0$$
$$\theta'(0) = 0, \quad \theta(1) = 0$$

for the steady-state temperature profile, $\theta(z)$, in a porous medium undergoing an exothermic reaction. The parameter B is the maximum possible temperature in the absence of natural convection, ϕ^2 is the ratio of the characteristic time for conduction to that for heat generation and γ is the dimensionless activation energy. For $B = 6.0$, $\phi^2 = 0.25$, and $\gamma = 30.0$, determine $\theta(z)$.

22. For values of l ranging from 0.1 to 5.0 in increments of 0.1, solve

$$n'' + 2\sqrt{n}(1 - \sqrt{n}) = 0, \quad n(0) = 0, \quad n'(l) = 0.$$

Use a step size of $\Delta x = 0.01$ for each l. Plot the maximum value of the population density as a function of l.

CHAPTER 9

Elliptic Partial Differential Equations

AN OVERVIEW

Fundamental Mathematical Problem

In this chapter, we will develop the finite difference method for approximating the solution of elliptic partial differential equations. The second-order linear partial differential equation

$$A(x,y)\frac{\partial^2 u}{\partial x^2}+2B(x,y)\frac{\partial^2 u}{\partial x \partial y}+C(x,y)\frac{\partial^2 u}{\partial y^2}+D(x,y)\frac{\partial u}{\partial x}+E(x,y)\frac{\partial u}{\partial y}+F(x,y)u=G(x,y)$$

is said to be *elliptic* over some region r in the x-y plane provided that

$$A(x,y)C(x,y)-[B(x,y)]^2>0$$

for all $(x,y)\in R$. The inequality $A(x,y)C(x,y)-[B(x,y)]^2>0$ is called the *ellipticity condition.*

Elliptic partial differential equations model time independent problems and often arise when determining the steady-state solutions to time-dependent problems. Elliptic problems are also the multidimensional analogue of the boundary value problems we studied in Chapter 8. There are two special cases of the general elliptic equation that we will investigate in detail in this chapter. The first is the Poisson equation

$$\frac{\partial^2 u}{\partial x^2}+\frac{\partial^2 u}{\partial y^2}=f(x,y).$$

The second arises when $f(x,y)=0$:

$$\frac{\partial^2 u}{\partial x^2}+\frac{\partial^2 u}{\partial y^2}=0.$$

This is the so-called Laplace equation.

We will focus our studies primarily on problems specified over rectangular domains. The left-hand part of the following figure illustrates a typical rectangular domain:

$$R=\{(x,y)|\,a<x<b,c<y<d\}.$$

The symbol ∂R denotes the boundary of r, which, in this case, is simply the rectangle itself. Elliptic partial differential equations can also be specified over nonrectangular, or irregular, domains, such as the one pictured below at the right. With the exception of circular domains, however, finite differences generally do not work well on nonrectangular domains. We will therefore consider special cases only.

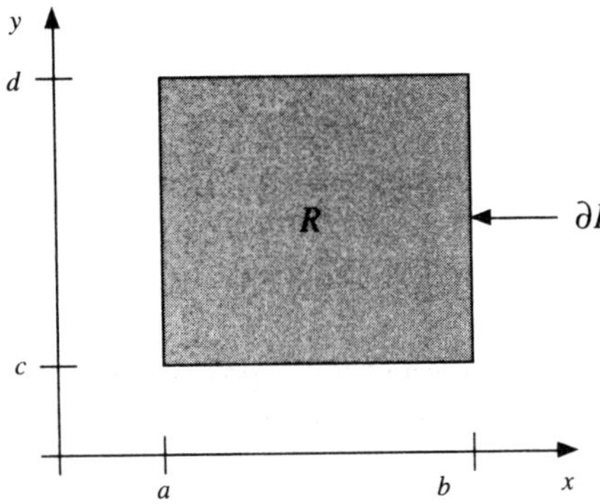

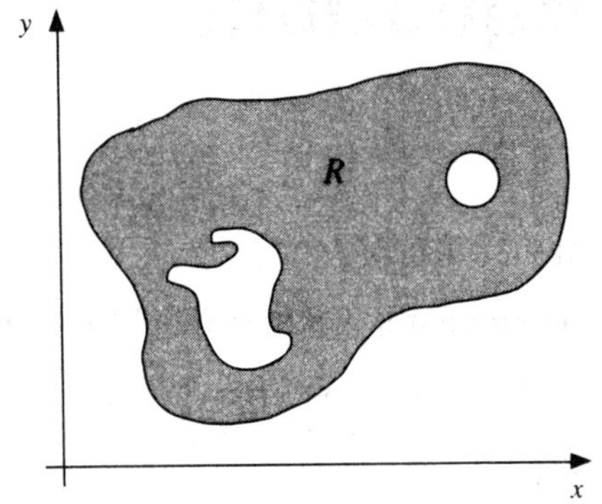

As for boundary conditions, we will discuss the same three types that were treated in Chapter 8:

Dirichlet:	$u(x,y) = r(x,y)$ on ∂R
Neumann:	$\frac{\partial u}{\partial n}(x,y) = r(x,y)$ on ∂R
Robin:	$\alpha(x,y)u(x,y) + \beta(x,y)\frac{\partial u}{\partial n}(x,y) = r(x,y)$ on ∂R

In general, different types of boundary conditions may be specified on different portions of the boundary. In the Neumann and Robin boundary condition, $\partial/\partial n$ refers to the derivative taken in the outward normal direction. The interpretation of this derivative for different domains is illustrated in Figure 9.1.

Flow Through a Contraction Duct

A fluid flows into the asymmetric contraction duct shown in Figure 9.2 with a uniform velocity of 3 meters/second and exits with a uniform velocity of 12 meters/second. The duct has a total length of three meters. The inlet section is one meter long and two meters high. Along the upper wall, the duct contracts at a 45° angle over a distance of one-half meter, while, along the lower wall, the duct contracts at a 45° angle over a distance of one meter. We will assume that the duct is of sufficient depth to warrant treating the flow as two dimensional.

Let $u(x,y)$ denote the flow velocity in the x-direction and $v(x,y)$ denote the flow velocity in the y-direction. Consider the representative fluid element in Figure 9.3, which has length Δx, height Δy, and unit depth. According to the law of conservation of mass, the rate of change of fluid mass within this representative element must be equal to the difference between the rate at which mass enters and the rate at which mass exits the element. If ρ denotes the mass density of the fluid,

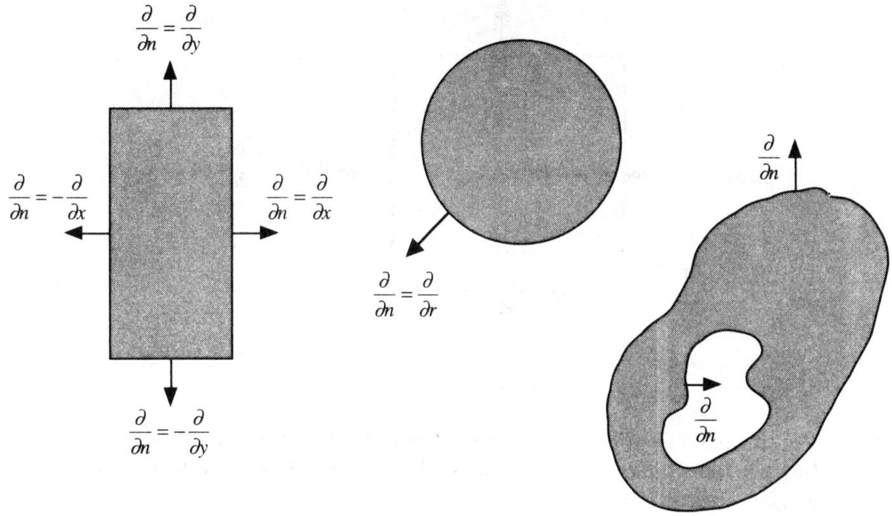

Figure 9.1 Interpretation of outward normal derivative for different domain configurations.

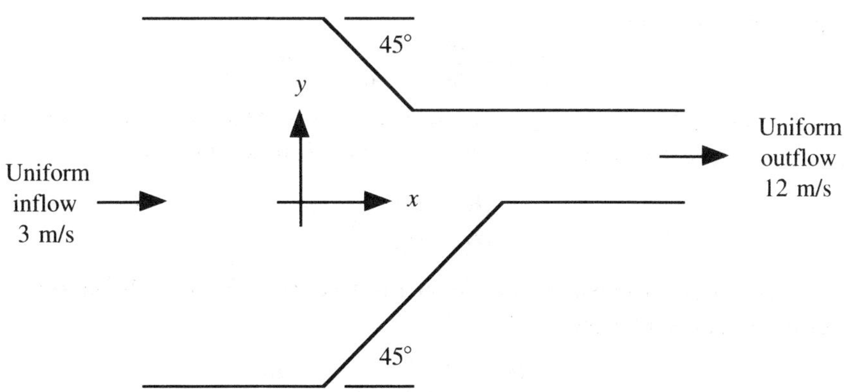

Figure 9.2 Geometry of an asymmetric contraction duct.

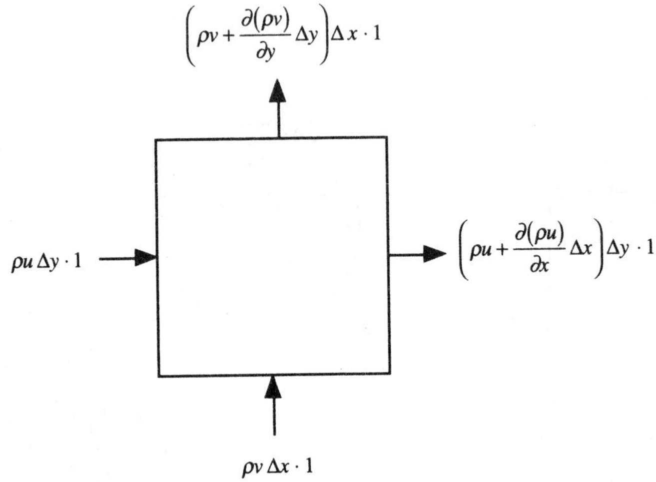

Figure 9.3 Representative fluid element for flow through the contraction duct of Figure 9.2. The mass flow rates across each face of the element are indicated.

then the element contains a total mass of $\rho\,\Delta x\,\Delta y \cdot 1$. The rate at which fluid mass crosses each face of the element is indicated in Figure 9.3. Therefore, conservation of mass requires

$$\frac{\partial(\rho\,\Delta x\,\Delta y)}{\partial t} = \rho u\,\Delta y + \rho v\,\Delta x - \left(\rho u + \frac{\partial(\rho u)}{\partial x}\Delta x\right)\Delta y - \left(\rho v + \frac{\partial(\rho v)}{\partial y}\Delta y\right)\Delta x,$$

or, upon simplification,

$$\frac{\partial \rho}{\partial t} + \frac{\partial(\rho u)}{\partial x} + \frac{\partial(\rho v)}{\partial y} = 0.$$

This is known as the *continuity equation.* If we assume that the fluid is incompressible (i.e., that ρ is constant), then the continuity equation becomes

$$\frac{\partial u}{\partial x} + \frac{\partial v}{\partial y} = 0. \tag{1}$$

At this point we introduce a clever mathematical device. Define the *stream function,* $\psi(x, y)$, such that

$$u = \frac{\partial \psi}{\partial y} \quad \text{and} \quad v = -\frac{\partial \psi}{\partial x}. \tag{2}$$

Notice that in terms of ψ

$$\frac{\partial u}{\partial x} + \frac{\partial v}{\partial y} = \frac{\partial}{\partial x}\left(\frac{\partial \psi}{\partial y}\right) + \frac{\partial}{\partial y}\left(-\frac{\partial \psi}{\partial x}\right) \equiv 0,$$

so equation (1) is satisfied identically. In addition to reducing the number of variables by one in such a way that conservation of mass is guaranteed, the stream function has an important geometric interpretation. Curves along which ψ is constant are the *streamlines* of the flow, where a streamline is defined as a curve that is everywhere tangent to the flow field at a given instant. For the problem under consideration, in which the flow is time independent, streamlines also coincide with the actual paths traveled by the fluid particles.

So how do we determine ψ? The expression

$$\frac{1}{2}\left(\frac{\partial v}{\partial x} - \frac{\partial u}{\partial y}\right) \tag{3}$$

defines the angular velocity of the flow, which is denoted by ω. Substituting (2) into (3) yields

$$\omega = \frac{1}{2}\left(\frac{\partial}{\partial x}\left(-\frac{\partial \psi}{\partial x}\right) - \frac{\partial}{\partial y}\left(\frac{\partial \psi}{\partial y}\right)\right) = -\frac{1}{2}\left(\frac{\partial^2 \psi}{\partial x^2} + \frac{\partial^2 \psi}{\partial y^2}\right),$$

or

$$\frac{\partial^2 \psi}{\partial x^2} + \frac{\partial^2 \psi}{\partial y^2} = -2\omega.$$

We will now make one last assumption, which is that the flow is irrotational. For an irrotational flow, $\omega = 0$ and so

$$\frac{\partial^2 \psi}{\partial x^2} + \frac{\partial^2 \psi}{\partial y^2} = 0. \tag{4}$$

We still need to specify appropriate boundary conditions for ψ. First, let's fix the origin of our coordinate system at the lower left corner of the duct. At the inlet, we know that $u(0, y) = 3$; therefore,

$$\frac{\partial \psi}{\partial y}(0, y) = 3 \quad \Rightarrow \quad \psi(0, y) = 3y, \tag{5}$$

where we've set the constant of integration to zero for convenience. Since the fluid cannot penetrate the walls of the duct, the fluid velocity normal to any wall must be zero. Using the definition of the stream function, it follows that the velocity normal to any surface is equal to the derivative of ψ taken along that surface. This leads us to conclude that ψ must be constant along the lower wall and along the upper wall of the duct. But $\psi(0, 0) = 0$ and $\psi(0, 2) = 6$, so the appropriate boundary conditions are

$$\psi(x, y) = 0 \text{ along the lower wall} \tag{6}$$

and

$$\psi(x, y) = 6 \text{ along the upper wall.} \tag{7}$$

Finally, at the exit of the duct, we find

$$\psi(3, y) = 12(y - 1). \tag{8}$$

Thus, to determine the flow through the contraction duct, we must solve the elliptic partial differential equation (4) subject to the Dirichlet boundary conditions (5)–(8).

Remainder of Chapter

We will begin our investigation of the numerical solution of elliptic partial differential equations with the Poisson problem on rectangular domains subject to Dirichlet boundary conditions. Emphasis will be placed upon the derivation of the discrete analogue of the Poisson equation and the organization of the discrete equations into matrix form. More general boundary conditions are treated in Section 9.2. Next, iterative strategies for solving the discrete equations will be considered. Collectively, these techniques are referred to as relaxation schemes. In the fourth section, an analysis of the convergence properties of relaxation schemes will be undertaken, and the notion of multigrid methods will be introduced. Elliptic equations over irregular domains will be treated in Section 9.5. The focus will be placed on circular domains.

Things to Come

As noted above, the partial differential operator

$$L = A(x,y)\frac{\partial^2}{\partial x^2} + 2B(x,y)\frac{\partial^2}{\partial x\,\partial y} + C(x,y)\frac{\partial^2}{\partial y^2} + D(x,y)\frac{\partial}{\partial x} + E(x,y)\frac{\partial}{\partial y} + F(x,y)$$

is said to be elliptic when $A(x,y)C(x,y) - [B(x,y)]^2 > 0$ for all $(x,y) \in R$. When $A(x,y)C(x,y) - [B(x,y)]^2 = 0$ for all $(x,y) \in R$, the partial differential equation is said to be parabolic. Such problems are often time dependent and frequently arise when modeling diffusive processes. When $A(x,y)C(x,y) - [B(x,y)]^2 < 0$, the partial differential equation is said to be hyperbolic. These problems arise in the study of wavelike phenomena. The numerical solution of parabolic equations will be treated in Chapter 10 and the numerical solution of hyperbolic equations in Chapter 11.

9.1 THE POISSON EQUATION ON A RECTANGULAR DOMAIN, I: DIRICHLET BOUNDARY CONDITIONS

We begin our study of the numerical solution of elliptic partial differential equations using the finite difference method with an investigation of the Poisson equation

$$\frac{\partial^2 u}{\partial x^2} + \frac{\partial^2 u}{\partial y^2} = f(x,y)$$

on the rectangular domain

$$R = \{\,(x,y)|\, a < x < b, c < y < d\}$$

subject to the Dirichlet boundary conditions $u(x,y) = g(x,y)$ for all $(x,y) \in \partial R$. Emphasis will be placed upon the derivation of the discrete analogue to the Poisson

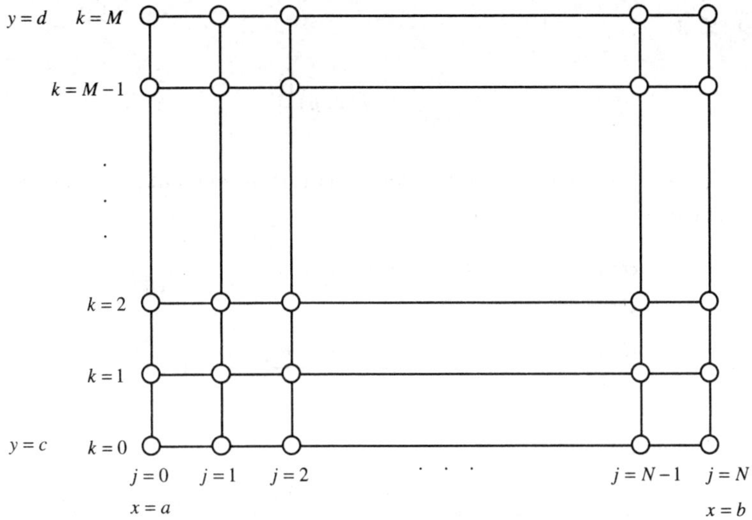

Figure 9.4 Computational grid for rectangular domain.

equation and the organization of the discrete equations into matrix form. Non-Dirichlet boundary conditions will be treated in Section 9.2.

Discrete Analogue to the Poisson Equation

The first step in determining the finite difference approximation to the solution of the Poisson problem is to introduce a computational grid. Divide the intervals $[a, b]$ and $[c, d]$ into N and M equal-sized subintervals, respectively, where N and M are positive integers. For convenience, we will assume that

$$\frac{b-a}{N} = \frac{d-c}{M} = h.$$

Note that this last condition imposes a restriction on the domain. In particular, the aspect ratio of the rectangle (i.e., the ratio of the height to the length) must be a rational number. We will remove this restriction in the exercises.

With the number of subintervals in each coordinate direction selected, the computational grid then consists of those points (x_j, y_k), where $x_j = a + jh$ and $y_k = c + kh$ for $j = 0, 1, 2, \ldots, N$ and $k = 0, 1, 2, \ldots, M$ (see Figure 9.4). To simplify the expressions for the discrete equations, we will adopt our usual subscript notation for function values:

$$u_{j,k} \equiv u(x_j, y_k), \quad f_{j,k} \equiv f(x_j, y_k), \quad g_{j,k} \equiv g(x_j, y_k).$$

With the computational grid defined, next evaluate the governing partial differential equation at each interior point (i.e., those for which $j = 1, 2, 3, \ldots, N-1$

and $k = 1, 2, 3, \ldots, M-1$)

$$\left[\frac{\partial^2 u}{\partial x^2} + \frac{\partial^2 u}{\partial y^2} = f(x,y)\right]\Bigg|_{(x_j,y_k)}$$

and substitute the following second-order central difference approximations for the partial derivatives:

$$\left.\frac{\partial^2 u}{\partial x^2}\right|_{(x_j,y_k)} = \frac{u_{j-1,k} - 2u_{j,k} + u_{j+1,k}}{h^2} + O(h^2)$$

$$\left.\frac{\partial^2 u}{\partial y^2}\right|_{(x_j,y_k)} = \frac{u_{j,k-1} - 2u_{j,k} + u_{j,k+1}}{h^2} + O(h^2).$$

This yields the equation

$$\frac{u_{j-1,k} - 2u_{j,k} + u_{j+1,k}}{h^2} + O(h^2) + \frac{u_{j,k-1} - 2u_{j,k} + u_{j,k+1}}{h^2} + O(h^2) = f_{j,k}.$$

Now, drop the truncation error terms and replace each $u_{j,k}$ (value of the exact solution) by $w_{j,k}$ (approximate value). Finally, multiply through by $-h^2$ and collect like terms to obtain

$$-w_{j-1,k} - w_{j+1,k} - w_{j,k-1} - w_{j,k+1} + 4w_{j,k} = -h^2 f_{j,k}. \tag{1}$$

For $j = 1$, $j = N-1$, $k = 1$ or $k = M-1$, at least one of the first four terms in equation (1) will be associated with a boundary grid point. Since Dirichlet boundary conditions have been specified, we can set these w values equal to the value of the function $g(x,y)$ at the corresponding grid point. This provides us with the set of boundary condition equations

$$w_{j,k} = g_{j,k} \quad (j = 0, j = N, k = 0, k = M). \tag{2}$$

Combining equation (1) for each interior grid point with equation (2) for each boundary grid point produces the discrete analogue of the Poisson problem subject to Dirichlet boundary conditions:

$$\begin{aligned} -w_{j-1,k} - w_{j+1,k} - w_{j,k-1} - w_{j,k+1} + 4w_{j,k} &= -h^2 f_{j,k} \\ &\quad \left(\begin{array}{c} 1 \le j \le N-1 \\ 1 \le k \le M-1 \end{array}\right) \\ w_{j,k} = g_{j,k} \quad (j = 0, j = N, k = 0, k = M) \end{aligned}$$

Since we arrived at this finite difference approximation by dropping second-order truncation error terms, we expect that this numerical method will be second order. We will verify this fact experimentally in both the worked examples and the exercises.

As in the case of one-dimensional boundary value problems, it is useful to think of equation (1) as a *computational template* which is to be applied at each

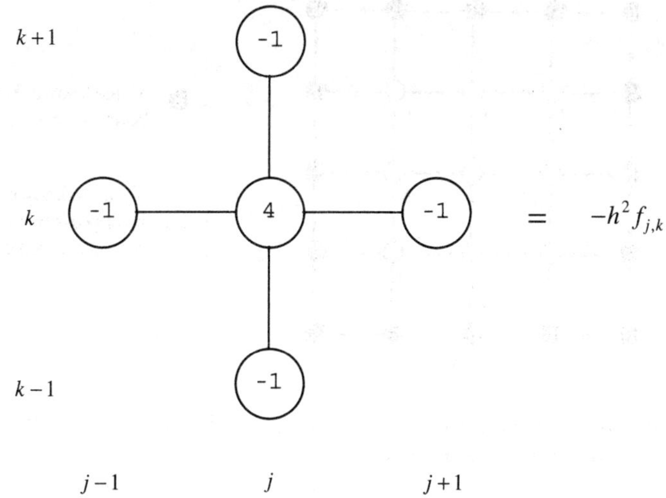

Figure 9.5 Five-point star approximation to the Laplacian operator.

grid point where the value of the approximate solution is unknown. Given the two-dimensional geometry of the problem, it is perhaps best to visualize this template as in Figure 9.5. Since the template uses values of the approximate solution from five neighboring points within the grid, it is often referred to as the five-point star approximation to the Laplacian operator

$$L = \frac{\partial^2}{\partial x^2} + \frac{\partial^2}{\partial y^2}.$$

Organizing the Equations

The system of algebraic equations embodied by equations (1) and (2) is linear in the unknowns; hence, the system can be written in matrix form. What is the structure of the coefficient matrix for the system? The answer to this question depends heavily upon the numbering scheme used to order the unknowns.

One of the most common numbering schemes is *lexicographic ordering.* Here, we start at the bottom left corner of the computational grid and number the unknowns consecutively from left to right and from bottom to top. For example, suppose the problem domain is a square and we take $N = M = 4$, as shown in Figure 9.6. The number written next to the location of each unknown is the lexicographic number associated with that unknown.

Placing the five-point star "over" the first unknown produces the equation

$$-w_{1,0} - w_{0,1} + 4w_{1,1} - w_{2,1} - w_{1,2} = -h^2 f_{1,1}.$$

The first two terms, $w_{1,0}$ and $w_{0,1}$, are associated with boundary nodes, so these values are given by the boundary conditions. Substituting $w_{1,0} = g_{1,0}$ and $w_{0,1} =$

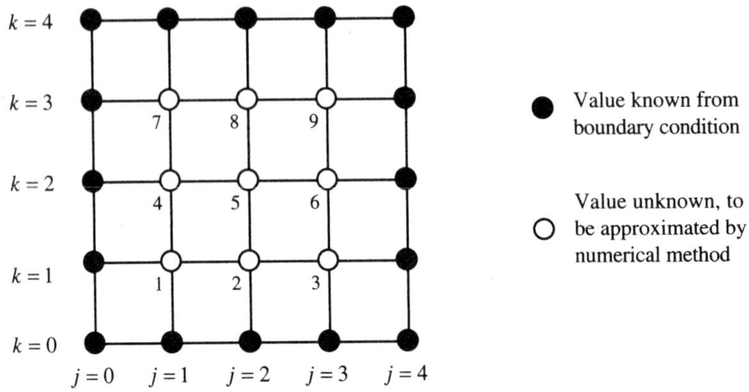

Figure 9.6 5×5 grid placed over a square domain. The unknowns are numbered in lexicographic order.

$g_{0,1}$ and rearranging so that only terms involving an unknown appear on the left-hand side, we arrive at

$$4w_{1,1} - w_{2,1} - w_{1,2} = -h^2 f_{1,1} + g_{1,0} + g_{0,1}.$$

Moving the five-point star to the second unknown, we obtain

$$-w_{2,0} - w_{1,1} + 4w_{2,1} - w_{3,1} - w_{2,2} = -h^2 f_{2,1}$$

or, since $w_{2,0} = g_{2,0}$ from the boundary conditions,

$$-w_{1,1} + 4w_{2,1} - w_{3,1} - w_{2,2} = -h^2 f_{2,1} + g_{2,0}.$$

Continuing to move the five-point star from grid point to grid point, we develop one equation at a time, until, after placing the template over the last unknown, we arrive at the system of nine equations shown in Figure 9.7. Notice how the coefficient matrix is organized into blocks, with each block associated with the unknowns along one row of the computational grid. Overall, we say that A is a block tridiagonal matrix since the non-zero entries are clustered into the blocks along the main diagonal and in the primary sub- and superdiagonals. For this problem, we see that whether we move across or down the coefficient matrix, there are three blocks. This comes about because the computational grid (recall Figure 9.6) has three rows of unknowns. Further, each block is square with side dimension equal to 3, the number of unknowns along each row. The right-hand side vector contains $-h^2$ times the values of the function f, the nonhomogeneous term from the partial differential equation, plus the values of the boundary value function, g, when the grid point is adjacent to the boundary.

$$A = \left[\begin{array}{c|c|c} D & -I & 0 \\ \hline -I & D & -I \\ \hline 0 & -I & D \end{array}\right] \quad \text{with} \quad D = \begin{bmatrix} 4 & -1 & 0 \\ -1 & 4 & -1 \\ 0 & -1 & 4 \end{bmatrix}$$

$$
\left.
\begin{array}{rcl}
4w_{1,1} - w_{2,1} - w_{1,2} & = & -h^2 f_{1,1} + g_{1,0} + g_{0,1} \\
-w_{1,1} + 4w_{2,1} - w_{3,1} - w_{2,2} & = & -h^2 f_{2,1} + g_{2,0} \\
-w_{2,1} + 4w_{3,1} - w_{3,2} & = & -h^2 f_{3,1} + g_{3,0} + g_{4,1} \\
\hline
-w_{1,1} + 4w_{1,2} - w_{2,2} - w_{1,3} & = & -h^2 f_{1,2} + g_{0,2} \\
-w_{2,1} - w_{1,2} + 4w_{2,2} - w_{3,2} - w_{2,3} & = & -h^2 f_{2,2} \\
-w_{3,1} - w_{2,2} + 4w_{3,2} - w_{3,3} & = & -h^2 f_{3,2} + g_{4,2} \\
\hline
-w_{1,2} + 4w_{1,3} - w_{2,3} & = & -h^2 f_{1,3} + g_{0,3} + g_{1,4} \\
-w_{2,2} - w_{1,3} + 4w_{2,3} - w_{3,3} & = & -h^2 f_{2,3} + g_{2,4} \\
-w_{3,2} - w_{2,3} + 4w_{3,3} & = & -h^2 f_{3,3} + g_{4,3} + g_{3,4}
\end{array}
\right\} \quad A\mathbf{w} = \mathbf{b}
$$

Figure 9.7 Organization of the discrete Poisson equations, with Dirichlet boundary conditions, for a 5 × 5 grid using lexicographic ordering of the unknowns.

$$\mathbf{w} = \left[\begin{array}{ccc|ccc|ccc} w_{1,1} & w_{2,1} & w_{3,1} & w_{1,2} & w_{2,2} & w_{3,2} & w_{1,3} & w_{2,3} & w_{3,3} \end{array}\right]^T$$

For a more general grid, containing N subdivisions along the horizontal direction and M subdivisions along the vertical direction, the coefficient matrix will have dimension $(N-1)(M-1) \times (N-1)(M-1)$. The structure of the matrix remains block tridiagonal. In particular,

$$A = \begin{bmatrix} D & -I & & & & \\ -I & D & -I & & & \\ & \cdot & \cdot & \cdot & & \\ & & \cdot & \cdot & \cdot & \\ & & & \cdot & \cdot & \cdot \\ & & & -I & D & -I \\ & & & & -I & D \end{bmatrix} \tag{3}$$

with

$$D = \begin{bmatrix} 4 & -1 & & & & \\ -1 & 4 & -1 & & & \\ & \cdot & \cdot & \cdot & & \\ & & \cdot & \cdot & \cdot & \\ & & & \cdot & \cdot & \cdot \\ & & & -1 & 4 & -1 \\ & & & & -1 & 4 \end{bmatrix}.$$

There are $M-1$ blocks both across and down the coefficient matrix, and each submatrix, D and I, has dimension $(N-1) \times (N-1)$.

Solving the Equations

Let's start by establishing that the coefficient matrix, A, for the finite difference equations, as given by equation (3), is nonsingular. On each row of A, the diagonal element is equal to 4 and there are at most four additional non-zero elements, each with magnitude equal to 1. Consequently, A is diagonally dominant. Furthermore, certain rows of A, specifically those for which at least one point of the computational template falls along the boundary of the domain, are strictly diagonally dominant. These conditions, however, are not sufficient to guarantee that a matrix is nonsingular (see Exercise 2).

Fortunately, A has another important property—the matrix is *irreducible.*

Definition. The $n \times n$ matrix A is REDUCIBLE if there exists a permutation matrix P such that

$$PAP^T = \begin{bmatrix} A_{1,1} & A_{1,2} \\ 0 & A_{2,2} \end{bmatrix},$$

where $A_{1,1}$ is an $r \times r$ matrix, $A_{1,2}$ is an $r \times (n-r)$ matrix and $A_{2,2}$ is an $(n-r) \times (n-r)$ matrix for some positive integer r. Note that the submatrices along the diagonal are square. A matrix is IRREDUCIBLE if no such permutation matrix exists.

Essentially, if the coefficient matrix of a system of equations is reducible, the solution may be found by breaking the original system into two smaller systems. When the coefficient matrix is irreducible, no such reduction in problem size is possible. It can be shown that a matrix which is irreducible, diagonally dominant and strictly diagonally dominant in at least one row is nonsingular (see Horn and Johnson [1]). Thus, A is nonsingular, and for any nonhomogeneous term f and any Dirichlet boundary data g, the finite difference equations for the Poisson problem on a rectangular domain have a unique solution.

Having established that the finite difference equations have a unique solution, how do we go about computing that solution? When N and M are both small, it is feasible in terms of both storage and computational cost to solve the finite difference equations using direct methods. For example, we can show that A is not only nonsingular, it is also symmetric positive definite (Exercise 1). Therefore, we might compute a Cholesky factorization or an LDL^T factorization of A. Alternatively, we can exploit the block tridiagonal structure of the matrix. See Varga [2] and Lindzen and Kuo [3] for specific algorithms.

In practice, however, neither N nor M will be small. When N and M are even of moderate size, it is recommended that the finite difference equations be solved using iterative methods. We will explore this issue in detail in Sections 9.3 and 9.4.

A Worked Example

EXAMPLE 9.1 A Sample Problem

To demonstrate the application of the finite difference method to the Poisson problem, consider

$$\frac{\partial^2 u}{\partial x^2} + \frac{\partial^2 u}{\partial y^2} = -[2 + \pi^2 x(1-x)]\cos(\pi y)$$

on the domain $R = \{(x, y) | 0 < x < 1, 0 < y < 1/2\}$. The Dirichlet boundary conditions are given by

$$\begin{aligned} u(0, y) &= u(1, y) = 0 \\ u(x, 0) = x(1-x), &\quad u(x, 1/2) = 0. \end{aligned}$$

Note that $f(x, y) = -[2 + \pi^2 x(1-x)]\cos(\pi y)$ and g is a piecewise function defined for $(x, y) \in \partial R$ by

$$g(x, y) = \begin{cases} 0, & x = 0 \\ 0, & x = 1 \\ x(1-x), & y = 0 \\ 0, & y = 1/2 \end{cases}.$$

Since the aspect ratio (the ratio of the height to the length) of R is one-half, we must choose $N = 2M$ to achieve the same grid spacing in both directions. Let's start with $M = 2$ and $N = 4$. This implies that the mesh spacing will be $h = 1/4$.

The computational grid, with the unknowns numbered in lexicographic order, is shown below.

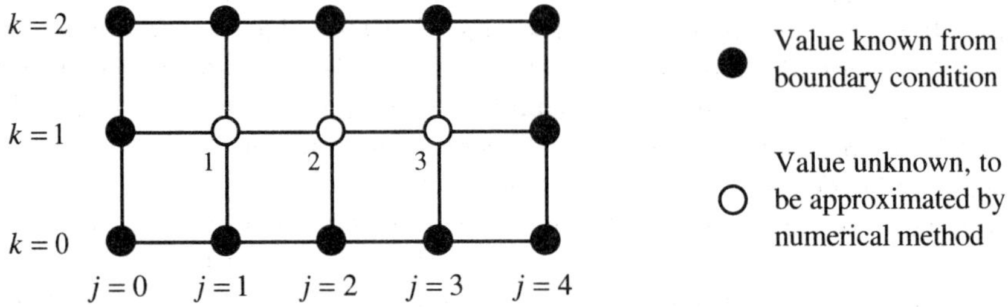

Successively placing the five-point star at the first, second, and third unknowns produces the system of equations

$$\begin{array}{rcrcrcl} 4w_{1,1} & - & w_{2,1} & & & = & -h^2 f_{1,1} + g_{1,0} + g_{0,1} + g_{1,2} \\ -w_{1,1} & + & 4w_{2,1} & - & w_{3,1} & = & -h^2 f_{2,1} + g_{2,0} + g_{2,2} \\ & & -w_{2,1} & + & 4w_{3,1} & = & -h^2 f_{3,1} + g_{3,0} + g_{4,1} + g_{3,2}, \end{array}$$

where

$$\begin{aligned} f_{1,1} &= -[2 + \pi^2 \cdot 0.25(0.75)] \cos(0.25\pi) \\ f_{2,1} &= -[2 + \pi^2 \cdot 0.5(0.5)] \cos(0.25\pi) \\ f_{3,1} &= -[2 + \pi^2 \cdot 0.75(0.25)] \cos(0.25\pi) \end{aligned}$$

and

$$\begin{aligned} g_{1,0} = 0.25(0.75), &\quad g_{0,1} = g_{1,2} = 0 \\ g_{2,0} = 0.5(0.5), &\quad g_{2,2} = 0 \\ g_{3,0} = 0.75(0.25), &\quad g_{4,1} = g_{3,2} = 0. \end{aligned}$$

Evaluating the entries of the right-hand side vector, the system can be written in matrix form as

$$\begin{bmatrix} 4 & -1 & 0 \\ -1 & 4 & -1 \\ 0 & -1 & 4 \end{bmatrix} \begin{bmatrix} w_{1,1} \\ w_{2,1} \\ w_{3,1} \end{bmatrix} = \begin{bmatrix} 0.357672 \\ 0.447433 \\ 0.357672 \end{bmatrix}.$$

The entries in the right-hand side vector have been displayed to six decimal places for convenience.

The solution to this system is listed in the third column of the following table. The values of the exact solution, $u(x, y) = x(1 - x)\cos(\pi y)$, at the corresponding grid points are listed in the fourth column. Notwithstanding the crudeness of the grid, the approximate solution is in error by only slightly more than 1%.

x_j	y_k	$w_{j,k}$	$u_{j,k}$	Absolute Error
0.25	0.25	0.134151	0.132583	0.001568
0.50	0.25	0.178934	0.176777	0.002157
0.75	0.25	0.134151	0.132583	0.001568

To investigate the order of convergence of the numerical method, the sample problem was solved on successively finer grids. The root mean square (rms) error,

$$\sqrt{\frac{\sum_{j=1}^{N-1}\sum_{k=1}^{M-1}(u_{j,k}-w_{j,k})^2}{(N-1)(M-1)}},$$

and the ratio of the rms error to h^2 were calculated for each grid. Results are summarized in the following table. The values in the third column of this table suggest that the approximation error is roughly a constant times h^2; that is, the approximation error is $O(h^2)$. Hence, we have numerical verification that the method is second-order accurate.

Step Size, h	rms Error	rms Error / h^2
1/4	1.786739×10^{-3}	0.02859
1/6	6.716379×10^{-4}	0.02418
1/8	3.493104×10^{-4}	0.02236
1/10	2.137212×10^{-4}	0.02137
1/12	1.441616×10^{-4}	0.02076
1/14	1.037838×10^{-4}	0.02034
1/16	7.827723×10^{-5}	0.02004

References

1. R. A. Horn and C. A. Johnson, *Matrix Analysis*, Cambridge University Press, Cambridge University, 1985.

2. R. S. Varga, *Matrix Iterative Analysis*, Prentice Hall, Englewood Cliffs, NJ, 1962.

3. R. S. Lindzen and H.-L. Kuo, "A Reliable Method for the Numerical Integration of a Large Class of Ordinary and Partial Differential Equations," *Monthly Weather Review*, **97**, 732–734, 1969.

EXERCISES

1. Let A be the matrix given in equation (3).

(a) Using the Gerschgorin Circle Theorem, show that $0 \le \lambda \le 8$, where λ is any eigenvalue of A.

(b) Use the fact that A is nonsingular to conclude that all of the eigenvalues of A are positive.

(c) Use Exercise 20 of Section 4.1 to conclude that A is symmetric positive definite.

2. Consider the matrix

$$M = \begin{bmatrix} 5 & 2 & 1 & 1 \\ 0 & -1 & 0 & 1 \\ 1 & 1 & 4 & 1 \\ 0 & 2 & 0 & -2 \end{bmatrix}.$$

(a) Show that M is diagonally dominant with at least one row that is strictly diagonally dominant.
(b) Show that M is singular.
(c) Show that M is reducible.

3. (a) Let $A(x)$ be any antiderivative of $a(x)$. Show that the change of variable $w(x,y) = e^{A(x)}u(x,y)$ transforms the equation

$$\frac{\partial^2 u}{\partial x^2} + \frac{\partial^2 u}{\partial y^2} + a(x)\frac{\partial u}{\partial x} = f(x,y) \quad \text{into} \quad \frac{\partial^2 w}{\partial x^2} + \frac{\partial^2 w}{\partial y^2} = e^{A(x)}f(x,y).$$

(b) Let $B(y)$ be any antiderivative of $b(y)$. Show that the change of variable $w(x,y) = e^{B(y)}u(x,y)$ transforms the equation

$$\frac{\partial^2 u}{\partial x^2} + \frac{\partial^2 u}{\partial y^2} + b(y)\frac{\partial u}{\partial y} = f(x,y) \quad \text{into} \quad \frac{\partial^2 w}{\partial x^2} + \frac{\partial^2 w}{\partial y^2} = e^{B(y)}f(x,y).$$

(c) What change of variable will transform

$$\frac{\partial^2 u}{\partial x^2} + \frac{\partial^2 u}{\partial y^2} + a(x)\frac{\partial u}{\partial x} + b(y)\frac{\partial u}{\partial y} = f(x,y)$$

into

$$\frac{\partial^2 w}{\partial x^2} + \frac{\partial^2 w}{\partial y^2} = \tilde{f}(x,y)?$$

How is $\tilde{f}(x,y)$ related to $f(x,y)$?

In Exercises 4–6:

(a) Using second-order central differences to approximate all partial derivatives, construct the computational template corresponding to the indicated elliptic partial differential equation. Assume uniform spacing, h, in both coordinate directions.
(b) Assuming a rectangular domain and Dirichlet boundary conditions, specify sufficient conditions for the finite difference equations to have a unique solution.

4. $\dfrac{\partial^2 u}{\partial x^2} + \dfrac{\partial^2 u}{\partial y^2} + a(x,y)u(x,y) = f(x,y)$

5. $\dfrac{\partial^2 u}{\partial x^2} + \dfrac{\partial^2 u}{\partial y^2} + a(x,y)\dfrac{\partial u}{\partial x} + b(x,y)\dfrac{\partial u}{\partial y} = f(x,y)$

6.

$$c_1(x,y)\frac{\partial^2 u}{\partial x^2} + 2c_2(x,y)\frac{\partial^2 u}{\partial x \partial y} + c_3(x,y)\frac{\partial^2 u}{\partial y^2} + c_4(x,y)\frac{\partial u}{\partial x} + c_5(x,y)\frac{\partial u}{\partial y} + c_6(x,y)u = f(x,y)$$

7. Consider the Poisson problem

$$\frac{\partial^2 u}{\partial x^2} + \frac{\partial^2 u}{\partial y^2} = 0 \text{ on } R = \{(x,y) \,|\, 0 < x < 1, 0 < y < 1\}$$

$$u(x,0) = 0, \quad u(x,1) = \frac{1}{(1+x)^2+1}, \quad u(0,y) = \frac{y}{1+y^2}, \quad u(1,y) = \frac{y}{4+y^2},$$

whose exact solution is

$$u(x,y) = \frac{y}{(1+x)^2+y^2}.$$

(a) Taking $N = M = 4$, set up and solve the corresponding system of finite difference equations.

(b) Numerically verify the second-order accuracy of the numerical method.

8. Repeat Exercise 7 (but use $N = M = 3$ in part (a)) for the Poisson problem

$$\frac{\partial^2 u}{\partial x^2} + \frac{\partial^2 u}{\partial y^2} = x^2 + y^2 \text{ on } R = \{(x,y) \,|\, 0 < x < 1, 0 < y < 1\}$$

$$u(x,0) = 0, \quad u(x,1) = \frac{1}{2}x^2, \quad u(0,y) = \sin(\pi y), \quad u(1,y) = e^{\pi}\sin(\pi y) + \frac{1}{2}y^2.$$

The exact solution for this problem is $u(x,y) = e^{\pi x}\sin(\pi y) + \frac{1}{2}(xy)^2$.

9. Repeat Exercise 7 for the Poisson problem

$$\frac{\partial^2 u}{\partial x^2} + \frac{\partial^2 u}{\partial y^2} = -52\cos(4x+6y) \text{ on } R = \{(x,y) \,|\, 0 < x < 1, 0 < y < 1\}$$

$$u(x,0) = \cos(4x), \quad u(x,1) = \cos(4x+6),$$
$$u(0,y) = \cos(6y), \quad u(1,y) = \cos(6y+4).$$

The exact solution for this problem is $u(x,y) = \cos(4x+6y)$.

10. Repeat Exercise 7 for the Poisson problem

$$\frac{\partial^2 u}{\partial x^2} + \frac{\partial^2 u}{\partial y^2} = 0 \text{ on } R = \{(x,y) \,|\, 1 < x < 2, 0 < y < 1\}$$

$$u(x,0) = 2\ln x, \quad u(x,1) = \ln(x^2+1), \quad u(1,y) = \ln(y^2+1), \quad u(2,y) = \ln(y^2+4).$$

The exact solution for this problem is $u(x,y) = \ln(x^2+y^2)$.

11. **(a)** Over a computational grid with a spacing of Δx in the x-direction and Δy in the y-direction (with Δx not necessarily equal to Δy), construct the computational template corresponding to the Poisson problem. Use second-order central differences to approximate all partial derivatives.

(b) Assuming a rectangular domain and Dirichlet boundary conditions, construct the coefficient matrix for the system of finite difference equations corresponding to the template from part (a).

(c) Under what conditions is the coefficient matrix from part (b) guaranteed to be nonsingular?

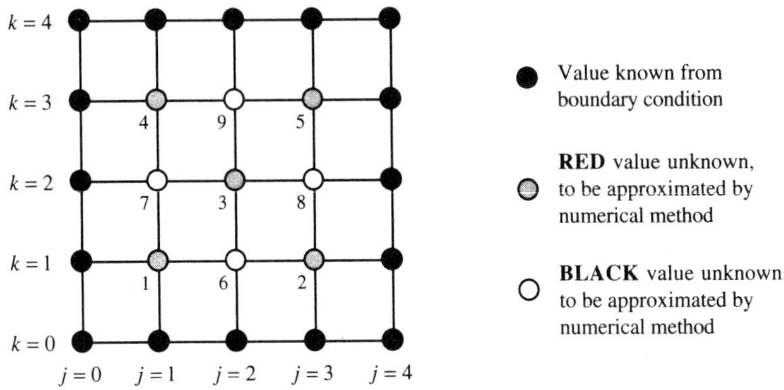

Figure 9.8 Red-black ordering of unknowns on a 5×5 grid placed over a square domain.

12. Using the results of Exercise 11,

(a) Repeat Exercise 7(a) with $N = 5$ and $M = 3$.

(b) Repeat Exercise 8(a) with $N = 4$ and $M = 5$.

(c) Repeat Exercise 9(a) with $N = 3$ and $M = 5$.

(d) Repeat Exercise 10(a) with $N = 5$ and $M = 4$.

13. A second popular numbering scheme for the unknowns is the *red-black ordering.* In this scheme, the unknowns are divided into two categories: red unknowns and black unknowns. An unknown, $w_{j,k}$, is considered red whenever $j + k$ is even and is considered black whenever $j + k$ is odd. The red unknowns are numbered first, in lexicographic order, followed by the black unknowns, also in lexicographic order. The red-black ordering of the unknowns from Figure 9.6 is shown in Figure 9.8.

(a) Write out the system of equations corresponding to the computational grid of Figure 9.8.

(b) What type of structure can you identify in the coefficient matrix? Generalize this structure to a grid with N subdivisions in the horizontal direction and M subdivisions in the vertical direction.

9.2 THE POISSON EQUATION ON A RECTANGULAR DOMAIN, II: NON-DIRICHLET BOUNDARY CONDITIONS

In this section we will examine how to handle non-Dirichlet boundary conditions for the Poisson equation on rectangular domains. Unlike one-dimensional boundary value problems, where the boundary conditions affect only the first and last row of the coefficient matrix, when working in two dimensions, the boundary conditions can affect the coefficient matrix in a variety of ways. There are too many combinations of different types of boundary conditions specified over different portions of the boundary to cover them all. Instead, we will discuss two specific cases in detail. Other cases will be treated in the exercises.

Neumann Boundary Condition along Bottom Edge

Suppose we are attempting to solve the Poisson equation on a rectangular domain with the Dirichlet boundary condition $u(x,y) = g(x,y)$ specified along the sides and the top of the rectangle, but with the Neumann condition $\partial u/\partial n = \alpha(x)$ along the bottom. Along the bottom of the rectangle the outward normal vector points in the negative y direction, so $\partial/\partial n = -\partial/\partial y$. Hence, in terms of the coordinate directions, the boundary condition along the bottom of the rectangle reads $-\partial u/\partial y = \alpha(x)$.

The Neumann boundary condition along the bottom of the rectangle produces an extra row of unknowns in the computational grid. To handle these unknowns, we have exactly the same three options open to us as when we were working with one-dimensional boundary value problems. First, we could derive a new template using a second-order forward difference to approximate $\partial^2 u/\partial y^2$ for application along the lower boundary. This, however, would destroy the block tridiagonal structure of the coefficient matrix. To maintain the matrix structure, we could use a first-order difference formula, but this would lower the overall order of approximation. This leaves us with the third alternative: the use of fictitious nodes. This approach maintains both the structure of the coefficient matrix (block tridiagonal, each block tridiagonal) and the order of approximation.

What effect will this have on the structure of the coefficient matrix? Let's consider a 5×5 grid placed over a square domain—see Figure 9.9. With the Neumann condition along the bottom boundary, there are four rows of unknowns with three unknowns per row (compare this with the grid shown in Figure 9.6 of Section 9.1). Therefore, the coefficient matrix will be 12×12, organized as four blocks across and down, each block being 3×3.

Taking the standard five-point star template and moving it along the first row of unknowns produces the set of equations

$$\begin{array}{cccccccc} -w_{F1} & +4w_{1,0} & -w_{2,0} & & -w_{1,1} & = & -h^2 f_{1,0} + g_{0,0} \\ -w_{F2} & -w_{1,0} & +4w_{2,0} & -w_{3,0} & -w_{2,1} & = & -h^2 f_{2,0} \\ -w_{F3} & & -w_{2,0} & +4w_{3,0} & -w_{3,1} & = & -h^2 f_{3,0} + g_{4,0} \end{array}. \qquad (1)$$

To eliminate the fictitious nodes from these equations, we turn to the Neumann boundary condition. Evaluating the boundary condition at $(x_1, 0)$ yields

$$-\left.\frac{\partial u}{\partial y}\right|_{(x_1,0)} = \alpha(x_1).$$

Replacing the derivative with a second-order central difference formula, we find

$$-\frac{w_{1,1} - w_{F1}}{2h} = \alpha(x_1) \quad \Rightarrow \quad w_{F1} = w_{1,1} + 2h\alpha(x_1).$$

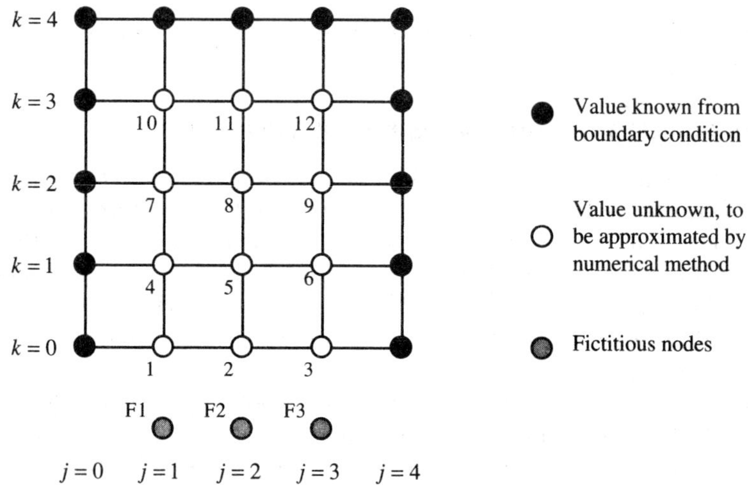

Figure 9.9 Computational grid for Poisson equation on a square domain with a Neumann boundary condition along the bottom side and Dirichlet boundary conditions on the other three sides. Note the placement of the fictitious nodes.

Repeating the process at the boundary points $(x_2, 0)$ and $(x_3, 0)$ gives

$$\begin{aligned} w_{F2} &= w_{2,1} + 2h\alpha(x_2) \\ w_{F3} &= w_{3,1} + 2h\alpha(x_3), \end{aligned}$$

and substituting these last three equations into (1) yields

$$\begin{array}{rrrrcl} +4w_{1,0} & -w_{2,0} & & -2w_{1,1} & = & -h^2 f_{1,0} + g_{0,0} + 2h\alpha(x_1) \\ -w_{1,0} & +4w_{2,0} & -w_{3,0} & -2w_{2,1} & = & -h^2 f_{2,0} + 2h\alpha(x_2) \\ & -w_{2,0} & +4w_{3,0} & -2w_{3,1} & = & -h^2 f_{3,0} + g_{4,0} + 2h\alpha(x_3) \end{array}. \tag{2}$$

Let

$$D = \begin{bmatrix} 4 & -1 & 0 \\ -1 & 4 & -1 \\ 0 & -1 & 4 \end{bmatrix},$$

let I denote the 3×3 identity matrix, and introduce the notation

$$\mathbf{w}_{Rj} = \begin{bmatrix} w_{1,j} & w_{2,j} & w_{3,j} \end{bmatrix}^T$$

for the unknowns along the jth row of the computational grid. Equation (2) can then be written in the form

$$D\mathbf{w}_{R0} - 2I\mathbf{w}_{R1} = \begin{bmatrix} -h^2 f_{1,0} + g_{0,0} + 2h\alpha(x_1) \\ -h^2 f_{2,0} + 2h\alpha(x_2) \\ -h^2 f_{3,0} + g_{4,0} + 2h\alpha(x_3) \end{bmatrix} = \mathbf{b}_{R0}. \tag{3}$$

Working the template across the remaining three rows of the grid, and taking into account values known from the Dirichlet boundary conditions where appropriate, generates the sets of equations:

$$-I\mathbf{w}_{R0} + D\mathbf{w}_{R1} - I\mathbf{w}_{R2} = \begin{bmatrix} -h^2 f_{1,1} + g_{0,1} \\ -h^2 f_{2,1} \\ -h^2 f_{3,1} + g_{4,1} \end{bmatrix} = \mathbf{b}_{R1} \qquad (4)$$

$$-I\mathbf{w}_{R1} + D\mathbf{w}_{R2} - I\mathbf{w}_{R3} = \begin{bmatrix} -h^2 f_{1,2} + g_{0,2} \\ -h^2 f_{2,2} \\ -h^2 f_{3,2} + g_{4,2} \end{bmatrix} = \mathbf{b}_{R2} \qquad (5)$$

$$-I\mathbf{w}_{R2} + D\mathbf{w}_{R3} = \begin{bmatrix} -h^2 f_{1,3} + g_{0,3} + g_{1,4} \\ -h^2 f_{2,3} + g_{2,4} \\ -h^2 f_{3,3} + g_{4,3} + g_{3,4} \end{bmatrix} = \mathbf{b}_{R3}. \qquad (6)$$

Combining (3), (4), (5) and (6) into a single system yields

$$\begin{bmatrix} D & -2I & & \\ -I & D & -I & \\ & -I & D & -I \\ & & -I & D \end{bmatrix} \begin{bmatrix} \mathbf{w}_{R0} \\ \mathbf{w}_{R1} \\ \mathbf{w}_{R2} \\ \mathbf{w}_{R3} \end{bmatrix} = \begin{bmatrix} \mathbf{b}_{R0} \\ \mathbf{b}_{R1} \\ \mathbf{b}_{R2} \\ \mathbf{b}_{R3} \end{bmatrix}. \qquad (7)$$

Thus, the Neumann condition along the bottom of the rectangle has resulted in the coefficient matrix having an extra block across and down and the first superdiagonal block being $-2I$ rather than $-I$. Note that the coefficient matrix in (7) is irreducible and diagonally dominant with at least one row being strictly diagonally dominant; hence (7) is guaranteed to have a unique solution.

A Worked Example

EXAMPLE 9.2 Handling a Neumann Boundary Condition

Consider the partial differential equation

$$\frac{\partial^2 u}{\partial x^2} + \frac{\partial^2 u}{\partial y^2} = -[2 + \pi^2 x(1-x)]\cos(\pi y)$$

on the domain $R = \{(x, y) | 0 < x < 1, 0 < y < 1/2\}$. The boundary conditions are given by

$$u(0, y) = u(1, y) = 0$$

$$\frac{\partial u}{\partial n}(x, 0) = 0, u(x, 1/2) = 0.$$

Note that this is the same problem that we solved in Section 9.1, but with the Dirichlet condition along the bottom of the domain replaced by a Neumann condition.

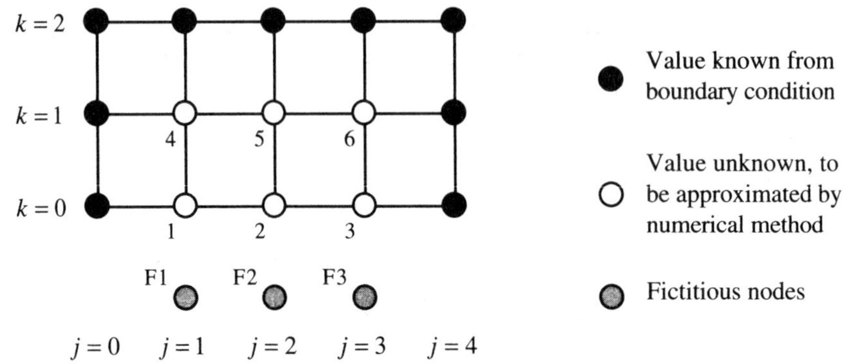

The first grid we will use for this problem is shown above. We have taken $N = 4$ and $M = 2$. The mesh spacing is $h = 1/4$. Note there are six unknowns, organized by rows into two groups of three.
Comparing the present problem with the model problem, we find

$$f(x, y) = -[2 + \pi^2 x(1 - x)] \cos(\pi y)$$
$$g(x, y) = 0 \quad \text{and} \quad \alpha(x) = 0.$$

The blocks which compose the right-hand-side vector are therefore

$$\mathbf{b}_{R0} = \begin{bmatrix} -h^2 f_{1,0} + g_{0,0} + 2h\alpha(x_1) \\ -h^2 f_{2,0} + 2h\alpha(x_2) \\ -h^2 f_{3,0} + g_{4,0} + 2h\alpha(x_3) \end{bmatrix} = \begin{bmatrix} \frac{1}{16}[2 + \pi^2(0.25)(0.75)] + 0 + 0 \\ \frac{1}{16}[2 + \pi^2(0.5)(0.5)] + 0 \\ \frac{1}{16}[2 + \pi^2(0.75)(0.25)] + 0 + 0 \end{bmatrix}$$

and

$$\mathbf{b}_{R1} = \begin{bmatrix} -h^2 f_{1,1} + g_{0,1} + g_{1,2} \\ -h^2 f_{2,1} + g_{2,2} \\ -h^2 f_{3,1} + g_{4,1} + g_{3,2} \end{bmatrix} = \begin{bmatrix} \frac{1}{16}[2 + \pi^2(0.25)(0.75)] \cos(\frac{\pi}{4}) + 0 + 0 \\ \frac{1}{16}[2 + \pi^2(0.5)(0.5)] \cos(\frac{\pi}{4}) + 0 \\ \frac{1}{16}[2 + \pi^2(0.75)(0.25)] \cos(\frac{\pi}{4}) + 0 + 0 \end{bmatrix},$$

so the complete system of equations we arrive at for this grid is

$$\left[\begin{array}{ccc|ccc} 4 & -1 & 0 & -2 & 0 & 0 \\ -1 & 4 & -1 & 0 & -2 & 0 \\ 0 & -1 & 4 & 0 & 0 & -2 \\ \hline -1 & 0 & 0 & 4 & -1 & 0 \\ 0 & -1 & 0 & -1 & 4 & -1 \\ 0 & 0 & -1 & 0 & -1 & 4 \end{array}\right] \left[\begin{array}{c} w_{1,0} \\ w_{2,0} \\ w_{3,0} \\ \hline w_{1,1} \\ w_{2,1} \\ w_{3,1} \end{array}\right] = \left[\begin{array}{c} 0.240659 \\ 0.279213 \\ 0.240659 \\ \hline 0.170172 \\ 0.197433 \\ 0.170172 \end{array}\right].$$

The entries in the right-hand-side vector have been displayed to six decimal places for convenience.

Here is a table listing the solution to the above system and comparing this approximate solution to the exact solution, $u(x, y) = x(1 - x) \cos(\pi y)$. The errors are roughly two and a half times larger than the errors introduced when Dirichlet conditions were specified around the entire boundary, but the accuracy is still quite good considering the coarseness of the grid.

x_j	y_k	$w_{j,k}$	$u_{j,k}$	Absolute Error
0.25	0.00	0.192371	0.187500	0.004871
0.50	0.00	0.256771	0.250000	0.006771
0.75	0.00	0.192371	0.187500	0.004871
0.25	0.25	0.136027	0.132583	0.003444
0.50	0.25	0.181564	0.176777	0.004787
0.75	0.25	0.136027	0.132583	0.003444

To demonstrate that the numerical method retains its second-order rate of convergence, the test problem was solved using successively finer grids. For each grid, the rms error in the approximate solution and the ratio of the rms error to h^2 were computed. Examining these latter values, we find strong evidence that the numerical scheme is $O(h^2)$.

Step Size, h	rms Error	rms Error / h^2
1/4	4.829513×10^{-3}	0.07727
1/6	1.889341×10^{-3}	0.06802
1/8	9.983822×10^{-4}	0.06390
1/10	6.156164×10^{-4}	0.06156
1/12	4.170598×10^{-4}	0.06006
1/14	3.010464×10^{-4}	0.05901
1/16	2.274558×10^{-4}	0.05823

Neumann Condition along the Right Side and Robin Condition along the Top Edge

As a second case, suppose that the Neumann boundary condition,

$$\frac{\partial u}{\partial n} = \alpha(y),$$

is specified along the right side of the domain and the Robin boundary condition,

$$p(x)\frac{\partial u}{\partial n} + q(x)u = r(x),$$

is specified along the top edge. Dirichlet boundary conditions are provided along the other two sides of the rectangle. To examine the effect of these boundary conditions on the finite difference equations, we will once again consider a 5×5 grid. Values along the left side and the bottom edge of the grid are given by the Dirichlet boundary conditions. With a non-Dirichlet boundary condition along the top of the grid, there will be four rows of unknowns. The non-Dirichlet condition along the right side dictates that each of these rows will contain four unknowns.

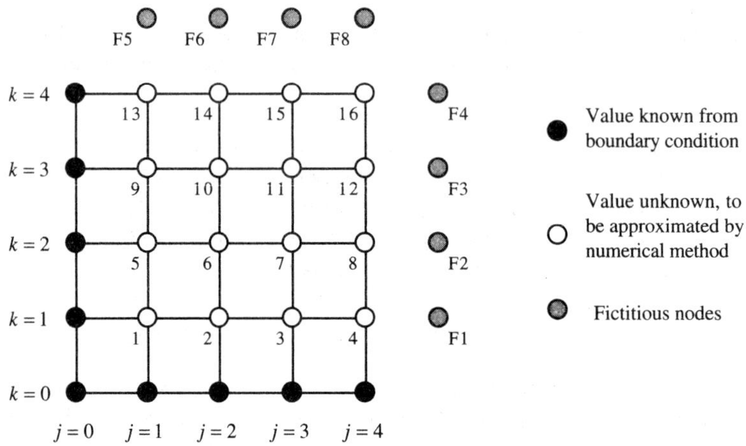

Figure 9.10 Computational grid for Poisson equation on a square domain with a Neumann boundary condition along the right side, a Robin boundary condition along the top, and Dirichlet boundary conditions on the remaining two sides. Note the placement of the fictitious nodes.

There will also be a total of eight fictitious nodes needed, four across the top and four down the right side. The complete computational grid, with fictitious nodes and lexicographic ordering of the unknowns, is shown in Figure 9.10.

Moving the five-point star across the first row of unknowns produces the set of equations

$$\begin{array}{ccccccccc} 4w_{1,1} & -w_{2,1} & & & -w_{1,2} & & = & -h^2 f_{1,1} + g_{1,0} + g_{0,1} \\ -w_{1,1} & +4w_{2,1} & -w_{3,1} & & -w_{2,2} & & = & -h^2 f_{2,1} + g_{2,0} \\ & -w_{2,1} & +4w_{3,1} & -w_{4,1} & -w_{3,2} & & = & -h^2 f_{3,1} + g_{3,0} \\ & & -w_{3,1} & +4w_{4,1} & -w_{4,2} & -w_{F1} & = & -h^2 f_{4,1} + g_{4,0} \end{array} \quad (8)$$

The value w_{F1} must be eliminated from the last equation in (8). This is accomplished using the Neumann boundary condition. Along the right side of the domain, the outward normal vector points in the positive x direction; hence,

$$\frac{\partial u}{\partial n} = \alpha(y) \quad \text{is equivalent to} \quad \frac{\partial u}{\partial x} = \alpha(y).$$

Using the second-order central difference formula for $\partial u/\partial x$, it follows that

$$w_{F1} = w_{3,1} + 2h\alpha(y_1).$$

With this expression for w_{F1}, (8) can be written in the form

$$D'\mathbf{w}_{R1} - I\mathbf{w}_{R2} = \begin{bmatrix} -h^2 f_{1,1} + g_{1,0} + g_{0,1} \\ -h^2 f_{2,1} + g_{2,0} \\ -h^2 f_{3,1} + g_{3,0} \\ -h^2 f_{4,1} + g_{4,0} + 2h\alpha(y_1) \end{bmatrix} = \mathbf{b}_{R1}, \quad (9)$$

where

$$D' = \begin{bmatrix} 4 & -1 & 0 & 0 \\ -1 & 4 & -1 & 0 \\ 0 & -1 & 4 & -1 \\ 0 & 0 & -2 & 4 \end{bmatrix}.$$

Proceeding in a similar manner, the sets of equations derived from the second, third, and fourth rows of unknowns are

$$-I\mathbf{w}_{R1} + D'\mathbf{w}_{R2} - I\mathbf{w}_{R3} = \begin{bmatrix} -h^2 f_{1,2} + g_{0,2} \\ -h^2 f_{2,2} \\ -h^2 f_{3,2} \\ -h^2 f_{4,2} + 2h\alpha(y_2) \end{bmatrix} = \mathbf{b}_{R2}, \tag{10}$$

$$-I\mathbf{w}_{R2} + D'\mathbf{w}_{R3} - I\mathbf{w}_{R4} = \begin{bmatrix} -h^2 f_{1,3} + g_{0,3} \\ -h^2 f_{2,3} \\ -h^2 f_{3,3} \\ -h^2 f_{4,3} + 2h\alpha(y_3) \end{bmatrix} = \mathbf{b}_{R3}, \tag{11}$$

and

$$-I\mathbf{w}_{R3} + D'\mathbf{w}_{R4} - I\mathbf{w}_{RF} = \begin{bmatrix} -h^2 f_{1,4} + g_{0,4} \\ -h^2 f_{2,4} \\ -h^2 f_{3,4} \\ -h^2 f_{4,4} + 2h\alpha(y_4) \end{bmatrix}. \tag{12}$$

In equation (12), $\mathbf{w}_{RF}$ is a column vector that contains the values at the fictitious nodes along the top of the grid. To eliminate this vector, we invoke the Robin boundary condition, which, along the top edge of the rectangle, is equivalent to

$$p(x)\frac{\partial u}{\partial x} + q(x)u = r(x).$$

After evaluating this equation at (x_1, y_4), replacing the partial derivative by its second-order central difference approximation, replacing $u_{1,4}$ by $w_{1,4}$ and solving for w_{F5}, we find

$$w_{F5} = w_{1,3} - \frac{2hq(x_1)}{p(x_1)} w_{1,4} + \frac{2hr(x_1)}{p(x_1)}.$$

Similar expressions are obtained for w_{F6}, w_{F7} and w_{F8}. Substituting these expressions into (12), the final set of finite difference equations becomes

$$-2I\mathbf{w}_{R3} + D''\mathbf{w}_{R4} = \begin{bmatrix} -h^2 f_{1,4} + g_{0,4} + 2hr(x_1)/p(x_1) \\ -h^2 f_{2,4} + 2hr(x_2)/p(x_2) \\ -h^2 f_{3,4} + 2hr(x_3)/p(x_3) \\ -h^2 f_{4,4} + 2h\alpha(y_4) + 2hr(x_4)/p(x_4) \end{bmatrix} = \mathbf{b}_{R4}, \tag{13}$$

where

$$D'' = \begin{bmatrix} 4+2h\dfrac{q(x_1)}{p(x_1)} & -1 & 0 & 0 \\ -1 & 4+2h\dfrac{q(x_2)}{p(x_2)} & -1 & 0 \\ 0 & -1 & 4+2h\dfrac{q(x_3)}{p(x_3)} & -1 \\ 0 & 0 & -2 & 4+2h\dfrac{q(x_4)}{p(x_4)} \end{bmatrix}.$$

Combining (9), (10), (11), and (13), the entire system of finite difference equations takes the form

$$\begin{bmatrix} D' & -I & & \\ -I & D' & -I & \\ & -I & D' & -I \\ & & -2I & D'' \end{bmatrix} \begin{bmatrix} \mathbf{w}_{R1} \\ \mathbf{w}_{R2} \\ \mathbf{w}_{R3} \\ \mathbf{w}_{R4} \end{bmatrix} = \begin{bmatrix} \mathbf{b}_{R1} \\ \mathbf{b}_{R2} \\ \mathbf{b}_{R3} \\ \mathbf{b}_{R4} \end{bmatrix}. \tag{14}$$

Another Worked Example (An Application)

EXAMPLE 9.3 Temperature Distribution in a Column Supporting an Industrial Furnace

A long column of fireclay brick, which has a square cross-section with side length of one meter, supports a large industrial furnace. During steady-state operation, three faces of the column are maintained at 500 K, while the fourth face is exposed to convective heat transfer with heat transfer coefficient $h = 10\ \text{W/m}^2\cdot\ \text{K}$ and ambient temperature $T_\infty = 300$ K. The thermal conductivity of fireclay brick is $k = 1\ \text{W/m}\cdot\text{K}$.

Let's orient the square in the first quadrant, with one corner at the origin and the top edge as the convective surface. Taking into account the symmetry of the problem about the vertical line $x = 1/2$, the temperature in the column, $T(x, y)$, satisfies the Poisson problem

$$\frac{\partial^2 T}{\partial x^2} + \frac{\partial^2 T}{\partial y^2} = 0, \quad 0 < x < 1/2, 0 < y < 1$$

$$T(0, y) = T(x, 0) = 500, \quad \frac{\partial T}{\partial x}\left(\frac{1}{2}, y\right) = 0, \quad \frac{\partial T}{\partial y}(x, 1) + 10T(x, 1) = 3000.$$

Consider the 5×3 grid shown in Figure 9.11. Note that the mesh spacing is $h = 1/4$. Since the grid has four rows of unknowns, the system of finite difference equations will be organized into four sets of equations. Each set will contain two equations since there are two unknowns per row.

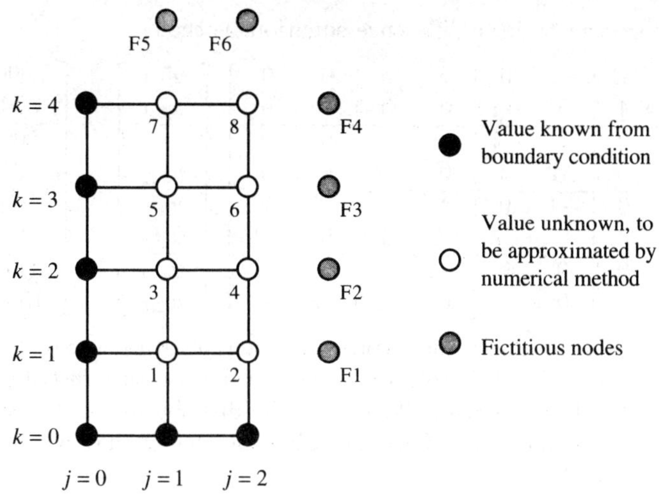

Figure 9.11 5 × 3 grid for Example 9.3.

Comparing the current problem with our model problem, we see that

$$\begin{gathered} f(x,y) = 0, \quad g(x,y) = 500, \\ \alpha(y) = 0, \quad p(x) = 1, \\ q(x) = 10 \text{ and } r(x) = 3000. \end{gathered}$$

Thus, the matrix D'' will take the form

$$\begin{aligned} D'' &= \begin{bmatrix} 4 + 2hq(x_1)/p(x_1) & -1 \\ -2 & 4 + 2hq(x_2)/p(x_2) \end{bmatrix} \\ &= \begin{bmatrix} 4 + 2(0.25)(10)/1 & -1 \\ -2 & 4 + 2(0.25)(10)/1 \end{bmatrix}, \end{aligned}$$

while the components of the right-hand side vector will be

$$\mathbf{b}_{R1} = \begin{bmatrix} -h^2 f_{1,1} + g_{1,0} + g_{0,1} \\ -h^2 f_{2,1} + g_{2,0} + 2h\alpha(y_1) \end{bmatrix} = \begin{bmatrix} 0 + 500 + 500 \\ 0 + 500 + 0 \end{bmatrix}$$

$$\mathbf{b}_{R2} = \begin{bmatrix} -h^2 f_{1,2} + g_{0,2} \\ -h^2 f_{2,2} + 2h\alpha(y_2) \end{bmatrix} = \begin{bmatrix} 0 + 500 \\ 0 + 0 \end{bmatrix}$$

$$\mathbf{b}_{R3} = \begin{bmatrix} -h^2 f_{1,3} + g_{0,3} \\ -h^2 f_{2,3} + 2h\alpha(y_3) \end{bmatrix} = \begin{bmatrix} 0 + 500 \\ 0 + 0 \end{bmatrix}$$

$$\mathbf{b}_{R4} = \begin{bmatrix} -h^2 f_{1,4} + g_{0,4} + 2hr(x_1)/p(x_1) \\ -h^2 f_{2,4} + 2h\alpha(y_4) + 2hr(x_2)/p(x_2) \end{bmatrix} = \begin{bmatrix} 0 + 500 + 2(0.25)(3000)/1 \\ 0 + 0 + 2(0.25)(3000)/1 \end{bmatrix}.$$

The complete system of finite difference equations is then

$$\left[\begin{array}{cc|cc|cc|cc} 4 & -1 & -1 & 0 & 0 & 0 & 0 & 0 \\ -2 & 4 & 0 & -1 & 0 & 0 & 0 & 0 \\ \hline -1 & 0 & 4 & -1 & -1 & 0 & 0 & 0 \\ 0 & -1 & -2 & 4 & 0 & -1 & 0 & 0 \\ \hline 0 & 0 & -1 & 0 & 4 & -1 & -1 & 0 \\ 0 & 0 & 0 & -1 & -2 & 4 & 0 & -1 \\ \hline 0 & 0 & 0 & 0 & -2 & 0 & 9 & -1 \\ 0 & 0 & 0 & 0 & 0 & -2 & -2 & 9 \end{array}\right] \left[\begin{array}{c} w_{1,1} \\ w_{2,1} \\ \hline w_{1,2} \\ w_{2,2} \\ \hline w_{1,3} \\ w_{2,3} \\ \hline w_{1,4} \\ w_{2,4} \end{array}\right] = \left[\begin{array}{c} 1000 \\ 500 \\ \hline 500 \\ 0 \\ \hline 500 \\ 0 \\ \hline 2000 \\ 1500 \end{array}\right].$$

Solving this system produces the approximate temperature values listed in Figure 9.12. Remember that this solution gives the temperature distribution in just the left half of the brick column. To obtain the distribution on the other half of the square, we simply need to reflect the values on the left half across the axis of symmetry at $x = 1/2$.

Figure 9.13 displays a contour plot of the temperature distribution within the brick column. A 17×9 grid was used both to improve the accuracy of the approximate solution and to increase the resolution of the contours. The approximate solution was computed for the left half of the square, and the output values were then reflected across $x = 1/2$ to create the full plot.

EXERCISES

1. What conditions must the functions p and q satisfy for the coefficient matrix in equation (14) to be diagonally dominant?

In Exercises 2–6, write out the system of finite difference equations corresponding to the Poisson problem on a rectangular domain with the indicated boundary conditions. Use a 4×4 grid.

2. Dirichlet conditions along the top edge and right side of the domain and Neumann conditions along the left side and bottom edge. To be specific, suppose

$$\begin{aligned} u(x,y) &= g(x,y) && \text{along top edge and right side,} \\ \frac{\partial u}{\partial n} &= \alpha(y) && \text{along left side, and} \\ \frac{\partial u}{\partial n} &= \beta(x) && \text{along bottom edge.} \end{aligned}$$

3. Dirichlet boundary conditions along the top and bottom edges, a Neumann boundary condition along the left side, and a Robin condition along the right side. To be specific, suppose

$$\begin{aligned} u(x,y) &= g(x,y) && \text{along top and bottom edges,} \\ \frac{\partial u}{\partial n} &= \alpha(y) && \text{along left side, and} \\ p(y)\frac{\partial u}{\partial n} + q(y)u &= r(y) && \text{along right side.} \end{aligned}$$

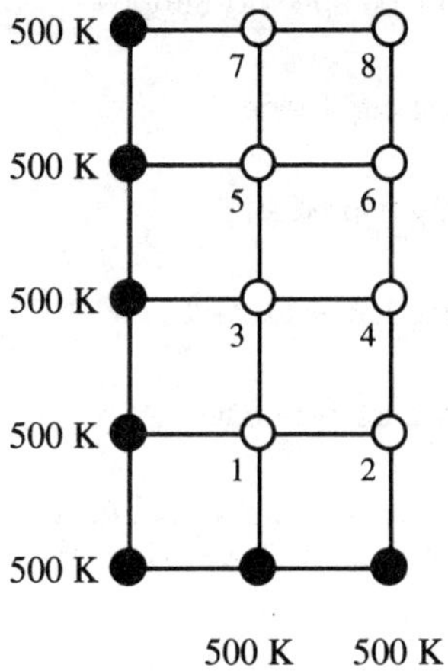

$T_7 \approx w_{1,4} = 356.99$ K $\qquad T_8 \approx w_{2,4} = 339.05$ K

$T_5 \approx w_{1,3} = 436.95$ K $\qquad T_6 \approx w_{2,3} = 418.74$ K

$T_3 \approx w_{1,2} = 472.07$ K $\qquad T_4 \approx w_{2,2} = 462.01$ K

$T_1 \approx w_{1,1} = 489.30$ K $\qquad T_2 \approx w_{2,1} = 485.15$ K

Figure 9.12 Approximate temperatures in a column supporting an industrial furnace.

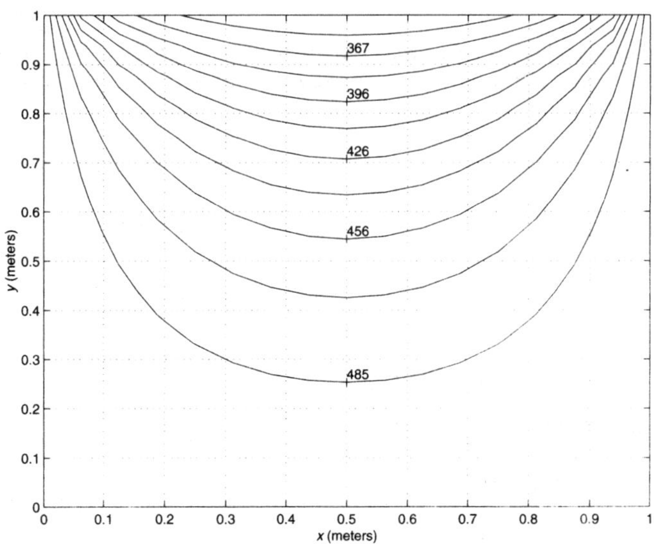

Figure 9.13 Temperature contour lines for the fireclay brick support column of an industrial furnace. Temperature values are in Kelvin, and the lower edge and sides are maintained at 500 K.

4. Neumann conditions along the top edge and right side of the domain and Robin conditions along the left side and bottom edge. To be specific, suppose

$$\begin{aligned} \frac{\partial u}{\partial n} &= \alpha(y) && \text{along right side,} \\ \frac{\partial u}{\partial n} &= \beta(x) && \text{along top edge,} \\ p_l(y)\frac{\partial u}{\partial n} + q_l(y)u &= r_l(y) && \text{along left side, and} \\ p_b(x)\frac{\partial u}{\partial n} + q_b(x)u &= r_b(x) && \text{along bottom edge.} \end{aligned}$$

5. Dirichlet boundary condition along the bottom edge, Neumann boundary conditions along the left side and the top edge, and a Robin condition along the right side. To be specific, suppose

$$\begin{aligned} u(x,y) &= g(x,y) && \text{along bottom edge,} \\ \frac{\partial u}{\partial n} &= \alpha(y) && \text{along left side,} \\ \frac{\partial u}{\partial n} &= \beta(x) && \text{along top edge, and} \\ p(y)\frac{\partial u}{\partial n} + q(y)u &= r(y) && \text{along right side.} \end{aligned}$$

6. Dirichlet boundary condition along the bottom edge, Neumann boundary conditions along the left side and the right side, and a Robin condition along the top edge. To be specific, suppose

$$\begin{aligned} u(x,y) &= g(x,y) && \text{along bottom edge,} \\ \frac{\partial u}{\partial n} &= \alpha(y) && \text{along left side,} \\ \frac{\partial u}{\partial n} &= \beta(y) && \text{along right side, and} \\ p(x)\frac{\partial u}{\partial n} + q(x)u &= r(x) && \text{along top edge.} \end{aligned}$$

7. Consider the Laplace equation

$$\frac{\partial^2 u}{\partial x^2} + \frac{\partial^2 u}{\partial x^2} = 0 \text{ on } R = \{(x,y)\,|\,0 < x < 1, 0 < y < 1\}$$

subject to the boundary conditions

$$u(x,0) = 0, \quad u(0,y) = \frac{y}{1+y^2}$$

$$\frac{\partial u}{\partial x}(1,y) = -\frac{4y}{(4+y^2)^2}, \quad \frac{\partial u}{\partial y}(x,1) + \frac{2}{(1+x)^2+1}u(x,1) = \frac{1}{(1+x)^2+1}.$$

Numerically verify that the finite difference method is second-order accurate for this problem. The exact solution is

$$u(x,y) = \frac{y}{(1+x)^2 + y^2}.$$

8. Repeat Exercise 7 for the Poisson problem

$$\frac{\partial^2 u}{\partial x^2} + \frac{\partial^2 u}{\partial y^2} = x^2 + y^2 \text{ on } R = \{(x,y) \,|\, 0 < x < 1, 0 < y < 1\}$$

$$\frac{\partial u}{\partial y}(x,0) = \pi e^{\pi x}, \quad u(x,1) = \frac{1}{2}x^2$$

$$u(0,y) = \sin(\pi y), \quad u(1,y) = e^{\pi} \sin(\pi y) + \frac{1}{2}y^2.$$

The exact solution for this problem is $u(x,y) = e^{\pi x}\sin(\pi y) + \frac{1}{2}(xy)^2$.

9. Repeat Exercise 7 for the Poisson problem

$$\frac{\partial^2 u}{\partial x^2} + \frac{\partial^2 u}{\partial y^2} = -52\cos(4x+6y) \text{ on } R = \{(x,y) \,|\, 0 < x < 1, 0 < y < 1\}$$

$$u(x,0) = \cos(4x), \quad \frac{1}{6}\frac{\partial u}{\partial y}(x,1) + u(x,1) = \cos(4x+6) - \sin(4x+6)$$

$$\frac{\partial u}{\partial x}(0,y) = -4\sin(6y), \quad \frac{\partial u}{\partial x}(1,y) = -4\sin(6y+4).$$

The exact solution for this problem is $u(x,y) = \cos(4x+6y)$.

10. Repeat Exercise 7 for the Poisson problem

$$\frac{\partial^2 u}{\partial x^2} + \frac{\partial^2 u}{\partial y^2} = 0 \text{ on } R = \{(x,y) \,|\, 1 < x < 2, 0 < y < 1\}$$

$$\frac{\partial u}{\partial y}(x,0) = 0, \quad u(x,1) = \ln(x^2+1), \quad \frac{\partial u}{\partial x}(1,y) = \frac{2}{y^2+1}, \quad u(2,y) = \ln(y^2+4).$$

The exact solution for this problem is $u(x,y) = \ln(x^2+y^2)$.

11. A long bar has a rectangular cross section, 0.4 meters by 0.6 meters on a side. The top and bottom edges of the bar are maintained at a temperature of 200° C, while the left side is insulated. The right side is subjected to convective heat transfer with a fluid whose temperature is 30° C. The temperature distribution within the bar, $T(x,y)$, satisfies the Poisson problem

$$\frac{\partial^2 T}{\partial x^2} + \frac{\partial^2 T}{\partial y^2} = 0 \text{ on } R = \{(x,y) \,|\, 0 < x < 0.4, 0 < y < 0.6\}$$

$$T(x,0) = T(x,0.6) = 200^\circ \text{ C}$$

$$\frac{\partial T}{\partial n} = 0 \text{ along } x = 0, \quad k\frac{\partial T}{\partial n} + hT = 30h \text{ along } x = 0.4.$$

Approximate $T(x,y)$ using the finite difference method with a uniform mesh spacing of 0.1 meters. Take $k = 1.5$ W/m · K for the thermal conductivity of the bar and $h = 50$ W/m^2· K for the convective heat transfer coefficient.

12. A solid bar having a rectangular cross-section one inch wide and two inches high is subjected to a constant rate of twist along its axis. The stresses established in the bar can be related to the nondimensionalized Prandtl stress function, $\psi(x, y)$, which, for the geometry of this problem, satisfies the Poisson problem

$$\frac{\partial^2 \psi}{\partial x^2} + \frac{\partial^2 \psi}{\partial x^2} = -1 \text{ on } R = \{(x, y) | 0 < x < 1/2, 0 < y < 1\}$$

$$\psi(x, 1) = \psi(1/2, y) = 0, \quad \left.\frac{\partial \psi}{\partial n}\right|_{(0,y)} = \left.\frac{\partial \psi}{\partial n}\right|_{(x,0)} = 0.$$

Approximate $\psi(x, y)$ using the finite difference method with a uniform mesh spacing of $h = 0.10$ inches.

13. The temperature distribution, $T(x, y)$, within a long bar of rectangular cross-section satisfies the Poisson problem

$$\frac{\partial^2 T}{\partial x^2} + \frac{\partial^2 T}{\partial y^2} = 0 \text{ on } R = \{(x, y) | 0 < x < 0.4, 0 < y < 0.3\}$$

$$T(x, 0) = 200°C$$

$$\frac{\partial T}{\partial n} = 0 \text{ along } x = 0 \text{ and } y = 0.3$$

$$k\frac{\partial T}{\partial n} + hT = 30h \text{ along } x = 0.4.$$

Approximate $T(x, y)$ using the finite difference method with a uniform mesh spacing of 0.05 meters. Take $k = 1.5$ W/m $\cdot$ K for the thermal conductivity of the bar and $h = 50$ W/m$^2\cdot$ K for the convective heat transfer coefficient.

14. Consider the Poisson problem

$$\frac{\partial^2 u}{\partial x^2} + \frac{\partial^2 u}{\partial y^2} = f(x, y) \text{ on } R = \{(x, y) | a < x < b, c < y < d\} \tag{15}$$

$$\frac{\partial u}{\partial n} = \alpha(x, y) \text{ for } (x, y) \in \partial R.$$

Note this problem has Neumann conditions specified around the entire boundary.

(a) Show that in order for (15) to have a solution, the functions f and α must satisfy the consistency condition

$$\iint_R f(x, y)\, dA = \int_{\partial R} \alpha(x, y)\, dl.$$

(b) Show that if $u(x, y)$ is a solution to (15) then $u(x, y) + k$ is a solution for any constant k. Hence, when (15) has a solution, it is not unique.

(c) Construct the system of finite difference equations associated with (15). Show that the coefficient matrix is singular.

(d) What condition must f and α satisfy in order for the system of finite difference equations developed in part (c) to have a solution?

9.3 SOLVING THE DISCRETE EQUATIONS: RELAXATION METHODS

We have seen that approximating the solution of the Poisson equation using the finite difference method requires the solution of a linear system of algebraic equations. If the computational grid that is placed over the rectangular domain has N subdivisions in one coordinate direction and M subdivisions in the other and Dirichlet boundary conditions are specified, then the coefficient matrix for the linear system will have dimensions

$$(N-1)(M-1) \times (N-1)(M-1).$$

With non-Dirichlet boundary conditions, the matrix is even larger. When N and M are small, direct techniques for the solution of linear systems, such as Gaussian elimination or special factorization algorithms, are viable choices. It was precisely these types of techniques that were used in Sections 9.1 and 9.2.

When N and M are even of moderate size, however, iterative techniques for the solution of linear systems are often more efficient alternatives than direct techniques. First, with an iterative technique, there is no need to store the coefficient matrix. All that must be known is the structure of the equations; that is, the structure of the computational template. Second, even though multiple iterations must be performed, an iterative solution typically requires fewer total operations. An added bonus is that iterative techniques are generally insensitive to roundoff error.

Any time an iterative method is used to solve the discrete Poisson equations, there are actually three solutions floating around of which we must be aware. The first of these is the true solution of the partial differential equation, u. The other two solutions are related to the linear system of equations: w^h is the true solution and $\tilde{w}^h$ the approximate solution. w^h and $\tilde{w}^h$ are referred to as *grid functions* since they consist of the values of some function at all points of the computational grid. The superscript, h, is used to identify the mesh spacing of that grid. Among these three solutions, there are also two different types of errors with which we must be concerned. Between u and w^h, there is the second-order discretization error that was discussed in the previous sections. Toward the end of this section, the iteration error between w^h and $\tilde{w}^h$ will be investigated.

In this section, we will consider the three iterative techniques presented in Section 3.8, namely, the Jacobi method (simultaneous relaxation), the Gauss-Seidel method (successive relaxation) and the SOR method (successive overrelaxation). These three methods, together with a host of others, are collectively known as *point relaxation schemes.* Given an approximate solution, $\tilde{w}^h$, to the finite difference equations, a point relaxation scheme sweeps through the grid updating the values of $\tilde{w}_{j,k}$ one by one to form a new approximate solution $\bar{w}^h$. Throughout this and the next section, tildes (˜) will be used to denote before sweep values and overbars to denote after sweep values.

Dirichlet Boundary Conditions

Let's start by considering the Poisson problem with Dirichlet conditions specified around the entire boundary. In this case, the generic finite difference equation

$$-w_{j-1,k} - w_{j+1,k} - w_{j,k-1} - w_{j,k+1} + 4w_{j,k} = -h^2 f_{j,k} \tag{1}$$

applies at every grid point. From Exercise 1 of Section 9.1, we know that the coefficient matrix for the complete system of finite difference equations is symmetric positive definite. Consequently, the Jacobi method, the Gauss-Seidel method and the SOR method are guaranteed to converge for any choice of initial approximation $\tilde{w}^h$.

Applying the Jacobi method to equation (1), we find that the update equation for the unknown at the (j,k) location of the grid is given by

$$\bar{w}_{j,k} = \frac{1}{4}\left(\tilde{w}_{j-1,k} + \tilde{w}_{j+1,k} + \tilde{w}_{j,k-1} + \tilde{w}_{j,k+1} - h^2 f_{j,k}\right). \tag{2}$$

Before we can write out the update equation for the Gauss-Seidel method, we have to specify the precise order in which the unknowns will be processed. If we work in lexicographic order, then by the time the sweep reaches location (j,k) of the grid, the unknowns at locations $(j,k-1)$ and $(j-1,k)$ have already been updated. Thus, the Gauss-Seidel equation for $\bar{w}_{j,k}$ takes the form

$$\bar{w}_{j,k} = \frac{1}{4}\left(\bar{w}_{j-1,k} + \tilde{w}_{j+1,k} + \bar{w}_{j,k-1} + \tilde{w}_{j,k+1} - h^2 f_{j,k}\right). \tag{3}$$

Finally, the update equation for the SOR method is given by

$$\bar{w}_{j,k} = (1-\omega)\tilde{w}_{j,k} + \frac{\omega}{4}\left(\bar{w}_{j-1,k} + \tilde{w}_{j+1,k} + \bar{w}_{j,k-1} + \tilde{w}_{j,k+1} - h^2 f_{j,k}\right), \tag{4}$$

where ω is the relaxation parameter. For the Poisson problem on a rectangular domain with Dirichlet boundary conditions, the optimal value for the relaxation parameter is given by

$$\omega_{opt} = \frac{4}{2 + \sqrt{4 - [\cos(\pi/N) + \cos(\pi/M)]^2}} \tag{5}$$

(see Ortega [1]).

Examining equations (2), (3), and (4), we see that the Jacobi method and the Gauss-Seidel method use exactly the same number of arithmetic operations per grid point, while the SOR method requires two additional operations per point. In terms of storage, with the Jacobi method, the value of $\tilde{w}_{j,k}$ will be needed to update other values in the grid, so $\tilde{w}_{j,k}$ cannot be overwritten by $\bar{w}_{j,k}$. We will therefore have to maintain two storage structures: one for $\tilde{w}^h$ and one for $\bar{w}^h$. In contrast, for both the Gauss-Seidel method and the SOR method, $\tilde{w}_{j,k}$ can be overwritten by $\bar{w}_{j,k}$; hence, only one storage structure need be maintained. As a final remark, we note that when using the Jacobi method on a vector or parallel machine, all entries in $\bar{w}^h$ can be computed simultaneously. With lexicographic ordering, neither the Gauss-Seidel method nor the SOR method is vectorizable; however, with red-black ordering of the unknowns (see Exercise 13 of Section 9.1), all of the red unknowns can be computed simultaneously followed by all of the black unknowns.

EXAMPLE 9.4 Point Relaxation Schemes in Action

Consider the sample Poisson problem

$$\frac{\partial^2 u}{\partial x^2} + \frac{\partial^2 u}{\partial y^2} = 0$$

over the rectangular domain $R = \{(x, y) | 1 < x < 2, 0 < y < 1\}$ and subject to the Dirichlet boundary conditions

$$u(x,0) = 2\ln x, \quad u(x,1) = \ln(x^2+1)$$
$$u(1,y) = \ln(y^2+1), \quad u(2,y) = \ln(y^2+4).$$

The exact solution to this problem is $u(x, y) = \ln(x^2 + y^2)$.

For various values of N, an $N \times N$ computational grid is placed over the domain. For each grid and each relaxation scheme, $\tilde{w}^h$ is initialized to zero at all interior grid points, and sweeps are terminated when

$$\max_{1\le j\le N-1, 1\le k\le N-1} |\bar{w}_{j,k} - \tilde{w}_{j,k}|$$

falls below the tolerance 5×10^{-11}. The optimal value of the SOR method relaxation parameter, as given by equation (5), is used in each case. The number of sweeps needed to achieve convergence is tabulated below.

N	Jacobi	Gauss-Seidel	SOR ($\omega = \omega_{\text{opt}}$)
5	107	57	23
10	422	220	47
20	1592	828	91
40	5940	3088	179

We make two important observations from the data in this table. First, on each grid, the Jacobi method requires roughly twice the number of sweeps to achieve convergence as does the Gauss-Seidel method, and the SOR method requires significantly fewer sweeps than the Gauss-Seidel method. Second, each time the grid is refined, the number of sweeps needed to achieve convergence increases for all of the relaxation schemes. In particular, each time N is doubled, the number of sweeps needed by the Jacobi method and the Gauss-Seidel method increases by roughly a factor of four, while the number of sweeps needed by the SOR method roughly doubles. We will examine these issues in more detail toward the end of the section.

Non-Dirichlet Boundary Conditions

To handle problems with non-Dirichlet boundary conditions, remember that each grid point gives rise to its own finite difference equation and corresponding update equation. We just have to be careful with our bookkeeping to make sure that the correct equation is applied at each grid point.

To illustrate this process, consider the Poisson problem

$$\frac{\partial^2 u}{\partial x^2} + \frac{\partial^2 u}{\partial y^2} = f(x,y)$$

on the rectangular domain $R = \{(x,y) | \, a < x < b, c < y < d\}$ subject to the boundary conditions

$$u(a,y) = g_1(y), \quad u(x,c) = g_2(x), \quad \frac{\partial u}{\partial y}(x,d) = \alpha(x)$$

and

$$p(y)\frac{\partial u}{\partial x}(b,y) + q(y)u(b,y) = r(y).$$

The computational grid for this problem is displayed in Figure 9.14. Note that the grid points have been separated into four distinct types. Grid points of the same type give rise to finite difference equations of identical structure; grid points of different types give rise to different equations. It is important to realize that the grid points along the right edge of the domain and those along the top edge have been classified as different types because of the location of the fictitious node within the computational template, not because one edge has a Robin condition and the other a Neumann condition.

Since we have identified four types of grid points, there will be four fundamentally different update equations with which we will have to work. Here, we will write out the update equations for the Jacobi method only. The equations for the Gauss-Seidel method and the SOR method will be considered in the exercises. When the computational template is placed at any Type I grid point, we obtain equation (1). Accordingly, the update equation at any Type I grid point is equation (2). Along the right edge of the domain, we have to take into account the Robin boundary condition. The resulting update equation at the Type II grid points is

$$\bar{w}_{N,k} = \left[4 + 2h\frac{q(y_k)}{p(y_k)}\right]^{-1}\left(2\tilde{w}_{N-1,k} + 2h\frac{r(y_k)}{p(y_k)} + \tilde{w}_{N,k-1} + \tilde{w}_{N,k+1} - h^2 f_{N,k}\right). \tag{6}$$

Taking into account the Neumann condition specified along the top of the domain, we find

$$\bar{w}_{j,M} = \frac{1}{4}\left(\tilde{w}_{j-1,M} + \tilde{w}_{j+1,M} + 2\tilde{w}_{j,M-1} + 2h\alpha(x_j) - h^2 f_{j,M}\right) \tag{7}$$

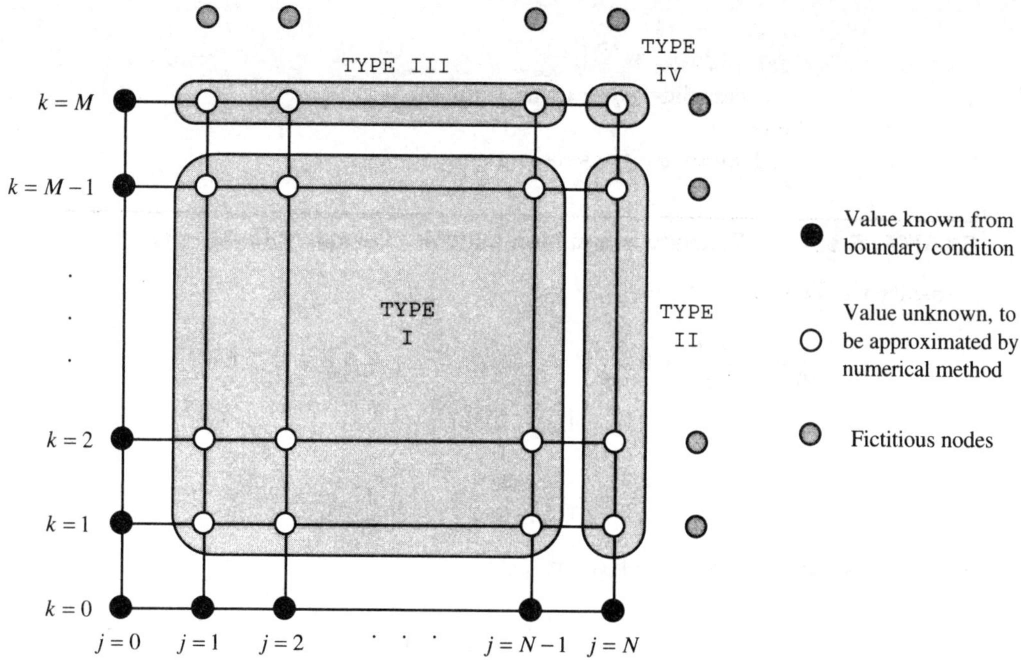

Figure 9.14 Computational grid for a problem with non-Dirichlet boundary conditions specified along the top edge and the right side of the domain.

at the Type III grid points. Finally,

$$\bar{w}_{N,M} = \left[4 + 2h\frac{q(y_M)}{p(y_M)}\right]^{-1}\left(2\tilde{w}_{N-1,M} + 2h\frac{r(y_M)}{p(y_M)} + 2\tilde{w}_{N,M-1} + 2h\alpha(x_N) - h^2 f_{N,M}\right). \tag{8}$$

Having determined the appropriate update equations, how should we organize the calculations to perform each relaxation sweep? Perhaps the best way to proceed is to take our cue from the grid itself. The first $M-1$ rows of unknowns all contain $N-1$ Type I grid points followed by a Type II grid point. The top row of unknowns then starts with $N-1$ Type III grid points and is terminated by a Type IV grid point. Thus the structure of the grid suggests the algorithm

for $k = 1$ to $M-1$
  for $j = 1$ to $N-1$
    calculate $\bar{w}_{j,k}$ using equation (2)
  end
  calculate $\bar{w}_{N,k}$ using equation (6)

```
end
for j = 1 to N − 1
    calculate w̄_{j,M} using equation (7)
end
calculate w̄_{N,M} using equation (8)
```

EXAMPLE 9.5 Relaxation and Non-Dirichlet Boundary Conditions

Consider the Poisson problem

$$\frac{\partial^2 u}{\partial x^2} + \frac{\partial^2 u}{\partial y^2} = 0 \text{ on } R = \{(x,y) | 0 < x < 1, 0 < y < 1\}$$

$$u(x,0) = 0, \quad u(0,y) = \frac{y}{1+y^2}$$

$$\frac{\partial u}{\partial y}(x,1) = \frac{x(2+x)}{((1+x)^2+1)^2}, \quad \frac{\partial u}{\partial y}(1,y) + \frac{4}{4+y^2}u(1,y) = 0.$$

The exact solution for this problem is

$$u(x,y) = \frac{y}{(1+x)^2+y^2}.$$

For various values of N, an $N \times N$ computational grid is placed over the domain. For each grid and each relaxation scheme, $\tilde{w}^h$ is initialized to zero at all grid points not along the left side or the bottom edge, and sweeps are terminated when

$$\max_{1 \le j \le N, 1 \le k \le N} |\bar{w}_{j,k} - \tilde{w}_{j,k}|$$

falls below the tolerance 5×10^{-11}. There is no formula, like equation (5), which gives the optimal value of the SOR method relaxation parameter in terms of the number of grid subintervals N and M for problems involving non-Dirichlet boundary conditions. Though there are algorithms for estimating ω_{opt} as the sweeps are being carried out (see, for example, Thomas [2] and Hageman and Young [3]), here, we will simply run the SOR method with

$$\omega = \omega_1 = \frac{2}{1+\sin\frac{\pi}{N}} \quad \text{and} \quad \omega = \omega_2 = \frac{2}{1+\sin\frac{\pi}{2N}}.$$

Note that ω_1 is the value obtained from (5) for an $N \times N$ grid. For this problem, the performance of the SOR method with $\omega = \omega_2$ turns out to be nearly optimal.

	Sweeps to Achieve Convergence			
N	Jacobi	Gauss-Seidel	SOR ($\omega = \omega_1$)	SOR ($\omega = \omega_2$)
5	306	157	89	37
10	1146	591	178	72
20	4249	2202	346	145
40	15621	8138	667	287

From the data in the above table, we can make exactly the same observations as we made earlier. On each grid, the Jacobi method requires roughly twice the number of sweeps to achieve convergence as does the Gauss-Seidel method, and the SOR method (even with far from optimal performance) requires significantly fewer sweeps than the Gauss-Seidel method. Also, each time N is doubled, the number of sweeps needed by the Jacobi method and the Gauss-Seidel method increases by roughly a factor of four, while the number of sweeps needed by the SOR method roughly doubles.

Convergence Analysis

We've observed in two examples that all three of our relaxation schemes converge toward the true solution of the finite difference equations more slowly when the mesh size decreases! This behavior has an extremely important practical consequence. A smaller mesh size corresponds to smaller truncation error. Thus, there is an inherent tradeoff between truncation error and the convergence rate of relaxation schemes. Let's see if we can establish a theoretical basis for this phenomenon.

When assessing the performance of a relaxation scheme, the main issue is the amount by which a single sweep through the grid reduces the difference between the true solution of the finite difference equations and the approximate solution (i.e., the iteration error). This quantity is measured by the *asymptotic convergence rate*, μ. If we let

$$\tilde{v}^h = w^h - \tilde{w}^h \quad \text{and} \quad \bar{v}^h = w^h - \bar{w}^h$$

denote the iteration error prior to commencing a sweep and upon completion of that sweep, respectively, then

$$\mu \equiv \text{asymptotic value of } \frac{\|\bar{v}^h\|}{\|\tilde{v}^h\|}.$$

In this context, "asymptotic value" refers to the value approached by the indicated ratio as the iteration nears convergence. Any appropriate norm can be applied to the grid functions $\bar{v}^h$ and $\tilde{v}^h$. The ideal situation is $\mu \ll 1$, which indicates that each sweep produces a substantial reduction in the iteration error. When $\mu \approx 1$, little error reduction is generated by each sweep as convergence nears.

Although throughout this section we have not formulated any of the methods in this way, we know (from Section 3.8) that the Jacobi method, the Gauss-Seidel method and the SOR method can all be written in the form

$$\bar{w}^h = T\tilde{w}^h + \mathbf{c},$$

for some iteration matrix t and some vector $\mathbf{c}$. We also know that each method converges linearly with an asymptotic error constant equal to $\rho(T)$, the spectral radius of the iteration matrix. Thus, as the iteration nears convergence,

$$\frac{\|\bar{v}^h\|}{\|\tilde{v}^h\|} \to \rho(T).$$

Given our definition of the asymptotic convergence rate, it follows that $\mu = \rho(T)$.

To carry our analysis further, let's consider the specific case of the Poisson problem on the unit square (i.e., $R = \{(x,y)|\, 0 < x < 1, 0 < y < 1\}$) subject to Dirichlet boundary conditions. Further, suppose we place an $N \times N$ grid over the domain. For this problem, it can be shown (see Stoer and Bulirsch [4]) that

$$\rho(T_{\text{jac}}) = \cos\frac{\pi}{N} \approx 1 - \frac{1}{2}\left(\frac{\pi}{N}\right)^2 \tag{9}$$

and

$$\rho(T_{\text{gs}}) = \left(\cos\frac{\pi}{N}\right)^2 \approx 1 - \left(\frac{\pi}{N}\right)^2, \tag{10}$$

where T_{jac} and T_{gs} are the Jacobi and Gauss-Seidel iteration matrices, respectively. For the SOR method,

$$\omega_{\text{opt}} = \frac{2}{1 + \sin\frac{\pi}{N}}.$$

With $\omega = \omega_{\text{opt}}$,

$$\rho(T_{\text{sor}}) = \omega_{\text{opt}} - 1 = \frac{1 - \sin\frac{\pi}{N}}{1 + \sin\frac{\pi}{N}} \approx 1 - 2\frac{\pi}{N}, \tag{11}$$

where T_{sor} is the SOR iteration matrix.

So how do we relate this information to the number of sweeps? Asymptotically, we know that one sweep reduces the norm of the error by the factor $\rho(T)$. Therefore, if we wish to reduce the error by a factor ϵ, say, we will need to perform s sweeps, where $[\rho(T)]^s \approx \epsilon$. Solving for s gives

$$s \approx \frac{\ln\epsilon}{\ln\rho(T)}. \tag{12}$$

Substituting (9) and (10) into (12), we find

$$s_{\text{jac}} \approx \frac{-\ln\epsilon}{\frac{1}{2}\left(\frac{\pi}{N}\right)^2} = \frac{-2\ln\epsilon}{\pi^2}N^2$$

and

$$s_{\text{gs}} \approx \frac{-\ln\epsilon}{\left(\frac{\pi}{N}\right)^2} = \frac{-\ln\epsilon}{\pi^2}N^2.$$

From these last two expressions we see that we should expect the Jacobi method to require twice as many sweeps as the Gauss-Seidel method, and that the number of sweeps needed by both methods is $O(N^2)$. In contrast, upon substituting (11) into (12), we obtain

$$s_{\text{sor}} \approx \frac{-\ln\epsilon}{2\pi}N,$$

so that the number of sweeps needed by the SOR method is $O(N)$.

Some Thoughts on the Programming of Relaxation Methods

When using a relaxation scheme to solve the finite difference equations associated with a Poisson problem, the code for the routine can be greatly simplified and the efficiency vastly improved by keeping two simple ideas in mind. First, dimension the matrix for storing the values of $w_{j,k}$ to be of size $(N+1) \times (M+1)$, even if Dirichlet conditions are specified along a portion of the boundary. Place any known values of the solution into the appropriate locations in the matrix. At the expense of at most $2(N+M+1)$ memory locations, the need to treat grid points that are adjacent to a Dirichlet boundary condition separately from all of the other interior grid points is removed.

Second, outside the iteration loop, construct a matrix which contains the value of the nonhomogeneous term, $f(x,y)$, at each grid point where the solution is not known from a Dirichlet boundary condition. Then, inside the iteration loop, use a table lookup, rather than a function evaluation. This approach will save $O(NM)$ function evaluations per sweep. Additional operations can be saved by multiplying each $f_{j,k}$ by h^2 when the matrix is constructed. How does this save operations? Take a close look at the update equations (2), (3), (4), (6), (7), and (8). Each requires the value $h^2 f_{j,k}$, not just $f_{j,k}$. The same will be true of any update equation. Hence, if the value $h^2 f_{j,k}$ can be looked up, we will reduce the computational cost of our relaxation scheme by one arithmetic operation per point per sweep.

An Application Problem: Electric Potential with Unusual Boundary Conditions

Consider the thin metal plate of length L and height h shown in Figure 9.15. The top edge of the plate is maintained at an electric potential V_0 while the bottom edge is maintained at an electric potential $-V_0$. No charge is allowed to enter or leave the plate along the other two sides. A uniform magnetic field of intensity B_0 acts normal to the plate and into the page. Our objective is to approximate the steady-state electric potential, $V(x,y)$, throughout the plate.

At steady state, there will be no volume charge density within the plate, so from Maxwell's equations, we have

$$\frac{\partial E_x}{\partial x} + \frac{\partial E_y}{\partial y} = 0, \tag{13}$$

where E_x and E_y are the x and y components of the electric field. The electric potential is related to the electric field by

$$E_x = -\frac{\partial V}{\partial x} \quad \text{and} \quad E_y = -\frac{\partial V}{\partial y}. \tag{14}$$

Substituting (14) into (13) yields

$$\frac{\partial^2 V}{\partial x^2} + \frac{\partial^2 V}{\partial y^2} = 0.$$

Along the top and bottom of the plate, we have the Dirichlet boundary conditions

$$V(x,H) = V_0 \quad \text{and} \quad V(x,0) = -V_0.$$

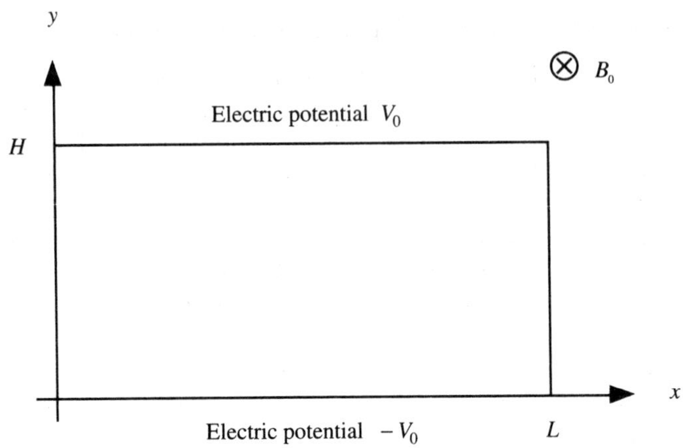

Figure 9.15 Geometry for "Electric Potential with Unusual Boundary Conditions" application problem.

Moelter et al., [5] have shown that the appropriate boundary condition along the other two sides of the plate is

$$\frac{\partial V}{\partial x} = \lambda \frac{\partial V}{\partial y},$$

where λ is a parameter which depends upon B_0.

Given the location of the non-Dirichlet boundary conditions, this problem will involve three types of grid points: those along the left side of the plate, those interior to the plate, and those along the right side of the plate. The finite difference equation for the interior grid points is just the generic equation (1). For the grid points along the left side and the right side of the plate, the corresponding finite difference equations are

$$-2w_{1,k} - (1+\lambda)w_{0,k-1} - (1-\lambda)w_{0,k+1} + 4w_{0,k} = -h^2 f_{0,k} \tag{15}$$

and

$$-2w_{N-1,k} - (1-\lambda)w_{N,k-1} - (1+\lambda)w_{N,k+1} + 4w_{N,k} = -h^2 f_{N,k}, \tag{16}$$

respectively. Let's now take $L = 2$, $H = 1$, $V_0 = 1$ and $\lambda = 0.25$, and place a uniform 20×10 grid over the domain. With $\omega = 1.64$, 49 iterations of the SOR method produce $\|\bar{w}^h - \tilde{w}^h\|_\infty < 5\times10^{-8}$. The resulting equipotential curves (curves along which V is constant) are displayed in Figure 9.16.

References

1. J. M. Ortega, *Numerical Analysis—A Second Course*, Academic Press, New York, 1972.

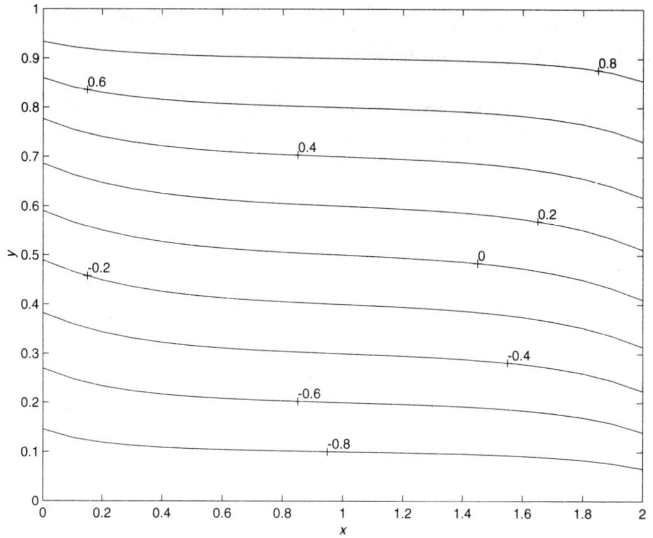

Figure 9.16 Equipotential curves for $L = 2$, $H = 1$, $V_0 = 1$ and $\lambda = 0.25$.

2. J. W. Thomas, *Numerical Partial Differential Equations: Conservation Laws and Elliptic Equations*, Texts in Applied Mathematics, volume 33, Springer-Verlag, New York, 1999.
3. L. A. Hageman and D. M. Young, *Applied Iterative Methods*, Academic Press, New York, 1981.
4. J. Stoer and R. Bulirsch, *Introduction to Numerical Analysis*, Springer-Verlag, New York, 1980.
5. M. J. Moelter, J. Evans, G. Elliott, and M. Jackson, "Electric Potential in the Classical Hall Effect: An Unusual Boundary-Value Problem," *American Journal of Physics,* **66**, 668–677, 1998.

EXERCISES

1. **(a)** Derive equations (6), (7), and (8).

(b) Assuming lexicographic ordering of the unknowns, what are the analogues of equations (6), (7), and (8) for the Gauss-Seidel method and the SOR method?

2. **(a)** Derive equations (15) and (16).

(b) What are the update equations corresponding to equations (15) and (16) for the SOR method?

In Exercises 3–7, the boundary conditions for a Poisson problem over a rectangular domain are provided. For each set of boundary conditions

(a) identify the different types of grid points;

(b) identify the finite difference equation for each type of grid point; and

(c) construct pseudocode to perform one relaxation sweep.

3. Dirichlet conditions along the top edge and right side of the domain and Neumann conditions along the left side and bottom edge. To be specific, suppose

$$\begin{aligned} u(x,y) &= g(x,y) && \text{along top edge and right side,} \\ \frac{\partial u}{\partial n} &= \alpha(y) && \text{along left side, and} \\ \frac{\partial u}{\partial n} &= \beta(x) && \text{along bottom edge.} \end{aligned}$$

4. Neumann conditions along the top edge and right side of the domain and Robin conditions along the left side and bottom edge. To be specific, suppose

$$\begin{aligned} \frac{\partial u}{\partial n} &= \alpha(y) && \text{along right side,} \\ \frac{\partial u}{\partial n} &= \beta(x) && \text{along top edge,} \\ p_l(y)\frac{\partial u}{\partial n} + q_l(y)u &= r_l(y) && \text{along left side, and} \\ p_b(x)\frac{\partial u}{\partial n} + q_b(x)u &= r_b(x) && \text{along bottom edge.} \end{aligned}$$

5. Dirichlet boundary conditions along the top and bottom edges, a Neumann boundary condition along the left side, and a Robin condition along the right side. To be specific, suppose

$$\begin{aligned} u(x,y) &= g(x,y) && \text{along top and bottom edges,} \\ \frac{\partial u}{\partial n} &= \alpha(y) && \text{along left side, and} \\ p(y)\frac{\partial u}{\partial n} + q(y)u &= r(y) && \text{along right side.} \end{aligned}$$

6. Dirichlet boundary condition along the bottom edge, Neumann boundary conditions along the left side, and the top edge and a Robin condition along the right side. To be specific, suppose

$$\begin{aligned} u(x,y) &= g(x,y) && \text{along bottom edge,} \\ \frac{\partial u}{\partial n} &= \alpha(y) && \text{along left side,} \\ \frac{\partial u}{\partial n} &= \beta(x) && \text{along top edge, and} \\ p(y)\frac{\partial u}{\partial n} + q(y)u &= r(y) && \text{along right side.} \end{aligned}$$

7. Dirichlet boundary condition along the bottom edge, Neumann boundary conditions along the left side, and the right side and a Robin condition along the top edge. To be specific, suppose

$$\begin{aligned} u(x,y) &= g(x,y) && \text{along bottom edge,} \\ \frac{\partial u}{\partial n} &= \alpha(y) && \text{along left side,} \\ \frac{\partial u}{\partial n} &= \beta(y) && \text{along right side, and} \\ p(x)\frac{\partial u}{\partial n} + q(x)u &= r(x) && \text{along top edge.} \end{aligned}$$

In Exercises 8–11, demonstrate that the Jacobi method and the Gauss-Seidel method require $O(N^2)$ sweeps and the SOR method $O(N)$ sweeps to achieve a given level of convergence. Regardless of the type of boundary conditions, use equation (5) to compute ω for the SOR method.

8. $\dfrac{\partial^2 u}{\partial x^2} + \dfrac{\partial^2 u}{\partial y^2} = 0$ on $R = \{\,(x,y)\,|\,0 < x < 1, 0 < y < 1\}$
$u(x,0) = 0, \quad u(x,1) = \dfrac{1}{(1+x)^2+1}, \quad u(0,y) = \dfrac{y}{1+y^2}, \quad u(1,y) = \dfrac{y}{4+y^2}$

9. $\dfrac{\partial^2 u}{\partial x^2} + \dfrac{\partial^2 u}{\partial y^2} = x^2 + y^2$ on $R = \{\,(x,y)\,|\,0 < x < 1, 0 < y < 1\}$
$\dfrac{\partial u}{\partial y}(x,0) = \pi e^{\pi x}, \ u(x,1) = \dfrac{1}{2}x^2, \ u(0,y) = \sin(\pi y), \ u(1,y) = e^{\pi}\sin(\pi y) + \dfrac{1}{2}y^2$

10. $\dfrac{\partial^2 u}{\partial x^2} + \dfrac{\partial^2 u}{\partial y^2} = -52\cos(4x+6y)$ on $R = \{\,(x,y)\,|\,0 < x < 1, 0 < y < 1\}$
$u(x,0) = \cos(4x), \quad u(x,1) = \cos(4x+6), \quad u(0,y) = \cos(6y), \quad u(1,y) = \cos(6y+4)$

11. $\dfrac{\partial^2 u}{\partial x^2} + \dfrac{\partial^2 u}{\partial y^2} = 0$ on $R = \{\,(x,y)\,|\,0 < x < 1, 0 < y < 1\}$
$u(x,0) = 0, \quad u(0,y) = \dfrac{y}{1+y^2}, \quad \dfrac{\partial u}{\partial x} = -\dfrac{4y}{(4+y^2)^2}$ for $x = 1$,
$\dfrac{\partial u}{\partial y} + \dfrac{2}{(1+x)^2+1}u = \dfrac{1}{(1+x)^2+1}$ for $y = 1$

12. A solid bar having a rectangular cross section one inch wide and two inches high is subjected to a constant rate of twist along its axis. The stresses established in the bar can be related to the nondimensionalized Prandtl stress function, $\psi(x,y)$, which, for the geometry of this problem, satisfies the Poisson problem

$$\frac{\partial^2 \psi}{\partial x^2} + \frac{\partial^2 \psi}{\partial y^2} = -1 \text{ on } R$$
$$\psi(x,y) = 0 \text{ on } \partial R,$$

where $R = \{\,(x,y)\,|\,0 < x < 1, 0 < y < 2\}$. Approximate $\psi(x,y)$ using the finite difference method with a uniform mesh spacing of $h = 0.10$ inches. Use the SOR method to solve the finite difference equations.

13. Rework the "Electric Potential with Unusual Boundary Conditions" application problem first with $\lambda = 0.05$ and then with $\lambda = 0.5$.

14. In the "Electric Potential with Unusual Boundary Conditions" application problem, determine the equipotential curves when $L = 1$, $H = 3$, $V_0 = 2$ and $\lambda = 0.3$.

In Exercises 15–22, we consider Jacobi and Gauss-Seidel *line relaxation schemes.* In a line relaxation scheme, the values of all of the unknowns along either a row or a column of the grid are updated simultaneously. For more detail on these and other line relaxation schemes consult Thomas [2].

15. For Jacobi y-line relaxation, we rewrite the finite difference equation (1) as

$$-\bar{w}_{j-1,k} + 4\bar{w}_{j,k} - \bar{w}_{j-1,k} = \tilde{w}_{j,k-1} + \tilde{w}_{j,k+1} - h^2 f_{j,k}$$

and then solve for all of the unknowns along the kth row of the grid simultaneously. This is referred to as y-line relaxation because the kth row of the grid corresponds to a fixed value of y. Jacobi x-line relaxation corresponds to solving for all of the unknowns along the jth column of the grid at the same time.

(a) Construct an algorithm to perform Jacobi y-line relaxation.

(b) Construct an algorithm to perform Jacobi x-line relaxation.

(c) How many operations per point does Jacobi line relaxation require? How does this compare with the number of operations per point for Jacobi point relaxation?

16. (a) Apply Jacobi x-line relaxation and Jacobi y-line relaxation to the problem

$$\frac{\partial^2 u}{\partial x^2} + \frac{\partial^2 u}{\partial y^2} = 0 \quad \text{on} \quad R = \{(x, y) | \, 1 < x < 2, 0 < y < 1\}$$
$$u(x, 0) = 2 \ln x, \quad u(x, 1) = \ln(x^2 + 1)$$
$$u(1, y) = \ln(y^2 + 1), \quad u(2, y) = \ln(y^2 + 4),$$

which was considered in the text. Use $N \times N$ grids with $N = 5$, 10, 20 and 40, and a convergence tolerance of 5×10^{-11}.

(b) How does the number of sweeps needed to achieve convergence vary with N?

(c) How does the number of sweeps needed by line relaxation compare with the number of sweeps needed by point relaxation?

17. Repeat Exercise 16 for the Poisson problem in Exercise 8.

18. (a) Apply Jacobi x-line relaxation and Jacobi y-line relaxation to the problem

$$\frac{\partial^2 u}{\partial x^2} + \frac{\partial^2 u}{\partial y^2} = -[2 + \pi^2 x(1 - x)] \cos(\pi y)$$
$$u(x, 0) = u(x, 10) = x(1 - x)$$
$$u(0, y) = u(1, y) = 0.$$

Take $h = \frac{1}{10}$ and use a convergence tolerance of 5×10^{-7}. Which scheme performs better?

(b) Apply Jacobi x-line relaxation and Jacobi y-line relaxation to the problem

$$\frac{\partial^2 u}{\partial x^2} + \frac{\partial^2 u}{\partial y^2} = -[2 + \pi^2 x(1-x)]\cos(\pi y)$$
$$u(x,0) = x(1-x), \quad u(x,1) = x(x-1)$$
$$u(0,y) = 0, \quad u(10,y) = -90\cos(\pi y).$$

Take $h = \frac{1}{10}$ and use a convergence tolerance of 5×10^{-7}. Which scheme performs better?

(c) Comment on the results of parts (a) and (b).

19. (a) Construct an algorithm to perform Gauss-Seidel y-line relaxation.

(b) Construct an algorithm to perform Gauss-Seidel x-line relaxation.

(c) How many operations per point does Gauss-Seidel line relaxation require? How does this compare with the number of operations per point for Gauss-Seidel point relaxation?

20. Repeat Exercise 16 using Gauss-Seidel line relaxation.

21. Repeat Exercise 17 using Gauss-Seidel line relaxation.

22. Repeat Exercise 18 using Gauss-Seidel line relaxation.

9.4 LOCAL MODE ANALYSIS OF RELAXATION AND THE MULTIGRID METHOD

In the previous section, it was first observed numerically and then established theoretically that the asymptotic convergence rate, μ, for the Jacobi method, the Gauss-Seidel method and the SOR method increases when the mesh spacing, h, decreases. This result implies an inherent, and unfortunate, tradeoff between truncation error and the convergence rate of relaxation schemes. In this section, we will perform what is known as a local mode analysis of these relaxation schemes. In a local mode analysis, we decompose the iteration error into components which have different frequencies and then quantify the amount by which a single relaxation sweep reduces each of the components. Our objective is to obtain a better understanding of the dependence of μ upon h. Once this has been done, an algorithm for improving the convergence of relaxation schemes will be developed.

Before we begin our discussions, it should be noted that the material in this section is due to Achi Brandt, who first presented the ideas in the paper "Multilevel Adaptive Solutions to Boundary Value Problems" [1].

Error Evolution Equation

The first step in performing a local mode analysis is to derive the error evolution equation for the particular relaxation scheme. To be specific, let's consider Gauss-Seidel relaxation with lexicographic ordering of the unknowns. With this choice, the equation

$$\frac{\bar{w}_{j-1,k} + \tilde{w}_{j+1,k} + \bar{w}_{j,k-1} + \tilde{w}_{j,k+1} - 4\bar{w}_{j,k}}{h^2} = f_{j,k}$$

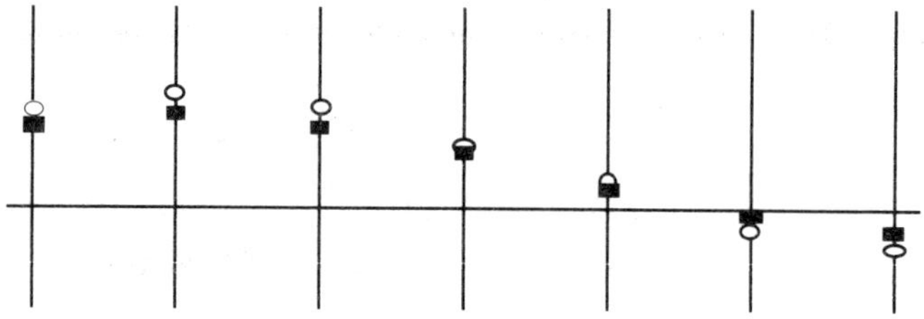

Figure 9.17 Example of a smooth error component. Open circles denote before sweep errors, and filled rectangles denote after sweep errors.

defines the after sweep value $\bar{w}_{j,k}$. The corresponding finite difference equation is

$$\frac{w_{j-1,k} + w_{j+1,k} + w_{j,k-1} + w_{j,k+1} - 4w_{j,k}}{h^2} = f_{j,k}.$$

Let $\tilde{v}^h = w^h - \tilde{w}^h$ and $\bar{v}^h = w^h - \bar{w}^h$ denote the before sweep error and the after sweep error, respectively, in the relaxation solution to the finite difference equations. Subtracting the Gauss-Seidel equation from the finite difference equation and rearranging terms yields

$$\bar{v}_{j,k} = \frac{1}{4}\left(\bar{v}_{j-1,k} + \tilde{v}_{j+1,k} + \bar{v}_{j,k-1} + \tilde{v}_{j,k+1}\right). \tag{1}$$

This is the *error evolution equation* for Gauss-Seidel relaxation with lexicographic ordering. It indicates that the new error at location (j, k) of the grid is merely the average of the current errors at the horizontally and vertically adjacent grid points.

To assess the importance of this statement, we will distinguish between two types of error components: smooth and nonsmooth. An error component which is either slowly oscillating, or non-oscillating, relative to the computational grid, will be considered smooth. An example is illustrated in Figure 9.17. The open circles indicate the before sweep error, and the filled rectangles indicate the after sweep error. Very little reduction in the amplitude of the error has occurred. On the other hand, a non-smooth error component, one that oscillates rapidly relative to the grid, is illustrated in Figure 9.18. Again, open circles indicate before sweep error, and filled rectangles indicate after sweep error. Here, there has been a significant reduction in the amplitude of the error.

From these figures we conclude that relaxation reduces the amplitude of smooth error components slowly, but reduces the amplitude of non-smooth error components more rapidly. Hence, relaxation is efficient at smoothing the error, but inefficient at solving the underlying equations.

Transformation to Fourier Space

To turn this qualitative conclusion into a quantitative one, we need to transform the problem from the physical space into the Fourier, or frequency, space. In

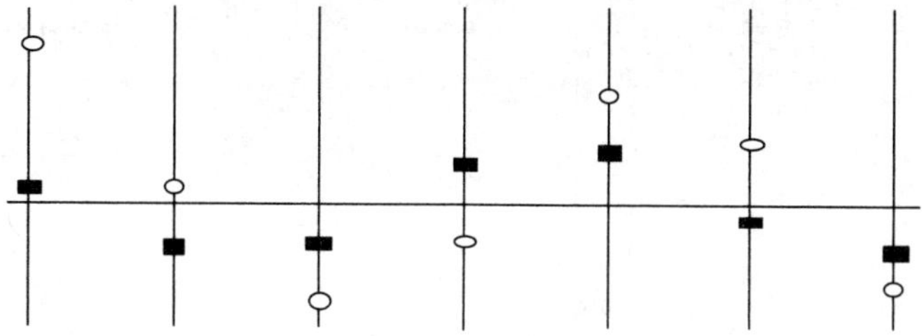

Figure 9.18 Example of a nonsmooth error component. Open circles denote before sweep errors, and filled rectangles denote after sweep errors.

the Fourier space, the dependence of the error on the frequency of oscillation is explicitly shown. Using frequency as a parameter will also provide a simple means for classifying error components as smooth or nonsmooth.

Over the rectangular domain $R = \{ (x, y) | \, a \le x \le b, c \le y \le d\}$, the continuous Fourier basis functions are given by

$$E_{p,q}(x, y) = \exp\left\{2\pi i \left[p\left(\frac{x-a}{b-a}\right) + q\left(\frac{y-c}{d-c}\right)\right]\right\}.$$

The wavenumbers $p, q = 0, \pm 1, \pm 2, \ldots$ indicate the frequency of oscillation in the x-direction and the y-direction, respectively. Evaluating this basis function at the grid point (x_j, y_k), where

$$\begin{aligned} x_j &= a + jh \text{ for some } j = 0, 1, 2, \ldots, N, \\ y_k &= c + kh \text{ for some } k = 0, 1, 2, \ldots, M \text{ and} \\ h &= \frac{b-a}{N} = \frac{d-c}{M}, \end{aligned}$$

yields

$$E_{p,q}(x_j, y_k) = \exp\left\{2\pi i\left(\frac{pj}{N} + \frac{qk}{M}\right)\right\}.$$

If we now define the *discrete Fourier mode* $\Theta = (\theta_1, \theta_2)$, where

$$\theta_1 = \frac{2\pi p}{N} \quad \text{and} \quad \theta_2 = \frac{2\pi q}{M},$$

then the *discrete Fourier basis functions* relative to the computational grid are given by

$$E_\Theta(x_j, y_k) = \exp\{i(j\theta_1 + k\theta_2)\}.$$

Since the complex exponential $e^{i\theta}$ is a 2π-periodic function, it follows that the discrete Fourier basic functions E_Θ, $E_{\Theta\pm(2\pi,0)}$, $E_{\Theta\pm(0,2\pi)}$, and $E_{\Theta\pm(2\pi,2\pi)}$ all produce identical values at every grid point. This is a phenomena known as *aliasing.*

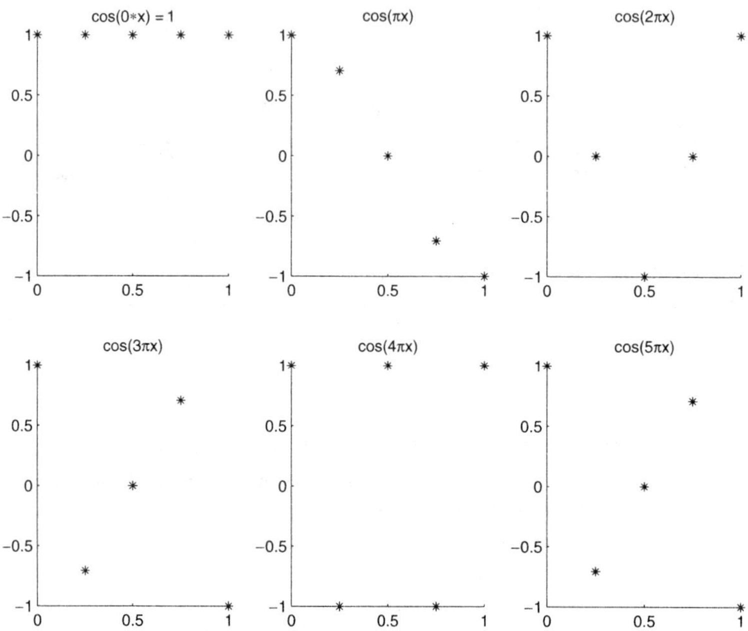

Figure 9.19 Demonstration of the aliasing phenomena over a uniform partition of the interval $[0, 1]$ into four subintervals. Note that, on this grid, the functions $\cos(3\pi x)$ and $\cos(5\pi x)$ cannot be differentiated.

Every computational grid has a maximum frequency that can be resolved. Any function that oscillates more rapidly than this maximum frequency, when evaluated at the grid points, will be indistinguishable from a grid function with a lower frequency. For our analysis, we will therefore restrict attention to the discrete frequency range $-\pi < \theta_1, \theta_2 < \pi$.

As a concrete example of aliasing, consider Figure 9.19. Here, we have a uniform partition of the interval $[0, 1]$ into four subintervals. The maximum frequency that this grid can resolve is 2–a function that alternates between maximum and minimum at successive grid points. Evaluated at the grid points, the functions $1 = \cos(0\pi x)$, $\cos(\pi x)$, $\cos(2\pi x)$, $\cos(3\pi x)$, and $\cos(4\pi x)$, which have frequencies of 0, 1/2, 1, 3/2, and 2, appear as distinct grid functions. However, the function $\cos(5\pi x)$, which has a frequency of 5/2, is identical to the function $\cos(3\pi x)$. The grid cannot differentiate these functions.

Local Mode Analysis

As the name implies, local mode analysis uses only local information. All boundary conditions are ignored. Therefore, this type of analysis is useful for local processes, like relaxation schemes. The outcomes from a local mode analysis are insight into the mechanisms at work and quantitatively correct convergence information.

Suppose that the before sweep and after sweep errors can be written as

$$\tilde{v}^h = \tilde{A}_\Theta E_\Theta \quad \text{and} \quad \bar{v}^h = \bar{A}_\Theta E_\Theta,$$

for each discrete Fourier mode $\Theta = (\theta_1, \theta_2)$ with $-\pi < \theta_1, \theta_2 < \pi$. The amplitudes, $\tilde{A}_\Theta$ and $\bar{A}_\Theta$, are associated with the discrete Fourier mode Θ and are independent of the spatial variables. We are interested in how well one sweep of the relaxation scheme reduces the amplitude of each component of the error, which is measured by the ratio of $\bar{A}_\Theta$ to $\tilde{A}_\Theta$.

Definition. The CONVERGENCE FACTOR, $\mu(\Theta)$, is given by

$$\mu(\Theta) = \left|\bar{A}_\Theta / \tilde{A}_\Theta\right|$$

and measures the reduction in error amplitude produced by a single sweep of the relaxation scheme as a function of the discrete Fourier mode, Θ.

Based on the analysis associated with Figures 9.17 and 9.18, we expect $\mu(\Theta)$ to be small for high-frequency modes but close to unity for low-frequency modes.

For Gauss-Seidel relaxation with lexicographic ordering, substitution of the discrete Fourier representation of the errors into the error evolution equation, (1), yields

$$\bar{A}_\Theta e^{i(j\theta_1 + k\theta_2)} = \frac{1}{4}\left[\bar{A}_\Theta e^{-i\theta_1} + \tilde{A}_\Theta e^{i\theta_1} + \bar{A}_\Theta e^{-i\theta_2} + \tilde{A}_\Theta e^{i\theta_2}\right] e^{i(j\theta_1 + k\theta_2)}.$$

Thus, the convergence factor for this scheme is

$$\mu_{\rm GS}(\Theta) = \left|\frac{\bar{A}_\Theta}{\tilde{A}_\Theta}\right| = \left|\frac{e^{i\theta_1} + e^{i\theta_2}}{4 - e^{-i\theta_1} - e^{-i\theta_2}}\right|.$$

A contour plot of this function is shown in Figure 9.20. Note that around the outside of the plot, where at least one of the frequencies of the Fourier mode is high, $\mu_{\rm GS} \approx 0.1 - 0.4$. Hence, error reduction for these modes is good. Unfortunately, when both frequencies are low, $\mu_{\rm GS} \approx 1$. In particular, for the smallest frequencies

$$(\theta_1, \theta_2) = \left(\pm\frac{2\pi}{N}, \pm\frac{2\pi}{M}\right) = \left(\pm\frac{2\pi h}{b-a}, \pm\frac{2\pi h}{d-c}\right),$$

expanding the exponential functions in the formula for $\mu_{\rm GS}(\Theta)$ in Taylor series leads to the estimate $\mu_{GS}(\Theta) \approx 1 - O(h^2)$.

In summary, local mode analysis has shown us that Gauss-Seidel relaxation with lexicographic ordering does an excellent job reducing the amplitude of the high-frequency components of the error. Error reduction is poor only for the low-frequency components of the error, with the convergence factor $\mu_{\rm GS}(\Theta) \approx 1 - O(h^2)$ for the lowest frequencies. Local mode analysis of the Jacobi method and the SOR method is left for the exercises.

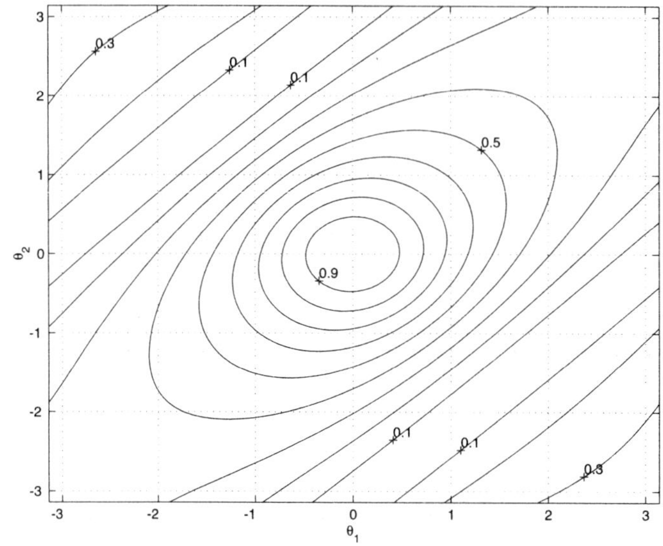

Figure 9.20 Contour plot of convergence factor for Gauss-Seidel relaxation with lexicographic ordering.

A Two-Grid Method

Although relaxation does a poor job reducing the amplitude of the low frequency error components, the key observation to make is that these smooth error components do not need a mesh spacing of h to be accurately resolved. They can be approximated just as well on a coarser mesh, one with a larger mesh spacing, with much less work involved. Furthermore, on a coarser mesh, some of the frequencies which are low relative to the fine grid will be high relative to the coarse grid.

This suggests working with two grids, rather than just one. Let the first grid have a mesh spacing of h. We will refer to this as the *fine grid.* The fine grid will be used to approximate the solution of the partial differential equation, so the mesh size should be selected to provide the desired resolution. After several relaxation sweeps on the fine grid, the error, $\bar{v}^h = w^h - \bar{w}^h$, will be smooth. At this point, we introduce a second grid, one with a mesh spacing of $2h$. This *coarse grid* will be used to approximate the smooth error from the fine grid, and this approximation will, in turn, serve as a correction to the fine grid solution.

Now, if we are going to use the coarse grid to approximate $\bar{v}^h$, we have to know the equation that $\bar{v}^h$ satisfies. Let L^h denote the discrete Poisson operator, which is defined by the relation

$$\left(L^h u^h\right)_{j,k} = \frac{u_{j-1,k} + u_{j+1,k} + u_{j,k-1} + u_{j,k+1} - 4u_{j,k}}{h^2}$$

for any grid function u^h. Applying L^h to the equation $\bar{v}^h = w^h - \bar{w}^h$ yields

$$L^h\bar{v}^h = L^h w^h - L^h\bar{w}^h = f^h - L^h\bar{w}^h, \tag{2}$$

where we have used the fact that w^h is the true solution of the discrete Poisson equation, so $L^h w^h = f^h$. Defining $r^h = f^h - L^h \bar{w}^h$, equation (2) becomes

$$L^h \bar{v}^h = r^h.$$

The grid function r^h is called the *residual.*

The basic two-grid solution algorithm consists of four steps. First, relax n_1 sweeps on the fine grid to produce an approximate solution $\bar{w}^h$. Second, construct the coarse grid problem to approximate $\bar{v}^h$, the error associated with $\bar{w}^h$. The analogue of the equation $L^h \bar{v}^h = r^h$ for a grid that has a mesh spacing of $2h$ is

$$L^{2h} \bar{v}^{2h} = I_h^{2h} r^h,$$

where I_h^{2h} represents the transfer of the residual from the fine grid to the coarse grid. Third, by some means, solve the coarse grid problem for $\bar{v}^{2h}$. Finally, correct the fine grid solution by adding $I_{2h}^h \bar{v}^{2h}$ to $\bar{w}^h$, where I_{2h}^h represents the transfer of the correction from the coarse grid to the fine grid. Repeat these four steps until convergence is achieved on the fine grid.

In the second step of the algorithm, we need to transfer the residual from the fine grid to the coarse grid. The easiest scheme for performing this task is known as *injection*: evaluate r^h on the fine grid and copy values to the coarse grid at points that are common to both grids. Another grid transfer operator is needed in the fourth step. The transfer of the correction from the coarse grid to the fine grid is generally carried out using bilinear interpolation. Taking into account the uniform mesh spacing of the grids, the operator I_{2h}^h is given by

$$\left(I_{2h}^h \bar{v}^{2h}\right)_{j,k} =$$
$$\begin{cases} \bar{v}^{2h}_{j/2,k/2} & j \bmod 2 = k \bmod 2 = 0 \\ 0.5(\bar{v}^{2h}_{(j-1)/2,k/2} + \bar{v}^{2h}_{(j+1)/2,k/2}) & j \bmod 2 = 1, \quad k \bmod 2 = 0 \\ 0.5(\bar{v}^{2h}_{j/2,(k-1)/2} + \bar{v}^{2h}_{j/2,(k+1)/2}) & j \bmod 2 = 0, \quad k \bmod 2 = 1 \\ \quad 0.25(\bar{v}^{2h}_{(j-1)/2,(k-1)/2} + \bar{v}^{2h}_{(j+1)/2,(k-1)/2} + & \\ \qquad \bar{v}^{2h}_{(j-1)/2,(k+1)/2} + \bar{v}^{2h}_{(j+1)/2,(k+1)/2}) & j \bmod 2 = k \bmod 2 = 1 \end{cases}$$

where j and k run over the subscripts on the fine grid.

What is the overall effect of this two-grid algorithm on the iteration error? To answer this question we must first recognize that because the mesh spacing on the coarse grid is double that of the fine grid, the error modes on the fine grid are divided into two groups: those for which $|\Theta| < \pi/2$ and those for which $\pi/2 \leq |\Theta| \leq \pi$, where $|\Theta| = \max(|\theta_1|, |\theta_2|)$—see Figure 9.21. The former group consists of those modes which can be resolved on the coarse grid, while the latter group contains those modes whose frequencies are too high to be resolved on the coarse grid.

The modes with $|\Theta| < \pi/2$ have their amplitudes reduced little by the relaxation sweeps on the fine grid but are subsequently eliminated when the coarse grid problem is solved. This, of course, assumes that the grid transfers introduce no errors. As for the modes with $\pi/2 \leq |\Theta| \leq \pi$, if we let

$$\bar{\mu} = \max_{(\pi/2) \leq |\Theta| \leq \pi} \mu(\Theta),$$

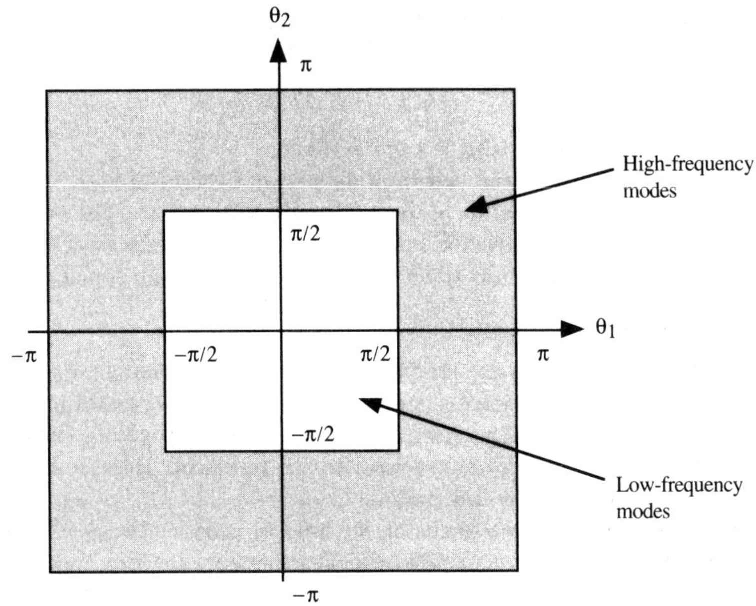

Figure 9.21 Designation of Fourier modes as high frequency versus low frequency.

then n_1 relaxation sweeps on the fine grid reduce the amplitudes of these modes by at least a factor of $\bar{\mu}^{n_1}$ per cycle. Hence, the result of the two-grid process is the reduction of all error mode amplitudes by at least $\bar{\mu}^{n_1}$ per cycle, *independent* of h.

The quantity $\bar{\mu}$ is called the *smoothing factor* and measures the efficiency with which a relaxation scheme reduces the amplitude of high frequency error modes. For Gauss-Seidel relaxation with lexicographic ordering, if we were to zoom in on the center of Figure 9.20, we would find that the 0.5 contour touches the boundary between low and high frequencies ($|\Theta| = \pi/2$) and is the highest valued contour to do so. Therefore, $\bar{\mu}_{\text{GS}} = 1/2$.

Multigrid Method

There is still one detail of the two-grid method which must be addressed, and that is the solution of the coarse grid problem in the third step. One choice is to repeat the fine grid procedure. Perform n_1 relaxation sweeps on the coarse grid to reduce the amplitude of the error components that are high frequency relative to the coarse grid. Next, transfer the residual to a still coarser grid, one with a mesh spacing of $4h$, to approximate the error components that remain low frequency with respect to the mesh spacing of $2h$. We can then continue to introduce coarser and coarser grids, doubling the mesh spacing each time. Note that with each doubling of h, we reduce the frequency resolution of the grid by a factor of two. Hence, half of the low frequencies on one grid become high frequencies relative to the next grid in the sequence.

If the problem domain is square and the finest grid has a number of subintervals that is a power of 2, eventually the coarsening process will produce a grid with only two subdivisions. One relaxation sweep over this grid (which has only one unknown) is equivalent to solving the finite difference equation on that grid. Even if the domain is not square, but the number of subdivisions in each direction on the finest grid is still a power of 2, we will eventually reach a grid which has either one row or one column of unknowns. The system of equations on that grid can be easily solved using direct techniques.

Once the problem on the coarsest grid has been solved, we work backward from the coarsest grid to the finest grid. Each step in this process requires interpolating the correction from the coarser of the two grids and adding it to the approximation on the finer grid. To achieve further reduction of the high-frequency error amplitudes, we can then perform n_2 relaxation sweeps on the finer grid. Figure 9.22 provides a schematic for this multiple grid algorithm. For obvious reasons, this approach is known as the *multigrid V-cycle.* The multigrid method uses repeated applications of the V-cycle until convergence has been achieved.

The cost of the V-cycle must be assessed in terms of both storage and work requirements. Suppose the finest grid contains P points. Since the coarse grids have been constructed by doubling the mesh spacing, each successive grid will contain one-quarter the number of points as the previous grid. The total number of points among all of the grids is then

$$\begin{aligned} P + \frac{P}{4} + \frac{P}{16} + \cdots + \frac{P}{4^{ng-1}} &= P\left(1 + \frac{1}{4} + \frac{1}{16} + \cdots + \frac{1}{4^{ng-1}}\right) \\ &< \frac{4}{3}P, \end{aligned}$$

where ng denotes the number of grids. Therefore, storage for all of the grids requires less than $4/3$ times the storage needed for the fine grid problem alone.

To measure the amount of work used by the multigrid algorithm, we first make the following definition.

> **Definition.** One WORK UNIT is the computational cost of one sweep over the finest grid.

Since the cost of one sweep over a grid is proportional to the number of points on the grid, the total work required to complete one V-cycle is

$$n + \frac{n}{4} + \frac{n}{16} + \cdots + \frac{n}{4^{ng-1}} = n\left(1 + \frac{1}{4} + \frac{1}{16} + \cdots + \frac{1}{4^{ng-1}}\right) < \frac{4}{3}n$$

work units, where $n = n_1 + n_2$ is the total number of sweeps performed on each grid. For example, if $n_1 = 2$ and $n_2 = 1$, then $n = 3$, and the total

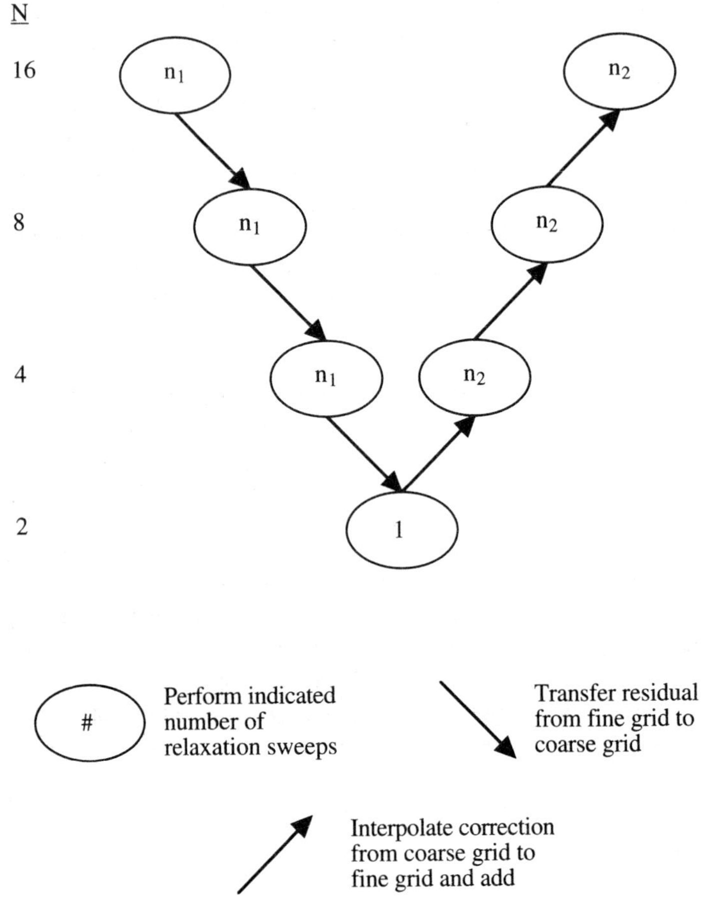

Figure 9.22 Schematic for operation of the multigrid V-cycle.

work per V-cycle is less than 4 work units. Thus, all of the sweeps over all of the additional grids amounts to less than one additional sweep over the finest grid.

What do we get in return for these modest increases in storage and work? Well, the lowest frequency components of the error on the finest grid are eliminated by solving the problem on the coarsest grid. Every other component of the error on the finest grid is high frequency on one of the grids, so the relaxation sweeps reduce the amplitude of every other error component by at least $\bar{\mu}^n$. Therefore, all components of the error are reduced by at least $\bar{\mu}^n$ per V-cycle, *independent* of h. Thus, even though each V-cycle requires a modest increase in the number of work units used, to achieve convergence, we expect to use significantly fewer work units than if we had worked only with relaxation over the finest grid.

EXAMPLE 9.6 Multigrid Performance

Consider the Poisson problem

$$\frac{\partial^2 u}{\partial x^2} + \frac{\partial^2 u}{\partial y^2} = -52\cos(4x+6y)$$

over the unit square, $R = \{(x,y) \,|\, 0 < x < 1, 0 < y < 1\}$, with the Dirichlet boundary conditions

$$\begin{aligned} u(x,0) &= \cos(4x), \quad u(x,1) = \cos(4x+6) \\ u(0,y) &= \cos(6y), \quad u(1,y) = \cos(6y+4). \end{aligned}$$

The number of sweeps required by the Gauss-Seidel method and the SOR method for three different mesh spacings is summarized below. A convergence tolerance of 5×10^{-8} was applied to $\|\bar{w}^h - \tilde{w}^h\|_\infty$. The optimal value of the relaxation parameter was used in each case for the SOR method.

	$h = 1/8$	$h = 1/16$	$h = 1/32$
Gauss-Seidel	91	320	1100
SOR with $\omega = \omega_{\text{opt}}$	28	52	103

Note that each time the mesh spacing is cut in half, the number of work units required by the Gauss-Seidel method increases by roughly a factor of 4. We now know that this is due to the fact that $\mu_{\text{GS}} \approx 1 - O(h^2)$ for the lowest frequency error components. On the other hand, the number of work units required by the SOR method increases by a factor of 2. In Exercise 2, we will see that $\mu_{\text{SOR}} \approx 1 - O(h)$ for the lowest frequency error components.

For the same three values of h and the same convergence tolerance applied to $\|\bar{w}^h - \tilde{w}^h\|_\infty$ after the final sweep on the finest grid, the convergence results for the multigrid method are summarized below. In each case $n_1 = 2$ relaxation sweeps were performed when moving from the finest grid to the coarsest grid, and $n_2 = 1$ sweep was performed when working back to the finest grid. The performance measures listed in the tables of Table 9.1 are defined as

$$\begin{matrix}\text{convergence} \\ \text{per V-cycle}\end{matrix} = \frac{\|\bar{w}^h - \tilde{w}^h\|_\infty}{\|\bar{w}^h - \tilde{w}^h\|_\infty \text{ from previous cycle}},$$

$$\begin{matrix}\text{convergence} \\ \text{per work unit}\end{matrix} = \left(\begin{matrix}\text{convergence} \\ \text{per V-cycle}\end{matrix}\right)^{1/\text{work units per V-cycle}}$$

and

$$\begin{matrix}\text{total number} \\ \text{of work units}\end{matrix} = \left(\begin{matrix}\text{number} \\ \text{of cycles}\end{matrix}\right)\left(\begin{matrix}\text{work units} \\ \text{per V-cycle}\end{matrix}\right),$$

where

$$\begin{matrix}\text{work units} \\ \text{per V-cycle}\end{matrix} = (n_1+n_2)\left(1 + \frac{1}{4} + \frac{1}{16} + \cdots + \frac{1}{4^{ng-1}}\right)$$

and ng denotes the number of grids.

$h = 1/8$—Three Grids

Cycle	$\|\bar{w}^h - \tilde{w}^h\|_\infty$ After Final Sweep	Convergence Factors Per V-Cycle	Per Work Unit
1	1.550×10^{-1}		
2	9.158×10^{-3}	0.059100	0.487554
3	3.349×10^{-4}	0.036568	0.431591
4	2.424×10^{-5}	0.072391	0.513329
5	1.036×10^{-6}	0.042746	0.449046
6	6.149×10^{-8}	0.059339	0.488054
7	3.838×10^{-9}	0.062407	0.494342

total number of work units used = 27.562500

$h = 1/16$—Four Grids

Cycle	$\|\bar{w}^h - \tilde{w}^h\|_\infty$ After Final Sweep	Convergence Factors Per V-Cycle	Per Work Unit
1	1.533×10^{-1}		
2	9.459×10^{-3}	0.061689	0.497011
3	5.090×10^{-4}	0.053816	0.480268
4	3.595×10^{-5}	0.070621	0.514167
5	3.064×10^{-6}	0.085225	0.539006
6	2.497×10^{-7}	0.081508	0.533007
7	1.583×10^{-8}	0.063400	0.500435

total number of work units used = 27.890625

$h = 1/32$—Five Grids

Cycle	$\|\bar{w}^h - \tilde{w}^h\|_\infty$ After Final Sweep	Convergence Factors Per V-Cycle	Per Work Unit
1	1.477×10^{-1}		
2	1.136×10^{-2}	0.076905	0.526280
3	5.001×10^{-4}	0.044012	0.457680
4	3.420×10^{-5}	0.068394	0.511058
5	2.657×10^{-6}	0.077701	0.527638
6	1.740×10^{-7}	0.065483	0.505525
7	1.573×10^{-8}	0.090366	0.547956

total number of work units used = 27.972656

TABLE 9.1: Tables for Example "Multigrid performance".

The important observation to make is that, regardless of h, the convergence per V-cycle and per work unit remains relatively constant. For less work than SOR needs to compute the solution with $h = 1/8$, the multigrid method computes the solution with $h = 1/32$.

We have barely scratched the surface on the topic of multigrid methods. The reader interested in more detail should consult one of the following references. As noted at the start of this section, the seminal work on multigrid methods is the paper "Multilevel Adaptive Solutions to Boundary Value Problems" by Brandt [1]. This paper presents many of the fundamental concepts in a very readable format. Fulton, Ciesielski, and Schubert [2] also provide a very readable review of the basic multigrid concepts. The books by Briggs, Henson and McCormick [3], Hackbusch and Trottenberg [4], and McCormick [5] provide material on various aspects of multigrid theory and application.

References

1. A. Brandt, "Multilevel Adaptive Solutions to Boundary Value Problems," *Mathematics of Computation,* **31**, 333–390, 1977.
2. S. Fulton, P. Ciesielski, and W. Schubert, "Multigrid Methods for Elliptic Problems: A Review," *Monthly Weather Review,* **114** (5), 943–959, 1986.
3. W. Briggs, V.E. Henson and S. McCormick, *A Multigrid Tutorial, Second Edition,* SIAM, Philadelphia, 2000.
4. W. Hackbusch and U. Trottenberg, *Multigrid Methods,* Springer-Verlag, Berlin, 1982.
5. S. McCormick, *Multigrid Methods,* SIAM Frontiers Series, volume 3, SIAM, Philadelphia, 1987.
6. S. Timoshenko and S. Woinowsky-Krieger, *Theory of Plates and Shells,* 2nd edition, McGraw-Hill, New York, 1959.

EXERCISES

1. Recall that the update equation for Jacobi relaxation is given by

$$\frac{\tilde{w}_{j-1,k} + \tilde{w}_{j+1,k} + \tilde{w}_{j,k-1} + \tilde{w}_{j,k+1} - 4\bar{w}_{j,k}}{h^2} = f_{j,k}.$$

(a) Determine the error evolution equation for Jacobi relaxation.
(b) Determine the formula for the convergence factor $\mu_{\text{JAC}}(\Theta)$.
(c) Show that $\mu_{JAC}(\Theta) \approx 1 - O(h^2)$ for the lowest non-zero frequencies

$$(\theta_1, \theta_2) = \left(\pm\frac{2\pi}{N}, \pm\frac{2\pi}{M}\right) = \left(\pm\frac{2\pi h}{b-a}, \pm\frac{2\pi h}{d-c}\right).$$

2. The update equation for the SOR method can be written in the form

$$\frac{\omega\bar{w}_{j-1,k} + \omega\tilde{w}_{j+1,k} + \omega\bar{w}_{j,k-1} + \omega\tilde{w}_{j,k+1} - 4(\omega-1)\tilde{w}_{j,k} - 4\omega\bar{w}_{j,k}}{\omega h^2} = f_{j,k}.$$

Take $\omega = \omega_{\text{opt}}$, which, for the case of a uniform mesh over the unit square, is

$$\omega_{\text{opt}} = \frac{2}{1+\sin(\pi h)}.$$

(a) Determine the error evolution equation for the SOR method.

(b) Determine the formula for the convergence factor $\mu_{\text{SOR}}(\Theta)$.

(c) Show that $\mu_{\text{SOR}}(\Theta) \approx 1 - O(h)$ for the lowest nonzero frequencies

$$(\theta_1, \theta_2) = \left(\pm\frac{2\pi}{N}, \pm\frac{2\pi}{M}\right) = \left(\pm\frac{2\pi h}{b-a}, \pm\frac{2\pi h}{d-c}\right).$$

(d) Estimate the smoothing factor $\bar{\mu}_{SOR}$.

In Exercises 3–8, apply the multigrid method to each of the following Poisson problems over the unit square. Use mesh spacings of $h = 1/8$, $h = 1/16$, and $h = 1/32$. Take $n_1 = 2$ and $n_2 = 1$ and apply a convergence tolerance of 5×10^{-8} to $\|\bar{w}^h - \tilde{w}^h\|_\infty$. Compute the convergence factor per V-cycle, the convergence factor per work unit, and the total number of work units expended. Compare the total number of work units to the number of iterations of the Gauss-Seidel method and the SOR method (using $\omega = \omega_{opt}$).

3. $\dfrac{\partial^2 u}{\partial x^2} + \dfrac{\partial^2 u}{\partial y^2} = -2\pi^2 \sin(\pi x)\sin(\pi y)$, exact solution: $u(x,y) = \sin(\pi x)\sin(\pi y)$
$u(x,0) = u(x,1) = u(0,y) = u(1,y) = 0$

4. $\dfrac{\partial^2 u}{\partial x^2} + \dfrac{\partial^2 u}{\partial y^2} = 0$, exact solution: $u(x,y) = \dfrac{y}{(1+x)^2 + y^2}$
$u(x,0) = 0$, $u(x,1) = \dfrac{1}{(1+x)^2+1}$, $u(0,y) = \dfrac{y}{1+y^2}$, $u(1,y) = \dfrac{y}{4+y^2}$

5. $\dfrac{\partial^2 u}{\partial x^2} + \dfrac{\partial^2 u}{\partial y^2} = x^2 + y^2$, exact solution: $u(x,y) = e^{\pi x}\sin(\pi y) + \dfrac{1}{2}(xy)^2$
$u(x,0) = 0$, $u(x,1) = \dfrac{1}{2}x^2$, $u(0,y) = \sin(\pi y)$, $u(1,y) = e^{\pi}\sin(\pi y) + \dfrac{1}{2}y^2$

6. $\dfrac{\partial^2 u}{\partial x^2} + \dfrac{\partial^2 u}{\partial y^2} = -[2 + \pi^2 x(1-x)]\cos(\pi y)$,
exact solution: $u(x,y) = x(1-x)\cos(\pi y)$
$u(x,0) = x(1-x)$, $u(x,1) = x(x-1)$, $u(0,y) = u(1,y) = 0$

7. $\dfrac{\partial^2 u}{\partial x^2} + \dfrac{\partial^2 u}{\partial y^2} = (x^2+y^2)e^{-xy}$, exact solution: $u(x,y) = e^{-xy}$
$u(x,0) = u(0,y) = 1$, $u(x,1) = e^{-x}$, $u(1,y) = e^{-y}$

8. $\dfrac{\partial^2 u}{\partial x^2} + \dfrac{\partial^2 u}{\partial y^2} = -\dfrac{x^2+y^2}{x^2y^2}$, exact solution: $u(x,y) = \ln(xy)$
$u(x,1) = \ln x$, $u(x,2) = \ln(2x)$, $u(1,y) = \ln y$, $u(2,y) = \ln(2y)$

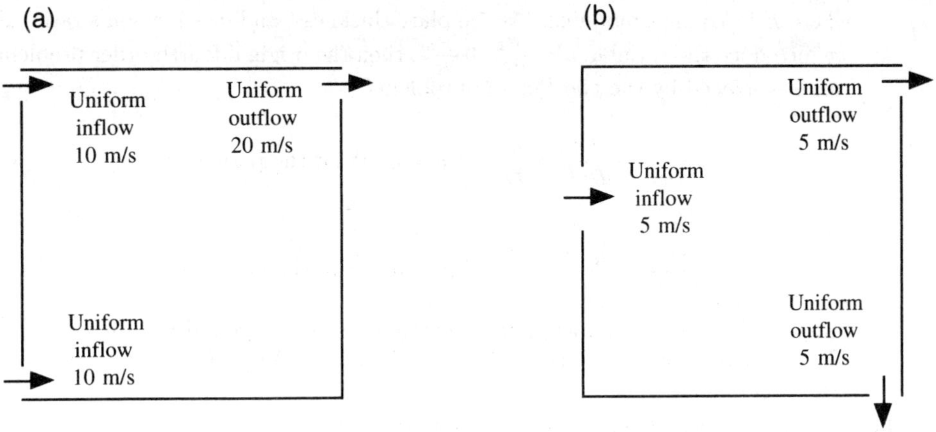

Figure 9.23 (a) Flow chamber for Exercise 9. (b) Flow chamber for Exercise 10.

9. Consider incompressible, irrotational flow through the chamber shown in Figure 9.23(a). The chamber is a square, one meter on a side, and the flow openings all have a width of 0.1 meters.

(a) Set up the boundary value problem for the stream function for this flow using the "Flow Through a Contraction Duct" problem capsule as a guide (see page 726).

(b) Solve the boundary value problem from part (a) and plot the resulting streamlines.

10. Repeat Exercise 9 for incompressible, irrotational flow through the chamber in Figure 9.23(b). The chamber is a square with side length of one meter. The opening on the left side has a width of 0.2 meters, and the bottom edge of the opening is at the midpoint of the side. The outflow openings each have a width of 0.1 meters.

11. A thin square plate is placed in a horizontal position and simply supported along its perimeter. A distributed load q is applied to the upper surface. The deflection, $w(x, y)$, of the plate from the horizontal satisfies (see Timoshenko and Woinowsky-Krieger [6])

$$\frac{\partial^4 w}{\partial x^4} + 2\frac{\partial^4 w}{\partial x^2 y^2} + \frac{\partial^4 w}{\partial y^4} = \frac{q}{D}$$

subject to the boundary conditions

$$w = 0 \quad \text{and} \quad \frac{\partial w}{\partial n} = 0 \quad \text{along the perimeter.}$$

The constant D is known as the flexural rigidity and is given by

$$D = \frac{Et^3}{12(1 - \sigma^2)},$$

where E is Young's modulus, t is the plate thickness, and σ is Poisson's ratio. If we introduce the variable $u = \frac{\partial^2 w}{\partial x^2} + \frac{\partial^2 w}{\partial y^2}$, then the original fourth-order problem can be replaced by the two Poisson problems

$$\frac{\partial^2 u}{\partial x^2} + \frac{\partial^2 u}{\partial y^2} = \frac{q}{D}, \qquad u = 0 \text{ along the perimeter}$$

and

$$\frac{\partial^2 w}{\partial x^2} + \frac{\partial^2 w}{\partial y^2} = u, \qquad w = 0 \text{ along the perimeter.}$$

Suppose the plate is 60 inches on a side and $q = 1$ psi (pound per square inch), $E = 30 \times 10^6$ psi, $\sigma = 0.27$, and $t = 0.15$ inches. With $N = 64$ on the finest grid, approximate $w(x, y)$.

12. Repeat Exercise 11 but replace the uniform distributed load by

$$q(x, y) = \frac{1}{360000} x(60 - x)y(60 - y).$$

13. Gauss-Seidel relaxation with red-black ordering of the unknowns (recall Exercise 13 of Section 9.1) can be programmed as follows:

```
for pass = 1 to 2
    ks = pass
    for j = 1 to N
        for k = ks to N by 2
            compute w̄(j, k)
        end
        ks = 3 - ks
    end
end
```

This code relaxes the red unknowns on the first pass and then the black unknowns on the second pass.

In the multigrid method, replace Gauss-Seidel relaxation with lexicographic ordering by Gauss-Seidel relaxation with red-black ordering. Also replace injection of the residual by half injection, in which the residual is multiplied by one-half before being copied to the coarser grid. With these modifications, resolve the Poisson problem from Exercise 3 with $h = 1/32$. What is the convergence factor per V-cycle? Use this value to estimate the smoothing factor for Gauss-Seidel relaxation with red-black ordering. Recall that in theory,

$$\begin{matrix}\text{convergence} \\ \text{per V-cycle}\end{matrix} = \bar{\mu}^n.$$

9.5 IRREGULAR DOMAINS

Thus far, we have worked exclusively with problems defined over rectangular domains. Real problems, unfortunately, often involve irregularly shaped domains. To finish out this chapter, we will therefore discuss the solution of the Poisson problem over nonrectangular domains. No attempt will be made to consider every possible configuration. Rather, several specific cases will be examined. These few examples, however, should be sufficient to allow for the investigation of a wide range of different domain geometries. The section will conclude with a general treatment of circular domains.

Sloped Boundaries

EXAMPLE 9.7 Flow through a Contraction Duct

In the Chapter 9 Overview (see page 726), we showed that the stream function, $\psi(x, y)$, for incompressible, irrotational flow through the asymmetric contraction duct in Figure 9.1 satisfies the Laplace equation

$$\frac{\partial^2\psi}{\partial x^2} + \frac{\partial^2\psi}{\partial y^2} = 0,$$

subject to the boundary conditions

$$\begin{aligned}\psi(0, y) &= 3y, \quad 0 \le y \le 2\\ \psi(3, y) &= 12(y-1), \quad 1 \le y \le \frac{3}{2}\\ \psi(x, y) &= 6, \quad \text{along the upper wall}\\ \psi(x, y) &= 0, \quad \text{along the lower wall.}\end{aligned}$$

Because the walls of the duct slope at an angle of 45°, it is possible to select the mesh spacing so that the grid follows the walls exactly. For instance, any mesh spacing of the form $h = \frac{1}{2N}$ for some positive integer N will suffice. The grid corresponding to $N = 2$ is shown in Figure 9.24.

The major difference between this problem and those which were treated in earlier sections is that here, the rows of the grid do not contain the same number of unknowns. In particular, with $h = \frac{1}{2N}$, the computational grid will have $4N-1$ rows of unknowns. With the sloped lower boundary, the rows numbered $k = 1$ through $k = 2N$ each contain $2N + k - 1$ unknowns. The rows numbered $k = 2N + 1$ through $k = 3N - 1$ each have $6N - 1$ unknowns. Finally, with the sloped upper boundary, the rows numbered $k = 3N$ through $k = 4N - 1$ each have $6N - k - 1$ unknowns.

To approximate $\psi(x, y)$, a mesh spacing of $h = 0.1$ meters was selected. The system of finite difference equations was solved using Gauss-Seidel relaxation with a convergence tolerance of 5×10^{-7}. The streamlines for the flow are plotted in Figure 9.25.

For any angle of inclination other than 45°, the only way for the computational grid to follow the entire boundary is to allow for a different mesh spacing in the x-direction than in the y-direction. This, of course, means that we must rederive

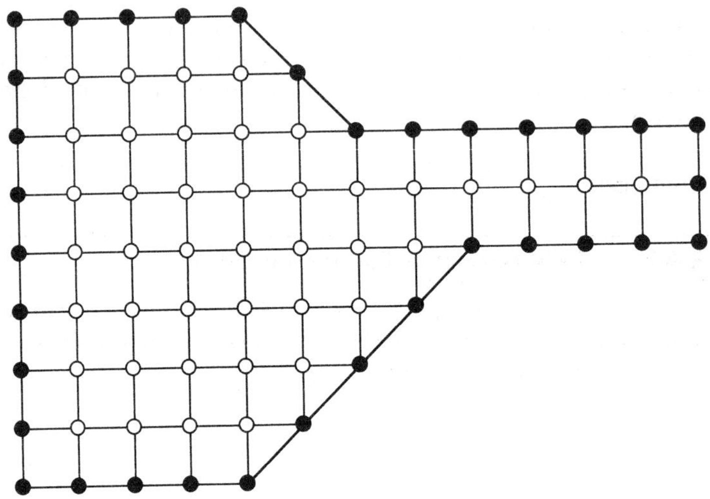

Figure 9.24 Grid corresponding to $N = 2$ for Example 9.7.

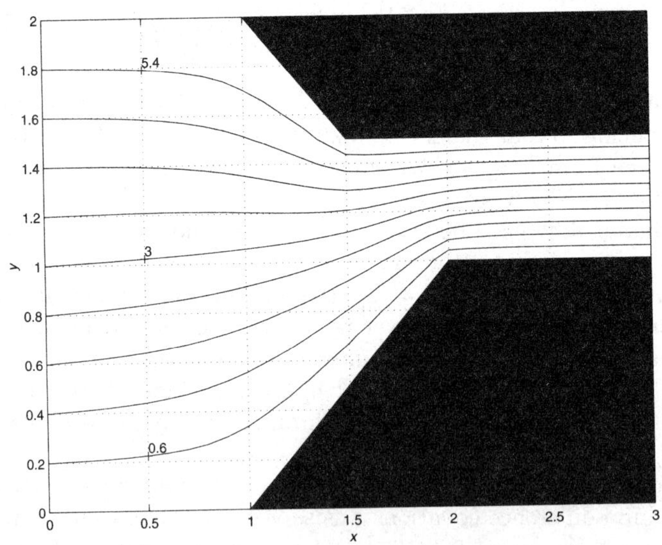

Figure 9.25 Streamlines for flow in an asymmetric contraction duct with 45° contraction angles. The values of three of the streamlines are given, and the increment between streamlines is $\Delta\psi = 0.6$.

the template for the discrete Poisson problem to account for a non-uniform mesh. To do this, let Δx denote the spacing between grid points in the x-direction and Δy denote the spacing between grid points in the y-direction. At the arbitrary interior grid point (x_j, y_k),

$$\frac{\partial^2 u}{\partial x^2} \approx \frac{w_{j-1,k} - 2w_{j,k} + w_{j+1,k}}{(\Delta x)^2}$$

and

$$\frac{\partial^2 u}{\partial y^2} \approx \frac{w_{j,k-1} - 2w_{j,k} + w_{j,k+1}}{(\Delta y)^2},$$

so the finite difference equation takes the form

$$\frac{w_{j-1,k} - 2w_{j,k} + w_{j+1,k}}{(\Delta x)^2} + \frac{w_{j,k-1} - 2w_{j,k} + w_{j,k+1}}{(\Delta y)^2} = f_{j,k}.$$

The truncation error associated with this equation is second-order in both Δx and Δy. Multiplying through by $-(\Delta x)^2$ and defining $\lambda = (\Delta x)^2/(\Delta y)^2$ yields

$$-w_{j-1,k} - w_{j+1,k} - \lambda w_{j,k-1} - \lambda w_{j,k+1} + 2(1+\lambda)w_{j,k} = -(\Delta x)^2 f_{j,k}. \quad (1)$$

We will demonstrate the use of this formula in the next example, where we also include a non-Dirichlet boundary condition along the sloped portion of the boundary.

EXAMPLE 9.8 Temperature Distribution in a Bar of Trapezoidal Cross Section

A long metallic bar has a trapezoidal cross section. The two parallel sides are maintained at constant temperatures, while the remaining sides are insulated. The temperature distribution within the bar, $T(x, y)$, satisfies Laplace's equation

$$\frac{\partial^2 T}{\partial x^2} + \frac{\partial^2 T}{\partial y^2} = 0.$$

The specific geometry of the domain and the boundary conditions are shown in Figure 9.26. Note that the angled top edge of the domain has a rise of 20 mm over a run of 30 mm. Hence, the slope is 2/3. For the computational grid to follow along this portion of the boundary, we will need to select Δy and Δx in the same ratio.

At all interior grid points, the appropriate finite difference equation is the one we derived previously:

$$-w_{j-1,k} - w_{j+1,k} - \lambda w_{j,k-1} - \lambda w_{j,k+1} + 2(1+\lambda)w_{j,k} = -(\Delta x)^2 f_{j,k},$$

where $\lambda = (\Delta x)^2/(\Delta y)^2$. With a non-Dirichlet boundary condition along the bottom edge of the domain, a separate equation must be derived for the unknowns along the first row of the grid. When the computational template, equation (1), is placed along the first row, each $w_{j,k-1}$ will be a fictitious node. The Neumann

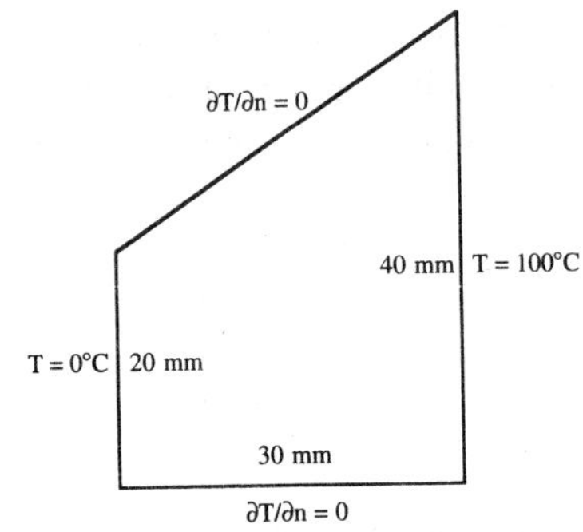

Figure 9.26 Geometry and boundary conditions for Example 9.8.

boundary condition along the bottom of the bar implies that, to second-order, $w_{j,k-1} = w_{j,k+1}$; therefore, the equation for the unknowns along the first row of the grid is

$$-w_{j-1,k} - w_{j+1,k} - 2\lambda w_{j,k+1} + 2(1+\lambda)w_{j,k} = -(\Delta x)^2 f_{j,k}.$$

A separate equation is also needed for the unknowns along the top edge of the domain. Consider the following diagram.

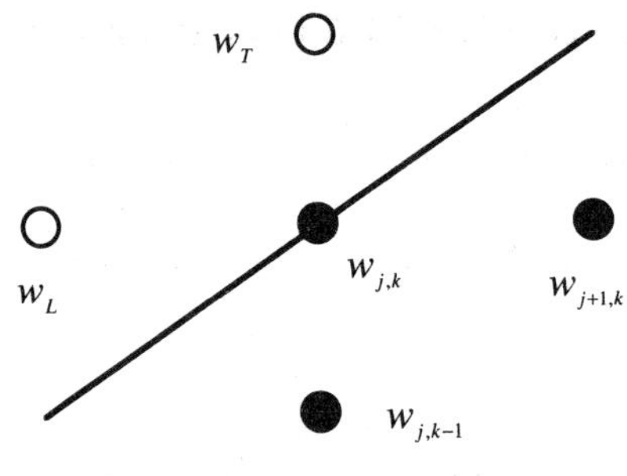

Placing the template at (x_j, y_k) produces the equation

$$-w_L - w_{j+1,k} - \lambda w_{j,k-1} - \lambda w_T + 2(1+\lambda)w_{j,k} = -(\Delta x)^2 f_{j,k},$$

where w_L and w_T are fictitious nodes. Since the slope of the boundary is $\Delta y/\Delta x$ and the outward normal points up and to the left, it follows that

$$\mathbf{n} = -\Delta y \cdot \mathbf{i} + \Delta x \cdot \mathbf{j} \quad \text{and} \quad \frac{\partial}{\partial n} = -\Delta y \frac{\partial}{\partial x} + \Delta x \frac{\partial}{\partial y}.$$

The boundary condition along the top edge of the domain then yields

$$-\Delta y \frac{w_{j+1,k} - w_L}{2\Delta x} + \Delta x \frac{w_T - w_{j,k-1}}{2\Delta y} = 0.$$

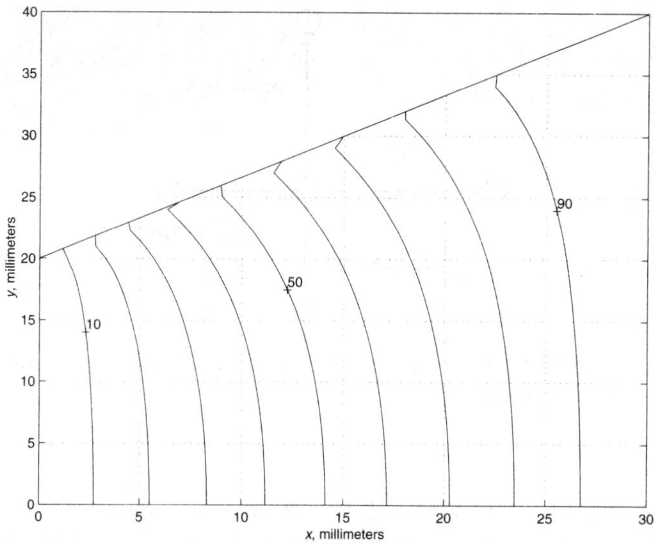

Figure 9.27 Temperature contours for a metallic bar with trapezoidal cross section. The values of three of the contours are given, and the increment between contours is $\Delta T = 10°$ C.

Multiplying through by $-2(\Delta x/\Delta y)$ and rearranging terms gives

$$-w_L - \lambda w_T = -w_{j+1,k} - \lambda w_{j,k-1}.$$

Therefore, for the unknowns along the upper boundary, the appropriate finite difference equation is

$$-2w_{j+1,k} - 2\lambda w_{j,k-1} + 2(1+\lambda)w_{j,k} = -(\Delta x)^2 f_{j,k}.$$

A contour plot of the temperature distribution within the bar is shown in Figure 9.27. For this calculation, the spacing between grid points was taken to be $\Delta x = 1.5$ mm and $\Delta y = 1.0$ mm. The system of finite difference equations was solved using Gauss-Seidel relaxation. A convergence tolerance of 5×10^{-6} applied to $\|\bar{w}^h - \tilde{w}^h\|_\infty$ was used to terminate iteration.

Curved Boundaries

Whereas sloped boundaries can generally be handled with some extra bookkeeping, curved boundaries are particularly messy with finite differences. To keep matters as simple as possible, we will only consider the case where Dirichlet conditions are specified along a curved boundary. Near the boundary, there is no way to maintain a uniform grid spacing. In the worst case, the distances from $w_{j,k}$ to its

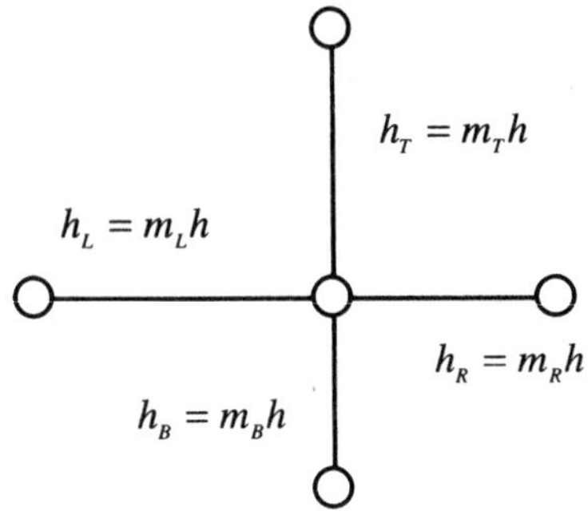

Figure 9.28 Example where distance from four neighboring grid points are all different.

four neighboring grid points would all be different, as indicated in Figure 9.28. To be able to handle this situation, we must determine the form of the five-point star template.

If we expand $u_{j-1,k}$ and $u_{j+1,k}$ into Taylor series with the appropriate increment in x, as indicated in the diagram, this leads to the approximation

$$\frac{\partial^2 u}{\partial x^2} \approx \frac{2}{h^2}\left[\frac{1}{m_L(m_L+m_R)}w_{j-1,k} - \frac{1}{m_L m_R}w_{j,k} + \frac{1}{m_R(m_L+m_R)}w_{j+1,k}\right]. \quad (2)$$

Proceeding in a similar fashion with $u_{j,k-1}$ and $u_{j,k+1}$, we find

$$\frac{\partial^2 u}{\partial y^2} \approx \frac{2}{h^2}\left[\frac{1}{m_B(m_B+m_T)}w_{j,k-1} - \frac{1}{m_B m_T}w_{j,k} + \frac{1}{m_T(m_B+m_T)}w_{j,k+1}\right]. \quad (3)$$

Combining these last two expressions, the generic interior grid point computational template for a nonuniform grid is found to be

$$2\left[\frac{1}{m_L(m_L+m_R)}w_{j-1,k} + \frac{1}{m_R(m_L+m_R)}w_{j+1,k} + \frac{1}{m_B(m_B+m_T)}w_{j,k-1} + \frac{1}{m_T(m_B+m_T)}w_{j,k+1} - \left(\frac{1}{m_L m_R} + \frac{1}{m_B m_T}\right)w_{j,k}\right] = h^2 f_{j,k}. \quad (4)$$

Not only is this formula cumbersome, it is only first order in h. Only when $m_R = m_L = m_B = m_T = 1$ does this formula reduce to the standard template and regain second-order accuracy.

EXAMPLE 9.9 Torsion of a Bar with a Curved Boundary

Suppose that a long bar whose cross section is indicated in Figure 9.29 is subjected to a constant rate of twist along its length. The stresses established in the bar can be related to the nondimensionalized Prandtl stress function, $\psi(x, y)$, which satisfies the Poisson problem

$$\frac{\partial^2 \psi}{\partial x^2} + \frac{\partial^2 \psi}{\partial y^2} = -1,$$

with Dirichlet boundary condition $\psi(x, y) = 0$ along the entire boundary.

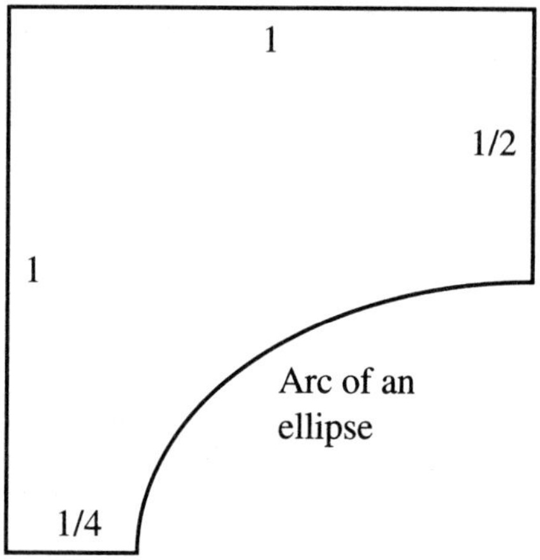

Figure 9.29 Cross section of long bar with curved boundary for Example 9.9.

Since the smallest dimension of the domain is 1/4, let's choose $h = 1/(4N)$ for some positive integer N. The computational grid will then contain $4N - 1$ rows of unknowns. With Dirichlet conditions specified around the entire boundary, every unknown is located at an interior grid point. Therefore, the system of finite difference equations will be constructed from the standard five-point star template and equation (4), only. No other finite difference equations will have to be developed.

Assuming lexicographic ordering of the unknowns, rows numbered $k = 1$ through $k = 2N - 1$ will contain at least N unknowns, though the actual number will vary from row to row and will depend on the x-coordinate of the intersection between the elliptical arc and $y = y_k$. If the origin is located at the bottom left corner of the domain, then the equation of the ellipse is

$$16(x - 1)^2 + 36y^2 = 9,$$

and the intersection between the ellipse and $y = y_k$ occurs at

$$x_{\text{ell}} = 1 - \frac{1}{4}\sqrt{9 - 36(kh)^2}.$$

The number of unknowns on row k, for $1 \leq k \leq 2N - 1$, is then

$$N + \left\lfloor \left(\frac{3}{4} - \frac{1}{4}\sqrt{9 - 36(kh)^2} \right) /h \right\rfloor .$$

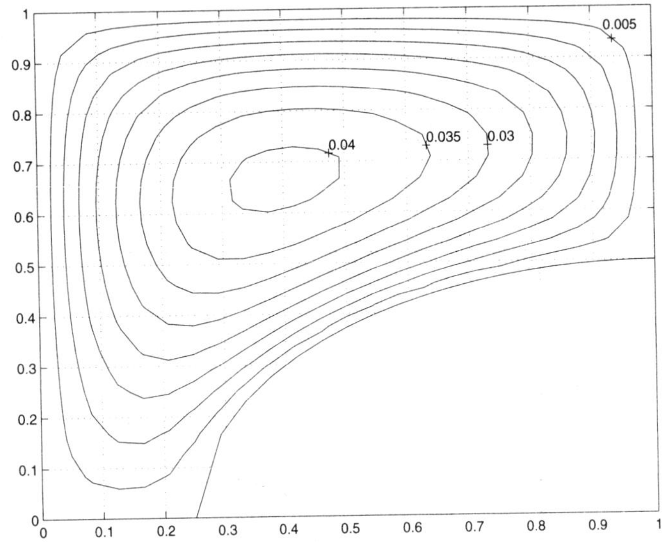

Figure 9.30 Contours of the Prandtl stress function for a bar with an elliptical arc as part of its boundary. The values associated with several contours are given, and the increment between contours is 0.005.

Rows numbered $k = 2N$ through $k = 4N-1$ of the grid all have the full complement of unknowns $(4N - 1)$.

Which finite difference equation is used for which unknowns? For $k > 2N$ and all j, as well as for $k \leq 2N$ and $j < N$, which correspond to grid points that are guaranteed not to interact with the elliptical boundary, the standard five-point star template is applied. For all other unknowns, we apply equation (4). From the geometry of the domain, we see that m_L and m_T are always equal to one, but m_R and m_B may be less than one. In particular, m_R and m_B are given by the formulas

$$m_R = \min(1, (x_{ell} - x_j)/h) \quad \text{and} \quad m_B = \min(1, (y_k - y_{ell})/h),$$

where

$$y_{ell} = \frac{1}{6}\sqrt{9 - 16(jh-1)^2}.$$

The approximate Prandtl stress function was computed for $N = 6$; that is, with $h = 1/24$. The contours of ψ are plotted in Figure 9.30. The maximum value of the stress function was 0.041307, which occurred at $x = 5/12$ and $y = 2/3$. The finite difference equations were solved using the SOR method with a relaxation parameter of $\omega = 1.7$. A convergence tolerance of 5×10^{-8} was applied to $\|\bar{w}^h - \tilde{w}^h\|_\infty$ to terminate iteration.

Circular Domains

As the previous three examples have illustrated, applying the finite difference method to solve a partial differential equation over a non-rectangular domain, especially one with curved boundaries, requires significantly more effort and/or bookkeeping than solving an equation defined over a rectangular domain. The one exception to this rule is a partial differential equation on a circular domain. For a problem with a circular domain, we can use polar coordinates to ensure that the grid follows the entire boundary.

The first thing that must be done is to convert the Laplacian operator, $L = \partial^2/\partial x^2 + \partial^2/\partial y^2$, into polar coordinates. Recall that the Cartesian coordinates x and y are related to the polar coordinates r and θ by the equations

$$\begin{array}{c} x = r\cos\theta \\ y = r\sin\theta \end{array} \quad \text{and} \quad \begin{array}{c} r = \sqrt{x^2+y^2} \\ \theta = \arctan(y/x) \end{array}.$$

After some tedious manipulation involving the chain rule, it can be shown that

$$L = \frac{\partial^2}{\partial r^2} + \frac{1}{r}\frac{\partial}{\partial r} + \frac{1}{r^2}\frac{\partial^2}{\partial \theta^2}.$$

Therefore, the Poisson problem in polar coordinates is

$$\frac{\partial^2 u}{\partial r^2} + \frac{1}{r}\frac{\partial u}{\partial r} + \frac{1}{r^2}\frac{\partial^2 u}{\partial \theta^2} = f(r,\theta).$$

To formulate the computational template, suppose the domain is $r_{\text{inner}} < r < r_{\text{outer}}$ and $\theta_{\text{start}} < \theta < \theta_{\text{end}}$. Let $\Delta r = (r_{\text{outer}} - r_{\text{inner}})/N$ and $\Delta\theta = (\theta_{\text{end}} - \theta_{\text{start}})/M$ for some positive integers N and M, and let $r_j = r_{\text{inner}} + j\Delta r$ and $\theta_k = \theta_{\text{start}} + k\Delta\theta$. Following the standard procedure for developing finite difference equations, we arrive at the generic interior grid point equation

$$\frac{w_{j-1,k} - 2w_{j,k} + w_{j+1,k}}{(\Delta r)^2} + \frac{1}{r_j}\frac{w_{j+1,k} - w_{j-1,k}}{2\Delta r} + \frac{1}{r_j^2}\frac{w_{j,k-1} - 2w_{j,k} + w_{j,k+1}}{(\Delta\theta)^2} = f_{j,k},$$

which is second order in both Δr and $\Delta\theta$. If we multiply through by $(\Delta r)^2$, define $\lambda = (\Delta r)^2/(\Delta\theta)^2$ and group like terms, the template becomes

$$\left(1 - \frac{\Delta r}{2r_j}\right) w_{j-1,k} + \left(1 + \frac{\Delta r}{2r_j}\right) w_{j+1,k} + \frac{\lambda}{r_j^2} w_{j,k-1} + \frac{\lambda}{r_j^2} w_{j,k+1} - 2\left(1 + \frac{\lambda}{r_j^2}\right) w_{j,k} = (\Delta r)^2 f_{j,k}. \quad (5)$$

This is the *polar five-point star.*

EXAMPLE 9.10 Torsion of Quarter Round Molding

A strip of quarter-round molding is subjected to a constant rate of twist along its length. The nondimensional Prandtl stress function, $\psi(r,\theta)$, satisfies the Poisson

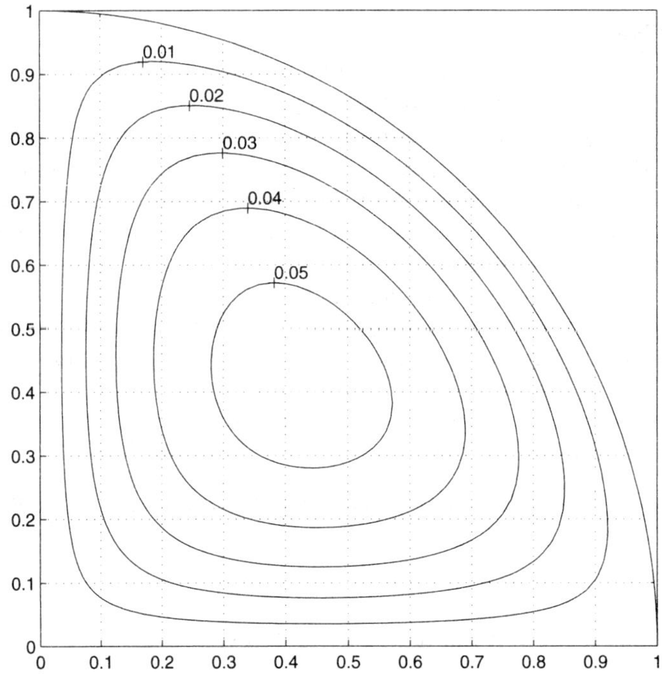

Figure 9.31 Contour plot of nondimensional Prandtl stress function for a strip of quarter-round molding.

problem

$$\frac{\partial^2 \psi}{\partial r^2} + \frac{1}{r}\frac{\partial \psi}{\partial r} + \frac{1}{r^2}\frac{\partial^2 \psi}{\partial \theta^2} = -1, \quad 0 < r < 1, 0 < \theta < \frac{\pi}{2}$$
$$\psi(0,\theta) = \psi(1,\theta) = 0$$
$$\psi(r,0) = \psi(r,\pi/2) = 0.$$

Since this problem has Dirichlet conditions around the entire boundary, every unknown is an interior grid point. The system of finite difference equations then consists solely of equation (5) for $j = 1,2,3,\ldots, N-1$ and $k = 1,2,3,\ldots,M-1$. Let's take $N = M = 64$ and use the SOR method with $\omega = 1.9$ and a convergence tolerance of 5×10^{-7} applied to $\|\bar{w}^h - \tilde{w}^h\|_\infty$. The value for ω was selected by trial and error. To simplify programming, we will represent the single point $r = 0$ by the collection of boundary grid points $(0, \theta_k)$ for $k = 0,1,2,\ldots,M$, all initialized with the value zero. A contour plot of the computed approximate stress function is shown in Figure 9.31.

EXAMPLE 9.11 An Annular Region with a Robin Boundary Condition

Consider the Poisson problem

$$\frac{\partial^2 T}{\partial r^2} + \frac{1}{r}\frac{\partial T}{\partial r} + \frac{1}{r^2}\frac{\partial^2 T}{\partial \theta^2} = -1$$
$$T(0.5, \theta) = 20$$
$$(1 + 0.5\cos 3\theta)T(2, \theta) + \frac{\partial T}{\partial r}(2, \theta) = 0.$$

This problem could be used to model the temperature distribution within a hollow, thick-shelled tube with internal heat generation. Fluid flowing through the inner core of the tube maintains an inner surface temperature 20 degrees higher than the ambient temperature surrounding the tube. The convective heat transfer coefficient varies about its mean value around the circumference of the tube.

This problem has two twists. First, note that there are no boundary conditions specified for θ. Periodicity of the solution in θ is implied. Suppose the interval $[0, 2\pi]$ is divided into M uniformly sized pieces. When the template, equation (5), is applied for $k = 0$, periodicity requires that we use the value $w_{j,M-1}$ for $w_{j,-1}$. Similarly, when the template is applied for $k = M - 1$, we use the value $w_{j,0}$ in place of $w_{j,M}$.

The second twist is the Robin boundary condition specified at the outer radius of the annulus. Suppose the radial interval is divided into N uniformly sized pieces. When the template is applied for $j = N$, the value $w_{N+1,k}$ will be a fictitious node, say w_F. The finite difference equation associated with the Robin boundary condition is

$$(1 + 0.5\cos 3\theta_k)w_{N,k} + \frac{w_F - w_{N-1,k}}{2\Delta r} = 0.$$

Solving this equation for w_F and substituting the result into (5) yields

$$2w_{N-1,k} + \frac{\lambda}{r_N^2}w_{N,k-1} + \frac{\lambda}{r_N^2}w_{N,k+1}$$
$$- 2\left(1 + \frac{\lambda}{r_N^2} + \left(\Delta r + \frac{(\Delta r)^2}{2r_N}\right)(1 + 0.5\cos 3\theta_k)\right)w_{N,k} = (\Delta r)^2 f_{N,k}$$

as the finite difference equation for the unknowns along the outer boundary.

To approximate $T(r, \theta)$, let's take $N = 12$ and $M = 32$ and use the SOR method to solve the complete system of finite difference equations. The value $\omega = 1.8$ was selected for the relaxation parameter by trial and error. A convergence tolerance of 5×10^{-7} was used to terminate relaxation sweeps. A contour plot of the computed temperature distribution is shown in Figure 9.32.

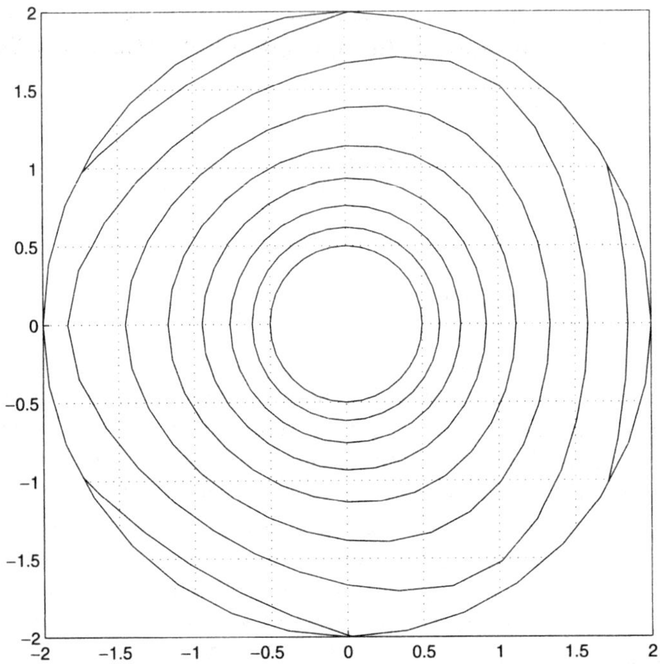

Figure 9.32 Temperature contours for annular region with a Robin boundary condition along the outer surface. The inner radius corresponds to the contour $T = 20$. Radiating outward, the contours decrement by 2.

EXERCISES

1. Derive equations (2) and (3), and show that each is only first-order accurate in the parameter h.
2. Approximate the solution of the Poisson problem

$$\frac{\partial^2 u}{\partial r^2} + \frac{1}{r}\frac{\partial u}{\partial r} + \frac{1}{r^2}\frac{\partial^2 u}{\partial \theta^2} = -\frac{\cos\theta}{r^2}, \quad 1 < r < 3, \quad 0 < \theta < \frac{\pi}{2}$$

$$u(1,\theta) = 0, \quad u(3,\theta) = \frac{2}{3}\cos\theta, \quad u(r,0) = \frac{r-1}{r}, \quad u(r,\pi/2) = 0.$$

3. Approximate the solution of

$$\frac{\partial^2 u}{\partial r^2} + \frac{1}{r}\frac{\partial u}{\partial r} + \frac{1}{r^2}\frac{\partial^2 u}{\partial \theta^2} = 0, \quad 1 < r < 2, \quad 0 < \theta < \pi$$

$$u(1,\theta) = 1 - \cos\theta, \quad u(2,\theta) = u(r,0) = 0, \quad u(r,\pi) = 4 - 2r.$$

4. A coaxial cable consists of a 0.1-inch-square inner conductor, maintained at a potential of 0 volts, and a 0.5 inch square outer conductor, maintained at a

potential of 110 volts. The potential, $\phi(x, y)$, within the cross section of the cable is governed by the partial differential equation

$$\frac{\partial^2 \phi}{\partial x^2} + \frac{\partial^2 \phi}{\partial y^2} = 0.$$

Given the symmetry of the cable and the potentials on the inner and outer conductors, ϕ can be determined by solving the above equation over the nonrectangular domain shown in Figure 9.33. The relevant boundary conditions are indicated in the diagram.

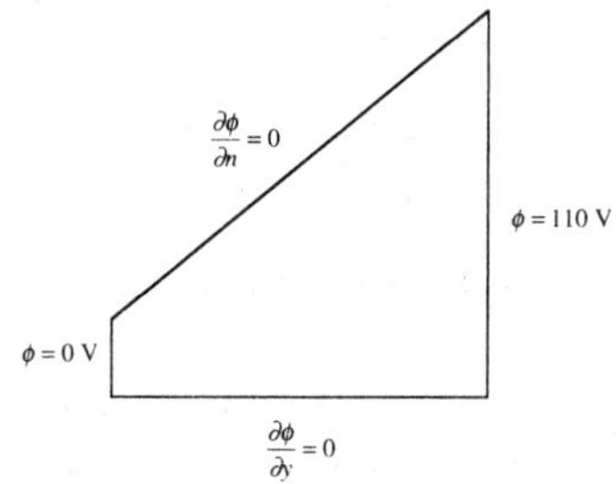

Figure 9.33 Diagram for Exercise 4.

5. A metallic plate has uniformly spaced V-shaped grooves milled into its upper surface. The upper surface, including the grooves, is maintained at a temperature of 200° C, while the bottom surface is held at 20° C. The temperature distribution within the plate satisfies the Laplace equation. Taking into account all available symmetries, the domain and the appropriate boundary conditions are as indicated in Figure 9.34.

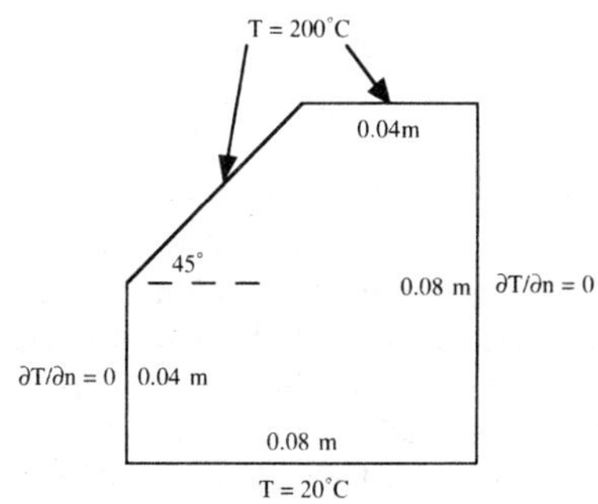

Figure 9.34 Diagram for Exercise 5.

6. A uniform flow of velocity 6 meters/second enters a contraction duct that contracts symmetrically along both walls and exits as a uniform flow with a velocity of 15 meters/second. The duct has a one-meter-long entry section with a width

of two meters, a one-half-meter contraction section that contracts 0.6 meters on each wall, and a one-meter-long exit section. The stream function, $\psi(x,y)$, satisfies the Laplace equation with boundary conditions

$$\begin{aligned}\psi(x,y) &= 0 \text{ along the bottom wall,}\\ \psi(x,y) &= 12 \text{ along the upper wall,}\\ \psi(0,y) &= 6y, \quad 0<y<2, \text{ and}\\ \psi(2.5,y) &= 15(y-0.6), \quad 0.6<y<1.4.\end{aligned}$$

7. A rod with a cross section in the shape of an equilateral triangle is subjected to a constant rate of twist along its length. The nondimensional Prandtl stress function, $\psi(x,y)$, satisfies the Poisson problem

$$\frac{\partial^2\psi}{\partial x^2}+\frac{\partial^2\psi}{\partial y^2}=-1$$

on the domain as shown in Figure 9.36 subject to the indicated boundary conditions.

8. A strip of half-round molding is subjected to a constant rate of twist along its length. The nondimensional Prandtl stress function, $\psi(r,\theta)$, satisfies the Poisson problem

$$\frac{\partial^2\psi}{\partial r^2}+\frac{1}{r}\frac{\partial\psi}{\partial r}+\frac{1}{r^2}\frac{\partial^2\psi}{\partial\theta^2}=-1, \quad 0<r<1, 0<\theta<\frac{\pi}{2}$$
$$\psi(0,\theta)=\psi(1,\theta)=\psi(r,0)=\frac{\partial\psi}{\partial\theta}(r,\pi/2)=0.$$

Note that symmetry has been applied to cut the domain at $\theta=\pi/2$.

9. A rod with a cross section as indicated in Figure 9.37 is subjected to a constant rateof twist along its length. The nondimensional Prandtl stress function, $\psi(x,y)$, satisfies the Poisson problem

$$\frac{\partial^2\psi}{\partial x^2}+\frac{\partial^2\psi}{\partial y^2}=-1$$

with $\psi(x,y)=0$ around the entire boundary.

10. A 100-mm-thick slab of granite has uniformly spaced heating pipes running through it. The pipes have a diameter of 50 mm and are centered in the thickness of the slab. The exposed surfaces of the slab are maintained at 300 K, and the fluid passing through the heating pipes maintains a temperature of 400 K at the interface between the pipes and the granite. Taking into account all symmetries, the temperature distribution within the slab satisfies

$$\frac{\partial^2 T}{\partial x^2}+\frac{\partial^2 T}{\partial y^2}=0$$

over the domain indicated in Figure 9.38. The appropriate boundary conditions are also shown in the diagram.

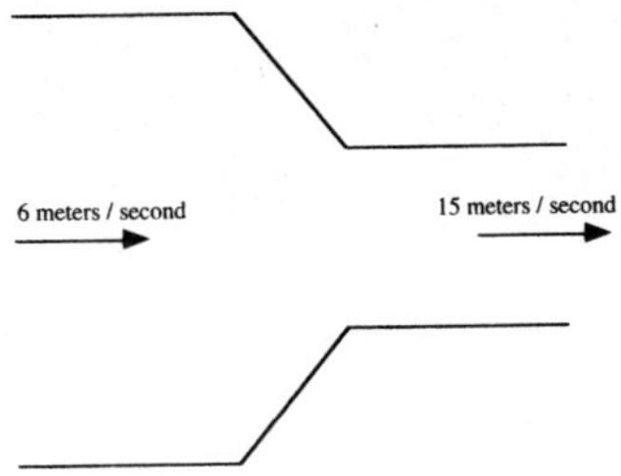

Figure 9.35 Diagram for Exercise 6.

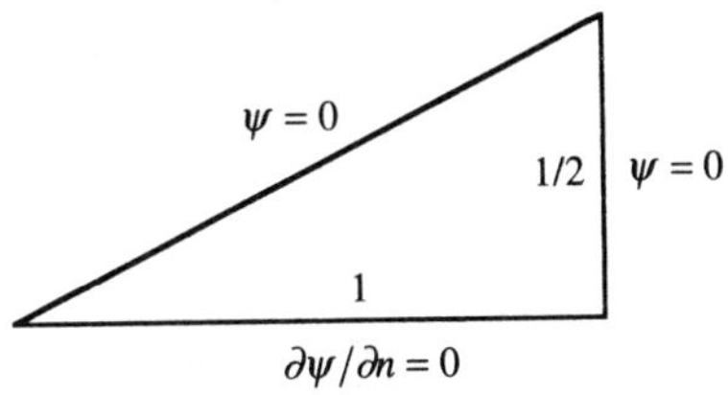

Figure 9.36 Diagram for Exercise 7.

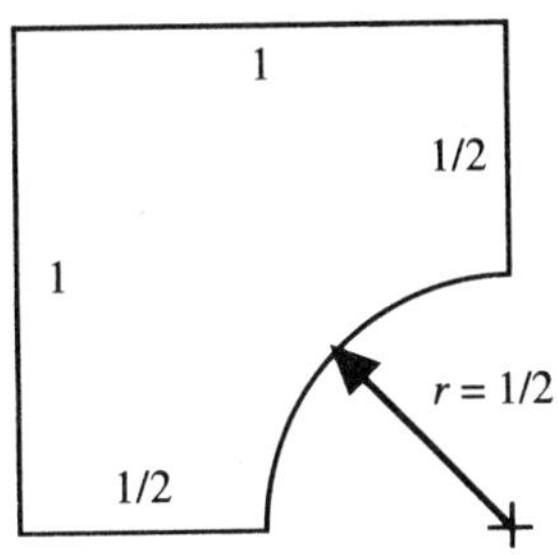

Figure 9.37 Diagram for Exercise 9.

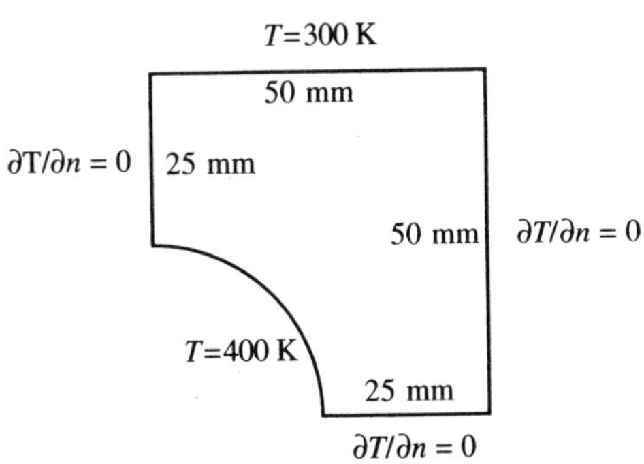

Figure 9.38 Diagram for Exercise 10.

CHAPTER 10

Parabolic Partial Differential Equations

AN OVERVIEW

Fundamental Mathematical Problem

In this chapter we will develop the finite difference method to approximate solutions of equations of the form

$$\rho\frac{\partial u}{\partial t} = \frac{\partial}{\partial x}\left(\alpha\frac{\partial u}{\partial x}\right) - \beta u + s$$

in one spatial dimension, or

$$\rho\frac{\partial u}{\partial t} = \frac{\partial}{\partial x}\left(\alpha\frac{\partial u}{\partial x}\right) + \frac{\partial}{\partial y}\left(\alpha\frac{\partial u}{\partial y}\right) - \beta u + s$$

in two spatial dimensions. Here, the parameter ρ represents a storage capacity, α measures conductivity or transmissivity, β is a decay rate, and s is a source term. In general, each parameter could be a function of any or all of the independent variables.

Equations of the type indicated above are called *parabolic* partial differential equations. The most basic parabolic partial differential equation is the heat equation

$$\frac{\partial u}{\partial t} = D\frac{\partial^2 u}{\partial x^2},$$

where D is called the coefficient of diffusion. As we will see throughout this chapter, despite its name, the heat equation appears as a model for many different physical phenomena.

The "Rise in the Water Table due to the Spring Thaw" problem capsule from the Chapter 1 Overvew (see page 8) is one application that gives rise to a parabolic partial differential equation. Here is another application.

Time-Dependent Temperature in a Fin

Recall that, in thermodynamic terms, a fin is any surface that extends from a larger object and is intended to enhance the dissipation of heat from that object. In the Chapter 8 Overview, we developed a model for the steady-state temperature within a fin of circular cross section and variable cross-sectional area (see page 656).

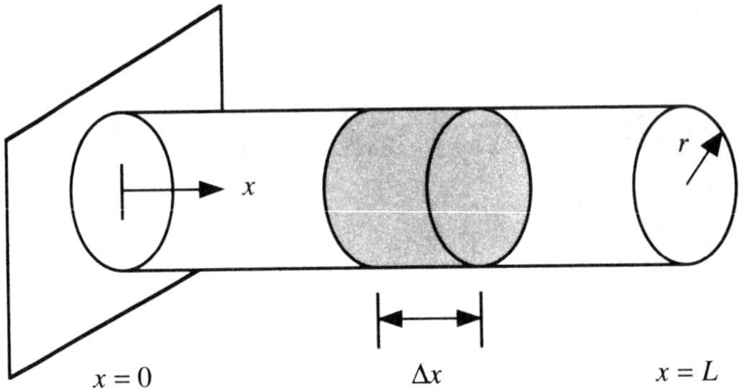

Figure 10.1

Here, we will develop a model for the time-dependent temperature within a fin with constant cross-sectional area.

The geometry for our problem is depicted in Figure 10.1. The fin has length L and a circular cross section with constant radius r. Distance along the axis of the fin will be measaured by the variable x, with $x = 0$ corresponding to the location at which the fin is attached to the larger object. We will denote the temperature at any location x and at any time t by $T(x, t)$.

To proceed with the development of our model, consider the arbitrary slice of the fin indicated in the diagram. Conservation of energy (specifically thermal energy for the current problem) requires that

$$\frac{dE_{\text{st}}}{dt} = E_{\text{in}} - E_{\text{out}}, \tag{1}$$

where E_{st} is the energy stored in the slice and E_{in} and E_{out} are the rates at which energy enters and exits the slice, respectively. The total energy stored in the slice is given by

$$E_{\text{st}} = \rho c_p V T,$$

where ρ is the mass density and c_p is the specific heat of the material from which the fin is constructed and $V = \pi r^2 \Delta x$ is the volume of the slice. If we assume ρ and c_p are constant, then

$$\frac{dE_{\text{st}}}{dt} = \rho c_p \pi r^2 \Delta x \frac{\partial T}{\partial t}. \tag{2}$$

Following the procedure described in the Chapter 8 Overview, we find that

$$E_{\text{in}} - E_{\text{out}} = \pi r^2 k \frac{\partial^2 T}{\partial x^2} \Delta x - 2\pi r \, \Delta x \, h(T - T_\infty), \tag{3}$$

where k is the thermal conductivity of the fin, h is the convection heat transfer coefficient between the fin and the surrounding air, and T_∞ is the temperature of the air. The first term in (3) accounts for conduction along the axis of the fin, while the second term accounts for convection from the lateral surface.

Substituting (2) and (3) into (1) and then dividing by $\rho c_p \pi r^2 \Delta x$ yields

$$\frac{\partial T}{\partial t} = \frac{k}{\rho c_p}\frac{\partial^2 T}{\partial x^2} - \frac{2h}{\rho c_p r}(T - T_\infty). \tag{4}$$

We will assume that the temperature throughout the fin is initially equal to T_∞ and that the temperature at $x = 0$ is known for all time; that is,

$$T(x,0) = T_\infty \qquad \text{and} \qquad T(0,t) = f(t) \tag{5}$$

for some function f. Any heat conducted to the exposed tip at $x = L$ is dissipated by convection, so

$$-k\left.\frac{\partial T}{\partial x}\right|_{x=L} = h(T(L,t) - T_\infty). \tag{6}$$

Equations (4)–(6) comprise what is called an *initial boundary value problem* for determining $T(x,t)$.

Remainder of the Chapter

We will begin our investigation of the numerical solution of parabolic partial differential equations with the heat equation in one dimension subject to Dirichlet boundary conditions. Three separate finite difference techniques will be developed. Section 10.2 will focus on the important issue of stability. A detailed analysis of the schemes developed in Section 10.1 will be presented. Parabolic problems more general than the heat equation will be treated in Section 10.3, and non-Dirichlet boundary conditions will be considered in Section 10.4. In Section 10.5, parabolic partial differential equations in polar coordinates will be discussed. The alternating direction implicit (ADI) scheme for problems in two dimensions will be presented in the final section.

10.1 THE HEAT EQUATION WITH DIRICHLET BOUNDARY CONDITIONS

Consider the initial boundary value problem

$$\text{IBVP}\quad\begin{cases}\dfrac{\partial u}{\partial t} = D\dfrac{\partial^2 u}{\partial x^2}, & A \le x \le B, t > 0\\ u(A,t) = u_A(t) \\ u(B,t) = u_B(t) \\ u(x,0) = f(x). \end{cases} \tag{1}$$

The diffusion coefficient, D, is assumed to be constant. This problem can serve as a model for heat conduction, soil consolidation, groundwater flow, and so on. Based on the geometry of the domain, as pictured in Figure 10.2, this problem is often referred to as a one-end open problem. In this section, three separate formulations of the finite difference method will be developed for equation (1).

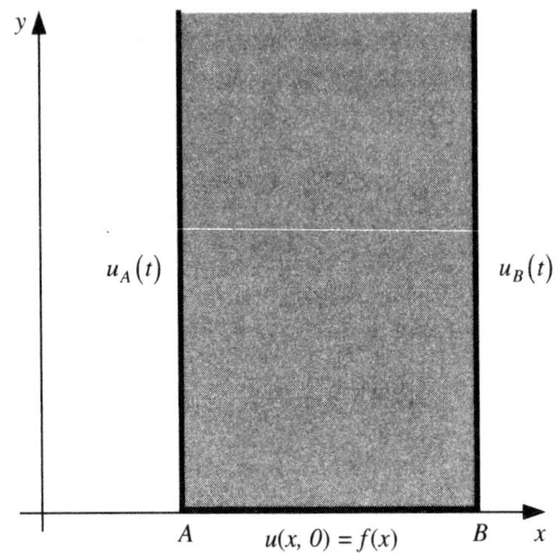

Figure 10.2 Geometry of domain for model initial boundary value problem.

Spatial Discretization

The discretization of (1) proceeds in two stages. First we discretize the space variable, then the time variable. To achieve the spatial discretization, let $\Delta x = (B - A)/N$ and $x_j = A + j\Delta x$ for $j = 0, 1, 2, \ldots, N$, where N is some positive integer. Next, evaluate the partial differential equation at the arbitrary interior grid point $x = x_j$:

$$\left\{ \frac{\partial u}{\partial t} = D\frac{\partial^2 u}{\partial x^2} \right\}\Bigg|_{x=x_j}.$$

Replace the derivative on the right-hand side of the equation by its second-order central difference approximation, and then drop the truncation error term. If we let $v_j(t)$ denote the *semidiscrete approximation* to $u(x_j, t)$, the functions $v_j(t)$ satisfy

$$\frac{dv_j(t)}{dt} = D\frac{v_{j-1}(t) - 2v_j(t) + v_{j+1}(t)}{(\Delta x)^2} \tag{2}$$

for $1 \leq j \leq N - 1$. These equations are supplemented by the boundary conditions $v_0(t) = u_A(t)$ and $v_N(t) = u_B(t)$. The initial values for the $v_j(t)$ are given by $v_j(0) = f(x_j)$.

The equations for the $v_j(t)$ can easily be written in matrix form. Let

$$\begin{aligned}
\mathbf{v}(t) &= \begin{bmatrix} v_1(t) & v_2(t) & v_3(t) & \cdot & \cdot & \cdot & v_{N-1}(t) \end{bmatrix}^T, \\
\mathbf{f} &= \begin{bmatrix} f(x_1) & f(x_2) & f(x_3) & \cdot & \cdot & \cdot & f(x_{N-1}) \end{bmatrix}^T, \\
\mathbf{b}(t) &= \begin{bmatrix} -u_A(t) & 0 & \cdot & \cdot & \cdot & 0 & -u_B(t) \end{bmatrix}^T
\end{aligned}$$

and

$$A = \begin{bmatrix} 2 & -1 & & & & & \\ -1 & 2 & -1 & & & & \\ & -1 & 2 & -1 & & & \\ & & \cdot & \cdot & \cdot & & \\ & & & \cdot & \cdot & \cdot & \\ & & & & \cdot & \cdot & \cdot \\ & & & & -1 & 2 & -1 \\ & & & & & -1 & 2 \end{bmatrix}.$$

With these definitions, equation (2) becomes

$$\frac{d\mathbf{v}(t)}{dt} = -\frac{D}{(\Delta x)^2}\left[A\mathbf{v}(t) + \mathbf{b}(t)\right], \quad \mathbf{v}(0) = \mathbf{f}. \tag{3}$$

This is a semidiscrete system of initial value problems that are an order $O\left((\Delta x)^2\right)$ approximation to the original partial differential equation.

Temporal Discretization

There are many techniques available for solving the equations represented by (3). Since (3) is a system of initial value problems, any of the methods that were developed in Chapter 7 could be used. For example, we might use the classical fourth-order Runge-Kutta method, a predictor/corrector technique or the RKF45 adaptive scheme. Instead, we will focus on the use of finite difference schemes for solving equation (3). Even under this heading, there are several possibilities available, of which we will develop three.

To complete the discretization of equation (1), we divide the time axis into uniform steps of length Δt. Let $t_n = n\,\Delta t$ for $n = 0, 1, 2, \ldots$ and so on. Values for the approximate solution will be obtained at these discrete time levels. For the fully discrete approximation, we will use the notation

$$w_j^{(n)} \approx v_j(t_n) \approx u(x_j, t_n).$$

Note the subscript on w indicates spatial location along the grid, while the superscript indicates the time level.

For our first method, let's evaluate equation (3) at the nth time level, $t = t_n$, and use a first-order forward difference approximation for the time derivative. Alternatively, we can integrate both sides of equation (3) from $t = t_n$ to $t = t_{n+1}$ and use a left endpoint approximation for the integral on the right-hand side. Following either procedure, we arrive at

$$\frac{\mathbf{w}^{(n+1)} - \mathbf{w}^{(n)}}{\Delta t} = -\frac{D}{(\Delta x)^2}\left[A\mathbf{w}^{(n)} + \mathbf{b}^{(n)}\right]$$
$$\mathbf{w}^{(0)} = \mathbf{f},$$

where

$$\mathbf{w}^{(n)} = \begin{bmatrix} w_1^{(n)} & w_2^{(n)} & w_3^{(n)} & \cdot & \cdot & \cdot & w_{N-1}^{(n)} \end{bmatrix}^T$$

and $\mathbf{b}^{(n)} = \mathbf{b}(t_n)$. Since a first-order difference formula was used for the time derivative, this equation is first order accurate in Δt. The overall truncation error between $w_j^{(n)}$ and $u(x_j, t_n)$ is then $O\left((\Delta x)^2 + \Delta t\right)$, or first order in time and second order in space.

Solving the previous equation for $\mathbf{w}^{(n+1)}$ yields

$$\mathbf{w}^{(n+1)} = (I - \lambda A)\mathbf{w}^{(n)} - \lambda \mathbf{b}^{(n)}, \tag{4}$$

where $\lambda = D\Delta t/(\Delta x)^2$. This is an explicit equation for computing the approximate solution at one time level from the values of the approximate solution at the previous time level. With $\mathbf{w}^{(0)}$ known from the initial conditions, we can march forward one increment of Δt at a time. The matrix $I - \lambda A$ is called the *evolution matrix* for the numerical method. This explicit method for solving the model initial boundary value problem, equation (1), is known as the forward in time/central in space, or FTCS, method.

EXAMPLE 10.1 Demonstration of the FTCS Method

Consider the initial boundary value problem

$$\begin{aligned} \frac{\partial u}{\partial t} &= \frac{1}{16}\frac{\partial^2 u}{\partial x^2} \\ u(0,t) &= u(1,t) = 0 \\ u(x,0) &= 2\sin(2\pi x). \end{aligned}$$

For this problem, $D = 1/16$, $u_A(t) = u_B(t) = 0$ and $f(x) = 2\sin(2\pi x)$. To construct our finite difference approximation, let's take $\Delta x = \Delta t = 0.25$. Then

$$\lambda = \frac{(1/16)(0.25)}{(0.25)^2} = 0.25.$$

With this value for λ, the evolution matrix for the FTCS method is

$$I - \lambda A = \begin{bmatrix} 0.5 & 0.25 & 0 \\ 0.25 & 0.5 & 0.25 \\ 0 & 0.25 & 0.5 \end{bmatrix}.$$

Since $u_A(t) = u_B(t) = 0$, the vector $\mathbf{b}^{(n)} = \mathbf{0}$ for all n. Hence, equation (4) reduces to

$$\mathbf{w}^{(n+1)} = \begin{bmatrix} 0.5 & 0.25 & 0 \\ 0.25 & 0.5 & 0.25 \\ 0 & 0.25 & 0.5 \end{bmatrix} \mathbf{w}^{(n)}.$$

The spatial grid consists of the points $x_0 = 0$, $x_1 = 0.25$, $x_2 = 0.5$, $x_3 = 0.75$, and $x_4 = 1$. Our calculations therefore start from the vector

$$\mathbf{w}^{(0)} = \begin{bmatrix} f(x_1) & f(x_2) & f(x_3) \end{bmatrix}^T = \begin{bmatrix} 2 & 0 & -2 \end{bmatrix}^T.$$

For the first time step we compute

$$\mathbf{w}^{(1)} = \begin{bmatrix} 0.5 & 0.25 & 0 \\ 0.25 & 0.5 & 0.25 \\ 0 & 0.25 & 0.5 \end{bmatrix} \begin{bmatrix} 2 \\ 0 \\ -2 \end{bmatrix} = \begin{bmatrix} 1 \\ 0 \\ -1 \end{bmatrix}.$$

The second time step produces

$$\mathbf{w}^{(2)} = \begin{bmatrix} 0.5 & 0.25 & 0 \\ 0.25 & 0.5 & 0.25 \\ 0 & 0.25 & 0.5 \end{bmatrix} \begin{bmatrix} 1 \\ 0 \\ -1 \end{bmatrix} = \begin{bmatrix} 0.5 \\ 0 \\ -0.5 \end{bmatrix},$$

while the third and fourth yield

$$\mathbf{w}^{(3)} = \begin{bmatrix} 0.5 & 0.25 & 0 \\ 0.25 & 0.5 & 0.25 \\ 0 & 0.25 & 0.5 \end{bmatrix} \begin{bmatrix} 0.5 \\ 0 \\ -0.5 \end{bmatrix} = \begin{bmatrix} 0.25 \\ 0 \\ -0.25 \end{bmatrix}$$

and

$$\mathbf{w}^{(4)} = \begin{bmatrix} 0.5 & 0.25 & 0 \\ 0.25 & 0.5 & 0.25 \\ 0 & 0.25 & 0.5 \end{bmatrix} \begin{bmatrix} 0.25 \\ 0 \\ -0.25 \end{bmatrix} = \begin{bmatrix} 0.125 \\ 0 \\ -0.125 \end{bmatrix}.$$

The exact solution for this problem is $u(x,t) = 2e^{-(\pi^2/4)t}\sin(2\pi x)$. The following table compares the values of the approximate solution at $t = 1$, $w_j^{(4)}$, with the values of the exact solution, $u(x_j, 1) = u_j(1)$. These errors are not unreasonable given the crudeness of the discretization in both time and space.

x_j	Approximate Solution, $w_j^{(4)}$	Exact Solution, $u_j(1)$	Absolute Error
0.00	0.000	0.000000	
0.25	0.125	0.169610	0.044610
0.50	0.000	0.000000	0.000000
0.75	−0.125	−0.169610	0.044610
1.00	0.000	0.000000	

A numerical verification of the second order spatial accuracy of the scheme is presented in the next table. Both the maximum absolute error and the root mean square (rms) error in the approximate solution at $t = 1$, as a function of the number of subintervals, N, into which the spatial axis has been divided, are displayed. The rms error was computed according to the formula

$$\text{rms error} = \sqrt{\frac{1}{N}\sum_{j=1}^{N-1}(w_j^{t=1} - u_j^{t=1})^2}.$$

To guarantee that we observe the full effect on the approximation error each time we change Δx, Δt must be chosen small enough so that the temporal truncation error is much less than the spatial truncation error. The value $\Delta t = 10^{-4}$ was used for all calculations. Note that with each doubling of N, the step size was cut in half, and the approximation error was reduced by roughly a factor of 4, which is what one would expect from a second-order numerical method.

N	Maximum Absolute Error	Error Ratio	rms Error	Error Ratio
4	0.101006		0.071422	
8	0.022388	4.511733	0.015830	4.511733
16	0.005384	4.158237	0.003807	4.158237
32	0.001296	4.152887	0.000917	4.152887
64	0.000285	4.553531	0.000201	4.553531

Numerically verifying the first-order temporal accuracy of the FTCS method is very difficult. To guarantee that we observe the full effect of any change in Δt, we want to choose Δx small enough so that the spatial truncation error is much less than the temporal truncation error. For the FTCS method, however, we must choose $\Delta t < (\Delta x)^2/(2D)$. The reason for this will be discussed in the next section. Unfortunately, with the time step restricted in this manner, the temporal truncation error will always be of the same order or smaller than the spatial truncation error.

The backward in time/central in space, or BTCS, method is obtained by evaluating equation (3) at the $(n+1)$-st time level, $t = t_{n+1}$, and using a first-order backward difference approximation for the time derivative. Alternatively, we can integrate both sides of equation (3) from $t = t_n$ to $t = t_{n+1}$ and use a right endpoint approximation for the integral on the right-hand side. Following either procedure leads to

$$\frac{\mathbf{w}^{(n+1)} - \mathbf{w}^{(n)}}{\Delta t} = -\frac{D}{(\Delta x)^2}\left[A\mathbf{w}^{(n+1)} + \mathbf{b}^{(n+1)}\right]$$

$$\mathbf{w}^{(0)} = \mathbf{f},$$

where

$$\mathbf{w}^{(n)} = \begin{bmatrix} w_1^{(n)} & w_2^{(n)} & w_3^{(n)} & \cdots & w_{N-1}^{(n)} \end{bmatrix}^T$$

and $\mathbf{b}^{(n+1)} = \mathbf{b}(t_{n+1})$. Like the FTCS method, this scheme is first order in time and second order in space.

Rearranging terms in the evolution equation, we obtain

$$(I + \lambda A)\mathbf{w}^{(n+1)} = \mathbf{w}^{(n)} - \lambda \mathbf{b}^{(n+1)}, \tag{5}$$

where $\lambda = D\Delta t/(\Delta x)^2$. From here, we see that the BTCS method defines $\mathbf{w}^{(n+1)}$ implicitly in terms of $\mathbf{w}^{(n)}$. Each time step therefore requires the solution of a linear system of equations. Since $I + \lambda A$ is a tridiagonal matrix, computing $\mathbf{w}^{(n+1)}$ from equation (5) requires only about twice as many algebraic operations as computing $\mathbf{w}^{(n+1)}$ from equation (4). The benefit derived from the additional effort will be explored in the next section.

EXAMPLE 10.2 Demonstration of the BTCS Method

Let's once again consider the initial boundary value problem

$$\frac{\partial u}{\partial t} = \frac{1}{16}\frac{\partial^2 u}{\partial x^2}$$
$$u(0,t) = u(1,t) = 0$$
$$u(x,0) = 2\sin(2\pi x).$$

With $\Delta x = \Delta t = 0.25$, the method parameters become $\lambda = 1/4$ and

$$I + \lambda A = \begin{bmatrix} 1.5 & -0.25 & 0 \\ -0.25 & 1.5 & -0.25 \\ 0 & -0.25 & 1.5 \end{bmatrix}.$$

Since $u_A(t) = u_B(t) = 0$, the vector $\mathbf{b}^{(n+1)} = \mathbf{0}$ for all n. Hence, equation (5) reduces to

$$\begin{bmatrix} 1.5 & -0.25 & 0 \\ -0.25 & 1.5 & -0.25 \\ 0 & -0.25 & 1.5 \end{bmatrix} \mathbf{w}^{(n+1)} = \mathbf{w}^{(n)}.$$

Starting from $\mathbf{w}^{(0)} = \begin{bmatrix} 2 & 0 & -2 \end{bmatrix}^T$, in the first time step, we must solve

$$\begin{bmatrix} 1.5 & -0.25 & 0 \\ -0.25 & 1.5 & -0.25 \\ 0 & -0.25 & 1.5 \end{bmatrix} \mathbf{w}^{(1)} = \begin{bmatrix} 2 \\ 0 \\ -2 \end{bmatrix} \quad \Rightarrow \quad \mathbf{w}^{(1)} = \begin{bmatrix} 4/3 \\ 0 \\ -4/3 \end{bmatrix}.$$

Thus, in the second time step, we are faced with the linear system

$$\begin{bmatrix} 1.5 & -0.25 & 0 \\ -0.25 & 1.5 & -0.25 \\ 0 & -0.25 & 1.5 \end{bmatrix} \mathbf{w}^{(2)} = \begin{bmatrix} 4/3 \\ 0 \\ -4/3 \end{bmatrix} \quad \Rightarrow \quad \mathbf{w}^{(2)} = \begin{bmatrix} 8/9 \\ 0 \\ -8/9 \end{bmatrix}.$$

The third and fourth time steps produce the results

$$\mathbf{w}^{(3)} = \begin{bmatrix} 16/27 \\ 0 \\ -16/27 \end{bmatrix} \quad \text{and} \quad \mathbf{w}^{(4)} = \begin{bmatrix} 32/81 \\ 0 \\ -32/81 \end{bmatrix}.$$

Comparing the values of the approximate solution at $t = 1$ with the exact solution, we see that the approximation error for the BTCS method, for this problem, is roughly five times that of the FTCS method.

x_j	Approximate Solution, $w_j^{(4)}$	Exact Solution, $u_j(1)$	Absolute Error
0.00	0.000000	0.000000	
0.25	0.395062	0.169610	0.225452
0.50	0.000000	0.000000	0.000000
0.75	−0.395062	−0.169610	0.225452
1.00	0.000000	0.000000	

The next table demonstrates the second-order spatial accuracy of the BTCS method. All calculations were performed with $\Delta t = 5 \times 10^{-5}$. Note that each time Δx was cut in half, both the maximum absolute error and the rms error were reduced by roughly a factor of 4.

Δx	Maximum Absolute Error	Error Ratio	rms Error	Error Ratio
1/4	0.101088		0.071480	
1/8	0.022467	4.499466	0.015886	4.499466
1/16	0.005462	4.113411	0.003862	4.113411
1/32	0.001374	3.975153	0.000972	3.975153
1/64	0.000362	3.793634	0.000256	3.793634

In contrast to the FTCS method, the BTCS method places no restrictions on the size of Δt. It is therefore an easy task to verify the first-order temporal accuracy of the scheme. With $\Delta x = 1/256$, we obtain the following results. The first-order nature of the approximation is apparent.

Δt	Maximum Absolute Error	Error Ratio	rms Error	Error Ratio
0.008	0.004147		0.002933	
0.004	0.002085	1.988911	0.001474	1.988911
0.002	0.001053	1.979563	0.000745	1.979563
0.001	0.000537	1.960657	0.000380	1.960657
0.0005	0.000279	1.924624	0.000197	1.924624

The third, and final, method that we will develop is the Crank-Nicolson scheme. Integrate both sides of equation (3) from $t = t_n$ to $t = t_{n+1}$, and approximate the integral on the right-hand side with the trapezoidal rule. This yields

$$\mathbf{w}^{(n+1)} - \mathbf{w}^{(n)} = -\frac{D\Delta t}{2(\Delta x)^2}\left[(A\mathbf{w}^{(n)} + \mathbf{b}^{(n)}) + (A\mathbf{w}^{(n+1)} + \mathbf{b}^{(n+1)})\right],$$

or

$$(I + \lambda A)\mathbf{w}^{(n+1)} = (I - \lambda A)\mathbf{w}^{(n)} - \lambda(\mathbf{b}^{(n)} + \mathbf{b}^{(n+1)}), \tag{6}$$

where $\lambda = D\Delta t/[2(\Delta x)^2]$. This is another implicit method, requiring the solution of a tridiagonal linear system at each time step. The computational cost associated with implementing equation (6) is roughly 50% larger than that associated with equation (5). However, since the trapezoidal rule is a second-order numerical integration scheme, the overall truncation error for the Crank-Nicolson method is $O\left((\Delta x)^2 + (\Delta t)^2\right)$—second order in both time and space.

EXAMPLE 10.3 Demonstration of the Crank-Nicolson Scheme

We once again turn to our sample initial boundary value problem

$$\begin{aligned}\frac{\partial u}{\partial t} &= \frac{1}{16}\frac{\partial^2 u}{\partial x^2}\\ u(0,t) &= u(1,t) = 0\\ u(x,0) &= 2\sin(2\pi x).\end{aligned}$$

With $\Delta x = \Delta t = 0.25$,

$$\lambda = \frac{(1/16)(1/4)}{2(1/4)^2} = 1/8,$$

so

$$I + \lambda A = \begin{bmatrix} 1.25 & -0.125 & 0 \\ -0.125 & 1.25 & -0.125 \\ 0 & -0.125 & 1.25 \end{bmatrix}$$

and

$$I - \lambda A = \begin{bmatrix} 0.75 & 0.125 & 0 \\ 0.125 & 0.75 & 0.125 \\ 0 & 0.125 & 0.75 \end{bmatrix}.$$

The evolution equation for the approximate solution is then

$$\begin{bmatrix} 1.25 & -0.125 & 0 \\ -0.125 & 1.25 & -0.125 \\ 0 & -0.125 & 1.25 \end{bmatrix} \mathbf{w}^{(n+1)} = \begin{bmatrix} 0.75 & 0.125 & 0 \\ 0.125 & 0.75 & 0.125 \\ 0 & 0.125 & 0.75 \end{bmatrix} \mathbf{w}^{(n)},$$

since both $\mathbf{b}^{(n)}$ and $\mathbf{b}^{(n+1)}$ are zero for all n.

Starting from the initial condition $\mathbf{w}^{(0)} = \begin{bmatrix} 2 & 0 & -2 \end{bmatrix}^T$, in the first time step, we must solve the system

$$\begin{aligned} \begin{bmatrix} 1.25 & -0.125 & 0 \\ -0.125 & 1.25 & -0.125 \\ 0 & -0.125 & 1.25 \end{bmatrix} \mathbf{w}^{(1)} &= \begin{bmatrix} 0.75 & 0.125 & 0 \\ 0.125 & 0.75 & 0.125 \\ 0 & 0.125 & 0.75 \end{bmatrix} \begin{bmatrix} 2 \\ 0 \\ -2 \end{bmatrix} \\ &= \begin{bmatrix} 1.5 \\ 0 \\ -1.5 \end{bmatrix}, \end{aligned}$$

which yields $\mathbf{w}^{(1)} = \begin{bmatrix} 1.2 & 0 & -1.2 \end{bmatrix}^T$. The next three time steps produce

$$\begin{aligned} \mathbf{w}^{(2)} &= \begin{bmatrix} 0.72 & 0 & -0.72 \end{bmatrix}^T, \\ \mathbf{w}^{(3)} &= \begin{bmatrix} 0.432 & 0 & -0.432 \end{bmatrix}^T \text{ and} \\ \mathbf{w}^{(4)} &= \begin{bmatrix} 0.2592 & 0 & -0.2592 \end{bmatrix}^T. \end{aligned}$$

Comparing the values of the approximate solution at $t = 1$ with the exact solution, we see that the approximation error for the Crank-Nicolson scheme, for this problem, is roughly twice that of the FTCS method.

x_j	Approximate Solution, $w_j^{(4)}$	Exact Solution, $u_j(1)$	Absolute Error
0.00	0.000000	0.000000	
0.25	0.259200	0.169610	0.089590
0.50	0.000000	0.000000	0.000000
0.75	−0.259200	−0.169610	0.089590
1.00	0.000000	0.000000	

The next table presents a numerical verification of the second-order spatial accuracy of the scheme. For all calculations, $\Delta t = 5 \times 10^{-2}$ was used. Note that for each Δx, the errors achieved by the Crank-Nicolson scheme are comparable with those of both the FTCS and the BTCS method, even with a time step that is two orders of magnitude larger.

Δx	Maximum Absolute Error	Error Ratio	rms Error	Error Ratio
1/4	0.101060		0.071460	
1/8	0.022439	4.503678	0.015867	4.503678
1/16	0.005435	4.128699	0.003843	4.128699
1/32	0.001347	4.034036	0.000953	4.034036
1/64	0.000336	4.015649	0.000237	4.015649

The Crank-Nicolson scheme also places no restrictions on the size of Δt. With $\Delta x = 1/1000$, we obtain the following results, which clearly verify the second-order temporal accuracy of the method.

Δt	Maximum Absolute Error	Error Ratio	rms Error	Error Ratio
1/5	0.008591		0.006074	
1/10	0.002128	4.037015	0.001505	4.037015
1/20	0.000530	4.016473	0.000375	4.016473
1/40	0.000131	4.033626	0.000093	4.033626
1/80	0.000032	4.130451	0.000022	4.130451

Application Problem: Rise in the Water Table due to the Spring Thaw

In the Chapter 1 Overview (see page 8), we developed the initial boundary value problem

$$\frac{\partial h}{\partial t} = a\frac{\partial^2 h}{\partial x^2},$$

$$h(x,0) = h_0(x), \qquad h(0,t) = h_L(t) \quad \text{and} \quad h(800,t) = h_R(t)$$

for determining the water table, $h(x,t)$, in an aquifer. The aquifer is situated between two monitoring wells located 800 meters apart. The constant a is called the hydraulic diffusivity, which has been experimentally determined to be

$$a = 0.0059 \text{ m}^2/\text{s} = 509.76 \text{ m}^2/\text{day}.$$

Suppose that during the spring thaw measurements made at the well at the left edge of the aquifer indicate

$$h_L(t) = \begin{cases} 2(1 - e^{-t/5}), & t, 30 \\ 2, & t \geq 30, \end{cases}$$

while the measurements made at the other well indicate

$$h_R(t) = 0.3 h_L(t).$$

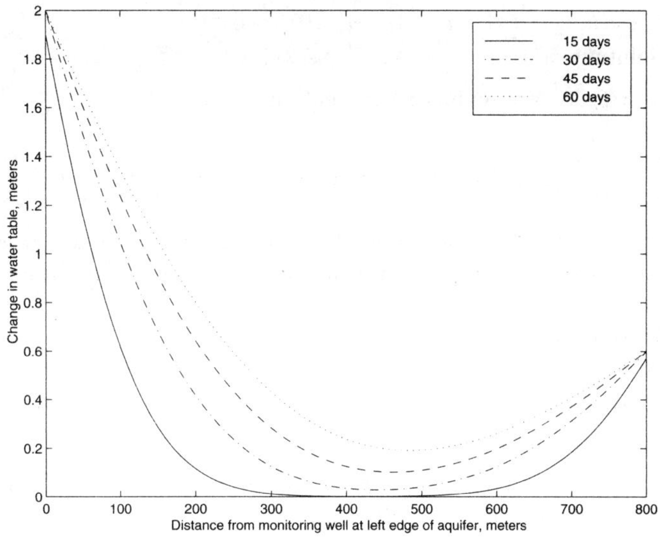

Figure 10.3 Change in the water table in an aquifer during the spring thaw.

Let's take $h_0(x) = 0$ so we can determine the change in the water table due to the spring thaw. Figure 10.3 displays the resulting water table profiles after 15 days, 30 days, 45 days, and 60 days. All solutions were obtained using the Crank-Nicolson scheme with $\Delta x = 20$ meters and $\Delta t = 0.2$ days.

EXERCISES

In Exercises 1–4, numerically verify that

(a) the FTCS method is second-order accurate in space;

(b) the BTCS method is first-order accurate in time and second-order accurate in space; and

(c) the Crank-Nicolson scheme is second-order accurate in both time and space

by approximating the solution of

$$\frac{\partial u}{\partial t} = \frac{\partial^2 u}{\partial x^2}$$

subject to the indicated initial and boundary conditions.

1. $u(0,t) = 1, \quad u(1,t) = 0, \quad u(x,0) = 1 - x - \frac{1}{\pi}\sin(2\pi x)$
exact solution: $u(x,t) = 1 - x - \frac{1}{\pi}e^{-4\pi^2 t}\sin(2\pi x)$

2. $u(0,t) = u(\pi,t) = 0, \quad u(x,0) = \sin x$
exact solution: $u(x,t) = e^{-t}\sin x$

3. $u(0,t) = u(\pi,t) = 0, \quad u(x,0) = \sin^3 x$
exact solution: $u(x,t) = \frac{3}{4}e^{-t}\sin x - \frac{1}{4}e^{-9t}\sin(3x)$

4. $u(0,t) = u(\pi,t) = \frac{1}{2}(1+e^{-4t}), \quad u(x,0) = \cos^2 x$
exact solution: $u(x,t) = \frac{1}{2} + \frac{1}{2}e^{-4t}\cos(2x)$

5. Let $\Delta x = 1/20$. Approximate the solution of

$$\frac{\partial u}{\partial t} = \frac{\partial^2 u}{\partial x^2}, \qquad u(0,t) = 1, \quad u(1,t) = 0, \quad u(x,0) = 1 - x - \frac{1}{\pi}\sin(2\pi x)$$

at $t = 1$ using the FTCS method with the indicated number of time steps. In each case calculate the corresponding value of λ and plot the resulting approximate solution.
   (a) 800 time steps
   (b) 777 time steps
   (c) 776 time steps
   (d) 775 time steps

6. Plot $u(x,t)$ for $t = 0.05$, $t = 0.1$, $t = 0.25$, and $t = 1$, where $u(x,t)$ is the solution to the initial boundary value problem

$$\frac{\partial u}{\partial t} = \frac{\partial^2 u}{\partial x^2}, \qquad u(0,t) = 1, \quad u(1,t) = 5, \quad u(x,0) = 0.$$

Take $\Delta x = 1/20$ for all calculations.

7. Plot $u(x,t)$ for $t = 1$, $t = 2$, $t = 3$, and $t = 4$, where $u(x,t)$ is the solution to the initial boundary value problem

$$\frac{\partial u}{\partial t} = \frac{\partial^2 u}{\partial x^2}, \qquad u(0,t) = 1 - e^{-t}, \quad u(1,t) = 1 - \cos(\pi t), \quad u(x,0) = 0.$$

Take $\Delta x = 1/20$ for all calculations.

8. Consider flow between two flat parallel walls separated by a distance h. The fluid and both walls are initially at rest. At $t = 0$, flow is initiated by impulsively bringing the lower wall, corresponding to $y = 0$, to the constant velocity u_0. The velocity profile between the walls, $u(y,t)$, satisfies

$$\frac{\partial u}{\partial t} = \nu\frac{\partial^2 u}{\partial y^2}, \qquad u(y,0) = 0, \quad u(0,t) = u_0, \quad u(h,t) = 0,$$

where ν is the kinematic viscosity of the fluid. If we introduce the nondimensional variables

$$U = u/u_0, \quad Y = y/h, \quad \text{and} \quad T = t\nu/h^2,$$

the problem becomes

$$\frac{\partial U}{\partial T} = \frac{\partial^2 U}{\partial Y^2}, \qquad U(Y,0) = 0, \quad U(0,T) = 1, \quad U(1,T) = 0.$$

Plot $U(Y,T)$ for $T = 0.05$, $T = 0.1$, $T = 0.25$, and $T = 1$.

9. Soil consolidation is the hydrodynamic process by which water is expelled from saturated soil voids when the soil is compacted. If the layer underneath the compressible soil has a higher permeability than the soil, then as the water is expelled, the pore water pressure, $\phi(x,t)$, satisfies

$$\frac{\partial \phi}{\partial t} = C_v \frac{\partial^2 \phi}{\partial z^2}, \qquad \phi(z,0) = \Delta\phi, \quad \phi(-H_d,t) = \phi(H_d,t) = 0,$$

where C_v is the coefficient of consolidation, H_d is the maximum drainage path, and $\Delta\phi$ is the change in pressure due to the compacting force. Introducing the nondimensional variables

$$\Phi = \phi/\Delta\phi, \quad Z = z/H_d, \quad \text{and} \quad T = tC_v/H_d^2,$$

the problem becomes

$$\frac{\partial \Phi}{\partial T} = \frac{\partial^2 \Phi}{\partial Z^2}, \qquad \Phi(Z,0) = 1, \quad \Phi(-1,T) = \Phi(1,T) = 0.$$

Plot $\Phi(Z,T)$ for $T = 0.05$, $T = 0.1$, $T = 0.25$, and $T = 1$.

10. An aquifer is located between two rivers, and fluctuations in the water table are monitored at two wells located 1100 meters apart. During a flood, the rise in the water table as measured at both wells was found to be

$$r(t) = \begin{cases} (5/3)t, & t \le 3 \\ 5e^{-(t-3)/5}, & t > 3 \end{cases},$$

where r is measured in meters, and t is measured in days. The change in the water table, $h(x,t)$, as a result of the flood is modeled by the initial boundary value problem

$$\frac{\partial h}{\partial t} = a\frac{\partial^2 h}{\partial x^2}, \qquad h(x,0) = 0, \quad h(0,t) = h(1100,t) = r(t).$$

The hydraulic diffusivity of the soil has been experimentally determined to be

$$a = 0.0059 \text{ m}^2/\text{s} = 509.76 \text{ m}^2/\text{day}.$$

(a) Determine $h(x,t)$ at the peak of the flood, $t = 3$.

(b) Plot $h(x,t)$ for $t = 10$, $t = 15$, and $t = 20$.

11. **(a)** Assuming that u is sufficiently differentiable, show that the truncation error for the FTCS method is

$$\frac{D^2\Delta t}{2}\left(1 - \frac{1}{6\lambda}\right)\frac{\partial^4 u}{\partial x^4} + O((\Delta t)^2 + (\Delta x)^4),$$

where $\lambda = D\Delta t/(\Delta x)^2$.

(b) Show that for $\lambda = \frac{1}{6}$, the leading term in the truncation error for the FTCS method is $O((\Delta t)^2 + (\Delta x)^4)$.

(c) Numerically verify that the FTCS is fourth order in space when $\lambda = \frac{1}{6}$ by approximating the solution of the initial boundary value problem from Exercise 5.

10.2 ABSOLUTE STABILITY

When approximating the solution of an elliptic partial differential equation by the finite difference method, our main concern was finding an efficient procedure for solving the discrete system of equations. When approximating the solution of a parabolic partial differential equation, however, our main analysis issue will be the absolute stability of the time discretization scheme applied to the semidiscrete system of equations [e.g., equation (3) from Section 10.1]. Recall from Section 7.9 that a time discretization scheme is said to be absolutely stable when the asymptotic character of the solution it produces matches that of the analytical solution. For the differential equations treated in this chapter, this definition translates to the requirement that the component of the approximate solution attributable to the initial conditions should not increase as time steps are computed.

We will suppose the mesh spacing Δx has been selected small enough to resolve any spatial variations in the solution. Once this has been done, our objective will be to determine what restrictions, if any, need to be placed upon the choice of Δt to ensure absolute stability. The ideal situation would be that no restrictions were needed, so Δt could be chosen solely on the basis of accuracy considerations. Methods for which no restrictions are imposed on the choice of Δt are said to be *unconditionally stable.* Methods for which an upper bound is imposed upon the value of Δt are called *conditionally stable.* An *unconditionally unstable* method is one for which no value of Δt will maintain absolute stability.

There are three basic approaches to performing stability analysis. Suppose the time discretization scheme that has been applied to the semidiscrete system can be identified as one of the methods developed in Chapter 7. We can then examine the region of absolute stability for that method, together with the eigenvalues of the coefficient matrix for the semidiscrete system, to determine whether or not Δt needs to be restricted. Alternatively, we can work from the fully discrete system of equations and perform either a matrix stability analysis or a von Neumann stability analysis. Since we have been focusing on expressing each method in matrix form, we will develop the matrix approach first. The von Neumann approach will be presented toward the end of the section. Stability analysis based on methods developed in Chapter 7 will be considered in the exercises.

Matrix Stability Analysis

In principle, each of the methods we have developed—the FTCS method, the BTCS method, and the Crank-Nicolson method—can be written in the form

$$\mathbf{w}^{(n+1)} = E\mathbf{w}^{(n)} + \mathbf{c}^{(n)},$$

for some evolution matrix E. The vector $\mathbf{c}^{(n)}$ incorporates the boundary conditions. Let $\mathbf{w}^{(0)}$ denote the vector obtained from the initial conditions. Then

$$\begin{aligned}
\mathbf{w}^{(1)} &= E\mathbf{w}^{(0)} + \mathbf{c}^{(0)} \\
\mathbf{w}^{(2)} &= E\mathbf{w}^{(1)} + \mathbf{c}^{(1)} = E^2\mathbf{w}^{(0)} + (\mathbf{c}^{(1)} + E\mathbf{c}^{(0)})
\end{aligned}$$

$$\mathbf{w}^{(3)} = E\mathbf{w}^{(2)} + \mathbf{c}^{(2)} = E^3\mathbf{w}^{(0)} + (\mathbf{c}^{(2)} + E\mathbf{c}^{(1)} + E^2\mathbf{c}^{(0)})$$

$$\vdots$$

$$\mathbf{w}^{(m)} = E\mathbf{w}^{(m-1)} + \mathbf{c}^{(m-1)} = E^m\mathbf{w}^{(0)} + \hat{\mathbf{c}},$$

where $\hat{\mathbf{c}} = \mathbf{c}^{(m-1)} + E\mathbf{c}^{(m-2)} + E^2\mathbf{c}^{(m-3)} + \cdots + E^{m-1}\mathbf{c}^{(0)}$.

For absolute stability, $\|E^m\mathbf{w}^{(0)}\|$ must be less than or equal to $\|\mathbf{w}^{(0)}\|$ for any initial vector $\mathbf{w}^{(0)}$. From here, it follows that we need $\|E^m\| \leq 1$ for the corresponding natural matrix norm of E^m. Using the fact that $\rho(E^m) \leq \|E^m\|$ and that $\rho(E^m) = [\rho(E)]^m$, we can conclude that $\rho(E) \leq 1$ is a necessary condition for absolute stability. Hence, matrix stability analysis requires an examination of the eigenvalues of the evolution matrix.

The following special tridiagonal matrix will play a central role in the analysis of the FTCS method, the BTCS method, and the Crank-Nicolson scheme:

$$M = \begin{bmatrix} a & c & & & & & \\ b & a & c & & & & \\ & b & a & c & & & \\ & & \cdot & \cdot & \cdot & & \\ & & & \cdot & \cdot & \cdot & \\ & & & & \cdot & \cdot & \cdot \\ & & & & b & a & c \\ & & & & & b & a \end{bmatrix}.$$

This matrix is defined by just three numbers—the value a that appears at each location along the main diagonal, the value b that appears along the principal subdiagonal, and the value c that appears along the principal superdiagonal. Assuming that M is an $N \times N$ matrix, its eigenvalues are given by

$$\mu_k = a + 2\sqrt{bc}\cos\frac{k\pi}{N+1} \tag{1}$$

for $k = 1, 2, 3, \ldots, N$.

Forward in Time

The evolution matrix for the FTCS method is

$$E_{\text{FTCS}} = I - \lambda A = \begin{bmatrix} 1-2\lambda & \lambda & & & & & \\ \lambda & 1-2\lambda & \lambda & & & & \\ & \lambda & 1-2\lambda & \lambda & & & \\ & & \cdot & \cdot & \cdot & & \\ & & & \cdot & \cdot & \cdot & \\ & & & & \cdot & \cdot & \cdot \\ & & & & \lambda & 1-2\lambda & \lambda \\ & & & & & \lambda & 1-2\lambda \end{bmatrix},$$

where $\lambda = D\,\Delta t/(\Delta x)^2$. This matrix matches the pattern of the special matrix M with

$$a = 1 - 2\lambda \quad \text{and} \quad b = c = \lambda.$$

It follows from equation (1), then, that the eigenvalues of E_{FTCS} are given by

$$\begin{aligned}\mu_k &= 1 - 2\lambda + 2\lambda\cos\frac{k\pi}{N}\\ &= 1 - 2\lambda\left(1 - \cos\frac{k\pi}{N}\right)\\ &= 1 - 4\lambda\sin^2\frac{k\pi}{2N}\end{aligned} \tag{2}$$

for $k = 1, 2, 3, \ldots, N-1$. In going from the second line to the third, the half-angle formula for $\sin^2\theta$ has been used.

The FTCS method will be stable provided $\rho(E_{\text{FTCS}}) \le 1$. From equation (2), this will happen when the compound inequality

$$-1 \le 1 - 4\lambda\sin^2\frac{k\pi}{2N} \le 1$$

holds for all k. Since λ is strictly positive by construction, the inequality on the right side is trivially satisfied. For the inequality on the left, note that the sine function has a maximum value of 1, so the quantity between the inequalities has a minimum value of $1 - 4\lambda$. Requiring $1 - 4\lambda$ to be greater than or equal to -1 leads to $\lambda \le 1/2$. Hence, the FTCS method is *conditionally stable*. Once Δx has been selected, we must choose

$$\Delta t \le \frac{(\Delta x)^2}{2D}$$

to maintain absolute stability.

EXAMPLE 10.4 The Conditional Stability of the FTCS Method

Let's once again examine the initial boundary value problem

$$\frac{\partial u}{\partial t} = \frac{1}{16}\frac{\partial^2 u}{\partial x^2}, \qquad u(0,t) = u(1,t) = 0, \quad u(x,0) = 2\sin(2\pi x).$$

The problem fixes the value of D at 1/16. Suppose we choose $\Delta x = 1/40$. Stability then requires that Δt be chosen to satisfy

$$\Delta t \le \frac{(\Delta x)^2}{2D} = \frac{1/1600}{2(1/16)} = \frac{1}{200}.$$

Figure 10.4 displays the approximate solution to the test problem at $t = 1$, computed using the FTCS method with different values of Δt, and therefore with different values of λ. The solution in the top graph was computed with $\Delta t = 1/200$, while the solution in the middle graph was computed with $\Delta t = 1/182$. The

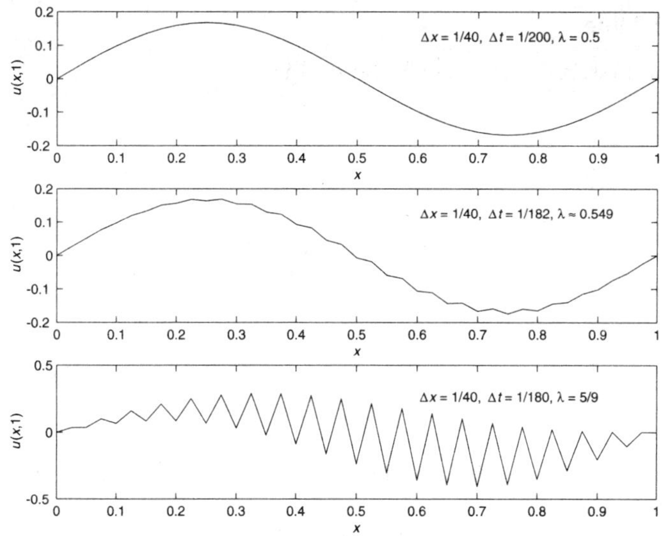

Figure 10.4 Approximate solutions to the initial-boundary value problem

$$\frac{\partial u}{\partial t} = \frac{1}{16}\frac{\partial^2 u}{\partial x^2},$$
$$u(0,t) = u(1,t) = 0, \quad u(x,0) = 2\sin(2\pi x)$$

computed using the FTCS method. Solutions at $t = 1$ are shown.

solution in the bottom graph was computed with $\Delta t = 1/180$. These time step values correspond to $\lambda = 1/2$, $\lambda \approx 0.549$ and $\lambda = 5/9$, respectively.

With the cutoff value of $\Delta t = 1/200$ ($\lambda = 1/2$), the computed approximate solution behaves as expected. The approximate solution compares favorably with the exact solution, $u(x,1) = 2e^{-\pi^2/4}\sin(2\pi x)$, and there are no signs of instability. Increasing the time step by less than 10%, to $\Delta t = 1/182$ ($\lambda \approx 0.549$), the solution begins to show the tell-tale signs of instability. Note the sawtooth-like oscillations in the middle graph. If additional time steps were performed with this value of Δt, the quality of the solution would degrade rapidly. Finally, taking $\Delta t = 1/180$ ($\lambda = 5/9$), we observe full-blown instability. (Compare the middle and bottom graphs of Figure 10.4 with the bottom graph of Figure 7.21.)

For $1/200 < \Delta t < 1/182$, the solution curves at $t = 1$ do not appear to indicate that the FTCS method is unstable, even though these values correspond to λ values which are larger than 0.5. This, however, is not a contradiction of our matrix analysis. With time step values in the range $1/200 < \Delta t < 1/182$, the amplification due to repeated multiplication by E_{FTCS} is small enough so that instability is not yet observed at $t = 1$. Were the solutions to be advanced beyond $t = 1$, eventually, the instability would become apparent. This will be explored in Exercise 1.

Backward in Time

The evolution matrix for the BTCS method is

$$E_{\text{BTCS}} = (I + \lambda A)^{-1} = \begin{bmatrix} 1+2\lambda & -\lambda & & & & & \\ -\lambda & 1+2\lambda & -\lambda & & & & \\ & -\lambda & 1+2\lambda & -\lambda & & & \\ & & \cdot & \cdot & \cdot & & \\ & & & \cdot & \cdot & \cdot & \\ & & & & \cdot & \cdot & \cdot \\ & & & & -\lambda & 1+2\lambda & -\lambda \\ & & & & & -\lambda & 1+2\lambda \end{bmatrix}^{-1},$$

where $\lambda = D\Delta t/(\Delta x)^2$. The matrix $I + \lambda A$ matches the pattern of the special matrix M with

$$a = 1 + 2\lambda \quad \text{and} \quad b = c = -\lambda.$$

It follows from equation (1), then, that the eigenvalues of $I + \lambda A$ are given by

$$\begin{aligned} \mu_k &= 1 + 2\lambda + 2\lambda \cos\frac{k\pi}{N} \\ &= 1 + 2\lambda\left(1 + \cos\frac{k\pi}{N}\right) \\ &= 1 + 4\lambda\cos^2\frac{k\pi}{2N} \end{aligned} \tag{3}$$

for $k = 1, 2, 3, \ldots, N - 1$.

From equation (3), we see that $\mu_k > 1$ for all k. Let τ_k denote the eigenvalues of E_{BTCS}. Since $E_{\text{BTCS}} = (I + \lambda A)^{-1}$, $\tau_k = 1/\mu_k$. With $\mu_k > 1$ for all k, it follows that $0 < \tau_k < 1$ for all k. Hence $\rho(E_{\text{BTCS}}) < 1$, regardless of the value of λ, and the BTCS method is *unconditionally stable.* This is the payoff we get in return for the extra work of solving a linear system of equations at each time step.

EXAMPLE 10.5 Unconditional Stability of the BTCS Method

Figure 10.5 displays approximate solutions to the initial boundary value problem

$$\frac{\partial u}{\partial t} = \frac{1}{16}\frac{\partial^2 u}{\partial x^2}, \qquad u(0,t) = u(1,t) = 0, \quad u(x,0) = 2\sin(2\pi x).$$

computed using the BTCS method with $\Delta x = 1/40$ and $\Delta t = 1/200$ (top graph), $\Delta t = 1/100$ (middle graph), and $\Delta t = 1/10$ (bottom graph). The corresponding values of λ are $\lambda = 1/2$, $\lambda = 1$, and $\lambda = 10$. Even with λ as large as 10, the computed solution shows no signs of instability.

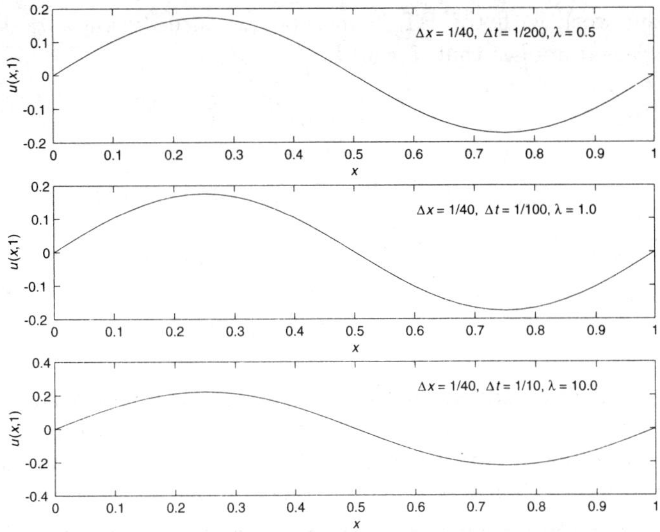

Figure 10.5 Approximate solutions to the initial boundary value problem

$$\frac{\partial u}{\partial t} = \frac{1}{16}\frac{\partial^2 u}{\partial x^2},$$

$$u(0,t) = u(1,t) = 0,$$
$$u(x,0) = 2\sin(2\pi x)$$

computed using the BTCS method. Solutions at $t = 1$ are shown.

Crank-Nicolson Scheme

The evolution matrix for the Crank-Nicolson scheme is given by

$$\begin{aligned} E_{\text{CN}} &= (I + \lambda A)^{-1}(I - \lambda A) \\ &= E_{\text{BTCS}} E_{\text{FTCS}}, \end{aligned}$$

where $\lambda = D\Delta t / \left[2(\Delta x)^2\right]$. To determine the eigenvalues of E_{CN}, note that

$$(I + \lambda A) + (I - \lambda A) = 2I.$$

Premultiplying this equation by $(I + \lambda A)^{-1}$ yields

$$I + (I + \lambda A)^{-1}(I - \lambda A) = 2(I + \lambda A)^{-1},$$

or

$$E_{\text{CN}} = -I + 2(I + \lambda A)^{-1}. \tag{4}$$

If τ_k are the eigenvalues of E_{CN} and μ_k are the eigenvalues of $I + \lambda A$, then equation (4) implies that

$$\tau_k = -1 + 2/\mu_k.$$

From our work with the BTCS method, we already know that $\mu_k > 1$ for all k. This relation implies that, for all k,

$$0 < \frac{1}{\mu_k} < 1,$$

and then

$$0 < \frac{2}{\mu_k} < 2.$$

Finally,

$$-1 < -1 + \frac{2}{\mu_k} = \tau_k < 1.$$

Hence $\rho(E_{\text{CN}}) < 1$, regardless of the value of λ. The Crank-Nicolson scheme, like the BTCS method, is *unconditionally stable.* With unconditional stability and second-order accuracy in both time and space, the Crank-Nicolson scheme is considered the method of choice for one-dimensional diffusion problems.

EXAMPLE 10.6 Unconditional Stability of the Crank-Nicolson Scheme

One last time we turn to the initial boundary value problem

$$\frac{\partial u}{\partial t} = \frac{1}{16}\frac{\partial^2 u}{\partial x^2}, \qquad u(0,t) = u(1,t) = 0, \quad u(x,0) = 2\sin(2\pi x).$$

Figure 10.6 displays approximate solutions to this problem computed using the Crank-Nicolson scheme with $\Delta x = 1/40$ and $\Delta t = 1/200$ (top graph), $\Delta t = 1/100$ (middle graph), and $\Delta t = 1/10$ (bottom graph). The corresponding values of λ are $\lambda = 1/2$, $\lambda = 1$ and $\lambda = 10$. Even with λ as large as 10, the computed solution shows no signs of instability.

---

von Neumann Stability Analysis

Von Neumann stability analysis is very similar to the local mode analysis we used in Chapter 9. We assume that the solution to the finite difference equations can be expressed in terms of the Fourier components

$$w_j^{(n)} = r^n e^{i(j\theta)}. \tag{5}$$

The amplitude coefficient, r, is called the *amplification factor.* As n increases (more time steps are computed), we want the amplitude of this mode of the solution to be bounded. In other words, we require the amplification factor satisfy

$$-1 \le r \le 1$$

for a real-valued amplitude, or

$$r\bar{r} \le 1$$

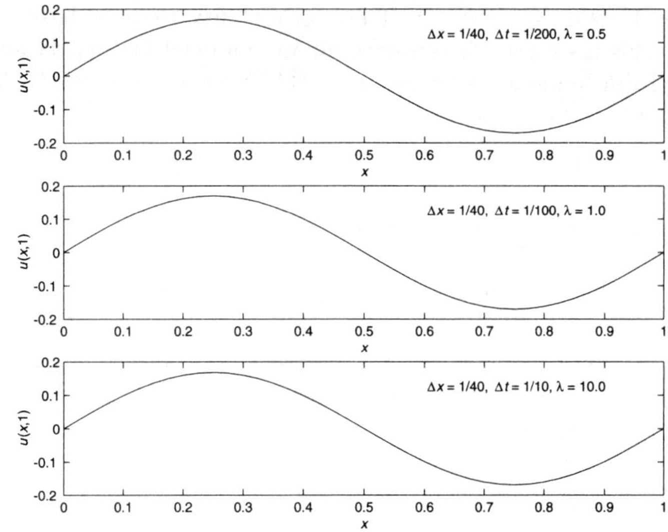

Figure 10.6 Approximate solutions to the initial-boundary value problem

$$\frac{\partial u}{\partial t} = \frac{1}{16}\frac{\partial^2 u}{\partial x^2},$$

$$u(0,t) = u(1,t) = 0,$$
$$u(x,0) = 2\sin(2\pi x)$$

computed using the Crank-Nicolson scheme. Solutions at $t = 1$ are shown.

for a complex-valued amplitude, where $\bar{r}$ denotes the complex conjugate of r. The equation for the amplification factor is obtained by substituting the Fourier component, (5), into the finite difference equation. This generally results in a polynomial in r.

To demonstrate von Neumann stability analysis, let's consider the FTCS method. The generic interior point finite difference equation for this method is

$$w_j^{(n+1)} - w_j^{(n)} = \lambda\left(w_{j-1}^{(n)} - 2w_j^{(n)} + w_{j+1}^{(n)}\right),$$

where $\lambda = D\Delta t/(\Delta x)^2$. Substitution of $w_j^{(n)} = r^n e^{i(j\theta)}$ into the difference equation yields

$$(r^{n+1} - r^n)e^{i(j\theta)} = \lambda r^n e^{i(j\theta)}(e^{-i\theta} - 2 + e^{i\theta}),$$

or, upon simplification,

$$r - 1 = \lambda(e^{-i\theta} - 2 + e^{i\theta}).$$

Making use of the identity $\cos\theta = (e^{-i\theta} + e^{i\theta})/2$ leads to $r = 1 - 2\lambda(1 - \cos\theta)$. r is clearly real-valued, so we require $-1 \le r \le 1$. Since $0 \le 1 - \cos\theta \le 2$, it follows that $1-4\lambda \le 1-2\lambda(1-\cos\theta) \le 1$. Thus, for stability, we must have $-1 \le 1-4\lambda$, or $\lambda \le 1/2$. This is exactly the same result we obtained via matrix analysis. The application of von Neumann analysis to the BTCS method and the Crank-Nicolson scheme will be left as exercises.

EXERCISES

1. Reconsider the initial boundary value problem

$$\frac{\partial u}{\partial t} = \frac{1}{16}\frac{\partial^2 u}{\partial x^2}, \qquad u(0,t) = u(1,t) = 0, \quad u(x,0) = 2\sin(2\pi x).$$

Using the FTCS method with $\Delta x = 1/40$ and the indicated value for Δt, determine the time t when instability becomes apparent in the approximate solution.
(a) $\Delta t = 1/185$
(b) $\Delta t = 1/190$
(c) $\Delta t = 1/195$

2. Use von Neumann stability analysis to show that the BTCS method is unconditionally stable. The generic interior grid point finite difference equation for the BTCS method is

$$w_j^{(n+1)} - w_j^{(n)} = \lambda\left(w_{j-1}^{(n+1)} - 2w_j^{(n+1)} + w_{j+1}^{(n+1)}\right),$$

where $\lambda = D\,\Delta t/(\Delta x)^2$.

3. Use von Neumann stability analysis to show that the Crank-Nicolson scheme is unconditionally stable. The generic interior grid point finite difference equation for the Crank-Nicolson scheme is

$$w_j^{(n+1)} - w_j^{(n)} = \lambda\left(w_{j-1}^{(n+1)} - 2w_j^{(n+1)} + w_{j+1}^{(n+1)} + w_{j-1}^{(n)} - 2w_j^{(n)} + w_{j+1}^{(n)}\right),$$

where $\lambda = D\,\Delta t/\left[2(\Delta x)^2\right]$.

4. (a) Show that if we apply Euler's method to the system of semidiscrete equations given in equation (3) of Section 10.1 we reproduce the FTCS method.
(b) Use the region of absolute stability for Euler's method to show that the FTCS method is conditionally stable and requires $\lambda \le \frac{1}{2}$.

5. (a) Show that if we apply the backward Euler method to the system of semidiscrete equations given in equation (3) of Section 10.1 we reproduce the BTCS method.
(b) Use the region of absolute stability for the backward Euler method to show that the BTCS method is unconditionally stable.

6. (a) Show that if we apply the trapezoidal method to the system of semidiscrete equations given in equation (3) of Section 10.1 we reproduce the Crank-Nicolson scheme.
(b) Use the region of absolute stability for the trapezoidal method to show that the Crank-Nicolson scheme is unconditionally stable.

7. **(a)** Suppose we apply the classical fourth-order Runge-Kutta method to the system of semidiscrete equations given in equation (3) of Section 10.1. What restriction must be placed upon Δt to guarantee absolute stability?

(b) Repeat part (a) for the two-step Adams-Moulton method.

8. The generic interior grid point finite difference equation for the Leapfrog method is

$$w_j^{(n+1)} - w_j^{(n-1)} = \lambda\left(w_{j-1}^{(n)} - 2w_j^{(n)} + w_{j+1}^{(n)}\right),$$

where $\lambda = 2D\,\Delta t/(\Delta x)^2$. Determine the stability of this scheme.

9. The generic interior grid point finite difference equation for the DuFort-Frankel method is

$$w_j^{(n+1)} - w_j^{(n-1)} = \lambda\left(w_{j-1}^{(n)} - w_j^{(n-1)} - w_j^{(n+1)} + w_{j+1}^{(n)}\right),$$

where $\lambda = 2D\,\Delta t/(\Delta x)^2$. Determine the stability of this scheme.

10. The FTCS method, the BTCS method, and the Crank-Nicolson scheme are special cases of the σ-general finite difference scheme. The σ-general scheme can be written in the form

$$A_I\mathbf{w}^{(n+1)} = A_E\mathbf{w}^{(n)}.$$

In arriving at this equation, homogeneous Dirichlet boundary conditions have been assumed. The $(N-1)\times(N-1)$ matrices A_I and A_E are of the form

$$\begin{aligned} A_I &= (1+2\sigma\lambda)I - \sigma\lambda X \\ A_E &= [1-2(1-\sigma)\lambda]\,I + (1-\sigma)\lambda X, \end{aligned}$$

where $\lambda = D\Delta t/(\Delta x)^2$ and

$$X = \begin{bmatrix} 0 & 1 & & & & \\ 1 & 0 & 1 & & & \\ & 1 & 0 & 1 & & \\ & & \ddots & \ddots & \ddots & \\ & & & 1 & 0 & 1 \\ & & & & 1 & 0 \end{bmatrix}.$$

(a) Show that $\sigma = 0$ corresponds to the FTCS method, $\sigma = 1$ corresponds to the BTCS method and $\sigma = 1/2$ corresponds to the Crank-Nicolson scheme.

(b) Show that the matrix A_I is non-singular.

(c) Show that the σ-general scheme is conditionally stable for $\sigma < 1/2$ and unconditionally stable for $\sigma \geq 1/2$.

(d) For $\sigma < 1/2$, show that the method is stable for

$$\lambda \leq \frac{1}{2(1-2\sigma)}.$$

11. Consider the FTCS method and recall that the eigenvalues of E_{FTCS} are

$$\mu_k = 1 - 4\lambda \sin^2 \frac{k\pi}{2N}$$

for $k = 1, 2, 3, \ldots, N-1$.

(a) Show that the vector $\mathbf{v}_k$ whose components are given by

$$\sin \frac{lk\pi}{N}$$

for $l = 1, 2, 3, \ldots, N-1$ is an eigenvector associated with μ_k.

(b) For $N = 10$, plot the components of $\mathbf{v}_9$ versus l. Repeat for $N = 40$ and $\mathbf{v}_{39}$.

(c) Using part (b) and the fact that μ_{N-1} is the most negative eigenvalue of E_{FTCS}, explain why instability manifests itself in the form of sawtooth oscillations.

10.3 MORE GENERAL PARABOLIC EQUATIONS

The previous two sections presented a detailed examination of three different finite difference methods for approximating the solution of the heat equation

$$\frac{\partial u}{\partial t} = D\frac{\partial^2 u}{\partial x^2} \tag{1}$$

with Dirichlet boundary conditions. The forward in time/central in space, or FTCS, method was found to be first-order accurate in the time discretization parameter, Δt, second-order accurate in the space discretization parameter, Δx, and conditionally stable, requiring $\lambda = D\,\Delta t/(\Delta x)^2 \leq 1/2$. The backward in time/central in space, or BTCS, method was also found to be first order in Δt and second order in Δx but was unconditionally stable. Finally, the Crank-Nicolson scheme was found to be second order in both Δt and Δx and to be unconditionally stable.

In this section we will address more general parabolic partial differential equations than equation (1). In particular, we will treat the equations

$$\frac{\partial u}{\partial t} = D\frac{\partial^2 u}{\partial x^2} + s(x,t)$$

and

$$\frac{\partial u}{\partial t} = D\frac{\partial^2 u}{\partial x^2} - \beta(x,t)u + s(x,t),$$

where $s(x,t)$ is a source term and $-\beta(x,t)u$ is a decay term. Note that we will continue to assume that the diffusion coefficient is constant. For each partial differential equation, we will develop the finite difference equations for the FTCS method, the BTCS method and the Crank-Nicolson scheme and will investigate any changes in stability.

Source Terms

Let's start generalizing the heat equation by including a source term only and continuing to specify Dirichlet boundary conditions. In other words, let's develop finite difference methods for approximating the solution of the initial boundary value problem

$$\text{IBVP} \quad \begin{cases} \dfrac{\partial u}{\partial t} = D\dfrac{\partial^2 u}{\partial x^2} + s(x,t), & A \le x \le B, t > 0 \\ u(A,t) = u_A(t) \\ u(B,t) = u_B(t) \\ u(x,0) = f(x). \end{cases} \tag{2}$$

In practice, a source term might represent the effect of precipitation in a groundwater flow problem, the effect of internal heat generation (possibly due to the passage of an electric current) in a heat conduction problem, or the creation of particles in a molecular diffusion problem.

To derive the finite difference approximation to (2), first introduce a uniform partition over the interval $A \le x \le B$, with mesh spacing $\Delta x = (B - A)/N$. Next, replace the second derivative term, $\partial^2 u/\partial x^2$, by its second-order finite difference formula. This produces the semidiscrete approximation

$$\frac{d\mathbf{v}(t)}{dt} = -\frac{D}{(\Delta x)^2}\left[A\mathbf{v}(t) + \mathbf{b}(t)\right] + \mathbf{s}(t), \quad \mathbf{v}(0) = \mathbf{f}. \tag{3}$$

Here

$$\begin{aligned} \mathbf{v}(t) &= \begin{bmatrix} v_1(t) & v_2(t) & v_3(t) & \cdots & v_{N-1}(t) \end{bmatrix}^T, \\ \mathbf{b}(t) &= \begin{bmatrix} -u_A(t) & 0 & \cdots & 0 & -u_B(t) \end{bmatrix}^T, \\ \mathbf{s}(t) &= \begin{bmatrix} s(x_1,t) & s(x_2,t) & s(x_3,t) & \cdots & s(x_{N-1},t) \end{bmatrix}^T, \\ \mathbf{f} &= \begin{bmatrix} f(x_1) & f(x_2) & f(x_3) & \cdots & f(x_{N-1}) \end{bmatrix}^T \end{aligned}$$

and $v_j(t) \approx u(x_j,t)$. The $(N-1) \times (N-1)$ matrix A takes the form

$$A = \begin{bmatrix} 2 & -1 & & & & & \\ -1 & 2 & -1 & & & & \\ & -1 & 2 & -1 & & & \\ & & \ddots & \ddots & \ddots & & \\ & & & \ddots & \ddots & \ddots & \\ & & & & -1 & 2 & -1 \\ & & & & & -1 & 2 \end{bmatrix}.$$

Finally, we obtain the FTCS method by evaluating (3) at $t = t_n$ and replacing the time derivative by its first-order forward difference approximation. This procedure yields

$$\mathbf{w}^{(n+1)} = (I - \lambda A)\mathbf{w}^{(n)} - \lambda\mathbf{b}^{(n)} + \Delta t\mathbf{s}^{(n)}.$$

The BTCS method,

$$(I+\lambda A)\mathbf{w}^{(n+1)} = \mathbf{w}^{(n)} - \lambda\mathbf{b}^{(n+1)} + \Delta t\mathbf{s}^{(n+1)},$$

and the Crank-Nicolson scheme,

$$(I+\lambda A)\mathbf{w}^{(n+1)} = (I-\lambda A)\mathbf{w}^{(n)} - \lambda\left(\mathbf{b}^{(n)} + \mathbf{b}^{(n+1)}\right) + \frac{\Delta t}{2}\left(\mathbf{s}^{(n)} + \mathbf{s}^{(n+1)}\right),$$

are obtained by evaluating (3) at $t = t_{n+1}$ and replacing the time derivative by its first-order backward difference approximation and by integrating (3) from $t = t_n$ to $t = t_{n+1}$ and using the trapezoidal rule to estimate the integral on the right-hand side, respectively. The vector

$$\mathbf{w}^{(n)} = \begin{bmatrix} w_1^{(n)} & w_2^{(n)} & w_3^{(n)} & \cdots & w_{N-1}^{(n)} \end{bmatrix}^T$$

is the fully discrete approximation to the solution of (2); that is, $w_j^{(n)} \approx v_j(t_n) \approx u(x_j, t_n)$, $\mathbf{b}^{(n)} = \mathbf{b}(t_n)$, and $\mathbf{s}^{(n)} = \mathbf{s}(t_n)$. For the FTCS and the BTCS methods, $\lambda = D\Delta t/(\Delta x)^2$, while for the Crank-Nicolson scheme, $\lambda = D\Delta t/\left[2(\Delta x)^2\right]$. The most important observation is that the inclusion of the source term in the partial differential equation has absolutely no effect upon the evolution matrix of the numerical method. Hence, there is no change to our previously established stability results. The FTCS method still requires $\lambda \leq 1/2$, and the BTCS method and the Crank-Nicolson scheme are still unconditionally stable.

EXAMPLE 10.7 A Test Problem with a Source Term

Let's consider the initial boundary value problem

$$\frac{\partial u}{\partial t} = \frac{1}{16}\frac{\partial^2 u}{\partial x^2} - t + 8x^2, \qquad u(0,t) = 0, \quad u(1,t) = 8t, \quad u(x,0) = 2\sin(2\pi x),$$

whose exact solution is $u(x,t) = 2e^{-(\pi^2/4)t}\sin(2\pi x) + 8x^2 t$. The source term for this problem is $s(x,t) = -t + 8x^2$, so the vector $\mathbf{s}^{(n)}$ is given by

$$\mathbf{s}^{(n)} = \left[-t_n + 8(\Delta x)^2 \; -t_n + 8(2\,\Delta x)^2 \; -t_n + 8(3\,\Delta x)^2 \cdots -t_n + 8(1-\Delta x)^2\right]^T.$$

The function $u(x, 10)$ is plotted in the upper left panel of Figure 10.7. The other panels of Figure 10.7 display the approximate solutions computed using our three finite difference methods. For each method, $\Delta x = 1/20$ was used. The upper right panel contains two approximate solutions computed using the FTCS method. With $\Delta x = 1/20$, stability requires

$$\Delta t \leq \frac{(1/20)^2}{2(1/16)} = \frac{1}{50}.$$

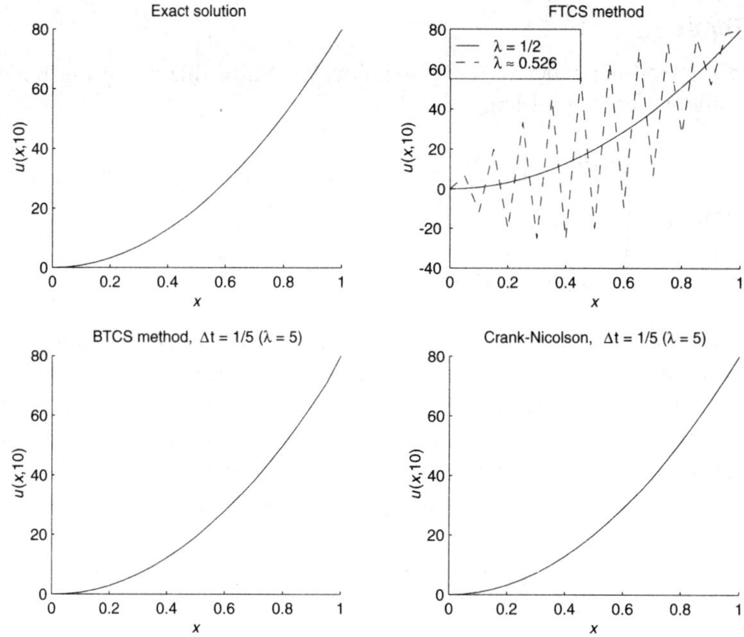

Figure 10.7 Exact solution to the initial boundary value problem

$$\frac{\partial u}{\partial t} = \frac{1}{16}\frac{\partial^2 u}{\partial x^2} - t + 8x^2,$$

$$u(0,t) = 0, \quad u(1,t) = 8t,$$
$$u(x,0) = 2\sin(2\pi x),$$

and approximate solutions computed using the FTCS method, the BTCS method, and the Crank-Nicolson scheme. All solutions are displayed at $t = 10$.

The solid curve corresponds to $\Delta t = 1/50$. The dashed curve was computed with $\Delta t = 2/95$, which corresponds to

$$\lambda = \frac{(1/16)(2/95)}{(1/20)^2} = \frac{10}{19} \approx 0.526.$$

The instability in this solution is clear. The approximate solutions computed using the BTCS method and the Crank-Nicolson scheme are shown in the lower left and lower right panels, respectively. Each of these curves was obtained with $\Delta t = 1/5$ ($\lambda = 5$), and neither displays any signs of instability.

Decay Terms

Next, let's introduce a decay term and develop finite difference methods for the initial boundary value problem

$$\text{IBVP}\quad \begin{cases} \dfrac{\partial u}{\partial t} = D\dfrac{\partial^2 u}{\partial x^2} - \beta(x,t)u + s(x,t), \quad A \le x \le B, t > 0 \\ u(A,t) = u_A(t) \\ u(B,t) = u_B(t) \\ u(x,0) = f(x), \end{cases} \tag{4}$$

where $\beta(x,t) \ge 0$ for all $x \in [A,B]$ and all $t \ge 0$. Like the source term, the decay term can model a variety of different phenomena. For example, in a heat conduction problem, the term $-\beta(x,t)u$ arises when there is convective heat loss from a lateral surface, while in a molecular diffusion problem, the decay term models the absorption of particles with a mean absorption rate of $\beta(x,t)$.

The semidiscrete approximation associated with problem (4) is

$$\frac{d\mathbf{v}\,(t)}{dt} = -\frac{D}{(\Delta x)^2}\left[A\mathbf{v}(t) + \mathbf{b}(t)\right] - B(t)\mathbf{v}(t) + \mathbf{s}(t), \quad \mathbf{v}(0) = \mathbf{f},$$

where the vectors $\mathbf{v}(t)$, $\mathbf{b}(t)$, $\mathbf{s}(t)$, and $\mathbf{f}$, and the matrix A, are the same as given above. B is the diagonal matrix

$$B = \operatorname{diag}(\ \beta(x_1,t),\ \ \beta(x_2,t),\ \ \beta(x_3,t),\ \ \ldots,\ \ \beta(x_{N-1},t)\ ).$$

Working from the semidiscrete approximation and following the basic procedures described earlier, we obtain the FTCS method

$$\mathbf{w}^{(n+1)} = (I - \lambda A - \Delta t\, B^{(n)})\mathbf{w}^{(n)} - \lambda\mathbf{b}^{(n)} + \Delta t\mathbf{s}^{(n)},$$

the BTCS method

$$(I + \lambda A + \Delta t\, B^{(n+1)})\mathbf{w}^{(n+1)} = \mathbf{w}^{(n)} - \lambda\mathbf{b}^{(n+1)} + \Delta t\mathbf{s}^{(n+1)},$$

and the Crank-Nicolson scheme

$$\left(I + \lambda A + \frac{\Delta t}{2}B^{(n+1)}\right)\mathbf{w}^{(n+1)} = \left(I - \lambda A - \frac{\Delta t}{2}B^{(n)}\right)\mathbf{w}^{(n)} - \lambda\left(\mathbf{b}^{(n)} + \mathbf{b}^{(n+1)}\right) + \frac{\Delta t}{2}\left(\mathbf{s}^{(n)} + \mathbf{s}^{(n+1)}\right).$$

The presence of the decay term in the partial differential equation has resulted in a change to the evolution matrix of each method. To determine whether these changes have an effect upon stability, we will have to investigate further. Unfortunately, $\beta(x,t)$ is generally not constant, so the evolution matrices no longer match the special form given in Section 10.2. As a result, we will turn to von Neumann stability analysis.

The generic finite difference equation for the FTCS method is

$$w_j^{(n+1)} = \lambda w_{j-1}^{(n)} + \left(1 - 2\lambda - \Delta t\, \beta_j^{(n)}\right) w_j^{(n)} + \lambda w_{j+1}^{(n)} + \Delta t\, s_j^{(n)}.$$

The term $\Delta t s_j^{(n)}$ is independent of w, so it will have no effect on the amplification factor for the discrete Fourier mode. We will therefore set this term to zero when we substitute $w_j^{(n)} = r^n e^{i(j\theta)}$. Upon substituting the Fourier mode and simplifying, we find

$$\begin{aligned} r &= \left(1 - 2\lambda - \Delta t\, \beta_j^{(n)}\right) + 2\lambda \cos\theta \\ &= 1 - \Delta t\, \beta_j^{(n)} - 2\lambda(1 - \cos\theta). \end{aligned}$$

Note that r is always less than $1 - \Delta t\, \beta_j^{(n)}$, which for $\beta_j^{(n)} > 0$ is always less than 1. As a function of θ, the smallest value taken on by r is $1 - \Delta t\, \beta_j^{(n)} - 4\lambda$. Hence, we need $1 - \Delta t\, \beta_j^{(n)} - 4\lambda \geq -1$ for stability. Recalling that $\lambda = D\,\Delta t/(\Delta x)^2$, after some algebraic manipulation, we arrive at the inequality

$$\Delta t \leq \frac{(\Delta x)^2}{2D + (\Delta x)^2 \beta_j^{(n)}/2}.$$

Since we desire stability for all time steps and at all locations of the grid, the final stability condition is

$$\Delta t \leq \frac{(\Delta x)^2}{2D + (\Delta x)^2 \left(\max_{1 \leq j \leq N-1, n \geq 0} \beta_j^{(n)}\right)/2}.$$

Hence, the FTCS is still conditionally stable, and a slightly smaller value for Δt is required to maintain stability.

For the BTCS method, we substitute $w_j^{(n)} = r^n e^{i(j\theta)}$ into

$$-\lambda w_{j-1}^{(n+1)} + \left(1 + 2\lambda + \Delta t\, \beta_j^{(n+1)}\right) w_j^{(n+1)} - \lambda w_{j+1}^{(n+1)} = w_j^{(n)}.$$

The term $\Delta t\, s_j^{(n+1)}$ has once again been dropped since it will not affect the amplification factor. Following simplification, we find

$$\begin{aligned} r &= \left[\left(1 + 2\lambda + \Delta t\, \beta_j^{(n+1)}\right) - 2\lambda \cos\theta\right]^{-1} \\ &= \left[1 + \Delta t\, \beta_j^{(n+1)} + 2\lambda(1 - \cos\theta)\right]^{-1}. \end{aligned}$$

It is clear that $1 + \Delta t \beta_j^{(n+1)} + 2\lambda(1 - \cos\theta) \geq 1$ for all θ and all λ; therefore, $0 \leq r = \left[1 + \Delta t \beta_j^{(n+1)} + 2\lambda(1 - \cos\theta)\right]^{-1} \leq 1$, and the method is unconditionally stable. A similar result holds for the Crank-Nicolson scheme. The details are left as an exercise.

EXAMPLE 10.8 A Test Problem with a Decay Term

Let's consider the initial boundary value problem

$$\frac{\partial u}{\partial t} = \frac{1}{16}\frac{\partial^2 u}{\partial x^2} - 4x(1-x)u + e^{-t}(8x^2 - t)$$

$$u(0,t) = 0, \quad u(1,t) = 8te^{-t}, \quad u(x,0) = 2\sin(2\pi x).$$

The source term for this problem is $s(x,t) = e^{-t}(8x^2 - t)$, and the decay coefficient is $\beta(x,t) = 4x(1-x)$. Note that

$$\max_{0\le x\le 1, t>0} \beta(x,t) = 1.$$

Figure 10.8 displays the approximate solution to this problem computed using our three finite difference methods. For each method, $\Delta x = 1/20$ was used. The upper left panel contains two approximate solutions computed using the FTCS method. With $\Delta x = 1/20$, stability requires

$$\Delta t \le \frac{(1/20)^2}{2(1/16) + (1/20)^2(1)/2} = \frac{2}{101} \approx 0.0198.$$

The solid curve corresponds to $\Delta t = 2/101$, while the dashed curve was computed with $\Delta t = 10/499$. The instability in the latter solution is clear. Note the difference six time steps make. The approximate solutions computed using the BTCS method and the Crank-Nicolson scheme are shown in the upper right and lower left panels, respectively. Each of these curves was obtained with $\Delta t = 1/5$, and neither displays any signs of instability.

Application Problem 1: Including the Effect of Rainfall

Let's reconsider the application problem from Section 10.1, "Rise in the Water Table Due to the Spring Thaw." Suppose that during the spring thaw, the average daily rainfall is given by the function $w(t)$. To incorporate the effect of this rainfall into our model, we need to add the term $w(t)\,\Delta x\,\Delta t$ to equation (18) from the Chapter 1 Overview. Carrying through with the remainder of the derivation, we find that our revised model is

$$\frac{\partial h}{\partial t} = a\frac{\partial^2 h}{\partial x^2} + \frac{w(t)}{S},$$

where S is the storativity of the soil.

As in Section 10.1, we take $a = 509.76$ m^2/day,

$$h(x,0) = 0, \quad h(0,t) = \begin{cases} 2(1-e^{-t/5}), & t < 30 \\ 2, & t \ge 30, \end{cases} \quad \text{and} \quad h(800,t) = 0.3h(0,t).$$

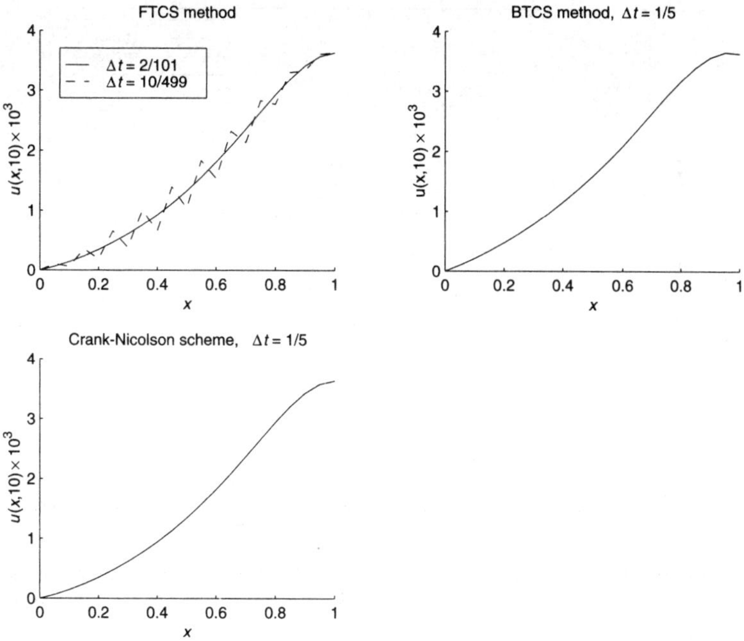

Figure 10.8 Approximate solutions to the initial boundary value problem

$$\frac{\partial u}{\partial t} = \frac{1}{16}\frac{\partial^2 u}{\partial x^2} - 4x(1-x)u + e^{-t}(8x^2 - t),$$

$$u(0,t) = 0, \quad u(1,t) = 8te^{-t}, \quad u(x,0) = 2\sin(2\pi x),$$

computed using the FTCS method, the BTCS method, and the Crank-Nicolson scheme. All solutions are displayed at $t = 10$.

Further, suppose $S = 0.2$ and

$$w(t) = \begin{cases} 0.004 \text{ m/day}, & t < 30 \\ 0, & t \geq 30 \end{cases}.$$

Figure 10.9 displays the resulting change in the water table along the aquifer after 30 days and after 60 days. Note how the water level diffuses once the rainfall has ended. Both solution profiles were computed using the Crank-Nicolson scheme with $\Delta x = 20$ meters and $\Delta t = 0.2$ days.

Application Problem 2: One-Dimensional Model for Color Photograph Development

When photographic film is exposed to light, chemical reactions cause the silver halide grains in the film to acquire latent image sites. During the development of this image, a chemical in the developer solution known as a reduced developer

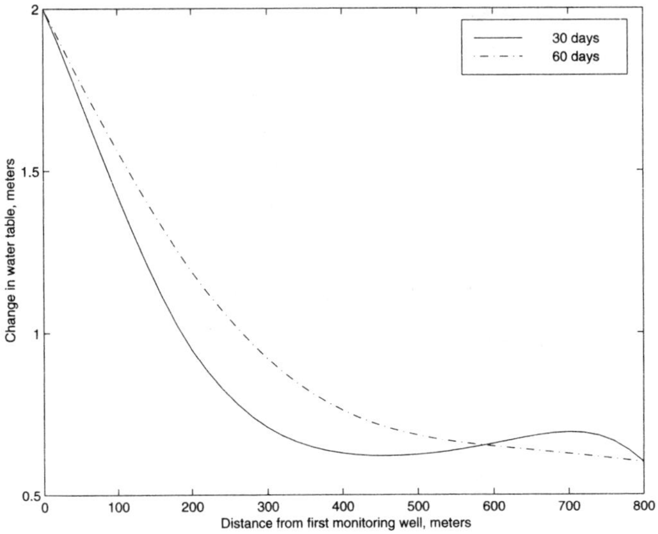

Figure 10.9 Change in the water table in an aquifer during the spring thaw, including the effect of average daily rainfall during the first thirty days.

deposits electrons at the latent image sites and becomes an oxidized developer. The oxidized developer diffuses and reacts with a dye-forming coupler to form an immobile dye and an inhibitor. The inhibitor then diffuses, and some of it adsorbs to the surface of the silver grain blocking the dissociation of the halides. The entire process is extremely complex, and modeling is not yet fully understood.

Friedman and Littman (in Chapter 4 of *Industrial Mathematics: A Course in Solving Real-World Problems*, SIAM, Philadelphia, 1994) present a simplified, one-dimensional model of the development process by considering only the density of the oxidized developer, $T(x)$, and the coupler, $C(x)$. The model consists of an initial boundary value problem for the oxidized developer plus an initial value problem for the coupler:

$$\frac{\partial T}{\partial t} = D\frac{\partial^2 T}{\partial x^2} - kTC + \gamma E(x), \qquad \frac{\partial C}{\partial t} = -kTC.$$

The initial conditions for this system are $T(x,0) = 0$ and $C(x,0) = C_0(x)$, while the boundary conditions are $T(0,t) = T(L,t) = 0$, where L is the length of the film. $E(x)$ is the exposure function which indicates those regions of the film which were exposed to light.

We will simulate three minutes (180 seconds) of development time, taking as parameter values

$$D = 100\,\mu\text{m}^2/\text{s}, k = 6.6\times 10^{12}\mu\text{m}/\text{moles}\cdot\text{s}, \gamma = 7.5\times 10^{-12}\text{ moles}/\mu\text{m}\cdot\text{s}$$
$$L = 1.5\times 10^5\mu\text{m} \text{ and } C_0 = 1.125\times 10^{-11}\text{ moles}/\mu\text{m}.$$

To help balance the drastic differences in order of magnitude between the parameters, let's introduce the following nondimensional variables:

$$\bar{T} = \frac{T}{\gamma L^2/D}, \quad \bar{C} = \frac{C}{C_0}, \quad \bar{t} = t\frac{D}{L^2}, \quad \text{and} \quad \bar{x} = \frac{x}{L}.$$

In terms of these variables, the system of differential equations becomes

$$\frac{\partial \bar{T}}{\partial \bar{t}} = \frac{\partial^2 \bar{T}}{\partial \bar{x}^2} - k_1 \bar{T}\bar{C} + E(\bar{x}), \qquad \frac{\partial \bar{C}}{\partial \bar{t}} = -k_2 \bar{T}\bar{C}$$

where

$$k_1 = \frac{C_0 k L^2}{D} = 1.670625 \times 10^{10} \quad \text{and} \quad k_2 = \frac{k\gamma L^4}{D^2} = 2.5059375 \times 10^{18}.$$

Furthermore, note that $t = 180$ seconds translates to $\bar{t} = 8 \times 10^{-7}$. For the exposure function, we will assume

$$E(\bar{x}) = \begin{cases} 1, & 0.1 \le \bar{x} \le 0.2 \text{ or } 0.45 \le \bar{x} \le 0.55 \\ 0, & \text{elsewhere.} \end{cases}$$

So how do we solve this system of differential equations? For each time step, we will first compute $\bar{C}^{(n+1)}$ using a semi-implicit discretization of the coupler equation. That is, on the right-hand side of the coupler equation, evaluate $\bar{T}$ at time level n and $\bar{C}$ at time level $n+1$. This yields the finite difference equation

$$\frac{\bar{C}_j^{(n+1)} - \bar{C}_j^{(n)}}{\Delta \bar{t}} = -k_2 \bar{T}_j^{(n)} \bar{C}_j^{(n+1)} \quad \Rightarrow \quad \bar{C}_j^{(n+1)} = \frac{\bar{C}_j^{(n)}}{1 + \Delta \bar{t} k_2 \bar{T}_j^{(n)}}.$$

By using a semi-implicit discretization we avoid having to solve a nonlinear algebraic equation to compute $\bar{T}^{(n+1)}$. Since the method for advancing the coupler density is only first order in time, once $\bar{C}^{(n+1)}$ has been calculated, we will use the BTCS method for the $\bar{T}$ equation:

$$\frac{\bar{T}_j^{(n+1)} - \bar{T}_j^{(n)}}{\Delta \bar{t}} = \frac{\bar{T}_{j-1}^{(n+1)} - 2\bar{T}_j^{(n+1)} + \bar{T}_{j+1}^{(n+1)}}{(\Delta \bar{x})^2} - k_1 \bar{T}_j^{(n+1)} \bar{C}_j^{(n+1)} + E_j.$$

Density profiles computed in this manner with $\Delta \bar{x} = 1/100$ and $\Delta \bar{t} = 2 \times 10^{-9}$ (i.e., 400 time steps) are shown in Figure 10.10.

EXERCISES

In Exercises 1–3, numerically verify that

(a) the FTCS method is stable for $\lambda = 1/2$ but unstable for any $\lambda > 1/2$;
(b) the BTCS method is unconditionally stable; and
(c) the Crank-Nicolson method is unconditionally stable.

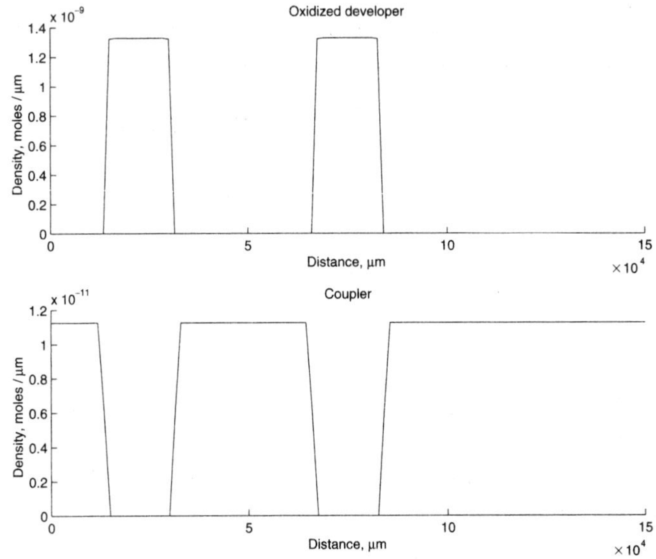

Figure 10.10 Oxidized developer and coupler density profiles after three minutes of development time.

1. $\dfrac{\partial u}{\partial t} = \dfrac{\partial^2 u}{\partial x^2}+2t+x(1-x), \quad u(0,t)=1, \quad u(1,t)=0, \quad u(x,0)=1-x-\dfrac{1}{\pi}\sin(2\pi x)$
2. $\dfrac{\partial u}{\partial t} = \dfrac{\partial^2 u}{\partial x^2}+x, \quad u(0,t)=0, \quad u(\pi,t)=\pi t, \quad u(x,0)=\sin x$
3. $\dfrac{\partial u}{\partial t} = \dfrac{\partial^2 u}{\partial x^2}-(x+t^2)e^{-xt}, \quad u(0,t)=1, \quad u(\pi,t)=e^{-\pi t}, \quad u(x,0)=1+\sin x$

In Exercises 4–6, numerically verify that

(a) the FTCS method is unstable for $\lambda = 1/2$, but stable for

$$\Delta t \le \frac{(\Delta x)^2}{2D+(\Delta x)^2(\max_{1\le j\le N-1, n\ge 0}\beta_j^{(n)})/2};$$

(b) the BTCS method is unconditionally stable; and

(c) the Crank-Nicolson method is unconditionally stable.

4. $\dfrac{\partial u}{\partial t} = \dfrac{\partial^2 u}{\partial x^2}-xu, \quad u(0,t)=1, \quad u(1,t)=0, \quad u(x,0)=1-x-\dfrac{1}{\pi}\sin(2\pi x)$
5. $\dfrac{\partial u}{\partial t} = \dfrac{\partial^2 u}{\partial x^2}-\dfrac{10}{1+xt}u, \quad u(0,t)=u(\pi,t)=0, \quad u(x,0)=\sin x$
6. $\dfrac{\partial u}{\partial t} = \dfrac{\partial^2 u}{\partial x^2}-e^{-x^2}u-(x+t^2)e^{-xt}, \quad u(0,t)=1, \quad u(\pi,t)=e^{-\pi t}, \quad u(x,0)=$ $1+\sin x$

7. Consider the partial differential equation

$$\frac{\partial u}{\partial t} = D\frac{\partial^2 u}{\partial x^2} - \beta(t)u + s(x,t),$$

where the coefficient on the decay term is a function of t only.

(a) Let $B(t)$ be any antiderivative of $\beta(t)$, and define $w(x,t) = e^{B(t)}u(x,t)$. Show that

$$\frac{\partial w}{\partial t} = D\frac{\partial^2 w}{\partial x^2} + e^{B(t)}s(x,t).$$

(b) What advantage is there to applying the FTCS method to the equation for w rather than the equation for u?

In Exercises 8–10, solve the indicated initial boundary value problem by first applying the technique of Exercise 7(a) to remove the decay term. Advance the solution to $t = 5$.

8. $\dfrac{\partial u}{\partial t} = \dfrac{\partial^2 u}{\partial x^2} - u, \quad u(0,t) = e^{-t}, \quad u(1,t) = 0, \quad u(x,0) = 1 - x - \dfrac{1}{\pi}\sin(2\pi x)$

9. $\dfrac{\partial u}{\partial t} = \dfrac{\partial^2 u}{\partial x^2} - \dfrac{2t}{1+t^2}u + x, \quad u(0,t) = 0, \quad u(\pi,t) = \dfrac{\pi t}{1+t^2}, \quad u(x,0) = \sin x$

10. $\dfrac{\partial u}{\partial t} = \dfrac{1}{16}\dfrac{\partial^2 u}{\partial x^2} - 2tu, \quad u(0,t) = u(1,t) = 0, \quad u(x,0) = 2\sin(2\pi x)$

11. A 15-cm-long piece of copper wire is initially at a uniform temperature of 20°C. At $t = 0$, a 10 amp current begins flowing through the wire generating heat, while the temperature at the ends of the wire is maintained at 20°C. The temperature within the wire satisfies the initial boundary value problem

$$\rho c_p \frac{\partial T}{\partial t} = k\frac{\partial^2 T}{\partial x^2} + \frac{4I^2R}{\pi D^2}, \qquad T(0,t) = T(0.15,t) = 20, \quad T(x,0) = 20.$$

The parameters in this problem are the mass density $\rho = 8933$ kg/m^3, the heat capacity $c_p = 385$ J/kg · °C, the thermal conductivity $k = 401$ J/m · s · °C, the diameter of the wire $D = 2.6$ mm, the current I, and the resistance per unit length

$$R = \frac{6.8 \times 10^{-8}\Omega \cdot \text{m}}{\pi D^2}.$$

Approximate the temperature profile along the length of the wire 30 seconds, 60 seconds, and 90 seconds after the current has begun to flow.

12. Rework the "Including the Effect of Rainfall" application problem with hydraulic diffusivity $a = 414.72$ m^2/day, storativity $S = 0.5$, initial and boundary conditions

$$h(0,t) = h(800,t) = 0, \quad h(x,0) = 0,$$

and rainfall

$$w(t) = \begin{cases} 0.01 \text{ m/day}, & t \le 20 \\ 0, & t > 20. \end{cases}$$

Determine the change in the water table in 10-day increments, up to $t = 60$ days.

13. Two chambers are connected by a hollow tube 0.4 meters long and contain pools of ethyl alcohol that are maintained at 30°C. At $t = 0$, the valves at each end of the tube are opened, and alcohol vapors diffuse into the tube. At 30°C, the diffusion coefficient of the alcohol vapors is $D = 1.19 \times 10^{-5}$ m^2/s and 10% alcohol vapor is present in the air. The tube contains a filter with a mean absorption rate of $\mu = 0.0069(\text{s})^{-1}$. The percent of alcohol vapor within the tube, $u(x,t)$, satisfies

$$\frac{\partial u}{\partial t} = D\frac{\partial^2 u}{\partial x^2} - \mu u, \qquad u(0,t) = u(0.4,t) = 10, \quad u(x,0) = 0.$$

Approximate $u(x,t)$ at $t = 100$ seconds, $t = 200$ seconds, and $t = 300$ seconds.

14. A circular commercial bronze rod of radius $r = 5$ cm and length 1 meter initially has a uniform temperature distribution 50 K above ambient temperature. At $t = 0$ the temperature at the right end of the rod is lowered to ambient. Let θ denote the difference between the temperature of the rod and ambient temperature. If the surface of the rod is exposed to convective heat transfer, then θ satisfies the initial boundary value problem

$$\rho c_p\frac{\partial \theta}{\partial t} = k\frac{\partial^2 \theta}{\partial x^2} - \frac{2h}{r}\theta, \qquad \theta(0,t) = 50, \quad \theta(1,t) = 0, \quad \theta(x,0) = 50,$$

where $\rho = 8800$ kg/m^3 is the mass density, $c_p = 420$ J/kg · K is the heat capacity, and $k = 52$ W/m · K is the thermal conductivity of the rod. $h = 25$ W/m^2 · K is the convective heat transfer coefficient. Determine the temperature profile along the rod after 600 seconds, 1200 seconds, and 1800 seconds.

15. Rework the "One-Dimensional Model for Color Photograph Development" application problem using the parameter values

$$D = 550\ \mu\text{m}^2/\text{s}, \quad k = 2.8 \times 10^{13}\ \mu\text{m/moles} \cdot \text{s},$$
$$\gamma = 3.2 \times 10^{-10}\ \text{moles}/\mu\text{m} \cdot \text{s},$$
$$L = 1.5 \times 10^5\ \mu\text{m} \quad \text{and} \quad C_O = 8.3 \times 10^{-8}\ \text{moles}/\mu\text{m}.$$

Take

$$E(\bar{x}) = \begin{cases} 1, & 0.1 \le \bar{x} \le 0.2 \text{ or } 0.35 \le \bar{x} \le 0.50 \text{ or } 0.62 \le \bar{x} \le 0.69 \\ 0, & \text{elsewhere} \end{cases}$$

as the exposure function and simulate five minutes of development time.

16. Show that the Crank-Nicolson scheme

$$\left(I + \lambda A + \frac{\Delta t}{2}B^{(n+1)}\right)\mathbf{w}^{(n+1)} = \left(I - \lambda A - \frac{\Delta t}{2}B^{(n)}\right)\mathbf{w}^{(n)} - \lambda\left(\mathbf{b}^{(n)} + \mathbf{b}^{(n+1)}\right) + \frac{\Delta t}{2}\left(\mathbf{s}^{(n)} + \mathbf{s}^{(n+1)}\right)$$

for approximating the solution of

$$\text{IBVP} \begin{cases} \dfrac{\partial u}{\partial t} = D\dfrac{\partial^2 u}{\partial x^2} - \beta(x,t)u + s(x,t), & A \le x \le B, t > 0 \\ u(A,t) = u_A(t) \\ u(B,t) = u_B(t) \\ u(x,0) = f(x) \end{cases}$$

is unconditionally stable.

10.4 NON-DIRICHLET BOUNDARY CONDITIONS

Consider the parabolic partial differential equation

$$\frac{\partial u}{\partial t} = D\frac{\partial^2 u}{\partial x^2} - \beta(x,t)u + s(x,t) \tag{1}$$

over the domain $A \le x \le B$ and $t \ge 0$. We will assume throughout the section that $\beta(x,t) \ge 0$ on the entire domain. The objective of this section is to investigate the numerical solution of equation (1) with non-Dirichlet boundary conditions. The appropriate finite difference equations for the FTCS method, the BTCS method, and the Crank-Nicolson scheme will be developed, and the effect of the boundary conditions upon stability will be determined.

Before turning to the boundary conditions, recall that the generic semidiscrete equation associated with equation (1) is

$$\frac{dv_j(t)}{dt} = D\frac{v_{j-1}(t) - 2v_j(t) + v_{j+1}(t)}{(\Delta x)^2} - \beta(x_j,t)v_j(t) + s(x_j,t). \tag{2}$$

In this equation $v_j(t) \approx u(x_j,t)$, $\Delta x = (B-A)/N$ for some positive integer N, and $x_j = A + j\Delta x$ for $j = 0, 1, 2, \ldots, N$.

A Model Problem

For argument's sake, let's develop the FTCS method, the BTCS method, and the Crank-Nicolson scheme for approximating the solution of the initial boundary value problem

$$\text{IBVP}\quad\begin{cases} \dfrac{\partial u}{\partial t} = D\dfrac{\partial^2 u}{\partial x^2} - \beta(x,t)u + s(x,t), \quad A \le x \le B, t > 0 \\[2ex] \dfrac{\partial u}{\partial n}(A,t) = \alpha(t) \\ p(t)u(B,t) + q(t)\dfrac{\partial u}{\partial n}(B,t) = r(t) \\ u(x,0) = f(x). \end{cases}$$

Note this problem has a Neumann condition at the left end of the domain and a Robin condition at the right end. One of each type of non-Dirichlet boundary condition has been selected so we can investigate any changes to stability criteria with a single problem.

We will handle non-Dirichlet boundary conditions here the same way we handled them in Chapters 8 and 9—by using fictitious nodes. The Neumann boundary condition at $x = A$ leads to the semidiscrete equation

$$\frac{dv_0(t)}{dt} = D\frac{-2v_0(t) + 2v_1(t) + 2\,\Delta x\,\alpha(t)}{(\Delta x)^2} - \beta(x_0,t)v_0(t) + s(x_0,t), \tag{3}$$

while the Robin condition at $x = B$ produces

$$\frac{dv_N(t)}{dt} = D\frac{2v_{N-1}(t) - 2\left[1 + \Delta x\frac{p(t)}{q(t)}\right]v_N(t) + 2\Delta x\frac{r(t)}{q(t)}}{(\Delta x)^2} - \beta(x_N,t)v_N(t) + s(x_N,t). \tag{4}$$

Combining equations (2), (3), and (4) yields the complete semidiscrete approximation to our model problem. In matrix form, we have

$$\frac{d\mathbf{v}(t)}{dt} = -\frac{D}{(\Delta x)^2}\left[A'\mathbf{v}(t) + \mathbf{b}'(t)\right] - B'(t)\mathbf{v}(t) + \mathbf{s}'(t)$$

$$\mathbf{v}(0) = \mathbf{f}',$$

where

$$\begin{aligned}
\mathbf{v}(t) &= \begin{bmatrix} v_0(t) & v_1(t) & v_2(t) & \cdots & v_N(t) \end{bmatrix}^T, \\
\mathbf{b}'(t) &= \begin{bmatrix} -2\Delta x\,\alpha(t) & 0 & \cdots & 0 & -2\,\Delta x\, r(t)/q(t) \end{bmatrix}^T, \\
\mathbf{s}'(t) &= \begin{bmatrix} s(x_0,t) & s(x_1,t) & s(x_2,t) & \cdots & s(x_N,t) \end{bmatrix}^T, \\
\mathbf{f}' &= \begin{bmatrix} f(x_0) & f(x_1) & f(x_2) & \cdots & f(x_N) \end{bmatrix}^T, \\
B'(t) &= \operatorname{diag}(\beta(x_0,t), \beta(x_1,t), \beta(x_2,t), \ldots, \beta(x_N,t))
\end{aligned}$$

and

$$A' = \begin{bmatrix}
2 & -2 & & & & & & \\
-1 & 2 & -1 & & & & & \\
 & -1 & 2 & -1 & & & & \\
 & & -1 & 2 & -1 & & & \\
 & & & \ddots & \ddots & \ddots & & \\
 & & & & -1 & 2 & -1 & \\
 & & & & & -1 & 2 & -1 \\
 & & & & & & -2 & 2\left(1 + \Delta x\frac{p(t)}{q(t)}\right)
\end{bmatrix}.$$

Note that the vectors each have $N+1$ components, and the matrices are $(N+1)\times(N+1)$.

From the semidiscrete approximation, following standard procedures, we obtain the formulation for the FTCS method

$$\mathbf{w}^{(n+1)} = \left(I - \lambda A' - \Delta t\, B'^{(n)}\right)\mathbf{w}^{(n)} - \lambda\mathbf{b}'^{(n)} + \Delta t\,\mathbf{s}'^{(n)},$$

the BTCS method

$$\left(I + \lambda A' + \Delta t\, B'^{(n+1)}\right)\mathbf{w}^{(n+1)} = \mathbf{w}^{(n)} - \lambda\mathbf{b}'^{(n+1)} + \Delta t\,\mathbf{s}'^{(n+1)}$$

and the Crank-Nicolson scheme

$$\left(I + \lambda A' + \frac{\Delta t}{2} B'^{(n+1)}\right) \mathbf{w}^{(n+1)} = \left(I - \lambda A' - \frac{\Delta t}{2} B'^{(n)}\right) \mathbf{w}^{(n)} - \lambda(\mathbf{b}'^{(n)} + \mathbf{b}'^{(n+1)}) + \frac{\Delta t}{2}(\mathbf{s}'^{(n)} + \mathbf{s}'^{(n+1)}).$$

For the FTCS and BTCS methods, $\lambda = D\,\Delta t/(\Delta x)^2$, while for the Crank-Nicolson scheme, $\lambda = D\,\Delta t/\left[2(\Delta x)^2\right]$.

Stability Analysis

We will start our stability analysis with the FTCS method. With a decay term in the partial differential equation and non-Dirichlet boundary conditions, the evolution matrix, $E_{\text{FTCS}} = I - \lambda A' - \Delta t B'^{(n)}$, does not match the pattern of the special matrix we used extensively in Section 10.2. Therefore, we cannot compute the spectral radius exactly. Since we have different finite difference equations for different locations on the grid, we also cannot use von Neumann stability analysis. Fortunately, however, we can obtain an estimate for the spectral radius of E_{FTCS}.

First, note that the off-diagonal elements of the tridiagonal evolution matrix are all of the same sign. In this case, they are all positive. This implies that the eigenvalues of the evolution matrix are all real. See Smith [1] for a proof of this result. To actually estimate the locations of the eigenvalues, we will use the Gerschgorin Circle Theorem (see the Chapter 4 Overview, page 263). Knowing that the eigenvalues are real, the Gerschgorin circles all reduce to intervals along the real line. For stability, we will have to guarantee that all of these Gerschgorin intervals are contained in the closed interval $[-1, 1]$.

For rows $j = 2$ through $j = N$ of E_{FTCS}, the diagonal element is $1 - 2\lambda - \Delta t\,\beta_j^{(n)}$ and the sum of the absolute values of the off-diagonal elements is 2λ. Hence, the Gerschgorin interval is given by

$$-2\lambda \le z - \left(1 - 2\lambda - \Delta t\,\beta_j^{(n)}\right) \le 2\lambda,$$

or

$$1 - 4\lambda - \Delta t\,\beta_j^{(n)} \le z \le 1 - \Delta t\,\beta_j^{(n)}.$$

Under the assumption that $\beta_j^{(n)} \ge 0$, the right endpoint of this interval is always less than or equal to 1. Stability then requires that the left endpoint be greater than or equal to -1, or

$$-1 \le 1 - 4\lambda - \Delta t\,\beta_j^{(n)}.$$

This is precisely the condition we derived in Section 10.3.

Along the first row of the evolution matrix, the diagonal element is $1 - 2\lambda - \Delta t\,\beta_0^{(n)}$ and the sum of the absolute values of the off-diagonal elements is again 2λ. The corresponding Gerschgorin interval is then given by

$$-2\lambda \le z - \left(1 - 2\lambda - \Delta t\,\beta_0^{(n)}\right) \le 2\lambda.$$

This leads to the condition

$$-1 \leq 1 - 4\lambda - \Delta t\, \beta_0^{(n)},$$

which is identical to the condition derived from the interior grid points. Hence, a Neumann boundary condition has no effect on the stability of the FTCS method.

What about the Robin condition? Along the last row of the matrix, the diagonal element is $1-2\lambda-\mu-\Delta t\,\beta_N^{(n)}$, where $\mu = 2\lambda\,\Delta x\, p^{(n)}/q^{(n)}$, and the sum of the absolute values of the off-diagonal elements is once again 2λ. The Gerschgorin interval obtained from this row is then

$$-2\lambda \leq z - \left(1 - 2\lambda - \mu - \Delta t\, \beta_N^{(n)}\right) \leq 2\lambda.$$

This leads to the stability condition

$$-1 \leq 1 - 4\lambda - \mu - \Delta t\, \beta_N^{(n)},$$

which is slightly stronger than the previous conditions. A Robin boundary condition therefore tightens the bound for stability for the FTCS method.

Putting all of our results together, we find that the most restrictive condition is the one obtained from the Robin condition. Recalling that $\lambda = D\,\Delta t/(\Delta x)^2$, the time step for the FTCS method must satisfy

$$\Delta t \leq \frac{(\Delta x)^2}{2D + D\,\Delta x\, \max_{n\geq 0}(p^{(n)}/q^{(n)}) + (\Delta x)^2 \max_{1\leq j\leq N, n\geq 0} \beta_j^{(n)}/2}.$$

Since we have only estimated the spectral radius, it is possible that this bound is overly restrictive. The accuracy of the estimate will be tested in the second example.

When it comes to the BTCS method and the Crank-Nicolson scheme, neither the Neumann boundary condition nor the Robin boundary condition affects stability. Both methods remain unconditionally stable. The details of each analysis are left as an exercise.

EXAMPLE 10.9 A Test Problem with a Neumann Boundary Condition

Let's consider the initial boundary value problem

$$\begin{aligned} \frac{\partial u}{\partial t} &= \frac{1}{16}\frac{\partial^2 u}{\partial x^2} - 4x(1-x)u + e^{-t}(8x^2 - t) \\ &\frac{\partial u}{\partial n}(0,t) = 0, \quad \frac{\partial u}{\partial n}(1,t) = 8te^{-t} \\ &u(x,0) = 2\sin(2\pi x). \end{aligned} \tag{5}$$

The source term for this problem is $s(x,t) = e^{-t}(8x^2 - t)$, and the decay coefficient is $\beta(x,t) = 4x(1-x)$. The boundary conditions match the pattern of our model problem with

$$\alpha(t) = 0$$

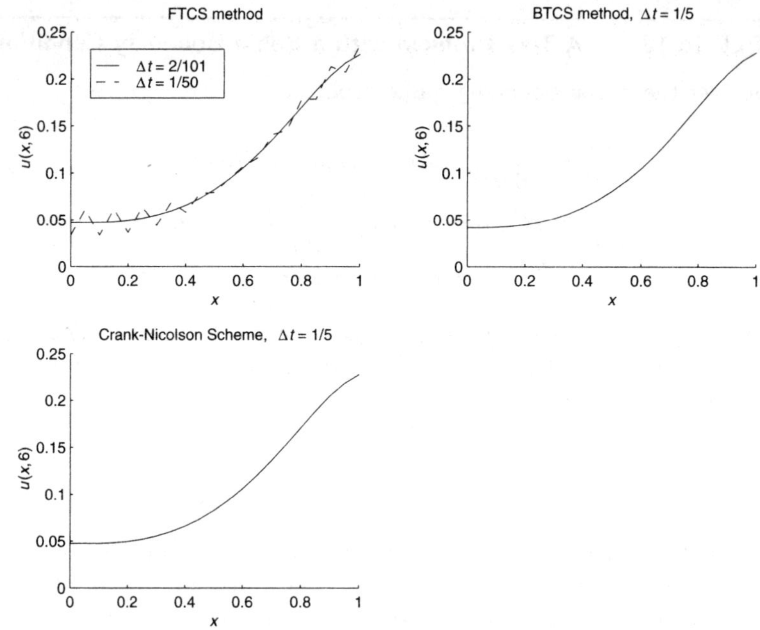

Figure 10.11 Approximate solutions to the initial boundary value problem given in equation (5) computed using the FTCS method, the BTCS method, and the Crank-Nicolson scheme. All solutions are displayed at $t = 6$.

and

$$p(t) = 0, \quad q(t) = 1, \quad \text{and} \quad r(t) = 8te^{-t}.$$

Figure 10.11 displays the approximate solution to this problem computed using our three finite difference methods. For each method, $\Delta x = 1/20$ was used. The upper left panel contains two approximate solutions computed using the FTCS method. With $\Delta x = 1/20$, stability requires

$$\begin{aligned} \Delta t &\leq \frac{(1/20)^2}{2(1/16) + (1/20)^2(1)/2} \\ &= \frac{2}{101} \approx 0.0198. \end{aligned}$$

The solid curve corresponds to $\Delta t = 2/101$, while the dashed curve was computed with $\Delta t = 1/50$. The instability in the latter solution is clear. The approximate solutions computed using the BTCS method and the Crank-Nicolson scheme are shown in the upper right and lower left panels, respectively. Each of these curves was obtained with $\Delta t = 1/5$, and neither displays any signs of instability.

EXAMPLE 10.10 A Test Problem with a Robin Boundary Condition

Let's consider the initial boundary value problem

$$\begin{gathered}\frac{\partial u}{\partial t}=\frac{1}{16}\frac{\partial^2 u}{\partial x^2}-4x(1-x)u+e^{-t}(8x^2-t)\\ \frac{\partial u}{\partial n}(0,t)=0,\quad 2u(1,t)+\frac{\partial u}{\partial n}(1,t)=8te^{-t}\\ u(x,0)=2\sin(2\pi x).\end{gathered}\tag{6}$$

The source term for this problem is $s(x,t)=e^{-t}(8x^2-t)$, and the decay coefficient is $\beta(x,t)=4x(1-x)$. The boundary conditions match the pattern of our model problem with

$$\alpha(t)=0$$

and

$$p(t)=2,\quad q(t)=1,\quad \text{and}\quad r(t)=8te^{-t}.$$

Figure 10.12 displays the approximate solution to this problem computed using our three finite difference methods. For each method, $\Delta x=1/20$ was used. The upper left panel contains two approximate solutions computed using the FTCS method. With $\Delta x=1/20$, stability requires

$$\Delta t\le\frac{(1/20)^2}{2(1/16)+(1/16)(1/20)(2)+(1/20)^2(1)/2}=\frac{1}{53}\approx 0.01887.$$

The solid curve corresponds to $\Delta t=1/53$, while the dashed curve was computed with $\Delta t=2/101$. The instability in the latter solution is clear, suggesting that our estimate for the spectral radius was quite accurate. The approximate solutions computed using the BTCS method and the Crank-Nicolson scheme are shown in the upper right and lower left panels, respectively. Each of these curves was obtained with $\Delta t=1/5$, and neither displays any signs of instability.

An Application Problem: Time-Dependent Temperature in a Fin

In the Overview to this chapter (see page 803), we showed that the time dependent temperature, $T(x,t)$, within a fin of circular cross section and constant cross-sectional area satisfies the initial boundary value problem

$$\frac{\partial T}{\partial t}=\frac{k}{\rho c_p}\frac{\partial^2 T}{\partial x^2}-\frac{2h}{\rho c_p r}(T-T_\infty)$$

$$T(0,t)=f(t),\quad T(x,0)=T_\infty,\quad -k\left.\frac{\partial T}{\partial x}\right|_{x=L}=h(T(L,t)-T_\infty).$$

Here, r is the radius and L the length of the fin; ρ is the mass density, c_p the specific heat, and k the thermal conductivity of the material from which the fin is

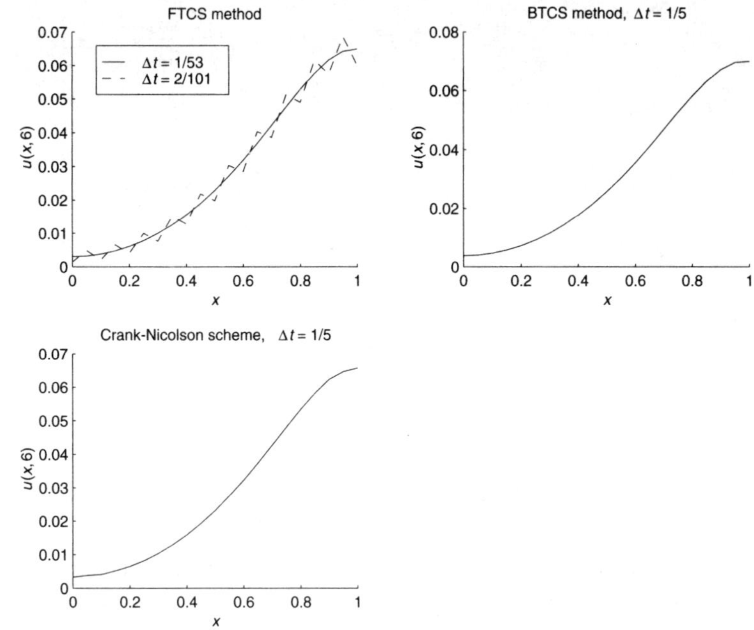

Figure 10.12 Approximate solutions to the initial-boundary value problem given in equation (6) computed using the FTCS method, the BTCS method, and the Crank-Nicolson scheme. All solutions are displayed at $t = 6$.

constructed; h is the convection heat transfer coefficient between the fin and the surrounding air; and T_∞ is the temperature of the air. The function f gives the time-varying temperature of the object to which the fin is attached.

Suppose the fin has a radius $r = 1.5$ cm, a length $L = 10$ cm, and is made from aluminum. For aluminum, $\rho = 2702$ kg/m^3, $c_p = 903$ J/kg $\cdot$ K, and $k = 237$ W/m $\cdot$ K. With $h = 20$ W/m^2 $\cdot$ K, $T_\infty = 20°$C and

$$f(t) = 20 + 80\left(1 - e^{-t/10}\right),$$

we obtain the temperature profiles shown in Figure 10.13. All calculations were performed using the Crank-Nicolson scheme with $\Delta x = 0.005$ meters and $\Delta t = 0.2$ seconds. Note that the temperature within the fin has essentially reached steady-state by $t = 300$ seconds.

Recall that a fin is designed to increase the heat dissipation rate from a larger object. Suppose the fin with which we've been working is attached to a square surface with side length $s = 0.25$ meters. Without the fin, the heat dissipation rate from this surface would be

$$s^2 h(f(t) - T_\infty) = 100\left(1 - e^{-t/10}\right).$$

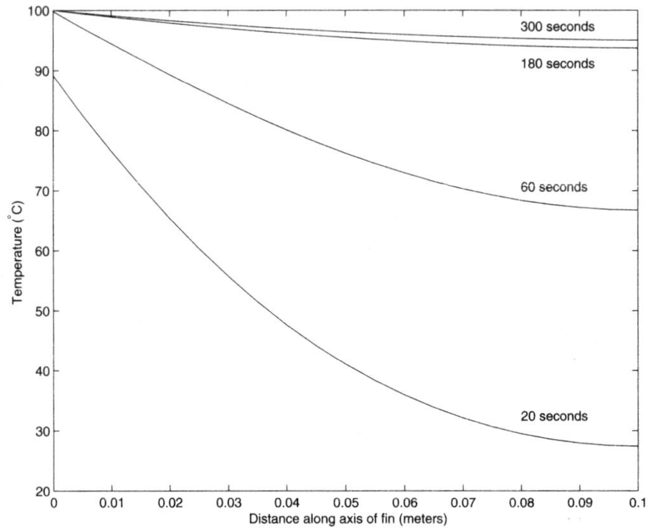

Figure 10.13 Time-dependent temperature distribution along length of a fin with uniform cross-sectional area and experiencing convective heat loss from its lateral surface.

The heat dissipation rate with the fin is

$$(s^2 - \pi r^2)h(f(t) - T_\infty) - \pi r^2 k \frac{\partial T}{\partial x}(0, t),$$

where the second term accounts for the heat dissipated by the fin itself. Figure 10.14 displays the two dissipation rates as functions of time. To obtain this figure, the second-order forward difference approximation for the first derivative was used to calculate $\frac{\partial T}{\partial x}(0,t)$. For $t \leq 50$ seconds, the heat dissipation rate with the fin is at least two to three times larger than the dissipation rate without the fin. At steady state, even with just one fin, the heat dissipation rate has been increased by nearly 14.6%.

References

1. G. D. Smith, *Numerical Solution of Partial Differential Equations: Finite Difference Methods*, 3rd edition, Oxford Applied Mathematics and Computing Science Series, Oxford University Press, Oxford, 1985.

EXERCISES

In Exercises 1–3, numerically verify that

(a) the FTCS method is stable for $\lambda = 1/2$ but unstable for any $\lambda > 1/2$;

(b) the BTCS method is unconditionally stable; and

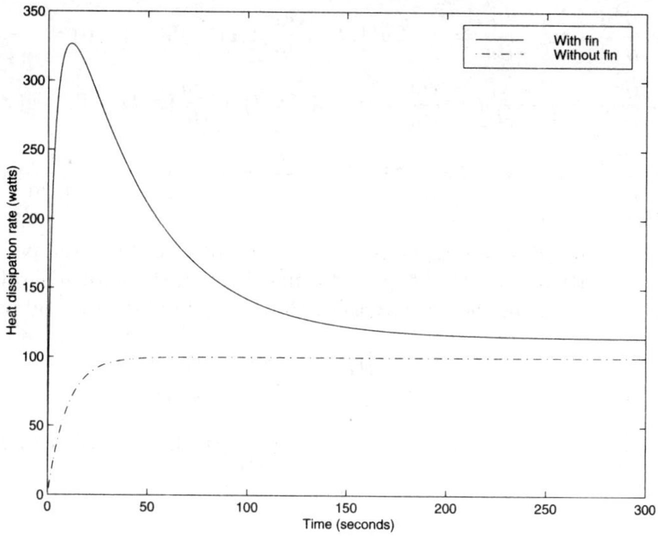

Figure 10.14 Heat dissipation rate with and without a single fin.

(c) the Crank-Nicolson method is unconditionally stable.

1. $\dfrac{\partial u}{\partial t} = \dfrac{\partial^2 u}{\partial x^2} + 2t + x(1-x), \quad u(x,0) = 1 - x - \dfrac{1}{\pi}\sin(2\pi x)$

$\dfrac{\partial u}{\partial n}(0,t) = 1 + 2e^{-4\pi^2 t} - t, \quad \dfrac{\partial u}{\partial n}(1,t) = -1 - 2e^{-4\pi^2 t} - t$

2. $\dfrac{\partial u}{\partial t} = \dfrac{\partial^2 u}{\partial x^2} + x, \quad \dfrac{\partial u}{\partial n}(0,t) = -e^{-t} - t, \quad \dfrac{\partial u}{\partial n}(\pi,t) = -e^{-t} + t, \quad u(x,0) = \sin x$

3. $\dfrac{\partial u}{\partial t} = \dfrac{\partial^2 u}{\partial x^2} - (x+t^2)e^{-xt}, \quad \dfrac{\partial u}{\partial n}(0,t) = -e^{-t} + t, \quad \dfrac{\partial u}{\partial n}(\pi,t) = -e^{-t} - te^{-\pi t},$
$u(x,0) = 1 + \sin x$

In Exercises 4–6, numerically verify that

(a) the FTCS method is unstable for

$$\Delta t \le \frac{(\Delta x)^2}{2D + (\Delta x)^2(\max_{1\le j\le N, n\ge 0}\beta_j^{(n)})/2}$$

but stable for

$$\Delta t \le \frac{(\Delta x)^2}{2D + D\Delta x \max_{n\ge 0}(p^{(n)}/q^{(n)}) + (\Delta x)^2 \max_{1\le j\le N, n\ge 0}\beta_j^{(n)}/2};$$

(b) the BTCS method is unconditionally stable; and

(c) the Crank-Nicolson method is unconditionally stable.

4. $\dfrac{\partial u}{\partial t} = \dfrac{\partial^2 u}{\partial x^2} - xu, \quad \dfrac{\partial u}{\partial n} = 1, \quad 2u(1,t) + \dfrac{\partial u}{\partial n}(1,t) = 0, \quad u(x,0) = 1 - x - \dfrac{1}{\pi}\sin(2\pi x)$

5. $\dfrac{\partial u}{\partial t} = \dfrac{\partial^2 u}{\partial x^2} - \dfrac{2}{1+xt}u, \quad \dfrac{\partial u}{\partial n} = 0, \quad 3u(\pi,t) + \dfrac{\partial u}{\partial n}(\pi,t) = 0, \quad u(x,0) = \sin x$

6. $\dfrac{\partial u}{\partial t} = \dfrac{\partial^2 u}{\partial x^2} - e^{-x^2}u - (x+t^2)e^{-xt}, \quad \dfrac{\partial u}{\partial n} = 1, \quad u(\pi,t) + \dfrac{1}{\pi}\dfrac{\partial u}{\partial n}(\pi,t) = e^{-\pi t},$
$u(x,0) = 1 + \sin x$

7. A 15-cm-long piece of copper wire is initially at a uniform temperature of 20°C. At $t = 0$, a 10 amp current begins flowing through the wire generating heat. The temperature within the wire satisfies the initial boundary value problem

$$\rho c_p \frac{\partial T}{\partial t} = k\frac{\partial^2 T}{\partial x^2} + \frac{4I^2R}{\pi D^2}$$

$$k\frac{\partial T}{\partial x}(0,t) = h[T(0,t) - T_\infty], \qquad -k\frac{\partial T}{\partial x}(0.15,t) = h[T(0.15,t) - T_\infty],$$

$$T(x,0) = 20.$$

The parameters in this problem are the mass density $\rho = 8933$ kg/m^3, the heat capacity $c_p = 385$ J/kg $\cdot\,^\circ$C, the thermal conductivity $k = 401$ J/m $\cdot$ s $\cdot^\circ$C, the convective heat transfer coefficient $h = 50$J/m$^2 \cdot s \cdot\,^\circ$C, the diameter of the wire $D = 2.6$ mm, the current I, and the resistance per unit length

$$R = \frac{6.8 \times 10^{-8}\Omega \cdot \text{m}}{\pi D^2}.$$

Approximate the temperature profile along the length of the wire 30 seconds, 60 seconds, 90 seconds, and 120 seconds after the current has begun to flow.

8. Two chambers are connected by a hollow tube 0.4 meters long and contain pools of ethyl alcohol that are maintained at 30°C. At $t = 0$, the valves at each end of the tube are opened, and alcohol vapors diffuse into the tube. At 30°C, the diffusion coefficient of the alcohol vapors is $D = 1.19 \times 10^{-5}$ m^2/s and 10% alcohol vapor is present in the air. The tube contains a filter with a mean absorption rate of $\mu = 0.0069(\text{s})^{-1}$. The percent of alcohol vapor within the tube, $u(x,t)$, satisfies

$$\frac{\partial u}{\partial t} = D\frac{\partial^2 u}{\partial x^2} - \mu u, \qquad u(0,t) = 10, \quad \frac{\partial u}{\partial x}(0.2,t) = 0, \quad u(x,0) = 0.$$

Approximate $u(x,t)$ at $t = 100$ seconds, $t = 200$ seconds, and $t = 300$ seconds.

9. A circular commercial bronze rod of radius $r = 5$ cm and length 1 meter initially has a uniform temperature distribution 50 K above ambient temperature. At $t = 0$ the temperature at the left end of the rod is lowered to ambient. Let θ denote the difference between the temperature of the rod and ambient temperature. If the surface of the rod is exposed to convective heat transfer, then θ satisfies the initial boundary value problem

$$\rho c_p \frac{\partial \theta}{\partial t} = k\frac{\partial^2 \theta}{\partial x^2} - \frac{2h}{r}\theta, \qquad \theta(0,t) = 0, \quad h\theta(1,t) + k\frac{\partial \theta}{\partial x}(1,t) = 0, \quad \theta(x,0) = 50,$$

where $\rho = 8800$ kg/m^3 is the mass density, $c_p = 420$ J/kg $\cdot$ K is the heat capacity, and $k = 52$ W/m $\cdot$ K is the thermal conductivity of the rod. $h = 25$ W/m$^2 \cdot$ K is the convective heat transfer coefficient. Determine the temperature profile along the rod after 600 seconds, 1200 seconds, and 1800 seconds.

10. Soil consolidation is the hydrodynamic process by which water is expelled from saturated soil voids when the soil is compacted. If the layer underneath the compressible soil is impermeable, then, as the water is expelled, the pore water pressure, $\phi(z,t)$, satisfies

$$\frac{\partial \phi}{\partial t} = C_v \frac{\partial^2 \phi}{\partial z^2}, \qquad \phi(z,0) = \Delta\phi, \quad \phi(0,t) = 0, \quad \frac{\partial \phi}{\partial z}(H_d, t) = 0,$$

where C_v is the coefficient of consolidation, H_d is the maximum drainage path (which, for this problem, is equal to the thickness of the soil layer) and $\Delta\phi$ is the change in pressure due to the compacting force. Introducing the nondimensional variables

$$\Phi = \phi/\Delta\phi, \qquad Z = z/H_d, \qquad \text{and} \qquad T = tC_v/H_d^2,$$

the problem becomes

$$\frac{\partial \Phi}{\partial T} = \frac{\partial^2 \Phi}{\partial Z^2}, \qquad \Phi(Z,0) = 1, \quad \Phi(0,T) = 0, \quad \frac{\partial \Phi}{\partial Z}(1,T) = 0.$$

Plot $\Phi(Z,T)$ for $T = 0.05$, $T = 0.1$, $T = 0.25$, and $T = 1$.

11. Rework the "Time-Dependent Temperature in a Fin" application problem for a copper fin with $\rho = 8933$ kg/m^3, $c_p = 385$ J/kg $\cdot$ K, and $k = 401$ W/m $\cdot$ K.

12. Rework the "Time-Dependent Temperature in a Fin" application problem for a stainless steel fin with $\rho = 8055$ kg/m^3, $c_p = 480$ J/kg $\cdot$ K, and $k = 15.1$ W/m$\cdot$K.

13. **(a)** Suppose a fin has a square cross section with side length s. Derive the partial differential equation for the temperature $T(x,t)$ within the fin.

(b) Let $s = 2.65$ cm. Using the parameter values, initial condition and boundary conditions from the "Time-Dependent Temperature in a Fin" application problem, calculate $T(x,t)$ for $t = 20$, $t = 60$, $t = 180$, and $t = 360$ seconds. How does the heat dissipation rate from the square fin compare with that of the circular fin?

14. Derive equations (3) and (4).

15. Use the Gerschgorin Circle Theorem to show that the BTCS method,

$$\left(I + \lambda A' + \Delta t B'^{(n+1)}\right) \mathbf{w}^{(n+1)} = \mathbf{w}^{(n)} - \lambda \mathbf{b}'^{(n+1)} + \Delta t \mathbf{s}'^{(n+1)},$$

and the Crank-Nicolson scheme,

$$\left(I + \lambda A' + \frac{\Delta t}{2} B'^{(n+1)}\right) \mathbf{w}^{(n+1)} = \left(I - \lambda A' - \frac{\Delta t}{2} B'^{(n)}\right) \mathbf{w}^{(n)} - \lambda(\mathbf{b}'^{(n)} + \mathbf{b}'^{(n+1)}) + \frac{\Delta t}{2}(\mathbf{s}'^{(n)} + \mathbf{s}'^{(n+1)})$$

are unconditionally stable.

10.5 POLAR COORDINATES

In this section we will develop finite difference methods for the following form of the heat equation in polar coordinates:

$$\frac{\partial u}{\partial t} = \frac{D}{r}\frac{\partial}{\partial r}\left(r\frac{\partial u}{\partial r}\right) - \beta(r,t)u + s(r,t).$$

As in previous sections, $-\beta(r,t)u$ is a decay term, and $s(r,t)$ is a source term. Equations of this type can arise from problems in cylindrical coordinates with axial symmetry and in which radial, rather than longitudinal, flow is important.

Starting Off Basic

Consider the initial boundary value problem

$$\begin{cases} \dfrac{\partial u}{\partial t} = D\left(\dfrac{\partial^2 u}{\partial r^2} + \dfrac{1}{r}\dfrac{\partial u}{\partial r}\right), & 0 \le r \le R, t > 0 \\[2ex] \dfrac{\partial u}{\partial r}(0,t) = 0 \\ u(R,t) = u_R(t) \\ u(r,0) = f(r). \end{cases}$$

This problem has neither a decay term nor a source term. We will add these later. The Neumann boundary condition at $r = 0$ is a consequence of the symmetry of the problem and allows us to handle the artificial singularity in the differential equation.

For the spatial discretization of the above problem, let $\Delta r = R/N$ for some positive integer N, and then define $r_j = j\Delta r$ for $j = 0, 1, 2, \ldots, N$. Evaluate the partial differential equation at an arbitrary interior grid point $r = r_j$ ($j = 1, 2, 3, \ldots, N-1$), and use second-order central difference formulas to approximate the space derivatives. Dropping the truncation error terms yields the semidiscrete equation

$$\frac{dv_j(t)}{dt} = D\left[\frac{v_{j-1}(t) - 2v_j(t) + v_{j+1}(t)}{(\Delta r)^2} + \frac{1}{r_j}\frac{v_{j+1}(t) - v_{j-1}(t)}{2\,\Delta r}\right], \tag{1}$$

where $v_j(t) \approx u(r_j, t)$. Taking into account the definition of $r_j = j\,\Delta r$, equation (1) becomes

$$\frac{dv_j(t)}{dt} = D\left[\frac{v_{j-1}(t) - 2v_j(t) + v_{j+1}(t)}{(\Delta r)^2} + \frac{1}{2j}\frac{v_{j+1}(t) - v_{j-1}(t)}{(\Delta r)^2}\right]. \tag{2}$$

Before evaluating the partial differential equation at $r_0 = 0$, we have to handle the artificial singularity. Taking the limit as $r \to 0$, we find that the appropriate partial differential equation is

$$\frac{\partial u}{\partial t}(0,t) = 2D\frac{\partial^2 u}{\partial r^2}(0,t).$$

We once again use a second-order central difference formula to approximate the space derivative. The fictitious node in the resulting equation is eliminated by applying the Neumann boundary condition at $r = 0$. The final semidiscrete equation is then

$$\frac{dv_0(t)}{dt} = 4D\left(\frac{v_1(t) - v_0(t)}{(\Delta r)^2}\right). \tag{3}$$

For the temporal discretization, we will start with the FTCS method. Recall that this method arises by evaluating the semidiscrete equations at time level $t = t_n$ and then replacing the time derivative with its first-order forward difference formula. Applying this procedure to equation (2) produces the finite difference equation

$$w_j^{(n+1)} = \lambda\left(1 - \frac{1}{2j}\right)w_{j-1}^{(n)} + (1-2\lambda)w_j^{(n)} + \lambda\left(1 + \frac{1}{2j}\right)w_{j+1}^{(n)}, \tag{4}$$

for $j = 1, 2, 3, \ldots, N-1$, where $\lambda = D\Delta t/(\Delta r)^2$. The finite difference equation obtained from equation (3) is

$$w_0^{(n+1)} = (1-4\lambda)w_0^{(n)} + 4\lambda w_1^{(n)}. \tag{5}$$

Before turning to the development of the other finite difference methods, let's investigate the stability of the FTCS method. From equations (4) and (5), we find that the evolution matrix, E_{FTCS}, takes the form

$$\begin{bmatrix} 1-4\lambda & 4\lambda & & & & & \\ \lambda/2 & 1-2\lambda & 3\lambda/2 & & & & \\ & 3\lambda/4 & 1-2\lambda & 5\lambda/4 & & & \\ & & \ddots & \ddots & \ddots & & \\ & & & \ddots & \ddots & \ddots & \\ & & & & \frac{(2N-5)\lambda}{2N-4} & 1-2\lambda & \frac{(2N-3)\lambda}{2N-4} \\ & & & & & \frac{(2N-3)\lambda}{2N-2} & 1-2\lambda \end{bmatrix}.$$

Since this is a tridiagonal matrix with all positive off-diagonal elements, the eigenvalues are guaranteed to be real. To estimate the locations of these eigenvalues, we consider the Gerschgorin intervals associated with each row. For the FTCS method to be stable, we must have all of these intervals contained in the closed interval $[-1, 1]$.

The first row of the evolution matrix produces the interval

$$|z - (1-4\lambda)| \le 4\lambda,$$

or, equivalently, $1 - 8\lambda \le z \le 1$. To guarantee this interval is contained in the closed interval $[-1, 1]$ requires $\lambda \le 1/4$. Rows 2 through $N-1$ produce the same Gerschgorin interval:

$$|z - (1-2\lambda)| \le 2\lambda,$$

or $1-4\lambda \le z \le 1$. This interval will be a subset of $[-1,1]$ provided $\lambda \le 1/2$. From the last row of the matrix, we obtain the Gerschgorin interval

$$|z-(1-2\lambda)| \le \frac{(2N-3)\lambda}{2N-2},$$

which leads to the condition

$$\lambda \le \frac{4N-4}{6N-7} \approx \frac{2}{3}.$$

The most restrictive condition on λ therefore comes from the first row. In conclusion, the FTCS method is once again conditionally stable, requiring $\lambda \le 1/4$, or

$$\Delta t \le \frac{(\Delta r)^2}{4D}.$$

Values of Δt that are less than or equal to $(\Delta r)^2/(4D)$ will definitely be sufficient to guarantee the stability of the FTCS method. However, since this bound was obtained using estimates for the eigenvalues of E_{FTCS} based on the Gerschgorin Circle Theorem, it is possible that the bound may be overly restrictive. We will examine this issue further in the examples and exercises.

Next, we will develop the BTCS method. Recall that this method arises by evaluating the semidiscrete equations at time level $t = t_{n+1}$ and then replacing the time derivative with its first-order backward difference formula. Applying this procedure to equations (2) and (3) yields the finite difference equations

$$-\lambda\left(1-\frac{1}{2j}\right)w_{j-1}^{(n+1)} + (1+2\lambda)w_j^{(n+1)} - \lambda\left(1+\frac{1}{2j}\right)w_{j+1}^{(n+1)} = w_j^{(n)},$$

for $j = 1,2,3,\ldots,N-1$, and

$$(1+4\lambda)w_0^{(n+1)} - 4\lambda w_1^{(n+1)} = w_0^{(n)}.$$

As for the FTCS method, $\lambda = D\,\Delta t/(\Delta r)^2$.

Each time step of the BTCS method requires the solution of a linear system of equations with the tridiagonal coefficient matrix

$$\begin{bmatrix} 1+4\lambda & -4\lambda & & & & \\ -\lambda/2 & 1+2\lambda & -3\lambda/2 & & & \\ & -3\lambda/4 & 1+2\lambda & -5\lambda/4 & & \\ & & \ddots & \ddots & \ddots & \\ & & & -\frac{(2N-5)\lambda}{2N-4} & 1+2\lambda & -\frac{(2N-3)\lambda}{2N-4} \\ & & & & -\frac{(2N-3)\lambda}{2N-2} & 1+2\lambda \end{bmatrix}.$$

The inverse of this matrix is the evolution matrix E_{BTCS}. It is easy to show, using the Gerschgorin Circle Theorem, that the eigenvalues of the above matrix are all greater than or equal to 1 in magnitude. Hence the eigenvalues of the evolution matrix all lie between 0 and 1, and the BTCS method is unconditionally stable.

Finally, let's develop the Crank-Nicolson scheme. Integrating both sides of equations (2) and (3) from $t = t_n$ to $t = t_{n+1}$ and using the trapezoidal rule to approximate the integral on the right-hand side yields

$$-\lambda\left(1-\frac{1}{2j}\right)w_{j-1}^{(n+1)} + (1+2\lambda)w_j^{(n+1)} - \lambda\left(1+\frac{1}{2j}\right)w_{j+1}^{(n+1)} =$$
$$\lambda\left(1-\frac{1}{2j}\right)w_{j-1}^{(n)} + (1-2\lambda)w_j^{(n)} + \lambda\left(1+\frac{1}{2j}\right)w_{j+1}^{(n)},$$

for $j = 1, 2, 3, \ldots, N-1$, and

$$(1+4\lambda)w_0^{(n+1)} - 4\lambda w_1^{(n+1)} = (1-4\lambda)w_0^{(n)} + 4\lambda w_1^{(n)}.$$

Here, $\lambda = D\,\Delta t/\left[2(\Delta r)^2\right]$. Like the BTCS method, each time step of the Crank-Nicolson scheme requires the solution of a linear system with a tridiagonal coefficient matrix.

To investigate the stability of the Crank-Nicolson scheme, note that as in Section 10.2, we can express the evolution matrix as

$$E_{\text{CN}} = E_{\text{BTCS}}E_{\text{FTCS}} = -I + 2E_{\text{BTCS}}.$$

Having just established that the eigenvalues of E_{BTCS} all lie between 0 and 1, it then follows that the eigenvalues of E_{CN} all lie between ± 1. Hence, the Crank-Nicolson scheme is unconditionally stable.

EXAMPLE 10.11 Demonstration of Stability

Let's approximate the solution of the initial boundary value problem

$$\begin{cases} \dfrac{\partial u}{\partial t} = \dfrac{1}{10}\left(\dfrac{\partial^2 u}{\partial r^2} + \dfrac{1}{r}\dfrac{\partial u}{\partial r}\right), & 0 \le r \le 1, t > 0 \\[2ex] \dfrac{\partial u}{\partial r}(0,t) = 0 \\[2ex] u(1,t) = 0 \\[2ex] u(r,0) = J_0(kr). \end{cases}$$

J_0 is the zeroth-order Bessel function of the first kind, and $k \approx 2.404825558$ is the smallest positive root of J_0. The exact solution of this problem is $u(r,t) = e^{-(k^2/10)t}J_0(kr)$.

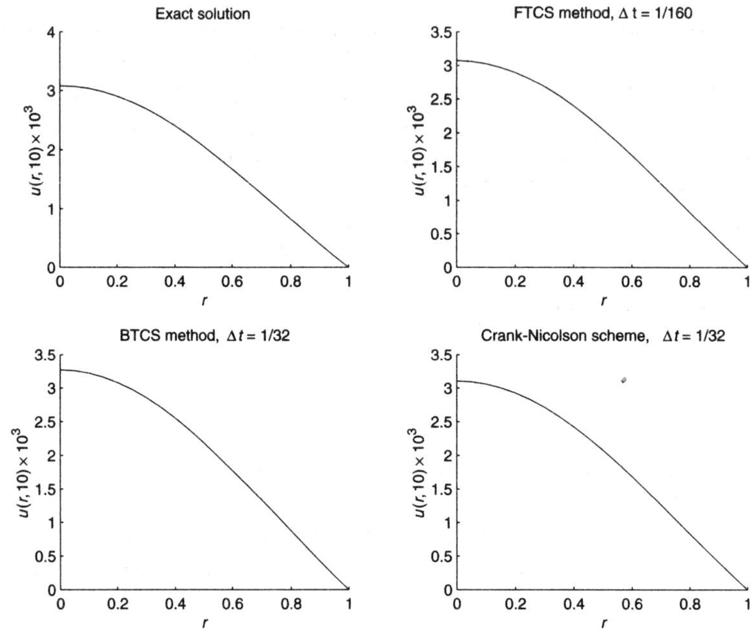

Figure 10.15 Exact solution to the initial boundary value problem

$$\frac{\partial u}{\partial t} = \frac{1}{10}\left(\frac{\partial^2 u}{\partial r^2} + \frac{1}{r}\frac{\partial u}{\partial r}\right),$$

$$\frac{\partial u}{\partial r}(0,t) = 0, \quad u(1,t) = 0,$$
$$u(r,0) = J_0(kr),$$

and approximate solutions computed using the FTCS method, the BTCS method, and the Crank-Nicolson scheme. All solutions are displayed at $t = 10$.

Figure 10.15 displays the exact solution and the approximate solutions computed using the FTCS method, the BTCS method, and the Crank-Nicolson scheme in the upper left, upper right, lower left, and lower right panels, respectively. All solutions are plotted at $t = 10$, and each numerical method used $\Delta r = 1/20$. The time step for the FTCS method was selected as the largest value allowed by our stability analysis:

$$\Delta t = \frac{(\Delta r)^2}{4D} = \frac{(1/20)^2}{4(1/10)} = \frac{1}{160}.$$

$\Delta t = 1/32$ was used for both the BTCS method and the Crank-Nicolson scheme. There are no signs of instability in any of the numerical approximations.

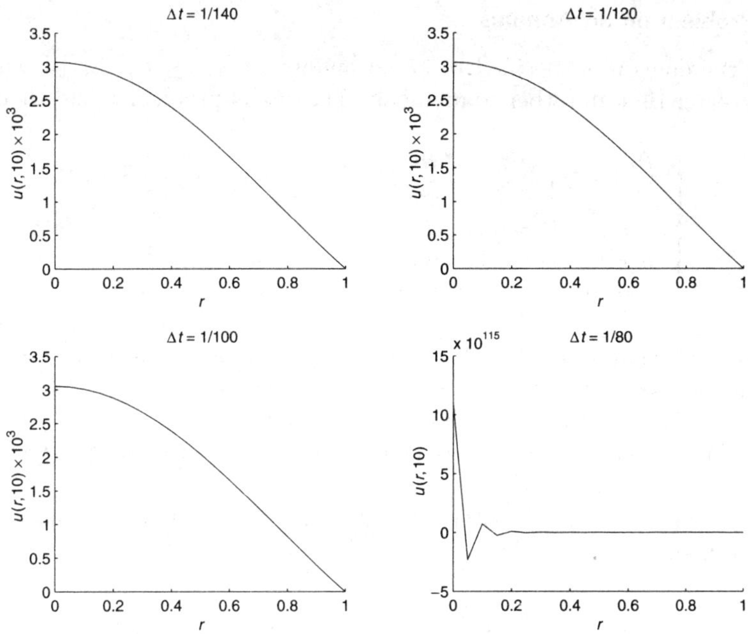

Figure 10.16 Approximate solutions to the initial-boundary value problem

$$\frac{\partial u}{\partial t} = \frac{1}{10}\left(\frac{\partial^2 u}{\partial r^2} + \frac{1}{r}\frac{\partial u}{\partial r}\right),$$

$$\frac{\partial u}{\partial r}(0,t) = 0, \quad u(1,t) = 0,$$
$$u(r,0) = J_0(kr),$$

computed using the FTCS method with different time steps. All solutions are displayed at $t = 10$.

Since our stability analysis was based not on analytical expressions for the eigenvalues of E_{FTCS}, but rather on estimates obtained from the Gerschgorin Circle Theorem, we should investigate the accuracy of the stability condition, $\lambda \le 1/4$, which was obtained for the FTCS method. Figure 10.16 displays approximate solutions obtained from the FTCS method with $\Delta t = 1/140$, $\Delta t = 1/120$, $\Delta t = 1/100$, and $\Delta t = 1/80$. These time steps correspond to $\lambda = 2/7$, $\lambda = 1/3$, $\lambda = 2/5$, and $\lambda = 1/2$, respectively. The solutions with the first three time steps appear to be stable, while the solution computed with $\Delta t = 1/80$ is clearly unstable. Hence, for this problem at least, the condition $\lambda \le 1/4$, though clearly sufficient to guarantee stability, is far from being a necessary condition. The actual stability condition for the FTCS method is somewhere between $\lambda \le 2/5$ and $\lambda < 1/2$. By computing the eigenvalues of E_{FTCS} for Δt ranging from 1/10 to 1/120, the condition $\lambda \le 0.41$ appears to be both necessary and sufficient to guarantee $\rho(E_{\text{FTCS}}) \le 1$.

Model Problem on an Annulus

Suppose the domain is specified to be an annulus, $r_{\text{inner}} \leq r \leq r_{\text{outer}}$ with r_{inner} strictly greater than 0, rather than a disk. The model problem would then be

$$\begin{cases} \dfrac{\partial u}{\partial t} = D\left(\dfrac{\partial^2 u}{\partial r^2} + \dfrac{1}{r}\dfrac{\partial u}{\partial r}\right), & r_{\text{inner}} \leq r \leq r_{\text{outer}}, t > 0 \\ u(r_{\text{inner}}, t) = u_{\text{inner}}(t) \\ u(r_{\text{outer}}, t) = u_{\text{outer}}(t) \\ u(r, 0) = f(r). \end{cases}$$

The Neumann condition at the left "boundary" has been replaced by a Dirichlet condition because there is no reason, in general, to expect that the solution will be symmetric about $r = r_{\text{inner}}$. Since $r = 0$ is no longer in the domain, the problem no longer has an artificial singularity. Equation (1) is therefore representative of all of the semidiscrete equations.

From equation (1), we obtain the following finite difference equation for the FTCS method:

$$w_j^{(n+1)} = \left(\lambda - \frac{\mu}{2r_j}\right) w_{j-1}^{(n)} + (1 - 2\lambda) w_j^{(n)} + \left(\lambda + \frac{\mu}{2r_j}\right) w_{j+1}^{(n)}.$$

Here $r_j = r_{\text{inner}} + j\,\Delta r$, $\lambda = D\,\Delta t/(\Delta r)^2$ and $\mu = D\,\Delta t/\Delta r$. For the BTCS method, the finite difference equation is

$$-\left(\lambda - \frac{\mu}{2r_j}\right) w_{j-1}^{(n+1)} + (1 + 2\lambda) w_j^{(n+1)} - \left(\lambda + \frac{\mu}{2r_j}\right) w_{j+1}^{(n+1)} = w_j^{(n)},$$

with the same expressions for λ and μ. Finally, the finite difference equation for the Crank-Nicolson scheme is found to be

$$\begin{aligned} -\left(\lambda - \frac{\mu}{2r_j}\right) w_{j-1}^{(n+1)} + (1 + 2\lambda) w_j^{(n+1)} - \left(\lambda + \frac{\mu}{2r_j}\right) w_{j+1}^{(n+1)} &= \\ \left(\lambda - \frac{\mu}{2r_j}\right) w_{j-1}^{(n)} + (1 - 2\lambda) w_j^{(n)} &+ \left(\lambda + \frac{\mu}{2r_j}\right) w_{j+1}^{(n)}, \end{aligned}$$

where $\lambda = D\,\Delta t/\left[2(\Delta r)^2\right]$ and $\mu = D\,\Delta t/(2\Delta r)$.

If we continue to use the Gerschgorin Circle Theorem to localize the eigenvalues of the evolution matrices, it is straightforward to establish that the BTCS method and the Crank-Nicolson scheme remain unconditionally stable. As for the FTCS method, recall that the condition $\lambda \leq 1/4$ was derived from the finite difference equation associated with $r = 0$, which is now no longer present. The remaining equations lead to the relaxed stability condition $\lambda \leq 1/2$.

EXAMPLE 10.12 Demonstration of Stability for Model Problem on an Annulus

Let's approximate the solution of the initial boundary value problem

$$\begin{cases} \dfrac{\partial u}{\partial t} = \dfrac{1}{10}\left(\dfrac{\partial^2 u}{\partial r^2} + \dfrac{1}{r}\dfrac{\partial u}{\partial r}\right), & 0.1 \le r \le 1, t > 0 \\ u(0.1,t) = e^{-(k^2/10)t}J_0(0.1k) \\ u(1,t) = 0 \\ u(r,0) = J_0(kr). \end{cases}$$

J_0 is the zeroth-order Bessel function of the first kind, and $k \approx 2.404825558$ is the smallest positive root of J_0. The exact solution of this problem is $u(r,t) = e^{-(k^2/10)t}J_0(kr)$.

$u(r,10)$ is plotted in the upper left panel of Figure 10.17. The other panels display the approximate solutions computed using the FTCS method, the BTCS method and the Crank-Nicolson scheme. For each numerical method, the spatial discretization parameter was $\Delta r = 1/20$. The upper right panel contains two approximate solutions computed using the FTCS method. With $\Delta r = 1/20$, the stability condition $\lambda \le 1/2$ requires

$$\Delta t \le \frac{(1/20)^2}{2(1/10)} = \frac{1}{80}.$$

The solid curve corresponds to this value of Δt. The dashed curve was computed with $\Delta t = 1/79$. This latter solution clearly exhibits the classical signs of instability. Hence, when the problem is defined over an annulus, the Gerschgorin intervals lead to a necessary and sufficient stability condition.

The approximate solutions obtained from the BTCS method and the Crank-Nicolson scheme are displayed in the lower left and lower right panels of Figure 10.17, respectively. Each of these solutions was computed with $\Delta t = 1/8$, which corresponds to $\lambda = 5$. As expected, neither curve exhibits any signs of instability.

More General Problems

Source terms, decay terms, and non-Dirichlet boundary conditions can be incorporated into the finite difference methods that have just been presented following exactly the same procedures that were described in Sections 10.3 and 10.4. For example, the BTCS method for the problem

$$\begin{cases} \dfrac{\partial u}{\partial t} = D\left(\dfrac{\partial^2 u}{\partial r^2} + \dfrac{1}{r}\dfrac{\partial u}{\partial r}\right) - \beta(r,t)u + s(r,t), & 0 \le r \le R, t > 0 \\ \\ \dfrac{\partial u}{\partial r}(0,t) = 0 \\ u(R,t) = u_R(t) \\ u(r,0) = f(r) \end{cases}$$

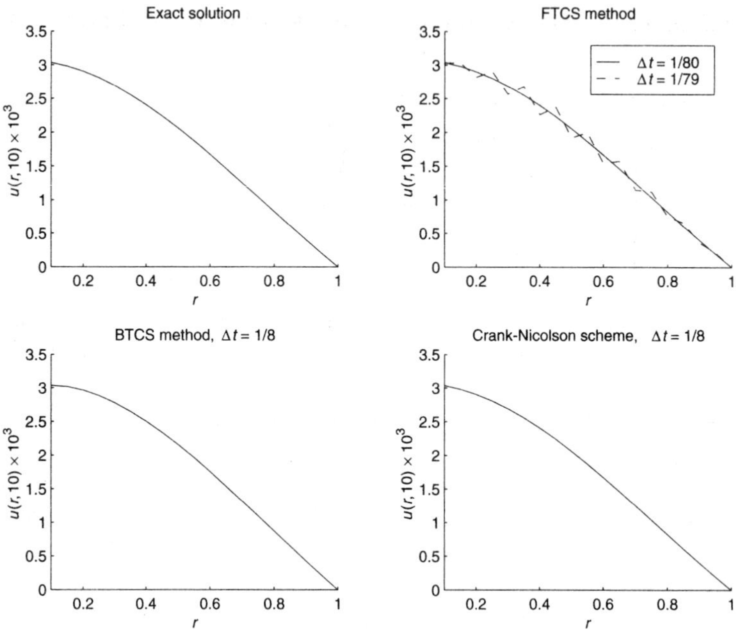

Figure 10.17 Exact solution to the initial boundary value problem

$$\begin{aligned}\frac{\partial u}{\partial t} &= \frac{1}{10}\left(\frac{\partial^2 u}{\partial r^2} + \frac{1}{r}\frac{\partial u}{\partial r}\right),\\ u(0.1, t) &= e^{-(k^2/10)t} J_0(0.1k),\\ u(1, t) &= 0,\\ u(r, 0) &= J_0(kr),\end{aligned}$$

and approximate solutions computed using the FTCS method, the BTCS method, and the Crank-Nicolson scheme. All solutions are displayed at $t = 10$.

consists of the finite difference equations

$$\left(1 + 4\lambda + \Delta t\, \beta_0^{(n+1)}\right) w_0^{(n+1)} - 4\lambda w_1^{(n+1)} = w_0^{(n)} + \Delta t\, s_0^{(n+1)} \tag{6}$$

and

$$-\lambda\left(1 - \frac{1}{2j}\right) w_{j-1}^{(n+1)} + \left(1 + 2\lambda + \Delta t\, \beta_j^{(n+1)}\right) w_j^{(n+1)} - \lambda\left(1 + \frac{1}{2j}\right) w_{j+1}^{(n+1)} = w_j^{(n)} + \Delta t\, s_j^{(n+1)}, \tag{7}$$

where the latter equation holds for $j = 1, 2, 3, \ldots, N-1$. As a second example, consider the FTCS method for approximating the solution of

$$\begin{cases} \dfrac{\partial u}{\partial t} = D\left(\dfrac{\partial^2 u}{\partial r^2} + \dfrac{1}{r}\dfrac{\partial u}{\partial r}\right) - \beta(r,t)u + s(r,t), & r_{\text{inner}} \le r \le r_{\text{outer}}, t > 0 \\ \dfrac{\partial u}{\partial r}(r_{\text{inner}}, t) = \alpha(t) \\ p(t)u(r_{\text{outer}}, t) + q(t)\dfrac{\partial u}{\partial r}(r_{\text{outer}}, t) = \rho(t) \\ u(r,0) = f(r) \end{cases}$$

At $r = r_{\text{inner}}$ and $r = r_{\text{outer}}$, the finite difference equations are

$$w_0^{(n+1)} = \left(1 - 2\lambda - \Delta t\, \beta_0^{(n)}\right) w_0^{(n)} + 2\lambda w_1^{(n)} + \Delta t\, s_0^{(n)} - 2\,\Delta r\, \alpha^{(n)}\left(\lambda - \frac{\mu}{2r_0}\right) \quad (8)$$

and

$$w_N^{(n+1)} = 2\lambda w_{N-1}^{(n)} + \left[1 - 2\lambda - \Delta t\, \beta_N^{(n)} - 2\Delta r \frac{p^{(n)}}{q^{(n)}}\left(\lambda + \frac{\mu}{2r_N}\right)\right] w_N^{(n)} + \Delta t\, s_N^{(n)} + 2\,\Delta r \frac{\rho^{(n)}}{q^{(n)}}\left(\lambda + \frac{\mu}{2r_N}\right), \quad (9)$$

respectively. The equation at each of the interior grid points ($j = 1, 2, 3, \ldots, N-1$) is

$$w_j^{(n+1)} = \left(\lambda - \frac{\mu}{2r_j}\right) w_{j-1}^{(n)} + \left(1 - 2\lambda - \Delta t\, \beta_j^{(n)}\right) w_j^{(n)} + \left(\lambda + \frac{\mu}{2r_j}\right) w_{j+1}^{(n)} + \Delta t\, s_j^{(n)}. \quad (10)$$

With regard to stability, the BTCS method and the Crank-Nicolson scheme remain unconditionally stable for all forms of the differential equation and for all combinations of boundary conditions. For the FTCS method, neither a source term nor a Neumann boundary condition produces any change in stability, but a decay term will reduce the maximum allowable time step. If the problem domain is an annulus, a Robin boundary condition will also reduce the maximum allowable time step. However, if the problem domain is a circle, a Robin boundary condition will produce no change in stability.

Application Problem: Transient Temperature in a Heat Pack

In Section 8.2 (see page 681), we developed a model for the steady-state temperature distribution in a heat pack. The corresponding model for the transient temperature distribution, $T(r,t)$, is the initial boundary value problem

$$\frac{\partial T}{\partial t} = \frac{k}{\rho c_p}\frac{1}{r}\frac{\partial}{\partial r}\left(r\frac{\partial T}{\partial r}\right) - \frac{2h}{\rho c_p w}(T - T_\infty) + \frac{\dot{q}}{\rho c_p},$$

$$\frac{\partial T}{\partial r}(0,t) = 0, \quad hT(R,t) + k\frac{\partial T}{\partial r}(R,t) = hT_\infty, \quad T(r,0) = T_\infty.$$

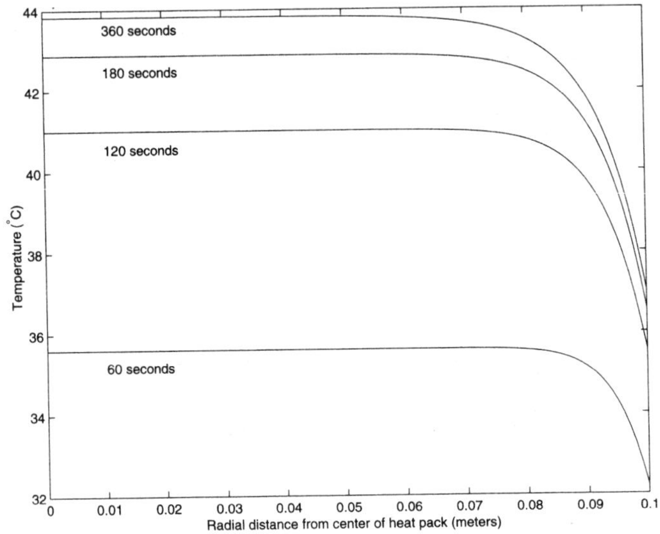

Figure 10.18 Transient temperature distribution in a heat pack.

Here, $\rho = 1591$ kg/m^3 is the mass density, $c_p = 237$ J/kg $\cdot$ K the specific heat, and $k = 0.4$ W/m $\cdot$ K the thermal conductivity of the material inside the pack; $h = 20$ W/m$^2 \cdot$ K is the convective heat transfer coefficient between the surface of the heat pack and the surrounding air and $T_\infty = 20°$C is the temperature of the air; $\dot{q} = \frac{500000}{\pi}$ W/m^3 is the rate of heat generation per unit volume within the pack; and $R = 10$ cm is the radius and $w = 6$ mm the thickness of the pack.

The Crank-Nicolson scheme was used to compute the approximate temperature distribution within the heat pack with $\Delta r = 1$ mm and $\Delta t = 0.2$ seconds. Figure 10.18 displays the temperature at $t = 60$, $t = 120$, $t = 180$, and $t = 360$ seconds. The temperature distribution has essentially reached steady state after six minutes.

EXERCISES

In Exercises 1–3, numerically verify that

(a) the FTCS method is stable for $\lambda = 1/2$ but unstable for any $\lambda > 1/2$;
(b) the BTCS method is unconditionally stable; and
(c) the Crank-Nicolson method is unconditionally stable.

1. $\dfrac{\partial u}{\partial t} = \dfrac{1}{r}\dfrac{\partial}{\partial r}\left(r\dfrac{\partial u}{\partial r}\right)$, $\quad u(1,t) = 0,\ \ u(3,t) = 1,\ \ u(r,0) = (r-1)/2$

2. $\dfrac{\partial u}{\partial t} = \dfrac{1}{r}\dfrac{\partial}{\partial r}\left(r\dfrac{\partial u}{\partial r}\right)$, $\quad u(1,t) = 20(1-e^{-t}),\ \ \dfrac{\partial u}{\partial r}(2,t) = 0,\ \ u(r,0) = 0$

3. $\dfrac{\partial u}{\partial t} = \dfrac{1}{r}\dfrac{\partial}{\partial r}\left(r\dfrac{\partial u}{\partial r}\right) + \dfrac{rt}{1+t^2}, \quad u(1/2,t) = -\dfrac{1}{4}\ln 2, \quad u(1,t) = 0, \quad u(r,0) = r^2 \ln r$

In Exercises 4–6,

(a) numerically verify that the FTCS method is stable for $\lambda = 1/4$ but unstable for $\lambda = 1/2$;

(b) experimentally determine the largest value of λ which will guarantee $\rho(E_{\text{FTCS}}) \leq 1$;

(c) numerically verify that the BTCS method is unconditionally stable; and

(d) numerically verify that the Crank-Nicolson method is unconditionally stable.

4. $\dfrac{\partial u}{\partial t} = \dfrac{0.01}{r}\dfrac{\partial}{\partial r}\left(r\dfrac{\partial u}{\partial r}\right), \quad \dfrac{\partial u}{\partial r}(0,t) = 0, \quad u(1,t) = 1, \quad u(r,0) = r^2$

5. $\dfrac{\partial u}{\partial t} = \dfrac{1}{r}\dfrac{\partial}{\partial r}\left(r\dfrac{\partial u}{\partial r}\right), \quad \dfrac{\partial u}{\partial r}(0,t) = 0, \quad \dfrac{\partial u}{\partial r}(1,t) = -1, \quad u(r,0) = 1 - \dfrac{1}{2}r^2$

6. $\dfrac{\partial u}{\partial t} = \dfrac{1}{r}\dfrac{\partial}{\partial r}\left(r\dfrac{\partial u}{\partial r}\right) + e^{-rt}, \quad \dfrac{\partial u}{\partial r}(0,t) = 0, \quad u(1,t) + \dfrac{\partial u}{\partial r}(1,t) = 0, \quad u(r,0) = 0$

In Exercises 7–9, numerically verify that

(a) the FTCS method is second-order accurate in space;

(b) the BTCS method is first-order accurate in time and second-order accurate in space; and

(c) the Crank-Nicolson scheme is second-order accurate in both time and space by approximating the solution of the indicated initial boundary value problem.

7. $\dfrac{\partial u}{\partial t} = \dfrac{0.1}{r}\dfrac{\partial}{\partial r}\left(r\dfrac{\partial u}{\partial r}\right), \quad \dfrac{\partial u}{\partial r}(0,t) = 0, \quad u(1,t) = 0, \quad u(r,0) = J_0(kr)$

exact solution: $u(r,t) = e^{-(k^2/10)t}J_0(kr), \quad k \approx 2.404825558$

8. $\dfrac{\partial u}{\partial t} = \dfrac{1}{r}\dfrac{\partial}{\partial r}\left(r\dfrac{\partial u}{\partial r}\right) - e^{-t}\left(r + \dfrac{1}{r}\right), \quad u(1/2,t) = \dfrac{1}{2}e^{-t} - \ln 2, u(1,t) = e^{-t},$

$u(r,0) = r + \ln r, \quad$ exact solution: $u(r,t) = \ln r + re^{-t}$

9. $\dfrac{\partial u}{\partial t} = \dfrac{1}{r}\dfrac{\partial}{\partial r}\left(r\dfrac{\partial u}{\partial r}\right) - 4r^2t^4 + r^4t^3, \quad \dfrac{\partial u}{\partial r}(0,t) = 0, \quad 4u(1,t) + \dfrac{\partial u}{\partial r}(1,t) = 4 + 2t^4,$

$u(r,0) = 1, \quad$ exact solution :$u(r,t) = 1 + \frac{1}{4}r^4t^4$

10. **(a)** Derive equations (6) and (7).

(b) Derive equations (8), (9), and (10).

11. The transient temperature distribution within an annular fin in a heating system satisfies the initial boundary value problem

$$\rho c_p \frac{\partial T}{\partial t} - \frac{1}{r}\frac{\partial}{\partial r}\left(kr\frac{\partial T}{\partial r}\right) + \frac{2h}{w}T = \frac{2h}{w}T_\infty,$$

$$T(r_{\text{inner}},t) = f(t), \quad hT(r_{\text{outer}},t) + k\frac{\partial T}{\partial r}(r_{\text{outer}},t) = hT_\infty, \quad T(r,0) = T_\infty.$$

For each of the following materials, determine the temperature distribution in the fin after 600 seconds. In each case, take $h = 50$ W/m$^2 \cdot$ K, $w = 5$ mm, $T_\infty = 20°$C, $f(t) = 20 + 40(1 - e^{-t/120})$, $r_{\text{inner}} = 2$ cm, and $r_{\text{outer}} = 5$ cm.

(a) pure copper fin: $\rho = 8933$ kg/m^3, $c_p = 385$ J/kg $\cdot$ K, $k = 401$ W/m $\cdot$ K

(b) stainless steel fin: $\rho = 8055$ kg/m^3, $c_p = 480$ J/kg $\cdot$ K, $k = 15.1$ W/m $\cdot$ K

(c) commercial bronze fin: $\rho = 8800$ kg/m^3, $c_p = 420$ J/kg $\cdot$ K, $k = 52$ W/m$\cdot$K

12. Consider the following simplified model for insect dispersal. Let $n(r,t)$ denote the insect population as a function of location, r, and time, t. Starting from an initial population distribution $n_0(r)$, the insects disperse randomly with a constant coefficient of diffusion, D. The mortality of the insects is proportional to population size, and there exists a finite interval which can sustain insect life. These assumptions lead to the initial boundary value problem

$$\frac{\partial n}{\partial t} = \frac{D}{r}\frac{\partial}{\partial r}\left(r\frac{\partial n}{\partial r}\right) - \lambda n, \qquad n(r,0) = n_0(t), \quad \frac{\partial n}{\partial r}(0,t) = n(L,t) = 0$$

for $n(r,t)$. Assume that the insect population is measured in thousands. Take $L = 1$ meter, $D = 0.05$ m^2/day and $\lambda = 0.1$ (day)$^{-1}$. The initial population distribution is given by

$$n_0(r) = \begin{cases} e^{-1000r^2}, & r \le 0.1 \\ 0, & \text{otherwise.} \end{cases}$$

Simulate the dispersal of the insect population over a four-day period.

13. Water is pumped from a well, one meter in diameter, at a daily rate of Q cubic meters. The well water is drawn from a confined aquifer that extends for a radius of 100 meters from the center of the well. The drawdown on the water level within the aquifer, $h(r,t)$, satisfies the initial boundary value problem

$$\frac{\partial h}{\partial t} = \frac{T}{S}\left(\frac{\partial^2 h}{\partial r^2} + \frac{1}{r}\frac{\partial h}{\partial r}\right), \qquad \left.\frac{\partial h}{\partial r}\right|_{r=1/2} = \frac{Q}{\pi T}, \quad \left.\frac{\partial h}{\partial r}\right|_{r=100} = 0, \quad h(r,0) = 0,$$

where T is the transmissivity and S the storativity of the aquifer.

(a) Suppose $T = 150$ m^2/day, $S = 0.2$, and $Q = 100$ m^3/day. Determine the drawdown within the aquifer after 15 days of pumping.

(b) Suppose that during the pumping period, average daily rainfall is w meters. To account for precipitation, add the term w/S to the right-hand side of the differential equation for $h(r,t)$. Using the same parameter values listed in part (a) and $w = 0.003$ m/day, determine the drawdown within the aquifer after 15 days of pumping.

14. The finite difference equations for the FTCS method applied to the initial boundary value problem

$$\begin{cases} \dfrac{\partial u}{\partial t} = D\left(\dfrac{\partial^2 u}{\partial r^2} + \dfrac{1}{r}\dfrac{\partial u}{\partial r}\right) - \beta(r,t)u + s(r,t), & r_{\text{inner}} \le r \le r_{\text{outer}}, t > 0 \\ \dfrac{\partial u}{\partial r}(r_{\text{inner}}, t) = \alpha(t) & \\ p(t)u(r_{\text{outer}}, t) + q(t)\dfrac{\partial u}{\partial r}(r_{\text{outer}}, t) = \rho(t) & \\ u(r,0) = f(r) & \end{cases}$$

are presented in the text.

(a) Derive the stability condition associated with these equations.

(b) Numerically verify the stability condition found in part (a) for the problem

$$\begin{cases} \dfrac{\partial u}{\partial t} = \left(\dfrac{\partial^2 u}{\partial r^2} + \dfrac{1}{r}\dfrac{\partial u}{\partial r} \right) - u, \quad 1/2 \le r \le 2, t > 0 \\ \dfrac{\partial u}{\partial r}(1/2, t) = 2e^{-t} \\ u(2,t) + \dfrac{\partial u}{\partial r}(2,t) = e^{-t}\left(\dfrac{1}{2} + \ln 2\right) \\ u(r,0) = \ln r. \end{cases}$$

The exact solution for this problem is $u(r,t) = e^{-t}\ln r$.

10.6 PROBLEMS IN TWO SPATIAL DIMENSIONS

To conclude our treatment of parabolic partial differential equations, we will consider problems in two space dimensions. We will restrict attention to the rectangular domain $R = \{(x,y) \,|\, a \le x \le b, c \le y \le d\}$, and our model problem will be the heat equation

$$\frac{\partial u}{\partial t} = D\left(\frac{\partial^2 u}{\partial x^2} + \frac{\partial^2 u}{\partial y^2}\right)$$

subject to Dirichlet boundary conditions. Source and decay terms will be included toward the end of the section. Non-Dirichlet boundary conditions will be considered in the exercises.

The FTCS Method

Let N_x be the number of subintervals along the x-axis and N_y be the number of subintervals along the y-axis. Define the spacing parameters

$$\Delta x = \frac{b-a}{N_x} \quad \text{and} \quad \Delta y = \frac{d-c}{N_y}$$

and the gridlines $x_j = a + j\Delta x$ and $y_k = c + k\Delta y$. For simplicity, we will assume throughout the section that $\Delta x = \Delta y = \Delta$. Next, evaluate the heat equation at the arbitrary interior grid point (x_j, y_k), and approximate each space derivative by its second-order central difference formula. Finally, drop the truncation error terms to produce the semidiscrete approximation

$$\frac{dv_{j,k}(t)}{dt} = D\left(\frac{v_{j-1,k}(t) - 2v_{j,k}(t) + v_{j+1,k}(t)}{\Delta^2} + \frac{v_{j,k-1}(t) - 2v_{j,k}(t) + v_{j,k+1}(t)}{\Delta^2}\right). \tag{1}$$

Here $v_{j,k}(t) \approx u(x_j, y_k, t)$.

To obtain the FTCS method, evaluate the semidiscrete template at time level $t = t_n$ and then replace the time derivative with its first-order forward difference approximation. Solving the resulting fully discrete equation for the value of the approximate solution at time level $t = t_{n+1}$ yields

$$w_{j,k}^{(n+1)} = \lambda\left(w_{j-1,k}^{(n)} + w_{j+1,k}^{(n)} + w_{j,k-1}^{(n)} + w_{j,k+1}^{(n)}\right) + (1-4\lambda)w_{j,k}^{(n)}, \tag{2}$$

where $w_{j,k}^{(n)} \approx v_{j,k}(t_n) \approx u(x_j, y_k, t_n)$ and $\lambda = D\,\Delta t/\Delta^2$.

What is the maximum allowable value for λ that will produce a stable approximate solution? Let's perform a von Neumann stability analysis. In two space dimensions the discrete Fourier mode takes the form

$$w_{j,k}^{(n)} = r^n e^{i(j\theta_1 + k\theta_2)}.$$

Substituting this expression into equation (2) gives

$$r^{n+1}e^{i(j\theta_1+k\theta_2)} = \left[\lambda\left(e^{-i\theta_1} + e^{i\theta_1} + e^{-i\theta_2} + e^{i\theta_2}\right) + 1 - 4\lambda\right] r^n e^{i(j\theta_1+k\theta_2)}.$$

Dividing out $r^n e^{i(j\theta_1+k\theta_2)}$ and using the identity $e^{i\theta} + e^{-i\theta} = 2\cos\theta$, we find

$$\begin{aligned} r &= 2\lambda(\cos\theta_1 + \cos\theta_2) + 1 - 4\lambda \\ &= 1 - 2\lambda(2 - \cos\theta_1 - \cos\theta_2). \end{aligned}$$

Since $0 \le 2 - \cos\theta_1 - \cos\theta_2 \le 4$ for all values of θ_1 and θ_2, it follows that $1 - 8\lambda \le r \le 1$. To guarantee $|r| \le 1$, we must therefore choose λ so that $1 - 8\lambda \ge -1$, which requires $\lambda \le 1/4$. Hence, the stability condition for the FTCS method in two space dimensions is even more restrictive than in one space dimension. Recalling the definition of λ, the restriction on the time step is given by

$$\Delta t \le \frac{\Delta^2}{4D}.$$

One programming note needs to be made before moving on to an example. When equation (1) is applied with $j = 1$, $j = N_x - 1$, $k = 1$ or $k = N_y - 1$, at least one of the w values will be obtained from the Dirichlet boundary conditions. The best approach for handling these cases is to dimension the w matrix to $(N_x+1)\times(N_y+1)$, rather than $(N_x - 1)\times(N_y - 1)$, and load the boundary condition values into appropriate locations of the matrix. At the expense of a few memory locations, the code for the algorithm will be simplified tremendously.

EXAMPLE 10.13 Stability of the FTCS Method in Two Space Dimensions

Consider the initial boundary value problem

$$\begin{aligned} \frac{\partial u}{\partial t} &= \frac{1}{16}\left(\frac{\partial^2 u}{\partial x^2} + \frac{\partial^2 u}{\partial y^2}\right), \quad 0 \le x \le 1, 0 \le y \le 1, t > 0 \\ u(0,y,t) &= u(1,y,t) = 0 \\ u(x,0,t) &= u(x,1,t) = 0 \\ u(x,y,0) &= \sin(2\pi x)\sin(2\pi y). \end{aligned}$$

The exact solution to this problem is $u(x,y,t) = e^{-(\pi^2/2)t}\sin(2\pi x)\sin(2\pi y)$. Two approximations to $u(x,y,5)$, obtained using the FTCS method with different size time steps, are displayed in Figure 10.19. Spatial resolution was set at $\Delta = \Delta x = \Delta y = 1/20$ for all calculations. With $\Delta = 1/20$, the maximum allowable time step, based on stability considerations, is

$$\Delta t_{\text{max}} = \frac{\Delta^2}{4D} = \frac{(1/20)^2}{4(1/16)} = \frac{1}{100}.$$

The approximate solution in the top graph of Figure 10.19 was calculated with a step size of Δt_{max}. The plotted surface exhibits no signs of instability. The approximate solution in the bottom graph, which is clearly unstable, was computed with a time step of $\Delta t = 1/98$. This time step corresponds to $\lambda \approx 0.255$.

Alternating Direction Implicit (ADI) Method

The stability condition $\lambda \le 1/4$ generally makes the FTCS method too expensive for practical use. In one space dimension, we found that the BTCS method and the Crank-Nicolson scheme were unconditionally stable and required little additional computational effort over the FTCS method. For the two-dimensional heat equation, the BTCS method is given by the finite difference equation

$$-\lambda\left(w_{j-1,k}^{(n+1)} + w_{j+1,k}^{(n+1)} + w_{j,k-1}^{(n+1)} + w_{j,k+1}^{(n+1)}\right) + (1+4\lambda)w_{j,k}^{(n+1)} = w_{j,k}^{(n)}, \qquad (3)$$

where $\lambda = D\Delta t/\Delta^2$, and the Crank-Nicolson scheme is given by

$$\begin{aligned} -\lambda\left(w_{j-1,k}^{(n+1)} + w_{j+1,k}^{(n+1)} + w_{j,k-1}^{(n+1)} + w_{j,k+1}^{(n+1)}\right) + (1+4\lambda)w_{j,k}^{(n+1)} = \\ \lambda\left(w_{j-1,k}^{(n)} + w_{j+1,k}^{(n)} + w_{j,k-1}^{(n)} + w_{j,k+1}^{(n)}\right) + (1-4\lambda)w_{j,k}^{(n)}, \qquad (4) \end{aligned}$$

where $\lambda = D\Delta t/\left[2\Delta^2\right]$. Von Neumann stability analysis easily establishes that both of these methods are unconditionally stable (see Exercise 1); however, the coefficient matrix for the system of equations that must be solved during each time step is no longer tridiagonal. Fortunately, equations (3) and (4) are trivial modifications of the discrete Poisson problem we treated in Chapter 9. The multigrid method can therefore be used to efficiently solve the corresponding systems. We leave it to the interested reader to explore this issue in more detail.

The generally preferred approach for dealing with the two-dimensional heat equation is to use one of a class of alternative differencing schemes that results in tridiagonal coefficient matrices. The first such scheme was proposed by Peaceman and Rachford [1]. Each time step of the scheme is carried out in two half-steps. First, the approximate solution is advanced from time level $t = t_n$ to the intermediate level $t = t_{n+1/2}$. The time derivative on the left-hand side of equation (1) is replaced by its first-order forward difference formula. On the right-hand side of

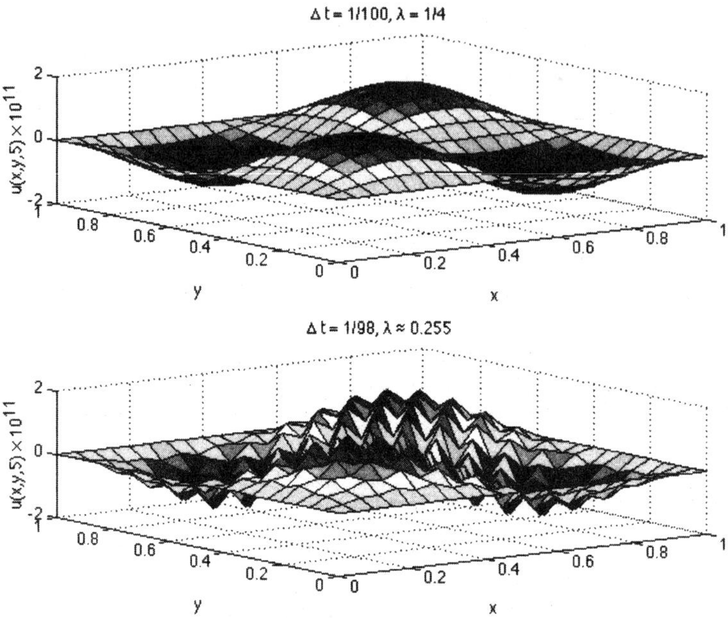

Figure 10.19 Two approximations to $u(x, y, 5)$, where $u(x, y, t)$ is the solution to the initial boundary value problem

$$\begin{aligned}\frac{\partial u}{\partial t} &= \frac{1}{16}\left(\frac{\partial^2 u}{\partial x^2} + \frac{\partial^2 u}{\partial y^2}\right),\\ u(0, y, t) &= u(1, y, t) = 0,\\ u(x, 0, t) &= u(x, 1, t) = 0,\\ u(x, y, 0) &= \sin(2\pi x)\sin(2\pi y).\end{aligned}$$

Approximate solutions were computed with the FTCS method with different step sizes.

equation (1), we evaluate the terms from one of the space derivatives at $t = t_n$ and the terms from the other space derivative at $t = t_{n+1/2}$; that is, one derivative is treated explicitly, the other implicitly. It does not matter which derivative is handled in which way, for in advancing the solution from $t = t_{n+1/2}$ to $t = t_{n+1}$, the treatment of the terms will be reversed. Thus, if we choose to treat the x derivative terms implicitly first, then the values at the intermediate time level are defined by

$$\begin{aligned}-\lambda\left(w_{j-1,k}^{(n+1/2)} + w_{j+1,k}^{(n+1/2)}\right) &+ (1+2\lambda)w_{j,k}^{(n+1/2)}\\ &= \lambda\left(w_{j,k-1}^{(n)} + w_{j,k+1}^{(n)}\right) + (1-2\lambda)w_{j,k}^{(n)}, \qquad (5)\end{aligned}$$

where $\lambda = D\Delta t/(2\Delta^2)$. The approximate solution at $t = t_{n+1}$ is ultimately computed by handling the y derivative terms implicitly:

$$-\lambda\left(w_{j,k-1}^{(n+1)}+w_{j,k+1}^{(n+1)}\right)+(1+2\lambda)w_{j,k}^{(n+1)} = \lambda\left(w_{j-1,k}^{(n+1/2)}+w_{j+1,k}^{(n+1/2)}\right)+(1-2\lambda)w_{j,k}^{(n+1/2)}. \quad (6)$$

Since this scheme alternately treats the x derivative and then the y derivative implicitly, it is known as the alternating direction implicit, or ADI, scheme.

It is important to note that equation (5) actually represents a set of $N_y - 1$ disjoint tridiagonal systems of equations, one for the values along each row of the computational grid. In matrix notation, each of these systems can be written as

$$E_x\mathbf{w}_{R(k)}^{(n+1/2)} = \lambda\left(\mathbf{w}_{R(k-1)}^{(n)}+\mathbf{w}_{R(k+1)}^{(n)}\right)+(1-2\lambda)\mathbf{w}_{R(k)}^{(n)}+\mathbf{b}_{R(k)}^{(n+1/2)},$$

where the subscript $R(k)$ denotes the kth row of the grid (for $k = 1, 2, 3, \ldots, N_y-1$), E_x is the $(N_x - 1) \times (N_x - 1)$ tridiagonal evolution matrix

$$\begin{bmatrix} 1+2\lambda & -\lambda & & & & \\ -\lambda & 1+2\lambda & -\lambda & & & \\ & -\lambda & 1+2\lambda & -\lambda & & \\ & & \ddots & \ddots & \ddots & \\ & & & -\lambda & 1+2\lambda & -\lambda \\ & & & & -\lambda & 1+2\lambda \end{bmatrix},$$

and $\mathbf{b}_{R(k)}^{(n+1/2)} = \begin{bmatrix} \lambda w_{0,k}^{(n+1/2)} & 0 & \cdots & 0 & \lambda w_{N_x,k}^{(n+1/2)} \end{bmatrix}^T$. The vector $\mathbf{b}_{R(k)}^{(n+1/2)}$ contains values along the kth row that are known from the boundary conditions. In a similar manner, equation (6) actually represents a set of $N_x - 1$ tridiagonal linear systems, one for the values along each column of the grid. These systems can be written as

$$E_y\mathbf{w}_{C(j)}^{(n+1)} = \lambda\left(\mathbf{w}_{C(j-1)}^{(n+1/2)}+\mathbf{w}_{C(j+1)}^{(n+1/2)}\right)+(1-2\lambda)\mathbf{w}_{C(j)}^{(n+1/2)}+\mathbf{b}_{C(j)}^{(n+1)},$$

where the subscript $C(j)$ denotes the jth column of the grid (for $j = 1, 2, 3, \ldots, N_x-1$), the evolution matrix E_y has the same tridiagonal structure as E_x but is of dimension $(N_y - 1) \times (N_y - 1)$, and

$$\mathbf{b}_{C(j)}^{(n+1)} = \begin{bmatrix} \lambda w_{j,0}^{(n+1)} & 0 & \cdots & 0 & \lambda w_{j,N_y}^{(n+1)} \end{bmatrix}^T.$$

Taking into account the computation of the right-hand-side vectors and the solution of the tridiagonal systems, each full time step of the ADI scheme uses roughly 24 algebraic operations per grid point. This is four times as much work as the FTCS method, which requires 6 operations per point. Fortunately, the

ADI scheme is unconditionally stable, so fewer time steps can be performed. To establish this result, we substitute the two-dimensional discrete Fourier mode into equation (5) to obtain the partial amplification factor

$$r_x = \frac{1 - 2\lambda + 2\lambda\cos\theta_2}{1 + 2\lambda - 2\lambda\cos\theta_1}.$$

From equation (6), we obtain

$$r_y = \frac{1 - 2\lambda + 2\lambda\cos\theta_1}{1 + 2\lambda - 2\lambda\cos\theta_2}.$$

Therefore, the total amplification factor per time step is

$$r = r_x r_y = \frac{(1 - 2\lambda + 2\lambda\cos\theta_2)(1 - 2\lambda + 2\lambda\cos\theta_1)}{(1 + 2\lambda - 2\lambda\cos\theta_1)(1 + 2\lambda - 2\lambda\cos\theta_2)},$$

which is always less than or equal to one in magnitude. Although we used a first-order difference formula for the time derivative, the overall truncation error of the ADI scheme is second-order in both time and space (see Morton and Mayers [2] for details). This is a second factor that allows for fewer time steps to be taken with the ADI scheme.

EXAMPLE 10.14 The ADI Scheme in Action

Consider the initial boundary value problem

$$\begin{aligned} \frac{\partial u}{\partial t} = \frac{1}{16}\left(\frac{\partial^2 u}{\partial x^2} + \frac{\partial^2 u}{\partial y^2}\right), \quad & 0 \le x \le 1, 0 \le y \le 1, t > 0 \\ u(0,y,t) = u(1,y,t) &= 0 \\ u(x,0,t) = u(x,1,t) &= 0 \\ u(x,y,0) &= \sin(\pi x)\sin(\pi y). \end{aligned}$$

Let's approximate the solution to this problem at $t = 1$ using the ADI scheme. We will take $\Delta = 1/4$ and $\Delta t = 1$. Then

$$\lambda = \frac{D\Delta t}{2\Delta^2} = \frac{(1/16)\cdot 1}{2(1/4)^2} = \frac{1}{2}.$$

The value of Δ implies that $N_x = N_y = 4$. Therefore, each half time step will require the solution of three 3×3 tridiagonal systems of equations. Furthermore, the evolution matrices, E_x and E_y, will be identical:

$$\begin{bmatrix} 2 & -1/2 & 0 \\ -1/2 & 2 & -1/2 \\ 0 & -1/2 & 2 \end{bmatrix}.$$

With $\lambda = 1/2$, $1 - 2\lambda = 0$ and one of the terms from the right-hand side of each of the systems vanishes. Since the problem has homogeneous boundary conditions specified around the entire boundary, all of the vectors $\mathbf{b}_{R(k)}^{(n+1/2)}$ and $\mathbf{b}_{C(j)}^{(n+1)}$ will also be zero.

From the initial condition, we find

$$\mathbf{w}_{R(1)}^{(0)} = \begin{bmatrix} 1/2 \\ \sqrt{2}/2 \\ 1/2 \end{bmatrix}, \quad \mathbf{w}_{R(2)}^{(0)} = \begin{bmatrix} \sqrt{2}/2 \\ 1 \\ \sqrt{2}/2 \end{bmatrix}, \quad \mathbf{w}_{R(3)}^{(0)} = \begin{bmatrix} 1/2 \\ \sqrt{2}/2 \\ 1/2 \end{bmatrix}.$$

Therefore, the first half time step requires the solution of the systems

$$\begin{bmatrix} 2 & -1/2 & 0 \\ -1/2 & 2 & -1/2 \\ 0 & -1/2 & 2 \end{bmatrix} \mathbf{w}_{R(1)}^{(1/2)} = \frac{1}{2}\left(\begin{bmatrix} 0 \\ 0 \\ 0 \end{bmatrix} + \begin{bmatrix} \sqrt{2}/2 \\ 1 \\ \sqrt{2}/2 \end{bmatrix} \right) = \begin{bmatrix} \sqrt{2}/4 \\ 1/2 \\ \sqrt{2}/4 \end{bmatrix},$$

$$\begin{bmatrix} 2 & -1/2 & 0 \\ -1/2 & 2 & -1/2 \\ 0 & -1/2 & 2 \end{bmatrix} \mathbf{w}_{R(2)}^{(1/2)} = \frac{1}{2}\left(\begin{bmatrix} 1/2 \\ \sqrt{2}/2 \\ 1/2 \end{bmatrix} + \begin{bmatrix} 1/2 \\ \sqrt{2}/2 \\ 1/2 \end{bmatrix} \right) = \begin{bmatrix} 1/2 \\ \sqrt{2}/2 \\ 1/2 \end{bmatrix},$$

and

$$\begin{bmatrix} 2 & -1/2 & 0 \\ -1/2 & 2 & -1/2 \\ 0 & -1/2 & 2 \end{bmatrix} \mathbf{w}_{R(3)}^{(1/2)} = \frac{1}{2}\left(\begin{bmatrix} 0 \\ 0 \\ 0 \end{bmatrix} + \begin{bmatrix} \sqrt{2}/2 \\ 1 \\ \sqrt{2}/2 \end{bmatrix} \right) = \begin{bmatrix} \sqrt{2}/4 \\ 1/2 \\ \sqrt{2}/4 \end{bmatrix}.$$

The solutions of these systems are found to be

$$\mathbf{w}_{R(1)}^{(1/2)} = \begin{bmatrix} (1+2\sqrt{2})/14 \\ (4+\sqrt{2})/14 \\ (1+2\sqrt{2})/14 \end{bmatrix}, \quad \mathbf{w}_{R(2)}^{(1/2)} = \begin{bmatrix} (4+\sqrt{2})/14 \\ (1+2\sqrt{2})/7 \\ (4+\sqrt{2})/14 \end{bmatrix},$$

and

$$\mathbf{w}_{R(3)}^{(1/2)} = \begin{bmatrix} (1+2\sqrt{2})/14 \\ (4+\sqrt{2})/14 \\ (1+2\sqrt{2})/14 \end{bmatrix}.$$

From these solutions, we construct

$$\mathbf{w}_{C(1)}^{(1/2)} = \begin{bmatrix} (1+2\sqrt{2})/14 \\ (4+\sqrt{2})/14 \\ (1+2\sqrt{2})/14 \end{bmatrix}, \quad \mathbf{w}_{C(2)}^{(1/2)} = \begin{bmatrix} (4+\sqrt{2})/14 \\ (1+2\sqrt{2})/7 \\ (4+\sqrt{2})/14 \end{bmatrix},$$

and

$$\mathbf{w}_{C(3)}^{(1/2)} = \begin{bmatrix} (1+2\sqrt{2})/14 \\ (4+\sqrt{2})/14 \\ (1+2\sqrt{2})/14 \end{bmatrix},$$

and the systems

$$\begin{bmatrix} 2 & -1/2 & 0 \\ -1/2 & 2 & -1/2 \\ 0 & -1/2 & 2 \end{bmatrix} \mathbf{w}_{C(1)}^{(1)} = \frac{1}{2}\left(\begin{bmatrix} 0 \\ 0 \\ 0 \end{bmatrix} + \begin{bmatrix} (4+\sqrt{2})/14 \\ (1+2\sqrt{2})/7 \\ (4+\sqrt{2})/14 \end{bmatrix} \right)$$
$$= \begin{bmatrix} (4+\sqrt{2})/28 \\ (1+2\sqrt{2})/14 \\ (4+\sqrt{2})/28 \end{bmatrix},$$

$$\begin{bmatrix} 2 & -1/2 & 0 \\ -1/2 & 2 & -1/2 \\ 0 & -1/2 & 2 \end{bmatrix} \mathbf{w}_{C(2)}^{(1)} = \frac{1}{2}\left(\begin{bmatrix} (1+2\sqrt{2})/14 \\ (4+\sqrt{2})/14 \\ (1+2\sqrt{2})/14 \end{bmatrix} + \begin{bmatrix} (1+2\sqrt{2})/14 \\ (4+\sqrt{2})/14 \\ (1+2\sqrt{2})/14 \end{bmatrix} \right)$$
$$= \begin{bmatrix} (1+2\sqrt{2})/14 \\ (4+\sqrt{2})/14 \\ (1+2\sqrt{2})/14 \end{bmatrix},$$

$$\begin{bmatrix} 2 & -1/2 & 0 \\ -1/2 & 2 & -1/2 \\ 0 & -1/2 & 2 \end{bmatrix} \mathbf{w}_{C(3)}^{(1)} = \frac{1}{2}\left(\begin{bmatrix} 0 \\ 0 \\ 0 \end{bmatrix} + \begin{bmatrix} (4+\sqrt{2})/14 \\ (1+2\sqrt{2})/7 \\ (4+\sqrt{2})/14 \end{bmatrix} \right)$$
$$= \begin{bmatrix} (4+\sqrt{2})/28 \\ (1+2\sqrt{2})/14 \\ (4+\sqrt{2})/28 \end{bmatrix},$$

which must be solved to complete the time step. The solutions of these systems are

$$\mathbf{w}_{C(1)}^{(1)} = \begin{bmatrix} (9+4\sqrt{2})/98 \\ (8+9\sqrt{2})/98 \\ (9+4\sqrt{2})/98 \end{bmatrix}, \quad \mathbf{w}_{C(2)}^{(1)} = \begin{bmatrix} (8+9\sqrt{2})/98 \\ (9+4\sqrt{2})/49 \\ (8+9\sqrt{2})/98 \end{bmatrix},$$

and

$$\mathbf{w}_{C(3)}^{(1)} = \begin{bmatrix} (9+4\sqrt{2})/98 \\ (8+9\sqrt{2})/98 \\ (9+4\sqrt{2})/98 \end{bmatrix}.$$

The following table compares the approximate solution at $t = 1$ with the exact solution, $u(x, y, 1) = e^{-(\pi^2/8)} \sin(\pi x) \sin(\pi y)$.

x	y	$w^{(1)}$	Exact	Absolute Error
0.25	0.25	0.149560	0.145606	0.003953
0.25	0.50	0.211509	0.205919	0.005591
0.25	0.75	0.149560	0.145606	0.003953
0.50	0.25	0.211509	0.205919	0.005591
0.50	0.50	0.299119	0.291213	0.007907
0.50	0.75	0.211509	0.205919	0.005591
0.75	0.25	0.149560	0.145606	0.003953
0.75	0.50	0.211509	0.205919	0.005591
0.75	0.75	0.149560	0.145606	0.003953

EXAMPLE 10.15 Convergence and Stability Properties of the ADI Scheme

Let's reconsider the initial boundary value problem

$$\frac{\partial u}{\partial t} = \frac{1}{16}\left(\frac{\partial^2 u}{\partial x^2} + \frac{\partial^2 u}{\partial y^2}\right), \quad 0 \le x \le 1, 0 \le y \le 1, t > 0$$
$$u(0, y, t) = u(1, y, t) = 0$$
$$u(x, 0, t) = u(x, 1, t) = 0$$
$$u(x, y, 0) = \sin(\pi x)\sin(\pi y)$$

whose exact solution is $u(x, y, t) = e^{-(\pi^2/8)t}\sin(\pi x)\sin(\pi y)$. The following two tables provide a numerical verification of the second-order accuracy of the ADI scheme in both time and space. The first table summarizes the error in the approximate solution at $t = 1$ for different values of Δ. For all calculations a time step of $\Delta t = 1/100$ was used. Each time Δ is cut in half, the error decreases by a factor of four. The second table presents the error in the approximate solution at $t = 1$ for different values of Δt, with Δ fixed at 1/100. For each reduction in the step size by a factor of two, the error drops by a factor of four.

The unconditional stability of the ADI scheme is shown in Figure 10.20. Here, the approximate solution at $t = 10$ is plotted. This solution was computed with ten time steps; that is, the step size was taken to be $\Delta t = 1$. Spatial resolution was set at $\Delta = 1/25$. Thus

$$\lambda = \frac{D\,\Delta t}{2\,\Delta^2} = \frac{(1/16)}{2(1/25)^2} = \frac{625}{32} \approx 19.53.$$

Even with a value of λ this large, the approximate solution exhibits no signs of instability.

Second-Order Accuracy in Space

Δ	Maximum Absolute Error	Error Ratio	rms Error	Error Ratio
1/4	0.0186651851		0.0093325926	
1/8	0.0046286252	4.032555	0.0023143126	4.032555
1/16	0.0011539171	4.011229	0.0005769585	4.011229
1/32	0.0002874745	4.013980	0.0001437373	4.013980
1/64	0.0000710048	4.048662	0.0000355024	4.048662

Second-Order Accuracy in Time

Δt	Maximum Absolute Error	Error Ratio	rms Error	Error Ratio
1	0.0118110448		0.0059055224	
1/2	0.0028450761	4.151398	0.0014225381	4.151398
1/4	0.0006840195	4.159349	0.0003420098	4.159349
1/8	0.0001485288	4.605300	0.0000742644	4.605300

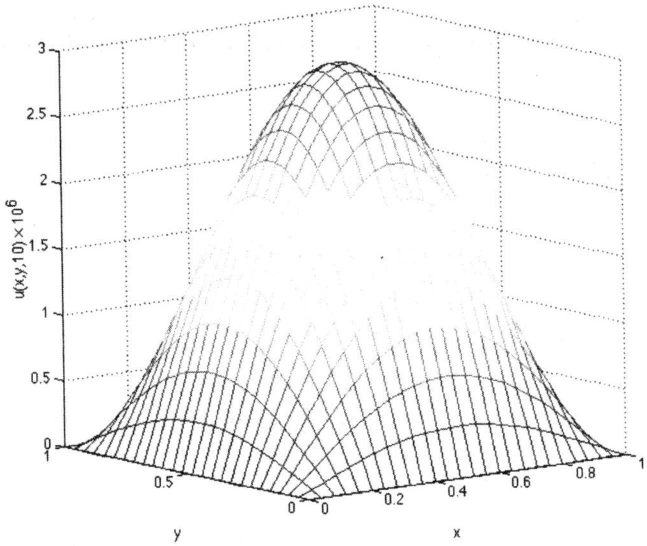

Figure 10.20 Demonstration of unconditional stability of the ADI scheme. Approximate solution was computed with $\lambda \approx 19.53$.

Source and Decay Terms in the ADI Scheme

Let's now discuss the inclusion of source and decay terms into the ADI scheme. In particular, we will consider the partial differential equation

$$\frac{\partial u}{\partial t} = D\left(\frac{\partial^2 u}{\partial x^2} + \frac{\partial^2 u}{\partial y^2}\right) - \beta(x,y,t)u + s(x,y,t).$$

As in previous sections, we will assume that $\beta(x,y,t) \geq 0$ for all $(x,y) \in R$, where $R = \{(x,y) | \, a \leq x \leq b, c \leq y \leq d\}$, and for all $t > 0$. The spatial discretization of the additional terms is straightforward—simply evaluate each term at the grid point (x_j, y_k). To maintain unconditional stability and second-order accuracy in time, the time discretization of the source term and the decay term must be carried out by averaging the values at the start and end of each half time step. Thus, to the right-hand side of equation (5), we must add

$$-\frac{\Delta t}{4}\beta_{j,k}^{(n)} w_{j,k}^{(n)} + \frac{\Delta t}{4}\left[s_{j,k}^{(n)} + s_{j,k}^{(n+1/2)}\right],$$

while to the left-hand side we add

$$\frac{\Delta t}{4}\beta_{j,k}^{(n+1/2)} w_{j,k}^{(n+1/2)}.$$

As for equation (6),

$$-\frac{\Delta t}{4}\beta_{j,k}^{(n+1/2)}w_{j,k}^{(n+1/2)}+\frac{\Delta t}{4}\left[s_{j,k}^{(n+1/2)}+s_{j,k}^{(n+1)}\right]$$

is added to the right-hand side, and

$$\frac{\Delta t}{4}\beta_{j,k}^{(n+1)}w_{j,k}^{(n+1)}$$

is added to the left.

EXAMPLE 10.16 Convergence and Stability of the ADI Scheme with Source and Decay Terms

Let's approximate the solution of the initial boundary value problem

$$\frac{\partial u}{\partial t}=\frac{1}{16}\left(\frac{\partial^2 u}{\partial x^2}+\frac{\partial^2 u}{\partial y^2}\right)-\frac{t}{t^2+1}u+\frac{8x^2y^2-t(x^2+y^2)}{\sqrt{t^2+1}},$$
$$0\le x\le 1, 0\le y\le 1, t>0$$

$$u(0,y,t)=0,\quad u(1,y,t)=8ty^2/\sqrt{t^2+1}$$
$$u(x,0,t)=0,\quad u(x,1,t)=8tx^2/\sqrt{t^2+1}$$
$$u(x,y,0)=\sin(\pi x)\sin(\pi y).$$

The source term for this problem is

$$s(x,y,t)=\frac{8x^2y^2-t(x^2+y^2)}{\sqrt{t^2+1}},$$

and the decay coefficient is

$$\beta(x,y,t)=\frac{t}{t^2+1}.$$

The exact solution is

$$u(x,y,t)=\frac{e^{-(\pi^2/8)t}\sin(\pi x)\sin(\pi y)+8tx^2y^2}{\sqrt{t^2+1}}.$$

The following two tables provide a numerical verification of the second-order accuracy of the ADI scheme in both time and space. The first table summarizes the error in the approximate solution at $t=1$ for different values of Δ. For all calculations a time step of $\Delta t=1/100$ was used. Each time Δ is cut in half, the error decreases roughly by a factor of four. The second table presents the error in the approximate solution at $t=1$ for different values of Δt, with Δ fixed at $1/100$. For each reduction in the step size by a factor of two, the error again drops by roughly a factor of four.

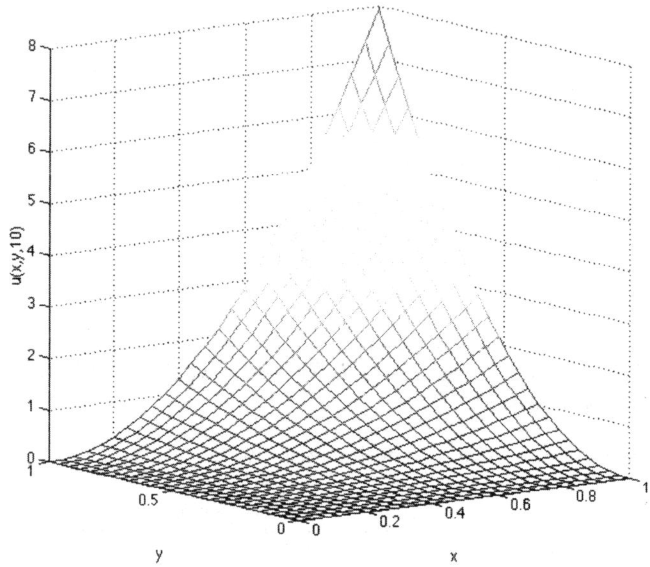

Figure 10.21 Demonstration of unconditional stability of the ADI scheme in the presence of source and decay terms. Approximate solution was computed with $\lambda \approx 19.53$.

The unconditional stability of the ADI scheme is shown in Figure 10.21. Here, the approximate solution at $t = 10$ is plotted. This solution was computed with $\Delta t = 1$. Spatial resolution was set at $\Delta = 1/25$. Thus

$$\lambda = \frac{D\Delta t}{2\Delta^2} = \frac{(1/16)}{2(1/25)^2} = \frac{625}{32} \approx 19.53.$$

Even with a value of λ this large, the approximate solution exhibits no signs of instability.

Second-Order Accuracy in Space

Δ	Maximum Absolute Error	Error Ratio	rms Error	Error Ratio
1/4	0.0131985079		0.0065994016	
1/8	0.0032730175	4.032520	0.0016367104	4.032113
1/16	0.0008159919	4.011090	0.0004082126	4.009456
1/32	0.0002033157	4.013423	0.0001018836	4.006657
1/64	0.0000502552	4.045661	0.0000253737	4.015320

Second-Order Accuracy in Time

Δt	Maximum Absolute Error	Error Ratio	rms Error	Error Ratio
1	0.1378577860		0.0311540693	
1/2	0.0096845702	14.234786	0.0031767904	9.806775
1/4	0.0026971364	3.590686	0.0007784414	4.080963
1/8	0.0006497332	4.151144	0.0001965845	3.959832
1/16	0.0001559938	4.165122	0.0000497160	3.954147

Application Problem: Two-Dimensional Model for Color Photograph Development

In Section 10.3, we considered a one-dimensional model for color photograph development. The corresponding two-dimensional model (see Friedman and Littman in Chapter 4 of *Industrial Mathematics: A Course in Solving Real-World Problems*, SIAM, Philadelphia, 1994) is

$$\frac{\partial T}{\partial t} = D\left(\frac{\partial^2 T}{\partial x^2} + \frac{\partial^2 T}{\partial y^2}\right) - kTC + \gamma E(x,y), \qquad \frac{\partial C}{\partial t} = -kTC,$$

where $T(x,y)$ is the density of the oxidized developer and $C(x,y)$ is the density of the coupler. The initial conditions for this system are $T(x,y,0) = 0$ and $C(x,y,0) = C_0(x,y)$, while the boundary conditions are $T(0,y,t) = T(L,y,t) = T(x,0,t) = T(x,L,t) = 0$, where L is the length of one side of the film. $E(x,y)$ is the exposure function that indicates those regions of the film that were exposed to light.

We will simulate three minutes (180 seconds) of development time, taking as parameter values

$$D = 100\mu\text{m}^2/\text{s}, k = 6.6\times 10^{12}\mu\text{m}/\text{moles}\cdot\text{s}, \gamma = 7.5\times 10^{-12}\ \text{moles}/\mu\text{m}\cdot\text{s}$$
$$L = 1.5\times 10^5\mu\text{m}\ \text{ and }\ C_0 = 1.125\times 10^{-11}\ \text{moles}/\mu\text{m}.$$

The exposure pattern is shown in Figure 10.22. To help balance the drastic differences in order of magnitude between the parameters, let's introduce the following nondimensional variables:

$$\bar{T} = \frac{T}{\gamma L^2/D}, \quad \bar{C} = \frac{C}{C_0}, \quad \bar{t} = t\frac{D}{L^2}, \quad \bar{x} = \frac{x}{L}, \quad \text{and} \quad \bar{y} = \frac{y}{L}.$$

In terms of these variables, the system of differential equations becomes

$$\frac{\partial \bar{T}}{\partial \bar{t}} = \frac{\partial^2 \bar{T}}{\partial \bar{x}^2} + \frac{\partial^2 \bar{T}}{\partial \bar{y}^2} - k_1\bar{T}\bar{C} + E(\bar{x},\bar{y}), \qquad \frac{\partial \bar{C}}{\partial \bar{t}} = -k_2\bar{T}\bar{C},$$

where

$$k_1 = \frac{C_0 k L^2}{D} = 1.670625\times 10^{10} \quad \text{and} \quad k_2 = \frac{k\gamma L^4}{D^2} = 2.5059375\times 10^{18}.$$

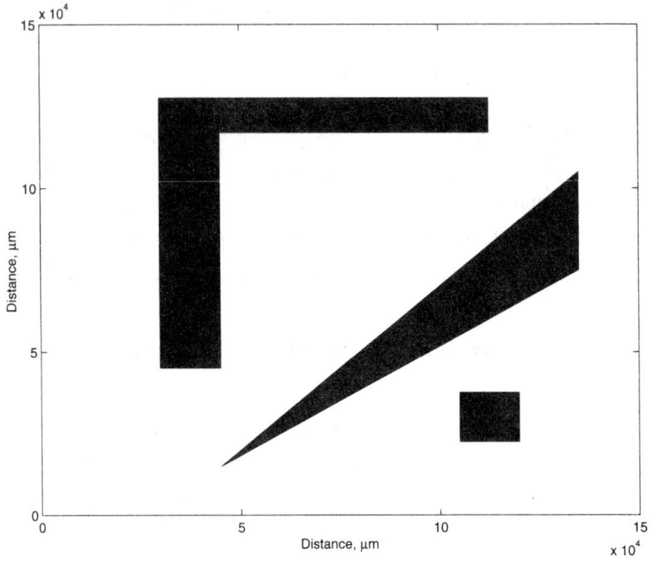

Figure 10.22 Exposure pattern for two-dimensional color development example. Shaded areas represent an exposure level of 1; nonshaded areas represent an exposure level of 0.

Furthermore, note that $t = 180$ seconds translates to $\bar{t} = 8 \times 10^{-7}$.

The oxidized developer equation is solved using the ADI scheme, while a semi-implicit discretization (with the coupler density treated implicitly and the oxidized developer density treated explicitly) is used for the coupler equation. The resulting finite difference equations for each half-step of the coupler equation are

$$\bar{C}_{j,k}^{(n+1/2)} = \frac{\bar{C}_{j,k}^{(n)}}{1 + \Delta\bar{t}k_2\bar{T}_{j,k}^{(n)}/2} \quad \text{and} \quad \bar{C}_{j,k}^{(n+1)} = \frac{\bar{C}_{j,k}^{(n+1/2)}}{1 + \Delta\bar{t}k_2\bar{T}_{j,k}^{(n+1/2)}/2}.$$

Density profiles computed in this manner, with $\Delta\bar{x} = \Delta\bar{y} = 1/100$ and $\Delta\bar{t} = 8 \times 10^{-9}$, are shown in Figure 10.23.

References

1. D. W. Peaceman and H. H. Rachford, "The Numerical Solution of Parabolic and Elliptic Differential Equations," *Journal of the Society for Industrial and Applied Mathematics*, **3**, 28–41, 1955.

2. K. W. Morton and D. F. Mayers, *Numerical Solution of Partial Differential Equations*, Cambridge University Press, Cambridge, 1994.

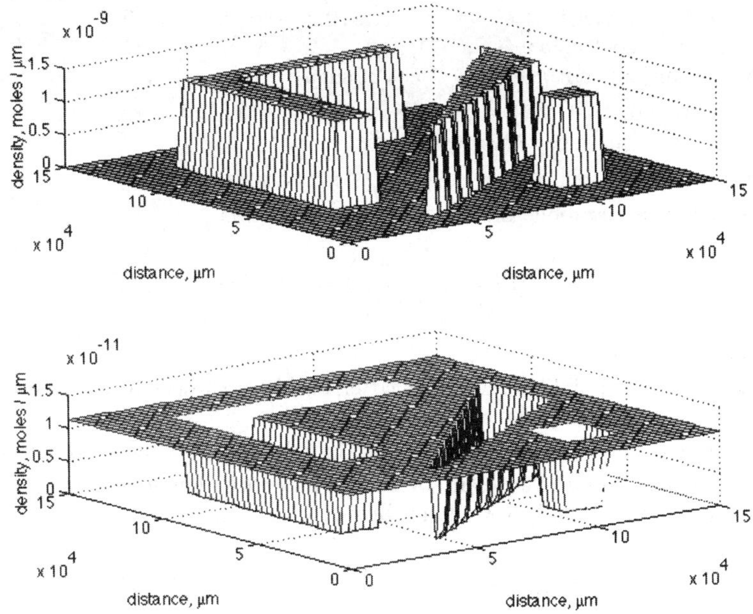

Figure 10.23 Oxidized developer (top graph) and coupler density (bottom graph) profiles after three minutes of development time.

EXERCISES

1. **(a)** Show that the BTCS method for the two-dimensional heat equation is unconditionally stable.

(b) Show that the Crank-Nicolson scheme for the two-dimensional heat equation is unconditionally stable.

In Exercises 2–5, numerically verify that

(a) the FTCS method is stable for $\lambda = 1/4$ but unstable for any $\lambda > 1/4$;

(b) the ADI scheme is second-order accurate in both time and space; and

(c) the ADI scheme is unconditionally stable by approximating the solution of

$$\frac{\partial u}{\partial t} = \frac{\partial^2 u}{\partial x^2} + \frac{\partial^2 u}{\partial y^2}$$

subject to the indicated initial and boundary conditions.

2. $u(0, y, t) = 1, \quad u(1, y, t) = 0, \quad u(x, 0, t) = u(x, 1, t) = 1 - x,$
$u(x, y, 0) = 1 - x - \frac{1}{\pi^2}\sin(2\pi x)\sin(2\pi y)$
exact solution: $u(x, y, t) = 1 - x - \frac{1}{\pi^2}e^{-8\pi^2 t}\sin(2\pi x)\sin(2\pi y)$

3. $u(0, y, t) = u(\pi, y, t) = u(x, 0, t) = u(x, \pi, t) = 0, \quad u(x, y, 0) = \sin x \sin y$
exact solution: $u(x, y, t) = e^{-2t}\sin x \sin y$

4. $u(0, y, t) = u(\pi, y, t) = e^{-5t} \sin y, \quad u(x, 0, t) = u(x, \pi, t) = 0,$
$u(x, y, 0) = \cos 2x \sin y$
exact solution: $u(x, y, t) = e^{-5t} \cos 2x \sin y$

5. $u(0, y, t) = 0, \quad u(1, y, t) = y, \quad u(x, 0, t) = 0, \quad u(x, 1, t) = x,$
$u(x, y, 0) = xy - \sin \pi x \sin \pi y$
exact solution: $u(x, y, t) = xy - e^{-2\pi^2 t} \sin \pi x \sin \pi y$

6. Consider the parabolic partial differential equation

$$\frac{\partial u}{\partial t} = D\left(\frac{\partial^2 u}{\partial x^2} + \frac{\partial^2 u}{\partial y^2}\right) - \beta(x, y, t)u + s(x, y, t),$$

where $\beta(x, y, t) \geq 0$.
   (a) Develop the FTCS method for this problem, subject to Dirichlet boundary conditions along the entire boundary.
   (b) Derive the stability condition for the FTCS method developed in part (a).
   (c) Demonstrate the necessity of the stability condition derived in part (b) using the initial boundary value problem

$$\frac{\partial u}{\partial t} = \frac{1}{16}\left(\frac{\partial^2 u}{\partial x^2} + \frac{\partial^2 u}{\partial y^2}\right) - u + e^{-t}\left[8x^2y^2 - t(x^2 + y^2)\right],$$
$$0 \leq x \leq 1, 0 \leq y \leq 1, t > 0$$

$$u(0, y, t) = 0, \quad u(1, y, t) = 8te^{-t}y^2$$
$$u(x, 0, t) = 0, \quad u(x, 1, t) = 8te^{-t}x^2$$
$$u(x, y, 0) = \sin(\pi x)\sin(\pi y).$$

7. Rework the "Two Dimensional Model for Color Photograph Development" application problem using the parameter values

$$D = 550\,\mu\text{m}^2/\text{s}, k = 2.8 \times 10^{13}\mu\text{m/moles} \cdot \text{s}, \gamma = 3.2 \times 10^{-10} \text{ moles}/\mu\text{m} \cdot \text{s}$$
$$L = 1.5 \times 10^5 \mu\text{m} \text{ and } C_0 = 8.3 \times 10^{-8} \text{ moles}/\mu\text{m}.$$

Take

$$E(\bar{x}, \bar{y}) = \begin{cases} 1, & \text{inside } S_1, \text{ but outside } S_2 \\ 0, & \text{elsewhere} \end{cases}$$

as the exposure function, where S_1 is a square of side length 0.25 centered at $(0.5, 0.5)$ and S_2 is a square of side length 0.1 centered at $(0.45, 0.55)$. Simulate five minutes of development time.

8. Consider flow through a square slot bounded by four flat plates. Each pair of parallel plates is separated by a distance h. The fluid and all four walls are initially at rest. At $t = 0$, flow is initiated by impulsively bringing the lower wall, corresponding to $y = 0$, to the constant velocity u_0. The velocity profile within the slot, $u(x, y, t)$, satisfies

$$\frac{\partial u}{\partial t} = \nu\left(\frac{\partial^2 u}{\partial x^2} + \frac{\partial^2 u}{\partial y^2}\right)$$

$$u(x, y, 0) = 0, \quad u(x, 0, t) = u_0, \quad u(x, h, t) = 0, \quad u(0, y, t) = u(h, y, t) = 0,$$

where ν is the kinematic viscosity of the fluid. If we introduce the nondimensional variables

$$U = u/u_0, \quad X = x/h, \quad Y = y/h, \quad \text{and} \quad T = t\nu/h^2,$$

the problem becomes

$$\frac{\partial U}{\partial T} = \frac{\partial^2 U}{\partial X^2} + \frac{\partial^2 U}{\partial Y^2}$$

$$U(X,Y,0) = 0, \quad U(X,0,T) = 1, \quad U(X,1,T) = 0, \quad U(0,Y,T) = U(1,Y,T) = 0.$$

Plot $U(X,Y,T)$ for $T = 0.05$, $T = 0.1$, $T = 0.25$, and $T = 1$.

9. Suppose we need to solve the two-dimensional heat equation (with no source term and no decay term) over a rectangular domain with Dirichlet conditions specified along the bottom and left edges and Neumann conditions specified along the top and right edges. To be specific, suppose

$$\begin{aligned} u(A,y,t) &= u_A(y,t) && \text{along the left edge} \\ u(x,C,t) &= u_C(x,t) && \text{along the bottom edge} \\ \frac{\partial u}{\partial n} &= \alpha(y,t) && \text{along the right edge, and} \\ \frac{\partial u}{\partial n} &= \beta(x,t) && \text{along the top edge.} \end{aligned}$$

(a) Develop the ADI scheme to solve this problem.

(b) Using the scheme developed in part (a), solve

$$\frac{\partial h}{\partial t} = 457.92\left(\frac{\partial^2 h}{\partial x^2} + \frac{\partial^2 h}{\partial y^2}\right)$$

$$h(0,y,t) = 1 - e^{-t/5}, \quad h(x,0,t) = 1 - \cos\left(\frac{\pi t}{15}\right)$$

$$\frac{\partial h}{\partial n}(400,y,t) = \frac{\partial h}{\partial n}(x,500,t) = 0$$

$$h(x,y,0) = 0.$$

This initial boundary value problem models the change in the water table of an aquifer that is bounded on two sides by impermeable surfaces. The variables h, x and y are measured in meters, while t is measured in days. Determine the change in the water table after 15 days.

10. Suppose we need to solve the two-dimensional heat equation (with a source term, but no decay term) over a rectangular domain with a Robin condition specified along the top edge and Neumann conditions specified along the other three edges.

To be specific, suppose

$$
\begin{aligned}
\frac{\partial u}{\partial n} &= \alpha_A(y,t) && \text{along the left edge} \\
\frac{\partial u}{\partial n} &= \beta(x,t) && \text{along the bottom edge} \\
\frac{\partial u}{\partial n} &= \alpha_B(y,t) && \text{along the right edge, and} \\
p(x,t)u + q(x,t)\frac{\partial u}{\partial n} &= r(x,t) && \text{along the top edge.}
\end{aligned}
$$

(a) Develop the ADI scheme to solve this problem.

(b) Using the scheme developed in part (a), solve

$$
\begin{aligned}
\rho c_p \frac{\partial T}{\partial t} &= k\left(\frac{\partial^2 T}{\partial x^2} + \frac{\partial^2 T}{\partial y^2}\right) + Q \\
\frac{\partial T}{\partial n}(0,y,t) = \frac{\partial T}{\partial n}&(0.024,y,t) = \frac{\partial T}{\partial n}(x,0,t) = 0 \\
hT(x,0.006,t) + k\frac{\partial T}{\partial n}&(x,0.006,t) = hT_\infty \\
T(x,y,0) &= T_\infty .
\end{aligned}
$$

This initial boundary value problem models heat conduction in a ceramic plate with internal heat generation. Take the following values for the parameters: $\rho = 2320$ kg/m^3, $c_p = 835$ J/kg · K, $k = 2$ W/m · K, $Q = 50$ W/m^3, $h = 100$ W/m^2· K, and $T_\infty = 30°$C. Determine the temperature distribution after 60 seconds, 120 seconds, 180 seconds, and 240 seconds.

CHAPTER 11

Hyperbolic Equations and the Convection-Diffusion Equation

AN OVERVIEW

Fundamental Mathematical Problem

The problems presented below, and throughout the remainder of the chapter, all involve either the transport of some material property with a flow or the analysis of some wavelike phenomena. From these physical problems, three fundamental mathematical problems can be identified:

The advection (or convection) equation

$$\frac{\partial u}{\partial t} + a(x,t,u)\frac{\partial u}{\partial x} = g(x,t,u)$$

The convection-diffusion equation

$$\frac{\partial u}{\partial t} + v(x,t)\frac{\partial u}{\partial x} = D\frac{\partial^2 u}{\partial x^2}$$

The wave equation

$$\frac{\partial^2 u}{\partial t^2} = c^2\frac{\partial^2 u}{\partial x^2}$$

The advection equation and the wave equation are the canonical examples of *hyperbolic* partial differential equations. Numerical methods for all three of these model problems will be developed in this chapter.

Pollution Transported by Groundwater Flow

Pollution seeps into the ground from a chemical plant and is then transported by groundwater flow. Once in the ground, the pollutant breaks down naturally, but very slowly. Let $C(x,t)$ denote the concentration (mass per unit volume) of pollutant in the ground x meters from the seepage site and t days after the initial contamination. To model the evolution of the concentration profile, consider the control volume shown below, which has length Δx and cross-sectional area A perpendicular to the direction of groundwater flow.

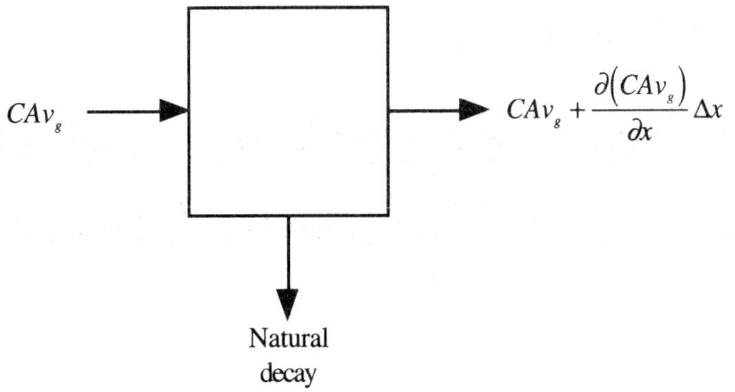

The total mass of pollutant within the control volume is $CA\,\Delta x$. Due to groundwater flow, pollutant enters and exits the control volume at the rate

$$CAv_g \quad \text{and} \quad CAv_g + \frac{\partial(CAv_g)}{\partial x}\Delta x,$$

respectively, where v_g is the velocity of the groundwater. The rate of natural breakdown is proportional to the amount of pollutant present, with constant of proportionality α. Applying the principle of conservation of mass to the control volume, it follows that

$$\frac{\partial(CA\,\Delta x)}{\partial t} = CAv_g - CAv_g - \frac{\partial(CAv_g)}{\partial x}\Delta x - \alpha CA\,\Delta x.$$

If A and v_g are constant, this simplifies to

$$\frac{\partial C}{\partial t} + v_g\frac{\partial C}{\partial x} = -\alpha C. \tag{1}$$

We will assume that the ground is initially pollutant free and that the concentration at the seepage site is constant; that is,

$$C(x,0) = 0 \quad \text{and} \quad C(0,t) = C_0 \tag{2}$$

for some constant C_0. If we now divide (1) and (2) by C_0 and define the normalized pollutant concentration

$$c(x,t) = \frac{C(x,t)}{C_0},$$

we arrive at our final model: the initial boundary value problem

$$\frac{\partial c}{\partial t} + v_g\frac{\partial c}{\partial x} = -\alpha c, \qquad c(x,0) = 0, \quad c(0,t) = 1. \tag{3}$$

Traffic Flow

Traffic on a one-lane highway through the mountains is stuck behind a slow moving truck. Eventually, the truck turns off the highway, and the vehicles which were behind the truck begin to spread out. We would like to model the evolution of the traffic pattern, which is described by the vehicle density, $\rho(x,t)$. Note that ρ is measured in cars/mile.

Let $J(x,t)$ denote the rate at which vehicles pass the arbitrary location x at time t. The function J is called the flux density or the flow rate. Requiring that the time rate of change of the number of vehicles along a representative section of the highway be equal to the net flux of vehicles into that section leads to the relation

$$\frac{\partial \rho}{\partial t} = -\frac{\partial J}{\partial x}. \tag{4}$$

To proceed further, we must specify a form for the flux density.

The simplest form we can choose for the flux density is

$$J = v(x,t)\rho(x,t),$$

where v is the traffic velocity. Using this expression for J implies that drivers react only to local conditions; however, drivers generally pay attention to the traffic ahead of them. All other things being equal, it therefore seems reasonable to expect that the flux density will be lower when travelling into a region of higher density and higher when travelling into a region of lower density. Accordingly, we include the term $-D\frac{\partial \rho}{\partial x}$ into our flux density function, where D is some positive constant. Thus, we take

$$J = v\rho - D\frac{\partial \rho}{\partial x}. \tag{5}$$

Substituting (5) into (4) yields

$$\frac{\partial \rho}{\partial t} + \frac{\partial (v\rho)}{\partial x} = D\frac{\partial^2 \rho}{\partial x^2}.$$

Empirical evidence suggests that traffic velocity is a function of vehicle density. Consequently, we will assume that $v = v(\rho)$ and define the function $f(\rho) = v(\rho)\rho$. Our final model for the vehicle density is then the convection-diffusion equation

$$\frac{\partial \rho}{\partial t} + \frac{\partial f(\rho)}{\partial x} = D\frac{\partial^2 \rho}{\partial x^2}. \tag{6}$$

The Transmission Line Equations

A simple transmission line consists of a cable stretched over an extended distance. When grounded, the cable acts much like an electric circuit with inductance per unit length l and capacitance per unit length c. We will also assume that the cable has resistance per unit length r and experiences current leakage proportional to the voltage, with constant of proportionality (per unit length) g. Our objective is to model the voltage, $V(x,t)$, and the current, $I(x,t)$, along the length of the transmission line.

Consider a representative section of the cable of length Δx. The resistance of the cable produces a voltage drop along this section of $Ir\,\Delta x$, while the inductance of the cable produces a voltage drop of $l\,\Delta x\frac{\partial I}{\partial t}$. Therefore,

$$V(x+\Delta x) = V(x) - Ir\,\Delta x - l\,\Delta x\frac{\partial I}{\partial t}.$$

Bringing all of the terms to the left side of the equation, dividing through by Δx and taking the limit as $\Delta x \to 0$, we obtain

$$l\frac{\partial I}{\partial t} + \frac{\partial V}{\partial x} + rI = 0. \tag{7}$$

In terms of the current flowing along this section of the cable, leakage produces a drop of $g\,\Delta\,x\,V$. The capacitance of the cable produces an additional drop of $c\,\Delta\,x\frac{\partial V}{\partial t}$. Thus,

$$I(x+\Delta x) = I(x) - g\,\Delta x\,V - c\,\Delta x\frac{\partial V}{\partial t},$$

which leads to

$$c\frac{\partial V}{\partial t} + \frac{\partial I}{\partial x} + gV = 0. \tag{8}$$

Equations (7) and (8) form a system of advection equations for determining $V(x,t)$ and $I(x,t)$.

If the resistance of the cable and current leakage are negligible, then the equations for the current and voltage can be decoupled as follows. First, set $r = g = 0$. Next, differentiate (7) with respect to t, multiply the resulting equation by c and subtract the result of differentiating (8) with respect to x. These steps yield

$$lc\frac{\partial^2 I}{\partial t^2} - \frac{\partial^2 I}{\partial x^2} = 0. \tag{9}$$

On the other hand, differentiating (8) with respect to t, multiplying the resulting equation by l and subtracting the result of differentiating (7) with respect to x yields

$$lc\frac{\partial^2 V}{\partial t^2} - \frac{\partial^2 V}{\partial x^2} = 0. \tag{10}$$

Thus, when resistance and leakage can be neglected, current and voltage along the transmission line satisfy the separate wave equations (9) and (10).

Characteristics

In keeping with previous chapters, we will work exclusively with finite difference methods. No treatment of hyperbolic equations is complete, however, without at least some discussion of characteristics. Consider the advection equation

$$\frac{\partial u}{\partial t} + a(x,t,u)\frac{\partial u}{\partial x} = g(x,t,u).$$

Suppose we can find a family of curves in the x-t plane, $x(t)$, for which

$$\frac{dx}{dt} = a(x, t, u).$$

The curves defined by this ordinary differential equation are called the *characteristics* of the partial differential equation. Along any characteristic, the solution to the partial differential equation, $u(x, t)$, satisfies

$$\frac{\partial u}{\partial t} + \frac{\partial u}{\partial x}\frac{dx}{dt} = g(x, t, u), \quad \text{or} \quad \frac{du}{dt} = g(x, t, u).$$

If the value of the solution is specified along any curve, C, which is not a characteristic, then the value of the solution away from C can be approximated by solving the system of initial value problems

$$\begin{aligned} x' &= a(x, t, u), & x(t_0) &= x_0 \\ u' &= g(x, t, u), & u(x_0, t_0) &= u_0 \end{aligned}$$

for a collection of points $(x_0, t_0) \in C$. This technique is known as the *method of characteristics.*

For a more detailed discussion of characteristics, in particular their use in determining analytical solutions to hyperbolic partial differential equations, consult any standard partial differential equations text. Morton and Mayers [1] and Smith [2] present a more detailed development of the method of characteristics as a numerical technique, with worked examples contained in the latter reference. Gerald and Wheatley [3] discuss the use of characteristics for the wave equation.

Remainder of the Chapter

The material in this chapter is organized as follows. In Section 11.1, the upwind finite difference method for the advection equation is presented. A detailed analysis of the stability and the error of this method is presented. A second scheme for approximating the solution of the advection equation, the MacCormack method, is presented in Section 11.2. Techniques for the convection-diffusion equation are treated in Section 11.3. The final section addresses the wave equation.

References

1. K. W. Morton and D. F. Mayers, *Numerical Solution of Partial Differential Equations*, Cambridge University Press, Cambridge, 1994.

2. G. D. Smith, *Numerical Solution of Partial Differential Equations: Finite Difference Methods*, 3rd edition, Oxford Applied Mathematics and Computing Science Series, Oxford University Press, Oxford, 1985.

3. C. F. Gerald and P. O. Wheatley, *Applied Numerical Analysis*, 5th edition, Addison-Wesley Publishing Company, Reading, 1994.

11.1 ADVECTION EQUATION, I: UPWIND DIFFERENCING

The advection equation arises naturally as a model for conservation laws. Suppose a fluid is flowing in the x-direction. Let $u(x,t)$ denote the time-dependent distribution of some physical property of the fluid (such as density, temperature or momentum) or the concentration of a substance dissolved in the fluid. Consider an infinitesimal control volume. Balancing the flows into and out from the volume, as well as any sources and sinks, with the time rate of change of $u(x,t)$ within the volume leads to

$$\frac{\partial u}{\partial t} + \frac{\partial (au)}{\partial x} = g(x,t,u). \tag{1}$$

Here, $a(x,t,u)$ is the velocity of the flow, and $g(x,t,u)$ represents the source and sink information. An alternative form for a conservation law is

$$\frac{\partial u}{\partial t} + \frac{\partial f(u)}{\partial x} = g(x,t,u), \tag{2}$$

where $f(u)$ does not depend explicitly on either of the independent variables and represents the flux of the quantity u.

The objective of this section is to develop the upwind finite difference method for the advection equation. We will start with a form that is simpler than either of equations (1) or (2) and then generalize. A detailed analysis of the stability and the error of the upwind scheme will also be presented. Toward the end of the section, the application of the upwind scheme to systems of advection equations will be briefly discussed. A second finite difference method for the advection equation will be presented in the next section.

The Basic Method

Consider the advection equation

$$\frac{\partial u}{\partial t} + a(x,t,u)\frac{\partial u}{\partial x} = g(x,t,u). \tag{3}$$

Let Δx denote the uniform spacing in the x-direction, and suppose that $t_{n+1} = t_n + \Delta t$. To approximate the solution at time level $t = t_{n+1}$ knowing the values at time level $t = t_n$, first evaluate equation (3) at (x_j, t_n). Next, replace the time derivative with its first-order forward difference approximation. To handle the space derivative, we will take our cue from the characteristics of the partial differential equation. If $a_j^{(n)}$ is positive, then the characteristic through (x_j, t_n) travels from left to right. This implies that information about the solution is transported from left to right and suggests the use of a backward difference formula for $\partial u/\partial x$. The resulting finite difference equation is

$$w_j^{(n+1)} - w_j^{(n)} + \lambda(w_j^{(n)} - w_{j-1}^{(n)}) = \Delta t\, g_j^{(n)}, \tag{4}$$

where $\lambda = a_j^{(n)}\Delta t/\Delta x$. By a similar thought process, if $a_j^{(n)}$ is negative, then the characteristic through (x_j, t_n) travels from right to left and suggests the use of a forward difference formula for $\partial u/\partial x$.

$$w_j^{(n+1)} - w_j^{(n)} + \lambda(w_{j+1}^{(n)} - w_j^{(n)}) = \Delta t\, g_j^{(n)} \tag{5}$$

is the corresponding finite difference equation. Combining equations (4) and (5) yields the *upwind finite difference method:*

$$w_j^{(n+1)} = \begin{cases} (1-\lambda)w_j^{(n)} + \lambda w_{j-1}^{(n)} + \Delta t\, g_j^{(n)}, & a_j^{(n)} > 0 \\ (1+\lambda)w_j^{(n)} - \lambda w_{j+1}^{(n)} + \Delta t\, g_j^{(n)}, & a_j^{(n)} < 0 \end{cases}. \tag{6}$$

EXAMPLE 11.1 A Simple Advection Equation

Consider the advection equation

$$\frac{\partial u}{\partial t} + 2t\frac{\partial u}{\partial x} = 0, \qquad -\infty < x < \infty, t > 0$$

subject to the initial condition

$$u(x,0) = u_0(x) = \begin{cases} 1, & -1 \le x \le 0 \\ 0, & \text{elsewhere.} \end{cases}$$

Comparing the current problem with equation (3), we see that $a(x,t,u) = 2t$ and $g(x,t,u) = 0$. Since the domain consists of $t > 0$, it follows that $a_j^{(n)}$ is always positive and the characteristics always travel from left to right. Hence, the upwind difference method will always use a backward difference for the space derivative.

Since we cannot compute the approximate solution over the entire real line, we will have to truncate the domain to the closed interval $[A, B]$ for some constants A and B. For the left endpoint of the domain, note that the initial data is zero for all $x \le -1$ and, as stated earlier, the characteristics all travel to the right. Hence, it would be appropriate to choose any $A \le -1$ and introduce the numerical boundary condition $u(A,t) = 0$. All calculations below were performed with $A = -2$. The appropriate selection of B depends upon how far forward in time the solution is desired—we must make sure that the solution does not interact with the artificial boundary. For this demonstration problem, we will advance the solution to $t = 3$, and $B = 11$ is found to be sufficiently large to be beyond the leading edge of the solution. No numerical boundary condition is needed at this point because we always use a backward difference formula.

Figure 11.1 displays the approximate solution at times $t = 1$, $t = 2$, and $t = 3$ (the solid curves) computed using the upwind difference scheme. The top graph was computed with $\Delta x = 1/20$ and $\Delta t = 1/200$, while the bottom graph was computed with $\Delta x = 1/40$ and $\Delta t = 1/400$. The exact solution of the partial differential equation, $u(x,t) = u_0(x - t^2)$, is indicated by the dotted curves in each graph. The upwind scheme propagates the initial condition to the right at roughly the correct speed: the center of mass of the true solution is at 0.5, 3.5, and 8.5 at times $t = 1$, $t = 2$, and $t = 3$, respectively, while the center of mass for the

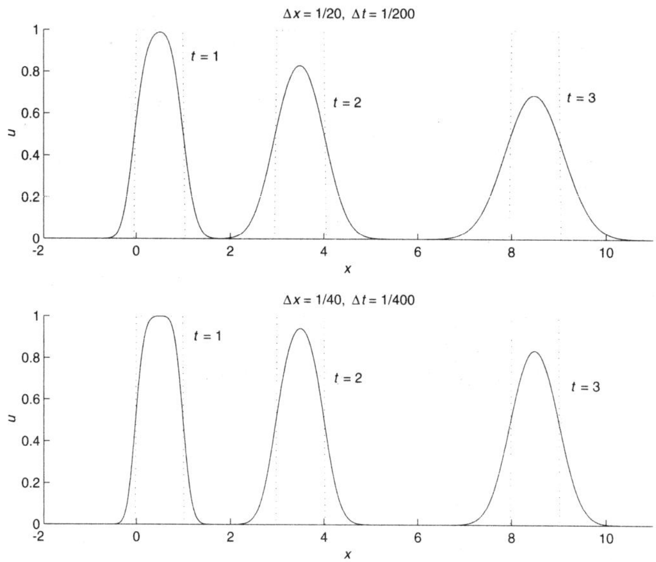

Figure 11.1 Solution of an advection equation at times $t = 1$, $t = 2$, and $t = 3$. Approximate solutions, computed with the upwind difference method, are indicated by the solid curves, and the exact solutions are indicated by the dotted curves.

corresponding approximate solutions is at 0.495, 3.49, and 8.48497. However, the upwind scheme also rounds over the corners of the original square wave, continually reduces the amplitude and spreads out the support of the pulse.

Increasing the resolution (that is, reducing the step sizes) improves the accuracy of the approximation to the extent one would expect from a first-order method. The amplitude errors have roughly been cut in half, as have the errors in the location of the center of mass. In particular, the centers of the approximate solutions in the bottom graph, from left to right, are located at 0.4975, 3.495, and 8.4925. There is also less spreading of the solution with the smaller step sizes.

Analysis of Upwind Differencing

Let's start with a stability analysis. In 1928, Courant, Friedrichs, and Lewy [1] developed a necessary condition for the stability of any difference scheme for the advection equation. The condition, which has become known as the CFL condition, states that the domain of dependence of the partial differential equation, as determined by the characteristic that passes through the point (x_j, t_{n+1}), must be contained in the domain of dependence of the finite difference equation used to compute the value at (x_j, t_{n+1}). For a computational grid with spacings of Δt in time and Δx in space, and assuming a is constant, the characteristic through (x_j, t_{n+1}) intersects the line $t = t_n$ at $x = x_j - a\,\Delta t$. For this to remain inside the domain of dependence of the finite difference equation of the upwind scheme, we must have $|a\,\Delta t| \le \Delta x$, or $|\lambda| \le 1$. This condition has an interesting physical interpretation: it dictates that once Δx has been selected, Δt must be chosen small enough so that material is not transported more than one grid space in a single time step.

To determine whether the condition $|\lambda| \le 1$ is sufficient for the upwind scheme to be stable, we perform a von Neumann stability analysis. Here, we will take the discrete Fourier mode to be of the form $w_j^{(n)} = r^n e^{ij(k\,\Delta x)}$, where k is the wavenumber. Representing the frequency of the mode in terms of the wavenumber and the grid spacing will simplify the error analysis that is presented below. Let's start with the case of $a_j^{(n)}$ being positive. For the time being, we will assume that g does not depend on u. This implies that the term $\Delta t\, g_j^{(n)}$ in equation (4) will have no effect on stability, so can be set to zero. When, in the worked examples, g does depend on u, we will reexamine the stability condition on a case-by-case basis. Substituting the Fourier mode into equation (4) produces

$$r - 1 + \lambda(1 - e^{-ik\,\Delta x}) = 0,$$

or

$$r = 1 - \lambda + \lambda e^{-ik\,\Delta x} = [1 - \lambda + \lambda\cos(k\,\Delta x)] - i\lambda\sin(k\,\Delta x).$$

After some algebraic manipulation, we arrive at

$$r\bar{r} = 1 - 2\lambda(1-\lambda)\,[1 - \cos(k\,\Delta x)] = 1 - 4\lambda(1-\lambda)\sin^2(k\,\Delta x)/2, \tag{7}$$

which is less than or equal to 1 provided $\lambda \le 1$. Similar analysis for the case when $a_j^{(n)}$ is negative leads to the requirement $\lambda \ge -1$ (recall that λ is negative when $a_j^{(n)}$ is negative). Thus $|\lambda| \le 1$ is necessary and sufficient for stability. Recalling the definition of λ, the restriction on the time step is

$$\Delta t \le \frac{\Delta x}{\max_j \left|a_j^{(n)}\right|}.$$

If we carry our Fourier analysis a little further, we can begin to quantify the errors introduced by the upwind scheme. Consider the function $f(x,t) = e^{i(kx+\omega t)}$. This function is a solution of equation (3)—with $a(x,t,u)$ constant and $g(x,t,u) = 0$—provided $\omega = -ak$. In one time step, the amplitude of f remains constant, and the phase changes by

$$\omega\,\Delta t = -ak\,\Delta t = -\lambda k\,\Delta x.$$

In contrast, equation (7) indicates that, for all $\lambda \ne 1$, the discrete Fourier solution to the finite difference equations experiences a reduction in amplitude on the order of $(k\,\Delta x)^2$ in a single time step. This explains both the amplitude decay and the rounding of the sharp corners (higher-frequency components experience a larger reduction in amplitude) observed in Figure 11.1. To account for the discrepancy in the speed of propagation, we examine the phase shift of the numerical mode, which is given by

$$\arctan\frac{\operatorname{Im} r}{\operatorname{Re} r} = -\arctan\frac{\lambda\sin(k\,\Delta x)}{1-\lambda+\lambda\cos(k\,\Delta x)}$$
$$\sim -\lambda k\,\Delta x\left[1-\frac{1}{6}(1-\lambda)(1-2\lambda)(k\,\Delta x)^2+\cdots\right].$$

Note this is also in error by an amount on the order of $(k\,\Delta x)^2$ in a single time step.

To explain the spreading out of the solution, we will turn to an analysis tool known as the *modified equation.* Essentially, this is the partial differential equation which is actually being solved when the numerical method is applied. We will continue to assume that $a(x,t,u)$ is constant, and, without loss of generality, will further assume that a is positive. The upwind difference equation is then

$$\frac{w_j^{(n+1)}-w_j^{(n)}}{\Delta t}+a\frac{w_j^{(n)}-w_{j-1}^{(n)}}{\Delta x}=g_j^{(n)}. \tag{8}$$

Substitute the true solution to the original partial differential equation, $u(x,t)$, into equation (8), and expand the terms for $u_j^{(n+1)}$ and $u_{j-1}^{(n)}$ in Taylor series about (x_j,t_n):

$$u_j^{(n+1)} = u_j^{(n)}+\Delta t\left.\frac{\partial u}{\partial t}\right|_{(x_j,t_n)}+\frac{1}{2}(\Delta t)^2\left.\frac{\partial^2 u}{\partial t^2}\right|_{(x_j,t_n)}+\cdots$$
$$u_{j-1}^{(n)} = u_j^{(n)}-\Delta x\left.\frac{\partial u}{\partial x}\right|_{(x_j,t_n)}+\frac{1}{2}(\Delta x)^2\left.\frac{\partial^2 u}{\partial x^2}\right|_{(x_j,t_n)}-+\cdots$$

This procedure, upon dropping the notation for evaluation at (x_j,t_n), yields

$$\frac{\partial u}{\partial t}+a\frac{\partial u}{\partial x}+\frac{1}{2}\Delta t\frac{\partial^2 u}{\partial t^2}-\frac{1}{2}a\,\Delta x\frac{\partial^2 u}{\partial x^2}+\cdots=g.$$

The second derivative in time can be replaced by a second derivative in space as follows:

$$\frac{\partial^2 u}{\partial t^2}=\frac{\partial}{\partial t}\left(\frac{\partial u}{\partial t}\right)=\frac{\partial}{\partial t}\left(a\frac{\partial u}{\partial x}\right)$$
$$=\frac{\partial}{\partial x}\left(a\frac{\partial u}{\partial t}\right)=\frac{\partial}{\partial x}\left(a^2\frac{\partial u}{\partial x}\right)=a^2\frac{\partial^2 u}{\partial x^2}.$$

Hence, the modified equation corresponding to the upwind scheme is

$$\frac{\partial u}{\partial t}+a\frac{\partial u}{\partial x}=g+\frac{1}{2}a\,\Delta x(1-\lambda)\frac{\partial^2 u}{\partial x^2}+\cdots.$$

The second derivative term introduces *numerical* (or *false*) *diffusion* and causes the spreading of the solution in Figure 11.1.

More General Formulations of the Advection Equation

What about the more general formulations of the advection equation given by equations (1) and (2)? First, we will consider equation (1):

$$\frac{\partial u}{\partial t} + \frac{\partial (au)}{\partial x} = g.$$

Expanding the space derivative

$$\frac{\partial (au)}{\partial x} = a\frac{\partial u}{\partial x} + u\frac{\partial a}{\partial x},$$

we see that the sign of a again determines the direction in which the characteristics travel. Thus, if $a_j^{(n)} > 0$, we backward difference the space derivative and the resulting finite difference equation becomes

$$w_j^{(n+1)} = w_j^{(n)} + \frac{\Delta t}{\Delta x}\left(a_{j-1}^{(n)}w_{j-1}^{(n)} - a_j^{(n)}w_j^{(n)}\right) + \Delta t\, g_j^{(n)}. \tag{9}$$

When $a_j^{(n)} < 0$, we forward difference the space derivative, leading to

$$w_j^{(n+1)} = w_j^{(n)} + \frac{\Delta t}{\Delta x}\left(a_j^{(n)}w_j^{(n)} - a_{j+1}^{(n)}w_{j+1}^{(n)}\right) + \Delta t\, g_j^{(n)}. \tag{10}$$

For advection equations of the form

$$\frac{\partial u}{\partial t} + \frac{\partial f(u)}{\partial x} = g,$$

we proceed in a similar manner. Applying the chain rule to the space derivative, we find

$$\frac{\partial f(u)}{\partial x} = \frac{\partial f}{\partial u}\frac{\partial u}{\partial x};$$

hence, the direction in which characteristics travel is determined by the sign of $\partial f/\partial u$. The resulting upwind finite difference scheme is then given by

$$w_j^{(n+1)} = \begin{cases} w_j^{(n)} + \frac{\Delta t}{\Delta x}\left[f(w_{j-1}^{(n)}) - f(w_j^{(n)})\right] + \Delta t\, g_j^{(n)}, & \partial f/\partial u > 0 \\ w_j^{(n)} + \frac{\Delta t}{\Delta x}\left[f(w_j^{(n)}) - f(w_{j+1}^{(n)})\right] + \Delta t\, g_j^{(n)}, & \partial f/\partial u < 0 \end{cases}. \tag{11}$$

EXAMPLE 11.2 An Advection Equation of the Form (1)

Consider the partial differential equation

$$\frac{\partial u}{\partial t} - \frac{\partial (xu)}{\partial x} = 0, \quad 0 \le x \le 1, t > 0$$

subject to the initial and boundary conditions

$$u(x,0) = \frac{x}{1+x} \quad \text{and} \quad u(1,t) = \frac{e^{2t}}{1+e^t}.$$

Note that $a(x,t,u) = -x$ and $g(x,t,u) = 0$. No boundary condition has been specified along $x = 0$ because $x = 0$ happens to be a characteristic of the partial differential equation. Consequently, we are not free to specify arbitrary values for $u(0,t)$. Rather, applying the method of characteristics, we find that $u(0,t)$ must satisfy the initial value problem

$$\frac{du}{dt} = u, \quad u(0,0) = 0.$$

Hence, the advection equation together with the specified initial condition determine $u(0,t) = 0$ for all t.

Figure 11.2 displays the approximate solution to this problem at $t = 1$ and $t = 2$ (the solid curves). The dotted curves correspond to the exact solution:

$$u(x,t) = \frac{xe^{2t}}{1+xe^t}.$$

A mesh spacing of $\Delta x = 1/20$ was selected, and the time step was chosen as the maximum allowed by stability considerations:

$$\Delta t = \Delta t_{\text{max}} = \frac{\Delta x}{\max_j |a_j^{(n)}|} = \frac{\Delta x}{\max_{x\in[0,1]} x} = \frac{1/20}{1} = \frac{1}{20}.$$

EXAMPLE 11.3 An Advection Equation of the Form (2)

Next, consider the partial differential equation

$$\frac{\partial u}{\partial t} + \frac{\partial (u^2/2)}{\partial t} = 0, \quad -\infty < x < \infty, t > 0$$

subject to the initial condition

$$u(x,0) = u_0(x) = \begin{cases} 1-\cos(2\pi x), & 0 \le x \le 1 \\ 0, & \text{elsewhere.} \end{cases}$$

With $f(u) = u^2/2$, it follows that $f'(u) = u$. Figure 11.3 displays the approximate solution at various times computed with $\Delta x = 1/100$ and $\Delta t = 1/200$. The time step was selected based on a maximum propagation speed of 2, which is the largest value taken on by the initial condition function, u_0. Note that the initial profile tilts toward the right as time advances. Eventually (around $t = 0.2$), the right side of the profile becomes vertical, and a jump discontinuity is produced in the solution. This phenomena is known as a shock and arises due to the crossing of characteristics. Once the shock has formed, it propagates to the right with a slowly decreasing amplitude.

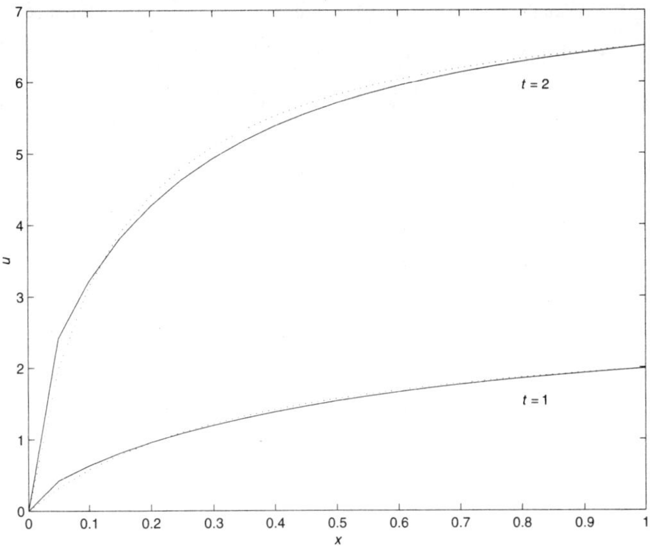

Figure 11.2 Solution of an advection equation at times $t = 1$ and $t = 2$. Approximate solutions, computed with the upwind difference method, are indicated by the solid curves, and the exact solutions are indicated by the dotted curves.

Systems of Equations

Consider the system of advection equations

$$\frac{\partial \mathbf{U}}{\partial t} + \frac{\partial \mathbf{F}(\mathbf{U})}{\partial x} = \mathbf{G},$$

where $\mathbf{U}$ is a vector of dependent variables and $\mathbf{F}(\mathbf{U})$ and $\mathbf{G}$ are vector-valued functions. Applying the chain rule to the space derivative yields

$$\frac{\partial \mathbf{U}}{\partial t} + J(\mathbf{U})\frac{\partial \mathbf{U}}{\partial x} = \mathbf{G},$$

where $J(\mathbf{U})$ is the Jacobian matrix for the function $\mathbf{F}(\mathbf{U})$. In general, it is a nontrivial task to use the upwind scheme on such a problem. For each time step and at each point of the computational grid, the eigenvalues and eigenvectors of the Jacobian matrix must be computed. The approximate solution must then be expressed as a linear combination of the eigenvectors. The appropriate finite difference formula, as determined by the sign of the corresponding eigenvalue, is next applied to each eigenvector. Finally, the eigenvectors must be recombined to form the solution at the next time step. It is only in the simplest cases, when the Jacobian is diagonal, that upwind differencing is a viable alternative for a system of equations. An example is presented below.

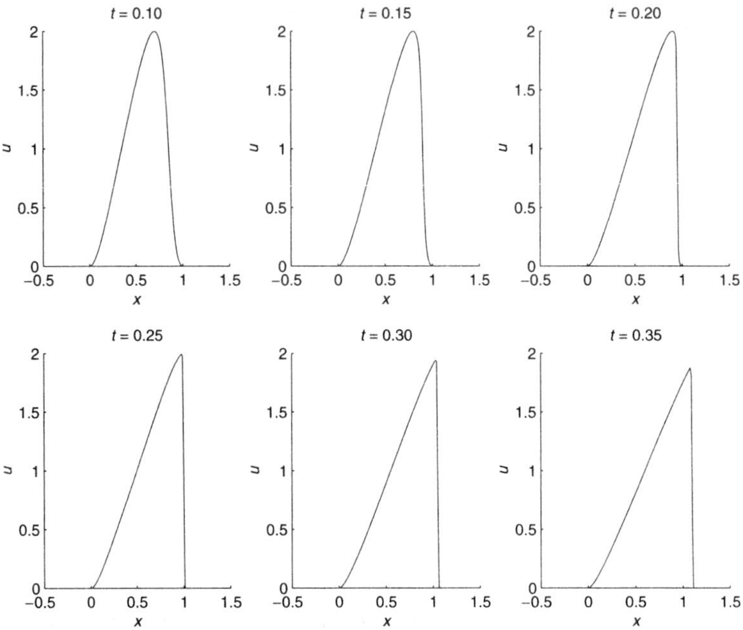

Figure 11.3 Approximate solution to the partial differential equation

$$\frac{\partial u}{\partial t} + \frac{\partial (u^2/2)}{\partial t} = 0$$

with one period of a vertically shifted cosine function as the initial condition.

Application Problem 1: Pollution Transported by Groundwater Flow

Pollution seeps into the ground from a chemical plant and is then transported by groundwater flow toward a river one kilometer away. Once in the ground, the chemical pollution breaks down naturally, but very slowly. In the Overview to this chapter (see page 883), we found that the normalized concentration of pollutant in the ground, $c(x,t)$, satisfies the initial boundary value problem

$$\frac{\partial c}{\partial t} + v_g \frac{\partial c}{\partial x} = -\alpha c, \qquad c(x,0) = 0, \quad c(0,t) = 1.$$

Here, x measures distance from the seepage site toward the river, t measures time from the initial contamination, $v_g = 10$ meters/day is the groundwater velocity, and $\alpha = 3.6 \times 10^{-3}$ day^{-1} is the natural decay rate.

Note the right-hand side of the above advection equation contains the dependent variable, so we must reexamine stability. Since $v_g = 10 > 0$, the appropriate finite difference equation is

$$w_j^{(n+1)} - w_j^{(n)} + \lambda(w_j^{(n)} - w_{j-1}^{(n)}) = -\alpha \, \Delta t \, w_j^{(n)},$$

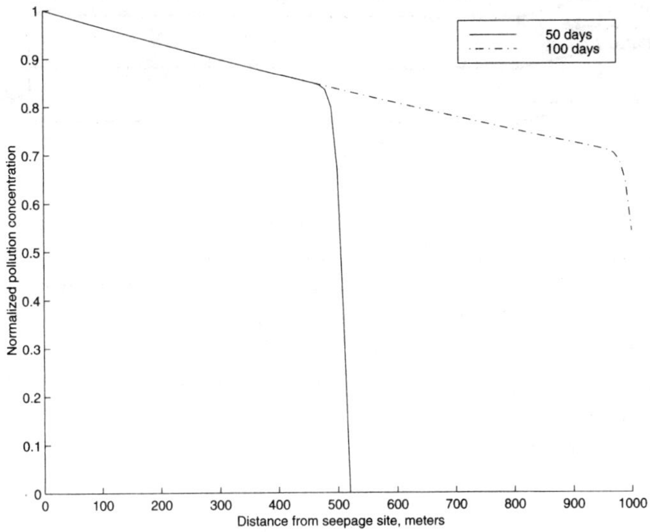

Figure 11.4 Normalized pollution concentration profile, as function of distance from seepage site, after 50 days and after 100 days.

where $\lambda = v_g \Delta t / \Delta x$. The amplification factor for this finite difference equation is

$$r = [1 - \alpha\,\Delta t - \lambda + \lambda \cos(k\,\Delta x)] - i\lambda \sin(k\,\Delta x);$$

hence, a sufficient condition for the magnitude of r to be less than or equal to one is

$$\Delta t \le \frac{\Delta x}{\alpha\,\Delta x + v_g}.$$

Figure 11.4 displays the normalized pollution concentration profile after 50 days and after 100 days. For this calculation $\Delta x = 10$ meters and $\Delta t = 50/51 \approx 0.980$ days, which is slightly below the maximum allowable time step

$$\Delta t_{\max} = 10/(0.036 + 10) \approx 0.9964.$$

Application Problem 2: Air-Air Countercurrent Heat Exchanger

A countercurrent air-air heat exchanger in a home uses the hot exhaust air from the furnace to heat cold air drawn from the outside. Within the exchanger, the two air flows travel in opposite directions, hence the designation as countercurrent. Consider the duct from the heat exchanger shown in the figure following. The duct has length L, width w, and is separated horizontally into two chambers, each of height d. Cold air flows from left to right through the upper chamber with uniform velocity v_C, while hot air flows through the lower chamber with uniform velocity v_H.

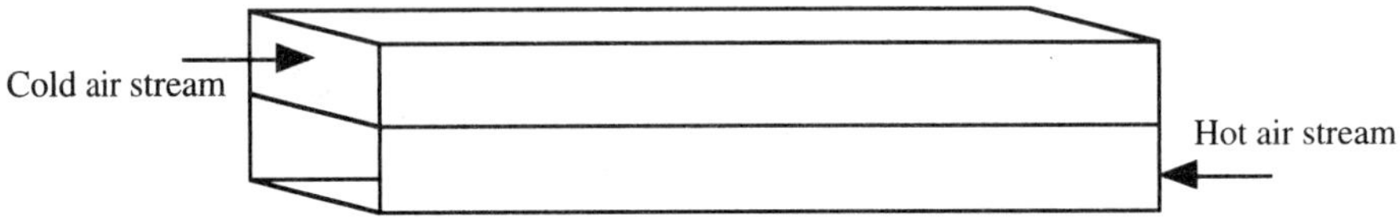

Let $C(x,t)$ and $H(x,t)$ denote the temperature of the cold air and the hot air streams, respectively. Orient the coordinate system so that $x=0$ corresponds to the cold air inlet. The heat flux (i.e., the rate at which thermal energy is transported by the air flow) at any location within the cold air stream is $\rho c_p w\, dv_C\, C$, where ρ is the mass denisty and c_p the specific heat of the air. The heat flux within the hot air is $-\rho c_p w\, dv_H\, H$. Assuming that diffusion effects are negligible and that heat is transferred from the hot air to the cold air by convection with an overall heat transfer coefficient hg, a control volume analysis leads to the system of advection equations

$$\frac{\partial C}{\partial t} + v_C \frac{\partial C}{\partial x} = \frac{h}{\rho c_p d}(H - C)$$

$$\frac{\partial H}{\partial t} - v_H \frac{\partial H}{\partial x} = \frac{h}{\rho c_p d}(C - H).$$

For simulation purposes, suppose $L = 3$ meters, $w = 0.4$ meters, and $d = 0.25$ meters. The initial temperatures are $C(x,0) = H(x,0) = 20°\text{C}$, and the boundary conditions are $C(0,t) = 5°\text{C}$ and $H(3,t) = 60°\text{C}$. For the remaining parameters, let's take $v_C = v_H = 0.1$ meters/second, $h = 20\ \text{W/m}^2 \cdot \text{K}$, $\rho = 1\ \text{kg/m}^3$, and $c_p = 1007\ \text{J/kg} \cdot \text{K}$. Though this is a system, it is clear that the $\partial C/\partial x$ term should be backward differenced and the $\partial H/\partial x$ term should be forward differenced. Figure 11.5 displays the approximate temperature profiles within the heat exchanger in ten second increments. The solid curves represent the cold air stream, and the dotted curves represent the hot air stream. All calculations were performed with $\Delta x = 1/100$ meters and $\Delta t = 1/20$ seconds.

As in the case of the groundwater pollution problem earlier, the presence of decay terms on the right-hand sides of the partial differential equations requires that the stability condition be recalculated. Following standard procedures, we find that the maximum allowable time step is given by

$$\Delta t_{\text{max}} = \frac{\Delta x}{h\Delta x/\rho c_p d + \max(v_C, v_H)}.$$

For the specified parameter values, $\Delta t \approx 0.0992$. A smaller value has been chosen to reduce the truncation error.

References

1. R. Courant, K. O. Friedrichs, and H. Lewy, "Uber die partiellen differenzengleichungen de mathematischen Physik," *Mathematische Annalen*, **100**, 32–74, 1928.

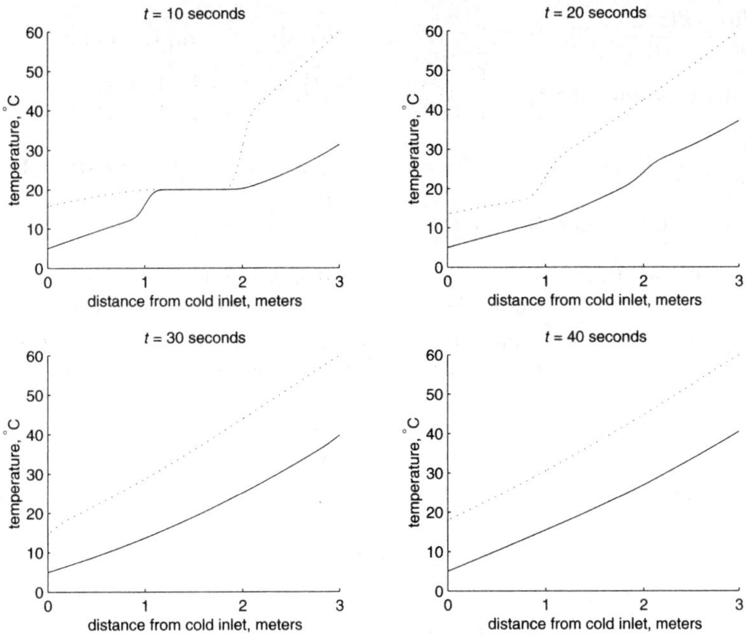

Figure 11.5 Temperature profiles of the cold (solid curves) and hot (dotted curves) air streams in an air-air countercurrent heat exchanger.

2. R. A. Gentry, R. E. Martin, and B. J. Daly, "An Eulerian Differencing Method for Unsteady Compressible Flow Problems," *Journal of Computational Physics*, **1**, 87–118, 1966.

EXERCISES

In Exercises 1–4, apply the upwind scheme to the given advection equation. Truncate the domain and introduce numerical boundary conditions where appropriate. Compare the approximate solution to the indicated exact solution.

1. $$\frac{\partial u}{\partial t} + \frac{1+x^2}{(1+x^2)^2 + 2tx}\frac{\partial u}{\partial x} = 0, \quad -\infty < x < \infty,\ t > 0,$$

$$u(x,0) = u_0(x) = \begin{cases} 1, & 0.5 < x < 1 \\ 0, & \text{elsewhere} \end{cases}$$

exact solution: $u(x,t) = u_0\left(x - \dfrac{t}{1+x^2}\right)$

2. $$\frac{\partial u}{\partial t} + (x+t)\frac{\partial u}{\partial x} = -1, \quad x > 0, t > 0, \quad u(x,0) = u_0(x) = \begin{cases} x, & 0 < x < 1 \\ 2-x, & 1 < x < 2 \\ 0, & \text{elsewhere} \end{cases}$$

exact solution: $u(x,t) = \begin{cases} u_0\left[e^{-t}(x+1+t) - 1\right] - t, & x > e^t - 1 - t \\ 0, & x \le e^t - 1 - t \end{cases}$

3. $\dfrac{\partial u}{\partial t} - \dfrac{\partial (x^2 u)}{\partial x} = 0, \quad 0 < x < 4, t > 0, \quad u(x,0) = 0, \quad u(4,t) = t^2$

exact solution: $u(x,t) = \begin{cases} \frac{16}{x^2}\left(\frac{1}{4} - \frac{1}{x} + t\right)^2, & x > 4/(4t+1) \\ 0, & x \le 4/(4t+1) \end{cases}$

4. $\dfrac{\partial u}{\partial t} + \dfrac{\partial}{\partial x}\left[(1-x^2)u\right] = -2xu, \quad -1 < x < 1, t > 0, \quad u(x,0) = 1 + \cos(\pi x)$

exact solution: $u(x,t) = u_0\left(\dfrac{(1+x)e^{-2t} - (1-x)}{1 - x + (1+x)e^{-2t}}\right)$

5. Repeat the "Pollution Transported by Groundwater Flow" application problem with $\Delta t = 1$ day. Observe the instability in the resulting solution.

6. Repeat the "Air-Air Countercurrent Heat Exchanger" application problem with $\Delta t = 0.1$ seconds. Observe the instability in the resulting solution.

7. Show that the restriction

$$\Delta t \le \frac{\Delta x}{\alpha \Delta x + v_g}$$

is sufficient to guarantee that the magnitude of r is less than or equal to one, where

$$r = [1 - \alpha\,\Delta t - \lambda + \lambda\cos(k\,\Delta x)] - i\lambda\sin(k\,\Delta x)$$

and $\lambda = v_g \Delta t/\Delta x$.

8. Consider the advection equation

$$\frac{\partial u}{\partial t} + a\frac{\partial u}{\partial x} = g,$$

where both a and g are constant.

(a) Derive a finite difference method for this problem using a first-order forward difference for the time derivative and a second-order central difference for the space derivative.

(b) What restriction on Δt is required for the method developed in part (a) to satisfy the CFL condition?

(c) Use von Neumann stability analysis to show that the method developed in part (a) is unconditionally unstable.

9. In a particular heat exchanger, steam is used to heat water as it flows through a four meter long pipe, also known as a heat exchanger tube. The steam maintains a constant temperature, $T_s = 100°$C, along the tube. The time-dependent temperature distribution of the fluid, $T(x,t)$, satisfies

$$\frac{\partial T}{\partial t} + v\frac{\partial T}{\partial x} = \frac{2h}{\rho c_p r}(T_s - T),$$

where it has been assumed that there is perfect thermal coupling between the fluid and the heat exchanger tube. Here, $v = 0.1$ m/s is the velocity of the water in the tube, $h = 1000$ W/m$^2 \cdot °$C is the heat transfer coefficient, $\rho = 989$ kg/m^3 is the density of the water, $c_p = 4180$ J/kg $\cdot °$C is the heat capacity of the water, and $r = 30$ mm is the radius of the tube. Initially, the inlet temperature of the water is $T(0,t) = 25°$C and the temperature distribution along the tube is

$$T(x,0) = 100 - 75\exp\left(-\frac{2hx}{\rho c_p r v}\right).$$

If the inlet water temperature is abruptly increased to $T(0,t) = 35°\text{C}$, approximate $T(x,t)$ at 10 second intervals from $t = 0$ through $t = 100$.

10. A 2-meter-long air-air heat exchanger in a home uses the hot exhaust air from the furnace to heat cold air drawn from the outside. Within the exchanger, the two air flows travel in the same direction. Let $C(x,t)$ and $H(x,t)$ denote the temperature of the incoming cold air and the exhausted hot air, respectively. These functions satisfy the system of equations

$$\frac{\partial C}{\partial t} + v_C \frac{\partial C}{\partial x} = \frac{h}{\rho c_p d}(H - C)$$

$$\frac{\partial H}{\partial t} + v_H \frac{\partial H}{\partial x} = \frac{h}{\rho c_p d}(C - H)$$

subject to the initial condition $C(x,0) = H(x,0) = 20°\text{C}$ and the boundary conditions $C(0,t) = 5°\text{C}$ and $H(0,t) = 60°\text{C}$. $v_C = v_H = 0.1$ meters/second is the velocity of the air through the heat exchanger, $h = 40\ \text{W/m}^2\cdot\text{K}$ is the overall heat transfer coefficient between the two air streams, $\rho = 1\ \text{kg/m}^3$ is the density, and $c_p = 1007\ \text{J/kg}\cdot\text{K}$ the heat capacity of the air, and $d = 0.25$ meters is the height of each chamber of the heat exchanger. Approximate the temperature profiles in each air stream at five second intervals from $t = 0$ through $t = 25$.

Exercises 11–13 deal with the "Donor Cell" method [2], which is a variation on the upwind scheme. For the form of the advection equation given by (1), the corresponding finite difference equation is

$$w_j^{(n+1)} = w_j^{(n)} + \frac{\Delta t}{\Delta x}(a_L w_L - a_R w_R) + \Delta t\, g_j^{(n)},$$

where

$$a_L = \frac{a_j^{(n)} + a_{j-1}^{(n)}}{2}, \quad a_R = \frac{a_j^{(n)} + a_{j+1}^{(n)}}{2}, \quad w_L = \begin{cases} w_{j-1}^{(n)}, & a_L > 0 \\ w_j^{(n)}, & a_L < 0 \end{cases}$$

and

$$w_R = \begin{cases} w_j^{(n)}, & a_R > 0 \\ w_{j+1}^{(n)}, & a_R < 0. \end{cases}$$

11. If a is constant, show that the donor cell method reduces to the upwind scheme.

12. Apply the donor cell method to the initial value problem in Exercise 3 and compare the approximate solution to that obtained using the upwind scheme.

13. Repeat Exercise 12 for the initial value problem in Exercise 4.

11.2 ADVECTION EQUATION, II: MACCORMACK METHOD

Though the upwind scheme has the benefit of being based on the analytical concept of characteristics, the method is only first-order accurate and difficult to implement for systems of equations. In this section, the MacCormack method for approximating the solution of the advection equation will be presented. This is a second-order method that is straightforward to extend from a single equation to systems of equations.

MacCormack Method

The MacCormack method is a predictor-corrector scheme for approximating the solution of advection equations. Suppose the solution is known at time level $t = t_n$ and we want to advance to time level $t = t_{n+1}$. With the MacCormack method we first replace both $\partial u/\partial t$ and $\partial u/\partial x$ by first-order forward difference formulas to predict the value of $w_j^{(n+1)}$. Denote this predicted value by w_j^*. In the corrector step, we again use a first-order forward difference formula for $\partial u/\partial t$. The space derivative is discretized as the average of a forward difference approximation computed using values at $t = t_n$ and a backward difference approximation computed using the values obtained from the predictor. The right-hand side of the differential equation is also handled as an average of values at the old and new time levels.

To be specific, consider the scalar advection equation

$$\frac{\partial u}{\partial t} + a\frac{\partial u}{\partial x} = g.$$

The predictor step takes the form

$$\frac{w_j^* - w_j^{(n)}}{\Delta t} + a_j^{(n)}\frac{w_{j+1}^{(n)} - w_j^{(n)}}{\Delta x} = g_j^{(n)},$$

or

$$w_j^* = (1+\lambda)w_j^{(n)} - \lambda w_{j+1}^{(n)} + \Delta t g_j^{(n)}, \tag{1}$$

where $\lambda = a_j^{(n)}\Delta t/\Delta x$. The corrector step is then

$$\frac{w_j^{(n+1)} - w_j^{(n)}}{\Delta t} + \frac{1}{2}\left[a_j^{(n)}\frac{w_{j+1}^{(n)} - w_j^{(n)}}{\Delta x} + a_j^*\frac{w_j^* - w_{j-1}^*}{\Delta x}\right] = \frac{1}{2}(g_j^{(n)} + g_j^*). \tag{2}$$

a_j^* denotes the function $a(x,t,u)$ evaluated at $x = x_j$, $t = t_{n+1}$ and $u = w_j^*$; g_j^* is similarly defined. If we solve the predictor equation for $w_{j+1}^{(n)}$, substitute the result into equation (2), solve for $w_j^{(n+1)}$ and simplify, the corrector equation can be written as

$$w_j^{(n+1)} = \frac{1}{2}\left[w_j^{(n)} + w_j^* - \lambda^*(w_j^* - w_{j-1}^*)\right] + \frac{1}{2}\Delta t\, g_j^*, \tag{3}$$

where $\lambda^* = a_j^*\,\Delta t/\Delta x$.

For the more general advection equation

$$\frac{\partial u}{\partial t} + \frac{\partial (au)}{\partial x} = g,$$

the predictor step of the MacCormack method becomes

$$w_j^* = w_j^{(n)} - \frac{\Delta t}{\Delta x}\left(a_{j+1}^{(n)} w_{j+1}^{(n)} - a_j^{(n)} w_j^{(n)}\right) + \Delta t\, g_j^{(n)}. \tag{4}$$

The corresponding corrector equation can be written as

$$w_j^{(n+1)} = \frac{1}{2}\left\{w_j^{(n)} + w_j^* - \frac{\Delta t}{\Delta x}\left(a_j^* w_j^* - a_{j-1}^* w_{j-1}^*\right)\right\} + \frac{1}{2}\Delta t\, g_j^*. \tag{5}$$

To arrive at equation (5), we must perform the same manipulations that were required to obtain equation (3). Finally, for an advection equation of the form

$$\frac{\partial u}{\partial t} + \frac{\partial f(u)}{\partial x} = g,$$

the predictor and corrector equations are

$$w_j^* = w_j^{(n)} - \frac{\Delta t}{\Delta x}\left(f_{j+1}^{(n)} - f_j^{(n)}\right) + \Delta t\, g_j^{(n)} \tag{6}$$

and

$$w_j^{(n+1)} = \frac{1}{2}\left[w_j^{(n)} + w_j^* - \frac{\Delta t}{\Delta x}(f_j^* - f_{j-1}^*)\right] + \frac{1}{2}\Delta t\, g_j^*, \tag{7}$$

respectively.

Although we have used first-order finite difference formulas in both the predictor and the corrector, the contributions to the truncation error cancel and produce a scheme which is second-order in both Δt and Δx. What about stability, amplitude and phase errors and numerical diffusion? For our additional analysis of the MacCormack method, we will assume that a is constant [alternatively, that $f(u) = au$ for some constant a] and that g is identically zero. With these assumptions, upon eliminating the intermediate values w_j^* from equations (1) and (3), we find

$$\begin{aligned}
w_j^{(n+1)} &= \frac{1}{2}\Big[w_j^{(n)} + (1+\lambda)w_j^{(n)} - \lambda w_{j+1}^{(n)} \\
&\quad -\lambda\left((1+\lambda)w_j^{(n)} - \lambda w_{j+1}^{(n)} - (1+\lambda)w_{j-1}^{(n)} + \lambda w_j^{(n)}\right)\Big] \\
&= \frac{1}{2}\left[\lambda(1+\lambda)w_{j-1}^{(n)} + 2(1-\lambda^2)w_j^{(n)} + \lambda(\lambda-1)w_{j+1}^{(n)}\right] \\
&= \frac{\lambda}{2}(1+\lambda)w_{j-1}^{(n)} + (1-\lambda^2)w_j^{(n)} + \frac{\lambda}{2}(\lambda-1)w_{j+1}^{(n)}.
\end{aligned} \tag{8}$$

where $\lambda = a\,\Delta t/\Delta x$. We arrive at the same end result using equations (4) and (5) or equations (6) and (7).

Since the characteristic through (x_j, t_{n+1}) intersects $t = t_n$ at $x = x_j - a\,\Delta t$ and equation (8) uses data from x_{j-1}, x_j and x_{j+1} to compute $w_j^{(n+1)}$, we must have $|a\,\Delta t| \le \Delta x$, or $|\lambda| \le 1$, to satisfy the CFL condition. Next, we substitute the discrete Fourier mode $w_j^{(n)} = r^n e^{ij(k\Delta x)}$ into (8). The resulting formula for the amplification factor is

$$\begin{aligned} r &= 1 - \lambda^2 + \lambda^2 \cos(k\,\Delta x) - i\lambda \sin(k\,\Delta x) \\ &= 1 - 2\lambda^2 \sin^2 \frac{k\,\Delta x}{2} - i\lambda \sin(k\,\Delta x), \end{aligned}$$

from which we calculate

$$\begin{aligned} r\bar{r} &= \left[1 - 2\lambda^2 \sin^2 \frac{k\,\Delta x}{2}\right]^2 + \lambda^2 \sin^2(k\,\Delta x) \\ &= 1 - 4\lambda^2(1-\lambda^2)\sin^4 \frac{k\,\Delta x}{2}. \end{aligned} \tag{9}$$

Equation (9), together with the CFL condition, indicates that $|\lambda| \le 1$ is necessary and sufficient for stability. From (9), we also find that one time step of the MacCormack method introduces an amplitude error on the order of $(k\,\Delta x)^4$—significantly smaller than the amplitude error of the upwind scheme. One time step of the MacCormack method also changes the phase of the Fourier mode by

$$\begin{aligned} \arctan \frac{\operatorname{Im} r}{\operatorname{Re} r} &= -\arctan \frac{\lambda \sin(k\,\Delta x)}{1 - 2\lambda^2 \sin^2(k\,\Delta x/2)} \\ &\sim -\lambda k\,\Delta x \left[1 - \frac{1}{6}(1-\lambda^2)(k\,\Delta x)^2 + \cdots\right], \end{aligned}$$

which is in error by $O\left((k\,\Delta x)^2\right)$. The modified equation associated with (6) is

$$\frac{\partial u}{\partial t} + a\frac{\partial u}{\partial x} = -\frac{1}{6}a(\Delta x)^2(1-\lambda^2)\frac{\partial^3 u}{\partial x^3} + \cdots.$$

Since this contains no second-derivative term, the method does not produce numerical diffusion. On the other hand, the presence of the third derivative term introduces what is known as *numerical dispersion*. This phenomena manifests itself as oscillations, or wiggles, before and after sharp wave fronts. The details of the derivation of the modified equation are left as an exercise.

EXAMPLE 11.4 Two Sample Advection Equations

Consider the advection equation

$$\frac{\partial u}{\partial t} + 2t\frac{\partial u}{\partial x} = 0, \qquad -\infty < x < \infty, t > 0$$

subject to the initial condition

$$u(x,0) = u_0(x) = \begin{cases} 1, & -1 \le x \le 0 \\ 0, & \text{elsewhere.} \end{cases}$$

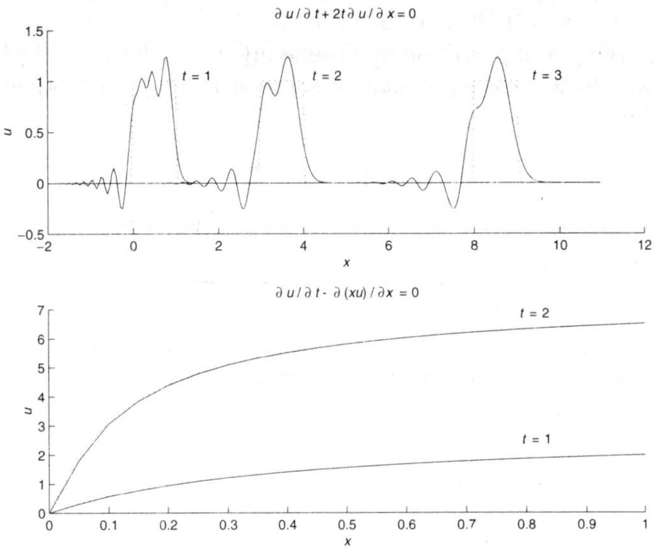

Figure 11.6 Performance of the MacCormack method on two sample advection equations. In each graph, the approximate solutions are plotted as solid curves and the exact solutions as dotted curves.

Here $a(x, t, u) = 2t$ and $g(x, t, u) = 0$. As in Section 11.1, we will advance the approximate solution to $t = 3$ and truncate the domain to the closed interval $[-2, 11]$. Though the resulting mathematical problem requires a boundary condition at $x = -2$ only, the MacCormack method needs a boundary condition at each end of the truncated domain. Examination of the initial condition and the characteristics indicates that the appropriate boundary conditions are $u(-2, t) = u(11, t) = 0$.

The top graph in Figure 11.6 displays the approximate solution at times $t = 1$, $t = 2$ and $t = 3$ (the solid curves) computed using the MacCormack method with $\Delta x = 1/20$ and $\Delta t = 1/200$. The exact solution of the partial differential equation, $u(x, t) = u_0(x - t^2)$, is indicated by the dotted curves. With fourth order amplitude errors and no numerical diffusion, the MacCormack method maintains both the height and the width of the initial pulse much better than does the upwind scheme (compare with Figure 11.1). However, numerical dispersion introduces oscillations which trail behind each discontinuity in the solution. Thus, on this problem, the performance of neither the upwind scheme nor the MacCormack method is entirely satisfactory.

The performance of the MacCormack method on this second example problem is, on the other hand, clearly superior to that of the upwind scheme. The bottom graph in Figure 11.6 displays the approximate solution (solid curves) to the initial boundary value problem

$$\frac{\partial u}{\partial t} - \frac{\partial (xu)}{\partial x} = 0, \quad 0 \leq x \leq 1, t > 0 \quad u(x, 0) = \frac{x}{1 + x}, \qquad u(1, t) = \frac{e^{2t}}{1 + e^t}$$

computed using $\Delta x = 1/20$ and $\Delta t = 1/20$. At $x = 0$, the advection equation together with the initial condition determine $u(0,t) = 0$ for all t. To the resolution of the graphics device, the approximate solution is almost indistinguishable from the exact solution,

$$u(x,t) = \frac{xe^{2t}}{1 + xe^t},$$

shown by the dotted curves. Compare this with the performance of the upwind scheme in Figure 11.2.

EXAMPLE 11.5 Resolution of a Developing Shock

In the previous section we found that the characteristics of the initial value problem

$$\frac{\partial u}{\partial t} + \frac{\partial (u^2/2)}{\partial x} = 0, \qquad -\infty < x < \infty, t > 0$$

$$u(x,0) = u_0(x) = \begin{cases} 1 - \cos(2\pi x), & 0 \le x \le 1 \\ 0, & \text{elsewhere} \end{cases}$$

intersect, producing a shock and introducing a jump discontinuity into the solution. Figure 11.7 displays the approximate solution to this initial value problem at various times, computed with $\Delta x = 1/100$ and $\Delta t = 1/200$. This problem demonstrates one of the main reasons that the MacCormack method has become popular in practice: for nonlinear partial differential equations, the MacCormack method provides high resolution of propagating fronts.

Systems of Equations

Another major advantage of the MacCormack method is the simplicity with which it can be extended to systems of equations. Essentially, little more than a change in notation is needed. In particular, consider the system of advection equations

$$\frac{\partial \mathbf{U}}{\partial t} + \frac{\partial \mathbf{F}(\mathbf{U})}{\partial x} = \mathbf{G},$$

where $\mathbf{U}$ is a vector of dependent variables and $\mathbf{F}(\mathbf{U})$ and $\mathbf{G}$ are vector-valued functions. For conservation laws of this type, equations (6) and (7) become

$$\mathbf{w}_j^* = \mathbf{w}_j^{(n)} - \frac{\Delta t}{\Delta x}\left(\mathbf{F}_{j+1}^{(n)} - \mathbf{F}_j^{(n)}\right) + \Delta t\,\mathbf{G}_j^{(n)}$$

and

$$\mathbf{w}_j^{(n+1)} = \frac{1}{2}\left[\mathbf{w}_j^{(n)} + \mathbf{w}_j^* - \frac{\Delta t}{\Delta x}\left(\mathbf{F}_j^* - \mathbf{F}_{j-1}^*\right)\right] + \frac{1}{2}\Delta t\,\mathbf{G}_j^*.$$

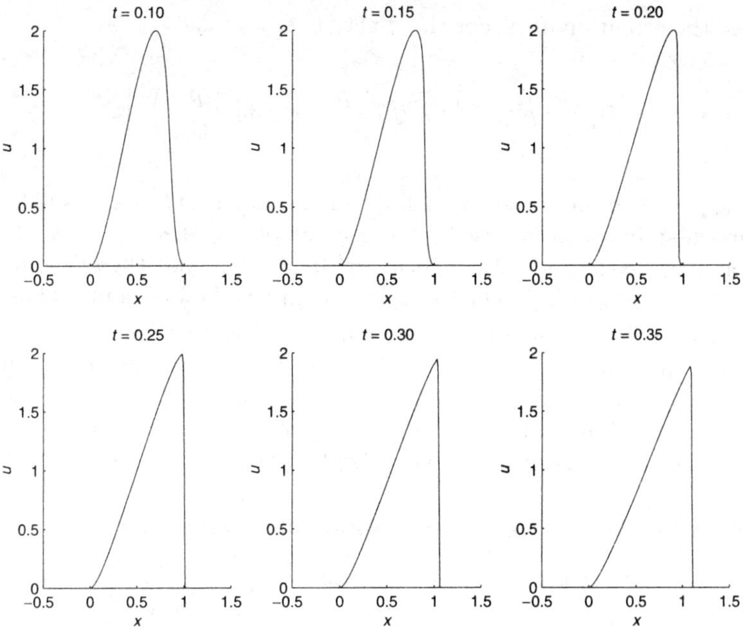

Figure 11.7 Resolution of a shock front in a nonlinear advection equation.

Application Problem 1: Stuck Behind a Slow-Moving Truck—A Simplified Model

In the Overview to this chapter (see page 885), we developed a model for the vehicle density, $\rho(x,t)$, on a stretch of highway. That model consisted of the convection-diffusion equation

$$\frac{\partial \rho}{\partial t} + \frac{\partial f(\rho)}{\partial x} = D\frac{\partial^2 \rho}{\partial x^2},$$

where $f(\rho) = v(\rho)\rho$, $v(\rho)$ is the density-dependent traffic velocity and D is a constant. Here, we will consider the case $D = 0$; that is, we will work with the simplified model

$$\frac{\partial \rho}{\partial t} + \frac{\partial f(\rho)}{\partial x} = 0.$$

Specifically, suppose

$$\rho(x,0) = \begin{cases} 40, & x \le -1 \\ 100 + 60\cos(\pi x), & -1 < x < 1 \\ 40, & x \ge 1, \end{cases}$$

which we use as an approximation to the density profile surrounding a slow moving truck located at $x = 0$. At $t = 0$, the truck exits the roadway, and we want to determine the evolution of ρ. For the function f, we take

$$f(\rho) = \rho V_{\max}\left[1 - 0.7\frac{\rho}{\rho_{\max}} - 0.3\left(\frac{\rho}{\rho_{\max}}\right)^2\right],$$

where $V_{\max} = 50$ mph is the speed limit on the road and $\rho_{\max} = 400$ cars/mile. This particular function is based on the assumptions that the flow of cars goes to zero at both extremes of car density, $f'(0) = V_{\max}$ and there is some density, $0 < \rho_* < \rho_{\max}$, for which f achieves a maximum (see Fowkes and Mahoney [1]).

Truncating the domain to the interval $[-2, 5]$ and choosing mesh spacings of $\Delta x = 0.01$ miles and $\Delta t = \Delta x/V_{\max} = 2 \times 10^{-4}$ hours, we obtain the density profiles displayed in Figure 11.8. Profiles are shown 0.6 minutes, 1.8 minutes, 3.0 minutes, and 4.2 minutes after the truck has exited the highway. Note the formation of a shock and the numerical dispersion trailing behind the jump discontinuity.

Application Problem : Free Surface Motion of Water in a Basin

A rectangular water basin, L meters long and W meters wide, opens on one end to a reservoir (see Figure 11.9). The water level in the reservoir changes over time, which causes the water level within the basin to change. By treating the basin dynamics as quasi-one dimensional flow and considering conservation of mass and conservation of momentum, Roberson and Crowe (*Engineering Fluid Mechanics*, 3rd edition, Houghton Mifflin Company, Boston, 1985) develop the system of partial differential equations

$$\frac{\partial h}{\partial t} + \frac{\partial(hV)}{\partial x} = 0$$

$$\frac{\partial(hV)}{\partial t} + \frac{\partial(hV^2 + gh^2/2)}{\partial x} = -\frac{c_f V|V|}{2}\frac{P}{W}$$

for the basin water surface profile, h, and the horizontal velocity of the water in the basin, V. The additional parameters in these equations are the acceleration due to gravity, g, the shear-stress coefficient, c_f, and the wetted perimeter $P = W + 2h$. The initial conditions are given by $h(x, 0) = h_0$ and $V(x, 0) = 0$. At the open end of the basin, $h(L, t)$ is known, while at the closed end of the basin $V(0, t) = 0$.

To apply the MacCormack method to this problem, we must first express the system in vector form. Since the system contains time derivatives of h and hV, we let

$$\mathbf{U} = \begin{bmatrix} h \\ hV \end{bmatrix} = \begin{bmatrix} u_1 \\ u_2 \end{bmatrix}.$$

Then

$$\mathbf{F}(\mathbf{U}) = \begin{bmatrix} hV \\ hV^2 + gh^2/2 \end{bmatrix} = \begin{bmatrix} hV \\ (hV)^2/h + gh^2/2 \end{bmatrix} = \begin{bmatrix} u_2 \\ u_2^2/u_1 + gu_1^2/2 \end{bmatrix}$$

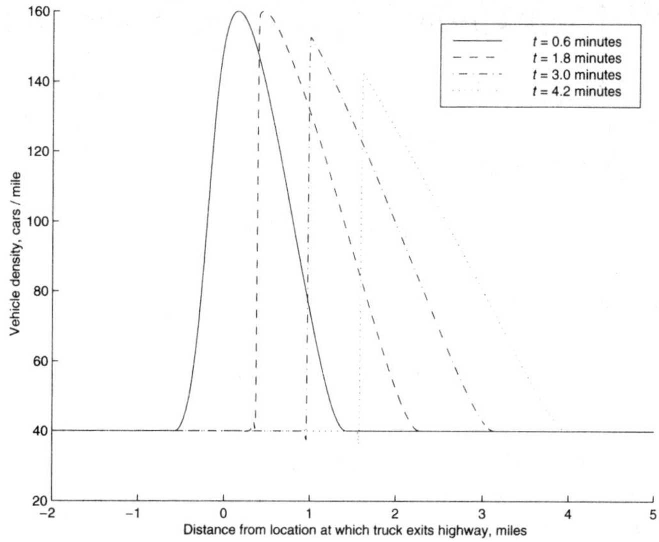

Figure 11.8 Vehicle density profiles after a slow moving truck exits the highway.

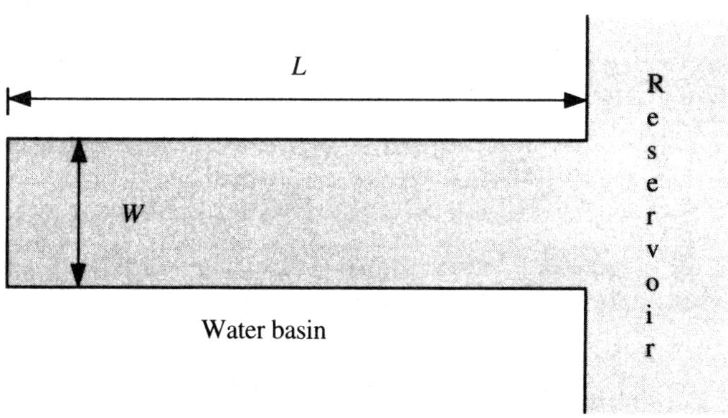

Figure 11.9 Rectangular water basin open at one end to a reservoir.

and

$$\mathbf{G} = \begin{bmatrix} 0 \\ -c_f V|V|P/2W \end{bmatrix} = \begin{bmatrix} 0 \\ -c_f (hV/h)|(hV/h)|P/2W \end{bmatrix}$$
$$= \begin{bmatrix} 0 \\ -c_f (u_2/u_1)|u_2/u_1|P/2W \end{bmatrix}.$$

The initial condition becomes

$$\mathbf{U}(x,0) = \begin{bmatrix} h_0 \\ 0 \end{bmatrix}$$

and the boundary conditions are given by

$$\mathbf{U}(L,t) = \begin{bmatrix} \text{known} \\ ? \end{bmatrix} \quad \text{and} \quad \mathbf{U}(0,t) = \begin{bmatrix} ? \\ 0 \end{bmatrix}.$$

Note that at each end of the domain we need to specify an additional boundary condition. The simplest scheme for doing so is to set the first difference of the unknown quantity to zero; that is, after computing the values of the approximate solution at all interior grid points, set

$$u_2(L,t) = u_2(L - \Delta x, t) \quad \text{and} \quad u_1(0,t) = u_1(\Delta x, t).$$

Suppose the basin is 60 meters long and 15 meters wide. The shear-stress coefficient is $c_f = 0.01$, the acceleration due to gravity is $g = 9.81$ and the initial, uniform depth of the water in the basin is $h_0 = 5$ meters. Take

$$h(L,t) = 5 + 0.5\sin(0.1t) \text{ meters}$$

as the varying depth at the open end of the basin and $\Delta x = 0.6$ meters. The maximum allowable time step for this problem is

$$\Delta t_{\text{max}} = \frac{\Delta x}{|V_{\text{max}}| + \sqrt{g h_{\text{max}}}},$$

where $|V_{\text{max}}| + \sqrt{g h_{\text{max}}}$ is the spectral radius of the Jacobian associated with $\mathbf{F}(\mathbf{U})$. Conservatively estimating $|V_{\text{max}}| = 1$ and $h_{\text{max}} = 10$ yields

$$\Delta t_{\text{max}} = \frac{0.6}{1 + \sqrt{98.1}} \approx 0.055.$$

Therefore, let's use $\Delta t = 1/20$ seconds. Simulation results are displayed in Figure 11.10. The top graph shows the free surface profiles after 20, 40, and 60 seconds. The bottom graph indicates the depth of the water, as a function of time, at the open and closed ends of the basin.

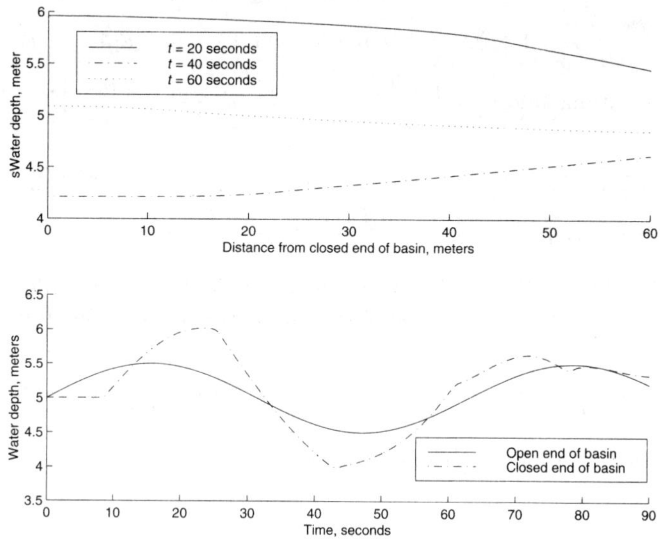

Figure 11.10 Motion of free surface of water in a basin which is open at one end to a reservoir. (Top graph) Free surface profiles at various times. (Bottom graph) Water depth as a function of time at the open and closed ends of the basin.

References

1. N. D. Fowkes and J. J. Mahoney, *An Introduction to Mathematical Modelling*, John Wiley & Sons, Chichester, 1994.

EXERCISES

1. Derive equations (5) and (7).

2. Suppose that a is constant. Show that equations (1) and (3) are identical to equations (4) and (5), which are, in turn, identical to equations (6) and (7).

For Exercises 3–6, apply the MacCormack method to the given advection equation. Truncate the domain and introduce numerical boundary conditions where appropriate. Compare the approximate solution to the indicated exact solution.

3. $\dfrac{\partial u}{\partial t} + \dfrac{1+x^2}{(1+x^2)^2+2tx}\dfrac{\partial u}{\partial x} = 0, \quad -\infty < x < \infty, t > 0, \quad u(x,0) = u_0(x) = \begin{cases} 1, & 0.5 < x < 1 \\ 0, & \text{elsewhere} \end{cases}$ exact solution: $u(x,t) = u_0\left(x - \dfrac{t}{1+x^2}\right)$

4. $\dfrac{\partial u}{\partial t} + (x+t)\dfrac{\partial u}{\partial x} = -1, \quad x > 0, t > 0, \quad u(x,0) = u_0(x) = \begin{cases} x, & 0 < x < 1 \\ 2-x, & 1 < x < 2 \\ 0, & \text{elsewhere} \end{cases}$

exact solution: $u(x,t) = \begin{cases} u_0\left[e^{-t}(x+1+t) - 1\right] - t, & x > e^t - 1 - t \\ 0, & x \le e^t - 1 - t \end{cases}$

5. $\dfrac{\partial u}{\partial t} - \dfrac{\partial(x^2u)}{\partial x} = 0, \quad 0 < x < 4, t > 0, \quad u(x,0) = 0, \quad u(4,t) = t^2$

exact solution: $u(x,t) = \begin{cases} \frac{16}{x^2}\left(\frac{1}{4} - \frac{1}{x} + t\right)^2, & x > 4/(4t+1) \\ 0, & x \le 4/(4t+1) \end{cases}$

6. $\dfrac{\partial u}{\partial t} + \dfrac{\partial}{\partial x}\left[(1-x^2)u\right] = -2xu, \quad -1 < x < 1, t > 0, \quad u(x,0) = 1 + \cos(\pi x)$

exact solution: $u(x,t) = u_0\left(\dfrac{(1+x)e^{-2t} - (1-x)}{1 - x + (1+x)e^{-2t}}\right)$

7. Numerically verify that the MacCormack method is second-order in Δx and Δt by approximating the solution of the initial boundary value problem

$$\frac{\partial u}{\partial t} - \frac{\partial(xu)}{\partial x} = 0, \quad 0 \le x \le 1, t > 0, \quad u(x,0) = \frac{x}{1+x}, \qquad u(1,t) = \frac{e^{2t}}{1+e^t}.$$

The exact solution for this problem is

$$u(x,t) = \frac{xe^{2t}}{1+xe^t}.$$

8. Repeat Exercise 7 using the initial value problem in Exercise 6.

9. Numerically verify that the MacCormack method is better than first order, but not second order, in Δx and Δt for the initial boundary value problem in Exercise 5. Why is the method not second order for this problem?

10. Reconsider the "Stuck Behind a Slow-Moving Truck—A Simplified Model" application problem.

(a) Estimate the amount of time needed for the peak of the car density profile to drop to 100 cars/mile.

(b) If the initial density profile is taken to be

$$\rho(x,0) = \begin{cases} 70, & x \le 0 \\ 100 - \frac{60}{2-x/2}\cos(\pi x), & 0 < x < 2 \\ 40, & x \ge 2, \end{cases}$$

calculate the density profile after 0.6, 1.8, 3.0, and 4.2 minutes. Estimate the amount of time needed for the peak of the profile to drop to 110 cars/mile.

11. For the function $\mathbf{F}(\mathbf{U})$ as defined in the "Free Surface Motion of Water in a Basin" application problem:

(a) Compute the Jacobian matrix $J(\mathbf{U})$.

(b) Determine the eigenvalues of $J(\mathbf{U})$.

(c) Show that $\rho(J) = |V_{\max}| + \sqrt{gh_{\max}}$.

12. Repeat the "Free Surface Motion of Water in a Basin" application problem for a basin that is 50 meters long and 20 meters wide. Take $h(L,t) = 5 + 0.75\sin(0.06t)$ and leave all other parameter values unchanged.

(a) Determine the free surface profiles after 25, 50, 75, and 100 seconds.

(b) Determine the water depth as a function of time at both the open and the closed end of the basin for $t \le 150$ seconds.

13. If the bottom of the basin is sloped, then the equations for the free surface motion become

$$\frac{\partial h}{\partial t} + \frac{\partial (hV)}{\partial x} = 0, \qquad \frac{\partial (hV)}{\partial t} + \frac{\partial (hV^2 + gh^2/2)}{\partial x} = Sgh - \frac{c_f V|V|}{2}\frac{P}{W}$$

where S is the slope of the bottom surface. Repeat the previous exercise with $S = 0.04$, assuming that the initial depth of the water is 3 meters at the closed end of the basin and 5 meters at the open end.

14. In the Overview to this chapter (see page 885), we showed that the voltage, $V(x,t)$, and the current, $I(x,t)$, along a transmission line satisfy the system of advection equations

$$l\frac{\partial I}{\partial t} + \frac{\partial V}{\partial x} + rI = 0$$

$$c\frac{\partial V}{\partial t} + \frac{\partial I}{\partial x} + gV = 0,$$

where l, r, c, and g are the inductance, resistance, capacitance, and current leakage per unit length of the transmission line, respectively. Suppose the line is 100 meters long with $l = 1.2$ henries/meter, $c = 0.3$ farads/meter, $g = 0.0038\,\mathrm{s}^{-1}$, and $r = 0.1$ ohms/meter. The initial conditions are

$$I(x,0) = 5.5\sin\frac{\pi x}{100} \quad \text{and} \quad V(x,0) = 100\sin\frac{\pi x}{100},$$

while the boundary conditions are

$$V(0,t) = V(100,t) = 0 \quad \text{and} \quad I(0,t) = I(100,t) = 0.$$

Determine the voltage along the line after 30, 60, 90, and 120 seconds.

15. Show that the modified equation associated with the finite difference equation

$$w_j^{(n+1)} = \frac{\lambda}{2}(1+\lambda)w_{j-1}^{(n)} + (1-\lambda^2)w_j^{(n)} + \frac{\lambda}{2}(\lambda-1)w_{j+1}^{(n)}$$

is

$$\frac{\partial u}{\partial t} + a\frac{\partial u}{\partial x} = -\frac{1}{6}a(\Delta x)^2(1-\lambda^2)\frac{\partial^3 u}{\partial x^3} + \cdots.$$

Exercises 16–18 deal with the Lax-Wendroff method, which for the advection equation

$$\frac{\partial u}{\partial t} + \frac{\partial f(u)}{\partial x} = 0,$$

is given by

$$w_j^{(n+1)} = w_j^{(n)} - \frac{\lambda}{2}(f_{j+1}^{(n)} - f_{j-1}^{(n)}) + \frac{\lambda^2}{2}\left[d_{j+1/2}(f_{j+1}^{(n)} - f_j^{(n)}) - d_{j-1/2}(f_j^{(n)} - f_{j-1}^{(n)})\right],$$

where

$$\lambda = \frac{\Delta t}{\Delta x}, \quad \text{and} \quad d_{j\pm1/2} = f'\left(\frac{1}{2}(w_{j\pm1}^{(n)} + w_j^{(n)})\right).$$

16. Suppose $f(u) = au$ for some constant a.

(a) Perform a von Neumann stability analysis on the Lax-Wendroff method.

(b) What are the amplitude and phase errors associated with the Lax-Wendroff method?

(c) What is the modified equation associated with the Lax-Wendroff method? Do you expect the method to produce numerical diffusion and/or numerical dispersion? Explain.

17. Apply the Lax-Wendroff method to the initial value problem

$$\frac{\partial u}{\partial t} + \frac{\partial (u^2/2)}{\partial x} = 0, \qquad -\infty < x < \infty, t > 0$$

$$u(x,0) = u_0(x) = \begin{cases} 1 - \cos(2\pi x), & 0 \le x \le 1 \\ 0, & \text{elsewhere.} \end{cases}$$

Truncate the domain to $[-0.5, 1.5]$ and use $\Delta x = 1/100$ and $\Delta t = 1/200$. Compare the solutions at $t = 0.10$, $t = 0.15$, $t = 0.20$, $t = 0.25$, $t = 0.30$, and $t = 0.35$ to those obtained from the MacCormack method (see Figure 11.7).

18. Apply the Lax-Wendroff method to the initial value problem in the "Stuck Behind a Slow-Moving Truck—A Simplified Model" application problem. How do the solutions compare to those obtained by the MacCormack method?

11.3 CONVECTION-DIFFUSION EQUATION

For the physical phenomena that can be modeled by the advection equation, such as contaminant transport, traffic flow, and heat flow, it is not uncommon for diffusion effects to be as important as convective (transport) effects. The objective of this section is to extend both the upwind differencing scheme and the MacCormack method to handle such convection-diffusion problems.

Upwind Scheme

Consider the partial differential equation

$$\frac{\partial c}{\partial t} + \frac{\partial (uc)}{\partial x} = D\frac{\partial^2 c}{\partial x^2}, \tag{1}$$

which includes both a convection term, $\partial(uc)/\partial x$, and a diffusion term, $\partial^2 c/\partial x^2$. u is the velocity of the underlying flow, which need not be constant, but is assumed to be independent of t. For the upwind scheme, we apply a first-order forward difference approximation for the time derivative and a second-order central difference approximation for the diffusion term. For the convection term, we apply either a first-order forward difference or a first-order backward difference approximation, depending on the sign of u.

When $u > 0$, c is being transported from left to right by the flow, so a backward difference is the appropriate choice for discretizing the convection term. The resulting finite difference equation is

$$\frac{c_j^{(n+1)} - c_j^{(n)}}{\Delta t} + \frac{(uc)_j^{(n)} - (uc)_{j-1}^{(n)}}{\Delta x} = D\frac{c_{j-1}^{(n)} - 2c_j^{(n)} + c_{j+1}^{(n)}}{(\Delta x)^2},$$

or

$$c_j^{(n+1)} = c_j^{(n)} + \frac{\Delta t}{\Delta x}\left[(uc)_{j-1}^{(n)} - (uc)_j^{(n)}\right] + \mu\left(c_{j-1}^{(n)} - 2c_j^{(n)} + c_{j+1}^{(n)}\right), \qquad (2)$$

where $\mu = D\Delta t/(\Delta x)^2$. When $u < 0$, a forward difference is the appropriate choice for discretizing the convection term, leading to the finite difference equation

$$c_j^{(n+1)} = c_j^{(n)} + \frac{\Delta t}{\Delta x}\left[(uc)_j^{(n)} - (uc)_{j+1}^{(n)}\right] + \mu\left(c_{j-1}^{(n)} - 2c_j^{(n)} + c_{j+1}^{(n)}\right). \qquad (3)$$

Equations (2) and (3), together, comprise the upwind differencing scheme for the convection-diffusion problem given by (1).

To assess the stability of the upwind differencing scheme, we will assume that u is constant and first consider the case when $u > 0$. With the assumption of constant flow velocity, equation (2) becomes

$$c_j^{(n+1)} = c_j^{(n)} + \lambda\left[c_{j-1}^{(n)} - c_j^{(n)}\right] + \mu\left(c_{j-1}^{(n)} - 2c_j^{(n)} + c_{j+1}^{(n)}\right), \qquad (4)$$

where $\lambda = u\Delta t/\Delta x$. Substituting the discrete Fourier mode $c_j^{(n)} = r^n e^{i(j\theta)}$ into equation (4) leads to

$$\begin{aligned} r &= 1 + \lambda\left(e^{-i\theta} - 1\right) + 2\mu(\cos\theta - 1) \\ &= 1 - \lambda - 2\mu + (\lambda + 2\mu)\cos\theta - i\lambda\sin\theta. \end{aligned}$$

Note that as θ ranges from $-\pi$ to π, the amplification factor traces out an ellipse in the complex plane. The center of this ellipse is located at the point $(1-\lambda-2\mu, 0)$, the horizontal semiaxis of the ellipse is of length $\lambda + 2\mu$ and the vertical semiaxis is of length λ. For stability, the ellipse must lie entirely inside the unit disk. This requires $\lambda \le 1$ so that the ellipse will not go beyond the top or bottom of the disk and $-1 \le 1 - 2\lambda - 4\mu$, or $\lambda + 2\mu \le 1$, so the ellipse will not pass the left side of the disk. The latter inequality poses the more restrictive condition, hence we need $\lambda + 2\mu \le 1$ for stability. If we follow precisely the same steps for the case when $u < 0$ and combine our results with the $u > 0$ case, we obtain the final stability requirement:

$$|\lambda| + 2\mu \le 1,$$

or, in terms of the time step,

$$\Delta t \le \left(\frac{|u|}{\Delta x} + \frac{2D}{(\Delta x)^2}\right)^{-1}.$$

The modified equation associated with (4) is given by

$$\frac{\partial c}{\partial t} + u\frac{\partial c}{\partial x} = \left[D + \frac{1}{2}u\Delta x(1-\lambda)\right]\frac{\partial^2 c}{\partial x^2} + \cdots$$

(see Exercise 1). In other words, the upwind scheme produces an effective diffusion coefficient of

$$D + \frac{1}{2}u\,\Delta x(1-\lambda).$$

Since we need $\lambda + 2\mu \le 1$ for stability, it follows that $1 - \lambda \ge 2\mu$. The increase in the diffusion coefficient is then at least

$$\frac{1}{2}u\,\Delta x \cdot 2\mu = \mu u\,\Delta x = \frac{D\,\Delta t}{(\Delta x)^2}u\,\Delta x = D\frac{u\,\Delta t}{\Delta x} = D\lambda.$$

Thus, λ provides a lower bound on the percentage increase in the diffusion coefficient.

MacCormack Method

The MacCormack method can also be easily extended from the advection equation to the convection-diffusion equation. Working from equation (1), the predictor step uses a first-order forward difference approximation for the convection term and a second-order central difference approximation for the diffusion term. This yields

$$\frac{c_j^* - c_j^{(n)}}{\Delta t} + \frac{(uc)_{j+1}^{(n)} - (uc)_j^{(n)}}{\Delta x} = D\frac{c_{j-1}^{(n)} - 2c_j^{(n)} + c_{j+1}^{(n)}}{(\Delta x)^2},$$

or

$$c_j^* = c_j^{(n)} + \frac{\Delta t}{\Delta x}\left[(uc)_j^{(n)} - (uc)_{j+1}^{(n)}\right] + \mu\left(c_{j-1}^{(n)} - 2c_j^{(n)} + c_{j+1}^{(n)}\right). \qquad (5)$$

c_j^* is used to denote the predicted value for $c_j^{(n+1)}$ and, again, $\mu = D\,\Delta t/(\Delta x)^2$. For the corrector step, the convection term is discretized as the average of a forward difference approximation computed using values at $t = t_n$ and a backward difference approximation computed using the values obtained from the predictor. The diffusion term is handled in exactly the same manner, thus producing

$$\frac{c_j^{(n+1)} - c_j^{(n)}}{\Delta t} + \frac{1}{2}\left[\frac{(uc)_{j+1}^{(n)} - (uc)_j^{(n)}}{\Delta x} + \frac{(uc)_j^* - (uc)_{j-1}^*}{\Delta x}\right] =$$
$$\frac{D}{2}\left(\frac{c_{j-1}^{(n)} - 2c_j^{(n)} + c_{j+1}^{(n)}}{(\Delta x)^2} + \frac{c_{j-1}^* - 2c_j^* + c_{j+1}^*}{(\Delta x)^2}\right).$$

Eliminating the term

$$\frac{(uc)_{j+1}^{(n)} - (uc)_j^{(n)}}{\Delta x}$$

by making use of the predictor equation, the corrector equation can be expressed as

$$c_j^{(n+1)} = \frac{1}{2}\left\{c_j^{(n)} + c_j^* - \frac{\Delta t}{\Delta x}\left[(uc)_j^* - (uc)_{j-1}^*\right] + \mu\left(c_{j-1}^* - 2c_j^* + c_{j+1}^*\right)\right\}. \qquad (6)$$

Hoffman [1] has determined that this method is stable provided $|\lambda| < 0.9$ and $\mu < 0.5$, where $\lambda = u\,\Delta t/\Delta x$. As with the advection equation, the MacCormack method does not introduce numerical diffusion into the approximate solution for the convection-diffusion equation, but may still introduce numerical dispersion (i.e., oscillations in the vicinity of a propagating front).

A Worked Example

EXAMPLE 11.6 A Sample Problem

Consider the convection-diffusion problem

$$\frac{\partial c}{\partial t} + 0.1\frac{\partial c}{\partial x} = D\frac{\partial^2 c}{\partial x^2}, \qquad c(0,t) = 1, \quad c(2,t) = 0, \quad c(x,0) = 0.$$

For all calculations, we will use $\Delta x = 1/20$. Let's start with $D = 0.025 = 1/40$. For the upwind scheme, we will take the maximum allowed time step:

$$\Delta t_{\text{up}} = \left(\frac{1/10}{1/20} + \frac{2(1/40)}{1/400}\right)^{-1} = (2+20)^{-1} = \frac{1}{22}.$$

With this value for Δt, it follows that

$$\lambda = \frac{(1/10)(1/22)}{1/20} = \frac{1}{11},$$

so there will be less than a 10% increase in the diffusion coefficient. The time step for the MacCormack method must satisfy

$$\Delta t_{\text{Mac},\mu} \le \frac{1/400}{2(1/40)} = \frac{1}{20} \quad \text{and} \quad \Delta t_{\text{Mac},\lambda} \le \frac{(9/10)(1/20)}{1/10} = \frac{9}{20};$$

we will therefore choose $\Delta t_{\text{Mac}} = 1/20$. The top graph in Figure 11.11 displays the approximate solutions at $t = 4$. Note there is only a slight increase in the diffusion of the front as calculated by the upwind scheme.

Next, suppose $D = 0.0025 = 1/400$. The maximum allowed time step for the upwind scheme is then

$$\Delta t_{\text{up}} = \left(\frac{1/10}{1/20} + \frac{2(1/400)}{1/400}\right)^{-1} = (2+2)^{-1} = \frac{1}{4},$$

for which

$$\lambda = \frac{(1/10)(1/4)}{1/20} = \frac{1}{2}.$$

In this case, there will thus be a 50% increase in the effective diffusion coefficient. For the MacCormack method, we must have

$$\Delta t_{\text{Mac},\mu} \le \frac{1/400}{2(1/400)} = \frac{1}{2} \quad \text{and} \quad \Delta t_{\text{Mac},\lambda} \le \frac{(9/10)(1/20)}{1/10} = \frac{9}{20}.$$

Since we still want to advance the solution to $t = 4$ in an integer number of time steps, we will use $\Delta t_{\text{Mac}} = 4/9$ (i.e., we will take 9 time steps). The approximate solutions with $D = 0.0025 = 1/400$ are shown in the bottom graph of Figure 11.11. As expected, the upwind scheme produces substantially more diffusion of the front in this case.

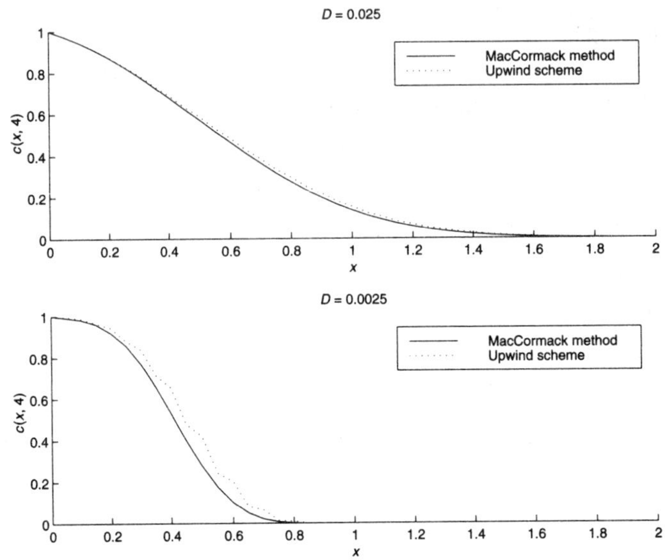

Figure 11.11 Approximate solution to a sample convection-diffusion problem computed with the upwind scheme and the MacCormack method for different values of the diffusion coefficient.

Application Problem 1: Passive Contaminant Transport

A smoke stack releases a passive contaminant (i.e., a chemical that does not react with air) into the air. Once released, the contaminant is transported to the surrounding area by prevailing wind currents and by diffusion. Suppose we consider only one-dimensional transport, and let $c(x,t)$ denote the normalized contaminant level as a function of distance from the smoke stack, x, and time, t. Performing a control volume analysis similar to the one used for the groundwater pollution problem in the Chapter 11 Overview, we find that $c(x,t)$ satisfies the convection-diffusion equation

$$\frac{\partial c}{\partial t} + \frac{\partial (uc)}{\partial x} = D\frac{\partial^2 c}{\partial x^2},$$

where u is the wind velocity and D is the coefficient of diffusion.

Let's take $D = 1.3 \times 10^{-4}$ miles^2/hour and $u = 2.5\left[1 + e^{-x/5}\cos(1.7\pi x/5)\right]$ miles/hour. Note that the velocity attains a maximum of 5 miles/hour at the smoke stack and decreases to a minimum of roughly 1.09 miles/hour at $x \approx 2.77$ miles. We will assume that the air is initially free from contamination ($c(x,0) = 0$) and that the contaminant level at the smoke stack is constant ($c(0,t) = 1$).

Figure 11.12 displays contaminant profiles at $t = 30$ minutes, $t = 1$ hour and $t = 2$ hours computed using both the upwind scheme (top graph) and the MacCormack method (bottom graph). Δx was selected as 0.01 miles. The time

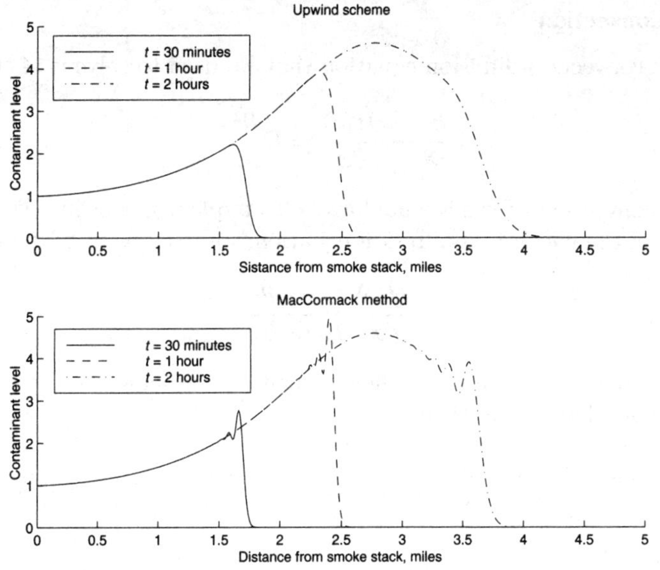

Figure 11.12 Contamination profiles for passive contaminant released into the air from a smoke stack.

steps, in hours, were taken as

$$\Delta t_{\text{up}} = \frac{1}{504} < \left(\frac{5}{0.01} + \frac{2.6 \times 10^{-4}}{1.0 \times 10^{-4}} \right)^{-1}$$

for the upwind scheme, and

$$\Delta t_{\text{Mac}} = \frac{1}{560} < \frac{(0.01)(0.9)}{5} = \Delta t_{\text{Mac},\lambda}$$

for the MacCormack method. For this problem, the stability condition for the MacCormack method as derived from the convection term is significantly more restrictive than the condition derived from the diffusion term.

Note that the peak contamination level at $t = 2$ is between four-and-a-half and five times the contamination released into the air. This build-up occurs because the wind velocity initially decreases as we move away from the smoke stack. The extent of the build-up can be expected to decrease with either an increase in the diffusion coefficient or a reduction in the difference between the maximum and minimum wind velocity. Comparing the numerical methods, we see that numerical dispersion introduces oscillations behind the propagating contamination front into the MacCormack method solution. This complicates the determination of the precise peak level of contamination. On the other hand, the numerical diffusion in the upwind scheme solution makes the localization of the front itself less precise than in the MacCormack method solution.

Nonlinear Convection

Suppose the convection-diffusion equation that we need to solve is of the form

$$\frac{\partial c}{\partial t} + \frac{\partial f(c)}{\partial x} = D\frac{\partial^2 c}{\partial x^2}; \tag{7}$$

that is, the convection effects are nonlinear. To implement upwind differencing for this version of the convection-diffusion equation, we note, as in Section 11.1, that

$$\frac{\partial f(c)}{\partial x} = \frac{\partial f}{\partial c}\frac{\partial c}{\partial x}.$$

Thus, $\partial f/\partial c$ plays the role of the flow velocity. The analogues of equations (2) and (3) for equation (7) are then

$$c_j^{(n+1)} = c_j^{(n)} + \frac{\Delta t}{\Delta x}\left[f_{j-1}^{(n)} - f_j^{(n)}\right] + \mu\left(c_{j-1}^{(n)} - 2c_j^{(n)} + c_{j+1}^{(n)}\right)$$

when $\partial f/\partial c > 0$ and

$$c_j^{(n+1)} = c_j^{(n)} + \frac{\Delta t}{\Delta x}\left[f_j^{(n)} - f_{j+1}^{(n)}\right] + \mu\left(c_{j-1}^{(n)} - 2c_j^{(n)} + c_{j+1}^{(n)}\right)$$

when $\partial f/\partial c < 0$. The finite difference equations for the MacCormack method applied to (7) are

$$\begin{aligned}
&\text{Predictor:} \quad && c_j^* = c_j^{(n)} + \tfrac{\Delta t}{\Delta x}\left[f_j^{(n)} - f_{j+1}^{(n)}\right] + \mu\left(c_{j-1}^{(n)} - 2c_j^{(n)} + c_{j+1}^{(n)}\right)\\
&\text{Corrector:} \quad && c_j^{(n+1)} = \tfrac{1}{2}\left\{c_j^{(n)} + c_j^* - \tfrac{\Delta t}{\Delta x}\left[f_j^* - f_{j-1}^*\right] + \mu\left(c_{j-1}^* - 2c_j^* + c_{j+1}^*\right)\right\}
\end{aligned}$$

Note that we only needed to replace each term of the form $(uc)_j^{(n)}$ from equations (5) and (6) by $f_j^{(n)}$. In each of these finite difference equations (for both the upwind scheme and the MacCormack method), we have used $f_j^{(n)}$ as a shorthand for $f\left(c_j^{(n)}\right)$.

Application Problem 2: Stuck Behind a Slow-Moving Truck

In the Chapter 11 Overview (see page 885), we considered the problem of traffic flow along a highway and developed the model

$$\frac{\partial \rho}{\partial t} + \frac{\partial f(\rho)}{\partial x} = D\frac{\partial^2 \rho}{\partial x^2}$$

for the evolution of the vehicle density profile, $\rho(x,t)$. Here, $f(\rho) = v(\rho)\rho$, $v(\rho)$ is the density-dependent traffic velocity and D is a constant. In the previous section, we worked with the simplified advection equation model corresponding to $D = 0$; we will now treat the full convection-diffusion model.

As in Section 11.2, we take

$$\rho(x,0) = \begin{cases} 40, & x \le -1 \\ 100 + 60\cos(\pi x), & -1 < x < 1 \\ 40, & x \ge 1 \end{cases}$$

as an approximation to the density profile surrounding a slow-moving truck that is located at $x = 0$ and that exits the highway at $t = 0$. For simplicity, we assume this initial condition extends out to infinity in both directions. Further, we will continue to use

$$f(\rho) = \rho V_{\text{max}} \left[1 - 0.7\frac{\rho}{\rho_{\text{max}}} - 0.3\left(\frac{\rho}{\rho_{\text{max}}}\right)^2\right],$$

where $V_{\text{max}} = 50$ mph is the speed limit on the road and $\rho_{\text{max}} = 400$ cars/mile. Finally, let $D = 0.114$ miles^2/hour.

With $\Delta x = 0.01$ miles, the maximum allowable time step for the upwind scheme is

$$\Delta t_{\text{up}} \le \left(\frac{V_{\text{max}}}{\Delta x} + \frac{2D}{(\Delta x)^2}\right)^{-1} = \left(\frac{50}{1/100} + \frac{0.228}{1/10000}\right)^{-1} = \frac{1}{7280}.$$

For convenience, $\Delta t_{\text{up}} = 1/7300$ hours is used. $\Delta t_{\text{Mac}} = \Delta x/V_{\text{max}} = 2 \times 10^{-4}$ hours is selected for the MacCormack method. Figure 11.13 displays the computed density profiles 0.6 minutes, 1.8 minutes, 3.0 minutes, and 4.2 minutes after the truck has exited the highway.

There are three observations that can be made from this figure. First, there was not enough diffusion present to prevent the formation of a shock. Second, comparing the bottom graph in Figure 11.13 with Figure 11.8, we see that the presence of diffusion has reduced the effects of dispersion in the MacCormack method solutions. Third, the MacCormack method maintains a steeper profile on the jump discontinuity and causes less rounding of the peak in the profile than does the upwind scheme.

References

1. J. D. Hoffman, *Numerical Methods for Engineers and Scientists*, McGraw-Hill, New York, 1992.

EXERCISES

1. Show that the modified equation associated with the finite difference equation

$$c_j^{(n+1)} = c_j^{(n)} + \lambda\left(c_{j-1}^{(n)} - c_j^{(n)}\right) + \mu\left(c_{j+1}^{(n)} - 2c_j^{(n)} + c_{j-1}^{(n)}\right)$$

is

$$\frac{\partial c}{\partial t} + u\frac{\partial c}{\partial x} = \left[D + \frac{1}{2}u\Delta x(1-\lambda)\right]\frac{\partial^2 c}{\partial x^2}.$$

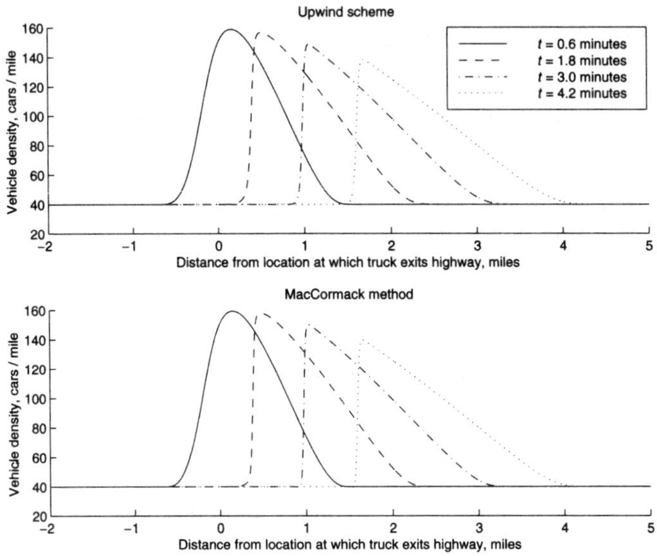

Figure 11.13 Vehicle density profiles after a slow moving truck exits the highway.

2. Derive the modified equation for the MacCormack method.
3. Use both the upwind scheme and the MacCormack method to approximate the solution of the partial differential equation

$$\frac{\partial c}{\partial t} + \frac{\partial (c^2/2)}{\partial x} = 0.1\frac{\partial^2 c}{\partial x^2}$$

subject to the initial condition

$$c(x,0) = \begin{cases} 1 - \cos(2\pi x), & 0 \le x \le 1 \\ 0, & \text{elsewhere} \end{cases}$$

at $t = 0.1$, $t = 0.2$, $t = 0.3$, and $t = 0.4$. Remember to truncate the domain and introduce appropriate numerical boundary conditions.

4. Use both the upwind scheme and the MacCormack method to approximate the solution of the initial boundary value problem

$$\frac{\partial c}{\partial t} + \frac{\partial [(1-x^2)c]}{\partial x} = 0.1\frac{\partial^2 c}{\partial x^2}, \quad c(x,0) = 1 + \cos(\pi x), \quad c(-1,t) = c(1,t) = 0$$

at $t = 1$, $t = 2$, $t = 3$, and $t = 4$.

5. Use both the upwind scheme and the MacCormack method to approximate the solution of the initial boundary value problem

$$\frac{\partial c}{\partial t} + \frac{\partial (xc)}{\partial x} = \frac{1}{2}\frac{\partial^2 c}{\partial x^2}, \quad x > 0, t > 0,$$

$$c(x,0) = \begin{cases} 1-\cos(2\pi x), & 0 \le x \le 1 \\ 0, & \text{elsewhere}, \end{cases} \qquad c(0,t) = 0$$

at $t = 0.5$, $t = 1$, $t = 1.5$, and $t = 2$. Remember to truncate the domain and introduce appropriate numerical boundary conditions.

6. In the "Stuck Behind a Slow-Moving Truck" application problem, we found that a diffusion coefficient of $D = 0.114$ miles^2/hour was not sufficient to prevent the formation of a shock. Estimate how large the diffusion coefficient must be to prevent the formation of a shock, at least up to $t = 4.2$ minutes.

7. Rework the "Stuck Behind a Slow-Moving Truck" application problem for each of the following initial vehicle density profiles.

(a) $\rho(x,0) = \begin{cases} 160, & x \le 0 \\ 100 + 60\cos(\pi x), & 0 < x < 1 \\ 40, & x \ge 1 \end{cases}$

(b) $\rho(x,0) = \begin{cases} 70, & x \le 0 \\ 100 - \frac{60}{2-x/2}\cos(\pi x), & 0 < x < 2 \\ 40, & x \ge 2 \end{cases}$

(c) $\rho(x,0) = \begin{cases} 160, & x \le 0 \\ 20 - 20\cos(\pi x/2), & 0 < x < 2 \\ 40, & x \ge 2 \end{cases}$

8. Rework the "Passive Contaminant Transport" application problem replacing the boundary condition $c(0,t) = 1$ with $c(0,t) = 1 - e^{-20t}$. Does the MacCormack method still produce oscillations trailing behind the moving front?

9. Reconsider the "Passive Contaminant Transport" application problem. Estimate the peak contamination level and the location of the contamination front after 30 minutes, 1 hour and 2 hours for the following combinations of diffusion coefficient and wind velocity.

(a) $D = 1.3 \times 10^{-3}$ miles^2/hour, $u = 2.5\left[1 + e^{-x/5}\cos(1.7\pi x/5)\right]$ miles/hour

(b) $D = 1.3 \times 10^{-4}$ miles^2/hour, $u = 2\left[1 + 0.8e^{-x/5}\cos(1.7\pi x/5)\right]$ miles/hour

(c) $D = 1.3 \times 10^{-3}$ miles^2/hour, $u = 2\left[1 + 0.8e^{-x/5}\cos(1.7\pi x/5)\right]$ miles/hour

10. Repeat Exercise 9 replacing the boundary condition $c(0,t) = 1$ with $c(0,t) = 1 - e^{-20t}$.

11. Consider the convection-diffusion equation

$$\frac{\partial c}{\partial t} + \frac{\partial (uc)}{\partial x} = D\frac{\partial^2 c}{\partial x^2}.$$

(a) Develop the forward in time, central in space (FTCS) method for this problem.

(b) Assuming that u is constant, determine the stability condition associated with the FTCS method.

(c) Determine the modified equation associated with the FTCS method.

(d) Approximate the solution of the convection-diffusion equation

$$\frac{\partial c}{\partial t} + 0.1\frac{\partial c}{\partial x} = D\frac{\partial^2 c}{\partial x^2}, \quad c(0,t) = 1, \quad c(2,t) = 0, \quad c(x,0) = 0$$

at $t = 4$ for $D = 1/40$ and $D = 1/400$. Compare the solution with that of the upwind scheme and the MacCormack method.

12. Consider the convection-diffusion equation

$$\frac{\partial c}{\partial t} + \frac{\partial f(c)}{\partial x} = D\frac{\partial^2 c}{\partial x^2}.$$

(a) Develop the FTCS method for this problem.

(b) Use the results of part (a) to rework Exercise 3. How does the performance of the FTCS method compare with that of the upwind scheme and the MacCormack method?

11.4 THE WAVE EQUATION

To finish this chapter, we will develop a finite difference method for the second-order hyperbolic partial differential equation

$$\frac{\partial^2 u}{\partial t^2} = c^2\frac{\partial^2 u}{\partial x^2}.$$

This is known as the *wave equation* and models such diverse phenomena as the vibration of a piano or guitar string, the oscillation of pressure within the pipe of an organ or within a blood vessel and the propagation of a signal along a transmission line. Being second order in both the time and the space variable, the wave equation requires two initial conditions and two boundary conditions. The initial conditions take the form

$$u(x, t_0) = f(x) \quad \text{and} \quad \frac{\partial u}{\partial t}(x, t_0) = g(x).$$

In general, the wave equation can be defined over the entire real line (an infinite interval), a semi-infinite interval or a finite interval, and a variety of different boundary conditions can be specified. Here, we will restrict attention to the finite interval $A \le x \le B$ with Dirichlet boundary conditions

$$u(A, t) = \alpha(t) \quad \text{and} \quad u(B, t) = \beta(t).$$

Generalizations to the differential equation and to the boundary conditions will be considered in the exercises.

The Method

Let $x_j = A + j\Delta x$ for $j = 0, 1, 2, \ldots, N$ for some positive integer N, where $\Delta x = (B - A)/N$. Further, let $t_n = t_0 + n\Delta t$ for some uniform time step Δt. Evaluate the wave equation at the arbitrary grid point (x_j, t_n). Next, replace each derivative by its second-order central difference approximation and drop the truncation error terms. This yields

$$\frac{w_j^{(n+1)} - 2w_j^{(n)} + w_j^{(n-1)}}{(\Delta t)^2} = c^2\frac{w_{j-1}^{(n)} - 2w_j^{(n)} + w_{j+1}^{(n)}}{(\Delta x)^2},$$

where $w_j^{(n)} \approx u(x_j, t_n)$. Solving for $w_j^{(n+1)}$, we arrive at the explicit finite difference equation

$$w_j^{(n+1)} = \lambda w_{j-1}^{(n)} + 2(1-\lambda)w_j^{(n)} + \lambda w_{j+1}^{(n)} - w_j^{(n-1)}, \tag{1}$$

where $\lambda = (c\Delta t/\Delta x)^2$. Equation (1) is a three time-level finite difference equation: values of the approximate solution at $t = t_{n+1}$ are determined by the values at both $t = t_n$ and $t = t_{n-1}$. Hence, this equation only applies for $n \geq 1$.

So how do we compute the approximate solution at $t = t_1$? We can handle this situation in precisely the same manner with which we have treated non-Dirichlet boundary conditions in the previous three chapters: use fictitious nodes. Evaluating equation (1) at $n = 0$ gives

$$w_j^{(1)} = \lambda w_{j-1}^{(0)} + 2(1-\lambda)w_j^{(0)} + \lambda w_{j+1}^{(0)} - w_j^{(-1)}, \tag{2}$$

where $w_j^{(-1)}$ is associated with a fictitious node. To eliminate this term, evaluate the derivative initial condition at (x_j, t_0) and replace $\partial u/\partial t$ by its second-order central difference approximation:

$$\frac{w_j^{(1)} - w_j^{(-1)}}{2\Delta t} = g_j \quad \Rightarrow \quad w_j^{(-1)} = w_j^{(1)} - 2\Delta t g_j. \tag{3}$$

Substituting equation (3) into equation (2) and solving for $w_j^{(1)}$ yields

$$w_j^{(1)} = \frac{1}{2}\lambda w_{j-1}^{(0)} + (1-\lambda)w_j^{(0)} + \frac{1}{2}\lambda w_{j+1}^{(0)} + \Delta t\, g_j.$$

Finally, taking into account the initial condition on u, we have

$$w_j^{(1)} = \frac{1}{2}\lambda f_{j-1} + (1-\lambda)f_j + \frac{1}{2}\lambda f_{j+1} + \Delta t\, g_j. \tag{4}$$

To summarize, the initial condition $u(x, t_0) = f(x)$ provides the values of the approximate solution to the wave equation at $t = t_0$. Next, the values at $t = t_1$ are computed from equation (4). Equation (1) is then used to advance the approximate solution to $t = t_2$, $t = t_3$, $t = t_4$ and so on. The resulting approximation is second-order accurate in both Δt and Δx.

Stability Analysis

When we use the numerical method that has just been developed, and select Δx to resolve the anticipated spatial variations in the solution, what restriction, if any, is placed on Δt? Let's answer this question by performing a von Neumann stability analysis. Substitution of the discrete Fourier mode, $w_j^{(n)} = r^n e^{i(j\theta)}$, into equation (1) leads to

$$r^{n+1}e^{i(j\theta)} = r^n e^{i(j\theta)}\left[\lambda e^{-i\theta} + 2(1-\lambda) + \lambda e^{i\theta}\right] - r^{n-1}e^{i(j\theta)}. \tag{5}$$

Dividing equation (5) by $r^{n-1}e^{i(j\theta)}$, using the identity $e^{i\theta} + e^{-i\theta} = 2\cos\theta$, and transposing all terms to the left-hand side of the equation, we find that the amplification factor satisfies the quadratic equation

$$r^2 - 2\left[(1-\lambda) + \lambda\cos\theta\right] + 1 = 0. \tag{6}$$

For the numerical method to be stable, both roots of this quadratic must be less than or equal to one in magnitude.

The first thing to note about equation (6) is that the leading coefficient and the constant term are both 1. This implies that the product of the roots is 1. If these roots are real and distinct, then one of them must be strictly larger than one in magnitude, and the method will be unstable. Thus, for the method to be stable, (6) must have either a double real root or complex conjugate roots. In other words, the discriminant,

$$4\left[(1-\lambda) + \lambda\cos\theta\right]^2 - 4,$$

must be less than or equal to zero. This will happen when

$$\left[(1-\lambda) + \lambda\cos\theta\right]^2 \le 1,$$

or

$$-1 \le (1-\lambda) + \lambda\cos\theta \le 1.$$

The inequality on the right holds for all λ, but the inequality on the left requires $\lambda \le 1$. Hence, our numerical method for the wave equation is conditionally stable and requires $\Delta t \le \Delta x / c$.

Worked Examples

EXAMPLE 11.7 Our Numerical Method in Action

Consider the wave equation

$$\frac{\partial^2 u}{\partial t^2} = \frac{1}{25}\frac{\partial^2 u}{\partial x^2}, \qquad 0 < x < 1, t > 0$$

with boundary and initial conditions given by

$$u(0,t) = -\sin(t/5), \quad u(1,t) = \sin(1 - t/5)$$

$$u(x,0) = \sin x, \quad \frac{\partial u}{\partial t}(x,0) = -\frac{1}{5}\cos x.$$

Comparing the current problem with the model wave equation, we find that $c^2 = 1/25$, or $c = 1/5$. For all of our calculations, we will choose $\lambda = 1$; that is, we will use the maximum time step allowed by stability. Given that $c = 1/5$, this means that Δt and Δx must satisfy the relation $\Delta t = 5\,\Delta x$.

Let's start with $\Delta x = 1/5$ and $\Delta t = 1$ and advance the solution from $t = 0$ to $t = 2$. Using the value of the initial condition $u(x, 0) = \sin x$ at the interior grid points and the boundary conditions at the boundary grid points, it follows that

$$\begin{aligned}\mathbf{w}^{(0)} &= \begin{bmatrix} 0 & \sin 0.2 & \sin 0.4 & \sin 0.6 & \sin 0.8 & \sin 1 \end{bmatrix}^T \\ &\approx \begin{bmatrix} 0 & 0.198669 & 0.389418 & 0.564642 & 0.717356 & 0.841471 \end{bmatrix}^T.\end{aligned}$$

Values have been displayed to six significant figures for convenience. With $\lambda = 1$ and $\Delta t = 1$, equation (4) becomes

$$w_j^{(1)} = \frac{1}{2}(f_{j-1} + f_{j+1}) + g_j.$$

Thus, for j =1, 2, 3, and 4,

$$w_j^{(1)} = \frac{1}{2}\left(\sin\frac{j-1}{5} + \sin\frac{j+1}{5}\right) - \frac{1}{5}\cos\frac{j}{5}.$$

The boundary conditions supply $w_0^{(1)} = -\sin 0.2$ and $w_5^{(1)} = \sin 0.8$. Therefore, after the first time step

$$\mathbf{w}^{(1)} = \begin{bmatrix} -0.198669 & -0.00130414 & 0.197444 & 0.388320 & 0.563715 & 0.717356 \end{bmatrix}^T.$$

For the second time step, note that with $\lambda = 1$, equation (1) becomes

$$w_j^{(n+1)} = w_{j-1}^{(n)} + w_{j+1}^{(n)} - w_j^{(n-1)}.$$

Applying this equation for j =1, 2, 3, and 4 and obtaining $w_0^{(2)} = -\sin 0.4$ and $w_5^{(2)} = \sin 0.6$ from the boundary conditions yields

$$\mathbf{w}^{(2)} = \begin{bmatrix} -0.389418 & -0.199895 & -0.00240239 & 0.196516 & 0.388320 & 0.564642 \end{bmatrix}^T.$$

The following table compares the values of the approximate solution with the values of the exact solution, $u(x, t) = \sin(x - t/5)$, at $t = 2$.

x_j	$w_j^{(2)}$	$u(x_j, 2)$	Absolute Error
0	-0.389418	-0.389418	0.000000
0.2	-0.199895	-0.198669	0.001226
0.4	-0.00240239	0.000000	0.002402
0.6	0.196516	0.198669	0.002153
0.8	0.388320	0.389418	0.001098
1.0	0.564642	0.564642	0.000000

A numerical verification of the second-order accuracy of the scheme is presented in the next table. Both the maximum absolute error and the root mean-square (rms) error in the approximate solution at $t = 1$, as a function of the number of subintervals along the x-axis, NX, and the number of time steps taken to reach $t = 1$, NT, are displayed. Clearly, each time Δt and Δx are cut in half, the corresponding error drops by a factor of 4.

NT	NX	Maximum Error	Error Ratio	rms Error	Error Ratio
1	5	0.0013041444		0.0010265081	
2	10	0.0003248943	4.014058	0.0002482593	4.134823
4	20	0.0000811525	4.003504	0.0000615365	4.034345
8	40	0.0000202837	4.000875	0.0000153510	4.008628
16	80	0.0000050706	4.000219	0.0000038357	4.002160

EXAMPLE 11.8 Better than Expected Performance

As a second example, consider the wave equation

$$\frac{\partial^2 u}{\partial t^2} = \frac{\partial^2 u}{\partial x^2}, \quad 0 < x < 1, t > 0$$

subject to the boundary conditions $u(0,t) = u(1,t) = 0$ and the initial conditions

$$u(x,0) = \sin \pi x, \quad \frac{\partial u}{\partial t}(x,0) = 0.$$

The two tables given below compare the approximate solution, computed with $\Delta x = \Delta t = 1/10$, to the exact solution, $u(x,t) = \sin \pi x \cos \pi t$, for two different values of t. In each case, the absolute error in the approximate solution is much smaller than one would expect from a second-order method. In fact, it appears as if we have obtained the exact solution, to within roundoff errors. That we have, indeed, obtained the exact solution is established following the end of this example.

	$t = 2$ (20 time steps)			$t = 3.2$ (32 time steps)		
x_j	$w_j^{(20)}$	$u(x_j, 2)$	Absolute Error	$w_j^{(32)}$	$u(x_j, 3.2)$	Absolute Error
0.0	0.000000	0.000000	0.000000	0.000000	−0.000000	0.000000
0.1	0.309017	0.309017	5.551115×10^{-17}	−0.250000	−0.250000	2.775558×10^{-17}
0.2	0.587785	0.587785	0.000000	−0.475528	−0.475528	7.771561×10^{-16}
0.3	0.809017	0.809017	0.000000	−0.654508	−0.654508	0.000000
0.4	0.951057	0.951057	3.330669×10^{-16}	−0.769421	−0.769421	7.771561×10^{-16}
0.5	1.000000	1.000000	0.000000	−0.809017	−0.809017	0.000000
0.6	0.951057	0.951057	3.330669×10^{-16}	−0.769421	−0.769421	7.771561×10^{-16}
0.7	0.809017	0.809017	0.000000	−0.654508	−0.654508	0.000000
0.8	0.587785	0.587785	0.000000	−0.475528	−0.475528	7.771561×10^{-16}
0.9	0.309017	0.309017	5.551115×10^{-17}	−0.250000	−0.250000	2.775558×10^{-17}
1.0	0.000000	−0.000000	0.000000	0.000000	0.000000	0.000000

How can we explain the performance in this second example? First, consider equation (1). If the exact solution of the wave equation is infinitely differentiable, then the truncation error associated with equation (1) is given by

$$\left(\frac{(\Delta t)^2}{4!}\frac{\partial^4 u}{\partial t^4} - c^2\frac{(\Delta x)^2}{4!}\frac{\partial^4 u}{\partial x^4}\right) + \left(\frac{(\Delta t)^4}{6!}\frac{\partial^6 u}{\partial t^6} - c^2\frac{(\Delta x)^4}{6!}\frac{\partial^6 u}{\partial x^6}\right) + \left(\frac{(\Delta t)^6}{8!}\frac{\partial^8 u}{\partial t^8} - c^2\frac{(\Delta x)^6}{8!}\frac{\partial^8 u}{\partial x^8}\right) + \cdots \quad (7)$$

Using the wave equation, we can establish that

$$\frac{\partial^4 u}{\partial t^4} = \frac{\partial^2}{\partial t^2}\left(\frac{\partial^2 u}{\partial t^2}\right) = \frac{\partial^2}{\partial t^2}\left(c^2\frac{\partial^2 u}{\partial x^2}\right) = c^2\frac{\partial^2}{\partial x^2}\left(\frac{\partial^2 u}{\partial t^2}\right) = c^4\frac{\partial^4 u}{\partial x^4}.$$

In a similar manner, it can be shown that

$$\frac{\partial^{2m} u}{\partial t^{2m}} = c^{2m}\frac{\partial^{2m} u}{\partial x^{2m}}.$$

Converting all of the t-derivatives in equation (7) to x-derivatives yields

$$\frac{c^2(\Delta x)^2}{4!}(\lambda - 1)\frac{\partial^4 u}{\partial x^4} + \frac{c^2(\Delta x)^4}{6!}(\lambda^2 - 1)\frac{\partial^6 u}{\partial x^6} + \frac{c^2(\Delta x)^6}{8!}(\lambda^3 - 1)\frac{\partial^8 u}{\partial x^8} + \cdots.$$

Thus, with $\lambda = 1$, the truncation error associated with the second, third, fourth, and so on time steps is identically zero.

What about the error associated with the first time step; that is, with equation (4)? Recall that when $\lambda = 1$, equation (4) reduces to

$$w_j^{(1)} = \frac{1}{2}(f_{j-1} + f_{j+1}) + \Delta t\, g_j. \quad (8)$$

This is to be compared with the d'Alembert solution to the wave equation

$$u(x,t) = \frac{1}{2}[f(x - ct) + f(x + ct)] + \frac{1}{2c}\int_{x-ct}^{x+ct} g(\xi)\, d\xi. \quad (9)$$

Evaluating (9) at $x = x_j$ and $t = \Delta t$ yields

$$u(x_j, \Delta t) = \frac{1}{2}(f_{j-1} + f_{j+1}) + \frac{\Delta t}{2\,\Delta x}\int_{x_{j-1}}^{x_{j+1}} g(\xi)\, d\xi, \quad (10)$$

where we have made repeated use of the relation $c\,\Delta t = \Delta x$ (which follows from $\lambda = 1$). Note that equation (8) can be viewed as an approximation to (10) obtained by estimating the value of the integral in (10) with the midpoint rule. The first time step will therefore introduce some error unless the midpoint rule integrates g exactly, which will happen when g is constant or linear.

Putting all of this together, we can expect our numerical method to produce the exact solution when

(1) the exact solution is infinitely differentiable;

(2) calculations are made with $\lambda = 1$; and

(3) g is constant or linear.

It is easy to verify that the problem in the second example satisfies all three conditions. Thus the only errors in the second example are roundoff errors.

An Application Problem: Transmission Line

Consider a 100-meter-long transmission line with inductance per unit length $L = 1.2$ henries/meter and capacitance per unit length $C = 0.3$ farads/meter. Let $V(x,t)$ denote the voltage along the line. If the resistance of the cable and current leakage are negligible, then, from the Chapter 11 Overview (see page 885), we know that $V(x,t)$ satisfies the wave equation

$$\frac{\partial^2 V}{\partial t^2} = \frac{1}{LC}\frac{\partial^2 V}{\partial x^2}.$$

Suppose that the line is originally dead; that is,

$$V(x,0) = \frac{\partial V}{\partial t}(x,0) = 0,$$

and the right end of the line is grounded, $V(100,t) = 0$. At the left end of the line, the following voltage is impressed:

$$V(0,t) = \begin{cases} 110\sin(\pi t/50), & 0 \le t \le 100 \\ 0, & t > 100 \end{cases}.$$

Figure 11.14 displays the voltage along the transmission line at $t = 30$, $t = 60$, $t = 90$, and $t = 120$ seconds. All calculations were performed with $\Delta x = 1$ meter and $\Delta t = 0.6$ seconds.

EXERCISES

In Exercises 1–4, approximate the solution of the given wave equation. In each case, compare the approximate solution with the indicated exact solution and, where possible, numerically verify the second-order accuracy of the finite difference scheme. Explain any unusual findings.

1. $\dfrac{\partial^2 u}{\partial t^2} = 9\dfrac{\partial^2 u}{\partial x^2}$, $0 < x < 1, t > 0$, $u(0,t) = u(1,t) = 0$,
$u(x,0) = 0$, $\dfrac{\partial u}{\partial t}(x,0) = 3\pi\sin\pi x$; exact solution: $u(x,t) = \sin\pi x \sin 3\pi t$

2. $\dfrac{\partial^2 u}{\partial t^2} = 4\dfrac{\partial^2 u}{\partial x^2}$, $0 < x < 1, t > 0$, $u(0,t) = \cosh 2t$, $u(1,t) = e\cosh 2t + t$,
$u(x,0) = e^x$, $\dfrac{\partial u}{\partial t}(x,0) = x$; exact solution: $u(x,t) = e^x\cosh 2t + xt$

3. $\dfrac{\partial^2 u}{\partial t^2} = \dfrac{1}{4}\dfrac{\partial^2 u}{\partial x^2}$, $0 < x < 1, t > 0$, $u(0,t) = 2\sinh(t/2)$,
$u(1,t) = 2e\sinh(t/2) + 1$, $u(x,0) = x$, $\dfrac{\partial u}{\partial t}(x,0) = e^x$
exact solution: $u(x,t) = 2e^x\sinh(t/2) + x$

4. $\dfrac{\partial^2 u}{\partial t^2} = \dfrac{\partial^2 u}{\partial x^2}$, $0 < x < 1, t > 0$, $u(0,t) = t^3/3$, $u(1,t) = t + t^3/3$,
$u(x,0) = 0$, $\dfrac{\partial u}{\partial t}(x,0) = x^2$; exact solution: $u(x,t) = x^2 t + t^3/3$

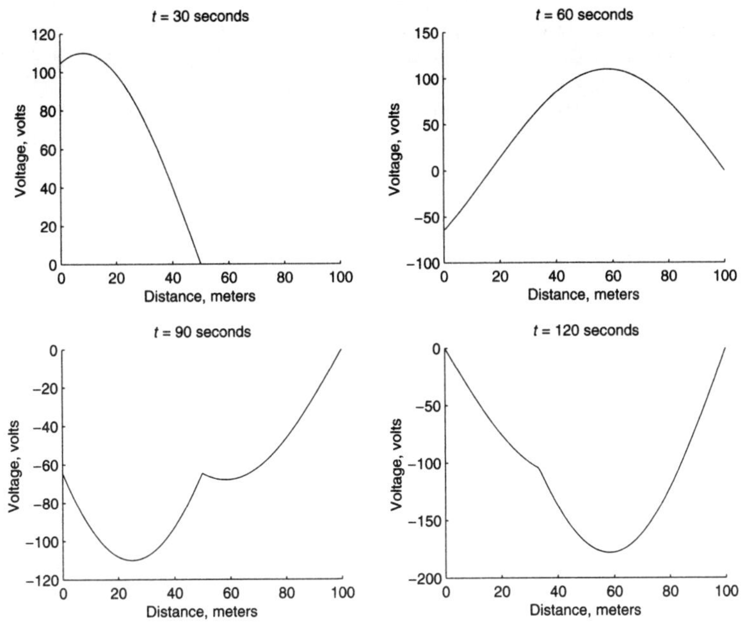

Figure 11.14 Voltage along 100 meter transmission line after 30, 60, 90, and 120 seconds.

5. Let $y(x,t)$ denote the lateral displacement of a vibrating string. If T is the tension in the string, w is the weight per unit length, and g is acceleration due to gravity, then y satisfies the equation

$$\frac{\partial^2 y}{\partial t^2} = \frac{Tg}{w}\frac{\partial^2 y}{\partial x^2}.$$

Suppose a particular string is 6 feet long and is fixed at both ends. Taking $T = 32$ pounds, $w = 0.01$ pounds/foot, and $g = 32$ feet/second^2, use the finite difference method to estimate the period of oscillation of the string. The initial conditions are

$$y(x,0) = \begin{cases} x/6, & 0 \le x \le 3 \\ (6-x)/6, & 3 < x \le 6 \end{cases} \quad \text{and} \quad \frac{\partial y}{\partial t}(x,0) = x(x-6).$$

6. Consider the wave equation with damping term:

$$\frac{\partial^2 u}{\partial t^2} + \mu\frac{\partial u}{\partial t} = c^2\frac{\partial^2 u}{\partial x^2}, \quad 0 < x < L, t > 0,$$

where μ and c are constant.

(a) Develop a second-order finite difference method for this equation. Assume that Dirichlet boundary conditions are specified at both ends of the domain.

(b) Use the method developed in part (a) to solve the problem:
Consider a 100-meter-long transmission line with inductance per unit length $L = 1.2$ henries/meter and capacitance per unit length $C = 0.3$ farads per meter. Let $V(x,t)$ denote the voltage along the line. If the resistance of the cable is negligible but current leakage is assumed proportional to voltage, then $V(x,t)$ satisfies

$$\frac{\partial^2 V}{\partial t^2} + G\frac{\partial V}{\partial t} = \frac{1}{LC}\frac{\partial^2 V}{\partial x^2}.$$

Take $G = 0.0038\ (\mathrm{s})^{-1}$ and suppose that the line is originally dead; that is,

$$V(x,0) = \frac{\partial V}{\partial t}(x,0) = 0.$$

The right end of the line is grounded, $V(100,t) = 0$. At the left end of the line, the following voltage is impressed:

$$V(0,t) = \begin{cases} 110\sin(\pi t/50), & 0 \le t \le 100 \\ 0, & t > 100 \end{cases}.$$

Determine the voltage after 30 seconds, 60 seconds, 90 seconds, and 120 seconds.

7. Consider the wave equation with source term:

$$\frac{\partial^2 u}{\partial t^2} = c^2\frac{\partial^2 u}{\partial x^2} + s(x,t), \quad 0 < x < L, t > 0,$$

where c is constant.

(a) Develop a second-order finite difference method for this equation. Assume that Dirichlet boundary conditions are specified at both ends of the domain.

(b) Use the method developed in part (a) to solve the problem:
Let $y(x,t)$ denote the lateral displacement of a vibrating string. If T is the tension in the string, w is the weight per unit length, g is acceleration due to gravity, and F is an applied force per unit length, then y satisfies the equation

$$\frac{\partial^2 y}{\partial t^2} = \frac{Tg}{w}\frac{\partial^2 y}{\partial x^2} + \frac{gF(x,t)}{w}.$$

Suppose a particular string is 6 feet long and is fixed at both ends. Taking $T = 32$ pounds, $w = 0.01$ pounds/foot, $g = 32$ feet/second^2, and $F(x,t) = 5\sin(12\pi t)$ pounds/foot, determine the profile of the string after 10 seconds, 20 seconds, and 30 seconds, assuming that the string starts from rest.

8. Consider the wave equation:

$$\frac{\partial^2 u}{\partial t^2} = c^2\frac{\partial^2 u}{\partial x^2}, \quad 0 < x < L, t > 0,$$

where c is constant, subject to the boundary conditions

$$u(0,t) = \alpha(t), \quad \frac{\partial u}{\partial x}(L,t) = \beta(t).$$

(a) Develop a second-order finite difference method for this equation.

(b) Use the method developed in part (a) to solve the problem:
Consider a 100-meter-long transmission line with inductance per unit length $L = 1.2$ henries/meter and capacitance per unit length $C = 0.3$ farads per meter. Let $V(x,t)$ denote the voltage along the line. If the resistance of the cable and current leakage are negligible, then $V(x,t)$ satisfies

$$\frac{\partial^2 V}{\partial t^2} = \frac{1}{LC}\frac{\partial^2 V}{\partial x^2}.$$

Suppose that the line is originally dead; that is,

$$V(x,0) = \frac{\partial V}{\partial t}(x,0) = 0.$$

The right end of the line is open, so that $\partial V/\partial x = 0$. At the left end of the line, the following voltage is impressed:

$$V(0,t) = \begin{cases} 110\sin(\pi t/50), & 0 \le t \le 100 \\ 0, & t > 100. \end{cases}$$

Determine the voltage after 30 seconds, 60 seconds, 90 seconds, and 120 seconds.

(c) Use the method developed in part (a) to solve the problem:
A 1-meter-long organ pipe is closed at the top end. The pressure, $p(x,t)$, in the pipe satisfies

$$\frac{\partial^2 p}{\partial t^2} = \frac{\partial^2 p}{\partial x^2},$$

subject to the initial conditions

$$p(x,0) = p_{\text{atm}}\cos(3\pi x), \quad \frac{\partial p}{\partial t}(x,0) = 0$$

and the boundary conditions

$$p(0,t) = p_{\text{atm}}, \quad \frac{\partial p}{\partial x}(1,t) = 0.$$

Take $p_{\text{atm}} = 1.05$, and determine the pressure at $t = 5$, $t = 10$, and $t = 15$.

APPENDIX A

Important Theorems from Calculus

The following theorems play an important role in the development and the analysis of many of the numerical methods presented in this text. Proofs for these theorems can be found in most elementary calculus or real analysis/advanced calculus textbooks.

Rolle's Theorem
If f is continuous on $[a,b]$ and differentiable on (a,b) with $f(a)=f(b)=0$, then there exists a number $c \in (a,b)$ with $f'(c)=0$.

Generalized Rolle's Theorem
If f is continuous on $[a,b]$, has n continuous derivatives on (a,b) and is equal to zero at $n+1$ distinct points in $[a,b]$, then there exists a number $c \in (a,b)$ with $f^{(n)}(c)=0$.

Mean Value Theorem
If f is continuous on $[a,b]$ and differentiable on (a,b), then there exists a number $c \in (a,b)$ such that

$$f'(c) = \frac{f(b)-f(a)}{b-a}.$$

Weighted Mean Value Theorem for Integrals
If f is continuous on $[a,b]$, g is integrable on $[a,b]$ and $g(x)$ does not change sign on $[a,b]$, then there exists a number $c \in (a,b)$ such that

$$\int_a^b f(x)g(x)\,dx = f(c)\int_a^b g(x)\,dx.$$

Extreme Value Theorem
If f is continuous on $[a,b]$, then there exist numbers $c_1, c_2 \in [a,b]$ with $f(c_1) \le f(x) \le f(c_2)$ for all $x \in [a,b]$.

NOTE: If f is differentiable on (a,b), then c_1 and c_2 occur either at the endpoints of $[a,b]$ or where $f'(c)=0$.

Intermediate Value Theorem
If f is continuous on $[a,b]$ and k is ANY number between $f(a)$ and $f(b)$, then there exists a number $c \in (a,b)$ with $f(c)=k$.

Taylor's Theorem
Suppose f is continuous on $[a, b]$, has n continuous derivatives on (a, b) and $f^{(n+1)}$ exists on $[a, b]$. Let $x_0 \in [a, b]$. For every $x \in [a, b]$ there exists $\xi(x)$ between x and x_0 such that

$$f(x) = P_n(x) + R_n(x),$$

where

$$P_n(x) = \sum_{k=0}^{n} \frac{f^{(k)}(x_0)}{k!}(x - x_0)^k$$

and

$$R_n(x) = \frac{f^{(n+1)}(\xi(x))}{(n+1)!}(x - x_0)^{n+1}.$$

APPENDIX B

Algorithm for Solving a Tridiagonal System of Linear Equations

Many of the techniques developed in this text require the solution of a linear system of equations with a tridiagonal coefficient matrix. These include the computation of a cubic spline interpolant, the solution of a one-dimensional two-point boundary value problem using the finite difference method and the solution of parabolic partial differential equations using the finite difference method.

Suppose the system that must be solved can be written in the form

$$\begin{bmatrix} a_1 & b_1 & & & & & \\ c_2 & a_2 & b_2 & & & & \\ & c_3 & a_3 & b_3 & & & \\ & & \cdot & \cdot & \cdot & & \\ & & & \cdot & \cdot & \cdot & \\ & & & & \cdot & \cdot & \cdot \\ & & & & c_{n-1} & a_{n-1} & b_{n-1} \\ & & & & & c_n & a_n \end{bmatrix} \begin{bmatrix} w_1 \\ w_2 \\ w_3 \\ \cdot \\ \cdot \\ \cdot \\ w_{n-1} \\ w_n \end{bmatrix} = \begin{bmatrix} f_1 \\ f_2 \\ f_3 \\ \cdot \\ \cdot \\ \cdot \\ f_{n-1} \\ f_n \end{bmatrix},$$

or $\mathbf{Aw} = \mathbf{f}$. The solution algorithm consists of three distinct parts. First, an LU decomposition is performed on the matrix $\mathbf{A}$. This process factors the coefficient matrix into a lower triangular matrix, $\mathbf{L}$, and an upper triangular matrix, $\mathbf{U}$, such $\mathbf{LU} = \mathbf{A}$. The LU decomposition transforms the original problem into the form $\mathbf{LUw} = \mathbf{f}$. Let $\mathbf{z}$ denote the vector $\mathbf{Uw}$. The solution to the original system can now be obtained by applying forward substitution to $\mathbf{Lz} = \mathbf{f}$, followed by backward substitution applied to $\mathbf{Uw} = \mathbf{z}$.

ALGORITHM

*** first, the LU decomposition ***

$L_1 = a_1$
$U_1 = b_1/a_1$
for $i = 2, 3, 4, \ldots, n-1$
  $L_i = a_i - c_i U_{i-1}$
  $U_i = b_i/L_i$
$L_n = a_n - c_n U_{n-1}$

*** now, the forward substitution ***

$z_1 = f_1/L_1$
for $i = 2, 3, 4, \ldots, n$
$\qquad z_i = (f_i - c_i z_{i-1})/L_i$

*** finally, the backward substitution ***

$w_n = z_n$
for $i = n-1, n-2, n-3, \ldots, 1$
$\qquad w_i = z_i - U_i w_{i+1}$

NOTE: It is possible to code this algorithm with no additional storage by storing each value of L in place of a, each value of U in place of b, each value of z in place of f, and finally each value of w in place of z.

References

O. Aberth, *Precise Numerical Methods using C++*, Academic Press, San Diego, 1998.

F. S. Acton, *Numerical Methods That Work*, Harper and Row, New York, 1970.

R. Aiken, ed., *Stiff Computation*, Oxford University Press, Oxford, England, 1985.

S. Alessandrini, "A Motivational Example for the Numerical Solution of the Algebraic Eigenvalue Problem," *SIAM Review*, **40** (4), 935–940, 1998.

J. G. Andrews and R. R. McLone, eds., *Mathematical Modeling*, Butterworths, 1976.

H. Anton and C. Rorres, *Elementary Linear Algebra with Applications*, John Wiley & Sons, New York, 1987.

N. S. Asaithambi, *Numerical Analysis: Theory and Practice*, Saunders College Publishing, Fort Worth, 1995.

K. E. Atkinson, *An Introduction to Numerical Analysis*, John Wiley & Sons, New York, 1978.

S. V. Babu and E. Barouch, "Exact Solution of Dill's Model Equations for Positive Photoresist Kinetics," *IEEE Electron Device Lett.*, **EDL-7**, 252–253, 1986.

R. K. Beatson, "On the Convergence of Some Cubic Spline Interpolation Schemes," *SIAM Journal on Numerical Analysis*, **23**, 903–912, 1986.

A. Beiser, *Concepts of Modern Physics*, McGraw-Hill, New York, 1981.

J. Bernasconi, "Hopping Transport in One-Dimensional Percolation Model: A Comment," *Physical Reviews* B, **25**, 1394–1395, 1982.

G. Birkhoff and C. de Boor, "Error Bounds for Spline Interpolation," *Journal of Mathematics and Mechanics*, **13**, 827–836, 1964.

E. Boyd and C.-D. Wang, "Optimizing Cell Production in a Chemostat," in Proceedings of *Mathematical Modeling in the Undergraduate Curriculum*, H. Skala, editor, University of Wisconsin/La Crosse, 1994.

A. Brandt, "Multilevel Adaptive Solutions to Boundary Value Problems," *Mathematics of Computation*, **31**, 333–390, 1977.

S. C. Brenner and L. R. Scott, *The Mathematical Theory of Finite Element Methods*, Springer-Verlag, New York, 1994.

R. P. Brent, "An Algorithm with Guaranteed Convergence for Finding a Zero of a Function," *Computer Journal*, **14**, 422–425, 1971.

W. Briggs, V. E. Henson and S. McCormick, *A Multigrid Tutorial, Second Edition*, SIAM, Philadelphia, 2000.

K. M. Brown, "A Quadratically Convergent Newton-Like Method Based upon Gaussian Elimination," *SIAM Journal on Numerical Analysis*, **6**, 560–569, 1969.

C. G. Broyden, "A Class of Methods for Solving Nonlinear Simultaneous Equations," *Mathematics of Computation*, **19**, 577–593, 1965.

C. G. Broyden, "Quasi-Newton Methods and Their Application to Function Minimization," *Mathematics of Computation*, **21**, 368–381, 1967.

C. G. Broyden, "A New Taxonomy of Conjugate Gradient Methods," *Computers and Mathematics with Applications,* **31**, 7–17, 1996.

R. L. Burden and J. D. Faires, *Numerical Analysis*, 5th edition, PWS Publishing Company, Boston, 1993.

J. C. Butcher, *The Numerical Analysis of Ordinary Differential Equations*, John Wiley & Sons, New York, 1987.

J. C. Butcher, "The Non-Existence of Ten-Stage Eighth-Order Runge-Kutta Methods," *BIT*, **25**, 521–542, 1985.

G. D. Byrne and C. A. Hall, eds., *Numerical Solution of Systems of Nonlinear Algebraic Equations*, Academic Press, New York, 1973.

R. M. Cassie, "Relationship Between Plant Pigments and Gross Primary Production in *Skeletonema costatum*," *Limnology and Oceanography*, **8**, 433–439, 1963.

G. Caughley, "Parameters for Seasonally Breeding Populations," *Ecology*, **48 (5)**, 834–839, 1967.

E. A. Coddington, *An Introduction to Ordinary Differential Equations*, Dover Publications, New York, 1989.

R. Courant, K. O. Friedrichs, and H. Lewy, "Uber die Partiellen Differenzengleichungen de Mathematischen Physik," *Mathematische Annalen*, **100**, 32–74, 1928.

D. T. Crouse, L. B. Crowder, and H. Caswell, "A Stage-Based Population Model for Loggerhead Sea Turtles and Implications for Conservation," *Ecology*, **68** (5), 1412–1423, 1987.

N. Curle, "Liquid Flowing from a Container," in *Mathematical Modeling*, J. G. Andrews and R. R. McLone, eds., Butterworths, pp. 39–55, 1976.

M. B. Cutlip and M. Shacham, *Problem Solving in Chemical Engineering with Numerical Methods*, Prentice Hall PTR, Upper Saddle River, NJ, 1999.

D. Cvetkovic, M. Doob and H. Sachs, *Spectra of Graphs: Theory and Application*, Academic Press, New York, 1979.

G. Dahlquist, "A Special Stability Property for Linear Multistep Methods," *BIT*, **3**, 27–43, 1963.

P. J. Davis and P. Rabinowitz, *Methods of Numerical Integration*, 2nd edition, Academic Press, New York, 1984.

C. de Boor, *A Practical Guide to Splines*, Springer-Verlag, New York, 1978.

C. de Boor, "Convergence of cubic spline interpolation with the not-a-knot condition," Mathematics Research Center preprint, University of Wisconsin, Madison, 1984.

K. Dekker and J. G. Verwer, *Stability of Runge-Kutta Methods for Stiff Nonlinear Differential Equations*, CWI Monographs **2**, North-Holland, Amsterdam, 1984.

J. Dennis and J. J. Moré, "Quasi-Newton Methods: Motivation and Theory," *SIAM Review*, **19**, 46–89, 1977.

J. E. Dennis and R. E. Schnabel, *Numerical Methods for Unconstrained Optimization and Nonlinear Equations*, Prentice Hall, Englewood Cliffs, New Jersey, 1983.

A. J. DeSanti, "A Model for Predicting Aircraft Altitude Loss in a Pull-Up from a Dive," *SIAM Review*, **30** (4), 625–628, 1988.

DeSantis, Gironi and Marelli, "Vector-Liquid Equilibrium from a Hard-Sphere Equation of State," *Industrial and Engineering Chemistry Fundamentals*, **15**, 182–189, 1976.

F. H. Dill, W. P. Homberger, P. S. Hauge and J. M. Shaw, "Characterization of Positive Photoresist," *IEEE Trans. Electron Devices*, **ED-22**, 445–452, 1975.

A. Douglass, *Introduction to Mathematical Analysis*, Addison-Wesley, Massachusetts, 1996.

G. H. Dunteman, *Principal Components Analysis*, Sage University Press series on Quantitative Applications in the Social Sciences, 07-069, Sage Publications, Beverly Hills, 1989.

A. C. Eisenstat and H. Walker, "Globally Convergent Inexact Newton Methods," *SIAM Journal of Optimization*, **4**, 393–422, 1994.

H. Eyring and D. Henderson, eds., *Theoretical Chemistry: Advances and Perspectives, Volume 2*, Academic Press, New York, 1976.

N. D. Fowkes and J. J. Mahoney, *An Introduction to Mathematical Modelling*, John Wiley & Sons, Chichester, 1994.

L. Fox and I. Parker, *Chebyshev Polynomials in Numerical Analysis*, Oxford University Press, Oxford, England, 1968.

J. G. F. Francis, "The QR Transformation I, II," *Computer Journal*, **4**, 265–271, 1961–1962.

R. Freund and N. Nachtigal, "QMR: A Quasi-Minimal Residual Method for Non-Hermitian Linear Systems," *Numerische Mathematik*, **60**, 315–339, 1991.

R. Freund and N. Nachtigal, "An Implementation of the QMR Method Based on Two Coupled Two-Term Recurrences," Technical Report 92.15, RIACS, NASA Ames, 1992.

A. Friedman and W. Littman, *Industrial Mathematics: A Course in Solving Real-World Problems*, SIAM, Philadelphia, 1994.

S. Fulton, P. Ciesielski, and W. Schubert, "Multigrid Methods for Elliptic Problems: A Review," *Monthly Weather Review*, **114** (5), 943–959, 1986.

C. W. Gear, "The Automatic Integration of Stiff Ordinary Differential Equations," *Proceedings of the IP68 Conference*, North-Holland, Amsterdam, 1969.

C. W. Gear, *Numerical Initial Value Problems in Ordinary Differential Equations*, Prentice Hall, Englewood Cliffs, NJ, 1971.

R. A. Gentry, R. E. Martin, and B. J. Daly, "An Eulerian Differencing Method for Unsteady Compressible Flow Problems," *Journal of Computational Physics*, **1**, 87–118, 1966.

C. F. Gerald and P. O. Wheatley, *Applied Numerical Analysis*, 5th edition, Addison-Wesley, Massachusetts, 1994.

M. L. Glasser, "The Electron Gas in a Magnetic Field: Nonrelativistic Ground State Properties," in *Theoretical Chemistry: Advances and Perspectives, Volume 2*, H. Eyring and D. Henderson, eds., Academic Press, New York, 1976, pp. 67–129.

M. L. Glasser, "Two Definite Integrals Arising in Light Transmission through a Crystal: Problem 93-4," *SIAM Review*, **35** (1), 136, 1993.

J. H. Goldthorpe, *Social Mobility and Class Structure in Modern Britain*, Clarendon Press, Oxford, 1987.

G. Golub and D. O'Leary, "Some History of the Conjugate Gradient and Lanczis Methods," *SIAM Review*, **31**, 50–102, 1989.

G. H. Golub and J. M. Ortega, *Scientific Computing and Differential Equations: An Introduction to Numerical Methods*, Academic Press, Boston, 1992.

G. Golub and C. van Loan, *Matrix Computations, Third Edition*, Johns Hopkins Press, Baltimore, 1996.

H. P. W. Gottlieb, "Simple Nonlinear Jerk Functions with Periodic Solutions," *American Journal of Physics*, **66** (10), 903–906, 1998.

W. Hackbusch and U. Trottenberg, *Multigrid Methods*, Springer-Verlag, Berlin, 1982.

L. A. Hageman and D. M. Young, *Applied Iterative Methods*, Academic Press, New York, 1981.

E. Hairer, S. P. Norsett and G. Wanner, *Solving Ordinary Differential Equations I: Nonstiff Problems*, Springer-Verlag, Berlin, 1991.

E. Hairer and G. Wanner, *Solving Ordinary Differential Equations II: Stiff Problems and Differential-Algebraic Equations*, Springer-Verlag, Berlin, 1991.

C. Hall and W. Meyer, "Optimal Error Bounds for Cubic Spline Interpolation," *Journal of Approximation Theory*, **16**, 105–122, 1976.

H. H. Harman, *Modern Factor Analysis*, The University of Chicago Press, Chicago, 1960.

W. Harrison, "Total Energies in the Tight-Binding Theory," *Physical Reviews B*, **23**, 5230–5245, 1981.

K. Hastings, "Reliability and the Cost of Guarantees," in *Applications of Calculus (Resources for Calculus, volume 3)*, P. Straffin, editor, MAA Notes #29, The Mathematical Association of America, pp. 152–166, 1993.

P. Henrici, *Discrete Variable Methods in Ordinary Differential Equations*, John Wiley & Sons, New York, 1962.

F. Hickey, "Death and Reproductive Rate of Sheep in Relation to Flock Culling and Selection," *New Zealand J. Agricultural Research*, **3**, 332–344, 1960.

F. Hickey, "Sheep Mortality in New Zealand," New Zealand Agriculturalist, **15**, 1–3, 1963.

F. B. Hildebrand, *Advanced Calculus for Applications*, 2nd edition, Prentice Hall, New Jersey, 1976.

A. J. Hoffman, "On eigenvalues and colorings of graphs," in *Graph Theory and Its Applications*, B. Harris, ed., Academic Press, 1970.

J. D. Hoffman, *Numerical Methods for Engineers and Scientists*, McGraw-Hill, New York, 1992.

R. A. Horn and C. A. Johnson, *Matrix Analysis*, Cambridge University Press, Cambridge University, 1985.

A. S. Householder, *The Numerical Treatment of a Single Nonlinear Equation*, McGraw-Hill, New York, 1970.

L. Howle, D. Schaeffer, M. Shearer and P. Zhong, "Lithotripsy: The Treatment of Kidney Stones with Shock Waves," *SIAM Review*, **40** (2), 356–371, 1998.

F. P. Incropera and D. P. DeWitt, *Fundamentals of Heat and Mass Transfer, Second Edition*, John Wiley & Sons, New York, 1985.

E. Isaacson and H. B. Keller, *Analysis of Numerical Methods*, John Wiley & Sons, New York, 1966.

A. Iserles, *A First Course in the Numerical Analysis of Differential Equations*, Cambridge Texts in Applied Mathematics, Cambridge University Press, Cambridge, 1996.

M. A. Jenkins, "Algorithm 493–Zeros of a Real Polynomial," *ACM Transactions on Mathematical Software*, **1**, 178–189, 1975.

M. A. Jenkins and J. F. Traub, "A Three-Stage Algorithm for Real Polynomials using Quadratic Iteration," *SIAM Journal on Numerical Analysis*, **7**, 545–566, 1970.

M. A. Jenkins and J. F. Traub, "Algorithm 419—Zeros of a Complex Polynomial," *Communications of the ACM*, **15**, 97–99, 1972.

C. Johnson, *Numerical Solution of Partial Differential Equations by the Finite Element Method*, Cambridge University Press, Cambridge, 1987.

R. A. Johnson and D. W. Wichern, *Applied Multivariate Statistical Analysis*, Prentice Hall, Englewood Cliffs, 1982.

H. B. Keller, *Numerical Methods for Two-Point Boundary Value Problems*, Blaisdell, Waltham, 1968.

J. Keller, "Probability of a Shutout in Racquetball," *SIAM Review*, **26**, 267–268, 1984.

W. O. Kermack and A. G. McKendrick, "Contributions to the Mathematical Theory of Epidemics, I.," *Proc. Roy. Soc.*, **115A**, 700, 1927.

P. K. Khosla and S. G. Rubin, "A Conjugate Gradient Iterative Method," *Computational Fluids*, **9**, 109–121, 1981.

I. Kollias, V. Hatzitaki, G. Papaiakovou, and G. Giatsis, "Using Principal Components Analysis to Identify Individual Differences in Vertical Jump Performance," *Research Quarterly for Exercise and Sport*, **72**, 63–66, 2001.

A. S. Kronrod, *Nodes and Weights of Quadrature Formulas*, Consultants Bureau, New York, 1965.

J. D. Lambert, *Numerical Methods for Ordinary Differential Equations*, John Wiley & Sons, London, 1991.

D. C. Lay, *Linear Algebra and Its Applications, Second Edition*, Addison-Wesley, Reading, MA, 1997.

C. Le Cunff and A. Zebib, "Nonlinear Spatially Developing Görtler Vortices in Curved Wall Jet Flow," *Phys. Fluids*, **8**, 2375–2384, 1996.

S. J. Leon, *Linear Algebra with Applications*, 6th edition, Prentice Hall, Upper Saddle River, NJ, 2002.

R. S. Lindzen and H.-L. Kuo, "A Reliable Method for the Numerical Integration of a Large Class of Ordinary and Partial Differential Equations," *Monthly Weather Review*, **97**, 732–734, 1969.

D. L. Logan, *A First Course in the Finite Element Method,* 2nd edition, PWS-Kent, Boston, 1992.

H. Longuet-Higgins, "Some Studies in Molecular Orbital Theory, I. Resonance Structures and Molecular Orbitals in Unsaturated Hydrocarbons," *The Journal of Chemical Physics*, **18** (3), 265–275, 1950.

D. Ludwig, D. G. Aronson and H. F. Weinberger, "Spatial Patterning of the Spruce Budworm," *Journal of Mathematical Biology*, **8**, 217–258, 1979.

S. McCormick, *Multigrid Methods*, SIAM Frontiers Series, Volume 3, SIAM, Philadelphia, 1987.

M. J. Moelter, J. Evans, G. Elliott, and M. Jackson, "Electric Potential in the Classical Hall Effect: An Unusual Boundary-Value Problem," *American Journal of Physics*, **66**, 668–677, 1998.

B. Mohar and S. Poljak, "Eigenvalues in Combinatorial Optimization," in *Combinatorial and Graph-Theoretical Problems in Linear Algebra*, R. A. Brualdi, S. Friedland and V. Klee, eds., IMA Volumes in Mathematics and its Applications, volume 50, Springer-Verlag, Berlin, 1993.

R. E. Moore, editor, *Reliability in Computing*, Academic Press, San Diego, 1988.

J. J. Moré and M. Y. Cosnard, "Numerical Solution of Nonlinear Equations," *ACM Transactions on Mathematical Software*, **5**, 64–85, 1979.

K. W. Morton and D. F. Mayers, *Numerical Solution of Partial Differential Equations*, Cambridge University Press, Cambridge, 1994.

E. R. Muller, "Ice Breaking with an Air Cushion Vehicle," in *Mathematical Modeling: Classroom Notes in Applied Mathematics*, M. S. Klamkin, editor, SIAM, Philadelphia, pp. 29–36, 1987.

J. D. Murray, *Mathematical Biology*, Springer-Verlag, Berlin, 1993.

D. N. Naik and R. Khattree, "Revisiting Olympic Track Records: Some Practical Considerations in the Principal Components Analysis," *The American Statistician*, **50**, 140–144, 1996.

R. M. Nisbet, E. McCauley, W. S. C. Gurney, W. W. Murdoch, and A. M. de Roos, "Simple Representations of Biomass Dynamics in Structured Populations," in *Case Studies in Mathematical Modeling—Ecology, Physiology and Biology*, H. G. Othmer, F. R. Adler, M. A. Lewis, and J. C. Dallon, eds., Prentice Hall Publishing, New Jersey, pp. 61–79, 1997.

J. T. Oden and J. N. Reddy, *An Introduction to the Mathematical Theory of Finite Elements*, John Wiley & Sons, New York, 1976.

J. Ortega, *Numerical Analysis—A Second Course*, Academic Press, New York, 1972.

J. Ortega and W. Rheinboldt, *Iterative Solution of Nonlinear Equations in Several Variables*, Academic Press, New York, 1970.

H. G. Othmer, F. R. Adler, M. A. Lewis, and J. C. Dallon, eds., *Case Studies in Mathematical Modeling—Ecology, Physiology and Biology*, Prentice Hall Publishing, New Jersey, 1997.

T. N. L. Patterson, "The Optimum Addition of Points to Quadrature Formulae," *Mathematics of Computation*, **22**, 847–856, 1968.

D. W. Peaceman and H. H. Rachford, "The Numerical Solution of Parabolic and Elliptic Differential Equations," *Journal of the Society for Industrial and Applied Mathematics*, **3**, 28–41, 1955.

R. H. Perry and D. W. Green, *Perry's Chemical Engineer Handbook*, 7th edition, McGraw-Hill, New York, 1997.

D. A. Peters, "Optimum Spring-Damper Design for Mass Impact," *SIAM Review*, **39** (1), 118–122, 1997.

G. Peters and J. Wilkinson, "Practical Problems Arising in the Solution of Polynomial Equations," *Journal of the Institute for Mathematics and Its Applications*, **8**, 16–35, 1971.

C. Philipsen, S. Markvorsen, and W. Kleim, "Modelling the Stem Curve of a Palm in a Strong Wind," *SIAM Review*, **38** (3), 483–484, 1996.

M. J. D. Powell, *Approximation Theory and Methods*, Cambridge University Press, Cambridge, 1981.

W. H. Press, B. P. Flannery, S. A. Teukolsky, and W. T. Vetterling, *Numerical Recipes: The Art of Scientific Computing*, Cambridge University Press, Cambridge, 1992.

A. Ralston, *A First Course in Numerical Analysis*, McGraw-Hill, New York, 1965.

A. Ralston and P. Rabinowitz, *A First Course in Numerical Analysis*, 2nd edition, McGraw-Hill, New York, 1978.

W. F. Ramirez, *Computational Methods for Process Simulation*, Butterworth, Boston, 1989.

S. Rinaldi, "Laura and Petrarch: An Intriguing Case of Cyclical Love Dynamics," *SIAM Journal of Applied Mathematics*, **58** (4), 1205–1221, 1998.

T. J. Rivlin, *The Chebyshev Polynomials*, John Wiley & Sons, New York, 1974.

T. J. Rivlin, *An Introduction to the Approximation of Functions*, Dover Publications, Inc., New York, 1981.

J. A. Roberson and C. T. Crowe, *Engineering Fluid Mechanics*, 3rd edition, Houghton Mifflin Company, Boston, 1985.

S. M. Rump, "Algorithms for Verified Inclusions: Theory and Practice," in *Reliability in Computing*, R. E. Moore, editor, Academic Press, San Diego, 1988.

Y. Saad, *Numerical Methods for Large Eigenvalue Problems*, Halsted Press, New York, 1992.

Y. Saad and M. Schultz, "GMRES: A Generalized Minimal Residual Algorithm for Solving Nonsymmetric Linear Systems," *SIAM Journal on Scientific and Statistical Computing*, **7**, 856–869, 1986.

M. H. Schultz, *Spline Analysis*, Prentice Hall, Englewood Cliffs, NJ, 1973.

L. Shampine and C. Gear, "A User's View of Solving Stiff Ordinary Differential Equations," *SIAM Review*, **21**, 1–17, 1979.

L. Shampine and M. Gordon, *Computer Solution of Ordinary Differential Equations*, W.H. Freeman, San Francisco, 1975.

T. Shifrin and M. R. Adams, *Linear Algebra: A Geometric Approach*, W. H. Freeman, New York, 2002.

G. D. Smith, *Numerical Solution of Partial Differential Equations: Finite Difference Methods*, 3rd edition, Oxford Applied Mathematics and Computing Science Series, Oxford University Press, Oxford, 1985.

P. Sonneveld, "CGS: A Fast Lanczos-tpye Solver for Nonsymmetric Linear Systems," *SIAM Journal on Scientific and Statistical Computing*, **10**, 36–52, 1989.

G. W. Stewart, *Introduction to Matrix Computations*, Academic Press, New York, 1973.

J. Stoer and R. Bulirsch, *Introduction to Numerical Analysis*, Springer-Verlag, New York, 1980.

P. D. Straffin, "Linear Algebra in Geography: Eigenvectors of Networks," *Mathematics Magazine*, **53**, 269–276, 1980.

P. Straffin, editor, *Applications of Calculus (Resources for Calculus, volume 3)*, MAA Notes #29, The Mathematical Association of America, Washington, DC, 1993.

S. Subramanian and V. Balakotaiah, "Convective Instabilities Induced by Exothermic Reactions Occurring in a Porous Medium," *Phys. Fluids*, **6** (9), 2907–2922, 1994.

P. D. Taylor and A. Sauer, "The Selective Advantage of Sex-Ratio Homeostasis," *American Naturalist*, **116**, 305–310, 1980.

K. G. TeBeest, "Numerical and Analytical Solutions of Volterra's Population Model," *SIAM Review*, **39**, 484–493, 1997.

J. W. Thomas, *Numerical Partial Differential Equations: Conservation Laws and Elliptic Equations*, Texts in Applied Mathematics, volume 33, Springer-Verlag, New York, 1999.

S. P. Timoshenko and J. M. Gere, *Theory of Elastic Stability*, McGraw-Hill, New York, 1961.

S. Timoshenko and S. Woinowsky-Krieger, *Theory of Plates and Shells*, 2nd edition, McGraw-Hill, New York, 1959.

C. Ueberhuber, *Numerical Computation 1: Methods, Software and Analysis*, Springer-Verlag, Berlin, 1997.

C. Ueberhuber, *Numerical Computation 2: Methods, Software and Analysis*, Springer-Verlag, Berlin, 1997.

M. B. Usher, "Studies on a Wood-Feeding Termite Community in Ghana, West Africa," *Biotropica*, **7**, 217–233, 1975.

M. B. Usher, T. J. Crawford, and J. L. Banwell, "An American Invasion of Great Britain: The Case of the Native and Alien Squirrel Species," *Conservation Biology*, **6**, 108–115, 1992.

H. A. van den Vorst, "Bi-CGSTAB: A Fast and Smoothly Converging Variant of Bi-CG for the Solution of Nonsymmetric Linear Systems," *SIAM Journal on Scientific and Statistical Computing*, **13**, 631–644, 1992.

J. H. van Lint and R. M. Wilson, *A Course in Combinatorics*, Cambridge University Press, Cambridge, 1992.

G. J. Van Wylen and R. E. Sonntag, *Fundamentals of Classical Thermodynamics, Third Edition*, John Wiley & Sons, New York, 1985.

R. S. Varga, *Matrix Iterative Analysis*, Prentice Hall, Englewood Cliffs, NJ, 1962.

R. Wait and A. R. Mitchell, *Finite Element Analysis and Applications*, John Wiley & Sons, New York, 1985.

B. Wendroff, *Theoretical Numerical Analysis*, Academic Press, New York, 1966.

F. M. White, *Fluid Mechanics, Second Edition*, McGraw-Hill, New York, 1986.

J. H. Wilkinson, *Rounding Errors in Algebraic Processes*, Prentice Hall, Englewood Cliffs, NJ, 1963.

J. H. Wilkinson, *The Algebraic Eigenvalue Problem*, Oxford University Press, Oxford, 1965.

J. Wilkinson, "The perfidious polynomial," in G. Golub, ed., *Studies in Numerical Analysis*, Mathematical Association of America, Washington, D.C., 1984.

J. H. Wilkinson and C. Reinsch, *Handbook for Automatic Computation. Volume 2: Linear Algebra*, Springer-Verlag, Berlin, 1971.

D. Winter, "On the Stem Curve of a Tall Palm in a Strong Wind," *SIAM Review*, **35** (4), 567–579, 1993.

J. F. Wittenberger and M. B. Dollinger, "The Effect of Acentric Colony Location on the Energetics of Avian Coloniality," *American Naturalist*, **124**, 189–204, 1984.

D. Young, *Iterative Solution of Large Linear Systems*, Academic Press, New York, 1971.

Index

RELATED TITLES OF INTEREST FROM PRENTICE HALL

Differential Equations, 2/E (0-13-143738-0)
John Polking, Al Boggess, David Arnold

Differential Equations with Boundary Value Problems, 2/E (0-13-186236-7)
John Polking, Al Boggess, David Arnold

Differential Equations: Computing and Modeling, 3/E (0-13-067667-4)
C. Henry Edwards and David E. Penney

Differential Equations and Boundary Value Problems: Computing and Modeling (0-13-065245-8)
C. Henry Edwards and David E. Penney

Elementary Differential Equations, 5e (0-13-145773-X)
C. Henry Edwards and David E. Penney

Elementary Differential Equations with Boundary Value Problems, 5/E (0-13-145774-8)
C. Henry Edwards and David E. Penney

Differential Equations: A Systems Approach (0-13-046026-5)
Bruce Conrad

Differential Equations and Boundary Value Problems: A Systems Approach (0-13-093419-4)
Bruce Conrad

Differential Equations with Boundary Value Problems (0-13-015927-1)
Selwyn Hollis

Differential Equations with Graphical and Numerical Methods (0-13-084376-8)
Bernard Banks

Elementary Differential Equations, 8/E (0-13-508011-8)
Rainville/Bedient/Bedient